# Animal Physiology

SECOND EDITION
# Animal

# Physiology

## MECHANISMS AND ADAPTATIONS

### Roger Eckert

UNIVERSITY OF CALIFORNIA, LOS ANGELES

*With Chapters 13 and 14 by*

### David Randall

UNIVERSITY OF BRITISH COLUMBIA

**W. H. FREEMAN AND COMPANY**
*New York*

*Project Editor:* Pearl C. Vapnek
*Manuscript Editor:* Liese Hofmann
*Interior Designer:* Paula Schlosser
*Cover Designer:* Sharon Helen Smith
*Production Coordinator:* William Murdock
*Illustration Coordinator:* Richard Quiñones
*Artists:* Jill Leland, Edna Indritz, Georg Klatt,
   Doreen Masterson, Donna Salmon
*Compositor:* Typothetae, Inc.
*Printer and Binder:* Kingsport Press


**Library of Congress Cataloging in Publication Data**


Eckert, Roger.
   Animal physiology.

   Includes bibliographies and index.
   1. Physiology.   I. Randall, David J., 1938–
II. Title.
QP31.2.E24 1983      591.1      82-18372
ISBN 0-7167-1423-X (cloth only)

Printed in the United States of America

   4 5 6 7 8 9 0    KP    10 8 9 8 7 6 5

*To our loved ones*

# Contents

# Preface

The diversity and adaptations of the several million species that make up the animal kingdom provide endless fascination and delight to those who love nature. Not the least of this pleasure derives from a consideration of how the bodies of animals function. At first it might appear that with so many kinds of animals adapted to such a variety of life-styles and environments, the task of even beginning to understand and appreciate the physiology of animals would be overwhelming. Fortunately (for scientist and student alike), the concepts and principles that provide a basis for understanding animal function are relatively few, for evolution has been conservative as well as inventive. The basic principles and mechanisms of animal physiology form the central theme of this book.

A beginning course in physiology is a challenge for both teacher and student because of the interdisciplinary nature of the subject. Not all students, even by their junior and senior years, have had exposure to all the chemical, physical, and biological subject matter required for an adequate background. On the other hand, most students are eager to come to grips with the subject and get on with the more exciting levels of modern scientific insight. For this reason, this book has been organized to present the essential background material in a way that allows students to review it on their own and go on quickly to the substance of animal function and to an understanding of its experimental elucidation.

*Animal Physiology* develops the major ideas in a simple and direct manner, stressing principles and mechanisms over the compilation of information and illustrating the functional strategies that have evolved within the bounds of chemical and physical possibility. Examples are selected from the broad spectrum of animal life, ranging from the protozoa at one end to our species and other vertebrates at the other end. Common principles, rather than exceptions, are emphasized. Thus, the more esoteric and peripheral details intentionally receive only passing attention, or none at all, so as not to distract from central ideas. Math is used where essential, but priority is given to the development of a qualitative and intuitive understanding.

The ideas developed in the text are illuminated and augmented by liberal use of illustrations and parenthetical "boxes." Other pedagogical aids are an 1100-word glossary and various chapter-end materials, including summaries, exercises, suggested readings, and lists of literature cited. References to the literature within the body of the text and in figure legends have been made unobtrusively, but with enough frequency so that the student can become aware of the role of scientists and their literature in the development of the subject. The text in places uses the device of a narrative describing actual, composite, or thought experiments to provide a feeling for methods of investigation while presenting information.

The chapters can be grouped into five sections. The first two chapters are intended primarily as an introduction and for review of the essential physical and chemical background not covered in later chapters. Chapters 3 and 4 are devoted to a survey of cell energetics and regulation of the intracellular milieu. Chapters 5 through 8 deal with excitable membranes, nerve signals, sensory mechanisms, and the function of the nervous system. Chapters 9 and 10 cover contractility and motility, and Chapters 11 through 16 describe the systems responsible for the homeostasis and supply of the internal environment.

I was fortunate to have been able to enlist David Randall, of the University of British Columbia, to write the two excellent chapters on respiratory and circulatory physiology.

The appreciative response to the first edition of *Animal Physiology* has encouraged us to bring out this second edition. The original 14 chapters were painstakingly reworked, reorganized, and selectively expanded to stay abreast of new developments, taking into con-

sideration hundreds of suggestions from colleagues in North America and abroad. New emphasis has been given to comparative organismic physiology and to environmental adaptations of animals, building on the tissue-level mechanistic approach that we feel is fundamental to an understanding of animal function.

Two entirely new chapters—Chapter 15, "Feeding, Digestion, and Absorption," and Chapter 16, "Animal Energetics and Temperature"—are the most important additions to this edition, for they make for a far more balanced coverage of organ systems and whole-animal physiology. As in the first edition, we have striven for clarity of exposition to help the reader grasp basic concepts and have given major consideration to student comprehension. We remain grateful for constructive criticism and suggestions.

October 1982                                        *Roger Eckert*

# *Acknowledgments*

Numerous persons helped us in important ways in preparing the second edition of *Animal Physiology*. Special recognition is due Melissa Derfler (University of California, Irvine), Luke Nwoye (University of California, Los Angeles), and Abigail Hafer (University of California, Los Angeles) for researching major portions of Chapters 15 and 16, to Frederick W. Munz (University of Oregon) for generous input to Chapter 16, and to Deanne Kuper for intelligent and accurate preparation of the typescript.

The second edition has also benefited significantly from detailed reviews in whole or in part by Mary Ann Baker (University of California, Riverside), Michael J. Berridge (Cambridge University), Charles J. Brokaw (California Institute of Technology), Carol Deutsch (University of Pennsylvania), David H. Evans (University of Miami), Gordon L. Fain (University of California, Los Angeles), Roger A. Gorski (University of California, Los Angeles), Harold T. Hammel (University of California, San Diego), Merrill B. Hille (University of Washington), Sarah E. Hitchcock-DeGregori (Carnegie-Mellon University), David E. Hornung (St. Lawrence University), Leonard B. Kirschner (Washington State University), Michael LaBarbera (University of Chicago), Hans Machemer (Ruhr-Universität Bochum), Kenneth A. Nagy (University of California, Los Angeles), Peter M. Narins (University of California, Los Angeles), Lee D. Peachey (University of Pennsylvania), Jane Peterson (University of California, Los Angeles), Guillermo Pilar (University of Connecticut), Robert Blake Reeves (State University of New York, Buffalo), Allen M. Scher (University of Washington), John D. Steeves (University of British Columbia), Larry C. Stoner (State University of New York, Syracuse), Malcolm H. Taylor (University of Delaware), Stephen C. Wood (University of New Mexico), Ernest Wright (University of California, Los Angeles), and Robert S. Zucker (University of California, Berkeley).

Of great value, too, were additional suggestions made by Bruce E. Beekman (California State University, Long Beach), J. K. Bowmaker (University of London), Annabelle Cohen (City University of New York), L. K. Ewing (Kent State University), Hector R. C. Fernandez (Wayne State University), Richard W. Heninger (Brigham Young University), D. J. Horvath (West Virginia University), August W. Jaussi (Brigham Young University), Eve Marder (Brandeis University), Michael A. Moskowitz (Massachusetts Institute of Technology), R. E. Ritzmann (Case Western Reserve University), Allen I. Selverston (University of California, San Diego), and Steven R. Vigna (University of Oregon).

Our associates at the University of California, Los Angeles, and University of British Columbia exercised commendable patience and provided inspiration as well as valuable comments while we were preoccupied with this project. Our families, too, were patient and understanding during the thousands of hours that went into the book. Finally, thanks are due for the high standards applied by the staff at W. H. Freeman and Company, especially to John H. Staples for editorial guidance, to Edna Indritz, Georg Klatt, Jill Leland, Doreen Masterson, and Donna Salmon for their excellent rendition of the figures, to Paula Schlosser for the design, to Sarah Elias for page layout, and to Lieselotte Hofmann and Pearl C. Vapnek for their very able and conscientious work as manuscript editor and project editor, respectively.

October 1982

*Roger Eckert*
*David Randall*

# CHAPTER 1

# *The Meaning of Physiology*

*A*nimal physiology can be defined as the study of the function of animals and their constituent parts. The ultimate goal of this subject is to understand, in physical and chemical terms, the mechanisms that operate in living organisms at all levels, ranging from the subcellular to the integrated whole animal. This goal is indeed an ambitious one, for each living organism, even a single cell, is incredibly complex. For this reason it has proved convenient to divide the subject of physiology into a number of subspecialties. Among these are general and cell physiology; organ, organismic, and environmental physiology; respiratory, circulatory, digestive, endocrine, developmental, neuro-, behavioral, sensory, and reproductive physiology. In spite of these somewhat contrived divisions, there are innumerable areas of overlap and common principles that recur throughout, providing a thread of continuity. It has also become apparent that these principles arise from the properties of matter and energy.

In an elementary textbook, it is necessary to introduce the specialties somewhat arbitrarily as separate chapters so as to avoid overwhelming the student with complexity. You should find it helpful, however, to remember that the various body functions require the coordinated activity of a number of tissues and organs. The brain, for example, cannot function without a constant supply of blood, carrying oxygen and glucose, provided by the pumping of the heart. The heart cannot survive more than minutes without oxygen supplied via the circulating blood by the lungs. The lungs cannot function without neural commands to the respiratory muscles from the brain. Similar examples abound also at subcellular levels of function. Thus, many chemical reactions in the cell require the integrity and metabolic activity of biological membranes, whereas the latter depend on some of these same reactions for production of energy-donor molecules required for the regulatory functions of the membrane.

## *Why* Animal *Physiology?*

From a biological standpoint, the human species is part of the animal kingdom, sharing with all other species a common evolutionary history, a common planet, and the same laws of physics and chemistry. The same principles and mechanisms of Mendelian and molecular genetics hold for us that operate for all creatures. Moreover, the fundamental biological processes that, in sum total, are termed "life" are shared in common by all animal species. Thus, the processes that give rise to the beating of the heart in the human body are fundamentally no different from those that underlie cardiac functions in a fish, frog, snake, bird, or ape. Likewise, the molecular and electrical events that produce a nerve impulse in the human brain are fundamentally no different from those that produce an impulse in the nerve of a squid, crab, or rat. In fact, most of what we have learned about the function of human cells, tissues and organs has first been learned on various species of vertebrate and invertebrate animals.

The first step in physiological research is to ask a question—for example, "Which ion carries the current that terminates the nerve impulse?" The next steps include the choice of tissue in which to investigate the problem. In studying nerve cells, it is extremely helpful to use one that is large, to facilitate certain procedures. For that reason, the major findings on nerve function have been made in work done on the giant axons of the cuttlefish and squid. Subsequent experiments, done with newer methods and with the benefit of the groundwork laid by the research on the squid, have confirmed that the nerves of humans and all other animals function in basically the same way. Our purpose, in this book, is to emphasize those processes that are basic to all animal groups and to see how they have been elucidated. In addition, we will note special evolutionary adaptations that serve to illuminate the ways in which environmental challenges have been met by the selection of functional specializations.

## Physiology and Medicine

There are several reasons why an understanding of body functions is relevant to our daily existence. First and most obvious is that physiology, especially as it applies to the human body, is the cornerstone of scientific medical practice. Throughout the ages, as in today's primitive societies, the approach to disease and malfunction has been almost entirely empirical—that is, by trial and error. Because this process has been applied over such long periods of time, human societies have found that certain ailments improve in response to certain treatments, be they herbs, hot water baths, acupuncture, or even the psychological treatments of witch doctors. In fact, the medicinal effects of many modern drugs—aspirin, for example—have been discovered by purely empirical means, and the primary actions of some important medicines remain unknown. As our understanding of function and malfunction of living tissue grows, it is becoming more feasible to develop effective, scientifically sound treatments for human ills. A physician who understands body function is better equipped to make intelligent and insightful diagnoses and decisions for effective treatment, and is less likely to embark on a course of treatment that is disruptive of the body's physiological balance. Physicians who are unfamiliar with physiological principles are, in effect, modern versions of the medicine man, dispensing drugs with little more understanding than that obtained from the advertising brochures of the pharmaceutical firms.

## Physiology and the Human Experience

Besides satisfying a natural curiosity as to how our bodies respire, move, metabolize food, and procreate—basic physical manifestations of life—the study of physiology is of great philosophical interest in helping us understand the nervous and sensory systems—those biological substrata of the human spirit in which resides all subjective experience: consciousness, awareness, thought, memory, learning, language, perception, and intellect—the sum of what is most specially human.

All animals, including humans, depend entirely on their sense organs and nervous system for information about the environment and the internal status of their bodies. Sensory input, together with the genetically inherited organization and properties of the nervous system, is responsible for all "knowledge" and determines how each animal behaves. (Some have claimed that there are channels of sensory input that bypass the physiological senses, but evidence in support of extrasensory perception has been equivocal and totally unconvincing to the scientific community at large.) Our ultimate dependence on our sense organs and the very personal nature of sensory perception become profoundly evident when we contemplate the problem of communicating subjective experience. For example, how would you explain to a person who has been totally blind since birth the visual sensations we term "red" or "green," or even "light" and "dark"?

It is difficult to say to what extent physiological and biochemical studies will eventually be able to explain higher mental experiences and answer such questions as "How does the brain 'remember' past experiences?" or "How is a mental image of a visual scene generated from past or present input to the eyes?" Questions about the origin of subjective experience may or may not be entirely answerable, but they are of such fundamental importance to human self-knowledge that the quest for answers must certainly equal or exceed in philosophical importance any other intellectual endeavor. The elucidation of the molecular basis of heredity and the exploration of the moon were the great scientific and technological adventures of the past 30 years. The elucidation of the mechanisms that give rise to human behavior and higher brain functions will undoubtedly be the great scientific adventure of the coming decades.

## Central Themes in Physiology

As in all other fields, certain principles recur throughout the study of animal function. We will consider a few of these here. You will doubtless discover more as you go on to later chapters.

### FUNCTION IS BASED ON STRUCTURE

The movement of an animal during locomotion depends on the structure of muscles and skeletal elements (e.g., bones). The movement produced by a contracting muscle depends on how it is attached to these elements and how they articulate with each other. In such a relatively familiar example, the relation between structure and function is obvious. The dependence of function on structure becomes more subtle, but no less real, as we direct our attention to the lower levels of organization—tissue, cell, organelle, and so on (Figure 1-1). One of the most intensively studied examples of functional dependence on structure is the contractile machinery of skeletal muscle. Our understanding of how a muscle contracts rests largely on an understanding of the ultrastructure of the contractile machinery as well as on its chemical properties.

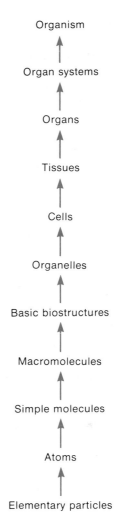

1-1  Structural hierarchy in a metazoan animal. In a protozoan, the cell is also the organism. "Basic biostructures" include membranes, microtubules, and filaments. At each level, function depends on the structural organization of that level and those below.

The principle that structure is the basis of function applies to the biochemical events as well. The interaction of an enzyme with its substrates, for example, depends on the configurations and electron distributions of the interacting molecules. Changing the shape of an enzyme molecule (i.e., denaturing it) by heating it above $40°C$ is generally sufficient to render it biologically nonfunctional by altering its shape.

### GENETICS AND PHYSIOLOGY

It is generally agreed among scientists that the information content of a *deoxyribonucleic acid* (*DNA*) molecule is the result of many generations of natural selection. Those spontaneous alterations (mutations) in the base sequence of the germ-line DNA that enhance the survival of the organism to reproduce are thereby statistically retained and increase in frequency of occurrence in the population of organisms. Conversely, those alterations in the base sequence of the DNA that render the organisms less well adapted to their environment will lessen the chances of reproduction and thus will be statistically suppressed and perhaps eliminated. Though it is common knowledge that Darwinian evolution has determined the basic structural details of all living species, it should be evident that function (which, as we have noted, is closely tied to structure) also has evolved through natural selection. Nevertheless, since evolutionary pressure can work only within the confines of chemical and physical laws, the nature and function of living systems are ultimately limited by the fundamental chemical and physical properties of the constituent elements and molecules.

Toward the end of the nineteenth century, August Weismann elaborated a theory of heredity in which he postulated the *continuity of the germ plasm*—namely, that genetic material passed on from metazoan parents to their offspring is contained in a line of *germ cells* that, in each generation, are derived directly from parent germ cells, creating an uninterruped lineage.

This germ plasm is hereditarily independent of the somatic cells, which arise from the germ cells and die off at the end of each generation. DNA is the molecular equivalent of Weismann's germ plasm and can be viewed as a continuous lineage of replicating strands that are passed from generation to generation within a species (Figure 1-2).

The blind process of evolution is centered on the survival of the germ-line DNA, for it is the informational content of the DNA that encodes a species, and once that is lost, the species becomes immediately and irreversibly extinct. Every somatic structure and function outside the continuous, generation-to-generation lineage of germ-line DNA is subservient to the survival of the germ line.

There exists, then, a symbiotic relationship between the germ-line DNA and the rest of the organism. Neither can survive without the other. The soma owes its existence to the DNA, and the DNA cannot survive without the somatic functions concerned with the short-term survival of the organism. The philosophical loop is closed, so to speak, with the realization that the structure and function of a species, and even its behavior, have evolved through natural selection for one ultimate "purpose"—to enhance the probability of the survival of the germ-line DNA of the species. To this end, the somatic functions of an organism are all directly or indirectly concerned with the acquisition and conversion of energy and matter from the environment. All else is frosting on the biological cake.

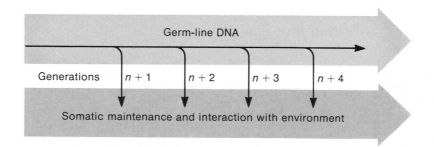

**1-2**   Concept of germ-line continuity. The germ line is preserved by the physiological activities of each succeeding generation. Natural selection favors those physiological processes that enhance the probability of reproduction and thus the transfer of the DNA to the next somatic generation.

## THE PRINCIPLE OF HOMEOSTASIS

The nineteenth-century French pioneer of modern physiology, Claude Bernard, was the first to enunciate the importance of *homeostasis* in animal function when he noted the ability of mammals to regulate the condition of their internal environment within rather narrow limits. This ability is familiar to all of us from routine clinical measurements on human blood and measurements of our body temperatures. "Constancy of the internal milieu," as Bernard phrased it, has been found to be a nearly universal phenomenon in living systems, allowing animals and plants to survive in stressful and varying environments (Figure 1-3). The evolution of homeostasis is believed to have been the essential factor that allowed animals to venture from physiologically friendly environments and invade environments hostile to life processes. One fascination of physiology is discovering how different groups of animals have dealt, in the evolutionary sense, with given environmental challenges.

Regulation of the internal milieu applies to unicellular organisms as well as to the most complex vertebrates. In the latter and in other metazoans, the composition of the fluid surrounding the cells of all tissues is subjected to constant regulation so that its composition (and even temperature in birds and mam-

mals) is kept within a narrow range. Single-celled animals, the protozoa, have been able to invade fresh water and other osmotically stressful environments because the concentrations of salts, sugars, amino acids, and other solutes in the cytoplasm are regulated by selective membrane permeability, active transport, and other mechanisms that maintain these concentrations within limits both favorable to the metabolic requirements of the cell and quite different from the extracellular environment. The same is true for the individual cells of the metazoan organism, which also regulate their intracellular milieu.

The regulatory processes of cells and multicellular organisms depend in many instances on the principle of *feedback* (Box 1-1). A man-made physical system such as a computer or a ballistic guidance system can be made so accurate that it will produce a predicted result under normal conditions. Living systems, however, must be able to function under the variable conditions to which they are subjected by the vagaries of nature. In the face of the finite accuracy of genetic and metabolic mechanisms—to say nothing of external perturbations—regulation requires continuous sampling and correction. For example, suppose that an experienced driver is placed in a car on an absolutely straight, 10-mile-long stretch of traffic-free highway, is allowed to position the car, is blindfolded, and is then required to drive the 10 miles without deviating from his lane. The slightest asymmetry in the neuromotor or sensory systems of the driver or in the steering mechanism of the car—not to mention wind or unevenness of the road surface—makes this an impossible task. On the other hand, if the blindfold is removed, the driver will employ visual information with negative-feedback principles to stay in his lane. As he perceives a gradual drift to one side of the lane or the other, due to whatever internal or external perturbations, he will simply correct by a compensatory motor output applied to the steering wheel. This can be summarized in the terms used in Box 1-1. The visual system of the driver acts as the sensor in this case, while his neuromotor system, by causing a correctional movement in the direction opposite to the perceived error, acts as the inverting amplifier that corrects for deviations from the *set point* (i.e., the center of the lane in this case).

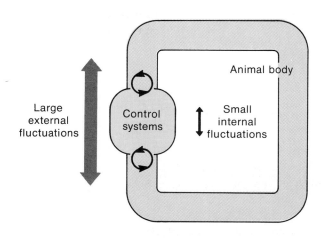

**1-3** Physiological regulatory systems maintain internal conditions within a relatively small range of fluctuation in the face of large variations in the external environment.

Large external fluctuations

Control systems

Small internal fluctuations

Animal body

***Box 1-1*** *The Feedback Principle*

Feedback is widely employed in biological systems, as well as in engineering, to maintain a preselected state. Feedback can be either positive or negative, each producing profoundly different effects.

*Positive Feedback*

In the model system shown in part A of the figure, an applied disturbance or perturbation acting on the *controlled system* is detected by a *sensor* and introduced at the *input* of an *amplifier*. Suppose the signal is amplified but that its sign (plus or minus) remains unchanged. In that case the *output* of the amplifier, when fed back to the controlled system, has the same effect as the original perturbation, reinforcing the perturbation of the controlled system. This configuration, which is called *positive feedback,* is highly unstable, because the output becomes progressively stronger as it is fed back

and reamplified. A familiar example occurs in public address systems when the output of the loudspeaker is inadvertently picked up by the microphone and reamplified, generating a loud squeal. Thus, a tiny perturbation at the input can cause a much larger effect at the output. The output of the system is usually limited in some way; for example, in the public address system, the intensity of the output is limited by the power of the audio amplifier and speakers or by saturation of the microphone signal. In biological systems, the response may be limited by the amount of energy or substrate available. It should be noted that positive feedback is generally used to produce a regenerative, explosive, or autocatalytic effect. In biological systems, it is often utilized for generating the rising phase of a cyclic event. An important example of this is seen in the upstroke of the nerve action potential.

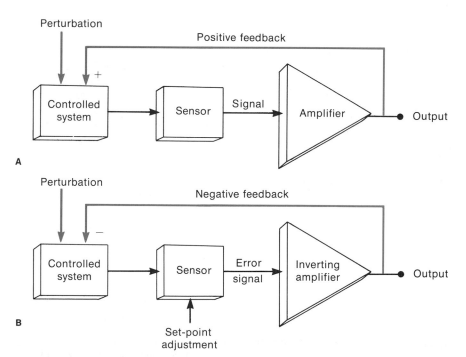

Basic elements of feedback systems. (A) Positive feedback occurs when a perturbation acting on a system is amplified and the amplifier output is "fed back" to the system without sign inversion. (B) Negative feedback occurs if somewhere in the feedback loop there is a "sign" inversion. In that case, the inverted signal stabilizes the controlled system at the set point. A change in sensitivity of the sensor is one way in which the set point can be adjusted.

*Negative Feedback*

Imagine an amplifier in which the "sign" of the output is opposite to that of the input (e.g., + changed to −, or vice versa). Such signal inversion provides the basis for *negative feedback* (see part B of the figure), which can be used to regulate a certain parameter (e.g., length, temperature, voltage, concentration) of the controlled system.

When the sensor detects a change in state (e.g., change in length, temperature, voltage, concentration) of the controlled system, it produces an *error signal* proportional to the difference between the *set point* to which the system is to be held and the actual state of the system. The error signal is then both amplified and inverted (i.e., changes in sign). The inverted output of the amplifier, fed back to the system, counteracts the perturbation. The inversion of sign is the most fundamental feature of negative feedback control. The inverted output of the amplifier, by counteracting the per-

turbation, reduces the error signal, and the system tends to stabilize near the set point.

A hypothetical negative-feedback loop with infinite amplification would hold the system precisely at the set point, because the slightest error signal would result in a massive output from the amplifier to counteract the perturbation. Since no amplifier—either electronic or biological—produces infinite amplification, negative feedback only approximates the set point during perturbation. The less amplification the system has, the less accurate is its control.

Finally, it should be noted that the elements of the physical example of feedback described here occur in a number of variations. For example, sensor and amplifier functions are, in some instances, carried out by a single element, and in others the inversion of sign may take place at the sensor. Nonetheless, the principles remain the same.

---

Another example of the regulation by negative feedback can be demonstrated with a thermostatic device that maintains the temperature of a hot water bath at or near the set point (Figure 1-4). As long as the water temperature is below the set point, the sensor maintains the heater switch "on." As soon as the set-point temperature is achieved, the heater switch opens, and further heating ceases until the temperature again drops below the set point. The "thermostat" that controls mammalian body temperature (situated in the brain) is set for about 37°C. Toxins produced by certain pathogens change the set point of this thermostat to a higher temperature, and a fever develops.

Examples of physiological feedback systems occur in intermediary metabolism (Chapter 3), neural control of muscle (Chapter 8), endocrine control (Chapter 11), circulatory control (Chapter 13), respiration (Chapter 14), and thermoregulation (Chapter 16).

## *The Literature of Physiological Sciences*

Our understanding of a science such as physiology is based on the work of individual scientists as described in published articles called "papers." These original papers, which include descriptions of the experimental methods, a summary of the results, and some discussion of them, are published in scientific

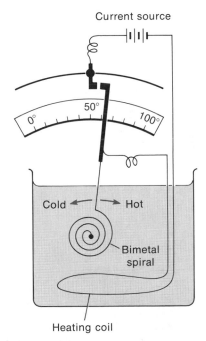

1-4 Example of a regulated system. A bimetal spiral, fixed at its center, winds slightly as the temperature of the water bath drops. The circuit is completed as the contacts touch, allowing electric current to flow through the heating coil. As the water warms, the coil unwinds slightly and the contacts separate. The desired temperature set point is adjusted by positioning the contact of the thermostat.

specialty journals. Before publication, the journal's editor sends the submitted manuscript to two or more other scientists expert in that field of study for their review and critical comments. The reviewers recommend acceptance or rejection on grounds of scientific quality and often make suggestions for improvement of acceptable papers. When the paper is published, members of the scientific community are free to test its conclusions by repeating key experiments and to accept or reject individually the conclusions stated in the paper. Healthy skepticism and attempts to improve on other scientists' work are of central importance to the self-correcting nature of an experimental science such as physiology.

Examples of widely read journals that publish original papers covering broad areas of physiology include the venerable *Journal of Physiology,* published in London; *Pflügers Archiv für gesammte Physiologie,* more recently titled *European Journal of Physiology;* and the *American Journal of Physiology.* Each of these includes all levels of study from cell to organ system. The *Journal of General Physiology* is devoted primarily to biophysical studies at the cell or subcellular level. The *Journal of Comparative Physiology* and *Comparative Physiology and Biochemistry* as well as the *Journal of Experimental Biology* cover many different areas and lean toward studies on lower vertebrates and invertebrates.

The weekly scientific news journals *Science* and *Nature* also publish some preliminary reports on physiological research that the editors feel will capture the general interest of the scientific community.

Specialty journals in physiology indicate by their names the disciplines they cover. Among these periodicals are the *Journal of Cell Physiology,* the *Journal of Neurophysiology,* the *Journal of Neurobiology,* the *Journal of Membrane Biology, Neuroscience, Brain Research, Journal of Endocrinology, Gastroenterology,* and *Nephron.*

Articles reviewing specialized areas of physiology, summarizing and evaluating the findings published in dozens to hundreds of original research papers, are published in periodicals devoted entirely to review articles. These include *Physiological Reviews, Annual Review of Physiology, Annual Review of Neuroscience,* and *Federation Proceedings.* The ultimate attempt to review the state of physiological knowledge is in an encyclopedic series, the *Handbook of Physiology,* published by the American Physiological Society as a growing set of volumes, each devoted to a major subdiscipline.

The periodicals noted above are only a small sampling of relevant publications. There are more than any one person can ever hope to digest, even with the aid of computerized retrieval systems. Nonetheless, you may wish to gain a glimpse into the vast literature of physiology by browsing casually through some recent issues of physiological journals in a university library. This experience should give ample indication that the information presented in this textbook is necessarily greatly condensed and greatly simplified. The contents of this book are the merest scratch through the surface of the rapidly growing body of knowledge in the physiological sciences.

## Summary

Animal physiology is concerned with the physical and chemical processes that take place in tissues and organs and form the bases of organismic function. The field is subdivided into many areas that are often interdependent and are interrelated by common genetic, physical, and chemical principles. Medicine, the practical application of physiology, is constantly evolving from a trial-and-error,

empirical approach toward practice founded on a rational understanding of cell and tissue function.

Besides practical application, physiology has philosophical value for us. This is apparent, for example, in a consideration of subjective human experience. All that we experience in life depends on the properties of our sensory and nervous systems. An understanding of how living organisms function helps us understand "what we are" and enhances our appreciation of our place in this world.

Three major ideas in physiology are:

*1.* Function is based on structure at all levels, beginning with atoms, molecules, and cell organelles.

*2.* Regulation of the intra- or extracellular environment (or both) provides the required constancy of conditions necessary for reliable and coordinated chemical and physical processes.

*3.* Cell and tissue functions have arisen through Darwinian evolution and are genetically determined.

Most of the information in this volume had its beginnings in research papers published in specialty journals devoted to the various subdisciplines of physiology. Some of these journals are listed, along with journals devoted to reviews or summaries of physiological advances.

## Exercises

*1.* Give an example of a simple structure–function relationship in physiology, and describe its conditions of operation.

*2.* What evolutionary advantage does successful maintenance of internal homeostasis confer on an animal species?

*3.* Why is negative rather than positive feedback required for maintenance of a constant state (i.e., homeostasis)?

*4.* Continuous regulation (i.e., homeostasis) of the internal environment is achieved through negative feedback. Describe an example and explain how the principle of negative feedback achieves regulation in the example.

*5.* Positive feedback is much rarer than negative feedback in biological systems. Give one example and explain the effect of positive feedback in that example.

*6.* Why can it be said that there exists a symbiotic relationship between the germ-line DNA and the somatic portions of the cell and organism?

*7.* Visit the university or college library and examine some recent issues of journals that publish original research articles presenting new findings in physiology.

## Suggested Reading

Bayliss, L. E. 1966. *Living Control Systems.* San Francisco: W. H. Freeman and Company.

Bernard, C. 1872. *Physiologie Générale.* Paris: Hatchette.

Fenn, W. O., and H. Rahn, eds. 1962. *Handbook of Physiology.* Washington, D.C.: American Physiological Society.

Grodius, F. S. 1963. *Control Theory and Biological Systems.* New York: Columbia University Press.

# CHAPTER 2

# Physical and Chemical Concepts

The living organisms found on our planet form a vast and varied array, ranging from viruses, bacteria, and protozoa to flowering plants, invertebrates, and the "higher" animals. In spite of this immense diversity, all forms of life as we know it are similar in certain fundamentals. Thus, all animals, plants, and microorganisms on our planet consist of the same chemical elements and of similar types of organic molecules. Moreover, all life processes take place in a milieu of water and depend on the physical–chemical properties of this ubiquitous and very special solvent. That all living organisms share a common biochemistry is one of the powerful evidences in support of their evolutionary kinship, the common thread that runs through all areas of biological study.

Biologists generally agree that life arose through processes of chance and natural selection under appropriate environmental conditions on the primitive Earth. Experiments first performed by Stanley Miller in 1953 show that certain molecules essential for primitive life (e.g., amino acids, peptides, nucleic acids) are formed by the action of lightninglike electric discharges on an atmosphere of methane, ammonia, and water—a simple mixture believed to be similar in composition to that of the primitive atmosphere about 4 billion years ago. The early atmosphere is believed to have been modified during subsequent eons by photosynthetic plants, which added the immense quantities of oxygen now present and which are capable of taking up nitrogen compounds for incorporation into nitrogenous biological compounds. The experimental formation of simple organic molecules under conditions similar to those that may have prevailed in the primeval atmosphere suggests that such molecules may have accumulated in ancient shallow seas, forming an organic "soup" in which life may then have undergone its first evolutionary stages of organization.

To what degree did the origin of life depend on the "right" conditions? Would life of another sort have appeared on Earth if the chemical and physical environment had been quite different? Suppose there had been no carbon atom? As we shall see shortly, the occurrence of life as we know it and can imagine it depends heavily on the "fitness of the environment." That is, life would be either nonexistent or vastly different if some of the fundamental properties of matter had been different.

A controversy once raged between the vitalists, who believed life was based on special "vital" principles not found in the inanimate world, and the mechanists, who maintained that life can ultimately be explained in physical and chemical terms. Until the early part of the nineteenth century, students of the natural world supposed that the chemical composition of living matter differed fundamentally from that of inanimate minerals. The vitalist view held that "organic" substances can be produced only by living organisms, setting them apart in a mysterious way from the inorganic world. This concept met its end in 1828, when Friedrich Wöhler synthesized the organic molecule urea (Figure 2-1) from lead cyanate and ammonia.

$$\underset{NH_2-\overset{\displaystyle\overset{O}{\|}}{C}-NH_2}{}$$

2-1   Structure of urea.

Wöhler's reagents were obtained from mineral sources. His successful organic synthesis set the stage for modern chemical and physical studies aimed at elucidating the mechanisms of life processes. It is now possible to duplicate in isolated cell-free systems nearly every synthetic and metabolic reaction normally performed by living cells.

The majority of biologists now agree that the bio-chemical and physiological processes of the living organism are based solely on the physical and chemical properties of the elements and compounds that constitute the living system. At first glance, this may appear to be a gross oversimplification, for certainly the properties of living systems seem far too marvelous and complex to be explained by a mere mixture of elements and compounds. Therein, of course, lies part of the answer. Living systems are not simple chemical "soups" but are highly organized structures composed of complex molecules. Macromolecules of many kinds participate in the regulation and direction of chemical activities within living cells. Organelles, such as the cell membrane, lysosomes, and mitochondria (Figure 2-2), lend structural organization to the living system by separating it into compartments and subcompartments. They also hold molecules in functionally important spatial relations to one another. Cells are organized

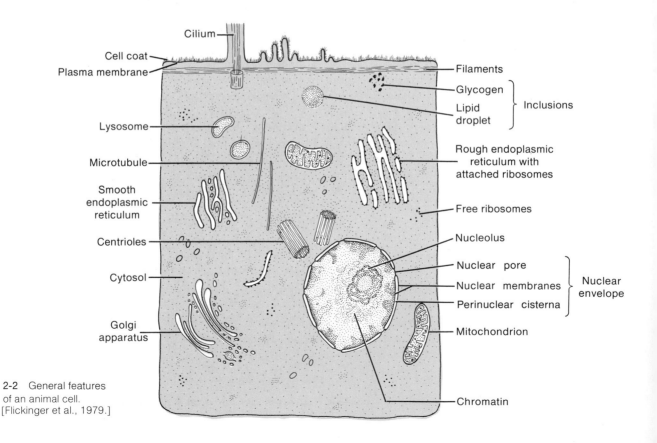

2-2   General features of an animal cell. [Flickinger et al., 1979.]

into tissues, tissues into organs, and those into interacting systems. Thus, the organism consists of an organizational hierarchy (Figure 1-1), with each higher level imparting further functional complexity to the whole. In this book we will begin with the most elementary, the chemical level, and progress to the more complex levels of organization.

## Atoms, Bonds, and Molecules

Figure 2-3 shows the periodic table of the chemical elements, of which all matter is composed. Table 2-1 lists the major components of the Earth's mineral crust, the human body, and seawater in their order of abundance in each. As the table shows, about 99% of the

| | | | | | | | | | | | | | | | | | |
|---|---|---|---|---|---|---|---|---|---|---|---|---|---|---|---|---|---|
| First shell | 1 H | | | | | | | | | | | | | | | | 2 He |
| Second shell | 3 Li | 4 Be | | | | | | | | | | 5 B | 6 C | 7 N | 8 O | 9 F | 10 Ne |
| Third shell | 11 Na | 12 Mg | | | | | | | | | | 13 Al | 14 Si | 15 P | 16 S | 17 Cl | 18 Ar |
| Fourth shell | 19 K | 20 Ca | 21 Sc | 22 Ti | 23 V | 24 Cr | 25 Mn | 26 Fe | 27 Co | 28 Ni | 29 Cu | 30 Zn | 31 Ga | 32 Ge | 33 As | 34 Se | 35 Br | 36 Kr |
| Fifth shell | 37 Rb | 38 Sr | 39 Y | 40 Zr | 41 Nb | 42 Mo | 43 Tc | 44 Ru | 45 Rh | 46 Pd | 47 Ag | 48 Cd | 49 In | 50 Sn | 51 Sb | 52 Te | 53 I | 54 Xe |
| Sixth shell | 55 Cs | 56 Ba | 57 La | 72 Hf | 73 Ta | 74 W | 75 Re | 76 Os | 77 Ir | 78 Pt | 79 Au | 80 Hg | 81 Tl | 82 Pb | 83 Bi | 84 Po | 85 At | 86 Rn |
| Seventh shell | 87 Fr | 88 Ra | 89 Ac | 104 | 105 | 106 | | | | | | | | | | | | |

| | | | | | | | | | | | | | |
|---|---|---|---|---|---|---|---|---|---|---|---|---|---|
| 58 Ce | 59 Pr | 60 Nd | 61 Pm | 62 Sm | 63 Eu | 64 Gd | 65 Tb | 66 Dy | 67 Ho | 68 Er | 69 Tm | 70 Yb | 71 Lu |
| 90 Th | 91 Pa | 92 U | 93 Np | 94 Pu | 95 Am | 96 Cm | 97 Bk | 98 Cf | 99 Es | 100 Fm | 101 Md | 102 No | 103 Lw |

2-3 The periodic table. Note that each row corresponds to a different orbital shell.

**TABLE 2-1** Comparison of the chemical composition of the human body with that of seawater and the Earth's crust. Values are percentages of total numbers of atoms.

| Human body | | Seawater | | Earth's crust | |
|---|---|---|---|---|---|
| H | 63 | H | 66 | O | 47 |
| O | 25.5 | O | 33 | Si | 28 |
| C | 9.5 | Cl | 0.33 | Al | 7.9 |
| N | 1.4 | Na | 0.28 | Fe | 4.5 |
| Ca | 0.31 | Mg | 0.033 | Ca | 3.5 |
| P | 0.22 | S | 0.017 | Na | 2.5 |
| Cl | 0.03 | Ca | 0.006 | K | 2.5 |
| K | 0.06 | K | 0.006 | Mg | 2.2 |
| S | 0.05 | C | 0.0014 | Ti | 0.46 |
| Na | 0.03 | Br | 0.0005 | H | 0.22 |
| Mg | 0.01 | | | C | 0.19 |
| All others < 0.01 | | All others < 0.1 | | All others < 0.1 | |

*Note:* Because figures have been rounded off, totals do not amount to 100.
*Source: Biology: An Appreciation of Life.*

human body is made up of hydrogen, oxygen, nitrogen, and carbon. This holds true for all living organisms. Is the preponderance of these elements in living systems simply a matter of chance, or is there a mechanistic explanation for their uniform prevalence in the great diversity of organisms that have evolved over the past 3 billion years?

George Wald, who has contributed much to our understanding of the chemical basis of vision, has argued that the biological predominance of hydrogen, oxygen, nitrogen, and carbon is not at all a matter of chance, but is the inevitable result of certain fundamental atomic properties of these elements—properties that render them especially suited for the chemistry of life. We will review briefly the factors that influence the chemical behavior of atoms, and then return to Wald's thesis.

Atomic structure is far more complex and subtle than can be fully described here; for our purposes, we need consider only the very simplest features. Each atom consists of a dense nucleus of protons and neutrons surrounded by a "cloud" of electrons equal in number to the protons in the nucleus. Since the negatively charged electrons are equal in number to the positively charged protons, each atom in its elemental state carries no net electric charge. The electrons do not occupy fixed orbits, but their statistical distribution is such that they occupy some positions with greater probabilities than others. This distribution is quite systematic, so that in an atom with one or two electrons, the orbital paths are virtually confined to a single "shell" around the nucleus, as in the hydrogen and helium atoms (Figure 2-4). In atoms

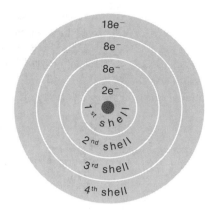

2-5  The first four orbital shells, showing the number of electrons necessary to fill each shell.

with three to ten electrons (Li, Be, B, C, N, O, F, Ne), this first shell is complete when occupied by two electrons; the remaining electrons occupy a second shell outside the first. The second shell can contain up to eight electrons. In atoms with more than ten protons and ten electrons, a third shell is formed, which can accommodate up to eight electrons, and so on (Figure 2-5).

When the outermost shell of an atom contains the maximum number of electrons possible in that shell (two in the first shell, eight in both the second and the third, eighteen in the fourth, etc.)—that is, when it cannot accommodate additional electrons—the atom is highly stable and resists reactions with other atoms. This is true of all the noble gases, such as helium and neon, which appear at the far right of the periodic table. Most elements, however, have incomplete outer electron shells and are therefore reactive with certain other atoms. Hydrogen, for example, has one rather than two electrons in its only shell, and oxygen has only six, instead of eight, electrons in its outer shell (Figure 2-6). Thus, the hydrogen atom and the oxygen atom both have a tendency to share electrons so as to fill their respective outer shells and bring them into more stable configurations.

Although the number of electrons in the outer shell has an important influence on the physical characteristics and reactivity of an atom, other physical features are also important in determining its chemical proper-

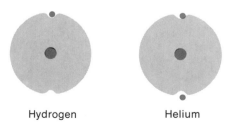

Hydrogen            Helium

2-4  The first shell of helium is filled (two electrons), but that of hydrogen is not.

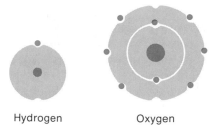

2-6  Both hydrogen and oxygen have unfilled outer shells.

ties. One of these is the size (or weight) of the atom. The heavier an atom (i.e., the more protons and neutrons in its nucleus), the more electrons surround the nucleus. As the number of electrons exceeds ten and a third shell of electrons exists, the valence electrons—that is, those of the outermost shell—are correspondingly more distant from the compact nucleus, and hence less strongly attracted by it than the valence electrons of atoms with only two shells (Figure 2-7). (Recall that electrostatic interactions between monopoles diminish with the square of distance.) Thus, chlorine, with seven electrons in its third and outer shell, is less reactive than fluorine, which has seven electrons in its second and outer shell. Both atoms have a tendency to gain one electron to complete the outermost shell, but this tendency is greater in fluorine, since its outermost shell feels a stronger electrostatic pull from its nucleus than

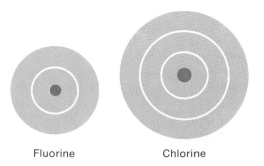

Fluorine          Chlorine

2-7  The outer shell of chlorine is more distant from its nucleus than is that of fluorine. Thus, the outermost electrons of chlorine are less strongly attracted by the nucleus than those of fluorine.

the larger chlorine atom. As a result, with all other things being equal, a small atom forms stronger and hence more stable bonds with other atoms than does a large atom.

## The Fitness of H, O, N, and C for Life

Now we can return to Wald's thesis that certain elements lend themselves especially well to the chemistry of living systems. Examination of the periodic table (Figure 2-3) reveals that of the common elements of the Earth's crust, only H, O, N, and C have two or fewer electron shells. Helium and neon are virtually inert, rare gases, while boron and fluorine form relatively rare salts. The metals lithium and beryllium form easily dissociated ionic bonds. In contrast, H, O, N, and C will form strong covalent bonds by sharing 1, 2, 3, and 4 electrons, respectively, to complete their outer electron shells.

Why are strong bonds important? Consider, for example, the biological chaos that would result if the chemical bonds in the hereditary material were easily dissociated. DNA, made up of H, O, N, C, and P, seldom undergoes alterations (i.e., mutations) during replication.† Although occasional mutations are essential for the process of evolution, it is important for the short-term integrity of each organism and each species that stable bonds reliably hold together the structures of DNA and other macromolecules.

Three of the four major biologically important elements (O, N, C) are among the very few to form double or triple bonds, which greatly increases the variety of molecular configurations that can be formed by reaction of these elements. Oxygen, for example, can oxidize carbon to form carbon dioxide.

$$O=C=O$$

Since the two double bonds satisfy the tendencies of the three atoms of this molecule to react, the $CO_2$ molecule is relatively inert and is therefore able to diffuse readily from its source of production to become available for recycling through the photosynthetic process of green plants.

---

†On the average, less than once per gene in every 10,000 replications.

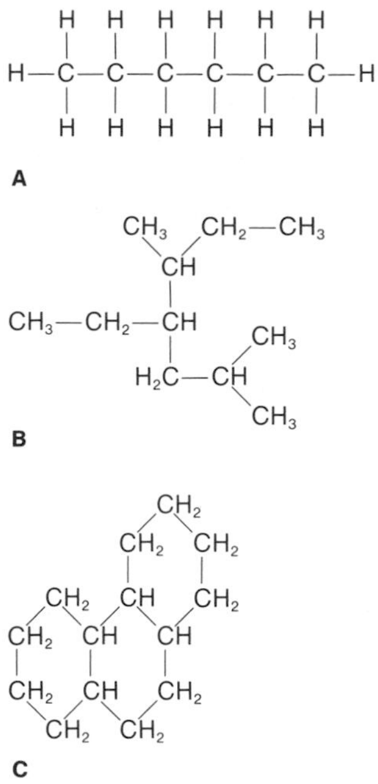

**A**

**B**

**C**

2-8  Examples of the diversity of molecular structures possible with the carbon atom.

The ability of the carbon atom to form four single or two double bonds endows it with great potential for varieties of atomic combinations with itself and other atoms. The carbon atom can form straight or branched chains and ring structures (Figure 2-8), and together with other atoms provides an almost infinite diversity of molecular structures and configurations.

Silicon, which is in the same column and just below carbon in the periodic table, has some properties similar to those of carbon. Unlike carbon, however, it is larger and does not form double bonds. Therefore, it combines with two atoms of oxygen by two single bonds only:

$$O-Si-O$$

leaving incomplete outer shells in all three atoms of silicon dioxide. Since the tendency to form bonds remains unfulfilled, the silicon dioxide molecule readily bonds with others of its kind, forming the huge polymeric molecules that make up silicate rocks and sand.

Thus, it is evident that silicon, even though it has some properties similar to those of carbon, is eminently more suited for the formation of stone than it is for large-scale participation in the organization of biological molecules.

Besides its important role in combining with hydrogen to form water, oxygen acts as the final electron acceptor in the sequence of oxidations through which chemical energy is released by cell metabolism. This important ability to oxidize (accept electrons from) other atoms and molecules is due to the oxygen atom's incomplete outer electron shell and relatively low atomic weight.

In addition to the four major biological elements, other elements participate in cell chemistry, though in lesser numbers. These include phosphorus and sulfur; the ions of four metallic elements ($Na^+$, $K^+$, $Mg^{2+}$, and $Ca^{2+}$); and the chloride ion ($Cl^-$). We will return to those later.

## Water

Water is directly and intimately involved in all details of animal physiology, but because it is so common, water is often regarded with indifference, as an inert space filler in living systems. The truth is, of course, that water is a highly reactive substance—quite different both physically and chemically from most other liquids. Water possesses a number of unusual and special properties of great importance for living systems. Indeed, life as we know it would be impossible if water did not have these properties. The first living systems presumably arose in the aqueous environment of shallow seas. It is therefore not surprising that the living organisms of the present are intimately adapted at the molecular level to the special properties of water. Today even terrestrial animals consist 75% or more of water, and much of their physiological effort is devoted to the conservation of body water and the regulation of the chemical composition of the internal aqueous environment.

The special properties of water so important to life stem directly from its molecular structure; therefore, we should begin with a brief consideration of the water molecule.

## THE WATER MOLECULE

Water molecules are held together by *polar covalent bonds* between one oxygen and two hydrogen atoms. The polarity (i.e., uneven charge distribution) of the covalent bonds results from the high electronegativity of the oxygen atom relative to hydrogen—an expression of the strong tendency of the oxygen atom to acquire electrons from other atoms, such as hydrogen. This high electronegativity causes the electrons of the two hydrogen atoms in the water molecule to occupy positions statistically closer to the oxygen atom than to the parent hydrogen atoms. The bond is therefore about 40% ionic in character, and the following partial charge distributions exist ($\delta$ represents the local partial charge of each hydrogen atom):

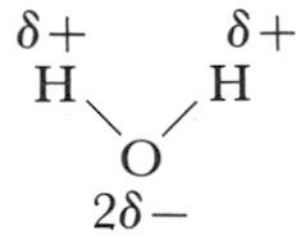

$$
\begin{matrix}
\delta + & & \delta + \\
H & & H \\
& \diagdown \; \diagup & \\
& O & \\
& 2\delta - &
\end{matrix}
$$

The water molecule can also be depicted in terms of molecular orbitals, as in Figure 2-9. The angle between the two oxygen–hydrogen bonds, rather than being 90°, as predicted for purely covalent bonding, is found to be 104.5°. The increased angle can be ascribed to the mutual repulsion of the two positively charged hydrogen nuclei, which tends to force them apart. In the hydride of sulfur, $H_2S$, the S — H bonds are purely covalent; there is no asymmetrical charge distribution as in $H_2O$. Thus, the bond angle in $H_2S$ is closer to 90°. Because of the semipolar nature of H — O bonds, $H_2O$ differs greatly, both chemically and physically, from $H_2S$ and other related hydrides. Why is this?

The water molecule's uneven distribution of electrons, due to the semipolar nature of the H — O bond, causes the water molecule to act like a *dipole*. That is, it behaves somewhat like a bar magnet, but instead of having two opposite magnetic poles, it has two opposite electric poles (+ and −) (Figure 2-9). As a result, it tends to align with an electrostatic field. The *dipole moment* is the turning force exerted on the molecule by an external field. The high dipole moment of water (4.8 debyes) is the most important physical feature of

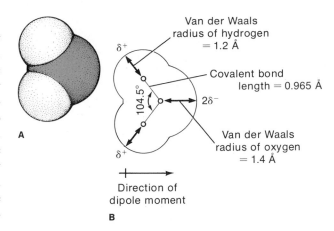

2-9  Diagrams of the water molecule, showing the relationships between the oxygen and hydrogen atoms. **(A)** Space-filling model. **(B)** Bond angles and lengths. [Lehninger, 1975.]

the molecule and accounts for many of its rather special properties.

The most important chemical feature of water is its ability to form *hydrogen bonds* between the nearly electron-bare, positively charged protons (hydrogen atoms) of one water molecule and the negatively charged electron-rich oxygens of neighboring water molecules (Figure 2-10). In each water molecule, four

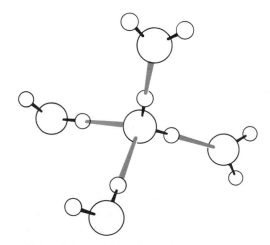

2-10  Tetrahedral nature of hydrogen bonding between water molecules.

of the eight electrons in the outer shell of the oxygen atom are covalently bonded with the two hydrogens. This leaves two pairs of electrons free to interact electrostatically (i.e., to form hydrogen bonds) with the electron-poor hydrogen atoms of neighboring water molecules. Since the angle between the two covalent bonds of water is about 105°, groups of hydrogen-bonded water molecules form tetrahedral arrangements. This arrangement is the basis for the crystalline structure of the most common form of ice (Figure 2-11).

## PROPERTIES OF WATER

The hydrogen-bonded structure of water is highly labile and transient, for the lifetime of a hydrogen bond in liquid water is only about $10^{-10}-10^{-11}$ s. This transience is due to the relatively weak nature of the hydrogen bond. It takes only 4.5 kcal/mol to break the hydrogen bond, whereas the covalent O—H bond within the water molecule has an energy of 110 kcal/ mol. As a result, no specific groups of $H_2O$ molecules remain bonded for more than a brief instant, yet a statistically constant fraction of the population is joined together by hydrogen bonding at all times at a given temperature.

In spite of its modest strength, the hydrogen bond increases the total energy (i.e., heat) required to separate individual molecules from the rest of the population. For this reason, the *melting* and *boiling points* and the *heat of vaporization* of water are much higher than those of other common hydrides of elements related to oxygen (e.g., $NH_3$, HF, $H_2S$). Of the common hydrides, only water has a boiling point (100°C) far above temperatures common to the surface of the Earth. The statistical loose bonding between water molecules also endows water with an unusually high *surface tension* and *cohesiveness.*

It is widely agreed that the oceans and lakes would have turned to solid ice except at the surface if ice were denser (heavier) than water and formed from the bottom up. Ice is less dense than water because it has an open crystalline lattice, whereas water, with its more random molecular organization, has a more closely packed, dense molecular population.

## WATER AS A SOLVENT

The medieval alchemists, looking for the universal solvent, were never able to find a more effective and "universal" solvent than water. The solvent characteristics of water are due largely to its high dielectric constant,† a manifestation of its electrostatic polarity. This is illustrated especially well by the *ionic compounds,* or *electrolytes,* which include salts, acids, and bases, and which all share the property of dissociating into ions when dissolved in water. (Solutes that undergo no dissociation, and therefore do not increase the conduc-

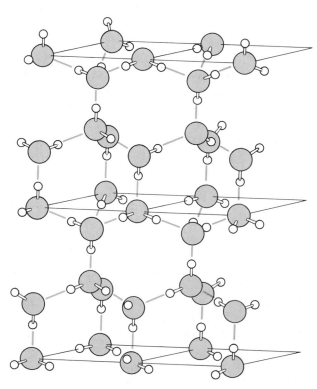

2-11    Crystalline structure of ice. [Loewy and Siekevitz, 1969.]

---

†The dielectric constant is a measure of the effect that water or any polar dielectric substance has in diminishing the electrostatic force between two charges separated by water or another dielectric medium. That force is given by the relation

$$f = \frac{q_1 q_2}{\varepsilon d^2}$$

where $f$ is the force (in dynes) between the two electrostatic charges $q_1$ and $q_2$ (in electrostatic units), $d$ is the distance (in centimeters) between the charges, and $\varepsilon$ is the dielectric constant.

tivity of a solution, are termed *nonelectrolytes*. Common examples of nonelectrolytes are the sugars, alcohols, and oils.) Figure 2-12 illustrates the arrangement of the ions $Na^+$ and $Cl^-$ in a sodium chloride crystal. The highly structured array is held together firmly by the electrostatic attraction between the positively charged sodium ions and the negatively charged chloride ions. A *nonpolar* liquid, such as hexane, cannot dissolve the crystal, because no source of energy exists in the nonpolar solvent to break an ion away from the rest of the crystal. Water, however, can dissolve the NaCl crystal, just as it can dissolve most other ionic compounds (e.g., salts, acids, bases), because the dipolar water molecules can overcome the electrostatic interactions

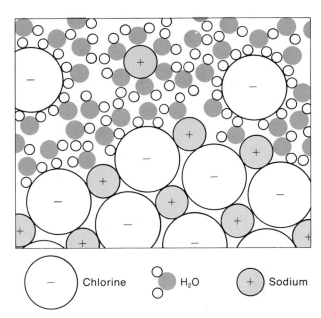

2-13   Hydration of a salt by water molecules. The oxygens of the water molecules are attracted to the cations, and the hydrogens of the water molecules are attracted to the anions. [*Biology Today.*]

between the individual ions, as shown in Figure 2-13. The partial negative charge of the oxygen causes weak electrostatic binding with the positively charged *cation* ($Na^+$ in this case), and the partial positive charge on the hydrogen causes weak electrostatic binding with the negatively charged *anion* ($Cl^-$ in this case). The clustering of water molecules about individual ions and polar molecules is called *solvation* or *hydration*.

The water molecules surround the ions, orienting themselves so that their positive poles face anions and their negative poles face cations, thereby further reducing the electrostatic attraction between the dissolved cations and anions of an ionic compound. In a sense, the $H_2O$ molecules act as insulators. The first shell of water molecules attracts a second shell of less tightly bound, oppositely oriented water molecules. The second shell may even attract more water in a third shell. Thus, the ion may carry a considerable quantity of *water of hydration.* The effective diameter of the hydrated ion varies inversely with the diameter of the ion of a given charge. For example, the ionic radii of $Na^+$ and

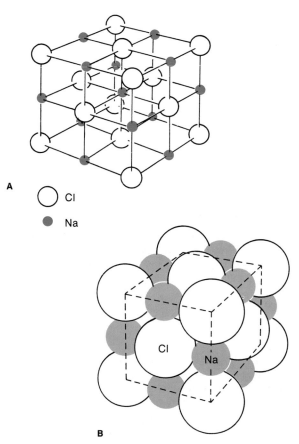

2-12   (A) Internal structure of the sodium chloride crystal lattice. (B) Representation of relative ionic sizes of $Na^+$ and $Cl^-$. [Mahan, 1966.]

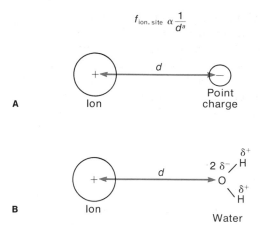

$$f_{\text{ion, site}} \propto \frac{1}{d^a}$$

**A**    Ion    Point charge

**B**    Ion    Water

**C**

2-14   Effect of distance on interactions between ions and charged sites. As indicated in the equation, the electrostatic force, *f*, between an ion and a site of opposite charge varies inversely with the distance *d* raised to some power, *a*. (A) For a point charge, or monopole, the exponent *a* equals 2.0, so that the force drops inversely with the square of the distance. (B) For a dipole such as the water molecule, the value of *a* can be as high as 4.0. (C) The drop in electrostatic force as a function of distance is illustrated for these two values of *a*. In the case of water and a point positive charge, the volume is closer to 3.0.

$K^+$ are 0.095 and 0.133 nm, respectively, whereas their effective hydrated radii are 0.24 and 0.17 nm, respectively. The reason for this inverse relationship is that the electrostatic force between the nucleus of the ion and the dipolar water molecule decreases markedly with distance between the water molecule and the nucleus of the ion (Figure 2-14). Thus, the smaller ion binds water more strongly and thereby carries a large number of water molecules with it.

Water also dissolves certain organic substances (e.g., alcohols and sugars) that do not dissociate into ions in solution, but these small organic molecules do have polar properties. In contrast, water does not dissolve or dissolve in compounds that are completely nonpolar, such as fats and oils, for it cannot react with them by hydrogen bonding. It does, however, react partially with *amphipathic* compounds, which have a polar group and a nonpolar group. A good example is a soap molecule (Figure 2-15), which has a *hydrophilic* (water-attracting) polar head and a *hydrophobic* (water-repelling) nonpolar tail. If a mixture of water and sodium oleate is shaken, the water will disperse the latter into minute droplets. The sodium oleate molecules in such a droplet, or *micelle*, are arranged in the configuration shown in Figure 2-16, with their hydrophobic, nonpolar tail groups huddled in the center and

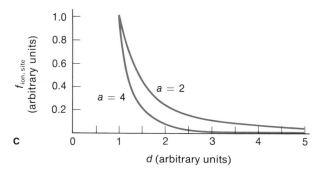

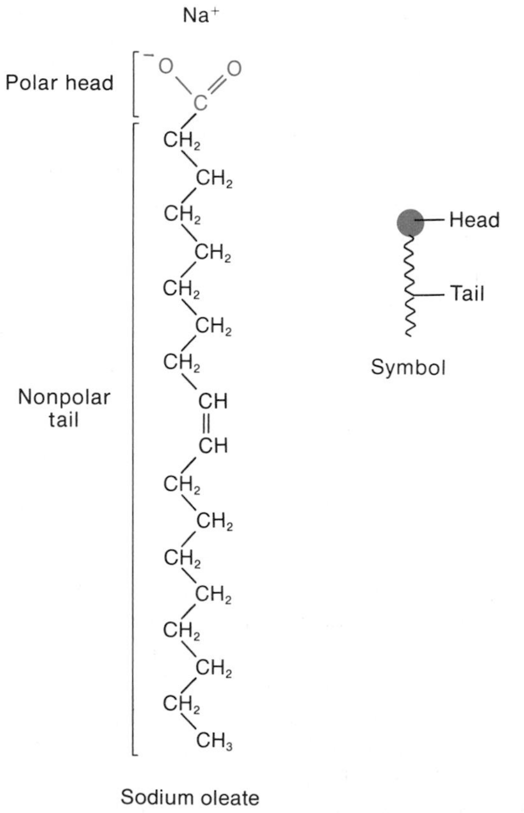

2-15   Structure of a polar lipid molecule, sodium oleate.

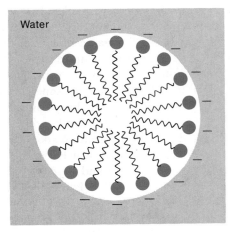

Water

Sodium oleate micelle

**2-16**  Behavior of a polar lipid in a polar solvent, such as water. The hydrophobic ends of the molecule tend to avoid contact with the polar solvent by grouping at the center of the micelle. [Lehninger, 1975.]

their hydrophilic, polar head groups arranged around the perimeter—facing outward, so as to interact with the water. The same behavior is exhibited by phospholipid molecules (pp. 91, 92), which also consist of hydrophobic and hydrophilic groups. This tendency of water to cause amphipathic molecules to form micelles is important in the formation of biological membranes in living cells, and may have provided the basis for the first cell-like organization of a living system in the organic-rich shallow seas in which life is believed to have undergone its first evolutionary stages.

## Solutions and Their Colligative Properties

It is a convention in chemistry to express the quantity of any pure substance in *moles* (abbreviated *mol*). This unit is defined as the molecular weight expressed in grams. Thus, 12.00 g of the pure nuclide $^{12}C$ is 1 mol of $^{12}C$, or $6.022 \times 10^{23}$ (Avogadro's number) carbon atoms. Conversely, there are $6.022 \times 10^{23}$ molecules in 2.00 g (or 1 mol) of $H_2$, in 28 g (or 1 mol) of $N_2$, and in 32 g (or 1 mol) of $O_2$.

For some purposes, it is necessary to express the amount of solute in terms of molality—the number of moles of solute in 1000 g of solvent (*not* solution). If 1 mol of a soluble substance (e.g., 342.3 g of sucrose) is dissolved in 1000 g of water, the resulting solution is said to be a *one molal* solution. Although 1 L of water equals 1000 g, the total volume of 1000 g of water plus 1 mol of the substance will be somewhat greater or lesser than 1 L by some unpredictable amount. Molality, therefore, is generally an inconvenient way of stating concentration. A more useful measure of concentration in physiology is *molarity*. A *one molar* (1 M) solution is one in which 1 mol of solute is dissolved in a *total final volume* of 1 L. This is written 1 mol/L or 1 M. In the laboratory, a 1 M solution is made by simply adding enough water to 1 mol of the solute to bring the volume of the final solution up to 1 L. A millimolar (mM) solution contains 1/1000 mol/L and a micromolar (μM) solution contains $10^{-6}$ mol/L. If a solution contains *equimolar* concentrations of two solutes, then the number of molecules of one solute equals the number of molecules of the other solute per unit volume of solution.

Since the molarity of a solution describes the number of individual *particles* dispersed in a given volume of solution, the concept becomes somewhat more complex for electrolytes than for nonelectrolytes because of ionic dissociation. As an example, 1 mol of NaCl dissolved in $H_2O$ produces nearly twice as many particles as a mole-equivalent weight of glucose, since the salt dissociates into $Na^+$ and $Cl^-$. Because of electrostatic interaction between the cations and anions of the dissolved electrolyte, there is a statistical probability that at any instant some $Na^+$ will be associated with $Cl^-$. The electrolyte therefore behaves as if it were not 100% dissociated. Because the electrostatic force between ions decreases with the square of the distance between them, the electrolyte will effectively become more dissociated if the solution is more dilute. Thus, the *activity* (i.e., effective free concentration) of an ion depends on its tendency to dissociate in solution, as well as on its total concentration. Table 2-2 lists the *activity coefficients*† of some common electrolytes. Those that dissociate to a large extent (i.e., have a large activity

† The activity coefficient is defined by the relation $\gamma = a/m$, in which $m$ is the molal concentration and $a$ the activity, which is defined as the effective concentration of the substance as indicated by the properties in solution.

TABLE 2-2   Activity coefficients of representative electrolytes at various molal concentrations.

| Substance | Molalities | | | | |
|---|---|---|---|---|---|
| | 0.01 | 0.05 | 0.10 | 1.00 | 2.00 |
| KCl | 0.899 | 0.815 | 0.764 | 0.597 | 0.569 |
| NaCl | 0.903 | 0.821 | 0.778 | 0.656 | 0.670 |
| HCl | 0.904 | 0.829 | 0.796 | 0.810 | 1.019 |
| CaCl$_2$ | 0.732 | 0.582 | 0.528 | 0.725 | 1.555 |
| H$_2$SO$_4$ | 0.617 | 0.397 | 0.313 | 0.150 | 0.147 |
| MgSO$_4$ | 0.150 | 0.068 | 0.049 | — | — |

*Source:* West, 1964.

coefficient) are termed *strong electrolytes* (e.g., KCl, NaCl, HCl), and those that dissociate only slightly are termed *weak electrolytes.*

Solute particles, irrespective of their chemical nature, impart to the solution a set of physical properties, including depression of the freezing point, elevation of the boiling point, and depression of the water vapor pressure. The solute particles also give the solution an osmotic pressure. These *colligative properties* are intimately related to one another, and are all quantitatively related to the number of solute particles dissolved in a given volume of solvent—that is, the molality. Thus, 1 mol of an ideal solute—that is, one in which the particles neither dissociate nor associate—dissolved in 1000 g of water at standard pressure (760 torr) depresses the freezing point by 1.86°C and elevates the boiling point by 0.54°C. In an ideal apparatus for determining osmotic pressure, such a solution will exhibit an osmotic pressure of 22.4 atm at standard temperature (0°C). Since colligative properties depend on the total number of solute particles in a given volume of solvent, the colligative properties of a 10 mM NaCl (strongly dissociating electrolyte) solution will be nearly equivalent to those of a 20 mM sucrose solution—more precisely, equivalent to an 18 mM sucrose solution, since the activity coefficient of NaCl at a concentration of 10 mM is 0.9 (Table 2-2).†

---

†The activity coefficients are given for various *molal* concentrations rather than *molar* concentrations. At low concentrations, however, molarity and molality are nearly identical.

## Solutions of Electrolytes

### IONIZATION OF WATER

The dynamic nature of hydrogen bonding between molecules of water can be depicted as in Figure 2-17, which shows how the covalent and hydrogen bonds may exchange places from one instant to the next. Because of the ever-changing nature of the bonding relations between adjacent water molecules, there is a finite probability that three hydrogen atoms will associate with one oxygen atom to form a *hydronium* ion, H$_3$O$^+$, leaving another oxygen atom with only one hydrogen to form a *hydroxyl* ion, OH$^-$. This probability is actually quite small. At any given time, a liter of pure water at 25°C contains only 1.0 × 10$^{-7}$ mol of H$_3$O$^+$ and an equal number of hydroxyl ions. The positively charged hydrogen atoms of the hydronium ion attract the electronegative (oxygen) ends surrounding nondissociated water molecules to form a stable hydrated hydronium ion (Figure 2-18).

The dissociation of water is conveniently written as

$$H_2O \rightleftharpoons H^+ + OH^-$$

Nevertheless, the reader should bear in mind that the proton (H$^+$) is not, in fact, free in solution but be-

2-17   Dynamic nature of the hydrogen bond in water. Resonance can cause separation of charges, producing hydronium and hydroxyl ions. [Dowben, 1969.]

2-18   The hydronium ion in solution, surrounded by three hydrogen-bonded water molecules. [Lehninger, 1975.]

2-19 Proton migration between water molecules. Each molecule in turn exists briefly as a hydronium ion. [Lehninger, 1975.]

comes part of the hydronium ion. A proton can, however, migrate to a surrounding $H_2O$ molecule, converting it to $H_3O^+$ or displacing one of its protons to another water molecule (Figure 2-19). A sequence of such migrations and displacements can, in the fashion of falling dominoes, conduct over relatively long distances, with any one proton traveling but a short distance. There is some evidence that such proton conduction may play an important role in some biochemical processes, such as photosynthesis and respiratory chain phosphorylation.

### ACIDS AND BASES

According to J. N. Brønsted's definition of acid and base, $H_3O^+$ is *acidic* and $OH^-$ is *basic*. That is, the former can donate a proton, and the latter can accept a proton. Thus, any substance that can form a hydrogen ion, $H^+$, is termed an acid, and any substance that combines with a hydrogen ion is termed a base. An acid–base reaction always involves such a *conjugate acid–base pair*—the proton donor and the proton acceptor, $H_3O^+$ and $OH^-$ in the example of water. Water is said to be *amphoteric* because it acts either as acid or as base. Amino acids also have amphoteric properties.

Examples of acids are:

$$\text{hydrochloric acid} \quad HCl \rightleftharpoons H^+ + Cl^-$$
$$\text{carbonic acid} \quad H_2CO_3 \rightleftharpoons H^+ + HCO_3^-$$
$$\text{ammonium} \quad NH_4^+ \rightleftharpoons H^+ + NH_3$$
$$\text{water} \quad H_2O \rightleftharpoons H^+ + OH^-$$

Examples of bases are:

$$\text{ammonia} \quad NH_3 + H^+ \rightleftharpoons NH_4^+$$
$$\text{sodium hydroxide} \quad NaOH + H^+ \rightleftharpoons Na^+ + H_2O$$
$$\text{phosphate} \quad HPO_4^{2-} + H^+ \rightleftharpoons H_2PO_4^-$$
$$\text{water} \quad H_2O + H^+ \rightleftharpoons H_3O^+$$

The dissociation of water into acid and base is an equilibrium process and can be described by the *Law of Mass Action*, which states that the rate of a chemical reaction is proportional to the active masses of the reacting substances. For example, the equilibrium constant for the reaction

$$H_2O \rightleftharpoons H^+ + OH^-$$

is given by

$$K_{eq} = \frac{[H^+][OH^-]}{[H_2O]} \tag{2-1}$$

The concentration of water remains virtually unaltered by its partial dissociation into $H^+$ and $OH^-$, since the concentration of each of the dissociated products is only $10^{-7}$ M $(10^{-7}$ mol/L), whereas the concentration of water in a liter of pure water (equal to 1000 g) is 1000 g/L divided by the gram molecular weight of water, 18 g/mol, or 55.5 M (55.5 mol/L). Equation 2-1 can thus be simplified to

$$55.5\, K_{eq} = [H^+][OH^-]$$

Recall that a consequence of the Law of Mass Action is the reciprocal relation between the concentrations of two compounds in an equilibrium system. This reciprocity is apparent in the constant $K_{eq}$, which may be lumped with the molarity of water (55.5) into a constant that will be termed the *ion product* of water, $K_w$. At 25°C this has a value of $1 \times 10^{-14}$:

$$K_w = [H^+][OH^-] = 10^{-14}$$

This equation follows from the fact, noted above, that $[H^+]$ and $[OH^-]$ each equal $10^{-7}$ mol/L. If $[H^+]$ for some reason increases, as when an acid substance is dissolved in water, [OH⁻] will decrease so as to keep $K_w = 10^{-14}$. This reaction is, of course, the basis for the *pH scale,* the standard for acidity and basicity, measured as concentration of H⁺ (actually $H_3O^+$). It will be noted in Table 2-3 that the pH scale is logarithmic and typically ranges from 1.0 M H⁺ to $10^{-14}$ M H⁺. The term pH is a sort of logarithmic shorthand. It is defined as

$$\text{pH} = \log_{10}\frac{1}{[\text{H}^+]}$$

or

$$\text{pH} = -\log_{10}[\text{H}^+]$$

Thus, a $10^{-3}$ M solution of a strong acid, such as HCl, which dissociates completely in water, has a pH of 3.0. A solution in which $[\text{H}^+] = [\text{OH}^-] = 10^{-7}$ M has a pH of 7.0, and so forth. A solution with a pH of 7 is said to be neutral—that is, neither acidic nor basic. The pH of a solution can be conveniently measured as the voltage produced by H⁺ diffusing through the proton-selective glass envelope of an electrode immersed in the solution (Figure 2-20).

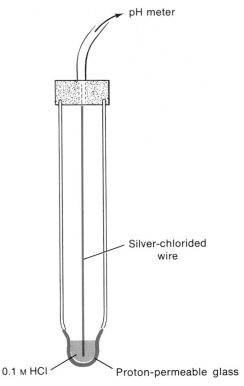

**2-20**  When the pH electrode is immersed in a solution of different [H⁺], the potential difference set up across the envelope of proton-selective glass is proportional to the log of the ratio of H⁺ concentrations on the two sides of the H⁺-selective glass.

TABLE 2-3   The pH scale.

| pH | [H⁺] (mol/L) | [OH⁻] (mol/L) |
|---|---|---|
| 0 | $10^{0}$ | $10^{-14}$ |
| 1 | $10^{-1}$ | $10^{-13}$ |
| 2 | $10^{-2}$ | $10^{-12}$ |
| 3 | $10^{-3}$ | $10^{-11}$ |
| 4 | $10^{-4}$ | $10^{-10}$ |
| 5 | $10^{-5}$ | $10^{-9}$ |
| 6 | $10^{-6}$ | $10^{-8}$ |
| 7 | $10^{-7}$ | $10^{-7}$ |
| 8 | $10^{-8}$ | $10^{-6}$ |
| 9 | $10^{-9}$ | $10^{-5}$ |
| 10 | $10^{-10}$ | $10^{-4}$ |
| 11 | $10^{-11}$ | $10^{-3}$ |
| 12 | $10^{-12}$ | $10^{-2}$ |
| 13 | $10^{-13}$ | $10^{-1}$ |
| 14 | $10^{-14}$ | $10^{0}$ |

### THE BIOLOGICAL IMPORTANCE OF pH

The hydrogen ion and hydroxyl ion concentrations are important in biological systems because protons are free to move from the $H_3O^+$ to associate with and thereby neutralize negatively charged groups, and hydroxyl ions are available to neutralize positively charged groups. This ability to neutralize is especially important in amino acids and proteins, which contain both carboxyl (i.e., —COOH) and amino (i.e., —NH₂) groups. In solution, amino acids normally exist in a dipolar configuration termed a *zwitterion:*

$$
\begin{array}{cc}
\text{NH}_2 & \text{NH}_3{}^+ \\
| & | \\
\text{R}-\text{C}-\text{COOH} & \text{R}-\text{C}-\text{COO}^- \\
| & | \\
\text{H} & \text{H} \\
\text{Undissociated} & \text{Zwitterion}
\end{array}
$$

Note that at a certain pH of the solution (the *isoelectric point*) the net charge of the amino acid in both undissociated and zwitterion forms is zero. In the event that the acidity of the solution is increased, the $H^+$ concentration of the solution will increase. As a result, the probability of a proton neutralizing a carboxyl group will be greater than the probability of a hydroxyl ion removing the extra proton from the amino group. A large proportion of the amino acid molecules will then bear a net positive charge:

$$NH_3^+ \qquad\qquad NH_3^+$$
$$R-\underset{H}{\overset{|}{C}}-COO^- + H^+ \rightleftharpoons R-\underset{H}{\overset{|}{C}}-COOH$$

Raising the pH will, of course, have the opposite effect. Each species of zwitterion has a characteristic isoelectric point—namely, a pH value at which it has a net charge (statistically) of zero. A number of amino acids have no amphoteric groups other than the alpha $-COOH$ and alpha $-NH_3$ groups, which enter into peptide bonds. Others, however, have additional carboxyl or amino side groups that can become acidic or basic. Dissociable side groups in a macromolecule will determine to a large extent the electrical properties of the molecule and will render it sensitive to the pH of its environment. This sensitivity is most dramatically evident in altered properties of the active site of an enzyme. Since the binding of a substrate to the active site of an enzyme generally includes electrostatic interactions, the formation of the enzyme–substrate complex is highly pH-dependent, showing highest probability at an optimum pH.

### THE HENDERSON–HASSELBALCH EQUATION

Some acids, such as HCl, dissociate completely, whereas others, such as acetic acid, dissociate only partially. For the generalized expression for the dissociation of an acid,

$$HA \rightleftharpoons H^+ + A^-$$

in which $A^-$ is the anion of the acid HA, the dissociation constant is derived from the Law of Mass Action:

$$K' = \frac{[H^+][A^-]}{[HA]} \qquad (2\text{-}2)$$

It is convenient to use the logarithmic transformation of $K'$, namely $pK'$, which is analogous to pH. Thus,

$$pK' = -\log_{10} K'$$

Hence, $pK' = 11$ means $K' = 10^{-11}$. A low $pK'$ indicates a strong acid, a high $pK'$ a weak acid.

Acid–base problems can be simplified by rearranging Equation 2-2. Taking the log of both sides, we obtain

$$\log K' = \log[H^+] + \log\frac{[A^-]}{[HA]} \qquad (2\text{-}3)$$

Rearranging gives us

$$-\log[H^+] = -\log K' + \log\frac{[A^-]}{[HA]} \qquad (2\text{-}4)$$

Substituting pH for $-\log [H^+]$ and $pK'$ for $-\log K$, we obtain

$$pH = pK' + \log\frac{[A^-]}{[HA]} \qquad (2\text{-}5)$$

Thus,

$$pH = pK' + \log\frac{[\text{proton acceptor}]}{[\text{proton donor}]}$$

This is the *Henderson–Hasselbalch equation*, which permits the calculation of the pH of a conjugate acid–base pair, given the $pK'$ and the molar ratio of the pair. Conversely, it permits the calculation of the $pK'$, given the pH of a solution of known molar ratio.

### BUFFER SYSTEMS

Because of the effect of pH on the ionization of basic and acidic groups in proteins and other biological molecules, the pH of intra- and extracellular fluids must be held within the narrow limits in which the enzyme systems have evolved. Deviations of one pH unit or more generally disrupt the orderly functioning of a living system. This sensitivity to the acidity of the aqueous intracellular milieu exists in part because reaction rates of different enzyme systems become mismatched and uncoordinated. For example, the pH of human blood is maintained at 7.4 by means of natural pH *buffers*. A buffered system is one that undergoes little change in pH over a certain pH range upon

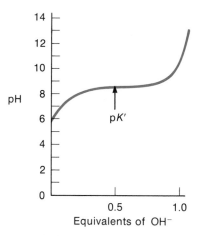

2-21    The greatest buffering capacity of a conjugate acid–base system is obtained when pH = pK. On the graph, this point corresponds to the part of the curve with the shallowest slope (small pH changes with large amounts of OH⁻ added).

addition of relatively large amounts of an acid or a base.

A buffer must contain an acid (HA) to neutralize added bases and a base (A⁻) to neutralize added acids. (We have already seen that HA is an acid because it acts as an H⁺ donor and that A⁻ is a base because it acts as an H⁺ acceptor.) The greatest buffering capacity of such a conjugate acid–base pair therefore occurs when [HA] and [A⁻] are both large and equal. Referring to the Henderson–Hasselbalch equation (Equation 2-5), we see that this situation exists when $pH = pK'$ (since $\log_{10} 1 = 0$). This is also apparent from that portion of a titration curve (Figure 2-21) along which there is relatively little change in pH.

The most effective buffer systems are combinations of weak acids and their salts. The former dissociate only slightly, thus ensuring a large reservoir of HA, whereas the latter dissociate completely, providing a large reservoir of A⁻. Added H⁺ therefore combines with A⁻ to form HA, and added OH⁻ combines with H⁺ to form HOH. As H⁺ is thereby removed, it is replaced by dissociation of HA. The most important inorganic buffer systems in the body fluids are the bicarbonates and phosphates. Amino acids, peptides, and proteins, because of their weak-acid side groups, form an important class of organic buffers in the cytoplasm and extracellular plasma.

## Electric Current in Aqueous Solutions

Water conducts electric current far more readily than do oils or other nonpolar liquids, and is therefore said to exhibit a higher *conductivity* than nonpolar liquids. Conductivity in aqueous solutions is defined as the rate of charge transfer caused by the migration of ions under a given potential. Thus, the conductivity of water depends entirely on the presence of charged atoms or molecules (ions) in solution. Electrons, which carry electric current in metals and semiconductors, play no direct role in the flow of electric current in aqueous solutions. Pure water contains the ions $H_3O^+$ and $OH^-$. Since they are present in low concentrations ($10^{-7}$ M at $25°C$), the electrical conductivity of pure water, though far higher than that of nonpolar liquids, is still relatively low. Conductivity is greatly enhanced by the addition of electrolytes, which, instead of dissolving as molecules, dissociate into cations and anions.

The role of ions in conducting electric current is illustrated in Figure 2-22, which shows two electrodes immersed in a solution of potassium chloride and connected by wires to a source of electromotive force, or emf, the two terminals marked + and −. The emf causes a current (i.e., a unidirectional displacement of positive electric charge) to flow through the electrolyte solution from one electrode to the other.

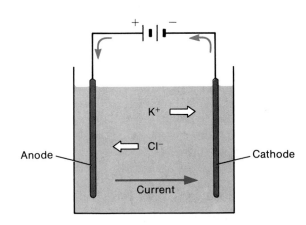

2-22    Flow of current through an electrolyte solution. Solid arrows indicate direction of current flow. Open arrows indicate direction of ion flow.

What does the electric current consist of? In the wire, it consists of the displacement of electrons from the outer shell of one metal atom to another, then to another, and so on. In the KCl solution, electric charge is carried primarily by $K^+$ and $Cl^-$ (and by displacement of $OH^-$, $H_3O^+$, and $H^+$, but this contribution is so small that it will be ignored). When a potential difference (voltage) is applied to an electrolyte solution, the cations, $K^+$, migrate toward the *cathode* (electrode with the negative potential) and the anions, $Cl^-$, migrate toward the *anode* (electrode with positive potential). The rate at which each species of ion migrates in solution is termed its electrical *mobility*. This mobility is determined by the ion's hydrated mass and the amount of charge (monovalent, divalent, or trivalent) that it bears. The mobilities of some common ions are given in Table 2-4. It is well to remember that ionic current is crudely analogous to a wave of falling dominoes, in which each domino (ion) is displaced just enough to cause a displacement of the next domino. Instead of interacting mechanically, like falling dominoes, ions influence each other through electrostatic interactions, with like charges repelling each other.

The current in a solution is said by convention to flow in the direction of cation migration. Anions flow in the opposite direction. The rate with which positive charges are displaced past a given point in the solution, plus the rate at which negative charges are displaced in the opposite direction, determines the intensity of the electric *current*. Current is a measure of the number of unit charges flowing past a point in 1 s, and

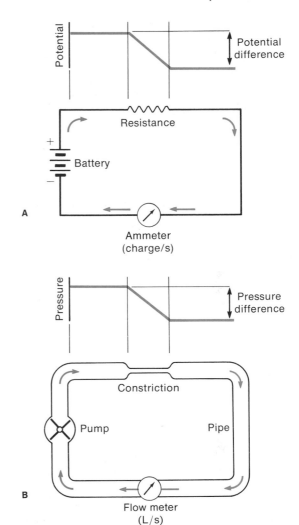

2-23  Analogy between (A) the flow of electrons in a wire and (B) the flow of water in a pipe.

**TABLE 2-4**  The electrical mobilities of some ions at 25°C extrapolated to infinite dilution.

| Ion | Mobility ($10^{-4}$ cm²/Vs) |
| --- | --- |
| $H^+$ | 36.3 |
| $Na^+$ | 5.2 |
| $K^+$ | 7.62 |
| $NH_4^+$ | 7.60 |
| $Mg^{2+}$ | 5.4 |
| $Li^+$ | 4.0 |

*Source:* Lehninger, 1970.

is thus analogous to the volume of water that flows per second past a point in a pipe (Figure 2-23B).

An electric current always meets some resistance to its flow, just as water meets a mechanical resistance owing to such factors as friction during its flow through a pipe. In order for the charges to flow through an electrical *resistance,* there must be an electrostatic force acting on the charges. This force (analogous to hydrostatic pressure in plumbing) is the difference in electric pressure, or *potential, E,* between the two ends of the

## *Box 2-1*    *Electrical Terminology and Conventions*

*Charge* ($q$) is measured in units of *coulombs* (C). To convert 1 g equivalent weight of a monovalent ion to its elemental form (or vice versa) requires a charge of 96,500 C (one *faraday*). Thus, in loose terms, a coulomb is equivalent to 1/96,500 g equivalent of electrons. The charge on one electron is $-1.6 \times 10^{-19}$ C. If this is multiplied by Avogadro's number, the total charge is one faraday or $-96,487$ coulombs/mole (C/mol).

*Current* ($I$) is the flow of charge. A current of 1 C/s is called an *ampere* (A). By convention, the direction of current flow is the direction that a positive charge moves (i.e., from the anode to the cathode).

*Voltage* ($V$ or $E$) is the electromotive force (emf) or electric potential expressed in volts. When the work required to move 1 C of charge from one point to a point of higher potential is 1 joule (J), or 1/4.184 cal, the potential difference between these points is said to be 1 *volt* (V).

*Resistance* ($R$) is the property that hinders the flow of current. The unit is the *ohm* ($\Omega$), defined as the resistance that allows exactly 1 A of current to flow when a potential drop of 1 V exists across the resistance. It is equivalent to the resistance of a column of mercury 1 mm² in cross-sectional area and 106.3 cm long. $R = \rho \times$ length/cross-sectional area.

*Resistivity* ($\rho$) is the resistance of a conductor 1 cm in length and 1 cm² in cross-sectional area.

*Conductance* ($G$) is the reciprocal of resistance, $G = 1/R$. The unit is the *siemens* (S) (formerly the *mho*).

*Conductivity* is the reciprocal of resistivity.

*Ohm's Law* states that current is proportional to voltage and inversely proportional to resistance:

$$I = \frac{E}{R} \quad \text{or} \quad E = IR$$

Thus, a potential of 1 V across a resistance of 1 $\Omega$ will result in a current of 1 A. Conversely, a current of 1 A flowing through a resistance of 1 $\Omega$ produces a potential difference across that resistance of 1 V.

*Capacitance* ($C$). A capacitor (or condenser) consists of two plates separated by an insulator. If a battery is connected in parallel with the two plates, charges will move up to one plate and away from the other until the potential difference between the plates is equal to the emf of the battery, or until the insulation breaks down. No charges move "bodily" across the insulation between the plates in an ideal capacitor, but charges of one sign accumulating on one plate electrostatically repel similar charges on the opposite plate. The *capacity,* or charge-storing ability, of the capacitor is given in *farads* (*fd*). If a potential of 1 V is applied across a capacitor and 1 C of positive charge is thereby accumulated by one plate and lost by the other plate, the capacitor is said to have a capacity of 1 fd:

$$C = \frac{q}{E} = \frac{1 \text{ C}}{1 \text{ V}} = 1 \text{ fd}$$

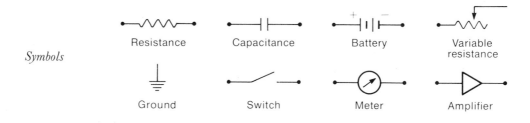

*Symbols*

Resistance    Capacitance    Battery    Variable resistance

Ground    Switch    Meter    Amplifier

resistive pathway (Figure 2-23A). A *potential,* or *voltage, difference* exists between separated negative ($-$) and positive ($+$) charges. This potential difference, or emf, is related to the current, $I$, and resistance, $R$, as described by *Ohm's Law* (Box 2-1). To force a given current (number of charges moving past a point per unit time) through a pathway of twice the resistance re-

quires twice the voltage (Figure 2-24A). Similarly, the current will be reduced to half its value if the resistance it encounters is doubled while the voltage is kept constant (Figure 2-24B).

Three major factors determine the resistance to current flow in a solution:

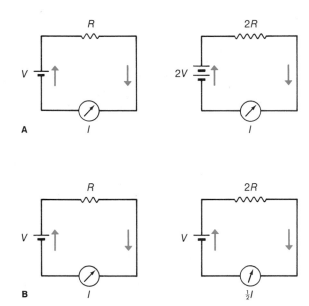

**A**

**B**

2-24  Representation of electrical relationships described by Ohm's Law. **(A)** Current remains unchanged if voltage and resistance are both doubled. **(B)** Current drops by half if resistance alone is doubled.

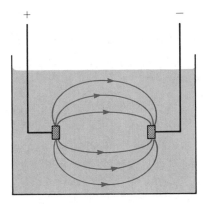

2-25  Current flow through a volume of electrolyte solution spreads so as to decrease current density.

*1.* Charge carriers are available in the solution (i.e., the ion concentration). The more dilute an electrolyte solution, the higher its resistance, and thus the lower its *conductivity* (Box 2-1). This makes sense, since fewer ions are available to carry current.

*2.* The smaller the cross-sectional area of the solution in a plane perpendicular to the direction of current flow, the higher the resistance encountered by the current. This, again, is analogous to the effect of the cross-sectional area of a pipe carrying water.

*3.* The total resistance to current flow increases with the distance traversed in solution by the current. The resistance encountered by a current passing through a distance of 2 cm of electrolyte solution is twice the resistance encountered in traversing 1 cm of the same solution.

The ions carrying current are distributed more or less evenly throughout a solution. Since the emf (i.e., voltage) applied between the two electrodes sets up a widespread electric field, the current does not follow a straight path. Instead, it arches out in curved paths between the two electrodes (Figure 2-25). It does so because more ions are brought into play than are present in a direct path between the two electrodes, thus providing a lower effective resistance to the flow of electric current.

The widespread importance of electrical phenomena in animal physiology will become apparent in later chapters. Familiarity with basic concepts of electricity is also of immense value in understanding laboratory instruments.

## Ion Selectivity

Ions free in solution inside or outside living cells will interact electrostatically with one another and with a variety of ionized or partially ionized portions of molecules such as proteins. These *ionic sites,* or *ion-binding sites,* carry electric charges. Their interactions with free inorganic ions are based on the same principles that determine ion exchange at sites on such nonbiological materials as soil particles, glass, and certain plastics. *Ion–site* interactions are highly important in certain physiological mechanisms, such as enzyme activation and ion selectivity of membrane channels and carriers, in which selective binding of ions plays a prominent role.

The energetic basis for interaction between an ion and an ion-binding site is the electrostatic attraction between the two and is identical in principle to the interactions that occur between anions and cations in free solution. Thus, a site with a negative charge or a partial negative charge (recall the partial charge on the oxygen of the water molecule) attracts cations,

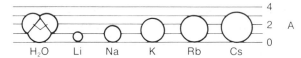

**2-26**   Scale drawings of the water molecule and unhydrated alkali-metal ions to show relative sizes. [Hille, 1972.]

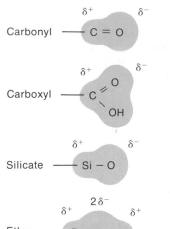

**2-27**   Electron-cloud distributions of molecular side groups.

and a site with a positive charge attracts anions. Two or more species of cations in solution will compete with each other to bind electrostatically to an anionic (i.e., electronegative) site. The negatively charged site will show an order of binding preference among cation species, ranging from those that bind most strongly to those that bind least strongly. This order to preference is termed the *affinity sequence,* or *selectivity sequence,* of the site.

The most extensively studied selectivity sequences are those of the five alkali-metal ions (lithium, sodium, potassium, rubidium, cesium). Of the 120† possible sequences for these ions, only 11 have been observed in nature. Sequence I ($Cs^+$, $Rb^+$, $K^+$, $Na^+$, $Li^+$) corresponds to the order of decreasing ionic radius (Figure 2-26), hence increasing degree of hydration in water. Sequence XI, the reverse of sequence I, corresponds to the order of increasing ionic radius, and hence decreasing degree of hydration.

Cation-binding sites on organic molecules are generally oxygen atoms in such groups as silicates ($-SiO^-$), carbonyls ($R-C=O$), carboxylates ($R-COO^-$), and ethers ($R_1-O-R_2$). As was noted earlier, the oxygen atom is strongly electron-hungry, and draws electrons from surrounding atoms in the molecule. In such neutral groups as the carbonyls or ethers, these sites can be treated as having a partial negative charge due to the statistically higher number of electrons around the oxygen atoms. (Since the group itself is neutral, there must also, of course, be partial positive charges on the other atoms, as shown in Figure 2-27.) The oxygens of silicates and carboxylates carry a full negative charge when ionized.

The energetics of electrostatic interaction of a site with an ion are expressed in terms of potential energy— namely, the energy $U$ of bringing together two charges,

$q_+$ and $q_-$, in a vacuum from a separation of infinity to the new distance of separation $d$:‡

$$U = \frac{(q_+ q_-)}{d^a} \qquad (2\text{-}6)$$

The exponent $a$, in the case of two monopoles each carrying a full charge (i.e., a monovalent anion and a monovalent cation), is equal to 1. For a dipolar molecule, such as water—that is, one in which there are centers of both negative and positive charge (but no net charge)—the energy of interaction falls off more rapidly with distance (i.e., $a$ in Equation 2-6 is greater than unity). This plays an important role in an electrostatic tug-of-war experienced by an ion dissolved in water attracted to a site of opposite charge.

Since the nucleus of a small atom can more closely approach another atom than can the nucleus of a large atom, it is evident that *small* monovalent cations can interact more strongly with an electronegative site than *large* monovalent cations. As a consequence, *in a vacuum* (i.e., without interference from water mole-

---

†$5! = 5 \times 4 \times 3 \times 2 \times 1 = 120$ possible rankings of the five cations.

‡The form of Equation 2-6 should not be confused with Coulomb's Law, in which the electrostatic force, $f$, of two monopoles varies inversely with the dielectric constant, $\varepsilon$, and the *square* of the distance, $d$, between them:

$$f = \frac{q_+ q_-}{\varepsilon d^2}$$

cules), the order of binding strength (from strongest to weakest) between an electronegative site and the alkali-metal cations should be $Li^+ > Na^+ > K^+ > Rb^+ > Cs^+$, the order of increasing atomic size. Thus, in a vacuum, the lithium ion will bind more strongly to an anionic site than will the cesium ion, because it carries the same unit charge but has a smaller *distance of closest approach.*

In an aqueous environment (i.e., in a solution as opposed to a vacuum), the coulombic relation (Equation 2-6) between atomic radius and affinity for a given electronegative fixed site is modified by the electrostatic interaction of the cation with water. This interaction is due to the dipole configuration of the water molecule. The cation, attracted to both the electron-rich oxygen atom of the fixed monopolar site and the electron-rich oxygen atom of the dipolar water molecule, is caught between the tendency to bind electrostatically to water and the tendency to bind electrostatically to the electronegative fixed site. The energetic tug-of-war between water and site will determine whether the site is more successful than the water in binding one species of ion or another. The more successfully it competes with water for a given ionic species, the greater the "selectivity" of the site for that ionic species. The selectivity sequence of a site for a group of different ions will be determined by the field strength and the polar/multipolar distribution of electrons near the site (Figure 2-28).

In addition to the principles of electrostatic interaction touched on here, there are steric constraints on the binding of ions with some sites. If, for example, a site is situated so that an interacting ion must squeeze into a narrow depression or hollow in or between mole-

cules, the hydrated size of the ion will, of course, also have an effect on the total energy required to reach and interact with the site.

## Biological Molecules

The precise molecular composition of an organism more complex than a virus has never been fully determined. This state of ignorance results largely from the incredible number and complexity of molecular species in an organism, even a unicellular one. This complexity is further compounded by the fact that no two animal species have the same molecular composition. In fact, the molecular composition of no individual of a species is identical to that of any other in the same species except in those reproduced by cellular fission (e.g., the two daughter cells of an amoeba or monozygotous mammalian twins). Such biochemical diversity is a major factor in evolution, for it provides an enormous number of variables in a population of organisms and acts as the raw material, so to speak, on which natural selection operates. This diversity is in part made possible by the great potential for structural variability exhibited by the carbon atom, with its ability to form four highly stable bonds. In fact, carbon is the "backbone" molecule for the four major classes of organic compounds found in living organisms—lipids, carbohydrates, proteins, and nucleic acids. We will review the chemical structures of these substances briefly as major classes and consider some properties important to their roles in physiology. More specialized texts, such as Lehninger's and Stryer's biochemistry books, may be consulted for further details.

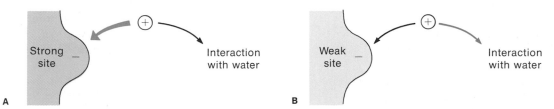

2-28   (A) The force of attraction of a small monovalent cation to a *strong* anionic point site is greater than its attraction to water (and vice versa for a large monovalent cation). (B) The force of attraction of a small monovalent cation to a *weak* point site is less than its attraction to water (and vice versa for a large monovalent cation).

## LIPIDS

Lipids are among the simplest of the biological molecules. The best-known are the fats. Each fat molecule consists of a glycerol molecule connected through ester bonds with three fatty acid chains. These fat molecules are therefore termed *triglycerides*. When they are *hydrolyzed* (i.e., digested) by the insertion of $H^+$ and $OH^-$ into the ester bonds, the fats break down into glycerol and the individual *fatty acids*, each of which contains an even number of carbon atoms (Figure 2-29). Some of the carbon atoms of the fatty acid chain may be linked by single bonds and some by double bonds, or the entire chain may be *saturated* with hydrogens, leaving no double bonds between the carbons. The degree of saturation and the length of the fatty acid chains (i.e., number of carbon atoms) determine the physical properties of the fat.

Fats with short chains and unsaturated fatty acids have low melting points (Table 2-5). These are oils or soft fats at room temperatures. Conversely, fats with long chains and saturated fatty acids are solids at room temperatures. That is why the process of hydrogenation (saturating the fatty acid chains with hydrogens and thereby breaking the double bonds) converts oily peanut butter into smooth, greasy peanut butter, and vegetable oil into Crisco. It is generally felt that saturated fatty acids are more readily converted by metabolic processes into the *steroid* cholesterol (Figure 2-30), and thus may be more likely to contribute to cardiovascular disease in humans.

Lipids act as energy stores, and are typically accumulated in the fat vacuoles of specialized adipose cells in vertebrates. Because of their low solubility in water, these energy-rich molecules can be stored in large

2-29   Hydrolysis of a triglyceride. R represents a fatty acid radical.

TABLE 2-5   Melting points of various fatty acids. Unsaturated bonds lower the melting point of a molecule, as can be seen by a comparison of saturated and unsaturated molecules of equivalent chain length.

| Fatty acid | Structure | Melting point (°C) |
|---|---|---|
| Saturated | | |
| Lauric acid | $CH_3(CH_2)_{10}COOH$ | 44 |
| Palmitic acid | $CH_3(CH_2)_{14}COOH$ | 63 |
| Arachidic acid | $CH_3(CH_2)_{18}COOH$ | 75 |
| Lignoceric acid | $CH_3(CH_2)_{22}COOH$ | 84 |
| Unsaturated | | |
| Oleic acid | $CH_3(CH_2)_7CH \overset{cis}{=\!=} CH(CH_2)_7COOH$ | 13 |
| Lineolic acid | $CH_3(CH_2)_4(CH\!=\!CHCH_2)_2(CH_2)_6COOH$ | −5 |
| Arachidonic acid | $CH_3(CH_2)_4(CH\!=\!CHCH_2)_4(CH_2)_2COOH$ | −50 |

*Source:* Dowben, 1971.

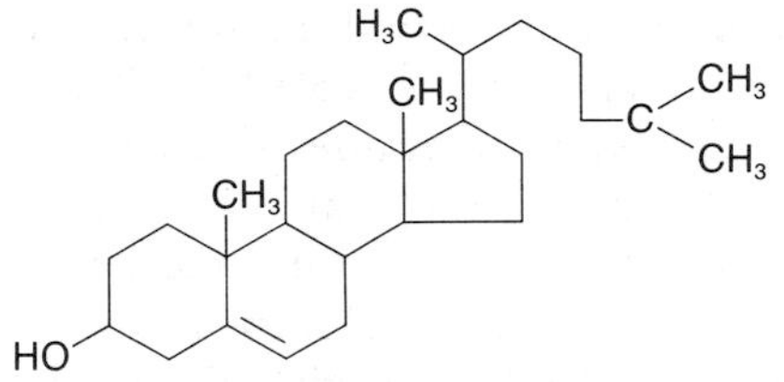

Cholesterol

2-30 Structure of the steroid cholesterol.

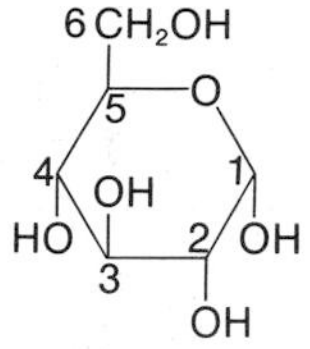

**A** A monosaccharide

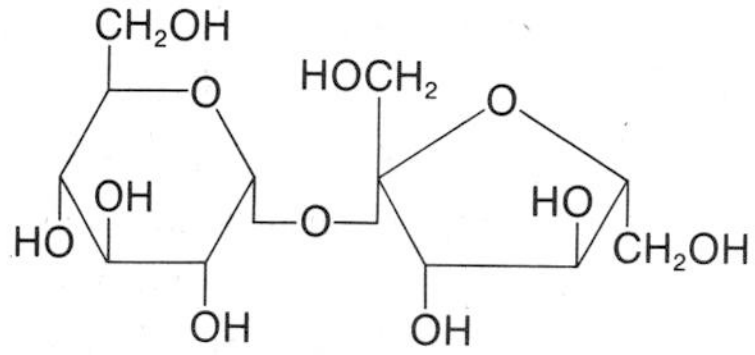

**B** A disaccharide

2-31 Structures of (**A**) a monosaccharide, glucose, and (**B**) a disaccharide, sucrose.

concentrations in the body without requiring large quantities of water as a solvent. Triglyceride energy stores are also rendered highly compact by the relatively high proportions of hydrogen and carbon and low proportions of oxygen in the molecule. Thus, 1g of triglyceride will yield about two times as much energy upon oxidation as 1g of carbohydrate (Table 2-6).

In *phospholipids* one of the outer fatty acid chains of the triglyceride is replaced with a phosphate-containing group. As is explained in Chapter 4, the phospholipids are important in the formation of biological membranes, because they exhibit *hydrophilic* (i.e., water-soluble) and *lipophilic* (i.e., lipid-soluble) properties at the head and tail ends, respectively, of their molecules. This allows a layer of oriented phospholipid molecules to form the transition between an aqueous phase and a lipid phase. Other groups of lipids are the waxes, glycolipids, sterols, and sphingolipids.

TABLE 2-6 The energy-producing capacity of the three foodstuffs.

| Substrate | Energy content (kcal/g) |
| --- | --- |
| Carbohydrates | 4.0 |
| Proteins | 4.5 |
| Fats | 9.5 |

### CARBOHYDRATES

Carbohydrate molecules have the ratio of one carbon to two hydrogens to one oxygen atom. In organic chemical terms, they are polyhydroxyl aldehydes and ketones. They include the mono- and disaccharide

*sugars* (Figure 2-31) and the polymerized sugars known as *starches* (Figure 2-32). Monosaccharides are typically ring structures containing five or six carbons, with one carbon outside the ring and an oxygen atom completing the ring. The hexose (six-carbon) sugar glucose is manufactured in green plants from $H_2O$ and $CO_2$ by the process of photosynthesis (Figure 2-31A). All the energy trapped by photosynthesis and transmitted as chemical energy to the living world (i.e., all plant and animal tissues) is channeled through such six-carbon sugars as glucose. As is noted in the next chapter, glucose is completely or partially degraded to water and $CO_2$ during cell respiration, with the release of the chemical energy that was stored in its molecular structure during photosynthesis. Cells also contain synthetic mechanisms by which glucose is modified and/or built up to form other monosaccharides or di- and polysaccharides, such as sucrose (Figure 2-31B) or starch.

Cells store carbohydrates as starches—substances that consist of polymerized D-glucose (Figure 2-32). The form found in animal cells is a highly branched polymer termed *glycogen*, in which the glucose residues are joined in a carbon 1 to carbon 4 linkage. The branches are between carbons 1 and 6. Like fats, these high-molecular-weight carbohydrate polymers require a

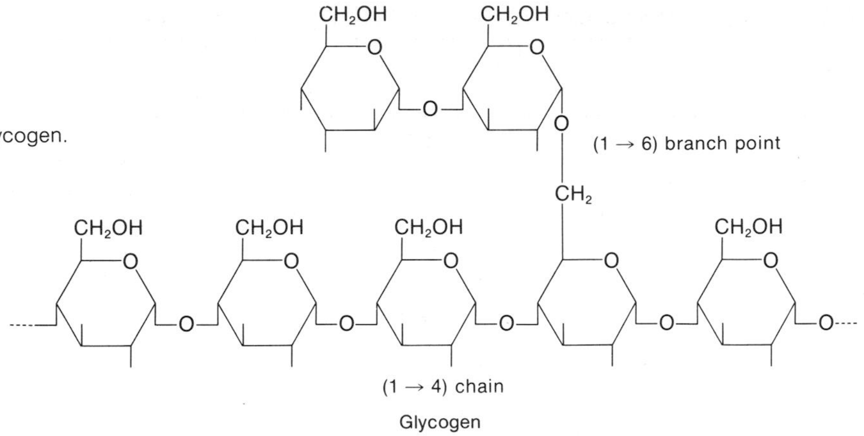

2-32 A portion of the polysaccharide glycogen. Branches occur every eight to ten glucose residues.

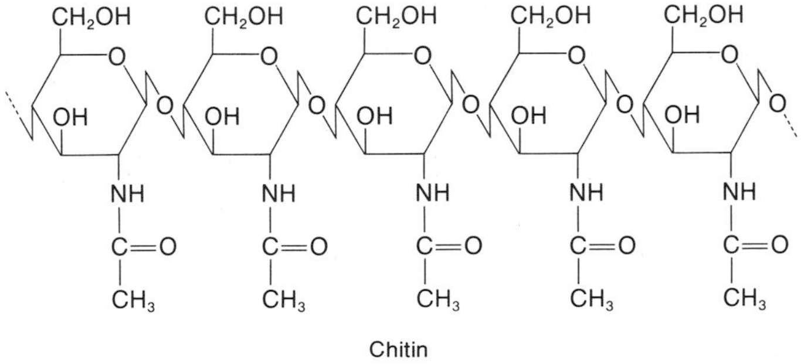

(1 → 6) branch point

(1 → 4) chain

Glycogen

2-33 Chitin, a polymer of *n*-acetyl glucosamine.

Chitin

minimum of water as a solvent, and constitute a concentrated form of food reserve in the cell. In vertebrates, glycogen is found in the form of minute intracellular granules, primarily in liver and muscle cells.

Carbohydrate polymers form structural substances.

Ribose

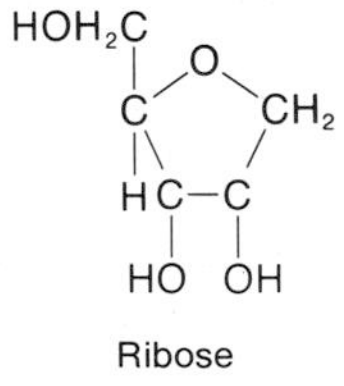

2-34 The pentose sugar ribose.

*Chitin,* which is a major constituent of the exoskeletons of insects and crustaceans, is a celluloselike polymer of an amino-acid-containing hexose termed D-glucosamine (Figure 2-33). Like the plant polymer cellulose, it is flexible and elastic and insoluble in water.

An important pentose (i.e., five-carbon) sugar is *ribose* (Figure 2-34), which occurs in the backbones of all nucleic acid molecules.

### PROTEINS

Proteins are the most complex and the most abundant organic molecules in the living cell, making up more than half the mass of a cell as measured by dry weight. Proteins are made up of linear chains of *amino acids* (Table 2-7). All the hereditary information encoded in a cell's genetic material is transmitted initially to

**TABLE 2-7**  Side groups or radicals (see Figure 2-35) of the 20 common α-amino acids.

| | | | | | |
|---|---|---|---|---|---|
| Glycine (Gly) | $-H$ | | Cysteine (Cys) | $-CH_2-SH$ | |
| Alanine (Ala) | $-CH_3$ | | Methionine (Met) | $-CH_2-CH_2-S-CH_3$ | 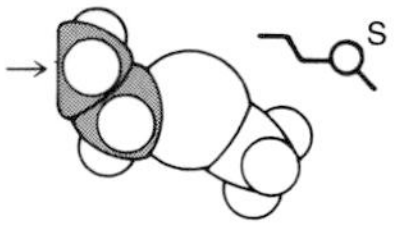 |
| Valine (Val) | $CH\begin{smallmatrix}CH_3\\CH_3\end{smallmatrix}$ | | Aspartic acid (Asp) | $-CH_2-C\begin{smallmatrix}O\\O^-\end{smallmatrix}$ | |
| Leucine (Leu) | $-CH_2-CH\begin{smallmatrix}CH_3\\CH_3\end{smallmatrix}$ | | Glutamic acid (Glu) | $-CH_2-CH_2-C\begin{smallmatrix}O\\O^-\end{smallmatrix}$ | |
| Isoleucine (Ile) | $-CH\begin{smallmatrix}CH_2-CH_3\\CH_3\end{smallmatrix}$ | | Asparagine (Asn) | $-CH_2-C\begin{smallmatrix}O\\NH_2\end{smallmatrix}$ | |
| Phenylalanine (Phe) | $-CH_2-C\begin{smallmatrix}CH=CH\\CH-CH\end{smallmatrix}CH$ | | Glutamine (Glu) | $-CH_2-CH_2-C\begin{smallmatrix}O\\NH_2\end{smallmatrix}$ | |
| Proline (Pro) | $O=C{-}CH{-}CH_2 / N{-}CH_2{-}CH_2$ | | Tyrosine (Tyr) | $-CH_2-C\begin{smallmatrix}CH=CH\\CH-CH\end{smallmatrix}C-OH$ | |
| Tryptophan (Trp) | $-CH_2-C=C$ ... indole ring ... $CH=C=CH$ / $CH=C=CH$ / $NH=CH$ | | Histidine (His) | $-CH_2-C=CH$ / $NH\ N$ / $CH$ | 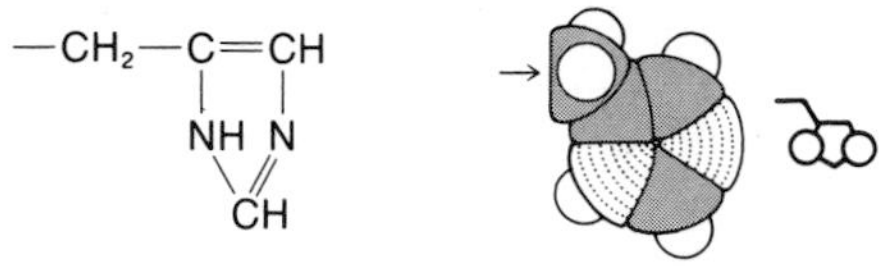 |
| Serine (Ser) | $-CH_2-OH$ | | Lysine (Lys) | $-CH_2-CH_2-CH_2-CH_2-\overset{+}{N}H_3$ | 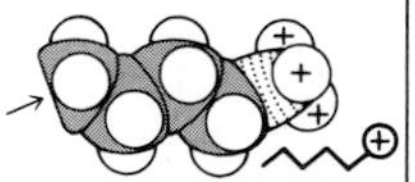 |
| Threonine (Thr) | $-CH\begin{smallmatrix}CH_3\\OH\end{smallmatrix}$ | | Arginine (Arg) | $-CH_2-CH_2-CH_2-NH_3-C\begin{smallmatrix}NH_2\\\overset{+}{N}H_2\end{smallmatrix}$ | 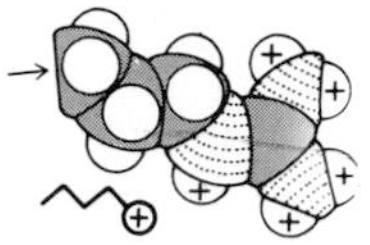 |

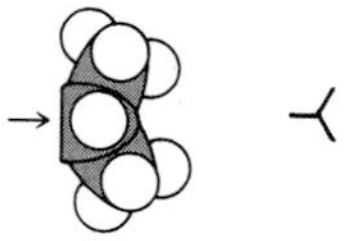

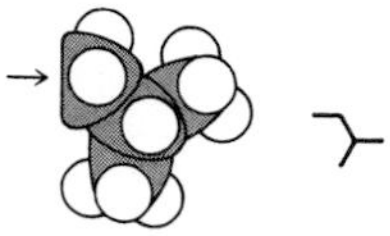

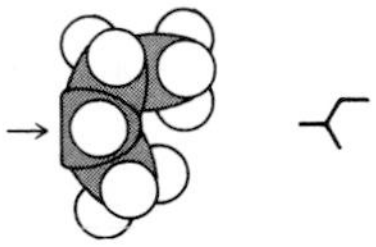

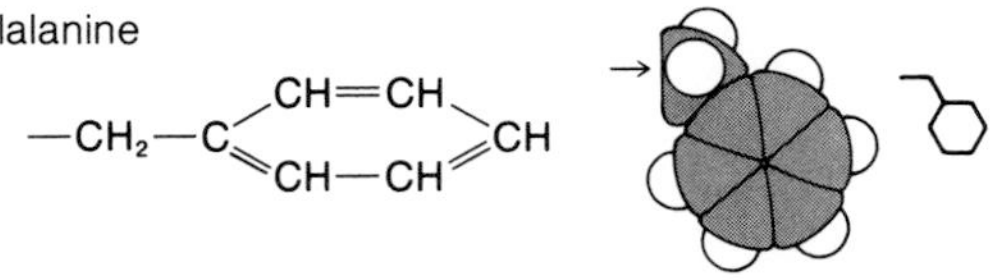

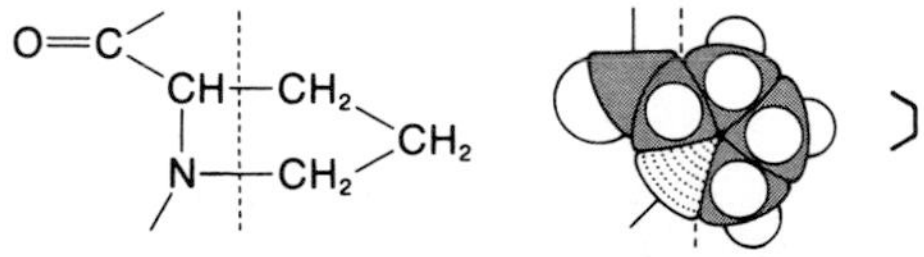

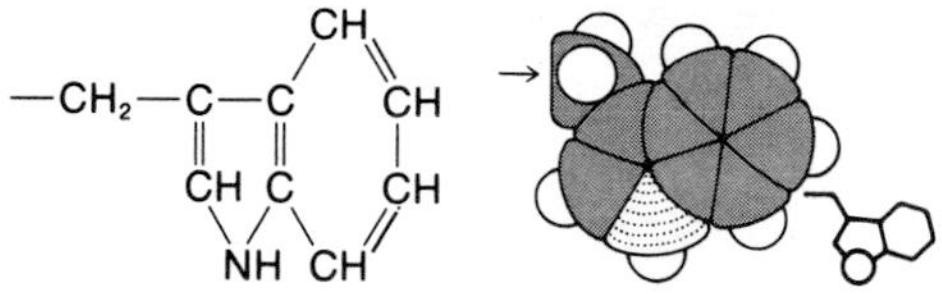

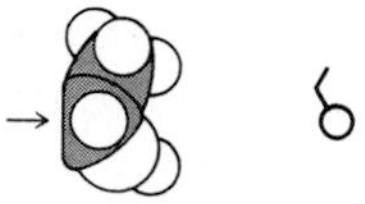

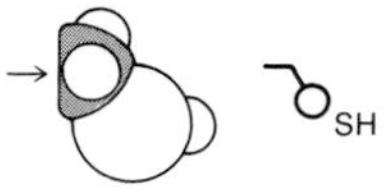

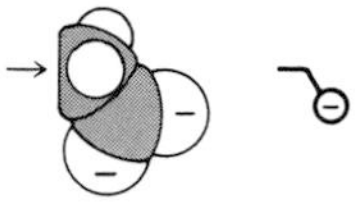

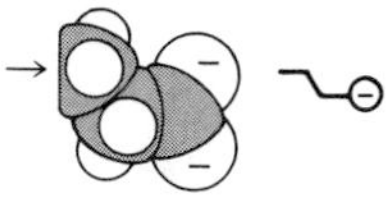

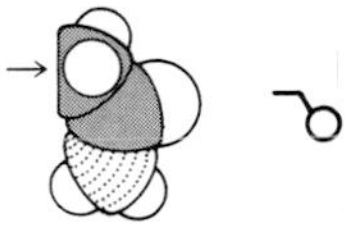

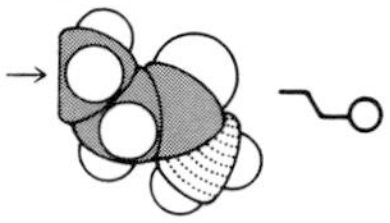

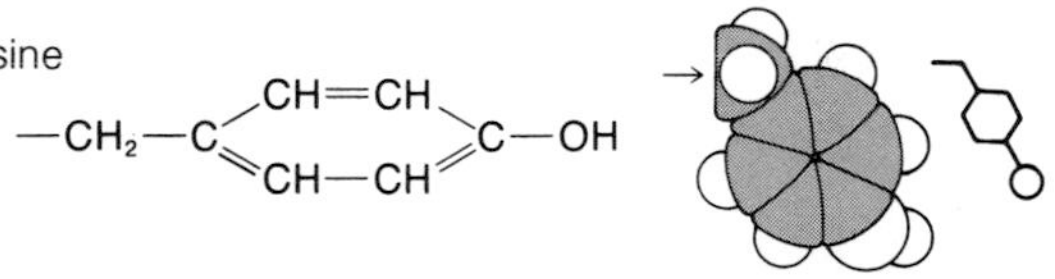

*Source:* Haggis et al., 1964.

**TABLE 2-8**  Classification of proteins according to biological function.

| Type and examples | Occurrence or function |
|---|---|
| **Enzymes** | |
|   Ribonuclease | Hydrolyzes RNA |
|   Cytochrome *c* | Transfers electrons |
|   Trypsin | Hydrolizes some peptides |
| **Regulatory proteins** | |
|   Calmodulin | Intracellular calcium-binding modulator |
|   Troponin C | Calcium-binding contraction regulator in muscle |
|   Tropomyosin | Contraction regulator in muscle |
| **Storage proteins** | |
|   Ovalbumin | Egg-white protein |
|   Casein | Milk protein |
|   Ferritin | Iron storage in spleen |
| **Transport proteins** | |
|   Hemoglobin | Transports $O_2$ in blood of vertebrates |
|   Hemocyanin | Transports $O_2$ in blood of some invertebrates |
|   Myoglobin | Transports $O_2$ in muscle |
|   Serum albumin | Transports fatty acids in blood |
| **Contractile proteins** | |
|   Myosin | Stationary filaments in myofibril |
|   Actin | Moving filaments in myofibril |
|   Dynein | Cilia and flagella |
| **Protective proteins in vertebrate blood** | |
|   Antibodies | Form complexes with foreign proteins |
|   Fibrinogen | Precursor of fibrin in blood clotting |
|   Thrombin | Component of clotting mechanism |
| **Toxins** | |
|   *Clostridium botulinum* toxin | Blocks neurotransmitter release |
|   Bungarotoxin | Agent in cobra venom that blocks neurotransmitter receptors |
| **Hormones** | |
|   Insulin | Regulates glucose metabolism |
|   Adrenocorticotrophic hormone | Regulates corticosteroid synthesis |
|   Growth hormone | Stimulates growth of bones |
| **Structural proteins** | |
|   Glycoproteins | Cell coats and walls |
|   $\alpha$-Keratin | Skin, feathers, nails, hoofs |
|   Sclerotin | Exoskeletons of insects |
|   Fibroin | Silk of cocoons, spider webs |
|   Collagen | Fibrous connective tissue (tendons, bone, cartilage) |
|   Elastin | Elastic connective tissue (ligaments) |

*Source:* Based on Lehninger, 1970.

protein molecules: the *amino acid sequence* laid down during protein synthesis is the expression of this information and is the primary determinant of the properties of any protein molecule. Since there are about 20 different amino acid building blocks, an impressive variety of different amino acid sequences is possible. Suppose, for example, that one were to construct a polypeptide molecule consisting of one of each of those 20 building blocks. How many different linear arrangements could one make without ever repeating the same sequence of amino acids? This is determined by multiplying $20 \times 19 \times 18 \times 17 \times 16 \times \cdot \cdot \cdot \times 2 \times 1$ (i.e., 20!), or $10^{18}$. Even this startling figure shrinks by comparison when one is reminded that a polypeptide containing only 20 amino acids is relatively small (molecular weight about 2400). For a more typical protein of molecular weight 35,000, containing only 12 kinds of amino acids, the number of possible sequences exceeds $10^{300}$. Thus, it is not at all surprising that the number of known enzymes (which are all proteins) exceeds 1000, with many unknown enzymes presumably still to be discovered.

Although enzymes constitute the largest functional group of proteins, there are many nonenzymatic proteins, all coded and transcribed by the same genetic mechanisms. Table 2-8 gives a classification of proteins on the basis of biological function, with a few examples.

The amino acids that make up proteins are all of the *alpha* type, since the *amino group* in each one arises from the alpha carbon atom of the molecule. The 20 common amino acids differ from one another in the structure of the side groups (Figure 2-35). The side groups are the letters in that protein alphabet, just as the four purine and pyrimidine bases are the molecular alphabet in DNA. The specific linear sequence of amino acid residues of a polypeptide is termed its *primary structure*. The *amino acid residues* of a polypeptide chain are linked to their neighbors by covalent *peptide bonds* to form a planar *amide group*. The peptide bond is formed by a condensation in which one molecule of water is removed. A protein molecule may consist of one, two, or several polypeptide chains, either covalently linked or held together by weaker bonding.

The primary structure of a polypeptide chain determines the three-dimensional conformation it will

Common to all
amino acids

$$R - \underset{\underset{H}{|}}{\overset{\overset{NH_2}{|}}{C}} - COOH$$

**A** General structure of α-amino acids

| R-group | Name of amino acid |
|---------|--------------------|
| H— | Glycine |
| $CH_6$— | Alanine |
| $HOCH_2$— | Serine |
| $\langle \rangle CH_2$— | Phenylalanine |

**B** Examples of side groups

2-35 (**A**) Generalized amino acid structure; R represents a side group (radical) of which four examples are shown (**B**). See also Table 2-7.

assume in a given environment. This structure depends on the nature and position of the side groups that project from the peptide backbone. In addition to the primary structure—that is, the amino acid sequence—there are additional levels of conformation, designated secondary, tertiary, and quaternary. *Secondary structure* refers to the conformation of the polypeptide chain; *tertiary structure* refers to the foldings of the chain to produce globular or rodlike molecules; and *quaternary structure* refers to the joining of two or more protein molecules to form dimers, trimers, and occasionally even larger aggregates.

Because of its semi-double-bond nature, the C—N peptide bond is not free to rotate; hence, the atoms of the amide group are confined to a single plane (Figure 2-36). This leaves two out of three bonds of the peptide backbone free to rotate. Linus Pauling and Robert Corey (1953), using precisely constructed atomic molecules, found that the simplest stable arrangement is a helix, as shown in Figure 2-37. In this so-called *alpha helix*, the plane of each amide group is parallel to the major axis of the helix, and there are 3.6 amino acid residues per turn. The side group of each amino acid

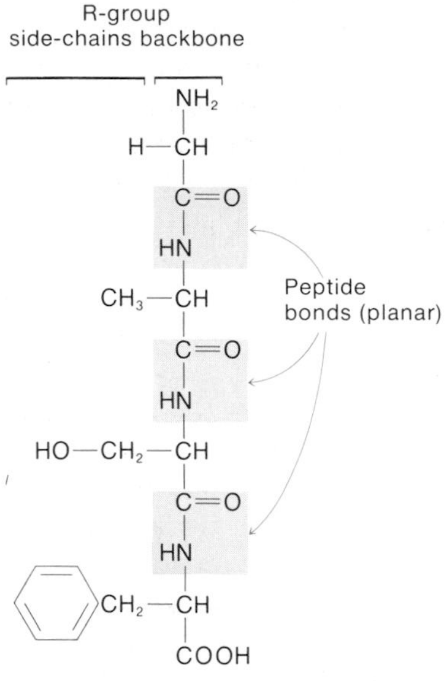

R-group
side-chains backbone

Glycylalanylserylphenylalanine

2-36  A tripeptide, showing the planar nature of the peptide bonds.

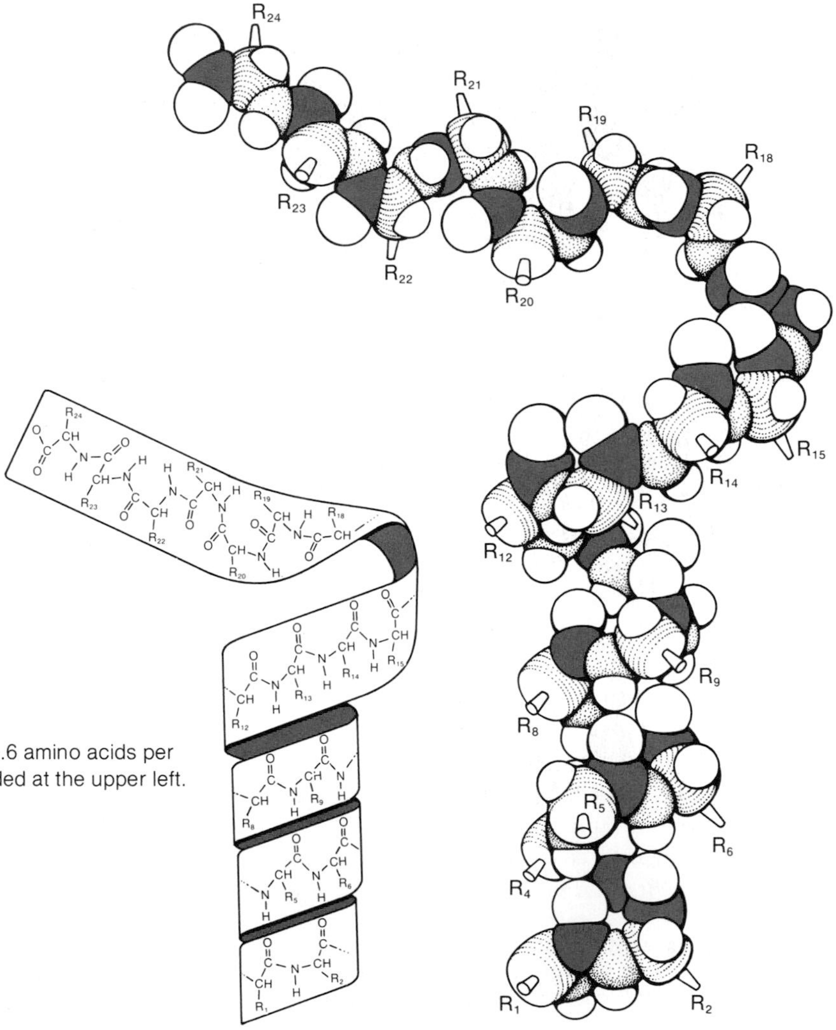

Peptide bonds (planar)

2-37  The protein α-helix, which contains 3.6 amino acids per turn. The polypeptide chain is shown extended at the upper left. [Haggis et al., 1964.]

residue extends outward from the helical backbone, free for interaction with other side groups or other molecules. The stability of the α-helix is enhanced substantially by hydrogen bonding between the oxygen of a carbonyl group and the hydrogen atom of the amide group four residues ahead (Figure 2-38). Thus, a peptide chain will assume the conformation of the α-helix spontaneously, provided that the side groups do not interfere. Long peptide chains with an uninterrupted α-helix conformation are characteristic of fibrous proteins, such as those that form hair, fingernails and claws, wool, horn, and feathers. Silk, which is produced by caterpillars, is an exception, having a secondary structure consisting of pleated sheets (*beta keratin*) rather than the α-helix.

Intracellular proteins that are not purely structural in function have a more or less globular conformation. These *globular proteins* also contain α-helices, but only in relatively short segments. In these proteins the polypeptide chain assumes a globular tertiary structure. The tertiary structure is in large part due to the properties of certain of the amino acid side groups. In the side group of the amino acid proline, for example (Figure 2-39), the N—C bond cannot rotate, even though it is not a peptide bond, because the nitrogen atom is part of a rigid ring. Thus, whenever proline (or hydroxyproline) occurs in a peptide chain, it interrupts the α helix, and the peptide backbone is bent at an angle, contributing to a globular conformation. Another factor that influences the tertiary structure of a protein is the electric charge on individual side groups and their resulting coulombic (electrostatic) interactions with other side groups. The side group of the

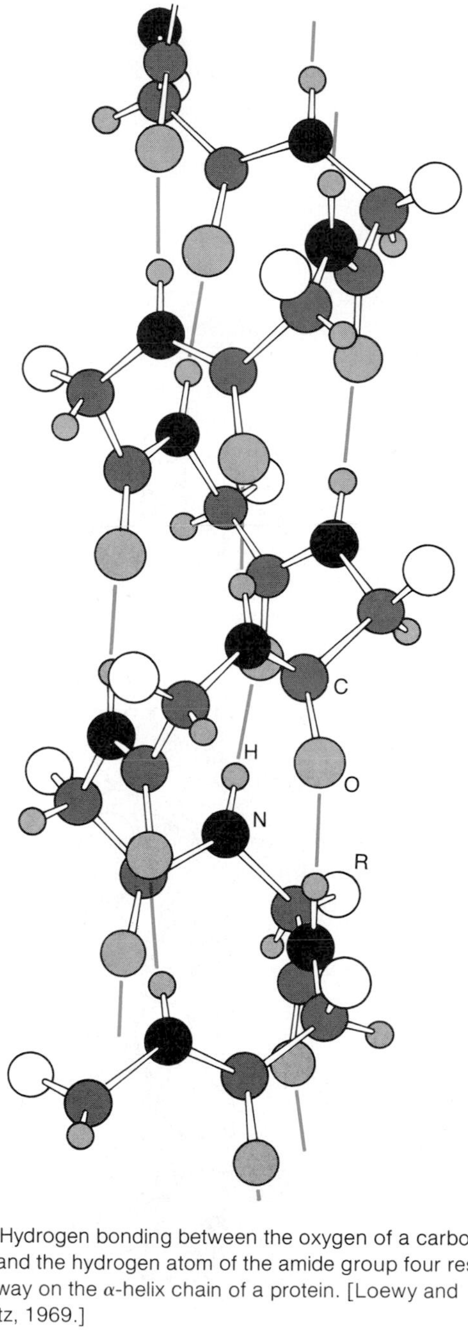

**2-38** Hydrogen bonding between the oxygen of a carbonyl group and the hydrogen atom of the amide group four residues away on the α-helix chain of a protein. [Loewy and Siekevitz, 1969.]

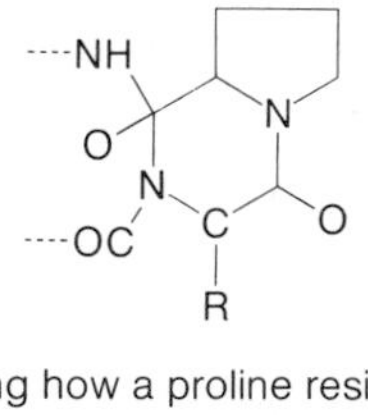

**2-39** Diagram showing how a proline residue can alter the direction of a polypeptide chain, "kinking" the α-helix formation and thereby contributing to the conformation of the protein molecule.

$$2\ SH-CH_2-\underset{\underset{NH_2}{|}}{\overset{\overset{H}{|}}{C}}-COOH \qquad \text{Cysteine}$$

$$COOH-\underset{\underset{NH_2}{|}}{\overset{\overset{H}{|}}{C}}-CH_2-S-S-CH_2-\underset{\underset{NH_2}{|}}{\overset{\overset{H}{|}}{C}}-COOH \qquad \text{Cystine}$$

Disulfide
linkage

**2-40**  Formation of a disulfide bond, important in the determination of the tertiary structure of a protein by linking together portions of polypeptide chains.

amino acid cysteine is worthy of special note. The sulfhydryl group (i.e., $-SH$) plays an important role in covalently cross-linking the protein structure (i.e., connecting two separate peptide chains or holding a chain in a folded position), because two cysteine residues can be joined by a disulfide linkage (Figure 2-40). Since the sulfhydryl group is highly reactive, it is not surprising that one or more cysteine residues frequently occupy the active sites of enzymes. The toxicity of mercury and other heavy metals is due in part to their reaction with the sulfur atom of cysteine, displacing the hydrogen atoms. This reaction can poison (i.e., render catalytically inoperative) the active site of an enzyme.

In addition to covalent $-S-S-$ bonding between cysteine residues, the tertiary structure of a protein depends directly on the conformation of certain residues, coulombic interactions, hydrogen bonding, and van der Waals forces. The last three interactions are relatively weak and heat-labile. This is why heating a protein tends to denature it—that is, cause it to unfold and alter its tertiary structure. In this way, high temperatures can render enzymes inactive and thus be fatal to living cells.

An important property of proteins is their ability to undergo *self-assembly*. The amino acid sequence of the peptide chain—and hence the positions of the various different amino acid side groups—not only determines the secondary and tertiary structure of the molecule, but also may produce sites of specific attractive interaction with certain other protein molecules, allowing several subunit molecules to assemble and to form stable quaternary complexes (Figure 2-41). These interactions can take place when the subunit molecules have complementary regions of interaction on their surfaces, so that (1) negatively charged groups of one subunit fit against positively charged groups of another subunit, and (2) hydrophobic, nonpolar side groups on the subunits meet to the mutual exclusion of water molecules. Some enzymes and other proteins are not single protein molecules, but consist of molecular subunits bound together in this way without covalent bonds. For example, the respiratory pigment hemoglobin consists of four subunits, two *alpha* and two *beta polypeptide molecules* (Figure 2-42A). These molecules will assemble themselves spontaneously if added separately to a solution and mixed. The manner in which they fit together is seen in Figure 2-42B.

**2-41**  Representation of the quaternary structure of the protein collagen, which consists of three $\alpha$-helixes, each twisted into a "superhelix." The superhelixes are held to each other by hydrogen bonding. [Dowben, 1971.]

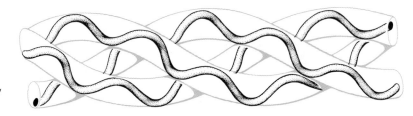

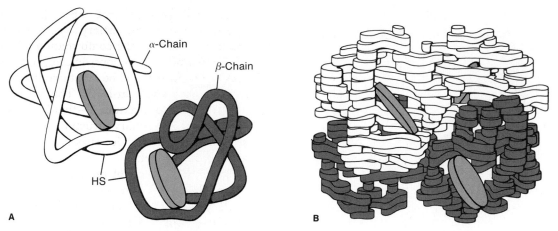

**2-42** Molecular structure of hemoglobin, which is composed of two α- and two β-chains. (A) The α- and β-chains are held together by disulfide bonds. [Dowben, 1971.] (B) Crystallographic reconstruction of the molecule. Disks represent heme groups, which bind oxygen. [From "The Hemoglobin Molecule" by M. F. Perutz. Copyright © 1964 by Scientific American, Inc. All rights reserved.]

## NUCLEIC ACIDS

*Deoxyribonucleic acid* was first isolated from white blood cells and fish sperm in 1869 by Friedrich Miescher. During the next decades the chemical composition of DNA was gradually worked out, and evidence slowly accumulated that implicated it in the mechanisms of heredity. It is now common knowledge that DNA, which is associated with the chromosomes, carries in its subunit sequence coded information that is passed from each cell to its daughter cells and from one generation of organisms to the next. It is the substance of Mendel's genes.

A second group of nucleic acids, *ribonucleic acid*

(*RNA*), was subsequently discovered; it is now known to be instrumental in translating the coded message of the DNA in sequences of amino acids in the synthesis of protein molecules.

The nucleic acids are polymers of *nucleotide* monomers, each of which consists of a *pyrimidine* or *purine* base (Figure 2-43), a *pentose* sugar, and a *phosphoric acid* residue (Figure 2-44). There are five major nucleotides: *adenine, thymine, guanine,* and *cytosine* are found in DNA, and *uracil* is found in place of thymine in RNA. The backbone of the polynucleotide chain (Figure 2-45) consists of the sugar portions of the nucleotides linked together by the phosphodiester bonds of phosphate

Adenine      Guanine      Cytosine      Uracil      Thymine

**2-43** Structures of the common purine and pyrimidine bases.

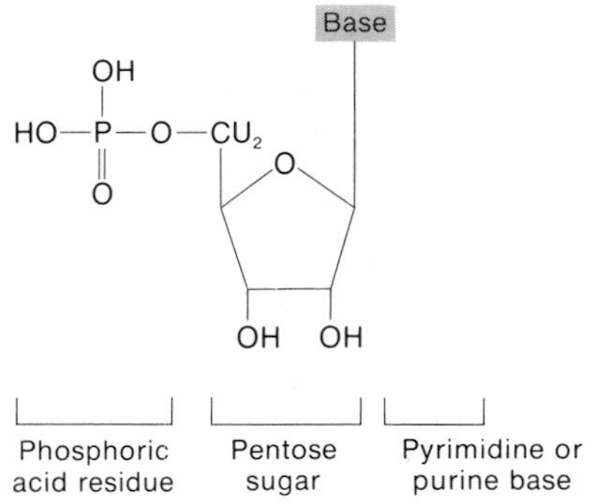

2-44   Structure of a nucleic acid.

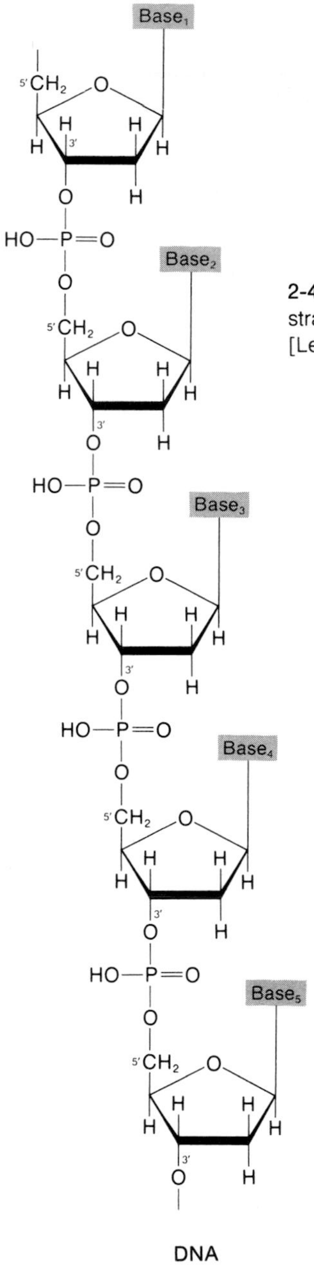

2-45   Structure of a single strand of DNA. [Lehninger, 1975.]

DNA

groups interpolated between the 3′ carbon of one pentose ring to the 5′ carbon of the next pentose. The purine and pyrimidine bases extend outward from the backbone and are not involved in the repetitive, nonvarying backbone. It is the sequence of the bases along this chain that codes the genetic information. The DNA chain (or strand) is coiled in the form of a helical staircase and is paired with a complementary strand through hydrogen bonding between thymine–adenine pairs and between cytosine–guanine pairs, as shown in Figure 2-46. The conformations and molecular dimensions are such that other base-pair combinations will not form. This allows either DNA strand, upon separating from the other, to act as a template for the formation of its complementary strand. The same principle allows a DNA strand to act as a template for *messenger RNA (mRNA)*, as shown in Figure 2-46.

After formation in the nucleus of a complementary base sequence of mRNA from the DNA template, the mRNA strand, which now contains the information for the amino acid sequence of a polypeptide chain, leaves the nucleus and enters the cytoplasm to be "read out" by a *ribosome*. In this process of transcription, certain sequences of three bases in the DNA code for certain amino acids. For example, GGU, GGC, CGA, and GGC all code for the amino acid glycine; and GCU, GCC, GCA and GCT code for the amino acid alanine. Thus, the genetic code consists of a four-letter alphabet (A, G, C, T) combined into three-letter words.

Once the primary structure (amino acid sequence) is formed, the new polypeptide chain curls and folds up by itself or together with other specific polypeptide chains to produce the characteristic secondary and tertiary structure of a protein molecule.

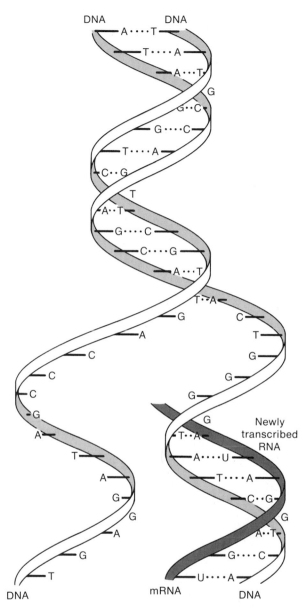

DNA          DNA

— A ···· T —
— T ···· A —
— A ··· T —
          G
        G·· C
— G ···· C —
— T ···· A —
— C ·· G
        T
— A··T —
— G ···· C —
— C ···· G —
— A ··· T —
          T··A —
        G          C —
      A          T —
    C          G —
  C          G —
  C        G —
 G        G —
A        G
T      T··A —
  A      A ···· U —
    G        T ···· A —
      G      C·· G —
        A        G
      G        A·T —
    T        G ···· C —
                U ···· A —

DNA          mRNA          DNA

Newly
transcribed
RNA

**2-46** Transcription: the formation of mRNA, complementary to the DNA.

More detailed accounts of the major steps relating protein synthesis to nucleic acids may be found in the references listed under Suggested Reading at the end of the chapter.

## Summary

Biologists generally accept the hypothesis that life on Earth arose spontaneously in shallow seas under special conditions that no longer exist. It is believed that organic molecules, synthesized in the primitive atmosphere by reactions energized by lightning discharges or radiation, accumulated in the water over long periods, providing the raw material for primordial living cells.

Living matter is composed primarily of carbon, nitrogen, oxygen, and hydrogen in stable (covalently bonded) associations. Carbon, nitrogen, and oxygen are capable of double and triple bonding, which greatly increases the structural variety of biological molecules.

The polarity of the $H_2O$ molecule is responsible for hydrogen bonding, which, besides linking hydrogen and oxygen atoms of adjacent water molecules, confers on water the many special properties that have profoundly shaped the evolution and survival of animal organisms. Water dissociates spontaneously into $H^+$ and $OH^-$. In 1 L of pure water, there is $10^{-7}$ mol of each ion. Many substances in solution contribute to an imbalance in $H^+$ and $OH^-$ concentration, giving rise to acid–base behavior (i.e., donation and acceptance of protons). Their concentrations are measured by the pH system. The pH of biological fluids influences the charges carried by amino acid side groups and hence the conformation and activity of proteins. Physiological buffering systems are necessary to maintain a narrow range of pH for the coordination of catalyzed reactions.

The electrostatic force attracting an ion to a site of opposite charge is determined by the distance of closest approach of the ion to the site. The ion selectivity of a site for different ion species depends on the relative success of the site in competing with the dipolar water molecules to bind those different ion species.

There are four major groups of organic molecules of which animal cells are composed. The lipids, which include triglycerides (fats), fatty acids, waxes, sterols, and phospholipids, are important as energy stores and as constituents of biological membranes. The carbohydrates include sugars, starches, and structural polymers such as chitin and cellulose. The sugars and starches are a major source of substrate for energy metabolism by cells. Proteins, made up of linearly arranged amino acid residues, form many structural materials, such as collagen, keratin, and subcellular fibrils and tubules. Enzymes are specialized proteins bearing catalytically active sites and are important in nearly all biological reactions. The nucleic acids DNA and RNA encode and translate the genetic information necessary for the orderly synthesis of all the protein molecules in the cell.

## Exercises

*1.* What evidence is there that the molecular building blocks of life might have arisen spontaneously on the primordial Earth?

*2.* What determines the reactivity of a given atom? Why?

*3.* What properties of C, H, O, and N make them especially well adapted for the construction of biological molecules?

*4.* Why is oxygen of such biological importance?

*5.* What important physical and chemical characteristics of $H_2O$ can be directly related to the dipole nature of the water molecule?

*6.* What is the pH of a 1 mM solution of an acid that is 10% dissociated?

*7.* Why is a weak acid rather than a strong one required for a pH buffer system?

*8.* What is the difference between molality and molarity?

*9.* How many grams does a mole of $CO_2$ weigh?

*10.* Approximately how many particles are in a 1 mM solution of NaCl?

*11.* What is the approximate boiling point of a 1 molal solution of NaCl?

*12.* Why do some liquids conduct electricity while others do not?

*13.* How many ions flow past a point (equivalents per second) at a current of 1 mA?

*14.* What are the primary factors that govern the binding of two cations, *a* and *b*, to an electronegative binding site? Write the expression that integrates these factors into a meaningful quantity.

*15.* For which does the force of attraction for a univalent ion fall off most rapidly with distance: (a) a monopolar binding site or (b) a multipolar site? Give the expression relating force and distance for each site.

*16.* (a) What determines primary protein structure? (b) Secondary? (c) Tertiary? (d) Quaternary?

*17.* What special characteristic does cysteine have that makes it a likely participant in the active sites of enzyme molecules?

*18.* Why do proteins become denatured (structurally disorganized) at elevated temperatures?

## Suggested Reading†

Baker, J. J. W., and G. E. Allen, 1974. *Matter, Energy, and Life.* 3rd ed. Reading, Mass.: Addison-Wesley.

Barry, J. M., and E. M. Barry. 1969. *An Introduction to the Structure of Biological Molecules.* Englewood Cliffs, N.J.: Prentice-Hall.

Calvin, M., and M. J. Jorgenson, eds. 1968. *Bio-organic Chemistry: Readings from Scientific American.* San Francisco: W. H. Freeman and Company.

Calvin, M., and W. A. Pryor, eds. 1973. *Organic Chemistry of Life: Readings from Scientific American.* San Francisco: W. H. Freeman and Company.

Lehninger, A. L. 1975. *Biochemistry.* 2nd ed. New York: Worth.

Oparin, A. I. 1953. *The Origin of Life.* 2nd ed. New York: Dover.

Oparin, A. I. 1974. *The Origin of Life and Evolutionary Biochemistry.* New York: Plenum.

Stryer, L. 1981. *Biochemistry.* 2nd ed. San Francisco: W. H. Freeman and Company.

Wald, G. 1954. The origin of life. *Scientific American* 191(2): 44–53.

Watson, J. D. 1965. *The Molecular Biology of the Gene.* Menlo Park, Calif.: Benjamin.

†Throughout this book, sources listed in Suggested Reading are not repeated in References Cited.

# References Cited

*Biology: An Appreciation of Life.* 1972. Del Mar, Calif.: CRM Books.

*Biology Today.* 1972. Del Mar, Calif.: CRM Books.

Dowben, R. M. 1969. *General Physiology: A Molecular Approach.* New York: Harper & Row.

Dowben, R. M. 1971. *Cell Biology.* New York: Harper & Row.

Flickinger, C. J., J. C. Brown, H. C. Kutchai, and J. W. Ogilvie. 1979. *Medical Cell Biology.* Philadelphia: Saunders.

Haggis, G. H., D. Michie, A. R. Muir, K. B. Roberts, and P. B. M. Walker. 1964. *Introduction to Molecular Biology.* London: Longman.

Hille, B. 1972. The permeability of the sodium channel to metal cations in myelinated nerve. *J. Gen. Physiol.* 59:637–658.

Lehninger, A. L. 1970. *Biochemistry.* New York: Worth.

Loewy, A. G., and P. Siekevitz. 1969. *Cell Structure and Function.* New York: Holt, Rinehart and Winston.

Mahan, B. H. 1966. *College Chemistry.* Reading, Mass.: Addison-Wesley.

Pauling, L., and R. B. Corey. 1953. Compound helical configurations of polypeptide chains: Structure of proteins of the L-keratin type. *Nature* 171:59–61.

Perutz, M. F. 1964. The hemoglobin molecule. *Scientific American* 211(5):64–76. Offprint 196.

West, E. S. 1964. *Textbook of Biophysical Chemistry.* New York: Macmillan.

# CHAPTER *3*

# *Enzymes and Energetics*

*A*nimals can be viewed as chemical machines. As in all machines, every event, no matter how trivial, is accompanied by an energy transaction. For a machine to do work, energy must be transferred from one part of the system to another, usually with the conversion of at least part of the energy from one form to another. This is true even when the parts of the system are as minute as reacting molecules.

Animals are fueled by the intake of organic food molecules, which are subsequently degraded by digestive and metabolic processes in which chemical energy inherent in their molecular structures is made available for the energetic needs of the organism. Besides the energy required for such overt activity as muscle contraction, ciliary movement, and the active transport of molecules by membranes, chemical energy is required for the synthesis of complex biological molecules from simple chemical building blocks and for the subsequent organization of these molecules into organelles, cells, tissues, organ systems, and complete organisms. Because living organisms require continual maintenance, they must take in fuel and continuously expend energy to maintain function and structure. If energy intake drops below the amount required for maintenance, the organism will consume its own energy stores. When these are exhausted, it no longer has any source of energy; hence, it cannot stave off the tendency to become disorganized, nor can it continue to perform the necessary energy-requiring functions. The result is death.

All the material and energy transactions that take place in an organism are lumped under the term *metabolism*. At the intracellular level, these transactions take place via intricate reaction sequences termed *metabolic pathways,* which in a single cell can involve thousands of different kinds of reactions. These reactions do not occur randomly, but in orderly sequences, regulated by a variety of genetic and chemical control mechanisms. Together with the organization of atoms and molecules into highly specific structures, it is *cell metabolism* that distinguishes living systems from the nonliving.

The processes of cell metabolism in animals are of two kinds:

*1.* The extraction of chemical energy from foodstuff molecules and the channeling of that energy into useful functions.

*2.* Chemical alteration and rearrangement of nutrient molecules into small precursors of other kinds of biological molecules.

An example of the first is the production of amino acids during the digestion of foodstuff proteins. The amino acids become available to the cell for the release of their chemical energy by oxidation to $CO_2$ and $H_2O$. An example of the second is the incorporation of amino acids in newly synthesized protein molecules in accord with the specifications of the genetic information of the cell. We are concerned here less with the biochemical details of cell metabolism than with the thermodynamic and chemical principles that underlie the transfer and utilization of chemical energy within the cell. Thus, we will consider the mechanisms by which chemical energy is extracted from foodstuff molecules and the manner in which it is made available for the energy-requiring processes discussed in subsequent chapters.

## Energy: Concepts and Definitions

Energy may be defined as the capacity to do work. Work, in turn, may be defined as the product of force times distance ($W = FD$); for example, when a force lifts a 1 kg weight a height of 1 m, the force is 1 kg, and the mechanical work done is 1 m-kg (Table 3-1 shows the symbols and units used in measuring the several forms of work). The energy expended to do this work (i.e., the useful energy, not including that expended in overcoming friction or expended as heat) is also 1 m-kg. Once the kilogram mass is raised to the height of 1 m, it possesses, by virtue of its position, a potential energy of 1 m-kg, which can be converted to kinetic energy (energy of movement) if the mass is allowed to drop. Thus, there are also different forms of energy. These include mechanical potential energy (e.g., a stretched spring or a lifted weight), chemical potential energy (e.g., gasoline, glucose), mechanical kinetic energy (e.g., a falling weight), thermal energy (actually kinetic energy at the molecular level), electrical energy, and radiant energy. We will be concerned in this chapter primarily with the potential energy stored in the structure of molecules—namely, *chemical energy*.

Before continuing, it will be useful to recall the First and Second Laws of Thermodynamics.

The *First Law* states that energy is neither created nor lost in the Universe. Thus, if one burns wood or coal to fuel a steam engine, this does not create new energy, but merely converts one form to another—in this example, chemical energy to thermal energy, thermal energy to mechanical energy, and mechanical energy to work.

The *Second Law* states that all the energy of the Universe will inevitably be degraded to heat and that the organization of matter will become totally randomized. In more formal terms, the Second Law states that the entropy of a closed system will progressively increase and that the amount of energy within the system capable of performing useful work will diminish. The term *entropy* refers to the degree of randomization of the system. A system that is ordered (nonrandom) contains energy in the form of its orderliness, because in becoming disordered (i.e., as a result of an increase in entropy), it can perform work. This is illustrated in Figure 3-1A, which shows gas molecules in thermal motion in a hypothetical system consisting of two compartments open to each other. Initially, the gas is confined almost entirely to compartment I, in which case the system possesses a certain degree of order. Clearly, this situation has a very small probability of occurring spontaneously if in the starting condition the gas molecules are evenly distributed between the two compartments. The gas molecules can all be forced into one compartment only by the expenditure of energy (e.g., a piston pushing the gas from one compartment to the other). As the gas is permitted to

TABLE 3-1  Various kinds of work, with symbols and units.

| Type of work | Driving force | Displacement variable |
|---|---|---|
| Expansion work | $-P$ (pressure) | Volume |
| Mechanical work | $F$ (force) | Length |
| Electrical work | $E$ (electric potential) | Electric charge |
| Surface work | $\Gamma$ (surface tension) | Surface area |
| Chemical work | $\mu$ (chemical potential) | Mole numbers |

*Source:* Dowben, 1971.

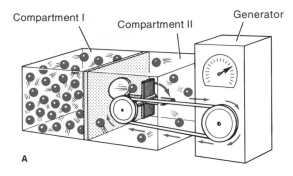

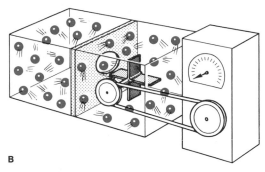

3-1 Mechanical analogy of low- and high-entropy states. In **(A)** the situation shown represents an organized, high-energy state in which nearly all the molecules are in compartment I. As they are allowed to diffuse, the molecules enter compartment II, thereby increasing the entropy and lowering the free energy of the system until equilibrium is reached **(B)**. The change from a low- to a high-entropy state releases free energy, which in this model is harnessed by the paddle wheel. The ability to do work approaches zero as the system comes into equilibrium. [Baker and Allen, 1965.]

Orderliness increases as the organism develops from a fertilized egg to the adult. It has therefore been correctly said that living systems defy the Second Law. It should be recalled, however, that the Second Law refers to a closed system (e.g., the Universe), and animals are not closed systems. Living organisms maintain a relatively low entropy at the expense of energy obtained from their environment. Thus, a rhinoceros eating, digesting, and metabolizing grass in quantities just sufficient to maintain constant weight ultimately increases the entropy of the matter it ingests (Figure 3-2). Highly ordered food molecules in the grass are eventually converted to $CO_2$, $H_2O$, and low-molecular-weight nitrogen compounds, releasing energy trapped in the organization of the larger molecules. For example, the $CO_2$ and water represent a less highly ordered state of the atoms in the carbohydrate, and the metabolic breakdown of cellulose in the grass therefore represents an increase in entropy. At the same time, the cells of the rhinoceros utilize for their own energy requirements a portion of the chemical energy originally stored in the molecular organization of foodstuff molecules.

Living systems must function at relatively uniform temperatures and pressures, for there can be only minor temperature or pressure gradients between the various parts of an organism. For this reason, biological systems can utilize only that component of the total energy capable of doing work under isothermal conditions. This component is termed the *free energy*, symbolized by the letter $G$. Changes in free energy are related to changes in heat and entropy by the equation

$$\Delta G = \Delta H - T\Delta S \qquad (3-1)$$

in which $H$ is the heat produced or taken up by the reaction (also termed the *enthalpy*), $T$ is absolute temperature, and $S$ is entropy (in units of calories/mole-K). From this equation, it is evident that in a reaction that produces no change in temperature ($\Delta H = 0$), there will be a decline in free energy (i.e., $^-\Delta G$) if there is a rise in entropy (i.e., $^+\Delta S$), and vice versa.

Since the direction of energy flow is toward increased entropy (Second Law), chemical reactions proceed spontaneously if they produce an increase in

escape from compartment I into compartment II, the entropy of the system increases (i.e., the system becomes more random). The movement of molecules from compartment I to compartment II is a form of useful energy that can be made to do work on an appropriate apparatus placed near the opening between the two compartments. Once the system is fully randomized (i.e., entropy is maximal), no further work can be extracted from the system, even though the gas molecules remain in constant thermal motion (Figure 3-1B).

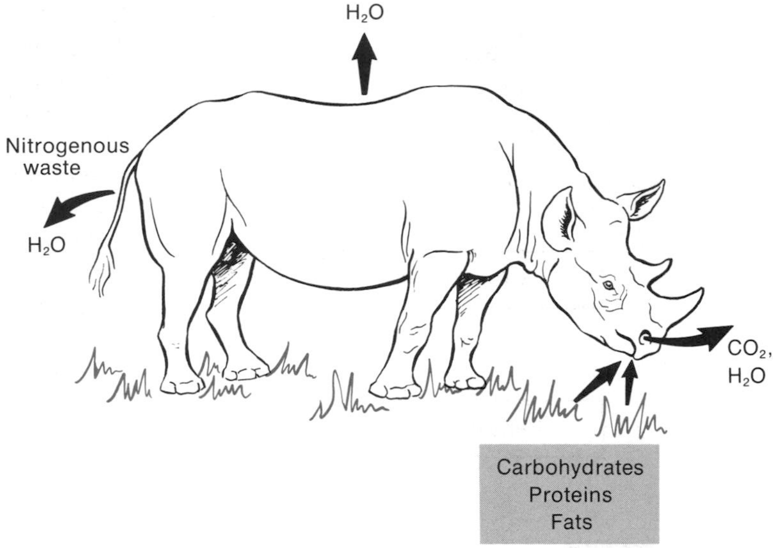

**3-2** Ingestion of food by an animal increases the entropy of the food molecules by breaking them down into smaller molecules of lower free-energy content. The free energy liberated is utilized by animal cells to drive energy-requiring reactions.

entropy (and thus a decrease in free energy). Stated otherwise, the reduction of free energy is the driving force in chemical reactions.

The inevitable trend toward increased entropy, with the inevitable degradation of useful chemical energy into useless thermal energy, requires that living systems must trap or capture new energy from time to time in order to maintain their structural and functional status quo. In fact, the ability to extract useful energy from their environment is one of the remarkable features that distinguish living systems from inanimate matter.

With the exception of chemotropic bacteria and algae, which obtain energy by the oxidation of inorganic compounds, all life on Earth depends on radiant energy from the sun. This electromagnetic energy (including visible light) has its origin in the process of nuclear fusion—a conversion of the energy of atomic structure to radiant energy. In this process, four hydrogen nuclei are fused to form one helium nucleus, with the release of an enormous amount of radiant energy. A very small fraction of this radiant energy reaches the planet Earth, and a small portion of that is absorbed by chlorophyll molecules in green plants and algae. The photically activated chlorophyll molecule

transfers this trapped energy, via activated electrons, to various electron acceptors. Eventually, these electrons become useful in the dark reactions of photosynthesis, in which the plant incorporates water and carbon dioxide to form glucose. The rearrangement of the atoms of water and carbon dioxide into carbohydrate molecules is brought about by the utilization of energy trapped by the chlorophyll. The chemical energy stored in the structure of glucose is available to the plant for controlled release during the processes of cell respiration. All animals depend ultimately on the photosynthetic process for their energy requirements, utilizing such organic compounds as carbohydrates, fats, and proteins manufactured by green plants. Herbivores (e.g., grasshoppers, cattle) obtain these energy-rich compounds by feeding directly on plant materials, whereas predators (e.g., spiders, cats) and scavengers (e.g., lobsters, vultures) obtain them at second, third, or fourth hand. The transfer of chemical energy between various *trophic levels* of the living world is diagrammed in Figure 3-3.

Later in this chapter we will consider the chemical pathways by which animal cells release energy through the oxidation of food molecules into $H_2O$ and $CO_2$. First, however, it will be useful to examine some gen-

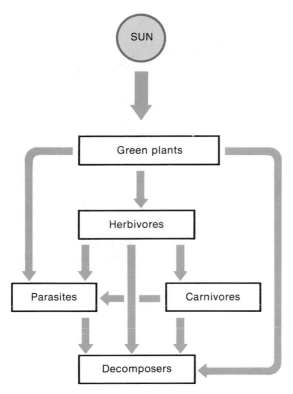

**3-3** The relation between trophic energy levels. Arrows indicate the direction of energy flow. Note the central position of green plants and herbivores. Bacterial decomposers are important in the recycling of organic matter. [Keeton, 1972.]

eral principles of energy transfer in chemical reactions, and also some features of enzymes—the biological catalysts that allow biochemical reactions to proceed rapidly at biological temperatures.

## Chemical Energy Transfer by Coupled Reactions

There are several categories of biochemical reactions, but the features of reaction rates and kinetics can be illustrated by a simple combination reaction in which two molecules react to form two new molecules:

$$A + B \rightleftharpoons C + D \qquad (3\text{-}2)$$

As the arrows indicate, this reaction is *reversible.* In theory, any chemical reaction is reversible—proceeding in either direction—provided that the products are not removed from the solution. In some reactions, however, the tendency for the reaction to go forward is so much greater than the tendency to go in reverse that for practical purposes they may be considered irreversible. A reaction tends to go forward if it liberates free energy (i.e., if the products contain less free energy than the reactants), and thus the reaction shows a negative free-energy change. In such a case, the reactants contain more potential energy than the products (Figure 3-4A), and the reaction is said to be *exergonic,* or *exothermic.* Such reactions typically liberate heat. An example is the oxidation of hydrogen:

$$2H_2 + O_2 \longrightarrow 2H_2O + \text{heat}$$

This reaction is reversed by the energy of chlorophyll-trapped quanta in the process of photosynthesis:

$$2H_2O \xrightarrow{\text{light quantum}} 2H_2 + O_2$$

This reaction, which requires the input of energy, is an example of an *endergonic,* or *endothermic,* "uphill" reaction (Figure 3-4B).

The amount of energy liberated or taken up by a reaction is related to the *equilibrium constant,* $K'_{eq}$, of the reaction. This is a constant of proportionality relating the concentrations of the products to the concentrations of the reactants when the reaction has reached equilibrium—that is, when the forward rate is equal to the reverse rate, and the concentration of reactants and products has stabilized:

$$K'_{eq} = \frac{[C][D]}{[A][B]} \qquad (3\text{-}3)$$

Here [A], [B], [C], and [D] are the equilibrium molar concentrations of the reactants and products in Equation 3-2. It is evident that the greater the tendency for the reaction in Equation 3-2 to go to the right, the higher the value of its $K'_{eq}$. This tendency depends on the difference in free energy, $\Delta G$, between the products C and D and the reactants A and B. The greater the drop in free energy, the more completely the reaction

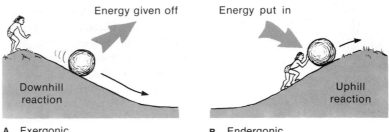

**A** Exergonic      **B** Endergonic

**3-4** **(A)** An exergonic (downhill) reaction is one in which the products have less potential energy than the reactants. **(B)** An endergonic (uphill) reaction requires energy input, as the products contain more potential energy than the reactants. [Baker and Allen, 1965.]

proceeds to the right and the higher its $K'_{eq}$. The equilibrium constant is related to the change in standard free energy, $\Delta G°$, of the system by the equation

$$\Delta G° = -RT \ln K'_{eq} \qquad (3\text{-}4)$$

It is evident from Equation 3-4 that if $K'_{eq}$ is greater than 1.0, $\Delta G°$ will be negative; and if less than 1.0, $\Delta G°$ will be positive. Exergonic reactions show a drop in free energy and hence have a negative $\Delta G°$. As a result, they occur spontaneously without the need of external energy to "drive" them. Endergonic reactions have a positive $\Delta G°$; that is, they require the input of energy from a source other than the reactants.

Some biochemical processes in living cells are exergonic and others endergonic. Exergonic processes, since they proceed on their own under the appropriate conditions, present relatively few problems in energetics. Endergonic processes, however, must be "driven." This is generally done in the cell by means of *coupled reactions,* in which a *common intermediate* serves to transfer chemical energy from a molecule of relatively high energy content to a reactant of lower energy content. As a result, the reactant is converted into a molecule of higher energy content and can undergo the required reaction by releasing some of this energy. The principle is illustrated in Figure 3-5 with compounds A, X, Y, and Z. The asterisk on A represents a group having a high energy content. This group is transferred, with the loss of some free energy, to X, transforming the latter to a high-energy common intermediate. As indi-

cated in the figure, this reaction, like Reactions 2 and 3, is exergonic. In the second reaction, the high-energy group is transferred from X* to reactant Y, again with the loss of some free energy. With its newly acquired high-energy group, reactant Y (now Y*) can react

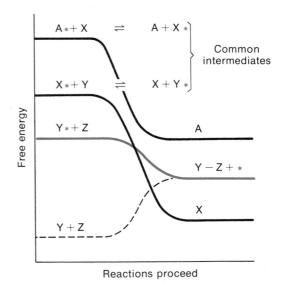

**3-5** Coupled reactions. A high-energy molecule, A*, transfers energy to a common intermediate, X, forming X*. This compound in turn reacts exergonically with Y, producing Y*, which then reacts exergonically with Z. Without the chemical energy contributed by X*, the reaction Y + Z, being endergonic, would not proceed.

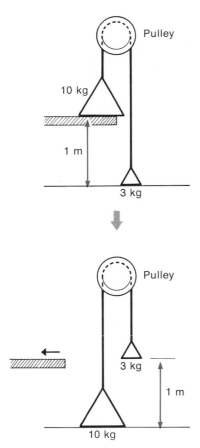

**3-6** Mechanical analogy of a coupled reaction: the fall of the 10 kg weight provides the energy required to lift the 3 kg weight.

weight, which initially had no potential energy of its own. It is evident that the falling weight can raise the other one only if it weighs more. Likewise, an exergonic reaction can "drive" an endergonic reaction only if the former liberates more free energy than the latter requires. As a consequence, some energy is lost, and the efficiency is, of necessity, less than 100%.

## ATP and the High-Energy Phosphate Group

The most ubiquitous energy-rich common intermediate is the nucleotide *adenosine triphosphate* (*ATP*), which can donate its terminal energy-rich phosphate group to any of a large number of organic acceptor molecules (e.g., sugars, amino acids, nucleotides). The *phosphorylation* raises the free-energy level of the acceptor molecule, allowing it to react exergonically in enzyme-catalyzed biochemical reactions.

The ATP molecule consists of an adenosine group, made up of the pyrimidine base adenine and the five-carbon sugar residue ribose, plus a triphosphate group (Figure 3-7). Much of the free energy of the molecule resides in the mutual electrostatic repulsion of the three phosphate units, with their positively charged phosphorus atoms and negatively charged oxygen atoms. The mutual repulsion of these phosphate units is analogous to the repulsion of bar magnets (Figure

spontaneously with Z to form the compound YZ, which has a higher free-energy content than Y + Z, but a lower energy content than Y* + Z. Thus, the chemical energy inherent in the A* molecule is utilized with the help of the common intermediates X* and Y* for the synthesis of YZ from Y and Z.

A mechanical analogy of a coupled reaction is seen in Figure 3-6. The 10 kg weight on the left can lose its potential energy of 10 m-kg by dropping a distance of 1 m, in which case it will lift the 3 kg weight on the right the same distance. Because the two weights are connected with a rope over a pulley, the fall of the 10 kg weight is coupled to the rise of the 3 kg

**3-7** Structural formula for ATP. The wavy bond symbols represent the structurally stored energy of the terminal and subterminal phosphate groups. [Baker and Allen, 1965.]

3-8), with their north and south poles aligned, held together by a sticky wax. If the wax, which is analogous to the oxygen–phosphorus bonds in ATP, is softened by warming, the energy stored by virtue of the proximity of the mutually repelling magnets is released as the magnets spring apart. Likewise, the breaking of the bonds between the phosphate units of ATP results in the release of free energy. Once the terminal phosphate group is removed by hydrolysis (Figure 3-9), the mutual repulsion of the two products, *adenosine diphosphate (ADP)* and inorganic phosphate ($P_i$), is such that the probability of their recombining is very low. That is, their recombination is highly endergonic. As Table 3-2 shows, the standard free energy for the

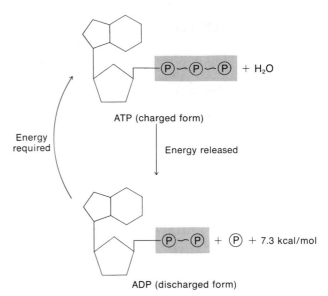

3-9 Hydrolysis of ATP to ADP and $P_i$ releases energy stored in the electrostatic repulsion between the terminal and subterminal phosphate groups, liberating about 7.3 kcal of free energy per mole of ATP. This reaction is conveniently monitored by measuring the concentration of inorganic phosphate.

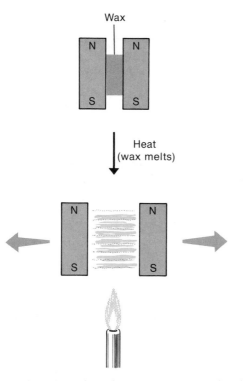

3-8 Magnetic analogy of the phosphate high-energy bond. Energy is stored in pushing the magnets together against bonding wax. When the wax melts, the magnets fly apart, releasing the energy. In this analogy, the flame supplies the activation energy for melting the wax.

TABLE 3-2 Standard free energy of hydrolysis of some phosphorylated compounds.

| | $\Delta G°$ (kcal) | Phosphate-group transfer potential† |
|---|---|---|
| Phosphoenolpyruvate | −14.8 | 14.8 |
| 3-Phosphoglyceroyl phosphate | −11.8 | 11.8 |
| Phosphocreatine | −10.3 | 10.3 |
| Acetyl phosphate | −10.1 | 10.1 |
| Phosphoarginine | −7.7 | 7.7 |
| ATP | −7.3 | 7.3 |
| Glucose 1-phosphate | −5.0 | 5.0 |
| Fructose 6-phosphate | −3.8 | 3.8 |
| Glucose 6-phosphate | −3.3 | 3.3 |
| Glycerol 1-phosphate | −2.2 | 2.2 |

†Defined as $-\Delta G° \times 10^{-3}$, where $G°$ is the standard free energy at pH 7.0.
*Source:* Lehninger, 1975.

hydrolysis of ATP under standard conditions is −7.3 kcal.

The role of ATP in driving otherwise endergonic reactions by means of coupled reactions is illustrated by the condensation of the two compounds X and Y.

$$X + ATP \rightleftharpoons X—phosphate + ADP$$

$$\Delta G° = -3.0 \text{ kcal/mol}$$

$$X—phosphate + Y \rightleftharpoons XY + P_i$$

$$\Delta G° = -2.3 \text{ kcal/mol}$$

The total free energy liberated in these two reactions (−5.3 kcal) will be equal to the sum of free energy changes of the two parent reactions:

$$ATP + HOH \rightleftharpoons ADP + P_i$$

$$\Delta G° = -7.3 \text{ kcal/mol}$$

$$X + Y \rightleftharpoons XY \qquad \Delta G° = \underline{+2.0 \text{ kcal/mol}}$$

$$-5.3 \text{ kcal/mol}$$

Although ATP and other nucleoside triphosphates, such as GTP, are responsible for the transfer of energy in many coupled reactions, it should be stressed that the mechanism of a common intermediate is widely employed in biochemical reaction sequences. Thus, portions of molecules—and even atoms, such as hydrogen—are transferred, along with chemical energy, from one molecule to another by common intermediates in consecutive reactions. The high-energy nucleotides are special only in that they act as *general energy currency* in a large number of energy-requiring reactions. In this role, ADP is the "discharged" form, and ATP is the "charged" form (Figure 3-9). As we shall see later on, there are various mechanisms in the biochemical machinery of the cell for channeling chemical energy into the formation of ATP. The ADP–ATP system is used to channel chemical energy from those phosphate compounds having higher free energies of hydrolysis than ATP (Table 3-2).

*Arginine phosphate* and *creatine phosphate* act as special reservoirs of chemical energy for the rapid phosphorylation of ADP to reconstitute ATP during vigorous muscle contraction. They are termed *phosphagens*. In

TABLE 3-3  Distribution of the two major phosphagens in the animal kingdom.

|  | Arginine phosphate | Creatine phosphate |
|---|---|---|
| Ciliata | + | − |
| Platyhelminthes | + | − |
| Annelida | − | + |
| Arthropoda | + | − |
| Mollusca | | |
| Lamellibranchiata | + | − |
| Cephalopoda | + | − |
| Echinodermata | | |
| Crinoidea | + | − |
| Asteroidea | + | − |
| Holothuroidea | + | − |
| Echinoidea | + | + |
| Ophiuroidea | − | + |
| Protochordata | | |
| Tunicata | − | + |
| Enteropneusta | + | + |
| Cephalochorda | − | + |
| Vertebrata (all classes) | − | + |

*Source:* Baldwin, 1964.

vertebrate muscle and in some invertebrates, the transphosphorylation takes the form

$$creatine\ phosphate + ADP \underset{enzyme}{\overset{transphosphorylase}{\rightleftharpoons}}$$

$$creatine + ATP$$

$$\Delta G° = 3.0 \text{ kcal/mol}$$

Creatine phosphate alone is found in the vertebrates, whereas arginine phosphate is distributed widely in invertebrate muscle (Table 3-3).

## Temperature and Reaction Rates

The rate at which a chemical reaction proceeds depends on the temperature. This is not surprising, since temperature is an expression of molecular motion. As temperature increases, so does the average molecular velocity. This greater velocity increases the number of collisions per unit time and thereby increases the probability of successful interaction of the reactant molecules. Furthermore, as their velocities increase, the molecules possess higher kinetic energies and thus are more likely to react on collision. The kinetic energy required to cause two colliding molecules to react is termed the free energy of activation, or *activation energy.* It is measured as the number of calories required to bring all the molecules in a mole of reactant at a given temperature to a reactive (or *activated*) state.

The requirement for activation applies to exothermic as well as endothermic reactions. Although a reaction may have the potential for liberating free energy, it will not proceed unless the reactant molecules possess the necessary energy. This situation can be compared to one in which it is necessary to push a boulder over a low ridge before it is free to roll down a large hill (Figure 3-10).

The relation between free energy and the progress of a reaction is shown in Figure 3-11. The reactants must first be raised to an energy state sufficient to activate them, allowing them to react. Since the reaction yields free energy, the energy state of the products is lower than that of the reactants. Note that the overall free-energy change of the reaction is independent of the activation energy required to produce the reaction.

In many industrial processes, both the reaction rate and the energy (i.e., the temperature) required to acti-

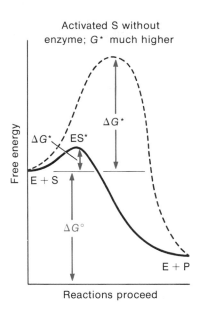

3-11   The activation energy, $\Delta G^{*}$, of a reaction is lowered by the catalytic action of an enzyme. Note that the overall free-energy change, $\Delta G^{\circ}$, is the same with or without the enzyme. E, enzyme; S, substrate; ES, activated enzyme–substrate complex.

vate reactants are significantly reduced by the use of *catalysts*—substances that are neither consumed nor altered by a reaction, but that offer the advantage of facilitating the interaction of the reactant particles. Reactions in the living cell are similarly aided by biological catalysts called *enzymes.* Figure 3-11 shows the effect of an enzyme lowering the activation energy of a reaction. Note that the presence of the enzyme has no effect on the overall free-energy change (and hence the equilibrium constant) of the reaction; it merely increases the rate of reaction.

The increase in reaction rates produced by enzymes is extremely useful biologically because it allows reactions that would otherwise proceed at imperceptible rates to proceed at useful rates at biologically tolerable temperatures. Figure 3-12A shows a normal distribution of particle velocities at a given temperature. Those molecules possessing the highest velocities (and hence kinetic energies) have sufficient energy to react (shaded area). The effect of adding a catalyst is shown in Figure 3-12B. The energy required for activation is reduced, so that a far larger number of molecules can react in a

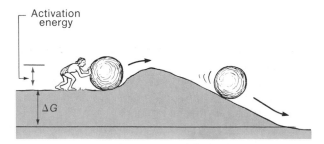

3-10   Activation energy is required to bring the reactants into position to interact. In this analogy, the potential energy of the rock cannot be liberated until some energy is expended to bring it into position at the crest of the hill.

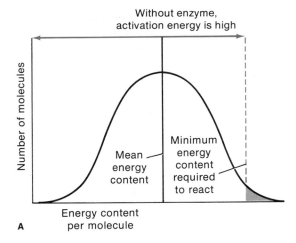

**Without enzyme, activation energy is high**

Number of molecules

Mean energy content

Minimum energy content required to react

Energy content per molecule

**A**

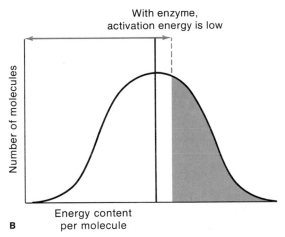

**With enzyme, activation energy is low**

Number of molecules

Energy content per molecule

**B**

**3-12** The energy distribution in a population of substrate molecules is illustrated by the bell-shaped curves. **(A)** Without enzyme, far fewer molecules (shaded area) have the required activation energy. **(B)** By lowering the activation energy, an enzyme makes possible a large increase in the fraction of substrate molecules that have sufficient kinetic energy for the reaction to proceed. [Reprinted by permission from Albert L. Lehninger, *Bioenergetics,* Second Edition, copyright © 1971, 1965 by W. A. Benjamin, Inc., Menlo Park, California.]

Another important advantage of catalyzed reactions is the possibility for regulating the rate of reaction by varying the concentration of catalyst. For example, when hydrogen and oxygen are burned noncatalytically, they explode in an uncontrolled manner because the heat released by the rapid combustion of the $H_2$ produces a rapid ignition of the remaining unburned $H_2$. When hydrogen is oxidized slowly at a low temperature with small quantities of the catalytic agent platinum, the release of heat is slowed enough so that no explosion occurs. The quantity of platinum relative to the fuel ($H_2$) and oxidant ($O_2$) regulates the rate of combustion. Likewise, most biological reactions are regulated by the quantity or the *activity* (i.e., catalytic effectiveness) of certain enzymes.

## Enzymes

Biological catalysts were first isolated from living cells in 1897 by the brothers Eduard and Hans Buchner by water extraction of yeast. These substances, which increase the rate of alcoholic fermentation, were found to be inactivated by heat, whereas the substrates were unaffected by heat. This finding was the first indication that enzymes are protein molecules. It was subsequently found that, without exception, each species of enzyme molecule is a protein of very specific amino acid composition and sequence. All of these proteins, or at least their enzymatically active portions, have a globular conformation (Figure 3-13). Each cell contains literally thousands of species of enzyme molecules, which catalyze just as many reactions. The work of molecular geneticists indicates that enzymes are the primary gene products, regulating all the synthetic and metabolic activities of the cell. By specifying the structure of each enzyme molecule that is produced, the genetic apparatus is indirectly responsible for all enzymatic reactions in a cell.

The catalytic effectiveness (i.e., activity) of an enzyme is measured in *turnover number,* which is the number of molecules of substrate per second with which 1 mol of the enzyme reacts to produce the product molecule(s). As we will discuss shortly, many enzymes require a cofactor—an ion or small molecule that combines with the protein molecule to form the active enzyme complex. If the cofactor is present in limiting

given time at the same temperature. The degrees of acceleration of reaction rates achieved by various enzymes range from $10^8$ to $10^{20}$ (!) times the rates of the uncatalyzed reactions.

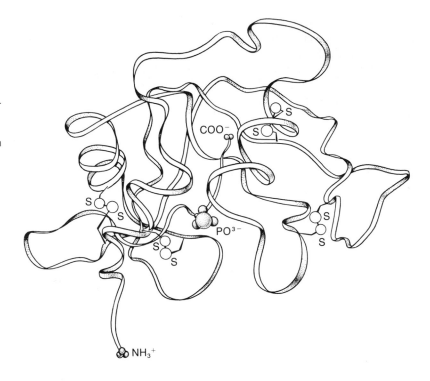

**3-13** The globular structure characteristic of enzymes is illustrated by this reconstruction of ribonuclease obtained from an ox. The phosphate ion is shown bound to the active site. [Barry and Barry, 1969.]

concentrations within the cell, the activity of the enzyme can be regulated by alterations in the concentration of the cofactor. In contrast, some enzymes do not become catalytically effective until an inhibitor molecule is removed from them.

### ENZYME SPECIFICITY

Each enzyme is, to some degree, specific for a certain *substrate* (reactant molecule). Some enzymes act at certain types of bonds and may therefore act on many different substrate molecules having such bonds. For example, the reaction catalyzed by *proteolytic enzymes* is the hydrolysis of peptide bonds. The proteolytic enzyme trypsin, found in the digestive tract, catalyzes the hydrolysis of any peptide bond in which the carbonyl group is part of an arginine or lysine residue, regardless of the position of those bonds in the polypeptide chain of a protein (Figure 3-14). Another intestinal proteolytic enzyme, carboxypeptidase, specifically catalyzes the hydrolysis of the peptide bond joining the terminal and subterminal amino acid residues in a polypeptide chain. Most enzymes are far more spe-

cific for their substrates. For example, the enzyme sucrase will catalyze only the hydrolysis of sucrose into glucose and fructose. Other disaccharides, such as lactose or maltose, are not attacked by that enzyme but are hydrolyzed by enzymes specific for them (lactase

**3-14** Specificity of the enzyme trypsin, which selectively hydrolyzes the C—N bond of the peptide linkage between two lysines, two arginines, or between a lysine and an arginine residue. [Baker and Allen, 1965.]

and maltase). Many enzymes differentiate between optical isomers—molecules that are chemically and structurally identical except that one is the mirror image of the other. For example, the enzyme L-amino oxidase catalyzes the oxidation of the L-isomer of an α-keto acid but is totally ineffective for the D-isomer of the molecule.

## CATALYTIC ACTIVITY

The steric specificity just noted is in agreement with the concept that the substrate molecule "fits" a special portion of the enzyme surface termed the *active site*. The enzyme molecule is made up of one or more peptide chains folded about so as to form the tertiary structure of a more or less globular protein of a specific conformation. The active site is thought to consist of the side groups of certain amino acid residues, which, in the tertiary structure, are brought into a conformation to which the substrate attaches through such attractive forces as electrostatic bonding, van der Waals forces, and hydrogen bonding, selectively fitting the active site.

The steric specificity of the active site has been well established by testing the reactivity of chemical analogs of substrate molecules (i.e., molecules similar to but slightly different from the substrate molecule). The active site becomes increasingly less effective as the analog molecules depart further from the optimum in terms of interatomic distances, number and position of charged groups, and bond angles.

The mechanisms of enzyme catalysis—the mechanics of how enzymes accelerate reaction rates—are based on organic reaction mechanisms. Lehninger (1975) lists four major factors that are believed to contribute to the enormous acceleration of reaction rates by enzymes:

*1.* Some enzymes may hold the reactants so that, with the appropriate orientation, the reacting groups are sufficiently close to one another to enhance the probability of reaction.

*2.* An enzyme may react with the substrate molecule to form an unstable intermediate that then readily undergoes a second reaction, forming the final products.

*3.* Groups within the enzyme may act as proton donors or acceptors to bring about general-acid or general-base reactions.

*4.* The binding of the enzyme to the substrate might cause it to undergo internal strain in the susceptible bond, increasing its probability of breaking.

In any event, formation of an *enzyme–substrate complex (ES)* increases the probability of reaction. On completion of the reaction, the enzyme and products separate, and the enzyme molecule is free to form a complex with a new substrate molecule. Since the ES persists for a finite time, all the enzyme can become tied up as ES if the substrate concentration is high enough relative to the enzyme concentration.

## TEMPERATURE AND REACTION RATES

Any factor that influences the conformation of the protein molecule, and hence the arrangement of amino acid side groups in the active site, will alter the activity of the enzyme. Temperature is one such factor. An increase in temperature increases the probability of denaturation (disruption of the tertiary structure of peptide chains), thereby rendering increasing numbers of enzyme molecules inactive. For this reason, enzyme-catalyzed reactions give a characteristic curve of reaction rate versus temperature (Figure 3-15). As temperature increases, the initial increase in reaction rate is due to the increased kinetic energy of substrate molecules. As temperature increases further, the rate

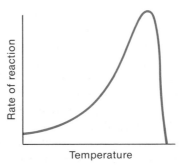

**3-15** Rate of an enzymatic reaction as a function of temperature.

of enzyme inactivation also increases owing to unfolding of the protein as hydrogen bonds and other bonds are weakened. At a certain temperature (the *optimum temperature*), the rate of enzyme destruction by heat just offsets the increase in enzyme–substrate reactivity, and the two effects of elevated temperature cancel. At that temperature, the reaction rate is maximal. At higher temperatures, enzyme destruction becomes dominant, and the rate of reaction rapidly decreases. The temperature sensitivity of enzymes and other protein molecules contributes to the lethal effects of excessive temperatures.

*pH Sensitivity*

Electrostatic bonds often participate in the formation of an ES. Since $H^+$ and $OH^-$ can act as counterions for electrostatic sites, a drop in pH exposes more positive sites for interaction with negative groups on a substrate molecule; conversely, a rise in pH facilitates the binding of positive groups to negative sites on the enzyme. Thus, it is not surprising that the activity of an enzyme will vary with the pH of the medium (Figure 3-16) and that each enzyme has an optimum pH range.

### MODULATION OF ENZYME ACTIVITY

Certain enzymes are subject to regulation of their activity by *modulator* or *regulator molecules,* which interact with a part of the enzyme molecule different from the active site. This part of the enzyme, termed the *allosteric site* when it combines with the modulator molecule, causes a change in the tertiary structure of the enzyme, changing the conformation of the active site (Figure 3-17) and thereby decreasing (or in some cases increasing) the affinity between the enzyme and its substrate. Allosteric enzymes operate at key points in metabolic pathways, and modulations of their activities play an important role in the regulation of these pathways.

*Cofactors*

As noted earlier, some enzymes require the participation of small molecules called *cofactors* in order to perform their catalytic function. In that case, the protein moiety is termed the *apoenzyme.* One class of cofactors consists of small organic molecules termed *coenzymes,* which *activate* their apoenzymes by accepting hydrogen atoms or protons from the ES. For example, the enzyme dehydrogenase requires the cofactor *NAD* (*nicotinamide adenine dinucleotide*) to accept hydrogen atoms from glutamate:

$$\text{glutamate} + \text{NAD}_{ox} \rightleftharpoons$$

$$\alpha\text{-ketoglutarate} + \text{NAD}_{red} + \text{NH}_3$$

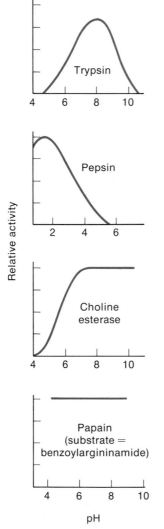

**3-16**   The effect of pH on the activity of several enzymes. [Lehninger, 1975.]

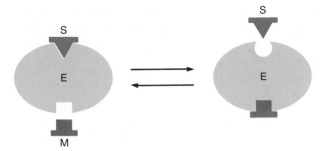

**A** Allosteric inhibition

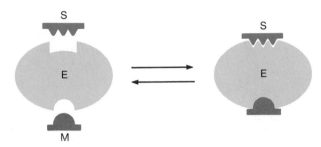

**B** Allosteric activation

**3-17** (A) An allosteric modulator molecule can indirectly alter the configuration of the active site of an enzyme, thereby rendering the enzyme inactive. Noncompetitive inhibitors act by this mechanism. (B) Conversely, an enzymatic site may be activated sterically by an allosteric modulator. S, substrate; E, enzyme; M, modulator.

**TABLE 3-4** Some enzymes and enzyme modulators that require or contain metal ions as cofactors.

$Zn^{2+}$
  Alcohol dehydrogenase
  Carbonic anhydrase
  Carboxypeptidase
$Ca^{2+}$
  Protein kinase
  Troponin

$Mg^{2+}$
  Phosphohydrolases
  Phosphotransferases
$Mn^{2+}$
  Arginase
  Phosphotransferases
$Fe^{2+}$ or $Fe^{3+}$
  Cytochromes
  Peroxidase
  Catalase
  Ferredoxin
$Cu^{2+}$ ($Cu^+$)
  Tyrosinase
  Cytochrome oxidase
$K^+$
  Pyruvate phosphokinase
  (also requires $Mg^{2+}$)
$Na^+$
  Plasma membrane ATPase
  (also requires $K^+$ and $Mg^{2+}$)

*Source:* Lehninger, 1975.

A number of coenzymes contain vitamins as part of the molecule. Since an apoenzyme cannot function without its coenzyme, it is not surprising that vitamin deficiencies can have profound pathological effects.

Other enzymes require monovalent or divalent metal ions as cofactors, generally in a highly selective manner. Some ion-activated enzymes are listed in Table 3-4 along with their cofactor ions. Especially interesting is the calcium ion, which differs from most of the other common physiologically important ions in being present within cells at very low concentrations (less than $10^{-6}$ M). Although other ions, such as $Mg^{2+}$, $Na^+$, $K^+$ and $Cl^-$, are generally present in nonlimiting concentrations, calcium is present in limiting concentrations for certain enzymes. The $Ca^{2+}$ concentration of the cytoplasm is regulated by the surface

membrane and by internal organelles, such as the mitochondria. In this way, the activity of calcium-activated enzymes can be regulated by the cell. Phenomena regulated by the concentration of calcium ions include muscle contraction, secretion of neurotransmitters and hormones, ciliary activity, assembly of microtubules, and amoeboid movement.

## ENZYME KINETICS

The rate at which a reaction proceeds depends on the concentrations of substrate, product, and active enzyme. For purposes of simplicity, let us imagine that the product is removed as fast as it is formed. The rate of reaction will be limited by the concentration of either the enzyme or the substrate. Suppose further that the enzyme is present in excess, so that the concentration of the substrate A determines the rate at which A is converted to the product P:

$$A \xrightarrow{k} P$$

The rate of conversion can be expressed as

$$\frac{-d[A]}{dt} = k[A] \qquad (3\text{-}5)$$

in which [A] is the instantaneous concentration of the substrate, $k$ is the *rate constant* of the reaction, and $-d[A]/dt$ is the rate at which A is converted to P. The disappearance of A and the appearance of P are plotted as functions of time in Figure 3-18. Note that as [A] decreases exponentially, [P] increases exponentially. An exponential time function is always generated when the rate of change of a quantity ($d[A]/dt$ in this example) is proportional to the instantaneous value of that quantity ([A] in this example). The kinetics of such a reaction is said to be *first order* (Figure 3-19B). The rate constant of a first-order reaction has the dimension of reciprocal time—that is, "per second," or $s^{-1}$. The rate constant can be inverted to yield the *time constant,* which has the dimension of time. Thus, a first-order reaction with a rate constant of $10/s$ has a time constant of $\frac{1}{10}$ s.

In a reaction with two substrates A and B, in the presence of enzyme,

$$A + B \xrightarrow{k} P$$

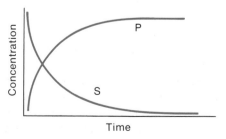

3-18 Changes in concentration of substrate S and product P during the reaction S ⟶ P.

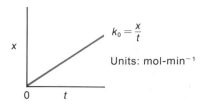

**A**   Zero order

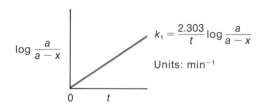

**B**   First order

3-19   (A) Zero-order and (B) first-order reaction kinetics plotted as straight lines. The symbol $x$ represents the amount of substrate S reacting within time $t$, and $a$ represents the initial amount of A at time zero. Note that the first-order plot is semilog, and therefore the straight line represents an exponential time course.

the rate of disappearance of A will be proportional to the product [A][B]. Thus,

$$\frac{d[A]}{dt} = k[A][B] \qquad (3\text{-}6)$$

This reaction will proceed with *second-order kinetics.*

It is noteworthy that the order of the reaction is not determined by the number of substrate species participating as reactants, but instead by the number of species present in rate-limiting concentrations. Thus, if B were present in great excess over A, the reaction A + B → P would become first-order, since its rate would be limited by only one substrate concentration.

The rate of reaction is independent of substrate concentrations when the enzyme is present in limiting concentrations and all the enzyme molecules are complexed with substrate (i.e., the enzyme is *saturated*). Such reactions proceed with *zero-order* kinetics (Figure 3-19A).

When the *initial* rate, $v_0$, of a reaction S → P is plotted as a function of substrate concentration, [S],

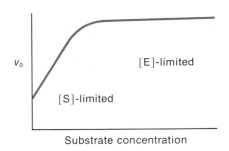

**3-20** At a given enzyme concentration, the initial rate, $v_0$, of the reaction S $\longrightarrow$ P rises linearly with increasing substrate concentration until all of the enzyme becomes saturated, at which time [E] becomes rate-limiting.

at a constant enzyme concentration, we see that at low substrate concentrations the reaction is first-order (i.e., $v_0 \propto$ [S]). At higher substrate concentrations, however, the reaction becomes zero-order, as all the enzyme is saturated with substrate, and the concentration of enzyme, not substrate, limits $v_0$ (Figure 3-20). In the living cell, all orders of reaction, as well as mixed-order reactions, occur.

*ENZYME–SUBSTRATE AFFINITY*

The maximum rate of any reaction, $V_{max}$, occurs when all the enzyme catalyzing that reaction is tied up, or saturated, with substrate—that is, when S is present in excess and [E] is rate-limiting (Figure 3-20). Each enzymatic reaction has a characteristic $V_{max}$. Although all enzymes can become saturated, they show great variation in the concentration of a given substrate that will produce saturation. The reason for this is that enzymes differ in affinity for their substrates. The greater the tendency for the enzyme and substrate to form a complex, ES, the higher the percentage of total enzyme, $E_t$, tied up as ES at any given concentration of substrate. Conversely, the higher this affinity, the lower the substrate concentration required to saturate the enzyme.

An intimate relationship exists between the affinity of S to E and the kinetics of the reaction

$$S \xrightarrow{\text{E}} P$$

The general theory of enzyme action and kinetics was

proposed by L. Michaelis and M. L. Menten in 1913, and later extended by G. E. Briggs and J. B. S. Haldane. Their derivation of $K_M$ can be found in most biochemistry texts. The *Michaelis–Menten equation* is the rate equation for a single enzyme-catalyzed substrate:

$$v_0 = \frac{V_{max} [S]}{K_M + [S]} \qquad (3\text{-}7)$$

where $v_0$ is the initial reaction rate at substrate concentration [S], $V_{max}$ is the reaction rate with excess substrate, and $K_M$ is the Michaelis–Menten constant. Let us take the special case in which $v_0 = \frac{1}{2} V_{max}$. Substituting for $v_0$ gives

$$\frac{V_{max}}{2} = \frac{V_{max} [S]}{K_M + [S]} \qquad (3\text{-}8)$$

Dividing by $V_{max}$ gives

$$\frac{1}{2} = \frac{[S]}{K_M + [S]} \qquad (3\text{-}9)$$

On rearranging, we obtain

$$K_M + [S] = 2[S] \qquad (3\text{-}10)$$

or

$$K_M = [S] \qquad (3\text{-}11)$$

Therefore, $K_M$ equals the substrate concentration at which the initial reaction rate is half what it would be if the substrate were present to saturation.

Thus, the Michaelis–Menten constant, $K_M$ (in units of moles per liter) depends on the affinity of the enzyme for a substrate. For a given enzyme and substrate, it is equal to the substrate concentration at which the initial reaction is $\frac{1}{2} V_{max}$. By inference, then, $K_M$ represents the concentration of substrate at which half the total enzyme present is combined with substrate in the ES; that is, $[E_t]/[ES] = 2$. The greater the affinity between an enzyme and its substrate, the lower the $K_M$ of the enzyme–substrate interaction. Stated inversely, $1/K_M$ is a measure of the affinity of the enzyme for its substrate. As illustrated by the plots for two enzyme concentrations in Figure 3-21, $K_M$ is independent of the enzyme concentration.

As Figure 3-21 shows, Equation 3-7 is a hyperbolic function, which cannot be accurately plotted without numerous data points. The equation can be algebraically transformed, however, into the Lineweaver–

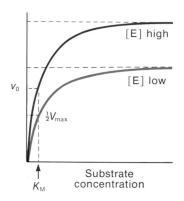

**3-21** The Michaelis–Menten constant, $K_M$, is equal to the substrate concentration at which the reaction rate is one-half maximum. The black and colored plots are for different enzyme concentrations. Note that $K_M$ is independent of [E].

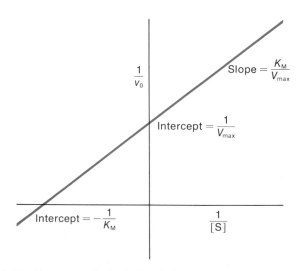

**3-22** Lineweaver–Burk plot, in which the reciprocal of reaction rate, $1/v_0$, is plotted against the reciprocal of the substrate concentration. The plot intercepts the *x*-axis where $-1/[S] = -1/K_M$, and the *y*-axis at $1/V_{max}$. [Patton, 1965.]

Burk form, which can be quickly and accurately plotted from few data points.

$$\frac{1}{v_0} = \frac{K_M}{V_{max}}\frac{1}{[S]} + \frac{1}{V_{max}} \qquad (3\text{-}12)$$

This gives a straight-line plot with a slope of $K_M/V_{max}$ and with intercepts of $1/V_{max}$ on the $1/v_0$ axis and of $-1/K_M$ on the $1/[S]$ axis (Figure 3-22). Thus, one need know only $V_{max}$ and one calculation for a given [S] to draw the curve. The intercept on the $1/[S]$ axis gives $K_M$.

### ENZYME INHIBITION

Certain molecules can inhibit the activity of an enzyme. Enzyme inhibition occurs in the living cell as a means of controlling enzymatic reactions. Through study of the molecular mechanisms of inhibition, enzymologists have discovered important features of the active site of an enzyme and of the mechanism of enzyme action.

Enzymes can be poisoned by agents that form highly stable covalent bonds with groups inside the active site and thereby interfere with the formation of the ES. Such interference can produce *irreversible inhibition.* More relevant to normal cell function, however, are two kinds of *reversible inhibition.* The first, *competitive inhibition,* can be reversed by an increase in substrate concentration, whereas the second, *noncompetitive inhibition,* cannot be reversed. Competitive inhibitors appear to react directly with the active site of the enzyme, whereas noncompetitive inhibitors appear to react with a part of the enzyme other than the active site. Competitive inhibitors (Figure 3-23) are generally analogs of the substrate molecule; noncompetitive inhibitors, however, can be completely different chemically from the substrate. A competitive inhibitor competes with substrate molecules for the active site. Thus, increasing the concentration of one reduces the binding probability of the other. Since the noncompetitive inhibitor complexes with a site other than the active site, it cannot be displaced from its binding site by the competition of substrate molecules, for the latter have no affinity for the allosteric site.

Competitive and noncompetitive inhibition are readily distinguishable in Lineweaver–Burk plots. The

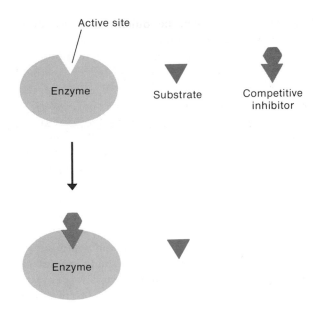

**3-23** Competitive inhibitor binds to the active site of an enzyme, thereby interfering with the binding of the substrate. It can be displaced, however, by a substrate molecule.

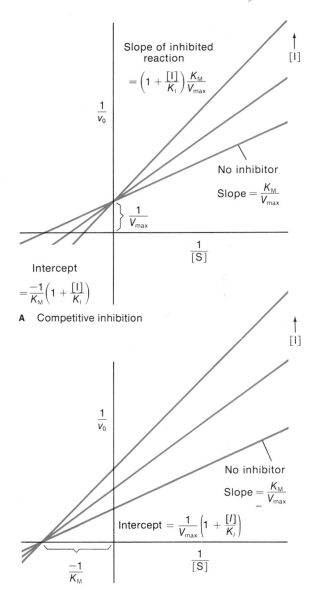

Slope of inhibited reaction

$$= \left(1 + \frac{[I]}{K_I}\right)\frac{K_M}{V_{max}}$$

No inhibitor

$$\text{Slope} = \frac{K_M}{V_{max}}$$

Intercept

$$= \frac{-1}{K_M}\left(1 + \frac{[I]}{K_I}\right)$$

**A** Competitive inhibition

No inhibitor

$$\text{Slope} = \frac{K_M}{V_{max}}$$

$$\text{Intercept} = \frac{1}{V_{max}}\left(1 + \frac{[I]}{K_I}\right)$$

**B** Noncompetitive inhibition

**3-24** Lineweaver–Burk plots of **(A)** competitive and **(B)** noncompetitive inhibition. Note that $K_M$ depends on the concentration of competitive inhibitor. The effect of the noncompetitive inhibitor is kinetically similar to a reduction in enzyme concentration, producing no change in $K_M$. I, inhibitor; S, substrate; $K_I$, dissociation constant of inhibitor–enzyme complex. [Lehninger, 1975.]

competitive inhibitor increases the slope of the plot, which corresponds to a decrease in reaction rate (Figure 3-24A). As the slope increases, however, the intercept on the $1/v_0$ axis remains the same; in other words, when extrapolated to infinite substrate concentration (i.e., $1/[S] = 0$), the rate of reaction in the presence of competitive inhibitor is the same as it is without the competitive inhibitor. The reason for this is that with increasing substrate concentration, the substrate competes successfully with the inhibitor for the active site, finally displacing the inhibitor completely in the hypothetical situation of infinite substrate concentration. The intercept with the $1/[S]$ axis indicates the $K_M$ value. This value is seen to shift toward a higher substrate concentration as the concentration of a competitive inhibitor is raised. This merely means that it takes a higher concentration of substrate in the presence of competing molecules (i.e., competitive inhibitor) to keep half of the enzyme molecules at any instant complexed with the substrate.

Noncompetitive inhibitors also produce an increase in slope in the Lineweaver–Burk plot (Figure 3-24B), but this is accompanied by a drop in reaction rate at infinite substrate concentration. The drop is seen as an increase in the value at which the plot crosses the $1/v_0$ axis. These kinetic effects result from the failure of increased substrate concentration to remove a noncompetitive inhibitor from the allosteric site of the enzyme. This situation has the same effect as destruction of those enzyme molecules complexed with the inhibitor. Concomitantly, the curves intersect the $1/[S]$ axis at the same point with or without the noncompetitive inhibitor, indicating no change in $K_M$ for the enzyme in the presence of the noncompetitive inhibitor. As noted earlier, the effect of the noncompetitive inhibitor is equivalent to lowering the enzyme concentration. It is not surprising, then, that the $K_M$ of an enzyme is unaltered by addition of a noncompetitive inhibitor, for, as noted above, $K_M$ is independent of $[E]$.

## Metabolic Regulatory Mechanisms

Without any regulation of reaction rates, cell metabolism would be uncoordinated and undirected. Growth, differentiation, and maintenance would be impossible, to say nothing of compensatory responses of the biological machine to externally imposed stresses. Most control is exerted through the quantity or activity of the various enzymes that catalyze nearly all biochemical reactions. The three major types of metabolic control are described next.

### GENETIC CONTROL OF ENZYME SYNTHESIS

The number of molecules of an enzyme present in a cell is a function of the rate of synthesis and the rate of destruction of enzyme molecules. Enzyme molecules are denatured by temperature increases, and they are broken down by the action of proteolytic enzymes. The rate of synthesis can be limited under special circumstances, such as inadequate diet, by the availability or unavailability of amino acid precursors, but the rate of synthesis of a particular enzyme is normally regulated by genetic mechanisms. *Structural genes* (i.e., sections of the DNA molecule coding the amino acid sequence of one or more peptide chains that make up the enzyme molecule) can be "turned off" by the action of *repressor proteins,* which are coded by a *regulator gene.* The repressor molecule interferes with the trans-

cription of the structural gene from DNA to RNA by binding to a third type of genetic locus called the *operator.* The latter controls the transcription into messenger RNA of one or more structural genes that constitute the *operon* subserved by that operator. The synthesis of enzymes coded by the operon is collectively controlled by the interaction of the regulator gene with the operator locus. This regulation is illustrated in Figure 3-25. Certain small molecules combine with the repressor protein, rendering it incapable of combining with the operator locus. They therefore halt repression of the structural genes and permit the synthesis of the enzymes that had previously been repressed. This scheme, proposed by Francois Jacob and Jacques Monod in 1961, explains the phenomenon of *enzyme induction,* in which cells synthesize certain enzymes only after they are exposed to the substrates (or related molecules) for those enzymes (Figure 3-26). This process is an example of metabolic economy, for inducible enzymes are synthesized only as needed. When the substrate is present, it combines with the repressor, halting repression (i.e., allowing transcription) of the respective operon. Thus, the segments of DNA that previously lay dormant become available for specifying the amino acid sequences of the induced enzymes.

The transcription of structural genes (and hence the synthesis of particular enzymes) is also regulated in some instances by the *end products* of a biosynthetic reaction sequence. But the repressor molecule synthesized by the regulator gene remains inactive until it combines with a small organic molecule—the *corepressor,* produced at the end of a biosynthetic reaction sequence (Figure 3-27). The repressor and corepressor together combine with the operator to prevent transcription of the structural gene for an enzyme that acts early in the biosynthetic pathway. In this way, the synthetic pathway and the rate of production of its end product are kept in check. If the end product begins to accumulate for any reason, such as a reduction of its incorporation into cell structures, the entire synthetic pathway is slowed by a drop in the rate of synthesis of the regulated enzyme.

Genetic regulatory mechanisms are of great importance in the development of an organism. Each somatic cell in an organism contains the same information coded in its DNA; cells in different tissues,

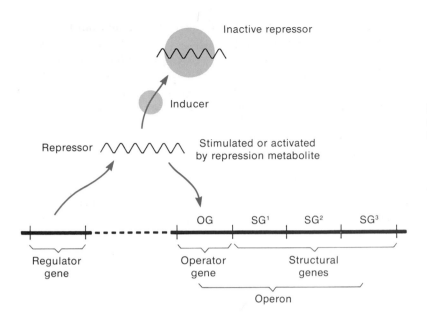

Inactive repressor

Inducer

Repressor

Stimulated or activated
by repression metabolite

OG SG¹ SG² SG³

Regulator
gene

Operator
gene

Structural
genes

Operon

**3-25** Operon model of regulation
of enzyme synthesis by the control
of gene expression. [Goldsby, 1967.]

however, contain widely divergent amounts of the
different enzymes coded by the genetic material. It
is evident that in any given tissue some genes are
turned on while others are turned off. This situation
may occur in part through mechanisms of enzyme induction and repression in response to differences in the
local chemical environments of different cells and tissues in the developing organism.

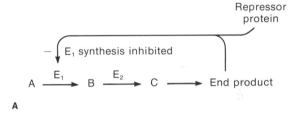

Repressor
protein

$-$ $E_1$ synthesis inhibited

$A \xrightarrow{E_1} B \xrightarrow{E_2} C \longrightarrow$ End product

**A**

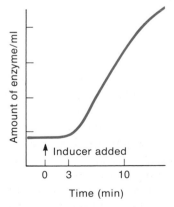

Amount of enzyme/ml

Inducer added

0 3 10

Time (min)

**3-26** Time course of induction of enzyme synthesis by the
addition of the substrate for the enzyme. [Dowben, 1971.]

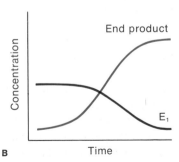

Concentration

End product

$E_1$

**B**  Time

**3-27** Repressive effect of an end-product synthesis of an
enzyme. **(A)** The negative feedback loop in which synthesis of $E_1$
is repressed by accumulation of a product several steps further
along the pathway. The end product binds to the repressor
molecule to inactivate the gene for $E_1$. **(B)** The level of $E_1$ drops
as the level of end product rises.

**3-28**  Allosteric inhibition of a rate-limiting enzyme by a product of the reaction sequence.

## METABOLIC FEEDBACK INHIBITION

Some metabolic pathways have built-in mechanisms for direct (i.e., nongenetic) self-regulation (Figure 3-28). In these reaction sequences, it is usually the first enzyme of the sequence that acts as a *regulatory enzyme,* because its activity is subject to modulation by the concentration of the end product of the sequence. This modulation usually takes the form of inhibition of the activity of that enzyme by some interaction with the end-product molecule. Such *end-product inhibition* limits the rate of accumulation of the end product by slowing the entire sequence. The interaction of the end-product molecule with the enzyme molecule has been shown to occur at a site other than the catalytically active site. Thus, the end product acts as a modulator molecule that reduces the activity of the enzyme molecule by an allosteric mechanism. Most regulatory enzymes catalyze reactions that are virtually irreversible under cellular conditions and therefore are not subject to mass-action kinetics, which would tend to slow the reaction as the concentration of product increases.

## ENZYME ACTIVATION

The requirement for cofactors exhibited by some enzymes provides the cell with another means of regulating the rate of biochemical reactions. The intracellular free concentration of certain ions depends on diffusion and active transport across membranes that separate the cytoplasmic compartment of the cell from the cell exterior and intracellular stores of certain ions. By regulating the levels of cofactor ions, the cell can modulate the concentration of cofactor ions and the activity of certain enzymes.

An important and common regulatory cofactor is the calcium ion, which is present within cells in lower concentrations than such common inorganic ions as $Mg^{2+}$, $Na^+$, $K^+$, and $Cl^-$. Changes in intracellular $Ca^{2+}$ concentration play an essential role in many physiological and biochemical functions. The special role of calcium as a widespread intracellular messenger and regulatory agent is related to its very low concentration (less than $10^{-6}$ M) in the *cytosol* (unstructured fluid phase of the cytoplasm). Minute changes in the net flux of $Ca^{2+}$ across the cell membrane or the membranes of cytoplasmic organelles can thereby produce substantial *percentage* changes in the intracellular free $Ca^{2+}$ concentration.

How do $Ca^{2+}$ and other intracellular regulatory agents exert their effects on intracellular events? There is no single answer, because there appear to be several ways in which biochemical regulation occurs. A number of enzymes exhibit allosteric regulation. For example, an ion such as $Ca^{2+}$, a metabolic end product, or a specific regulatory molecule may combine with a site on the enzyme molecule in such a way as to alter the tertiary conformation of the enzyme and thereby modify the catalytic efficiency of the active site on another part of the enzyme molecule.

## Metabolic Production of ATP

For the sake of argument, we can continue the analogy between an animal and a machine introduced at the beginning of this chapter. If we compare the energy utilization of an animal with that of an automobile, we note that both types of machines require the intermittent intake of chemical fuel to energize their activities. Their use of fuel differs, however, in at least one very important aspect. In the automobile engine, the organic fuel molecules in the gasoline are oxidized (ideally) to $CO_2$ and $H_2O$ in one explosive step. The heat generated by the rapid oxidation produces a great increase in the pressure of gases in the cylinder. In this way, the chemical energy of the fuel is converted to mechanical movement (kinetic energy). This conversion depends on the high temperatures produced by the burning gasoline, for the chemical energy of the gasoline is converted directly into heat, and heat can be used to do work only if there is a temperature and pressure difference between two parts of the machine.

Since living systems are capable of sustaining only

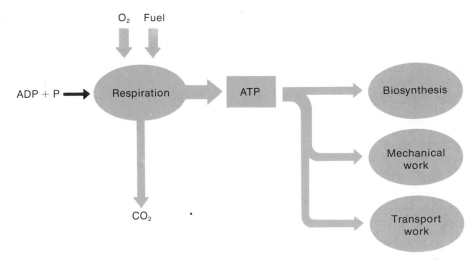

**3-29** Uses of ATP in biological systems. The ADP produced by hydrolysis is recycled to ATP by rephosphorylation energized by the oxidation of foodstuff molecules to $CO_2$ and $H_2O$. [Lehninger, 1965.]

small temperature and pressure gradients, the heat provided by the simple one-step combustion of fuel would be essentially useless for energizing the activities of a living system. For this reason, cells have evolved metabolic mechanisms for the stepwise conversion of chemical energy in a series of discrete reactions. The energy of foodstuff molecules is recovered for useful work through the formation of intermediate compounds of progressively lower energy content. At each exergonic step, some of the chemical energy is liberated as heat, while the rest is transferred as free energy to the reaction products. Chemical energy conserved and stored in the structure of intermediate compounds is then transferred to the general-purpose, high-energy intermediate ATP and to other high-energy intermediates through which chemical energy is made available for a wide variety of cellular processes (Figure 3-29).

Chemical energy is extracted primarily from three classes of food molecules: carbohydrates, lipids, and proteins. After digestion, these generally enter the circulatory system as five- or six-carbon sugars, fatty acids, and amino acids, respectively (Figure 3-30).

These small molecules then enter into the tissues and cells of the animal, where they may (1) be immediately broken down into smaller molecules for the extraction of chemical energy or for rearrangement and recombination into other types of molecules *or* (2) be built up into larger molecules, such as polysaccharides (e.g., glycogen), fats, or proteins. With few exceptions, these too will eventually be broken down and eliminated as $CO_2$, $H_2O$, and urea. Nearly all molecular constituents of a cell are in dynamic equilibrium, constantly being replaced by components newly synthesized from the simpler organic molecules.

Some simple organisms, including certain bacteria and yeasts, as well as a few invertebrate species, can live indefinitely under totally *anaerobic* (i.e., essentially oxygen-free) conditions. The anaerobes fall into two groups: (1) *obligatory anaerobes,* those that cannot grow in the presence of oxygen (e.g., botulism bacterium, *Clostridium botulinum*); and (2) *facultative anaerobes,* those, like the yeasts, that survive and reproduce well either in the absence or in the presence of oxygen. All vertebrates and most invertebrates require molecular oxygen for cell respiration and are therefore termed *aerobic*.

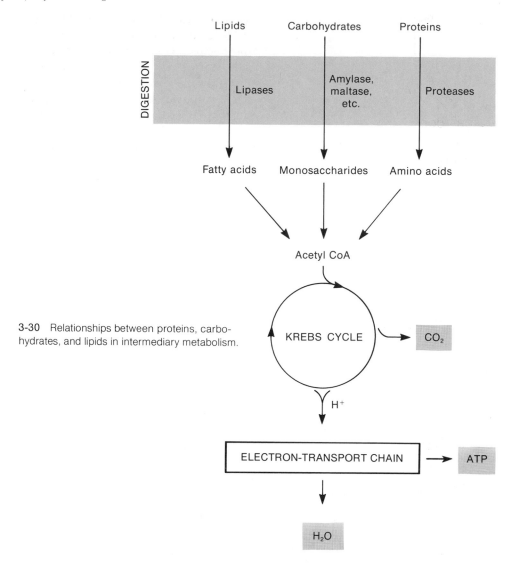

3-30   Relationships between proteins, carbohydrates, and lipids in intermediary metabolism.

Even these generally possess tissues that can metabolize anaerobically for periods of time, building up an *oxygen debt* that is repaid when sufficient oxygen becomes available.

As these observations suggest, there are two kinds of energy-yielding metabolic pathways in animal tissues: (1) *aerobic metabolism*, in which foodstuff molecules are finally oxidized completely to $CO_2$ and water by molecular oxygen; and (2) *anaerobic metabolism*, in which foodstuff molecules are oxidized incompletely to lactic acid (Figure 3-31). The energy yield per molecule of glucose in anaerobic metabolism is only a fraction of the energy yield for aerobic metabolism. For this reason, cells with high metabolic rates survive only briefly when their tissues are deprived of oxygen. The nerve cells of the mammalian brain are a familiar example. An oxygen deficiency lasting only a few minutes will lead to massive cell death and permanently impaired brain function.

In animal cells, aerobic metabolism is intimately

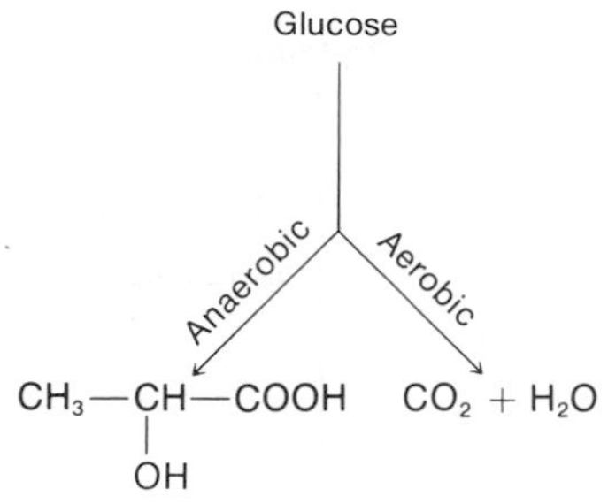

Glucose

*Anaerobic*  *Aerobic*

$CH_3—CH—COOH$     $CO_2 + H_2O$

$OH$

Lactic acid

**3-31** The two fates of glucose catabolism.

associated with the *mitochondria*. These organelles, just visible in the light microscope, had to await the electron microscope for their detailed description (Figure 3-32). They consist of an *outer membrane* and an *inner membrane,* which are not connected with each other. These two membranes carry out completely different functions. The inner membrane is thrown into folds termed *cristae,* which serve to increase the area of the inner membrane relative to the outer membrane. The space within the confines of the inner membrane is termed the *matrix compartment,* and the space between the two membranes is the *outer space.* As we shall see

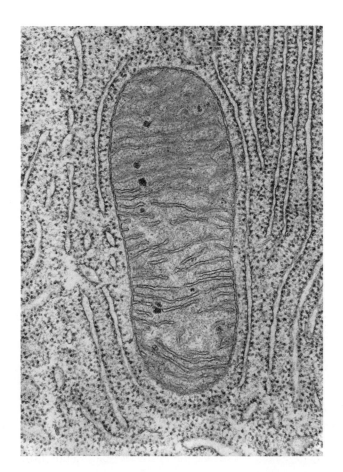

**3-32** Electron micrograph of a mitochondrion in a bat pancreas cell. Magnification 50,000×. [Courtesy of K. R. Porter.]

later on, the inner membrane is very important in the production of ATP during aerobic metabolism. The matrix compartment contains DNA, which is involved in replication of the mitochondria, ribosomes, and dense granules, the last consisting primarily of salts of calcium.

Mitochondria are quite numerous in most cells, with estimates ranging from 800 to 2500 for a liver cell. They also tend to congregate most densely in those portions of a cell that are most active in utilizing ATP.

## Oxidation, Phosphorylation, and Energy Transfer

Before going on to consider the biochemical pathways in cellular energy metabolism, we will see how chemical energy, liberated during metabolism, is conserved and channeled into high-energy intermediates. You will remember that when a complex organic molecule is taken apart, free energy is liberated, thus increasing the entropy (degree of randomness) of the constituent matter. This situation occurs when glucose is oxidized to carbon dioxide and water by combustion in the overall reaction

$$C_6H_{12}O_6 + 6O_2 \longrightarrow 6CO_2 + 6H_2O$$

$$\Delta G° = -686 \text{ kcal/mol}$$

The 686,000 cal liberated by the oxidation of 1 mol of glucose is the difference between the free energy incorporated into the structure of the glucose molecule during photosynthesis and the total free energy contained in the $CO_2$ and $H_2O$ produced. If 1 mol of glucose is oxidized to carbon dioxide and water in a one-step combustion (i.e., if it is burned), the free-energy change will appear simply as 686 kcal of heat. During cell respiration, however, a portion of this energy, instead of appearing as heat, is conserved as useful chemical energy and is channeled into ATP through the phosphorylation of ADP. The overall reaction for the metabolic oxidation of glucose by the cell can be written as

$$C_6H_{12}O_6 + 38P_i + 38ADP + 6O_2$$

$$\longrightarrow 6CO_2 + 6H_2O + 38ATP$$

$$\Delta G° = -420 \text{ kcal (as heat)}$$

Thus, 266 kcal (686 minus 420) is incorporated into 38 mol of ATP (7 kcal/mol ATP).

How is the free energy of the glucose molecule transferred to ATP? To understand this, we must first recall that oxidation of a molecule is most broadly defined as the transfer of electrons from that molecule to another molecule. In an oxidation–reduction reaction, the *reductant* (electron donor) is oxidized by the *oxidant* (electron acceptor). Together they form a *redox pair:*

$$\text{electron donor} \rightleftharpoons e^- + \text{electron acceptor}$$

or

$$\text{reductant} \rightleftharpoons ne^- + \text{oxidant}$$

where $n$ is the number of electrons transferred. Whenever electrons are accepted from a reductant by an oxidant, energy is liberated, for the electrons move into a more stable (higher-entropy) situation in transferring to the oxidant. This is akin to water dropping from one level to a lower level. It is the *difference* between the two levels that determines the energy liberated.

Thus, chemical energy is liberated when electrons are transferred from a compound of a given *electron pressure* (tendency to donate electrons) to one of lower electron pressure. If a molecule has a higher electron pressure than the molecule with which it undergoes a redox reaction, it is said to have a greater *reduction potential,* and will act as a reducing agent; if it has less electron pressure, it will act as an oxidant. The free-energy change in each reaction is proportional to the difference between the electron pressures of the two molecules of the redox pair.

In aerobic cell metabolism, electrons move to progressively lower energy levels from compounds of high electron pressure to compounds of lower electron pressure. The *final electron acceptor* in aerobic metabolism is molecular oxygen. Oxygen presumably became the universal final oxidant because of its very low electron pressure (strongly oxidizing nature) and its abundance on the Earth's surface as a result of photosynthesis. Since oxygen acts merely as an electron acceptor, it is possible in theory to support aerobic metabolism without $O_2$, provided that a suitable electron acceptor is supplied in place of oxygen.

In being transferred from glucose to oxygen, electrons undergo an enormous drop in both reduction potential and free energy. One of the functions of cell metabolism is to transport electrons gently from glucose to oxygen in a series of small steps instead of one large drop. This transport is implemented by two mechanisms found in all cells. First, as we noted earlier, the conversion of foodstuff molecules such as glucose to the fully oxidized end products (e.g., $CO_2$ and $H_2O$) is accomplished in a series of many small intermediate stages of molecular change and oxidation. Second, electrons removed from substrate molecules are passed to oxygen via a series of electron acceptors

and donors of progressively lower electron pressure. As we shall see shortly, this activity allows energy to be channeled into the synthesis of ATP in "packets" of appropriate size.

### THE ELECTRON-TRANSFERRING COENZYMES

Electrons, together with protons (i.e., hydrogen), are removed from substrate molecules during certain reactions along metabolic pathways by enzymes collectively termed *dehydrogenases*. These enzymes all function in conjunction with *pyridine* or *flavin* coenzymes. The most common are nicotinamide adenine dinucleotide (NAD), noted earlier, and *flavin adenine dinucleotide* (*FAD*). Their structural formulas are shown in Figure 3-33. These coenzymes act as electron acceptors in their

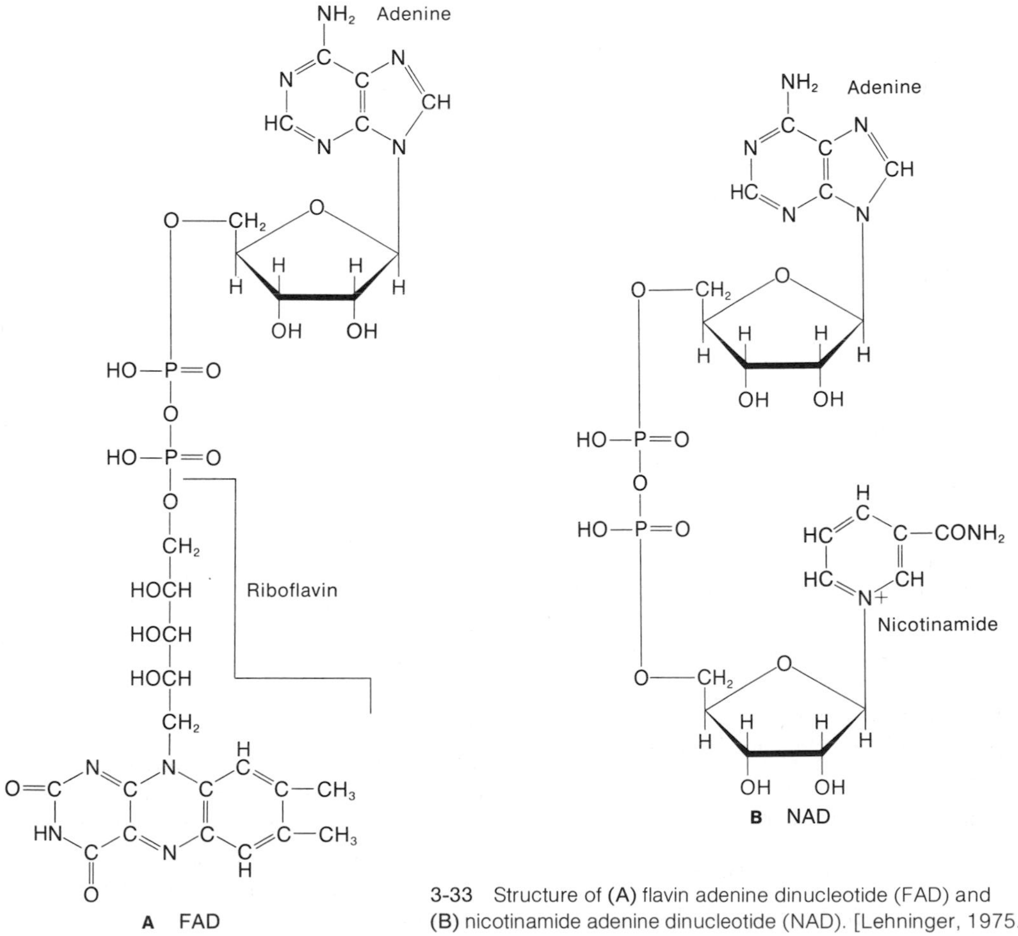

3-33 Structure of (**A**) flavin adenine dinucleotide (FAD) and (**B**) nicotinamide adenine dinucleotide (NAD). [Lehninger, 1975.]

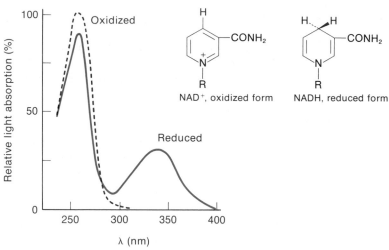

**3-34**  Absorption spectra of NAD+ and NADH. Since the difference in absorption is greatest at 340 nm, that wavelength is used to monitor the reduction of NAD+ to NADH. [Lehninger, 1975.]

oxidized form and as electron donors in their reduced form:

$$\text{reduced substrate} + NAD^+ \rightleftharpoons$$

$$NADH + H^+ + \text{oxidized substrate}$$

A very convenient feature for studies of these coenzymes is that their absorption spectra in the ultraviolet range differ for the reduced and oxidized forms (Figure 3-34). They also undergo a change in ultraviolet-excited fluorescence upon oxidation and reduction. These two features have permitted physiologists and biochemists to use photometric methods to monitor changes in the amount of reduced coenzyme under experimental conditions in living cells.

The energy level of the reduced coenzyme molecule, NADH or FADH, is very high relative to oxygen. As a result, the transfer of two electrons from NADH to $O_2$ produces a change in free energy of about $-52$ kcal/mol (Figure 3-35). This energy represents a large proportion of the 686 kcal of free energy available from the oxidation of glucose, for 1 mol of glucose yields 10 mol of reduced NAD and 2 mol of reduced

FAD. Multiplying $12 \times 52$ gives a total of 624 kcal. Thus 91% of the free energy of glucose is transferred to the electron-transferring coenzymes to be released in subsequent stages of electron transfer. As noted earlier, 266 kcal of this free energy is eventually retained by the synthesis of ATP.

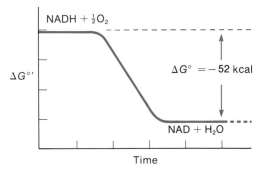

**3-35**  The oxidation of NADH leads to the release of free energy ($-52$ kcal).

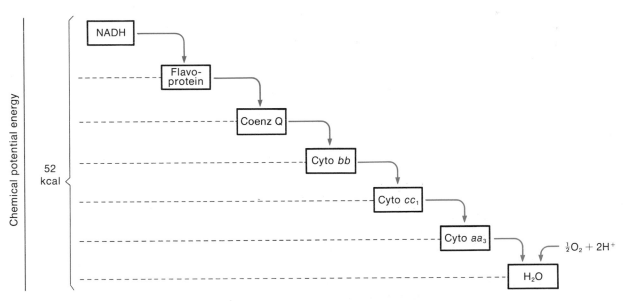

**3-36** Structure of heme A, which acts as the electron acceptor-donator group of cytochrome $aa_3$. At the center of the porphyrin ring is the iron atom that is oxidized or reduced during transport. [Lehninger, 1975.]

## The Electron-Transport Chain

It is significant that in spite of the large difference in electron pressure between NADH and $O_2$, there is no enzymatic mechanism by which NADH and FADH can be directly oxidized by oxygen. Instead, an elaborate *electron-transport chain,* or *respiratory chain,* has evolved in which the electrons move through about seven discrete steps—from the high reducing potential of NADH and FADH to the final electron acceptor, molecular oxygen. This sequence of electron transfers is the final common pathway for all electrons during aerobic metabolism. Its function, as we will see, is to utilize the energy of electron transfer efficiently for the phosphorylation of ADP to ATP.

The electron-transport chain consists of a sequence of enzymes termed *cytochromes,* each of which contains a deeply colored *heme* group. The heme group consists, essentially, of a *porphyrin* ring with an *iron* atom at its center (Figure 3-36); it is similar to the pigmented heme group in the hemoglobin molecule of vertebrate red blood cells. The functional order of the cytochrome sequence is diagrammed in Figure 3-37. From left to

**3-37** Electron pressure produced by a cascading of electrons during electron transport in the respiratory chain.

right, each successive cytochrome molecule has a lower electron pressure than its predecessor. As a result, electrons are transferred from NADH down the cytochrome chain in a series of seven coupled reactions, ending with the reduction of molecular oxygen. Only the last enzyme in the chain, cytochrome $a_3$, is able to transfer its electrons directly to oxygen.

The cytochromes show characteristic absorption spectra in their oxidized and reduced forms, absorbing more strongly in the red when reduced. This behavior led to the first discoveries of their function by David Keilin in 1925. Using a spectroscope, he discovered that the flight muscles of insects contain compounds that are oxidized and reduced during respiration. He named these compounds cytochromes and hypothesized that they transfer electrons from energy-rich substrates to oxygen. When the final step—electron transfer by cytochrome oxidase (composed of subunits of $a$ and $a_3$) to $O_2$—is blocked by cyanide, the effect on electron transport is identical to the removal of molecular oxygen. Electrons pile up, so to speak, be-

cause transport is interrupted along the chain, reducing all the cytochrome molecules above the point of the block. Another poison, *antimycin,* blocks the flow of electrons from cytochrome $b$ to $c$ (Figure 3-38), causing the cytochromes above the block to become fully reduced and those below the block to become completely oxidized. Such selective poisoning at various points along the respiratory chain has enabled biochemists to work out the order of electron transfer, using spectrophotometric methods to follow the oxidation or reduction of the cytochromes.

A great energetic advantage is provided by the cascading of electrons through a series of small discrete steps rather than by allowing direct reduction of oxygen by NADH. The "logic" of the electron-transport system becomes apparent when it is recalled that the standard currency in biological energy exchange is small compared with the total change in free energy produced by the transfer of electrons from NADH to oxygen. It requires a minimum of 7.3 kcal to synthesize ATP from ADP and inorganic phosphate, whereas

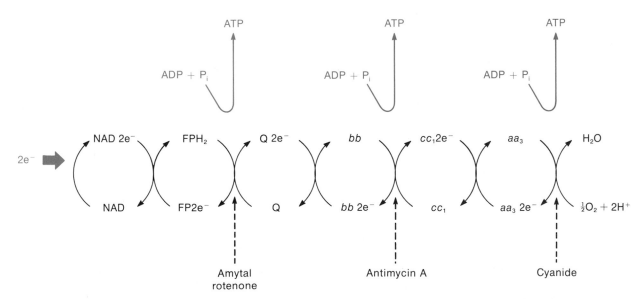

**3-38** Electron-transport chain. Broken arrows indicate where the chain can be uncoupled by respiratory poisons. One molecule of ADP is phosphorylated to ATP owing to the transport of one pair of electrons along the respiratory chain. FP, flavoprotein; Q, coenzyme Q. Symbols $b$, $c$, $c_1$, $a$, and $a_3$ refer to the respective cytochromes, shown working in pairs transporting pairs of electrons.

52 kcal would be released in a one-step oxidation of NADH, permitting the formation of only one molecule of ATP. This procedure would conserve only 14% (7.3 divided by 52) of the available free energy in the form of ATP, the rest being lost as heat. To avoid this loss, the large energy drop that would be experienced by the electrons in direct oxidation is subdivided into smaller steps of free-energy change. Thus, the electron-transport system is a mechanism for releasing quantities of energy in doses just large enough for efficient synthesis of ATP. As noted next, there are three stages of electron transfer along the *respiratory chain*, each with a drop in free energy adequate to drive the phosphorylation of ADP to form ATP (Figure 3-38).

The actual synthesis of ATP from ADP and inorganic phosphate during electron transport is termed *oxidative phosphorylation* or *respiratory chain phosphorylation*. The phosphorylation of ADP to ATP occurs as a consequence of the tranfer of electrons from (1) flavoprotein to coenzyme $Q$, (2) from cytochrome $b$ to cytochromes $c$ and $c_1$, and (3) from cytochromes $aa_3$ (cytochrome oxidase) to molecular oxygen. Thus, for each pair of electrons that passes along the entire chain, three molecules of ATP are generated from three molecules of ADP and three molecules of inorganic phosphate ($P_i$). Each pair of electrons finally reduces one-half molecule of $O_2$ to form one molecule of water:

$$2e^- + 2H^+ + \tfrac{1}{2}O_2 \longrightarrow H_2O$$

By comparing the amount of oxygen consumed (i.e., converted to water) and the amount of inorganic phosphate consumed (i.e., incorporated into ATP), we can establish the *P/O ratio* (ratio of inorganic phosphate to atomic oxygen). For example, if one oxidative phosphorylation occurs at each of the three steps noted above, 3 mol of inorganic phosphate will be incorporated into ATP for each mole of oxygen atoms ($\tfrac{1}{2}O_2$) consumed in the formation of $H_2O$. Thus, P/O = 3. Certain electron carriers, however, bypass the first stage of phosphorylation, reducing coenzyme $Q$ directly, in which case electron transport allows only two ATP phosphorylations per pair of electrons and P/O = 2.

Several theories have been developed to explain how the synthesis of ATP is *coupled* at the molecular level to the free energy liberated during electron transfer. The chemiosmotic theory of energy transduction is the most widely accepted of the three and will be discussed later. It is interesting to note here that oxidative phosphorylation becomes *uncoupled* from electron transport whenever anything happens that makes the inner mitochondrial membrane "leaky." If the membrane is broken or if a chemical agent is used to increase its permeability to $H^+$ or other cations, the production of ATP drops or ceases while both electron transport and the reduction of $O_2$ to $H_2O$ continue, all the released energy being given off as heat. Oxidative phosphorylation is also uncoupled from electron transport by certain drugs, such as dinitrophenol (DNP). Because this drug reduces the efficiency of energy metabolism, it was once prescribed by physicians to help patients lose weight. Its use as a weight-reducing drug was discontinued when it was found to produce pathological side effects.

## Glycolysis

The term *glycolysis*, which means "breakdown of sugar," refers to the pathway of reactions leading from glucose to pyruvic acid (Figures 3-39, 3-40). This sequence of reactions, the most fundamental in the energy metabolism of animal cells, is required for both anaerobic and aerobic release of energy from foodstuffs. The glycolytic pathway is also termed the *Embden–Meyerhof pathway* after the two German biochemists who worked out the details of glycolysis in the 1930s.

Glucose is first phosphorylated by ATP, either during the phosphorolysis of glycogen (Figure 3-41) or in the reaction

$$\text{glucose} + \text{ATP} \longrightarrow \text{glucose-6-phosphate} + \text{ADP}$$

After conversion of glucose (step 2 in Figure 3-39) to *fructose-6-phosphate*, the *hexose* (six-carbon sugar) is again phosphorylated to *fructose 1,6-diphosphate* at the expense of a second mole of ATP (step 3). At first glance, it would seem uneconomical for the cell to expend 2 mol of ATP to phosphorylate 1 mol of hexose,

**3-39** Glycolysis. Note the formation of two molecules of triose for each hexose at step 4, doubling molarities of reactants in the remainder of the pathway. The common energy intermediates, ATP and NADH, are shaded.

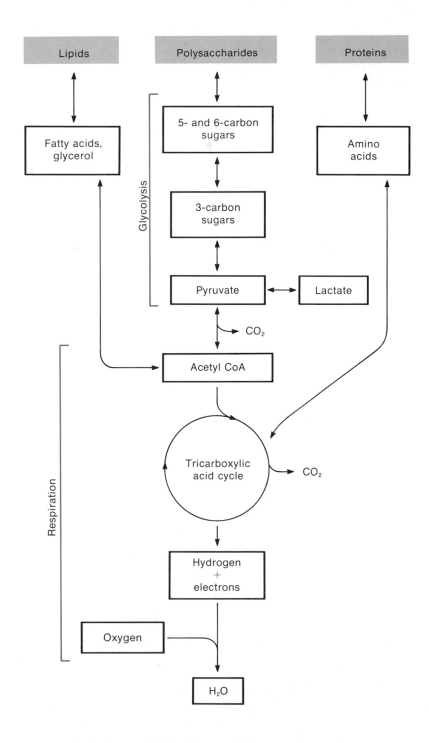

**3-40** Metabolic interrelations between fats, carbohydrates, and proteins. Each class of foodstuff can fuel the tricarboxylic acid cycle.

3-41   Phosphorolysis of glycogen to glucose-1-phosphate. [Lehninger, 1975.]

since the object of glycolysis is to produce ATP. On closer examination, however, the phosphorylation of glucose does make sense. As a result of phosphorylation, the hexose and *triose* (three-carbon sugar) phosphate molecules become ionized; and, as explained in Chapter 2, polar molecules have very low membrane permeabilities. Thus, although the unphosphorylated glucose is free to enter (or leave) the cell by diffusion through the surface membrane, the phosphorylated form is conveniently trapped along with its phosphorylated derivatives within the cell. The 2 mol of ATP expended in these so-called *priming phosphorylations* is, in fact, not really lost, for later in the glycolytic pathway these phosphate groups —and their intramolecular free energies—are transferred to ADP (step 10 in Figure 3-39), thereby conserving the energy of the phosphate groups utilized in the priming phosphorylations.

Fructose 1,6-diphosphate is cleaved in step 4 into two triose sugars, *glyceraldehyde 3-phosphate* and *dihydroxyacetone phosphate*. The latter molecule is enzymatically rearranged into the former, so that each mole of glucose yields 2 mol of glyceraldehyde 3-phosphate,

both of which follow the same pathway. This completes the *first stage of glycolysis,* which is concerned with the conversion of each mole of six-carbon sugar into 2 mol of the three-carbon sugar glyceraldehyde 3-phosphate (steps 1 to 5).

The *second stage of glycolysis* begins with the oxidation of glyceraldehyde 3-phosphate to *1,3-diphosphoglycerate* (step 6). This reaction is very important, because the addition of the second phosphate group to the triose molecule conserves the energy that would otherwise be released by oxidation of the aldehyde group. The elucidation of the mechanism of this reaction and of the following one (step 7), in which ADP is directly phosphorylated to ATP by the substrate, is considered among the most important contributions to modern biology. Through these discoveries, Otto Warburg and his colleagues provided, in the late 1930s, the first insight into a mechanism by which chemical energy of oxidation is conserved in the form of ATP. This is termed a *substrate-level phosphorylation* to distinguish it from respiratory chain phosphorylation.

In steps 8 to 10, 3-phosphoglyceric acid is converted to *2-phosphoglyceric acid,* water is removed to form *phosphoenolpyruvate,* and in step 10 the latter yields its phosphate group to ADP, forming ATP and *pyruvic acid.* Thus, the glycolytic pathway ends with 2 mol of pyruvic acid produced from each mole of glucose. The phosphorylation of each mole of hexose consumes 2 mol of ATP, and each mole of triose generates 2 mol of ATP (steps 7 and 10). Since each mole of glucose yields 2 mol of triose, the net gain per mole of glucose in anaerobic glycolysis is 2 mol of ATP (Figure 3-42).

In the absence of oxygen—that is, during *anaerobic glycolysis*—the reduction of pyruvic acid to *lactate* (step 11, Figure 3-39) or ethanol (in certain microorganisms such as yeast) serves the very important function of oxidizing the NADH formed in step 6 back to $NAD^+$. In this case, the electrons of NADH are accepted by pyruvate instead of oxygen. Without this *anaerobic oxidation* of the reduced coenzyme, there would be a depletion of the oxidized form of the coenzyme, and glycolysis would be blocked for lack of an electron acceptor at step 5 (the oxidation of 3-phosphoglyceraldehyde to 1,3-diphosphoglycerate) in the absence of

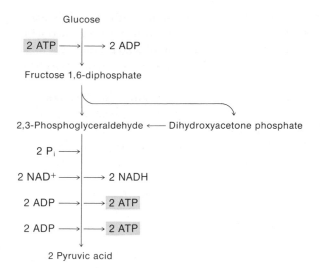

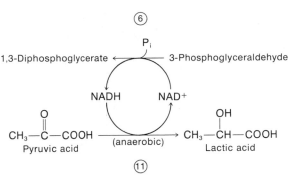

**3-43** The $NAD^+ \rightleftharpoons$ NADH cycle between steps 6 and 11 (Figure 3-39) in anaerobic glycolysis.

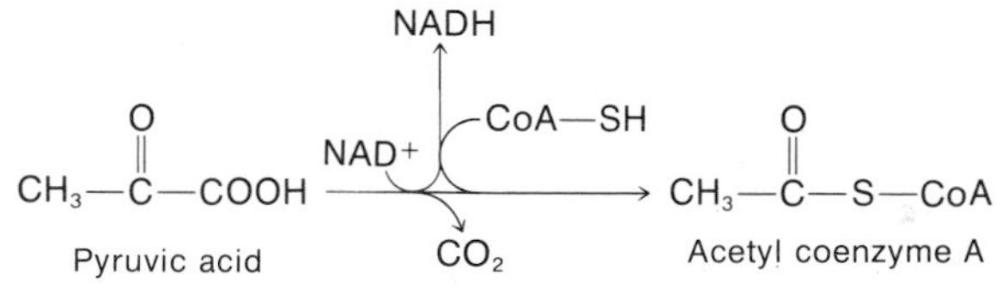

**3-42** Utilization and production of ATP during glycolysis. Note that net ATP production equals 2 mol of ATP per mole of glucose oxidized to pyruvic acid. [Vander et al., 1975.]

**3-44** Formation of acetyl CoA from pyruvic acid.

molecular oxygen. The anaerobic $NAD^+ \rightleftharpoons$ NADH cycle that operates between step 6 and 11 is shown in Figure 3-43. During *aerobic glycolysis,* this mole of NADH is oxidized with the concomitant production of 3 mol of ATP by molecular oxygen via the electron-transport system discussed earlier.

## *Tricarboxylic Acid (Krebs) Cycle*

Under aerobic conditions, pyruvic acid is decarboxylated; that is, 1 mol of $CO_2$ is removed, leaving a two-carbon *acetate* residue. The oxidized form of the coenzyme $NAD^+$ accepts one hydrogen atom from pyruvic acid and one from *coenzyme A (CoA).* This allows the

two-carbon residue from the pyruvic acid to be condensed with the coenzyme to form *acetyl coenzyme A* (*acetyl CoA*) (Figure 3-44). The coenzyme acts as a carrier for the acetate residue, transferring it to *oxaloacetic acid* in the next reaction, in which free CoA is released. CoA is not consumed, but repeatedly transfers acetate residues from pyruvic acid to oxaloacetate.

All the reactions of the glycolytic pathway up to the formation of pyruvic acid occur in free solution in the cytosol. The formation of acetyl CoA and $CO_2$ from pyruvic acid is followed by the eight major reactions of the *tricarboxylic acid* (*TCA*) *cycle* (Figure 3-45), in which each acetate residue is degraded to two additional molecules of $CO_2$ and two molecules of $H_2O$.

These reactions are all catalyzed by enzymes confined to the matrix compartment of the mitochondrion. The overall reaction is

$$2CH_3-\overset{\overset{\textstyle O}{\|}}{C}-COOH + 5O_2 \longrightarrow 6CO_2 + 4H_2O$$

The tricarboxylic acid cycle is also known as the *Krebs cycle* in honor of Hans Krebs, who in the early 1940s elucidated the major features of the reaction sequence and its cyclic nature. The two-carbon acetate residue of acetyl CoA is first condensed with the four-carbon oxaloacetic acid to form the six-carbon *citric acid* (step 2, Figure 3-45). In steps 5 and 6, two carboxyl groups of *isocitric acid* are removed to form the second and third molecules of $CO_2$. Moreover, four hydrogen atoms are transferred to $NAD^+$ to form two molecules of NADH. Step 7 takes place on the inner

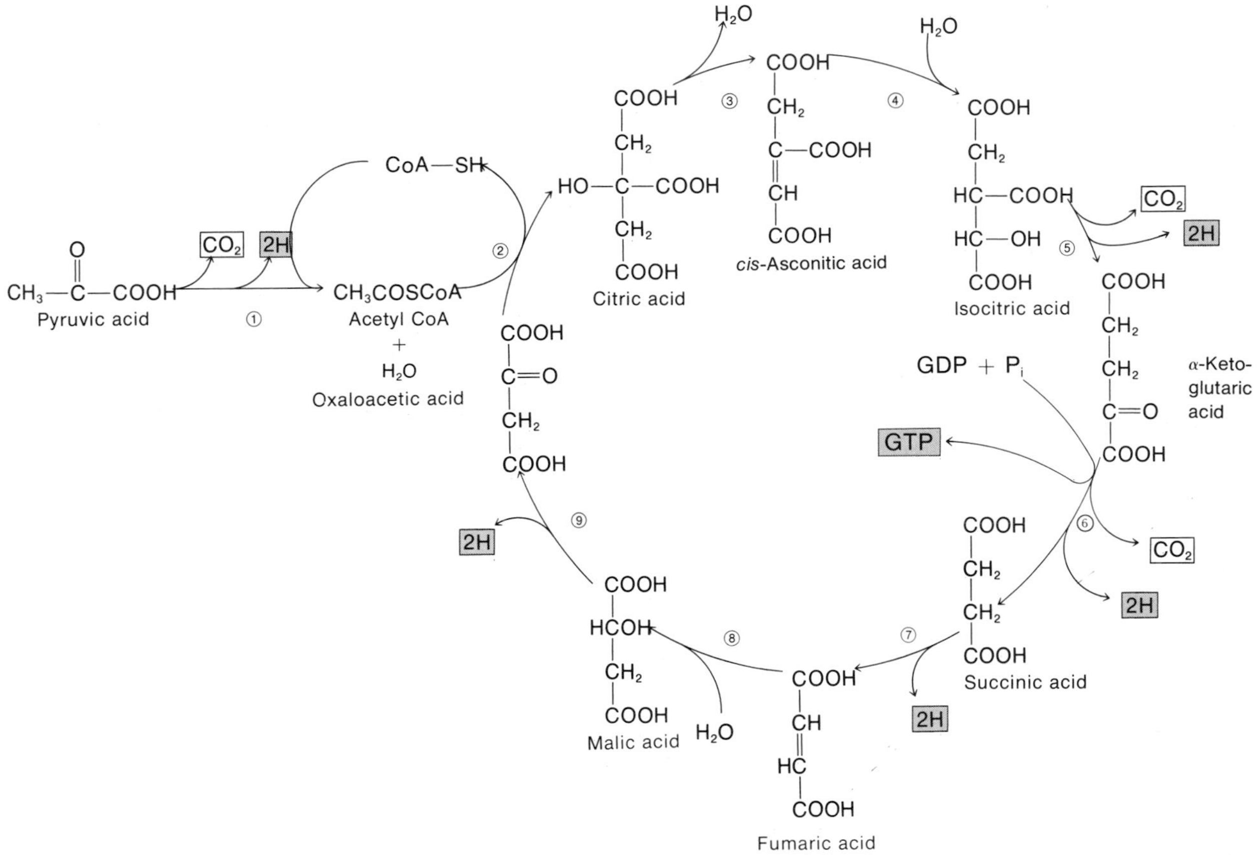

3-45  Tricarboxylic acid (Krebs) cycle. With each circuit, one acetate group is fed in at step 2. The equivalent in carbon, oxygen, and hydrogen atoms is removed in later steps as $CO_2$ and as $H^+ + e^-$ entering the electron-transport chain.

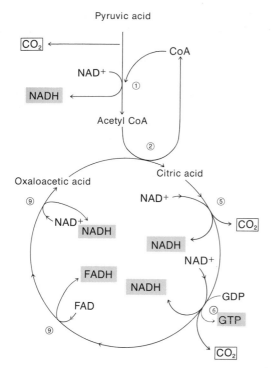

**3-46** Production of $CO_2$, NADH, FADH, and GTP in the tricarboxylic acid cycle. The circled numbers refer to the corresponding steps in Figure 3-45. [Vander et al., 1975.]

mitochondrial membrane, to which is bound *succinic dehydrogenase*. This enzyme, along with $FAD^+$, removes two hydrogen atoms from *succinic acid* to form *fumaric acid*. Another oxidation occurs at step 9 when *malic acid* is converted to *oxaloacetic acid* by the transfer of two hydrogen atoms to $NAD^+$. A new acetate residue is then condensed with the oxaloacetate to reconstitute the citric acid molecule, and thus the cycle repeats.

Each time one circuit of the TCA cycle is completed, two carbon atoms and four oxygen atoms are removed as two molecules of $CO_2$ (Figure 3-46), and with each circuit eight hydrogen atoms are removed, two at a time. These hydrogens (as electrons accompanied by protons) are oxidized to $H_2O$ by molecular oxygen via $NAD^+$, $FAD^+$, and the cytochrome of the

respiratory chain. The $CO_2$ leaves the mitochondrion and then the cell by simple diffusion, and is finally eliminated as a gas via the circulatory and respiratory systems.

## Efficiency of Energy Metabolism

The direct oxidation (burning) of glucose and its metabolic oxidation both liberate the same amount of free energy—namely, 686 kcal. If water is boiled by the heat of burning glucose to produce steam pressure for a steam engine, the mechanical output of the engine divided by the free-energy drop of 686 kcal represents the efficiency of the conversion of chemical to mechanical energy. Modern steam engines have attained efficiencies of approximately 30%. Now let us see how efficiently the living cell transfers chemical energy from glucose to ATP.

Under standard conditions, it takes about 7 kcal to phosphorylate 1 mol of ADP to form ATP. If the free energy of glucose were conserved with an efficiency of 100%, each mole of glucose could energize the synthesis of 98 (i.e., 686/7 = 98) mol of ATP from ADP and inorganic phosphate. As we shall see, only 38 mol of ATP is synthesized, giving an overall efficiency of about 42% or more.† The remaining free energy is liberated as metabolic heat, which accounts for a part of the heat that warms and thereby increases the metabolic rate of the tissue. Essentially all the energy incorporated into ATP and transferred to other molecules is eventually degraded to heat. The oxidation of fossil fuels represents a long-delayed return of stored energy to the original low-energy, high-entropy state of $CO_2$ and water.

---

†The 42% calculated here is for standard conditions. The efficiency of energy conservation may in fact be as high as 60%, for under intracellular conditions the free energy of hydrolysis of ATP has been estimated to be greater than that under standard conditions. The energetic efficiency of ATP production is therefore substantially better than that of a steam engine, or in fact better than any other method as yet devised by humans for converting chemical energy to mechanical energy.

$$CH_3-\underset{\underset{\text{Lactic acid}}{|}}{\overset{\overset{OH}{|}}{CH}}-COOH \xrightarrow[\underset{NAD^+ \quad\quad NADH}{}]{\text{Lactate dehydrogenase}} CH_3-\overset{\overset{O}{\|}}{C}-COOH$$
Pyruvic acid

**3-47** Regeneration of pyruvate from lactate by reduction of $NAD^+$ to NADH. $NAD^+$ is replenished during oxidative phosphorylation.

It is interesting to compare the efficiencies of anaerobic versus aerobic glucose metabolism, keeping in mind that since each mole of glucose yields 2 mol of the three-carbon derivatives, it is necessary to double all the molarities beyond step 5 of glycolysis. In anaerobic glycolysis, there is a net production of 2 mol of ATP per mole of glucose (Figure 3-42), since 2 of the 4 mol produced by substrate-level phosphorylation of ADP are consumed in the priming phosphorylations. The 2 mol of NADH produced by the oxidation of 3-phosphoglyceraldehyde (step 6, Figure 3-39) is oxidized again to $NAD^+$ when the two pairs of hydrogen atoms are transferred to 2 mol of pyruvic acid to form 2 mol of lactic acid under anaerobic conditions (Figure 3-43).

Under aerobic conditions, each of the 2 mol of NADH produced in glycolysis by the oxidation of 3-phosphoglyceraldehyde yields 3 mol of ATP during respiratory phosphorylation (Figure 3-38). Pyruvic acid goes on to fuel the TCA cycle, yielding a total of 10 pairs of hydrogen atoms for every 2 mol of pyruvic acid (Figure 3-45). Eight pairs are carried by $NAD^+$, yielding 24 mol of ATP, while two pairs are carried by the coenzyme $FAD^+$, yielding 4 mol of ATP. Finally, 2 mol of GTP is produced by substrate-level phosphorylation of guanosine diphosphate during the oxidation of $\alpha$-ketoglutaric acid to succinic acid (step 6 in the TCA cycle). This adds up to 38 mol of nucleotide triphosphate per mole of glucose during aerobic respiration. As noted earlier, only 2 mol is produced during anaerobic respiration. Thus, although aerobic respiration conserves about 42% of the free energy of the glucose molecule, anaerobic respiration conserves only about 2%. Stated differently, the energy conservation of glucose metabolism via aerobic glycolysis and the TCA cycle is about 20 times as efficient as that via anaerobic glycolysis. It is not surprising, then, that most animals carry on aerobic metabolism and require molecular oxygen for survival.

## Oxygen Debt

When animal tissue, such as active muscle, receives less $O_2$ than is required to produce adequate amounts of ATP by respiratory chain phosphorylation, some of the pyruvic acid, instead of going on to fuel the TCA cycle, is reduced to lactic acid. For every 2 mol of pyruvic acid reduced, 2 mol of NADH is oxidized (Figure 3-43), costing 6 mol of ATP that might have been synthesized by respiratory chain phosphorylation. If the oxygen deficiency is maintained, lactic acid concentrations rise, and some may be lost into the extracellular space and circulatory system. When the muscle stops its strenuous activity, the accumulated lactic acid is oxidized by $NAD^+$ and the enzyme lactate dehydrogenase back to pyruvic acid (Figure 3-47). Respiratory chain oxidation of the NADH produced in this reaction restores the ATP forfeited during the anaerobic formation of lactic acid. Moreover, the pyruvic acid regenerated from lactic acid goes on to fuel the TCA cycle. Thus, the oxygen-poor state of the tissue results in a switch to anaerobic glycolysis, in which ATP is formed with low efficiency. But the unused chemical energy is stored in the tissue as lactic acid, and later becomes available for aerobic metabolism when sufficient oxygen is available. With cessation of heavy exercise, the respiratory and circulatory systems continue for some time to supply large amounts of oxygen in order to "repay" the oxygen debt that was built up as an accumulation of lactic acid.

# References Cited

Baker, J. J. W., and G. E. Allen. 1965. *Matter, Energy, and Life*. Reading, Mass.: Addison-Wesley.

Baldwin, E. 1964. *Comparative Biochemistry*. New York: Cambridge University Press.

Barry, J. M., and E. M. Barry. 1969. *An Introduction to the Structure of Biological Molecules*. Englewood Cliffs, N. J.: Prentice-Hall.

Buchner, E. 1897. Alcoholic fermentation without yeast cells. In M. Gabriel and S. Fogel, eds., *Great Experiments in Biology*. Englewood Cliffs, N. J.: Prentice-Hall, 1955.

Dowben, R. M. 1971. *Cell Biology*. New York: Harper & Row.

Goldsby, R. A. 1967. *Cells and Energy*. New York: Macmillan.

Jacob, F., and J. Monod. 1961a. Genetic regulatory mechanisms in the synthesis of proteins. *J. Molec. Biol.* 3:318–356.

Jacob, F., and J. Monod. 1961b. On the regulation of gene activity. *Cold Spring Harbor Symp. Quant. Biol.* 26:193–209.

Keeton, W. T. 1972. *Biological Science*. 2nd ed. New York: Norton.

Lehninger, A. L. 1965. *Bioenergetics*. Menlo Park, Calif.: Benjamin.

Lehninger, A. L. 1970. *Biochemistry*. New York: Worth.

Michaelis, L., and M. L. Menten. 1913. Der Kinetik der Invertinwirkung. *Biochem. Z.* 49:333–369.

Patton, A. R. 1965. *Biochemical Energetics and Kinetics*. Philadelphia: Saunders.

Vander, A. J., J. H. Sherman, and D. S. Luciano. 1975. *Human Physiology*. New York: McGraw-Hill.

# CHAPTER *4*

# *Permeability and Transport*

*T*he membranes found at the surface of all living cells and within them perform functions of great importance to the integrity and activities of cells and tissues. So important, pervasive, and varied are these functions in nearly all areas of physiology that two chapters are required to provide an elementary understanding of biological membranes. We will begin here by considering membrane structure and transport functions. The electrical behavior of cell membranes follows in the next chapter.

The most obvious function of membranes is compartmentalization. Membranes are never seen to terminate with free ends; they always form enclosed compartments. The largest of these is, of course, formed by the *surface membrane,* also termed *cell membrane, plasma membrane,* or *plasmalemma.* This compartment encloses the cytosol (free cytoplasm) and all cell organelles and inclusions, such as mitochondria, vesicles, nucleus, and reticulum. Many of these constitute still smaller subcompartments, separated from the cytosol by their own surface membranes. Membranes act as barriers to free diffusion. The enormous retardation of free diffusion enables membranes, with the aid of metabolic mechanisms, to regulate the *net* movement and hence the concentrations of substances in cellular or subcellular compartments. The existence of *concentration gradients* across the membrane implies that the membrane actively participates in the transport of substances into or out of the cell. Indeed, the cell membrane regulates the cytoplasmic concentration of dissolved ions and molecules rather precisely, establishing an *intracellular milieu* conducive to the finely balanced metabolic and synthetic activities of the cell.

Experimental studies demonstrate that containment and regulation of the cell contents is only one function of membranes. Other functions include enzymatic activity carried out by enzyme molecules imbedded in membranes, as in the conversion of ATP to cyclic adenosine monophosphate (cAMP), the oxidation of succinic acid, electron transport and respiratory chain phosphorylation, and so on; enzymatic assembly of secretory products in Golgi apparatus membranes; transduction of environmental stimuli into electrical signals; conduction of bioelectric impulses; release of synaptic transmitter substances; and pinocytosis. Finally, it should not be overlooked that membranes perform the functions of compartmentalization and concentration of solutes within the various intracellular organelles, producing microenvironments that differ from the environment of the cytosol.

As recently as the 1930s, the existence of a differentiated membrane structure at the surface of the cell was still a matter of uncertainty and debate. Since there was little or no direct anatomical evidence for biological membranes at that time, their existence could only be inferred from physiological studies. The first observations on the diffusion-limiting properties of the cell surface were made in the mid-nineteenth century by Karl Wilhelm von Nägeli, who noticed that the cell surface acts as a barrier to free diffusion of dyes into the cell from the extracellular fluid. From this he deduced the presence of a "plasma membrane." He also discovered the osmotic behavior of cells, noting that cells swell when placed in dilute solutions and shrink in concentrated solutions. Several years later Wilhelm Pfeffer drew a parallel between the osmotic properties of artificial semipermeable membranes and the osmotic properties of living cells. This observation added to the growing evidence that living tissue obeys the laws of physics and chemistry.

Using red blood cells as osmometers (indicators of osmotic pressure), E. Overton discovered late in the nineteenth century that there is a close relation between the lipid solubility of a substance and its ability to penetrate the cell membrane: the greater its lipid solubility, the less effective it is in producing osmotic shrinkage of the cell. He correctly interpreted this to indicate that high lipid solubilities allow a substance to penetrate the cell membrane more readily. This finding provided the first evidence that membranes contain substantial quantities of lipids.

Morphological evidence for the existence of a discrete cell membrane came with the development of methods for ultrathin sectioning of tissues chemically fixed for electron microscopy. Seen at the surface of every cell type is a continuous layer (Figure 4-1) that takes on electron-opaque stains more strongly than the free cytoplasm. The range of membrane thickness is 60–120 Å.

More recently, freeze-fracture methods combined with scanning electron microscopy have permitted visualization of the fine structure of membrane surfaces (Figure 4-10).

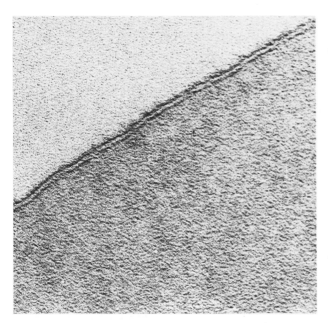

4-1   Electron micrograph of plasma membrane seen in cross section. The cell interior (lower right) is separated from the cell exterior by the surface membrane, which is seen as a dark-light profile about 100 Å thick. The dark-light-dark sandwichlike appearance is due to the differential staining of the "unit membrane" by an electron-opaque substance during preparation of the tissue. [Robertson, 1960.]

## Membrane Composition

Membranes consist almost entirely of lipids and proteins. The relative amounts of lipid and protein vary greatly among the membranes of different organelles. The enzymatic properties of membranes are, of course, due to *membrane proteins,* such as the flavoproteins and cytochromes of mitochondrial inner membranes, the ATPase associated with active transport mechanisms, and the adenylate cyclase that catalyzes the conversion of ATP to cAMP. Proteins other than enzymes have also been isolated from some membranes, including those of mitochondria. In one such nonenzymatic protein, alteration of the molecular structure by the substitution of a single amino acid residue leads to the leakage of cytochrome *a* from the mitochondrion. Thus, nonenzymatic proteins as well as enzymes appear to be functionally important in membranes. Examples of important nonenzymatic membrane proteins are those that make up specific ion channels and their gating mechanisms and those that comprise membrane receptors that bind hormones and other messenger molecules. Some of these proteins are intimately associated with lipid molecules because of lipophilic groups exposed on the surface of the protein molecule. The protein–lipid complex is termed a *lipoprotein*.

Membrane lipids, being smaller and simpler molecules than membrane proteins, are more completely known and can be conveniently divided into three major classes. The *phosphoglycerides* (e.g., phosphatidyl ethanolamine and phosphatidyl choline) are characterized by a glycerine backbone; the *sphingolipids* (e.g., the sphingomyelins) have backbones made of sphingosine bases. These two classes of molecules are *amphipathic;* that is, they have polar heads and nonpolar tails (Figure 4-2). The polar head groups are hydrophilic (water soluble), and the nonpolar tail groups are hydrophobic (water insoluble).

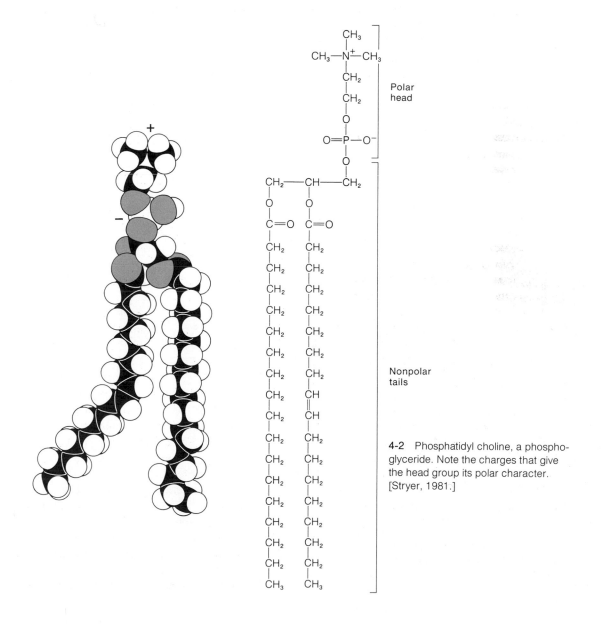

4-2   Phosphatidyl choline, a phosphoglyceride. Note the charges that give the head group its polar character. [Stryer, 1981.]

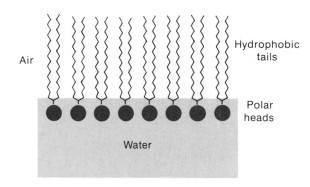

**4-3** Orientation of phospholipid molecules at an air–water interface. The polar heads of the molecules seek the company of water, and the hydrophobic tails project into the air.

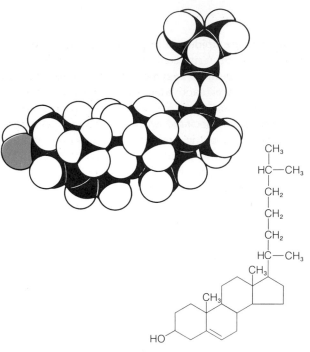

**4-4** Cholesterol, a sterol. [Lehninger, 1975.]

The dual nature of these membrane lipids, with their hydrophilic heads and hydrophobic tails, is important to the organization of biological membranes, for the polar heads of these molecules seek the company of water (Figure 4-3), and their nonpolar tails seek each other's company and are mutually attracted by van der Waals forces. Thus, these molecules are ideally suited to form an interface between a nonaqueous lipid environment ("phase") within the membrane and the aqueous intra- and extracellular phases in contact with the two membrane surfaces. This concept forms the basis of the most widely accepted models of membrane structure, as discussed in the next section.

The remaining major class of membrane lipids is the *sterols* (e.g., cholesterol) (Figure 4-4). The sterols are largely nonpolar and only slightly soluble in water. In aqueous solution they form complexes with proteins. The complexes are far more water soluble than the sterols are alone. Once in the membrane, the sterol molecule fits snugly between the hydrocarbon tails of the phospholipids and sphingolipids (Figure 4-5) and acts to increase the viscosity of the hydrocarbon core of the membrane.

The hydrophobic properties of the hydrocarbon tails of the phospholipids are believed to be responsible for the low permeability of membranes to polar substances (e.g., the inorganic ions and such polar nonelectrolytes as sucrose and inulin).

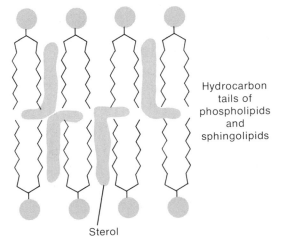

**4-5** The nonpolar sterols insert themselves between the hydrocarbon tails of the polar lipids in the membrane.

Membrane lipids are important to the activity of some membrane-bound enzyme molecules. For example, the highly organized enzyme systems of mitochondrial membranes are inactivated when extracted and isolated from membrane lipids. Likewise, certain transport enzymes associated with the surface membrane lose their enzymatic activity if deprived of specific lipids. The enzyme $\beta$-galactosidase, isolated from surface membranes, can be reactivated by the addition of phosphatidyl serine; other phospholipids, however, are ineffective. These examples illustrate the functional interrelations of the various molecules that make up a membrane.

## *Membrane Organization*

The precise molecular mechanisms involved in various membrane functions are poorly understood. To study these mechanisms, it is helpful to know how the molecules of the membrane are assembled into functioning units. But the structural and functional integrity of the membrane is lost when its components are isolated and purified, and this loss places severe limitations on studies of membrane organization and substructure. To make matters still more difficult, there appears to be a broad spectrum of membrane types. It is now apparent that many of the controversies among proponents of different theories of membrane organization arose because of this diversity.

Information on structure and organization of membranes has come principally from three approaches: (1) chemical dissection, (2) inferences based on the physical properties of membranes, and (3) preparation of artificial membranes into which selected molecules are introduced in order to study their functions.

### SIMPLE BILAYER MODELS

In 1925, E. Gorter and F. Grendel published the results of experiments that led to one of the major models of membrane organization. They dissolved the lipids from red blood cell ghosts† and allowed the extracted mem-

---

†A "ghost" is the empty membrane sac remaining after the red blood cell is hemolyzed by a hypotonic solution.

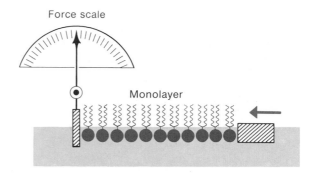

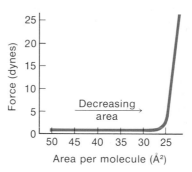

**4-6**   Gorter and Grendel's experiment, which led to the lipid bilayer model of the membrane. Lateral compression of the lipid film produces an abrupt increase in the measured force, indicating the point at which the molecules are crowded into a dense single layer.

brane lipids to spread out on the surface of water in a trough. Because of their asymmetry, lipid molecules become oriented so that their polar head groups form hydrogen bonds with the water, and their hydrophobic hydrocarbon chains stick up into the air, as in Figure 4-3. The film of dispersed lipid molecules at the air–water interface was gently compressed laterally, and the force required to compress the film was measured. This force was low as long as the lipid molecules were dispersed on the surface. The formation of a uniformly compact monolayer was signaled by a sharp increase in the force required for further lateral compression (Figure 4-6). The area of the monolayer of lipid was then measured and found to be about twice the calculated surface area of the cell membranes from which they were extracted. Gorter and Grendel concluded from this observation that the lipid of the cell membrane is arranged in a bilayer composed of two monomolecular layers of molecules oriented so that the two are apposed to one another, with the hydrophobic hydrocarbon ends of the ordered molecules facing inward, or toward each other, and the polar groups facing outward, or toward the aqueous phases adjacent to the surfaces of the double layer (Figure 4-7A). This arrangement is the essence of the *bimolecular leaflet* or *lipid bilayer* model of membrane structure. Evidence for this concept is listed in Box 4-1. Later studies showed that the lipid content of the red blood cell is only $1\frac{1}{2}$ times the surface area. The remainder is now known to be made up of proteins, as explained in the next subsection.

A similar model developed subsequently by J. F. Danielli is based on the same double layer of phospholipid molecules, but in addition it contains protein-lined pores and a layer of protein molecules on the surfaces of the lipid bilayer (Figure 4-7B). Danielli's original reason for adding surface proteins to the model was to account for the relatively low surface tension of cell membranes compared to that of an oil–water interface. The low surface tension is now understood to result from the hydrophilic properties of the phospholipid head groups of the bilayer molecules.

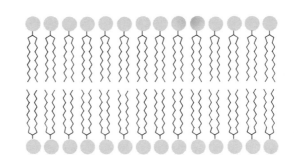

A

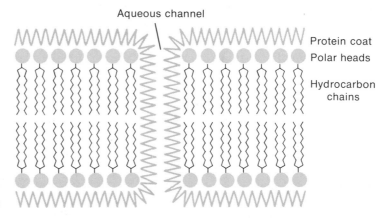

Aqueous channel

Protein coat
Polar heads

Hydrocarbon
chains

B

4-7   (A) Gorter–Grendel bimolecular leaflet. (B) Danielli membrane model, showing both lipid and protein components of the membrane.

## Box 4-1    *Evidences for the Lipid Bilayer Membrane*

*1.* The lipid content of membranes is consistent with a bilayer of oriented lipid molecules, as first shown by Gorter and Grendel.

*2.* The ease with which nonelectrolytes pass through the membrane corresponds to their partitioning between oil and water (Figure 4-24). That is, the greater the tendency for a molecule to leave an aqueous phase for a lipid phase, the more permeant the molecule. This situation suggests that substances crossing the membrane encounter a barrier layer of lipid. Moreover, certain lipid-insoluble substances must first be converted to a lipid-soluble form (by attachment of a lipid-soluble molecule) before they can cross the membrane.

*3.* The capacitance of biological membranes, typically $10^{-6}$ fd/cm$^2$, is the same as that of a layer of lipid the thickness of two phospholipid molecules placed end to end (i.e., 60–75 Å).

*4.* When fixed with permanganate, membranes appear as triple-layered profiles, a lightly staining central zone sandwiched between two electron-dense outer layers (Figure 4-1), with a total thickness of about 75 Å. In 1955 J. D. Robertson named this three-layered structure the *unit membrane*. The unit-membrane concept is consistent with a bimolecular layer of lipid between two layers of protein.

*5.* The thickness of a lipid bilayer calculated as twice the length of a single membrane lipid molecule agrees roughly with the dimensions of the inner zone (ca. 75 Å) of the unit membrane seen in electron micrographs.

*6.* Freeze-etch electron microscopy shows that membranes have a preferential plane of splitting down the middle, which is consistent with separation of a bilayer into two monolayers.

*7.* Reconstituted, or artificial, lipid bilayers (Box 4-2), similar in thickness and presumed structure to the bimolecular lipid core of the Danielli model, have permeabilities and electrical properties fundamentally similar to those of cell membranes. Those differences that exist can be attributed to special channels and carriers present in natural membranes.

### THE FLUID MOSAIC MODEL

As we have noted, chemical fractionation of membranes confirms that proteins are a component of membranes; moreover, the enzymatic properties of membranes exhibited in active transport and other metabolic functions require the participation of proteins. For some time it was speculated that these proteins have a globular conformation, since all other known enzymes are globular proteins or hàve active groups that are globular. Optical studies confirmed the globular nature of membrane proteins. It was also found that the protein molecules are largely free to diffuse laterally along the membrane, presumably because of the fluidity of the lipid matrix. In addition, labeling studies demonstrated that protein molecules or parts of molecules facing one side of the membrane differ from those facing the other side and that they normally do not flip-flop across the membrane. On the basis of such evidence, Singer and Nicolson (1972) proposed the *fluid mosaic* *model* of the membrane, in which globular proteins are integrated with the lipid bilayer, with some protein molecules penetrating the bilayer completely and others penetrating only partially (Figure 4-8). These *integral proteins* are thought to be amphipathic, their nonpolar portions buried in the hydrocarbon core of the bilayer and their polar portions protruding from the core to form a hydrophilic surface with charged amino acid side groups in the aqueous phase. Uncharged hydrophobic side groups, on the other hand, are associated with the hydrocarbon bilayer (Figure 4-9).

Morphological evidence for the mosaic arrangement of globular proteins in a lipid bilayer is seen in Figure 4-10, which shows three freeze-etch electron micrographs of the surface of a membrane. Globular units in the membrane are removed by progressive proteolytic digestion. The specificity of the protein-digesting enzyme used in this experiment indicates that these globular units are proteins.

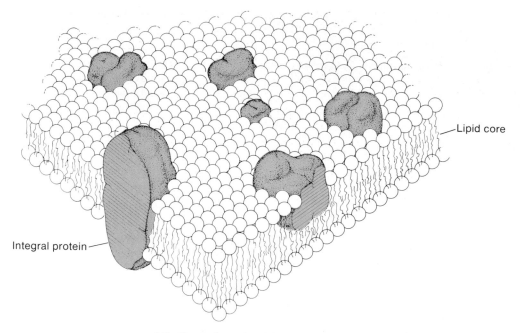

**4-8**   Three-dimensional diagram of the fluid mosaic model of the membrane, showing the globular integral proteins embedded in the lipid bilayer. [Singer and Nicolson, 1972.]

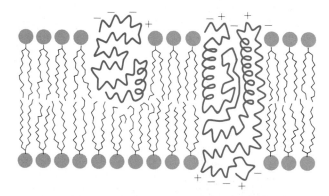

**4-9**   Cross-sectional view of the mosaic bilayer model, showing the charged hydrophilic amino acid side groups of the proteins projecting into the aqueous phase and the uncharged hydrophobic groups in the contact with the lipid phase of the bilayer. [Singer and Nicolson, 1972.]

The fluid mosaic bilayer model appears to offer the most acceptable structural picture of the surface membrane and most of the intracellular membranes. It incorporates both large areas of uninterrupted phospholipid bilayer, for which there is very strong evidence, and the globular proteins responsible for the metabolic activities of membranes. It is also consistent with the evidence for intimate interaction between membrane proteins and lipids in the form of lipoproteins.

*SUBUNIT MODELS*

In some membranes, such as those of mitochondria and visual receptor cells, the mosaic bilayer model appears to be modified so that the lipid bilayer is largely obliterated by identical repeating macromolecular units. In visual receptor cells, these units are molecules of

visual pigment, consisting primarily of protein. In mitochondrial membranes, which have an unusually high proportion of protein, there is specialization for intense enzymatic activity, and the subunits appear to be complexes of enzyme molecules. Electron micrographs suggest repeating units having a globular shape (Figure 4-11). Fractionation of mitochondrial membranes has yielded several components with specific enzymatic functions. When those components are allowed to reassemble, they regain the ability to perform the complete sequence of reactions characteristic of the intact membrane. On the other hand, mixtures containing all the dissociated components of the membrane are unable to perform the sequence of reactions. So apparently an important function of certain membranes is to bestow a highly ordered structural arrangement on enzymatic subunits involved in tightly coupled sequential reactions.

It now appears that at one end of the spectrum of membrane organization, there is the metabolically inert myelin sheath (Figure 4-12) of certain nerve cells, in which the lipid bilayer is largely uninterrupted. At

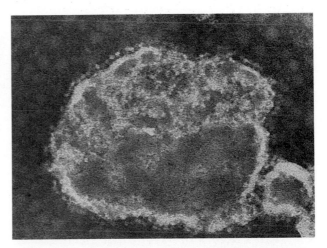

**4-10**  Freeze-etch electron micrographs showing morphological evidence for the mosaic membrane model. In each preparation, the membrane split along the middle of the bilayer, exposing membrane-embedded particles with diameters of 50–80 Å. Digestion with a proteolytic enzyme produced progressive loss of these particles, indicating that they are globular proteins inserted into the lipid phase of the membrane. (A) Control; (B) 45% of the particles digested; (C) 70% digested. Magnification 55,000×. [Courtesy of L. H. Engstrom and D. Branton.]

**4-11**  Electron micrograph of negatively strained fragment of the inner mitochondrial membrane prepared from mammalian heart muscle. Magnification 152,000×. Note the orderly arrangement of knoblike projections. [Courtesy of B. Tandler.]

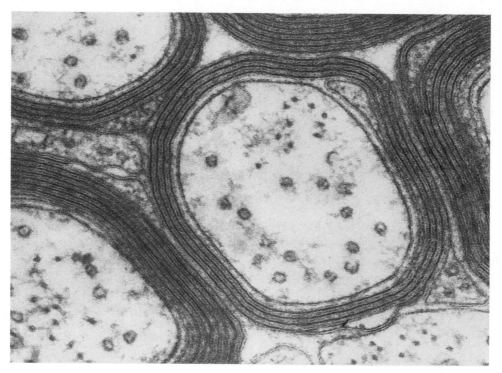

**4-12**    Electron micrograph of myelin sheath surrounding a nerve fiber cut in cross section. The sheath is formed during development by many layers of Schwann cell surface membrane wrapped spirally around the nerve fiber. Magnification 75,000×. [Peters and Vaughn, 1970.]

the other end of the spectrum, there is the rapidly metabolizing mitochondrial membrane, composed almost entirely of repeating subunits of ordered enzymatic aggregates. In between these extremes are the surface membrane and most intracellular membranes, in which the mosaic bilayer is interrupted here and there by lipoprotein molecules or aggregates of molecules. Thus, the basic structure of the bilayer appears to be modified as required for the functional specializations of different membranes.

## Physical Factors in Membrane Permeation

### DIFFUSION

Before we can properly consider the mechanisms by which substances cross membranes, it is necessary to review the physics of solute and solvent displacement in solution and across semipermeable membranes. This displacement is based on the concept of *diffusion*. Random thermal motion of suspended or dissolved molecules causes their gradual dispersion from regions of higher to regions of lower concentration. The dispersion is a very slow process when viewed on a large scale. A crystal of copper sulfate dissolves in water, spreading so slowly that it takes days for a liter of undisturbed water to become uniformly dyed by the salt. However, at the microscopic dimensions of cell function, times for diffusion are short, in some instances being measured in fractions of 1 millisecond (1/1000 s).

The rate of diffusion of a solute $s$ can be defined by the relationship:

$$\frac{ds}{dt} = D_s A_s \frac{dC}{dx} \qquad (4\text{-}1)$$

in which $ds/dt$ is the rate of diffusion (i.e., displacement of $s$ per unit time), $D_s$ is the *diffusion coefficient* of $s$, $A_s$ is the cross-sectional area through which $s$ is diffusing, and $dC/dx$ is the concentration gradient of $s$ (i.e., the change in concentration with distance). The gradient factor $dC/dx$ is clearly very important, since it will determine how fast $s$ diffuses down the gradient. $D_s$ varies with the nature and molecular weight of $s$ and the solvent, which in most physiological situations will, of course, be water.

## MEMBRANE FLUX

If a solute exists on both sides of a membrane through which it can diffuse, it will exhibit a *unidirectional flux* in each direction (Figure 4-13A). The flux can be expressed as the amount of the solute that penetrates

a unit area of membrane every second in the direction under consideration, or

$$J = \frac{ds}{dt} \qquad (4\text{-}2)$$

in which $J$ is the unidirectional flux, $s$ is the amount of solute in moles traversing a unit area of membrane, $t$ is the time in seconds, and $ds/dt$ is the amount of solute traversing a unit area of membrane per unit time (moles per square centimeter per second) in the direction under consideration. The flux in one direction (say, from cell exterior to cell interior) will be considered independent of the flux in the opposite direction. Thus, if the *influx* and *efflux* are equal, the net flux is zero. If the unidirectional flux is greater in one direction, there is said to be a *net flux,* which is the difference in the two unidirectional fluxes (Figure 4-13B).

The *permeability* of the membrane to a substance refers to the rate at which that substance penetrates the membrane under a given set of conditions. A greater permeability will be accompanied by a greater flux if other factors remain equal. If it is assumed that the membrane is a homogeneous barrier and that a continuous concentration gradient exists for a nonelectrolyte substance between the side of high concentration (I) and the side of low concentration (II), then

$$\frac{ds}{dt} = P(C_{\mathrm{I}} - C_{\mathrm{II}}) \qquad (4\text{-}3)$$

in which $ds/dt$ is the amount of the substance crossing a unit area of membrane per unit time, $C_{\mathrm{I}}$ and $C_{\mathrm{II}}$ are the respective concentrations (moles per cubic centimeter) of the substance on the two sides of the membrane, and $P$ is the *permeability constant* of the substance, with the dimension of velocity (centimeters per second).

The permeability constant incorporates all the factors inherent in the membrane and in the substance in question—factors that determine the probability with which a molecule of that substance will cross the membrane. This can be expressed formally as

$$P = \frac{D_m K}{x} \qquad (4\text{-}4)$$

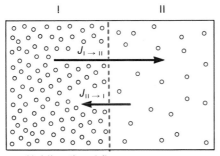

**A** Unidirectional flux

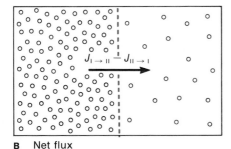

**B** Net flux

**4-13** Movement of a solute through a membrane. **(A)** The arrows represent the actual unidirectional fluxes of a substance between compartments I and II. **(B)** The single arrow indicates the resulting net flux.

in which $D_m$ is the diffusion coefficient of the substance within the membrane (the more viscous the membrane or the larger the molecule, the lower this value), $K$ is the partition coefficient of the substance, and $x$ is the thickness of the membrane.

Permeability constants for different membranes and different substances vary greatly. The permeability of red blood cells to different solutes ranges from $10^{-12}$ to $10^{-2}$ cm/s. Furthermore, the permeability of many membranes to given substances can be altered greatly by hormones and other molecules that react with receptor sites on the membrane and thereby influence channel size or carrier mechanisms. Antidiuretic hormone, for example, can increase the water permeability of the renal collecting duct in mammals by as much as 10 times. Similarly, neural transmitter substances, acting on specialized portions of the surface membranes of nerve and muscle cells, produce large increases in permeability to an ion such as $Na^+$, $K^+$, $Ca^{2+}$, or $Cl^-$.

## OSMOSIS

Osmotic pressure is the colligative property of greatest importance to living systems. It was first noted by Abbé Jean Antoine Nollet, in 1748, that if pure water is placed on one side of an animal membrane (e.g., bladder wall) and an aqueous solution is placed on the other side, the water passes through the membrane into the solution. This movement of water down its concentration gradient was termed *osmosis* (from Greek for "push"). It was later found that this phenomenon produces a hydrostatic pressure gradient. As can be seen in Figure 4-14, the pressure difference causes a rise in the level of the solution as water diffuses through the *semipermeable membrane* into the solution. The rise in the level of the solution continues until the net rate of water movement (the net *flux*) across the membrane becomes zero. This state occurs when the hydrostatic pressure of the solution in compartment II is sufficient to force water molecules back through the membrane from compartment II to compartment I at the same rate that osmosis causes water molecules to diffuse from I to II. The hydrostatic back pressure required to cancel the osmotic diffusion of water from compartment I to compartment II is termed the *osmotic pressure* of the solution in compartment II.

Pfeffer, in 1877, made the first quantitative studies of osmotic pressure. By depositing a "membrane" of

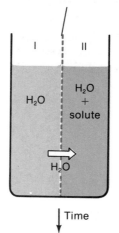

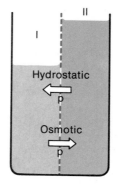

4-14  Osmotic pressure forces water to enter compartment II from compartment I until the hydrostatic pressure difference equals the opposing osmotic pressure difference. When the pressures are equal, the flux is zero. Compartment I contains pure water; compartment II, water with impermeant solute.

copper ferrocyanide on the surface of porous clay cups, he produced membranes that would allow water molecules to diffuse through them far more freely than sucrose molecules and that were also strong enough, because of the clay substratum, to withstand relatively high pressures without rupturing. Using these membranes, he was able to make the first direct measurements of osmotic pressure. Some of his results are shown in Table 4-1. Note in the table that the osmotic pressure was proportional to the solute concentration.

To demonstrate the role of a semipermeable membrane in osmosis, let us suppose that a 1.0 M aqueous solution of sucrose is carefully layered under an 0.01 M aqueous solution of sucrose. There would be net diffusion of water molecules from the solution of lower sucrose concentration (the 0.01 M solution) into the 1.0 M sucrose solution, and sucrose would show net

**TABLE 4-1** Results obtained by Pfeffer in experimental measurements of the osmotic pressure of sucrose solutions of various concentrations.

| Percentage of sucrose | Osmotic pressure (atm) | Ratio of osmotic pressure to percentage of sucrose |
|---|---|---|
| 1 | 0.70 | 0.70 |
| 2 | 1.34 | 0.67 |
| 4 | 2.74 | 0.68 |
| 6 | 4.10 | 0.68 |

*Source:* Getman and Daniels, 1931.

diffusion in the opposite direction. If we were to place a membrane between these two solutions that would allow water molecules, but not sucrose molecules, to pass through it, the water molecules would still show a net diffusion from the solution in which $H_2O$ is more concentrated (the 0.01 M sucrose solution) into the 1.0 M sucrose solution, in which the concentration of $H_2O$ is lower. The sucrose, however, would be prevented from diffusing because of the membrane barrier. The result would be a net diffusion of water (*osmotic flow*) through the membrane from the solution of lower solute concentration to the solution of higher solute concentration. As explained further on, osmosis provides the basis for net water movement across many biological membranes and epithelia.

Osmotic pressure, $P$, is proportional not only to the concentration, $C$, of the solute but also to the absolute temperature, $T$.

$$P = K_1 C \qquad (4\text{-}5)$$

and

$$P = K_2 T \qquad (4\text{-}6)$$

where $K_1$ and $K_2$ are constants of proportionality. Jacobus van't Hoff related these observations to the gas laws and showed that solute molecules in solution behave thermodynamically like gas molecules. Thus,

$$P = RTC$$

or

$$P = \frac{nRT}{V} \qquad (4\text{-}7)$$

where $n$ is the number of mole equivalents of solute, $R$ is the molar gas constant (0.082 L-atm/K-mol),† and $V$ is the volume in liters. Like the gas laws, however, this expression for osmotic pressure holds true only for dilute solutions; corrections must be made for concentrated solutions and for dissociation of electrolytes.

For purposes of illustration, we can calculate the osmotic pressure of a 0.1 M aqueous solution of NaCl. At 25°C the activity coefficient 0.1 M NaCl will be 0.78. Thus, the mole equivalents/liter of a bimolecular salt solution of this molarity will be

$$2 \times 0.1 \times 0.78 = 0.156 \text{ equiv/L}$$

According to Equation 4-7, then,

$$P = \frac{0.156\,RT}{1\text{ L}}$$

$$= (0.156 \text{ mol})(0.082 \text{ L-atm/K-mol})(298 \text{ K})$$

$$= 3.81 \text{ atm}$$

## OSMOLARITY AND TONICITY

Two solutions that exert the same osmotic pressure through a membrane permeable only to water are said to be *isosmotic* to each other. If one solution exerts less osmotic pressure than the other, it is *hypoosmotic* with respect to the other solution; if it exerts greater osmotic pressure, it is *hyperosmotic*. Osmolarity (or *osmoticity*) is thus defined on the basis of an ideal osmometer—one in which the osmotic membrane allows water to pass but completely prevents the solute from passing. All solutions with the same number of dissolved particles per unit volume have the same osmolarity and are thus defined as isosmotic.

In contrast, *tonicity* is defined in terms of the response of cells or tissues immersed in a solution. A solution is considered to be *isotonic* with a given cell or tissue if the cell or tissue immersed in it neither shrinks nor swells. If the tissue swells, the solution is said to be *hypotonic*, and if it shrinks the solution is said to be *hypertonic* to the tissue. These effects are due, of course, to movement of water across the cell membrane in response to osmotic

---

†$R$ is the constant of proportionality in the gas equation $pv/T = R$ when referring to 1 mol of a perfect gas, and it has the value of 1.985 cal/mol K; $p$ is in atmospheres and $v$ is in liters.

pressure differences between the cell interior and the extracellular solution.

If cells actually behaved as ideal osmometers, tonicity and osmolarity would be equivalent, but this is not generally true. For example, sea urchin eggs maintain a constant volume in a solution of NaCl that is isosmotic relative to seawater, but they swell if immersed in a solution of $CaCl_2$ that is isosmotic relative to seawater. The NaCl solution therefore behaves isotonically relative to the sea urchin egg, whereas the $CaCl_2$ solution behaves hypotonically for the sea urchin egg. The tonicity of a solution depends on the rate of intracellular accumulation of the solute in the tissues in question, as well as on the concentration of the solution. The more readily the solute accumulates, the lower the tonicity of a solution of given concentration or osmolarity. This is so because as the cell gradually loads up with the solute, water follows according to osmotic principles, causing the cell to swell (Figure 4-20). Thus, the terms *isotonic, hypertonic,* and *hypotonic* are meaningful only in reference to actual experimental determinations on living cells or tissues.

The sum of the concentrations of solutes in a solution is sometimes obtained by measuring the colligative properties of the solution (e.g., by measuring the depression of freezing point). Concentrations measured in this way are often expressed in *osmoles* or *milliosmoles.* These units are, in theory, equivalent to the molarity of a solution of an ideal nondissociating solute exhibiting the same colligative properties.

### ELECTRICAL INFLUENCES ON ION DISTRIBUTION

If a molecule bears an electric charge—that is, if it is present in ionized form—its net flux across a membrane will be determined not only by the permeability of the membrane to that ion and the concentration gradient of the ion, but also by the electric potential difference between the two sides of the membrane. This concept will be developed more extensively in Chapter 5, so the following brief observations will suffice here.

*1.* For charged molecules (e.g., $Na^+$, $K^+$, $Cl^-$, $Ca^{2+}$, amino acids) two forces act to produce net passive diffusion of the species across a membrane: (a) the chemical gradient arising from differences in the concentration of the substance on the two sides of the membrane, and (b) the electrical field (i.e., difference in potential

across the membrane) experienced by the ion as it enters the membrane. Clearly, a positively charged ion will tend to move in the direction of increasing negative potential. The sum of the combined forces of concentration gradient and electrical gradient acting on an ion determine the net *electrochemical gradient* acting on the ion.

*2.* It follows that there must be a potential difference just sufficient to balance and counteract the chemical gradient acting on an ion, so as to prevent a net transmembrane flux of the ion in question. The potential at which an ion is in electrochemical equilibrium is termed the *equilibrium potential.* The value of this potential depends on several factors, the most prominent being the ratio of concentrations of the ion in question. For a monovalent ion at 18°C, the equilibrium potential is equal to 0.058 V $\log_{10}$ of the extracellular concentration/intracellular concentration of the ion. Thus, a 58 mV potential difference across the membrane has the same effect on the net diffusion of that ion as a transmembrane concentration ratio of 10:1.

*3.* Passive diffusion of an ion species will therefore take place against its chemical concentration gradient if the electrical gradient (i.e., potential difference) across the membrane is in the opposite direction of and exceeds the concentration gradient. For example, if the interior of a cell is more negative than the equilibrium potential for $K^+$, potassium ions will diffuse into the cell even though the intracellular concentration of $K^+$ is much higher than the extracellular concentration.

Electrical forces do not act directly, of course, on uncharged molecules such as sugars.

### DONNAN EQUILIBRIUM

In 1911, the physical chemist Frederick Donnan examined the distribution of diffusible solutes separated by a membrane that is freely permeable to water and electrolytes but totally impermeable to one species of ion confined to one of two compartments. In this situation, as Donnan discovered, the diffusible solutes become unequally distributed among the two compartments.

To begin with, pure water is placed in both compartments, and some KCl is dissolved in the water in one of them. The dissolved salt ($K^+$ and $Cl^-$) will diffuse through the membrane until the system is in equilibrium—that is, until the concentrations of $K^+$

and $Cl^-$ become equal on both sides of the membrane (Figure 4-15A). If the potassium salt of a nondiffusible anion (a macromolecule, $A^-$, having multiple negative charges) is added to the solution in compartment I, the $K^+$ and $Cl^-$ become redistributed until a new equilibrium is established by movement of some $K^+$ and some $Cl^-$ from compartment I to compartment II (Figure 4-15B). *Donnan equilibrium* is characterized by a reciprocal distribution of the anion and cation such that

$$\frac{[K^+]_I}{[K^+]_{II}} = \frac{[Cl^-]_{II}}{[Cl^-]_I}$$

At equilibrium, the diffusible cation, $K^+$, is more concentrated in the compartment in which the nondiffusible anion, $A^-$, is confined than in the other, whereas the diffusible anion, $Cl^-$, becomes less concentrated in that compartment than in the other.

This equilibrium situation arises from the following physical requirements:

*1.* There must be electroneutrality in both compartments. That is, within each compartment, the total number of positive charges must *virtually* equal the total number of negative charges. Thus, in this example, $[K^+] = [Cl^-]$ in compartment II.

*2.* The diffusible ions $K^+$ and $Cl^-$ must, statistically, cross the membrane in pairs to maintain electrical neutrality. The probability that they will cross together is proportional to the product $[K^+][Cl^-]$.

*3.* At equilibrium the rate of diffusion of KCl in one direction through the membrane must equal the rate of KCl diffusion in the opposite direction. Thus, at equilibrium the product $[K^+][Cl^-]$ in one compartment must be equal to the same product in the other compartment. Letting $x, y,$ and $z$ represent the concentrations of the ions in compartments I and II, as shown in Figure 4-16, we can express the equilibrium condition (i.e., equality of the product $[K^+][Cl^-]$ in the two compartments) algebraically:

$$x^2 = y(y + z) \qquad (4\text{-}8)$$

This equation also holds, of course, if $A^-$ is not present, for in that case $K^+$ and $Cl^-$ are equally distributed, and $z = 0$ and $x = y$.

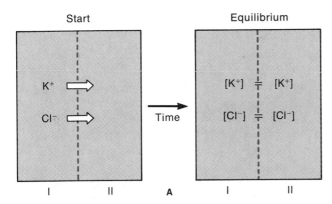

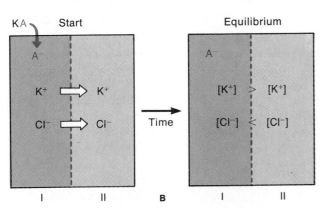

4-15 (A) When KCl is added to compartment I of a container divided by a permeable membrane, $K^+$ and $Cl^-$ diffuse across the membrane until the concentrations are equal on either side. (B) If the potassium salt of an impermeant anion is added to compartment I, some $K^+$ and $Cl^-$ diffuse into compartment II until electrochemical equilibrium is reestablished.

4-16 Algebraic description of the equilibrium condition established in Figure 4-15B after the salt of an impermeant anion is added to compartment I.

By rearranging Equation 4-8, we can see that, at equilibrium, the distributions of the diffusible ions in the two compartments are reciprocal:

$$\frac{x}{y} = \frac{y + z}{x} \qquad (4\text{-}9)$$

Thus, it is evident that as the concentration of the nondiffusible anion, *z,* is increased, the concentrations of the diffusible ions will become increasingly divergent. It is this unequal distribution of diffusible ions that is the outstanding feature of Donnan equilibrium.

This explanation of Donnan equilibrium uses an ideal set of conditions for the sake of simplicity. The living cell and its surface membrane are, of course, far more complex. The "nondiffusible anion" in this example represents various anionic side groups of proteins and other large molecules. The cell membrane is somewhat permeable to a variety of ions and molecules. Although the physical and mathematical principles recognized by Donnan play a role in regulating the distribution of electrolytes in living cells, it is evi-

dent that some *non*equilibrium mechanisms must modify the distribution of many substances across the cell membrane, as discussed in the next section. Thus, the cell cannot be considered a passive osmometer, and the distribution of substances across biological membranes cannot be predicted entirely by the Donnan principles alone.

## Osmotic Properties of Cells

With this physical background we can now turn to the properties of the cell membrane that are responsible for the different concentrations of ions inside and outside the cell (Figure 4-17) and for the regulation of cell volume.

### IONIC STEADY STATE

Although the intracellular concentrations of inorganic solutes (Table 4-2) differ somewhat among different cell types and different organisms, certain generalizations can be made. The most concentrated inorganic

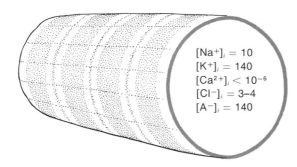

$[Na^+]_i = 10$
$[K^+]_i = 140$
$[Ca^{2+}]_i < 10^{-6}$
$[Cl^-]_i = 3\text{-}4$
$[A^-]_i = 140$

$[Na^+]_o = 120$
$[K^+]_o = 2.5$
$[Ca^{2+}]_o = 2.0$
$[Cl^-]_o = 120$

4-17   Representative permeabilities and concentrations of common ions on the inside and outside of a vertebrate skeletal muscle cell. The concentrations are in millimoles. The concentration given for intracellular $Ca^{2+}$ is for the free, unbound, and unsequestered ion in the myoplasm. As the list of ions is incomplete, the totals do not balance out perfectly. $[A^-]_i$ represents the molar-equivalent negative charges carried by various impermeant anions.

TABLE 4-2   Internal and external concentrations of some electrolytes in some nerve and muscle tissues.

| Tissue | Internal concentrations (mM) | | | External concentrations (mM) | | | Ratios, inside/outside | | |
|---|---|---|---|---|---|---|---|---|---|
| | Na+ | K+ | Cl− | Na+ | K+ | Cl− | Na+ | K+ | Cl− |
| Squid nerve | 49 | 410 | 40-100 | 440 | 22 | 560 | 1/9 | 19/1 | 1/14−1/6 |
| Crab leg nerve | 52 | 410 | 26 | 510 | 12 | 540 | 1/10 | 34/1 | 1/21 |
| Frog sartorius muscle | 10 | 140 | 4 | 120 | 2.5 | 120 | 1/12 | 56/1 | 1/30 |

ion in the cytosol is $K^+$, which is typically 10 to 30 times as concentrated in the cytosol as in the extracellular fluid. Conversely, the internal concentrations of free $Na^+$ and $Cl^-$ are typically less (ca. one-tenth or less) than the external concentrations. Another important generalization is that the intracellular concentration of $Ca^{2+}$ is maintained several orders of magnitude below the extracellular concentration. This situation is due in part to active transport of $Ca^{2+}$ out across the cell membrane and in part to the *sequestering* of this ion within such organelles as the mitochondria and reticulum. As a result, the activity of $Ca^{2+}$ in the cytosol is generally well below $10^{-6}$ M.

Cell membranes are typically far more permeable (ca. 30 times) to $K^+$ than to $Na^+$. Membrane permeability to chloride varies. In some cells, it is similar to potassium, while in others it is lower. The permeability of the cell membrane to $Na^+$ is low; but it is not low enough to prevent sodium from leaking steadily into the cell.

In light of the general leakiness of the cell membrane, let us consider to what degree Donnan equilibrium contributes to the steady-state ionic distributions between cell interior and cell exterior. Three related factors are involved:

*1.* A preponderance of net negative charge resides in the form of anionic sites such as carboxyls on peptide and protein molecules that are nonpermeant and thus trapped within the cell. These charges must be balanced by positively charged counterions such as $Na^+$, $K^+$, $Mg^{2+}$, and $Ca^{2+}$.

*2.* Because such "immobile" anionic sites are trapped within the cell by the inability of the parent peptides and proteins to cross the surface membrane, we have a natural situation similar in some respects to the artificial situation shown at the bottom right of Figure 4-15B and illustrating Donnan equilibrium—namely, a higher concentration of the diffusible cation in the cell relative to the extracellular concentration. If $K^+$ and $Cl^-$ were the only diffusible ions, an equilibrium situation in the cell would indeed develop similar to that in Figure 4-15B. However, the cell membrane is leaky to $Na^+$ and other inorganic ions, and with time the cell would load up with these ions if they were simply allowed to accumulate. This, in turn, would cause

osmotic movement of water into the cell, causing it to swell.

*3.* Such osmotic disasters are avoided by the ability of the cell membrane to pump out $Na^+$, $Ca^{2+}$, and some other ions at the same rate they leak in, keeping the intracellular $Na^+$ concentration about an order of magnitude lower than the extracellular concentration. This active pumping is discussed further later on. Suffice it to note here that this activity confers on the membrane an *effective impermeability* to $Na^+$ and $Ca^{2+}$. As a result, the concentrations of these ions are not allowed to come into equilibrium, and the cell in fact behaves very much on the surface *as if* it were in a state of Donnan equilibrium. In spite of this resemblance, the unequal distribution of ions represents a *steady state* requiring the continual expenditure of energy (to pump ions) rather than a true equilibrium.

Since $K^+$ and $Cl^-$ are by far the most concentrated and most permeant ions in the tissue, they distribute themselves in a way similar to that in an ideal Donnan equilibrium; that is, the *KCl product* ($[K^+] \times [Cl^+]$) of the cell interior will approximately equal the KCl product of the extracellular solution (Figure 4-18), providing the membrane permeabilities of chloride and potassium are both high relative to those of other ions present.

### CELL VOLUME

Plant and bacterial cells possess rigid walls secreted by the cell membrane. These walls place an upper limit on the size of the cell, allowing the osmotic build-up of turgor pressure in these cells. In contrast, animal cells have no rigid walls and therefore cannot build up

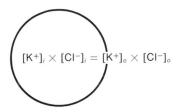

**4-18**  The KCl product. The distribution of $K^+$ and $Cl^-$ will be governed by Donnan equilibrium, provided the membrane is permeable to $Cl^-$ and does not actively transport this ion.

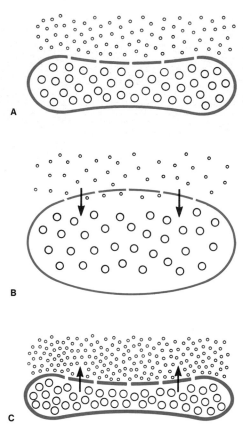

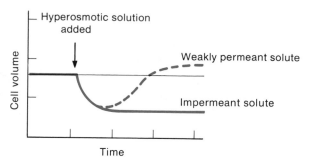

**4-20**   A hyperosmotic solution in which the solute is completely impermeant causes maintained shrinkage of the cell. In that instance, the solution is also hypertonic. If the solute slowly enters the cell, it is followed by the osmotic flow of water and eventually produces swelling. In that instance, the solution is hypotonic in spite of being hyperosmotic.

**4-19**   Osmotic changes in the volume of a red blood cell. (A) In an isotonic solution, the cell volume remains unchanged. (B) In a more dilute solution, water (arrows) enters the cell because of the higher osmoticity of the cytoplasm with respect to the solution, producing swelling. (C) Water leaves the cell in a more concentrated medium, causing shrinkage.

any large intracellular pressure. As a result, they will change size when placed in different concentrations of impermeable substances dissolved in water. This shrinkage or swelling is due to osmotic movement of water (Figure 4-19).

The osmotically active intracellular solutes consist of colloidal molecules of proteins and peptides as well as smaller molecules and diffusible ions. The concentration of impermeant solutes is higher inside the cell than outside; thus, the cell faces the problem of osmotic swelling by entry of water and permeant solutes. There

are two ways in which the surface membrane might prevent osmotic swelling. One is to pump water out as fast as it leaks in; there is no evidence that this occurs, although a similar effect is achieved by the contractile vacuole of certain protozoas. The other, which appears to be the major mechanism for regulation of cell volume, is the active extrusion of solutes that leak into the cell (Figure 4-20). Thus, at steady state, $Na^+$, the major osmotic constituent outside the cell, is extruded from the cell by active transport as rapidly as it leaks in. In effect, there is no net entry. The situation is osmotically equivalent to complete sodium impermeability, with a relatively fixed concentration of $Na^+$ trapped in the cell. Since $Na^+$ is not allowed to further accumulate in the cell, there is no compensatory further osmotic influx of water.

The low intracellular (relative to extracellular) sodium concentration is important in balancing the other osmotically active solutes in the cytoplasm. The importance of active transport in maintaining the sodium gradient, and thereby the osmolarity of the cell and the cell volume, is seen when the energy metabolism of the cell is interrupted by metabolic poisons (Figure 4-21). Without ATP to energize uphill extrusion of $Na^+$, the sodium ion, together with its chloride counterion, leaks into the cell, and water follows osmotically, causing the cell to swell.

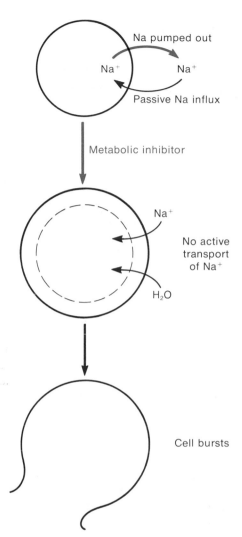

**4-21** Addition of a metabolic inhibitor results in an inability of the cell to pump out the Na+ that steadily leaks in. As a result, [Na+]$_i$ rises, water follows osmotically, and the cell bursts owing to massive swelling.

## *Mechanisms of Passive Permeation*

There are three major routes by which substances are believed to cross membranes without the aid of metabolically energized processes (Figure 4-22). In the first, the solute molecule remains in the aqueous phase and diffuses through *aqueous channels,* that is, water-filled

pores in the membrane. In the second, the molecule leaves the aqueous phase on one side of the membrane, dissolves directly in the lipid layer of the membrane, diffuses across the thickness of the lipid or protein layer, and finally enters the aqueous phase on the opposite side of the membrane. In the third route, the solute molecule combines with a *carrier* molecule dissolved in the membrane. This carrier "mediates" or "facilitates" the movement of the solute molecule across the membrane. In those instances in which the permeating molecules are inorganic ions, the channels or carriers by which they cross the membrane are termed *ionophores.* Because of their lipid solubility, the

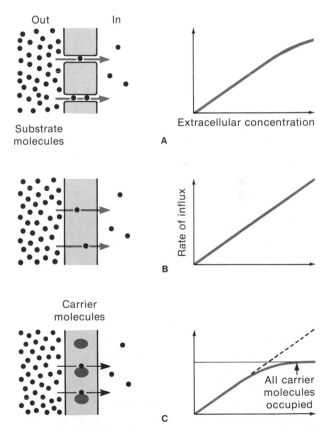

**4-22** Three major mechanisms by which substances cross membranes. (**A**) Diffusion through labile or fixed aqueous channels. (**B**) Dissolving in lipid phase. (**C**) Carrier-mediated transport (either facilitated or active transport).

## *Box 4-2*  *Artificial Bilayers*

Many of our present concepts of how molecules and ions pass across membranes have grown out of experiments and observations on artificial bilayers that are similar to the bimolecular leaflet that forms the basis of the cell membrane. Artificial bilayers are extremely useful in studies of permeation mechanisms because they can be made from chemically defined mixtures of lipids. Selected substances can be added to test their effects on permeability. Channel-forming substances, such as the antibiotic ionophores (molecules that facilitate the diffusion of ions across membranes, Box 4-3) and membrane channel components of excitable tissues, have been incorporated into artificial bilayers, allowing their properties to be studied in isolation under the highly controlled conditions shown in the accompanying figure.

The principle of bilayer formation is shown in the figure.

The most stable configuration attained consists of two layers of lipid molecules whose hydrophobic, lipophilic hydrocarbon tails are loosely associated to form a liquid-lipid phase sandwiched between the hydrophilic, polar ends of the molecules, which are directed outward toward the aqueous medium. The thickness of the lipid film is easily determined from the interference color of light reflected from the two surfaces of the film. Membranes with thicknesses of approximately 70 Å (black interference color) are most commonly used. These membranes have electrical conductances (ion permeabilities) and capacitances consistent with their thickness and lipid composition. Although their permeability to ions is much lower than that of cell membranes, the addition of certain ionophores increases it to values that are characteristic of cell membranes.

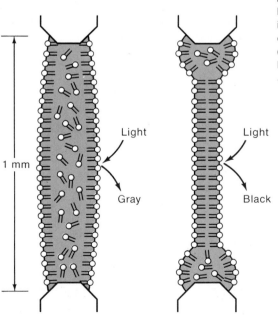

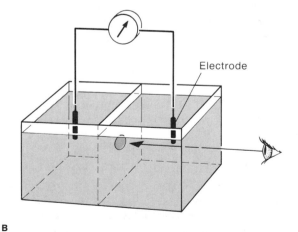

Bilayer of lipid formed in a 1 mm opening between two chambers. (A) A small amount of the lipid in a solvent such as hexane is placed in the opening. As it assumes the more stable bilayer configuration, the interference color changes from gray to black. (B) The chambers are filled with the test medium. The permeability to electrolytes can be measured electrically. [Kotyk and Janáček, 1970.]

A

B

carriers allow the solute molecules to diffuse more readily across the membrane, down their concentration or electrochemical gradient. This is termed *carrier-mediated* (or *facilitated*) *transport* and takes several forms to be discussed below.

### DIFFUSION THROUGH MEMBRANE CHANNELS

Water and certain other polar substances can pass through special channels that penetrate the lipid bilayer. Since water has a very low solubility in long-chain hydrocarbons, it can be predicted that its diffusion through a lipid bilayer will be very slow. It is significant that the permeability of artificial lipid bilayers is as predicted, whereas the permeabilities of natural membranes to water are far higher than predicted for truly uninterrupted lipid films. The few water molecules that do penetrate lipid bilayers may pass through temporary channels between lipid molecules. In natural membranes, aqueous channels formed by hydrophilic portions of protein molecules, as in Figure 4-7B, provide the main pathway for water permeation and most likely account for the relatively high water permeability of these membranes.

The selective permeability of the surface membrane to certain inorganic ions such as $K^+$ and $Cl^-$, and under certain conditions $Na^+$ and $Ca^{2+}$, also suggests that the membrane contains specific water-filled channels through which these ions can diffuse. Again, the existence of such channels is supported by studies of artificial lipid bilayer membranes (Box 4-2). Such membranes have very low permeabilities to inorganic ions, which, being highly polar and hydrophilic, avoid entry into the lipid phase of the membrane. However, the addition to the membrane of small amounts of channel-producing ionophore molecules produces a dramatic increase in ion permeability, so that the artificial membrane can exhibit ion permeability similar in many respects to that of a natural cell membrane.

The introduction of channel-producing ionophores in an artificial bilayer causes the appearance of discrete pulses of current carried by ions from one side of the membrane to the other. These *unitary currents* are due to the sudden opening of individual channels that allow hundreds or thousands of ions to stream down their gradients and across the membrane. Similar unitary events are now observed in cell membranes, providing

strong evidence for ion permeation through discrete channels (Box 4-3).

Studies of the permeabilities of cell membranes to other polar substances give an estimated 7 Å for the *equivalent pore size*—the pore diameter that would account for the rate of diffusion across the membrane. Thus, membrane channels presumably have diameters of less than 10 Å, close to the practical limits of resolution of contemporary electron microscopes and fixation methods.

Note that only a very small percentage of membrane area need be occupied by ionic and water channels to account for the observed rate of diffusion of polar substances through the membrane. For example, rod-shaped molecules of the antibiotic nystatin applied to both sides of a membrane aggregate to form channels through either natural or artificial membranes. The pores thus formed permit the passage of water, urea, and chloride, all of which are less than 4 Å in diameter. Larger molecules cannot penetrate the channels. Cations are excluded, presumably because there are fixed positive sites along the channel walls (Figure 4-23). Incorporation of nystatin into artificial membranes produces an insignificant increase in membrane area occupied by fixed pores (0.000–0.01%), but

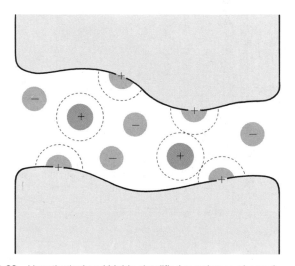

**4-23** Hypothetical and highly simplified membrane channel lined with positive charges. These allow anions to pass but retard the diffusion of cations through the channel.

## Box 4-3    *Antibiotic Ionophores*

Our understanding of membrane permeation has been advanced by the discovery of naturally occurring antibiotic molecules from soil-dwelling fungi. These molecules, some of which are polypeptides, facilitate the movement of certain ions across membranes. Some of these *ionophores*, like valinomycin and monactin, are doughnut-shaped; others, like gramicidin and nystatin, are rod-shaped. These molecules enter the membrane because they have hydrophobic surfaces. As the concentration of the ionophore in the membrane is increased, there is an increase in permeability to ions. The rod-shaped ionophores cooperate in small groups to form *channels,* assembling like the staves of a barrel so as to line a water-filled pore through which ions can pass without having to enter the hydrophobic environment of the lipid bilayer.

The doughnut-shaped ionophores act as membrane *carriers* by sequestering an ion in the "hole" of the cyclic molecule (part A of figure). These molecules often have no net charge but have inwardly directed electronegative oxygens with which such cations as $K^+$ form electrostatic bonds. Thus, a cation, sitting snugly within the polar core of the carrier-type ionophore, can be safely transported through the unfriendly nonpolar interior of the lipid bilayer. These ionophores are produced by various fungi as antibiotics, which destroy bacterial cells by upsetting their ionic balance.

(A) Transport of an inorganic ion across the lipid phase of the membrane by a carrier-type ionophore. The oxygen atoms, shown as open circles, are electron-rich, thus providing an energetic environment for cations similar to that produced by the oxygen atoms of water. The potassium ion is not drawn to scale. (B) Relationship between ionophore concentration and membrane permeability to ions carried by the ionophore. The ionophore enters the membrane from the solution in amounts proportional to its concentration in solution. [Läuger, 1972.]

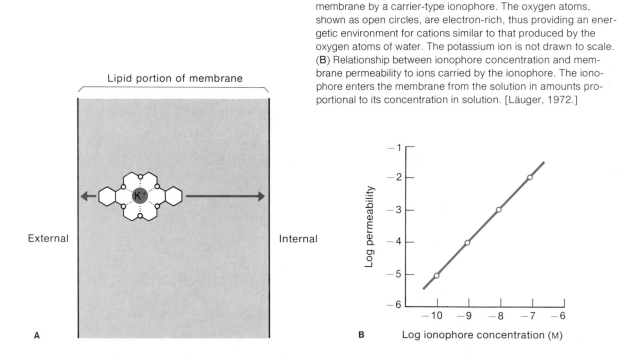

it produces a 100,000-fold increase in membrane permeability to chloride ions. In principle, then, very little membrane area need be devoted to pores to account for the ion permeabilities of natural membranes. This

conclusion is supported by the fact that the electric capacitance of the cell membrane remains relatively unchanged during large changes in the membrane's permeability.

## SIMPLE DIFFUSION THROUGH THE LIPID BILAYER

If a solute molecule comes into contact with the lipid layer of the membrane, it may enter into the lipid phase by virtue of its thermal energy and cross the lipid layer, finally emerging into the aqueous phase again on one side of the membrane or the other. To leave the aqueous phase and enter the lipid phase, a solute must first break all its hydrogen bonds with water. This activity requires kinetic energy in amounts of about 5 kcal per hydrogen bond. Moreover, the solute molecule crossing the lipid phase of the membrane must dissolve in the lipid bilayer. Its lipid solubility will therefore play a role in determining whether or not it will cross the membrane. Thus, it is evident that those molecules with a minimum of hydrogen bonding with water will most readily enter the lipid bilayer, whereas the probability is extremely low that polar molecules such as water and inorganic ions will dissolve in the bilayer.

Factors such as molecular weight and molecular shape have a modifying effect on the mobility (Equation 4-4) of nonelectrolytes within the membrane, but the primary overall property that determines the diffusion of a nonelectrolyte across the lipid bilayer is its lipid–water *partition coefficient*. The test substance is shaken in a closed tube containing equal amounts of water and olive oil, and the coefficient, $K$, is determined from the relative solubilities in water and oil at equilibrium by the equation

$$K = \frac{\text{solute concentration in lipid}}{\text{solute concentration in water}} \quad (4\text{-}10)$$

Overton, at the end of the nineteenth century, suggested that nonelectrolyte permeability is correlated with the lipid–water partition coefficient of the solute. R. Collander (1937) systematically tested this idea in the giant algal cell *Chara* by plotting the permeability coefficient (Equation 4-4) against the partition coefficient. His results are graphed in Figure 4-24, which shows a nearly linear relationship between lipid solubility and permeability of a substance. Nonelectrolytes exhibit a wide range of partition coefficients. For example, the value for urethan is 1000 times that for glycerol (Figure 4-24). The reason for these differences can be illustrated by considering the two molecules shown in Figure 4-25. The molecular structures are quite similar except that hexanol contains only one

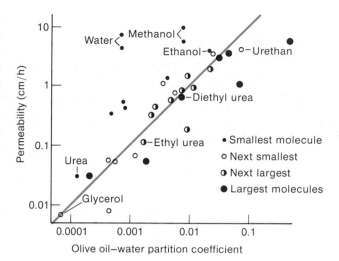

**4-24** Membrane permeability of nonelectrolytes plotted against their respective oil–water partition coefficients. Note that the permeability of nonelectrolytes is independent of molecular size. [Collander, 1937.]

|  | |
|---|---|
| H₂C—OH | CH₂—OH |
| HCH | HO—CH |
| HCH | HO—CH |
| HCH | HC—OH |
| HCH | HC—OH |
| H | H₂C—OH |
| Hexanol | D-Mannitol |

**4-25** Structures of two 6-carbon molecules—hexanol and mannitol. Note the difference in the number of hydroxyl groups. Hexanol is poorly soluble in water and highly soluble in lipids, whereas mannitol is highly soluble in water and poorly soluble in lipids owing to its hydrogen-bonding capacity.

—OH group and mannitol contains six. The —OH groups facilitate hydrogen bonding to water and therefore decrease lipid solubility. In fact, each additional hydrogen bond results in a 40-fold decrease in the partition coefficient. This decrease, in turn, is reflected in a decrease in permeability (Figure 4-26). Hexanol, therefore, diffuses across membranes much more readily than mannitol.

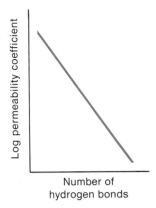

**4-26** Graph showing the relationship between permeability coefficient and number of possible hydrogen bonds in a molecule. Hydrogen bonds greatly decrease the lipid solubility and hence the permeability. [Stein, 1967.]

Simple diffusion through the lipid bilayer exhibits *nonsaturation kinetics* (Figure 4-22B). The rate of influx increases more or less in proportion to the concentration of the solute in the extracellular fluid because the net rate of influx is determined by the difference in the number of ions impinging on the two sides of the cell membrane. This proportionality between external concentration and rate of influx over a large range of concentrations distinguishes simple diffusion from channel permeation or carrier-mediated transport mechanisms (Figure 4-22).

### FACILITATED TRANSPORT

Some solutes exhibit *saturation kinetics;* that is, the rate of influx reaches a plateau beyond which a further increase in solute concentration produces no further increase in the rate of influx (Figure 4-22C). Such saturation kinetics are taken as evidence that there is a rate-limiting step in the permeation process. This step may indicate either (1) a binding of a permeating molecule to a site in or near the channel through which it passes, so that the rate of permeation is limited by the kinetics of the binding step, or (2) the transport of the permeating solute through the membrane by carrier molecules that are free to diffuse within the membrane from one side to the other so as to facilitate the passage of the solute molecules through the membrane. Since the number of carrier molecules and the rate at which they

react with and traverse the membrane must be finite, the rate of carrier-mediated transport will reach a maximum when all the carrier molecules are occupied by the solute.

The carrier hypothesis assumes the formation of a carrier–substrate complex similar in concept to an enzyme–substrate complex; that is, the carrier and the solute molecule temporarily form a complex based on bonding and/or steric specificity. This hypothesis is consistent with the observation that carrier-mediated transport exhibits Michaelis–Menten kinetics, in which the number of interactions with a carrier or enzyme reaches a maximum when the carrier or enzyme molecules are all occupied by substrate molecules. Also characteristic is inhibition by certain chemical analogs of the substrate. Addition of these analogs reduces the rate of transport of the substrate across the membrane at a given substrate concentration (Figure 4-27A). The two curves in the Lineweaver–Burk plot (Figure 4-27B)

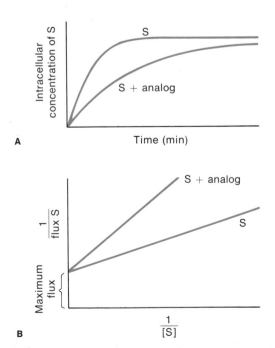

**4-27** Effect of a competitive inhibitor on the transport kinetics of a particular substrate molecule (S) passing through a membrane. The inhibitor (the analog) reduces the rate of transport. (**A**) Accumulation of S slowed by the analog. (**B**) A Lineweaver–Burk plot shows that the inhibition is competitive.

intercept the ordinate at the same point for an infinite concentration of the substrate (i.e., when $1/[S] = 0$), indicating that the inhibition is a competitive one, rather than an irreversible poisoning of the transport system.

## Active Transport

Both simple diffusion through channels or lipid and facilitated transport are passive processes in which no energy is released other than the potential energy that may exist in the form of differences in solute concentrations on opposite sides of the membrane. As diffusion proceeds, the solute concentrations in the two compartments approach equilibrium (Figure 4-28). When the concentrations on the two sides of the membrane are in equilibrium, no further net diffusion occurs, although equal and opposite unidirectional fluxes continue. It should be recalled that if the molecule has a net electric charge and if there is a potential difference between the two sides of the membrane, the concentrations will, of course, be unequal at equilibrium.

Most solutes distributed across the surface membrane of a living cell are not in equilibrium. Active processes of the membrane maintain the transmem-

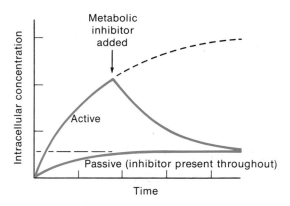

**4-29** The active transport of ions requires a source of metabolic energy. If that energy supply is interrupted by the use of a metabolic inhibitor, active transport stops. In this hypothetical case, the cell was presented with a transportable substance at time zero.

brane concentrations of such substances well out of equilibrium by the continual expenditure of chemical energy, generally in the form of ATP. The poorly understood mechanisms that actively transport substances against a gradient are loosely termed *membrane pumps*. When the source of energy for such pumps is cut off, active uphill transport ceases; passive diffusion then determines the distribution of substances to which the membrane is permeable, and the concentrations of these substances gradually redistribute toward equilibrium (Figure 4-29).

Sodium ions, for example, are actively transported outward across the cell membrane at the same rate as they leak in. This process is attributed to the *sodium pump*, an ATP-requiring enzyme system that operates in the cell membrane. In the steady state, the number of sodium ions pumped, or transported, out of the cell is equal to the number of sodium ions that leak in. Thus, even though there is a continual turnover of $Na^+$ (and other ion species) across the membrane, the net sodium flux over any period of time is zero. There are two factors that determine how large a $Na^+$ concentration gradient will be built up between the cell interior and cell exterior. These are (1) the rate of active transport of $Na^+$ and (2) the rate at which it can leak (i.e., diffuse passively) back into the cell. The rate at which the membrane allows sodium to leak back

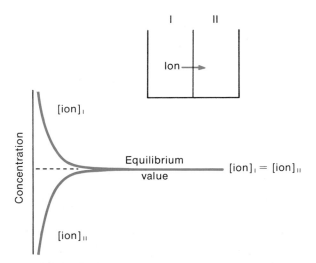

**4-28** Equilibration in concentration by diffusion of a solute across a membrane with time. At time zero, the solute was confined to compartment I. At infinite time, the concentrations in I and II are equal.

into the cell determines, of course, the rate at which the sodium pump has to work in order to maintain a given ratio of extracellular to intracellular Na⁺. There is evidence that an increase in the intracellular concentration of sodium leads to an increase in the rate of Na⁺ extrusion by the pump (which may merely be due to the increased availability of intracellular Na⁺ to the carrier molecules in the membrane).

No concrete picture of the mechanism of active transport exists at this time, but several basic features can be recognized.

*1. Transport can take place against substantial concentration gradients.* The most commonly studied membrane pump is the one that transports Na⁺ from the cell interior to the external fluid. The cytoplasm typically has a free Na⁺ concentration about one-tenth that of the extracellular fluid.

*2. The active-transport system generally exhibits a high degree of selectivity.* The sodium pump, for example, fails to transport lithium ions, which have properties very similar to those of sodium ions.

*3. ATP or other sources of chemical energy are required.* Metabolic poisons that stop the production of ATP bring active transport to a halt (Figure 4-29).

*4. Certain membrane pumps exchange one kind of molecule or ion from one side of the membrane for another kind of molecule or ion from the other side.* This feature is illustrated by the active outward transport of Na⁺ concomitant with the inward transport of K⁺ by the sodium pump. This process involves the obligatory exchange of two potassium ions from outside the cell for three sodium ions† from inside the cell (Figure 4-30). When external K⁺ is removed, the sodium ions that normally would have been exchanged for potassium ions are no longer pumped out.

*5. Some pumps can perform electrical work by producing a net flux of charge.* For example, the Na⁺–K⁺ exchange pump just mentioned produces a net outward movement of one positive charge per cycle in the form of three Na⁺ exchanged for only two K⁺. Ionic pumps

---

†The precise ratio of K⁺ to Na⁺ is difficult to measure and is not fully settled at this time. It may in fact be 1 K⁺/2 Na⁺ in some or all sodium pumps.

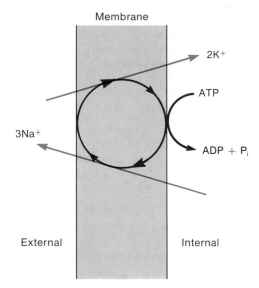

**4-30**   Na⁺–K⁺ exchange pump. An ATPase harnesses the energy of hydrolysis of ATP. A possible mechanism for active transport is given in Figure 4-31.

that produce net charge movement are said to be *rheogenic* because they produce a transmembrane electric current. If the current produces a measurable effect on the voltage across the membrane, the pump is also said to be *electrogenic*.

*6. Active transport can be selectively inhibited by certain metabolic blocking agents.* When the cardiac glycoside *ouabain* is applied to the extracellular surface of the membrane, it blocks the potassium-dependent active extrusion of Na⁺ from the cell by the Na⁺–K⁺ pump.

*7. Energy for active transport is released by the hydrolysis of ATP by enzymes (ATPases) present in the membrane.* Active transport exhibits Michaelis–Menten kinetics and competitive inhibition by analog molecules, behaviors that are characteristic of enzymatic reactions. Calcium-activated ATPases have been associated with calcium-pumping membranes. Associated with the Na⁺–K⁺ pump are sodium- and potassium-activated ATPases isolated from red blood cell membranes and other tissues. These enzymes catalyze the hydrolysis of ATP into ADP and inorganic phosphate only in the presence of Na⁺ and K⁺, and they bind the specific sodium pump inhibitor ouabain. Since ouabain binds to the membrane and blocks the Na⁺–K⁺ pump, this evi-

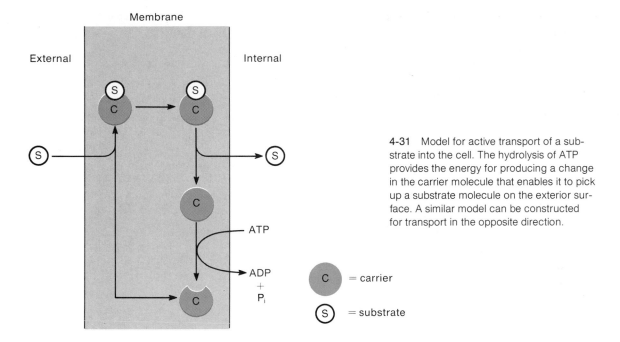

Membrane

External　　　　　　　　Internal

**4-31**　Model for active transport of a substrate into the cell. The hydrolysis of ATP provides the energy for producing a change in the carrier molecule that enables it to pick up a substrate molecule on the exterior surface. A similar model can be constructed for transport in the opposite direction.

C　= carrier

S　= substrate

dence indicates that these ATPases are involved in active transport of sodium and potassium, and they are believed to be integral parts of the $Na^+$–$K^+$ pump.

Although the molecular mechanism for active transport remains a matter of speculation, several theories exist. According to one scheme (Figure 4-31), a carrier molecule that has a specific site binds the transport substrate, S (e.g., $Na^+$). In its activated state, this site binds the substrate on one side of the membrane and releases it on the other side, producing a net flux in one direction. Once the substrate is released, the site must be reactivated by a conformational change or phosphorylation of the carrier molecule by ATP. The ATPase, which hydrolyzes the ATP, may be part of the carrier molecule or may be separate from it. The carrier, once primed by the conformational change, accepts another molecule of substrate, and the cycle repeats. This scheme is intended only as a generalization and is largely hypothetical.

The actual process of metabolically energized transport takes place across the cell membrane, pumping molecules either into or out of the cell. However, the organization of cells into an epithelial sheet makes possible the active transport of substances from one side of the epithelial sheet to the other because the cell surfaces at each side are asymmetrical in their transport properties. Thus, the epithelia of amphibian skin and bladder, fish gills, the vertebrate cornea, kidney tubules, the intestine, and many other tissues transport salts and other substances.

## Ion Gradients as Sources of Cell Energy

We noted in Chapter 3 that organic phosphagens such as creatine phosphate and ATP act as common currency for cellular energy transactions. Another important source of immediate free energy in living cells exists in the form of *electrochemical gradients* across biological membranes. That is, the energy-rich separation of ion concentrations by the membrane is produced by membrane pumps at the expense of chemical bond energy (as discussed in the preceding section) or in photosynthetic membranes by radiant energy. The free energy stored in an electrochemical gradient depends on the ratio of ion concentrations or, more accurately, the ratio of the chemical activities of an ion species on the two sides of the membrane. The energy is released when the ions are allowed to flow down their gradient across the membrane.

Among the major biological phenomena that utilize the free energy of biological gradients are the following three:

***1.*** *Production of electric signals.* Electrochemical energy is stored across the membrane primarily as $Na^+$ and $Ca^{2+}$ gradients. The release of this electrical energy is under the control of "gated" channels. These channels are normally closed. In response to certain chemical or electrical signals, they switch to an open state in which they exhibit selective permeability to specific ions. These ions then flow passively across the membrane down their electrochemical gradients. Because of the charges it carries, the movement of an ion species produces an electric current and changes the potential difference that exists across the membrane. Such bioelectric phenomena are the subject of Chapter 5.

***2.*** *Chemiosmotic energy transduction.* This concept is illustrated by a widely accepted theory, proposed by Peter Mitchell (1966) in 1961, for the conversion of the energy of oxidation released by the electron-transport chain into the chemical bond energy stored during the oxidative phosphorylation of ADP to ATP that takes place in mitochondria. Two central ideas are put forth in this theory: (a) Owing to a highly specific orientation of the redox enzymes within the inner membrane of the mitochondrion, the electron-transport system of the respiratory chain pumps hydrogen ions from inside the mitochondrion to the cytoplasmic space outside (Figure 4-32). It is proposed that the inner mitochondrial membrane has a low permeability to $H^+$, so that this pumping produces an excess of $OH^-$ (e.g., high pH) within the mitochondrion and an excess of $H^+$ (e.g., low pH) outside the mitochondrion. (b) The energy-rich proton gradient set up in this way provides the free energy that removes HOH from ADP + $P_i$ as required for the production of ATP:

$$ADP + P_i \longrightarrow ATP + H_2O$$
$$\Delta G^{0\prime} = +7.3 \text{ kcal/mol}$$

This reaction, according to Mitchell's theory, also depends on the physical orientation of the ATPase complex in the inner mitochondrial membrane so as to take advantage of the $H^+/OH^-$ separation across the membrane. The $H^+$ that is enzymatically removed

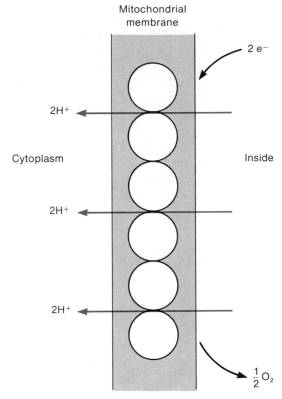

**4-32**  Part I of Mitchell's chemiosmotic theory of energy transduction. In this simplified scheme, protons are shown to be transported out of the mitochondrion by the electron-transport chain by the energy of oxidation. The unidirection displacement of $H^+$ requires a high degree of structural specificity in the organization of the respiratory chain molecules (circles) in the inner mitochondrial membrane. The consequence of this activity is an excess of $H^+$ on the cytoplasmic side of the membrane and an excess of $OH^-$ on the mitochondrial side.

from ADP is thought to be siphoned off, so to speak, into the $OH^-$-rich mitochondrial interior to form HOH (Figure 4-33). The $OH^-$ removed from the inorganic phosphate molecule is similarly shunted outside the mitochondrion to react with the excess $H^+$ to form HOH. Thus, the $H^+/OH^-$ gradient provides the energy needed to remove the water during the phosphorylation. Following the dehydration, phosphate bond formation,

$$ADP^- + P_i^+ \longrightarrow ATP$$

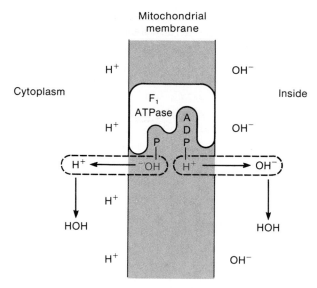

Mitochondrial
membrane

Cytoplasm

Inside

HOH

HOH

**4-33** Part II of Mitchell's chemiosmotic theory of energy transduction. With the catalytic aid of $F_1$ ATPase located in the inner mitochondrial membrane, ADP and $P_i$ have $H^+$ and $OH^-$, respectively, stripped away by high $OH^-$ in the mitochondrial interior and by the relatively high concentration of $H^+$ in cytoplasm. This process allows $P_i^+$ to condense with $ADP^-$ to form ATP.

goes forward on the active site of the ATPase without further need for energy input.

Chemiosmotic energy transduction similar to that proposed for oxidative phosphorylation in mitochondria has also been implicated as the mechanism for energy transduction during photosynthesis in chloroplasts and photosynthetic bacteria. In addition, there is evidence that the $Na^+$–$K^+$ pump, which normally utilizes ATP to produce the $Na^+$ gradient, can under special circumstances be run in reverse so that the movement of $Na^+$ down its gradient will cause the pump to synthesize ATP from ADP and $P_i$.

*3. Uphill transport of other molecules.* Movement of some molecules *up* a concentration gradient is driven by movement of another substance *down* its concentration gradient. Thus, the ubiquitous $Na^+$ gradient is used to carry certain sugars and amino acids along through the membrane by a *cotransport* mechanism and to drive $Ca^{2+}$ out of the cell by a *countertransport* mechanism. These forms of transport are discussed next.

## COTRANSPORT

Figure 4-34A shows the time course of the accumulation (uptake by cell) of the amino acid alanine in the presence and absence of extracellular sodium. In the presence of $Na^+$, the amino acid is taken up by the cell until the internal concentration is 7 to 10 times that of the external concentration. In the absence of $Na^+$, the intracellular concentration of alanine merely approaches the extracellular concentration. As the Lineweaver–Burk plots show (Figure 4-34B), the maximum rate of alanine influx approaches the same maximum (intercept on ordinate) with or without extracellular $Na^+$. In both cases the rate of influx shows saturation kinetics, indicating a carrier mechanism. The different

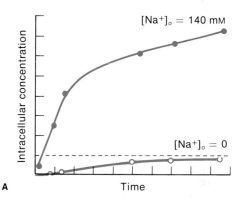

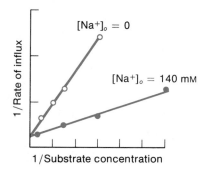

**4-34** Effect of $Na^+$ concentration on uptake of the amino acid alanine. (A) Intracellular content as a function of time with and without extracellular $Na^+$ present. (B) Lineweaver–Burk plots with and without extracellular $Na^+$. The common intercept indicates that at infinite concentration of alanine, the rate of transport is independent of $[Na^+]_o$. [Schultz and Curran, 1969.]

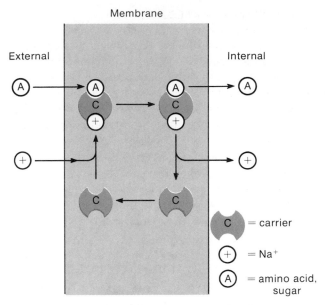

4-35  Hypothetical mechanism of sodium-mediated cotransport of amino acids (or sugars). The carrier must bind both the $Na^+$ and the amino acid before it will transport either. Net transport is inward because of the $Na^+$ gradient.

slopes of the two plots indicate that the effect of extracellular $Na^+$ is to enhance the activity of the alanine carrier. Increasing the intracellular $Na^+$ concentration by blocking the sodium pump with ouabain has the same effect as decreasing the extracellular $Na^+$ concentration. Thus, it appears to be the sodium gradient that is important for inward alanine transport, and not merely the presence of sodium ions in the extracellular fluid.

The transport of amino acids and sugars appears to be *coupled* to inward $Na^+$ leakage by means of a *common carrier*. The carrier molecule apparently binds both $Na^+$ and the organic substrate molecule before it can transport either (Figure 4-35). The tendency for $Na^+$ to diffuse down its concentration gradient is what drives this carrier system. Anything that reduces the concentration gradient of $Na^+$ (low extracellular $Na^+$ or increased intracellular $Na^+$) reduces the inwardly directed driving force and thereby reduces the coupled transport of amino acids and sugars into the cell. If the direction of the sodium gradient is experimentally

reversed, the direction of transport of these molecules is also reversed. The carrier-mediated transport of $Na^+$ in this case also depends on the presence of amino acids and sugars. In the absence of amino acids and sugars, the common carrier will transport $Na^+$ only very weakly, and as a result, the inward leakage of $Na^+$ is reduced.

The common carrier appears to shuttle between the two sides of the membrane passively, without *direct* utilization of metabolic energy. The coupled uphill transport of organic molecules derives its energy from the downhill diffusion of sodium ions, but the potential energy stored in the sodium gradient is, of course, derived from metabolic energy that drives the sodium pump (Figure 4-36). The sodium concentration gradient can be thought of as an intermediate form of common energy currency that is used to drive several energy-requiring processes in the membrane.

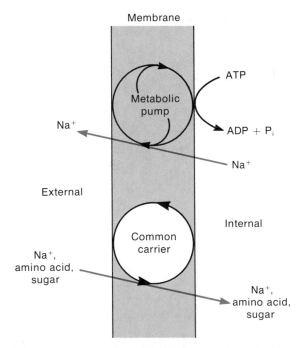

4-36  The transport of amino acids and sugars depends ultimately on the chemical energy of ATP derived from cell metabolism. The $Na^+$ concentration gradient acts as an intermediate form of potential energy utilized to drive the organic molecules against their concentration gradients.

## COUNTERTRANSPORT

The sodium concentration gradient also plays a role in the maintenance of a very low intracellular calcium concentration in certain cells. In most cells, if not all, the intracellular calcium concentration is several orders of magnitude below the extracellular concentrations (less than $10^{-6}$ M), and certain cell functions are regulated by changes in the intracellular calcium concentration. Efflux of $Ca^{2+}$ from cells is reduced when extracellular $Na^+$ is removed. This reduction suggests that $Ca^{2+}$ is extruded from the cell in exchange for $Na^+$ leaking in, with the opposing movements of these two ions coupled to each other by an *exchange carrier*. One view is that $Ca^{2+}$ and $Na^+$ both compete for the carrier, but that $Ca^{2+}$ competes more successfully inside the cell than on the outer surface, so that there is a net efflux of $Ca^{2+}$. Again, the immediate source of energy is the $Na^+$ gradient, which ultimately depends on the ATP-energized active transport of $Na^+$. This exchange pump is particularly important when intracellular calcium becomes abnormally high in concentration. There is evidence that $Ca^{2+}$ is also transported independent of the sodium gradient by an ATP-energized calcium pump, which is the major form of $Ca^{2+}$ extrusion under normal conditions.

## Membrane Selectivity

Each kind of membrane transport displays selectivity, and the selectivity generally differs in a given membrane for different transport systems. For example, when the sodium in a physiological saline solution used to bathe a nerve cell is replaced with lithium ions, the $Li^+$ readily passes through the sodium channels, which open during electrical excitation of the nerve cell membrane. The other alkali-metal cations, $K^+$, $Rb^+$, and $Cs^+$, are essentially impermeant through these channels. On the other hand, the ATPase of the sodium pump in the same membrane is highly specific for intracellular $Na^+$ and is not activated by lithium ions. Lithium ions will therefore gradually accumulate in the cell. This is an example of electrolyte selectivity. We will now consider mechanisms of selectivity for both electrolytes and nonelectrolytes.

## SELECTIVITY FOR ELECTROLYTES

The most commonly encountered example of selectivity is the ability of the resting membrane to distinguish $K^+$ from the other major monovalent cation, $Na^+$. The resting nerve cell is about 30 times more permeable to $K^+$ than to $Na^+$. At first glance, one might conclude that these ions are distinguished on the basis of their hydrated size, with $K^+$ passing freely through channels that are too small for $Na^+$. The *sieving hypothesis* is not adequate, however, to explain most other mechanisms of membrane selectivity. For example, during the excitation of nerve or muscle membrane, the $Na^+$ permeability of the membrane increases about 300-fold to a value about 10 times greater than the $K^+$ permeability at rest. If, during excitation, the membrane were suddenly to develop channels that pass the $Na^+$ ion on the basis of size alone, the sieving hypothesis would predict a simultaneous increase in permeability to $K^+$ through the same channels. Since this increase does not occur, the membrane's selectivity must rest on mechanisms other than those based on size alone.

One factor that influences permeation of ions is illustrated in Figure 4-23, which shows fixed charges producing a net positivity on the walls of a hypothetical membrane channel. Such a channel would preferentially pass anions of appropriate size but would hinder passage of cations by electrostatic repulsion.

J. M. Diamond and E. M. Wright reviewed the selectivity exhibited by various membranes and found that nearly all experimentally demonstrated selectivity sequences for the alkali cations ($Ca^{2+}$, $Rb^+$, $K^+$, $Na^+$, $Li^+$) fall into one of the 11 predicted ion sequences. Some membranes show preferences that correspond to the order of decreasing ion diameter, whereas others show a selectivity sequence that corresponds to increasing size. Since these are only 2 of the 11 observed membrane selectivity sequences, correlations between ion diameters, hydrated or nonhydrated, appear to be secondary to more subtle interactions.

The thermodynamic basis for interactions between ions and electrostatic sites on enzymes and membranes has been discussed in Chapter 2. The probability of binding between an ion and a polar site on the membrane is determined by the differences between the electrostatic attraction of the ion to water and to specific electrostatically charged sites on the membrane.

The greater the electrostatic attraction of an ion to the site relative to its attraction to water, the more successfully it competes with other ions for the site. This theory does not specify the mechanism of permeation, but it does suggest why certain ions are able to gain access to a channel or to bind to the channel interior or to a carrier molecule with a higher probability than other ion species.

### SELECTIVITY FOR NONELECTROLYTES

For nonelectrolytes, as shown in Figure 4-24, there is a somewhat linear relationship between permeability and partition coefficient. This is one evidence that those substances cross the membrane by dissolving in (i.e., entering) the lipid bilayer and simply diffusing across it. In relation to Equation 4-4, this means that the permeability to those nonelectrolytes is limited primarily by the partition coefficient, $K$. Those few nonelectrolytes that deviate from the linear relation between partition coefficient and permeability (Figure 4-24) all do so in the direction of greater than predicted permeability. One explanation for these deviations is that these substances cross the membrane by carrier-mediated transport and hence show higher permeabilities than can be accounted for by simple diffusion through the lipid layer. Another reason may be that small molecules such as ethyl alcohol, methyl alcohol, and urea can cross through both the lipid layer and water-filled channels. All the deviants are small and water-soluble regardless of their relative solubilities in water versus lipid (i.e., their partition coefficients).

## Pinocytosis and Exocytosis

Membranes sometimes transport substances by the transfer of small bulk quantities of material into or out of the cell. The membrane can take up bulk quantities into the cell by first trapping the material in a vesicle formed from a tiny invagination. The vesicle then pinches off, isolating the material on the cytoplasmic side of the membrane (Figure 4-37). Digestion or rupture of the vesicle liberates the contents into the cytosol. This process is termed *pinocytosis* if fluid is ingested, and *phagocytosis* if solids are ingested.

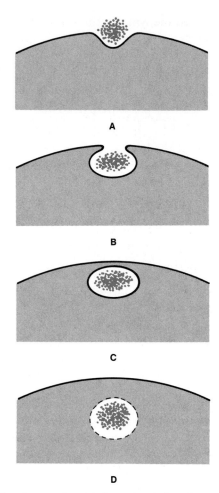

4-37   Stages of vesicle formation during pinocytosis. (A) Invagination begins. (B) Invagination continues. (C) Vesicle pinches off. (D) Vesicle membrane disintegrates, releasing engulfed material.

A related process, called *exocytosis*, plays an important role in the endocrine and nervous systems. For example, the presynaptic terminals of nerve cells contain many membrane-limited internal vesicles about 500 Å in diameter, which contain the neural transmitter substance. It now appears that these vesicles can coalesce with the surface membrane of the nerve terminal and release their contents to the cell exterior. This activity occurs with greatly enhanced probability when the terminal is invaded by a nerve impulse and serves to release the synaptic transmitter that interacts with the postsynaptic membrane. Similar mechanisms are involved in the secretion of hormones.

Implicit in the theory of exocytosis is that the membrane of the vesicle is incorporated into the surface membrane, allowing all of the free contents—hormones and accessory molecules—to diffuse away into the interstitial space. But what keeps the surface area of the plasma membrane from continually growing with the addition of new vesicle membranes? The answer appears to be that new *microvesicles* are formed by pinocytotic budding from the surface membrane, as suggested in Figure 11-6. Evidence favoring the formation of new secretory vesicles by this kind of pinocytotic process comes to us from experiments in which a large electron-opaque molecule, *horseradish peroxidase,* is introduced into the extracellular fluid and its progress followed with electron-microscopical methods. It subsequently appears inside the cell only within vesicles. Since the large size of the horseradish peroxidase molecule prevents its penetration by direct passage across biological membranes, it must be taken up in bulk by a process similar to pinocytosis during the formation of the microvesicles—that is, while they are budding off from the plasmalemma into the cytoplasm.

The calcium ion has been closely linked to the exocytotic secretion of neurotransmitter substances from nerve cells and of hormones from endocrine cells. Although the precise role of $Ca^{2+}$ in initiating secretion is unknown, it appears that an elevation of intracellular $Ca^{2+}$ somehow enhances the probability of exocytotic activity—perhaps the coalescence of vesicles with the inner surface of the membrane. The membrane regulates exocytotic activity by regulating the intracellular accumulation of $Ca^{2+}$. As enhanced calcium influx allows $Ca^{2+}$ levels to rise, the rate of exocytotic secretion increases (Figure 4-38). Thus, $Ca^{2+}$ is said to act as a *secretogogue.*

The vesicle membrane itself may participate actively in the initial steps leading to exocytosis. The secretory granules of the adrenal medulla have been found to be rich in an unusual phospholipid, *lysolecithin,* which facilitates the fusion of membranes and thus may help the vesicle membrane fuse with the surface membrane. Before fusion of the two mem-

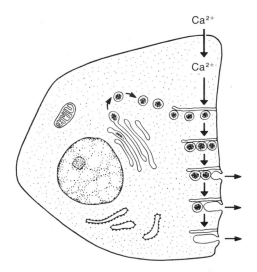

**4-38** Possible role of microtubules in secretion, as hypothesized for pancreatic beta cells. The microtubules, often seen at secretory sites with vesicles attached in series, may propel the vesicles to the membrane in an energy-requiring process regulated by calcium ions. Different stages are shown from top to bottom. [Lacy, 1972.]

branes can take place, the secretory granule (or vesicle) must come into contact with the plasmalemma. Release of secretory products from glandular secretory cells can be blocked by *colchicine,* an antimitotic agent that leads to the disassembly of microtubules, or by *cytochalasin,* an agent that disrupts microfilaments. This pharmacological evidence has led to the suggestion that microtubules or microfilaments participate in the movement of secretory granules toward sites of exocytotic release on the inner side of the surface membrane (Figure 4-38). Although there is some doubt as to the specificity of these pharmacological agents, secretory vesicles are often seen associated with microtubules in the electron microscope.

An exciting lead in the search for the mechanism of exocytosis is the identification of two proteins in nerve endings isolated from mammalian brain. One, termed *stennin,* is associated with the synaptic vesicles, has properties similar to the muscle protein myosin, and exhibits ATPase activity. The other, termed *neurin,* is a filamentous protein similar to muscle actin and

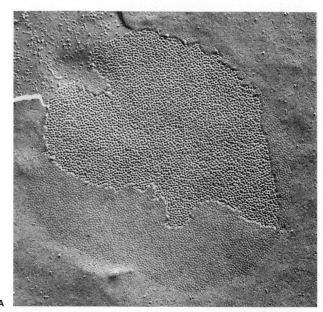

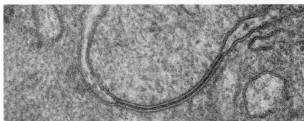

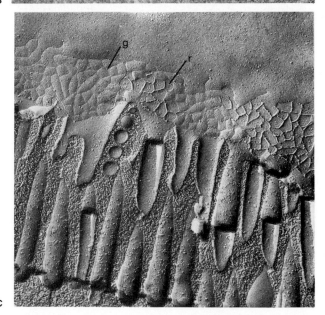

**4-39** Electron micrographs showing junctions between
membranes of neighboring cells. (**A**) Freeze-fracture preparation of a gap junction between pancreatic cells in guinea
pig, showing characteristic densely packed particles. Note
that a fragment of the uppermost membrane is fractured
away, giving a view of the particles that connected it with
the membrane of the adjacent cell below, which is now visible. Pits associated with the particles can be seen in the
surrounding portions of the upper membrane. Magnification
78,500×. (**B**) Transverse thin section showing gap junction
between two mouse liver cells. Note close apposition of unit
membranes and the connecting particles. Magnification
130,000×. (**C**) Freeze-fracture preparation from rat small
intestine showing lacework of tight junctions connecting
two epithelial cells, one above and one below the plane of
cleavage. The grooves (g) belong to the upper cell, and the
corresponding ridges (r) to the lower cell. The cylindrical
projections at the bottom are the microvilli of the epithelial
cells. Magnification 55,000×. [Courtesy of N. B. Gilula and
D. S. Friend.]

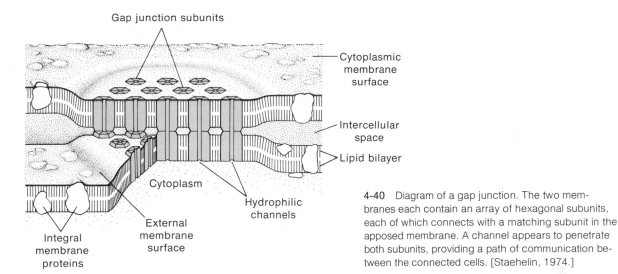

Gap junction subunits

Cytoplasmic
membrane
surface

Intercellular
space

Lipid bilayer

Cytoplasm

Hydrophilic
channels

External
membrane
surface

Integral
membrane
proteins

**4-40** Diagram of a gap junction. The two membranes each contain an array of hexagonal subunits, each of which connects with a matching subunit in the apposed membrane. A channel appears to penetrate both subunits, providing a path of communication between the connected cells. [Staehelin, 1974.]

appears to be attached to the inner surface of the plasmalemma. In the presence of $Ca^{2+}$, these two proteins interact as *neurostennin*, and it has been suggested that this interaction brings the secretory vesicle into contact with the inner surface of the plasmalemma as a key step in the secretory process.

## Cell–Cell Junctions

To this point, we have considered the cell membranes of cells in isolation. Most cells occur in multicellular *tissues* in which their membranes are in some way held together, with a thin space between them largely filled with the extracellular aqueous solution. In certain tissues, including epithelium, smooth muscle, cardiac muscle, central nervous tissues, and many embryonic tissues, neighboring cells are connected by specializations of apposed surface membranes. These specializations are of two major types, termed *gap junctions* and *tight junctions* (Figures 4-39 and 4-42). Intimate contacts made between such neighboring cells are the basis for (1) cell–cell communication through minute water-filled channels that connect adjacent cells through gap junctions and (2) transepithelial transport of substances by cells that are more or less intimately "sewn" together into a sheet by tight junctions.

### GAP JUNCTIONS

The distance between two membranes of a gap junction is only 20 Å, and the membranes are connected by a hexagonal array of subunits about 50 Å in diameter (Figure 4-40) that resemble miniature doughnuts whose hollow centers form passageways between the interior of one cell and the interior of the neighboring cell. The gap junction permits the passage of ions and molecules between cells. The continuity of the cell–cell passageways through the gap junction has been demonstrated by the injection of a fluorescent dye, such as fluorescein (molecular weight 332) and procion yellow (molecular weight 500), into one cell and following its diffusion into the neighboring cells (Figure 4-41). This continuity has been corroborated for direct exchange of ions by the finding that electric current readily passes directly from one cell into another if gap junctions are present. Since the intercellular channels in these junctions appear to pass molecules of at least molecular weight 500, it is possible that they may also pass long, thin informational molecules such as messenger RNA in cell–cell communication.

Gap junctions are labile and are uncoupled by any treatment that increases the intracellular level of $Ca^{2+}$ or the intracellular $H^+$ concentration. Uncoupling of cells from their neighbors can be demonstrated by directly injecting $Ca^{2+}$ or $H^+$ into a coupled cell, by lowering temperature, or by using poisons that inhibit energy metabolism. These experiments indicate uncoupling by the loss of electrical transmission between cells. Thus, gap junctions are maintained intact only if the metabolic activity of the surface membrane maintains sufficiently low concentrations of intracellular free $Ca^{2+}$ and $H^+$.

### TIGHT JUNCTIONS

In tight junctions, the outer surfaces of the two apposing cell membranes make intimate contact (Figure 4-39C), fully occluding the extracellular space between the adjacent cells at the points of contact. Tight junctions are found most commonly in epithelial tissues in the form of a *zonula occludens,* which completely encircles each cell, restricting paracellular passage between the two sides of an epithelial tissue. The zonula occludens is important in transport epithelia, for it controls leakage past the cells situated between the two sides of an epithelium. In some tissues, these zonulae are not fully continuous and thus not really very

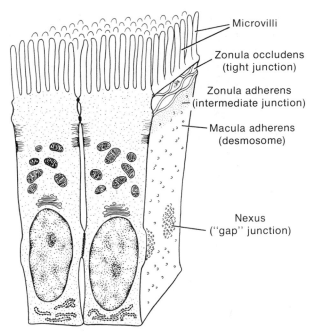

4-42 Artist's reconstruction of intercellular junctions found between adjacent epithelial cells like those that line the mammalian small intestine. The membranes and associated structures are drawn disproportionately large. [Weinstein and McNutt, 1972.]

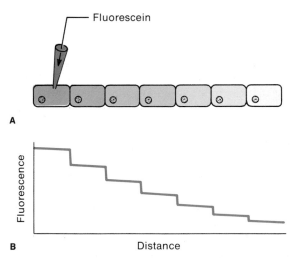

4-41 Injection of fluorescent dye in one cell of a group of coupled epithelial cells of insect salivary gland. Subsequent diffusion of the dye into neighboring cells without loss into extracellular space indicates that there are direct pathways from the cytoplasm of one cell to the cytoplasm of the adjacent cell. Gap junctions can invariably be found between such coupled cells.

"tight." These occur in the mammalian small intestine, gallbladder, and proximal tubule of the nephron. These tissues have such a marked *paracellular leak* that, unlike frog skin, they do not produce a transepithelial potential difference, even though their cells contain ion pumps like those found in frog skin. Unlike gap junctions, tight junctions appear to have no special channels for cell–cell communication.

Two other types of cell junction, shown in Figure 4-42, are the *zonula adherens* and the *macula adherens* (more commonly termed *desmosome*). These junctions serve primarily to aid the structural bonding of neighboring cells.

## Epithelial Transport

The osmoregulatory activities of animals are carried out by a variety of specialized tissues and organs, some of which we will consider in Chapter 12. Fundamental to osmoregulatory process is the transport of water and solutes across epithelia. Osmoregulatory tissues have several features in common. First, they occur at surfaces that separate the internal space of the organism from the environment. This occurrence is not always evident, because the external space may extend in a tortuous way deep into the body, as in the lumen of the intestine. Second, the cells forming the outermost layer of the epithelium are generally sealed together by tight junctions, which to varying degrees in different epithelia obliterate *paracellular pathways* between the *serosal* (inner) and *mucosal* (external) sides of the epithelium (Figure 4-43). In some epithelia, such as the endothelium of capillary walls, the junctions are leaky. As a consequence, water and solute molecules can cross these epithelia by diffusing within the passages that exist between the epithelial cells. Since diffusion of material through paracellular pathways is not coupled to any metabolically energized transport mechanism, such passages allow only passive diffusional movements of water and ions. Substances that are actively transported across an epithelium must follow *transcellular pathways* in which the cell membrane partici-

pates. Such substances must cross the cell membrane first on one side of the cell and then on the other. As discussed in the next section, the functional properties of the surface membrane of an epithelial cell are dissimilar in some respects on the serosal and mucosal surfaces of the cell. This asymmetry is important to epithelial active transport.

### ACTIVE SALT TRANSPORT ACROSS AN EPITHELIUM

Metabolically energized transport of ions from one side of an epithelium to the other has been demonstrated in a number of epithelial tissues including amphibian skin and urinary bladder, the gills of fishes and aquatic invertebrates, insect and vertebrate intestine, and vertebrate kidney tubule and gallbladder. Much of the initial work on epithelial active transport was done on the skin of the frog. In amphibians, the skin acts as a major osmoregulatory organ. Salt is actively transported from the mucosal side (i.e., facing the pond water) to the serosal side of the skin to compensate for the salt that leaks out of the skin into the fresh water surrounding the frog. Similar uptake occurs in the gut. Water that enters the skin due to the osmotic gradient between the hypotonic pond water and the more concentrated internal fluid is eliminated in the form of a copious dilute urine that is hypotonic relative to the body fluids.

Frog skin was developed as a preparation for the study of epithelial transport in the 1930s and 1940s by the German physiologist Ernst Huf and the Danish physiologist Hans Ussing. A piece of abdominal skin several square centimeters in area is removed from an anesthetized and decapitated frog and placed between two halves of an *Ussing chamber* (Figure 4-44). The dissection is very simple, since the skin of the frog lies largely unattached over an extensive lymph space. Once the skin is gently clamped between the two half chambers, a test solution—for example, frog Ringer (Table 4-3)—is introduced, with the frog skin acting as a partition between the two compartments. The compartment facing the mucosal side of the skin can be designated as the outside compartment and the one facing the serosal side as the inside compartment. Air is bubbled through the two solutions to keep them well oxygenated.

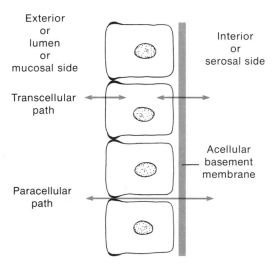

**4-43** Two pathways, paracellular and transcellular, for passage of substances across an epithelial layer. Active transport takes place only across cell membranes; hence all actively transported molecules follow the transcellular pathway.

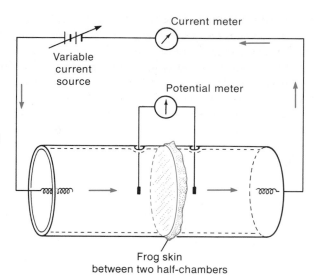

4-44 Ussing chamber. The frog skin separates the two half-chambers, each filled with Ringer's or other test solution. The current source is adjusted until the potential difference across the skin is zero. Under those conditions, the current flowing through the circuit (and thus through the skin) is equivalent to the rate of charge transferred by the active movement of sodium ions across the skin.

TABLE 4-3  Physiological salines (concentrations in millimoles per liter).

| | NaCl | KCl | $CaCl_2$ | $MgCl_2$ | $NaHCO_3$ | Other constituents |
|---|---|---|---|---|---|---|
| **Marine** | | | | | | |
| Artificial seawater | 470 | 10.0 | 10.0 | 25.0 | 2.0 | |
| Lobster | 455 | 13.5 | 16.5 | 4.0 | 4.0 | $H_3BO_3$ + NaOH buffer |
| Squid | 475 | 10.3 | 10.6 | 53.1 | 25.7 | Phosphate to pH 7.6 |
| Dogfish | 224 | 1.35 | 1.8 | 1.1 | 2.38 | Glucose 15<br>Urea 333 |
| **Freshwater** | | | | | | |
| Artificial pond water | 0.5 | 0.05 | 0.4 | | 0.2 | |
| Leech | 115 | 4 | 1.8 | | | Glucose 10<br>Tris buffer 10 |
| Crayfish | 207 | 5.4 | 13.6 | 2.64 | 2.4 | |
| Freshwater fish | 101.8 | 3.38 | 1.36 | | 2.5 | $MgSO_4$ 1.19 |
| Frog | 112 | 1.9 | 1.1 | | 2.4 | Glucose 11.1<br>$NaH_2PO_4$ 0.7 |
| **Terrestrial** | | | | | | |
| Cockroach | 210 | 3.1 | 1.8 | | | Phosphate buffer to pH 7.2 |
| Lizard | 116 | 3.2 | 1.2 | | 2.0 | $NaH_2PO_4$ 0.3<br>Glucose 1.7<br>$MgSO_4$ 1.4 |
| Bird | 117 | 2.33 | 5.8 | | 28 | $MgSO_4$ 2.12 |
| Mammal Tyrode | 138 | 2.7 | 1.84 | 1.06 | 11.9 | $NaH_2PO_4$ 0.5 |
| Mammal Krebs | 119 | 5.0 | 2.5 | 1.0 | 2.5 | $NaH_2PO_4$ 0.5<br>Glucose 11.0 |

*Source:* Prosser, 1973.

In 1947, Ussing reported the first experiments in which two isotopes of the same ion were used to measure bidirectional fluxes (i.e., the simultaneous movements of that ionic species in opposing directions across the epithelium). The Ringer solution in the outside compartment was prepared with the isotope $^{22}Na^+$, and the Ringer in the inside compartment was prepared with $^{24}Na^+$. The appearance of each of the two isotopes on the opposite side of the skin was followed as a function of time. The two isotopes were switched around in other experiments of the same type to rule out any effects due to possible (but unlikely) differences in transport rates inherent in the isotopes themselves. In all experiments, it was found that $Na^+$ shows a net movement across the skin from the outside compartment to the inside one. That is, more ions move in that direction than in the opposite direction. For the following reasons, it was concluded that the net movement of sodium ions is apparently the result of active transport:

*1.* It occurs without any concentration gradient, even *against* an electrochemical gradient.

*2.* It is inhibited by general metabolic inhibitors, such as cyanide and iodoacetic acid, and by specific transport inhibitors, such as ouabain.

*3.* It displays a strong temperature dependence.

*4.* It exhibits saturation kinetics.

*5.* It shows chemical specificity.

The initial experiments of Ussing led to several important questions. For example, does the inward active transport of $Na^+$ require cotransport of an anion? If not, the transport of $Na^+$ should produce an electric current (because of the movement of positive charge in the form of the sodium ion). Most important, how is transport carried out across a layer of cells?

The electrical correlates of active sodium transport were reported by Ussing and K. Zerahn in 1951. They reasoned that if $Na^+$ alone is actively transported across the epithelium, there should be quantitative agreement between the number of sodium ions transported across a unit area of skin per second and the strength of the resulting current (i.e., charges crossing the membrane per second). Under ordinary conditions, this comparison is hard to make, since the measured current would be reduced by any passive movement of

$Cl^-$ or other ions that would move across the skin down a voltage gradient set up by the movement of $Na^+$. That is, as soon as the inside compartment gains a few excess cations, that side would become more positive, and $Cl^-$, for example, would be electrostatically drawn across the skin. To avoid any electrochemical gradients, Ussing and Zerahn neutralized any charge difference, and hence any potential difference, by drawing off positive charges with an external electrical network at the same rate as they appeared across the skin (Figure 4-44). The adjustment to zero potential difference accomplished two important things. First, $Na^+$ would not be hindered in its movement by the buildup of a potential that would counteract its net movement by active transport. Second, the current flowing through the external circuit (identical, of course, to the current flowing through the skin) could be compared to the net amount of sodium ions transported through the skin. If the current were in quantitative agreement with the net sodium transport measured by isotopic means, it could be concluded that $Na^+$ transport can conveniently be determined from the skin current.

Indeed, Ussing and Zerahn found a close agreement between the skin current and the isotopically measured sodium flux (Figure 4-45). Both the current and

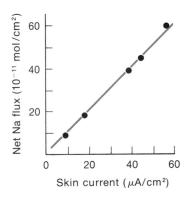

**4-45**  Agreement between frog skin current and the net sodium flux across a unit area of skin. Relating the current (coulombs per second) to the flux (equivalents per second) through Faraday's constant (96,500 C/g-equiv) shows a close agreement between the net transport of sodium and the skin current. The latter is therefore considered to be primarily a sodium current in the isolated frog skin. [Ussing, 1954.]

the transport of $Na^+$ are reduced or abolished by ouabain (the sodium-pump inhibitor), by metabolic inhibitors, and by sodium-free solutions in the external compartment. This finding provided further evidence that, in the isolated frog skin, active sodium transport seems to predominate over any other active ion transport.

When the external short-circuiting current is removed, the frog skin quickly develops a potential difference because of the active transport of $Na^+$ to the serosal side. The potential difference between the two sides leads, in turn, to a passive inward movement of chloride down its electrochemical gradient. Thus, the active transport of $Na^+$ alone is sufficient to produce a net influx of NaCl under the conditions in which both sides of the skin are bathed in frog Ringer's solution. However, under more natural conditions (e.g., with the external surface of the skin bathed in a dilute solution more closely approximating pond water), the skin is not leaky to $Cl^-$. Instead, at external $Na^+$ concentrations below 3–10 mM, chloride is actively taken up by a transport system that is independent of sodium transport. The uptake of $Cl^-$ occurs in exchange for $HCO_3^-$ and therefore produces no electric current. Thus, a frog sitting in its pond actively takes up both $Na^+$ and $Cl^-$ to replenish salt lost to the dilute environment.

We can now turn to the problem of how an active movement of ions is produced across an epithelium. It will be recalled (Figure 4-43) that adjacent cells of a transport epithelium are intimately tied together with tight junctions. Let us assume for the sake of simplicity that this closeness eliminates all extracellular passageways for the diffusion of ions between the two sides of the epithelium (in reality, there are minute residual paracellular pathways). This would force all substances that cross the epithelium to traverse the epithelial cell membrane twice, first traversing the membrane on one side of the cell and then leaving through the membrane on the other side. Active transport by this route requires that the surface membrane of each epithelial cell be differentiated, so that the portion of the cell membrane facing the serosal side of the epithelium differs in functional properties from that portion of the membrane facing the mucosal side. Experiments on frog skin have provided several lines of evidence to support this hypothesis. For example:

*1.* Ouabain, which blocks the $Na^+$–$K^+$ pump, inhibits transepithelial sodium transport only when applied to the inner (serosal) side of the epithelium. It is ineffective on the outer (mucosal) side. Conversely, the drug amiloride, a powerful inhibitor of passive carrier-facilitated transport, blocks sodium movements across the skin only when applied to the outer side of the skin.

*2.* Potassium must be present in the solution on the inner side for active sodium transport to take place but is not required on the outer side.

*3.* Transport of sodium exhibits saturation kinetics as a function of $Na^+$ concentration in the outer solution; it is unaffected by $Na^+$ concentration in the inner solution.

Such evidence led to the model of epithelial sodium transport shown in Figure 4-46. According to this model, a $Na^+$–$K^+$ exchange pump (plus $Na^+$–$H^+$ and $Na^+$–$NH_4^+$ exchange pumps in the intact animal) is located in the membrane of the serosal side of the epithelial cell. This membrane behaves in the manner typical of many cell membranes, pumping $Na^+$ out in exchange for $K^+$, thus maintaining a high intracellular potassium concentration and a low intracellular so-

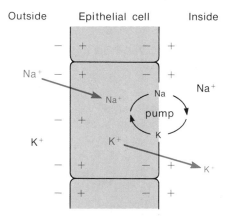

**4-46** Model devised by V. Koefoed-Johnsen and Hans Ussing for transepithelial sodium transport in isolated frog skin bathed on both sides in Ringer's solution. Sodium diffuses passively down its concentration gradient into the cell from the mucosal solution. $K^+$ diffuses out of the cell into the serosal space as it is displaced by $Na^+$ influx. In the face of these leaks, a $Na^+$–$K^+$ exchange pump in the serosal membrane of the cell maintains the high internal $K^+$ and low internal $Na^+$ concentrations. [Koefoed-Johnsen and Ussing, 1958.]

dium concentration. The outward diffusion of potassium ions across the membrane on this side of the cell produces an inside negative resting potential. The situation on the mucosal side is different. The cell membrane on this side of the cell is relatively impermeable to potassium. Moreover, a net *inward* diffusion of sodium across this membrane (apparently facilitated by carriers or through channels in the membrane) replaces the $Na^+$ pumped out of the cell on the serosal side. This model explains why sodium-pump inhibitors exert an effect only from the serosal side of the epithelium and why only changes in the concentration of $K^+$ on that side influence the rate of sodium transport.

Thus, there is a net flow of $Na^+$ across the frog skin from the mucosal side to the serosal side as a result of the functional asymmetries of the membranes on the two sides. The driving force is none other than the active transport of $Na^+$ that is common to cell membranes of all tissues.

The frog skin has served as a model system for the general problem of epithelial salt transport. Although details may differ from one type of epithelial tissue to another, the major features, listed below, are probably common to all transport epithelia.

*1.* To varying degrees, tight junctions obliterate paracellular pathways. As a result, transport through transcellular pathways assumes major importance in epithelial transport.

*2.* Mucosal and serosal portions of the cell membranes exhibit functional differences, being asymmetrical in both pumping activity and membrane permeabilities.

*3.* The active transport of cations across an epithelium is typically accompanied by transport (passive or active) of anions in the same direction or by exchange for another species of cation, minimizing the buildup of electric potentials. The converse applies to actively transported anions.

*4.* Epithelial transport is not limited to the pumping of sodium and chloride ions. Various epithelia are known to transport $H^+$, $HCO_3^-$, $K^+$, and other inorganic ions.

### TRANSPORT OF WATER

A number of epithelia absorb or secrete fluids; for example, the stomach secretes gastric juice, the choroid plexus secretes cerebrospinal fluid, the gallbladder and intestine transport water, and the kidney tubules of birds and mammals absorb water from the glomerular filtrate. In some of these tissues, water moves across an epithelium in the absence of or against an osmotic gradient between the bulk solutions on either side of the epithelium. A number of possible explanations for the uphill movement of water have been given, but all these hypotheses can be placed in one of two major categories:

*1.* Water is transported by a specific water-carrier mechanism driven by metabolic energy.

*2.* Water is transported secondarily as the consequence of solute transport.

The latter includes classic osmosis, in which water undergoes a net diffusion in one direction owing to concentration gradients built up by solute transport. So far, there has been no convincing evidence to indicate that water is actively transported by a primary water-carrier pump.

The osmotic hypothesis of water transport received a boost when Peter Curran pointed out in 1965 that an osmotic gradient produced by active salt transport from one subcompartment of the epithelium into the other could, in theory, result in a net flow of water across the epithelium (Figure 4-47). Biological correlates of Curran's model were subsequently found in

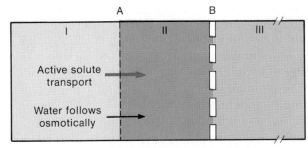

**4-47** Curran's model for solute-linked water transport. Solute (e.g., $Na^+$) is pumped by barrier A from compartment I to compartment II. Semibarrier B slows diffusion of solute into compartment III and thereby keeps the osmolarity high in II. The rise in osmolarity in compartment II causes water to be drawn from I into II. In the steady state, both water and solutes diffuse into compartment III at the same rate at which they appear in II. Compartment III is much larger than II, as indicated by the breaks. [Curran, 1965.]

the epithelium of mammalian gallbladder by J. M. Diamond and J. McD. Tormey in 1966, which led to the *standing-gradient hypothesis* of *solute-coupled water transport* of Diamond and W. H. Bossert. A simplified schematic version is shown in Figure 4-48. Two anatomical features are of major importance: (1) the tight junctions near the luminal (mucosal) surface, which obliterate extracellular passages through the epithelium, and (2) the lateral intercellular spaces, or intercellular *clefts,* between adjacent cells. These clefts are restricted at the luminal ends by the tight junctions and are freely open at the basal ends.

The basis for the standing-gradient hypothesis is the active transport of salt across the portions of the epithelial cell membranes facing the intercellular clefts. The membranes bordering the lateral clefts have been shown to be especially active in pumping sodium out of the cell. It is suggested that as salt is transported out of

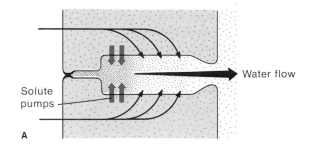

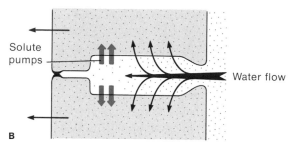

4-49   The standing-gradient flow system operating in (**A**) "forward" and (**B**) "backward" directions. The direction of water flow depends on the direction of salt pumping. The density of the dots shows the relative osmolarity. [Diamond, 1971.]

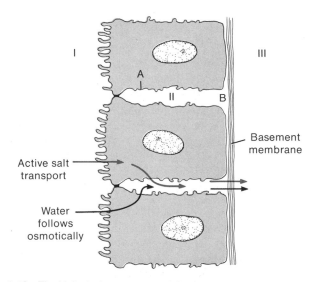

4-48   The biological counterpart of Curran's three-compartment model for solute-coupled water transport. The compartments corresponding to those in Figure 4-47 are numbered I, II, and III. Salt transported actively into the intercellular clefts produces a high osmolarity within the clefts. Water flows osmotically into the clefts across the cell, and the bulk solution flows through the freely permeable basement membrane and into the bulk fluid of the interstitium. The barriers A and B are analogous to A and B in Figure 4-47. [Diamond and Tormey, 1966.]

the cell into these long, narrow clefts, the salt concentration will set up an osmotic gradient between the extracellular spaces on either side of the tight junctions that join the epithelial cells. There may also be an osmotic gradient within the cleft such that the salt concentration will be highest near the closed ends of the clefts, diminishing toward the open ends of the clefts, where it comes into equilibrium with the bulk phase (Figure 4-49A). As a consequence of the high extracellular osmolarity in the clefts, water is osmotically drawn into the cleft across the "not so tight" tight junction, or possibly from within the cell across the cell membrane into the intercellular space. The water leaving the cell would have to be replaced by water drawn osmotically into the cell at the mucosal surface. The water that enters the clefts gradually moves, together with solute, out into the bulk phase. In this way, the steady, active extrusion of salt by one surface membrane of the cell produces an elevated concentration in the narrow intercellular spaces. This, in turn, results in a steady osmotic flow of water from one side of the epithelium to the other. Solute-coupled water transport can occur in the opposite direction if the active salt transport takes place in that direction (Figure 4-49B).

The general applicability of the standing-gradient mechanism of solute-coupled water transport is supported by ultrastructural studies showing that the necessary cellular geometry—namely, narrow intercellular spaces closed off at the luminal end by tight junctions—is present in all the water-transporting epithelia that have been examined. In epithelia fixed during conditions that produce water transport, these spaces are dilated; in epithelia fixed in the absence of water transport, the intercellular clefts are largely obliterated.

## Summary

Membranes form one class of fundamental structures of the cell. They not only form the surface but serve as the structural bases of numerous cellular organelles. They are responsible for:

*1.* Cellular and subcellular compartmentalization.

*2.* Maintenance of the intracellular milieu by virtue of selective permeability and transport mechanisms.

*3.* Regulation of cell metabolism by determining the concentrations of enzyme cofactors and substrates.

*4.* Metabolic activities carried out by enzymes that are often present in ordered arrays in or on the membrane.

*5.* Electric signals that conduct messages and/or regulate the transport of substances across the membrane.

*6.* Endo- and exocytosis of bulk material.

The foundation structure of membranes is a lipid bilayer in which the hydrophilic heads of the phospholipid molecules face outward and the lipophilic tails face inward, toward the center of the bilayer. The most widely accepted model of membrane structure has a mosaic of globular proteins, including enzymes, penetrating the bilayer.

Because of an unequal distribution of solutes between cell interior and exterior, water enters the cell owing to its tendency to flow from a region of lower to a region of higher osmotic pressure. Osmotic pressure is equal to the hydrostatic pressure necessary to balance osmotic flow (water movement across a semipermeable membrane) down a concentration gradient at equilibrium. The concept of tonicity describes the osmotic effects that a solution has on a given tissue, whereas osmolarity describes the number of dissolved particles per volume of solvent, as well as the behavior of a solution in an ideal osmometer.

Permeability is a measure of the ease with which a substance traverses a membrane. There are several ways in which substances cross the membrane. Water and some small polar molecules diffuse very slowly through transient aqueous channels created by thermal motion. Nonpolar molecules can diffuse more readily through the lipid phase of the membrane. There is a great deal of evidence for the existence of fixed channels that are more or less specific for certain ions and molecules. Diffusion across the membrane of some substances can occur via carrier molecules that complex with the substance, facilitating its transport across the membrane while the carrier shuttles within the lipid phase of the membrane.

Active transport of a substance occurs by means of carriers and requires metabolic energy, usually provided by ATP. It is responsible for the movement of a substance across a membrane against a concentration gradient. The most familiar active transport system is the sodium–potassium pump, which maintains the intracellular $Na^+$ concentration below that of the cell exterior. The energy stored in the form of an extracellular/intracellular $Na^+$ concentration gradient is utilized to drive the uphill movement of a number of other substances, such as calcium ions, amino acids, and sugars, by means of exchange diffusion and coupled transport. The $Na^+$ and $K^+$ gradients are also important for the production of electric signals, such as nerve impulses.

Another important function of active transport is to

compensate for the tendency of certain substances such as sodium ions to leak into cells and thereby cause uncontrolled increases in osmotic pressure and subsequent swelling of the cell. Continual removal of $Na^+$ by the $Na^+$–$K^+$ pump is therefore a major factor in controlling cell volume.

Transepithelial transport depends on an asymmetry in the permeability and pumping activities of the mucosal and serosal portions of epithelial cell membranes. On the serosal side of the cell, ions are actively transported across the membrane against an electrochemical gradient; on the mucosal side, ions cross the membrane by diffusion or facilitated transport. Diffusion of ions back through the epithelial layer is slow, because the spaces between cells are restricted by tight junctions. Water is transported across some epithelia by being drawn osmotically down a standing salt-concentration gradient built up by active salt transport between the epithelial cell interior and the intercellular clefts. There is no evidence for true active transport of water.

## Exercises

*1.* What are some of the physiological functions of membranes?

*2.* What is the evidence for the existence of membranes as real physical barriers?

*3.* What is the evidence for the lipid bilayer model of the membrane?

*4.* What is the evidence for a mosaic of globular proteins set into the lipid bilayer of the membrane?

*5.* Explain the meanings of *isotonic* and *isosmotic*. How can a solution be isosmotic but not isotonic to another solution?

*6.* What factors determine the permeability of the membrane to a given electrolyte? Nonelectrolyte?

*7.* Describe the probable mechanisms by which water and other small (less than 10 Å diameter) polar molecules pass through the membrane.

*8.* Why do nonpolar substances diffuse more easily than polar substances through the membrane?

*9.* Active transport and facilitated transport both show saturation kinetics. What does this tell us about the mechanisms underlying these two kinds of transport?

*10.* How does facilitated transport differ from simple diffusion?

*11.* What factors influence the rate of facilitated transport of ions across a membrane?

*12.* How does active transport differ from facilitated transport?

*13.* Why can the sodium concentration gradient be considered a common cellular energy currency?

*14.* What are some parameters by which the membrane discriminates between ions of the same charge?

*15.* Explain the osmotic consequences of poisoning the metabolism of a cell.

*16.* How does the cell maintain a higher concentration of $K^+$ inside the cell than in the extracellular fluid?

*17.* What are the morphological and functional distinctions between gap and tight junctions?

*18.* A given cell is 40 times as permeable to $K^+$ and $Cl^-$ as to any other ions present. If the inside/outside ratio of $K^+$ is 25, what would the approximate inside/outside ratio of $Cl^-$ be?

*19.* Given that cell membranes can transport substances only into or out of a cell, explain how substances are transported *through* cells.

*20.* Describe the experiments that first demonstrated active transport of $Na^+$ across an epithelium.

*21.* What is some of the evidence that active transport of $Na^+$ and $K^+$ occurs only across the serosal membranes of epithelial cells?

*22.* There is no convincing evidence for direct active transport of water. Explain one way in which water is moved by epithelia against a concentration gradient—that is, from a concentrated salt solution to a more dilute salt solution.

## Suggested Reading

Cereijido, M., and C. A. Rotunno. 1970. *Introduction to the Study of Biological Membranes.* New York: Gordon & Breach.

Green, D. E., and J. F. Danielli, eds. 1972. *Membrane Structure and Its Biological Applications.* New York: New York Academy of Sciences.

Jain, M. K. 1972. *The Bimolecular Lipid Membrane: A System.* New York: Van Nostrand Reinhold.

Kyte, J. 1981. Molecular considerations relevant to the mechanism of active transport. *Nature* 292:201–204.

Levine, Y. K. 1972. Physical studies of membrane structure. *Progr. Biophys. Molec. Biol.* 24:1–74.

Lightfoot, E. N. 1974. *Transport Phenomena and Living Systems.* New York: Wiley.

Lockwood, A. P. M. 1971. *The Membranes of Animal Cells.* London: Arnold.

Scott, W. N., and D. B. P. Goodman, eds. 1981. *Hormonal Regulation of Epithelial Transport of Ions and Water.* New York: New York Academy of Sciences. Vol. 372.

Singer, S. J., and G. L. Nicolson. 1972. The fluid mosaic model of the structure of cell membranes. *Science* 175:720–731.

Sleigh, M. A., and D. H. Jennings. 1974. *Transport at the Cellular Level.* New York: Cambridge University Press.

Solomon, A. K. 1962. Pumps in the living cell. *Scientific American* 207(2):100–118. Also available as Offprint 131.

Vanderkooi, G., and D. E. Green. 1971. New insights into biological membrane structure. *BioScience* 21(9):409–415.

Weissmann, G., and R. Clairborne, eds. 1975. *Cell Membranes.* New York: Hospital Practice Publishing Co.

## References Cited

Collander, R. 1937. The permeability of plant protoplasts to non-electrolytes. *Trans. Faraday Soc.* 33:985–990.

Curran, P. F. 1965. Ion transport in intestine and its coupling to other transport processes. *Fed. Proc.* 24:993–999.

Diamond, J. M. 1971. Standing-gradient model of fluid transport in epithelia. *Fed. Proc.* 30:6–13.

Diamond, J. M., and W. H. Bossert. 1967. Standing-gradient osmotic flow: A mechanism for coupling of water and solute transport in epithelia. *J. Gen. Physiol.* 50:2061–2083.

Diamond, J. M., and J. McD. Tormey. 1966. Studies on the structural basis of water transport across epithelial membranes. *Fed. Proc.* 25:1458–1463.

Diamond, J. M., and E. M. Wright. 1969. Biological membranes: The physical basis of ion and nonelectrolyte selectivity. *Ann. Rev. Physiol.* 31:581–646.

Donnan, F. G. 1911. Theorie der Membrangleichgewichte und Membranpotentiale bei Vorhandensein von nicht dialysierenden Elektrolyten: Ein Beitrag zur physikalisch-chemischen Physiologie. *Z. Elektrochem.* 17:572–573.

Getman, F. H., and F. Daniels. 1931. *Outlines of Theoretical Chemistry.* New York: Wiley.

Gorter, E., and F. Grendel. 1925. On bimolecular layers of lipoids on the chromocytes of blood. *J. Exp. Med.* 41:439–443.

Huf, E. G. 1935. Versuche über den Zusammenhang zwischen Stoffwechsel, Potentialbildung und Funktion der Froschhaut. *Pflügers Arch. Ges. Physiol.* 235:655–673.

Koefoed-Johnsen, V., and H. H. Ussing. 1958. The nature of frog skin. *Acta Physiol. Scand.* 42:298–308.

Kotyk, A., and K. Janáĉek. 1970. *Cell Membrane Transport.* New York: Plenum.

Lacy, P. E. 1972. Microtubule–microfilament system in beta cell secretion. In S. Taylor, ed., *Proceedings of the 3rd International Symposium on Endocrinology.* London: Heinemann.

Läuger, P. 1972. Carrier-mediated ion transport. *Science* 178:24–30.

Lehninger, A. L. 1975. 2nd ed. *Biochemistry.* New York: Worth.

Mitchell, P. 1966. Chemiosmotic coupling in oxidative and photosynthetic phosphorylation. *Biol. Rev.* 41:445–502.

Nägeli, K. W. von. 1884. *Mechanisch-physiologische Theorie der Abstammungslehre.* Munich: Oldenbourg.

Overton, E. 1902. Beiträge zur allgemeinen Muskel und Nervenphysiologie. *Pflügers Arch. Ges. Physiol.* 92:115–280.

Peters, A., and J. E. Vaughn. 1970. Morphology and development of the myelin sheath. In D. N. Davison and A. Peters, eds., *Myelination.* Springfield, Ill.: Thomas.

Pfeffer, W. F. P. 1899. *Osmotische Unterschungen.* Leipzig: Engleman.

Prosser, C. L. 1973. *Comparative Animal Physiology.* Vol. 1. Philadelphia: Saunders.

Robertson, J. D. 1960. The molecular structure and contact relationships of cell membranes. *Progr. Biophys. Mol. Biol.* 10:343–418.

Schultz, S. G., and P. F. Curran. 1969. The role of sodium in non-electrolyte transport across animal cell membranes. *Physiologist* 12:437–452.

Staehelin, L. A. 1974. Structure and function of intercellular junctions. *Int. Rev. Cytol.* 39:191–283.

Stein, W. D. 1967. *The Movement of Molecules Across Cell Membranes.* New York: Academic.

Stryer, L. 1981. *Biochemistry.* 2nd ed. San Francisco: W. H. Freeman and Company.

Ussing, H. H. 1954. Ion transport across biological membranes. In H. T. Clarke and D. Nachmansohn, eds., *Ion Transport Across Membranes.* New York: Academic.

Ussing. H. H., and K. Zerahn. 1951. Active transport of sodium as the source of electric current in the short-circuit isolated frog skin. *Acta Physiol. Scand.* 23:110–127.

Weinstein, R. S., and N. S. McNutt. 1972. Cell junctions. *New England J. Med.* 286:521–524.

# CHAPTER 5

# Ions and Excitation

The origin of electrochemical theory and the discovery that tissues produce electric currents can both be traced to observations made late in the eighteenth century by Luigi Galvani, a professor of anatomy at Bologna, Italy. Working with a nerve–muscle preparation from a frog leg, Galvani noted that muscles contract when dissimilar metals in contact with each other are brought into contact with the tissue, one metal touching the muscle and the other touching the nerve. Galvani and his nephew Giovanni Aldini, a physicist, ascribed this response to a discharge of "animal electricity" delivered by the nerves and stored in the muscle. They postulated that an "electric fluid" passed from the muscle through the metal and back into the nerve and that the discharge of electricity from the muscle triggered the contraction. This interpretation, published in 1791, was largely incorrect in retrospect; nevertheless, this work stimulated many inquisitive amateur and professional scientists of that revolutionary age to investigate two new and important areas of science, the physiology of excitation in nerve and muscle and the chemical origin of electricity.

Alessandro Volta, a physicist at Pavia, Italy, quickly took up Galvani's experiments and in 1792 proposed that the electric stimulus leading to contraction in Galvani's experiments came not from a discharge of current from the tissue, as claimed by Galvani and Aldini, but was in fact generated outside the tissue by the contact of dissimilar metals with the saline fluids of the tissue. It took several years for Volta to demonstrate unequivocally the electrolytic origin of electric currents from dissimilar metals, for there was no physical instrument available at that time sufficiently sensitive to detect weak currents. Indeed, the nerve–muscle preparation from the frog leg was probably the most sensitive indicator of electric current in use at that time.

In his search for a means of producing stronger sources of electricity, Volta found that he could increase the electricity produced electrolytically by placing metal–saline cells in series. The fruit of his labor was the so-called Voltaic pile—a stack of alternating silver and zinc plates separated by saline-soaked papers. This first "wet-cell" battery produced proportionately higher voltages than can be produced by a single silver–zinc cell.

Although Galvani's original experiments did not really prove the existence of "animal electricity," they did demonstrate the sensitivity of excitable tissues to minute electric currents. In 1840, Carlo Matteucci used the action current of a contracting muscle to stimulate a second nerve–muscle preparation (Figure 5-1). His experiment was the first recorded demonstration that excitable tissue produces electric current.

Since the nineteenth century, it has become evident that the production and processing of signals in the nervous system and the contraction of muscle depend on the electrical properties of cell membranes.

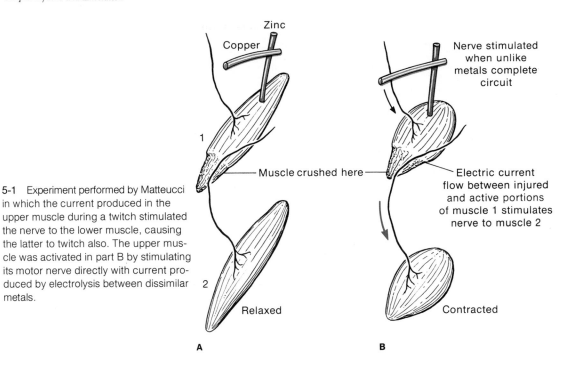

5-1   Experiment performed by Matteucci in which the current produced in the upper muscle during a twitch stimulated the nerve to the lower muscle, causing the latter to twitch also. The upper muscle was activated in part B by stimulating its motor nerve directly with current produced by electrolysis between dissimilar metals.

## Membrane Excitation

Electrical phenomena† in living tissues can be detected by placing two electrodes in the tissue to measure the field set up by electric currents flowing through the extracellular fluids. Since these currents originate across cell membranes, a more direct and quantitative approach is to measure electrical events across the membrane of a single cell. This measurement is done by comparing the electric potential (voltage) of one side of the membrane with that of the other side. Subtracting one from the other gives the *potential difference* (*p.d.*), which is commonly called the *membrane potential*, or $V_m$. One sensing electrode is placed in electrical continuity with one side of the membrane, and another in electrical continuity with the other side of the membrane, and the potential difference is electronically amplified for display on a recording instrument, such as an oscilloscope (Figure 5-2). This procedure was

---

†A review of the electrical definitions and conventions given in Box 2-1 may be useful for this chapter.

impossible to do satisfactorily except in certain very large cells until G. Ling and R. W. Gerard (1949) perfected the glass capillary *microelectrode* (Figure 5-3A). Because of their minute tip diameter, microelectrodes can be inserted into cells with negligible damage to their membranes.

Once the fine capillary tip has penetrated the membrane, the cytoplasm is in continuity with the wire leading to the amplifier via a fine column of electrolyte that fills the inside of the capillary electrode (e.g., a 3 M solution of KCl). The membrane potential is always given as the intracellular potential relative to the extracellular potential, which is arbitrarily defined as zero.

A simple stimulating and recording arrangement is shown in Figure 5-3B. A cell is immersed in a physiological saline solution that is in contact with a reference electrode. Before the tip of the recording microelectrode enters the cell, the microelectrode and the reference electrode are at the same potential. The potential difference between the two electrodes is then zero (Figure 5-4A). As the tip of the microelectrode is advanced,

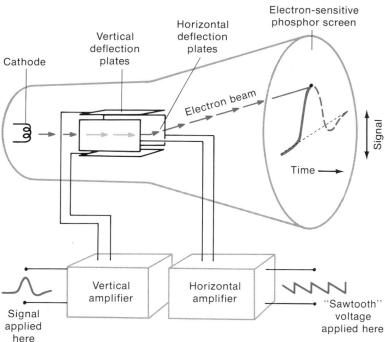

**5-2** The oscilloscope. A beam of electrons "writes" on the phosphor screen while driven from left to right by a gradually changing "sawtooth" voltage. Signals applied to the oscilloscope input are amplified and fed to the vertical deflection plates as a changing voltage. Vertical deflection of the beam plots the signal against time.

a negative voltage, or potential, suddenly appears as a downward shift of the voltage trace, indicating penetration of the surface membrane (Figure 5-4B; in electrophysiological recordings, negative potentials are conventionally shown as downward displacements on the oscilloscope screen). The steady negativity recorded by the electrode tip in the cytoplasm is the *resting potential*, $V_{rest}$, of the cell membrane and is given in *millivolts* (mV, thousandths of a volt). All cells that have been investigated have a negative resting potential. The size of this potential in various cells ranges as high as $-100$ mV.

The potential sensed by the intracellular electrode does not change as the tip is advanced further into the cell. Thus, the entire potential difference between the cell interior and cell exterior exists across the surface membrane and in the regions immediately adjacent to the inner and outer membrane surfaces.

The electrical properties of the cell membrane can be examined by causing a pulse of current to pass through the membrane so as to produce a perturbation in the membrane potential. A second microelectrode,

the current electrode, can deliver such a current (Figure 5-3B, left). This current can be made to flow across the membrane in either the inward (bath to cytoplasm) or the outward (cytoplasm to bath) direction,† depending on the polarity of the electric current applied to the electrode.

When current pulses are passed so that positive charge is removed from inside the cell via the current electrode, the potential difference across the membrane increases (*hyperpolarizes*). That is, the intracellular negative potential is increased (e.g., from $-60$ to $-70$ mV). With hyperpolarization, the membrane (with some exceptions) produces no response other than a passive potential change due to the applied current (Figure 5-5).

If a current pulse is passed from the electrode into the cell, positive charge will be added to the inner

---

†As was noted in Chapter 2, all the current carried in solution and through the membrane is in the form of migrating ions. By convention, the flow of ionic current is from a region of relative positivity to one of relative negativity and corresponds to the direction of cation migration.

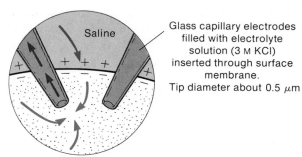

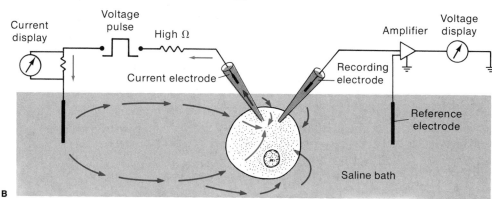

5-3 (A) Glass capillary microelectrodes inserted through the membrane of a cell. The electrode at the left passes current into or out of the cell. (B) Current flows in a circuit through the wires, bath, electrodes, and cell membrane. The high-ohm resistor has far greater resistance than the other resistances in the stimulating circuit and therefore maintains the constancy of the stimulating current. The recording amplifier has a very high input resistance, preventing any appreciable current from leaving the cell through the recording electrode.

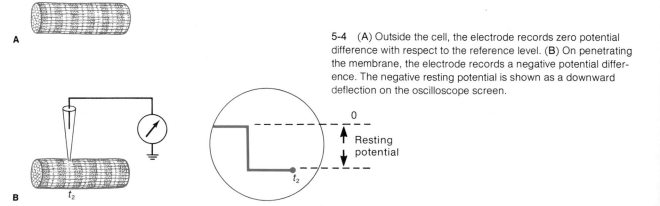

5-4 (A) Outside the cell, the electrode records zero potential difference with respect to the reference level. (B) On penetrating the membrane, the electrode records a negative potential difference. The negative resting potential is shown as a downward deflection on the oscilloscope screen.

surface of the cell membrane. This charge causes the potential difference across the membrane to decrease, and the cell is then said to become *depolarized.* That is, the intracellular negative potential decreases (e.g., from $-60$ to $-50\,mV$). As the strength of the outward current pulse is intensified, depolarization will increase. *Excitable cells,* such as nerve, muscle, and many receptor cells, exhibit a *threshold potential* at which the membrane will produce a strong active response, the *action potential* (Figure 5-5). The action potential is caused by the activation of membrane channels permeable to sodium, which themselves are activated by the reduction in voltage difference between the two sides of the cell membrane. The opening of the sodium channels in response to depolarization and the resulting flow of

sodium ions into the cell provide an example of *membrane excitation.* The mechanisms underlying the action potential and other instances of membrane excitation are considered in more detail later.

As just illustrated, cell membranes respond to stimuli with two quite different classes of electrical behavior—passive and active:

*1.* A *passive electrical response* is always produced when an electric current is forced across a biological membrane because of the elementary electrical properties of the membrane—capacitance and conductance (or resistance), which are described in the next section. Passive responses occur independent of any molecular changes such as opening or closing of membrane channels.

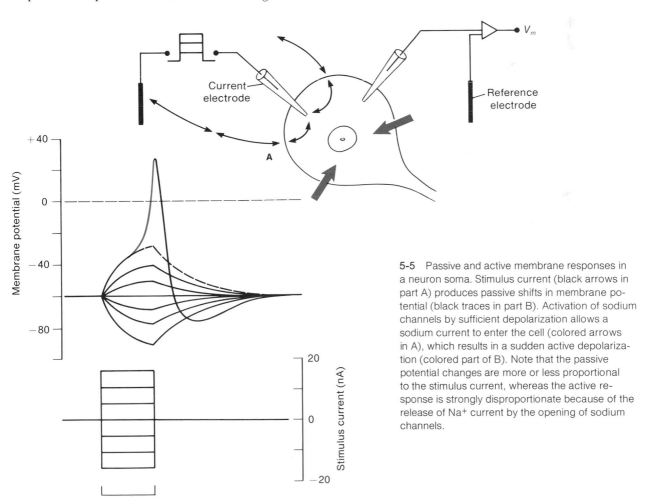

**5-5** Passive and active membrane responses in a neuron soma. Stimulus current (black arrows in part A) produces passive shifts in membrane potential (black traces in part B). Activation of sodium channels by sufficient depolarization allows a sodium current to enter the cell (colored arrows in A), which results in a sudden active depolarization (colored part of B). Note that the passive potential changes are more or less proportional to the stimulus current, whereas the active response is strongly disproportionate because of the release of $Na^+$ current by the opening of sodium channels.

*2.* An *active electrical response* (i.e., membrane excitation), which is found in excitable tissues such as nerve, muscle, and sensory receptors, depends on the opening and/or closing of numerous and minute *ion channels* (also called *membrane channels*) in response to a stimulus. Some ion channels are *gated* (i.e., opened and shut) by changes in voltage across the membrane, while others are opened by the binding of transmitter or messenger molecules, and still others, primarily in sensory receptor cells, are activated by specific stimulus energies such as light (photoreceptors) or mechanical strain (mechanoreceptors). When a certain group of channels, selectively permeable to a certain species of ion, is opened, a current can be carried across the membrane. As in the case of the sodium channels noted earlier, such a current normally produces a voltage signal across the membrane. The gating of ion channels, as we will see later, is the immediate cause for nearly all electrical activity in living tissue.

## Passive Electrical Properties of Membranes

Before we can go on to gain an understanding of active electrical processes in nerve cell function, it is essential that we consider the passive electrical properties of the cell membrane. Two structural elements in the membrane give rise to two corresponding electrical properties:

*1.* The ion-impermeant lipid bilayer can separate charge. This bilayer conveys the property of *capacitance*.

*2.* Ion channels, as noted above, provide a pathway for inorganic ions to carry electric charge across the membrane. These channels convey the property of *conductance*.

These two properties account entirely for the passive electrical behavior of cell membranes.

### MEMBRANE CAPACITANCE
The rate at which ions traverse a membrane is as little as $10^{-8}$ times the average rate at which they diffuse across an equivalent distance (50–100 Å) of cytoplasm or extracellular fluid. Since electric current in an aqueous solution is carried by ions, the relatively low mobility of ions through the lipid bilayer results in a high electrical resistance of the bilayer.

Because they are very thin (less than 100 Å) and are

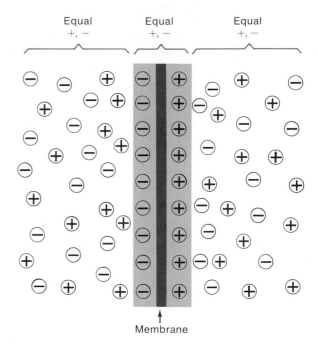

5-6   The cell membrane is able to separate charges because it acts as a capacitor. The charges form a diffuse layer on each side and interact electrostatically across the thin barrier.

virtually impermeable to ions over most of their surface areas, cell membranes are able to violate the principle of electroneutrality at the microscopic level. That is, they can separate ionic charge on a microscopic scale. Negative charges accumulated at or near one surface of a membrane will interact electrostatically over the short distance of membrane thickness with positive charges on the other side (Figure 5-6). The ability of the cell membrane to accumulate and separate electric charge is called its capacitive property (Box 5-1). In simple terms, the capacitance of a membrane is a measure of the amount of charge it will separate for a given potential difference across the membrane.

Capacitance increases in proportion to the dielectric constant of the material separating the charges; and it decreases with increasing distance between charges. Most membranes appear to contain a layer of lipid about 50 Å thick. If we assume that the lipid has a dielectric constant of 3, which is about that of an 18-carbon fatty acid, the membrane capacitance can be calculated to be about 1 microfarad (1 μfd = $10^{-6}$ fd)

## Box 5-1 Capacitance and Time Constant

When a sudden step of current is forced across a membrane, the current divides itself between the membrane capacitance and resistance in a manner that changes with time, so that initially most of the current flows through the membrane capacitance. Let us consider a constant current, I, applied in an abrupt, step pulse to a capacitor in parallel with a resistor (Figure 5-7). The potential, $V_c$ (in volts), developed across the capacitor is proportional to the charge, q (in coulombs), accumulated by a capacitor. The proportionality constant, C (in farads), is an indication of the ability of the capacitor to accumulate and store charge. Thus,

$$q = V_c C \qquad (1)$$

The *capacitive current*, $i_c$, is defined as the charge accumulated per unit time:

$$i_c = \frac{dq}{dt} \qquad (2)$$

Differentiation of Equation 1 shows that the rate of charge accumulation is proportional to the rate of change of potential across the capacitor, since C remains constant throughout:

$$\frac{dq}{dt} = C\frac{dV_c}{dt} \qquad (3)$$

Finally, by substitution into Equation 2, we obtain

$$i_c = C\frac{dV_c}{dt} \qquad (4)$$

That is, the current at any instant through a capacitor is proportional to the rate of change of the potential at that instant.

As current passes through a capacitor, the accumulated charge causes the potential across the capacitor to increase. The voltage produced across the capacitor repels new charges, and the rate of charging drops. Hence, the capacitive current, $i_c$, varies inversely with $V_c$ (part B of figure) and diminishes with an exponential time course.

The relationship between potential and time during the charging of the capacitance is given by the equation

$$V_t = V_\infty(1 - e^{-t/RC}) \qquad (5)$$

where $V_\infty$ is the potential across the capacitor at $t = \infty$ produced by a constant current applied to the network, t is the time in seconds after the beginning of the current pulse, R is the resistance of the circuit in ohms, C is the capacitance of the circuit in farads, and $V_t$ is the potential across the capacitor at time t.

When t is equal to the product RC, then $V_t = V_\infty$ $(1 - 1/e) = 0.63\ V_\infty$. The value of t (in seconds) that equals RC is termed the *time constant* ($\tau$) of the process. Note that it is independent of both $V_\infty$ and current strength and is the time required for the voltage across a charging capacitor to reach 63% of the asymptotic value $V_\infty$ (Figure 5-7B).

The current applied to the membrane flows initially through the capacitance of the membrane. As the capacitance charges (or discharges) exponentially, an increasing fraction of the total current passes through the resistive component of the membrane (part A of figure). After several time constants have elapsed, the potential closely approaches an asymptote, and all the applied current flows through the membrane resistance. When the applied current is terminated, the charge stored across the capacitor leaks back through the resistor, $R_m$, and the potential returns to the rest level with an exponential time course.

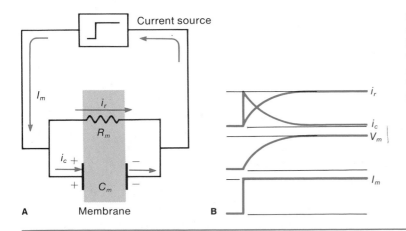

(A) The equivalent circuit for a cell membrane across which an abrupt pulse of constant current is passed. The high resistance is used to supply an unvarying current. (B) Time courses of resistive current, $i_r$, capacitive current, $i_c$; membrane potential, $V_m$ (i.e., the potential across both the membrane resistance and capacitance), and the total membrane current, $I_m$.

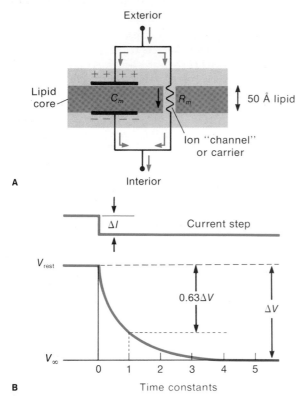

**A**

**B**

**5-7** (A) Simplest equivalent circuit for a cell membrane, showing membrane capacitance, $C_m$, and resistance, $R_m$. Arrows indicate direction of current. (B) Electrotonic response of a cell membrane to an applied step of inward-going current, $\Delta I$. The time required for the voltage to reach 63% of its asymptotic value is proportional to the product of the resistance and the capacitance of the membrane. This product is termed the time constant, $\tau$, of the membrane.

per square centimeter. Measured values for membranes range from 1 to 3 $\mu$fd/cm$^2$.

It is helpful to conceptualize the properties of conductance and capacitance in the form of an *equivalent circuit* (Figure 5-7A) in which a capacitor is wired in parallel with a resistor. The resistor represents the conductance conferred on the membrane by its ion channels or carriers. The capacitor represents the major area of lipid bilayer, which is essentially impermeable to ions.

The properties of membrane conductance and capacitance can be illustrated with the experimental setup

shown in Figure 5-3B. Consider a current of constant intensity, $\Delta I$ (amperes), passed across the membrane from the reference electrode in the bath to the current electrode in the cell. This current is applied as a square pulse with an abrupt onset. All the current must pass through the membrane to make a complete circuit. While crossing the membrane, the current distributes itself between the parallel conductance and capacitance of the membrane (Figure 5-7A).

The passive potential change of the membrane in response to an applied pulse of current, shown in Figure 5-7B, is termed an *electrotonic potential*. This potential change is produced by the applied current flowing through the capacitance and resistance of the membrane.

### MEMBRANE CONDUCTANCE

The conductance of a membrane is a measure of its permeability to ions. The greater the conductance, the more ions will cross the membrane per unit time under a given potential difference. When a step pulse of steady current is applied across the membrane, the membrane potential shifts with an exponential time course toward a steady-state level (Box 5-1). The displacement, $\Delta V_m$, of the membrane potential from the resting value to the asymptotic value is a function of both the magnitude of the applied current ($\Delta I$) and the input conductance ($G_{\text{input}}$) that the current encounters as it passes across the membrane of a cell. The relationship between applied current, conductance, and recorded steady-state voltage is described by Ohm's Law, which states that *the voltage across a membrane is proportional to the current passed through the membrane and inversely proportional to the conductance of the membrane.* Thus,

$$\Delta V_m = \Delta I / G_{\text{input}} \qquad (5\text{-}1)$$

or

$$G_{\text{input}} = \frac{\Delta I}{\Delta V_m}$$

It will be recalled that the reciprocal of conductance (in units of siemens) is resistance ($R$, in units of ohms):

$$R = \frac{1}{G}$$

Consider two spherical cells, one small, the other large, both with membranes having the same *specific resistance*, $R_m$, to electric current (i.e., the same resistance for a square centimeter of membrane). For a given increment of current, $\Delta I$, the large cell will show a smaller increment of voltage, $\Delta V_m$, because the same current will flow through a larger area of membrane; the current density will therefore be smaller across the membrane of the large cell than across the membrane of the smaller cell. This principle is illustrated by the fact that a current passing through two equal parallel resistors produces half the potential drop as the same current passing through only one of these resistors. Thus, if all else is equal, a large cell will have a lower input resistance than a small cell (Figure 5-8). Because the *input resistance, R,* of a cell (i.e., the total resistance encountered by current flowing into or out of the cell) is a function of both membrane area, $A$, and specific resistance, $R_m$, of the membrane, it is useful when comparing membranes of different cells to correct for the effect of membrane area on the current density. Thus, the specific resistance is calculated as

$$R_m = RA$$

Since

$$R = \frac{\Delta V_m}{\Delta I} \tag{5-2}$$

then

$$R_m = \frac{\Delta V_m}{\Delta I} A$$

Since $\Delta V_m / \Delta I$ has the units of ohms, and area is in square centimeters, $R_m$ is in units of ohms $\times$ centimeters squared. Note that membrane area and input resistance, $R$, are reciprocally related. The specific resistance, $R_m$, of the membrane is, of course, purely a property of the membrane structure. Specific resistances of various membranes range from hundreds to tens of thousands of ohms $\times$ centimeters squared.

The reciprocal of the specific resistance of a membrane is the specific *membrane conductance, $G_m$* (in units of siemens per square centimeter). Conductance is related to the ionic permeability of the membrane, but conductance and permeability are not synonymous. Conductance to a given species of ion is defined by Ohm's Law as the current carried by that species of ion divided by the electrical force acting on that species. Thus, membrane conductance for species $X$ is defined as

$$g_X = \frac{I_X}{\text{emf}_X} \tag{5-3}$$

in which $g_X$ is the membrane conductance for ion species $X$, $I_X$ is the current carried by that species, and

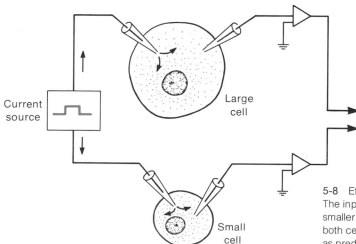

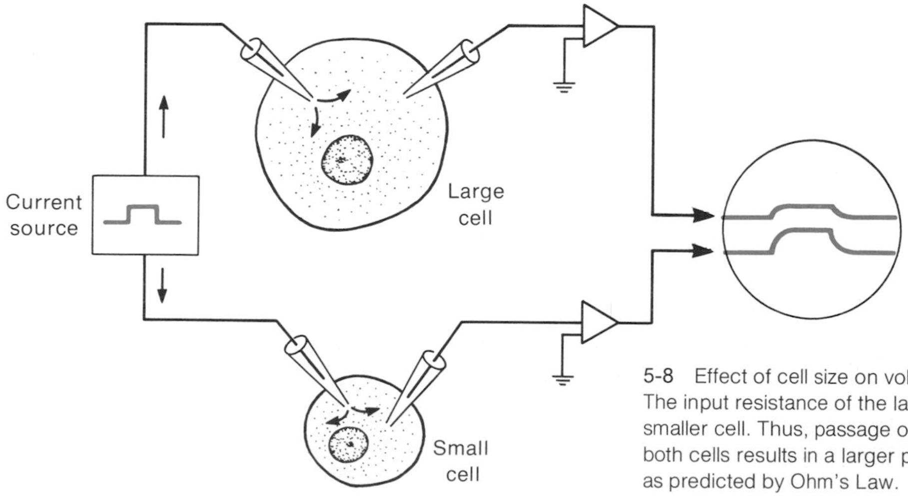

5-8  Effect of cell size on voltage response to a given current. The input resistance of the larger cell is lower than that of the smaller cell. Thus, passage of the same amount of current into both cells results in a larger potential change in the smaller cell, as predicted by Ohm's Law.

$\text{emf}_X$ is the electromotive force (in volts) acting on that species. Although $\text{emf}_X$ varies with membrane potential, it is not identical with membrane potential, as noted in the next section.

Even though a membrane may be permeable to ion X, the conductance, $g_X$, depends on the presence and concentration of this species in the solution; unless it is present, it cannot carry current. It is also evident that permeability of nonelectrolytes does not contribute any conductance, since a nonelectrolyte does not carry a charge and hence cannot carry current.

## Equilibrium Potentials

Now that we have considered the two basic electrical properties of cell membranes, capacitance and conductance, we can go on to consider the origin of *electrochemical potentials*. It is the electrical energy of these transmembrane potentials that is responsible for nearly all the electrical phenomena that occur in the animal body. These potentials originate, as you will see, from two features of biological membranes: (1) asymmetrical distribution of ions between the intracellular and the extracellular compartment, and (2) the selective permeability of the membrane.

To begin with, consider the chamber in Figure 5-9. It is divided into two compartments by a selectively permeable membrane, one that is permeable only to potassium ions. At the start of the experiment, both compartments contain 0.01 M KCl. Electrodes inserted one each into the two compartments record no potential across the membrane. Since this hypothetical membrane is permeable to $K^+$ but *not* to $Cl^-$, $K^+$ can diffuse across the membrane without its counterion. On the average, for each potassium ion that passes in one direction through the membrane, another will pass in the opposite direction under these conditions. As long as the two compartments contain the same concentration of KCl, the net $K^+$ flux is zero and the p.d. between the two sides of the membrane remains zero (Figure 5-9A). We then add KCl to compartment I to produce an instantaneous concentration of 0.1 M (10 times that of compartment II, Figure 5-9B). As $K^+$ now exhibits net diffusion, through potassium-selective channels in the membrane, from compartment I to II, a p.d. will quickly develop across the membrane and

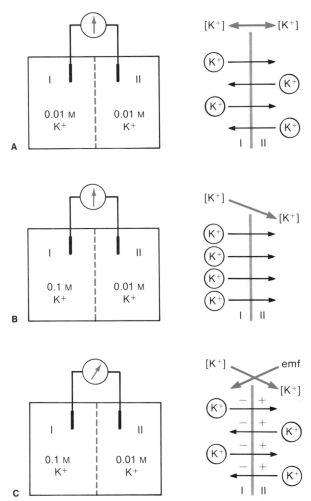

5-9 Electrochemical equilibrium. A hypothetical membrane permeable to only $K^+$ separates compartments I and II, containing the concentrations of KCl indicated. The net movement of $K^+$ across the membrane is zero when the electromotive force acting on $K^+$ balances the concentration gradient. See text for details.

the voltmeter will indicate that compartment I is more negative than compartment II (Figure 5-9C). This p.d. will be maintained indefinitely, provided there is no leakage of $Cl^-$ across the membrane.

How do we explain the steadily maintained p.d.? For every $K^+$ that is statistically available for diffusion through the membrane from compartment II to I,

10 potassium ions in I are available to pass through the membrane to compartment II. That is, the difference in K$^+$ concentration gradients represents a *chemical p.d.,* causing an initial net diffusion through the membrane from I to II (Figure 5-9B). Each additional K$^+$ that diffuses from I to II adds its positive charge to that side, and Cl$^-$ is left behind, since it cannot accompany K$^+$ across this hypothetical membrane. As potassium ions accumulate in compartment II and chloride ions are left behind in compartment I, the p.d. across the membrane quickly rises, since the membrane then separates a *slight* excess of positive charges on one side from a *slight* excess of negative charges on the other, as in Figure 5-6. As K$^+$ leaks into compartment II, it builds up positive potential in that compartment and leaves behind net negative charge. It is repelled by the former and attracted back to the latter. Thus, each K$^+$ now entering the membrane has two forces acting on it: a chemical p.d. favoring net K$^+$ flux from I to II, and an *electrical p.d.* favoring net K$^+$ flux from II to I (Figure 5-9C). These two opposite forces come into equilibrium and remain balanced, with the electrostatic force of the electrical p.d. precisely offsetting the tendency for K$^+$ to diffuse down its concentration gradient. The potassium ion is then said to be in *electrochemical equilibrium;* the p.d. that is established in this way is termed the *equilibrium potential* for the ion in question (in this case, the potassium equilibrium potential, $E_K$). When an ion is in electrochemical equilibrium, it undergoes no net flux across the membrane (even if the membrane is freely permeable to that ion).

To illustrate the equilibrium state between ionic concentration gradient and the resulting electric potential gradient, a simple analogy is given in Figure 5-10. A mass is gently lowered from a spring. As gravity pulls the mass down, tension develops in the spring as the mass stretches the spring. This tension holds the mass up with a force equal and opposite to the force of gravity acting to pull down the mass; the system is therefore in equilibrium, with the mass suspended on the stretched spring. The gravity pulling on the mass is analogous to the chemical gradient, and the tension developed in the spring is analogous to the potential developed across the membrane. The gravity acting on the mass produces the tension in the spring by stretching it, and the tension develops until it just balances

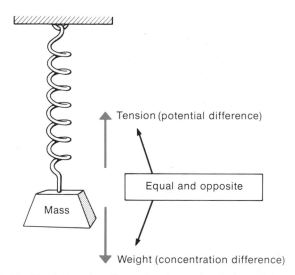

**5-10**  Physical analogy illustrating electrochemical equilibrium. The tension in the spring is analogous to the potential difference produced by diffusion of an ion across a semipermeable membrane. The weight of the mass is analogous to the concentration difference responsible for that diffusion. Note that the tension is produced by, and is equal to, the weight pulling on the spring.

the pull of gravity and keeps the mass suspended. Likewise, movement of charge from compartment I to II produces the electrical "tension" (p.d.); the p.d., in turn, prevents further movement of charge and thereby balances the unequal ionic concentrations.

If for any reason the membrane potential, $V_m$, is not at the equilibrium potential, $E_X$, for an ion X, there will exist an emf acting on that ion, emf$_X$, equal to the difference between $V_m$ and $E_X$:

$$\text{emf}_X = V_m - E_X \qquad (5\text{-}4)$$

Clearly, when $V_m = E_X$, ion X will experience no emf and will be in electrochemical equilibrium across the membrane.

The equivalent electrical circuit for the development of the membrane potential in Figure 5-9C is given in Figure 5-11. Positive charge (in the form of potassium ions), driven by the emf acting on potassium (i.e., $V_m - E_K$), leaks through the potassium conductance (i.e., $R_K$) of the membrane so as to accumulate on the other side of the membrane. When the voltage

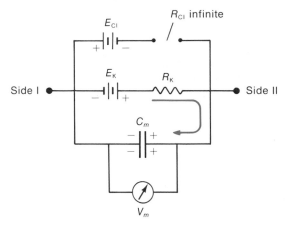

**5-11** Equivalent electrical circuit for development of a potential across the membrane in Figure 5-9. $E_K$ supplies the emf for $K^+$ to carry current through the membrane's potassium channels, $R_K$. This transport causes the buildup of positive charge on side II of the membrane capacitance, $C_m$.

across the capacitance of the membrane equals the potassium equilibrium potential (i.e., when $V_m - E_K = 0$), net $K^+$ diffusion ceases and the system is at equilibrium, side II positive with respect to side I. Although the electrochemical gradient for the chloride ion is in the opposite direction, it has no effect, because our hypothetical membrane is impermeable to chloride ions.

As explained in Box 5-2, the amount of $K^+$ crossing the membrane to produce the potential difference is so small that there is *virtually no change in concentration* of that ion in the two compartments.

## THE NERNST RELATION

It seems right, intuitively, that the equilibrium potential of an ion should increase in value with an increase in its concentration gradient across the membrane, just as the tension developed in a spring (Figure 5-10)

---

### Box 5-2  *Charge Separation by Membranes*

Very few ions actually diffuse across 1 cm² of the membrane in Figure 5-9 before the membrane potential equals $E_K$. The actual number of excess ions that cross the membrane is easy to calculate for a system with a single diffusible ion. The number of excess potassium ions accumulated in compartment II (and excess chloride ions left behind in compartment I) depends on two factors: (1) the potassium equilibrium potential and (2) the capacitance of the membrane. You will recall that the charge, $q$ (in coulombs), accumulated across a capacitor is proportional to both the capacitance, $C$, of the capacitor and the voltage, $V$, developed across the capacitor (Equation 1, Box 5-1). Biological membranes typically have capacitances of about 1 $\mu$fd ($10^{-6}$ fd) per square centimeter. We can calculate the coulombs of charge that diffuse across 1 cm² of membrane when the membrane separates a 10-fold difference in the concentrations of a diffusible monovalent cation (i.e., a potential difference of 58 mV after equilibrium is achieved):

$$q = CV$$
$$= (10^{-6} \text{ fd/cm}^2)(5.8 \times 10^{-2} \text{ V})$$
$$= 5.8 \times 10^{-8} \text{ C/cm}^2$$

There is 96,500 C of charge (1 $F$) in 1 g-equiv wt, or 1 mol, of a monovalent ion. Thus, the amount of $K^+$ in moles required

to transfer $5.8 \times 10^{-8}$ C across 1 cm² of membrane is calculated by dividing by the number of coulombs in 1 $F$:

$$\frac{5.8 \times 10^{-8} \text{ C/cm}^2}{9.65 \times 10^4 \text{ C/mol K}^+}$$
$$= 6 \times 10^{-13} \text{ mol K}^+/\text{cm}^2$$

The number of potassium ions accumulated in compartment II of the membrane in Figure 5-9 is found by multiplying the number of moles by Avogadro's number ($6 \times 10^{23}$ molecules/mol):

$$(6 \times 10^{-13})(6 \times 10^{23}) = 3.6 \times 10^{11} \text{ K}^+/\text{cm}^2$$

An equal number of chloride ions remains in excess in compartment I of the membrane. This number is more than 10,000,000 smaller than the number of potassium ions in a cubic centimeter of solution II ($6 \times 10^{18}$ potassium ions). Thus, the concentrations in compartments I and II are virtually unchanged as a result of the charge separation across the membrane. Even though there is a slight separation of anions from cations across the membrane, the segregation exists only on a microscopic scale, separated by about the thickness of the membrane (Figure 5-6). The rule of electroneutrality—that positive charges must equal negative charges—remains essentially unviolated on the macroscopic scale.

## Box 5-3   *Derivation of the Nernst Equation*

The Nernst equation is probably the most widely used mathematical relation in physiology and is essential for an understanding of bioelectric phenomena. Its derivation is based on the concept of a thermodynamic equilibrium between the osmotic work that is required to move a given number of ions across a membrane in one direction and the electrical work required to move the same number of charges back across the membrane in the opposite direction. The osmotic work required to transfer 1 mol-equiv of an ion, X, from a concentration $[X]_I$ to a concentration 10 times higher, $[X]_{II}$, can be derived from the gas laws as

$$W = RT \ln \frac{[X]_I}{[X]_{II}} \qquad (1)$$

in which $W$ is mechanical (or osmotic) work (equals force times distance).

This, then, represents the thermodynamic work that would be required to establish an e-fold difference in the concentration across a membrane barrier of 1 mol of ion X, starting with equal concentrations on both sides of the membrane.

If the membrane is permeable to this species, these ions will tend to diffuse back toward the low concentration, until the resulting equilibrium potential just balances this tendency. Note that we can relate osmotic work to electrical work through the equality

$$W = EFz \qquad (2)$$

in which the potential difference, $E$, is multiplied by the Faraday constant ($F$), namely, the charge per mole of univalent ion. The valence, $z$, of the ion corrects for multivalent species. Substituting Equation 2 in Equation 1, we obtain

$$EFz = RT \ln \frac{[X]_I}{[X]_{II}} \qquad (3)$$

or

$$E = \frac{RT}{Fz} \ln \frac{[X]_I}{[X]_{II}} \qquad (4)$$

which is the general form of the Nernst equation.

---

increases with mass. In other words, a greater chemical p.d. across the membrane should develop a greater electric p.d. in order to offset the tendency for the ions to diffuse down their concentration gradient. The equilibrium potential, in fact, is proportional to the *logarithm* of the ratio of the concentrations in the two compartments. The relation between concentration ratios and membrane potential was derived in the latter part of the nineteenth century by Walther Nernst from the gas laws (Box 5-3). The equilibrium potential depends on the absolute temperature, the valence of the diffusible ion, and, of course, the ratio of concentrations on the two sides of the membrane:

$$E_X = \frac{RT}{Fz} \ln \frac{[X]_I}{[X]_{II}} \qquad (5\text{-}5)$$

Here $R$ is the gas constant, $T$ is the absolute temperature (Kelvin); $F$ is the Faraday constant (96,500 C/g-equiv charge); $z$ is the valence of ion X; $[X]_I$ and $[X]_{II}$ are the concentrations (more accurately, activities) of ion X on sides I and II of the membrane; and $E_X$ is the equilibrium potential for ion X (potential

of side II minus side I). At a temperature of 18°C and with a monovalent ion, and converting from ln to log, the Nernst equation can be reduced to

$$E_X = 0.058 \text{ V} \log \frac{[X]_I}{[X]_{II}} \qquad (5\text{-}6)$$

Note that $E_X$ will be positive if X is a cation and the ratio of $[X]_I/[X]_{II}$ is greater than unity. The sign will become negative if the ratio is less than 1. Likewise, the sign will be reversed if X is an anion rather than a cation, because $z$ will be negative. The Nernst relation predicts a rise in potential difference of 58 mV every time the concentration ratio of the permeant ion is increased by a factor of 10. When the potential is plotted as a function of log $[K^+]_I/[K^+]_{II}$, the relation has a slope of $-58$ mV per 10-fold increase in the ratio (Figure 5-12).

Recall that in living cells the electrical potential inside the cell, $V_i$, is described in reference to the potential outside the cell, $V_o$. That is, the membrane potential, $V_m$, is given as $V_i - V_o$, so that the potential of the cell exterior is arbitrarily defined as zero. For this reason,

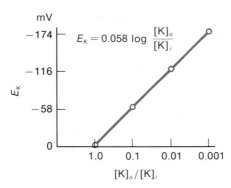

**5-12**   Relationship between the equilibrium potential of a monovalent ion, such as $K^+$, and the ratio of concentrations of that ion on the two sides of a membrane.

when determining the equilibrium potential, we place the extracellular concentration of the ion in question in the numerator and the intracellular concentration in the denominator of the concentration ratio. Applying the Nernst relation (Equation 5-5), we can calculate the potential at which potassium will be in equilibrium, $E_K$, in a hypothetical cell in which $[K]_o = 0.01$ M and $[K]_i = 0.1$ M:

$$E_K = 0.058 \text{ V} \times \log \frac{[K]_o}{[K]_i}$$

$$= 58 \text{ mV} \times \log \frac{0.01}{0.1}$$

$$= 58 \text{ mV}(-1)$$

$$= -58 \text{ mV}$$

Note that $E_K$ has a negative sign, since intracellular negativity will result when a minute amount of $K^+$ leaks out of the cell owing to its high intracellular and low extracellular concentrations. It should also be apparent from Equation 5-8 that if the ion in question is a divalent cation (i.e., $z = +2$), the slope of the relation becomes $-29$ mV per 10-fold increase in concentration ratio.

## The Resting Potential

### PASSIVE CHARGE DISTRIBUTION

The principle of an equilibrium potential was illustrated in the preceding section with a simplified, ideal system in which only one ionic species was diffusible. This principle will now be applied to biological mem-

branes, which are permeable in varying degrees to all of the inorganic ions present.

It is evident that the electrochemical gradient of an ion species has no effect on the membrane potential if the membrane is impermeable to that species. After all, nonpermeant ions cannot carry charge from one side of the membrane to the other. It follows that a species to which the membrane is only slightly permeable will have a smaller effect on the membrane potential than another species that can diffuse across the membrane more freely. It is, in fact, the relative ease with which different ions can cross the membrane that determines their relative contributions to the potential they produce in diffusing across the membrane. On this basis, and by making the assumption that there is a uniform gradient of potential in going from one side of the membrane to the other side, D. E. Goldman (1943) derived an equation that is related to the Nernst equation and that takes into consideration the relative permeability of each species of ion:

$$V_m = \frac{RT}{F} \ln \frac{P_K[K^+]_o + P_{Na}[Na^+]_o + P_{Cl}[Cl^-]_i}{P_K[K^+]_i + P_{Na}[Na^+]_i + P_{Cl}[Cl^-]_o} \quad (5-7)$$

in which $P_K$, $P_{Na}$, and $P_{Cl}$ are respective permeability constants of the major ion species in the intra- and extracellular compartments.

Thus, the probability that the ions of one species will cross the membrane is proportional to the product of their concentration (more accurately, their thermodynamic *activity*) on that side and the permeability of the membrane to that species. As a result, the contribution of an ion species to the membrane potential diminishes as its concentration is reduced. This point is illustrated in Figure 5-13, in which the membrane potential of a living cell is plotted as a function of the concentration of extracellular $K^+$, the ion most important in determining the resting potential. At the higher $K^+$ concentrations, the slope of the plot is about $+58$ mV per 10-fold increase in potassium concentration. At lower $K^+$ concentrations, the curve deviates from this theoretical slope for $E_K$ because $Na^+$ becomes a more important contributor, in spite of its low permeability, as the product $P_{Na}[Na^+]_o$ approaches $P_K[K^+]_o$.

R. D. Keynes (1954) determined the permeability constants for the major ions in frog muscle with the

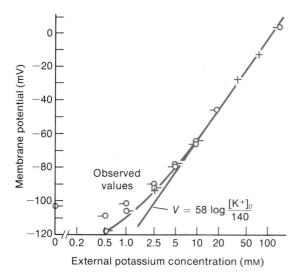

**5-13** Resting potential of a frog muscle cell plotted against the extracellular $K^+$ concentration. The theoretical 58 mV change for every 10-fold increase in the ratio of $[K^+]_o:[K^+]_i$, as predicted by the Nernst equation, is shown as a straight line. The measured values are shown by the plotted points. Curved line was calculated from Equation 5-7, using $P_{Na} = 0.01\ P_K$. $[K^+]_i$ was taken as 140 mM. [Hodgkin and Horowicz, 1960.]

use of radioisotopes. The permeability of sodium was found to be about 0.01 times that of potassium. Chloride is in electrochemical equilibrium across the muscle membrane (i.e., $E_{Cl} = V_m$) because it is not actively transported and can therefore be ignored. Thus, for muscle cell membranes, the Goldman equation can be simplified to

$$V_m = \frac{RT}{F} \ln \frac{[K^+]_o + 0.01[Na^+]_o}{[K^+]_i + 0.01[Na^+]_i}$$

$$= 0.058 \log \frac{2.5 + (0.01)(120)}{140 + (0.01)(10)}$$

$$= -92\ \text{mV}$$

Microelectrode measurements of the resting potential in frog skeletal muscle cells range from $-90$ to $-100$ mV.

Resting potentials of muscle, nerve, and other cells are far more sensitive to changes in the extracellular potassium level than to changes in the concentrations of other cations because of the predominant permeability of most cell membranes to $K^+$. Large changes in

extracellular $Na^+$, for example, have little effect on the resting potential because of the low permeability of the cell membrane to sodium ions.

To summarize, the negative resting potential of cells is due to their high internal potassium concentrations relative to extracellular potassium concentrations and to the tendency of $K^+$ to leak out of the cell, leaving behind a net negative charge. Other ions such as $Na^+$ contribute little to the resting potential, because the resting membrane has a relatively low permeability to those ions.

### THE ROLE OF ACTIVE TRANSPORT

An idealized membrane (Figures 5-9 and 5-11) permeable to only one ion species will maintain a constant membrane potential indefinitely without expenditure of energy, because it is in a state of thermodynamic equilibrium. Real membranes, however, are leaky to varying degrees to all inorganic ions.

Take the case of $Na^+$. If we take the concentrations of extracellular $Na^+$ and intracellular $Na^+$ in frog muscle (Figure 4-17) as 120 and 10 mM/L, respectively, we can calculate the sodium equilibrium potential as follows:

$$E_{Na} = 0.058\ \text{V} \times \log \frac{120}{10} = +63\ \text{mV}$$

Since $V_m$ in frog muscle ranges from $-90$ to $-100$ mV, the sodium ions are over 150 mV (i.e., $V_m - E_{Na}$) out of equilibrium. Even with a small resting membrane permeability to $Na^+$, there will be a steady influx of $Na^+$ driven by the emf acting on that ion. If it were not removed from the cell interior at the same rate at which it leaks in, $Na^+$ would accumulate in the cell, displacing internal $K^+$ that would leak out in response to intracellular sodium accumulation. The high internal $[K^+]$ and low internal $[Na^+]$ result from the continual active transport of $Na^+$ out of the cell. During active transport of $Na^+$ there is an obligatory uptake of $K^+$, usually two potassium for three sodium ions. Even without coupled $K^+$ uptake, however, potassium ions, which are relatively permeant, would move into the cell passively to replace the $Na^+$ pumped out, maintaining electroneutrality of the cytoplasm.

When metabolically energized transport is eliminated by an inhibitor of oxidative metabolism, such as cyanide or azide, or by a specific inhibitor of sodium transport, such as ouabain, $Na^+$ exhibits a net influx, internal $K^+$ is gradually displaced, and the resting potential shows a corresponding *slow* decay as the ratio $[K^+]_i / [K^+]_o$ gradually decreases. Over the long term, it is the metabolically energized extrusion of $Na^+$ that keeps the $Na^+$ and $K^+$ concentration gradients from running downhill to equilibrium. By continual maintenance of the potassium concentration gradient, the sodium pump plays an important *indirect* role in determining the resting potential (Figure 5-14).

Active transport has been shown to contribute *directly,* as well as indirectly, to the resting potential of some cells. This situation occurs because the amount of $Na^+$ extruded from the cell per unit time exceeds the rate of $K^+$ uptake by the pump (Figure 5-14). The pump is said to be *electrogenic* because it is directly responsible for a steady transfer of net positive charge out of the cell, which contributes to the internal negative potential. The size of the potential contributed by an electrogenic pump depends on the rate at which charge, generally in the form of $K^+$ entry or $Na^+$ entry, can leak back into the cell. This direct contribution of the

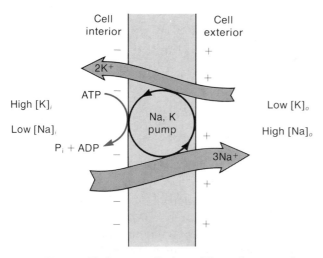

**5-14** Direct and indirect contributions of the sodium pump to the resting potential. Because of the 3Na:2K exchange ratio, the pump may contribute *directly* to the resting potential by removing positive charge from the cell interior. By maintaining a high internal potassium concentration, the pump *indirectly* contributes to the resting potential.

sodium pump to the resting potential seldom exceeds several millivolts, since the tendency for positive charge to leak back into the cell increases as the resting potential becomes more negative than the equilibrium potentials of the relatively highly permeant ions $K^+$ and $Cl^-$.

## *Release of Stored Electrical Energy*

The electrical energy stored across the cell membrane by the activity of metabolic pumps is selectively released in the form of ionic currents to produce the action potential and other electrical signals in living tissue. To understand this, it is again helpful to think of the membrane as having the properties of a capacitance shunted by a conductance (Figure 5-15A). This time, however, we will include a separate conductance in the equivalent circuit for each species of ion being considered, plus some batteries to represent the equilibrium potential of each species.

In the resting state, the membrane is predominantly permeable to $K^+$; hence the resting membrane potential is in the vicinity of $E_K$, and potassium ions are nearly in electrochemical equilibrium across the resting membrane. As sodium is more concentrated outside the cell than inside, both its electrochemical gradient and the polarity of its equilibrium potential are opposite those of potassium. The sodium potential, $E_{Na}$, is positive, whereas $E_K$ is negative (Figure 5-15B). Because the sodium equilibrium potential is far from the resting potential, there is a large emf acting on $Na^+$ equal to the difference between the membrane potential and the sodium equilibrium potential (i.e., $V_m - E_{Na}$). This emf is a substantial source of potential energy, which is released in some cells by stimuli that increase the sodium permeability of the membrane so as to permit a transient *inward sodium current* (i.e., net influx of $Na^+$) through the membrane.

Suppose that in the example of Figure 5-15B the membrane potential, $V_m$, is initially $-90$ mV, while $E_{Na} = +63$ mV. The emf acting on sodium is the difference between these two values:

$$\begin{aligned} \text{emf}_{Na} &= V_m - E_{Na} \\ &= -90 \text{ mV} - 63 \text{ mV} \\ &= -153 \text{ mV} \end{aligned}$$

The negative sign means that sodium current will be

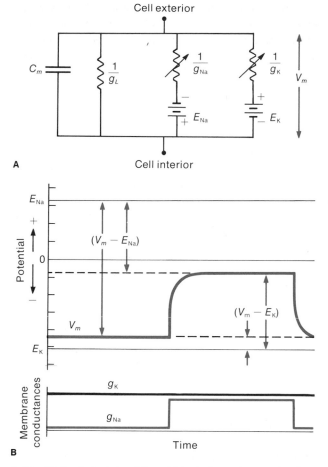

Cell exterior

Cell interior

**A**

**B**

5-15 (A) Equivalent circuit for a biological membrane, showing the electrically excited conductances for Na+ and K+. The remaining conductances are lumped as $1/g_L$. The concentration batteries $E_{Na}$ and $E_K$ are connected in series with the respective conductances. Note that these batteries have opposite polarities. (B) Changes in membrane potential, $V_m$ (top), produced by changes in sodium conductance, $g_{Na}$ (bottom). Note the changes in the terms $(V_m - E_{Na})$ and $(V_m - E_K)$ produced by the change in $V_m$. Note also that $E_K$ and $E_{Na}$ remain constant throughout.

directed into the cell. Assigning a value of $-102$ mV to $E_K$ in this hypothetical example (Figure 5-15B), we can calculate

$$\begin{aligned} \text{emf}_K &= V_m - E_K \\ &= -90 \text{ mV} - (-102 \text{ mV}) \\ &= +12 \text{ mV} \end{aligned}$$

and we see from the positive value that the current carried by K+ will flow out of the cell.

The effect of a temporary increase in the sodium conductance, $g_{Na}$, is illustrated in Figure 5-15B. When this conductance increases, Na+ will flow through it into the cell, adding positive charge to the inside surface of the capacitor, $C_m$ (Figure 5-15A), until a new steady-state potential is attained between $E_K$ and $E_{Na}$, and K+ (which is now out of equilibrium) carries charges out of the cell at the same rate as Na+ carries them in. Thus, the membrane will undergo a *depolarization* in response to an increase in its sodium conductance. When $g_{Na}$ is restored to its previous low level, the membrane potential returns to its original value near $E_K$. Should there be a conductance increase to an ion such as K+ that has its equilibrium potential more negative than the prevailing membrane potential, $V_m$ will shift in the *hyperpolarizing* direction toward the equilibrium potential of that ion (toward $E_K$, for example).

To summarize, the production of an electrical signal by a cell membrane depends on two factors:

*1.* The existence of an electrochemical gradient due to unequal distribution of ions across the membrane.

*2.* The opening of ion channels to allow the ionic current to flow across the membrane. When this opening occurs, a gradient that is out of electrochemical equilibrium provides the emf that drives the ions across the membrane to produce an electric current.

## *Ionic Basis of the Action Potential*

### *GENERAL PROPERTIES*

Action potentials† are produced by the membranes of neurons and muscle cells, as well as by some receptor cells, secretory cells, and protozoa. They perform two major functions:

*1.* Rapid transmission of information over long distances, in nerve and muscle fibers.

---

†The appearance of the recorded action potential has led to the term *spike* in neurophysiological jargon. *Spike, impulse,* and *action potential* are generally used interchangeably.

*2.* Control of effector responses, such as muscle contraction and the release of neurotransmitters and hormones.

To demonstrate some of the properties of action potentials, we will continue with the experiment begun on page 137. Current is passed outward across the membrane of a nerve cell as a short pulse (Figure 5-16, lower trace), producing a passive depolarization until the current delivered is strong enough to depolarize the membrane to the threshold potential. This potential then causes the production of an action potential. If the depolarization is just short of reaching threshold, there often appears an aborted, nonpropagated excitation termed a *local response.* It is simply the beginning of an action potential that died out before it was irreversibly under way.

The intensity of stimulating current just sufficient to displace the membrane to the threshold potential, and thus to elicit the action potential, is defined as the *threshold current.* No consistent values can be assigned to either the threshold current intensity or the threshold potential, because these depend on several factors, such as the condition of the membrane and the conditions of its environment, the duration of the current pulse, and the resistance of the membrane.

Once the threshold potential is attained, the upstroke of the action potential becomes *regenerative* (i.e., explosive; self-perpetuating). The cell interior continues to lose negative potential until the potential passes zero (i.e., reverses polarity), increasing in positive potential until it reaches a peak of $+30$ to $+50$ mV (Figure 5-16). The portion of the action potential on the positive side of zero is called the *overshoot.*

It is the regenerative property of the action potential that permits it to propagate itself along a nerve cell axon or a muscle fiber for long distances without any decline in amplitude. Propagation of the action potential is discussed in Chapter 6.

Because there are no responses (under a given set of conditions) that are intermediate in size between a full response and a minute, abortive response, the action potential is said to be *all-or-none.* Although the amount of overshoot can change if the condition of the membrane or the compositions of the intra- or extracellular solutions change, this does not contradict the all-or-none principle, which is simply that the amplitude of the response (action potential) is independent of the intensity of the stimulus. An analogy can be drawn with the flushing of a toilet. Once initiated, the flush generally continues until the tank is emptied, independent of the pressure applied to the lever to initiate the operation.

Another feature that is characteristic of an action potential is the rapid repolarization from the peak of the overshoot to the resting level (Figure 5-16). Action potentials range in duration from less than a millisecond in some nerve fibers (axons) to nearly a half-second in heart muscle cells.

When the interval between two action potentials is reduced, the second one becomes progressively smaller, and it fails completely if the stimulus is delivered too soon after the end of the first action potential (Figure 5-17A). The interval at which the second spike fails depends on the stimulus strength. No stimulus, how-

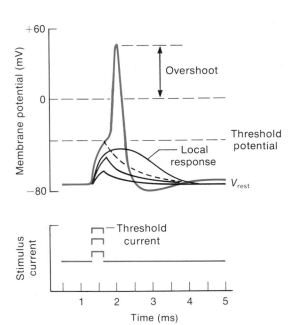

5-16  Action potential produced by a nerve cell membrane in response to a depolarizing stimulus that brought the membrane to the threshold potential. Smaller polarizations failed to evoke the all-or-none response.

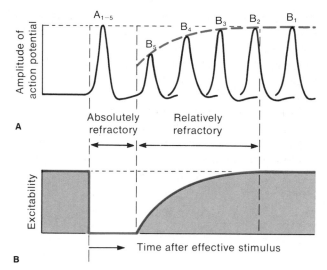

**A**

**B**

5-17 (**A**) Refractoriness following an action potential. Five pairs of stimuli ($A_1$, $B_1$, to $A_5$, $B_5$) were delivered with progressively shorter intervals between the first stimulus (A) and the second (B) in each pair. During the relatively refractory period, spike B became smaller. During the absolutely refractory period, no second spike can be elicited. (**B**) Time course of change in excitability.

5-18 Accommodation. The current required to reach threshold increases with time during subthreshold stimulation. Thus, slowly increasing currents (*b* to *c*) are less effective in reaching thresholds than an abruptly increasing current (*a*). If it rises slowly enough, as in *d*, the current will never reach threshold. [Davson, 1964.]

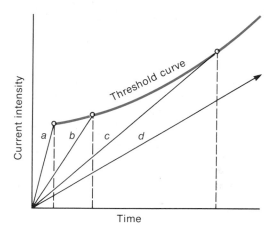

ever, is sufficient to evoke a second action potential during the *absolute refractory period,* which exists during and persists for a short time after an action potential. Following this period, the excitable cell membrane enters the *relative refractory period.* This is characterized by (1) a threshold that is higher than before the action potential (i.e., more current is required to elicit an action potential) and (2) a reduced amplitude of the action potential (i.e., the overshoot is smaller). The threshold potential progressively decreases (i.e., excitability increases) during this period to the level characteristic of the resting membrane before stimulation (Figure 5-17B). Diminished excitability (*refractoriness*) during and immediately after the action potential prevents fusion of impulses, but permits the propagation of discrete impulses.

The membrane undergoes a time-dependent decrease in excitability (i.e., threshold increase) during subthreshold depolarizations. This can be demonstrated by depolarizing the membrane gradually with a cur-

rent of steadily increasing intensity rather than with an abrupt steplike stimulus current. Greater depolarizations are required to elicit an action potential with such a slowly increasing current (Figure 5-18). This characteristic of excitable membranes—a result of time-dependent changes in the sensitivity of membrane channels to depolarization—is termed *accommodation.*

Accommodation of excitable membranes also occurs during the passage of a constant-intensity current. Some nerve cell membranes accommodate rapidly and generate only one or two spikes at the beginning of a prolonged constant-current stimulus; others accommodate more slowly and therefore fire repetitively with gradually decreasing frequency in response to a maintained current of constant intensity (Figure 5-19). It will be seen in Chapter 7 that accommodation is important in sensory physiology, for it is one of several factors that determine whether a maintained stimulus will elicit a transient or maintained discharge in a sensory neuron. The reduction in discharge frequency during a sustained stimulus is termed *adaptation* (Figure 5-19B).

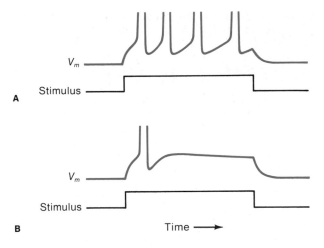

A

B

Time ⟶

**5-19** Adaptation of impulse discharge during a constant-stimulus current. (A) Some membranes show little accommodation except for a progressive lengthening of the interspike interval. (B) Other membranes produce only one or two impulses in response to a similar stimulus. Tops of action potentials are cut off.

## THE SODIUM HYPOTHESIS

Much of what we know about electrically excitable membranes began with a report in 1936 by the British zoologist J. Z. Young that certain longitudinal structures in squids and cuttlefish (Figure 5-20) previously thought to be blood vessels were really extraordinarily large axons. He quickly recognized the possible usefulness of these *giant axons* for the study of membrane physiology, for their unusually large diameters (up to 1 mm) allow electrode wires to be inserted longitudinally for stimulation and recording (Figure 5-21A).

The first major discoveries made with the squid axon were published separately in 1939 by K. S. Cole and H. J. Curtis, working at Woods Hole, Massachusetts, and by A. L. Hodgkin and A. F. Huxley, working in Plymouth, England. Cole and Curtis demonstrated an increase in membrane conductance (without any significant change in capacitance) concomitant with the passage of the action potential (Figure 5-22). Hodgkin and Huxley found that the membrane potential does not simply go to zero during the action potential but instead reverses sign during the impulse (Figure 5-21B). This important detail was at odds with the earlier

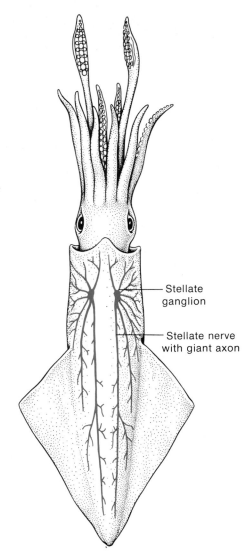

**5-20** The squid *Loligo*. The stellate nerve contains the giant axon. The axon, because of its large size, conducts rapidly, ensuring relatively synchronous activation of the muscles of the mantle. These muscles are responsible for the production of the water jet that propels the squid backward when the animal is startled. [Keynes, 1958.]

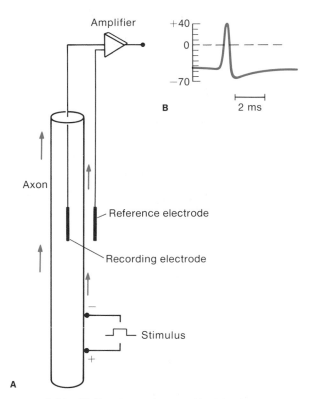

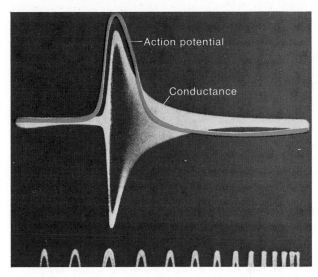

**5-22** Change in membrane conductance during the action potential in squid axon. This classic display from Cole and Curtis (1939) includes the action potential and an envelope of high-frequency oscillation, which indicates by its increase in width the increased membrane conductance during and briefly following the action potential. By repeating this with different frequencies of oscillation, Cole and Curtis found the membrane capacitance to remain constant. Time markers are 1 ms apart.

**5-21** (**A**) Experimental setup with which Hodgkin and Huxley (1939), using the squid giant axon, demonstrated that the membrane potential reverses sign during the action potential (**B**). The arrows show propagation of the impulse past the recording electrodes.

assumption that the permeability increase during excitation was nonspecific and that it resulted in a simple collapse of membrane polarization.

Hodgkin and B. Katz (1949) subsequently found that the action potential fails to develop if the extracellular $Na^+$ is removed. When extracellular $Na^+$ is present in reduced concentration, the rate of depolarization decreases, and the amplitude of the action potential diminishes (Figure 5-23). This finding led them to propose the *sodium hypothesis*—namely, that the upstroke and overshoot of the action potential results from an influx of $Na^+$ through a transiently increased membrane permeability to sodium.

The sodium hypothesis receives support from the following considerations and observations.

*1.* The extracellular concentration of $Na^+$ exceeds the intracellular concentration by a factor of about 10, so that $E_{Na}$ is about $+50$ to $+60$ mV. Thus, the emf will tend to drive $Na^+$ through the membrane into the cell. The emf acting on $Na^+$ is $V_m - E_{Na}$.

*2.* Since it carries a positive charge, $Na^+$ on entering the cell will produce the observed positive shift in intracellular potential.

*3.* The overshoot of the action potential approaches the sodium equilibrium potential calculated from the 10:1 ratio of external to internal $Na^+$ concentration:

$$E_{Na} = \frac{RT}{Fz}\ln\frac{[Na^+]_o}{[Na^+]_i} = +0.058 \text{ V log } 10$$

$$= +58 \text{ mV}$$

## Box 5-4  *The Voltage-Clamp Method*

An important method of studying the behavior of voltage-gated membrane channels is to "clamp" the voltage difference across the membrane by electronic feedback, subjecting the membrane to precisely controlled steplike changes in voltage. The currents that flow across the membrane through channels activated by the voltage change are measured at the same time.

There are several reasons for clamping the membrane potential in an abrupt step from a steady holding potential to a new stimulus potential:

*1.* Total membrane current ($I_m$) is made up of resistive current, $i_r$, and capacitive current, $i_c$. Capacitive current is proportional to both the rate of change of membrane potential and the membrane capacitance, $C_m$:

$$i_c = \frac{dV_m}{dt}C_m \tag{1}$$

Since

$$I_m = i_r + i_c \tag{2}$$

by substitution

$$I_m = i_r + \frac{dV_m}{dt}C_m \tag{3}$$

With the membrane potential clamped at a constant value, $dV_m/dt$ equals zero, and

$$I_m = i_r \tag{4}$$

Thus, all the membrane current after the abrupt step in voltage passes through the membrane conductances. The current delivered to charge up the membrane capacitance to this level is over in a fraction of a millisecond and can therefore be ignored.

*2.* In the unclamped membrane, conductance and potential interact with each other to produce the Hodgkin cycle (Figure 5-30), with resulting complex voltages and currents. By clamping the potential at a controlled value, one major variable, $V_m$, is held constant, the Hodgkin cycle is broken, and the complexity of the analysis is greatly reduced.

*3.* By clamping the membrane to different voltages and measuring the currents produced, it is possible to determine the behavior of the ion channels through which these currents pass.

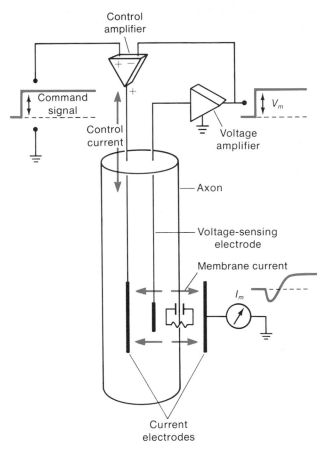

Basic circuit for voltage-clamping a squid axon. The control amplifier compares the membrane potential, $V_m$, with the command signal. Any difference produces a current at the output of the amplifier, which is fed back across the membrane, bringing $V_m$ into agreement with the command potential. The currents, $I_m$, depend on the conductance changes that take place in the membrane.

In voltage-clamping, the potential recorded from inside the axon is compared by a control amplifier with an electronically generated "command" potential (see figure). The voltage difference between the membrane potential and the command signal is amplified so that the output of the amplifier supplies a control current that passes across the membrane in the direction that brings the membrane potential into agreement with the command signal. This automatic adjustment is complete within a fraction of a millisecond after a steplike command potential is presented. Thus, when the membrane becomes permeable to Na⁺, and that ion carries a current across the membrane into the axon, the entering charge is immediately removed from within the axon by the control system so as to keep the membrane voltage constant. The current supplied or removed by the amplifier to maintain the selected membrane potential is equal, of course, to the current crossing the membrane, $I_m$.

***

*4.* The overshoot varies, as noted above, with the extracellular Na⁺ concentration as predicted for the change in $E_{Na}$.

After World War II, which interrupted their work on the squid axon, Hodgkin and Huxley (1952a, 1952b) obtained further evidence in support of the sodium hypothesis with a powerful new electronic technique termed *voltage clamping* (Box 5-4). In a nutshell, this method, first applied to the squid giant axon, employs a feedback system that abruptly changes and maintains ("clamps") the membrane voltage constant at any preselected value while the ionic current that flows across the membrane is measured. This approach includes fewer uncontrolled variables than are present when the membrane is permitted to produce an action potential in response to a pulse of depolarizing current (Figures 5-5 and 5-16). Moreover, it can be used to investigate the behavior of the all-important ion channels that are responsible for the production of bioelectric signals.

Figure 5-24 shows that a *hyper*polarizing voltage step (trace *a*) results in a very small and constant inward membrane current (trace *a'*) during maintained hyperpolarization. A *de*polarizing potential step of the same magnitude (trace *b*) is accompanied by a stronger and more complex sequence of membrane currents (trace *b'*). The initial transient downward deflection of trace b' indicates an early surge of inward current. This subsides in 1 or 2 ms and is followed by a more slowly developing, *delayed outward current* (upward deflection of the trace). The *early inward current* is especially interesting, since it represents an influx of positive charge into the cell and could therefore be related to the upstroke of the action potential believed to be due to an influx of Na⁺.

By replacing the Na⁺ in the extracellular medium with choline ion, an organic cation to which the membrane is essentially impermeable, Hodgkin and Huxley tested the hypothesis that sodium carries the early inward membrane current. The result, seen in Figure 5-25, was the disappearance of the early inward current, with a small early outward current appearing

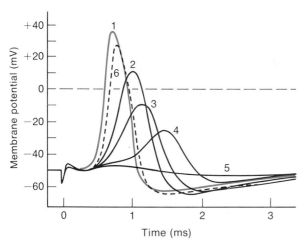

**5-23** Dependence on Na⁺ of the overshoot of the action potential in the squid giant axon. Trace I shows the control action potential in the normal solution (i.e., seawater). Traces 2 to 5 show the progressive change with time after the replacement with artificial seawater containing choline chloride in place of NaCl. Trace 6 was made after replacement of normal seawater at the end of the experiment. [Hodgkin and Katz, 1949.]

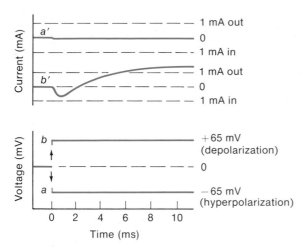

**5-24** Results of a voltage-clamp experiment. A hyperpolarization (trace *a*) results in a small, constant membrane current (*a'*). A depolarization produces an initial inward current (downward deflection in trace *b'*) followed by a slowly developing, delayed outward current (upward in trace *b'*). [Hodgkin, Huxley, and Katz, 1952.]

in its place. The delayed outward current was unaffected. Returning the axon to the normal sodium-containing bathing solution restored the early inward current. This reappearance indicated that the early inward current is produced by a transient influx of Na+ across the membrane into the cell. According to this interpretation, a sudden depolarization causes the sodium channels to open briefly, which allows Na+ to flow into the axon and produce a sodium current. Thus, according to Equation 5-3,

$$I_{Na} = g_{Na}(V_m - E_{Na}) \qquad (5\text{-}8)$$

In the normal extracellular environment, the electrochemical gradient ($V_m - E_{Na}$) drives Na+ into the cell when $g_{Na}$ rises. When extracellular Na+ is replaced by an impermeant ion such as choline, the electrochemical gradient of Na+ becomes reversed, and hence the direction of the sodium current is reversed.

Hodgkin and Huxley proceeded to separate the time course of the early inward current from the delayed outward current. After recording the current in normal saline, they lowered the external sodium concentration, so that Na+ would be in equilibrium during the stimulus

potential (Figure 5-25). Since Na+ was in equilibrium at the stimulus voltage (i.e., $V_m - E_{Na} = 0$), no sodium current flowed in response to the depolarizing potential step (Figure 5-25B), leaving the delayed outward current, which was later shown to be carried by K+. The delayed current was subtracted from the complex current sequence obtained with normal saline. The difference between these two currents (shaded area, Figure 5-25B) is taken as the time course of the current carried inward through the membrane by Na+, as plotted in Figure 5-25C.

What does the time course of sodium current tell us about the behavior of the membrane? It is neces-

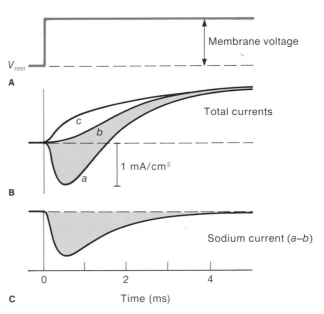

**5-25** Separation of the time course of the sodium current from that of the delayed outward current. (**A**) Voltage change, a 60 mV depolarization. (**B**) Trace *a*, current carried by both Na+ and K+ recorded in normal seawater; trace *b*, current carried by K+ alone recorded in low-sodium seawater, in which $V_m$ equaled the sodium equilibrium potential during the stimulus potential, preventing sodium current flow through the open sodium channels. (**C**) Trace *b* subtracted from trace *a* gives time course of the sodium current. Trace *c* was obtained after *all* the extracellular sodium was replaced with an impermeant cation, choline. In that case there was an outward current (upward bump) carried by Na+ leaving the cell during the opening of the sodium channels. [Hodgkin and Huxley, 1952a.]

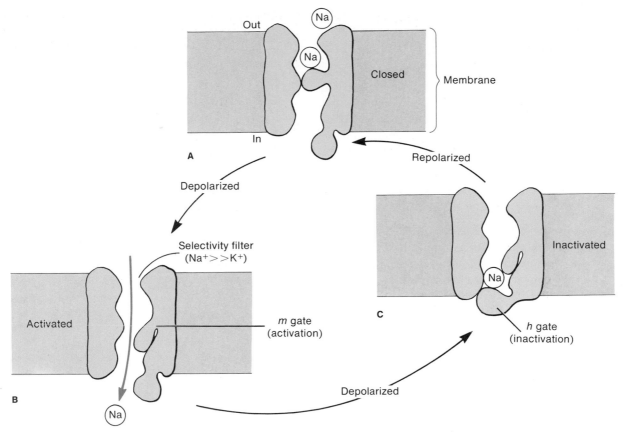

**5-26** Major states of the sodium channel. **(A)** Before depolarization, the channel is nonconducting because the *m* gate is closed. **(B)** Depolarization causes the sodium channel to become activated and conducting owing to the opening of the *m* gate. The conductance of the channel is now largely determined by its selectivity filter, which blocks all anions and has a much greater selectivity for Na+ than for K+ or Ca²+. **(C)** While still under depolarization, the *h* gate on the inner end of the channel subsequently closes, causing inactivation of the channel. Repolarization to the resting potential opens the *h* gate and closes the *m* gate. The channel is now again ready for activation by a new depolarization.

sary, first, to note in Equation 5-8 that, according to Ohm's Law, the sodium current, $I_{Na}$, is determined by two factors: (1) the conductance, $g_{Na}$, of the membrane to Na+ and (2) the electrochemical driving force, or emf, acting on Na+ (i.e., $V_m - E_{Na}$). This means that when the membrane is suddenly shifted to a steady depolarized level, the time course of the sodium current must be proportional to the time course of the underlying increase in the sodium conductance in response to the depolarization. The time course of

increased $I_{Na}$ in Figure 5-25C can therefore be equated with the time course of the increased sodium conductance. Note that although $V_m$ is held constant at the depolarized potential, the sodium conductance, after reaching a maximum within 1 ms, rapidly subsides to its low prestimulus level (Figure 5-25C). Thus, there are two separate processes. The increase in sodium conductance evoked by a depolarization is termed *activation,* and its subsequent time-dependent return to the base level is termed *inactivation.*

# The Sodium Channel

A number of observations indicate that the sodium current passes through a finite number of channels in the excited membrane (Figure 5-26). These channels, which become *activated* (i.e., opened up to carry a current) in response to a depolarization, exhibit high selectivity for $Li^+$ and $Na^+$ relative to other ions (Figure 5-27). This property resides in a portion of the channel that is called the *selectivity filter*. Since lithium is normally absent, sodium normally carries the current through this channel, while little or no $K^+$ passes through.

The way in which the sodium channel is gated (i.e., the process by which it is turned ON and OFF) by membrane depolarization is not fully understood. There appears to be a charge-bearing mechanical barrier to sodium ions in the channel at rest that shifts to a new position so as to unblock the channel when the membrane is depolarized (Figure 5-26B). The major evidence favoring such a physical reordering or conformational change is the detection of *gating currents* associated with the opening and closing of the sodium channels. These minute currents are related to the activation of the sodium channel and can be observed when the much larger ionic current through the sodium channel is pharmacologically blocked. The gating currents are believed to arise from movements of a charged group associated with the opening of the *activation gate* (*m* gate) during channel activation. Similar gating currents have been detected for the potassium and calcium channels in several tissues.

During maintained depolarization, the sodium channel becomes *inactivated*. Since this inactivation can be prevented by proteolytic enzymes applied to the cytoplasmic side of the membrane, it is widely believed to be due to a proteinaceous *inactivation gate* (*h* gate) located near the inner end of the sodium channel (Figure 5-26C). The closing of the *h* gate occurs within

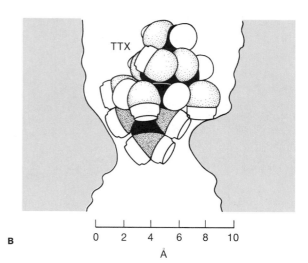

**A**   Tetrodotoxin

**B**

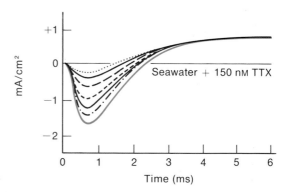

**C**

| Li | Na | K | Rb | Cs | Choline |
|----|----|----|----|----|---------|
| 1.1 | 1 | $\frac{1}{12}$ | $\frac{1}{40}$ | $\frac{1}{60}$ | $\frac{1}{73}$ |

**5-27** The relative selectivity of the sodium channel of squid axon is indicated by the ratios of permeability coefficients of ions that pass through the sodium channel. [Taylor, 1967.]

**5-28** The puffer-fish poison, tetrodotoxin. (A) Molecular structure of TTX. (B) Space-filling model of the toxin as it is believed to fit into the sodium channel. [Hille, 1975.] (C) Effect of TTX on the early inward current. Traces are recorded at 15 s intervals after the application of $1.5 \times 10^{-7}$ M TTX to squid axon. Note reduction in size of early current, whereas late current remains unchanged. [Moore and Narahashi, 1967.]

several milliseconds after opening of the *m* gate and appears to be related to and dependent on the *m* gate being in the open state.

*Tetrodotoxin* (TTX), a compound extracted from the viscera of the Japanese puffer fish, blocks the sodium channel by inserting itself into the channel (Figure 5-28). Experiments on several types of nerves have shown that less than 100 molecules of TTX bind to the sodium-selective sites in 1 $\mu m^2$ of axon membrane, completely blocking the increase in sodium conductance that normally accompanies depolarization (Table 5-1). The kinetics of the blocking indicate that each TTX molecule blocks a single sodium channel. Thus, the number of channels per square micrometer must be less than 100. If all these channels are simul-

taneously open, the total cross-sectional area occupied by all of them (assuming a channel diameter of 5 Å, Figure 5-28B) will be less than 1/50,000th of the membrane surface area. That this minute fraction of the surface is occupied by channels is consistent with the Danielli and Singer membrane models (pp. 94, 95), in which large areas of uninterrupted lipid bilayer account for the low permeability of the membrane to polar molecules. The invariant capacitance of the membrane during the large conductance changes characteristic of membrane excitation is also consistent with the infinitesimal fraction of membrane area occupied by ion-conducting channels.

F. Sigworth and E. Neher (1980) proved it possible to record the current carried through single sodium

**TABLE 5-1** Major nonsynaptic ion channels of excitable membranes.

| Channel | Current | Characteristics | Blocked by | Function |
|---|---|---|---|---|
| Sodium channel | $I_{Na}$ | Rapid activation by depolarization, followed by voltage-dependent inactivation | Tetrodotoxin (TTX) | Current for upstroke of action potential; impulse conduction |
| Calcium channel | $I_{Ca}$ | Slower activation by depolarization and lower single-channel conductance than sodium channel; inactivates as function of $[Ca]_i$ and/or membrane potential | Verapamil, D600, $Co^{2+}$, $Cd^{2+}$, $Mn^{2+}$, $Ni^{2+}$, $La^{3+}$ | Slow depolarizations; calcium acts as "messenger molecule" to cell interior |
| Potassium channel, delayed rectifying | $I_{K(V)}$ | Delayed activation by depolarization; inactivates slowly and incompletely under steady depolarization | Intra- and extra-cellular tetraethyl-ammonium (TEA); amino pyridine | Rapid repolarization to terminate action potential |
| Potassium channel, calcium-activated | $I_{K(Ca)}$ | Activation by $[Ca^{2+}]_i$. Remains activated until $[Ca]_i$ is lowered. Activation by calcium is enhanced by depolarization | Extracellular TEA | Repolarization of calcium action potential; production of outward current to balance inward calcium current, thereby limiting depolarization due to $I_{Ca}$ |
| Potassium channel, early transient | $I_{K(A)}$ | Rapid activation followed by inactivation during depolarization; inactivated at low resting potentials | Amino pyridine | Reduces rebound excitation at end of hyperpolarizing inhibitory potentials |
| Potassium channel, resting | $I_{K(leak)}$ | Responsible for potassium leakiness of resting cell | Partially by TEA | Largely responsible for resting potential |
| Inward rectifying | | Carries potassium current during hyperpolarization; carries little current during depolarization; may be identical to resting potassium channel in some tissues | Extracellular $Cs^+$ | |

channels that are activated by depolarization. The currents, which are recorded with $\frac{1}{2}$ μm suction pipettes, occur as all-or-none pulses having a square shape—indicating abrupt opening and closing of the channel. They are of similar amplitude for different individual channels and have durations that vary widely and at random. The mean channel *open time,* which is less than 1 ms, depends on membrane voltage, becoming shorter as the stimulus voltage is made more positive. The conductance of a single sodium channel is about 10 pS (i.e., $10 \times 10^{-12}$ S, or $10^{11}$ Ω resistance). By using Ohm's Law, Faraday's constant, and Avogadro's number, it can be calculated that the activated sodium channel carries $Na^+$ at a rate of about six ions per microsecond if the driving force ($V_m - E_{Na}$) is $-100$ mV (approximately what it is as the action potential gets under way).

### THE HODGKIN CYCLE

By subjecting the membrane to sudden changes in potential of different values, Hodgkin and Huxley demonstrated that the peak intensity of the early in-

---

### Box 5-5   *Current–Voltage Relations*

When the peak intensity of the early sodium current (figure, part A) is plotted against stimulus voltage, a current–voltage plot results (figure, part B) that is characteristic of the regenerative current of electrically excitable membranes. To generate the current–voltage plot, a series of voltage steps are applied, each beginning at the holding potential, $V_{rest}$, and stepping abruptly to a selected potential along the voltage axis. The inward current (figure, part A) recorded in response to each potential step is then plotted as a function of the test voltage, $V_m$.

At small voltage steps, the current–voltage plot (figure, part B) shows a region of positive slope in which a positive potential displacement is accompanied by an increase in outward current, as predicted for a resistor behaving according to Ohm's Law. At potentials somewhat more positive, there begins a region of negative (descending) slope, along which each increment in positive potential produces an increment in inward current.

The region of negative slope in the sodium current curve (figure, part B) results from the progressive increase in $g_{Na}$, which occurs with increased depolarization along that segment of the voltage axis (figure, part C). At the potential at which $g_{Na}$ reaches a maximum (figure, part C), the transient inward sodium current is at a maximum (bottom of curve in figure, part B) and begins to decrease at more positive potentials. This drop occurs because the emf, acting on $Na^+$ (i.e., $V_m - E_{Na}$), drops linearly with further increase in positivity. Thus, the current–voltage plot crosses the voltage axis (i.e., zero current) at $V_m = E_{Na}$.

The inward sodium current can be blocked with application of tetrodotoxin (broken line).

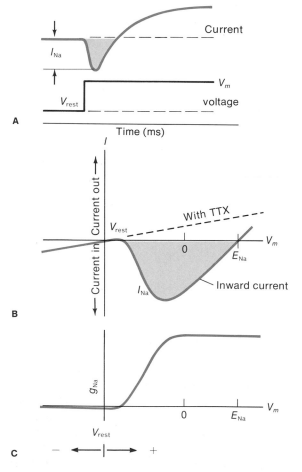

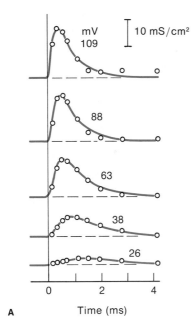

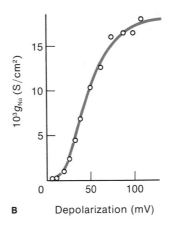

**5-29** (A) Plots of $g_{Na}$ against time at different amplitudes of depolarization given in millivolts. [Hodgkin, 1958.] (B) Maximum values of $g_{Na}$ at each potential plotted against amplitude of depolarization. This curve is similar to the idealized curve in part C of the figure in Box 5-5, but it is based on actual measurements. [Hodgkin and Huxley, 1952b.]

ward current depends on the value of the stimulating potential, $V_m$. As the figure in Box 5-5 shows, the peak current increases with moderate positive-going steps but then decreases with greater positive potential. To understand the decrease, it is necessary to recall that $I_{Na} = g_{Na} (V_m - E_{Na})$. Thus, as $V_m$ approaches the potential, $E_{Na}$, at which sodium is in electrochemical equilibrium (about +55 mV), $I_{Na}$ must approach zero even though $g_{Na}$ is very high. Figure 5-29A shows plots of sodium conductance against time at various stimulus potentials. The plots were calculated with Equation 5-8 from data similar to those in Box 5-5. In Figure 5-29B it can be seen that $g_{Na}$ increases sigmoidally with depolarization, reaching a plateau when the membrane is depolarized approximately 100 mV.

The increase in sodium conductance produced by depolarization is the basis of the regenerative behavior necessary for the production of an all-or-none action potential. When the membrane is allowed to respond freely to a stimulus without voltage-clamp control, as in Figure 5-16, an increase in $g_{Na}$ produced by a depolarizing stimulus allows $Na^+$ to flow inward across the membrane, carrying its positive charge into the cell

and thus producing further depolarization. The additional depolarization due to $Na^+$ influx leads to a further increase in the sodium conductance (Figure 5-30), and this permits a further increase in the rate of $Na^+$ influx. The circular relation between membrane potential and sodium conductance is known as the *Hodgkin cycle*. This is a rare example, in a biological system, of positive feedback, which is inherently unstable and explosive, so to speak. The rapidity of the upstroke of the action potential is due in part to the regenerative, explosive nature of the Hodgkin cycle. The extent of the upstroke is limited, however, by its dependence on the driving force, $V_m - E_{Na}$. There-

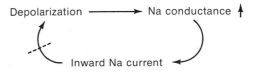

**5-30** The Hodgkin cycle. The broken line indicates where the voltage clamp interrupts the positive-feedback loop. With the loop closed, the system exhibits positive feedback and accounts for the upstroke of the action potential.

fore, even though there is strong positive feedback (Box 1-1) between depolarization and increased sodium conductance, the upstroke of the action potential cannot exceed the sodium equilibrium potential. For a further description of the relations between membrane voltage and membrane current, see Box 5-5.

It may seem confusing that the current that flows *into* the cell through open sodium channels depolarizes the membrane, whereas depolarization also occurs when an electrotonic current flows *out* of the cell (Figure 5-5). It may be helpful to consider that, in both instances, the depolarization results from increased positive charge added to the cell interior. In the first case, it arrives as $Na^+$ through the Na channels, and in the second case, it is delivered by positive current from an intracellular electrode or from an excited region of membrane nearby.

## The Potassium Current

The regenerative increase in sodium conductance is responsible, as we have seen, for the upstroke of the action potential. How, then, does the membrane potential subsequently return to the resting level? Hodgkin and Huxley suspected that the delayed outward current in Figure 5-25 corresponds to the efflux of positive charge that helps shift the membrane potential from the peak of the overshoot to the resting potential. The intensity of the delayed current during maintained potential steps is plotted in Figure 5-31 against membrane potential. The slope of the curve (conductance) increases as the membrane potential is made more positive. Another way of stating this is that in the steady state, several milliseconds after a potential change, the membrane passes current more readily (i.e., rectifies) in the outward direction than in the inward direction. This behavior is known as *delayed rectification*.

It was logical to suppose that the delayed outward current evoked by a depolarization is carried by potassium ions, because the emf acting on $K^+$ (Equation 5-4) increases as the interior is made progressively more positive than $E_K$. Thus, the tendency for $K^+$ to leak out and carry positive charge out of the cell

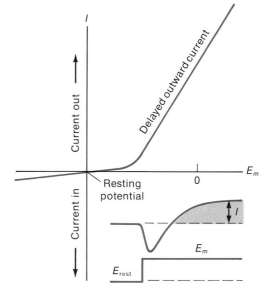

**5-31**  Current–voltage relations of the delayed outward current. The method used to obtain this plot is the same as that for part B of the figure in Box 5-5, except that the current was measured about 5 ms after $E_m$ was applied, as shown in the inset. Note that, after this delay, the membrane passes current more easily in the outward than in the inward direction.

increases as the membrane potential becomes more positive. At the peak of the action potential, the emf tending to drive $K^+$ out of the cell and restore the resting potential is at a maximum. In addition, it can be reasoned that $K^+$ efflux is enhanced by the delayed activation of the potassium conductance seen in the delayed development of the outward current (Figure 5-25).

To test this idea, Hodgkin and Huxley (1953) used radioactive potassium to measure the movement of $K^+$ across the membrane during the passage of steadily applied current. The results showed that potassium efflux was small during inward current and large during outward current. Moreover, the efflux agreed quantitatively with the amount of charge carried by the outward current (Figure 5-32). This finding provided evidence that the delayed outward current is carried by $K^+$. The increase in membrane conductance to $K^+$ on depolarization is termed *potassium activation*.

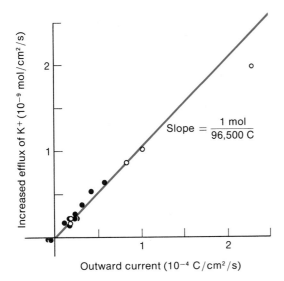

**5-32** Efflux of radioactively labeled K+ plotted against membrane current during steady electrical depolarization of squid axon. The proportionality provides strong evidence that K+ carries the late outward current. [Hodgkin, 1958.]

*1.* Stimulating current of threshold intensity is passed from an electrode inside the cell across the membrane to the exterior. This current applies positive charge to the intracellular side of the membrane, partially reducing the resting potential and causing the depolarization from point a to b in Figure 5-33.

*2.* As the membrane potential approaches the threshold voltage, the sodium channels of the membrane begin to open, allowing Na+ current to flow into the cell. Below threshold, K+ efflux is sufficient to cancel the charge carried by Na+ influx. The potential at which Na+ influx begins to exceed K+ efflux is the threshold (point b).

*3.* The net inward current causes the membrane to depolarize further. The depolarization becomes regenerative, since each increment of positive potential causes new sodium channels to open, and this further hastens the inflow of Na+, producing another increment in

The delayed increase in potassium permeability can be suppressed by some agents. Procaine and xylocaine, both local anesthetics, diminish potassium activation as well as sodium activation, and thereby block impulse conduction in nerves. Tetraethylammonium (TEA) ions, when injected into the squid axon, interfere only with potassium activation (Table 5-1). As a result, repolarization is delayed and the action potential is prolonged.

In addition to the delayed voltage-dependent potassium current, $I_{K(V)}$, described by Hodgkin and Huxley in the squid axon, other potassium currents have been discovered more recently. The most important of these are a calcium-dependent potassium current, $I_{K(Ca)}$, and a transient voltage-activated potassium current, $I_{K(A)}$ (Table 5-1 and p. 169).

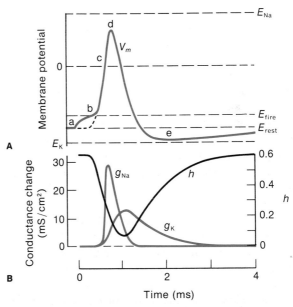

**5-33** An action potential (**A**), with underlying changes in Na+ and K+ conductances plotted (**B**). The value *h* (right-hand scale) represents the ability of the sodium channels to become activated (i.e., opened) by a depolarization. Note that the decrease in *h* far outlasts the increase in $g_{Na}$. [Hodgkin and Huxley, 1952c.]

## *SUMMARY OF IONIC EVENTS*
## *DURING THE ACTION POTENTIAL*

The sequence of events that produce the action potential can be outlined with reference to the graphs in Figure 5-33.

positive potential (Hodgkin cycle). This process produces the upstroke.

*4.* As the membrane potential approaches $E_{Na}$, the emf $(V_m - E_{Na})$ acting on $Na^+$ becomes progressively smaller. This decrease causes the rate of potential change to slow progressively from point c until the overshoot reaches a maximum somewhat short of $E_{Na}$ (point d). The peak of the overshoot is about $+120$ mV from the resting potential. Thus, the initial passive depolarization to threshold (ca. $+20$ mV) produced by the stimulating current is *amplified* five or six times by the regenerative depolarization of the membrane.

*5.* The open sodium channels inactivate—that is, close. This behavior in itself would cause the action potential to gradually subside as the resting potential reestablishes itself. As we see in the next step, the process of repolarization is speeded up by the opening of potassium channels.

*6.* Potassium channels begin to open, allowing a small amount of $K^+$ to diffuse rapidly outward in accord with the large emf $(V_m - E_K)$ acting on it. This efflux of $K^+$ removes positive charge from the cell and repolarizes the membrane rapidly to the resting level (points d to e).

On completion of the repolarization, the number of potassium ions having left the cell equals the number of sodium ions that initially entered the axon, restoring the membrane to its resting potential. $Na^+$ inactivation and high potassium conductance persist for a number of milliseconds after the action potential, producing the depressed excitability characteristic of the absolute and relative refractory periods. In the absolute refractory period, sufficient numbers of sodium channels cannot be activated to produce enough inward current to keep ahead of $K^+$ efflux. During the relative refractory period, strong depolarization will activate sufficient numbers of sodium channels to produce an action potential in the face of the elevated potassium conductance. Because of the higher $g_K$, the action potential peak will not get as close to $E_{Na}$ as normal, and so the overshoot will be smaller.

Sodium inactivation and potassium activation also contribute to the increase in threshold that accompanies a subthreshold depolarization (accommodation, Figure 5-18). The different degrees of accommodation observed in different membranes result from quantitative differences in time- and voltage-dependencies of sodium inactivation and potassium activation.

## CHANGES IN ION CONCENTRATION DURING EXCITATION

It is important to note that the ionic movements responsible for the potential changes of a single action potential are extremely small and *do not appreciably change the intracellular ionic concentrations* in any but the smallest cells or axons.

It can be calculated from the principles discussed in Box 5-2 that about $10^{-12}$ mol of $Na^+$ crossing 1 cm² of membrane is sufficient to produce an action potential of 100 mV amplitude. That is, only 160 sodium ions per square micrometer are needed. But because of the partial canceling effect of some simultaneous $K^+$ outflow, the actual number is closer to 500 sodium ions per square micrometer per impulse. Extending the calculation, it becomes evident that a single action potential changes the internal $Na^+$ concentration of a squid axon 1 mm in diameter by only 1 part in more than 100,000. For this reason, a squid axon can generate thousands of impulses with the sodium pump incapacitated by a metabolic poison. Eventually, of course, the concentrations, and hence the equilibrium potentials, of $Na^+$ and $K^+$ in a poisoned cell will show significant shifts.

In small axons, the increased surface/volume ratio that attends decreasing diameter of a cylinder leads to more significant change in axoplasmic concentration, even as the result of a single action potential. For example, in mammalian "C" fibers (axons), which have diameters of only $10^{-3}$ mm, the flux from a single impulse changes the internal $Na^+$ and $K^+$ concentrations by about 1%. The result is a drop in resting potential of about 0.3 mV and, for 10 action potentials in close succession, a depolarization of 2 mV. Thus, in axons of small diameter, it is important that the intracellular resting concentrations of $Na^+$ and $K^+$ be restored rapidly by active transport before cumulative ion fluxes produce significant changes in ion gradients.

Note that *metabolic pumping of ions across the cell mem-*

*brane does not enter directly into the production or recovery of an action potential,* but serves to maintain the ionic concentration gradients required for the production of membrane currents.

## Calcium and Membrane Excitability

### THE CALCIUM CHANNEL

Since the work of Hodgkin and Huxley that gave rise to the ionic hypothesis, it has become apparent that calcium-selective channels activated by membrane depolarization are equally as ubiquitous and important as sodium channels. Calcium ions carry all or part of the regenerative depolarizing current through such channels in crustacean muscle fibers, in smooth muscle cells, in the cell bodies and the dendrites and terminals of many nerve cells, in embryonic cells, and in ciliates like *Paramecium.* In some of these membranes, $Ca^{2+}$ carries the inward current together with $Na^+$, while in others it carries all the current by itself. The calcium current is generally weaker than the sodium current, which results in part from the lower conductance of calcium channels. For this reason the calcium current usually is not strong enough to produce an all-or-none action potential without help from the sodium current or without agents that block the potassium current. The calcium current therefore generally produces a graded response, the size of which depends on the stimulus depolarization. In most membranes exhibiting a calcium current, the upstroke of an all-or-none action potential is generated largely by a strong sodium current that is primarily responsible for rapidly depolarizing the membrane. The calcium channels are opened by the depolarization, and $Ca^{2+}$ enters the cell, often to act as a messenger, as in activating the release of a transmitter substance from the presynaptic terminals of a neuron (p. 121). Calcium current is notably absent from most axons, where the inward current is carried solely by the more rapidly acting sodium channels, and purely for the purpose of rapid conduction of impulses along the axon. During embryonic development, calcium channels typically appear first, and sodium channels become functional at later stages. This behavior, together with the greater prevalence of calcium channels in lower organisms such as the proto-

zoa, suggests that sodium channels may be a more recent evolutionary specialization for impulse conduction, whereas calcium channels, controlling the entry of "messenger" $Ca^{2+}$ into a great number of cell types, may have had a more primitive origin.

Calcium channels are rendered defective or are eliminated by certain mutations in *Paramecium,* as can be seen in Figure 5-34. Failure of the channels to open in response to a depolarizing stimulus renders the membrane inexcitable, and the response is the simple exponential change in voltage similar to that produced when a pulse of current is passed across a capacitance in parallel with a simple resistance.

The calcium channels of metazoan excitable cells are blocked to greater or lesser extents by certain divalent and trivalent cations, most notably $Co^{2+}$, $Cd^{2+}$, $Mn^{2+}$, $Ni^{2+}$, and $La^{3+}$ (Table 5-1). These ions compete with $Ca^{2+}$ at anionic sites associated with the channel but do not pass through the channel as readily as $Ca^{2+}$. The ions $Sr^{2+}$ and $Ba^{2+}$, on the other hand, compete with $Ca^{2+}$ but are equally as capable, or more so, of carrying current through the calcium channels to produce action potentials. This capability is illustrated by the large all-or-none action potentials produced by *Paramecium* in a barium-containing solution compared to the small graded responses produced by $Ca^{2+}$ entry alone (Figure 5-34B, left).

Some calcium channels differ markedly from the sodium channel in failing to fully inactivate (i.e., close down) even under maintained depolarization. Instead, these channels become blocked by elevation of intracellular free $Ca^{2+}$. They inactivate during maintained depolarization in response to an action of intracellular calcium on the channel as intracellular $Ca^{2+}$ increases during the flow of calcium current. Since the calcium that enters is "buffered" by the cytoplasm (p. 471), any buildup of $Ca^{2+}$ near the membrane is limited. As $Ca^{2+}$ entry leads to some inactivation of the calcium current, the rate of entry drops to a level at which the $Ca^{2+}$ entry, the elevation of $[Ca^{2+}]_i$, and the calcium permeability are in a state of balance. For these reasons, $Ca^{2+}$ entry can persist at a low steady level under prolonged depolarization (Figure 5-35), whereas the sodium current exhibits rapid and complete inactivation during depolarization (Figure 5-25).

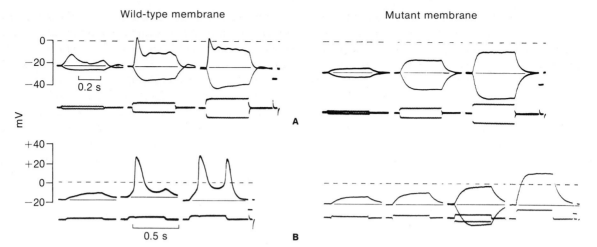

Wild-type membrane          Mutant membrane

**5-34** Electrical excitability of a wild-type *Paramecium* (left) compared with the inexcitability of the membrane mutant "pawn," which has defective calcium channels (right). **(A)** Bath solution contained 1 mM $CaCl_2$. The wild type shows graded calcium responses to depolarizing current, while the "pawn" mutant shows pure electrotonic responses. **(B)** Addition of 4 mM $BaCl_2$ to the solution. The wild type exhibits strong all-or-none barium action potentials, produced by $Ba^{2+}$ carrying current through calcium channels, and the "pawn" again shows only electrotonic responses. The lower set of traces in each record indicates the stimulus current. [Kung and Eckert, 1972.]

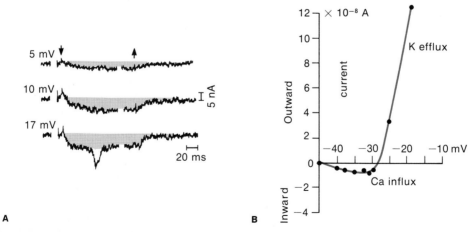

**5-35** Persistent calcium current in a snail neuron. **(A)** Inward current (downward deflection of trace) during prolonged ($\frac{1}{2}$ s) depolarizations of three amplitudes (shown at left). The depolarization began and ended as shown by arrows. The recording was interrupted at the gap for 250 ms. **(B)** Plot of the late current recorded 200 ms after onset of the voltage pulse against membrane potential during the pulse. Compare with part B of the figure in Box 5-5. Note the small inward current carried by $Ca^{2+}$ between −45 and −28 mV. This small inward current is obscured at more positive potentials by the strong outward potassium current. [Eckert and Lux, 1976.]

### THE CALCIUM-DEPENDENT POTASSIUM CHANNEL
Calcium ions are also known to affect the permeability of other channels, such as the sodium- and calcium-carrying channels that become activated in some invertebrate visual receptor cells by the action of visible light (p. 270). These channels are blocked by elevation of intracellular $Ca^{2+}$. More common, however, are a class of potassium channels that are found in many different tissues and are activated by an elevation of intracellular $Ca^{2+}$. Thus, $Ca^{2+}$ flowing into a cell through calcium channels and accumulating near the inner surface of the membrane causes an activation of the calcium-dependent potassium channels, as well as a partial inactivation of some calcium channels (Figure 5-36). The accumulation of $Ca^{2+}$ within the cell therefore fosters repolarization owing to both enhanced outward potassium current and diminished inward calcium current. The hyperpolarizing effect of the calcium-dependent potassium current can be seen in Figure 5-37. After the calcium current is eliminated

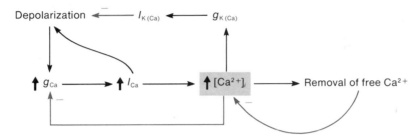

**5-36** Interrelations of calcium current and calcium-dependent potassium current common to many excitable tissues. The colored arrows indicate negative actions. The intracellular accumulation of free $Ca^{2+}$ during depolarization activates some of the calcium-dependent potassium channels and inactivates some of the calcium channels. Note that both these pathways beginning with elevated calcium and ending with depolarization include a colored arrow and therefore limit depolarization by negative feedback. This negative feedback counteracts the positive feedback that is associated with depolarization, elevated $g_{Ca}$ and $I_{Ca}$. The continual action of $Ca^{2+}$ removal mechanisms keep $[Ca]_i$ from building up over time in proportion to calcium entry.

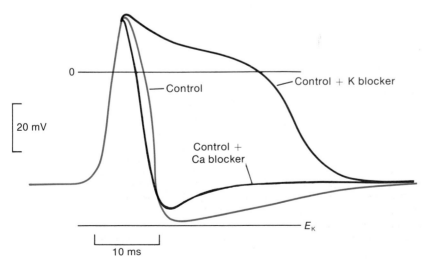

**5-37** Role of calcium currents and calcium-dependent potassium currents in excitable tissues. Normal action potential in molluscan neuron recorded in calcium-containing physiological saline is shown in color. Addition of tetraethylammonium (TEA), which blocks much of the repolarizing potassium current, reveals a prolonged plateau depolarization due to maintained calcium current. In control saline containing a calcium-blocking agent such as $Co^{2+}$, $Mn^{2+}$, $Ni^{2+}$, or $Cd^{2+}$, the prolonged undershoot caused by the calcium-dependent potassium current is eliminated, leaving the more transient undershoot due to voltage-dependent potassium current.

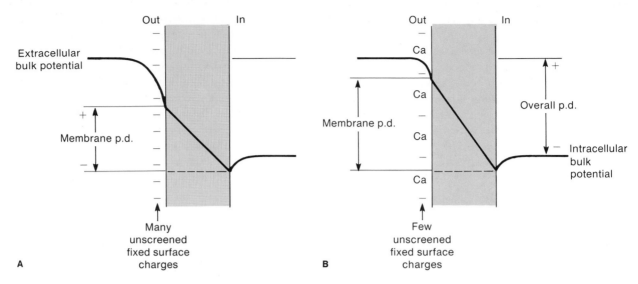

**5-38**    Surface charge screening. (**A**) Fixed negative charges distributed over the extracellular surface of the membrane produce a steep profile of negative surface potential. (**B**) Addition of divalent cations to the bath neutralizes some of the negative surface charge, and the membrane then experiences a steeper potential gradient even though the overall bulk potential difference (p.d.) remains unchanged. Membrane channels and gating mechanisms sense only the voltage gradient existing across the membrane and thus will "feel" a steeper intramembrane voltage gradient. The steeper this gradient, the more depolarization is required to activate electrically sensitive channels.

with a blocking agent, the strong persisting hyperpolarization toward $E_K$ is diminished, and the repolarization of the action potential can be ascribed purely to the voltage-activated population of potassium channels (Table 5-1).

It should be noted that axons, including the squid giant axon, are largely devoid of calcium channels and calcium-dependent potassium channels. In nerve cells, these two kinds of channels are commonly found in the dendrites, cell body, and axon terminals.

### SURFACE-CHARGE SCREENING BY Ca²⁺

Calcium plays several important roles in modifying the electrical properties of membranes. It had been known for some time that the concentration of extracellular $Ca^{2+}$ influences the threshold of firing in nerve and muscle membranes. B. Frankenhaeuser and Huxley (1957) had shown that the current–voltage curve of the early sodium current (Box 5-5) is displaced along the voltage axis toward more positive potentials in high $[Ca^{2+}]_o$ and toward more negative potentials in low $[Ca^{2+}]_o$—in either case, about 12 mV per e-fold change

in $[Ca^{2+}]_o$. This displacement is ascribed to a nonspecific *screening effect* of the divalent cation—that is, the neutralization of surface charge (Figure 5-38). The charges associated with fixed negative groups residing on the cell membrane cause a negative *surface potential* that extends from the membrane into the bulk phase of fluid decaying over a short distance in a steep profile. Figure 5-38A shows that because of this, the membrane itself experiences only a portion of the total potential drop between the extra- and intracellular bulk phases from which the electrodes record voltage. Increasing the cation concentration *screens* some of those surface charges; and divalent cations, especially $Ca^{2+}$, do so most effectively by associating with the negative surface sites. The result of such *surface-charge screening* is that the potential drop across the membrane itself is increased, while the overall potential difference usually remains unchanged (Figure 5-38B). The molecules comprising ionic channels within the membrane now "feel" a higher potential gradient, and they require a larger overall depolarization to feel the same stimulus voltage and respond. This situation accounts for the increase in threshold associated with increased extracellular $Ca^{2+}$ and, conversely, the increased excitability seen with a decreased extracellular $Ca^{2+}$ concentration.

## PACEMAKER POTENTIALS

The interactions between individual ionic mechanisms can lead to oscillations in membrane potential. If these oscillations are regular, are slow, and give rise to action potentials, they are termed *pacemaker potentials*. Although no precise single sequence of events explains all pacemaker activity, slowly inactivating inward current carried by calcium and/or sodium ions has been shown, along with calcium-activated potassium currents, to participate in pacemaker oscillations. The two most extensively studied pacemaker potentials are found in the vertebrate heart and in certain molluscan neurons. In both tissues, the pacemaker activity has been shown to be truly spontaneous and independent of any extrinsic signals or interactions.

The "bursting pacemaker" neurons of snails and other gastropod mollusks generate repeated trains (or "bursts") of impulses every several seconds. Each train arises from a spontaneous pacemaker wave (Figure 5-39). The ionic mechanisms generating the pacemaker waves are indicated in Figure 5-39 and appear to be as follows. Calcium accumulates during the depolarization owing to $Ca^{2+}$ entry through activated calcium channels. As the $Ca^{2+}$ accumulates, it activates many of the calcium-activated potassium channels and inactivates many calcium channels. Both effects contribute to carrying positive charge out of the cell and repolarizing the membrane toward $E_K$. The repolarization turns off the calcium channels, $Ca^{2+}$ entry ceases, and the accumulated free $Ca^{2+}$ is gradually taken out of circulation in the cytosol by calcium-buffering mechanisms. As the accumulated free $Ca^{2+}$ is removed, the calcium-dependent potassium conductance drops, and the membrane potential drifts away from $E_K$, slowly depolarizing and gradually activating the voltage-sensitive sodium and calcium channels.

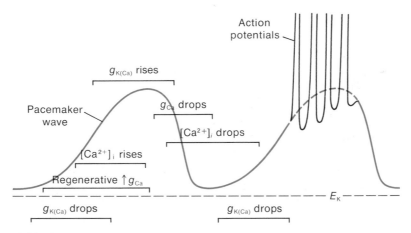

**5-39**   Proposed events responsible for pacemaker waves in spontaneously "bursting" neurons. During the slow depolarization, the calcium conductance, $g_{Ca}$, is turned on, allowing an influx of $Ca^{2+}$. As the $[Ca^{2+}]_i$ rises, it gradually activates the calcium-dependent potassium conductance, $g_{K(Ca)}$. Thereupon the membrane repolarizes toward $E_K$, causing the calcium conductance to turn off. As a result of the calcium-buffering activity of cytoplasm, $[Ca^{2+}]_i$ drops, $g_{K(Ca)}$ drops, and the membrane potential slowly shifts away from $E_K$ (i.e., depolarizes), turning on $g_{Ca}$ and initiating a new cycle. The slow depolarizing pacemaker waves give rise to trains of action potentials as the wave reaches and exceeds the firing level. For the sake of simplicity, the wave at the left is shown without action potentials.

Entry of $Ca^{2+}$ feeds the depolarization of the membrane until a full-blown pacemaker wave is regenerated before the accumulated $Ca^{2+}$ again turns on the calcium-dependent potassium conductance and the cycle repeats itself. The time lags inherent in the accumulation and removal of free $Ca^{2+}$ from the inner surface of the cell membrane provide the time-dependent alternations essential for an oscillatory process.

In the experiment illustrated in Figure 5-40, the intracellular ionized calcium concentration was measured with a calcium-sensitive dye, arsenazo III, injected into a pacemaker neuron. It can be seen with photometric measurements that the ionized calcium level reaches a maximum prior to the repolarization ascribed to activation of calcium-dependent potassium channels and inactivation of calcium channels. The calcium signal then slowly subsides as free $Ca^{2+}$ is taken up by calcium-buffering mechanisms of the cell. Reduction in calcium-activated potassium conductance during removal of free $Ca^{2+}$ causes $V_m$ to shift away from $E_K$ and thus produces the slow pacemaker depolarization that leads into the next wave.

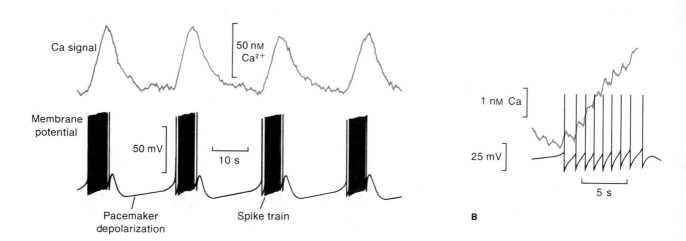

**5-40**  Experimental measurement of free $Ca^{2+}$ fluctuations in cytoplasm during spontaneous pacemaker activity described in Figure 5-39. Colored traces are absorbance signals from the calcium-sensitive metallochromic dye arsenazo III at 650 nM wavelength simultaneous with the recording of membrane potential below. Upward deflection of the colored trace denotes increased cytoplasmic free $Ca^{2+}$ activity. (**A**) Five pacemaker cycles showing buildup of $[Ca]_i$ during the pacemaker depolarization and its decay coincident with the slow depolarization. (**B**) High-sensitivity recording of the calcium signal showing $Ca^{2+}$ increments with successive action potentials. [Gorman and Thomas, 1978.]

## Summary

The electrical properties of the cell membrane can be related to the molecular makeup of the membrane. The lipid bilayer has the property of capacitance; that is, it does not readily allow charge carriers (i.e., ions) to pass through, but because it is very thin (ca. 50 Å) it can store electric charge by means of electrostatic interaction between cations and anions on opposite sides of the membrane. Channels that are composed of protein molecules imbedded in the lipid bilayer and that penetrate the lipid bilayer have the property of conductance, since they permit physical passage of ions across the membrane. These two properties, capacitance and conductance (reciprocal of resistance), determine the time course of voltage changes produced across electrically active cell membranes.

An asymmetrical distribution of ions in solution on the two sides of a membrane can produce an electric potential across the membrane, depending on the relative permeabilities of the membrane to the ions present. If it is permeable to one ion species only, the membrane will develop a potential difference proportional to the log of the ratio of concentrations of the ion on opposite sides of the membrane, the difference in potential resulting from the electrostatic charge carried across the membrane by the diffusible ion. The diffusional and electrostatic forces acting on that ion species will be equal and opposite at the so-called equilibrium potential, determined by the ratio of concentrations of that ion across the membrane. If the membrane shows significant permeability to more than one ion species, as is common in biological systems, the membrane potential reflects the contributed diffusion potentials of those ions. Since resting cell membranes are most permeable to $K^+$ and $Cl^-$, the resting potential is close to the equilibrium potentials of these two ions, which usually have similar but always oppositely directed concentration gradients. Since the free $K^+$ concentration of the cell interior is usually 10 to 60 times that of the cell exterior, the resting potential (inside voltage with respect to outside voltage) is $-60$ to $-100$ mV (i.e., $-0.6$ to $-1 \times 10^{-1}$ V). Active transport of $Na^+$ and $Ca^{2+}$ causes these ions to be more concentrated outside the cell than within the cytoplasm; they have a strong tendency to leak into the

cell and must be continually pumped out. Stimuli that increase the normally low permeability to either $Na^+$ or $Ca^{2+}$ lead to a strong influx of one or the other, which tends to decrease the negative potential of (i.e., depolarize) the cell interior. A transient opening of sodium channels, in fact, is responsible for the upstroke (depolarization) of the nerve impulse. Since this increase in sodium permeability is evoked by membrane depolarization, the upstroke of the nerve impulse is regenerative and causes the membrane potential to approach briefly the sodium equilibrium potential of $+50$ to $+60$ mV at the peak of the nerve impulse. A delayed rise in potassium permeability brought on by the voltage change across the membrane, together with rapid inactivation of the sodium permeability, brings the membrane back to the resting potential.

In some excitable membranes, calcium ions carry all or part of the inward depolarizing current. In some of those membranes, the $Ca^{2+}$ that enters turns on potassium channels, which hasten the repolarization process. The subsequent removal of intracellular $Ca^{2+}$, restoring $[Ca^{2+}]_i$ to the original low level, restores the resting potassium permeability. A rise in $[Ca^{2+}]_i$ also causes the closure of calcium channels that opened in response to membrane depolarization.

Spontaneous pacemaker depolarizations result from interdependent activation of two or more classes of membrane channels that produce opposing shifts in membrane potential. Thus, $Ca^{2+}$ entry causes depolarization but leads slowly to subsequent activation of potassium channels that carry $K^+$ out of the cell and repolarize it. Because of time delays between these opposing events, the membrane potential in pacemaker cells oscillates spontaneously.

## Exercises

*1.* Galvani and Volta each considered the other's view of "animal electricity" (i.e., electrophysiology) incorrect. Explain why they were, in fact, both correct in certain respects.

*2.* The cell membrane separates charge and therefore has a potential difference across it. Does this violate the rule of electroneutrality? Explain.

*3.* What is the structural basis for membrane capacitance? For conductance?

*4.* How is the time course of potential changes across the cell membrane related to the resistance and capacitance of the membrane?

*5.* A given intensity of injected current will produce a larger potential change in a small cell than in a large cell having similar membrane properties. Why?

*6.* Explain why the side of the membrane having the higher concentration of the permeable cation is electrically negative relative to the other side.

*7.* In an artificial system in which the membrane is totally impermeable to all ion species but one, why would you expect a permanently unequal distribution of the permeant ion across the membrane?

*8.* In a living cell that is permeable to ions other than $K^+$, what prevents the $K^+$ from gradually being displaced from the cell interior by other cations?

*9.* For a cell that is 100 times more permeable to $K^+$ than to any other ion, use the Nernst equation to determine the potential change produced by a doubling of the extracellular $K^+$ concentration.

*10.* What are the equilibrium potentials for each of the following ions of the given concentrations?

   (a) $[K^+]_o = 3$ mM, $[K^+]_i = 150$ mM
   (b) $[Na^+]_o = 100$ mM, $[Na^+]_i = 10$ mM
   (c) $[Ca^{2+}]_o = 10$ mM, $[Ca^{2+}]_i = 10^{-3}$ mM

*11.* In 1939, Cole and Curtis reported that membrane conductance increases but capacitance remains essentially unchanged during an action potential. Relate these findings to membrane structure and to changes in structure thought to occur during excitation.

*12.* Give two lines of evidence that $Na^+$ carries the inward current responsible for the upstroke of the action potential.

*13.* How did Hodgkin and Huxley show that the repolarization (return to resting potential) of the action potential is due in part to an increase in potassium permeability?

*14.* Does the sodium pump play a *direct* role in any part of the action potential?

*15.* How is the sodium pump *indirectly* important in the production of an action potential?

*16.* Explain the following classic phenomena in terms of the modern ionic hypothesis:

   (a) threshold potential
   (b) all-or-none overshoot
   (c) refractoriness
   (d) accommodation

*17.* Calculate the approximate number of sodium ions entering through each square centimeter of axon surface during an action potential having an amplitude of 100 mV. (Recall that 96,500 C is equivalent to 1 mol-equiv of charge; that membranes have a typical capacitance of $10^{-6}$ fd/cm$^2$; and that Avogadro's number is $6.022 \times 10^{23}$ atoms/mol.)

*18.* Why can it be said that an axon of large diameter undergoes essentially no change in ionic concentration during several action potentials, whereas the very thinnest axons undergo significant changes in concentration during several impulses?

*19.* The Hodgkin cycle is an example of positive feedback in a biological system. Since positive feedback is inherently unstable, how does one account for the limited amplitude of the upstroke?

## Suggested Reading

Aidley, D. J. 1978. *The Physiology of Excitable Cells.* 2nd ed. New York: Cambridge University Press.

Bezanilla, F. 1982. Gating charge movements and kinetics of excitable membrane proteins. In *Proteins in the Nervous System: Structure and Function.* New York: Alan R. Liss.

Bullock, T. H., R. Orkand, and A. D. Grinnell. 1977. *Introduction to Nervous Systems.* San Francisco: W. H. Freeman and Company.

Cooke, I., and M. Lipkin. 1972. *Cellular Neurophysiology.* New York: Holt, Rinehart and Winston.

Davison, P. F. 1967. Protein of nervous tissue: Specificity, turnover, and functions. In F. O. Schmitt, ed., *The Neurosciences: A Study Program.* New York: Rockefeller University Press.

Duncan, C. J., ed. 1976. *Calcium in Biological Systems.* New York: Cambridge University Press.

Hagiwara, S., and L. Byerly. 1981. Calcium channel. *Ann. Rev. Neurosci.* 4:69–125.

Hille, B. 1970. Ionic channels in nerve membranes. *Progr. Biophys. Molec. Biol.* 21:1–32.

Hodgkin, A. L. 1958. Ionic movements and electrical activity in giant nerve fibres. *Proc. Roy. Soc.* (London) Ser. B. 148:1–37.

Hodgkin, A. L. 1964. *The Conduction of the Nervous Impulse.* Springfield, Ill.: Thomas.

Junge, D. 1980. *Nerve and Muscle Excitation.* 2nd ed. Sunderland, Mass.: Sinauer.

Katz, B. 1966. *Nerve, Muscle and Synapse.* New York: McGraw-Hill.

Keynes, R. D., and D. J. Aidley. 1981. *Nerve and Muscle.* Cambridge: Cambridge University Pres.

# References Cited

Cole, K. S., and H. J. Curtis. 1939. Electric impedance of the squid giant axon during activity. *J. Gen. Physiol.* 22:649–670.

Davson, H. 1964. *A Textbook of General Physiology.* 3rd ed. Boston: Little, Brown.

Eckert, R., and H. D. Lux. 1976. A voltage-sensitive persistent calcium conductance in neuronal somata of *Helix. J. Physiol.* 254:129–151.

Frankenhaeuser, B., and A. F. Huxley. 1957. The action of calcium on the electrical properties of squid axons. *J. Physiol.* 137:218–244.

Goldman, D. E. 1943. Potential, impedance, and rectification in membranes. *J. Gen. Physiol.* 27:37–60.

Gorman, A. L. F., and M. V. Thomas. 1978. Changes in intracellular concentration of free calcium ions in a pacemaker neurone, measured with metallochromic indicator dye arsenazo III. *J. Physiol.* 275:357–376.

Hille, B. 1975. The receptor for tetrodotoxin and saxitoxin. *Biophys. J.* 15:615–619.

Hodgkin, A. L., and P. Horowicz. 1959. The influence of potassium and chloride ions on the membrane potential of single muscle fibres. *J. Physiol.* 148:127–160.

Hodgkin, A. L., and A. F. Huxley. 1939. Action potentials recorded from inside a nerve fibre. *Nature* 144:710–711.

Hodgkin, A. L., and A. F. Huxley. 1952a. Currents carried by sodium and potassium ions through the membrane of the giant axon of *Loligo. J. Physiol.* 166:449–472.

Hodgkin, A. L., and A. F. Huxley. 1952b. A quantitative description of membrane current and its application to conduction and excitation in nerve. *J. Physiol.* 117:500–544.

Hodgkin, A. L., and A. F. Huxley. 1952c. Properties of nerve axons: (I) Movement of sodium and potassium ions during nervous activity. *Cold Spring Harbor Symp. Quant. Biol.* 17:43–52.

Hodgkin, A. L., and A. F. Huxley. 1953. The mobility and diffusion coefficient of potassium in giant axons from *Sepia. J. Physiol.* 119:513–528.

Hodgkin, A. L., A. F. Huxley, and B. Katz. 1952. Measurement of current–voltage relations in the membrane of the giant axon of *Loligo. J. Physiol.* 116:424–448.

Hodgkin, A. L., and B. Katz. 1949. The effect of sodium ions on the electrical activity of the giant axon of the squid. *J. Physiol.* 108:37–77.

Keynes, R. D. 1954. The ionic fluxes in frog muscle. *Proc. Roy. Soc.* (London) Ser. B. 142:359–382.

Keynes, R. D. 1958. The nerve impulse and the squid. *Scientific American* 199(6):83–90. Also available as Offprint 58.

Kung, C., and R. Eckert. 1972. Genetic modification of electric properties in an excitable membrane. *Proc. Nat. Acad. Sci.* 69:93–97.

Ling, G., and R. W. Gerard. 1949. The normal membrane potential of frog sartorius fibres. *J. Cell. Comp. Physiol.* 34:383–396.

Moore, J. W., and T. Narahashi. 1967. Tetrodotoxin's highly selective blockage of an ionic channel. *Fed. Proc.* 26:1655–1663.

Sigworth, F. J., and E. Neher. 1980. Single Na channel currents observed in cultured rat muscle cells. *Nature* 287:447–449.

Taylor, R. E. 1967. The role of inorganic ions in the nerve impulse. In F. O. Schmitt, ed., *The Neurosciences: A Study Program.* New York: Rockefeller University Press.

Young, J. Z. 1936. The giant nerve fibers and epistellar body of cephalopods. *Quart. J. Microscop. Sci.* 78:367–386.

# CHAPTER 6

# *Propagation and Transmission of Signals*

Nervous systems are undoubtedly the most intricately organized structures to have evolved on Earth. The human nervous system contains $10^{10}$–$10^{11}$ functional units called *neurons* (nerve cells) plus as many or more inexcitable supportive satellite cells termed *glial cells* or *neuroglia*. During development, these units organize themselves quite marvelously into the interacting arrays that we call *neural circuits*. The question of how the nervous system gives rise to perception and behavior is undoubtedly one of the greatest of all challenges to science, for to unravel the functioning of the nervous system, the brain is called on to analyze and understand itself. Whether such subjective phenomena as perception and consciousness will ever be understood in physical and chemical terms can only be conjectured.

In spite of the profound complexity of the nervous system, a great deal has already been learned about its basic mechanisms of operation. We will consider a number of these and limit ourselves to those ideas and findings that are of general importance to neural function. Sensory reception will be discussed in Chapter 7, and neuronal integration and motor control in Chapter 8. In this chapter, we will begin by considering the electrical properties of neurons and the means they use to propagate signals and transmit information to one another.

## *Nerve Cells*

The complexity of the nervous system lies not in large numbers of fundamentally different signals, but in the number and intricacy of the interconnections made among its neurons. This is not to imply that all neurons are alike. Indeed, by variations of their anatomy and a few basic functions, neurons have generated a great variety of specialized functions derived from but a few basic themes. They occur in a diversity of shapes and sizes (Figure 6-1). There are several ways to classify neurons on the basis of their morphology. One major distinction is whether or not the neuron has an *axon* (long process; nerve fiber). The renowned anatomist Camillo Golgi termed those with axons *Type I* and those without axons *Type II*. The latter, obviously, must restrict their contacts to *local circuits* made with cells in their immediate vicinity.

Shown diagrammatically in Figure 6-2 is a vertebrate *motoneuron* (motor neuron), which originates in the spinal cord and innervates skeletal muscle fibers. This is a classic Type I neuron, specialized for conducting impulses over long distances, generally as part of a *through circuit* (Figure 6-3). In these neurons, the surface membrane of the *dendrites* (threadlike cytoplasmic extensions) and the *soma* (cell body) is innervated by the terminals of other nerve cells. The axon carries action potentials from the *spike-initiating zone* near the *axon*

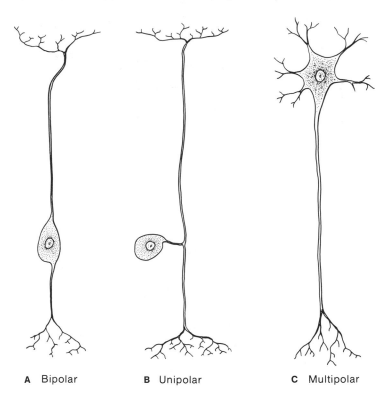

**A** Bipolar    **B** Unipolar    **C** Multipolar    **D**

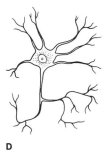

**6-1**  Three morphological types of neurons. There are wide variations within each type. The neurons in A, B, and C correspond to Golgi's Type I, while the neuron in D, which lacks an axon, is representative of his Type II. [Montagna, 1959.]

*hillock* to the *axon terminals,* which in this example innervate muscle cells. The dendrites and the axon are processes that grow out from the soma during development, and into which there is a slow but steady flow of proteins and other constituents synthesized in the soma. Once disconnected from the cell body, these processes gradually deteriorate and die within a few days or weeks. In mammals, regeneration (regrowth) of axons is limited to peripheral nerves; in the lower invertebrates, regeneration and reinnervation of muscles occur quite readily.

The physiological behavior of a neuron depends on its anatomical form and on the properties of its surface membrane. The membrane properties are not uniform over the neuron surface, but are specialized in different regions for different functions. Thus, the axon membrane is specialized for the conduction of impulses in most (notably Type I) neurons. The membrane of the neuron terminals is specialized to secrete a *transmitter substance* into the extracellular space for the purpose of communicating with the cells with which the terminals make special junctions termed *synapses.* The membrane of the dendrites and the soma has regions of special sensitivity to transmitter substances released by the terminals of other nerve cells. *Postsynaptic currents* produced in the dendrites and the soma in response to those transmitters are *integrated* (i.e., "put together"; summed algebraically) to produce *postsynaptic potentials* in the dendrites, soma, and axon hillock.

Membrane properties include passive electrical properties such as the membrane capacitance and resistance discussed in Chapter 5. In addition, there are the various gated ion channels that give rise to the active behavior of the membrane. The nature of these channels and their distribution in the neuron are both very important. For example, the axon membrane, specialized for rapid conduction, contains the fast-acting voltage-gated sodium channels. The cell body and dendrites contain transmitter-activated channels that carry the postsynaptic current. Thus, the various parts of the neuron are both anatomically and functionally specialized.

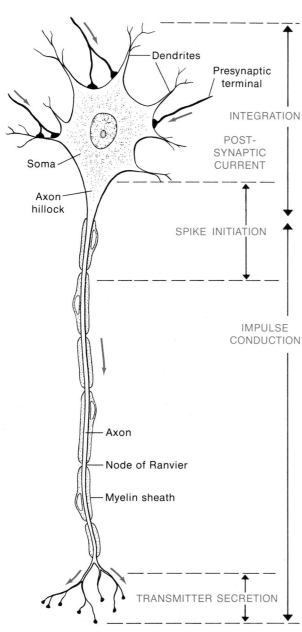

**6-2** Vertebrate spinal motoneuron, with the functions of different parts indicated. The flow of information is indicated by colored arrows. Axon and surrounding myelin sheaths are shown in longitudinal section.

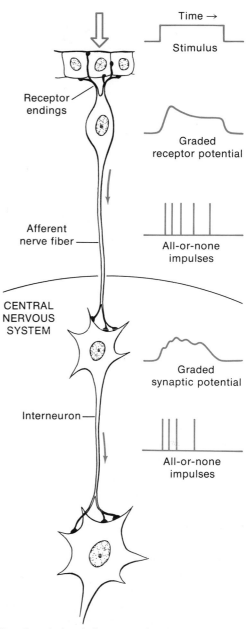

**6-3** Alternation of graded and all-or-none electrical signals in a neural "through circuit." The receptor potential produced in the sensory endings of the afferent (sensory) nerve cell is an analog approximation of the stimulus. This receptor potential spreads through the soma and sets up all-or-none propagated action potentials in the axon. On arriving at the terminals, the action potentials cause the release of a transmitter that sets up a graded synaptic potential in the next neuron. If the synaptic potential reaches the spike threshold, it will initiate a new action potential or a train of action potentials.

Glial cells are inexcitable cells of several types that occupy about half the volume of the nervous system in vertebrates (less in most invertebrates). They fill all the interneuronal space except for a very thin extracellular space (about 200 Å wide) between the glial and neuronal membranes.

Because the glial cells are electrically silent, their function had long been a puzzle. The glial cell membranes are highly permeable to $K^+$, and adjacent glial cells are often electrically coupled, allowing $K^+$ to flow between them. This flux permits glia to take up and redistribute high concentrations of $K^+$, which build up in the narrow extracellular space owing to $K^+$ efflux from electrically active neurons. Since the resting potential is highly potassium-dependent, electrical recordings from glial cells have provided a useful measure of changes in extracellular $K^+$ produced when surrounding neurons generate electric signals. The depolarizing shift in the glial cell membrane due to extracellular $K^+$ accumulation during neural activity subsides within several seconds, indicating the removal of the excess $K^+$ from the extracellular space, presumably through the highly potassium-permeable surface membrane. Such removal of $K^+$ by glial cells helps prevent $K^+$ accumulation in the extracellular space, thus providing an important service for the neurons, which otherwise would undergo significant depolarization by elevation of extracellular $K^+$, impairing such functions as synaptic transmitter release, attainment of the threshold potential, and pacemaker activity. Besides this potassium-buffering function, glial cells provide intimate structural and perhaps metabolic support for the neurons. As will be discussed later, two kinds of glia, Schwann cells and the oligodendrocytes, produce the myelin sheath around axons.

### THE TWO CLASSES OF SIGNALS EMPLOYED BY NEURONS

Even the commonly studied "simple" nervous systems of invertebrates consist of tens or hundreds of thousands of neurons. It should therefore come as a relief to learn that, regardless of their number, all neurons operate (with minor variations) on only two themes. That is, they employ only two basic classes of electrical signals—*graded potential changes* and *all-or-none impulses* (Figure 6-3).

Stimulus energy impinging on specialized receptor endings of neurons produces *receptor potentials* that are *graded* (i.e., amplitude-modulated) according to the stimulus energy intercepted and that generally continue for as long as the stimulus persists. Since the time course and intensity of the receptor potential are related to the intensity and time course of the stimulus, the receptor potential can be said to be an electrical analog of the stimulus. It spreads electrotonically in the sensory endings of the nerve cell and therefore decays progressively with distance from its site of origin. These amplitude-modulated analog signals from the receptor regions are therefore incapable of transmitting information over appreciable distances. If the information is to be transmitted over long distances to the central nervous system, it must be converted into action potentials, since these are conducted without decrement (i.e., without loss of signal strength).

The impulses arriving at the central terminals of a sensory neuron cause the release of a transmitter substance. The action of the transmitter (discussed later in this chapter) produces a potential change in the *postsynaptic* neuron. The amount of transmitter released and the amplitude of the postsynaptic response elicited by the transmitter are both a function of the impulse frequency in the *presynaptic* neuron (Figure 6-3). Within limits, the higher the presynaptic frequency, the greater the postsynaptic depolarization. The postsynaptic potentials are also analogs (although nonlinear and quite distorted) of the original stimulus. The postsynaptic depolarization, if large enough, then initiates a train of impulses in the postsynaptic neuron. Thus, graded, local, analog-type membrane potential changes typically alternate along the pathway with all-or-none, far-traveling, conducted impulses. Graded potentials occur at sensory and postsynaptic membranes, and impulses are confined largely to conducting structures, such as axons, that lie in between.

With minor exceptions, all the signals lumped under the two major categories—all-or-none impulses and graded potential changes—are generated by the gating of membrane channels. When activated, these channels carry ionic currents driven by simple electrochemical gradients. We will now see how these mechanisms produce impulse conduction within a neuron and transmission of signals from one neuron to another.

## Passive Spread of Electrical Signals

Membrane resistance and capacitance are properties that influence the way in which potentials and currents spread with distance within cells. In a spherical cell like that shown in Figure 5-3, potentials spread essentially uniformly with minimal decay. This is so because the resistance of the cell membrane is high compared to the total resistance a current encounters on passing through the cytoplasm across the relatively short diameter of the cell. The result is that a current injected into a spherical cell spreads out and passes through the membrane with relatively uniform density over the entire cell surface. However, real cells are seldom that simple. In long, thin, cylindrical elements such as an axon, dendrite, or muscle cell an *electrotonic* spread of signals occurs. That is, there is a strong decay of membrane potential with distance from the point of current injection because of the relatively high electrical resistance that occurs along the cytoplasm from one part within the cylinder to a distant point along the axis. Thus, the longitudinal flow of a current injected at one point undergoes a progressive decline with distance, as some of it leaks out across the cell membrane at each point along the cylinder.

Axons have certain similarities to an insulated submarine cable. The cytoplasm is analogous to the conducting core of the cable, and the membrane is analogous to the insulation. Moreover, the extracellular fluid bathing the cell is analogous to the seawater that surrounds the submarine cable. The *cable properties* of nerve and muscle cells play an important role in the spread of current and the conduction of impulses along the cell membrane, as we will see.

Current distributes itself along the axon according to the passive electrical properties described by the equivalent circuit shown in Figure 6-4. The components $R_m$ and $C_m$ are the same as in Figure 5-7A and represent the uniformly distributed passive resistance and capacitance of the inactive membrane. They are depicted as discrete elements only for the sake of convenience. We will first ignore the membrane capacitance and assume that current flows only through the resistance.

According to Kirchhoff's First Law, the sum of all the currents leaving a point must equal the sum of all the currents entering that point. Kirchhoff's Law, together with Ohm's Law, requires that the current distribute itself in inverse proportion to the resistances of the various routes open to it at each branching point. Thus, when the switch in the equivalent circuit of Figure 6-4 is closed, the pulse ($\Delta I$) of constant intensity current will distribute itself accordingly and flow across the "membrane" at all branching points (0, 1, 2, 3, 4). It will be diminished by each longitudinal resistor, $R_l$, encountered, because longitudinal resistance is cumulative. At each branching point, the current divides, with the same proportion passing through each $R_m$ and the remainder through the $R_l$. The transmembrane current through the $R_m$'s is therefore a decreasing exponential function of longitudinal distance from the point of current injection. Since all the $R_m$'s in the circuit are of the same resistance, Ohm's Law requires that the potentials developed across them also decrease exponentially with distance. Thus, the transmembrane steady-state potential ($\Delta V$) will diminish exponentially with distance along the axon (Figure 6-4).

The exponential decay of a steady-state potential with distance is described by the following equation, first applied to axons by A. L. Hodgkin and W. A. H. Rushton (1946):

$$V_x = V_0 e^{-x/\lambda} \qquad (6\text{-}1)$$

Here $V_x$ is the potential change measured at a distance $x$ from the point ($x = 0$) at which the current is injected, and $V_0$ is the potential change at the point $x = 0$. The symbol $\lambda$ is the *length* (or *space*) *constant*, which is related to the resistance of the axon by the expression

$$\lambda = \sqrt{\frac{r_m}{r_i + r_o}} = \sqrt{\frac{r_m}{r_l}} \qquad (6\text{-}2)$$

in which $r_m$ is the resistance of a unit length of axon membrane, and $r_l$ is the summed longitudinal internal and external resistances over a unit length ($r_i + r_o$). Equation 6-1 reveals that when $x = \lambda$,

$$V_x = V_0 e^{-1} = V_0 \frac{1}{e} = 0.37\, V_0$$

Therefore, $\lambda$ is defined as the distance over which a steady-state potential shows a 63% drop in amplitude (Figure 6-5).

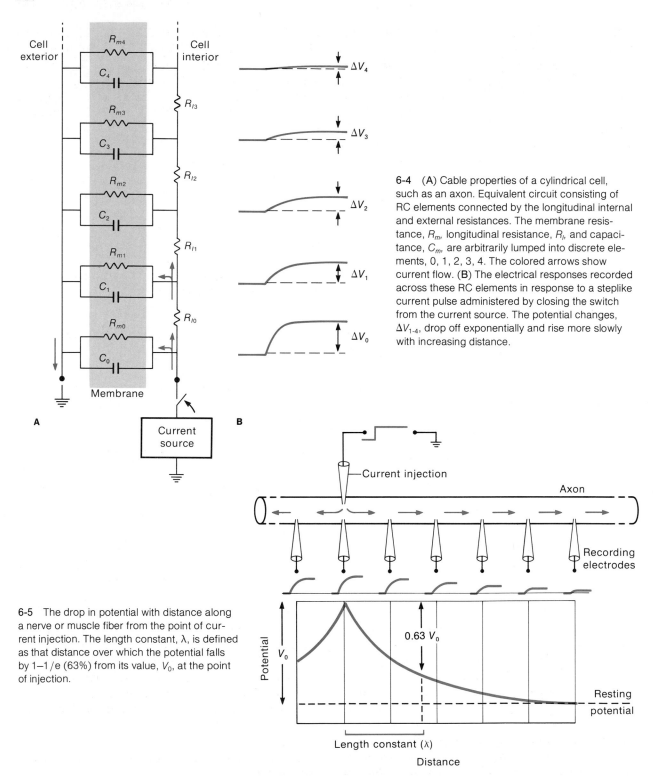

**6-4** (**A**) Cable properties of a cylindrical cell, such as an axon. Equivalent circuit consisting of RC elements connected by the longitudinal internal and external resistances. The membrane resistance, $R_m$, longitudinal resistance, $R_l$, and capacitance, $C_m$, are arbitrarily lumped into discrete elements, 0, 1, 2, 3, 4. The colored arrows show current flow. (**B**) The electrical responses recorded across these RC elements in response to a steplike current pulse administered by closing the switch from the current source. The potential changes, $\Delta V_{1-4}$, drop off exponentially and rise more slowly with increasing distance.

**6-5** The drop in potential with distance along a nerve or muscle fiber from the point of current injection. The length constant, $\lambda$, is defined as that distance over which the potential falls by $1-1/e$ (63%) from its value, $V_0$, at the point of injection.

Note in Equation 6-2 that the value of $\lambda$ is directly proportional to the square roots of both $r_m$ and $1/r_l$. This means that *spread of electric current along the interior of an axon is enhanced by a high membrane resistance and/or low longitudinal resistance.* We will see shortly that the velocity of conduction of an action potential is closely related to the effectiveness of current spread along the interior of the axon. Cable properties of nerve cells also are important in the information-processing of the nervous system, to be discussed in Chapter 8.

## Propagation of Nerve Impulses

Many Type II neurons as well as some small Type I neurons are sufficiently short relative to the length constant of their *neurites* (cell processes) that they carry out all or most of their normal electrical functions without propagated impulses. Such cells are typically incapable of producing action potentials and are therefore referred to as *nonspiking neurons* or *local circuit neurons.* Their graded signals are conducted electrotonically to their terminals without the aid of all-or-none impulses. The signals are attenuated by loss of amplitude along the way, but nevertheless remain sufficiently large at the terminals that they can modulate the release of the transmitter substance. Examples of such local circuit neurons are seen in the vertebrate retina and other portions of the central nervous system (CNS), the barnacle eye, the insect nervous system, and the crustacean stomatogastric ganglion. These cells are seldom more than a very few millimeters in overall length, and they are generally characterized by a high specific membrane resistance, which contributes to efficient, low-loss electrotonic spread of signals.

Communication between different parts of the nervous system separated by greater distances depends, however, on the propagation of nerve impulses (action potentials) along the axons of neurons. Similar propagation also occurs in muscle cells. As noted in the previous chapter, the action potential entails a potential change across the surface membrane that is about five times as large as the threshold depolarization. With this *safety factor,* the excited portion of an axon is able to excite the portions ahead of it and thereby cause the action potential to propagate along the axon.

To understand the way in which impulses propagate, it is necessary to recall from Chapter 5 that the

electrically excited membrane becomes permeable to $Na^+$, so that this ion carries a momentary strong current into the excited region of an axon. This current spreads longitudinally along the inside of the axon and leaks transversely outward across the membrane to complete the circuit of current flow. This electrotonic spread of *local circuit current*† is due to the cable properties of the axon (Figure 6-4). Thus, the influx of sodium, producing the upstroke of the action potential (Figure 6-6A), results in a current that spreads longitudinally, both forward and backward within the axon (Figure 6-6B).

Consider only that part of the current that flows forward longitudinally within the axon—that is, in the direction of impulse propagation (to the left in Figure 6-6B). To complete its circuit, this current must flow out across unexcited portions of the membrane ahead of the region of inward sodium flux and then back to the active region. Current flowing outward across the unexcited membrane partially depolarizes that part of the membrane to produce the *foot* of the propagated impulse (compare Figures 5-16 and 6-6A). Such electrotonic depolarization of the inactive membrane ahead of the active region was first clearly demonstrated with the experiment shown in Figure 6-7, done by Alan Hodgkin as a student in 1937.

As the membrane ahead of the impulse is depolarized by local circuit current, the sodium conductance of that region increases, initiating the regenerative sequence (i.e., the Hodgkin cycle) that produces the action potential. The newly excited region then generates the local circuit current that depolarizes and thereby excites portions of the axon ahead of it. Thus, local circuit current from each excited region depolarizes and excites in turn the next region ahead of it. The signal is continuously boosted and maintained at full strength in this manner as it travels along the axon.

Some of the current that enters an axon at the excited portion spreads backward longitudinally—that is, in the direction from which the impulse originated. This current does not stimulate, since the membrane just behind the advancing action potential is in a refractory state. The potassium channels in that region

---

†"Local circuit" in this usage is unrelated to the concept of local circuit neurons.

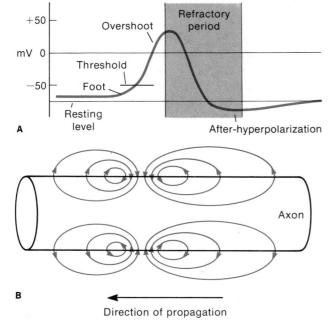

**A**

**B**

Direction of propagation

6-6   The action potential in a nerve (**A**) is accompanied by a local circuit current flowing across the membrane, as shown in part B. The "foot" is produced by outward current depolarizing the membrane ahead of the active (sodium-entering) region. Outward current in the inactive regions is carried across the membrane primarily by K+.

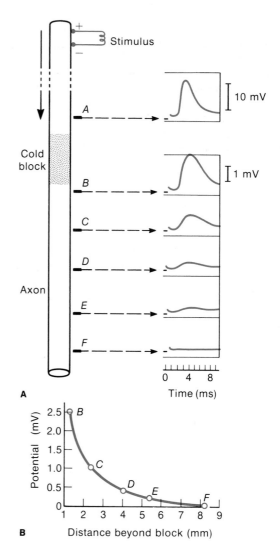

**A**

**B**   Distance beyond block (mm)

6-7   Experiment in which it was first demonstrated that the action potential generates current that spreads within the axon to produce an electrotonic depolarization of the membrane ahead of the active portion. (**A**) Section (stippled area) was blocked by cooling, and potentials recorded at points *B* to *F*. (**B**) The electrotonic potential was found to decrease exponentially with distance from the region of the block. [Hodgkin, 1937.]

are still open and carry the current out of the cell as an efflux of K+.

Propagation of an impulse depends on two factors:

*1.* The electrical excitability of the axon membrane, which results in a fivefold regenerative amplification of the electrotonic potential that reaches threshold.

*2.* The cable properties of the axon, which permit the electrotonic spread of local circuit current from the region of sodium influx to stimulate nearby regions of inactive membrane.

One might ask why extracellular action currents from one conducting axon do not excite other nearby axons, creating "cross talk" between the axons. The answer, in short, is that the resistances of the inactive membranes are so much greater than the resistance of the extracellular current path that only a minute fraction of the total current produced by an active membrane flows into a neighboring inactive axon.

Because of this extracellular shunting action, the currents developed by an active axon are normally too small to excite neighboring inactive neurons. However, extracellular currents generated by an action potential can be detected by extracellular electrodes (Box 6-1).

## Box 6-1  Extracellular Signs of Impulse Conduction

Nerve impulses can be recorded with a pair of extracellular recording leads (first figure). The electronics are designed so that a negative potential in lead $A$ will cause the oscilloscope beam to go up while a similar potential recorded by lead $B$ will make the beam go down; positive potentials do the opposite. An action potential passing along an axon is seen as a wave of negative potential, because the cell exterior becomes more negative than its surroundings when sodium flows into the cell to produce the overshoot. As a result, the action potential passing two electrodes makes a biphasic wave shape on the oscilloscope (part A of figure).

The recording is simplified by preventing invasion of the part of the axon that is in contact with electrode $B$. This is done by anesthetizing, cooling, or crushing that part of the

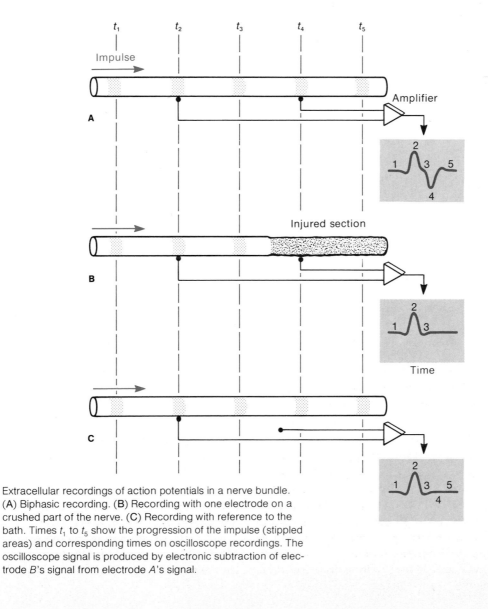

Extracellular recordings of action potentials in a nerve bundle. (A) Biphasic recording. (B) Recording with one electrode on a crushed part of the nerve. (C) Recording with reference to the bath. Times $t_1$ to $t_5$ show the progression of the impulse (stippled areas) and corresponding times on oscilloscope recordings. The oscilloscope signal is produced by electronic subtraction of electrode $B$'s signal from electrode $A$'s signal.

axon (part B of figure). A similar effect can be obtained by placing electrode *B* in the bath, at some distance from the axon (part C of figure).

Extracellular recordings are frequently made from nerve bundles or tracts made up of many axons. The summed activity of many axons gives a compound recording whose characteristics depend on the number of axons conducting and their relative timing and current strength. Larger axons generate larger extracellular currents because membrane current increases in proportion to membrane area. The size of the action potential recorded with extracellular electrodes is proportional to the amount of current flowing through the extracellular fluid. Thus, action potentials from axons of large diameter appear bigger when recorded extracellularly, even though the potential changes across their membranes are no larger than those of smaller nerve fibers. Because of these amplitude differences in extracellularly recorded action potentials, the signals originating in individual axons can often be distinguished by their size in such recordings (figure at the right).

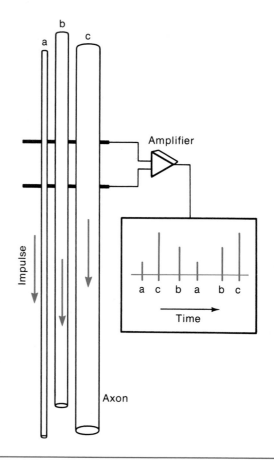

Extracellular recordings can distinguish between fibers of different diameter on the basis of differences in the sizes of action currents. A larger current flowing through the extracellular fluid produces a larger voltage between the recording electrodes.

### VELOCITY OF PROPAGATION

Johannes Müller declared in the 1830s that the velocity of the action potential would never be measured. He reasoned that the action potential, being an electrical impulse, must travel with a speed approaching that of light ($3 \times 10^{10}$ cm/s), and was therefore too fast to resolve over biological distances, even with the best instruments available at that time.

Within 15 years, one of his own students, Hermann von Helmholtz, had measured the velocity of impulse propagation in frog nerves by an elegantly simple method (Figure 6-8), which can easily be duplicated in a student laboratory course. The nerve is stimulated at each of two locations 3 cm apart, and the latency to the peak of the muscle twitch is determined. Suppose the latency decreases by 1 ms when the stimulating electrode is moved to the more distal (closer to the muscle) position. The velocity of propagation, $V_p$, is calculated as

$$V_p = \frac{\Delta d}{\Delta t} = \frac{3 \text{ cm}}{1 \text{ ms}} = 3 \times 10^3 \text{ cm/s}$$

This is seven orders of magnitude slower than the velocity of electric current flow in a copper wire or in an electrolyte solution. From this kind of experiment, Helmholtz correctly concluded that the nerve impulse is more complex than a simple longitudinal flow of current within the nerve fiber.

The velocity of impulse propagation in various axons ranges from 120 m/s in some large axons down

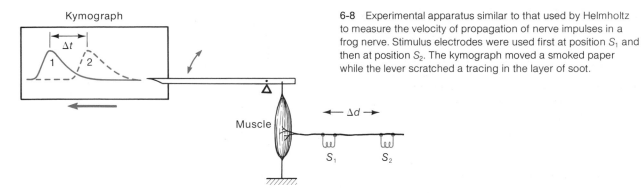

6-8 Experimental apparatus similar to that used by Helmholtz to measure the velocity of propagation of nerve impulses in a frog nerve. Stimulus electrodes were used first at position $S_1$ and then at position $S_2$. The kymograph moved a smoked paper while the lever scratched a tracing in the layer of soot.

to several centimeters per second in very thin axons. Differences in conduction velocity are illustrated in Table 6-1 and Figure 6-9.

The velocity of conduction depends in large part on the rate at which the membrane at any given distance ahead of the active region is depolarized to threshold by the local circuit currents. The greater the length constant, the farther the local circuit current flow, the more rapid the depolarization of the membrane in advance of the excited region, and hence the greater the velocity of propagation. The effect that decreasing the length constant has on conduction velocity can be demonstrated by placing an axon in oil or in air. This procedure leaves only a thin film of saline on the surface of the axon, and the length constant is decreased because the external longitudinal resistance ($r_o$ in Equation 6-2) is thereby raised. Under these conditions, the rate of conduction is slower than if the axon is immersed in saline.

The evolutionary process has exploited two means for increasing the length constants of axons and thereby the velocity of impulse propagation. One, typified by the giant axons of squid, arthropods, annelids, and teleosts, is an *increase in axonal diameter* and consequently a reduction in internal longitudinal resistance ($r_i$ in Equation 6-2). This is explained further in Box 6-2.

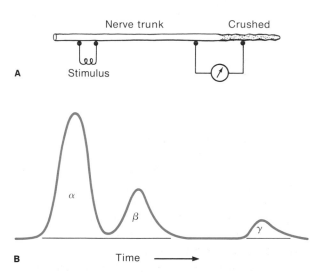

6-9 Conduction velocities of different axons in a frog nerve. (A) Experimental setup for stimulating and recording from the nerve bundle. (B) Externally recorded compound action potentials (the summed signals from all active fibers in the bundle). The $\alpha$-fibers have the largest diameter and highest conduction velocity. The $\gamma$-fibers have the smallest and slowest of those shown here (Table 6-1).

TABLE 6-1   Classification of frog nerve fibers on the basis of diameter and conduction velocity.

| Fiber class | | Fiber diameter ($\mu$m) | Velocity (m/s) |
|---|---|---|---|
| A | $\alpha$ | 18.5 | 42 |
| | $\beta$ | 14.0 | 25 |
| | $\gamma$ | 11.0 | 17 |
| B | | — | 4.2 |
| C | | 2.5 | 0.4–0.5 |

*Source:* Erlanger and Gasser, 1937.

## Box 6-2   *Axon Diameter and Conduction Velocity*

The transverse resistance, $r_m$, of a unit length, $l$, of axon membrane is *inversely proportional to the radius, $\rho$, of the axon*. The reason is that the area, $A_s$, of a cylindrical surface of unit length is equal to $2\pi\rho l$. The longitudinal resistance, $r_i$, of a unit length of axoplasm is inversely proportional to the cross-sectional area, $A_x$, of the axon. Since $A = \pi\rho^2$, the resistance $r_i$ is *inversely proportional to the square of the radius*. It follows that, with any given increase in radius, the drop in $r_i$ will be greater than the drop in $r_m$. Since the length constant $\lambda = \sqrt{r_m/r_i + r_o}$ (Equation 6-2), the disproportionate drop in longitudinal resistance $r_i$, which accompanies an increase in axon diameter, results in an increased length constant. Typically, $r_i \gg r_o$, so $\lambda \approx k\sqrt{\rho}$, in which $k$ is simply a constant of proportionality; that is, the length constant increases in proportion to the axon radius.

Because propagation velocity depends on the *rate* of depolarization at a given distance, $x$, ahead of the action potential, membrane capacitance cannot be ignored. Note that the time constant ($r_m \times c_m$) of a unit length of axon membrane remains constant with changes in axon diameter, because capacitance, $c_m$, *increases* in direct proportion, whereas the resistance, $r_m$, *decreases* in direct proportion to an increase in membrane area. The increase in length constant that accompanies increased axon diameter therefore occurs without an increase in the time constant of the membrane; and thus increase in diameter results in a greater outward membrane current at distance $x$ without an increase in membrane time constant. The increased rate of depolarization brings the membrane to threshold sooner at distance $x$ and results in an increase in velocity of conduction.

---

Giant axons have evolved in some species for rapid and synchronous activation of locomotor reflexes, as in the mantle of squid and the escape or withdrawal reflexes of certain arthropods (e.g., crayfish, cockroach) and annelids (e.g., earthworm).

### SALTATORY CONDUCTION

The other evolutionary means of increasing conduction velocity, developed only in the vertebrates, is the *insulation of segments of the axon with myelin*. This greatly increases the length constant of these segments, thus enhancing efficient longitudinal current spread. Myelin is laid down around axons of peripheral and central tracts in vertebrates during development by glial cells that grow around the axon, leaving behind a tightly wrapped, multilayered sheath of cell membrane. Myelin-depositing glial cells include *Schwann cells* in the peripheral nerves and *oligodendrocytes* (Figure 6-10) in the central nervous system. Cross sections of the sheath show a periodic 120 Å spacing, which represents the repeated layering of glial cell membrane.

Each unit membrane contributes to the high transverse resistance of the sheath. Because of its many layers of membrane, the myelin sheath exhibits far less capacitance than does a single unit membrane. The multilayered sheath is interrupted at regular intervals (*nodes of Ranvier*), directly exposing short sections of electrically excitable axon membrane to the extracellular fluid. Between nodes, the sheath is closely applied to the axon membrane, nearly obliterating the extracellular space of the axon membrane. Moreover, the internodal axon membrane appears to lack the channels that carry the sodium current.

The insulating properties of the myelin greatly increase the length constant of the axon, since the sheath has the same effect as increasing $r_m$ in Equation 6-2. Because of the high insulating resistance along the internode, the local circuit current in advance of the action potential leaves the axon almost exclusively at the nodes of Ranvier. Moreover, very little current is expended in discharging membrane capacitance along the internodes because of the low capacitance of the thick myelin sheath. The action potential at one node

Oligodendrocyte

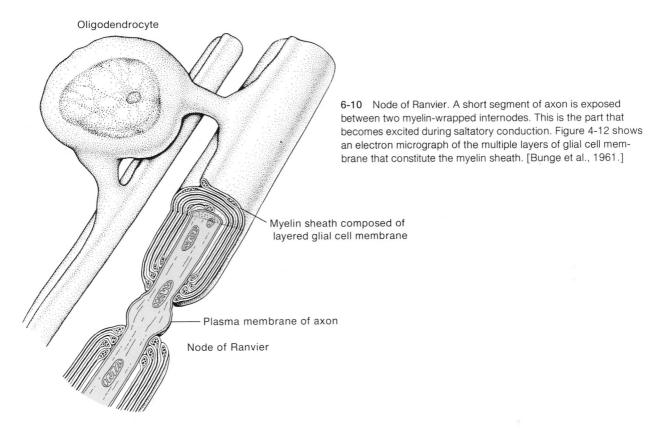

6-10  Node of Ranvier. A short segment of axon is exposed between two myelin-wrapped internodes. This is the part that becomes excited during saltatory conduction. Figure 4-12 shows an electron micrograph of the multiple layers of glial cell membrane that constitute the myelin sheath. [Bunge et al., 1961.]

Myelin sheath composed of layered glial cell membrane

Plasma membrane of axon

Node of Ranvier

electrotonically depolarizes the membrane at the next node, and thus the action potential does not propagate with continuity along the axon membrane as it does in nonmyelinated nerve fibers, such as the squid axon. Instead, it is produced only at the small areas of membrane exposed at the nodes of Ranvier. The result is *saltatory conduction*, a series of discontinuous action potentials, one at each node, as illustrated in Figure 6-11. The velocity of signal transmission is greatly enhanced, for the electrotonic spread of local circuit current occurs rapidly over internodal distances. Thus, the vertebrates "solved" the problem of high-velocity impulse conduction without need for cumbersome trunks of giant axons.

6-11  Saltatory conduction in a myelinated axon. (A) The action potentials occur only at the nodes of Ranvier, and current spreads longitudinally between nodes. The action potential jumps from node to node. (B) The large arrows indicate Na⁺ entry through activated sodium channels. The circles indicate the intracellular potentials present at each corresponding node at the instant shown in part A.

Direction of propagation

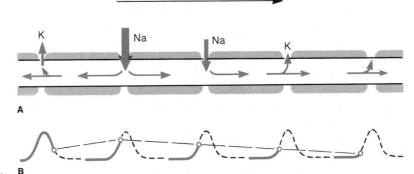

## Concept of the Synapse

The great histologist Santiago Ramon y Cajal, using Golgi's silver impregnation technique, showed early in this century that neurons appear histologically to be discrete units. In spite of this, some anatomists held the belief that neural tissue is made up of a continuous reticulum rather than morphologically separate neurons. Unequivocal evidence for cellular discontinuity between neurons and for special regions of neural interaction came with the development of electron microscopy in the 1940s.

In 1897, long before the ultrastructural basis for neuron–neuron interactions became known, the functional junction between two neurons was given the name *synapse* by Sir Charles Sherrington, who is regarded as the founder of modern neurophysiology. It was his conclusion that "the neurone itself is visibly a continuum from end to end, but continuity fails to be demonstrable where neurone meets neurone—at the synapse. There a different kind of transmission may occur" (Sherrington, 1906, p. 21). Although Sherrington had no direct information about the microstructure or microphysiology of these specialized regions of interaction between excitable cells, he displayed extraordinary insight, gained from his cleverly designed experiments on the spinal reflexes of cats and dogs. He deduced that some synapses are *excitatory,* leading to the initiation of action potentials, and that others are *inhibitory,* counteracting the initiation of action potentials.

It is now known that there are two major categories of synapses, *chemically transmitting* and *electrically transmitting.* Electrical synapses will be considered first, since they are the simpler of the two.

## Transmission at Electrical Synapses

At an *electrical synapse,* the pre- and postsynaptic membranes are in close apposition (Figure 6-12A) and form gap junctions through which electric current flows preferentially from one cell into the other. The injection of a subthreshold current pulse into cell *A* (Figure 6-13) elicits a change in the membrane potential of that cell. If a significant fraction of the current injected into cell *A* spreads through gap junctions into cell *B*, it will cause a detectable change in the membrane poten-

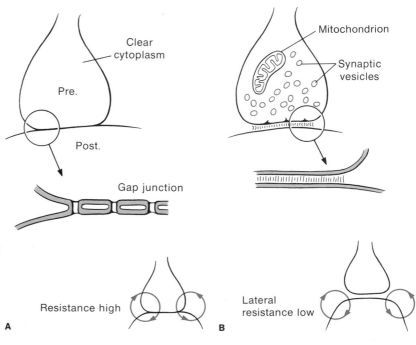

**6-12** Two kinds of synaptic transmission. (A) Electrical synapse. Gap junctions between pre- and postsynaptic membranes permit current to flow through fine intercellular channels (inset). The action current from the presynaptic cell simply flows into the postsynaptic cell, depolarizing it. (B) Chemical synapse. There is no intercellular continuity. Synaptic current flows only across the postsynaptic membrane in response to transmitter-activated opening of membrane channels. [Whittaker, 1968.]

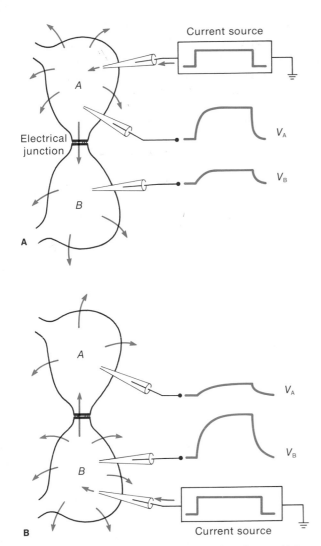

6-13 (A) In electrically coupled cells, injection of current into one cell elicits potential changes in both cells. (B) This coupling is generally symmetrical, current passing equally well in either direction. For the purpose of simplicity, the hypothetical cells used in this figure exhibit no membrane excitation, but passive potentials only.

corded in cell *A*. The gap junctions through which the current flows from one cell to the other are generally (but not always) symmetrical in resistance—that is, current generally meets the same resistance in either direction.

An electrical junction between two neurons will allow local circuit current from an action potential in one to spread into the other and depolarize it. Transmission of an action potential through an electrical synapse is basically no different from propagation within one cell, since both phenomena depend on the electrotonic spread of local circuit current ahead of the action potential to depolarize and excite new regions of membrane. As we have noted, the safety factor of an action potential is typically about 5; so the attenuation in amplitude from one cell to the next must be no greater than the safety factor if the electrotonic depolarization of the postsynaptic cell is to reach threshold and initiate an impulse. It would therefore be very difficult for a single action potential in a fine axon to singlehandedly supply enough local circuit current across an electrical synapse to elicit an action potential in a comparatively large cell, such as a muscle fiber, because of the enormous membrane area (and hence low input resistance) of the muscle fiber compared to that of the motor axon. This is undoubtedly one evolutionary reason why electrical synapses appear not to be as widespread as chemical synapses.

*Electrical transmission* between excitable cells was first demonstrated by E. J. Furshpan and D. D. Potter in 1959 in the crayfish. The synapse between the crayfish lateral giant nerve fiber and a large motor axon has the usual property of passing current preferentially in one direction (Figure 6-14). Since 1959, electrical transmission has been discovered between cells in the central nervous system, smooth muscle, cardiac muscle, receptor cells, and axons. Because current flows directly from the presynaptic cell into the postsynaptic cell without intervening events, transmission occurs with less delay at electrical synapses than at chemical synapses. Electrical transmission is therefore well suited for the synchronization of electrical activity in a group of nerve cells or for rapid transmission across a series of cell–cell junctions, such as those that occur in giant nerve fibers of the earthworm and in the myocardium of the vertebrate heart.

tial of cell *B* as well. Since there is a potential drop for the current crossing the gap junctions from cell *A* into cell *B*, the electrotonic potential change recorded across the membrane of cell *B* will always be less than that re-

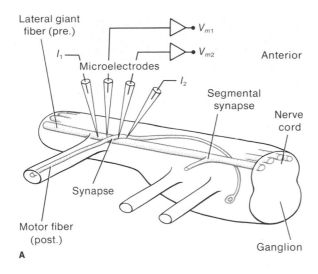

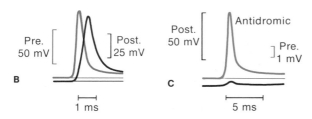

6-14  Giant electrical synapse in the crayfish. (**A**) The arrangement used for recording from and stimulating the pre- and
postsynaptic axons. (**B**) An action potential produced in the presynaptic (lateral giant) axon electrically excites an action potential in the postsynaptic (giant motor) axon. (**C**) An action potential in the postsynaptic axon does not produce a significant
potential change in the presynaptic axon. Injection of current
pulses shows that there is preferential flow of current from pre-
to postsynaptic cells. This is unusual for an electrical synapse.
[Furshpan and Potter, 1959.]

## *Transmission at Chemical Synapses*

Most synapses in the nervous system utilize a transmitter substance with which the presynaptic neuron
communicates with the postsynaptic cell. The sequence
of events that occurs during *chemical transmission* is outlined in Figure 6-15 as a preview for the greater detail
presented in the next sections. At the top, we see the
arrival of a presynaptic action potential. The depolar-

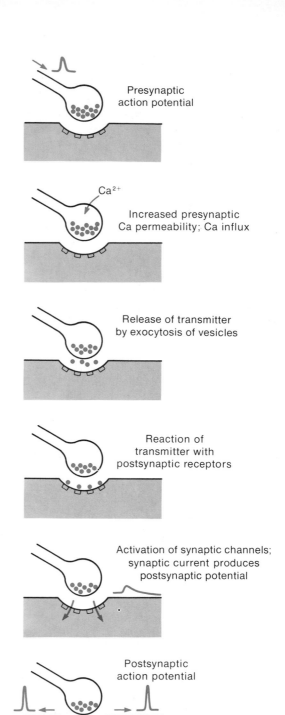

6-15  Sequence of events from presynaptic
to postsynaptic action potentials.

ization activates the calcium channels of the terminal, and $Ca^{2+}$ enters. The rise in $[Ca^{2+}]_i$ initiates the exocytosis of vesicles containing the transmitter substance. These spill their contents into the extracellular space, and the transmitter diffuses so that some of it binds to receptor molecules in the postsynaptic membrane. The binding of the transmitter activates ion channels associated with the receptor molecules, allowing the permeant ions to carry a postsynaptic current according to the prevailing electrochemical gradients. The postsynaptic current produces a postsynaptic potential. If the potential change is sufficient to exceed the threshold potential, it initiates an action potential. Generally speaking, chemical transmission is more flexible than electrical transmission, and it freely allows inhibitory as well as excitatory action. In addition, chemical transmitters allow small presynaptic fibers to excite large postsynaptic cells by chemically activating postsynaptic channels that carry strong currents.

The verity of chemical transmission and the existence of transmitter substances was debated during the first six decades of this century. The first direct evidence for a chemical transmitter substance was obtained by Otto Loewi (1921), who found that inhibition of one frog heart by stimulation of the vagus nerve produced a substance that could cause a second frog heart to beat more slowly. Loewi's finding led to the subsequent identification of *acetylcholine (ACh)* as the transmitter substance released by the postganglionic neurons in response to stimulation of the vagus nerve (Figure 8-13B) and by motoneurons innervating skeletal muscle in vertebrates. Since then, much more has been learned about the action of transmitter substances, and many additional neurotransmitters have been identified.

### MORPHOLOGY OF CHEMICAL SYNAPSES

Chemical transmission occurs across an extracellular *synaptic cleft,* a space of about 200 Å that separates the membranes of the pre- and postsynaptic cells (Figure 6-12B). The presynaptic terminal contains membrane-bound *synaptic vesicles* (Figure 6-16), about 400 Å in diameter, each containing $1 \times 10^4$ to $5 \times 10^4$ molecules of the transmitter substance. Presynaptic terminals contain thousands of these vesicles. For example, the branches of the nerve terminal innervating a single

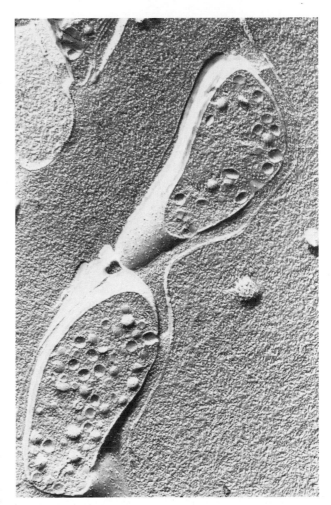

**6-16** Synaptic terminal as seen in transverse section in a freeze-etched specimen of an electric organ of the ray *Torpedo.* Synaptic vesicles can be seen in the terminal, with two vesicles frozen in place as they were opening into the synaptic cleft. Magnification 40,000×. [Nickel and Potter, 1970.]

frog muscle fiber typically contain a total of about $10^5$ synaptic vesicles. During synaptic transmission, the transmitter substance is released into the synaptic cleft and reaches the postsynaptic membrane by diffusion. The cleft is filled with a mucopolysaccharide that "glues" together the pre- and postsynaptic membranes, which usually show some degree of thickening at the synapse.

6-17   Frog motor endplate (neuro-muscular synapse). (**A**) Light micrograph of a whole mount preparation showing motor nerve approaching from above and branching to left and right along the surface of a muscle fiber (note striations). A histochemical reaction has stained the acetylcholinesterase in the postsynaptic (muscle) membrane black. (**B**) Electron micrograph of the endplate region. The muscle cell appears at the bottom, showing striated myofibrils. The muscle membrane shows extensive infoldings, termed *junctional folds*. The axon terminal is seen in longitudinal section above and contains pale synaptic vesicles grouped in bunches over regions of presynaptic membrane thickenings, forming the active zones. Above them are seen denser granules and mitochondria. The synaptic cleft is filled with an amorphous mucopolysaccharide. [McMahan et al., 1972.]

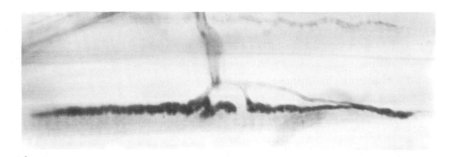

A

B

The most extensive studies of synaptic transmission have been done with the *motor terminal,* or *motor endplate* (i.e., neuromuscular synapse), of vertebrate skeletal muscle (Figure 6-17), especially with the endplates of the frog sartorius muscle.† Except for the identities of the transmitter substances and certain quantitative differences, excitatory synaptic transmission between neurons in the central nervous system is similar to transmission at a nerve–muscle synapse, such as the motor endplate. It will serve as the model system in much of what follows on chemical transmission.

An examination of Figure 6-17 in the light of the three-dimensional reconstruction of the frog motor endplate in Figure 6-18 shows that it consists of specializations of the postsynaptic membrane, the motor nerve terminal, and Schwann cells. The axon terminal bifurcates, and each approximately 2-$\mu$m-thick branch lies in a longitudinal depression along the surface of the muscle fiber. The muscle membrane lining the depression is thrown into transverse *junctional folds* at intervals of 1 to 2 $\mu$m. Directly above these folds within the nerve terminal are the *active zones,* transverse regions of slight thickening of the presynaptic membrane, above which are clustered the synaptic vesicles. There is evidence that the vesicles are released along the active zones by the process of exocytosis. Evidence of exocytosis can be seen in Figure 6-16. Transmitter release from the presynaptic terminal is

†Strictly speaking, the term *endplate* does not accurately describe the organization of the neuromuscular junction of amphibians, having been first applied to the neuromuscular junction of mammals, which is more platelike and less strung out than the amphibian terminal.

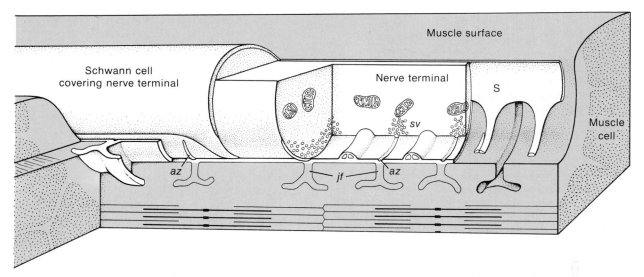

**6-18**   Three-dimensional reconstruction of the frog endplate based on electron-microscopical data. The nerve terminal lies in a longitudinal depression in the surface of the muscle fiber. The depression contains transverse *junctional folds* ( *jf* ). Overlying the junctional folds are *active zones* (*az*) in the nerve terminal, rich in synaptic vesicles (*sv*). A Schwann cell (S) overlies the terminal, sending slender fingers underneath the terminal. Compare with Figure 6-16. [Peper et al., 1974.]

triggered by the arrival of an action potential in the terminal.

Once released, the transmitter—acetylcholine in vertebrate neuromuscular synapses—is subject to hydrolysis by the enzyme *acetylcholinesterase*. The presence of this enzyme can be detected by histochemical means (Figure 6-17) and is found to be located in the synaptic folds of the frog motor terminal. Before it is completely hydrolyzed, the acetylcholine interacts with acetylcholine receptor molecules located in the postsynaptic membrane of the endplate, causing closely associated ion channels that are more or less selective for $Na^+$ and $K^+$ to open briefly.

### SYNAPTIC POTENTIALS

In 1942, S. W. Kuffler reported experiments on single fibers of frog muscle that showed depolarizations intimately associated with the motor endplate. These occurred in response to motor nerve impulses and preceded the action potential generated across the membrane of the muscle cell. These potential changes, which at that time were recorded with crude extracellular methods, were greatest in amplitude at the endplate and gradually disappeared with distance as

recordings were made farther away from the endplate region. They were consequently termed *endplate potentials*. It was correctly concluded that the propagated action potential in the muscle cell arises from a local depolarization of the postsynaptic membrane associated with the arrival of an action potential in the presynaptic terminal.

The development of the glass capillary microelectrode in the late 1940s made it possible to go far beyond these early exploratory experiments. Described next is a composite of intracellular studies—based largely on work from the laboratory of Bernard Katz—of synaptic transmission at the frog neuromuscular synapse.

A microelectrode inserted into a muscle fiber at a distance of several millimeters from the region of the motor terminal arborizations will record the resting potential, plus an all-or-none *muscle* action potential with a delay of several milliseconds after the arrival of an action potential in the terminals of the motor axon. Every time the motor axon is stimulated, a muscle action potential will be recorded, and the muscle fiber will respond with a twitch. If increasing concentrations of the South American arrow poison *curare*

## ***Box 6-3*** *Pharmacological Agents Useful in Synaptic Studies*

Studies of synaptic transmission have been greatly aided by the discovery and application of a number of agents that selectively interfere with or partially mimic certain steps in the process of transmission. Listed here are some of these agents.

*d-Tubocurarine* is the active principle of curare, the South American blow-dart poison. This molecule blocks transmission postsynaptically by competing with acetylcholine for the ACh-binding site of nicotinic receptors found at the motor endplate and autonomic ganglion cells. It binds competitively to the sites without opening the postsynaptic channels, thereby interfering with the generation of the postsynaptic current.

*α-Bungarotoxin (BuTX)* is isolated from the venom of the krait, a member of the cobra family. This protein molecule binds highly specifically and irreversibly to the ACh receptor of the postsynaptic membrane. Thus, by labeling the BuTX radioactively, it has been possible to determine the number of ACh receptors present in a membrane and also to isolate and purify the receptor.

*Eserine* (physostigmine), an anticholinesterase, blocks the action of acetylcholinesterase, the enzyme that destroys ACh after its release into the synaptic cleft. Use of this alkaloid has enabled the recovery and measurement of released ACh by preventing its enzymatic destruction. Partial doses accentuate the postsynaptic potential at cholinergic synapses.

*Decamethonium, carbachol, succinylcholine* are ACh analogs that, like curare, bind to the ACh receptor, except that instead of blocking, they activate the receptor and open the postsynaptic channels.

*Hemicholinium* compounds block the uptake of choline by the nerve endings, thereby retarding the resynthesis of ACh from choline and acetate within the endings.

*Botulinum toxin* from the food poisoning bacterium *Clostridium botulinum* is one of the most potent poisons known. It acts to prevent, even in minute doses, the release of ACh from the nerve terminal.

*Muscarine* and other muscarinic agents such as *pilocarpine* activate ACh receptors that are most prevalent at the visceral tissues innervated by the cholinergic axons of the parasympathetic innervation.

*Nicotine* and other nicotinic agents such as *lobeline* activate the ACh receptors of the muscle endplate and of autonomic ganglion cells.

*Atropine* (belladonna) is an alkaloid that blocks the muscarinic action of ACh agonists.

*Reserpine,* an alkaloid, causes the gradual depletion of catecholamines and serotonin from synaptic terminals until these transmitters are no longer released in response to presynaptic activity.

*Naloxone,* like other opiate antagonists, blocks opiate receptors in the nervous system.

---

(d-tubocurarine, Box 6-3) are applied, a concentration will be found at which there is a sudden (all-or-none) failure of the muscle action potential and a concomitant failure of the muscle fiber to contract. The action potentials of the motor axon, however, will remain unaffected, and so will the ability of the muscle fiber to give an action potential and contract in response to an electrical stimulus applied directly to it. Since the presynaptic and postsynaptic acton potentials remain unaffected by the poison, we can conclude that curare somehow interferes with synaptic transmission at the neuromuscular synapse.

If the same experiment is repeated with the microelectrode inserted in the muscle fiber close to (within less than 0.1 mm of) the endplate region (Figure 6-19B), the following observations are noted:

*1.* As the concentration of curare is gradually raised, the action potential can be seen to arise not abruptly from the resting potential, but instead from a depolarization that is distinctly slower in time course and lower in amplitude than the action potential. This slow potential wave is an endplate potential, or postsynaptic potential.

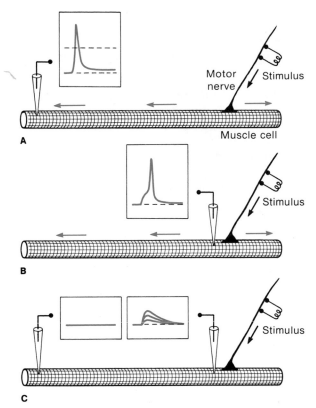

**A**

**B**

**C**

**6-19** Separation of endplate and action potentials. **(A)** All-or-none muscle action potential is recorded at a distance from the endplate region. **(B)** Recording close to the endplate shows the action potential arising out of the endplate potential. **(C)** Endplate potentials can be recorded without the superimposed action potential if they are reduced to an amplitude below the firing level by application of curare. In that case, only the undisturbed resting potential is recorded at a distance from the endplate region.

Motor nerve
Stimulus
Muscle cell

Stimulus

Stimulus

At a concentration of curare sufficient to reduce the size of the synaptic potential to just below the firing level, the action potential is eliminated, and the synaptic potential is revealed without the superimposed action potential (Figure 6-19C). If the recording electrode is now reinserted a number of times at positions along the muscle fiber at progressively greater distances from the motor endplate, it is found that the amplitude of the postsynaptic potential drops more or less exponentially (more accurately, as an error function) with distance from the endplate (Figure 6-20). In contrast to the action potential, which propagates in an unattenuated manner because of its regenerative nature, the synaptic potential spreads passively, decaying progressively with distance.

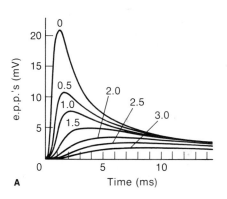

**A**

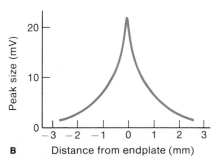

**B**

**6-20** **(A)** Recordings of endplate potentials at various distances from the endplate region. **(B)** The peak potential decreases more or less exponentially with increased distance from the endplate. [Fatt and Katz, 1951.]

*2.* The blocking agent, curare, reduces the amplitude of the postsynaptic potential.

*3.* The synaptic potential must reach a minimum level (the threshold potential or *firing level*) to trigger the muscle action potential—hence the abrupt failure of the action potential as the amplitude of the synaptic potential is reduced below the spike threshold by increased concentrations of curare.

### SYNAPTIC CURRENTS

In Chapter 5, it was shown that changes in membrane conductance to one or more species of ion shift the membrane potential toward a new level, in accord with Equation 5-7. The flow of ions that produces such a potential shift constitutes an electric current that transfers charge from one side of the membrane to the other through the synaptic channels opened by the action of the transmitter substance. Chemical transmission depends on the ability of the transmitter substance to produce a specific change in the ionic conductance of the postsynaptic membrane, for this determines the direction and extent of current flow across the membrane.

The ionic currents responsible for the endplate potential can be recorded independently of any membrane potential change by holding the postsynaptic potential constant by voltage-clamping (Box 5-4) the membrane in the region of the endplate (Figure 6-21A). The motor nerve is stimulated while the membrane potential is held constant by electronic feedback. The release of transmitter by the motor nerve ending is followed by a characteristic *endplate current,* or *synaptic current* (Figure 6-21B). This current represents the flow of ions down their electrochemical gradients through channels in the postsynaptic membrane opened by the action of the transmitter.

The ions carrying the motor endplate current were initially identified by changing the extracellular concentrations of ions and noting the effects on the synaptic current. In this way, it was found that the inward synaptic current at the endplate is carried by a large influx of $Na^+$, which is partially canceled by a simultaneous smaller efflux of $K^+$. It is now known that sodium and potassium ions both pass through the same postsynaptic ACh-activated channels. This behavior is, of course, in contrast to the high selectivity of the separate sodium and potassium channels activated by membrane depolarization (Table 5-1).

It can be seen in Figure 6-21 that the synaptic current is of considerably shorter duration than the synaptic potential. The duration of the synaptic channel opening is quite brief, as the transmitter is rapidly removed by enzymatic destruction and/or uptake of the transmitter by surrounding cells. Once the trans-

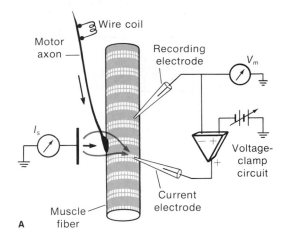

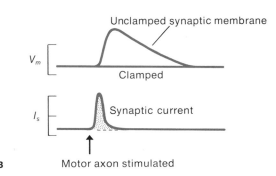

**6-21** Comparison of synaptic current with synaptic potential in frog motor endplates. (**A**) Voltage-clamping the muscle membrane holds the postsynaptic potential constant while the endplate current is recorded. (**B**) Stimulation of motor axon produces synaptic current (lower trace). When membrane is unclamped, an endplate potential is produced (upper trace), which decays much more slowly than the underlying endplate current. This is explained in the text.

mitter is removed, the synaptic current ceases to flow. The membrane potential then returns with some delay to the resting level, with a time course determined by the time constant of the membrane (p. 141).

### REVERSAL POTENTIAL

When a chemical transmitter or other stimulus activates membrane channels each of which is permeable to two ions, such as $Na^+$ and $K^+$, the resulting flow of

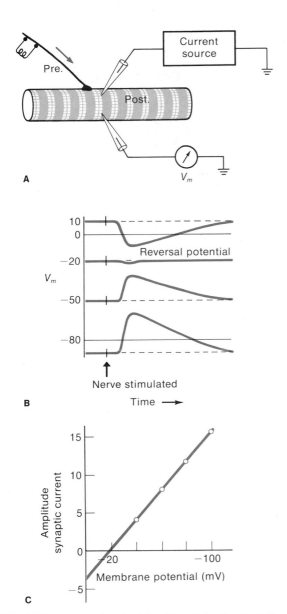

**A**

**B**

**C**

**6-22** (A) Experimental method for determining the reversal level of a synaptic potential. Polarizing current injected with one electrode is used to displace the membrane potential to differ-ent levels during synaptic stimulation via the motor nerve. The responses are recorded with the other electrode. (B) The synap-tic potential reverses sign at potentials more positive than the reversal potential. At the reversal potential, the open channels carry no net current and therefore produce no potential charge. (C) The amplitude of the synaptic potential is plotted against the membrane potential.

mixed ionic current shifts the membrane potential *toward* a new level somewhere between the equilibrium potentials of the two permeant ions. *The potential shift produced by the activation of those channels will not exceed this level regardless of the number of channels activated.* If the initial membrane potential for some reason were to be on the far side of that level, activation of the synaptic channels would produce a current in the opposite, or "reverse," direction. The membrane potential beyond which the synaptic current reverses its direction is termed the *reversal potential,* $E_{rev}$, for the current (and for the potential change) produced by activation of the channels. The reversal potential for the endplate potential is demonstrated in Figure 6-22. The presyn-aptic axon was stimulated so that transmitter release would activate the endplate channels (which carry a mixed $Na^+$–$K^+$ current), while the postsynaptic poten-tial was recorded intracellularly. A current was passed across the postsynaptic membrane so as to shift the prevailing potential to various selected levels. The amplitude of the synaptic potential became progres-sively smaller as the prevailing potential was progres-sively depolarized by the applied current. At the rever-sal potential, somewhere between $-10$ and $-20$ mV, the synaptic potential disappeared. Shifting the mem-brane potential to still more positive levels caused the synaptic potential to reappear, but with its polarity now reversed.

Let us consider, for a moment, the bi-ionic current that flows through the ACh-activated endplate chan-nel in frog muscle. If the membrane potential is elec-tronically "clamped" at $E_K$, the driving force acting on $K^+$ will be null, and all the current through the ACh-activated endplate channel will be carried by $Na^+$ moving into the cell, driven by the large electrochem-ical driving force, $V_m - E_{Na}$ (Figure 6-23, row *a*). Suppose, now, that $V_m$ were instead "clamped" at $E_{Na}$, and the endplate channels were activated by ACh. There would be no driving force acting on $Na^+$, but a large driving force acting on $K^+$, and so all the end-plate current would be carried by $K^+$ moving out through the activated channels (Figure 6-23, row *e*). It follows that somewhere between $E_{Na}$ and $E_K$ there must be a membrane potential at which the sodium and potassium partial currents through the endplate

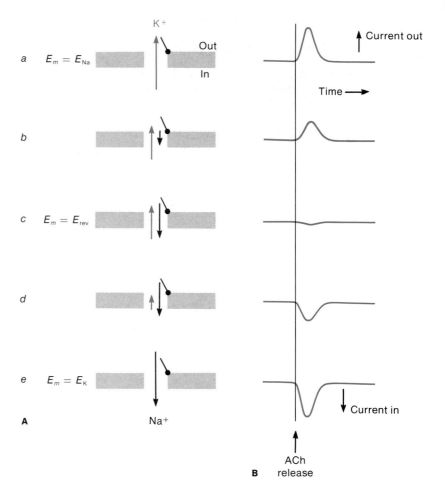

**6-23** Partial currents through endplate channel at different potentials. (**A**) Cartoons of sodium and potassium currents through ACh-activated channel at different membrane potentials beginning at $E_K$ (*e*). The channel gate is shown open, and relative magnitudes of sodium and potassium currents are represented by lengths of arrows. (**B**) Corresponding net current intensities shown against time. At a certain membrane potential, the partial currents will be equal and opposite (*c*) and the net current zero. This is the reversal potential, $E_{rev}$. See text and Box 6-4 for explanation.

channel will be equal and opposite, so that no net current will result when the channels open (Figure 6-23, row *c*). This level is defined as the reversal potential for the endplate current. In the case of the frog endplate channel, the conductances for the two permeant ions, $Na^+$ and $K^+$, are approximately equal, and therefore the reversal potential equals the algebraic mean of $E_{Na}$ and $E_K$. It will be apparent from Equation 5 in Box 6-4 that the reversal potential for a given synaptic current (or other active current) depends on two factors:

*1.* The relative conductances of the activated channel to the permeant ions. This depends on the ionic selectivity of the channel.

*2.* The equilibrium potentials for the permeant ions (based on their concentration gradients).

The concept of a reversal potential is useful for two major reasons. First, it is essential in distinguishing between inhibitory and excitatory synaptic actions (described in the next section). Second, the reversal potential of a current determined experimentally provides an indication of the ionic composition of the current. For example, if changing $[Na]_o$ shifts the reversal potential for a given synaptic current, this indicates that at least part of the current is carried by $Na^+$. The degree of the shift indicates what proportion of the current is carried by $Na^+$.

### POSTSYNAPTIC INHIBITION
A synaptic event that increases the probability of initiation of an action potential in the postsynaptic cell is termed excitatory; conversely, an event that reduces

## Box 6-4    *Calculation of Reversal Potential*

The value of the reversal potential depends on the relative conductances of the permeant ions as well as their equilibrium potentials. Assuming that only $Na^+$ and $K^+$ are permeant, the reversal level can be related to conductances by using Equation 5-3. The values $g_K$ and $g_{Na}$ represent the respective conductances to the two permeant ions.

$$I_K = g_K \ (V_m - E_K) \tag{1}$$

$$I_{Na} = g_{Na} \ (V_m - E_{Na}) \tag{2}$$

At the reversal potential, $I_K$ and $I_{Na}$ must be equal and opposite regardless of the relative conductances. Thus, when $E_m$ is at the reversal potential, $E_{rev}$,

$$-I_K = I_{Na} \tag{3}$$

Therefore, from Equations 1 and 2,

$$- g_K \ (V_m - E_K) = g_{Na} \ (V_m - E_{Na}) \tag{4}$$

It is obvious that if $g_K$ is greater than $g_{Na}$, then $E_m$ must be closer to $E_K$ than to $E_{Na}$, and vice versa.

Solving for $V_m \ (= E_{rev})$ gives

$$V_m = \left(\frac{g_K}{g_{Na} + g_K}\right)E_K + \left(\frac{g_{Na}}{g_K + g_{Na}}\right)E_{Na} \tag{5}$$

From this it is apparent that $E_{rev}$ will not simply be the algebraic sum of $E_{Na}$ and $E_K$, but will lie somewhere between the two, depending on the ratio $g_{Na}/g_K$. Thus, if $g_{Na}$ and $g_K$ become equal to each other (e.g., as during the activation of endplate channels by ACh in frog muscle), the membrane potential will shift toward a reversal potential that lies exactly halfway between $E_{Na}$ and $E_K$:

$$E_{rev} = \tfrac{1}{2}E_K + \tfrac{1}{2}E_{Na} = \tfrac{1}{2}(E_K + E_{Na})$$

For frog muscle, $E_K$ is about $-90$ mV, and $E_{Na}$ about $+60$. Hence during synaptic activation, $E_{rev} = \tfrac{1}{2}(-90 + 60) = -15$ mV.

To summarize, the reversal potentials of different membrane currents differ according to the species of ions that participate, the equilibrium potentials of those ions, and the relative conductances to each of the kinds of ions.

---

that probability is termed inhibitory. Thus, any postsynaptic current with a reversal potential more positive than the firing level is defined as excitatory (Figure 6-24A), and any postsynaptic current with a reversal potential on the negative side of the firing level is defined as inhibitory. Thus, excitatory current is carried through channels that are permeable to $Na^+$ or $Ca^{2+}$ and often to $K^+$ as well. The inhibitory synaptic current is carried by channels permeable to $K^+$ and $Cl^-$, since both of these ions typically have equilibrium potentials in the general vicinity of the resting potential, well below the firing level.

If the reversal potential for a transmitter action happens to be equal to the resting potential, no synaptic current and no potential change will result from the increase in postsynaptic conductance caused by the action of the inhibitory transmitter substance. Even though the conductance to $Cl^-$ or $K^+$ increases, the membrane potential in that special instance will remain constant at the rest level. Nevertheless, the trans-

mitter will have an inhibitory action, since it will tend to hold $V_m$ below spike threshold if there should be a simultaneous activation of excitatory current. If the reversal potential is more negative than the resting potential, the action of the transmitter will hyperpolarize the cell toward that level (Figure 6-24). If the reversal potential for the transmitter action is more positive than the resting potential but more negative than threshold, the transmitter will produce a depolarization (Figure 6-25B). However, if this transmitter acts simultaneously with an excitatory transmitter that, presented by itself, would cause depolarization to threshold, it will lead to a smaller depolarization than that produced by the excitatory transmitter alone (Figure 6-25).

There is, of course, nothing *inherently* "excitatory" or "inhibitory" about a transmitter substance. For example, acetylcholine is the excitatory transmitter at the motor endplate and at synapses of sympathetic ganglia, producing a predominant increase in sodium

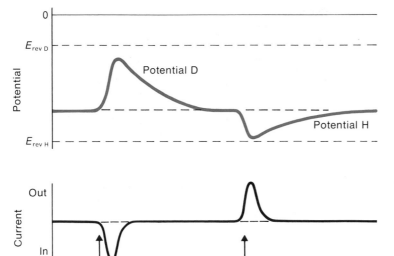

6-24 (A) Comparison of depolarizing and hyper-polarizing conductance changes. Transmitter D evokes a depolarizing postsynaptic potential because the conductance increase that it produces leads to a net inward current (carried primarily by $Na^+$) that adds positive charge to the cell interior. The increase in positive charge occurs because the current flow has a reversal potential less negative than the resting potential. Transmitter H produces a hyperpolarizing synaptic potential because the conductance increase that it produces is primarily to ions (e.g., $K^+$ or $Cl^-$) that will carry net current out of the cell (i.e., add negative charge to the cell interior). (B) The direction of current flow through the postsynaptic membrane is the opposite for currents D and H.

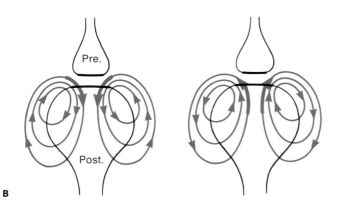

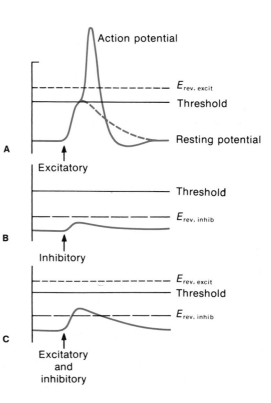

6-25 Inhibitory–excitatory interaction. (A) An action potential arises out of an excitatory postsynaptic potential if the latter exceeds threshold. (B) An inhibitory postsynaptic potential may depolarize, but its reversal potential is below threshold. (C) An inhibitory transmitter acting simultaneously with an excitatory transmitter can keep the membrane potential below threshold.

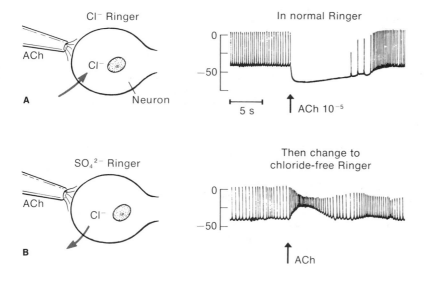

**6-26** The importance of ionic gradients in determining the direction of potential changes produced in response to a transmitter substance. (A) ACh applied to H cells in the snail brain produces a hyperpolarization because $Cl^-$ flows into the cell down its concentration gradient. (B) Exchange of the extracellular $Cl^-$ with $SO_4^{2-}$ causes the direction of potential change to reverse. Opening of the chloride channels by the transmitter now causes a depolarization because the new concentration gradient causes $Cl^-$ to flow out of the cell. The effect on impulse discharge is shown at the right. [Kerkut and Thomas, 1964.]

and potassium conductance in the postsynaptic membrane. In contrast, it is an inhibitory transmitter at parasympathetic endings in the heart and viscera, where it produces an increase in potassium and/or chloride conductance. The molecular properties of the ion channels in the postsynaptic membrane determine the ion specificity of the permeability increase that is set up across the postsynaptic membrane when these receptors react with the transmitter molecules. The relative ionic permeabilities and the electrochemical gradients of the permeant ions determine the reversal level of the synaptic potential, and thus those factors determine whether the postsynaptic effect is excitatory or inhibitory.

It follows that a transmitter that normally is inhibitory in its effect on a given cell can be made to have an excitatory action by an experimental redistribution of the appropriate ionic gradients across the postsynaptic membrane. This action has, in fact, been induced in neurons of the mammalian spinal cord and in the snail. In certain snail neurons, the effect of the natural transmitter (acetylcholine) is to increase the chloride conductance of the postsynaptic membrane. In one such group of cells (H cells, or hyperpolarizing cells), the intracellular $Cl^-$ concentration is relatively low, so that $E_{Cl}$ is more negative than the resting potential. The

neural transmitter acetylcholine produces a hyperpolarization when applied to the H cells by opening chloride channels, allowing $Cl^-$ to flow into the cell, and thus shifting the membrane potential toward $E_{Cl}$ (Figure 6-26A). When the extracellular chloride is replaced with sulfate, which cannot pass through the chloride channels, application of acetylcholine leads to an efflux of $Cl^-$, which now has an outwardly directed electrochemical gradient. This efflux of negative charge produces both a depolarization and an increase in the frequency of action potentials (Figure 6-26B). Thus, acetylcholine, the transmitter that is normally inhibitory in these cells, will produce excitation if the electrochemical gradient of the chloride ion is reversed.[†]

### PRESYNAPTIC INHIBITION

Neural inhibition can also result from the action of an inhibitory terminal ending on the presynaptic terminal of an excitatory axon (Figure 6-27), causing a

---

[†]It is interesting that other cells (D cells, or depolarizing cells) in the snail maintain a high intracellular $Cl^-$ concentration naturally by active chloride pumping. In these cells, ACh produces a similar increase in chloride permeability. Since the chloride gradient is normally outward in these D cells, they depolarize in response to the normal transmitter, ACh.

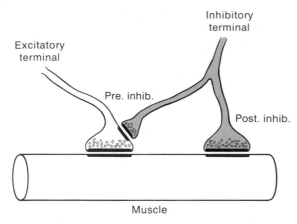

6-27  Anatomical basis for presynaptic inhibition in a crustacean muscle. An inhibitory neuron synapses with the presynaptic terminals of an excitatory axon. Release of inhibitory transmitter increases the potassium and chloride conductance of the excitatory terminal, reducing the size of the action potential in the terminal and hence the amount of transmitter released from the terminal. Postsynaptic inhibition occurs where the inhibitor terminates on the muscle fiber directly. [Lang and Atwood, 1973.]

reduction in the amount of transmitter released from the excitatory terminal. This effect is termed *presynaptic inhibition*. The inhibitory transmitter apparently increases the membrane permeability of the presynaptic terminals of the excitatory axon to $K^+$ or $Cl^-$. This increase in conductance reduces the size of the action potential invading the excitatory terminal, thereby reducing the amount of excitatory transmitter released. There is also growing evidence that in some cases there may be a partial inactivation of presynaptic calcium channels by the inhibitory transmitter. Reduced $Ca^{2+}$ entry causes reduced transmitter release. The end effect of presynaptic inhibition is that the postsynaptic cell receives less transmitter and thus produces a smaller postsynaptic potential.

Presynaptic inhibition has been found in the vertebrate central nervous system, where it plays an extensive role in synaptic integration, discussed later in this chapter. It has also been demonstrated in the crustacean neuromuscular apparatus (Figure 6-27), where branches of the axons that make inhibitory endings on the muscle fiber also send branches that end on the terminals of excitatory axons.

## Postsynaptic Receptors and Channels

The transmitter, as we have seen, interacts with the postsynaptic membrane to produce conductance changes to certain ions. This interaction must entail two major events:

*1.* The transmitter combines with a receptor molecule in the postsynaptic membrane.

*2.* Interaction of a receptor molecule with the transmitter molecule causes a previously closed ion channel associated with the receptor molecule to open transiently. In brief, the channel becomes "activated." The receptor may be a molecule separate from those that make up the channel, or it may be a component molecule of the channel.

When a channel opens in response to its activation by transmitter–receptor binding, a minute current (single-channel current) passes through the opened channel. A large number of such single-channel currents sum to produce the synaptic current in response to the tens or hundreds of thousands of transmitter molecules released from the presynaptic terminal in response to a presynaptic action potential.

### THE ACETYLCHOLINE RECEPTOR

The problem of isolating, identifying, and characterizing the molecule that binds the transmitter and transduces this binding into the opening of a channel is not a simple one, since the number of channel molecules relative to the total number of protein molecules in the membrane is rather small. In such studies, advantage has been taken of two fortuitous biological aids. One is the occurrence of an acetylcholine receptor in high densities on one surface of the *electroplax* (flattened cells originating embryonically from muscle tissue) found in the powerful electric organs of certain elasmobranch and teleost fishes. The high receptor content of these tissues provides useful yields during the chemical isolation of the receptor. The second aid is the use of *α-bungarotoxin (BuTX)* (Box 6-3). Since BuTX binds irreversibly to the ACh receptor, it can be isotopically labeled and then used as a tagging molecule for the identification and isolation of the receptor.

The ACh receptor consists of five protein subunits ranging in molecular weight from 40,000 to 65,000.

These form a tubular structure having a total molecular weight of about 250,000. This molecular weight agrees well with the size of channel structures seen with the electron microscope to penetrate the surface membrane of electroplax cells, with the funnel-shaped opening protruding outward from the cell surface.

The ACh receptor is located on the exterior side of the cell membrane, as deduced from the fact that ACh injected into a muscle cell near the endplate produces no electrical effect. Whether the molecule or molecules that constitute the ACh receptor are also a part of the ACh-activated channel is not yet known.

### ACh-ACTIVATED CHANNELS

As we have seen, the postsynaptic channels of the motor endplate of frog skeletal muscle become permeable to both $K^+$ and $Na^+$ when activated by ACh. This permeability permits the flow of an inward current with a reversal potential of about $-15$ mV. Normally, these channels and the associated ACh receptors are confined to the postsynaptic membrane in the endplate region. The density of ACh-activated channels in the postsynaptic membrane of the frog endplate has been determined to be about $10^4/\mu m^2$.

When a muscle is *denervated* by section of the motor nerve, ACh sensitivity gradually spreads from the endplate region over most or all of each muscle cell, indicating that receptors and channels, normally confined to the endplate region and suppressed elsewhere in the membrane, appear at extrajunctional sites. The normal suppression of *extrajunctional receptors* (and channels) is mediated in part by a poorly understood *trophic action* of the motoneuron that innervates the muscle cell, and in part by electrical and contractile activity of the innervated muscle cell. If the motor axon is allowed to reinnervate the muscle, the extrajunctional receptors disappear, and sensitivity to ACh again becomes confined to the junctional (i.e., endplate) region.

The sparse distribution of extrajunctional channels that develop in denervated muscle has been exploited in a study of the gating of the ACh-activated $Na^+$–$K^+$ channel in frog muscle by E. Neher and B. Sakmann (1976). The muscle membrane was voltage-clamped at a hyperpolarized potential to increase the driving force for inward current. A micropipette, with a smooth-polished tip having an opening of 10 $\mu m^2$, was filled

with Ringer solution containing a low concentration of ACh or one of its agonists (compound having a similar action). The surface of a denervated muscle fiber was explored with the tip of the pipette, which was connected to a highly sensitive current-recording amplifier (Figure 6-28A). Applied snugly to the surface of the muscle fiber, the pipette detected minute (less than $5 \times 10^{-12}$ A) pulses of inward current (Figure 6-28B) produced by the transient opening of the ACh-activated (or agonist-activated) channels. It is notable that those pulses are more or less rectangular in shape (i.e., they turn abruptly ON and OFF) and are all-or-none.

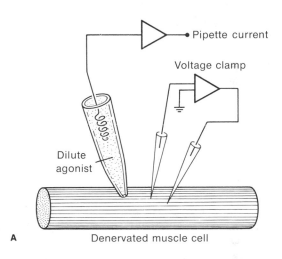

Pipette current

Voltage clamp

Dilute agonist

**A**     Denervated muscle cell

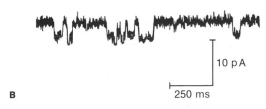

10 pA

250 ms

**B**

6-28 Currents recorded from a single ACh-sensitive channel in denervated frog muscle. (A) The muscle membrane is held at a high potential ($-120$ mV) by a voltage-clamp circuit while its surface is explored with a pipette containing a $2 \times 10^{-7}$ M solution of the ACh agonist suberylcholine in Ringer solution. (B) With the opening of the pipette placed snugly against the membrane, pulses of inward-going current are recorded by the pipette. These represent transient opening of channels as the agonist interacts with the receptor sites associated with individual channels. [Neher and Sakmann, 1976.]

This is, of course, to be expected if the channels exhibit two states, completely shut or completely open. Moreover, the different current pulses detected in this way are all of about the same size, indicating that the individual channels in the sample all have similar conductances. The current pulses are absent unless the pipette contains ACh or an agonist, and they occur with a frequency related to the concentration of the transmitter or agonist in the pipette. The conductance of a single open channel was calculated from the current to be about $2 \times 10^{-11}$ S (i.e., a resistance of $5 \times 10^{10}$ Ω).

Since these pioneering "patch-clamp" experiments of Neher and Sakmann, the ACh-activated channel as well as a variety of other chemically activated postsynaptic channels have been studied intensively with this method of recording single-channel currents. Statistical analysis of these minute currents indicate that the channels fluctuate between several "closed" states and at least one "open" state. It appears that binding of the appropriate transmitter molecule to the receptor site of the closed channel greatly increases the probability that it will enter into an open state, allowing ions to flow briefly through the channel before the channel again reverts to a closed state.

### DECREASED POSTSYNAPTIC CONDUCTANCES

In the type of synaptic activation just discussed, which is by far the most common, the transmitter substance causes an increase in the conductance of postsynaptic channels to one or more ions. Depending on the relation of the reversal potential for the synaptic current, the transmitter has an excitatory or inhibitory action. We should briefly note that there are a few known synapses at which the transmitter causes a *decrease* in postsynaptic conductance to one or more ions. Thus, the transmitter *serotonin (5-hydroxytryptamine, 5-HT)* has been shown in neurons of the sea hare *Aplysia californica* to produce either classic postsynaptic conductance increases or two kinds of conductance decrease—a decreased conductance to both $Na^+$ and $K^+$, producing hyperpolarization and thus inhibition, and a decreased conductance to $K^+$ alone, producing depolarization. Synaptically activated decrease in postsynaptic conductance has been found also in certain neurons of vertebrate autonomic ganglia. However, this type of mechanism seems to be far less common than the type utilizing conductance increases. Recent evidence sug-

gests that transmitter-evoked decreases in postsynaptic conductance are mediated by intracellular messenger molecules such as cAMP (p. 463).

## Transmitter Substances

One of the more difficult problems in neurophysiology is the unequivocal chemical identification of transmitters acting at various synapses. A limited number of transmitters have been identified with certainty (Table 6-2). For a putative (i.e., suspected and likely, but not completely established) transmitter substance to be considered an established transmitter in a given tissue it must meet the following criteria:

*1.* When applied directly to the postsynaptic membrane, it must elicit in the postsynaptic cell precisely the same physiological effects as does presynaptic stimulation.

*2.* It must be shown to be released during presynaptic activity of the presynaptic neuron.

*3.* Its action must be blocked by the same agents that block natural transmission.

Acetylcholine (Figure 6-29) is the most familiar of the established transmitter substances. It is released by the terminals of vertebrate motor axons, by preganglionic terminals of the vertebrate autonomic system (Figure 8-13), by postganglionic terminals of the parasympathetic division of the autonomic system, and by the presynaptic terminals of certain neurons of the vertebrate central nervous system. It also appears to act as the transmitter in a number of invertebrate neurons, including some cells of the molluscan central nervous system and sensory neurons of arthropods.

The ACh released from the terminals of a *cholinergic* neuron is hydrolyzed to choline and acetate by the enzyme acetylcholinesterase, which is present at the surface of the postsynaptic membrane (Figure 6-30). The enzymatic destruction of ACh terminates its effect on the postsynaptic membrane. The choline is actively

Acetylcholine

6-29  Structure of acetylcholine.

**TABLE 6-2** Some known and possible neurotransmitters and neuromodulators.

| Substance | Site of action | Type of action | Status† |
|---|---|---|---|
| Acetylcholine | Skeletal muscle; neuromuscular junction | Excitatory | Est. |
| | Autonomic nervous system: | | |
| |   preganglionic sympathetic | Excitatory | Est. |
| |   pre- and postganglionic parasympathetic | Excitatory or inhibitory | Est. |
| | Renshaw cell in CNS | Excitatory | Est. |
| | Other CNS | — | Poss. |
| Norepinephrine | Most postganglionic sympathetic | Excitatory or inhibitory | Est. |
| | CNS | — | Poss. |
| Glutamic acid | CNS | Excitatory | Poss. |
| | Crustacea, CNS and PNS | Excitatory | Est. |
| Aspartic acid | Vertebrate retina | — | Poss. |
| γ-Aminobutyric acid (GABA) | CNS | Inhibitory | Poss. |
| | Crustacea, CNS and PNS | Inhibitory | Est. |
| Serotonin (5-hydroxytryptamine) | Vertebrate and invertebrate CNS | — | Est. |
| Dopamine (3,4-dihydroxy-phenylethylamine) | CNS | — | Est. |
| Octopamine | Insect CNS | — | Est. |
| Substance P | CNS | Inhibitory modulation | Est. |
| Various peptides | Vertebrate and invertebrate CNS; gut | Various | Est. |

†Est. = established transmitter. Poss. = possible transmitter.

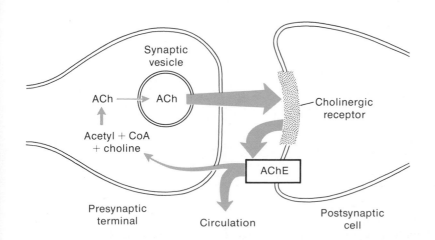

6-30 Transmitter chemistry at a cholinergic synapse. Acetylcholine released from the terminal is hydrolyzed by acetylcholinesterase (AChE) on the surface of the postsynaptic membrane. Choline is taken up by the presynaptic terminal and reacetylated to ACh. [Mountcastle and Baldessarini, 1968.]

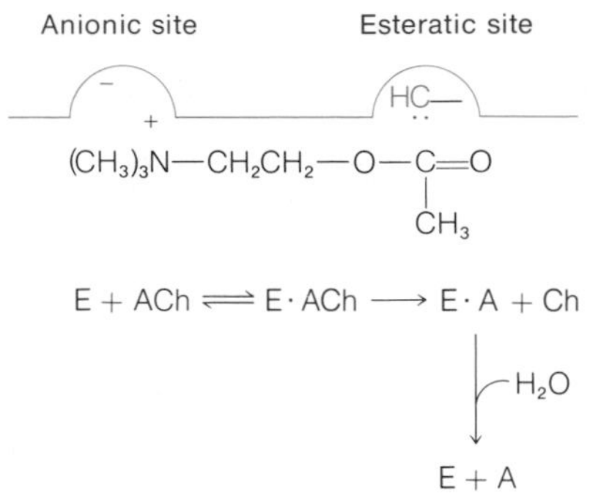

**6-31**  Steric relations between the active site on the acetylcholinesterase molecule and its substrate ACh. Hydrolysis of ACh is shown at bottom.

reabsorbed by the presynaptic endings, in which it is recycled by combination with acetyl coenzyme A to form a new molecule of acetylcholine.

The acetylcholinesterase molecule has two distinct sites (Figure 6-31): (1) an anionic site that binds the quaternary nitrogen of ACh, and (2) an esteratic site that can donate electrons to the acetate portion of the ACh molecule, cleaving it from the choline moiety. The actions at these sites allow the hydrolysis of ACh and terminate the postsynaptic action of the transmitter. The esterase is inactivated by certain nerve gases and insecticides. With the esterase blocked, ACh lingers and builds up in the synaptic regions. This situation either prevents repolarization of the postsynaptic membrane or, in many types of synapses, causes inactivation of the ACh receptors so that the postsynaptic channels remain closed despite presynaptic release of ACh. In either case, the normal functioning of the nervous and neuromuscular systems is disrupted, and death ensues very rapidly owing to paralysis of the respiratory muscles. The anticholinesterase *eserine* (physostigmine) is used in the laboratory when it is necessary to retard the hydrolysis of acetylcholine.

### BIOGENIC AMINES

The monoamines *norepinephrine, dopamine,* and *serotonin* (Figure 6-32) are closely related compounds. They are concentrated in some nerve terminals, and have also

been isolated in the synaptic vesicles of some nervous tissues. These substances can be detected in individual neurons, because they fluoresce in ultraviolet light after fixation with formaldehyde. They are found in some invertebrate neurons and in the central and autonomic nervous systems of vertebrates and are suspected of being neurotransmitters (Table 6-2). The evidence for dopamine as a vertebrate CNS transmitter is especially convincing.

Norepinephrine (also known as noradrenalin) is known to be the excitatory transmitter at postganglionic cells of the sympathetic system (Figure 8-13), such as the *chromaffin cells* of the adrenal medulla and the sympathetic neurons innervating the vertebrate heart. The chromaffin cells are derived embryologically from postganglionic neurons and secrete epinephrine (adrenalin) as well as norepinephrine. These two *catecholamines* have similar pharmacological actions. The chemistry of norepinephrine at an *adrenergic*† synapse is outlined briefly in Figure 6-33. The pathway for

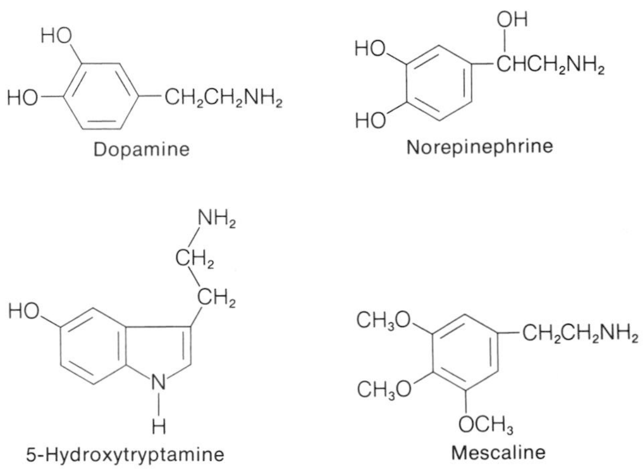

**6-32**  The monamines dopamine, norepinephrine, epinephrine (not shown), and 5-HT (serotonin) are a closely related group of transmitter substances found in vertebrate and invertebrate nervous systems. Mescaline, a psychoactive drug extracted from the peyote cactus, is believed to induce hallucinations by interfering with its analog norepinephrine at synapses in the central nervous system.

---

†The term *adrenergic* pertains to neurons or synapses that liberate epinephrine, norepinephrine, and other catecholamines.

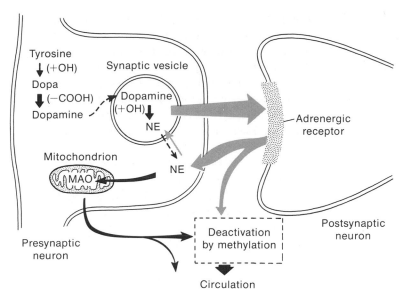

**6-33** Transmitter chemistry at an adrenergic synapse. Norepinephrine (NE) is synthesized from the amino acid phenylalanine via tyrosine and stored in synaptic vesicles. After release, some NE is taken back up into the presynaptic terminal and some is deactivated by methylation and carried away in the blood. Cytoplasmic NE is either taken up into a synaptic vesicle or degraded by monamine oxidase (MAO). [Mountcastle and Baldessarrini, 1968.]

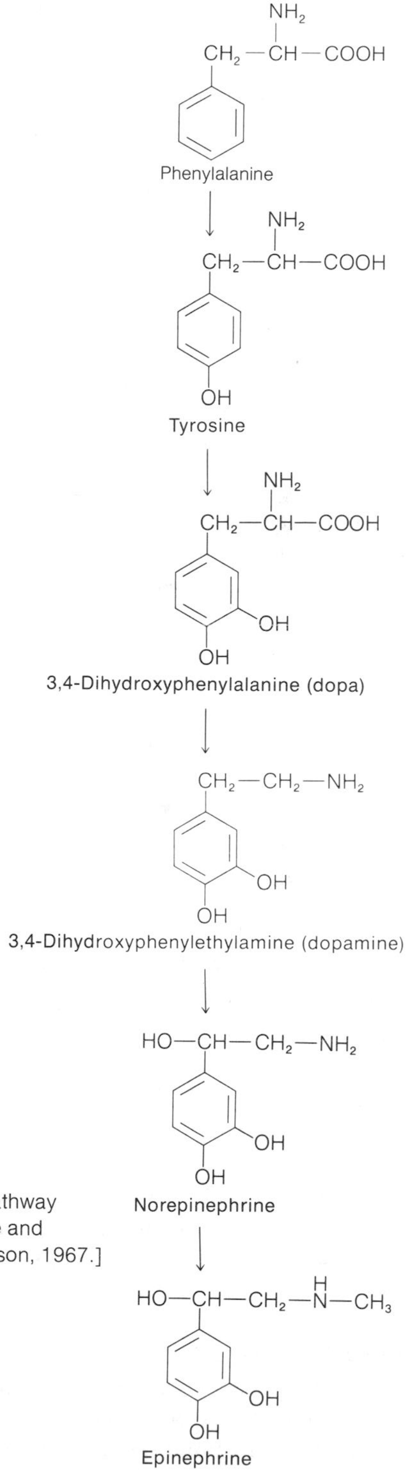

synthesis of norepinephrine from the amino acid phenylalanine (Figure 6-34) shows the close relationship between dopamine and norepinephrine. It is interesting to note the structural similarities between these amines and certain extremely potent psychoactive drugs, such as mescaline (Figure 6-32) and lysergic acid diethylamide (LSD).

### AMINO ACIDS

There is growing evidence that certain amino acids, such as *glutamic acid* (Figure 6-35), are released at excitatory synapses of the vertebrate CNS and at excitatory neuromuscular synapses (motor nerve terminals) in insects and crustaceans (Table 6-2). A putative amino acid transmitter found in the vertebrate retina is *aspartic acid. Gamma aminobutyric acid (GABA)* (Figure 6-35) is the transmitter at the inhibitory motor synapse in crustacean muscle, and there is evidence that it may be one of the inhibitory transmitters in the vertebrate CNS.

**6-34** Biosynthetic pathway leading to epinephrine and norepinephrine. [Eiduson, 1967.]

Phenylalanine

Tyrosine

3,4-Dihydroxyphenylalanine (dopa)

3,4-Dihydroxyphenylethylamine (dopamine)

Norepinephrine

Epinephrine

$$HOCO \quad COOH$$

Glutamic acid

$$NH_2-CH_2-CH_2-CH_2-COOH$$
γ-Aminobutyric acid

**6-35** Structures of glutamate and γ-aminobutyric acid (GABA). Glutamate, the neutral form of glutamic acid, has been identified as an excitatory transmitter in some central nervous systems and in the excitatory innervation of insect muscle. GABA is the inhibitory transmitter in arthropod muscle and at some synapses in central nervous systems.

## NEUROPEPTIDES

In addition to the relatively small, "classic" transmitter molecules already described, there is a growing list (over 20 by now) of polypeptides produced and released in the nervous system and known to influence neural activity. Interestingly, a number of these *neuropeptides* are also produced in various nonneural tissues. Thus, some are secreted by intestinal endocrine cells, by autonomic neurons, and by various sensory neurons and in various parts of the CNS. In fact, some neuropeptides were initially discovered in nonneural tissue. The gastrointestinal hormones glucagon, gastrin, and cholecystokinin are prime examples.

It is not yet clear how many of the 20-odd known neuropeptides are utilized as true neurotransmitters—mediators that act on closely apposed postsynaptic cells. Some, like the releasing hormones of the hypothalamus, are liberated from certain neurons as neurosecretory substances; that is, they are liberated from nerve terminals into the circulation so as to be carried to their target cells in the blood. There is evidence that the same peptide can be liberated as a transmitter from some neurons, as a neurosecretory substance from other neurons, and as a hormone from nonneuronal tissue. This release is not really all that novel, since the biogenic amine norepinephrine has long been known as both a hormone liberated by the adrenal medulla and as a transmitter liberated from certain nerve terminals. A rather unexpected recent finding has been

that a neuropeptide will frequently occur in the same nerve terminals along with a "classic" transmitter such as ACh, serotonin, or norepinephrine.

The first neuropeptide was discovered in 1931 by Ulf S. von Euler and J. P. Gaddum while they were assaying for ACh in extracts of rabbit brain and intestine. The extracts caused contraction of the isolated intestine much like ACh does, but the contractions were not blocked by ACh antagonists. This result led to the discovery that the contraction was produced in response to a polypeptide, which the researchers named *substance P*. Since then, substance P and a growing list of other neuropeptides have been discovered in various parts of the central, peripheral, and autonomic nervous systems of vertebrates and also in many invertebrate nervous systems. A recently developed means of locating individual peptides utilizes immunological labeling with fluorescent antibodies that can be detected in histological sections under the microscope.

Neuropeptides have been implicated in a number of neural and behavioral phenomena. To give only a few examples: The posterior pituitary neurosecretory hormones—antidiuretic hormone and oxytocin—seem to be involved in memory mechanisms; substance P, in primary sensory neurons, is active in neural mechanisms of pain, apparently as a synaptic transmitter; elevated brain levels of cholecystokinin cause satiety (and thus loss of appetite); angiotensin II stimulates drinking; and LH-releasing hormone stimulates sexual behavior.

## ENDOGENOUS OPIOIDS

A great deal of interest has focused on two groups of neuropeptides known as *endorphins* and *enkephalins*, which have natural analgesic (pain-lessening) and other opiatelike, pleasure- and euphoria-producing properties. For example, their level in the brain rises in response to eating and to certain other activities that are generally agreed to be pleasurable. Because of such properties, and because these neuropeptides bind to the same receptors in the nervous system to which *opiates* such as opium and its derivatives bind, they are referred to as *endogenous opioids*. Until recently, we had no idea why certain alkaloids, such as opium, morphine, and heroin, had such powerful effects on the nervous system. We now know that the surface membranes of certain neurons contain *opioid receptors*. This class of receptors normally binds the peptide enkephalins and endorphins produced within the nervous system. Only secondarily and coincidentally do they bind the narcotic opiates, which are alkaloid molecules derived from plants and are chemically unrelated to the normally occurring polypeptide opioids. The pleasurable effects they elicit have led humans to use opiate narcotics such as opium, morphine, and heroin to stimulate these receptors. The subjective sensations associated with such abnormally strong stimulation are reputedly very pleasurable—comparable to the orgasm but more protracted. Repeated doses of these opiates provoke compensatory changes in neuronal metabolism, such that removal of the opiate leaves the nervous system in a rebounded state in which the subject experiences extreme discomfort until the opiate is readministered. Such metabolically induced dependence is termed *addiction*.

A useful tool in studies on opioid peptide receptors is the drug *naloxone* (Box 6-3), which acts as a competitive blocker of the opioid receptor. Because naloxone interferes with the ability of either opiates or the opioid peptides to act on their target cells, it is very useful in determining whether a response is mediated by opioid receptors. It was found, for example, that naloxone largely blocks the analgesic effect produced by a *placebo* (inert substance given to patients with the suggestion that it will relieve pain). Apparently the act of believing that a medication or other treatment will relieve pain can induce the release of opioid peptides, which may provide the physiological basis for the *placebo effect*. Similarly, naloxone renders acupuncture ineffective in

relieving pain. This behavior has led to the suggestion that acupuncture stimulation causes the release of natural opioid peptides within the central nervous system.

The pain-lessening properties of the endogenous opioids, the enkephalins and the endorphins, may depend on the ability of these neuropeptides to block the release of transmitter from certain nerve endings. Consistent with this possibility is the presence of these agents in the dorsal horn of the vertebrate spinal cord, which lies in the pathway of sensory input to the spinal cord. The sensation of pain may be diminished by the release of neuropeptides that interfere with synaptic transmission along afferent pathways that signal noxious stimuli. It has been postulated that analgesic effects of acupuncture may involve the stimulated release of endogenous opioids.

## Release of Transmitter Substances

### QUANTAL NATURE OF RELEASE

P. Fatt and B. Katz (1952) reported miniature (less than 1 mV) spontaneous depolarizations recorded from the vicinity of the postsynaptic membrane of the motor endplate of the frog muscle (Figure 6-36). These spontaneous signals became progressively smaller as the intracellular recording electrode was inserted at greater distances from the endplate. Since these potentials have a shape, time course, and drug-sensitivity similar to those of the endplate potential, they were termed *miniature endplate potentials (m.e.p.p.'s)*. As we shall see below, the m.e.p.p.'s have played an important role in the analysis of transmitter release mechanisms.

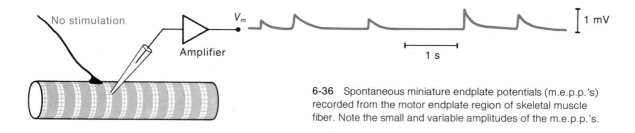

6-36 Spontaneous miniature endplate potentials (m.e.p.p.'s) recorded from the motor endplate region of skeletal muscle fiber. Note the small and variable amplitudes of the m.e.p.p.'s.

A progressive increase in $Mg^{2+}$ and/or decrease in $Ca^{2+}$ causes the normal endplate potential to become smaller until, at the appropriate concentrations of these cations, it attains an amplitude similar to that of a single spontaneously occurring m.e.p.p. By measuring postsynaptic responses to presynaptic motor nerve impulses under high $Mg^{2+}$ and low $Ca^{2+}$, Fatt and Katz found:

*1.* Some motor impulses produce no response at all (i.e., failures).

*2.* Some produce m.e.p.p.'s having approximately the same amplitudes as single spontaneous m.e.p.p.'s.

*3.* Some produce potentials whose amplitudes are integral multiples (e.g., $2\times, 3\times, 4\times$, etc.) of the mean amplitude of single spontaneous m.e.p.p.'s (Figure 6-37).

Subsequent studies revealed that an m.e.p.p. is produced by a unit release of the transmitter ACh. Further-

more, it was shown that the normal endplate potential results from the release of a large number of such units of ACh in unison in response to the arrival of the presynaptic action potential in the nerve terminal. At the frog motor endplate, the endplate potential consists of about 100–300 such units.

Cooling the nerve–muscle preparation slows the release of transmitter from the nerve terminal and therefore results in an asynchronous release of these units. The jagged, stepwise rise of the endplate potential under these conditions, each peak, presumably due to a separate unit of transmitter release, is further evidence that the endplate potential consists of many small units that are normally released in unison and that sum to produce a single, large depolarization. Since transmitter release seems to consist of discrete units, packets or "quanta" of transmitter, this process has been termed *quantal release*. The source of a single packet of transmitter is believed to be a single pre-

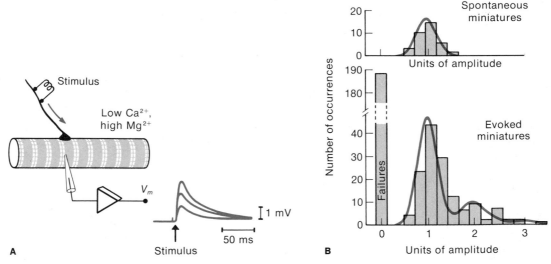

**6-37**  Quantal release at the motor endplate. (A) A motor nerve is stimulated in a solution containing $Ca^{2+}$ and high $Mg^{2+}$, which reduces the amount of transmitter released in response to stimulation. Evoked endplate potentials have variable miniature amplitudes. (B) Histograms record the number of m.e.p.p.'s of various amplitudes (vertical bars). The upper histogram plots the numbers of spontaneously occurring m.e.p.p.'s of different amplitudes recorded over a period of time. The lower histogram does the same for m.e.p.p.'s that were evoked by motor–nerve stimulation. Note that there were large numbers of failures. The largest number of evoked m.e.p.p.'s had a size distribution similar to those of single spontaneous m.e.p.p.'s. The continuous curves give the theoretical distributions obtained on the assumption that the evoked m.e.p.p.'s are made up of units corresponding to those that occur spontaneously. [Histograms from Del Castillo and Katz, 1954.]

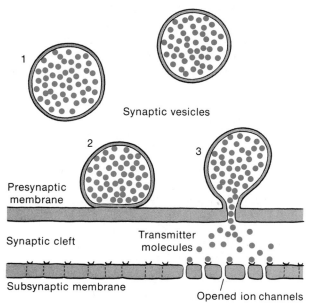

1

Synaptic vesicles

Presynaptic
membrane

2

3

Synaptic cleft

Transmitter
molecules

Subsynaptic membrane

Opened ion channels

**6-38**  Exocytosis of transmitter-containing synaptic vesicles. The vesicles release their contents into the synaptic cleft when their membranes fuse with the plasma membrane and an opening forms. The transmitter diffuses and interacts with receptor molecules on the postsynaptic membrane. [Eccles, 1965.]

synaptic vesicle releasing its contents by exocytosis (Figure 6-38). Each unit release of transmitter at the frog neuromuscular junction consists of 10,000–40,000 molecules of acetylcholine, and these activate about 2000 postsynaptic channels. Each channel is activated by the binding of two ACh molecules by its ACh receptor.

The quantal release of transmitter at the frog endplate has been studied by statistical analysis at great length. The picture that has emerged is that a fraction of the population of vesicles inside a nerve terminal are available for immediate release in response to a presynaptic action potential. Under a given set of physiological conditions (e.g., $Ca^{2+}$ and $Mg^{2+}$ concentrations, temperature), there is a certain probability that any one member of that population of "available" vesicles will be released. Lowering extracellular $Ca^{2+}$ or raising extracellular $Mg^{2+}$ reduces the entry of $Ca^{2+}$ into the terminal (essential for transmitter release) and thus lowers the probability of release of presynaptic vesicles. If the probability is sufficiently low, we have the condi-

tion illustrated in Figure 6-37, in which a presynaptic action potential leads to "failures" (i.e., no vesicles released) or the release of only one, two, or a few vesicles to produce endplate potentials with amplitudes corresponding to multiples of 1, 2, 3, . . . m.e.p.p.'s. With the normal *quantal content* of 100–300 units of release lowered to 0, 1, 2, 3, and so on, by reduced extracellular $Ca^{2+}$, it is possible to determine the number of vesicles released in response to each stimulus of a large series and to do a statistical analysis of those numbers. This type of analysis has shown that the probability of vesicle release follows a *Poisson distribution* (Figure 6-37), which is diagnostic of a random process.

## DEPOLARIZATION–RELEASE COUPLING

According to the statistical theory of quantal transmitter release, the probability that a given vesicle will undergo exocytosis and release its contents at any given instant is quite low if the presynaptic membrane potential is at the resting level. The occurrence of a spontaneous release is random and independent of other vesicles undergoing release, but when the presynaptic membrane is depolarized, and $Ca^{2+}$ entry occurs, the probability of quantal release dramatically increases. The relation between membrane potential and probability of release can be seen in the increased frequency of m.e.p.p.'s that accompanies a steady depolarization (Figure 6-39). The increased probability of vesicle release results in the simultaneous release of many vesicles in response to a presynaptic action potential.

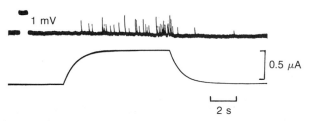

1 mV

0.5 μA

2 s

**6-39**  Electrically induced increase in the rate of m.e.p.p. occurrence. Electrotonic depolarization by application of current (below) to the presynaptic terminal increases the probability of transmitter release, as evidenced by the increase in frequency of m.e.p.p.'s seen in the upper trace. [Katz and Miledi, 1967b.]

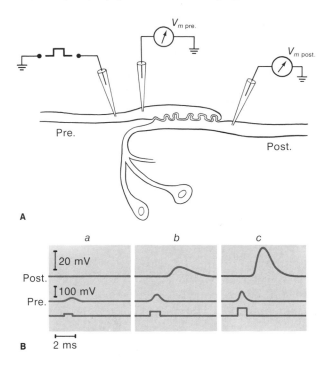

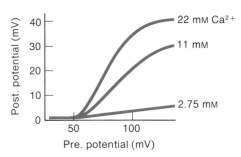

**6-40** Relation between presynaptic depolarization and transmitter release at the squid giant synapse. (**A**) The presynaptic membrane is depolarized with current from an intracellular electrode. (**B**) Pre- and postsynaptic potentials are recorded with two additional microelectrodes. The current used to produce the presynaptic depolarization was increased in going from *a* to *b* to *c*. (**C**) Increased presynaptic membrane depolarization for a given extracellular concentration of $Ca^{2+}$ results in increased transmitter release and hence increased amplitude of the postsynaptic potential. With decreasing $[Ca^{2+}]_o$, the postsynaptic amplitude drops. [Katz and Miledi, 1966 and 1970.]

The relation between presynaptic membrane potential and transmitter release was examined by Katz and R. Miledi in an unusually large synapse that occurs on the squid giant axon (Figure 6-40). Because of its large size, microelectrodes can be inserted into both the presynaptic terminal and the postsynaptic cell close to the synaptic region—a feat that is impossible in most other synapses because of the small diameter of most presynaptic terminals. Sodium activation was eliminated by the application of TTX, and potassium activation was blocked by the injection of TEA. This made it possible for Katz and Miledi to depolarize the presynaptic membrane in a graded fashion by passing a current into the terminal through an intracellular electrode (Figure 6-40). The postsynaptic potential, recorded with a third electrode, provided a highly sensitive microchemical assay of the transmitter released by the presynaptic membrane. The results (Figure 6-40B, C) show:

*1.* Depolarization of the presynaptic membrane results in transmitter release (detected as postsynaptic depolarization), even though the normal mechanism of the action potential was eliminated.

*2.* The amplitude of the postsynaptic potential (related to transmitter release) increases with increased depolarization of the presynaptic membrane.

Evidence has accumulated that $Ca^{2+}$ plays an important role in the process of transmitter release. One example is the reduction of transmitter release in response to a given presynaptic depolarization when the extracellular $Ca^{2+}$ concentration is reduced (Figure 6-40). When the presynaptic terminal is depolarized all the way to $E_{Ca}$, the strong intracellular positive potential prevents $Ca^{2+}$ from entering through the calcium channels present in the presynaptic terminal. Under those conditions, there is no transmitter release until the membrane potential is allowed to drop enough after the high depolarization to provide sufficient electrochemical gradient for $Ca^{2+}$ to enter the terminal. A relation between $Ca^{2+}$ entry and transmitter release has been demonstrated in the nerve terminal of the squid axon by light emission from injected aequorin, a calcium-sensitive protein extracted from the jellyfish *Aequorea*. Membrane potential was recorded pre- and postsynaptically, while the light emission, which resulted from the reaction of the aequorin with $Ca^{2+}$

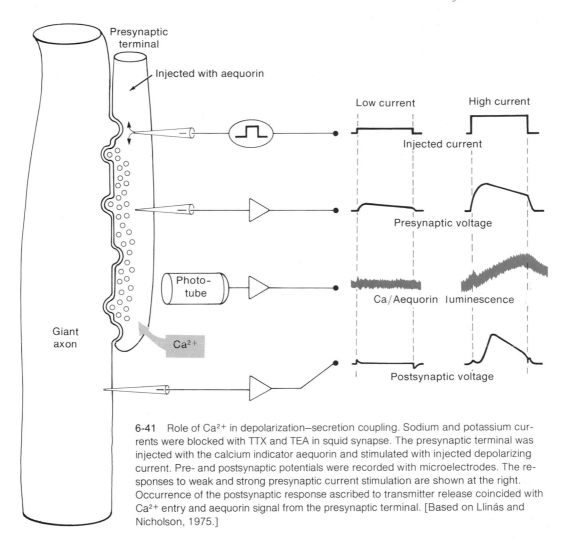

**6-41**  Role of $Ca^{2+}$ in depolarization–secretion coupling. Sodium and potassium currents were blocked with TTX and TEA in squid synapse. The presynaptic terminal was injected with the calcium indicator aequorin and stimulated with injected depolarizing current. Pre- and postsynaptic potentials were recorded with microelectrodes. The responses to weak and strong presynaptic current stimulation are shown at the right. Occurrence of the postsynaptic response ascribed to transmitter release coincided with $Ca^{2+}$ entry and aequorin signal from the presynaptic terminal. [Based on Llinás and Nicholson, 1975.]

entering the presynaptic terminal, was monitored by a phototube (Figure 6-41). When depolarizing current was injected, a postsynaptic potential was elicited only when there was also a signal from the aequorin to indicate $Ca^{2+}$ entry into the presynaptic terminal.

It has also been demonstrated that calcium is required for transmitter release during impulse propagation into the presynaptic terminal at the muscle endplate. For example, conditions that interfere with the entry of $Ca^{2+}$ into the axon (e.g., low extracellular $Ca^{2+}$; competing ions, such as $Mg^{2+}$ and $La^{3+}$) also prevent release of the transmitter in response to a presynaptic action potential. Finally, it has been shown

that microinjection of $Ca^{2+}$ into the squid presynaptic terminal elicits release of transmitter.

Thus, the evidence indicates that $Ca^{2+}$, which enters upon arrival of the action potential, is necessary for triggering the release of transmitter. The intracellular role of $Ca^{2+}$ in the release of transmitter is not fully understood. Perhaps $Ca^{2+}$ is required for the fusion of synaptic vesicles with the inner surface of the cell membrane of the nerve terminal.

Other lines of evidence indicate that transmitter release requires more than one calcium ion per reactive site and occurs in proportion to the $n$th power (with $n$ ranging from 2 to 4 depending on the tissue in question) of the number of calcium ions that enter the terminal. That is, where $n$ is equal to 4, as it is at the frog neuromuscular junction, release will increase 16-fold in response to a doubling of $[Ca]_i$ ($2^4 = 16$), and it will increase 81-fold in response to a tripling of $[Ca]_i$ ($3^4 = 81$), and so on. In light of this, it was proposed that two to four calcium ions must bind to a molecule X within the terminal to form an active complex, $Ca_nX$, that is essential for transmitter release.

## Synaptic Integration

The processing of neural signals is given the general term *integration*, defined as "combining into an integral whole." At the level of a single neuron, this means reacting to incoming information from various synaptic inputs so as to produce or fail to produce propagated impulses. Thus, each neuron integrates the various excitatory and inhibitory synaptic signals that impinge on it. This process depends largely on the passive electrical properties of those portions of the neuron that lie between the synapses and the spike-initiating zone. In addition, the sensitivity of the sodium and potassium channels to depolarization determines the threshold and rate of firing in response to depolarization.

Much of our knowledge about neuronal integration has been obtained from the large $\alpha$-*motoneurons* (Figure 6-42), which have their cell bodies in the ventral horn of the vertebrate spinal cord and which innervate groups of skeletal muscle fibers. Each motoneuron receives thousands of inhibitory and excitatory synaptic terminals on its dendrites and cell body. The frequency of firing (impulses per second) of a motoneuron in response to its synaptic input determines the strength of contraction of the muscle fibers it innervates. All the integrative activity of the motoneuron is centered on the production of action potentials (i.e., excitation) or the suppression of excitation (i.e., inhibition). Because only action potentials can carry information over distances exceeding a few millimeters, only those signals that lead to the production of action potentials can effect the contraction of muscle cells innervated by the motoneuron. Excitatory input that fails to reach threshold, either by itself or by summation with other inputs, is effectively discarded.

### SUMMATION

The action potentials generated in a motoneuron arise in the *initial segment* of the axon, which lies beyond the axon hillock. This region is more sensitive to depolarization and has a lower firing level than do the soma and dendrites (Figure 6-43). It therefore is the site of impulse initiation. Excitatory synaptic current must flow outward across the membrane of the impulse-

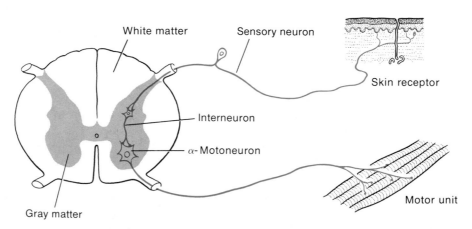

**6-42** An $\alpha$-motoneuron located in the ventral horn of gray matter. It is shown as part of a bisynaptic reflex arc in which a noxious stimulus applied to the skin results in excitation of the motoneuron via the interneuron. Activation of the motoneuron leads to contraction of its motor unit, the group of muscle fibers that it innervates.

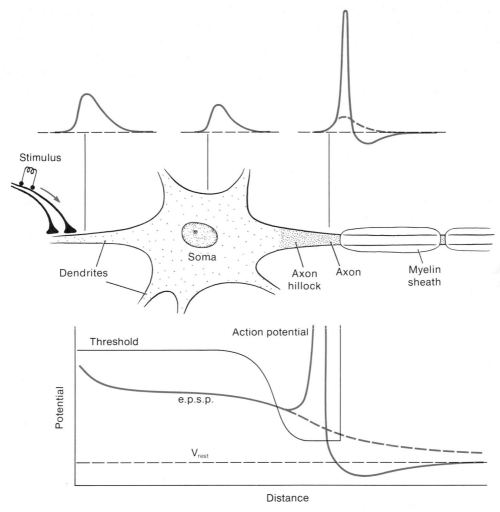

**6-43** Spatial decay of synaptic potential and impulse initiation. Excitatory postsynaptic potential (e.p.s.p.) originating in dendrites gets smaller with distance of electrotonic spread. Density of sodium channels (colored dots) determines threshold (colored traces) for generation of the action potential. Thus, even though synaptic potential (top of figure) gets smaller as it spreads toward the axon, the action potential is generated at the spike-initiating zone in the initial segment of the axon, where density of the sodium channel is high and threshold is low. The dashed line shows the course the synaptic potential would take if the action potential could be blocked.

initiating zone in order to depolarize the membrane toward the firing level.

Synaptic currents spread electrotonically from the synapses on dendrites and soma in accord with the cable properties of the neuron. As shown in Figure 6-43, synaptic potentials become smaller as they spread away from their sites of origin and toward the impulse-initiating zone characterized by a high density of sodium channels. Because of the cable loss that takes place when a nonregenerating electrical signal spreads down a cell process, a synaptic current set up at the end of a long slender dendrite will experience especially high decrement; hence it will exert a far smaller effect on the impulse-initiating zone than will synaptic current originating on the cell body close to the axon hillock. It is apparent, then, that the cable properties of the cell play a major role in the summation of synaptic currents originating in different parts of the neuron. Inhibitory synapses generally occur with highest density near the axon hillock, where they are most effective in preventing excitatory synaptic current from depolarizing the impulse-initiating zone to the firing level.

The integrative properties of motoneurons can be studied by first exposing several segments of the spinal cord of an anesthetized cat, so that a microelectrode can be lowered into the ventral horn of the gray matter and inserted into a single motoneuron soma. Stimulating electrodes are placed on bundles of afferent axons dissected from the dorsal root so that excitatory or inhibitory axons† can be made to fire as required for the experiment.

In the absence of presynaptic stimulation, the recording electrode inserted into the soma of the motoneuron records randomly occurring synaptic activity. Some of this is due to activity in presynaptic neurons not under the control of the investigator. But some of it results from spontaneous release of transmitter from inactive endings. This effect produces spontaneous *miniature postsynaptic potentials* with amplitudes of about 1 mV analogous to the m.e.p.p.'s recorded at the muscle endplate (Figure 6-36). Stimulation of a single presynaptic excitatory fiber shows that a single presynaptic ending releases only one to several units of transmitter in response to a presynaptic action potential. In this respect, the excitatory synapses ending on a motoneuron differ from the endplate potential in vertebrate skeletal muscle, in which the motoneuron terminals release about 100–300 units of transmitter and produce an endplate potential of 60 mV or more. The transmitter released from a single synaptic ending on the motoneuron depolarizes the neuron by only about 1 mV—far less than the amount required to shift the membrane potential to the firing level. The small effect

of a single ending is significant in the integrative behavior of the neuron, in that it requires that *many* presynaptic terminals be activated more or less simultaneously to reach the firing level and initiate a postsynaptic action potential. Thus, the "decision" to fire is a response to a population of active presynaptic inputs. This rather democratic sort of behavior prevents activation of the motoneuron by trivial input or spontaneous activity.

As the strength of the stimulus current applied to the presynaptic axons in the dorsal root is intensified, an increasing number of excitatory axons are *recruited*. As these fire in unison, the total amount of transmitter released rises accordingly to produce a larger postsynaptic potential (Figure 6-44). The increase in depolarization occurs because of the additive effect of synaptic currents originating at more than one synapse, and it is termed *spatial summation*. Depolarization to the firing level is suppressed when inhibitory transmitter is released simultaneously with excitatory transmitter (Figure 6-45). The opening of these "inhibitory" channels short-circuits the current carried in by $Na^+$ through the "excitatory" channels. That is, as charge is carried into the cell by $Na^+$ (depolarizing current), some of that charge is promptly carried out of the cell by $K^+$ or $Cl^-$ (repolarizing current). This minimizes the depolarization of the impulse-initiating zone and reduces the probability of impulse generation.

When a second postsynaptic potential is elicited within a short time after the first, it will ride piggyback upon the first (Figure 6-46). This effect is called *temporal summation*. The shorter the interval between two successive synaptic potentials, the higher the second rides on the first. Further summation can be achieved by additional stimuli, the third synaptic potential riding on the second, and so on. Under natural conditions, spatial and temporal summations generally occur together. For example, if different excitatory synapses on one motoneuron are active at slightly different times, their effects will sum both spatially and temporally.

Spatial and temporal summations both depend on the passive electrical properties of the neurons. Spatial summation occurs because synaptic currents originating at the same time, at different synapses, spread electrotonically in an additive manner to the impulse-initiating zone (Figure 6-43). Temporal summation,

---

†Whether these axons are excitatory or inhibitory depends on the transmitter substances they release and on the consequent conductance changes in the postsynaptic membrane.

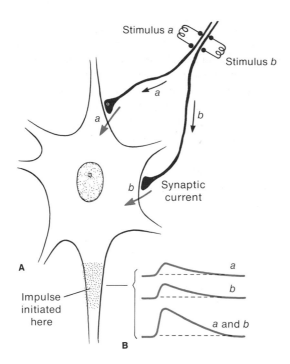

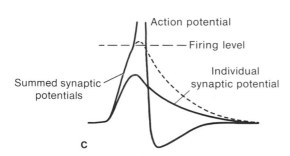

6-44  Spatial summation in a motoneuron. (**A**) Two excitatory synaptic potentials, *a* and *b,* arising at spatially separated synapses give rise to separate synaptic currents. (**B**) Synaptic potentials due to stimulation of either pathway *a* or *b,* or both, as they might be recorded at the impulse-initiating zone. (**C**) The summation of currents from many synapses is required to exceed the firing level.

by the inward synaptic current is removed slowly, and the potential returns to the resting level gradually after the synaptic current has ceased. The duration of the synaptic potential is therefore much longer than the duration of the synaptic currents, and if a second synaptic current flows before the resting potential is reattained, it will cause a second depolarization to add to the falling phase of the first, even though the two synaptic currents do not overlap in time. Thus, because of the membrane's resistance and charge-storing capacity, its time constant makes possible the interaction of brief synaptic currents separated in time. The longer the time constant of the membrane, the longer the synaptic potential takes to decay, and the more effective the summation of asynchronous synaptic inputs. The time constant is about 10 ms in spinal motoneurons and ranges from 1 ms to 100 ms in other neurons.

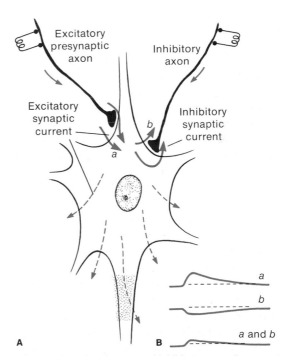

6-45  Summation of excitatory and inhibitory synaptic currents. (**A**) Stimulation of separate presynaptic pathways gives rise to excitatory, *a,* and inhibitory, *b,* synaptic currents. (**B**) Potentials produced by currents as they might be recorded from the impulse-initiating zone.

on the other hand, does not require the summation of synaptic *currents* (Figure 6-46). The first synaptic current partially discharges the resting potential of the cell membrane. Because of the time constant of the membrane, the positive charge carried into the neuron

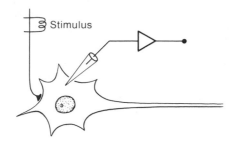

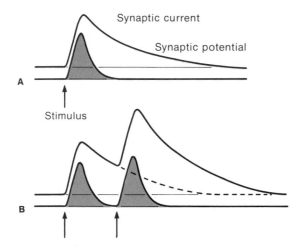

**6-46** Temporal summation. (**A**) A single stimulus evokes synaptic current (shaded signal) and more slowly decaying synaptic potential. (**B**) Summation of synaptic currents is not required for summation of synaptic potentials, because the time course of the potential change exceeds the time course of the synaptic current. Arrows indicate time of stimulus to presynaptic axon.

and causes the contraction of the muscle fibers it innervates.

The membrane in the impulse-initiating zone fails to accommodate completely to maintained depolarization. Therefore, a maintained, intense synaptic input to the motoneuron causes it to fire a sustained train of action potentials. The frequency of impulses in such a train rises with increased depolarization. The postsynaptic spike frequency thus reflects the intensity of excitatory synaptic input (Figure 6-47).

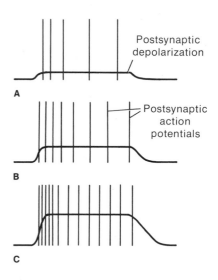

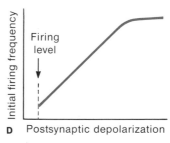

**6-47**  At the spike-initiating zone of a motoneuron, the initial frequency of impulses generated is approximately proportional to the amplitude of the synaptic depolarization. (**A–C**) Depolarizing current is injected into the soma with one electrode, and potentials are recorded with another. Increasing depolarization causes an increased rate of firing. (**D**) Steady-state firing frequency plotted against depolarization.

Under normal conditions, the motoneuron is almost never electrically silent, but always exhibits *synaptic noise* (irregular fluctuations in membrane potential) owing to a "background" level of activity in the presynaptic neurons. The result is a constantly changing, irregular membrane potential. Sufficient excitatory inputs will intermittently sum to trigger an action potential in the motoneuron, which in turn leads to an action potential and twitch in each one of the muscle fibers in its motor unit (p. 291). The result is a constant background of low-level tonus (maintained tension) in a muscle as first one and then another motoneuron fires

To summarize, the generation of impulses by a neuron requires that the low-threshold initial segment be depolarized to the firing level. Beyond that level, the frequency of firing rises as the depolarization increases up to a maximum firing frequency. The amount of depolarization depends on the timing and place of origin (relative to the spike-initiating zone) of excitatory and inhibitory synaptic currents.

## Synaptic Plasticity

One of the most interesting properties of nervous systems is *plasticity*—that is, the modification of neural function as a result of experience. Neural plasticity is evidenced in our lives by learning and by the development of reflexes, motor skills, and habits. This is an area about which very little concrete knowledge of physiological mechanisms exists, but it is widely suspected that it involves a change in *synaptic efficacy* (i.e., the effectiveness of a presynaptic impulse in producing a postsynaptic potential change). A change in synaptic efficacy is not the only way in which neural function might be modified, but at present it is the only one for which there is any experimental support. The evidence indicates that neural activity can modify the amount of transmitter released from presynaptic terminals. There are two major categories of presynaptic mechanisms. In one, it is the activity of the terminal itself that produces a use-dependent functional change, usually of short duration. In the other, the changes in presynaptic function, usually of longer duration, are induced by the action of a modulator substance released from another nerve terminal nearby. We will begin with use-dependent changes in presynaptic efficacy.

### FACILITATION

A change in synaptic efficacy can be illustrated by recording from the endplate region in a skeletal muscle fiber while applying two stimuli to the motor axon, the second stimulus following the first after varying intervals (Figure 6-48). If the second synaptic potential begins before the first has subsided, they will of course sum, but the second will reach an amplitude greater than can be accounted for by summation alone. If the second synaptic potential begins soon after the first

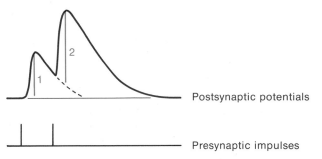

**6-48** Synaptic facilitation. The second synaptic potential sums with the falling phase of the first, but the amplitude is greater than can be accounted for by summation alone.

has completely subsided, it reaches a higher amplitude than the first one. This effect, which is termed synaptic *facilitation*, subsides in 100–200 ms at the motor endplate.

The best evidence indicates that facilitation is due to an increase in the intracellular free $Ca^{2+}$ concentration of the presynaptic terminal, and that this increase is related in some way to $Ca^{2+}$ entry into the terminal during the prior action potential, as first noted by Katz and Miledi (1968). They used a carefully positioned micropipette to supply pulses of $Ca^{2+}$ to the vicinity of the motor terminals of the endplate of a frog muscle immersed in a $Ca^{2+}$-free Ringer solution (Figure 6-49A). The size of the endplate potential depended on the presence of $Ca^{2+}$ (Figure 6-49B, C). Katz and Miledi also found that facilitation of the postsynaptic potential evoked by the second stimulus is greatest when a pulse of $Ca^{2+}$ coincides in time with the arrival of the first action potential at the motor endplate (Figure 6-49E). The first pulse of $Ca^{2+}$ does not significantly enhance facilitation if it is given after the first action potential arrives at the motor nerve terminals (Figure 6-49D). Thus, to be effective in facilitation, calcium must be available to enter the presynaptic terminal during the invasion of the terminal by an action potential. It is presumed that some of this $Ca^{2+}$ persists within the ending so as to sum with calcium entering in response to a second presynaptic action potential arriving after a short interval, and that the additional $Ca^{2+}$ results in the release of a larger amount of transmitter, and hence a larger postsynaptic response.

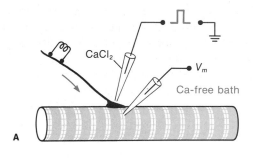

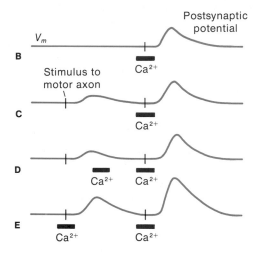

**6-49** Dependence of facilitation of a second synaptic potential on availability of extracellular calcium during the first presynaptic action potential. (**A**) Method of applying calcium pulses to the endplate region. (**B–E**) Black bars show timing of calcium pulses. Facilitation occurs only if calcium is available when the first presynaptic action potential reaches the endplate. [Katz and Miledi, 1968.]

There are several possible mechanisms by which calcium entry during the first presynaptic impulse might lead to a facilitation of the release of the transmitter in response to the second presynaptic action potential. For example:

*1.* The second presynaptic action potential may be accompanied by a greater influx of $Ca^{2+}$ than the first. This might result from (a) more extensive invasion of the fine presynaptic terminals by the second impulse, producing a larger depolarization, or (b) increased duration of the second action potential due to an effect on membrane conductances of the internal $Ca^{2+}$ from the first impulse. The longer period of depolarization might allow a more protracted influx of $Ca^{2+}$.

*2.* A higher concentration of *free* $Ca^{2+}$ could result during the second presynaptic impulse if the $Ca^{2+}$ entering during the first impulse loads a larger proportion of the calcium binding sites within the terminal. The increment of $Ca^{2+}$ entering during the second impulse would then produce a large increment of unbound ionized calcium available for activating the transmitter-release mechanism.

*3.* The $Ca^{2+}$ entering during the second depolarization should add to residual $Ca^{2+}$ remaining in the terminal from the first action potential. This addition should lead to a disproportionately increased release due to the nonlinear relation between transmitter release and $[Ca^{2+}]_i$. That is, a little extra $Ca^{2+}$ should produce a large increase in transmitter release. This possibility has received strong support from recent studies utilizing an intracellular calcium-sensitive indicator dye.

None of these possibilities is incompatible with present data, although the last two seem the most likely. It will require more study to determine which of these mechanisms actually are responsible for synaptic facilitation.

### POSTTETANIC POTENTIATION

When a presynaptic neuron is stimulated *tetanically* (i.e., at a high frequency), synaptic transmission is initially *depressed* following the tetanic stimulus. Test pulses applied at various times are then *potentiated* (i.e., increased in amplitude) for up to several minutes. How does past activity affect the behavior of the synapse? A tentative answer was obtained in the experiment on frog muscle shown in Figure 6-50. Curare (Box 6-3) was first applied to keep the endplate potentials from reaching the firing threshold. Endplate potentials were evoked initially by stimulation at a low control rate (e.g., 1 stimulus every 30 s). The rate was then increased to 50/s for a period of 20 s, after which a series of test stimuli were administered at the rate of 1 every 30 s. The results are plotted for a normal extracellular calcium concentration (1.8 mM) and for a low $Ca^{2+}$ concentration ($\frac{1}{8}$ times normal). In the

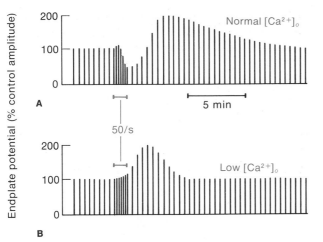

6-50 Posttetanic depression and posttetanic potentiation of frog endplate potentials. (A) When $[Ca^{2+}]_o$ is normal, a period of high-frequency motor stimulation is followed first by a depression of e.p.p.'s and delayed potentiation. (B) When $[Ca^{2+}]_o$ is low, only an early phase of potentiation is seen. See text for explanation. [Rosenthal, 1969.]

normal calcium concentration, a *posttetanic depression* of the endplate potential occurred initially after the tetanic stimulation. After the depression, there was a rapid increase in amplitude, termed *posttetanic potentiation (PTP)*, up to nearly twice the control amplitude. The amplitude of the endplate potential then returned to the control level in about 10 min. In Ringer solutions containing low calcium (Figure 6-50B), there is no depression, and the PTP subsides more rapidly.

These results have been interpreted as follows: During high-frequency stimulation in the normal extracellular calcium concentration (Figure 6-50A), the available transmitter units are released faster than they can be replaced, and the amount of transmitter available for release is depressed immediately after a period of high-frequency discharge. The units are then replenished during the posttetanic period, and the depression subsides. During tetanic stimulation, $Ca^{2+}$ entering the terminals as a consequence of excitation accumulates, loads up calcium-binding sites, and lingers on within the terminals as it is gradually pumped out of the terminals by active transport. It is believed that PTP

and its slow decay reflect an increase and subsequent reduction in intracellular calcium concentration. The increased $[Ca^{2+}]_i$ increases the probability of quantal release, and as the number of quanta available for release is replenished following the period of tetanic stimulation, the number released in response to a presynaptic impulse exceeds the control number (i.e., potentiation). In Ringer solutions containing low calcium (Figure 6-50B), the limited availability of $Ca^{2+}$ lowers the number of synaptic vesicles released; hence there is less depletion of available transmitter units. As a result, posttetanic depression is absent, and PTP is just as pronounced but decays more rapidly, perhaps because the presynaptic terminal is able to pump the accumulated calcium out more rapidly in the face of a lower extracellular calcium concentration.

### CHEMICAL MODULATION OF TRANSMITTER RELEASE

The release of transmitter from nerve terminals elicited by presynaptic action potentials can be influenced at some synapses by application of certain natural neurohumoral (neurotransmitter- or neurohormonelike) agents. Some of these putative *modulatory agents* are serotonin in mollusks and vertebrates, octopamine in insects, and norepinephrine and γ-aminobutyric acid (GABA) in vertebrates. All these agents are also proven or putative neurotransmitters (Table 6-2). In addition, several opiates (e.g., etorphine) and the endogenous opioid enkephalin have been shown to act as modulatory agents in vertebrate neurons. Such agents, released into the circulation or liberated by nerve endings near a synapse, are believed to modulate the release of transmitter from the preterminal of the synapse. When liberated near a presynaptic ending, such agents are said to act *heterosynaptically,* since transmission through the synapse is altered by an additional, third neuron releasing a modulatory agent that alters transmitter release by the presynaptic terminal. One class of heterosynaptic action noted previously is presynaptic inhibition; another is termed *heterosynaptic facilitation.*

In those instances in which the mechanism of modulation has been examined, the agent was found to alter the amount of ionized calcium that enters the nerve terminal in response to a presynaptic action potential. It is characteristic of synaptic modulators that they do not directly open (or close) channels to exert their modulatory action. Instead, synaptic modulatory agents typically influence (i.e. modulate) the response of certain membrane channels to another stimulus. Thus, they will increase or decrease the current carried through a population of channels that become activated by depolarization. Classic synaptic transmitter substances, on the other hand, act as chemical stimulants that directly open membrane channels.

With the exception of serotonin in molluscan ganglia, all the agents listed in the first paragraph of this section are reported to exhibit a depressing action on transmitter release and/or $Ca^{2+}$ entry. This behavior is illustrated by the action of enkephalin, serotonin, GABA, and norepinephrine on the neurons of the dorsal root ganglion grown in tissue culture. The action of these modulatory agents in the soma of the ganglion cell is to shorten the calcium action potential, and they do so without changing the resting potential or the resting membrane conductance. Voltage-clamp experiments have shown that the rate of $Ca^{2+}$ entry (i.e., the calcium current) and hence total $Ca^{2+}$ influx during depolarization of the soma are reduced by norepinephrine and GABA. This reduction explains the shortened calcium plateau of the action potential, and it indicates that the modulatory agent causes the channels to be less sensitive to depolarization of the membrane, so that the probability of channel opening is lessened. If the response of the soma membrane to the agent is indicative of what is occurring at the presynaptic terminal, reduced calcium-channel activation, due to the action of modulatory agents, will result in a smaller $Ca^{2+}$ entry during the depolarization and thus a smaller release of transmitter substance.

A decrease in transmitter release in response to those agents has, in fact, been demonstrated in dorsal root neurons. Mudge and co-workers (1979) found that the endogenous opioid enkephalin inhibits the release of substance P from the sensory neurons in the dorsal root ganglion. This effect was prevented by the drug naloxone, which also selectively blocks the familiar actions of opioids, such as their analgesic effects.

## Summary

The common unit of structure in the nervous system is the neuron, which communicates with other neurons or effector cells by electrical or chemical transmission across synapses. The neuron typically consists of:

*1.* A soma, which contains the nucleus.

*2.* A variable number of dendrites, which, along with the soma (in some neurons), are supplied with synaptic inputs from the terminals of other neurons.

*3.* An axon, which carries impulses to its presynaptic terminals.

Arrival of impulses in the terminals leads to a release of neurotransmitter substance.

The nervous system functions with two types of signals: (1) graded, nonpropagating potential changes and (2) all-or-none, propagated action potentials. These generally alternate in a pathway, the graded potentials occurring at sensory and postsynaptic membranes, and the action potentials being confined largely to axons and associated structures. Intensity is coded by amplitude in the graded potentials and by frequency in the firing of action potentials.

Propagation along an axon depends on two phenomena: (1) the spread of current longitudinally along the axon according to its cable properties and (2) the continual amplification of the signal as a result of excitation of new membrane channels by the local circuit current as it flows out of the axon ahead of the impulse. Information travels between different parts of the nervous system via action potentials, which are propagated in this manner without decrement along the axons. The velocity of propagation depends on axon diameter and on the presence (in some vertebrate axons) of myelin sheaths between bare sections termed nodes of Ranvier. In myelinated axons, saltatory conduction occurs from node to node, skipping the intimately sheathed portions in between and thereby increasing the velocity.

Synapses are of two major types—electrical and chemical. The principle of electrical synaptic transmission is essentially that of impulse propagation; the current flows from one cell into another through low-resistance gap junctions and depolarizes the second cell. In chemical transmission, the presynaptic ending liberates a transmitter substance that interacts with receptor molecules on the postsynaptic membrane. This interaction opens ion-specific channels in the postsynaptic membrane, resulting in an ionic current that produces a synaptic potential across the postsynaptic membrane. The action of the transmitter can be blocked by inhibitor substances that bind with the receptor molecule, as in the case of curare, a drug that blocks neuromuscular transmission in vertebrates.

A synapse can be either excitatory or inhibitory. In excitatory synapses, the action of the transmitter is to change the membrane conductance so as to shift the potential to or beyond the firing potential for the action potential. The action of the transmitter substance at an inhibitory synapse is to change the conductance so as to counteract depolarization to the firing level. The properties of excitation and inhibition are not inherent in the transmitter substances, but depend on the ion selectivity of the postsynaptic channels activated by the transmitters and by the reversal potentials of the currents carried through those channels.

Both excitatory and inhibitory substances are stored in and released from vesicles found in the nerve terminal. The arrival of an action potential depolarizes the presynaptic membrane, allowing $Ca^{2+}$ to enter the terminal. The calcium, in a manner as yet not understood, increases the probability that the vesicles will coalesce with the terminal membrane and release their contents by exocytosis into the synaptic cleft.

Temporal and spatial summations of synaptic potentials occur according to the passive electrical properties at the postsynaptic cell, and the integration of synaptic inputs is determined by the total effect of the synaptic currents in depolarizing the membrane of the spike-initiating zone.

Some changes in synaptic efficacy occur owing to prior activity. In some cases, it has been shown that these changes result from altered amounts of transmitter released in response to a presynaptic impulse. Modulatory agents released from endocrine tissues or from a third neuron alter the effectiveness of synaptic transmission by altering the amount of $Ca^{2+}$ entry during the presynaptic impulse and thereby modulate the amount of transmitter released.

## Exercises

*1.* Contrast the characteristics of the two basic types of electrical signals used in the nervous system.

*2.* Compare the encoding of intensity in graded signals, such as receptor potentials and synaptic potentials, with the encoding of intensity with all-or-none impulses.

*3.* Action potentials consist of electric currents. Explain why they do not propagate along axons with the velocity of an electric current in a wire.

*4.* If the action potential is not a simple linear flow of current, why can one draw an analogy between the behavior of a myelinated axon and that of a submarine cable?

*5.* How does an action potential propagate over long distances without loss of amplitude?

*6.* Explain why, if all else is equal, an axon of large diameter will conduct impulses at higher velocity than will an axon of small diameter.

*7.* Explain why interrupted myelination increases the velocity of impulse conduction.

*8.* How can one determine experimentally if a junction between two nerve cells is an electrical or a chemical synapse?

*9.* Compare the properties of electrical and chemical synapses. Outline the presynaptic events leading to the release of transmitter substance.

*10.* What determines if a postsynaptic conductance change will be excitatory or inhibitory?

*11.* What factors determine whether a transmitter depolarizes or hyperpolarizes the postsynaptic membrane?

*12.* Explain how a synapse can produce a depolarizing postsynaptic potential and still be inhibitory.

*13.* What prevents the acetylcholine released from the presynaptic terminal from persisting and interfering with subsequent synaptic transmission?

**14.** Of what possible significance are the structural similarities between the monoamine neurotransmitters and such psychotropic agents as LSD and mescaline in explaining the effects of these agents?

**15.** What is the evidence that an endplate potential is composed of smaller units termed miniature endplate potentials?

**16.** How might previous synaptic activity affect the amplitude of postsynaptic potentials?

**17.** Explain why the amplitude of a postsynaptic potential is limited in spite of simultaneous release of many transmitter quanta.

**18.** Why is it that impulses originate in the axon hillock of a vertebrate motoneuron rather than at the postsynaptic membranes of the dendrites and soma?

**19.** Since postsynaptic potentials decay as they spread, where on a neuron would you expect a synapse to exert the strongest influence?

**20.** What is meant by spatial summation and by temporal summation? What are the membrane properties responsible for each?

**21.** Discuss the role of $Ca^{2+}$ in each of the following: depolarization–release coupling, facilitation, posttetanic potentiation, heterosynaptic modulation of transmitter release.

## Suggested Reading

Aidley, D. J. 1978. *The Physiology of Excitable Cells.* New York: Cambridge University Press.

Bullock, T. H., R. Orkand, and A. D. Grinnell. 1977. *Introduction to Nervous Systems.* San Francisco: W. H. Freeman and Company.

Cooper, J. R., F. E. Bloom, and R. Roth. 1978. *The Biochemical Basis of Neuropharmacology.* 3rd ed. New York: Oxford University Press.

Eccles, J. C. 1965. The synapse. *Scientific American* 212(1)56–66. Also available as Offprint 1001.

Gainer, H., ed. 1977. *Peptides in Neurobiology.* New York: Plenum.

Hall, Z., J. Hildebrand, and E. Kravitz. 1974. *The Chemistry of Synaptic Transmission.* Newton, Mass.: Chiron.

Hodgkin, A. L. 1964. *The Conduction of the Nervous Impulse.* Springfield, Ill.: Thomas.

Kandel, E. R. 1976. *Cellular Basis of Behavior.* San Francisco: W. H. Freeman and Company.

Kandel, E. R., and J. H. Schwartz. 1982. Molecular biology of learning: Modulation of transmitter release. *Science* 218:433–443.

Katz, B. 1966. *Nerve, Muscle and Synapse.* New York: McGraw-Hill.

Katz, B. 1969. *The Release of Neural Transmitter Substances.* Springfield, Ill.: Thomas.

Kravitz, E. A., and J. E. Treherne, eds. 1980. *Neurotransmission, Neurotransmitters and Neuromodulators.* Cambridge: Cambridge University Press.

Kuffler, S. W., and J. Nicholls. 1976. *From Neuron to Brain.* Sunderland, Mass.: Sinauer.

Llinás, R. 1982. Calcium in synaptic transmission. *Scientific American* 247(4):56–65.

Roberts, A., and B. M. Bush, eds. 1980. *Neurones Without Impulses.* Cambridge: Cambridge University Press.

Schmitt, F. O., and F. G. Worden, eds. 1979. *The Neurosciences: Fourth Study Program.* Cambridge, Mass.: M.I.T. Press.

*The Synapse.* 1976. *Cold Spring Harbor Symp. Quant. Biol.* Vol. 40.

## References Cited

Bunge, M. B., R. P. Bunge, and H. Ris. 1961. Ultrastructural study of remyelination in an experimental lesion in adult cat spinal cord. *J. Biophys. Biochem. Cytol.* 10:67–94.

Del Castillo, J., and B. Katz. 1954. Quantal components of the endplate potential. *J. Physiol.* 124:560–573.

Eccles, J. C. 1965. The synapse. *Scientific American* 212(1): 56–66. Also available as Offprint 1001.

Eiduson, S. 1967. The biochemistry of behavior. *Science J.* 3:113–117.

Erlanger, J., and H. S. Gasser. 1937. *Electrical Signs of Nervous Activity.* Philadelphia: University of Pennsylvania Press.

Fatt, P., and B. Katz. 1952. An analysis of the endplate potential recorded with an intracellular electrode. *J. Physiol.* 115:320–370.

Fatt, P., and B. Katz. 1952. Spontaneous subthreshold activity at motor nerve endings. *J. Physiol.* 117:109–128.

Furshpan, E. J., and D. D. Potter. 1959. Transmission at the giant motor synapses of the crayfish. *J. Physiol.* 145:289–325.

Hodgkin, A. L. 1937. Evidence for electrical transmission in nerve. *J. Physiol.* 90:183–232.

Hodgkin, A. L., and W. A. H. Rushton. 1946. The electrical constants of a crustacean nerve fibre. *Proc. Roy. Soc.* (London) Ser. B. 133:444–479.

Katz, B., and R. Miledi. 1966. Input–output relation of a single synapse. *Nature* 212:1242–1245.

Katz, B., and R. Miledi. 1967. Tetrodotoxin and neuromuscular transmission. *Proc. Roy. Soc.* (London) Ser. B. 167:8–22.

Katz, B., and R. Miledi. 1968. The role of calcium in neuromuscular facilitation. *J. Physiol.* 195:481–492.

Katz, B., and R. Miledi. 1970. Further study of the role of calcium in synaptic transmission. *J. Physiol.* 207:789–801.

Kerkut, G. A., and R. C. Thomas. 1964. The effect of anion injection and changes in the external potassium and chloride concentration on the reversal potentials of the IPSP and acetylcholine. *Comp. Physiol. Biochem.* 11:199–213.

Kuffler, S. W. 1942. Further study on transmission in an isolated nerve–muscle fibre preparation. *J. Neurophysiol.* 6:99–110.

Lang. F., and H. Atwood. 1973. Crustacean neuromuscular mechanisms, functional morphology of nerve terminals and the mechanism of facilitation. *American Zool.* 13:337–338z

Llinás, R., and C. Nicholson. 1975. Calcium role in depolarization–secretion coupling: An aequorin study in squid giant synapse. *Proc. Nat. Acad. Sci.* 72:187–190.

Loewi, O. 1921. Über humorale Übertragbarkeit der Herznervenwirkung. *Pflügers Arch. Ges. Physiol.* 189:239–242.

McMahan, M., N. C. Spitzer, and K. Peper. 1972. Visual identification of nerve terminals in living isolated skeletal muscle. *Proc. Roy. Soc.* (London) Ser. B. 181:421–430.

Montagna, W. 1959. *Comparative Anatomy.* New York: Wiley.

Mountcastle, V. B., and R. J. Baldessarini. 1968. Synaptic transmission. In V. B. Mountcastle, ed., *Medical Physiology.* 13th ed. St. Louis: Mosby.

Mudge, A. W., S. E. Leeman, and G. D. Fischbach. 1979. Enkephalin inhibits release of substance P from sensory neurons in culture and decreases action potential duration. *Proc. Nat. Acad. Sci.* 76:526–530.

Neher, E., and B. Sakmann. 1976. Single channel currents recorded from membrane of denervated frog muscle fibers. *Nature* 260:799–802.

Nickel, E., and L. Potter. 1970. Synaptic vesicles in freeze-etched electric tissue of *Torpedo. Brain Res.* 23:95–100.

Peper, K., F. Dreyer, C. Sandri, K. Ackert, and H. Moor. 1974. Structure and ultrastructure of the frog endplate. *Cell Tissue Res.* 149:437–455.

Rosenthal, J. 1969. Post-tetanic potentiation at the neuromuscular junction of the frog. *J. Physiol.* 203:121–133.

Sherrington, C. S. 1906. *The Integrative Activity of the Nervous System.* New Haven: Yale University Press.

Whittaker, V. 1968. Synaptic transmission. *Proc. Nat. Acad. Sci.* 60:1081–1091.

# CHAPTER *7*

# *Sensory Mechanisms*

Sense organs provide the only channels of communication into the nervous system from the external world. Thus, sensory input—interacting with the organization and properties of the nervous system, inherited through genetic mechanisms, and organized during the ontogeny of the animal—provides the animal with its entire store of "knowledge." This concept was recognized by Aristotle when he said, "Nothing is in the mind that does not pass through the senses." An understanding of how environmental information is converted into neural signals and how these signals are then processed would seem to have considerable philosophical as well as scientific interest.

The processes of *sensory reception* begin in sense organs—more specifically, in the *receptor cells*. These receptor cells are tuned, so to speak, to specific *modalities* of stimuli. In the higher animals (extrapolating from human experience), a consequence of most sensory input is a subjective sensation identified with the stimulus. Thus, light of wavelength 650–700 nm falling on the eye is perceived as "red"; sugar on the tongue is perceived as "sweet." Sensations are subjective phenomena generated by unknown physical and chemical means within the nervous system, and they are in no way inherent in the sources of stimulation themselves.

The traditional categorization of senses—sight, hearing, touch, taste, and smell—is both subjective and incomplete, for it fails to include certain types of *intero-ceptive* (internal) senses of which we are largely unconscious, such as *proprioceptors* (positional receptors in muscles and joints) and certain sensory receptors that monitor the chemical and thermal conditions within the organism. A more fundamental classification of receptors is based simply on the form of energy to which each is specially sensitive. We will classify receptors as *chemoreceptors, mechanoreceptors, photoreceptors, thermoreceptors,* and *electroreceptors.*

During the course of evolution, sensory systems developed from single, independent receptor units into complex sense organs, such as the vertebrate eye, in which receptor cells are organized into a tissue associated with elaborate accessory structures. The architecture of the accessory structures and the organization of the receptor cells permit far more intricate and accurate sampling of the environment than is possible by independent, isolated receptor cells.

The difference in capability between a simple receptor and a highly organized association of receptor and accessory structures in the form of a sense organ can be illustrated by comparing the barnacle "eye," for example, with more elaborate eyes that possess optic mechanisms. The barnacle eye consists of three simple photoreceptor cells, which are not equipped with a lens and thus cannot process an image. The barnacle eye evidently evolved to carry out photoreception, but not vision. The photoreceptors of a barnacle merely signal changes in light intensity, allowing the nervous system to respond with appropriate protective reflexes when the shadow of a predator falls on the barnacle. The vertebrate eye, on the other hand, can form an optical image, certain parameters of which are encoded in the neural activity of the optic nerve. From this information, the central nervous system abstracts a complex neural counterpart of the image, which in turn gives rise to the subjective experience of "vision."

The first part of this chapter considers some of the general principles of sensory function, and the remainder deals in turn with the major senses and examines the physiology of selected sense organs. An understanding of the principles covered in Chapters 5 and 6 is assumed throughout.

## Receptor Cells as Sensory Transducers

As just noted, sensory processing begins in the *receptor cells*—more specifically, in specialized membranes of these cells. These specialized cells exhibit two major properties that are important in processing stimulus signals (Figure 7-1). First, they are *highly selective* for specific forms of stimulus energy, largely ignoring other forms. That is, they exhibit differential sensitivity, *transducing* certain forms of stimulus energy into biological signals composed of membrane currents and potential changes. With the exception of electroreceptors, in which the stimulus is already an electrical signal, each receptor cell transduces the stimulus into a change in membrane potential. A receptor cell thus may perform a role analogous to that of a microphone, which converts the mechanical energy of sound to modulated electrical signals, or a role analogous to that of a photocell, which converts changing light intensities into corresponding electrical signals.

For example, the visual cells of the eye are far more sensitive to light than to any other form of energy. Although extremely sensitive to light, these receptor cells are quite insensitive to other stimuli, such as mechanical energy. Conversely, the hair cells found in the vertebrate ear are extremely sensitive to mechanical energy and quite insensitive to light. Thus, each receptor cell type normally responds only to an *adequate stimulus*.

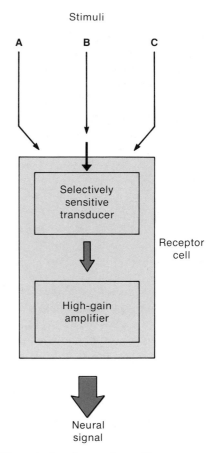

7-1  The major functions performed by sensory receptor cells in the processing of stimuli. Although many forms of stimulus energy may impinge on a receptor, only one form is intercepted at weak to moderate levels of stimulus energy. This form is transduced and amplified into a neural signal.

How is this selectivity achieved? Photoreceptor cells contain a visual pigment that consists of molecules that capture quanta of light. As the light energy is trapped, the molecular structure undergoes a transient alteration. The pigment is part of the cell membrane, so the change in molecular conformation may produce a change in membrane conductance, exciting the receptor cell. Excitation of the receptor cell gives rise to a neural signal. In contrast, a mechanoreceptor cell is equipped with a membrane that responds to a slight distortion or stretch. Thus, the adequate stimulus of a receptor cell is determined largely by the molecular mechanisms built into its membrane.

The second major property of receptor cells is closely related to their high degree of selectivity. Namely, it is the extraordinary ability to produce an electrical signal containing far greater energy than that contained in the stimulus itself. Some of the most impressive examples of such *amplification* of stimulus energy are exhibited by the receptors of the vertebrate eye and ear and by the olfactory receptors of certain insects. One photon of red light contains only $3 \times 10^{-19}$ J of radiant energy. However, the capture of a single photon by a visual receptor cell leads to a receptor current containing about $5 \times 10^{-14}$ J of electrical energy, which represents an amplification in energy of about $1.7 \times 10^5$ times. The receptor current contributes to the modulation of visual impulse frequency. Because of its extraordinary sensitivity, the dark-adapted human visual system permits the subject to sense, as a faint flash, as few as 5 to 10 photons delivered simultaneously over a small portion of the retina.

## STEPS BETWEEN STIMULUS INPUT AND SENSORY OUTPUT

A receptor cell that senses muscle length in the abdomen of crayfish and lobsters is diagrammed in Figure 7-2. Because it is relatively large, this *stretch receptor* has been useful for studies in which intracellular recording methods must be used. If the receptor is stimulated by a weak stretch applied to the muscle to which the dendrites attach, a steady sequence of impulses can be recorded from the axon. The frequency of firing depends on the amount of stretch applied. With a microelectrode inserted into the cell body, it is seen that a small stretch applied to the relaxed muscle leads to a small depolarization, the *receptor potential* (Figure 7-3A). A stronger stretch produces a larger receptor potential. Sufficiently large receptor potentials are accompanied by one or more action potentials.

What is the relation between stimulus, receptor potential, and action potential? The action potential (Figure 7-3B) can be eliminated by blocking the electrically excited sodium system with TTX. When this is done, the receptor potential remains, indicating that

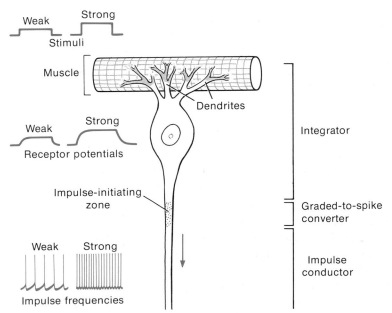

7-2  Stretch receptor in the abdomen of a crayfish, consisting of a sensory neuron with its stretch-sensitive dendrites embedded in a special muscle bundle. When the tail of the crayfish bends, the muscle is stretched and the receptor is activated. Recordings show its electrical responses to both weak and strong stimuli. The parts of the neuron are functionally differentiated as labeled at the right. Graded receptor currents from the stretch-sensitive membrane of the dendrites are converted to all-or-none impulses at the impulse-initiating zone.

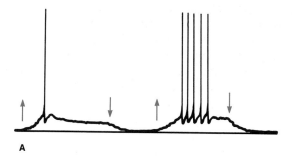

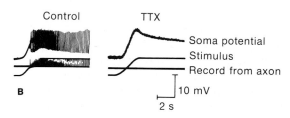

A

Control    TTX

Soma potential
Stimulus
Record from axon

B

10 mV

2 s

7-3  Intracellular records from the stretch receptors of a cray-fish. (A) Response of quickly adapting receptor to a weak (left) and a strong (right) stimulus. Arrows indicate ON and OFF of applied stretch. [Eyzaguirre and Kuffler, 1955.] (B) Slowly adapting receptor in normal saline (at left). With addition of tetrodotoxin (right), the action potentials are blocked, showing the underlying receptor potential. [Loewenstein, 1971.]

it is produced by a different mechanism from the one that generates the all-or-none upstroke of the action potential. It is also seen that the receptor potential is graded with stimulus strength (Figure 7-3B) rather than all-or-none (Figure 7-3A). In these respects, the receptor potential resembles the excitatory synaptic potential of muscle and nerve cells.

The stimulus and the receptor potential can be bypassed by injecting a depolarizing current into the receptor cell with a second electrode. When this is done, a steady train of impulses is evoked by the depolarization. The frequency of impulses is seen to be related more or less linearly to the amount of depolarization, showing a small drop in frequency with time.

Local stimulation of the receptor cell was used to test the ability of various parts of the cell membrane

to produce sustained trains (sequences) of impulses. As shown in Figure 7-4, a sustained stimulus current produces a sustained steady frequency of discharge only when the current depolarizes the low-threshold region where the impulses are normally initiated. Other portions of the cell membrane show much more rapid accommodation. Finally, measurements of membrane conductance have shown that the effect of stimulating the stretch receptor is an increase in conductance to $Na^+$ and $K^+$.

We can now put together a general sequence of steps leading from a stimulus to a train of impulses in a sensory neuron (Figure 7-5). The initial sensory transduction produces a change in membrane conductance in response to the stimulus. That is, it causes the opening or closing of a population of special membrane channels. How this occurs is not entirely clear.

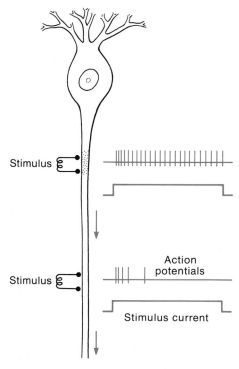

Stimulus

Stimulus

Action potentials

Stimulus current

7-4  Capacity of the spike-initiating zone of the slowly adapting crayfish stretch receptor to respond with a sustained discharge of impulses during sustained stimulation. Other areas of the cell show rapid adaptation to the steady stimulation. [Nakajima and Onodera, 1969.]

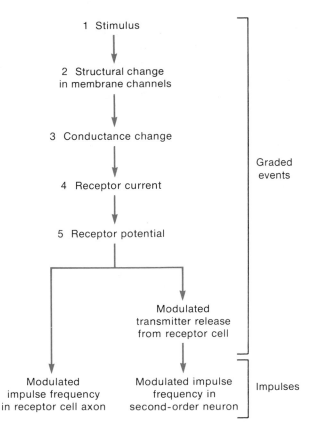

other than sodium (i.e., $Ca^{2+}$ and $K^+$) are also involved in several receptors that have been studied. The *receptor current* is produced by movement of ions through the membrane channels that are activated by the stimulus. In some instances (e.g., the vertebrate visual cells) the stimulus results in channel closing and thus a drop in current.

As the intensity of a stimulus increases, greater numbers of channels respond (i.e., open or close). The result is an increased (or decreased) conductance to certain ions, an increased change in the receptor current, and a larger receptor potential.

All steps leading up to and including the receptor potential are graded in amplitude. Unlike the sodium current of an action potential, the receptor current is *not* regenerative and can therefore spread only by passive electrotonic means. In order for sensory information to be propagated over long distances into the central nervous system, the information contained in the receptor potential must, of course, be converted into action potentials. This is done in one of two ways (Figure 7-6):

*1.* In some receptors, a depolarizing receptor potential spreads electrotonically from its site of origin in the *sensory zone* to a spike-initiating zone in the axon membrane, which then generates action potentials. The receptor zone may be part of the afferent neuron (Figure 7-6A and B) or part of a special receptor cell that is electrically coupled to the afferent neuron (not shown in figure). When the receptor potential spreads directly to the electrically excitable membrane and directly modulates the generation of action potentials, it is sometimes also termed a *generator potential.*

*2.* In other receptors, a depolarizing or hyperpolarizing receptor potential spreads electrotonically to the presynaptic portion of the receptor cell, modulating the release of a transmitter substance (Figure 7-6C). The postsynaptic action of the transmitter then modulates the frequency of impulses generated in the axon of an afferent nerve fiber. In this case, it is not necessary for the receptor cell itself to produce an action potential.

We see, then, that the receptor potential is the message with which the sensory portion of a receptor cell

**7-5** Sequence of events in a receptor cell from stimulus to sensory axon impulses. In some systems, the impulses are initiated in the receptor cell itself and travel to the central nervous system in the axon of that cell (left branch). In other systems, the receptor cell synaptically modulates impulse discharge in a second-order neuron (right branch).

In any case, a change in membrane conductance, such as an increase in sodium conductance, will cause a shift in the membrane potential in accord with the principles discussed in Chapter 5. If the stimulus produces an increase in membrane conductance to sodium, the membrane potential will shift toward the equilibrium potential for $Na^+$, and the cell interior will become more positive—that is, it will exhibit some degree of depolarization in accordance with Equation 5, Box 6-4. If the stimulus decreases the sodium conductance, the membrane potential will shift away from the sodium equilibrium potential, becoming more negative. Ions

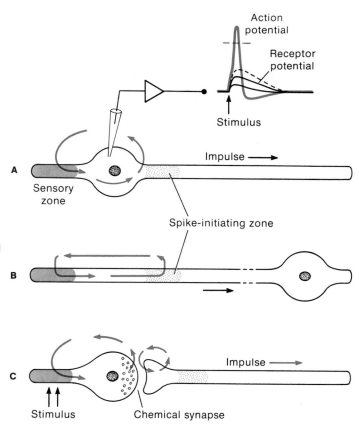

**7-6** Conversion of the receptor potential into propagated impulses. Compare with Figure 7-5. In A and B, the receptor current arising in the sensory zone spreads electrotonically to depolarize the spike-initiating zone directly. In both A and B, the receptor cell also provides the afferent sensory fiber, the only difference being the peripheral (A) or central (B) locations of the cell body. In C, a synapse intervenes between the sensory zone in the primary receptor cell and a secondary afferent fiber.

signals the secretory or impulse-generating regions, modulating the production of action potentials that signal the stimulus to the central nervous system.

## Intensity Coding

Individual action potentials carrying information from the different sense organs are essentially indistinguishable from one another. That is, the modality of the stimulus is *not* coded by any characteristics inherent in the individual action potentials. Rather, the modality perceived by the organism depends on the anatomical specificity with which the sensory neurons connect with the "higher" cognitive centers in the brain. This idea receives further consideration in Chapter 8.

The only way in which the information carried in a single nerve fiber can be coded other than by anatom-

ical specificity is by the temporal distribution of the impulses. Thus, a high frequency of impulses represents a strong stimulus, and a cessation of spontaneously occurring impulses means a reduction in the strength of the stimulus. There is no simple rule for sensory coding, because the relations between stimulus and sensory response differ in different kinds of receptors. Nevertheless, some generalizations can be made at the level of the receptor cell. As the intensity of a stimulus is increased, the receptor current is increased and a greater depolarization is produced. In many receptors, the impulse-initiating zone (Figure 7-2) shows little accommodation and in the face of a steady depolarization can produce a steady train of impulses. An important exception to the rule that increased stimulus strength leads to increased depolarization occurs in vertebrate visual cells, which are discussed at some length further on.

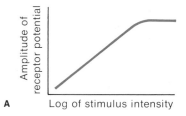
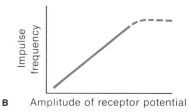
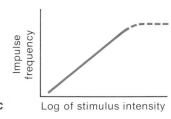

7-7  Input–output relations of sense organs. (A) In many receptors, the amplitude of the receptor potential is linearly related to the logarithm of stimulus intensity over a large range. (B) The frequency of sensory impulses is linearly related (i.e., proportional) to the amplitude of the receptor potential. (C) As a result of A and B, the impulse frequency in many sensory fibers varies linearly with the log of the stimulus intensity. The broken parts of the curves in B and C indicate the failure of impulses to follow due to refractoriness of the axon membrane.

## INPUT–OUTPUT RELATIONS

The range of stimulus intensities over which a receptor can respond without saturation—that is, over which the receptor can code intensity by producing increasingly larger signals for increasing stimulus energies—is termed the *dynamic range* of the receptor or sensory organ. The amplitude of the receptor potential in most receptor cells is approximately proportional to the logarithm of the stimulus intensity (Figure 7-7A). The frequency of sensory impulses is related more or less linearly to the amplitude of the receptor potential (Figure 7-7B) up to the point at which the refractory state of the axon membrane after each impulse limits the rate of firing. As a consequence of those two relations, the impulse frequency in a slowly adapting receptor (to be discussed shortly) is generally a function of the logarithm of the stimulus intensity (Figure 7-7C). The impulses, on reaching the central terminals of the sensory neurons, set up synaptic potentials that sum and facilitate as a function of the frequency of impulses. Thus, the postsynaptic potential produced at this point is graded as a function of the stimulus intensity, and it can be regarded as a very crude analog of the stimulus (Figure 6-3).

The logarithmic relation between stimulus energy and sensory impulse frequency seen in many sensory systems is significant in view of the enormous range of stimulus intensities encountered by sense organs. For example, the difference in intensity between moonlight and sunlight is about $10^9$ times. The human auditory system can perceive sounds that differ in energy by as much as $10^{12}$ times without significant distortion. The ability of sense organs to function over such enormous energy ranges is quite remarkable, and it depends not only on the large dynamic range of the initial transduction process, but also, as we will see, on processes of *adaptation* that take place with maintained exposure to the stimulus and on properties of the neural network that processes the sensory signals.

At low stimulus intensities, the receptor potential in the nonadapted receptor represents a very large amplification of energy. The amplification is progressively reduced as the stimulus energy increases. This logarithmic relation between stimulus intensity and amplitude of the receptor potential is predicted by the Goldman equation (p. 148), which states that the membrane potential varies with the log of the ion permeabilities multiplied by the concentrations of the ions distributed across the membrane. Thus, changes in the permeability to $Na^+$, for example, should produce changes in membrane potential that are proportional to the log of stimulus-induced changes in sodium permeability. Stimulus intensities normally encountered in the environment are generally within the logarithmic portion of the input–output curve (Figure 7-7B). Some receptors depart from the logarithmic relationship, exhibiting a power function in which the log of the response amplitude is proportional to the log of the stimulus intensity.

Three factors limit the maximum sensory response to intense stimuli:

*1.* A finite number of receptor current channels will place an upper limit on the maximum receptor current that can flow in response to a strong stimulus.

*2.* Regardless of stimulus intensity, the receptor potential will saturate (i.e., become limited) as it approaches a certain value (the reversal potential of the receptor current), which it cannot exceed. This is analogous to the reversal potential of a synaptic current.

*3.* Impulse frequencies in sensory axons are limited by refractoriness following each impulse; thus, most nerve fibers are limited to maximum impulse frequencies of several hundred or fewer per second.

The roughly logarithmic relationship between stimulus strength and receptor potential in many sense organs is interesting from an evolutionary point of view, because it makes far better adaptive sense than a simple proportional (i.e., linear) relationship. The logarithmic stimulus–response relationship (illustrated in Figure 7-7C) "compresses" the high-intensity end of the scale, thereby greatly extending the range of discrimination.

Another consequence of the logarithmic relationship between receptor-potential amplitude and stimulus intensity is that a given *percentage* of change in stimulus intensity evokes the same *increment* of change (i.e., same number of millivolts) in the receptor potential over a large range of intensities. In other words, a doubling of the stimulus intensity at the low end of the intensity range will evoke the same increase in receptor-potential amplitude as will a doubling of intensity at the high end of the range. So we have

$$\frac{\Delta I}{I} = K$$

where $I$ is stimulus intensity and $K$ is a constant. This relation is similar to that governing subjectively perceived changes in stimulus intensity, known in psychology as the *Weber–Fechner Law.*

This feature of sensory systems makes excellent adaptive sense. For example, the luminosities of objects within a scene viewed in a given light (e.g., at high noon) differ far less than do the changes in overall luminosity due to changing intensities of illumination (e.g., high noon versus dusk). Thus, detection of *relative* intensities of stimuli and *changes* in stimulus intensity within a given scene (for example, the spatial and temporal changes in luminosity produced by an object moving in the visual field) are far more important for the survival of an animal than are absolute measurements of luminosity.

### RANGE FRACTIONATION

The dynamic range of intensity discrimination exhibited by a multineuronal sensory system is typically much larger than that exhibited by any single receptor or afferent sensory fiber. The extended dynamic range of the entire system is possible because individual afferent fibers of a sensory system cover different portions of the

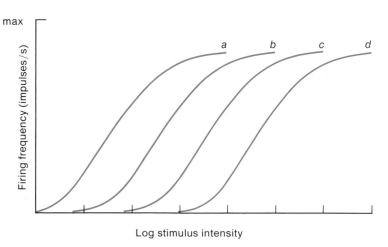

**7-8**  Range fractionation. The curves labeled *a* through *d* each represent the discharge frequencies of an individual sensory afferent plotted against stimulus intensity. In this hypothetical example, each sensory fiber has a dynamic range of about 3–4 log units of stimulus intensity, whereas the overall dynamic range of *a–d* covers 7 log units of intensity.

spectrum of sensitivity. Thus, at the lowest stimulus energies a few especially sensitive sensory fibers will respond weakly. Then, if the stimulus energy is increased, their discharge frequencies will increase, while new, less sensitive fibers will join in weakly. With still greater stimulus intensities, another, previously quiet lower-sensitivity population of afferents will join in. With ever-increasing stimulus intensity, the least sensitive sensory fibers will finally be *recruited,* and all will be firing maximally. At that point, the system will be saturated and therefore unable to detect further increases in intensity. Such *range fractionation,* in which individual receptors or sensory afferents cover only a fraction of the total dynamic range of the sensory system (Figure 7-8), enables the sensory processing centers of the central nervous system to discriminate stimulus intensities over a range much greater than that of any single sensory fiber.

## Determinants of Receptor Sensitivity

With what fidelity is sensory information conveyed to the central nervous system? How do sense organs compare in this respect with physical transducers such as thermometers, lightmeters, and strain gauges? From our own experiences, we know that biological sensory systems are not very trustworthy indicators of absolute energy levels. Moreover, many sensations change over time. For example, a pleasantly sunny day may seem painfully bright for a few minutes after one emerges from a dimly lit interior. For this reason, even an experienced photographer requires a lightmeter to make accurate judgments of camera exposure settings. When one dives into an unheated pool for a swim, the water initially feels colder than it does a minute or two later. Phenomena of this type are lumped under the general term *sensory adaptation.* Where does the adaptation take place? There is no simple answer. Some adaptation occurs in the receptor cell, some occurs as a result of time-dependent changes in accessory tissues, and some occurs in the central nervous system.

### RECEPTOR ADAPTATION

Receptors exhibit various degrees of adaptation. So far we have considered only *tonic* (i.e., slowly adapting) *receptors*—that is, those that continue to fire steadily in

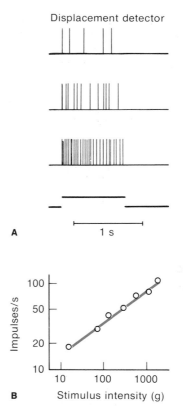

Displacement detector

**A**          1 s

**B**          Stimulus intensity (g)

7-9   Behavior of a displacement detector. This tonic (i.e., slowly adapting) mechano-receptor responds to a steady displacement with a relatively steady frequency of impulses. (A) Discharge at three stimulus intensities. (B) A plot of steady-state discharge frequency against grams of tension applied. [Schmidt, 1971.]

response to a constant stimulus. This behavior is illustrated by a mechanical displacement detector (Figure 7-9) that fires at an almost constant rate with constant stimulus. As the stimulus intensity is increased, the firing frequency rises as a power function of stimulus intensity (Figure 7-9B). Many receptors are of the *phasic* (i.e., quickly adapting) type, in which the firing subsides soon after the onset of a steady stimulus. In one class of phasic receptors, firing occurs only during changes in stimulus energy. For example, some mechanoreceptors fire only during the mechanical change, with the firing frequency a function of the velocity of displacement (Figure 7-10).

Velocity detector

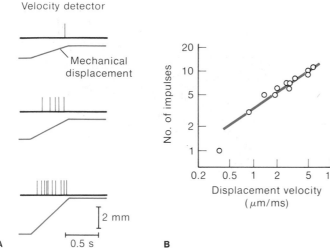

7-10 Behavior of a velocity detector. This rapidly adapting mechanoreceptor responds to the rate of change of position (i.e., velocity of displacement). (A) The higher the rate of change, the higher the frequency of the impulse discharge. (B) The number of impulses during a 0.5 s stretch varies with the log of the displacement velocity. [Schmidt, 1971.]

Adaptation of a receptor organ can occur in at least three ways:

*1.* The receptor molecules themselves may "run down" in the face of a constant stimulus. For example, the visual pigment of photoreceptors becomes bleached during exposure to light and requires metabolic regeneration before responding again to illumination.

*2.* The electrical properties of the receptor cell may change during sustained stimulation. In some receptors, activation of receptor channels diminishes with accumulation of $Ca^{2+}$ within the cell during sustained stimulation. $Ca^{2+}$ accumulation can also activate certain potassium channels so as to shift the membrane potential back toward the resting potential.

*3.* Accessory structures may show time-dependent changes, as with the pupil of the eye or the connective tissue associated with mechanoreceptors.

The second and third of these mechanisms of adaptation are well illustrated by the stretch receptors of crayfish and lobsters. These receptors occur in pairs, each pair composed of one phasic receptor and one tonic receptor. A step increase in receptor muscle-fiber length produces a transient response in the former (Figure 7-11A) and a sustained discharge in the latter

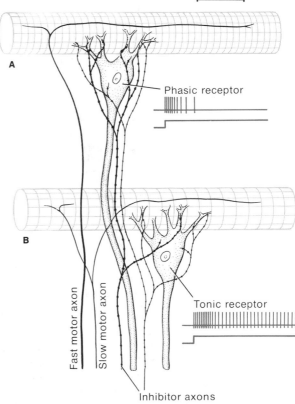

7-11 The phasic (A) and tonic (B) stretch receptors of the crayfish. The phasic receptor adapts quickly to a constant stretch, producing only a short train of impulses. The tonic receptor fires steadily during maintained stretch. [Horridge, 1968.]

(Figure 7-11B). When the receptor current is bypassed by direct injection of depolarizing current with a microelectrode, the two receptor cells retain some of their differences. That is, the phasic receptor responds to a sustained current with a more rapid drop in firing frequency than does the tonic receptor under similar conditions of stimulation. But in both the phasic and the tonic stretch receptors, firing slows down to some extent as a result of a slackening of tension of the receptor muscle fibers after a sudden increase in length. The slackening is due to viscous properties of the tissue, which is not perfectly elastic.

The filtering effects of accessory structures are also important in the rapid adaptation of the *Pacinian corpuscle,* a pressure and vibration receptor found in the skin, muscles, mesentery, tendons, and joints of mammals (Figure 7-12). It consists of a mechanically sensitive axon terminal, surrounded by concentric lamellae of connective tissue resembling the layers of an onion. When a corpuscle is deformed by something pressing on it, the disturbance is transmitted mechanically through these layers to the sensitive membrane of the axon terminal. The axon terminal normally responds with a brief, transient depolarization at both the ON and the OFF phases of the deformation (Figure 7-12B). When the layers of the corpuscle are peeled away, permitting a mechanical stimulus to be applied directly to the naked axon, the receptor potential obtained is a far better analog of the stimulus than that obtained in the undissected Pacinian corpuscle. Although the receptor potential still shows some degree of adaptation (the sag in Figure 7-12C), it gives only an ON response. The rapid adaptation of the undissected Pacinian corpuscle to maintained deformation can be ascribed in part to the properties of the corpuscle, which preferentially passes rapid *changes* in pressure. This behavior explains, in part, why we quickly lose awareness of moderate, steady pressure stimuli. You can demonstrate this by pressing a finger into the palm of your hand and noting how quickly the sensation of pressure subsides.

Regardless of its site or mechanism of origin, adaptation plays a major role in extending the dynamic range of sensory reception. Together with the logarithmic nature of the primary transduction process,

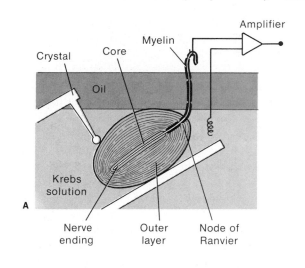

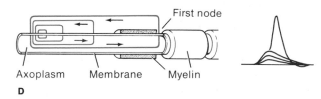

**7-12** Adaptation in the Pacinian corpuscle. (A) Experimental arrangement for tapping the receptor with a crystal-driven stylus. The electrical recording was made between the hook electrode and the oil–water interface. (B) Electrical response of the intact corpuscle. The ON and OFF of the stimulus (arrows) both produce transient depolarizations. (C) Electrical response was sustained during displacement after the lamellae were removed. (D) Receptor current flow in response to deformation of the sensory zone of the axon. The impulse arises out of the receptor potential (right) at the first node of Ranvier. [Loewenstein, 1960.]

sensory adaptation allows an animal to detect changes in stimulus energy against background intensity differences of many orders of magnitude.

## MECHANISMS FOR ENHANCED SENSITIVITY

Certain receptor cells exhibit spontaneous firing without any sensory input. This behavior has two important consequences. First, if the sensory fiber is in the process of firing spontaneously, any small increase in stimulus energy will produce an increase in its rate of firing. Small receptor currents in response to weak stimuli can modulate the impulse frequency by shortening the intervals between impulses (Figure 7-13). Such modulation of impulse frequency lends far greater sensitivity than would be possible if the receptor current had to depolarize to the firing level a completely inactive spike-initiating zone. The input–output relations of such a sensory fiber are described by the sigmoid curve in Figure 7-14. In the unstimulated condition, the firing frequency is on the steep portion of the curve, and even a small input will produce a significant increase in firing frequency. Second, in some spontaneously active sensory neurons, modulation of impulse frequency can take the form of either an *increase* or a *decrease,* which permits the receptor to respond differentially to different stimulus polarities or directions. An example is seen in the discharge of the electroreceptor

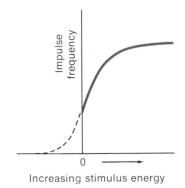

7-14   Typical input–output relations of a spontaneously active receptor cell or second-order sensory fiber. In the absence of any input above or below the norm (0 on the abscissa), the cell fires spontaneously. Because this norm lies on the steep portion of the stimulus-frequency curve, any deviation in stimulus energy from the norm will produce either a decrease or an increase in the rate of firing.

nerve of some fishes (Figure 7-34). The fibers in the nerve fire spontaneously in the absence of any electric field in the surrounding water. An electric field that causes current to flow into the receptor pit depolarizes the receptor and increases the firing frequency of the sensory nerve. A field that causes an outflow of current hyperpolarizes the receptor and decreases the afferent impulse frequency.

Another way in which the nervous system enhances its ability to differentiate between a "signal" (i.e., a change in activity attributable to a stimulus) and the "noise" (i.e., background activity in the sensory afferents) is to average the input from a number of receptor cells simultaneously. Thus, random signals in individual input channels are given less weight than an increase in simultaneous activity in several channels. Consequently, a human observer cannot reliably perceive a single photon absorbed by a single receptor cell; but if each of several receptors simultaneously absorbs a photon, the observer experiences the sensation of light.

## EFFERENT CONTROL OF RECEPTOR SENSITIVITY

The responsiveness of some sense organs is influenced by the central nervous system, which sends impulses through efferent (centrifugally conducting) axons that

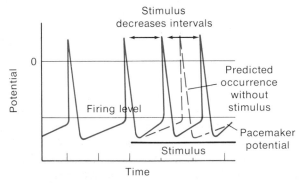

7-13   Impulse interval in a spontaneously firing receptor cell can be decreased by extremely small stimuli because of a slight increase in the slope of the pacemaker potential. (In reference to second-order sensory fibers, read "synaptic potential" instead of "pacemaker potential.")

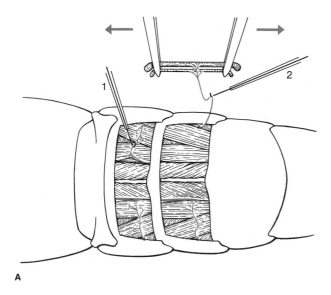

**A**

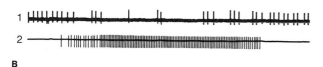

**B**

7-15 (A) Reflex inhibition of muscle stretch receptors (see Figure 7-11) in the tail of the crayfish. The muscle stretch receptors of one segment are removed with innervation intact, and recordings are made as shown with electrodes 1 and 2. (B) Electrode 1 records a steady train of action potentials from the tonic receptor in response to a steady stretch. When the isolated tonic receptor is stretched, electrode 2 records a train of sensory impulses and electrode 1 records a drop in frequency of firing in the undisturbed intact receptor. The activity recorded by electrode 2 results in the activation of an inhibitory axon, which inhibits the steady stretch-receptor output by electrode 1. [Eckert, 1961.]

innervate the sense organ. For example, the stretch receptors, or "muscle spindles," found in vertebrate skeletal muscle and in muscle receptors of crustaceans have a motor innervation to their receptor muscles. By controlling the length of the receptor muscles, this innervation controls the range of response of the muscle spindle to changes in overall muscle length. The sensory neurons of the abdominal stretch receptors of crayfish and lobsters are also innervated directly by efferent neurons that make inhibitory synapses on the

stretch-receptor cells (Figure 7-11). The size of the receptor potential is diminished by a transmitter substance released by the inhibitory terminals, causing either a cessation of the sensory impulses produced by the stretch-receptor cells or a reduction in their frequency. The inhibitory fibers are activated reflexly by sensory discharge from the same receptor cells and in similar stretch receptors in neighboring body segments (Figure 7-15). This activation indicates that the axon of the receptor cell must make synaptic connections in the CNS, either directly or via interneurons, with these inhibitory neurons. The central output—in this case, reflex—reduces the sensitivity of the receptor at the periphery to a given mechanical stimulus.

Generally speaking, efferent control of receptor sensitivity serves to extend the range of sensitivity of the receptor. In the case of the crustacean stretch receptor, for example, the sensory neuron has a maximum rate of discharge and saturates in response to strong stretch. Activation of the inhibitor reduces the sensitivity of the receptor, extending its dynamic range so that it can continuously code degrees of stretch that otherwise would saturate the firing frequency of the axon.

### FEEDBACK INHIBITION OF RECEPTORS

Closely related to efferent control of receptor sensitivity is the occurrence of *feedback inhibition*. In a number of receptor systems, activity in the receptors results in an inhibitory signal feeding back on the receptors. The crustacean abdominal stretch receptors again offer an illustration (Figure 7-11B). The afferent activity of the sensory neuron produces a reflex efferent output from the CNS in the inhibitory nerves leading back to the same sensory neuron (*autoinhibition*) and also to its neighbors (*lateral inhibition*). Thus, with strong stimulation, the output of the receptor is less likely to exceed its operating range (e.g., maximum rate of impulse discharge). The inhibitory feedback is said to extend the dynamic range of the receptor neuron. At low stimulus intensities, on the other hand, the feedback plays little or no role, since it takes a relatively strong sensory discharge to evoke a reflex discharge of the inhibitory neurons.

Another consequence of mutual inhibition is *contrast enhancement* by lateral inhibition. Although first discovered in visual systems, it occurs in a number of sensory systems. The stronger the output of a receptor, the stronger the resulting inhibition of its neighbors through mutual inhibitory connections. This situation causes a disproportionate suppression of output from the more weakly stimulated receptors, as discussed at greater length in Chapter 8. We shall simply note here that another outcome of the connections responsible for lateral inhibition, as well as autoinhibition, is the extension of dynamic range for the sensory organ.

## Chemoreception

The sensitivity of cells to specific molecules is widespread; this includes metabolic responses of tissues to chemical messengers, as well as the ability of lower organisms such as bacteria to detect certain substances in the environment. We will restrict ourselves to special *chemoreceptor* cells that include both *gustatory* (taste) receptors, which detect dissolved ions and molecules, and *olfactory* (smell) receptors, which detect airborne molecules. Because airborne molecules enter an aqueous layer covering the membrane of an olfactory receptor, any fundamental distinction between olfactory and gustatory reception disappears. Figure 7-16 illus-

trates vertebrate taste and olfactory receptors and an insect olfactory receptor.

Chemosensory systems can be extraordinarily sensitive. A case in point is the sensitivity of antennal chemoreceptors of the male silkworm moth (*Bombyx mori*) to *bombycol*, the female's sex-attractant pheromone (p. 417). The male responds behaviorally to concentrations as low as one molecule per $10^{17}$ molecules of air. The moth's receptors are highly specific, responding only to bombycol and a few of its chemical analogs. The zoological significance of this highly evolved stimulant–receptor system is obvious: it allows the male of the species to locate a single female at night from distances of up to several miles downwind.

Recordings of electrical responses of the antennal olfactory receptors of the moth have been used to investigate their sensitivity to bombycol. Only about 90 bombycol molecules impinging on a single receptor cell per second are required to produce a significant increase in the rate of firing of the cell. However, the male moth reacts behaviorally (i.e., flaps his wings excitedly) when only about 40 receptor cells (out of a total of 20,000 per antenna) each intercept one molecule per second. Because the change in discharge rate

**7-16**  Three types of chemoreceptor organs. (A) Vertebrate taste bud with secondary sensory axons innervating the primary receptor cells. [Murray and Murray, 1970.] Vertebrate (B) and insect (C) olfactory receptors send primary afferents to the central nervous system. Analogous structures are drawn similarly in B and C. In all three types, fine processes extend from the receptor cell into the mucous layer. In the insect, these are true dendrites. [Steinbrecht, 1969.]

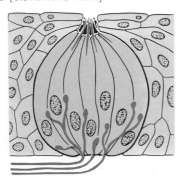

**A**   Vertebrate taste bud

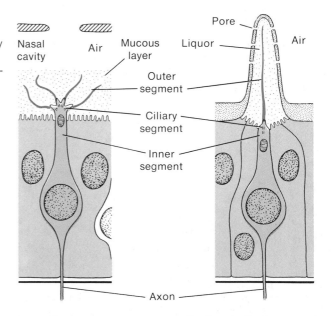

**B**   Vertebrate olfactory receptor    **C**   Insect olfactory receptor

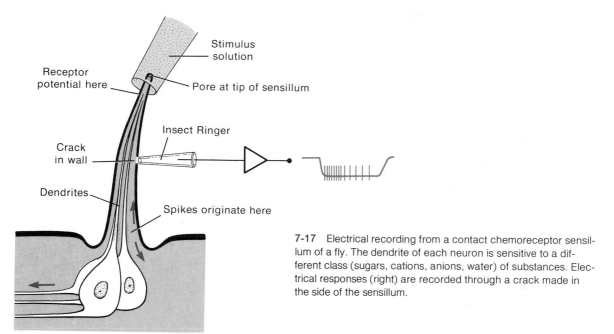

**7-17** Electrical recording from a contact chemoreceptor sensillum of a fly. The dendrite of each neuron is sensitive to a different class (sugars, cations, anions, water) of substances. Electrical responses (right) are recorded through a crack made in the side of the sensillum.

of a single cell in response to a single odorant molecule is not readily apparent in electrophysiological recordings, it is inferred that the moth's central nervous system is capable of sensing very slight average increases in impulse frequency arriving along numerous chemosensory channels.

The *contact chemoreceptors* (taste hairs) of insects have proven very useful in electrophysiological studies of single chemoreceptor cells. These receptor cells send fine dendrites to the tips of hollow hairlike projections of the cuticle, called *sensilla.* Each sensillum has a minute pore that provides access for stimulant molecules to the sensory cells (Figure 7-17). The sensillum contains several cells, each of which is sensitive to a different chemical stimulus (e.g., water, cations, anions, carbohydrates). In this case, a behavioral response can be evoked by appropriate stimulation of a single sensillum. A small drop of sugar solution applied to a sensillum on the foot causes the fly to respond by lowering its proboscis to feed. The effectiveness of various compounds in evoking this stereotyped behavior has been tested with this reflex. All compounds that release the feeding reflex also evoke electrical activity in the

*sugar receptor,* one of the receptor cells in the sensillum. This receptor cell responds only to certain carbohydrates. Carbohydrates, such as D-ribose, that do not release feeding behavior also fail to stimulate the sugar receptor. It is interesting that the sugar receptor of the fly shows the same sequence of sensitivity (fructose > sucrose > glucose) as do the sweetness receptors of the human tongue.

The electrical activity of the contact chemoreceptors of the fly can be recorded either through the stimulus solution applied to the tip or through a crack made in the side wall of the sensillum (Figure 7-17). Using the latter technique, it is possible to observe two components of the extracellularly recorded electrical response: (1) a receptor potential (slow downward deflection of trace), and (2) impulses. Even though the receptor potential under these recording conditions appears to be negative-going, it corresponds to a depolarization of the cell membrane. The receptor potential is produced at the ends of the dendrites that extend to the tip of the sensillum, whereas the action potentials originate near the cell body. The long distance between the tip of the sensillum, where the stimulus acts on

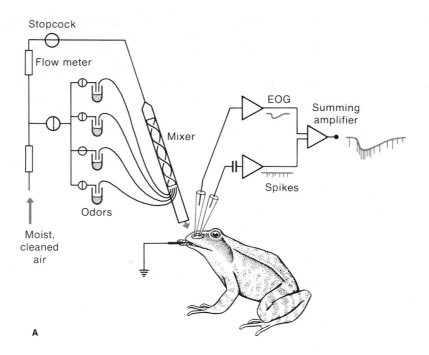

**A**

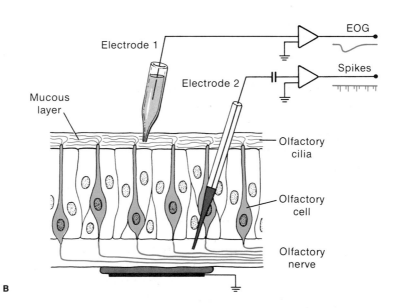

**B**

**7-18** Method for studying sensory coding in the olfactory epithelium of the frog. (**A**) Various odors can be applied to the nasal epithelium while the electro-olfactogram (EOG) and spikes are recorded and summed to give a composite recording (right). (**B**) Detail of tissue and electrodes. Electrode 1 records the overall EOG potential by virtue of its position. [Gesteland, 1966.]

the *receptor membrane,* and the site of impulse initiation requires electrotonic spread of the receptor potential along the length of the dendrite to the site of impulse initiation.

Olfactory coding has been examined electrically in the olfactory epithelium of the frog by recording the activity of single receptor axons (electrode 2, Figure 7-18B) along with the summed potential (*electro-olfactogram,* or *EOG*) of large numbers of olfactory receptors in the epithelium (electrode 1). Impulses from individual receptors were superimposed electronically on the EOG before display on the oscilloscope. This technique permitted comparison of the activity of single units with the total response of many receptors to selected odor stimuli, or odorants. The results indicate that stimulus coding in the vertebrate nose is far more complex than that in the contact chemoreceptors of insects. Different receptors respond differently to the same odorant. In some sensory axons, a particular odorant increases the impulse frequency (Figure 7-19). Perhaps most interesting is the finding that odorants smelling alike to humans have similar effects on some frog olfactory cells but have differing effects on other cells (Figure 7-19). It has not been possible to establish a one-to-one relation between certain smells and certain types of olfactory cells in the frog. Instead, each olfactory receptor cell may have a mosaic of odorant– receptor molecules with differing specificities. Depending on the proportions of receptor molecules having different olfactory specificities, a given olfactory cell may show various combinations of excitation or depression in response to a given odorant. The ability

of mammals to distinguish among a wide variety of odorants may therefore reside in the ability of the olfactory centers of the brain to "recognize" a large number of different combinations of enhanced and depressed impulse frequencies arriving from various olfactory cells in the nasal epithelium.

How does a taste or olfactory receptor cell transduce a chemical stimulus? In the context of this question, we can recall the behavior of the postsynaptic membrane of a chemical synapse, which can be regarded as the chemosensory membrane of the postsynaptic nerve or muscle cell. The postsynaptic receptor molecules of the vertebrate motor endplate, on binding with ACh, undergo a conformational change that produces an increase in the conductance of the membrane to $Na^+$ and $K^+$, with the result that a synaptic current (in the present context, read also "receptor current") flows into the cell, depolarizes it, and generates action potentials.

The ACh receptor molecule is known to be a protein (or rather, as discussed in Chapter 6, a group of protein molecules). This fact is not surprising, for protein molecules are known to undergo changes in shape in response to interactions with other molecules. So a change in shape on binding an odorant molecule is one way in which an olfactory receptor might cause the opening of channels for the passage of current through the membrane. Although no receptor molecules of chemosensory cells have yet been isolated, it is probable that they are proteins.

According to one classification, there are seven primary odors: *camphoraceous, musky, floral, pepperminty, ethereal, pungent,* and *putrid.* Humans can distinguish several hundred different compounds on the basis of their odors, but it is not reasonable to suppose that corresponding to each odorant there is a different receptor molecule or receptor cell responsive only to that odorant. More likely, as we proposed in the case of the frog, there is a distribution of receptor molecules in different olfactory cells that permits the coding of combinations of basic odor types.

## Mechanoreception

The simplest mechanoreceptors are morphologically undifferentiated nerve endings found in the connective tissue of skin (Figure 7-20). In many mechanoreceptors, there have evolved accessory structures whose function

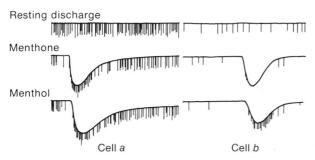

**7-19** Two substances that have similar effects on cell *a* have very different effects on cell *b*. Thus, cell *b* can differentiate some odors that *a* cannot. [Gesteland, 1966.]

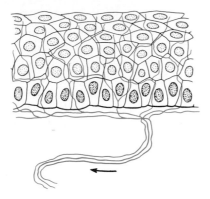

**7-20**  Mechanoreceptors in the connective tissue of skin are morphologically undifferentiated axon terminals. [From "Biological Transducers" by W. R. Loewenstein. Copyright © 1960 by Scientific American, Inc. All rights reserved.]

is the efficient transfer of mechanical energy to the receptor cells. These accessory structures are generally differentiated to filter the mechanical energy in some way. Examples are Pacinian corpuscles, in which the sensitive ending is covered by a capsule (Figure 7-12); muscle stretch receptors, in which the mechanical sensitive neuronal endings are associated with specialized muscle fibers (Figures 7-11 and 8-36); and the hairlike sensilla in the exoskeletons of arthropods (Figure 7-21). The most elaborate accessory structures, used to detect and analyze sound waves, are found in the vertebrate middle and inner ear (Figures 7-28 and 7-29).

The immediate stimulus thought to act on mechanoreceptors is a stretch or distortion of the surface membrane. Stretching the membrane of one of the large axons of the lobster has been shown to increase its permeability to sodium. One hypothesis for mechanoreceptor transduction is that stretching of the receptor

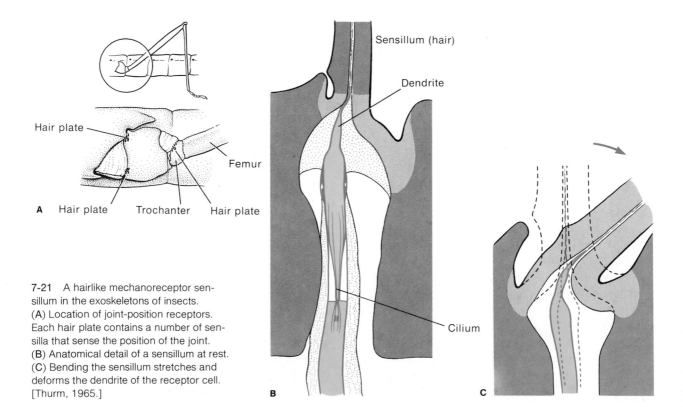

**7-21**  A hairlike mechanoreceptor sensillum in the exoskeletons of insects.
(A) Location of joint-position receptors. Each hair plate contains a number of sensilla that sense the position of the joint.
(B) Anatomical detail of a sensillum at rest.
(C) Bending the sensillum stretches and deforms the dendrite of the receptor cell. [Thurm, 1965.]

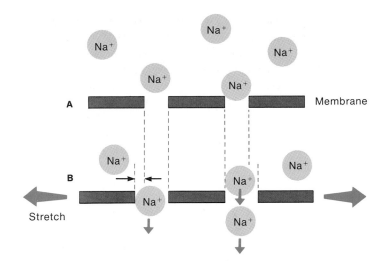

7-22   A simple hypothesis for explaining the increase in permeability to cations produced by stretching a mechanoreceptor membrane. The sodium ion is used for illustration. (A) If the diameter of the pores in the unstretched receptor is just below the diameter of the sodium ion, a small stretch (B) will increase the diameter enough to allow $Na^+$ to flow into the cell. It is not known if this is, in fact, the way a mechanoreceptor functions.

membrane slightly enlarges water-filled ion-selective channels in the membrane. Such an enlargement of channels that would otherwise be just below the critical size for the common physiological ions might cause a large increase in permeability to those ions (Figure 7-22). The very small amount of mechanical energy required to stretch the membrane just enough to produce an increased conductance to the ion would release a much larger amount of electrical energy stored as the concentration gradient of that ion. Supporting this hypothesis is the observation that some mechanoreceptor channels appear to be water-filled pores with a lower ion selectivity than many other channels. However, the size of the channel enlarged by stretch cannot by itself determine its ionic selectivity. For example, the mechanoreceptors at opposite ends of a *Paramecium* exhibit entirely different selectivities (Figure 8-47). A mechanical disturbance of the posterior end causes $K^+$-selective channels to open, while a similar disturbance of the anterior end causes $Ca^{2+}$-selective channels to open. The ion selectivity of these two kinds of mechanically activated channels must rely on properties other than size alone.

## HAIR CELLS

The hair cells of vertebrates (Figure 7-23A and B) are extraordinarily sensitive mechanoreceptors found in the lateral-line (sensory) systems of fishes and amphibians (Figure 7-24), in the vertebrate organs of hearing

(Figures 7-28 and 7-29), and in the organs of equilibrium (semicircular canals) (Figure 7-28). The name *hair cell* derives from the cilia that project from the apical end of each receptor cell—a single *kinocilium* and about two dozen *stereocilia*. The kinocilium has a "9 + 2" arrangement of microtubules similar to that of motile cilia, to be described in Chapter 10; however, the cilia of hair cells have not been observed to exhibit motility. The stereocilia contain many fine longitudinal actin filaments, and they are structurally and developmentally unrelated to the kinocilium. Although the kinocilium is present in the lateral-line and vestibular hair cells, it is absent from the hair cells of the adult mammalian cochlea and probably does not play a primary role in mechanotransduction. The stereocilia are arranged in an order of increasing length from one side of the cell to the other (Figure 7-23B).

The hair cells are sensitive to the direction of the mechanical displacement of the stereocilia. Bending of the cilia in the direction of the tallest cilia leads to depolarization of the hair cell, whereas bending in the opposite direction leads to hyperpolarization (Figure 7-23B). The receptor potential of the hair cell modulates the steady release of a transmitter substance from the basal end of the cell. This transmitter release, in turn, determines the rate of discharge of the sensory axons with which the hair cells make chemical synapses. Depending on the direction in which the cilia are bent, the rate of transmitter release and hence the

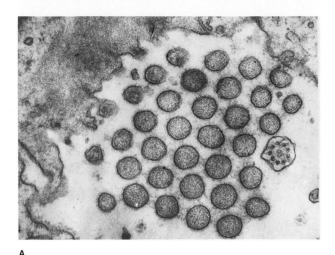

A

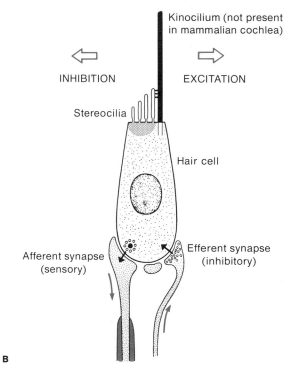

Kinocilium (not present in mammalian cochlea)

⇐ INHIBITION    ⇒ EXCITATION

Stereocilia

Hair cell

Afferent synapse (sensory)    Efferent synapse (inhibitory)

B

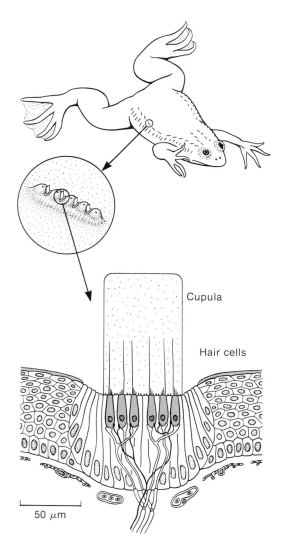

Cupula

Hair cells

50 μm

7-24  Hair cells in the lateral-line system of an amphibian. Water motion causes displacement of the cupula, which stimulates the hair cells.

7-23  (A) Electron micrograph of a cross section through the cilia of a hair cell. The large cilium with the 9 + 2 structure is the kinocilium, the others are stereocilia. [Flock, 1967. From *Lateral Line Detectors,* Phyllis H. Cahn, ed., © 1967 by Indiana University Press. Reprinted by permission of the publisher.] (B) Diagram showing anatomical relations of the hair cell, stimulus direction, and secondary afferent fibers. Depending on the direction in which the cilia are bent, the hair cell can either increase or decrease its rate of firing. [Harris and Flock, 1967.]

firing frequency are modulated above or below a spontaneous frequency corresponding to zero displacement. An important feature of the hair-cell response is that its input–output relations are markedly asymmetrical (Figure 7-25). That is, the potential change produced by a displacement of the stereocilia in the direction of the tallest cilia (depolarization) is larger than that

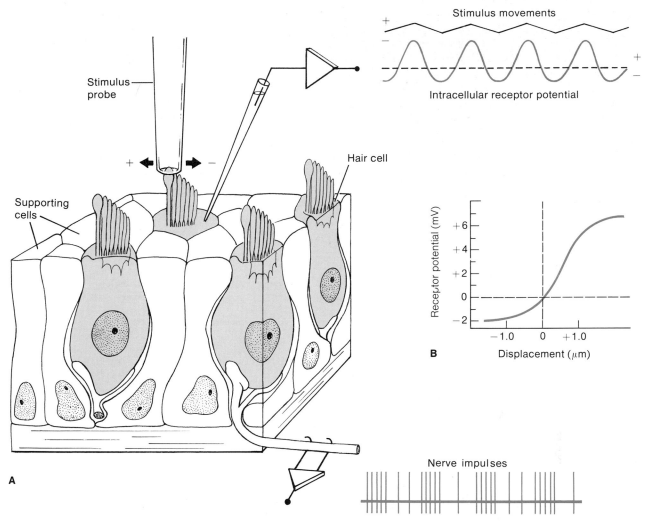

**7-25** Electrical signals recorded from hair cells in the bullfrog sacculus. (**A**) Stimulus pipette placed over cilia of hair cell was used to produce a linear back-and-forth motion at 10 Hz while intracellular potential was recorded with a microelectrode inserted into the hair cell. [Hudspeth and Corey, 1977.] Extracellular recording from the axon at bottom shows impulses arising in a branch (not shown) of the axon that innervates the cell being stimulated. (**B**) Amplitudes of receptor potentials plotted against amplitudes of stimulus pipette excursions. Note that depolarizing responses (+) produced by movement toward the kinocilium were stronger than hyperpolarizations (−) in response to movement away from the kinocilium. The significance of this is explained in the text. [Russell, 1980.]

produced by a similar displacement of the cilia in the opposite direction (hyperpolarization). This asymmetry is important because in hair cells subjected to symmetrical vibrations such as sound waves, the receptor potential can faithfully "follow" the alternating phases of the stimulus only up to frequencies of several hundred hertz (Hz). At higher frequencies, the vibrations lead to a steady depolarization, since the depolarizing direction of stimulus is more effective than the hyperpolarizing direction. The steady depolarization at such high frequencies produces steady transmitter release by the hair cell and hence high-frequency firing of the afferent nerve (Russell, 1980).

Hair cells present some remarkable problems in sensory transduction. One question that arises is how the directional sensitivity is achieved. Why does displacement of the cilia in one direction produce a depolarization, while displacement in the other direction produces hyperpolarization? One hypothesis, based on the premise that the transduction takes place in a stretch-sensitive region of membrane next to the tallest cilia, is that the tension on this part of the membrane is modulated higher or lower by tilting of the bundle of stereocilia. The changes in stress on the stretch-sensitive membrane may produce corresponding changes in ionic permeability, which would account for the depolarization and hyperpolarization produced when the cilia are displaced toward and away from the kinocilium, respectively.

### ORGANS OF EQUILIBRIUM

The simplest organ that has evolved to detect position with respect to gravity and to detect acceleration is the invertebrate *statocyst*. Forms of this organ are found in a number of animal groups, ranging from jellyfish to vertebrates. The organ consists of a hollow cavity lined with mechanoreceptor cells that are generally ciliated and that make contact with a *statolith* (sand grains, calcareous concretions, etc.). The statolith is either taken up from the animal's surroundings or secreted by the epithelium of the statocyst. In either case, the statolith has a higher specific gravity than the surrounding fluid. If a lobster, for example, is tilted to one side, the statolith stimulates the receptor cells on that side of the statocyst, causing a tonic (steady) discharge in the sensory fibers of the stimulated receptor cells

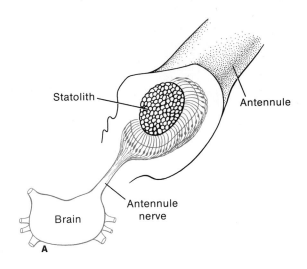

**A**

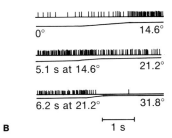

**B**

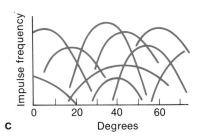

**C**

7-26  Statocyst of the lobster. **(A)** A statolith rests on the sensory processes of an array of neurons. **(B)** Electrophysiological records from the nerve fibers in response to the tilting of the lobster. The trace below the recording indicates tilting. **(C)** A plot of impulse frequency versus position of the animal shows that each cell responds with maximum discharge at an optimum position. [Horridge, 1968.]

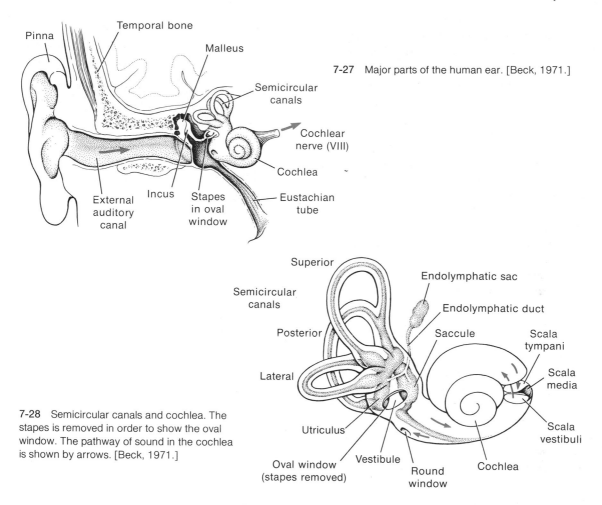

7-27　Major parts of the human ear. [Beck, 1971.]

7-28　Semicircular canals and cochlea. The stapes is removed in order to show the oval window. The pathway of sound in the cochlea is shown by arrows. [Beck, 1971.]

(Figure 7-26). Each cell has an optimum angle at which its discharge is maximal. The sensory discharge produced by tilting the lobster sets up reflex movements of the appendages, which right the animal.

The vertebrate organs of equilibrium reside in the *membranous labyrinth* that develops from the anterior end of the lateral-line system. It consists of the *sacculus* and the *utriculus,* two bony chambers. The utriculus gives rise to the three *semicircular canals* of the inner ear, which lie in three mutually perpendicular planes (Figures 7-27 and 7-28). Their function is to detect changes in the rate of rotation or translation of the head. As the head is accelerated in one of these planes, the inertia of the endolymph fluid in the corresponding canal results

in relative motion of the endolymph past a gelatinous projection, the *cupula.* Movement of the cupula stimulates the hair cells at its base. The hair cells are all oriented with the kinocilium on the same side. Thus, the hair cells in the cupula are all excited by acceleration of the fluid in one direction and inhibited by acceleration in the other direction. The three semicircular canals are admirably suited to detecting movements of the head in three-dimensional space.

Below the semicircular canals, the large bony chambers contain three more patches of hair cells. Associated with these *maculae* are mineralized concretions termed *otoliths.* The otoliths function very much like invertebrate statoliths, except that they do not appear to

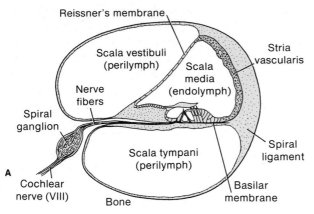

**7-29**  Details of mammalian cochlear anatomy. (**A**) Cross section through cochlear canal, showing the two outer chambers and the organ of Corti in the central canal attached to the basilar membrane. (**B**) Arrangement of the hair cells in the organ of Corti. Only the cilia of the outer hair cells are in contact with the tectorial membrane.

be important in signaling acceleration, which is taken care of by the semicircular canals. The otoliths signal position relative to the direction of gravity, and in the lower vertebrates are used to detect vibrations such as sound waves. The sensory signals from the semicircular canals are integrated with other sensory input in the cerebellum for the control of postural and other motor reflexes.

## The Mammalian Ear

### STRUCTURE AND FUNCTION OF THE COCHLEA

The hair cells of the mammalian ear are located in the *organ of Corti* in the *cochlea* (Figure 7-29). They resemble the hair cells of the lateral-line system of lower vertebrates except that the kinocilium is absent in the adult, leaving only the stereocilia. Most of the structures of the ear aid in the transformation of sound

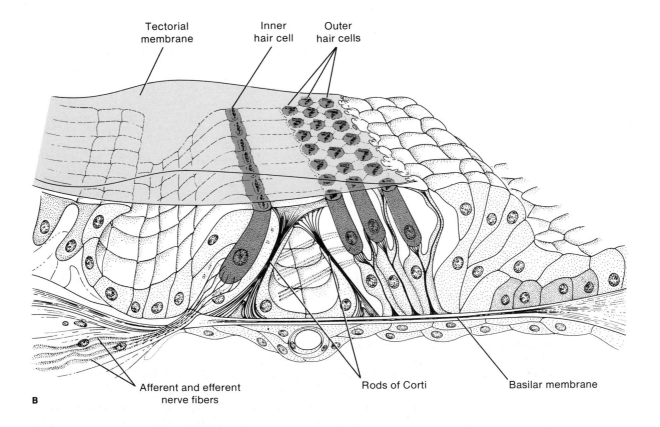

waves (airborne vibrations) into movements of the organ of Corti, which stimulate the hair cells; the hair cells, in turn, excite the sensory axons of the auditory nerve.

Among the vertebrates, only mammals possess a true cochlea, although birds and crocodilians have a nearly straight cochlear duct that contains some of the same features, including the basilar membrane and the organ of Corti. The other vertebrates have no cochlear ducts. Detection of sound waves in the lower vertebrates is carried out by hair cells associated with the otoliths of the utriculus and sacculus and with the *lagena* (one of the three maculae). We will now consider how frequency and intensity of sound are encoded by the mammalian cochlea and the sensory axons that innervate it.

The cochlea, a tapered tube encased in the mastoid bone, is coiled somewhat like the shell of a snail. It is divided internally into three longitudinal compartments (Figure 7-29). Continuity exists between the two outer compartments (*scala tympani* and *scala vestibuli*) via the *helicotrema*, an opening located near the apical (tip) end of the cochlea. This space is filled with an aqueous *perilymph*. Between these compartments, and bounded by the *basilar membrane* and *Reissner's membrane*, is another compartment, the *scala media*, filled with a fluid *endolymph*. The organ of Corti, which bears the hair cells, lies within the scala media and sits upon the basilar membrane.

The adult mammal has four rows of hair cells—one inner and three outer rows. The stereocilia of the three outer rows of outer hair cells make contact with the overlying *tectorial membrane*, but the inner row makes no such direct contact. The cilia of the inner row are believed to be stimulated by movements of the viscous mucus covering the tectorial membrane and therefore to be sensitive primarily to the velocity of displacement. The direct contact of the outer hair cells with the tectorial membrane allows sensitivity to the extent of deflection, and not merely to the velocity of deflection. The hair cells are arranged in orderly rows. The stereocilia of the outer rows are arranged in V-shaped arrays (Figure 7-29), with the tallest cilia at the apex of each array.

Airborne vibrations impinging on the *eardrum* are transmitted through the auditory ossicles and through the *oval window* to the perilymph (Figure 7-28). The ossicles amplify the pressure of vibrations set up in the eardrum by airborne vibrations. Because the area of the oval window is only about 1/25th the area of the eardrum, the sound energy is concentrated in a smaller area, and the force per unit area (pressure) is increased. These conditions are important, as the inertia of the fluid is greater than that of air.

Vibrations applied to the oval window pass through the cochlear fluids and the membranes (Reissner's and basilar) separating the cochlear compartments before dissipating their energy through the membrane-covered *round window*. The distribution of perturbations within the cochlea depends on the frequencies of vibrations entering the oval window. To visualize this, imagine a displacement of the eardrum transferred through the ossicles of the middle ear to the oval window. Very long-wave, low frequencies displace the incompressible perilymph along the scala vestibuli, through the helicotrema, and back through the scala tympani toward the round window. In contrast, short-wave, rapid displacements due to high sound frequencies show a greater tendency to take a shortcut from the scala tympani to the scala vestibuli by displacing the intervening membranes and endolymph of the scala media before traveling far from the base of the cochlea. This view is based on Georg von Békésy's observations of exposed cochleae. His studies showed:

*1.* The perturbations of the basilar membrane in response to a pure (sine wave) tone have the same frequency as the tone.

*2.* These perturbations, regardless of frequency, move as a traveling wave over the *whole length* of the basilar membrane.

*3.* The location along the basilar membrane of the maximum amplitude of the wave is a function of the frequency of the tone.

Thus, each point along the basilar membrane is displaced most effectively by some unique frequency.

### EXCITATION OF THE COCHLEAR HAIR CELLS

Electrical recordings from several places in the cochlea show fluctuations in electric potential that are similar in frequency, phase, and amplitude to the sound waves that produce them. These *cochlear microphonics,* as they are called, result from the summation of receptor currents from the numerous hair cells stimulated by movements of the basilar membrane. The receptor currents of the hair cells faithfully signal movements of the basilar membrane over the whole audible sound-frequency spectrum (about 20–20,000 Hz in young humans). The hair cells make synaptic contact with sensory axons of the *eighth cranial nerve,* axons that terminate in the cochlear nucleus. Synaptic output of the hair cells modulates the firing rate in the auditory nerve fibers. In response to *low* frequencies of sound, these afferent nerve fibers discharge impulses in synchrony with the receptor potentials produced in the hair cells. But at sound frequencies approaching 1000 Hz and beyond, the rate of discharge is limited by the time constant of the receptor cell membrane, as well as by the refractoriness of the axons. (The coding of tonal frequency is discussed in the next subsection.)

Serious gaps remain in our understanding of the way that movements of the basilar membrane produce receptor potentials in the hair cells. The initial transductional step appears to be a lateral shearing (i.e., relative sliding) component of motion between the organ of Corti and the tectorial membrane, to which the stereocilia of the outer hair cells are attached. The shearing component of the movement is a consequence of the arrangement of the pivot points of the basilar and tectorial membranes (Figure 7-30). Because of this arrangement, a perturbation of the basilar membrane applies a lateral displacement to the tips of the stereocilia. This displacement presumably increases or decreases the tension on the mechanoreceptor membrane near the attachments of the cilia at the apical end of the hair cell. According to this theory, an upward displacement of the basilar membrane produces a shearing movement that bends the stereocilia in the direction of the tallest cilia, causing an increase in conductance of the receptor membrane and a flow of current from the scala media through the receptor membrane into the hair cell (Figure 7-31). Conversely, a downward displacement of the basilar membrane will cause the stereocilia to be bent in the opposite direction, relieving some of the tension on the receptor membrane and reducing the resting-state depolarizing receptor current.

The emf responsible for the receptor current origi-

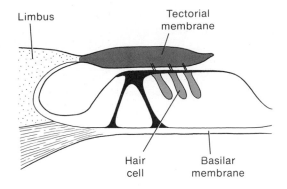

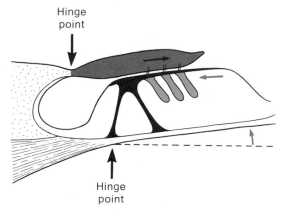

**7-30**  Diagram of the presumed shearing force applied to the cilia as a result of an upward movement of the basilar membrane. The tectorial membrane slides over the organ of Corti because the tectorial membrane and the basilar membrane have laterally displaced points of pivoting. The movements are greatly exaggerated. [Davis, 1968.]

nates in part from an electrogenic pumping of $K^+$ from the blood into the endolymph by the *stria vascularis* (Figure 7-31), a capillary bed specialized for active transport. The stria thereby sets up a battery, generating a large positive potential of about $+80$ mV in the scala media relative to the scala tympani. This potential is in series with the resting potential of the hair cell, about $-60$ mV, giving a total emf at the receptor membrane of about 140 mV. It is conjectured that stimulation of the hair cell changes the resistance of the receptor membrane at its apical end (variable resistance in Figure 7-31), and that this modulates the flow

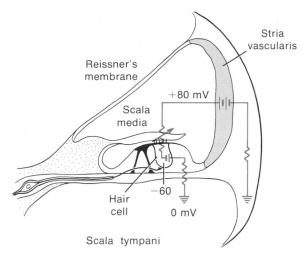

**7-31** Equivalent circuit for excitation of the hair cells. The stria vascularis is the source of a large positive potential (ca. +80 mV relative to the scala tympani) in the endolymph of the scala media. Modulation of the conductance of the sensory membrane at the apical end of the hair cell by movement of the cilia alters the receptor current flowing into the hair cell. [Davis, 1968.]

of current from the endolymphatic space through the hair cell to the perilymphatic space, producing receptor potentials in the hair cell. The receptor potentials modulate the entry of $Ca^{2+}$ at the basal end and hence the release of the synaptic transmitter that influences the discharge of the afferent fibers in the auditory nerve.

### FREQUENCY ANALYSIS BY THE COCHLEA

As we have noted, for frequencies exceeding about 1000 Hz, the time constant of the hair cell and the impulse capability of the axons of the auditory nerve prevent a 1:1 correspondence between sound waves and electrical signals. Thus, we must look for some mechanism other than the frequency of nerve impulses to inform the central nervous system of the sound frequency.

Hermann von Helmholtz noted in 1868 that the basilar membrane consists of many transverse bands increasing in length from the proximal end (round window) to the apical end of the basilar membrane. This reminded him of the strings of a piano and led to his *resonance theory,* which proposed that various parts

along the length of the basilar membrane vibrate to the exclusion of other parts of the membrane in resonance with a specific tonal frequency (just as the appropriate string of a piano will resonate in response to a tone from a tuning fork). This theory was later challenged by von Békésy (1960), who found (as noted earlier) that the movements of the basilar membrane are *not* standing waves like those of a piano (as Helmholtz suggested), but consist instead of *traveling waves* that move from the narrow end of the basilar membrane toward the broader apical end (Figure 7-28). These waves have the same frequency as the sound entering the ear, but their velocity is much less than that of sound in air.

A familiar example of a traveling wave is seen when a rope, secured at one end, is given a shake at the free end. The basilar membrane differs from the rope in that its mechanical properties change along its length. The progressive increase in mechanical compliance (i.e., looseness) of the basilar membrane—from its narrow end toward its broad end—results in changes in the amplitude of traveling waves as they move along the membrane (Figure 7-32). The position at which the displacement of the basilar membrane has its maximum amplitude (and hence maximum stimulation of hair cells and sensory axons) depends on the frequency of the traveling waves and, consequently, on the frequency of the stimulating sound. At high sound frequencies, the traveling waves produce maximum

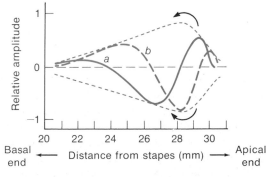

**7-32** Profile of a traveling wave moving along the basilar membrane, stopped at two times, *a* and *b.* The light broken lines show the envelope generated by the movements. Amplitudes are greatly exaggerated. [Von Békésy, 1960.]

displacements near the basal end of the cochlea. At low sound frequencies, the region of maximum displacement of the basilar membrane shifts toward the apical end of the cochlea. The extent of displacement at any point along the basilar membrane determines how strongly the hair cells are stimulated, and so it also determines the rate of discharge of sensory fibers arising from different parts of the basilar membrane. It is in this way that the basilar membrane and its receptor cells analyze the sound-frequency spectrum. As explained in Box 7-1, frequency discrimination, at least for sounds that exceed the frequency capabilities of individual nerve fibers, is further heightened by neural contrast-enhancing mechanisms in the central nervous system.

It should be stressed that the diagrams of traveling waves in Figure 7-32 exaggerate greatly the actual amplitudes of displacement of the basilar membrane. The loudest sounds produce displacements of the basilar membrane of only about 1 $\mu$m. It has been calcu-

---

### Box 7-1    *Von Békésy's Model of the Cochlea*

How can a pure tone of one frequency be resolved from other tones if traveling waves pass along the entire basilar membrane, mechanically stimulating all the hair cells along its entire length? Georg von Békésy provided an important insight by experimenting with a mechanical model of the basilar membrane (part A of figure). The model "basilar membrane," which is tapered along its length like its biological counterpart, consists of a thin metal membrane with a rib along its length. Traveling waves of selected frequencies are set up along the "basilar membrane" by sine wave perturbations produced by an electrically activated mechanical driver at one end. The driver is analogous to the stapes on the oval window. As the "sound" frequency is changed, the maximum excursion associated with the traveling wave changes to a new location along the forearm.

A subject's forearm is placed on the rib, so the tactile receptors in the skin of the arm can feel the movements of the "basilar membrane." The tactile receptors in this model are analogous to the cochlear hair cells along the length of the real basilar membrane.

The important finding made with this model is that vibrational waves traveling along the entire length of the forearm are felt in only one narrow transverse band (part B of figure). The band of sensation moves along the arm as the frequency of stimulation is altered and corresponds to the region at which the traveling wave in the artificial basilar membrane reaches its maximum amplitude. Presumably the subject "feels" the vibration only in the region of maximum displacement because the strong sensory signals arriving in the brain from the region of maximum displacement inhibit weaker signals arriving from surrounding regions (a form of lateral inhibition). It is believed that a form of lateral inhibition takes place in the central nervous system (specifically, in the cochlear nucleus), and that this accounts in part for the auditory system's ability to resolve tones sharply, even though traveling waves stimulate receptors along the entire length of the basilar membrane.

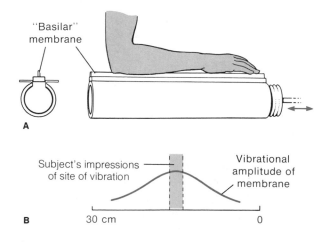

Von Békésy's mechanical model of the basilar membrane. A traveling wave moving along the artificial basilar membrane produces a tactile sensation at only one locus, although the entire skin touching the artificial basilar membrane is stimulated by traveling waves. The position of the sensation corresponds to the position of maximum traveling-wave vibration. [Von Békésy, 1955.]

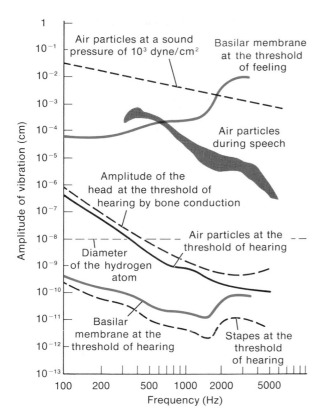

**7-33** Amplitudes of vibration of air particles and basilar membrane at different sound frequencies and intensities. Note the dimensions relative to the size of the hydrogen atom. [Von Békésy, 1962.]

lated that for sound levels that are just perceptible to the mammalian auditory system, the amplitude of vibration of the basilar membrane may be as small as $10^{-11}$ cm (Figure 7-33). It seems incredible that such minuscule movements (more than 1000 times smaller than the diameter of a hydrogen atom) should have any effect on the receptor cells. At this time, we can only assume that such movements produce small changes in the firing rates of a large number of sensory fibers, and that these changes are detected statistically in the auditory centers of the brain.

The most highly evolved auditory systems are found in animals that sense their surroundings by sending out sounds to produce echos from the objects around them. Bats and porpoises have evolved exquisite audi-

tory capabilities enabling them to find or avoid objects as small as fruit flies on the wing. This phenomenon, called *echolocation,* is discussed in Chapter 8.

## Electroreceptors

Hair cells located in the skin of certain fishes have lost their cilia and have become modified for the detection of electrical currents in the water. The sources of these currents are the electric organs found in *weak electric fishes* (such as the Mormyrids) or external currents that originate in the active tissues of other animals in the vicinity. The electroreceptors are distributed over the body in the lateral-line system (Figure 7-34A).

In weak electric fishes (as opposed to *strong electric fishes,* such as the electric eel), electrical pulses produced by modified muscle or nerve tissue at one end of the body reenter the fish over much of its surface, through pores in the epithelium. At the base of each pore, the current encounters an *electroreceptor cell* (Figure 7-34B), which makes synaptic contact with eighth-nerve axons that innervate the receptors of the lateral-line system. The receptor-cell membrane that faces the exterior has a lower electrical resistance than the membrane at the base of the cell. The result is that the membrane at the base limits the current flow through this cell and responds to a current flowing from outside the fish into the pore with a positive-going shift in membrane potential (depolarization). The depolarization of the membrane at the base of the receptor cell causes the release of synaptic transmitter to exceed the spontaneous rate of release and hence increases the frequency of firing in the sensory nerve fiber that innervates the receptor. Conversely, a current flowing out of the body of the fish hyperpolarizes the membrane at the base of the receptor cell and reduces the release of transmitter to less than the spontaneous rate. Thus, the firing frequency goes up or down depending on the direction of the current flowing through the electro-receptor cell (Figure 7-34B and C). The sensitivity of these receptors and their innervation, like that of the hair cells of the vertebrate ear, is truly remarkable. As seen in Figure 7-34C, changes in sensory nerve discharge occur in response to changes in receptor-cell membrane potential of as little as several microvolts (millionths of a volt).

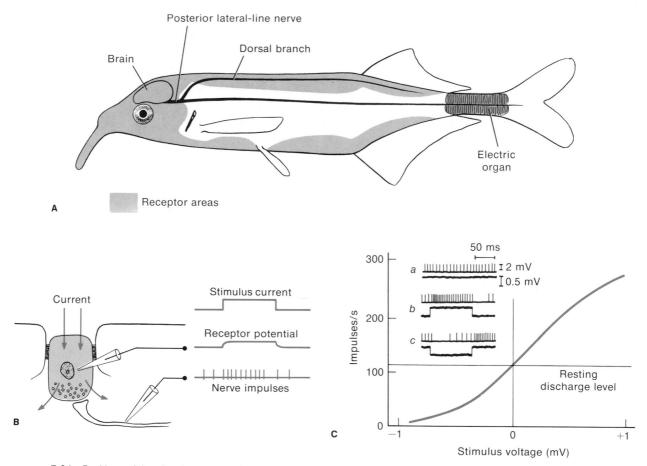

**7-34**    Positions of the electric organ and the lateral-line nerve trunk and the distribution of electroreceptor pores in the weak electric fish *Gnathonemus petersii*. **(B)** At the base of each pore is an electroreceptor cell whose apical membrane has a low electrical resistance compared to that of the basal membrane. **(C)** The spontaneously active receptor (*a*) is extremely sensitive to inward (*b*) and outward (*c*) current through the receptor pore. [Bennett, 1968.]

Electroreceptors capable of detecting minute fields are used extensively for orientation, communication, and detection of objects by weak electric fishes and by certain nonelectric fishes, such as catfishes and elasmobranchs. The latter can detect prey by the electric currents produced by the active tissues of the prey, even when the prey is buried in the bottom sand. The lateral-line system of elasmobranchs is developed extensively over the head. Each electroreceptor lies at the base of an *ampulla of Lorenzini* that connects to the surface through a long, narrow tube.

Certain eels, torpedoes, and other fishes produce a powerful discharge of current to stun enemies and prey. In contrast, the weak electric fishes produce a series of synchronous depolarizations in the cells of their electric organs. The currents of these cells sum to produce a train of current pulses that flow through the water from the posterior to the anterior end of the fish (Figure 7-35). Any object whose conductivity differs from that of the water will distort the lines of current flow. The lateral-line electroreceptors detect the distribution of current flowing back into the fish through the

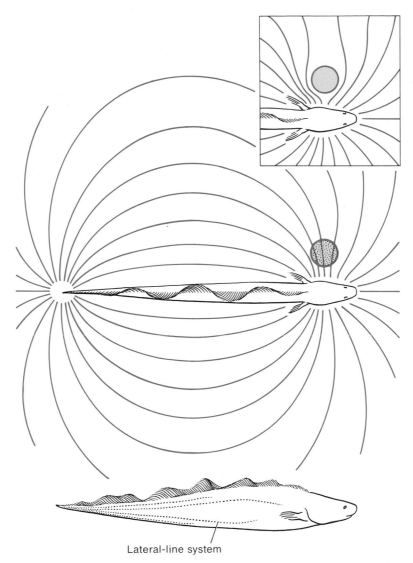

Lateral-line system

**7-35** Current flowing between the electric organ (in the tail) and receptor pores in the head. An object having a conductivity greater than the water deflects current toward the axis of flow. An object whose conductivity is lower than that of the water (inset) diverts the current away from the axis of flow. [From "Electric Location by Fishes" by H. W. Lissman. Copyright © 1963 by Scientific American, Inc. All rights reserved.]

lateral-line pores on the head and anterior end of the body, and this sensory information is then processed in the greatly enlarged cerebellum of the fish, enabling it to detect and locate objects in its immediate environment.

## Thermoreceptors

Certain nerve endings are specialized for detecting and signaling temperature and/or changes in temperature of the skin. Selected neurons in the central nervous

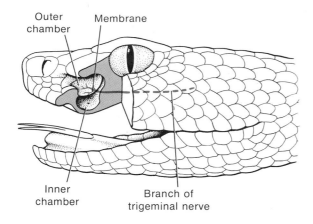

Outer chamber
Membrane
Inner chamber
Branch of trigeminal nerve

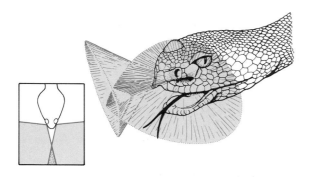

7-36 (A) Structure of the facial pit of the rattlesnake *Crotalus viridis.* (B) The positions of the facial pits lend directional sensitivity to the thermoreceptors within the pits. [Bullock and Diecke, 1956.]

systems of vertebrates are highly sensitive to absolute temperature. Especially important in the homeotherms (mammals and birds), they monitor the temperature of the internal organs and provide feedback to the temperature-regulating mechanisms of the body.

Relatively little is known about thermoreception other than some relations between stimulus and response. Some temperature receptors have remarkable sensitivity. The infrared detectors in the facial pits of rattlesnakes (Figure 7-36A) are warmth receptors that increase their firing rate transiently in response to temperature increases as small as 0.002 °C. This sensitivity

is also seen in behavioral tests. A rattlesnake can detect the radiant heat (infrared wavelengths) from a mouse 40 cm away if the mouse is 10°C above the ambient temperature. Because these receptors are located on a connective tissue membrane deep within the facial pits, the snake is able to detect the direction of the infrared source (Figure 7-36B). The infrared-receptor organ consists of the branching endings of sensory nerve fibers and shows no obvious structural specializations.

The mechanisms by which temperature changes alter receptor output are not known. Both the skin and the upper surface of the tongue in mammals have two kinds of temperature receptors, those that increase firing when warmed ("warm" receptors) and those that increase firing when cooled ("cold" receptors). These categories are distinguished on the basis of responses to temperature changes near normal body temperature (about 37°C), because both kinds of receptor show peaks beyond which the steady-state response to temperature changes is reversed (Figure 7-37). The transient response, however, always reliably signals a single direction of temperature change.

## Photoreceptors

All photoreceptors have in common the presence of photoexcitable pigments associated with the receptor membrane. These photopigment molecules, which are, of course, the primary sites of photoreception, are structurally altered by the absorption of light. In the invertebrates, the photopigment is present in profuse miniature evaginations of the surface membrane termed *microvilli* (Figure 7-38). In the vertebrates, the membrane elaborations containing the photopigment molecules are *lamellae* in the form of either flattened foldings of the surface membranes or such foldings that have pinched away from the surface membrane to become internal disks. The vertebrates and the invertebrates have also evolved different mechanisms by which the primary transduction process influences the conductance of the photoreceptor membrane and alters the membrane potential of the receptor cell. Let us first consider the simpler of the two classes of mechanisms.

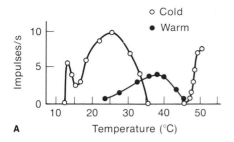

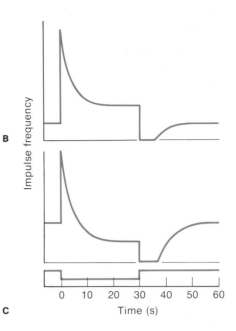

**7-37** (**A**) Responses of ''cold'' and ''warm'' receptors of the mammalian tongue. (**B**) Frequency changes of a ''cold'' fiber upon cooling, then rewarming, at temperatures above the point of maximum resting frequency. (**C**) Same as in B, except at temperatures below the point of maximum resting frequency. [Zotterman, 1959.]

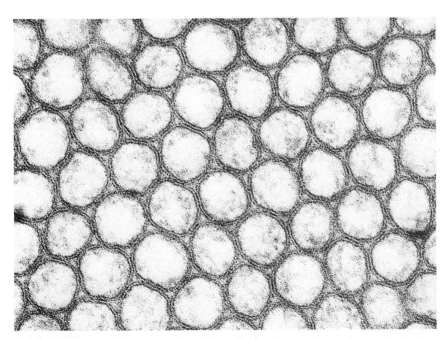

**7-38** Electron-microscopic view of a rhabdomere in *Limulus*. Note the profusion of microvilli (shown in cross section), the site of visual pigment concentration. Magnification 170,000 ×. [Courtesy of A. Lasansky.]

## INVERTEBRATE PHOTORECEPTOR CELLS

Invertebrate photoreceptors and their organization into eyes show great variety, but they appear to exhibit principles common to all. The most intensively studied invertebrate photoreceptors have been those in the *lateral eye* and the *ventral eye* of the horseshoe crab *Limulus polyphemus.*† Most of the early electrical recording from single visual units utilized the lateral eye, for this *compound eye* lends itself well to simple electrical recording techniques.

The receptor cells at the base of a single *ommatidium* in the lateral eye of the horseshoe crab are shown in Figure 7-39. Ommatidia are the individual functional units of the compound eye. Each ommatidium lies beneath a hexagonal facet of the cornea. A central dendrite arising from the *eccentric cell* is surrounded by about 12 *retinular cells,* the primary photoreceptors. Each retinular cell has a *rhabdomere,* consisting of the surface membrane thrown into a dense profusion of microvilli (Figure 7-38). Thus, the surface area of the cell membrane is greatly expanded in the rhabdomere. Light enters through the lens and is absorbed by molecules of the photopigment located in the receptor membrane forming the rhabdomere. Transient, random depolarizations of the membrane potential occur in the retinular cells of *Limulus* when the eye is exposed to very dim steady illumination. These "quantum bumps" on the recording increase in frequency when photons are allowed to impinge on the receptor with greater frequency. The transient depolarizations are evidently the electrical signals generated by the absorption of individual quanta of light by single photopigment molecules. A single photon captured by a single visual pigment molecule in *Limulus* leads to a receptor current of $10^{-9}$ A. The power amplification of this transduction represents between $10^5$ and $10^6$ times the energy of the light quantum. How does a single photon captured by a molecule of photopigment lead to the rapid release of so many times its own energy? One obvious possibility is that, as a consequence of the bleaching of the pigment molecule, an enzyme becomes activated that catalyzes the production of a

†This "living fossil" is an arachnid, closely related to the spiders; it is not a crustacean.

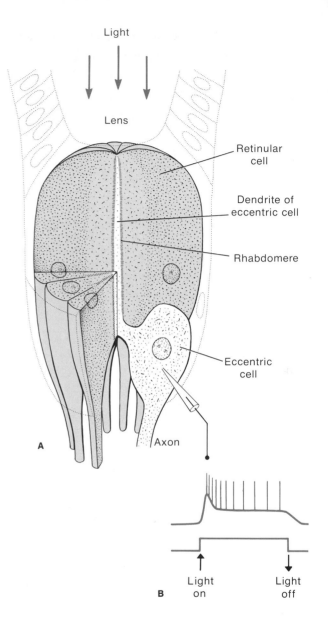

**7-39**   Ommatidium in the compound eye of the horseshoe crab *Limulus polyphemus.* (**A**) Light enters via the lens at the top and is intercepted by visual pigment in the retinular cells, which are arranged like the segments of an orange around the dendrite of the eccentric cell. [From "How Cells Receive Stimuli" by W. H. Miller, F. Ratliff, and H. K. Hartline. Copyright © 1961 by Scientific American, Inc. All rights reserved.] (**B**) Intracellularly recorded signal in response to a light stimulus. Receptor potential arising in the retinular cells spreads through gap junctions to the eccentric cell, which produces action potentials. [Fuortes, 1959.]

large number of intracellular transmitter molecules, each of which then causes the opening of one channel in the membrane for the diffusion of many ions down their electrochemical gradient.

In *Limulus,* the receptor current that flows through the light-activated channels is carried by $Na^+$ and $K^+$. This current leads to a depolarizing receptor potential by principles similar to those that operate when ACh activates the motor endplate channels in muscle. When the light goes off, these channels close again, and the membrane repolarizes.

Although in *Limulus* the photoreceptors in the lateral eye (the retinular cells) have axons, they apparently do not support action potentials. Instead, the receptor current spreads electrotonically through low-resistance gap junctions (electrical synapses) from the retinular cells into the dendrite of the eccentric cell. The depolarizing current spreads into the cell body and axon of the eccentric cell, and there the electrotonic depolarization leads to the generation of action potentials (Figure 7-39B), which are conducted in the optic nerve by the axon of the eccentric cell to the central nervous system. Thus, the one axon of the eccentric cell carries all the information for the ommatidium. This is just as well, for the retinular cells are all electrically coupled and act with the eccentric cell as one unit. In the next chapter, we will see how the neighboring ommatidia interact through lateral nerve branches in the initial stages of processing the visual information.

## CORRELATES OF VISUAL PHENOMENA

Studies on the *Limulus* eye were begun by H. K. Hartline in the mid-1930s. By recording impulses in the axons of the optic nerve, he was able to correlate primary receptor activity with stimulus parameters. One of the most significant generalizations that came from his work is that a number of the subjective phenomena of human vision parallel the electrical behavior of single visual cells. This suggests that some of the very simple features of visual perception originate in the behavior of the photoreceptor cells and are relatively unmodified by the nervous system. Examples follow.

Impulse frequencies recorded from the axons of single ommatidia are proportional to the logarithm of the stimulating light intensity (Figure 7-40, right-hand column). This logarithmic relation is also true of a human subject's judgment of intensity in comparing different light intensities.

The receptor's response to light flashes less than 1 s in duration is proportional to the total number of photons in the flash, irrespective of flash duration (Figure 7-40). That is, the number of impulses generated remains constant, provided that the product of intensity and duration of the flash are kept constant. This is reasonable, because the response should be determined (within limits) by the number of photopigment molecules altered by the photons impinging on the receptor. For flashes of short duration, a human

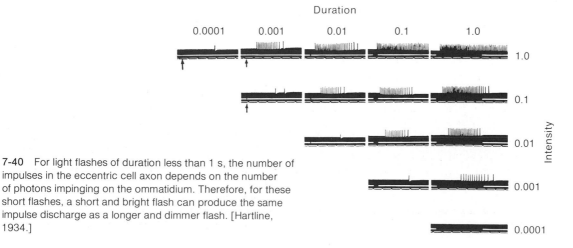

**7-40**  For light flashes of duration less than 1 s, the number of impulses in the eccentric cell axon depends on the number of photons impinging on the ommatidium. Therefore, for these short flashes, a short and bright flash can produce the same impulse discharge as a longer and dimmer flash. [Hartline, 1934.]

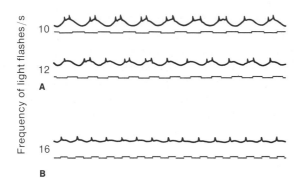

observer cannot tell the difference between reciprocal changes in intensity and duration of the flash.

The frequency of impulses in the eccentric cell is directly proportional to the amplitude of the receptor potential. If the receptor is stimulated with a flickering light, the membrane potential will follow the flashes up to nearly 10 Hz (Figure 7-41). Beyond this frequency, the receptor potential can no longer follow the flashes; hence the ripples in the membrane potential fuse into a steady level of depolarization. The impulses then no longer group with the flashes, but show a steady rate of discharge. When grouping of impulses fails, the central nervous system no longer receives information about the rate of flicker, and the light might as well be constant. This appears to be the reason why a human cannot tell the difference between a steady light and one (such as a lamp powered by a 60 Hz current) that flickers at a rate above the *critical fusion frequency.*

### PHOTORECEPTOR ADAPTATION IN LIMULUS

Photoreceptor cells exhibit *light adaptation:* they reduce their sensitivity (i.e., adapt), avoiding a saturation of their response over the upper end of the intensity range. The ventral photoreceptor of *Limulus* is much simpler in morphology than the lateral eye, and has proven more favorable than the compound eye in recent years for ionic and metabolic studies of receptor mechanisms. Using the ventral photoreceptor cells in voltage-clamp experiments, visual physiologists have found that light adaptation (i.e., loss of sensitivity during pro-

longed background illumination) depends in part on a reduction of the number of channels activated per photon absorbed by the photopigment. It was discovered that this effect depends on an elevation of intracellular ionized calcium. Experimentally raising $[Ca^{2+}]_i$ in the cytoplasm of the ventral photoreceptor by various means caused a reduction in the receptor current evoked by illumination (Figure 7-42A). Moreover, injection of the calcium-sensitive photoprotein aequorin revealed that illumination causes an elevation of intracellular $Ca^{2+}$ in the photoreceptor (Figure 7-42B). Injection of the calcium-chelating agent EGTA (p. 471) prevents this rise in $Ca^{2+}$ and also prevents the adaptation that accompanies illumination.

Other lines of evidence suggest that $Ca^{2+}$, released from intracellular storage sites by the action of light, produces adaptation (i.e., reduced sensitivity) by blocking one or more steps that lie between the absorption of light by the photopigment and the opening of the $Na^+–K^+$ channels in the surface membrane of the photoreceptor cells by the postulated intracellular transmitter agent. Thus, a light stimulus appears to have two major actions on the *Limulus* photoreceptor cell: (1) production of an intracellular messenger that activates membrane channels and thereby produces the receptor current; and (2) release of intracellular bound $Ca^{2+}$ that interferes with step 1 at some stage. This interference causes a time-dependent loss of sensitivity (i.e., adaptation) of the illuminated photoreceptor cell. In addition, strong light intensities can deplete the unbleached photopigment, causing additional loss of sensitivity and long-lasting light adaptation.

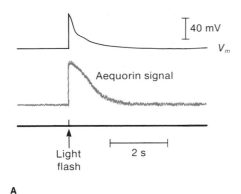

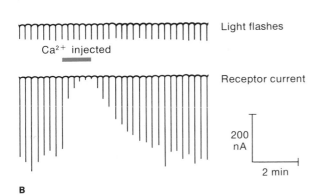

**7-42** Role of Ca²⁺ in adaptation of *Limulus* ventral photo-receptor. (A) The receptor was injected with the calcium-sensitive photoprotein aequorin. The light stimulus (lower trace) was followed by the receptor potential (top trace) and the calcium—aequorin signal (middle trace, indicating the appearance of free Ca²⁺ in the cytoplasm. (B) A different cell was stimulated repeatedly with a series of light flashes (top trace) while the receptor current across the cell membrane was monitored under a voltage clamp. Injection of Ca²⁺ into the cell caused precipitous transient reduction of the receptor current. [Brown and Blinks, 1975; Brown and Lisman, 1975.]

## VISUAL RECEPTOR CELLS OF VERTEBRATES

In mammals, birds, and many other vertebrates, the visual receptor cells (*rods* or *cones*) are distributed over the *retina* in a tightly packed mosaic. The *fovea*, or *area centralis*, is the small (1 mm²) central part of the retina specialized for highest *visual acuity* (angular discrimination). In humans, and certain other mammals with

color vision, the fovea, a pitlike depression, contains only cones, whereas the remainder of the retina contains rods as well as cones. In mammals, the cones are responsible for color vision, and the more sensitive rods are restricted to achromatic vision. This distinction between rods and cones does not pertain to all vertebrates. In fact, some retinas contain only rods but may nonetheless be capable of color vision (discussed in the next section).

The two types of vertebrate photoreceptor cells are structurally and functionally more uniform than the great variety of photoreceptors found in the invertebrates. Each receptor cell contains a rudimentary *cilium* that connects the *outer segment,* which contains the receptor membranes, to the inner segment, which contains the nucleus, mitochondria, synaptic contacts, and so on (Figures 7-43 and 7-44). The receptor membranes of vertebrate visual cells consist of flattened lamellae derived from the surface membrane near the origin of the outer segment. In the cones (Figure 7-45B) of mammals and some other vertebrates, the lumen of each lamella is open to the cell exterior. In rods, the lamellae pinch off completely during the continual growth of the outer segment so as to form flattened sacs, or disks (Figure 7-45A), which are stacked like hollow pancakes or pita bread on top of one another within the outer segment. The stack of disks is completely contained within the surface membrane of the visual cell. The *inside* surface of the membrane composing each sac corresponds to the *outside* surface of the cell membrane, because the disk is formed as an invagination of the outer membrane. The photopigment molecules have been shown to be intimately associated with these membranes. Because the disk membranes of the rod outer segment contain the photopigment, the primary step in photochemical transduction must take place on the disks rather than in the surface membrane. The earliest electrical signal recorded from the intact eye, a signal termed an electroretinogram (Box 7-2), is believed to be related to the process of phototransduction. This initial signal is followed by a complex wave of signals having their origin in different parts of the retina and pigment epithelium.

In vertebrate rods and cones, light produces a *hyperpolarizing* receptor potential (Figure 7-46) instead of

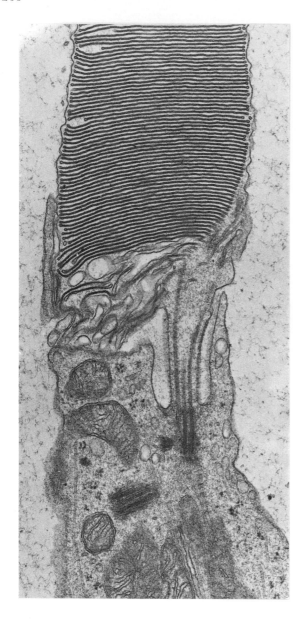

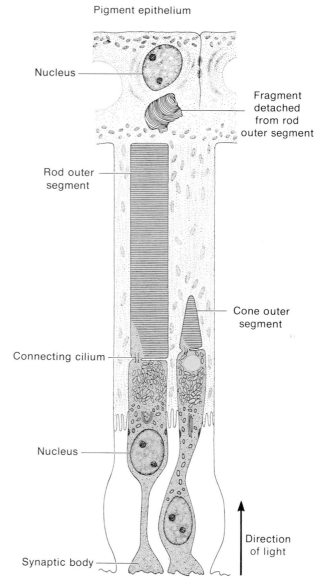

**7-43**  Electron micrograph of a portion of a visual rod. Inner (below) and outer (above) segments are attached by a modified cilium. The outer segment contains numerous flattened lamellar sacs, or disks. The inner segment contains the nucleus, mitochondria, and other common organelles. Magnification 28,000 ×. [Courtesy of T. Kuwabara.]

**7-44**  Rod and cone of the frog. Note that the outer segments are pointed away from the source of light, toward the pigment epithelium at the back of the eye. As the visual cells grow, fragments detach from the apical ends and a new membrane reseals the outer segment. [From "Visual Cells" by R. W. Young. Copyright © 1970 by Scientific American, Inc. All rights reserved.]

## *Box 7-2*    *The Electroretinogram*

It is sometimes useful to record the summed electrical activity of the eye, for this is much easier than recording from single cells with microelectrodes. The recording electrode (generally a saline wick) is placed on the cornea, and the indifferent electrode is attached to another part of the body. When a light is flashed on the eye, a complex waveform is recorded (part A of figure). This *electroretinogram (ERG)* results primarily from the summed activity of the visual cells and the neurons in the retina (Figure 8-25). After years of debate about the structural correlates of the various components of the ERG, it is now evident that the *a* wave is due to the receptor current produced by the visual receptor cells. The *a* wave is followed by the *b* wave, which results from electrical activity of the retinal neurons innervated by the receptor cells. The *c* wave is found only in vertebrates, and appears to be produced by the nonretinal pigmented epithelial cells against which the outer segments of the visual cells abut. In tadpoles, only the *a* wave is produced before synaptic contacts are established; the *a* wave is produced alone also in adult frog retinas after synaptic transmission is eliminated by means of a blocking agent.

When the stimulus is a very intense brief flash, another component can be recorded with minimum latency after the onset of the flash, before the ERG develops (part B of figure). This *early receptor potential (ERP)*, discovered in recent years, differs from the classic ERG in several fundamental respects. It increases linearly in amplitude with flash intensity (rather than logarithmically like the ERG) and saturates when the flash is intense enough to bleach all the visual pigment molecules. Fixation of the retina in gluteraldehyde abolishes the ERG, but leaves the early receptor potential intact. These lines of evidence, plus others, suggest that the early receptor potential represents a charge displacement that accompanies a change in the conformation of the pigment molecules of the receptor membrane. The visual pigment molecules are all aligned similarly so that, on

absorption of light quanta, their minute individual shifts in charge add up to generate the early receptor potential. The ERP is not the electrical event that leads to excitation of the receptor; instead, it occurs concomitantly with the structural change in the visual pigment, which is believed to be an early step in the transduction of light energy into a change in receptor-membrane conductance.

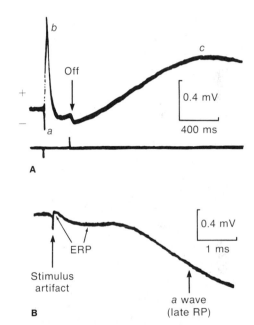

Components of the vertebrate electroretinogram (ERG).
(A) Low-gain, slow time base shows components *a*, *b*, and *c*.
(B) The early receptor potential (ERP) is seen before the *a* wave at high gain and high sweep speed, in response to a brief but intense flash. [Brown, 1974.]

a depolarization as in the photoreceptors of invertebrates. The responses of invertebrate and vertebrate visual receptor cells are compared in Figure 7-47. Membrane-conductance measurements before and during illumination have shown that the effect of light

in these vertebrate visual receptors is to *decrease* the sodium conductance of the outer segment of the membrane. In the dark, the surface membrane of the vertebrate photoreceptor is nearly equally permeable to $Na^+$ and $K^+$. The result is a resting potential about

**7-45** Formation of the receptor membranes in a rod (**A**) and cone (**B**) of the frog retina. The rod lamellar sacs become detached after folding in from the surface membrane at the base of the outer segment. [From "Visual Cells" by R. W. Young. Copyright © 1970 by Scientific American, Inc. All rights reserved.]

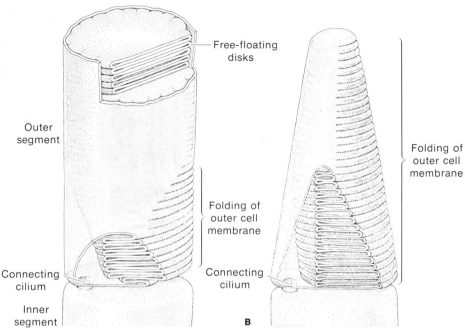

**7-46** Receptor potentials in a rod of the giant toad *Bufo marinus*. The vertical bar above and to the left of the response gives the duration of light flashes, the relative intensities of which are indicated by the numbers next to the traces. These are in units of log photons per square centimeter incident on the retina. [Fain, 1981.]

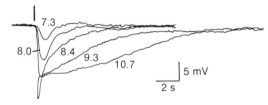

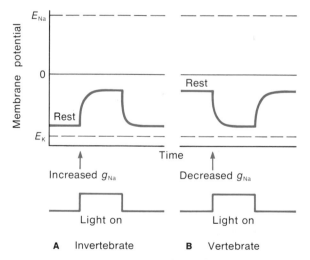

**7-47** Electrical responses to light by invertebrate and vertebrate visual cells. (**A**) Most invertebrate photoreceptors undergo an *increase* in sodium and potassium conductance in response to light. The membrane potential shifts toward $E_{Na}$, producing a depolarization. (**B**) Vertebrate photoreceptors respond with a decrease in sodium conductance. This response causes a shift away from $E_{Na}$ toward $E_K$, producing a hyperpolarization. [Toyoda et al., 1969.]

halfway between $E_K$ and $E_{Na}$. Sodium ions leak into the outer segment through sodium channels that are steadily open in the dark. This *dark current* carried into the outer segment by $Na^+$ is balanced by a current flowing out of the inner segment, carried presumably by $K^+$ (Figure 7-48A). The $Na^+$–$K^+$ pump maintains the concentration gradients in the face of the steady influx of $Na^+$ and efflux of $K^+$. The dark current is characteristic of vertebrate visual receptor cells.

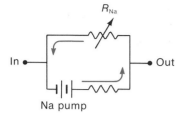

**7-48** Effect of illumination on the dark current of vertebrate rods. The sodium conductance of the rod outer segment is high in the dark (**A**) and becomes reduced in the light (**B**). For this reason, the dark current, carried by Na+ leaking into the outer segment, drops during illumination. In the equivalent circuit (top left), the battery represents the sodium pump, and the light-activated variable resistor represents the sodium conductance of the outer segment. [Hagins, 1972.]

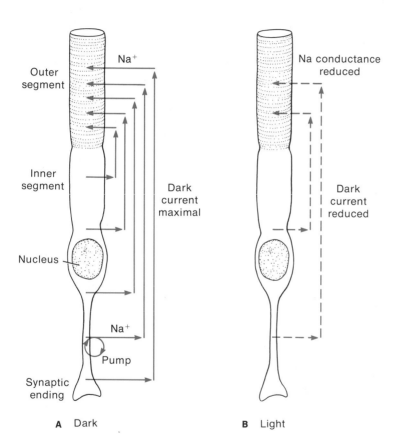

Following the light absorption by the photopigment, the sodium conductance of the outer segment decreases, the dark current decreases, and the membrane potential hyperpolarizes toward $E_K$ (Figure 7-48B), according to Equation 5 in Box 6-4. When the light ceases, the sodium conductance of the membrane increases to its high resting value and the membrane potential shifts in the positive direction back to its resting level between $E_{Na}$ and $E_K$.

How does illumination produce a *decrease* in sodium conductance? It was first proposed by W. A. Hagins and S. Yoshikami (1974) that the photochemical alteration of photopigment molecules in the membrane of a disk opens calcium channels in that membrane, and that calcium ions stored at a high concentration inside the disk then leak out into the cytoplasm of the rod outer segment and diffuse to the surface membrane of the rod, blocking the sodium channels through which the Na$^+$ current flows in the dark. Because many calcium ions can flow through a single photon-activated calcium channel, and each calcium ion can block the inflow through a sodium channel of many sodium ions, this mechanism could account for the large power amplification exhibited by the photoreceptor cell. The *calcium hypothesis* of visual excitation is supported by experiments showing that the local application of Ca$^{2+}$ or of certain drugs (theophylline, caffeine) that release internal stores of Ca$^{2+}$ mimics the effect of light in hyperpolarizing the photoreceptor cell.

If this hypothesis is correct, the function of the membrane-limited flattened disks of the outer segment is similar to that of the calcium-sequestering sarcoplasmic reticulum, which releases Ca$^{2+}$ into the myoplasm of the striated muscle cell to activate the contractile mechanism. In some vertebrate cones, the photosensitive membranes remain continuous with the surface membranes, so that the interior of each disk is continuous with the extracellular space (Figure 7-45). Calcium ions enter the cytoplasm of such cones through the invaginated membranes from the cell exterior, where the calcium concentration is high relative to the cell interior. Calcium is continually pumped out of the cytoplasm into the disks of rods and into the extracellular spaces of cones. This maintains a low cytoplasmic calcium concentration, so that in the dark, according to the calcium hypothesis, the sodium channels are open and Na$^+$ carries a strong dark current (Figure 7-49A).

In the next chapter, we see that illumination of a vertebrate visual receptor cell produces a hyperpolarization of the bipolar cell that it innervates. Thus, regardless of the mechanism by which the dark current is blocked after the action of light on the photopigment, it is an interesting fact that it is a hyperpolarization rather than a depolarization that occurs as a consequence of excitation in the vertebrate visual receptor cell. How is the hyperpolarization of the receptor cell used to signal the neurons with which it makes synaptic contacts? In the dark, the receptor cell, in its partially depolarized state produced by high sodium conductance, steadily secretes a transmitter. On hyperpolarization in response to illumination, the release of transmitter is slowed! In those cells, then, the signal for excitation is a *reduction* of transmitter release.

## Visual Pigments

The spectrum of electromagnetic radiation extends from gamma rays, with wavelengths as short as $10^{-12}$ cm, to radio waves, with wavelengths greater than $10^6$ cm (Figure 7-50). The portion of the electromagnetic spectrum between $10^{-8}$ cm and $10^{-2}$ cm is termed *light*. Only a small portion of this segment of the spectrum is visible to us, ranging from about 400 nm to about 740 nm. Below this range is the ultraviolet (UV) portion of the spectrum and above it the infrared, neither of which is visible to humans. There is nothing

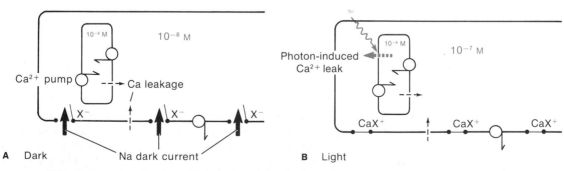

**7-49**  Proposed mechanism of light-activated drop in the sodium conductance of the outer segment. (A) In the dark, Ca$^{2+}$ is sequestered inside the disks, keeping the intracellular Ca$^{2+}$ concentration very low. (Ca$^{2+}$ values are shown in color.) Na$^+$ leakage across the surface membrane is high. (B) Light, acting on the visual pigment, causes calcium channels to open in the disk membranes. Calcium that leaks out of the disks blocks sodium channels in the surface membrane. [Hagins and Yoshikami, 1974.]

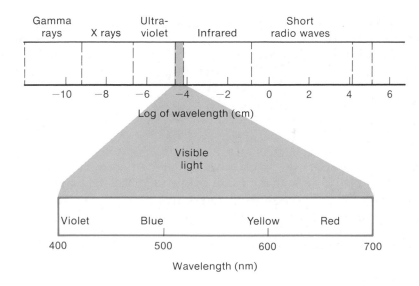

qualitatively special about those portions of the spectrum that renders them invisible to us. What we see depends on which wavelengths, on reaching the visual cells, are absorbed by *visual pigments.* For example, persons who have had their lenses (which absorb UV) removed because of cataracts can see into the UV range, which is invisible to the normal eye. The compound eyes of insects can see into the UV, and thus some flowers with patterns of UV-reflecting pigments will not look the same to insects as they do to mammals.

The energy in a quantum of radiation is equal to Planck's constant divided by the wavelength (in centimeters):

$$E = \frac{2.854}{\lambda} \text{ g-cal/mol}$$

Thus, the energy in a quantum increases as the wavelength of radiation decreases. Quanta with wavelengths below 1 nm have so much energy that they break bonds and atomic nuclei. Quanta with wavelengths greater than 1000 nm lack sufficient energy to affect molecular structure. The pigments evolved by living organisms for capturing the sun's radiant energy were selected for their maximal absorption between these limits. When a quantum of radiation is absorbed by a photopigment molecule, it raises the energy state of the molecule by increasing the orbital diameter of the electrons associated with a conjugated double bond. This process is the basis of the photosynthetic conversion of radiant energy into chemical energy in plants; it is also the basis of visual excitation in animals. It is interesting that all known organic pigments owe their photochemical properties to the presence of a carbon chain or ring having alternating single and double bonds.

## PHOTOCHEMISTRY OF VISUAL PIGMENTS

The concept that a pigment is essential for the process of photoreception originated with J. W. Draper, who concluded in 1872 that, in order to be detected, light must be absorbed by molecules in the receptor system. R. Boll found soon thereafter that the characteristic reddish purple color of the frog's retina fades (*bleaches*) with exposure to light. The light-sensitive substance *rhodopsin,* responsible for the color, was extracted in 1878 by W. Kühne, who also found that once the pigment is bleached by light, its reddish purple color can be restored by keeping the retina in the dark, provided that the receptor cells remain in contact with the pigment epithelium at the back of the eye.

Rhodopsin has since been shown to absorb light maximally at wavelengths of about 500 nm. It is found in the outer segments of rods in many vertebrate species and in the photoreceptors of many invertebrates. Rhodopsin molecules are present in the receptor membranes in high density—$5 \times 10^{12}$ molecules per square centimeter, which is equivalent to an intermolecular spacing in the membrane of about 5 nm.

All visual pigments that have been studied consist of two major components: one a protein moiety called an *opsin* and the other a prosthetic group consisting in all instances of either *retinal* (Figure 7-51; retinal is the aldehyde of the carotenoid vitamin $A_1$, which is the alcohol retinol) or *3-dehydroretinal* (the aldehyde of vitamin $A_2$, the alcohol 3-dehydroretinol). To be more specific, a visual pigment molecule is made up of (1) the prosthetic group, (2) a protein moiety to which is attached (3) a six-sugar polysaccharide chain and (4) a variable number (30 or more) of phospholipid molecules. The lipoprotein opsin (which includes 2 through 4 in the preceding list) appears to be part of the mosaic structure of the visual receptor membrane; the carotenoid prosthetic group exchanges between the visual receptor membrane and the pigment epithelium at the back of the retina during bleaching and regeneration of the visual pigment. The pigment in the epithelium is photochemically inactive and unrelated to the visual pigment. It keeps light from scattering and reflecting diffusely back toward the retina.

The effect of light on rhodopsin, representative of the effect of light on all the visual pigments, can be summarized as

$$\text{rhodopsin} \underset{\text{dark}}{\overset{\text{light}}{\rightleftharpoons}} \text{retinal} + \text{opsin}$$
$$\Updownarrow$$
$$\text{retinol}$$

In the absence of light, the opsin and the retinal are closely associated, with the retinal in the 11-*cis* configuration (Figure 7-51A). It is believed that the retinal fits snugly into a special site on the opsin (Figure 7-52). The absorption of a quantum of light isomerizes the 11-*cis*-retinal into the all-*trans* configuration, forming *lumirhodopsin* (Figure 7-53). The *cis–trans isomerization* is the only action of light on the visual pigment, and this initial step (the conversion of 11-*cis*-retinal to all-*trans*-retinal) is the only reaction in the sequence that does not proceed spontaneously. All the subsequent steps are energy-yielding reactions, and all proceed spontaneously at physiological (body) temperatures. The

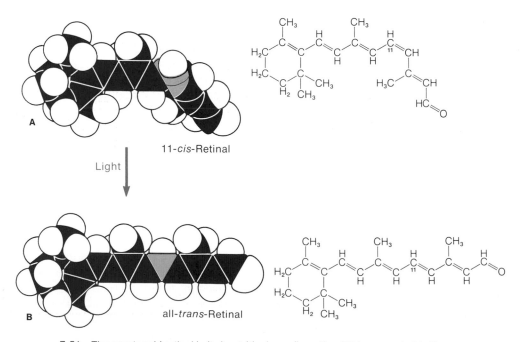

**7-51**  The carotenoid retinal in its bent 11-*cis* configuration (**A**) is converted to the straight all-*trans* configuration (**B**) by light. [From "Molecular Isomers in Vision" by R. Hubbard and A. Kropf. Copyright © 1967 by Scientific American, Inc. All rights reserved.]

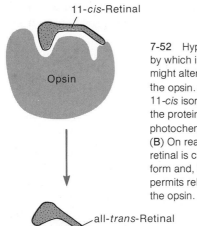

**7-52** Hypothetical mechanism by which isomerization of retinal might alter the conformation of the opsin. (**A**) In this scheme, the 11-*cis* isomer fits snugly into the protein molecule before photochemical isomerization. (**B**) On reacting with light, the retinal is converted to the all-*trans* form and, in straightening out, permits relaxation of a part of the opsin.

conversion from 11-*cis* to all-*trans* straightens out the conjugated chain of the retinal to form *metarhodopsin I*. As a result, the opsin undergoes a change in molecular conformation, because the all-*trans*-retinal no longer fits into the special site on the opsin (Figure 7-51B). It is the conformational change in the opsin that presumably results in the opening in the membrane of an ion channel, of which the opsin is believed to be a part. In this context, it is interesting that vertebrate visual pigment behaves in gel-filtration analysis as if it has undergone a 36% increase in volume on becoming bleached by exposure to light. It has therefore been suggested that the retinal group "locks" the opsin into a compact conformation (Figure 7-52A), which is altered to a less compact form when the retinal undergoes isomerization from the *cis* to the *trans* conformation by the action of light (Figure 7-52B).

Subsequent chemical permutations of the retinal (Figure 7-53) appear to be irrelevant to the excitation

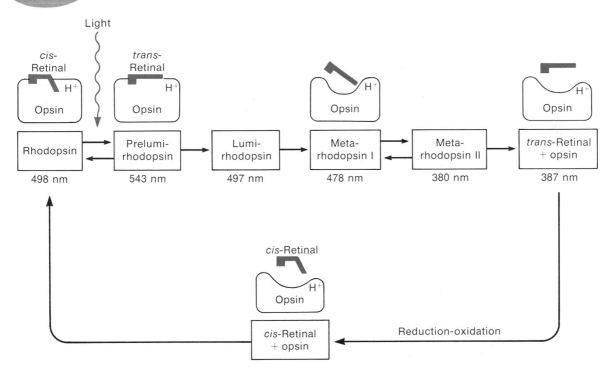

**7-53** The visual pigment cycle. A sequence of reactions beginning with photochemical isomerization of retinal leads to the dissociation of rhodopsin into retinal and the opsin. See text for details. [From "How Light Interacts with Living Matter" by S. B. Hendricks. Copyright © 1968 by Scientific American, Inc. All rights reserved.]

of the visual receptor cell, but are necessary for the regeneration of rhodopsin. *Metarhodopsin* is hydrolyzed spontaneously to retinal and opsin. The all-*trans*-retinal is then reisomerized to the 11-*cis*-retinal. This process requires the action of an enzyme and consumes chemical energy. The newly reformed 11-*cis*-retinal then combines spontaneously with the opsin to form rhodopsin. Retinal and the opsin are both reused in repeated cycles of bleaching and reconstitution. Any retinal that is lost or chemically degraded is replenished from vitamin $A_1$ (retinol) stored in the cells of the pigment epithelium, which actively take up the vitamin from the blood. A nutritional deficiency of vitamin $A_1$ is reflected in a decrease in the amount of retinal that is synthesized and hence a decrease in the amount of rhodopsin. The result is a lessened photosensitivity of the eye, commonly known as "night blindness."

Any opsin can combine either with 11-*cis*-retinal or with 11-*cis*-3-dehydroretinal to form a visual pigment molecule. For a given pigment, the *absorption spectrum* depends on the electronic substructure of the *chromophore*, the photon-absorbing portion of the pigment molecule, consisting of the carotenoid prosthetic group and closely associated portions of the opsin.

Retinal acts as the prosthetic group for three cone pigments in humans, in addition to the rod pigment, rhodopsin. Each of the cone pigments has a different absorption spectrum and is important in color vision, as described in the next subsection. The other carotenoid found in visual cells, 3-dehydroretinal—which differs from retinal by a single double bond in its ring structure—occurs in the *porphyropsins,* a class of pigments found in rods, and in other visual pigments present in cones.

The distribution of the rhodopsins, as compared with that of the porphyropsins, shows an interesting ecological trend. In land vertebrates, the visual receptor cells are supplied by the pigment epithelium with retinol only, and thus the visual pigments of land vertebrates are all rhodopsins. Rhodopsins are found in invertebrates such as *Limulus,* insects, and crustaceans. Porphyropsins are most common in freshwater and euryhaline fishes and in some amphibians. An interesting case is that of the grass frog, in which the tadpole at first has only porphyropsins, whereas the adult has only rhodopsins. During the course of metamorphosis,

dehydroretinal supplied in the bloodstream is gradually replaced by retinal. Fishes that migrate from fresh water to salt water similarly change from porphyropsin to rhodopsin in anticipation of the marine environment. There is a correlation between the presence of visual pigments with higher red-sensitivity (e.g., porphyropsins) and longer-wavelength light in the environment. Thus, it is suspected that the adaptive significance of the changeover from porphyropsin to rhodopsin during seaward migration of euryhaline fishes lies in the greater absorbance of rhodopsin for shorter-wavelength (blue) light, which is prevalent at ocean depths.

## COLOR VISION

In 1666, Sir Isaac Newton demonstrated that white light can be separated into a number of colors by passing it through a prism. Each *spectral color* is monochromatic—that is, it cannot be separated into other colors. It was already known, however, that a painter could match any spectral color (e.g., orange) by mixing two pure pigments (e.g., red and yellow), each reflecting a wavelength different from that of the spectral color. Thus, there seemed to be a paradox between Newton's demonstration that there are an infinite number of colors and the growing awareness of Renaissance painters that all colors can be produced by combinations of three primary pigments—red, yellow, and blue. This paradox was resolved by the suggestion of Thomas Young in 1802 that the receptors in the eye are selective for the three *primary colors:* red, yellow, and blue. Young was able in this way to reconcile the infinite variety of spectral colors that can be duplicated with the limited number of painter's pigments by proposing that each class of color receptor is excited to a greater or lesser degree by any wavelength of light, so that the "red" and "yellow" receptors would be stimulated, respectively, by separate monochromatic "red" and "yellow" wavelengths or that both would be stimulated to a lesser degree by monochromatic orange light. In other words, the sensation for "orange" is the result of the simultaneous excitation of "red" receptors and "yellow" receptors.

Young's *trichromacy theory* was supported by the extensive psychophysical investigations of James C. Maxwell and Helmholtz in the nineteenth century, and those of W. A. H. Rushton in more recent years, but the direct evidence for three classes of color receptors had to wait until 1964, when W. B. Marks—working with E. F. MacNichol, Jr.—made the first spectrophotometric (color-absorption) measurements on single cones of the goldfish retina (Figure 7-54). He found three classes of cones, each with a different maximum absorption wavelength corresponding to one of three visual pigments. Similar measurements on human and monkey retinas produced the same findings. It is a principle of photochemistry that light made up of different wavelengths energizes photochemical reactions in accord with the proportions of each wavelength absorbed. A photon that is not absorbed has no effect on the pigment molecule. Once absorbed, the photon transfers part of its energy to the molecule. This energy transfer suggests that a photoreceptor cell will be excited by different wavelengths (as recorded by the *action spectrum*) in proportion to the efficiency with which its pigments absorb those wavelengths (as recorded by the *absorption spectrum*).

The existence of three pigment classes of cones was corroborated by the demonstration that there are three electrophysiological classes with action spectra corresponding to the absorption spectra (Figure 7-55).

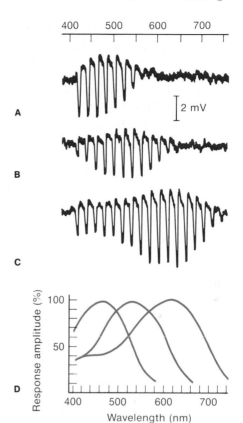

**7-55**  The action spectra of three classes of cones in the carp. (**A–C**) Electrical responses of single cones to flashes of different wavelength, as shown by the scale at the top. (**D**) When the responses were plotted, three classes of cones were found, each with an action spectrum approximating one of the absorption spectra in Figure 7-54. [Tomita et al., 1967.]

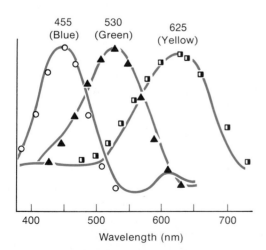

**7-54**  Microspectrophotometric measurements on the cones reveals three classes of absorption spectra, each representing a different pigment with a different absorption maximum. [Marks, 1965.]

Thus, we can now state Young's trichromacy theory in terms of cone pigments. There are three classes of cones, each of which contains a visual pigment that is maximally sensitive to either blue, green, or yellow light. The electrical output of each class depends on the number of quanta that excite the photopigment. The sensation of color apparently arises from the proportions of neural signal from the three classes of cones. According to this concept, excitation of the blue-absorbing cones alone will give the sensation of "blue," but, in combination with partial excitation of yellow-absorbing cones by red light, they will give "purple."

At this point you may be puzzled because the three classes of cone pigments—blue, green, and yellow—seem to disagree with the three primary colors of the painter's pigments—blue, yellow, and red. Even though the *absorption peaks* of the visual pigments do not coincide closely with the three primary colors, each class of color receptor has a broad absorption spectrum and is more sensitive to one of the primary colors than it is to the other two. Thus, red wavelengths stimulate the "yellow" cones far more effectively than they do the "blue" or "green" cones. Light of a given wavelength will produce a unique degree of relative excitation in each class of color receptor. This distinctive excitation provides the nervous system processing the input from a three-pigment visual system with sufficient information to identify the wavelength of any light within the visible spectrum.

What about the rods? In humans, and probably in other primates, color vision is mediated primarily by the cones. In the human retina, the rods are found entirely outside the fovea and are important primarily for vision at low light levels because they show greater convergence (Figure 8-15B) in their neural connections and thus produce greater summation of weak stimuli than do the cones. Since the rods make only a small contribution to color vision, we see primarily shades of black and gray in very dim light. As the fovea contains the less sensitive cone receptors, we are most sensitive to dim light that falls outside the cone-populated fovea. For example, a dim star appears brighter if its image falls just outside the fovea into a population of rods than it does if it falls on the fovea.

Color vision has been demonstrated in all classes of vertebrates. Few generalities can be made about the contribution of rods and cones to color vision. In general, retinas with cones are associated with color vision, but examples of color classes among rods have been found. For example, frogs have two kinds of rods, red (containing rhodopsin, which absorbs in the blue-green) and green (containing a pigment that absorbs in the blue), in addition to cones. Among the invertebrates, the insects have a well-developed ability to distinguish color, including ultraviolet.

## Optic Mechanisms

Primitive animals possess means of detecting the direction of light sources and changes in intensity; more highly evolved visual systems form images with optic mechanisms. In protozoa and flatworms, the direction of a light source is detected with the help of a screening pigment that casts a shadow on the photoreceptor. Some flagellates, for example, have a light-sensitive organelle near the base of the flagellum, shielded on one side by a pigmented "eye spot." This shielded organelle provides a crude indication of directionality. As the flagellate swims along, it rotates about its longitudinal axis roughly once per second. If it enters a beam of light shining from one side perpendicular to its path of locomotion, the eyespot is shaded each time the shielding pigment passes between the source of the light and the photosensitive part of the base of the flagellum. Each time this happens, the flagellum moves just enough to turn the flagellate slightly toward the side bearing the shielding pigment. Because this is in the same direction as the source of light, the flagellate turns by increments in the direction of the light source. This is an example of *positive phototaxis*. One group of dinoflagellate protozoa even has a subcellular lenslike organelle positioned to concentrate light on a "pigment cup." Although this organelle is capable of forming crude images, it seems unlikely that an image is utilized for visual discrimination by unicellular organisms. Vision also requires that the spatial details of an image formed by a lens be represented in neural activity that is processed by the central nervous system so as to abstract certain features of the optic image.

The formation of an image requires a lens; but this is not enough. There must be a spatial sampling by the receptor cells of the optical image formed by the lens. In most image-forming eyes, this sampling is done by a myriad of receptor cells arrayed on a retina, so that each individual receptor samples one portion of the image formed by the lens and conveys the temporal pattern of stimulation to the CNS. A very interesting and unusual mechanism of spatial sampling occurs in the eye of the copepod crustacean *Copilia,* which has only three receptor cells. A set of muscles moves these photoreceptors back and forth in the image plane to "scan" the image. The great disadvantage of this mechanism of optic sampling is the inability of a scanning system to form an instantaneous neural "image" of the optical image. The neural image produced by the scanning system of *Copilia* must of necessity be extremely coarse.

### COMPOUND EYES

The compound eyes of arthropods solve the problem of image formation by having many optic units (the ommatidia) aimed at different parts of the visual field (Figure 7-56). Each ommatidium samples an angular cone (about 2–3°) of the visual field. The visual acuity of such a compound eye is therefore less than that of the vertebrate eye, in which each receptor samples only 0.02° of the visual field. The *mosaic image* projected on the visual cells of an insect compound eye is therefore coarse compared to that of a vertebrate eye (Figure 7-57).

The detailed structure and optics of compound eyes of different arthropods are varied and complex, and will therefore not be considered in detail here. The ommatidium of *Limulus* (Figure 7-39) is one of the least complex and represents one of two variants of the compound eye. In *Limulus* and in some insects, such as bees, the rhabdomeres of the retinular cells are grouped at the center of the ommatidium, and all "see" the same part of the optic field (Figure 7-58A).

A

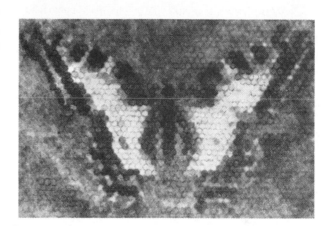

B

**7-57** Visual acuity of a compound eye. (A) Photograph of the butterfly *Papilio machaon*. (B) Mosaic image of that picture approximately as it would be perceived by a dragonfly at a distance of 10 cm. [Mazokhin-Porshnyakov, 1969.]

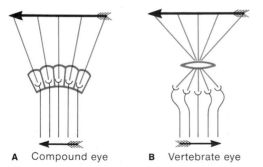

**A** Compound eye  **B** Vertebrate eye

**7-56** Compound and simple eyes. (A) In a compound eye, each ommatidium samples a different point in the field through a separate lens. (B) In the simple eye, each receptor cell samples part of the field through a common lens. [Kirschfeld, 1971.]

The other variant of the compound eye is represented in the flies, whose retinular cells are arranged in an open circle, often with one or two retinular cells in the center. In this arrangement, each retinular cell of an ommatidium "sees" a different part of the field, but the retinular cells of neighboring ommatidia channel into the neural network in such a way that the axons of all seven (or so) of those retinular cells, which "see" the same part of the field, converge on the same postsynaptic neurons (Figure 7-58B). The convergent

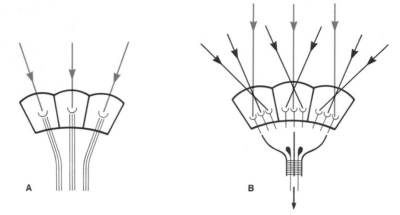

**7-58** (A) In one type of compound eye (e.g., that of *Limulus*), all of the retinular cells of one ommatidium "look at" about the same point in space. (B) In another type, one retinular cell in each of seven or eight ommatidia looks at the same point in the visual field. The outputs of these cells, each in a different ommatidium, converge on the same secondary cell. [B: after Kirschfeld, 1971.]

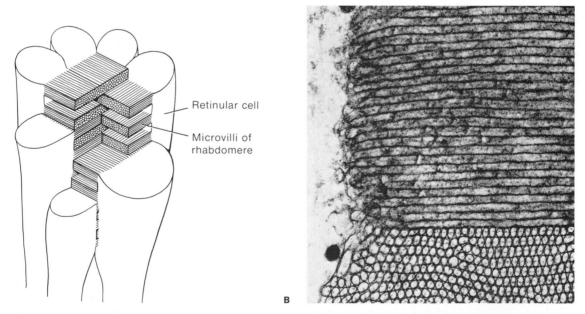

**7-59** The polarized-light detector of the crustacean eye. (A) The interdigitating rhabdomeres of separate retinular cells give rise to two sets of mutually perpendicular microvilli. [Horridge, 1968.] (B) Electron micrograph of a section of the rhabdome formed by two sets of microvilli, the upper set sectioned parallel to the axes of the microvilli, the lower set sectioned perpendicular to the axis of orientation. Magnification 24,400 ×. [Waterman et al., 1969.]

wiring of retinular cells receiving the same optic signal serves to maintain spatial coherence in the neural translation of the visual field.

Some insects and crustaceans can orient with respect to the sun even when the sun is blocked from their view. Sunlight is polarized differently over parts of the sky in relation to the position of the sun. It has been found that many arthropods can detect differences in the plane of the electric vector of polarized light entering the eye (Figure 7-59A). Measurements of birefringence (the ability of a substance to absorb light polarized in various planes) on retinular cells in the crayfish compound eye show that that the absorption of polarized light is maximum when the plane of the electric vector of the light is parallel to the longitudinal axis of microvilli that form the rhabdomere of the retinular cell (Figure 7-59). The rhabdomeres of the seven retinular cells of each ommatidium interdigitate, as shown in Figure 7-59A, and form two groups. The microvilli of each group are oriented at 90° to those of the other group. Polarized light, with its electric vector parallel to the microvilli of one group of receptor cells, is absorbed about six times as efficiently by that group of retinular cells as by the other group. If we assume that the photopigment molecules are oriented systematically in the microvilli, preferentially absorbing light with its electric vector parallel to the microvilli, this arrangement provides the anatomical basis for the detection of planes of polarized light by arthropods. Electrical recordings from single retinular cells (Figure 7-60) have shown that the electrical response to a given intensity of light depends on the plane of polarization of the stimulating light, consistent with the preferential absorption of light polarized parallel to the microvilli. Unlike the situation in the ommatidium of the horseshoe crab *Limulus* (Figure 7-39), each retinular cell in the ommatidia of crustaceans and insects independently sends sensory information along its own axon. Thus, the relative degree of excitation of the retinular cells whose microvilli are orientated in different directions can be utilized at higher centers in the visual system to analyze the relation between planes of polarized light incident upon the compound eye, and this information can in turn be used by the animal for orientation or navigation.

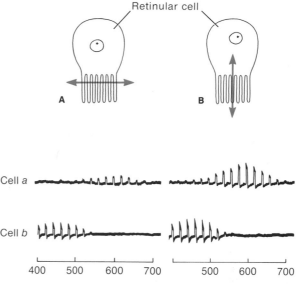

**7-60** Electrophysiological evidence of polarized-light analysis in the crayfish. Two cells, *a* and *b,* were presented with a series of equal-energy flashes of plane-polarized light of different wavelengths (lower scale). Cell *a* showed maximum response at about 600 nm; cell *b,* about 450 nm. When the plane of polarization was parallel to the microvilli, the responses were small (left). Responses of both cells were significantly enhanced when the plane of polarization was rotated so as to lie perpendicular to the microvilli (right). [Waterman and Fernández, 1970.]

### THE VERTEBRATE EYE

Vertebrate eyes have certain structural features similar to those of a camera (Figures 7-56B and 7-61). The lens focuses light that enters through the pupil and forms an inverted image on the retina. In a camera, the image is focused on the film by moving the lens along the optic axis. For close objects, the lens must be moved away from the film to keep the image focused. The eyes of certain bony fishes also focus by this method. The eyes of higher vertebrates are not equipped to move the lens along the optic axis; instead, the image is focused by changing the shape of the lens and thereby its *focal length.* The lens is suspended from radially oriented fibers of the *zonula.* These connective tissue fibers, which originate in the outer margin of the ciliary body, exert radial tension on the outer rims (equator)

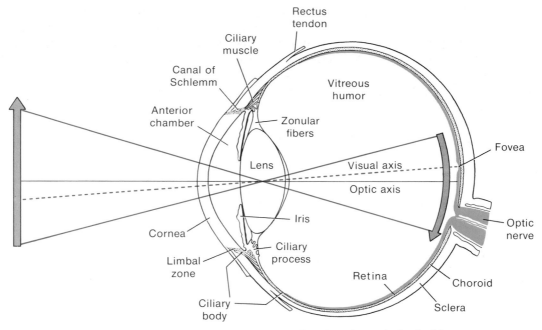

**7-61**   The mammalian eye. An inverted image is focused on the retina at the back of the eye. The refraction of light is diagrammatically simplified, for refraction of light at the air–cornea interface has not been shown. The ciliary body is anchored near the limbal zone; when it contracts, tension on the zonular fibers is reduced, allowing the lens to bulge elastically. The visual axis differs from the optic axis because the fovea does not lie precisely in the path of the optic axis.

of the lens. When the radially oriented muscles of the ciliary body are relaxed, the lens is somewhat flattened by the elastic tension exerted on the equator of the lens by the fibers of the zonula. *Accommodation* to closer objects is accomplished by contraction of the radially oriented smooth muscle fibers of the *ciliary body*. When these contract, the origins of the zonular fibers near the outer edge of the ciliary body are moved toward the lens. This relieves the tension exerted on the equator of the lens by the zonular fibers and allows the lens to bulge more and thereby to shorten its focal length, bringing closer objects into focus. Most of the refraction occurs as light passes from air into the corneal surface, for this is where the light encounters the greatest change in refractive index. However, modifications of the focal length of the lens by contraction of the ciliary body are superimposed on the basic lens effect of the corneal surface to accommodate to objects at

different distances. The ability to accommodate decreases with age in humans, hence the need for bifocal eyeglasses.

The most remarkable thing about accommodation is not the mechanism for altering the shape of the lens, but the neural mechanism by which the correct focus of a "selected" image on the retina is reflexly adjusted by nerve impulses to the ciliary muscle. This is related in principle, if not in detail, to the neural mechanisms responsible for *binocular convergence,* in which the left and right eyes are positioned so that the images formed fall on analogous portions of the two retinas, regardless of the distance and hence the angle formed by the object and the two eyes.

In a camera, the intensity of light admitted to the film is controlled by adjusting the aperture of a mechanical diaphragm through which light is admitted when the shutter opens. The eye has an opaque *iris diaphragm* analogous to the mechanical diaphragm of a camera. In the center of the iris, there is an aperture, the *pupil,* through which light enters the eye. When circular, smooth muscle fibers in the iris contract, the pupillary diameter decreases, and the proportion of light entering the eye is reduced. Contraction of radially oriented muscle fibers causes the pupil to enlarge. The contraction of these muscles, and hence the diameter of the pupil, is controlled by a central neural reflex originating in the retina. The *pupillary reflex* can be seen in a dimly lit room in response to sudden illumination of another person's eye with a flashlight.

Changes in pupillary diameter are transient. The pupil gradually returns to its average size after several minutes. Moreover, the area of the pupil can change only about fivefold, which makes it no match for the changes in intensity of illumination normally encountered by the eye; these changes amount to many orders of magnitude. The pupillary reflex can account for only a small component of visual adaptation of different ambient light levels; it is useful primarily for making rapid adjustments to moderate changes in light intensity. Bleaching and regeneration of visual pigment and neural adaptation are the more effective means by which the eye adapts to extremes of illumination. One consequence of pupillary constriction is an improved image on the retina, because the edges of the lens, which are optically less perfect than the center of the lens, and therefore introduce optical aberration, are screened off as the pupil becomes smaller. Furthermore, depth of focus increases with decreased pupillary diameter, just as it does in a camera when the aperture is reduced.

The structure and function of the vertebrate retina are included in the discussion of visual information processing in the next chapter.

## Summary

Receptor cells are highly sensitive to their specific modalities (i.e., qualities) of stimulus energy while relatively insensitive to other modalities, and they transduce the stimulus into an electrical signal, usually (but not always) a depolarization. The transduction process generally exhibits greatest sensitivity in the lower ranges of stimulus strength, producing receptor signals several orders of magnitude stronger than the stimulus energy. In most receptor cells, the primary sites of reception and transduction are in the cell membrane or intracellular membranes, which undergo changes in ionic permeability in response to the stimulus. The receptor potential, in turn, initiates action potentials by electrotonic spread to the impulse-initiating zone of the sensory neuron. Stimulus intensity is typically encoded in the frequency of impulses, which is roughly proportional to the log of the intensity up to a limiting frequency. The log relationship between stimulus and response magnitudes permits reception over a large dynamic range while retaining high sensitivity at low stimulus intensities. Parallel inputs from receptors that cover different parts of the intensity range enable the sensory system to handle a larger range of stimulus intensities than is possible by any one receptor alone. The time-dependent loss of sensitivity to a maintained stimulus is termed sensory adaptation. Some receptor cells adapt rapidly, others slowly. The mechanisms responsible for sensory adaptation vary, some taking place in the receptor cell, others in the nervous system. In at least one case (the *Limulus* photoreceptor), adaptation results in part from intracellular elevation of $Ca^{2+}$. The calcium blocks the light-dependent activation of $Na^+$–$K^+$ channels.

Some receptor cells occur individually, but others are organized into sensory tissues and organs, such as the nasal epithelium or the retina of the eye. The anatomical organization can play an important role in the functioning of the sensory organ. For example, the image formed by the visual system depends on the presence of a lens and a mosaic of many visual receptor cells in the retina.

Chemoreception (taste, smell) is poorly understood. One theory is that it depends on steric specificity—the ability of stimulus molecules to "fit" into protein receptor sites on the receptor-cell membrane. Mechanoreception occurs as a result of distortion or stretching of receptor membrane, producing changes in ion conductances. The deflection of the stereocilia of hair cells provides directional information by modulating, up or down, the frequency of spontaneously occurring impulses in the eighth-nerve sensory fibers. This function is important in the organs of equilibrium and hearing and in the lateral-line system of fishes and amphibians.

The vertebrate cochlea analyzes sound frequencies according to their effectiveness in displacing different portions of the basilar membrane, which bears hair cells. Traveling waves of the basilar membrane, set up by sound-stimulating movements of the eardrum and auditory ossicles, stimulate the hair cells, which in turn modulate their synaptic action on auditory nerve fibers.

Electroreceptors of fishes are modified hair cells that have lost their cilia, but retain a similar high degree of sensitivity. The electroreceptor responds with potential changes to exogenous currents channeled across the low-resistance apical membrane. These potential changes modulate the release of transmitter at the base of the receptor cells, and this release in turn determines the rate of firing of the eighth-nerve fibers, as in the cochlea.

Visual receptors employ photoexcitable membrane-bound pigment molecules that undergo a conformation change after absorbing a photon. This produces a change in the conductance of the receptor membrane. All visual pigments consist of one protein molecule (opsin) combined with one carotenoid chromophore moiety, either retinal or dehydro-retinal. Differences in the opsin moiety determine the absorption spectrum of the visual pigment. It is the *cis–trans* isomerization of the carotenoid that initiates all visual responses. The mechanisms of coupling between these photochemical events and the opening (in invertebrates) or closing (in vertebrates) of membrane channels in response to illumination remain obscure.

Some vertebrates have three types of cones in the fovea, each with a visual pigment maximally sensitive to a different part of the spectrum. The integration of input from each of these cones produces the sensation of color vision. Rods, which contain only one type of photopigment in humans, are present in greater densities in the periphery of the retina outside the fovea, are more sensitive than cones and also show much greater synaptic convergence. As a result, they exhibit low acuity and high sensitivity.

## Exercises

*1.* Visual receptor cells can be stimulated by pressure, heat, and electricity, as well as by light. How is this fact reconciled with the concept of receptor specificity?

*2.* Outline the steps from energy absorption in a receptor cell to the initiation of impulses in a sensory neuron innervating the receptor cell.

*3.* Why must receptor potentials be converted into action potentials?

*4.* All sensory information enters the central nervous system in the form of action potentials having similar properties. How, then, does the CNS differentiate among various stimulus modalities?

*5.* How are stimulus intensities that differ by up to many orders of magnitude all handled by the same receptor cells?

*6.* What is the typical relationship between stimulus intensity and impulse frequency? What advantage does this offer?

*7.* Discuss three mechanisms that contribute to sensory adaptation.

*8.* Discuss an example of CNS modulation of receptor sensitivity.

*9.* Chemoreception probably requires the presence of protein receptor molecules on or in receptor membranes. What properties of proteins are probably most important in chemoreception?

*10.* Von Békésy noted that sound waves produce traveling waves that propagate along the entire length of the basilar membrane. Explain how his theory accounts for frequency discrimination in the cochlea.

*11.* How does von Békésy's traveling-wave hypothesis of frequency discrimination differ from Helmholtz's resonance hypothesis?

*12.* How are movements of the basilar membrane converted into auditory nerve impulses?

*13.* How does spontaneous firing enhance the sensitivity of certain receptor systems—for example, lateral-line electroreceptors?

*14.* How is the presence of an object perceived by electroreceptors of the weak electric fish?

*15.* What features of receptor-cell behavior help determine the flicker-fusion frequency of the visual system?

*16.* What is the major difference between vertebrate and invertebrate photoreceptor cells in their electrical responses to illumination?

*17.* Outline the *photochemical* steps in the transduction of light energy in visual receptors.

**18.** What modern evidence substantiates Young's trichromacy theory of color vision?

**19.** Describe the various morphological and functional differences that exist between vertebrate rods and cones.

**20.** How is the orientation of polarized light detected in the visual systems of some arthropods?

**21.** Compare the ways in which mammalian and teleost lenses are focused.

**22.** Visible light is restricted to a narrow band between about 350 and 700 nm. Why should vision have evolved to use these wavelengths?

## Suggested Reading

Baylor, D. A., and P. M. O'Bryan. 1971. Electrical signaling in vertebrate photoreceptors. *Fed. Proc.* 30:79–83.

Békésy, G. von. 1960. *Experiments in Hearing.* New York: McGraw-Hill.

Békésy, G. von. 1970. Traveling waves as frequency analyzers in the cochlea. *Nature* 225:1207–1209.

Burtt, E. T. 1974. *The Senses of Animals.* New York: Springer-Verlag.

Fain, G. L., and J. E. Lisman. 1981. Membrane conductances of photoreceptors. *Progr. Biophys. Molec. Biol.* 37:91–147

Fein, A., and E. Z. Szuts. 1982. *Photoreceptors: Their Role in Vision.* Cambridge: Cambridge University Press.

Geldard, F. A. 1972. *The Human Senses.* New York: Wiley.

Hubbell, W. L., and D. Bownds. 1979. Mechanisms of visual transduction. *Ann. Rev. Neurosci.* 2:17–34.

Kennedy, D., ed. 1967. *From Cell to Organism: Readings from*

*Scientific American.* Pt. 4. San Francisco: W. H. Freeman and Company.

Loewenstein, W. R. 1971. *Handbook of Sensory Physiology: Principles of Receptor Physiology.* New York: Springer-Verlag.

McGough, J. L., N. M. Weinberger, and R. E. Whalen, eds. 1967. *Psychobiology: Readings from Scientific American.* San Francisco: W. H. Freeman and Company.

Mellon, D., Jr. 1968. *The Physiology of Sense Organs.* San Francisco: W. H. Freeman and Company.

Rushton, W. A. H. 1972. Pigments and signals in colour vision. *J. Physiol.* 220:1–31P.

Russell, I. J. 1980. The responses of vertebrate hair cells to mechanical stimulation. In A. Roberts and B. M. Bush, eds., *Neurones Without Impulses.* Cambridge: Cambridge University Press.

*Sensory Receptors.* 1965. *Cold Spring Harbor Symp. Quant. Biol.* Vol. 30.

## References Cited

Beck, W. S. 1971. *Human Design.* New York: Harcourt Brace Jovanovich.

Békésy, G. von. 1955. Human skin perception of traveling waves similar to those of the cochlea. *J. Acoust. Soc. Amer.* 27:830–841.

Békésy, G. von. 1962. The gap between the hearing of external and internal sounds. *Symp. Soc. Exp. Biol.* 16:267–288.

Bennett, M. V. L. 1968. Similarities between chemical and electrical mediated transmission. In F. D. Carlson, ed., *Physiological and Biochemical Aspects of Nervous Integration.* Englewood Cliffs, N.J.: Prentice-Hall.

Brown, J. E., and J. R. Blinks. 1975. Changes in intracellular free calcium concentration during illumination of inver-

tebrate photoreceptors: Detection with aequorin. *J. Gen. Physiol.* 64:643–665.

Brown, J. E., and J. E. Lisman. 1975. Intracellular Ca modulates both sensitivity and time scale in *Limulus* ventral photoreceptors. *Nature* 258:252–253.

Brown, K. T. 1974. Physiology of the retina. In V. B. Mountcastle, ed., *Medical Physiology.* 13th ed. St. Louis: Mosby.

Bullock, T. H., and F. P. J. Diecke. 1956. Properties of an infrared receptor. *J. Physiol.* 134:47–87.

Davis, H. 1968. Mechanisms of the inner ear. *Ann. Otol. Rhinol. Laryngol.* 77:644–655.

Draper, J. W. 1872. On the distribution of chemical force in the spectrum. *Phil Mag.* Ser. 4. 44:422–443.

Eckert, R. O. 1961. Reflex relationships of the abdominal stretch receptors of the crayfish. *J. Cell. Comp. Physiol.* 57:149–162.

Eyzaguirre, C., and S. W. Kuffler. 1955. Processes of excitation in the dendrites and in the soma of single isolated sensory nerve cells of the lobster and crayfish. *J. Gen. Physiol.* 39:87–119.

Fain, G. 1980. Integration by spikeless neurones in the retina. In A. Roberts and B. M. Bush, eds., *Neurones Without Impulses.* Cambridge: Cambridge University Press.

Flock, A. 1967. Ultrastructural and function in the lateral line organs. In P. H. Cahn, ed., *Lateral Line Detectors.* Bloomington: Indiana University Press.

Fuortes, M. C. F. 1959. Initiation of impulses in visual cells of *Limulus. J. Physiol.* 148:14–28.

Gesteland, R. C. 1966. The mechanics of smell. *Discovery* 27(2). London: Proprietors, Professional and Industrial Publishing Co.

Hagins, W. A. 1972. The visual process: Excitatory mechanisms in the primary receptor cells. *Ann. Rev. Biophys. Bioeng.* 1:131–158.

Hagins, W. A., and S. Yoshikami. 1974. Proceedings: A role for Ca$^{2+}$ in excitation of retinal rods and cones. *Exp. Eye Res.* 18:299–305.

Harris, G. G., and A. Flock. 1967. Spontaneous and evoked activity from *Xenopus laevis* lateral line. In P. H. Cahn, ed., *Lateral Line Detectors.* Bloomington: Indiana University Press.

Hartline, H. K. 1934. Intensity and duration in the excitation of single photoreceptor units. *J. Cell. Comp. Physiol.* 5:229–274.

Helmholtz, H. von. 1867. *Handbuch der Physiologischen Optik.* Leipzig: L. L. Voss.

Hendricks, S. B. 1968. How light interacts with living matter. *Scientific American* 219(3):175–186.

Horridge, G. A. 1968. *Interneurons.* San Francisco: W. H. Freeman and Company.

Hubbard, R., and A. Kropf. 1967. Molecular isomers in vision. *Scientific American* 216(6):64–76. Also available as Offprint 1075.

Hudspeth, A. J., and D. P. Corey. 1977. Sensitivity, polarity and conductance change in the response of vertebrate hair cells to controlled mechanical stimuli. *Proc. Nat. Acad. Sci.* 74:2407–2411.

Kirschfeld, K. 1971. *Verhandlungen der Gesellschaft Deutscher Naturforscher und Ärtze.* Berlin: Springer-Verlag.

Kühne, W. 1878. *On the Photochemistry of the Retina and on Visual Purple.* Ed. with notes by M. Foster. London: Macmillan.

Lehninger, A. L. 1965. *Bioenergetics.* Menlo Park, Calif.: Benjamin.

Lissman, H. W. 1963. Electric location by fishes. *Scientific American* 208(3):50–59. Also available as Offprint 152.

Loewenstein, W. R. 1960. Biological transducers. *Scientific American* 203(2):98–108.

Marks. W. B. 1965. Visual pigments of single goldfish cones. *J. Physiol.* 178: 14–32.

Mazokhin-Porshnyakov, G. A. 1969. *Insect Vision.* New York: Plenum.

Miller, W., F. Ratliff, and H. K. Hartline. 1961. How cells receive stimuli. *Scientific American* 205(3):223–238. Also available as Offprint 99.

Murray, R., and A. Murray. 1970. *Taste and Smell in Vertebrates.* London: Churchill.

Nakajima, S., and K. Onodera. 1969. Membrane properties of the stretch receptor neurones of crayfish with particular reference to mechanisms of sensory adaptations. *J. Physiol.* 200:161–185.

Schmidt, R. F. 1971. Möglichkeiten und Grenzen der Hautsinne. *Klin. Wochenschr.* 49:530–540.

Steinbrecht, R. A. 1969. Comparative morphology of olfactory receptors. In C. Pfaffman, ed., *Olfaction and Taste.* Vol. 3. New York: Rockefeller University Press.

Thurm, U. 1965. An insect mechanoreceptor. *Cold Spring Harbor Symp. Quant. Biol.* 30:75–82.

Tomita, T., A. Kaneko, M. Murakami, and E. L. Pautler. 1967. Spectral response curves of single cones in the carp. *Vision Res.* 7:519–531.

Toyoda, J., H. Nosaki, and T. Tomita. 1969. Light-induced resistance changes in single photoreceptors of *Necturus* and *Gekko. Vision Res.* 9:453–463.

Waterman, T., H. Fernández, and T. Goldsmith. 1969. Dichroism of photosensitive pigment in rhabdoms of the crayfish *Orconectes. J. Gen. Physiol.* 54:415–432.

Waterman, T. H., and H. R. Fernández. 1970. E-vector and wavelength discrimination by retinular cells of the crayfish *Procambarus. Z. Vergl. Physiol.* 68:157–174.

Young, R. W. 1970. Visual cells. *Scientific American* 223(4): 89–91. Also available as Offprint 1201.

Young, T. 1802. On the theory of light and colours. *Phil. Trans. Roy. Sci.* (London) 92:12–48.

Zotterman, Y. 1959. Thermal sensations. In H. W. Magoun, ed., *Handbook of Physiology* (Section 1, Neurophysiology, Vol. I). Baltimore: Williams and Wilkins.

CHAPTER *8*

# Neural Processing and Behavior

*F*ew features of the natural world have captured the human imagination more than the behavior of animals. The behavior of the human species is the basis of nearly all literature and religion, politics, and history, and for the past century it has been the subject of intensive study by behavioral scientists. In spite of this effort, human behavior remains one of the least understood areas of biology. Most behavioral scientists are of the opinion that an understanding of the function of the nervous system is essential for a solid understanding of human and other animal behaviors.

All behavioral acts are ultimately generated by the motor output of the nervous system, which controls the contraction of muscles. The motor output, in turn, is strongly influenced by sensory input. The behavior (i.e., sum total of movements) of an organism is constantly modified by stimuli from the environment. Some of these input–output relations are simple· and predictable reflexes. Other kinds of behavior are highly dependent on information stored from past experience, and therefore are less predictable to an observer.

The simplest neural network is a *reflex arc*. The primordial reflex arc may have consisted of a *receptor cell* directly innervating an *effector cell* (Figure 8-1A). As neural circuitry became more complex through the process of evolution, centralized nervous systems developed to permit compactness and economy as well as complexity of interconnections. Long sensory and motor axons became useful to connect the receptors and effectors at the periphery to the *central nervous system*. This gave rise to the *monosynaptic reflex arc* (Figure 8-1B), in which a sensory neuron makes synaptic connection in the CNS with a motoneuron that innervates a muscle. The result is a reflex excitation of the effector organ whenever a stimulus sets up sufficiently intense activity (impulses) in the sensory neuron. Polysynaptic pathways with *interneurons* connecting the sensory and motor neurons are actually more common (Figure 8-1C). Interneurons have become progressively more numerous through evolution, vastly increasing the behavioral potential of animals and gradually improving their ability to learn from experience and to associate combinations of stimuli. It is noteworthy that the most elementary components of the reflex arc—namely, the sensory input pathways and the *final common pathways* of motor output—remain essentially unchanged from the most primitive invertebrates to the most complex vertebrates.

The neurophysiological mechanisms of complex behavior—learning and memory—are very poorly understood, but it is agreed that the physical substratum for complex behavior is the vast array of neural circuitry that lies between the relatively simple afferent sensory pathways and efferent motor pathways. Symbolized by the central box in Figure 8-2, this enormously complex interface between the sensory and

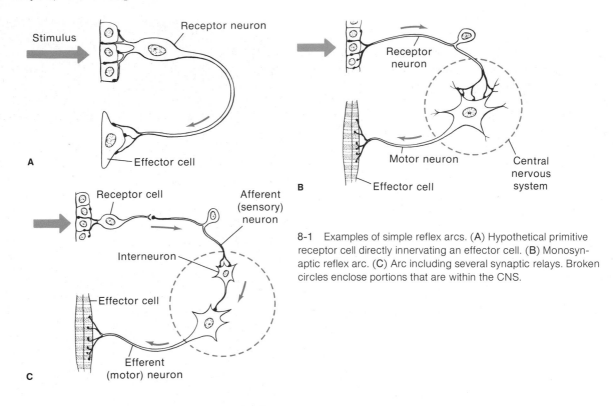

8-1 Examples of simple reflex arcs. (A) Hypothetical primitive receptor cell directly innervating an effector cell. (B) Monosynaptic reflex arc. (C) Arc including several synaptic relays. Broken circles enclose portions that are within the CNS.

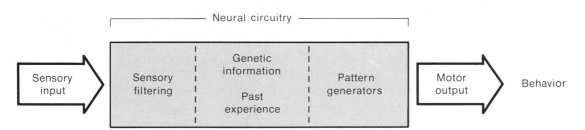

8-2 Major functional divisions of the CNS. Genetic determination and past experience both contribute to the organization of the CNS, which can be crudely divided, as shown, into parts that process sensory input, associative circuits that integrate various inputs and make decisions influenced by past sensory input, and networks that in turn generate appropriate neuromotor output.

motor sides of the nervous system is understood only in the most fragmentary sense. Thus, at the most complex levels of behavior, the nervous system must still be viewed as a "black box" that will produce a more or less predictable output, depending on past and present sensory input. This is in the domain of behavioral psychology. To begin to understand how the nervous system functions, we will examine some of the simpler neurophysiological correlates of animal behavior.

# Evolution of Nervous Systems

The evolution of the nervous system cannot be reconstructed directly from the fossil record because the soft neural tissues leave little trace. However, it is instructive to examine the progression of neural organization in present-day phyla. This progression provides some basis for speculation as to the routes by which the more complex nervous systems may have evolved.

At the cellular level of function, the nervous system appears to have undergone less modification during the course of evolution than perhaps any other tissue. The electrical and chemical properties of nerve cells in the vertebrates and invertebrates are remarkably similar. Many principles of neural function common to all nervous systems have in fact been elucidated first from studies of the simpler systems of invertebrates and lower vertebrates, which are more readily suited for some types of experiments than the more complex nervous systems of the higher vertebrates.

Those nervous systems that are anatomically the most simply organized consist of very fine nerve fibers (axons) criss-crossing as a diffuse network (Figure 8-3) and making synaptic contacts at points of intersection. Such *nerve nets,* found most extensively in the coelenterates, show little or no preference in the direction of conduction. A stimulus applied to one part of the organism produces a response that spreads to some degree from the point of stimulation. If the stimulus is repeated at brief intervals, the system "facilitates," spreading the response further. Very little is known about the synaptic mechanisms in diffuse nerve nets, because the nerve fibers are extremely fine and do not lend themselves readily to intracellular recording techniques. Coelenterates and ctenophores show the beginnings of an organization of neurons into reflex arcs.

A major early advance in the evolution of nervous systems was the organization of neurons into ganglia. Ganglia are first seen in the coelenterates and are common at all higher levels of the animal kingdom. A *ganglion* is a cluster of many neuron cell bodies organized around a tangle of nerve fibers termed *neuropil* (Figure 8-4). This mode of organization permits extensive interconnections among those neurons with an economy of collateral processes. The collaterals (side branches arising from the axons) form arborizations and make contacts in the neuropil. Although the neu-

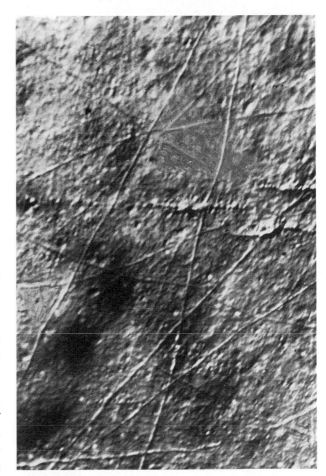

**8-3** Nerve net in subumbrellar surface of the jellyfish *Aurelia,* viewed in oblique light. The axons running in all directions innervate muscles that cause contraction of the umbrella. [Horridge, 1968.]

ropil gives the appearance of a confused tangle of fine processes, recent studies done with injected fluorescent dyes (Figure 8-5) indicate that these arborized collaterals differ greatly in fine detail from one specimen to another. However, the physiological evidence shows that connections in the neuropil are sufficiently orderly that identical synaptic interactions can be observed between homologous neurons in different individuals of the same species.

In segmented invertebrates, each body segment is equipped with a ganglion. A segmental ganglion usu-

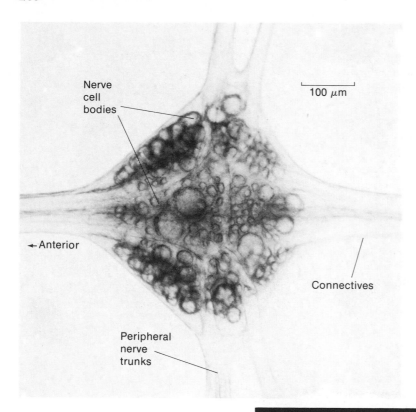

Nerve
cell
bodies

100 μm

← Anterior

Connectives

Peripheral
nerve
trunks

8-4   Photomicrograph of a living segmental ganglion of the leech *Hirudo,* showing the nerve cell bodies. The paired connectives at the left and right contain axons and connect the ganglion to similar ganglia in adjacent body segments. The peripheral nerve trunks emerging laterally carry motor and sensory axons to the viscera and muscle. [Van Essen, 1973.]

8-5   Sensory neuron in leech ganglion, injected with the fluorescent dye procion yellow before fixation. The dye remains within the cell, diffusing into all branches. Note the numerous small branches for synaptic contact with dendrites of other cells. The two large axons enter the peripheral nerve trunks at the left. [Van Essen, 1973.]

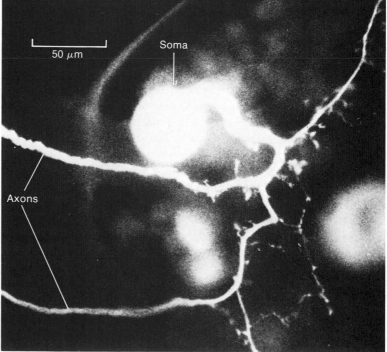

50 μm

Soma

Axons

ally serves the reflex functions of the segment it occupies, plus one or more adjacent body segments. The ganglia of successive body segments are joined by nerve fiber trunks called *connectives*. The result is a series of ganglia and connectives constituting the *ventral nerve cord* characteristic of annelids and arthropods (Figure 8-6), in which segmental organization is most clearly seen. The segmental ganglia of these animals provide convenient material for neurophysiological investigation because of the relatively small numbers of neurons in each segment and the redundancy of structure and function from one segment in the nerve cord to the next. Thus, an analysis of the interactions of the neurons of one segment provides the basic picture for all other segments of the nerve cord. This approach has been especially useful in the nervous system of the leech (Figure 8-5), in which the ganglia are highly redundant.

An important development in the evolution of complex behavior was the trend toward fusion of several of the more anterior ganglia into superganglia, or "brains." These are more complex than the segmental ganglia and exert varying degrees of control over them. The prominence of the brain over the more posterior portions of the CNS results, in part, from the relatively large amount of sensory input entering the brain from the large number of receptors at the anterior of an animal, and in part from the development of regulatory centers in the brain.

In contrast to worms and arthropods, which are segmentally structured and have bilateral symmetry, echinoderms typically have a nerve ring around an axis of secondary radial symmetry. Perhaps as a result of the radial symmetry, there is no brainlike ganglion in the echinoderms. The mollusks have nonsegmental nervous systems with several dissimilar ganglia connected by long nerve trunks. Opisthobranch mollusks, like the sea hare *Aplysia,* and nudibranchs, like *Tritonia,* have a number of neurons with extraordinarily large cell bodies; in *Aplysia,* some are over 1 mm in diameter. These neurons have become a favorite material for the study of cellular neurophysiology, for they can be recognized visually from preparation to preparation as individual neurons, and lend themselves well to long-term electrical recording, injection of experimental agents, isolation for microchemical analysis, and so forth. In mollusks, as well as in annelids and arthro-

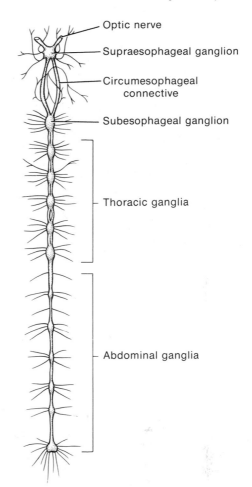

8-6   The ventral nerve cord of the lobster *Homarus* illustrates the segmented arrangement of the nervous system in many invertebrates. The roots emerging from the ganglia carry sensory and motor axons.

pods, it is now possible to select time and again a given identified cell and, under microscopic visual guidance, insert electrodes for physiological studies.

The most complex nervous system of all the invertebrates belongs to the octopus. The brain alone is estimated to contain about $10^8$ neurons. These are arranged in a series of highly specialized lobes and tracts that evidently evolved from the more dispersed ganglia of the lower mollusks. By invertebrate standards, the octopus is quite intelligent indeed.

The nervous systems of invertebrate animals, especially below the octopus, are significantly less complex than those of the vertebrates. The number of neurons in an invertebrate nervous system is far lower than in a vertebrate one. For this reason invertebrate nervous systems are often called "simple." However, superficial appearances can be deceiving, and the functional sophistication of even relatively "simple" animals and their parts becomes apparent on close examination.

## Vertebrate Nervous Systems

The organization of anterior ganglia into a multifunctional brain shows the highest development in the vertebrates, although a rudimentary segmentation remains in the form of the cranial and spinal nerve roots (Figure 8-7). In spite of its awesome structural complexity, the vertebrate nervous system offers certain advantages for experimental neurophysiology. One of these, expressed as the so-called *Bell–Magendie rule,*

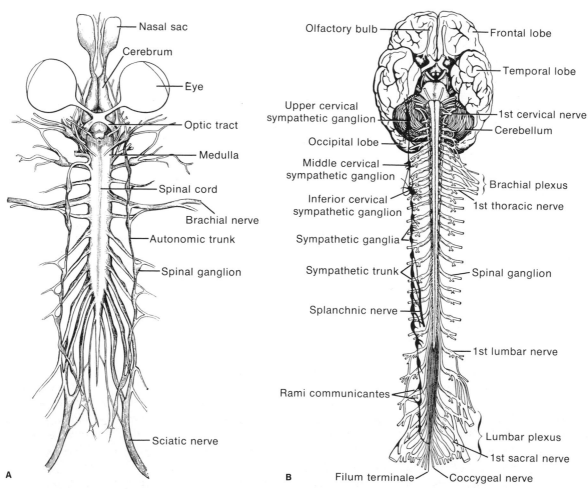

8-7   Brain and spinal cord of a frog (A) and human (B) in ventral aspect. Rudimentary segmentation remains in the form of cranial and spinal neurons. [Wiedersheim, 1907; Neal and Rand, 1936.]

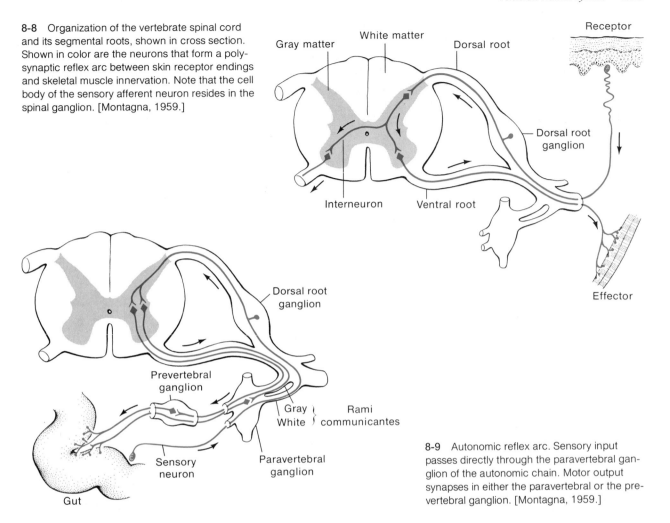

**8-8** Organization of the vertebrate spinal cord and its segmental roots, shown in cross section. Shown in color are the neurons that form a polysynaptic reflex arc between skin receptor endings and skeletal muscle innervation. Note that the cell body of the sensory afferent neuron resides in the spinal ganglion. [Montagna, 1959.]

**8-9** Autonomic reflex arc. Sensory input passes directly through the paravertebral ganglion of the autonomic chain. Motor output synapses in either the paravertebral or the prevertebral ganglion. [Montagna, 1959.]

enunciated over a century ago, is that the *afferent* (sensory) nerve fibers (i.e., axons) enter the CNS via the dorsal roots of the cranial and spinal nerves, whereas the *efferent* (motor) nerve fibers leave the CNS via the ventral roots (Figures 8-8 and 8-9). It was recently found, however, that exceptions to this rule do exist. Applebaum and colleagues (1976) found fine unmyelinated sensory afferents entering the spinal cord through at least some of the ventral roots in the cat. The motor fibers arise from nerve cell bodies in the *ventral horn* of the spinal cord. The afferent fibers arise from monopolar cell bodies in the *dorsal root* or *spinal ganglia* (Figure 8-9). The segregation of sensory and motor axons into dorsal

and ventral roots makes it possible selectively to stimulate, or to eliminate by transection, the sensory input to or motor output from the CNS.

Another convenience to neurophysiologists is the high degree of redundancy of neurons of any given type in the vertebrate CNS. In the arthropod nervous system, a single motoneuron may innervate essentially all the muscle fibers of one muscle, and in some cases more than one muscle of a limb. In vertebrates, each skeletal muscle typically is innervated by a large *pool* of several hundred motoneurons, each of which controls a *motor unit* consisting of up to 100 muscle fibers. Since the motoneurons of each pool are qualitatively similar in physiological properties, data obtained from one motoneuron are largely representative of the whole pool. If this redundancy were not the case, and if all neurons were substantially different in their functional properties from all others, attempts by neurophysiologists to analyze the functions of the vertebrate nervous system would most likely be utterly hopeless.

## MAJOR PARTS
## OF THE CENTRAL NERVOUS SYSTEM

The *spinal cord,* enclosed in the vertebral column (Figure 8-10), is the site of the cervical, thoracic, lumbar, and sacral segmental reflex connections. Seen in cross section (Figure 8-8), ascending (sensory) and descending (motor) interneurons located in the periphery of the spinal cord form well-defined tracts of *white matter,* which gets its color from the myelin sheathing of the axons. The more centrally located *gray matter* of the spinal cord contains the cell bodies, dendrites, and presynaptic terminals, which are all nonmyelinated. A fluid-filled central lumen, the spinal canal, is continuous with the *ventricles* of the brain.

The enlarged portion of the upper spinal cord forms the *medulla oblongata* (Figures 8-7 and 8-11). This contains centers for the control of respiration and cardiovascular reflexes. The spinal cord contains the reflex connections for locomotion and other limb movements and for such visceral functions as bladder control and erection of the penis or clitoris. Spinal reflexes are under descending control from various portions of the brain.

The *cerebellum,* which overlies the medulla, consists of a pair of hemispheres that are convoluted in the higher vertebrates. It integrates input from the semicircular canals and other proprioceptors (position and movement sensors) and from the visual and auditory systems. These inputs are compared in the cerebellum, and the resulting output helps coordinate the motor signals responsible for maintaining posture, for orientation of the animal in space, and for accurate limb movements.

The *hypothalamus* contains a number of centers that control visceral functions and concomitant emotional reactions having to do with feeding, drinking, sexual appetite, pleasure, rage, and temperature control. Hormone-secreting neurons in the hypothalamus control water and electrolyte balance and the secretory activity of the pituitary gland. Further discussion of the endocrine functions of the hypothalamus is found in Chapter 11.

The *cerebral hemispheres* show the most dramatic changes in the course of vertebrate evolution (Figure 8-11). In the higher mammals, the *cortex* (surface layer of gray matter) is thrown into prominent folds that greatly increase the cortical area. Some areas of the cerebral cortex are purely sensory, and others are purely motor. During some neurosurgical procedures, local electric stimulation of the *somatosensory cortex* evokes sensations in waking patients, who are able to tell

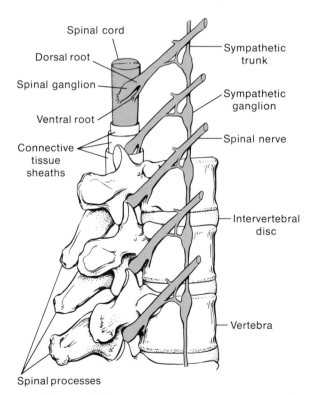

Spinal cord
Dorsal root
Spinal ganglion
Ventral root
Connective tissue sheaths

Sympathetic trunk
Sympathetic ganglion
Spinal nerve

Intervertebral disc

Vertebra

Spinal processes

**8-10**   Vertebral column and spinal cord. The spinal cord runs through openings in the vertebrae, the spinal nerves emerging between the vertebrae. Nervous tissue is shown in color.

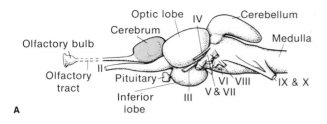

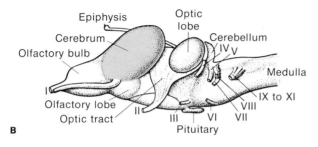

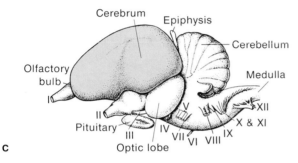

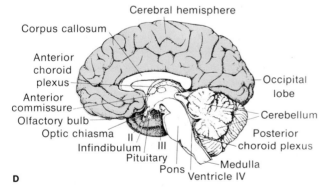

**8-11** Representative vertebrate brains. **(A)** Fish. **(B)** Frog. **(C)** Bird. **(D)** Human. Note the gradual increase in the size of the cerebrum with evolutionary development. The cerebellum, which has an important function in the coordination of movement, is highly developed in birds and mammals. The Roman numerals refer to the cranial nerves. [Romer, 1955.]

the surgeon to which part of the periphery the sensation projects (is felt). This is one line of evidence that *all sensation* takes place in the CNS—primarily in the sensory portions of the cerebral cortex. Peripheral stimulation produces the converse result; local points in the somatosensory cortex produce electrical signals in response to sensory input when specific areas of skin are stimulated. Both approaches have permitted the point-by-point construction of a somatosensory map (Figure 8-12, left).

The *auditory cortex* of the temporal lobe and the *visual cortex* of the occipital lobe are both purely sensory; direct electrical stimulation of these areas during neurosurgery evokes rudimentary auditory and visual sensations. Attempts are being made to develop neural prostheses in which electronic signals from artificial sensors are used to stimulate the respective areas of the brain so as to provide at least crude perceptions of sensory inputs. Although such prostheses are unlikely to mimic in detail the normal sensory mechanisms, they might enable the blind or the deaf to "see" and "hear" in a rudimentary manner.

Anterior to the central sulcus (groove) is the *motor cortex,* which also exhibits topographical representation of the periphery (Figure 8-12, right). Large *Betz cells,* with their somata in the motor cortex, send axons down the spinal cord and make synaptic connections on the motoneurons that innervate the skeletal muscles. These cells maintain a steady background of low-level synaptic input to the motoneurons. An increase in the activity of the Betz cells results in synaptic activation of the motoneurons and forceful movements of the limb. This behavior occurs naturally when a subject consciously generates a strong contraction of a muscle; it also occurs experimentally in anesthetized animals when the Betz cells are directly stimulated with electrodes.

In primitive mammals, essentially the entire cortex consists of sensory or motor areas, and the cortex is without the strong convolutions that increase its area in higher mammals. In humans especially, and in other higher mammals, large portions of the cortex are "unassigned" to either sensory or motor functions. Unassigned areas are involved in intersensory associations, memory, and (in humans) speech.

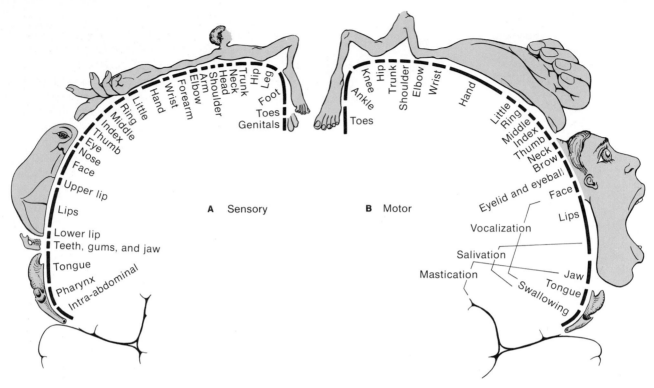

**8-12**  Sensory and motor homunculi in the human cortex. (A) Map of a transverse section of the somatosensory cortex depicts the point localization of areas corresponding to their peripheral projections—that is, the places in the periphery where the stimuli are subjectively "felt." Note that the region representing the genitalia is tucked discreetly out of sight into the intercerebral cleft. (B) Transverse section through the motor cortex, showing the projection of cortical regions to the skeletal musculature. [Penfield and Rasmussen, 1950.]

## THE AUTONOMIC NERVOUS SYSTEM

Visceral functions, largely without conscious control, are regulated in large part by the two divisions of the *autonomic nervous system* (Figure 8-13). The autonomic reflex arc is shown in Figure 8-9. The afferent (sensory) side of the arc is largely indistinguishable from the somatic reflex arc (Figure 8-8); in fact, the same sensory fiber may activate both somatic and autonomic reflexes. The two arcs differ primarily in the location of the effector neuron. In the autonomic system, where it is known as the *postganglionic neuron,* it lies completely outside the CNS. The somatic motoneuron lies, of course, within the gray matter of the spinal cord.

In the *sympathetic* or *thoracolumbar* division of the autonomic system, the preganglionic neurons arise from thoracic and lumbar divisions of the spinal cord (Figure 8-13A), and synapse with the postganglionic neurons in the *paravertebral chain* of autonomic ganglia. The *parasympathetic* or *craniosacral* division arises from cranial and sacral divisions of the CNS (Figure 8-13B). Parasympathetic interneurons characteristically synapse with postganglionic neurons in or near the visceral end organs rather than in separate ganglia.

The parasympathetic and sympathetic systems innervate the same end organs but, in general, liberate different neural transmitter substances and have generally opposing actions on visceral functions. Pacemaker activity of the heart, for example, is inhibited by

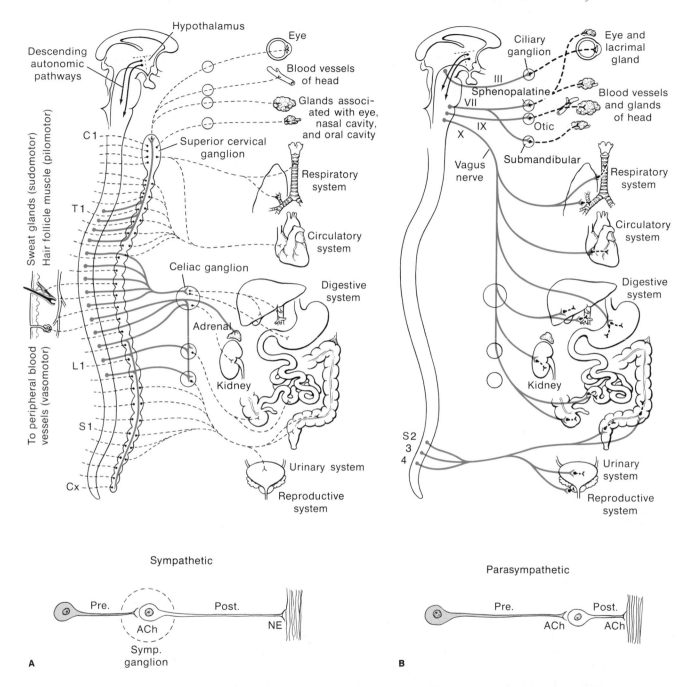

**8-13** Autonomic nervous system. **(A)** Sympathetic or thoracolumbar division. **(B)** Parasympathetic or craniosacral division. Postganglionic neurons represented by broken lines. The major differences in organization are shown at the bottom. The preganglionic endings liberate acetylcholine (ACh) in both divisions. The postganglionic endings of the parasympathetic division also liberate ACh, but the postganglionic endings of the sympathetic system liberate norepinephrine (NE). [Noback and Demarest, 1972.]

the liberation of the transmitter acetylcholine from parasympathetic neurons but is accelerated by norepinephrine liberated from the sympathetic innervation to the heart. These actions of the parasympathetic and sympathetic systems are reversed in the digestive tract, where parasympathetic cholinergic (ACh-liberating) innervation stimulates and sympathetic adrenergic (norepinephrine-liberating) innervation inhibits intestinal motility and digestive secretion. The sympathetic system responds rapidly to stress and danger, generally mobilizing the organism for strenuous or stressful activity, diverting resources from the viscera to the somatic musculature. Commonly termed the *fight-or-flight reaction,* this response is backed up, so to speak, by the secretion of epinephrine by the adrenal medulla into the general circulation.

The sympathetic and parasympathetic divisions of mammals are characterized by certain systematic differences in the synaptic connections (Figure 8-13) and synaptic chemistry. In both divisions, the preganglionic nerve cell bodies lie within the spinal cord and send myelinated axons out through the ventral roots. In the sympathetic division, the innervation of the postganglionic neurons by preganglionic terminals is made close to the spinal cord in the paravertebral and prevertebral ganglia (Figure 8-9). The transmitter at this synapse is ACh. The sympathetic postganglionic axons are much longer than the preganglionic, leading all the way to the visceral end organ, where they release norepinephrine as the primary transmitter. The parasympathetic preganglionic axons of the vagus nerve and sacral segments S2–4 are very long compared to the postganglionic neurons they innervate, because the latter are confined entirely to the visceral target tissues (Figure 8-13). The major transmitter liberated by both pre- and postganglionic parasympathetic neurons is ACh.

The synaptic chemistry of the autonomic nervous system is of interest in part because it illustrates that a given transmitter may have more than one kind of postsynaptic receptor. Thus, ACh and norepinephrine each have two kinds of receptors in the autonomic nervous system. In both cases these can be distinguished on pharmacological grounds—that is, by drugs that act as agonists, mimicking the natural transmitter, or as blockers. The two kinds of cholinergic (ACh-sensitive) receptors are termed *nicotinic* and *muscarinic*. These names derive from early studies in which it was found that nicotine acts as an agonist for some ACh-mediated responses, while muscarine acts as an agonist for others (Box 6-3). Curare blocks the action of ACh on nicotinic receptors (including those at the endplate of skeletal muscle); the muscarinic receptors are blocked by atropine.

The synapses between pre- and postganglionic neurons in both sympathetic and parasympathetic divisions are nicotinic. The postganglionic neurons of the parasympathetic division innervate primarily muscarinic receptors. These mediate inhibitory actions in some tissues (e.g., cardiac muscle) and excitatory action in other tissues (e.g., intestinal smooth muscle).

The adrenergic postganglionic neurons of the sympathetic division innervate two kinds of norepinephrine receptors designated *alpha* and *beta receptors*. The α-receptor can be distinguished from the β-receptor because it is more sensitive to norepinephrine than to the drug *isoproterenol*, and is selectively blocked by *phenoxybenzamine.* The β-receptor is more sensitive to isoproterenol than to norepinephrine, and is blocked selectively by the drug *propranolol.*

The α- and β-adrenergic receptors are distinguished also in some tissues, at least, by serving as mediators for separate but parallel intracellular regulatory pathways (Figure 11-42).

The postganglionic sites of transmitter release in the visceral tissues differ in their organization from classic neuroeffector synapses such as the endplate of skeletal muscle. The transmitter vesicles of postganglionic neurons lie in numerous *varicosities,* or swellings, along the length of the axon (Figure 9-43C). Release occurs from these varicosities, but there are no well-defined postsynaptic specializations limited to the area of the varicosities as there are at the motor endplate. Instead, the postsynaptic receptors are scattered about in the target tissue, and so the transmitter must diffuse over greater distances to reach the receptors.

## Neural Circuits

Several generalizations can be made about the organization and function of nervous systems. First, the anatomical substratum consists of specific connections

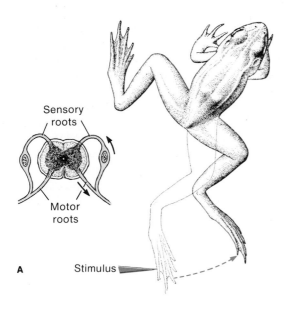

Sensory roots

Motor roots

**A**  Stimulus ◁▭

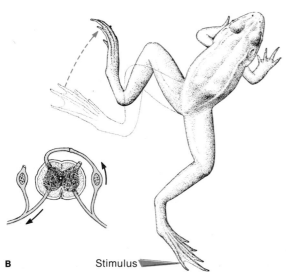

**B**  Stimulus ◁▭

8-14  Effect of rerouting sensory information. **(A)** The normal reflex of a frog to a noxious stimulus applied to the leg is withdrawal of that leg. **(B)** If the dorsal roots are cut and are allowed to regrow as shown, the frog will exhibit withdrawal of the contralateral leg. The reason for this reflex is that sensory input from the right leg has been channeled to the motor network of the left leg. [From "The Growth of Nerve Circuits" by R. W. Sperry. Copyright © 1959 by Scientific American, Inc. All rights reserved.]

between neurons composing the many *neural networks* of a nervous system. These neuronal circuits are established during the development of the nervous system; they are subsequently maintained and modified by use during the entire lifetime of the organism. Disuse typically leads to major loss of function. In kittens, for example, covering the eyes (or even placing translucent diffusers over the eyes) leads to incomplete development of function in the visual system during a critical period following birth. Once established, neural circuits are less labile.

The importance of proper connections for simple reflex acts is illustrated in Figure 8-14. In this experiment, the consequences of altering neural connections was tested in the frog by surgically disconnecting and incorrectly reconnecting the sensory fibers entering the spinal cord from one side to the dorsal root of the opposite side. A noxious stimulus to the leg of an unoperated frog causes reflex withdrawal of the leg (Figure 8-14A). In the case where the connections were surgically redirected (Figure 8-14B), the stimulus evoked an inappropriate movement of the other leg.

The importance of specificity of neural connections was recognized formally over a century ago by Johannes Müller when he stated that the modality of a sensation is determined ultimately by the central connections of the nerve fibers activated by a stimulus, rather than by the nature of the stimulus applied to the sense organ. There is widespread agreement that this is the major—perhaps the only—way in which sensory modality is conveyed to the CNS.

The various senses are represented centrally by "assigned" regions of the brain (Figure 8-12). Direct electrical stimulation of any one of those regions evokes sensations in the subject's consciousness that are more or less similar to those produced by stimulation of the corresponding sense organ.

The second generalization is that the synaptic and electrical properties of individual neurons determine the way each neuron responds to the sum total of the synaptic signals impinging on it. Each active neuron, by virtue of the connections it makes, influences in turn the activity of other neurons.

The third and most sweeping generalization we can make about the nervous system is that *its complexity and variety of functions are manifestations of the complexity and variety of neural circuits, and not of a large variety of different kinds of signals.* The two basic kinds of signals—propagated, all-or-none (action potentials), and nonpropagated, graded (synaptic and receptor potentials)—are illustrated in Figure 6-3. As discussed in Chapters 5, 6, and 7, bioelectrical signals of both major categories all have their origin in the flow of ionic currents driven by electrochemical gradients and regulated by the gating of ion channels.

Neural networks can be grouped into several categories (Figure 8-2). *Sensory filter* networks, discussed in the next section, are organized so as to pass on only certain features of a complex sensory input while ignoring other features. *Pattern-generating* networks are responsible for the production of stereotyped movements. The output of some pattern generators is cyclic—for example, those governing locomotion and respiration. That of others is noncyclic, such as those governing tongue ejection of frogs and toads when catching insects.

Superimposed on some *centrally programmed,* pattern-generating networks are reflex arcs, noted earlier, in which moment-to-moment changes in sensory input modulate the motor output. Some behaviors involve the participation of a filter network on the input side and a programmed pattern generator on the output side (Figure 8-2). An example is the feeding reflex of the frog, discussed in the next section. The simplest reflexes (e.g., the knee jerk, to be described later) involve neither a sensory nor a motor network.

The number of combinations with which neurons can form different circuits is enormous. A single neuron may receive thousands of presynaptic terminals from other neurons, some of which are excitatory, and others inhibitory. In addition, the neuron may itself branch many times and innervate many other neurons. *Divergence*—the repeated branching of a neuron—gives it a widespread influence on many postsynaptic neurons (Figure 8-15A). *Convergence* of inputs on a single neuron

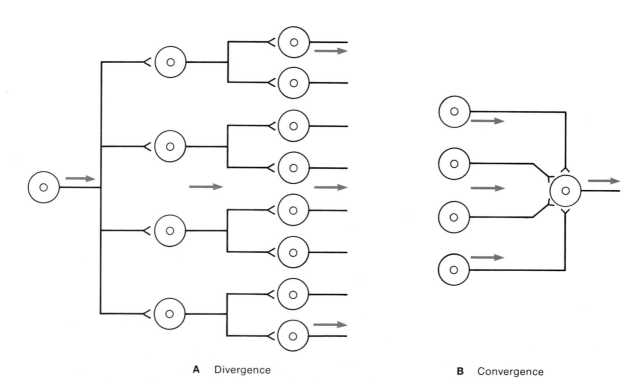

**A**    Divergence                                              **B**    Convergence

8-15    (A) Divergence is the branching of one neuron so as to innervate several others. (B) Convergence refers to the innervation by many cells of a single cell.

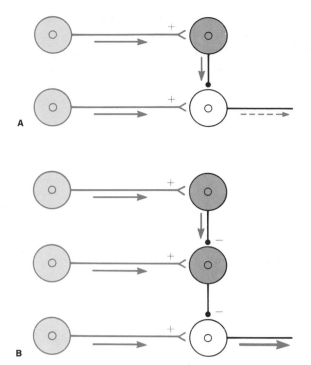

**A**

**B**

8-16   Two consequences of activity in an inhibitory interneuron. (A) With a single inhibitor in a sequence, the end effect is a reduced probability of firing in the terminal cell. (B) Disinhibition. If there are two inhibitors in the sequence of interneurons, and if the second one is tonically excited or spontaneously active, excitation of the first inhibitor will, in effect, increase the probability of firing in the terminal cell.

inhibitory synapses on a neuron, the greater the number of excitatory synapses that must be activated to depolarize the integrating neuron to the firing level. It is interesting, however, that with the appropriate circuitry, the end effect of activity in an inhibitory neuron can be an increase in the discharge frequency of another neuron (Figure 8-16B). This increase simply requires that the inhibitory neuron inhibit another inhibitory neuron, releasing ongoing inhibition exerted by the latter on a third neuron. This behavior is termed *disinhibition.*

Feedback is employed extensively in neural circuits. An example of positive feedback is shown in Figure 8-17A, for a hypothetical reverberating circuit in which some branches of a neuron excite interneurons that feed back upon that neuron to reexcite it and keep it firing for extended periods. In theory, the neuron could continue to fire indefinitely once excited by synaptic input. If the interneuron in such a circuit is inhibitory instead of excitatory (Figure 8-17B), the result is negative feedback, which reduces any tendency for the

(Figure 8-15B) allows that unit to integrate signals from numerous presynaptic neurons. Most neurons, such as the mammalian motoneuron, discussed earlier, are not depolarized to the firing level without considerable spatial and temporal summation of excitatory synaptic input. As a result, those neurons fire only in response to more or less simultaneous activity in a number of presynaptic neurons.

Excitation of a neuron will be suppressed if sufficient numbers of its inhibitory synapses are active at the same time as the excitatory synapses. The inhibitory synapses on a neuron can be thought of as a variable gain (amplification) control that modulates the ease with which excitatory inputs can excite a neuron (Figure 8-16A). Thus, the greater the number of active

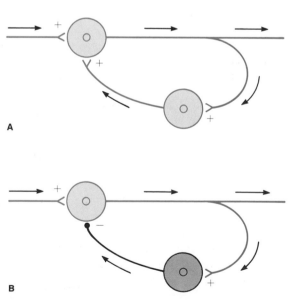

**A**

**B**

8-17   Two simple circuits in which a branch of a neuron excites an interneuron that feeds back on it. (A) Recurrent facilitation. If the interneuron is excitatory, the feedback is positive. (B) Recurrent inhibition. An inhibitory interneuron produces negative feedback.

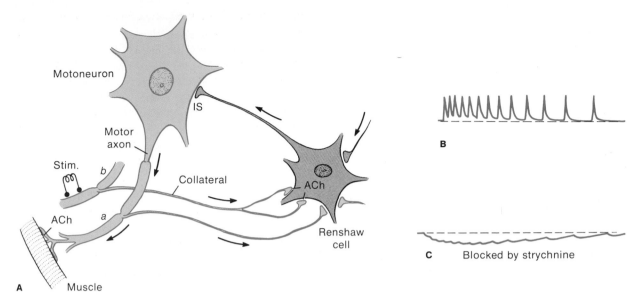

**8-18**   Inhibition of α-motoneurons by the Renshaw cells. (A) Simplified diagram of motoneuron axons with their collateral axons innervating a group of Renshaw cells (only one is shown). The Renshaw cells make inhibitory endings on the motoneurons (IS is inhibitory synapse). (B) Discharge of Renshaw cell in response to antidromically excited discharge in motoneuron *b*. (C) Inhibitory postsynaptic potentials in motoneuron *a* in response to the Renshaw cell discharge. These inhibitory potentials are blocked by strychnine. It is not known if the Renshaw cell feeds back on the same motoneurons that innervate it or only on others of the same pool of motoneurons. [Eccles, 1969.]

neuron to fire repetitively. A well-known instance of inhibitory (negative) feedback occurs in the motoneuron pool of the vertebrate spinal cord. Here the α-motoneurons (the neurons that activate the skeletal muscle fibers) have small branches that innervate short inhibitory interneurons, the *Renshaw cells* (Figure 8-18). These feed back upon the motoneurons, and the Renshaw cells are excited each time the motoneurons fire. The Renshaw cells respond with a high-frequency train of impulses (Figure 8-18B) that set up inhibitory postsynaptic potentials in the motoneurons. Although the reason for the existence of this circuit is not entirely clear, we do know that it serves to keep the motor discharge in check. Strychnine interferes with the inhibitory synapses made on the motoneurons by the Renshaw cells, and presumably with other inhibitory synapses. This explains why the poison produces con-

vulsions, spastic paralysis, and death due to loss of coordination of the respiratory center. These gruesome consequences of an inhibitory blocking agent demonstrate the pervading importance of synaptic inhibition in the nervous system.

## Sensory Filter Networks

The function of sensory filtering networks is to screen out all but certain features of sensory input—a function that should not be confused with energy-filtering properties of receptor cells and their accessory structures.

The concept of sensory filtering is well illustrated in the visual system. Unlike a television system, in which there is a point-to-point transfer of light–dark color information from the photocathode of the TV camera to the phosphor screen of the TV receiver, the visual

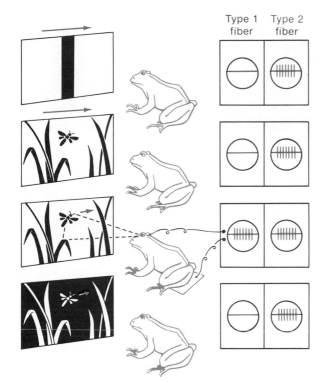

| Type 1 fiber | Type 2 fiber |

**8-19** Sensory filtering in the retina of the frog. The frog will respond to a small, moving object in the field but not to the movement of a background against the stationary small object or to other nonspecific visual stimuli. The frog's response is seen behaviorally as the ejection of its sticky tongue. Certain fibers (Type 1 in the figure) in the optic nerve become active only in response to the sight of a small, dark, sharply outlined object moving across a lighter background. Other fibers (Type 2) are activated by a wide variety of movement in the field. [Bullock and Horridge, 1965.]

their orientation, or their direction of movement, as described later in this chapter.

A classic example of sensory filtering is shown in Figure 8-19. Recordings from the optic nerve of a frog show that some of the optic nerve fibers respond only to certain specific features of the visual field. Some are more specific in this regard than others. One kind of optic fiber fires only in response to a small object, such as a fly, moving against a light background containing stationary objects. It does not respond if the entire scene moves or if the light is turned on or off. Because frogs catch flying insects, this class of optic fibers presumably tells the frog's brain that supper is on the wing, a bit of information undoubtedly more significant to the frog than most of the other details in the visual scene. Anyone who tries to maintain frogs in captivity soon learns that a frog does not recognize a dead insect as food. Apparently, to the frog, a dead fly does not appear at all similar to a moving fly. In fact, to the frog, the image of the dead fly does not include those features that activate the circuitry responsible for the feeding response.

## LATERAL INHIBITION

One common form of filtering that occurs in sensory systems serves to enhance visual contrast. Changes in stimulus energy or quality, either in time or in space, are generally the significant cues to an animal, and neural mechanisms have evolved for playing up such differences. Enhancement of visual contrast can be experienced by viewing Figure 8-20. The bands give the appearance of increasing luminosity (brightness) toward the side bordering the darker neighbor. Conversely, each band looks darker on the side bordering its lighter neighbor. This effect is an illusion, for the actual luminosity is uniform across the width of each band. (Use two sheets of paper to mask out all but one band to convince yourself.)

How does this illusion arise? Our first understanding came from the compound eye of *Limulus*, described in Chapter 7. Although the organization is much simpler in the *Limulus* eye than in the retina of the vertebrate eye, interactions do occur among the ommatidia. This was first noted in the laboratory of H. K. Hartline at Rockefeller University in the mid-1950s. A room

system selects certain features from the total input to the eye while largely ignoring the rest. Thus, an individual cell in the visual centers of the brain does not respond to the mere presence or absence of light impinging on a corresponding visual receptor cell in the retina, but reacts specifically to certain parameters of the visual input to a number of receptor cells. Each neuron abstracts certain features of the image falling on the retina, such as straight borders or lines,

**8-20** Enhanced visual contrast. Each band is actually uniform, but appears to be lighter near its dark neighbor and darker near its light neighbor. Edge-to-edge uniform luminosity can be demonstrated by covering the neighbors to either side of a band.

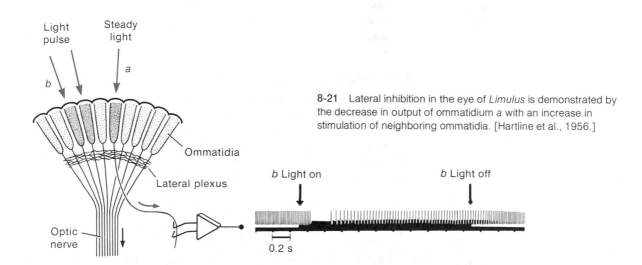

**8-21** Lateral inhibition in the eye of *Limulus* is demonstrated by the decrease in output of ommatidium *a* with an increase in stimulation of neighboring ommatidia. [Hartline et al., 1956.]

light was turned on while the activity of a single ommatidium was being recorded in response to a bright stimulus light focused on that ommatidium alone. The additional stimulus from the room light, rather than producing an increase in the frequency of discharge, caused the frequency in that unit to drop. It was subsequently determined that excitation of the surrounding units by the diffuse room light is what caused an inhibition (reduced frequency of firing) in the test ommatidium. This phenomenon, which was termed *lateral inhibition,* has since been observed in other visual systems as well as in some other sensory systems.

Evidence of the interaction of receptors in the lateral eye of *Limulus* is illustrated in Figure 8-21; the steady discharge of ommatidium *a* is inhibited during the illumination of ommatidium group *b*. The inhibi-

Stimulus

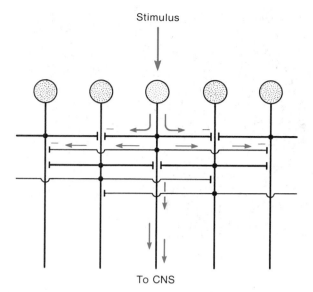

To CNS

8-22  Connections in the retina of *Limulus* responsible for
lateral inhibition. Each eccentric cell axon gives off branches
that make inhibitory synapses on the neighboring axons. These
connections become fewer with distance away from each axon;
hence the interaction becomes weaker between more distant
neighbors.

tory effects are, of course, reciprocal between interacting units. An important feature is that they fall off with the distance. Thus, inhibition is strongest for the nearest neighbors. Lateral inhibition in the *Limulus* eye is mediated through the so-called *lateral plexus*, which consists of collaterals from the eccentric cell axons, which form inhibitory synapses on one another. These connections are illustrated in greatly simplified form in Figure 8-22. Impulse activity in the eccentric cell collaterals causes release of inhibitory transmitter from synaptic terminals that end on neighboring eccentric cell axons. The effect of the transmitter on the postsynaptic axon is to reduce its probability of firing. Because the inhibition exerted by a unit on its neighbors increases as its activity (i.e., impulse frequency) increases, a strongly stimulated ommatidium will strongly inhibit neighboring, less strongly stimulated units and, in turn, will receive weak inhibition from those units. This enhances the contrast (increases the difference in response) between units subjected to different intensities of photostimulation (Figure 8-23). The contrast is greatest for units immediately opposite each other

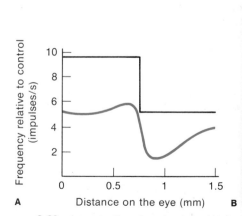

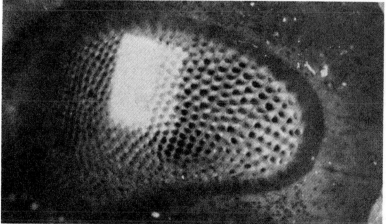

8-23  Accentuation of contrast at a border between dark and light. A light–dark border is moved across the compound eye (**B**) while the output of a single ommatidium is recorded (**A**). If all other ommatidia are masked, the output of the ommatidium will change in an abrupt, steplike manner (upper plot in A). If the light–dark border is allowed to pass over all the ommatidia, the output of the single ommatidium plotted against position of the border will generate a curve similar to the lower plot in A. This results from lateral inhibition, as explained in the text. [From ''How Cells Receive Stimuli'' by W. H. Miller, F. Ratliff, and H. K. Hartline. Copyright © 1961 by Scientific American, Inc. All rights reserved.]

across a bright–dim border, since the lateral effects diminish with distance. As a shadow slowly passes over a brightly lit compound eye, the impulse frequency recorded from one receptor unit increases as the edge of the shadow approaches that ommatidium. This higher frequency occurs because the inhibition from nearly half the units neighboring that ommatidium is reduced as they become nearly quiescent in the shadow. Then, as the shadow passes over that ommatidium, its discharge frequency drops abruptly to a low level. At that moment, the unit is subjected not only to a weaker stimulus, but to strong inhibitory influence from nearly half of its closest neighbors still exposed to the bright light. As the edge progresses further beyond the selected unit, shading the rest of the neighboring ommatidia, inhibitory input to that unit from its neighbors is reduced, and that ommatidium assumes a new, somewhat higher frequency of firing.

Although the impulse-frequency profile in Figure 8-23A was produced as a function of *time* by moving the edge of a shadow over the field of receptor units, it is obvious that a similar impulse frequency profile would result as a function of *distance*. If it were possible to record and to plot the discharge frequencies of all of the receptors along a straight line perpendicular to the edge of a stationary shadow, the result would be a frequency profile similar to that in Figure 8-23. Thus, lateral inhibition serves to sharpen visual edges by increasing contrast at borders between areas of different luminosities, as experienced by an observer viewing the bands in Figure 8-20.

### VISUAL PROCESSING IN THE VERTEBRATE RETINA

The vertebrate retina consists of the visual receptor cells and a highly ordered network of neurons. These neurons are part of the CNS even though their network is located at the periphery. It is not surprising, therefore, that the process of *visual abstraction* begins in the retina before the partially processed visual information is sent along the axons of the optic nerve for further processing in the lateral geniculate bodies and the visual cortex (in birds and mammals) or optic tectum (in lower vertebrates) (Figure 8-24). The signals in the optic nerve are carried by the axons of the *ganglion cells* located in the retina.

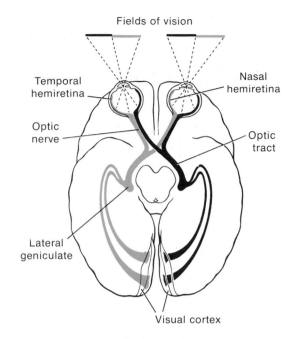

**A** Mammal

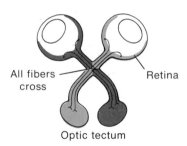

**B** Amphibian

8-24  Visual pathways in mammal and amphibian. **(A)** In the mammal, each side of the visual field is projected to the opposite side of the visual cortex. [Noback and Demarest, 1972.] **(B)** In the amphibian, the left and right optic tecta each receive projections of the entire field of view of the contralateral eye. [From ''Retinal Processing of Visual Images'' by C. R. Michael. Copyright © 1969 by Scientific American, Inc. All rights reserved.]

The visual receptor cells make indirect contact with the ganglion cells via the *bipolar cells* (Figure 8-25). We will consider the receptors as the first-order cells, the bipolars as second-order, and the ganglion cells as third-order in the afferent pathway. This is admittedly oversimplified, because two other types of retinal neurons—the *horizontal cells* and the *amacrine cells*—participate in these connections and are particularly important for lateral interactions in the retina. The horizontals receive input from the neighboring and moderately

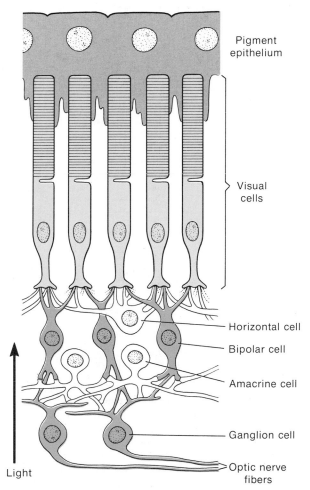

8-25 Cellular organization in the vertebrate retina. The bipolar cells carry signals from the visual receptor cells toward the ganglion cells, which give rise to the axons of the optic nerve. The horizontal and amacrine cells carry signals laterally. [From "Visual Cells" by R. W. Young. Copyright © 1970 by Scientific American, Inc. All rights reserved.]

distant receptor cells and innervate the bipolars, and the amacrines interconnect bipolars and ganglion cells. Recent studies combining intracellular recording techniques with the injection of fluorescent marker dyes have succeeded in identifying the electrical activity of each retinal cell type (Figure 8-26).

As noted in Chapter 7, the visual receptor cells of vertebrates produce hyperpolarizing (i.e., negative-going) potential shifts when illuminated and normally do not produce action potentials. They release synaptic transmitter continuously in the dark, and this release is reduced when the receptor hyperpolarizes in response to illumination. The horizontal cells, too, show only graded, nonspiking potential changes, which are also hyperpolarizing (Figure 8-26). The bipolar cells likewise fail to produce spike potentials, but produce graded potential changes of both polarities. Thus, the name "bipolar" can apply as well to the electrical behavior of this type of cell as it does to the morphological feature from which it received its name. The membrane potential of the ganglion cell changes with the same polarity as the bipolar cells innervating it. It becomes depolarized and fires action potentials when bipolar cells synapsing on it depolarize, and it becomes hyperpolarized and ceases spontaneous firing when its bipolar inputs hyperpolarize. The amacrine cells show transient ON and OFF effects in response to input from bipolar cells.

The bipolars generally connect more than one receptor to each ganglion cell, and they may also connect each receptor cell to several ganglion cells. Thus, convergence and divergence already occur between first- and third-order cells of the visual system; both are minimal in the *area centralis,* or fovea, where acuity is greatest and where there is a tendency for 1:1:1 connections between cones, bipolar cells, and ganglion cells. Outside the area centralis, ganglion cells receive input from many receptor cells, primarily rods, conveying on those ganglion cells a greater sensitivity to dim illumination but a lower degree of visual acuity.

Each ganglion cell is active spontaneously in the dark and responds to a spot of light directed at any point inside a more or less circular area of the retina. Depending on which receptor cells are illuminated by a small spot of light, a ganglion cell may give an ON

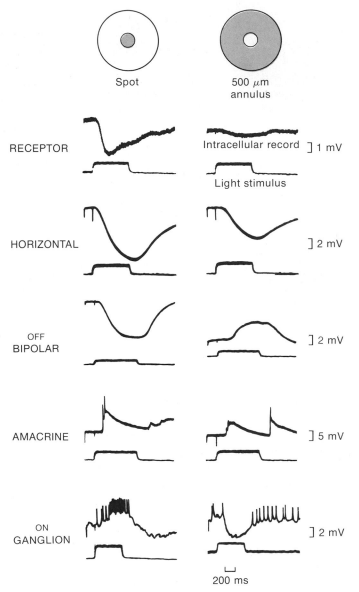

**8-26** Electrical responses of the retinal cells to a spot of light (left) and to an annulus (doughnut) of light (right). In this example, the ganglion cell has an ON-center receptive field. The duration of the light stimulus is indicated in the lower trace on each record. Note that the bipolar and ganglion cells produce responses of opposite polarities to spot and annulus. This effect is believed to be due to lateral inhibition similar to that in *Limulus*. It is important to note that the OFF bipolar and the ON ganglion cells are not synaptically connected. Figure 8-28 shows how ganglion cell responses are related to bipolar cell signals. [Werblin and Dowling, 1969.]

response (increased rate of firing when stimulus light is turned on) or an OFF response (increased rate of firing when stimulus light is turned off) (Figure 8-27). The region of the retina in which illumination causes a response (either increased or decreased rate of firing) in a nerve cell is termed the *receptive field* of that cell. The receptive field of a neuron is that portion of the sensory field (in this case the retina) within which a stimulus can influence the electrical activity of that neuron.

The receptive field of a ganglion cell is centered more or less on the ganglion cell and varies in extent from a few receptors at the center of the area centralis to several thousand receptor cells near the periphery

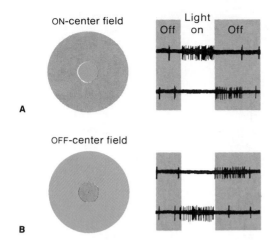

ON-center field

Light

Off | on | Off

OFF-center field

**8-27** Electrical behavior of ON-center and OFF-center ganglion cells in the mammalian retina. (**A**) An ON-center cell fires in response to light in the center of its receptive field (upper trace). If the stimulus spot is outside the center, the firing is suppressed (lower trace). (**B**) An OFF-center cell is inhibited by a spot of light in the center of its receptive field (upper trace) and excited by a spot of light falling on the surround (lower trace). [From "The Visual Cortex of the Brain" by D. H. Hubel. Copyright © 1963 by Scientific American, Inc. All rights reserved.]

of the retina where the receptive field of a ganglion cell can be up to 2 mm in diameter.

Two main types of ganglion cells can be recognized. One type, the *on-center* cell, responds to illumination of a region comprising the center of its receptive field with an ON response (Figure 8-27). Illumination of the receptive field in a concentric band (annulus) peripheral to the ON region of such a cell causes an OFF response in the cell. A weaker OFF response is elicited by a spot of light illuminating a part of the band. This band is termed the *inhibitory surround* of the receptive field. The other type, the *off-center* cell, exhibits the converse behavior, ceasing or reducing its firing when the center of its receptive field is illuminated and increasing its firing when the surround is illuminated.

It is characteristic of both ON- and OFF-center ganglion cells that they exhibit little or no response to uniform illumination over the entire receptive field. This is due to mutual cancellation by the antagonistic center and surround. Thus, a small spot of light is more effective than diffuse illumination in affecting activity

in the optic nerve. A small spot of light moved about on the retina will cause an increase in frequency of firing of an optic nerve fiber (axon of a ganglion cell) when the spot is in the ON portion of that cell's receptive field. At the same time, the spot will be in the OFF portions of the receptive fields of other fibers. As the spot is moved, it will turn fibers ON or OFF according to its position in their individual receptive fields.

The center–surround organization of receptive fields is, of course, an example of lateral inhibition similar to that which we encountered earlier in the compound eye of *Limulus*. Another feature of the organization of the vertebrate retina is the presence of the OFF-center receptive field in addition to the ON-center type.

Much of the lateral interaction in the retina takes place via the horizontal cells. These cells, which lie in the *outer plexiform layer* of the retina, have extensive lateral processes and are interconnected with their neighboring horizontal cells through electrotonic junctions. In addition, they make chemical synapses on bipolar cells and receive synaptic inputs from many receptor cells. Light that falls in the surround of a ganglion cell's receptive field exerts its effects on that cell via the lateral connections made by horizontal cells. Since the horizontal cells form an extensive syncytial network, communicating through low-resistance gap junctions that they make with one another, input from any receptor into a horizontal cell produces a hyperpolarizing signal that spreads electrotonically with distance in all directions from that receptor (Figure 8-28). Every bipolar cell gets input from surrounding receptor cells via the lateral network of horizontal cells, and this input falls off with distance because the graded hyperpolarizing potentials undergo cable loss during their electrotonic spread. The *indirect* input a bipolar cell receives from outlying receptors via the horizontal cell network opposes the *direct* input it receives via *direct-line* connections from overlying visual receptors. This arrangement is the basis of the center–surround organization of receptive fields. The local, direct-line pathway of receptor ⟶ bipolar cell ⟶ ganglion cell is responsible for the center response, while the lateral, *indirect-line* pathway—namely, receptor ⟶ horizontal cells ⟶ bipolar cells ⟶ ganglion cell—mediates the response to illumination of the surround.

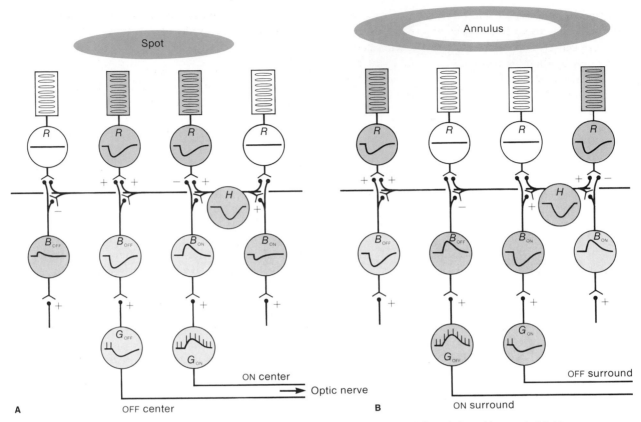

**8-28**   Retinal organization responsible for ON- and OFF-center receptive fields. (A) Stimulation with a spot of light. (B) Stimulation with an annulus. Two kinds of bipolar cells, $B_{ON}$ and $B_{OFF}$. exhibit reciprocal responses to direct input from the receptors, *R*, and indirect input carried laterally by the horizontal cells, *H*. The ON bipolars become depolarized during activation of their overlying receptor cells and weakly hyperpolarized in response to lateral input from horizontal cells. The OFF bipolars show the converse behavior. The amacrine cells have been omitted for simplicity. Direct-line pathway to the ganglion cell, *G*, is shown by colored shading. Indirect, lateral pathway via horizontal cells is shown in gray. The plus and minus signs indicate synaptic transfer that conserves ( + ) or inverts ( − ) the polarity of the signal.

A key factor in the origin of two kinds of receptive fields—ON-center and OFF-center—is the existence of two kinds of bipolar cells, the *on-bipolar* and the *off-bipolar,* which exhibit mutually inverse responses to synaptic input both from receptors and from horizontal cells (Figure 8-28).

The OFF-bipolar cells become hyperpolarized, whereas the ON bipolars become depolarized, by illumination (and hyperpolarization) of overlying receptors. In both types of bipolar cells, a light flashed onto the surround produces a response, mediated by horizontal cells, that is of the opposite electrical sign to that produced by illumination of the center. As just noted, each bipolar cell produces synaptic potential changes in its ganglion cell(s) that are of the same sign as the potential change occurring in the bipolar. Thus, the ganglion cells innervated by ON bipolars will have ON-center receptive fields, whereas those innervated by OFF bipolars will have OFF-center fields. An ON-center ganglion cell will be excited by direct receptor input to its bipolar cells and be inhibited via horizontal cells in response to illumination of the surround. The converse applies to the ganglion cells innervated by OFF bipolars.

The two opposing sets of responses mediated by the ON- and OFF-bipolar cells result from differences in the postsynaptic responses of the bipolars to the neurotransmitter released by the receptor cells and by a different neurotransmitter released by the horizontal cells. The ON bipolars are steadily *hyper*polarized in the dark by the action of the transmitter that is steadily secreted by the partially depolarized receptor cells in the dark. When a light stimulus causes hyperpolarization of the visual receptor cell, its secretion of transmitter diminishes and the ON bipolar is allowed to depolarize to its low (i.e., near zero) resting potential. This depolarization causes it to release an excitatory transmitter that depolarizes its ganglion cells, causing them to increase their rate of firing (i.e., ON response). In contrast, the OFF bipolars, having different postsynaptic channels with different ionic selectivity, are steadily *de*polarized in the dark by steady release of the visual cell transmitter. Illumination of the visual receptor and the concomitant drop in transmitter secretion by the receptor cause the OFF bipolar to hyperpolarize toward its relatively high (i.e., very negative) resting potential. This hyperpolarization is accompanied by a drop in the OFF bipolar's transmitter release and a hyperpolarization of its ganglion cells (i.e., OFF response).

In summary, then, the receptive field organization of the vertebrate retina depends on three basic features:

*1.* Two kinds of ganglion cells, having ON-center and OFF-center receptive fields, respectively, receive input from two kinds of bipolar cells.

*2.* Receptors in the antagonistic surround exert their effects on these two kinds of bipolar cells through a network of electrotonically interconnected horizontal cells.

*3.* The direct input to the bipolar cells from overlying receptors and the indirect input via the horizontal cell network oppose each other and thereby produce the contrasting center–surround effects exhibited by both the ON-center and OFF-center ganglion cells.

The retina teaches us several lessons about neural organization that appear to be of widespread importance in the rest of the central nervous system. First, we note that cells can signal each other without action potentials if they are small and do not require impulses for long-distance propagation. Such nonspiking neurons have the advantage of continuously graded signals more sensitive than those possible with intervening all-or-none activity. Moreover, it permits signal attenuation as a function of distance. Second, excitation does not necessarily mean depolarization; in some nerve cells (e.g., visual receptors, horizontal cells), hyperpolarization is a normal response that modulates synaptic transmission, namely by causing a drop in the steady release of transmitter. Third, the postsynaptic response of a neuron cannot be predicted a priori from the sign of the presynaptic potential change. Thus, a cell can be depolarized in response to hyperpolarization of the presynaptic cell, or vice versa. The postsynaptic response, it will be recalled, depends on the ionic currents produced in the postsynaptic cell in response to the modulated release of transmitter by the presynaptic neuron.

## INFORMATION PROCESSING IN THE VISUAL CORTEX

In mammals and birds, the axons of the ganglion cells distribute themselves at the optic chiasm between ipsilateral (same) and contralateral (opposite) sides of the brain, as shown in Figure 8-24A, whereas in vertebrates more primitive than birds, all optic fibers cross at the chiasm (Figure 8-24B). In the *geniculate body* of the thalamus, they synapse with fourth-order cells that go on to innervate fifth-order cortical neurons (simple cells) in Area 17 of the *visual cortex,* located at the rear of the cerebral hemispheres (Figure 8-11). The receptive fields of the fourth-order cells in the geniculate body somewhat resemble those of the ganglion cells, except that contrast between brightly and dimly lit regions of the retina receives greater emphasis.

Among the most fruitful studies in sensory processing have been those of D. Hubel and T. Wiesel during the 1960s on the visual centers of the brain. While

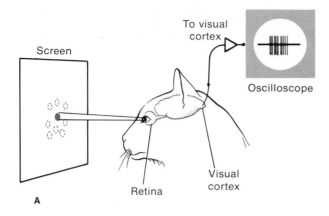

A

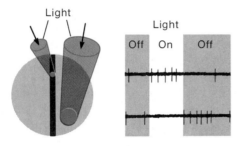

**8-29**  Experimental setup used to study neural responses in the visual cortex of a cat. (**A**) Electrode in cortex records responses to light stimuli projected onto a screen. [From "Cellular Communication" by G. S. Stent. Copyright © 1972 by Scientific American, Inc. All rights reserved.] (**B**) The receptive field of the cortical simple cells is bar-shaped. A spot of light anywhere along the ON portion of this receptive field produces a small excitation of the simple cell. A spot of light adjacent to the bar-shaped ON region causes inhibition of discharge. [From "The Visual Cortex of the Brain" by D. H. Hubel. Copyright © 1963 by Scientific American, Inc. All rights reserved.]

electrodes recorded the activity of individual neurons in the brain of an anesthetized cat, a simple illuminated silhouette was projected onto a screen positioned to cover the visual field of the cat. The responses of the neurons were correlated with the position, shape, and movement of the silhouette on the screen (Figure 8-29A). Several functional classes of cortical neurons were identified in Area 17. These were named *simple cells, complex cells,* and *hypercomplex cells.*

The simple cells, along with other neurons of the cortex, are arranged in an orderly array of vertical columns in the cortical tissue. They are believed to be the first way station for input to the visual cortex, and they have receptive fields very different from those of the retinal ganglion cells and the geniculate cells. Although the receptive fields of these neurons are of several types, they all have one thing in common. The ON region of the field always makes a straight border with the OFF region, rather than a circular border as in the ganglion and the geniculate cells. In some simple cells, the receptive field has a bar-shaped ON region surrounded by an OFF region (Figure 8-29B). Others have a receptive field consisting of an OFF bar surrounded by an ON region. Still others have a receptive field consisting of a straight border with an OFF region on one side and an ON region on the other side. As in ganglion and geniculate cells, the receptive field for a simple cell is in a fixed position on the retina.

The orientation and the location of the ON–OFF boundary differ from one simple cell to another, so that a bar of light moved horizontally or vertically on the retina activates one simple cell after another as it enters one receptive field after another. Figure 8-30 shows the effect of rotating a bar of light about the center of the ON area for a given simple cell. The bar has either no effect or an inhibitory effect on spontaneous activity in the simple cell when at right angles to the field, but it elicits a maximum discharge when in register with the ON region of the receptive field (this effect, of course, defines the receptive field). If the bar of light is then displaced so that it falls completely outside the ON region onto the OFF region, the cell is inhibited.

How do the simple cells come to be specifically sensitive to straight bars or borders of precise orientation and location projected onto the retina? Hubel and Wiesel suggest that each simple cell with a bar-type receptive field receives excitatory connections from lateral geniculate cells whose ON centers are arranged linearly on the retina (Figure 8-31A). Simple cells with border-type receptive fields are presumed to receive inputs as shown in Figure 8-31B. Thus, the simple cell responds most strongly when all the receptors included in the ON-center fields of the ganglion and the geniculate cells innervating that simple cell are

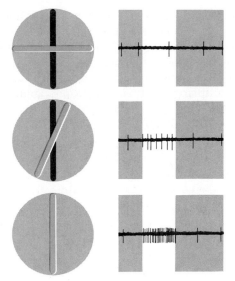

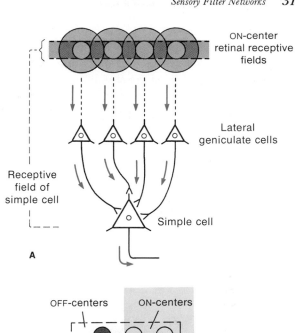

**A**

**8-30** Rotating a bar of light across the receptive field of a cortical simple cell produces a maximum discharge when the bar coincides with the ON region of the receptive field, and partial excitation depending on the degree of overlap between the bar of light and the ON region at other orientations. [From "The Visual Cortex of the Brain" by D. H. Hubel. Copyright © 1963 by Scientific American, Inc. All rights reserved.]

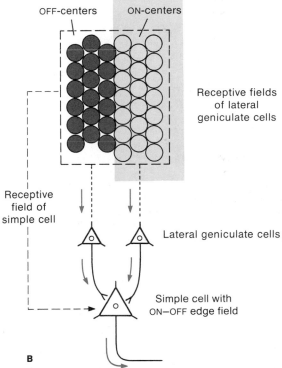

**B**

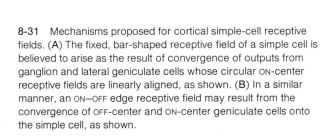

**8-31** Mechanisms proposed for cortical simple-cell receptive fields. (A) The fixed, bar-shaped receptive field of a simple cell is believed to arise as the result of convergence of outputs from ganglion and lateral geniculate cells whose circular ON-center receptive fields are linearly aligned, as shown. (B) In a similar manner, an ON–OFF edge receptive field may result from the convergence of OFF-center and ON-center geniculate cells onto the simple cell, as shown.

illuminated. Any additional illumination of the inhibitory surrounds of the ganglion cells serves only to reduce the response of the cortical cell.

The next order of complexity is seen in the complex cells. These are believed to be innervated by the simple cells, which would make them the sixth order in the visual hierarchy. Complex cells respond best to straight borders of specific angular orientation on the retina.

Unlike the simple cells, however, complex cells do not have fixed receptive fields. Appropriate stimuli presented within relatively large portions of the retina are equally effective, whereas general illumination over the whole receptive area proves ineffective. Some complex cells respond to bars of light of specific orientation (Figure 8-32A), and others give an ON response to a straight border with the light on one side, and an OFF response when the light is on the other side. Certain complex cells respond to a moving border progressing in one direction only (Figure 8-32B). Movement in the other direction evokes either no response at all or a weak one. This effect has been explained as follows. A complex cell appears to receive its input largely from the simple cells of one cortical column. The simple cells of one column all have their receptive fields in the same angular orientation, but systematically displaced with respect to one another. As a light–dark border moves through the corresponding receptive fields of these simple cells, they excite the complex cell in sequence as the light–dark edge passes in turn through each of the ON–OFF borders in the receptive fields of the simple cells. Directional sensitivity in the complex cell results from the inhibitory

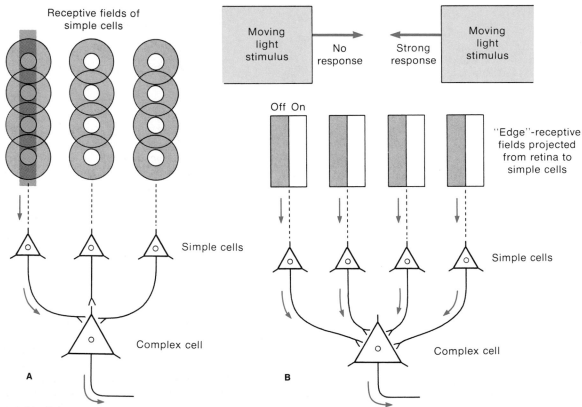

8-32    Origin of complex-cell responses. (A) Some complex cells respond to bars of light located anywhere within a large receptive field but having a specific angular orientation. This response may occur by convergence of many simple cells having similarly oriented bar-shaped receptive fields. (B) Some complex cells respond to edges of light moving in one direction only. This response could result from convergence of a population of simple cells sensitive to light–dark edges of the same orientation. Excitation of the complex cell occurs if the edge moves so as to illuminate the ON side of the simple-cell receptive fields before illumination of the OFF side. Movement in the opposite direction causes only inhibition.

effect of illuminating the OFF region of each simple cell's receptive field before illuminating the adjacent ON region (Figure 8-32B). Remember that illumination of the entire receptive field is relatively ineffective in stimulating a cortical cell. If the border moves so as to illuminate each ON region before the adjacent OFF region of the simple-cell receptive field, the complex cell will receive a sequence of excitation as one simple cell after another is excited.

The complex cells are believed to innervate the hypercomplex cells (seventh order), which respond specifically to such silhouettes as corners and moving "tongues." The organization of the receptive fields of the hypercomplex cells is based on the same principles of visual processing that hold for the lower-order cells.

In summary, the visual system appears to be organized primarily on the principle of convergence. Each cell receives inputs from numerous cells of lesser complexity of receptive field. The result is a hierarchy in which each complex cell receives input from a number of simple cells below it in the neuronal hierarchy and abstracts information from this input on the basis of their connections and synaptic mechanisms. The hypercomplex cell in turn receives input from numerous complex cells and makes a further abstraction. Although convergence is most evident in this scheme, a great deal of divergence also exists, for each receptor cell connects to many ganglion cells and ultimately to thousands of cortical cells. Thus, the visual cortex does not receive a simple one-to-one projection from the retina. Instead, the activity of each neuron represents an abstraction of some relatively simple feature of the optic image. These conclusions suggest that the neurons of the brain are organized during development with a high degree of anatomical specificity, with information flow channeled rather precisely between individual neurons (Box 8-1).

---

## Box 8-1 *Specificity of Neuronal Connections and Interactions*

Note in Figure 8-24A that the portions of the image falling on the temporal (ear-side) portion of one retina fall on the nasal portion of the other retina, and vice versa. The ganglion cells on the right side of each retina send their axons to the right side of the brain, and those on the left side send their axons to the left side of the brain. Thus, it is evident that the right side of the brain "sees" the left half of the field of view, and that the left side of the brain "sees" the right half.

D. H. Hubel and T. N. Wiesel, in their studies of visual processing in the brain, found that some individual cells of the right and left visual cortices have receptive fields in each of the two retinas, and that these fields are positioned so as to be optically in register. That is, cortical cells receive from both retinas input that arises from *corresponding parts of the image*. This means that these cortical cells receive extremely accurate neural projections from ganglion cells "seeing" the same part of the field but located in each of the two retinas. These findings confirm the suggestion of Johannes Müller over a century ago that information originating from analogous receptors (i.e., those "seeing" the same part of the visual field) on both right and left retinas converges on specific neurons in the brain. Such a high degree of morphological specificity is in direct contrast to the idea that the activity of the nervous system is diffusely organized, and that it is the patterning of electrical activity rather than the precise circuitry of neurons that is significant in coding neural messages. It is clear from recordings of single-unit activity in the visual cortex that each cell responds optimally to certain parameters of a stimulus. Furthermore, the behavior of that cell depends in large part on the behavior of cells from which it receives its input. A single impulse in a neuron is thus significant by virtue of the connections that neuron makes with other neurons.

The neurons of the visual cortex are arranged in a remarkably orderly manner. When a recording electrode is gradually advanced into the cortex perpendicular to its surface and past successive simple cells, it is seen that all of the cells of that vertical column have retinal receptive fields of the same orientation, but progressively displaced along the surface of the retina. Those cells forming the adjacent vertical column have receptive fields whose angles of orientation differ slightly from those of the first column, and so on. This is an example of the orderliness of the vast number of connections present in the CNS. One of the major problems of modern neurobiology is to discover how connections are made during the development of the nervous system with such specificity and precision.

## Neuromotor Networks

Although we now understand some of the elementary principles of sensory processing, a great deal remains obscure. As yet, we do not understand the neural mechanisms by which the nervous system gives rise in the mind's eye to the mental image from the optic image projected on the retina, nor for that matter how it gives rise to sensory perception of any kind. Likewise, the circuits that exist between sensory input and motor output constitute a highly complex and poorly understood no-man's land. These circuits, lumped together as a "black box" in Figure 8-33, receive filtered sensory input from past and present. They are also influenced by hormones acting on the genetically and developmentally determined neural networks. The end result of the central integration of past and present inputs is the coordinated motor output that we term the animal's behavior. We next consider some of the simpler examples of neuromotor control.

### THE MYOTATIC REFLEX

In its rudimentary form, the *myotatic* (or *stretch*) *reflex* of vertebrates depends on only two kinds of neurons: the *1a afferent sensory fiber* and the α-motoneuron (Figure 8-34). The 1a afferents have endings in the muscle that are sensitive to stretch. Centrally, they make synapses on the α-motoneurons to form a monosynaptic reflex arc. The sensory terminals of a 1a afferent fiber spiral around the central, noncontractile region of a *muscle spindle,* or *spindle organ.* This organ is a *stretch receptor* that contains a small bundle of special *intrafusal* muscle fibers, which are distinct from the mass of *extrafusal* muscle fibers that constitutes the bulk of the muscle. The extrafusal muscle fibers are innervated by the α-motoneurons. These muscle fibers are the ones that cause the muscle to develop tension and shorten. The intrafusal muscle fibers are present in far smaller numbers, and they make little direct contribution to overall muscle tension. They are innervated by a separate set of motoneurons, the *gamma motoneurons.* We will return to gamma control of the intrafusal fibers in the next subsection.

Changes in the overall length of a muscle are detected by the 1a afferent terminals as they sense the changes in the length of the central, noncontractile segments of intrafusal muscle fibers located in the midregion of each muscle spindle. It is important to note that the muscle spindles are arranged in *parallel*

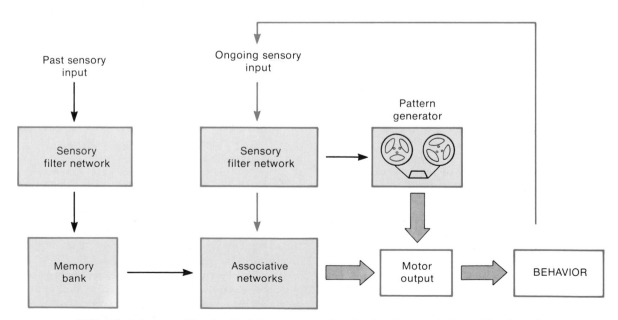

**8-33**  Block diagram of the circuitry between sensory input and motor output. See text for discussion.

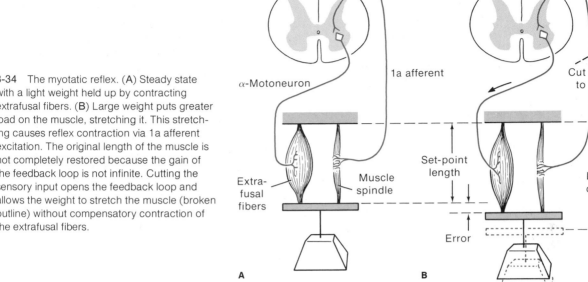

**8-34** The myotatic reflex. **(A)** Steady state with a light weight held up by contracting extrafusal fibers. **(B)** Large weight puts greater load on the muscle, stretching it. This stretching causes reflex contraction via 1a afferent excitation. The original length of the muscle is not completely restored because the gain of the feedback loop is not infinite. Cutting the sensory input opens the feedback loop and allows the weight to stretch the muscle (broken outline) without compensatory contraction of the extrafusal fibers.

with the extrafusal muscle fibers (Figure 8-34). Thus, they are stretched when the muscle is stretched by an external force (e.g., gravity acting on the skeleton) or by contraction of an antagonist muscle. Stretching of the central region of the spindle organ causes an increase in 1a afferent sensory discharge, and because of the synaptic junctions made by the 1a afferents with the $\alpha$-motoneurons, their discharge causes a reflex contraction of the extrafusal muscle fibers.

Consider this reflex arc as a *feedback loop* (Box 1-1). When the system is perturbed by a force that stretches the muscle, the increased 1a afferent discharge evokes an increased discharge of the $\alpha$-motoneurons (Figure 8-35) and thus greater contraction of the extrafusal muscle fibers innervated by those motoneurons. This reflex contraction counteracts the force that originally stretched the muscle, causing the muscle to return toward its initial length. By reducing the tension on the intrafusal fibers, this reflex lowers the sensory discharge of the 1a afferents until the system is in a steady state. Although the reflex shortening approximates the original muscle length, it does not fully compensate. To do so would require a muscle of infinite strength and a feedback loop of infinite sensitivity. The difference between the original length and the new length is termed the *error* (Figures 8-34 and 8-35).

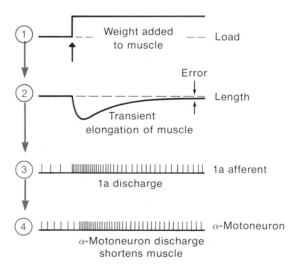

**8-35** Cause–effect sequence of the components of the myotatic reflex initiated by the sudden addition of a weight that stretches the muscle. The subsequent time-dependent shortening of the muscle is due to reflex activation of the $\alpha$-motoneurons.

From this simplified account, it is evident that the myotatic reflex constitutes a feedback system that operates to keep the muscle close to a preset length, in spite of the changes in load applied to the muscle. This mechanism, which operates without conscious control, is important in maintaining posture. Postural muscles must keep the skeleton erect in the face of gravitational pull. Extensors, especially, are subject to stretch by the force of gravity. Flexion of a limb by gravity applies a stretch to the spindle receptor organs in the extensors. In response to this stretch, the 1a fibers discharge and synaptically excite the α-motoneurons innervating the extrafusal muscle fibers of the extensors. The ensuing contraction opposes the force that tends to flex the limb. This reaction commonly occurs in passengers standing in a moving bus during a rough ride.

A familiar display of the stretch reflex is the knee jerk evoked by a tap applied to the tendons of the knee extensor muscles where they pass over the knee joint. The tap results in a sudden stretch of the muscle and its spindle organs. The reflex discharge of the α-motoneurons produces the brief extension of the lower leg.

The myotatic reflex loop can be "opened" for analysis by interrupting the 1a sensory feedback. This has been done in an anesthetized animal by sectioning the dorsal root through which the 1a afferents enter the spinal cord (Figure 8-34B). The operation causes the corresponding muscles to lose their tone, and they no longer contract reflexly when stretched. It was this behavior that led to C. S. Sherrington's discovery of the myotatic reflex in his pioneering neurophysiological studies in the early part of this century (Sherrington, 1906). It was interesting that the muscles should go limp even though the motor innervation remained intact and connected to the spinal cord. This reaction indicated to Sherrington that the afferent fibers provide continual synaptic input to the neurons innervating the muscles and are responsible in part for *muscle tone,* the state of partial contraction that exists in the absence of active movement.

### EFFERENT CONTROL THROUGH THE GAMMA LOOP

The account of the stretch reflex just given is simplified by omitting the mechanism that determines the *set point* of the sensory portion of the feedback system

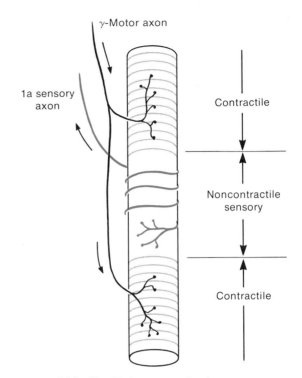

**8-36** Simplified structure of an intrafusal muscle, the 1a sensory axon, and γ-motoneuron innervation. The striated part of the muscle is contractile. The central region is noncontractile and is stretched either by γ-motoneuron activity or by elongation of the muscle within which it resides.

(Box 1-1). The set point is that length of the muscle spindle at which the 1a discharge produces just enough reflex α-motoneuron discharge (and hence extrafusal muscle contraction) to balance the forces tending to stretch the muscle. It is analogous to the thermostat setting on a temperature-controlling device (Figure 1-4).

What determines the set-point spindle length in the myotatic reflex? Before considering this, it should be recalled that the contractile portions at the ends of the intrafusal muscle fibers are arranged *in series* with the noncontractile sensory regions midway along these fibers (Figure 8-36). The contractile portions of the fibers are innervated by the γ-motoneurons, which have their cell bodies located in the gray matter of ventral horn of the spinal cord (Figure 8-37). These motoneurons

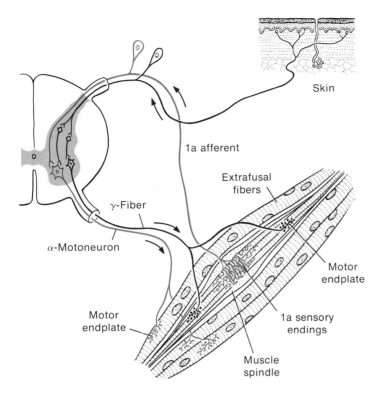

Skin

1a afferent

Extrafusal
fibers

γ-Fiber

α-Motoneuron

Motor
endplate

Motor
endplate

1a sensory
endings

Muscle
spindle

**8-37**   The γ- and α-systems shown together. Painful stimuli to the skin activate both systems simultaneously, shortening the muscle. Stretching of the muscle activates only the α-motoneuron.

are smaller than the α-motoneurons that innervate the extrafusal fibers, and their axons (γ-*efferents*) innervate only the intrafusal muscle fibers of the spindle organs. Contraction of the intrafusal muscle fibers of the spindle organ differs from that of the extrafusal fibers in three important ways:

*1.* It does not directly contribute significantly to muscle tension, because the intrafusal fibers are relatively few in number.

*2.* The intrafusal fibers contract only near their ends and not in the central sensory region.

*3.* The intrafusal fibers contract in response to impulses in the γ-efferent system (i.e., γ-motoneurons) that innervate them.

The γ-motoneurons and the intrafusal muscle fibers are collectively termed the *fusimotor system.* Activation of the fusimotor system by neural commands from the motor centers of the brain or from reflex connections in the spinal cord causes shortening of the terminal contractile portions of the intrafusal muscle fibers,

which *stretches the noncontractile sensory portion* of the intrafusal fibers.† This stretching causes an increase in 1a sensory activity and reflex shortening of the extrafusal muscle fibers until the spindle organs reach their new set-point length (Figure 8-38). A stretch applied to the muscle will now set up additional 1a sensory discharge, which will cause additional firing of α-motoneurons and maintenance of the new set-point length. Likewise, a reduction in the steady excitation of the fusimotor system by signals from the brain resets the system to maintain a greater length. The γ-system is also activated reflexly along with the α-motor system in response to painful stimuli to the skin (Figure 8-37), as well as during willful movement. This *coactivation* of α- and γ-motoneurons serves to keep the muscle spindle from going slack during the contraction of the extrafusal muscle fibers and maintains the sensitivity of the spindle organs to stretch at different muscle lengths.

---

†It is important to note that the α-motoneurons and γ-motoneurons have contrary actions on the length of the stretch-sensitive sensory region of the intrafusal muscle fiber.

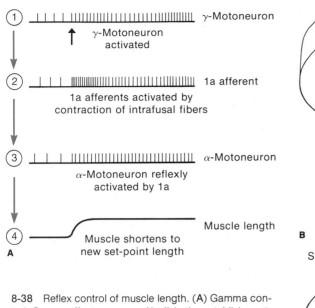

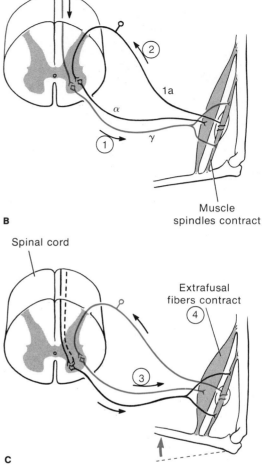

8-38  Reflex control of muscle length. **(A)** Gamma control. Cause—effect sequence (1—4) in the establishment of a new set point by the fusimotor system. Activation of the γ-motoneurons causes the intrafusal fibers to contract, stretching the sensory portion of the spindle. This stretching stimulates the 1a sensory axons to fire and activate the α-motoneurons. The muscle contracts until tension in the sensory portion of the spindle is reduced. **(B)** Diagram of steps 1 and 2 from part A. **(C)** Steps 3 and 4. The muscle spindle is drawn separate from the rest of the muscle for clarity. [From "How We Control the Contractions of Our Muscles" by P. A. Merton. Copyright © 1972 by Scientific American, Inc. All rights reserved.]

Relatively slow movements and postural attitudes may be controlled through adjustments of the set point by the γ-system and reflex activation of the α-system that maintains the muscle at any given set-point length. Rapid and forceful movements are mediated by direct activation of the α-motoneurons by nerve fibers descending from the cerebral cortex and brain stem (Figure 8-38C) and also via polysynaptic reflexes initiated from pain receptors.

### GOLGI TENDON REFLEX

A second kind of receptor in muscle is the *Golgi tendon organ,* formed by sensory arborizations of the so-called *1b afferent neurons* terminating in the tendons of skeletal muscle (Figure 8-39A). These endings in the tendon are sensitive to tension developed in the muscle during contraction. Because they are in series with the muscle fibers (rather than in parallel like the spindle organs), the Golgi tendon organs sense muscle tension rather than length. They indirectly innervate the α-motoneurons of the corresponding muscle through inhibitory interneurons (Figure 8-39B) and thus act to dampen the α-motor output. The role of the 1b reflex is not entirely clear, but it is conjectured that the tension-sensing negative feedback to the α-motoneurons helps correct for imperfections in the length-control mechanism of the myotatic reflex. Thus, as the α-motoneurons accommodate or the synapses fatigue during a main-

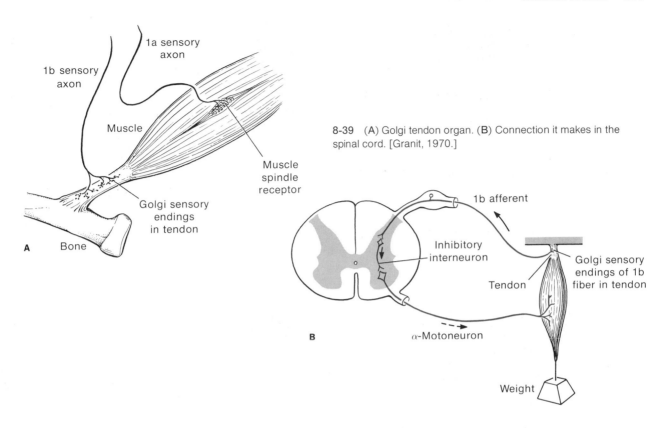

8-39 (A) Golgi tendon organ. (B) Connection it makes in the spinal cord. [Granit, 1970.]

tained 1a input, the 1b afferents will respond to the ensuing drop in muscle tension by slowing their firing rate. The reduced 1b inhibition on the α-motoneurons is responsible for a stabilization of the α-motoneuron discharge as the myotatic reflex weakens. It has also been proposed that unusually high and potentially damaging muscle tension might be prevented by 1b inhibition of the α-motoneurons.

We are aware of the positions of our limbs even when blindfolded. It is possible to touch one's nose on the first try with both eyes shut. However, it is neither the spindle organs nor the Golgi tendon organs that provide the major sensory input for conscious sensation of limb or joint position. Injection of a local anesthetic into a joint eliminates most conscious sensation of the position of the limb, demonstrating that mechanoreceptors located on the articulating surfaces of each joint monitor its position. The information thus obtained enters the consciousness as a sense of limb position.

### THE FLEXION REFLEX
### AND RECIPROCAL INNERVATION

Limb movement about a given joint requires the cooperation of the various muscles that act on that joint. Contraction in one set of muscles occurs in coordination with relaxation of its *antagonists* (i.e., those that have the opposite action), so that mutually antagonistic sets of muscles do not counteract each other.

Consider two muscles, *A* and *B*, that produce opposite movements about a joint (Figure 8-40). When muscle *A* is stretched, its 1a afferents activate reflexly the α-motoneurons that cause contraction. At the same time, branches of the 1a afferents of muscle *A* activate inhibitory interneurons ending on the α-motoneurons of muscle *B*. Thus, although a stretch applied to muscle *A* produces a reflex contraction in that muscle, it simultaneously produces relaxation in the antagonistic muscle. Conversely, a stretch applied to muscle *B* sets up a myotatic reflex in that muscle and reciprocally inhibits the stretch reflex of muscle *A*. Without this

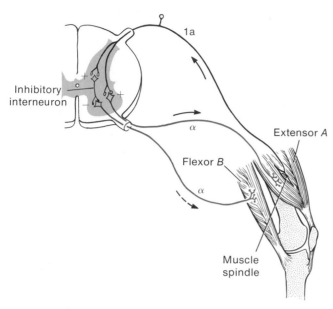

**8-40** Myotatic reflex connections of two antagonistic muscles of the same limb. When muscle *A* is reflexly activated by stretch, the motoneurons to its antagonist, *B*, receive input from inhibitory interneurons reflexly activated by the 1a afferents of muscle *A*. Excitatory and inhibitory synapses are labeled + and − respectively. [Eccles, 1960.]

*reciprocal inhibition,* the stretching and consequent activation of the 1a afferents of one muscle by its contracting antagonist would evoke a counterproductive reflex contraction of the stretched muscle.

Inhibitory circuitry is also important in the muscular coordination of movements between different limbs in vertebrates. This is most clearly demonstrated in *decerebrate* animals. Decerebration—cutting through the brain stem above the respiratory centers in the medulla to disconnect the pathways between the forebrain and the spinal cord—results in exaggerated spinal reflexes, because cerebral modification of spinal reflexes is eliminated. A noxious stimulus to the skin of limb *A* will cause a reflex withdrawal (flexion) of that limb. This *flexion reflex* is accompanied by inhibition of motoneurons to the antagonistic muscles of that limb, plus a reflexive *extension* of the contralateral limb. The latter, termed the *crossed extension reflex,* happens because, simultaneous with the activation of flexor motoneurons

and inhibition of the extensor neurons going to limb *A,* there is inhibition of flexor motoneurons and excitation of extensor neurons innervating muscles of limb *B* (Figure 8-41). It is evident that the flexion and crossed extension reflexes are adaptively related. Thus, if one foot steps on a thorn and reflexly withdraws, the crossed extension reflex assures that the opposite limb is extended to carry the animal's weight momentarily.

## Programmed Behavior

### CENTRALLY GENERATED MOTOR RHYTHMS

Locomotory and respiratory movements typically consist of rhythmic movements controlled by repetitive patterns of neuromotor discharge. Each phase of such a neuromotor cycle is both preceded and followed by characteristic motoneuron discharges that are consistently related to one another in time. There was a long-standing question about the generation of locomotor output: To what extent is it dependent on moment-to-moment sensory input, and to what extent does it arise from autonomous motor output of pattern-generating networks independent of sensory input? This question has been examined in a number of animals. As might be expected, both mechanisms have been found in some instances, and one or the other has been found to predominate strongly in others.

Central motor patterns have been most clearly demonstrated in the nervous systems of some invertebrates, such as the neuromotor control of rhythmic movements in locomotion. Thus, the various muscles that cause alternate up-and-down movements of the two pairs of wings in a grasshopper are controlled by an appropriate sequence of nerve impulses carried by several motor axons (Figure 8-42). The output patterns of the motoneurons continue to occur in proper phase relation to one another even after all sensory input from the muscles or joints of the wings is eliminated. This persistence suggests that the motor output pattern is generated entirely within the CNS by a network of neurons that interact to produce a coordinated motor output controlling rhythmic muscle contractions.

What role, if any, does sensory input play in the control of such centrally programmed motor output? Sensory feedback from stretch receptors at the bases of

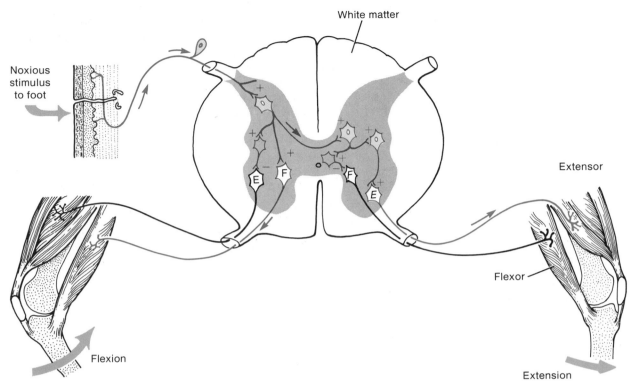

White matter

Noxious
stimulus
to foot

Extensor

Flexor

Flexion

Extension

**8-41** Pathways for flexor and crossed extension reflexes. Noxious stimulation of skin activates pain afferents that make various synaptic connections with the interneuron. This interneuron excites ipsilateral flexor (*F*) motoneurons, causing the flexor muscles to contract and the joint to flex. By virtue of the gray-shaded sign-reversing (i.e., inhibitory) interneurons, the ipsilateral extensor (*E*) motoneurons are inhibited (−). The crossed extension reflex, seen as the converse activation and inhibition of extensors and flexors on the other side, causes contralateral extension.

the locust's wings, stimulated by the movement of the wings, increases the frequency of the motor rhythm. In the absence of feedback from the wing joints (i.e., when the receptors are destroyed), the neuromotor output slows down to about half its normal frequency, but the phase relations of impulses in the different motoneurons are retained. The original frequency can be restored by electrical stimulation of the nerve roots containing the sensory axons of the wing-joint receptors. It is interesting that although the motor output rhythm increases in frequency when provided with sensory input, the timing of the motor output is independent of the timing of impulses in the sensory nerves. Randomly ordered stimulation of the sensory nerves is just as effective in speeding up the motor output as a regular train of stimuli. Thus, proprioceptive feedback is not required for the proper sequencing of motor impulses to the muscles producing wing movements. Nevertheless, once the central flight motor is turned on, sensory nerves from the wing-joint receptors provide feedback that reinforces the central output (Figure 8-43C).

What turns the flight motor on and off? When the grasshopper jumps off to fly, hair receptors on the head are stimulated by the passing air. This specific sensory input initiates the output of the flight motor. When the

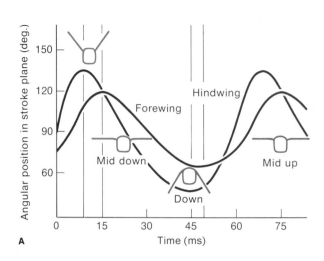

**A**

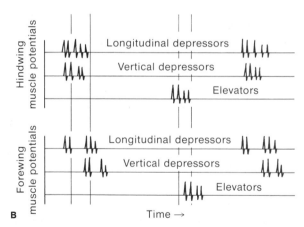

**B**

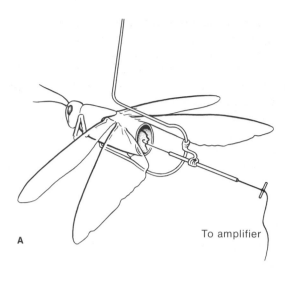

**A** To amplifier

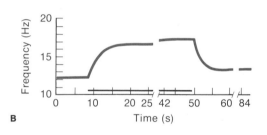

**B**

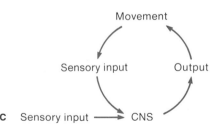

Movement

Sensory input          Output

**C** Sensory input → CNS

**8-42** Centrally generated motor-output rhythms (B) in flight muscles of grasshoppers are correlated with wing position (A) during one and one-half cycles of wing beat. Each deflection of the baseline represents time of firing of several motor units. The phase relations of firing of different muscles are determined in the CNS and are independent of sensory input. [Reprinted from "The Origin of the Flight-Motor Command in Grasshoppers," by Donald Melvin Wilson, in *Neural Theory and Modeling: Proceedings of the 1962 Ojai Symposium,* edited by Richard F. Reiss, with the permission of the publishers, Stanford University Press. © 1964 by the Board of Trustees of the Leland Stanford Junior University.]

**8-43** Role of proprioceptive feedback in grasshopper flight. (A) Experimental arrangement. An eviscerated grasshopper or locust is mounted so that it can flap its wings when stimulated by air blowing on facial receptor hairs. Electrodes for recording motor output and for stimulating receptor nerves are fixed in place. (B) The result of sensory input is to speed up the frequency of the endogenous rhythm of motor output. (C) Cyclic organization of behavior. External sensory input (puff of air on hair receptors) stimulates behavior (flying). The wing movements stimulate proprioceptors that provide further input that stimulates the flight motor. [Wilson, 1964, 1971.]

insect alights, the motor is turned off by signals originating in tarsal (i.e., foot) mechanoreceptors.

Endogenous pattern-generating networks have been shown to exist in a number of other invertebrate nervous systems. The cyclic motor output to the abdominal swimmerets of the crayfish persists in the isolated nerve cord and even in single ganglia. This intrinsic rhythm is initiated and maintained by activity of command interneurons that descend from the supraesophageal ganglion. Although the intrinsic bursting pattern requires maintained activity from one or perhaps several of the interneurons, there is no simple one-to-one relationship between the firing pattern of the interneuron and the pattern of the motor output to the swimmerets.

Varying degrees of central neural automaticity also occur in the vertebrates. Respiratory movements persist in mammals when sensory input from the thoracic muscles is eliminated by cutting the appropriate sensory roots. Toads in which all sensory roots except those of the cranial nerves have been cut still show some rudimentary coordinated walking movements, although these are hard to demonstrate owing to the loss of neuromotor tone that results from interruption of the myotatic reflex arc. Motor output to the swimming muscles of sharks continues to fire in the normal alternating way when segmental sensory input is eliminated. But the intersegmental sequencing (anterior-to-posterior waves) of motor output disappears in the absence of sensory input, and all the segments fire together, left and right motor outputs firing alternately with each other. This results in violent, alternating lateral flexions of the shark's entire body instead of the sinuous traveling waves that normally propel it through the water. Thus, sensory input appears to be required in the shark to coordinate the motor output in a spatial sequence that progresses in waves from anterior to posterior segments.

Walking movements have been investigated in spinal cats (in which the brain stem has been transected above the medulla oblongata) supported over a treadmill. Such studies reveal that the walking sequence can occur without input to the spinal neurons from the brain. Moreover, a rudimentary walking rhythm has been seen to continue even after the peripheral sensory input to the spinal cord has been essentially eliminated by transection of the dorsal roots.

## FIXED ACTION PATTERNS

The stimulation of appropriate neurons in the CNS can elicit coordinated movements of varying degrees of complexity. Electrical stimulation of one such *command interneuron,* or *trigger neuron,* in the nerve cord of the crayfish causes the animal to assume the defense posture, with open claws held high and body arched upward on extended forelegs. The appropriate sensory input excites this interneuron, and it (via its widespread processes) effects the excitation of some motoneurons and the inhibition of others, thereby eliciting the fixed defensive posture. It is characteristic of command interneurons of arthropods that they activate many muscles in a coordinated manner and produce reciprocal actions in a given body segment. That is, antagonists are inhibited while agonists are excited. Those command interneurons most effective in eliciting a coordinated motor response are generally least easily activated by simple sensory input. A number of command interneurons have been described in arthropods; some act as ON or OFF switches for internally generated, repetitive patterns of motor output.

Command interneurons must not be confused with motoneurons. As stated by Sherrington (1906) early in the century, motoneurons are the *final common pathways* of animal behavior. Each motoneuron activates its motor unit of muscle cells and thus produces only one component of movement in the complex sequence of movements that, together, constitute a behavioral act. This activation was illustrated by A. O. D. Willows in the nudibranch *Tritonia* (Figure 8-44). Stimulation of specific neurons in the brain of this mollusk produces specific movements in the semirestrained experimental animal. Activity in one neuron causes contraction of a set of muscles on one side, while activity in its counterpart on the opposite side causes contraction of homologous muscles on the other side. This is not surprising. What is more noteworthy is the ability of a command interneuron to control the orchestrated activation of motoneurons so as to produce a well-defined temporal pattern of swimming movements (Figure 8-45). This command interneuron is activated by sensory input when *Tritonia* is touched by its natural predator, a starfish. The central motor program of escape-swimming thus *released* by the appropriate sensory input is played out even in the isolated nervous system deprived of reflex feedback from receptor organs.

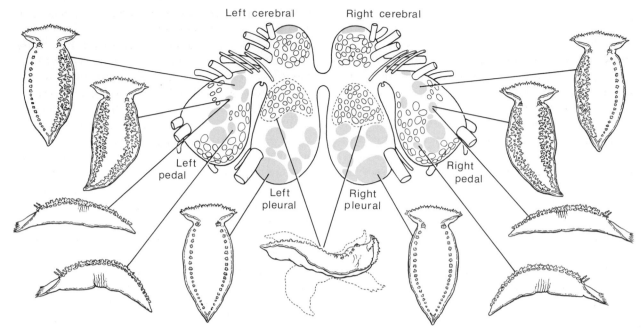

8-44 Stimulation of identified nerve cells in the brain of the marine nudibranch *Tritonia* results in stereotyped postures or behavior patterns. Stimulation of the small cells indicated in the pleural ganglion elicits the alternating movements of swimming. [From "Giant Brain Cells in Mollusks" by A. O. D. Willows. Copyright © 1971 by Scientific American, Inc. All rights reserved.]

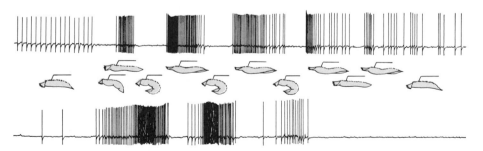

8-45 Stimulation of one command interneuron in *Tritonia* results in the activation of several motoneurons to produce a complicated sequence of muscle contractions termed the escape response. The upper and lower recordings show the discharge of two cells during the sequence of movements shown in outline. Firing of the lower cell is correlated with flexion, and firing of the upper cell with extension. [From "Giant Brain Cells in Mollusks" by A. O. D. Willows. Copyright © 1971 by Scientific American, Inc. All rights reserved.]

## Behavior in Animals Without Nervous Systems

Early claims that subcellular "neuromotor" systems exist in certain protozoa have not survived modern scrutiny, and it is evident that the unicellular organisms have nothing analogous to the circuitry of a metazoan nervous system. Nonetheless, protozoa are capable of detecting stimuli and giving adaptive locomotory responses. Some even exhibit a very rudimentary kind of memory. How is this possible without a nervous system?

When a ciliate, such as *Paramecium,* bumps into an obstacle, it reverses the beat of its cilia, swims backward a short distance, pivots about its posterior end, and resumes forward locomotion, usually at some angle to its previous path (Figure 8-46A). This behavior is termed the *avoiding reaction.* In contrast, a mechanical stimulus to the posterior end causes the paramecium to swim forward more rapidly—the *escape reaction.* Although these responses have the superficial appearance

of being purposive, electrophysiological studies show them to be purely mechanistic.

As explained in Chapter 10, the frequency and the direction of the ciliary beat in protozoa is controlled directly or indirectly by the membrane potential. Depolarization causes the cilia to reverse their beat, so that the ciliate swims backward, whereas hyperpolarization evokes an increased rate of beat in the forward-swimming direction. Electrical recordings were made from paramecia with intracellular microelectrodes while mechanical stimuli were applied to selected parts of the cell surface (Figure 8-47). Mechanical stimulation at the anterior end elicited a depolarization graded in intensity with the strength of the stimulus; stimulation of the posterior end elicited a graded hyperpolarization.

Why do the opposite ends of the paramecium give different electrical responses to the same kind of stimulation? Experiments using different concentrations of the external cations $K^+$ and $Ca^{2+}$ revealed that mechanical stimulation of the anterior end causes an increase in calcium permeability. Because calcium is

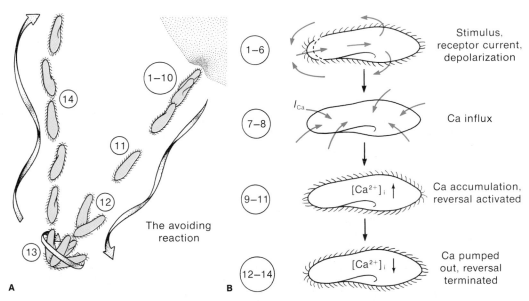

**8-46**  The avoiding reaction in *Paramecium.* (**A**) After colliding with an object, the ciliate backs up, rotates, and heads off in a new direction. [Grell, 1973.] (**B**) Ionic events that give rise to this behavior; the numbers refer to steps shown in A. See text for details. [Eckert, 1972.]

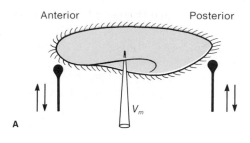

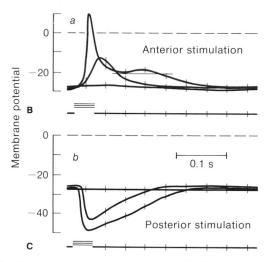

8-47    Receptor potentials in *Paramecium*. (**A**) The anterior or posterior ends of the ciliate were stimulated mechanically with a crystal-driven microprobe while an intracellular electrode recorded responses. (**B**) Stimulation of the anterior produced a depolarizing receptor potential, *a*, and (**C**) stimulation of the posterior produced a hyperpolarizing receptor potential, *b*. The lower trace in each set shows relative intensity and the duration of each stimulus. Note gradation of response amplitudes. [Naitoh and Eckert, 1969.]

very different electrical consequences of mechanical stimulation at anterior and posterior ends are due to differences in the ion-selective channels activated in the cell membrane by mechanical stimuli to opposite ends of the ciliate.

The depolarizing receptor potential elicited by a mechanical disturbance to the anterior end acts as an electrical stimulus for the entire cell membrane, causing the activation of calcium channels located in portions of the membrane covering the cilia. The inflow of $Ca^{2+}$ through these channels produces a graded regenerative depolarization (Figure 8-48), and the elevation of intraciliary $Ca^{2+}$ levels leads to a reversal in the beat direction of the cilia. The behavioral consequence of these events is a retreat of the swimming paramecium from obstacles that it bumps into (Figure 8-46).

The calcium response of the paramecium membrane is not all-or-none like that of a true action potential, but is instead graded with stimulus intensity.

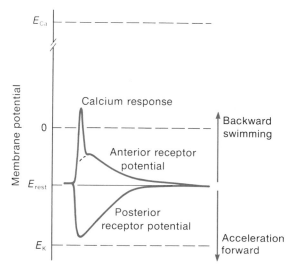

8-48    Control of swimming by membrane potential in *Paramecium*. The depolarizing receptor potential in response to stimulation of the anterior gives rise to a second potential, the spikelike $Ca^{2+}$ response. This response is graded with stimulus strength and evokes reversed beating of the cilia and hence backward swimming. The hyperpolarizing potential elicited by mechanical stimulation of the posterior causes the cilia to beat more rapidly in the forward-swimming direction. [Eckert, 1972.]

about 10,000 times more concentrated outside the cell than inside, the local increase in calcium permeability produced by the mechanical stimulus results in a transient influx of $Ca^{2+}$. This influx constitutes a depolarizing receptor current. The depolarization spreads electrotonically throughout the cell. A similar mechanical stimulus to the posterior surface leads to a local increase in potassium ion permeability, and the consequent efflux of $K^+$ constitutes a hyperpolarizing receptor current that spreads throughout the cell. Thus, the

An all-or-none response is unnecessary in this system because the cell is short enough so that electrotonic spread communicates the electrical signal to the entire cell. Although the overall mechanisms of the avoiding reaction in *Paramecium* have been worked out, it is still not clear how the hyperpolarization produced by mechanical stimulation of the hind end causes the cilia to beat faster to produce the escape reaction.

The trumpet-shaped ciliate *Stentor* spends most of its time attached to the substratum at its narrow base (Figure 8-49A). When a mechanical disturbance of sufficient strength is applied to the surface of the cell, it contracts longitudinally with the aid of an internal system of microfilaments. It is interesting that the probability of occurrence of this all-or-none contraction diminishes with repeated stimuli. The response to a carefully controlled mechanical stimulus decreases exponentially with number of trials (Figure 8-49B). This resembles a simple type of learning in metazoans termed *habituation,* in which the nervous system eventually ignores a stimulus after it is presented several times. Recordings made from *Stentor* with intracellular electrodes show that a mechanical stimulus elicits a receptor potential that in turn evokes an all-or-

none action potential (Figure 8-50A). Contraction follows the action potential. As the mechanical stimulus is applied repeatedly to the cell at 1 min intervals, the receptor potential becomes progressively smaller, until finally it is no longer large enough to evoke the all-or-none action potential (Figure 8-50B). Thus, the probability of a behavioral response to mechanical stimulation decreases as the size of the receptor potential decreases. It is not known why repeated stimulation produces a progressive reduction in the receptor potential of *Stentor.*

It is evident from these two examples that the surface membrane of a unicellular organism can perform the functions of sensory reception, simple integration of receptor signals, and control of the motor response—and that a rather simple modification of membrane function can give rise to an elementary form of memory. Thus, the rudiments of the sensory–neuromotor systems of metazoans are present already in the electrophysiological properties of the Protozoa. The similarity between the electrical properties of nerve cells and those of protozoa suggests that these properties, primarily related to the functioning of ion-selective membrane channels, are very general and very ancient.

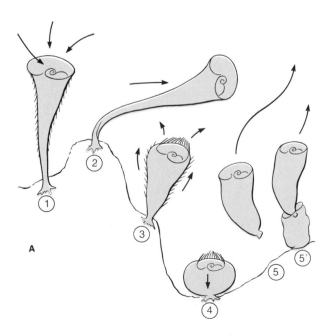

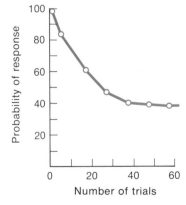

**8-49**  Habituation in *Stentor*. (A) Behavior in response to adverse stimuli: (1) posture is undisturbed; (2) cell bends away from stimulus; (3) cilia reverse; (4) cell contracts; (5, 5′) cell swims away. With repeated stimulation, the responses decrease, with the responses toward the left becoming more probable than those toward the right. [Jennings, 1906.] (B) Decay of probability of contraction as a function of number of trials in which *Stentor* is mechanically stimulated. [Wood, 1970–1971.]

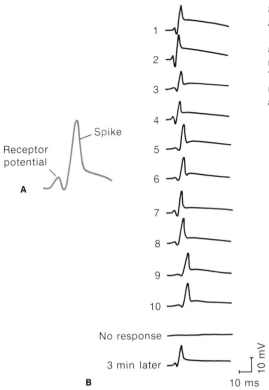

**A** Spike

Receptor potential

1
2
3
4
5
6
7
8
9
10

No response

3 min later

**B**

10 mV

10 ms

8-50   Receptor potentials and action potentials recorded from *Stentor*. **(A)** Control recording. **(B)** Series of potentials elicited at 1 min intervals by mechanical stimulation. Note the decrease in amplitude of receptor potentials with time. Eventually, the receptor potential fails to reach threshold for the action potential. The reduced probability of eliciting the action potential with repeated stimulation results in behavioral habituation in this animal. [Wood, 1970–1971.]

## Animal Orientation

### TAXES AND CORRECTIVE RESPONSES

A common nocturnal sight for many city dwellers is the scurrying of cockroaches as they abandon their tasty scraps and disappear into the woodwork when the kitchen light is switched on. This delightful behavior is an example of a *taxis* (movement in response to a stimulus), more specifically, *negative phototaxis*. Another example of negative phototaxis is illustrated by *Dugesia* and other flatworms that will turn away from the source of light on entering the light beam (Figure 8-51). An animal programmed to turn toward the light is said to exhibit *positive phototaxis*. Jacques Loeb (1918) suggested early in this century that simple taxes like these result from asymmetrical motor activation by asymmetrical sensory input. According to this view,

negative phototaxis occurs when the light impinging on one eye leads to a stronger ipsilateral motor output (i.e., on the same side as that eye), causing the animal to veer away from the source of the light. Positive phototaxis occurs if the light stimulus to an eye stimulates contralateral locomotor output, causing the animal to turn toward the source of light. The concept of a *sensorimotor servosystem* (Figure 8-52) as the basis for locomotor orientation is supported by the simple observation that positively phototaxic animals blinded in one eye will orient so that the intact eye points away from the light. It appears to apply in a number of cases of orientation toward or away from stimuli such as light, heat, odor, sound, and gravity.

Sensory information is used to correct structural or functional asymmetries of central pattern generators or of the structures (wings, legs, fins, etc.) that influence locomotion. A locust continues to fly in a straight line even after one of its four wings is partially or entirely removed, providing it can use its eyes to orient. In the dark, even an intact tethered locust (Figure 8-53A) will roll about its longitudinal axis when induced to fly. The roll is due to the slight asymmetry of the wings and of the centrally generated motor output (Figure 8-53B). The tethered locust ceases to roll when it is provided with a visual cue in the form of an artificial horizon. This stabilization is the result of a correction in the motor output to the wings (Figure 8-53C). Visual input provides information to the central flight motor, regulating the relative outputs to the left and right sets of wing muscles so as to maintain the horizon level.

329

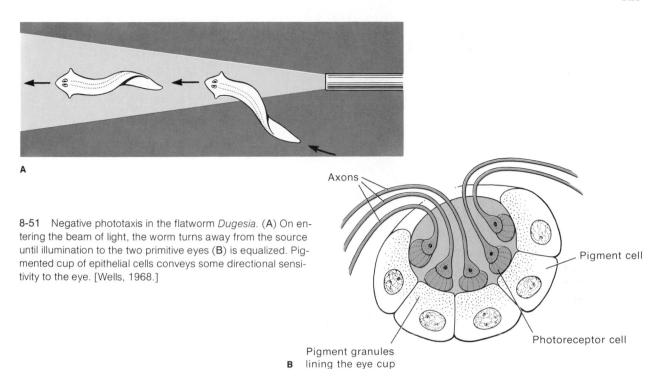

A

8-51 Negative phototaxis in the flatworm *Dugesia*. (A) On entering the beam of light, the worm turns away from the source until illumination to the two primitive eyes (B) is equalized. Pigmented cup of epithelial cells conveys some directional sensitivity to the eye. [Wells, 1968.]

Axons

Pigment cell

Photoreceptor cell

Pigment granules
B lining the eye cup

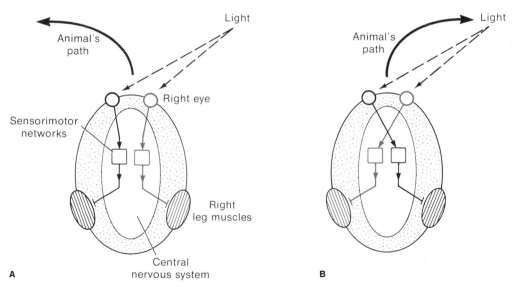

8-52 Jacques Loeb's hypothesis of negative and positive taxes. (A) Negative phototaxis. Stronger illumination of one eye leads to stronger locomotor movements of the appendages on the homolateral side. (B) Positive phototaxis. Stronger illumination of one eye causes stronger locomotor movements of the contralateral appendages. [From "Flight Orientation in Locusts" by J. M. Camhi. Copyright © 1971 by Scientific American, Inc. All rights reserved.]

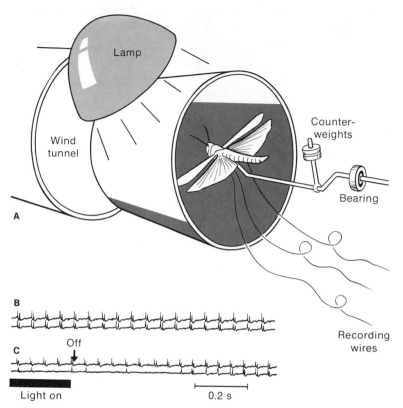

**8-53** Visual correction of an endogenous neuromotor asymmetry. **(A)** Experimental arrangement. The locust is tethered so as to be free to roll while flying in a wind tunnel. An artificial horizon is provided when the lamp is on. When the lamp is turned off, the flying locust has no visual clues as to flight orientation; consequently, it rolls slowly about its longitudinal axis because of natural asymmetry of the wing beat. **(B)** Electrical recordings of left and right flight muscles during flight of a locust that has had one hindwing removed. In this recording of flight with the lamp off, the output to both left and right wing muscles was equal, as in the normal animal. Because of the missing hindwing, this motor output produced severe rolling during flight. **(C)** When the light was turned on, output to the intact wing was reduced in order to compensate and prevent rolling. After the light was turned off, removing the visual reference, the output again became symmetrical and rolling resumed. [Wilson, 1968.]

The importance of sensory feedback correction of orientation and locomotion in humans is confirmed in our daily experience. For example, as noted in Chapter 1, the driver of a car continually makes minor steering adjustments: his eyes are the sensors in a feedback system in which his neuromuscular system coupled to the car's steering mechanism corrects any deviations from the center of the lane due to asymmetries in his neuromotor system, as well as to side winds or imperfections in the road and vehicle. Related to this is the finding that blindfolded human subjects, either walking or driving a car in an open flat field, will take a more or less circular course, the size and direction (right or left) of the average circle differing from one subject to another. Similar turning biases are seen in animals at all phylogenetic levels. Visual and other exteroceptive feedback compensate for such inherent locomotor biases, some of which are probably due to congenital asymmetries in the functioning of the neuromotor apparatus.

## VIBRATIONAL ORIENTATION

Many animals locate their prey from vibrations set up by the prey. Spiders are alerted to prey in their web, for example, by vibrations of the silken threads. These vibrations are detected by mechanoreceptor organs located in the legs of the spider. Another group of arachnids, the nocturnal desert scorpions, use sand-borne vibrations produced by movements of their prey to locate and orient toward such sources up to 0.5 m away. At distances up to 15 cm, the scorpion can deter-

mine the distance, as well as the direction, of the source of vibration. Besides the common mechanoreceptor sensilla, the scorpion possesses a specially sensitive vibration receptor on each of its eight legs, the *basitarsal compound slit sensillum (BCSS)* (Figure 8-54A). The sensitivity of the BCSS receptors to vibration is considerably less than that calculated for cochlear hair cells, but is nonetheless remarkable. Using a calibrated mechanical displacement stimulator, Brownell and Farley (1979a and b) found that the receptor fires in response to displacements of the tarsal segment of the leg of less than 1 Å. The ordinary mechanoreceptor sensilla of the leg are an order of magnitude less sensitive. Thus, the BCSS's appear to be the sensors used for detecting the direction of travel of sand-borne vibrations.

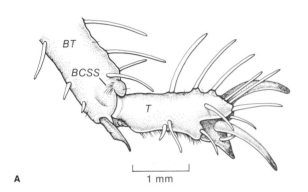

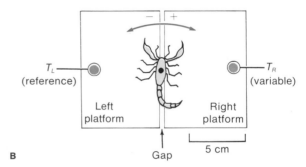

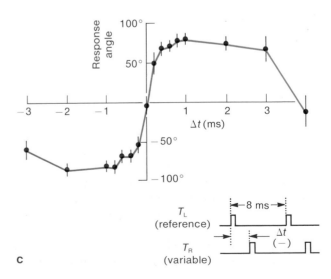

**8-54** Orientation response of the scorpion. (A) Lower part of leg showing basitarsus, *BT,* tarsus, *T,* and basitarsal compound slit sensillum, *BCSS.* (B) Apparatus used to vary time differences of platform movements between left and right sets of legs. The scorpion straddles a 2 mm wide gap between two movable platforms driven by left and right transducers, $T_L$, $T_R$. Input to right-hand transducer was varied in time with respect to the left-hand transducer. The angle of the scorpion's turn to left ($-$) or right ($+$) was measured in degrees. (C) Turning response toward legs receiving the earlier stimulation, with angle plotted against various time delays between brief displacements of the left and right platforms. The scorpion turned right ($+$) when the activation pulse to the right platform preceded the ($+\Delta t$) pulse to the left platform. When $\Delta t = 0$ or $+4$ ms (pulses 180° out of phase), there was no consistent turning direction. [Brownell and Farley, 1979a and b.]

For accurate orientation toward the source of vibration, the scorpion keeps all of its eight legs in contact with the substratum, in a precisely spaced circle. Clearly, the leg closest to the source will be the earliest to intercept the vibrational waves along the sandy substratum. Experiments were performed with the scorpion astride a split vibration platform, with half of its legs on each side of the platform (Figure 8-54B). The scorpion was allowed to orient in response to vibrations produced independently in the two sides of the platform. The responses indicated that directional discrimination by the scorpion depends on the timing of the vibrations reaching each leg (Figure 8-54C). The scorpion would orient toward the side that vibrated slightly ahead of the other side. In another set of experiments, the BCSS's of various legs were ablated to test the importance of these receptor organs for the integration of incoming stimuli. When the organs were ablated on all four legs of the left side, for example, the scorpion would lose proper directional discrimination for the defensive orientation and would turn to the right in each trial regardless of the direction of origin of the vibration. The intact scorpion, on the other hand, invariably oriented correctly to within a few degrees.

In producing the orientation response, the central neural apparatus appears to determine the direction of the vibrational waves by the timing of impulses arising in the BCSS receptors of all the legs. Those legs facing the stimulus source intercept the waves first, and those on the opposite side about 1 ms later (the waves travel 40–50 m/s in the sand). By integrating the delays of impulses from the different legs, the CNS appears to calculate the direction of the stimulus source and then produces the appropriate motor output for the orientation response.

More complex forms of vibrational orientation occur in some aquatic organisms. Among the invertebrates, the surface-swimming whirligig beetle *Dytiscus,* a common pond dweller, is reputed to detect the reflections, from nearby surface objects, of miniature surface waves that it produces in the course of swimming. Among the vertebrates, various aquatic amphibians and fishes, as well as marine fishes, can detect nearby objects from reflections of the pressure waves produced by their own swimming movements. The returning pressure "echoes" are detected by the exquisitely sensitive hair cells of the lateral lines. The timing and distribution of sensory signals, produced by the returning pressure waves, are utilized by the nervous system to determine the locations of nearby objects.

## ECHOLOCATION

The highly refined auditory mechanisms of mammals and birds have led to the evolution of a truly remarkable form of acoustical orientation in which the animal utilizes the return echoes of high-frequency emitted sound pulses to detect the direction, distance, size, and texture of objects in its environment. This sonarlike use of auditory signals, termed *echolocation,* is most highly developed in two groups of mammals: (1) the microchiropteran bats and (2) certain cetaceans, most notably porpoises and dolphins. Among birds, only two genera, the Asian cave swift *Collocalia* and the South American oil bird *Steatomis* appear to use echolocation, producing audible tongue clicks for this purpose.

The discovery that bats utilize echolocation had to wait until the end of the eighteenth century, when the Italian naturalist Lassaro Spallanzani became curious about the ability of bats to avoid obstacles in the air in total darkness, whereas his pet owl required at least a dim light to do the same. After some false leads, he confirmed a report published by Louis Jurine, a Swiss surgeon, that proper plugging of the bat's ears interferes with its remarkable ability to navigate in the dark. On the other hand, he found that blinded bats could find their way home to their favorite belfry in the cathedral at Pavia. Moreover, the blinded bats caught insects quite successfully, for their stomachs were stuffed with such morsels caught on the wing during their trip back to the belfry. Owing perhaps to a general ignorance of the properties of sound in those days, Spallanzani overlooked the possibility that the bats emitted sounds inaudible to humans, and he came to the erroneous conclusion that they navigated by the detection of echoes from the sounds of their wingbeats and located their prey by homing in on the buzzing of insect wings.

It was not until about 150 years later, in 1938, that Donald Griffin and Robert Galambos, both students at Harvard University, made use of recently developed acoustical equipment to determine that bats in fact emit ultrasonic cries to produce the echoes with which they "see" in the dark. Further studies by Griffin and his colleagues revealed some of the phenomenal echolocating capabilities of insectivorous bats. High-speed photography showed that bats could make two separate mosquito or fruitfly captures in about $\frac{1}{2}$ s. Equally amazing is the more recently discovered ability of a fish-eating bat in Trinidad to locate and capture its underwater prey by detection of surface ripples produced by the fish swimming near the surface.

A bat captures an insect via three phases of acoustical orientation. In the little brown bat *Myotis lucifugus,* the "cruising" phase, emitted during straight flights, consists of about a dozen pulsed sounds separated by silent periods of at least 50 ms. Each pulse of sound is *frequency-modulated (FM)*, sweeping downward through a frequency spectrum of about one octave somewhere between 100 kHz and 20 kHz. The second phase of emission appears to begin as the bat detects its prey. The pulses are then produced at shorter intervals, so that the cries occur up to 100/s. The third and final phase consists of a buzzlike emission during which the intervals become even shorter than 10 ms, the pulse duration shortens to about 0.5 ms, and the sound frequencies drop to about 25–30 kHz. Finally, the bat scoops up the insect with its wings or with the webbing between its hind legs, guiding the morsel to its mouth.

The sounds produced by the bat are remarkably energetic (even though inaudible to humans), reaching intensities in excess of 200 dynes/cm$^2$ (equal to $+120$ db sound pressure level) close to the mouth of the bat. Such a sound is equivalent in intensity to that of a jet plane during takeoff passing 100 m overhead, and it is 20 times as intense as the sound of a pneumatic jackhammer several yards distant. Nonetheless, the sound energy returning as an echo from a small object is very weak because, like other forms of radiating energy, sound intensity drops off as the square of distance. Since this relation holds for both the cries and the tiny fraction of the emitted energy that is reflected back from a small object (Figure 8-55), the bat faces a formidable neural task in differentiating the very weak and complex echoes from the far more powerful emitted sounds.

A number of morphological and neural modifications in echolocating bats contribute to these phenomenal abilities. The snout is provided with complex folds and a nostril spacing that make for a megaphone effect. The pinnae of the ears are greatly enlarged to help capture echoes. The eardrum and ear ossicles are especially small and light for high-frequency fidelity. Contraction of the muscles of the auditory ossicles during sound emissions briefly reduces sensitivity, a characteristic of the mammalian ear. Blood sinuses, connective tissue, and fatty tissue isolate the inner ear from the skull, reducing the direct transmission of sound from the mouth to the inner ear. Not surprisingly, the auditory centers of the brain are enormously enlarged relative to the small size of the brain.

A number of specializations of the auditory centers of bats are important in the neurophysiology of echolocation. One of these is *contralateral inhibition.* In this special form of lateral inhibition, the activation of auditory nerve fibers on one side of the head that are sensitive to certain wavelengths leads to an inhibition of cells in the auditory center of the opposite side, homologous to the areas innervated by the activated auditory fibers. The effect of the contralateral inhibition is to increase contrast between the intensities perceived on the two sides of the head, thus enhancing the ability to discriminate the direction of an echo. Two other important specializations are the auditory system's rapid recovery after a loud sound, and its short-term enhanced sensitivity to sounds having characteristics similar to the (emitted) sound just heard. Thus, for a period of 2–20 ms (the time it takes an echo to return from an object 34–340 cm away) following a given emission, the auditory apparatus is extrasensitive to a second sound (the echo) of similar frequency. Thus, the bat has a facilitated ability to perceive its echoes over this critical range of distances. Besides enhancing the accuracy of echolocation, this mechanism may help explain why bats flying in a swarm are not subject to jamming by the cries of their neighbors.

## Animal Navigation

The ability of certain animals to navigate in unfamiliar terrain and bodies of water, over long distances, is truly impressive. The navigational abilities of animals have become surrounded by an aura of mystery because of our limited understanding of the cues they use to sense

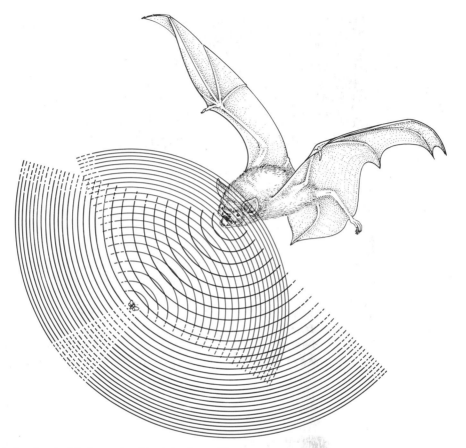

**8-55**   Echolocation by little brown bat *Myotis lucifugus*. The insect is located by detection of returning sound waves (colored waves). Spacing of curves represents changing wavelength of the emitted sound and echo. Note that only a minute fraction of the emitted sound is reflected, and only a small portion of that is intercepted by the bat. Not drawn to scale. [From ''More About Bat 'Radar''' by D. R. Griffin. Copyright © 1958 by Scientific American, Inc. All rights reserved.]

direction. An understanding of these sensory cues has developed slowly because of the redundancy of sensory systems that appear to be used in navigation by some animals. Several sensory systems are utilized by some homing or navigating species, one or more systems predominating when conditions for the other systems are less than favorable. This diversity of systems makes controlled experimentation, in which one deals with a single variable, extremely difficult to carry out. It is now evident, for example, that to varying degrees, birds use landmarks, visual cues (including polarized light), odors, sounds, sun and star positions, and even the Earth's magnetic field in homing and navigation.

## CLOCK-COMPASSES

Bees use the sun's position and the pattern of polarized light in the sky to keep track of the direction from hive to food sources, and they communicate this direction to their hive-mates with their now-famous waggle dances. Certain birds navigate over vast stretches of ocean devoid of landmarks. Some species of nocturnally migrating birds such as garden warblers, when exposed to the night sky in a planetarium, orient with respect to certain patterns of stars (Figure 8-56). As night progresses, and the projected constellation is moved across the dome to mimic the Earth's rotation, the birds orient themselves in the "correct" direction with

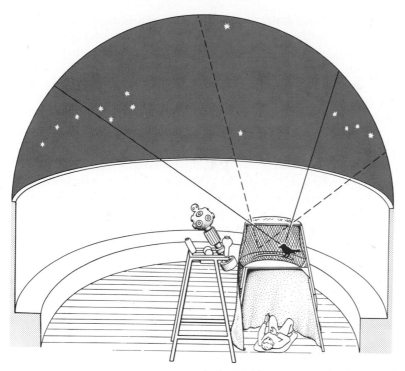

**8-56** Celestial orientation in birds. A planetarium is used to study the orientation of migratory birds to night sky constellations. As the projected sky is rotated, the warbler changes its perch to face a given direction with respect to certain constellations. The orientation is recorded by the observer below the bird. [From ''Celestial Navigation by Birds'' by E. G. F. Sauer. Copyright © 1958 by Scientific American, Inc. All rights reserved.]

respect to the projected sky, continually compensating for the time-dependent shift in position of the constellation about the Earth's axis. Arbitrary changes in the position of the projected sky produce corresponding changes in the orientation of the birds. It appears, then, that in order to compensate for the rotation of the Earth relative to celestial references, birds, bees, and other celestial navigators refer to *internal "clocks."* The poorly understood mechanism for time-compensated orientation to celestial markers is generally termed the *clock-compass.* If a bee or bird is put on a day–night schedule with "dawn" and "dusk" shifted by several hours, it will enter the incorrect time into

its internal clock-compass and will orient with a compass deviation equivalent to the artificial phase shift in the day–night cycle.

### GEOMAGNETIC CUES

It was suspected for a long time, but resisted unequivocal demonstration, that some animals utilize the Earth's magnetic field for orientation and navigation. Several lines of recent evidence have strengthened this view. Homing pigeons deprived of familiar landmarks can find their way home without sight of the sun on an overcast day. They normally head in the correct direction after a brief period of flight. However, they become disoriented if small magnets are attached to their heads, or if they are transported to their release points in containers that are impervious to the magnetic field. Local magnetic anomalies, due to iron deposits, also produce disorientation in newly released homing pigeons.

The cave salamander *Eurycea* finds its way "home" in complete darkness, and so its reliance on magnetic field was tested in the laboratory. After training in a given field, it was tested in cross-shaped passageways with different orientations of the field and was found to utilize the field for orientation. This finding is significant for two important reasons. First, since the salamanders move slowly (as compared to birds), their ability to detect the magnetic field appears to be a direct detection, rather than an indirect response to induced electric currents set up within the animal as a result of rapid movement through the field. Second, since the salamander moves through air, electric currents set up by water moving through a magnetic field cannot be the immediate cue, though it appears to be in certain fish, as we will see.

How can an animal detect the magnetic field directly? This question cannot be answered with certainty at this time. However, *magnetite,* a magnetic material of biological origin, has been found in a small black structure between the brain and skull of pigeons. Certain mollusks, bees, and mud bacteria have also been shown to contain magnetite. Bees give behavioral evidence of sensing the Earth's magnetic field, and so do the mud bacteria. Northern-latitude mud bacteria orient toward the North Pole, while those indigenous to the southern hemisphere orient toward the South Pole. These orientations enable them to swim downward at an angle (the angle with which the magnetic field enters the Earth at that location) into deeper mud. If placed in a drop of water in an artificial magnetic field, they collect at the corresponding side of the water drop and swim to the opposite side after the magnetic field is reversed.

The American eel (*Anguilla rostrata*) illustrates another method available to a marine organism for navigation based on geomagnetism. The larvae of this species migrate from spawning grounds in the Sargasso Sea to the Atlantic coast of North America, a distance of about 1000 km. Suggestions that they might employ the Earth's magnetic field were initially scoffed at because of the field's low density. However, eels have sensitive electroreceptors in their lateral-line system. The movement of the seawater in ocean currents acts as an enormous generator, for the water functions as a conductor moving through the Earth's magnetic field. The geoelectric fields set up in the sea by ocean currents, such as the Gulf Stream, reach intensities of about 0.5 $\mu$V/cm. This is equivalent to a 1.0 V drop over 20 km. Classic (Pavlovian) conditioning has been used to train eels to slow their heart rates in response to changes in the electric field. Once trained, the eels showed reduced heart rates in response to changes in dc fields as low as 0.002 $\mu$V/cm. Because the fields generated in the ocean are two or three orders of magnitude greater, it is entirely possible that the eel uses the orientation of a geoelectric field for navigational purposes.

## Genetics and Instinctive Behavior

The evidence for genetic influences on the physiology of animal behavior comes from several lines. To begin with, there are long-standing observations that animals at all phyletic levels are born or hatched with locomotory and behavioral patterns that are expressed *de novo* without the opportunity for practice and learning outside the womb or egg. These kinds of behavior, termed *instinctive,* suggest that complex neural functions are programmed in the genetic material in the form of information for the anatomical and physiological organization of the nervous system. Examples of some genetically fixed behaviors are shown in Figure 8-57. Such behaviors can often be modified through experience, but only to a limited extent.

A well-documented example of genetically programmed motor output is seen in the wing movements responsible for the chirps and trills in the courting song of the male cricket. The patterns of cricket songs are species-specific and are largely independent of environmental factors other than temperature. Moreover, the sound pattern is directly related to the pattern of motor impulses to the sound-producing muscles. $F_1$ hybrids of two species, one with a short trill (2 sound pulses) and the other with a long trill (about 10 sound

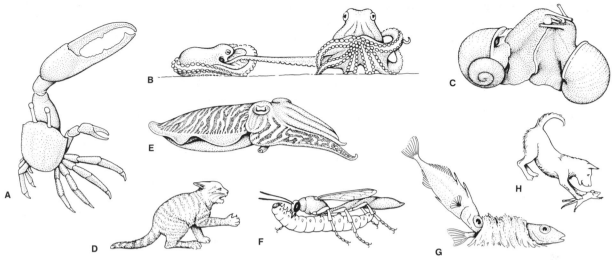

**8-57**  Stereotyped behaviors in several species. The patterns of action appear to be independent of learning. (A) Fiddler crab waving its claw. (B) Mating in octopus. (C) Mating in *Helix pomatia*. (D) European wildcat striking with its paw. (E) Male *Sepia officinalis* in sexual display. (F) The digger wasp with prey. (G) Male three-spined stickleback stimulating the female to spawn by quivering. (H) The mouse jump in the domestic dog. [A, C, E, F, and G after Tinbergen, 1951; B, Buddenbrock, 1956; D, Lindemann, 1955; H, Lorenz, 1954.]

pulses), produce trills with an intermediate number (about 4) of pulses. Backcrosses to produce various genetic combinations demonstrated that the neural network producing the song pattern is under rigid genetic control, sufficiently precise to specify differences as small as the exact number of impulses in the motor output to the chirp-producing muscles.

A similar finding, using hybridization, has helped confirm the genetic fixity of more complex vertebrate behavior. Closely related species of fowl with different behavior patterns were mated. The hybrid offspring exhibited hybrid behavior patterns with some components from one parent species and other components from the other parent species. Of course, care was taken to isolate the hybrids from either parent species to rule out any possibility of learning by exposure to the adults.

At birth, the calves of some ungulates walk about immediately and react to danger in much the same way as do adults of the species. Nevertheless, advocates of the idea that all, or at least most, behavior is learned (the "nurture" camp of the nature-versus-nurture controversy) have contended that the rudiments of feeding and locomotor behavior are learned gradually by the fetus through movements in the mammalian womb or the avian egg. This argument is difficult to disprove directly, because it is not practical to control behavior *in utero*. But observations on certain migratory birds give strong support for the idea that complex behavioral information can be genetically coded.

The ability to orient with respect to stellar patterns, complete with time compensation to account for the Earth's rotation, is fully present in birds that have been artificially hatched indoors and shielded from prior exposure to the sky. There is no way under those conditions for the birds to have learned from practice to orient properly in relation to the sky. It is difficult to avoid the conclusion that information necessary for celestial navigation resides in the genetic material, and that the nervous system of the bird is programmed to generate a certain behavior when it is presented with the appropriate visual input from the night sky.

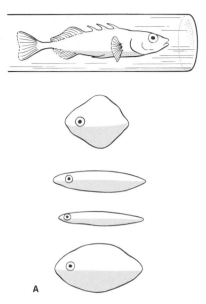

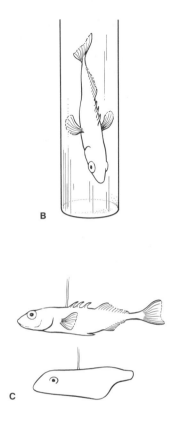

8-58   (A) Models used to study the aggressive behavior of male sticklebacks. Responses to these models indicate that the stimulus that releases the aggressive behavior is the horizontal red underside of the intruding male. The models in A were all effective releasers; those shown in B and C were not. (B) Vertically oriented males elicit no response. (C) The shape of the model fish is of little importance. [Tinbergen, 1951.]

A classic example of instinct is the release of aggressive behavior in the courting male of the three-spined stickleback by the appearance of another male. With models of various shapes and coloration, it was determined that the *sign stimulus* that releases this behavior is the red underside characteristic of the males (Figure 8-58A). The shape of the model male is relatively unimportant, but the red underside loses its effectiveness as a sign stimulus if the fish is in any orientation other than horizontal (Figure 8-58B). Thus, it is not simply the color red, but the red belly in the proper visual context, that acts to release the aggressive behavior.

The genetic origin of the innate releasing mechanism for the aggressive behavior of the male stickleback is evident in another stickleback species discovered in a mountain lake. The sign stimulus in this species is a *black* underbelly. Having evolved from a population of the red-bellied stock when the lake was

formed tens of thousands of years ago by glacial action, this isolated group, in which the males have black bellies, now is genetically programmed to react to underbellies that are black instead of red.

It is reasonable to say that behavior in higher animals has both genetic and learned components, the contributions of heredity and experience showing wide variations among the different behavioral acts and among the different species. Genetically programmed behavior is most dominant in animals with simple nervous systems. Examples of learned behavior are too familiar to enumerate here, and the problems of learning and memory too poorly understood to consider in depth in a textbook that stresses known mechanisms. It will suffice here to note that the larger (in numbers of neurons) and hence the more complex a nervous system, the greater its potential to learn from experience. This potential enables it to depart from a limited repertoire of inherited, fixed patterns of behavior.

## GENETIC DISSECTION OF NEURAL MECHANISMS

Neurobiologists have adopted the approach termed *genetic dissection* to examine the effects of single gene mutations on the functioning of a complex system. This approach was initially brought to neurobiology from molecular biology by S. Benzer at the California Institute of Technology. Hundreds of behavioral mutants of the fruitfly *Drosophila* were isolated that exhibit abnormal locomotor responses to visual stimuli. Wild-type (i.e., normal) fruitflies will walk toward a light, exhibiting positive phototaxis. If only one eye is blind, the fly walks with the defective eye toward the light, as predicted by Loeb's hypothesis (Figure 8-52). Mutants that are blind fail to show this behavior, of course, and are of interest primarily to students of the visual process. One line of mutants shows a normal electroretino-gram and a normal optokinetic response to moving patterns of alternating black and white stripes, which indicates that it can see, but it shows no phototaxis. The locomotor mutants include "Runner," which walks very quickly; "Hyperkinetic," which jumps when a shadow passes over; and "Easily Shocked," which goes into convulsions upon mechanical stimulation. There are also inbred strains of mice with genetically determined neurological deficiencies. In all cases, modification of the genetic material (i.e., mutation) has produced a structural or chemical modification in one or more loci of the nervous system. The major interest in such behavioral mutants arises from the possibilities they present to identify structural or chemical mechanisms involved in the altered behavior.

## Summary

During the course of evolution, primitive, distributed, structurally diffuse "nerve nets" characteristic of coelenterates became condensed into nerve cords and ganglia—first apparent in some jellyfish. In segmented animals, the anterior end, initially specialized as the sensory center, became differentiated as a superganglion, or brain.

Lying between the sensory inputs and motor outputs of the CNS are highly complex neural networks responsible for all reflexive and higher functions of the nervous system. The connections of these central networks appear to be preprogrammed genetically to a large extent, but during development and thereafter they are sustained and modified by use.

The integration of sensory input and subsequent activity depends primarily on two major sets of factors: (1) the organization of circuits and synapses formed by interacting neurons, and (2) the way in which the individual neurons process or integrate the incoming synaptic signals to produce their own action potentials. The integrative properties of neurons depend on the anatomical layout of the neuron and its connections and on the properties of the cell membrane. Simultaneous synaptic activity impinging on a neuron sums spatially because of the cable properties of the neuron. Synaptic activity also undergoes temporal summation as permitted by the time constant or signal decay time of the membrane. At the same time, there is interaction of excitatory and inhibitory input, the former tending to exceed the firing level and the latter counteracting depolarization to the firing level.

The simplest neural networks are monosynaptic reflex arcs, the most familiar of which is the stretch reflex of vertebrates. Elongation of a muscle stimulates its stretch receptors, which include the sensory endings of 1a afferent axons. These fibers enter the spinal cord and make direct synaptic connection with the $\alpha$-motoneurons that activate contraction of the muscle. This reflex contraction counteracts the force that produced the initial elongation of the muscle, causing it to shorten toward its original length. Simultaneously, collateral branches of the 1a afferent synaptically activate interneurons that make inhibitory synapses on the $\alpha$-motoneurons to the antagonist muscles. This reflex inhibition of contraction of the antagonist muscles eliminates their opposition to the reflex contraction of the protagonist muscle.

Locomotor movements of animals are based in part on central "motor scores," which program the sequence of muscle contractions to produce the coordinated locomotor movements. Feedback from proprioceptors can exert an influence on the strength and frequency of the motor output and, in some motor activity, plays an important role in its coordination.

The CNS "filters" sensory input—namely, it enhances certain features of the incoming stimuli and suppresses others. A classic example of neural filtering is lateral inhibition, which produces enhanced contrast in apparent luminosity at visual edges, in intensity of sound between adjacent frequencies in the audio spectrum, and in apparent vibration perceived between adjacent somatosensory loci. Electrical recording from cells in the visual cortex suggests that there is a hierarchical arrangement of neurons, and that the specificity of sensory features evoking activity in the neurons increases with each level, until only very specific features of the visual stimulus evoke responses at the higher levels.

Sensorimotor behavior, as well as a rudimentary form of learning termed habituation, occurs in single-celled organisms. These phenomena have been traced to electrophysiological properties of the cell membrane, which resemble closely the properties of nerve cell membranes.

Certain animals use highly developed sensory abilities to orient and navigate. Birds use star patterns, the sun, landmarks, and perhaps even the Earth's magnetic field. Bats become familiar with the sonar "image" of the landscape and can find their way back to their roosts as readily as if they were using visual images. Bees use the sun's position and the pattern of polarized light in the sky to keep track of the direction of the hive or a source of food. Birds and bees both possess internal clock-compasses that they use to correct for time-dependent changes in the position of celestial indicators (sun, stars).

## Exercises

*1.* Since all action potentials are fundamentally alike, how is the modality of input from the various sense organs recognized by the CNS?

*2.* Outline the general organization of the vertebrate brain and the spinal cord.

*3.* Contrast the sympathetic and the parasympathetic division of the autonomic nervous system.

*4.* What are muscarinic, nicotinic, alpha, and beta receptors, and how are they distinguished from one another?

*5.* How can increased rate of firing of an inhibitory interneuron result in an increased rate of firing of other neurons?

*6.* What is responsible for the continuous low-level synaptic input and slow tonic firing in motoneurons?

*7.* Explain the two means with which the nervous system controls the intensity of contraction of a muscle.

*8.* Each eye of a primate sees about the same field, but the right hemisphere of the brain "sees" the left half of the visual field, while the left hemisphere "sees" the right half of the field. How does this occur?

*9.* Why does the evening sky appear to have a lighter band outlining the silhouette of a mountain range?

*10.* What is meant by the "receptive field" of a neuron?

*11.* What is the neural organization proposed as the explanation for the sensitivity to a bar of light of specific orientation in a "simple cell" of the visual cortex?

*12.* What would happen to a subject if all of her muscle spindles ceased to function?

*13.* How is the set-point length of a muscle regulated by the nervous system?

*14.* What basic difference in arrangement is there between the muscle spindle and the Golgi tendon organ? What effect does this have on the respective functions of these two classes of muscle receptor?

*15.* Relate the answer for Exercise 14 to the differences in reflex function between the 1a and the 1b afferent fibers.

*16.* How does the mammalian nervous system sense the relative position of the limbs?

*17.* The ciliate *Paramecium* has no nervous system, yet it responds to collision with an object by reversing beat, backing off, and resuming forward locomotion in a new direction. How is this accomplished?

*18.* The nervous system is sometimes compared to a telephone system. Give some reasons why this analogy is a poor one.

*19.* Discuss some of the evidence that some complex behavior patterns are inherited and cannot be ascribed entirely to learning.

*20.* Cite observations that suggest that the major factor responsible for differences in the functioning of different nervous systems is neural circuitry and not the properties of single nerve cells.

*21.* What evidence is there that some nervous systems can detect the Earth's magnetic field?

*22.* Give several examples of how studies of invertebrates and lower vertebrates have provided fundamental insights into mechanisms of neural function.

## Suggested Reading

Bennett, M. V. L., and A. B. Steinbach. 1969. *Neurobiology of Cerebellar Evolution and Development.* Chicago: American Medical Association.

*The Brain.* 1979. September issue of *Scientific American.* San Francisco: W. H. Freeman and Company.

Bullock, T. H., R. Orkand, and A. D. Grinnell. 1977. *Introduction to Nervous Systems.* San Francisco: W. H. Freeman and Company.

Chalmers, N., R. Crawley, and S. P. R. Rose, eds. 1971. *The Biological Bases of Behavior.* New York: Harper and Row.

Dethier, V. G., and E. Stellar. 1970. *Animal Behavior.* Englewood Cliffs, N. J.: Prentice-Hall.

Eccles, J. C. 1973. *The Understanding of the Brain.* New York: McGraw-Hill.

Eckert, R., and P. Brehm. 1979. Ionic mechanisms of excitation in *Paramecium. Ann. Rev. Biophys. Bioeng.* 8:353–383.

Ewert, J.-P. 1980. *Neuroethology.* Berlin: Springer-Verlag.

Fain, G. L. 1980. Integration of spikeless neurones in the retina. In A. Roberts and B. M. Bush, eds., *Neurones Without Impulses.* Cambridge: Cambridge University Press.

Fentress, J. C. 1976. *Simpler Networks and Behavior.* Sunderland, Mass.: Sinauer.

Grillner, S. 1975. Locomotion in vertebrates: Control mechanisms and reflex interaction. *Physiol. Rev.* 55:247–304.

Hoyle, G., ed. 1977. *Identified Neurons and Behavior in Arthropods.* New York: Plenum.

Kandel, E. R. 1976. *Cellular Basis of Behavior.* San Francisco: W. H. Freeman and Company.

Kuffler, S. W., and J. Nicholls. 1976. *From Neuron to Brain.* Sunderland, Mass.: Sinauer.

Manning, A. 1972. *An Introduction to Animal Behavior.* Menlo Park, Calif.: Addison-Wesley.

Schmidt-Koenig, K., and W. T. Keeton, eds. 1978. *Animal Migration, Navigation, and Homing.* New York: Springer-Verlag.

Sherrington, C. S. 1906. *The Integrative Activity of the Nervous System.* 2nd ed., 1947. New Haven: Yale University Press.

Wechert, C. K. 1970. *Anatomy of the Chordates.* New York: McGraw-Hill.

## References Cited

Applebaum, M. L., G. L. Clifton, R. E. Coggeshall, J. D. Coulter, W. H. Vance, and W. D. Willis. 1976. Unmyelinated fibers in the sacral 3 and caudal 1 ventral roots of the cat. *J. Physiol.* 256:557–572.

Brownell, P., and R. D. Farley. 1979a. Detection of vibrations in sand by tarsal sense organs of the nocturnal scorpion *Paruroctonus mesaensis. J. Comp. Physiol.* 131:23–30.

Brownell, P., and R. D. Farley. 1979b. Orientation to vibrations in sand by the nocturnal scorpion *Paruroctonus mesaensis:* Mechanism to target location. *J. Comp. Physiol.* 131:31–38.

Buddenbrock, W. von. 1956. *The Love of Animals.* London: Muller.

Bullock, T. H., and G. A. Horridge. 1965. *Structure and Function in the Nervous Systems of Invertebrates.* San Francisco: W. H. Freeman and Company.

Camhi, J. M. 1971. Flight orientation in locusts. *Scientific American* 225(2):74–81. Also available as Offprint 1231.

Eccles, J. 1960. The nature of central inhibition. *Proc. Roy. Soc.* (London) Ser. B. 153:445–476.

Eccles, J. C. 1969. Historical introduction to central cholinergic transmission and its behavioral aspects. *Fed. Proc.* 28:90–94.

Eckert, R. 1972. Bioelectric control of ciliary activity. *Science* 176:473–481.

Granit, R. 1970. *The Basis of Motor Control.* New York: Academic.

Grell, K. G. 1973. *Protozoology.* New York: Springer-Verlag.

Griffin, D. R. 1958. More about bat "radar." *Scientific American* 199(1):40–44.

Hartline, H. K., and F. Ratliff. 1957. Inhibitory interaction of receptor units in the eye of *Limulus. J. Gen. Physiol.* 40:357–376.

Hartline, H. K., H. G. Wanger, and F. Ratliff. 1956. Inhibition in the eye of *Limulus. J. Gen. Physiol.* 39:651–673.

Horridge, G. A. 1968. *Interneurons.* San Francisco: W. H. Freeman and Company.

Hubel, D. H. 1963. The visual cortex of the brain. *Scientific American* 209(5):54–62. Also available as Offprint 168.

Jennings, H. S. 1906. *Behavior of the Lower Animals.* New York: Columbia University Press.

Lindemann, W. 1955. Über die Jugendentwicklung beim Luchs (*Lyns* 1. *lynx* Kerr.) und bei der Wildkatze (*Feliss. sylvestris* Schreb.). *Behavior* 8:1–45.

Loeb, J. 1918. *Force Movements, Tropisms, and Animal Conduct.* Philadelphia: Lippincott.

Lorenz, K. Z. 1954. *Man Meets Dog.* Trans. by M. K. Wilson. London: Methuen.

Merton, P. A. 1972. How we control the contractions of our muscles. *Scientific American* 226(5):30–37. Also available as Offprint 1249.

Michael, C. R. 1969. Retinal processing of visual images. *Scientific American* 205(5)104–114. Also available as Offprint 1143.

Miller, W. H., F. Ratliff, and H. K. Hartline. 1961. How cells receive stimuli. *Scientific American* 205(3):222–238. Also available as Offprint 99.

Montagna, W. 1959. *Comparative Anatomy.* New York: Wiley.

Naitoh, Y., and R. Eckert. 1969. Ionic mechanisms controlling behavioral responses of *Paramecium* to mechanical stimulation. *Science* 164:963–965.

Neal, H. V., and H. W. Rand. 1936. *Comparative Anatomy.* Philadelphia: Blakiston.

Noback, C. R., and R. J. Demarest. 1972. *The Nervous System: Introduction and Reviews.* New York: McGraw-Hill.

Penfield, W., and T. Rasmussen. 1950. *The Cerebral Cortex of Man.* New York: Macmillan.

Romer, A. S. 1955. *The Vertebrate Body.* Philadelphia: Saunders.

Sauer, E. G. F. 1958. Celestial navigation by birds. *Scientific American* 199(2):42–47. Also available as Offprint 133.

Sperry, R. W. 1959. The growth of nerve circuits. *Scientific American* 201(5):68–75.

Stent, G. S. 1972. Cellular communication. *Scientific American* 227(3):42–51. Also available as Offprint 1257.

Tinbergen, N. 1951. *The Study of Instinct.* Oxford: Clarendon.

Van Essen, D. 1973. The contribution of membrane hyperpolarization to adaptation and conduction block in sensory neurones of the leech. *J. Physiol.* 230:509–534.

Wells, M. 1968. *Lower Animals.* New York: McGraw-Hill.

Werblin, F. S., and J. E. Dowling. 1969. Organization of the retina of the mudpuppy, *Necturus maculosus:* II, Intracellular recording. *J. Neurophys.* 32:339–355.

Wiedersheim, R. E. 1907. *Comparative Anatomy of Vertebrates.* London: Macmillan.

Willows, A. O. D. 1971. Giant brain cells in mollusks. *Scientific American* 224(2):68–75.

Wilson, D. M. 1964. The origin of the flight-motor command in grasshoppers. In R. F. Reiss, ed., *Neural Theory and Modeling: Proceedings of the 1962 Ojai Symposium.* Stanford, Calif.: Stanford University Press.

Wilson, D. M. 1968. Inherent asymmetry and reflex modulation of the locust flight motor pattern. *J. Exp. Biol.* 48:631–641.

Wilson, D. M. 1971. Neural operations in arthropod ganglia. In F. O. Schmitt, ed., *The Neurosciences: Second Study Program.* New York: Rockefeller University Press.

Wood, D. C. 1970–1971. Electrophysiological correlates of the response decrement produced by mechanical stimuli in the protozoan *Stentor coeruleus. J. Neurobiol.* 2:1–11.

Young, R. W. 1970. Visual cells. *Scientific American* 223(4): 80–91. Also available as Offprint 1201.

CHAPTER *9*

# *Muscle and Movement*

*A*nimal movements—such as locomotion, eating, copulation, and nearly all forms of nonchemical communication—are expressions of neural commands translated into coordinated mechanical activity. Three fundamental mechanisms generate movement in animals: amoeboid movement, ciliary and flagellar bending, and muscle contraction. The contractions of muscles are the most apparent and dramatic macroscopic signs of animal life and have therefore excited the imagination since the times of the ancients. In the second century, Galen postulated that "animal spirits" flow from nerves into muscle, inflating the muscle so as to increase its diameter at the expense of its length. As recently as the 1950s, it was postulated that muscles shorten as the result of a shortening of linear molecules of "contractile proteins." It was suggested that such molecules possess helical configurations and that changes in the pitch of the helix produce changes in length. This hypothesis, however, was short-lived, and the development of new techniques subsequently led to a dramatic improvement in our knowledge of muscle function. As a result, our understanding of how muscles function is more complete and intellectually satisfying than are most other areas of physiological knowledge. Through evidence from electron microscopy, biochemistry, and biophysics, we have learned how the contractile mechanism of muscle is organized and how it produces shortening. It is also becoming clear how the process of contraction is initiated and controlled by the electrical activity of the muscle-cell membrane.

Muscles are classified on both morphological and functional grounds into two major types, *smooth* and *striated*. Because the most is known about the structure and function of vertebrate striated muscle (primarily frog and rabbit skeletal muscle), we use it as the model system in most of this chapter. Striated muscle itself can be subdivided into skeletal and cardiac muscle. This division, however, is not a basic one, since in both these muscle types the organization and function of the contractile mechanism are nearly identical, although the cellular organization shows some major differences.

## Structural Basis of Contraction

The hierarchy of organization of skeletal muscle tissue is shown in Figure 9-1. The cells (or *fibers*) of striated muscle typically extend from the tendon or other connective tissue attached to one bone to a tendon attached to another bone, and they thus function in parallel. Striated muscle fibers range from 5 to 100 $\mu$m in diameter and may be as much as several centimeters or more in length. Their extraordinary size is possible because of their syncytial origin; that is, they arise from single cells, the *myoblasts,* which fuse into *myotubes.* These, in turn, differentiate into the multinucleate membrane-limited *muscle fiber.* Each fiber is made up of numerous parallel subunits termed *myofibrils,* which consist of longitudinally repeated units called *sarcomeres,* which are bounded by *Z lines.* The sarcomere of a myofibril is the functional unit of striated muscle.

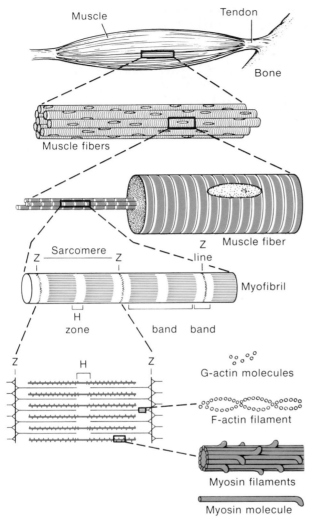

Muscle

Tendon

Bone

Muscle fibers

Muscle fiber

Z line

Sarcomere

Z                    Z

Myofibril

H
zone

band  band

Z          H          Z

G-actin molecules

F-actin filament

Myosin filaments

Myosin molecule

**9-1**  Hierarchy of skeletal muscle organization.
[Bloom and Fawcett, 1968.]

the protein *actin*. These interdigitate with *thick filaments* made up of the protein *myosin*. The myosin filaments make up the densest portion of the sarcomere, the *A band* (the "A" stands for "anisotropic," as this band strongly polarizes visible light). The light region in the center of the A band is called the *H zone*. In the middle of the H zone is the *M line*, which has been shown to be the locus of enzymes important in energy metabolism (e.g., creatine kinase). The portion of the sarcomere between two A bands is called the *I band* (the "I" stands for "isotropic," as these bands are relatively nonpolarizing).

A cross section through the I band reveals profiles of only the actin filaments (Figure 9-3), and a cross section through the H zone reveals profiles of only the myosin filaments. In the region of overlap, each myosin filament is surrounded by six actin (thin) filaments; it shares these with surrounding myosin (thick) filaments. Each actin filament is surrounded by three myosin filaments.

Close inspection with the electron microscope shows small projections, termed *cross bridges*, that extend outward from the myosin filaments (Figure 9-4A). These projections make contact with the actin filaments during contraction. The cross bridges are arranged in a two-stranded helical pattern along the myosin filament (Figure 9-4B). The cross bridges are spaced about 143 Å apart along the axis of one helix. The angular displacement around the filament between successive cross bridges is $120°$.

## SUBSTRUCTURE OF THE MYOFILAMENTS

It has been known since the work of Wilhelm Kühne in the mid-nineteenth century that different protein fractions can be extracted from the muscle by soaking minced muscle in water containing various concentrations of salts. Nonstructural soluble proteins, such as myoglobin, are extracted by distilled water. The actin and myosin filaments are solubilized by highly concentrated salt solutions, which break the bonds that hold together their respective monomers. Along with actin and myosin, certain other proteins are extracted in this way. Our present knowledge of muscle contraction rests in part on the isolation of actin and myosin filaments and on subsequent analyses of their structure and composition.

The myofibrils of one muscle fiber are lined up with sarcomeres in register, giving the fiber a banded or *striated* appearance in the light microscope.

The fine structure of striated muscle is an elegant example of structure as the basis of function. The electron micrograph in Figure 9-2 shows a longitudinal section of several myofibrils. The Z line contains α-actinin, one of the proteins found in all motile cells. Extending in both directions from the Z line of a myofibril are numerous *thin filaments* consisting largely of

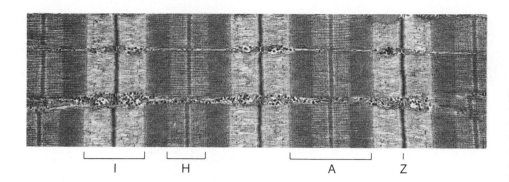

9-2 Electron micrograph of frog muscle in a longitudinal thin section that includes two whole and two half sarcomeres of three myofibrils. I, H, and A bands and Z line are labeled. Dark granules between fibrils are glycogen. Magnification 15,000×. [Courtesy of L. D. Peachey.]

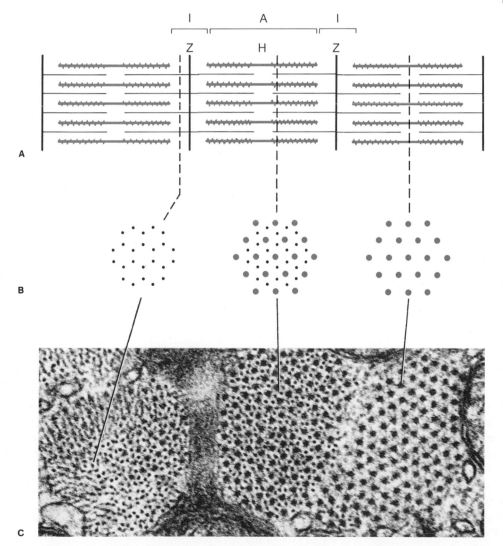

9-3 Organization of the myofibril. (A) Diagram of three sarcomeres, showing thick and thin myofilaments forming I, A, and H bands and Z lines. (B) Imaginary sections through the sarcomere at different levels show profiles of thin (left) and thick (right) filaments, and both types (center). (C) Electron micrograph of a cross section in which the sarcomeres of adjacent myofibrils are out of register and can thus be matched with the corresponding profiles shown above. Spider monkey extraocular muscle. Magnification 100,000×. [Courtesy of L. D. Peachey.]

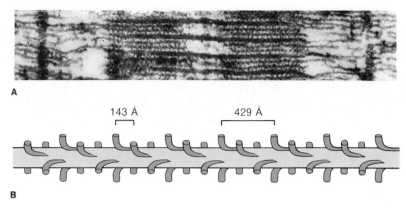

**A**

143 Å          429 Å

**B**

9-4    The myosin cross bridges. (**A**) Electron micrograph in which they can be seen as faint projections extending from the myosin toward the actin filaments. [Huxley, 1963.] (**B**) The two-stranded helical arrangement of cross bridges on the thick filaments, shown enlarged relative to A. [Huxley, 1969.]

Isolated fragments of myofibrils several sarcomeres in length can be prepared by homogenizing fresh muscle in a laboratory blender. Gentle homogenization in a *relaxing solution* of $Mg^{2+}$, ATP, and a calcium *chelating agent* such as EGTA (p. 354) prevents the formation of bonds between the myosin cross bridges and the actin filaments. As a result, the myofibril falls apart into its constituent actin and myosin filaments.

The *actin filament* resembles two strings of beads twisted into a two-stranded helix (Figure 9-5). Each "bead" is a monomeric molecule of G-actin, so called because of its globular shape. The molecules of G-actin (diameter 55 Å) are polymerized to form the long two-stranded helix of F-actin, so called because of its filamentous appearance. This polymerization can be produced in vitro with purified preparations of G-actin. The F-actin helix has a pitch of about 730 Å, so that its two strands cross over each other once every 365 Å. It should not be confused with the far smaller $\alpha$-helix of peptide chains. Actin filaments are about 1 $\mu$m long and 80 Å thick, and they are joined at one end to the material that constitutes the Z line. Positioned in the grooves of the actin helix are the filamentous molecules of the protein *tropomyosin*. Attached to each tropomyosin is a complex of globular protein molecules collectively called *troponin*. The troponin complexes are spaced out along the actin filament at intervals of

about 400 Å. As described further on, troponin and tropomyosin together play an important role in the control of muscle contraction.

The individual monomers that polymerize to form the *myosin filament* are long and thin, with an average length of 1500 Å and a width of about 20 Å (Figure 9-6). One end of the myosin molecule forms the globular double "head region," which is about 40 Å thick and 200 Å long. The long slender portion of the molecule constitutes its "neck" and "tail."

When the myosin molecule is treated with the proteolytic enzyme trypsin, it separates into two parts—the so-called *light meromyosin* (*LMM*) and *heavy meromyosin* (*HMM*). The LMM constitutes the major part of the tail region, and the HMM makes up the globular head and a thin segment of "neck." The head region is of special interest, for it contains all the enzymatic and actin-binding activity of the parent molecule. The myosin molecule consists of two peptide chains twisted about each other the full length of the straight portion. The head is actually a composite of the globular ends of the two peptide chains plus several (three or four depending on the species) so-called *myosin light chains*. These are short (thus "light") polypeptides of two kinds. They differ in different muscle types and influence the ATPase activity of the myosin.

Myosin molecules, like the molecules of G-actin,

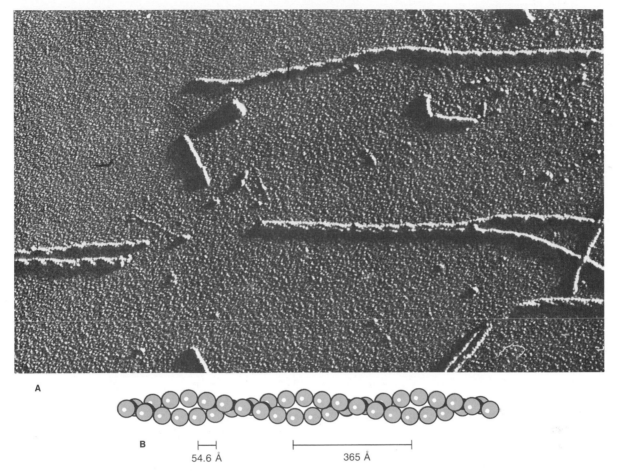

**9-5**    Actin filaments. (**A**) Electron micrograph of F-actin filaments. Note the two-stranded helical arrangement of globular monomers. The specimen was prepared for microscopy by shadowing with a thin film of metal. Magnification 150,000×. [Courtesy of R. V. Rice.] (**B**) Diagram showing the two-stranded helix of F-actin, made up of G-actin monomers. This structure has been deduced from electron micrographs (as in A) and from X-ray diffraction studies. [Huxley, 1969.]

will aggregate and polymerize in vitro so as to reconstitute myosin (A-band) filaments. This behavior occurs spontaneously in vitro when the ionic strength of a solution of myosin molecules is lowered. The first step in the formation of myosin filaments occurs when several myosin molecules aggregate with tails overlapping and heads pointing outward from the region of overlap in opposite directions (Figure 9-7). The result is a short filament with a central region bare of heads. The filament grows as molecules of myosin add onto each end,

with their tails pointing toward the center of the filament and overlapping with the tails of previously added molecules. With the addition of each myosin molecule, a new head projects laterally from the filament. Because the myosin molecules add on symmetrically to the growing ends, the heads on each half of the filament are oriented opposite those of the other half. Aggregation continues until the myosin filament is about 1.5 $\mu$m long and 120 Å thick. Why filaments stop growing at that length is not clear.

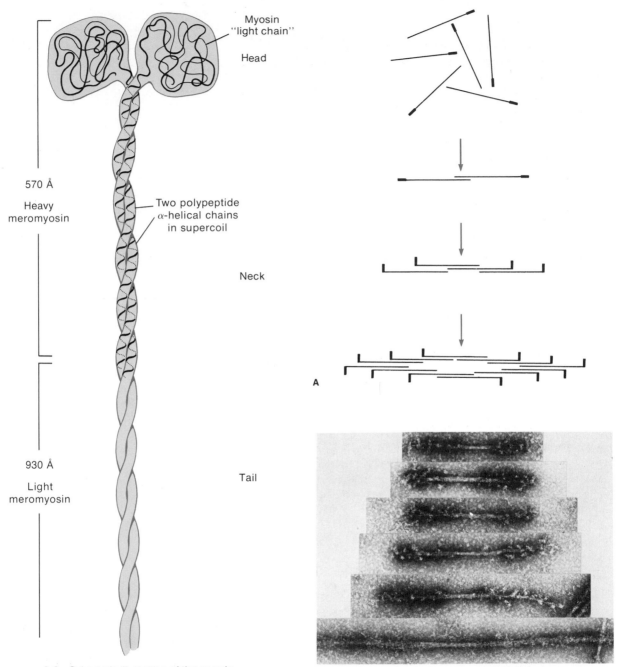

Myosin "light chain"

Head

570 Å

Heavy meromyosin

Two polypeptide α-helical chains in supercoil

Neck

930 Å

Light meromyosin

Tail

A

B

9-6   Schematic illustration of the myosin molecule, with globular head and long thin tail. Light and heavy meromyosin are differentiated on the basis of trypsin digestion, which breaks the myosin molecule into those two parts. [Lehninger, 1975.]

9-7   (A) The spontaneous in vitro formation of thick filaments from myosin molecules. (B) Electron micrographs of myosin molecules assembled into thick filaments of various lengths. Note the double-ended organization. [Huxley, 1969.]

# Sliding-Filament Theory

The bands that define the sarcomeres were first observed in the light microscope well over a century ago. It was also observed that the sarcomeres undergo changes in length during stretch or contraction of the muscle, and that these changes correspond to the change in muscle length. Using a specially built interference light microscope, which permitted more accurate measurement of the sarcomeres, A. F. Huxley and R. Niedergerke in 1954 confirmed earlier reports that the A bands maintain a constant width during shortening of the muscle, whereas the I bands and the H zone become narrower. When the muscle is stretched, the A band again maintains constant width, but the I bands and H zone broaden. That same year, H. E. Huxley and J. Hanson (1954) reported that the myosin (A-band) and actin (I-band) filaments seen in electron micrographs do not change their lengths when the sarcomere shortens or is stretched (Figure 9-8). Instead, the *extent of overlap* between actin and myosin filaments changes as the sarcomere changes length. Largely on the basis of these two points of evidence, H. E. Huxley and A. F. Huxley independently proposed in 1954 the *sliding-filament theory of muscle contraction.* This theory states that shortening of the sarcomere (and hence the muscle fiber) during contraction occurs as a consequence of active sliding of the thin (actin) filaments between the thick (myosin) filaments. Shortening results when the actin filaments are drawn farther in toward the center of the A band. When the muscle relaxes or is stretched, the overlap between thin and thick filaments is reduced.

The sliding-filament theory differed radically from earlier hypotheses of muscle contraction. Some workers had suggested that contraction might be due to a shortening of the protein molecules themselves, either as a result of increased folding of plaited molecules or as a result of changes in the helical pitch or diameter of helical molecules. In contrast, the sliding-filament theory holds that filaments of constant length slide past each other as a result of forces developed between the actin and myosin filaments. Evidence that cross bridges produce the sliding force is presented in the next subsection.

It was noted that the myosin monomers that make up one half of the thick filament are assembled with

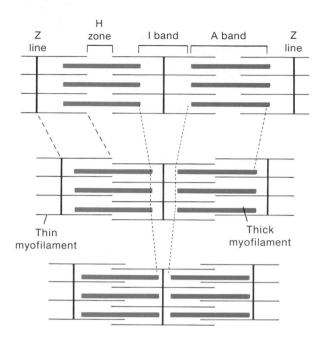

**9-8** Relations of the myofilaments during shortening of two sarcomeres. Note that the thick filaments and thin filaments retain a constant length during sliding. The I band narrows as the thin filaments slide toward the center of each A band.

their heads all pointing toward one Z line, whereas those that make up the other half are oriented with their heads toward the other Z line. This arrangement is crucial to the sliding-filament mechanism, as described below.

### LENGTH–TENSION CURVE

One of the strongest evidences for the sliding-filament theory is the relation between amount of overlap between actin and myosin filaments and the tension produced by the active sarcomere at various amounts of filament overlap. It was reasoned by Huxley and Niedergerke that if each myosin cross bridge that interacts with the actin filament provides an increment of tension, the total tension produced by the sarcomere should be proportional to the number of cross bridges that can interact with actin filaments. Because the number of myosin cross bridges that can interact with the actin filaments increases linearly with the distance

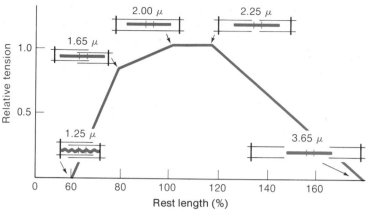

**9-9**   Length–tension relations for striated muscle sarcomere. The state of the sarcomere is shown at critical points on the curve. The theory is described in the text. [Gordon et al., 1966.]

of filament overlap, tension should be proportional to the amount of overlap. The sliding-filament theory also predicts that no *active* tension (i.e., beyond that due to the elasticity of the muscle fiber) will develop if the sarcomere is stretched so far that there is no longer any overlap between actin and myosin filaments.

To test these predicted relations between filament overlap and tension, single amphibian muscle fibers were stimulated to contract at different fixed sarcomere lengths, which, of course, are linearly related to the amount of overlap between the actin and myosin filaments (Figure 9-9). Sarcomere length was adjusted with the aid of an electromechanical system that controlled muscle-fiber tension so as to hold the sarcomeres at any desired constant length. The tension required to do so was measured and plotted as a function of sarcomere length. Tension was found to increase linearly with decreasing sarcomere length between 3.65 $\mu$m and 2.25 $\mu$m. When the sarcomere was allowed to shorten until the actin filaments overlapped completely with the segment of the myosin filament that bears the cross bridges, the tension was maximal. When the fiber was stretched until there was no overlap at all, stimulation produced no tension beyond the passive elastic tension of the resting state. At the other extreme, when the fiber was allowed to shorten until the actin filaments from the two halves of the sarcomere collided, tension decreased with further shortening. Tension decreased still more steeply when shortening was al-

lowed to proceed beyond the point at which the myosin filaments crumpled up against the Z line.

It is crucial in such experiments that the length measurements be made on small groups of sarcomeres of uniform behavior near the center of the muscle fiber. Measurements made earlier with less precise techniques yielded rounded curves, because the various sarcomeres of a whole muscle—and, indeed, of a single fiber—were in different stages of overlap at any instant. From the length–tension relation, it was predicted that muscles with longer actin and myosin filaments (long sarcomeres) should develop greater tensions per unit cross-sectional area. This prediction has since been confirmed.

## Cross-Bridge Function and the Production of Force

The most fundamental questions in muscle research are concerned with the precise functioning of the cross bridges. According to recent versions of the sliding-filament theory, the force for muscle contraction arises from the sequential binding of several sites on the myosin head to sites on the actin filament. The head of the cross bridge then separates from the actin filament, freeing it for another cycle of sequential binding farther along the actin filament. The following subsections explain this model in detail.

## CHEMISTRY OF CROSS-BRIDGE ACTIVITY

Clues to the chemistry of cross-bridge interaction with the actin filaments have their origin in studies begun several decades ago on crude and purified extracts of muscle. Semipurified solutions of actin and myosin—extracted from freshly minced rabbit muscle with concentrated salt solutions and subsequently precipitated by ammonium chloride—exhibited several interesting physical properties. Among them are:

*1.* The formation of an actin–myosin, or *actomyosin,* complex that has a distinctly higher viscosity than a dissociated mixture of actin and myosin.

*2.* The formation of a precipitate of the actomyosin gel, which contracts on addition of $Mg^{2+}$–ATP in the presence of small amounts of calcium (initially the calcium was present as a contaminant and was therefore overlooked).

The contraction of the gel squeezes water out from between its molecular lattice (Figure 9-10). This process, termed *syneresis* by polymer chemists, proved useful in muscle chemistry as a test-tube analog of muscle contraction.

When mixed together, the actin and myosin solutions form the actomyosin complex spontaneously without ATP. This explains the high viscosity that results when an actin solution is added to a myosin solution. But the actomyosin complex requires $Mg^{2+}$–ATP to undergo

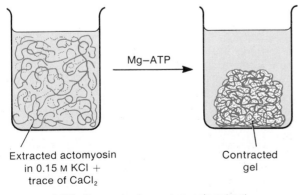

Extracted actomyosin in 0.15 M KCl + trace of $CaCl_2$

Mg–ATP

Contracted gel

**9-10** Syneresis of an actomyosin extract after addition of Mg–ATP.

syneresis. The ATPase activity (i.e., enzymatic hydrolysis of the terminal phosphate bond of ATP) of the myosin catalyzes the following reaction:

$$Mg–ATP + actomyosin \rightarrow actomyosin–Mg^{2+}–ATP$$
$$\rightarrow actomyosin–Mg^{2+}–ADP + P_i$$

Binding of myosin to actin can take place in the absence of ATP or when the ATPase site is poisoned. This suggests that binding of actin to myosin occurs at a site different from the ATPase site of the myosin. Even though different parts of the myosin appear to be involved, there is an interaction between the actin-binding site of the myosin and the enzymatic site for splitting of ATP, for the ATPase activity of myosin is greatly enhanced by the formation of the actomyosin complex. This is evident from the observation that the ATPase activity of myosin by itself is much lower than that of actomyosin. The formation of the actomyosin complex may activate the ATPase site by an allosteric mechanism. The head portion alone shows all the ATPase and actin-binding activity of the myosin. It is also the only part of the myosin molecule that comes in contact with the actin filament in the living muscle.

## CROSS-BRIDGE ACTIVITY AND MUSCLE CONTRACTION

It has proven possible to correlate tension with cross-bridge position in living contracting muscle by using X-ray diffraction technology. As early as the 1950s, H. E. Huxley studied the molecular organization of skeletal muscle with X-ray diffraction. It is possible to use this rather esoteric method on muscle because of the highly ordered, almost crystalline organization of the muscle filaments. To examine changes in the organization of the filaments during contraction, A. Miller and R. T. Tregear (1971) utilized the X-ray diffraction method on insect fibrillar flight muscle (to be described later). This muscle was used because it can be made to oscillate spontaneously between relaxation and contraction without nervous control after the ions and soluble molecules are extracted by low-temperature storage in a water–glycerol solution. The diffraction pattern was projected onto a lead screen (Figure 9-11A). The screen had a hole in the position corresponding to the spot produced by the inherent

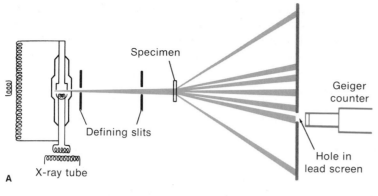

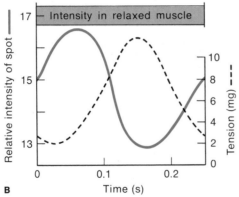

9-11    Use of an X-ray diffraction technique to correlate muscle tension with cross-bridge movement. (A) Experimental arrangement for X-ray diffraction studies. A Geiger counter behind a hole in the lead screen monitors the fluctuations in intensity of the X-ray spot produced by reflections from the actin. The intensity of the spot declines as cross bridges attach to the actin and obscure its intrinsic periodicity. (B) Tension in water beetle flight muscle undergoing cyclic contraction is plotted in broken curve. The intensity of the X-ray spot declines (i.e., the number of cross-bridge attachments increases; solid line) as the tension in the muscle rises. Note the high X-ray intensity for relaxed muscle. [Miller and Tregear, 1971.]

periodicity of the actin (Figure 9-5). Without going into the theory of X-ray diffraction, we can simply state that the intensity of the X rays of such a spot on the diffraction pattern is a function of the number of units in the specimen that produce the X-ray spot. Thus, as the cross bridges swing out from the myosin filament and attach to the actin, they scatter X rays away from the spot corresponding to the actin periodicity, and that spot then receives less intense bombardment of X rays. The intensity of X radiation producing that spot was monitored by means of a Geiger counter placed behind the opening in the lead screen. When the muscle was made to undergo oscillatory contractions and relaxations, the fluctuating intensity of the actin spot was correlated with fluctuating muscle tension; the relationship is plotted in Figure 9-11B. Increased muscle tension was correlated with decreased signal due to cross-bridge attachment to the actin filaments (Figure 9-12A and B), thus supporting the idea

that the sliding movement between the myosin and actin filaments results from forces produced by the cross bridges acting on the actin filaments.

The cross bridges must alternately attach to the actin filament, exert force, detach, and reattach at another locus. For maintained active contraction, the activity of the cross bridges must be asynchronous, so that at any instant some of them are attached to the actin while others are detached from it. Presumably, after detaching, the cross bridge reattaches to the actin filament farther toward the Z line, because there is continued sliding in that direction during shortening.

One of the major problems in cross-bridge function is to understand how chemical energy is transduced into mechanical energy. How do the cross bridges produce a sliding force between thick and thin filaments? There have been a number of suggestions. The most widely accepted view is that a *rotation* of the myosin head (Figure 9-12B and C) produces force, and that

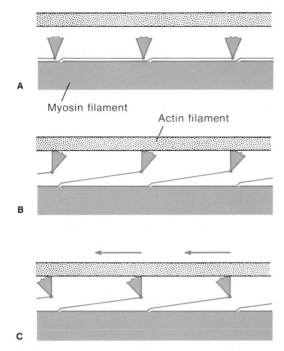

**A**

Myosin filament

Actin filament

**B**

**C**

9-12 Sequence of events in the attachment of cross bridges to actin filaments. (**A**) Relaxed state. (**B**) Myosin heads attach to actin. (**C**) Rotation of heads pulls on the actin filament, causing it to slide past the myosin filament. Although the cross bridges are shown acting in unison, they actually act out of synchrony. [Huxley, 1969; Huxley and Simmons, 1971.]

as the head makes a rotational movement against the actin filament (Figure 9-13A). According to their hypothesis, several sites on the head interact with sites on the actin filament. These sites are ordered in a sequence such that the actin–myosin affinity increases from $M_1$ to $M_2$, from $M_2$ to $M_3$ (left to right in Figure 9-13A), and so on. Thus, after site $M_1$ attaches, the tendency is for the head to rotate so that site $M_2$ attaches, then $M_3$, and so forth. This sequence causes point $b$ to move toward the right in Figure 9-13A.

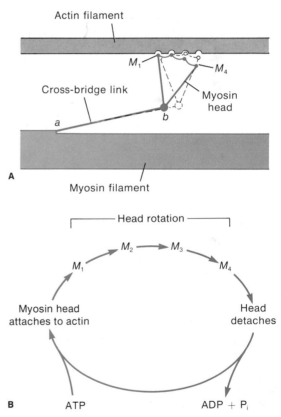

9-13 Mechanism proposed for cross-bridge function. (**A**) This schematized model shows four sites on the myosin head. These sites, $M_1$ to $M_4$, interact in sequence (left to right) with sites on the actin. The rocking motion produced in this manner causes the myosin head to pull on the elastic cross-bridge link, $a$ to $b$, stretching it. This tension pulls the actin filament toward the left, causing it to slide past the myosin filament. [Huxley and Simmons, 1971.] (**B**) Cross-bridge cycle summarized. [Keynes and Aidley, 1981.]

this force is transmitted to the thick filament through the neck of the myosin molecule, the neck forming the *cross-bridge link* between the head of the myosin molecule and the thick filament. In this hypothesis, the *link* acts as a connection between the myosin head and the thick filament, transmitting the force produced by rotation of the head on the actin filament.

Studies of the mechanical properties of contracting muscle by A. F. Huxley and R. M. Simmons have provided support for this view of cross-bridge function. They obtained evidence that much of the elasticity that exists *in series* with the contractile component (p. 367) of the muscle resides in the cross bridges themselves—presumably in the cross-bridge link. They postulated that the elastic stretching of the link serves the important function of storing mechanical energy

The elasticity of the link (segment *a* to *b* in Figure 9-13A) would allow the rotational movement to occur without abrupt changes in tension. Once stretched, the link would transmit its tension smoothly to the thick filament, contributing force to promote filament sliding. One of the major evidences in favor of this scheme is the finding by Huxley and Simmons that the series elasticity of a muscle fiber is proportional to the amount of thick- and thin-filament overlap and hence is proportional to the number of attached cross bridges. They also found that sudden small decreases in length are accompanied by very rapid recovery of tension, which they explain as being due merely to the rotation of the cross-bridge heads into their more stable positions of interaction with actin sites (i.e., from the $M_1$ to the $M_4$ position of interaction).

The details of cross-bridge function are not all rigorously established at this time. Our present understanding of the sequence of events in cross-bridge function (Figure 9-13B) is summarized as follows:

*1.* The head of the cross bridge attaches to the actin filament at the first of a sequence of stable sites. It then goes to the second, third, and successive sites, each of which exhibits a greater myosin–actin affinity than the previous one (i.e., a lower energy state).

*2.* This interaction produces a rocking, or rotation, of the myosin head (Figure 9-12), causing it to pull on the cross-bridge link connecting the myosin head to the thick filament. The elasticity of the link allows the steplike rocking of the head to progress without sudden large transients in tension.

*3.* Tension in the link is transmitted to the myosin filament, and sliding occurs, allowing the tension built up in the stretching of the link to subside.

*4.* When the rotation of the head is complete, the myosin head dissociates from the actin filament. $Mg^{2+}$–ATP attached to the ATPase site of the head region is necessary for this dissociation. The ATP is hydrolyzed at the ATPase site of the free myosin head. This hydrolysis is accompanied by a conformational change in the myosin head, leaving the head in an energized state. When the myosin head reattaches to the actin filament, the stored energy is used to rotate the head against the actin to produce active sliding. On subsequent detachment of the myosin from the actin filament, the cycle repeats itself a little farther along the actin filament. In this way, sliding is produced in small incremental steps of cross-bridge attachment, rotation, and detachment of the myriad of cross bridges in each sarcomere.

Two things are especially noteworthy in this sequence. First, ATP is not utilized directly to produce cross-bridge force, but is required to "prime" the myosin head so that it can attach and repeat the cycle. Second, cross-bridge attachment requires the presence of free intracellular $Ca^{2+}$ concentrations above $10^{-7}$ M. The role of calcium in controlling cross-bridge function is discussed next.

## Role of Calcium in Contraction

### CROSS-BRIDGE ACTIVATION

Only gradually did evidence accumulate that calcium ions play an important role in contractile activity of muscle. Calcium is active in such low ($10^{-7}$ M or above) sarcoplasmic concentrations that, before the discovery of calcium-chelating agents such as EDTA and EGTA,† it was not feasible to keep the $Ca^{2+}$ concentration of experimental solutions to such low levels, since even twice-distilled water contains more than $10^{-6}$ M $Ca^{2+}$. The earliest evidence for a physiological role of $Ca^{2+}$ came from the work of S. Ringer and D. W. Buxton (1887) in the late nineteenth century. They found that the isolated frog heart stops contracting if calcium is omitted from the bathing saline. (This was the origin of *Ringer's solution* and other physiological salines.) Later, when electrophysiological methods were available, it was found that contraction ceases in low $Ca^{2+}$, even though cardiac action potentials continue.

The possibility that $Ca^{2+}$ participates in the regulation of muscle contraction was tested by the introduction of various cations to the interior of muscle fibers by T. Kamada and H. Kinosita (1943), and later by L. V. Heilbrunn and F. J. Wierczinski. Of all the ions tested, only calcium was found to produce contraction at concentrations commensurate with those normally found in living tissue. It was subsequently found that skeletal muscle fails to contract in response

---

†EDTA = ethylenediamine tetraacetic acid; EGTA = ethylene-glycol-*bis* (β-aminoethylether)-N,N'-tetraacetic acid.

## Box 9-1  *Extracted Cell Models*

One of the key advances in cell physiology was made in the late 1940s when Albert Szent-Györgyi (1947) developed the glycerine-extracted muscle "model," which he and his followers used to elucidate certain facets of muscle activity. After a muscle fiber has been soaked in a solution made up of equal parts of glycerine and water at subzero temperatures for several days or weeks, the cell membrane is disrupted and all soluble substances are leached out, leaving the contractile mechanism intact. The glycerine acts as an antifreeze and also solubilizes membranes. The subfreezing storage temperature serves to keep the enzymes intact while slowing catabolic processes that would cause the cells to digest themselves. The *glycerinated fibers* can be made to contract and relax under the appropriate conditions. The significance of using *extracted, reactivated* cells is that they allow the investigator to control the composition of the

intracellular milieu without interference from the regulatory mechanisms of the cell membrane and other metabolic processes.

More recently, this method of producing extracted models of cells has been improved on by the use of nonionic detergents, such as the Triton X series. These agents, used at about 0°C, rapidly solubilize the lipid components of the cell membrane, allowing dissolved metabolites to diffuse out of the cell and substances in the extracellular medium to diffuse rapidly into the cell. Because this all takes place within minutes instead of days, much time, is saved and enzymatic activity remains higher. Many of the structurally fixed enzymes remain active, permitting the reactivation of motile cells such as amoebae, fibroblasts, and ciliates by external application of ATP and essential ions.

---

to depolarization after its internal calcium stores are depleted, and that extracted models of skeletal muscle fibers (Box 9-1) fail to contract in response to added ATP if $Ca^{2+}$ is absent.

The quantitative relation between sarcoplasmic free $Ca^{2+}$ concentration and muscle contraction has been determined more recently by peeling off the surface membrane and exposing the naked myofibrils to solutions of different calcium concentrations. As seen in Figure 9-14A, the tension rises sigmoidally from zero at a calcium concentration of about $10^{-8}$ M to a maximum at about $5 \times 10^{-6}$ M. This relation between tension and $Ca^{2+}$ concentration is closely paralleled by the relation seen between the rate of ATP hydrolysis (ATPase activity) of homogenized myofibrils and $Ca^{2+}$ concentration (Figure 9-14B). These lines of evidence suggested that $Ca^{2+}$ might act as cofactor for the ATPase activity of myosin, but this turned out *not* to be so.

In *semi*purified extracts of actomyosin, the hydrolysis of ATP and syneresis of the actomyosin require the presence of minute ($10^{-7}$ M or higher) concentrations of free $Ca^{2+}$. The low-level requirement for calcium was initially overlooked because calcium is normally

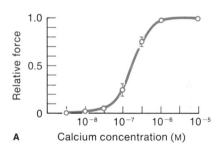

**A**

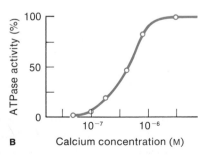

**B**

**9-14** Relationship between the calcium concentration and the force produced in a glycerinated muscle. [Hellam and Podolsky, 1967.] **(B)** ATPase activity in a suspension of isolated myofibrils as a function of calcium concentration. [Bendall, 1969.]

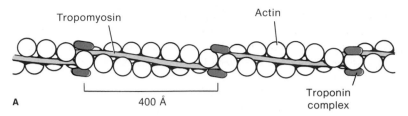

A        400 Å

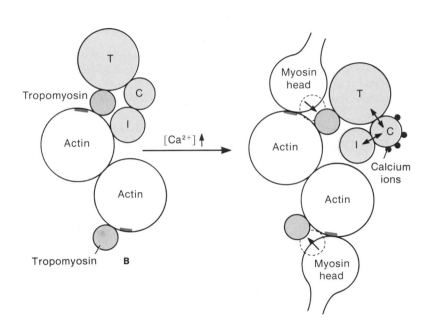

B

$[Ca^{2+}]\uparrow$

**9-15** Muscle regulatory proteins. (A) Arrangement of troponin and tropomyosin on the actin filament, as first postulated by Ebashi et al. (1969). (B) A more recent model, shown in cross section, of their manner of action. At left, under resting conditions of low $[Ca^{2+}]_i$, the troponin (Tn) complex, consisting of units C, T (1, 2), and I, bonds in such a way with the actin and tropomyosin as to cause the latter to interfere sterically with the attachment of the cross-bridge heads to the myosin-binding sites of the actin. At right, an increase in sarcoplasmic $Ca^{2+}$ allows TnC to bind $Ca^{2+}$, causing a change in subunit affinities and a consequent shift of the tropomyosin molecule away from the myosin-binding site. The cross-bridge activity can then proceed until the $Ca^{2+}$ is removed from the TnC. (C) Three-dimensional concept of the troponin complex in relation to tropomyosin (TM). The two portions of the T unit of troponin are labeled $T_1$ and $T_2$. [Ebashi et al., 1980.]

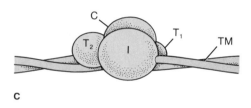

C

present as a contaminant of even the most pure off-the-shelf salts. As noted earlier, it was only after the introduction of calcium-chelating agents that this problem was recognized and overcome.

The calcium dependence of ATPase activity and syneresis disappeared when the actomyosin extracts were purified sufficiently to remove from the actin filaments the closely associated proteins troponin and tropomyosin. It then became evident that $Ca^{2+}$ is not directly required for the contractile activity of actin and myosin. Magnesium is, in fact, the only ion necessary for the ATPase activity of myosin. It became clear, then, that the calcium requirement for activation of the ATPase, and hence for contraction, is closely tied to the presence of troponin and tropomyosin.

How, then, does $Ca^{2+}$ induce contraction? As noted earlier, the ATPase activity of a pure myosin solution

is low, but shows a great increase when purified actin is added. This indicates that *the ATPase site of the myosin is activated when myosin binds to actin.* In the intact muscle, the ATPase site of the cross bridge is activated when the cross bridge attaches to the actin filament. Experiments done in S. Ebashi's laboratory indicate that troponin and tropomyosin, which lie along the actin helix (Figure 9-15), inhibit the attachment of cross bridges to the actin. Troponin is the only protein in the actin and myosin filaments of vertebrate striated muscle with a high binding affinity for $Ca^{2+}$, each troponin complex binding four calcium ions. The troponin complex occurs every 400 Å along the actin filament, attached to both the actin filament and the tropomyosin molecule. In the resting state, the tropomyosin lies in a position that interferes sterically with the binding of the myosin heads to the actin filament. On binding

$Ca^{2+}$, the troponin undergoes a change in conformation that moves the tropomyosin out of the way, permitting the myosin cross bridges to attach to actin sites (Figure 9-15B). Thus, binding of $Ca^{2+}$ to troponin removes a constant inhibition of the cross-bridge attachment. It is inferred from experiments like those shown in Figures 9-14 and 9-16 that release of cross-bridge inhibition occurs at free $Ca^{2+}$ concentrations above $10^{-7}$ M.

The aforegoing is an account of the role of $Ca^{2+}$ in regulating actin–myosin interaction in vertebrate skeletal and cardiac muscle. The role of $Ca^{2+}$ differs in most other muscles. At least two other calcium-dependent regulatory mechanisms controlling actin–myosin interactions exist. In most invertebrate striated muscles, calcium initiates contraction by binding to the

myosin light chains of the cross-bridge heads. In vertebrate smooth muscle and in nonmuscle actomyosin (p. 394), contraction depends on a calcium-dependent phosphorylation of the myosin head.

### CROSS-BRIDGE INACTIVATION AND MUSCLE RELAXATION

In the resting state, an internal system of membrane-limited sacs, the *sarcoplasmic reticulum* (described later), actively takes up $Ca^{2+}$. This process keeps the level of free $Ca^{2+}$ below $10^{-7}$ M. At those concentrations, very little calcium binds to the troponin, and the cross bridges remain inactive. Thus, removal of $Ca^{2+}$ from the sarcoplasm by the reticulum causes the muscle to relax after a contraction.

Because ATP supplies the energy for contraction, one might conclude that relaxation can also be produced by removal of ATP. This, it turns out, does not happen.

A muscle becomes rigid and inextensible when all of its ATP and phosphagen stores are depleted. This state, known as *rigor mortis,* is due to the failure of the cross bridges to detach from the actin filaments. That $Mg^{2+}$–ATP is required for relaxation has been known since the early experiments with glycerine-extracted models of muscle (Box 9-1). In the presence of $Ca^{2+}$, the extracted muscle contracts when $Mg^{2+}$–ATP is added (Figure 9-16A) and subsequently relaxes if $Ca^{2+}$ is removed. Relaxation, like contraction, occurs only if $Mg^{2+}$–ATP is present (Figure 9-16B). Normally, there is an adequate supply of ATP, and the cross bridges readily detach. The muscle then relaxes if the sarcoplasmic free $Ca^{2+}$ concentration is below the level required to allow cross-bridge attachment to the actin filaments.

Thus, muscle relaxation depends on the presence of $Mg^{2+}$–ATP to break the myosin–actin complex and on a calcium concentration sufficiently low to prevent reattachment of the cross bridges to the actin filaments.

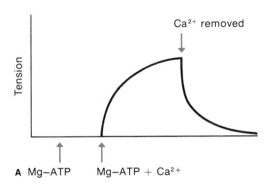

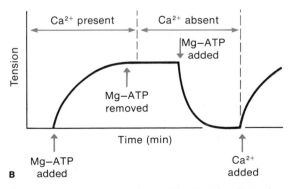

**9-16** Dependence of tension in glycerine-extracted muscle on Mg–ATP and $Ca^{2+}$. (A) Contraction requires $Ca^{2+}$. (B) Mg–ATP is required for both contraction and relaxation.

## Electromechanical Coupling

We will now consider how the contractile machinery in the muscle cell becomes activated. In vertebrate skeletal muscle, the arrival of an action potential at

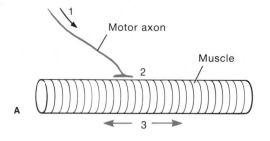

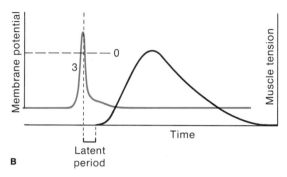

9-17    Excitation of muscle fiber. (**A**) Action potential in motor nerve (1) leads to postsynaptic potential (2), and this gives rise to a propagated muscle action potential (3). (**B**) The action potential in the muscle fiber is followed after a latent period by an all-or-none contraction, the muscle twitch.

## MEMBRANE POTENTIAL AND CONTRACTION

As noted in Chapter 5, replacing some of the $Na^+$ with $K^+$ in the extracellular fluid will depolarize the cell membrane. It is therefore a convenient way of depolarizing the muscle membrane to different levels, for the degree of depolarization depends on the $K^+$ concentration. Muscle fibers show a transient contraction (termed *contracture,* to differentiate from a normal contraction) in response to sudden, maintained depolarizations. In the setup shown in Figure 9-18A, single frog-muscle fibers were exposed to various concentrations of $K^+$ while membrane potential and muscle

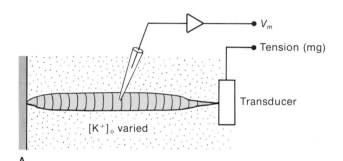

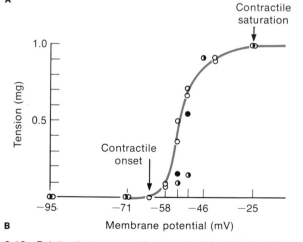

9-18    Relation between membrane potential and contractile tension. (**A**) Experimental setup for measuring tension and potential changes in an isolated muscle fiber in response to various external concentrations of KCl. (**B**) Membrane depolarized to varying degrees by adjusting KCl concentration. The tension was found to rise sigmoidally with depolarization. [Hodgkin and Horowicz, 1960.]

the terminals of the motor axon causes the release of the neurotransmitter acetylcholine. This leads to a postsynaptic potential at the muscle endplate, which in turn triggers an all-or-none action potential in the surface membrane of the muscle fiber. This action potential propagates away from the endplate in both directions, so as to excite the entire muscle-fiber membrane. The contractile mechanism responds with a latency of several milliseconds from the peak of the action potential. Whenever the action potential occurs, the muscle fiber responds with an all-or-none twitch (Figure 9-17). Somehow the action potential sets in motion the sequence of events leading to contraction. We will now consider the *coupling* of contraction to the electrical events that accompany the action potential.

tension were monitored. With depolarization, tension began to develop at about −60 mV (the mechanical threshold) (Figure 9-18B); with a further depolarization, tension increased sigmoidally, reaching a maximum at about −25 mV. This experiment demonstrates that the contractile system is capable of *graded contraction* in response to varying degrees of reduced membrane potential.

During the action potential, the membrane swings from a resting value of about −90 mV to an overshoot of about +50 mV, an excursion of 140 mV. The peak is as much as 75 mV more positive than the potential required to give a maximum contracture (Figure 9-19). Thus, the muscle twitch is all-or-none because the action potential is all-or-none, and the potential greatly exceeds the plateau (Figure 9-18B) at which mechanical activation during steady depolarization is saturated.

The direct physical influence of a potential difference across the surface membrane can extend only several micrometers at most from both surfaces of the membrane. Thus, a potential change across the surface membrane cannot exert a direct effect on the great bulk of myofibrils in a muscle fiber 50–100 $\mu$m in diameter. It therefore became necessary to look for a substance or a process that couples changes in surface membrane potentials to the myofibrils deep within the muscle fiber. The local circuit currents produced by a propagated action potential were ruled out because currents of physiological magnitude passed between two microelectrodes inserted in a muscle fiber produced no contraction.

During the 1930s and 1940s, L. V. Heilbrunn emphasized the importance of calcium in cellular processes, including muscle contraction. He proposed that the contraction of muscle is controlled by intracellular changes in calcium concentration (1940). We now know that this hypothesis is essentially correct. It was widely rejected at first, however, because (1) before the discovery of the sarcoplasmic reticulum it was assumed that, to initiate contraction, calcium would have to enter the myoplasm through the surface membrane, and (2) A. V. Hill (1948) pointed out that the rate of diffusion of an ion or a molecule from the surface membrane to the center of a muscle fiber 25–50 $\mu$m in radius is several times too slow to account for the short latent period (about 2 ms) between the action potential at the surface membrane and activation of the entire cross section of the muscle fiber. Hill correctly concluded that a *process* rather than a substance must communicate the surface signal to myofibrils deep within the muscle fiber in order to initiate contraction. As we shall see, the surface signal is conducted into the cell interior, where it causes the release of intracellular $Ca^{2+}$ from internal storage depots. The $Ca^{2+}$ released in this way permits the cross bridges to enter into action and produce active sliding.

### THE SARCOTUBULAR SYSTEM

Anatomical and physiological evidence for a process of intracellular communication linking the surface membrane to the internal myofibrils came to light about 10 years after Hill's calculation. In 1958, A. F. Huxley

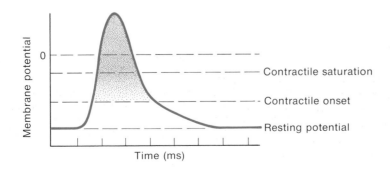

9-19  The muscle action potential greatly exceeds the potential that produces contractile saturation—that is, the potential that will produce maximum contraction as determined from Figure 9-18.

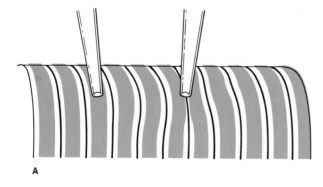

**A**

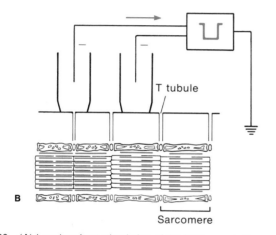

**B**

Sarcomere

9-20  (A) Local surface stimulation of the frog muscle fiber.
The opening of the pipette was placed over different loci on
the surface of the muscle fiber. [Peachey, 1965.] (B) Local
contraction occurred in response to negative pulses applied
with the pipette when its opening was placed over the minute
entrances to the T tubules, located in the plane of the
Z line. [Ashley, 1971.]

*1.* Local contractions occur only if the tip of the
electrode is positioned on the fiber surface with its
opening over a Z line.

*2.* These contractions occur only at certain loci
spaced around the perimeter of the fiber along each
Z line.

*3.* Contractions spread further inward as the intensity of stimulating current is raised.

*4.* Contractions are limited to both half-sarcomeres
immediately on either side of the Z line over which the
electrode is positioned; thus, there is inward spread,
but no longitudinal spread, of the graded contraction.

Electron micrographs of amphibian skeletal muscle
have provided the anatomical correlate of these findings. Running around the perimeter of each myofibril
at the level of the Z line is a membrane-limited *transverse tubule (T tubule)* less than 0.1 μm in diameter,
which branches so that it is continuous with similar
tubules surrounding neighboring myofibrils at the
same level (Figure 9-21). The anastomosing system
of tubules eventually reaches and connects with the
surface membrane. For some time it was uncertain
whether continuity exists between the lumen of the
T-tubule system and the cell exterior. Continuity was
finally confirmed by the demonstration that ferritin
and horseradish peroxidase (large electron-opaque
protein molecules) placed in the bath find their way
into the T tubules before the tissue is fixed for electron
microscopy. Because these molecules do not cross cell
membranes, the T tubules must be open to the extracellular space, arising as invaginations of the surface
membrane.

The T-tubule system provides the anatomical link
between the surface membrane and the myofibrils
deep inside the muscle fiber. With Huxley's stimulating pipette placed over the entrance to a T tubule at
the surface membrane (Figure 9-20), stimulating current spreads down the tubule to initiate contraction
deep within the muscle fiber. This conclusion is supported by comparative studies that correlate the location of the T-tubule system with surface sensitivity to
current. In muscles where the T tubules are located at
the ends of the A band instead of at the Z line—for example, in the crab (Figure 9-22) and the lizard—the

and R. E. Taylor stimulated the surface of single muscle fibers of the frog with tubular glass microelectrodes
(Figure 9-20). Externally applied pulses of current too
small to initiate a propagated action potential in the
surface membrane, but sufficient to produce depolarization of the membrane under the opening of the
microelectrode, were observed to produce local contractions near the surface of the fiber. Listed here are
Huxley and Taylor's most significant findings:

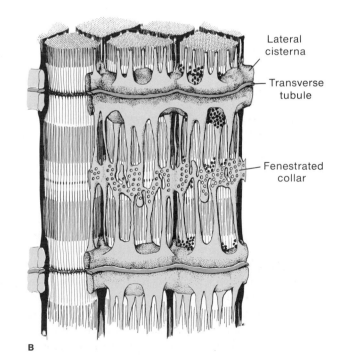

current-sensitive spots on the surface membrane are found at the edges of the A band.

The role of the T tubules in the process that couples activation of the sarcomeres to depolarization of the surface membrane was demonstrated by breaking the connections of the T tubules to the surface membrane through osmotic shock with a 50% glycerol solution. When the tubules are disconnected from the surface membrane in this way, membrane depolarization no longer evokes a contraction. Thus, physical uncoupling of the T-tubule system results in functional uncoupling of the contractile system from the surface membrane.

The inward spread of electrical signals down the T tubules was at first thought to be simply electrotonic, but it then was discovered that the spread of contraction to the center of a twitch fiber in response to membrane depolarization is reduced by adding TTX or by lowering the $Na^+$ concentration. Either treatment will reduce or eliminate sodium action potentials. It appears, then, that the action potential characteristic

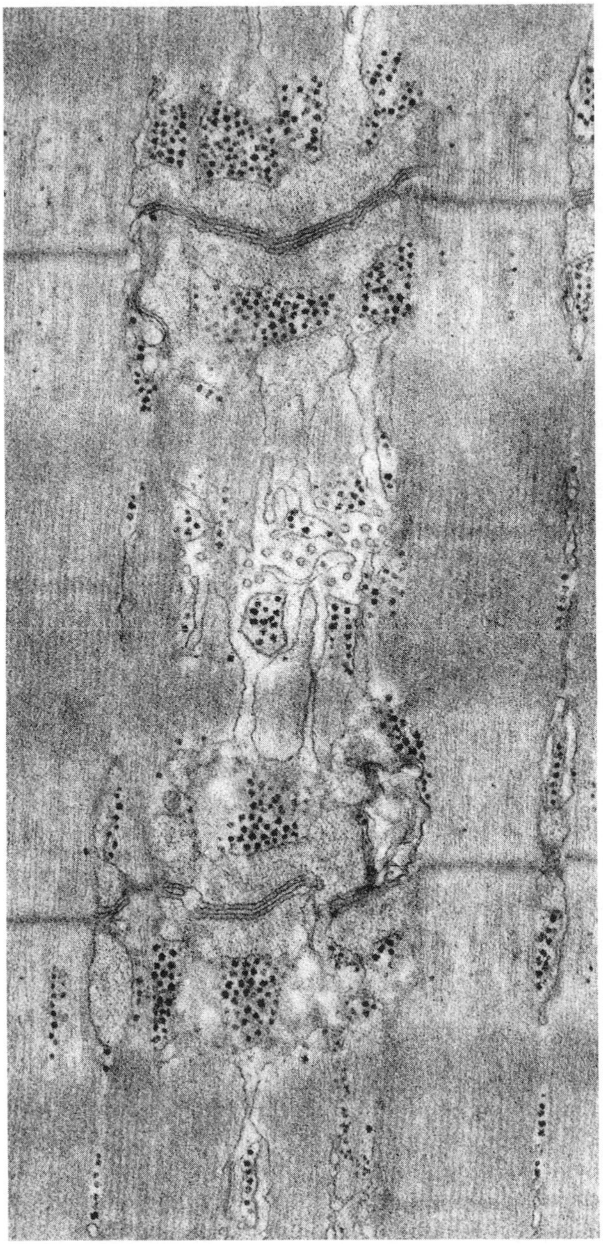

Lateral cisterna

Transverse tubule

Fenestrated collar

**9-21** Electron micrograph (A) and diagram (B) showing the sarcoplasmic reticulum (light color) and T tubules (dark color) of striated skeletal muscle of the frog. Dark spots are glycogen granules. [Peachey, 1965.]

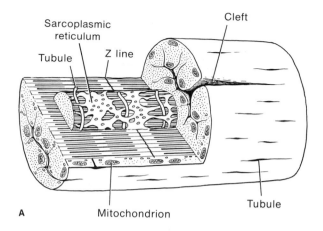

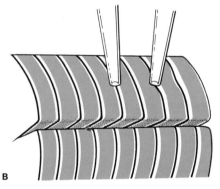

9-22   Crab muscle. (A) The structure of crab muscle fibers resembles that of vertebrate fibers, except for larger diameter, deep clefts, and location of T tubules. [Ashley, 1971.] (B) The current-sensitive spots are located at the edges of the A bands, which suggested that the T tubules open to the surface near the edges of the A bands, rather than at the Z lines as in vertebrate muscle. This hypothesis was subsequently confirmed with the electron microscope. [Peachey, 1965.]

of the surface membrane of a vertebrate twitch fiber is carried deep into the muscle fiber along the membranes of the T tubules. In muscle fibers that do not produce action potentials (e.g., many arthropod muscles), the T tubules carry passive electrotonic signals in keeping with the graded depolarizations of the surface membrane.

## THE SARCOPLASMIC RETICULUM

In addition to the system of T tubules, there is the *sarcoplasmic reticulum (SR),* which is wrapped separately as a hollow collar around each myofibril from one Z line to another (Figure 9-21). The SR surrounding each sarcomere consists of a membrane-enclosed compartment separate from the sarcoplasm. The *terminal cisternae* of the SRs of two adjacent sarcomeres make intimate contact with the T tubule, which is sandwiched between them. The manner in which the depolarization of the T tubules conveys a signal to the terminal cisternae of the SR remains unknown. There is speculation, but no direct evidence, that the region of contact between the T tubule and terminal cisternae of the SR may be specialized as a low-resistance pathway for the spread of electric current between the interior of the T tubule and the lumen of the SR.

When isolated by fractionation techniques, membranes of the SR form microscopic vesicles about 1 $\mu$m in diameter. These can take up calcium ions from the surrounding medium. If oxalate is present, a calcium oxalate precipitate is seen to form within the vesicles as the calcium concentration within the vesicles increases because of active calcium transport by the membrane of the reticulum. In unfractionated tissue, the calcium oxalate can be visualized in the terminal cisternae with the electron microscope. The calcium-sequestering activity of the SR is sufficiently powerful to keep the concentration of free $Ca^{2+}$ in the sarcoplasm of the resting muscle below $10^{-7}$ M, which is sufficient to remove the calcium bound to the troponin and prevent contraction.

## CALCIUM RELEASE
## BY THE SARCOPLASMIC RETICULUM

Once it became known that calcium ions are accumulated by the SR, it seemed likely that muscle contraction is initiated by release of $Ca^{2+}$ from the interior of the SR cisternae into the sarcoplasm. Direct evidence that free sarcoplasmic $Ca^{2+}$ increases in response to stimulation has come from a photometric method that utilizes the calcium-sensitive protein aequorin. When a molecule of aequorin combines with three calcium ions, it emits a photon of visible light. Aequorin was injected into a large, nonspiking muscle fiber of the

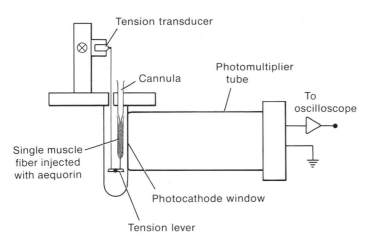

9-23   Method for monitoring calcium transients in muscle fibers with the photoprotein aequorin. A photomultiplier tube monitors the light emission brought about by $Ca^{2+}$ reacting with aequorin. The tension produced by muscle contraction is measured with a force transducer. [Ashley, 1971.]

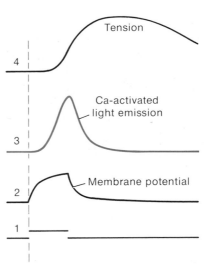

9-24   Relation of calcium transient to excitation and contraction in an aequorin-injected crustacean muscle. A pulse of outward-going current (1) caused a graded depolarization of the surface membrane (2). This elicited a signal from the calcium-sensitive aequorin (3) and development of tension (4). The aequorin signal (colored trace) is related to the square of the sarcoplasmic free $Ca^{2+}$ concentration. [Ashley and Ridgway, 1970.]

giant barnacle, and a photomultiplier was used to monitor changes in light emission (Figure 9-23) as calcium entered the muscle fiber through the membrane and was released from the SR during excitation. Depolarization of the muscle membrane was graded because of the absence of an all-or-none action potential. This grading permitted the investigators to produce larger or smaller depolarizations, depending on the amount of current injected. With sufficient depolarization, the onset of the depolarization was followed by light emission, which began before an increase in tension was recorded from the muscle fiber (Figure 9-24). Increased intensity or duration of depolarization produced greater light emission (indicating a greater calcium concentration) along with greater tension (Figure 9-25). In this way, tension developed in response to membrane depolarization was directly correlated with transients in the concentration of free $Ca^{2+}$ in the sarcoplasm.

Contraction is normally controlled electrically by membrane depolarization. A muscle fiber will nevertheless contract without any change in membrane potential if placed in a solution containing a small

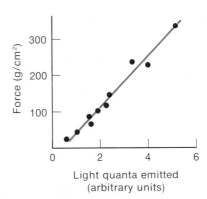

9-25   The tension developed in response to a depolarizing current in a crustacean muscle fiber is closely related to the total amount of light from the injected aequorin in the kind of experiment shown in Figures 9-23 and 9-24. [Ashley and Ridgway, 1970.]

amount of caffeine or theophylline,† methyl xanthines that produce contracture by causing the release of calcium sequestered in the SR. The increase in sarcoplasmic free $Ca^{2+}$ in response to caffeine has been demonstrated with a colorimetric method similar in principle to the photometric method depicted in Figure 9-23. After removal of caffeine, the calcium is taken up again by the SR, and the muscle relaxes.

There is an interesting correlation between structure of the sarcotubular system and muscle function. Muscles that contract and relax very rapidly have a highly developed SR and extensive systems of T tubules. Those that contract and relax slowly have either a rudimentary or a less well developed SR. These different rates of contraction and relaxation, of course, correlate with the efficiency of the SR in regulating the changes in calcium concentrations that turn the contractile system on and off.

### CONTRACTION AND RELAXATION SUMMARIZED

The events controlling contraction of skeletal muscle are summarized in Figure 9-26 as follows:

*1.* The surface membrane of the muscle fiber is depolarized by an action potential or, in some muscles, by synaptic potentials.

*2.* The signal is conducted deep into the muscle fiber along the T tubules.

*3.* The electrical signal spreads from the T tubules to the SR.

*4.* This signal brings about the release of calcium ions sequestered in the reticulum.

*5.* The free $Ca^{2+}$ concentration of the sarcoplasm increases from a resting value below about $10^{-7}$ M to an active level of about $10^{-6}$ M or higher. Calcium binds to the troponin, and a conformational change takes place.

*6.* This conformational change causes a change in the position of the tropomyosin molecule, eliminating steric inhibition of cross-bridge binding to the actin filaments.

*7.* Cross bridges attach to the actin filaments and undergo a sequential interaction of sites that causes

---

†Theophylline is the stimulant found in tea.

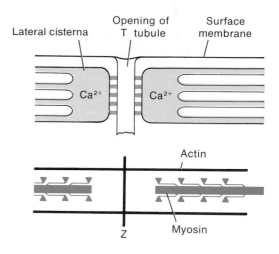

**A**    Relaxed

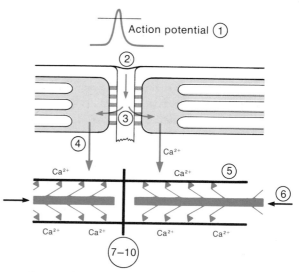

**B**    Contracting

**9-26**  Steps in the sequence of excitation–contraction coupling. (1) The action potential propagates along the surface membrane of the muscle fiber and conducts down the T tubule (2). The signal is communicated to the lateral cisternae of the SR (3), which thereupon release sequestered $Ca^{2+}$ (4). The calcium then relieves the inhibition of actin–myosin interaction by binding to troponin (5). The myosin cross bridges attach to the actin filaments and produce sliding of the filaments (6).

the myosin head to rock against the actin filament, pulling on the cross-bridge link.

*8.* This pulling produces active sliding of the actin filaments into the A band. The sarcomere shortens a small distance.

*9.* Prior to the next cycle of cross-bridge movement, ATP (bound to the ATPase site on the myosin head) is hydrolyzed, and the energy of hydrolysis is stored as a conformational change in the myosin molecule. The head is then free to attach to the next site along the actin filament and repeat the cycle described in items 7 and 8. During a single contraction, each cross bridge attaches, pulls, and detaches many times as it progresses along the actin filament toward the Z line.

*10.* Finally, the free $Ca^{2+}$ level of the sarcoplasm is again lowered through active calcium uptake by the SR; the tropomyosin again inhibits cross-bridge attachment, and the muscle relaxes until the next depolarization.

## Mechanical Properties of Contracting Muscle

Now that we have reviewed the sliding-filament and cross-bridge mechanisms, we can examine the mechanical and contractile properties of a whole muscle.

Many of the mechanical properties of contracting muscle were elucidated during the first half of this century, before the mechanism of contraction was understood. It is interesting to reexamine these classic findings in the light of the sliding-filament theory.

Contraction is expressed in two different ways: in terms of shortening or in terms of tension. Two corresponding means are used to measure activity of the contractile system. In one, changes in muscle length are measured during contraction while the muscle shortens against a load (Figure 9-27A). This is called *isotonic contraction*, because the tension remains constant. In the other method, the muscle is held at an essentially constant length while tension produced during contraction is measured with a strain gauge (Figure 9-27B). This is called *isometric contraction*. Although there is no appreciable external shortening during isometric contraction, there is a small amount of *internal* shortening (sliding of actin filaments into the A band) owing to stretching of intracellular components and extracellular connective tissue in series with the muscle fibers. The time course of isometric contraction differs greatly among muscles of different animals and among different muscles of the same animal (Figure 9-28).

One simple physical concept is essential to an understanding of the following paragraphs: the tension in one element of a linear series (e.g., one link in a

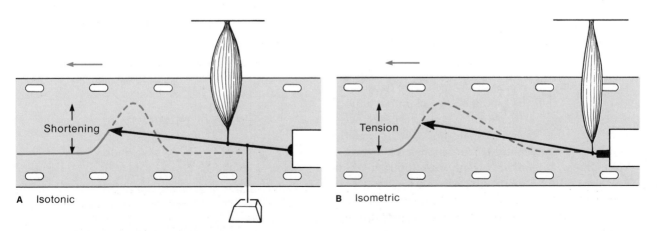

**A**  Isotonic      **B**  Isometric

9-27  Two basic ways of measuring contraction of a muscle. (A) In isotonic contractions, change in the length of a muscle is recorded while it contracts with constant tension. (B) In isometric contractions, tension change is recorded while the length of the muscle is held essentially constant. [Wilkie, 1968.]

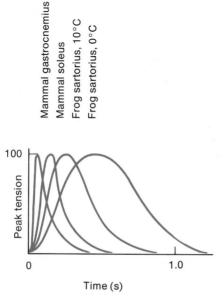

9-28    Time courses of isometric twitches in various muscles. The peak tensions are normalized to the same height. [Wilkie, 1968.]

chain) is equal to the tension in each of the other elements in the series. This concept holds true for tension generated by contraction or produced by an external weight pulling on the muscle.

### SARCOMERE LENGTH AND CONTRACTILE PROPERTIES

The velocity with which two points along a muscle fiber approach each other (i.e., the fiber shortens) under a given load is a linear function of the number of sarcomeres present *in series* between these two points. Assuming that all the sarcomeres of one muscle fiber shorten by the same increment per unit time, the overall reduction in length in that time will be directly proportional to the number of sarcomeres in series. What this means, in brief, is that the opposite ends of a long muscle approach each other with higher velocity than do those of a short muscle. This is undoubtedly an important factor in the evolution of skeletal morphology.

Likewise, it will be self-evident that maximum velocity of shortening and maximum force development

are reciprocally related. For a muscle of a given mass and cross-sectional area, long sarcomeres (i.e., large myofilament overlap) will convey high strength and low overall velocity of shortening, whereas short sarcomeres will convey high velocity with low force. It should be kept in mind, however, that just as a chain is no stronger than its weakest link, the tension developed by a myofibril cannot exceed the tension developed by its weakest sarcomere. Thus, the contractile tension produced by a muscle fiber limited by the number of myofibrils (or more precisely, actin and myosin filaments) working in parallel, and a thick muscle can lift more than an equivalent thin muscle, regardless of length. Physical exercise increases the number of myofibrils and mitochondria per muscle fiber, so there is an increase both in size and in strength of the muscle.

### LATENT PERIOD

The effect on the kinetics of the isotonic twitch of increasing the load on the muscle is seen in Figure 9-29. As the load is increased, the muscle takes more time to lift the weight clear of its support. The reason for this is that the muscle needs time to develop tension (described in the next subsection). The greater the weight, the longer the time required to build up the necessary tension. The muscle also shortens less with greater loads. Even with the lightest load, there is a latent period—the sum of all the delays between membrane excitation, propagation via the T tubules into the fiber,

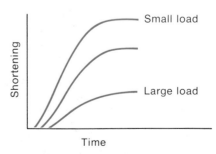

9-29    The effect on latent period and rate and extent of isotonic shortening of increasing the load on a muscle. In each case, tetanic stimulation was begun at time zero. [Wilkie, 1968.]

time for calcium release and diffusion, and activation of the cross bridges by attachment to the actin filaments. The time from peak of the action potential to the first sign of tension in frog muscle is about 2 ms.

## FORCE–VELOCITY RELATIONS

It also is apparent in Figure 9-29 that the initial velocity of shortening decreases as the load increases. Maximum velocity is attained when the load is zero—in which case, of course, the force (tension) developed by the muscle is also zero.

The relation between load and maximum velocity of shortening is plotted in Figure 9-30. The velocity drops as the force necessary to lift the load (i.e., weight of load) increases. If the load is made sufficiently heavy, there is no *external* shortening, and the contraction (by definition) is isometric. The dependence of velocity of shortening on load is easily understood in the light of the sliding-filament theory. The tendency for the actin filaments to slip backward against the production of force by cross-bridge activity should increase as the load increases.

A. V. Hill (1948) measured the velocity of shortening under various loads in different muscle and found that for each muscle the force–velocity relation could be consistently described by the hyperbolic equation

$$V = \frac{b(P_0 - P)}{P + a}$$

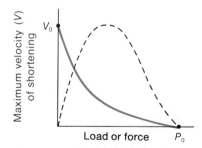

**9-30** Force–velocity relations in isotonic contractions. As the load to be lifted increases, the maximum shortening decreases. The broken line represents mechanical power (force × velocity) produced by the muscle under various loads. [Wilkie, 1968.]

where $V$ is initial velocity of shortening; $P$, the force (or load); $P_0$, maximum isometric tension of that muscle; $b$, a constant with dimensions of velocity; and $a$, a constant with dimensions of force. Unfortunately, although Hill's equation describes the force–velocity relation, it provides no real insight into the mechanism of muscle contraction. X. Aubert (1956) devised an exponential equation—entirely different from Hill's—that fits the force–velocity data equally well. An empirical equation is but a model of the physical process, and models often have no true relation to the real system other than that they describe its behavior. Our present knowledge of muscle physiology has grown out of many concrete morphological, chemical, and biophysical data.

## SERIES ELASTIC COMPONENTS

The muscle can be represented (Figure 9-31) as a contractile component *in parallel* with an elastic† component (sarcolemma, connective tissue, etc.) and *in series* with a second elastic component, the so-called *series elastic components* (*SEC*). Lumped into the category of the SEC are tendons, connective tissues linking muscle fibers to the tendons, and perhaps the Z-line material. An important component contributing to series elasticity appears to be the elasticity of the cross-bridge links. Thus, Figure 9-31 is greatly simplified.

As the contractile component shortens, it must stretch the SEC while tension is developed and transmitted to the external load (Figure 9-31A and B). When the tension developed in the SEC equals the weight of the load, the muscle begins to shorten externally and lift the load (Figure 9-31C). In Figure 9-31B the contraction is isometric, whereas in Figure 9-31C it becomes isotonic with a load. At maximum tension during an isometric contraction, the SEC stretch by an amount equivalent to about 2% of the muscle length. The contractile component must therefore shorten by an equivalent amount even though the external length of the muscle, under those conditions, does not change.

It takes time for the filaments to slide past each other by cross-bridge activity as the SEC become

---

†Elasticity is defined by Hooke's Law, which states that the length of an object with ideal elasticity increases in proportion to the force applied.

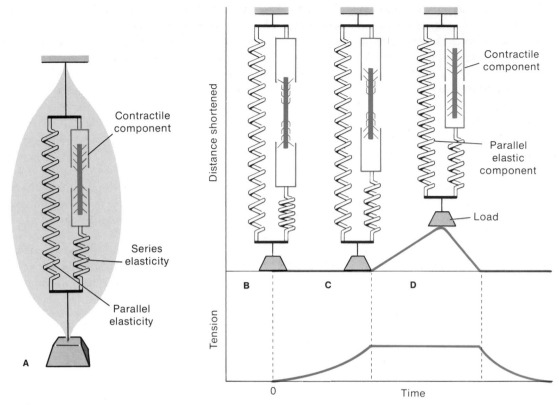

**9-31**    Mechanics of contraction. (A) A muscle represented as a contractile component in series and in parallel with elastic components. (B) The beginning of contraction: the weight rests on the substratum; sliding of filament begins to exert stretch on series elasticity. (C) Tension building up with progressive stretching of series elasticity: contraction is isometric to this point. (D) Once tension equals the weight of the load, the load is lifted and the contraction becomes isotonic. Note the progressive increase in overlap of filaments and in number of active cross bridges throughout the contraction. [Vander et al., 1975.]

stretched and tension builds up. Thus, the development of tension in a muscle progresses with time as a function of internal shortening. The effect of the series elasticity is to slow development of tension in the muscle and to smooth out abrupt changes in tension.

### ACTIVE STATE

Shortening or tension reaches a maximum within 10–100 ms, depending on the kind of muscle and the load. At first glance, this might suggest that the contractile mechanism is activated with a similar slowly rising time course. It is important, however, not to confuse the time course of tension developed by the

muscle with the time course of cross-bridge activity. It will be recalled that, on excitation and release of $Ca^{2+}$ from the SR, the cross bridges first attach to the actin filaments before active sliding begins. Moreover, sliding must first take up the slack in the SEC before full tension is developed (Figure 9-31).

The state of cross-bridge activity that prevails before the muscle has had a chance to develop full tension can be determined by application of *quick stretches* with a special apparatus at various times after stimulation before and during contraction. The object of applying a quick stretch to the muscle is to stretch the SEC and thereby eliminate the time normally required for the contractile mechanism to take up the slack in them. In this way, the state of cross-bridge activity can be measured with improved time resolution. The tension recorded by the sensing device during the quick stretch represents the tensile strength of the contractile mechanism at the instant of the quick stretch. This tensile strength depends on the holding strength of the cross bridges at the instant of stretch, because the cross bridges will slip and the filaments will simply slide past each other, preventing the development of any higher tension. Thus, the tension just necessary to make the thick and thin filaments slide apart approximates the load-carrying capacity of the muscle at the time of stretch. This tension should be proportional to the average number of active cross bridges per sarcomere.

In the relaxed state, muscle has very little resistance to stretch other than the compliance of connective tissue, sarcolemmae, and the like. The quick-stretch technique revealed that, after stimulation, the resistance to stretch rises steeply and reaches a maximum within 100 ms—about the time when external shortening or tension in the unstretched muscle is just getting under way (Figure 9-32). After a brief plateau, the load-carrying ability decreases to the low level characteristic of the relaxed muscle.

The increase in load-carrying ability of the muscle, as determined by quick-stretch experiments, is termed the *active state*. It corresponds to the formation of actomyosin complexes by the attachment of myosin cross bridges to actin filaments and by subsequent shortening activity of the cross bridges. Because cross-bridge activity is controlled by the concentration of free $Ca^{2+}$ in the sarcoplasm, the time course of the active state is believed to approximate the time course of increased calcium concentration in the sarcoplasm.

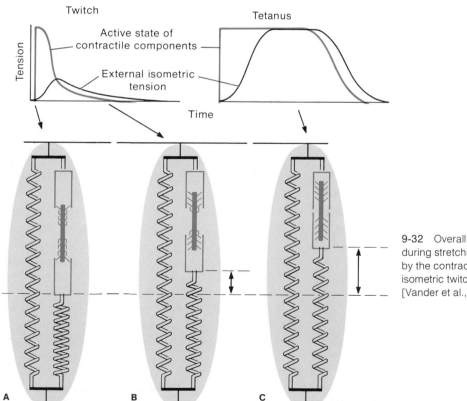

9-32  Overall length remains constant during stretching of the series elasticity by the contractile elements during an isometric twitch (**B**) and tetanus (**C**). [Vander et al., 1975.]

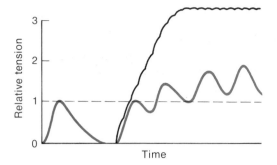

**9-33**  Varying degrees of tetanus are produced, depending on the frequency of muscle action potentials. A single twitch is seen at left. The higher stimulus frequency produces a higher and smoother tension curve.

### TWITCH AND TETANUS TENSIONS

An obvious question is raised by a comparison of active state and twitch tension in Figure 9-32. Why is the maximum tension produced by the muscle during a twitch so much lower than the active state—that is, the maximum tension of which the contractile mechanism is actually capable?

During a single twitch, the active state is rapidly terminated by the calcium-sequestering activity of the SR, which removes the $Ca^{2+}$ soon after it is released. Thus, the active state begins to decay before the filaments have had time to slide far enough to stretch the SEC to a fully developed tension (Figure 9-32). For this reason, the tension of which the contractile system is capable cannot be realized in a single twitch. Before the peak of twitch tension, the contractile elements store potential energy in the SEC by progressive stretching. If a second action potential follows the first before the SR can remove the previously released $Ca^{2+}$, the calcium level remains high in the sarcoplasm, and the active state is prolonged. With the active state prolonged, isometric tension continues to increase with time until the tension produced by internal shortening of the contractile components and stretching of the SEC is just sufficient to cause cross bridges to slip and prevent further shortening of the contractile components. The muscle has then reached full *tetanus tension.* (The prolongation of the active state by closely repeated action potentials is called *tetanus.*) Depending

on the repetition rate of muscle action potentials, varying degrees of twitch fusion and tetanus tension are produced (Figure 9-33).

### ENERGY, HEAT, AND WORK

The immediate source of energy for the contractile process is ATP:

$$ATP \rightleftharpoons ADP + P_i + \text{free energy}$$

The concentration of ATP in muscle, however, is only 2–4 mM; hence ATP is quickly used up in sustained contraction. The ATP is rapidly regenerated by rephosphorylation of ADP, by an energy-rich phosphagen compound (p. 55). In the vertebrates and several invertebrates, the phosphagen is creatine phosphate; in the muscles of most invertebrates, it is arginine phosphate (Figure 9-34). The enzyme creatine phosphokinase catalyzes the transphosphorylation:

$$PC + ADP \rightleftharpoons C + ATP$$

This reaction proceeds freely to supply ATP during exercise. The level of ATP in active muscle shows no substantial decline as long as the phosphagen has not been depleted.

During and after contraction, the heat production of the muscle rises above the basal (resting) level. There is a relatively fixed production of heat associated with the process of electrical excitation and mechanical activation; this is termed the *activation heat.* In addition, there is the *heat of shortening,* which is proportional to the distance through which the muscle is allowed to shorten during the contraction. It should be noted that

**9-34**  Two phosphagens responsible for the rephosphorylation of ADP produced from ATP during muscle contraction.

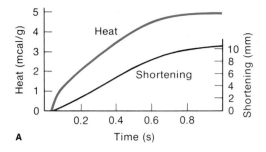

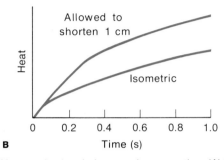

**9-35** Heat production during muscle contraction. (A) The heat produced is composed of two components: (1) the activation heat (a relatively constant amount presumed to be associated with the processes of excitation, release and reuptake of calcium, and activation), and (2) a variable heat of shortening (related to the distance through which the muscle shortens). (B) If allowed to shorten, the muscle produces more heat than if contracting isometrically. [Hill, 1938.]

when a muscle shortens as it lifts a load, the energy released in the muscle is the sum of the work done (weight × distance) and the heat produced. The energy liberated as heat is about five times as great as the energy converted into work. This level of efficiency is typical of biological processes. What is unusual about the energy utilization of muscle is its high degree of control. One example of this is seen in Figure 9-35. With increased shortening, there is a proportional increase in heat liberated. This is now understandable in terms of the sliding-filament mechanism. With increased overlap of actin and myosin filaments, the number of activated cross-bridge ATPase sites increases linearly with the degree of overlap. Those cross bridges that are out of reach of the actin filaments cannot attach and therefore cannot enzymatically hydrolyze ATP to do work and liberate heat.

# Metabolic Subtypes of Striated Muscle

The skeletal muscles of vertebrates consist of muscle fibers of more than one type. Some contain a high proportion of fibers specialized for slow *tonic* (steady) contraction, most useful for postural tone. Other muscles contain a higher percentage of fibers specialized for *phasic* (twitch) contraction, most useful for rapid movements of limbs. These different kinds of muscle fibers can be distinguished also on the basis of biochemical, metabolic, and histochemical criteria. Four major groups of vertebrate skeletal muscle are illustrated in Figure 9-36 (Goldspink, 1977).

*1. Tonic muscle fibers.* These very slowly contracting, nontwitching fibers are found in the postural muscles of amphibians, reptiles, and birds, in the muscle spindles, and in the extraocular muscle of mammals. They receive a multiterminal innervation very much like arthropod muscle receives, except that each muscle fiber is normally innervated by the branches of only one excitatory motor axon. Tonic fibers normally do not produce action potentials and, in fact, do not require an action potential to spread excitation, since that is done by the branches of motor nerve that make repeated synapses with the muscle fiber along its length. A single presynaptic impulse elicits an insignificant contraction. With a train of impulses, the postsynaptic potential exhibits temporal summation and facilitation so as to produce enhanced graded depolarization. In addition, the slow contractile system has time to develop tension, which occurs slowly by comparison with twitch muscle. One of the biochemical specializations of tonic fibers is an extremely low turnover number for the myosin ATPase, which allows them to maintain isometric tension very efficiently.

*2. Slow phasic fibers.* These slowly contracting, slowly fatiguing fibers occur in mammalian postural muscles. They generate all-or-none action potentials and therefore contract in response to motor nerve impulses with all-or-none slow twitches. Like all the phasic (i.e., twitch) fibers, they typically have one or only a few endplates, usually from a single motor axon. Like tonic fibers, slow phasic fibers are used for maintaining posture and for slow, repetitive movements. They fatigue very slowly, probably because they contain a large number of mitochondria and use ATP at a relatively

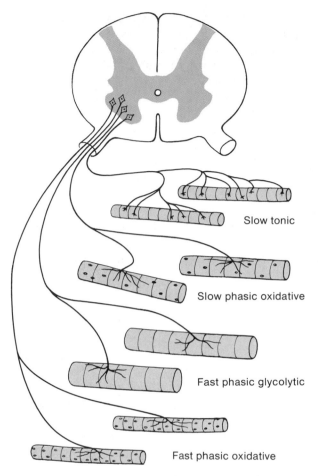

Slow tonic

Slow phasic oxidative

Fast phasic glycolytic

Fast phasic oxidative

**9-36** Major categories of muscle fiber types in vertebrate skeletal muscles. These types probably never all occur in the same muscle, and some muscles consist of only one fiber type. Note the multiterminal innervation of the slow tonic-fiber type, which cannot produce its own propagated muscle action potential. [Courtesy of G. Goldspink.]

slow rate owing to the low turnover number of the myosin ATPase. They are also characterized by a reddish color (hence the "dark" meat of fish and fowl) because of high concentrations of the oxygen-storage protein *myoglobin.*

*3. Fast phasic glycolytic fibers.* These powerful fibers contract rapidly but fatigue quickly. The rapid contraction can be ascribed in part to a myosin ATPase that has a high turnover number. These fibers are gen-

erally recruited when very rapid contraction is required. The ATP is produced by glycolysis, and the fibers contain few mitochondria. During contraction, they build up an energy debt that is subsequently replenished. A familiar example of this type makes up the white breast muscles of domestic fowl, which cannot sustain activity.

*4. Fast phasic oxidative fibers.* These quickly contracting twitch fibers fatigue slowly because, with their large numbers of mitochondria, they are able to produce ATP quickly by oxidative phosphorylation. Thus, they are specialized for rapid repetitive movements, as in sustained, strenuous locomotion. They are prominent in bird flight muscles.

These categories are somewhat arbitrary, since a gradient of intermediate types exist between the phasic-fiber categories, especially if comparisons are made between vertebrate species. However, in a given muscle, these types can be distinguished by their histological (i.e., staining) properties. These properties can be based, for instance, on the chemistry of the different myosin ATPases found in these fibers or on the abundance of an oxidative enzyme such as succinic dehydrogenase.

Three known factors, then, seem to determine the behavior of the different muscle fiber types:

*1.* The electrical properties of the membrane determine if the fiber will respond with an all-or-none twitch or a graded contraction.

*2.* The myosin ATPase activity will determine the intrinsic speed of contraction. These differences in the rate at which the ATPase site of the myosin head utilizes ATP (and hence differences in the rate of cross-bridge activity) result from differences in the myosin "light chains" that are part of the cross-bridge head.

*3.* The density of mitochondria and hence the rate of ATP production will determine the resistance to fatigue.

In addition, the rate of release and uptake of $Ca^{2+}$ from the SR will determine the duration of each twitch. Extremely fast muscle fibers, like those found in hummingbird wing muscles and those controlling the vocal cords of echolocating bats, contain highly developed sarcotubular systems.

# Neural Control of Muscle Contraction

Effective animal movements require that contractions of muscles be correctly timed with respect to one another. Their timing is, of course, regulated by the timing of motor impulses generated by the CNS. In addition, it is necessary that the degree or strength of contraction of each muscle be regulated by the nervous system. A motor system limited to all-or-none contractions of entire skeletal muscles would produce spastic behavior with a very limited repertoire of movement. Fine control of muscle contraction has been achieved in different organisms by various means during the course of evolution. Vertebrate and arthropod neuromuscular mechanisms lend themselves especially well to comparison because of the divergence of neuromotor organization that has evolved in these two groups for the control of movement.

## VERTEBRATE NEUROMOTOR ORGANIZATION

Vertebrate skeletal muscle is innervated by motoneurons whose cell bodies are located in the ventral horn of the gray matter of the spinal cord. The motor axon leaves the spinal cord by a ventral root, continues to the muscle via a peripheral nerve trunk, and finally branches repeatedly to innervate about 100 skeletal muscle fibers.

The motoneuron and the muscle fibers that it innervates form a *motor unit* (Figure 6-42). An action potential originating as a consequence of synaptic input to the motoneuron travels from its site of origin in the axon hillock along the axon toward the periphery, spreading into all the terminal branches of the axon to the endplates that innervate the muscle fibers of that motor unit. A postsynaptic potential is produced in each muscle fiber by the action of acetylcholine, the neuromuscular transmitter substance. In twitch muscle fibers, such as those that make up the bulk of the sartorius muscle of the frog, this depolarization generally exceeds the firing level for the muscle action potential (Figure 9-37A); therefore, each time the motoneuron fires (generates an action potential), all the muscle fibers of the motor unit are fully activated by the release of transmitter from all the motor terminals of that neuron. Whether the contractions consist of single twitches or of sustained tetanic contractions depends on the frequency of motor impulses generated by the synaptic input to the motoneuron.

The degree with which tension can be modulated in such an all-or-none motor unit is very small, because there is no gradation between inactivity and a twitch. With repeated motor impulses, the response is rather uneven except at firing rates high enough to elicit a full tetanic contraction (Figures 9-32C and 9-33). In the vertebrates, the problem of increasing overall muscle tension in a graded fashion is solved by increasing the number of motor units active at any moment (known as *recruitment* of motor units) and by varying the average frequency with which the motoneurons fire. If, for example, a small number of the motor units of a given muscle are maximally active, the muscle will contract with a small fraction of maximum tension. On the other hand, if all the motoneurons of the muscle are recruited to fire at a high rate, all the motor units that constitute the muscle are brought into a state of full tetanus, producing a maximal contraction of the muscle.

The advantage of the neuromotor organization of vertebrate twitch muscle is that it is able to produce rapid, powerful contractions as well as finely graded tensions. The cost of this flexibility is a large number of motoneurons serving each muscle.

As described earlier, the tonic muscle fibers of vertebrates receive a multiterminal innervation—that is, the motor nerve makes many synapses along the length of each fiber; they also exhibit strongly facilitating synaptic potentials, and it is the synaptic potentials that produce the graded depolarizations responsible for their slow, graded contractions. In these respects, tonic muscle fibers are similar to the muscle fibers of arthropods, as described in the next subsection. Tonic muscle fibers generally are found where slow, sustained contractions are required. The major electrical difference, then, between twitch and tonic muscle fibers is the presence or absence, respectively, of a muscle action potential (Figure 9-38).

Except for the extraocular muscles and intrafusal spindle fibers, mammals do not have multiterminal, nonspiking muscle fibers. But the skeletal muscles of these two classes of vertebrates contain several types of twitch muscle. Muscles occupied primarily with postural functions tend to contain a high percentage of slow phasic fibers that produce all-or-none twitches

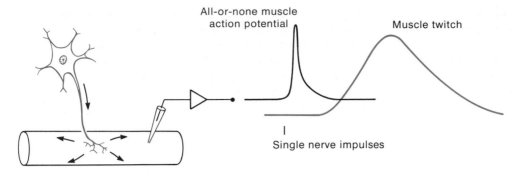

All-or-none muscle action potential

Muscle twitch

Single nerve impulses

**A** Vertebrate

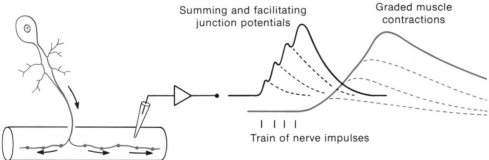

Summing and facilitating junction potentials

Graded muscle contractions

Train of nerve impulses

**B** Arthropod

9-37 Comparison of vertebrate and arthropod muscle excitation. (A) Vertebrate twitch muscle produces an all-or-none twitch in response to each all-or-none, conducted muscle action potential. (B) In many arthropod fibers, contractions are graded in response to graded synaptic potentials produced at the motor synapses. In those muscles, nerve fibers carry the impulses to distributed motor synapses.

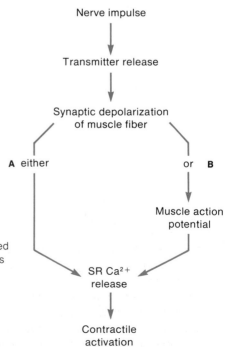

Nerve impulse

Transmitter release

Synaptic depolarization of muscle fiber

**A** either

or **B**

Muscle action potential

SR Ca²⁺ release

Contractile activation

9-38 Steps between nerve impulse and muscle contraction. (A) In graded or tonic muscle, the membrane potential is graded by summation and facilitation of synaptic potentials at synapses distributed along the length of the muscle fiber. (B) In twitch muscle, an action potential intervenes, giving the contraction all-or-none characteristics.

of slower time course and low fusion frequency (Figure 9-36, Box 9-2). Muscles concerned with rapid movements contain a high percentage of fast phasic fibers that exhibit high tetanus fusion frequencies. Between the extreme fast and slow twitch fibers, there exist muscle fibers of intermediate properties.

### ARTHROPOD NEUROMUSCULAR ORGANIZATION

Arthropod nervous systems consist of a relatively small number of neurons; hence large numbers of motor units are not available to produce fine gradations of movement by varying the recruitment of motor units. Moreover, many types of arthropod muscles do not produce action potentials or do so only in response to synaptic input from certain motor endings. In these muscles, contraction is controlled by *graded* depolarization of the muscle fiber membrane. As seen in Figure 9-18, the tension developed in a single muscle fiber is a function of membrane depolarization, the two being related by the release of $Ca^{2+}$ from the SR (Figure 9-26). This relation appears to be true for all muscle fibers. Muscle fibers that do not produce all-or-none action potentials produce a graded release of calcium—and, hence, graded tension—in response to graded changes in synaptic depolarization. The latter is determined in part by the frequency of impulses in the motor axon.

The vertebrate twitch muscle fiber is innervated at only one or two restricted regions—the endplates, where the motor axon forms minute presynaptic arborizations. The postsynaptic action potential originates at or near an endplate and spreads nondecrementally along the muscle fiber (Figure 9-37A). In contrast, the crustacean skeletal muscle fiber receives multiterminal innervation (Figure 9-37B). Thus, because the motor axon carries the message along the entire length of the muscle fiber, there is no need for a propagated action potential in the muscle fiber. Many crustacean muscle fibers do not, in fact, generate all-or-none action potentials, but produce graded responses (Figure 9-37B). The postsynaptic potentials exhibit a large degree of facilitation and temporal summation at the distributed neuromuscular synapses. The shorter the interval between excitatory synaptic potentials, the greater the depolarization of the muscle membrane. Because the coupling between membrane potential and tension

is graded (Figure 9-18), each muscle fiber can produce a wide range of contractions instead of being limited, like vertebrate twitch muscle, to all-or-none twitches or tetanus. For this reason, arthropod muscles can function quite well over a large range of tensions with very few motor units. In some arthropod muscles, a single motoneuron innervates all or most of the fibers of the muscle.

The flexibility of peripheral control of contraction is further enhanced in the crustaceans and other arthropods by *multineuronal innervation*—that is, each muscle fiber receives branches from several motor axons, including one or two inhibitory axons (Figure 9-39). Firing of the inhibitor counteracts the depolarization induced by the excitatory motor nerves. There is usually one excitor axon that produces larger excitatory synaptic potentials in the muscle fiber. This *fast excitor axon* can generate a strong contraction with less facilitation and summation than can a *slow excitor axon,* which must fire repeatedly at high frequency to produce similar levels of depolarizations, and hence contraction, in the muscle fiber.

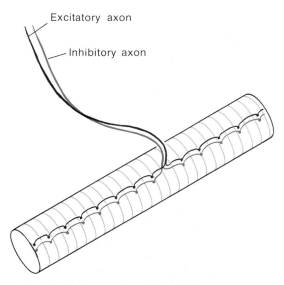

Excitatory axon

Inhibitory axon

**9-39**  Multineuronal and multiterminal innervation in arthropod muscle. There are usually several excitor axons and one or more inhibitors. One of each is shown here for simplicity.

## Box 9-2 *Trophic Effects of Nerves on Muscles*

Cross reinnervation experiments indicate that biochemical and physiological differences between fast and slow muscle fibers are regulated by the nerve fibers that innervate them. For example, when motor nerves that normally innervate slow muscle fibers are transplanted into denervated fast fibers, they will, on innervation, induce properties characteristic of slow fibers (see figure). One hypothesis is that different *trophic substances* released from the axon terminals influence the transcription of genetic information in the muscle fibers. The different myosin light chains in fast and slow muscle are, of course, products of different genes. In this regard, it is interesting that the changes in myosin ATPase activity that can be demonstrated as one of the consequences of cross-innervation can be directly ascribed to changes that occur in the composition of the myosin light chains in response to the cross-innervation. Trophic substances elaborated by nerves to regulate the molecular biology of target cells have not yet been demonstrated.

It was recently reported that the amount of contractile activity, dependent, of course, on the average frequency of impulses in the motor nerve, plays an important role in determining whether a muscle exhibits "slow" or "fast" biochemical and physiological properties, irrespective of the identity of the nerve innervating the muscle. Thus, the differences between slow and fast muscle may be due, at least in part, to differences in steady impulse activity of the motor nerves that innervate them. Some of the differences in biochemical and contractile properties may indeed arise within each muscle fiber as a consequence of the amount of contractile activity it experiences. Although the contraction itself may not influence the chemistry of the myoplasm, concomitant chemical changes, such as elevated $Ca^{2+}$ or altered intracellular pH, may influence the differentiation of the muscle fiber and differential genetic transcription.

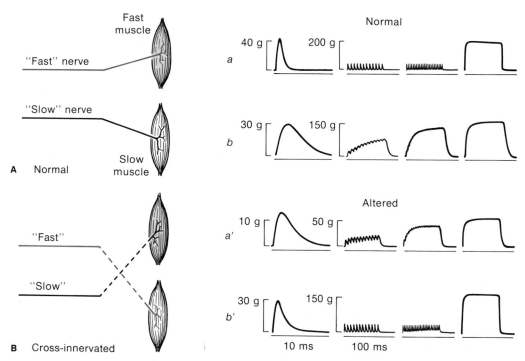

Crossed-innervation experiment. (**A**) Two muscles, *a* and *b*, have fast and slow contractile properties, as illustrated by the recordings of twitch and tetanus tensions at the right. (**B**) Cutting the nerves and crossing them, to allow reciprocal regeneration of the "incorrect" nerve to each of the two muscles, causes each muscle to assume the properties of the other. This experiment suggests that the contractile properties are determined to a large extent by the nerve innervating the muscle. The contractile changes have been shown to parallel biochemical change in the cross-innervated muscles. [Close, 1971.]

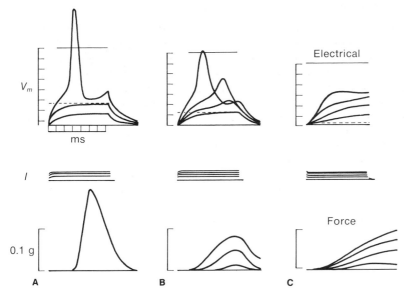

9-40   Summary of the variations in crustacean muscle-fiber properties. Membrane potentials shown at top, stimulating current in the middle, and tension at the bottom of each set of records. (A) All-or-none twitch fibers; these produce action potentials and fast twitches. (B) Intermediate graded fibers; these produce nonpropagating graded potentials and graded contractions. (C) Slow muscle fibers; these produce only very small and slow active depolarizations and contract very slowly. [Hoyle in *Invertebrate Nervous Systems,* C. A. G. Wiersma, ed. © 1967 by University of Chicago Press.]

The variety and complexity of peripheral motor organization is increased still further by the presence in most arthropod muscles of several types of muscle fibers exhibiting different electrical, contractile, and morphological properties. At one end of the spectrum are fibers with rapid all-or-none contractions, which resemble vertebrate twitch fibers. A series of intracellular current pulses (Figure 9-40A) evokes a series of electrotonic depolarizations until the firing level is exceeded and the membrane responds with an all-or-none action potential. This elicits an all-or-none fast twitch. At the opposite end of the spectrum of crustacean muscle-fiber types are the fibers in which the electrical responses show little sign of regenerative depolarization and the contractions are fully graded with the amount of depolarization (Figure 9-40C). Between these two extremes is a continuum of intermediate muscle-fiber types (Figure 9-40B). The differences in contractile behavior of these fiber types are correlated with morphological differences. The slowly contracting fibers have relatively fewer T tubules and less SR than the rapidly contracting fibers. The fast excitor tends to innervate fast, or twitch, fibers, while the slow excitor tends to innervate slow muscle fibers. This accounts for the different actions of the two types of motoneurons.

## ASYNCHRONOUS FLIGHT MUSCLES

One type of striated skeletal muscle is a notable exception to the rule that each contraction is evoked by a depolarization of the surface membrane. The flight muscles of most species of four orders of insects (bees and wasps, flies, beetles, and bugs) show no relation between the timing of the individual contractions of the muscles and the timing of the arrival of motor impulses. These muscles are termed *asynchronous* to distinguish them from muscles that contract in synchrony with each motor impulse. They are known also as *fibrillar muscles.* Although the timing of contraction has no relation to the timing of neural input to these muscles, a constant train of motor impulses and muscle depolarizations maintains the muscle in an active condition.

In species of small insects, the wing-beat frequency (and the frequency of wing-muscle contractions) far exceeds the maximum maintained discharge rates of which axons are capable. Wing-beat frequency increases with decrease in wing size. A tiny midge, for example, will have a wing-beat frequency of more than 1000 Hz.

How do the contractions of fibrillar muscle occur with a timing that is independent of membrane potentials? As in other muscles, the active state (cross-bridge activity) of fibrillar muscle requires that there be a sufficient $Ca^{2+}$ concentration in the sarcoplasm. The sarcoplasmic calcium concentration is maintained at a steady activating level as long as there is steady neural input to the muscle. But the active state is not initiated until the muscle is given a sudden stretch. Conversely, the active state is terminated if tension is released. The role that changes in length play in producing repetitive contractions was examined in glycerine-extracted (Box 9-1) fibrillar muscles lacking functional cell membranes. In the presence of constant levels of $Ca^{2+}$ above $10^{-7}$ M, the extracted muscle will contract (actively develop tension) in response to an applied stretch and oscillate between contractions and relaxations if provided with a resonant mechanical system (Figure 9-41). When the calcium levels are reduced to $10^{-9}$ M, the muscle fails to contract.

Insects having asynchronous muscles have musculoskeletal configurations with two stable states. Contractions of the antagonist flight muscles produce changes in the shape of the thorax, which favor only two wing positions—up or down (Figure 9-42B). As the *elevators* (muscles that elevate the wings by pulling down the roof of the thorax) contract, they cause the

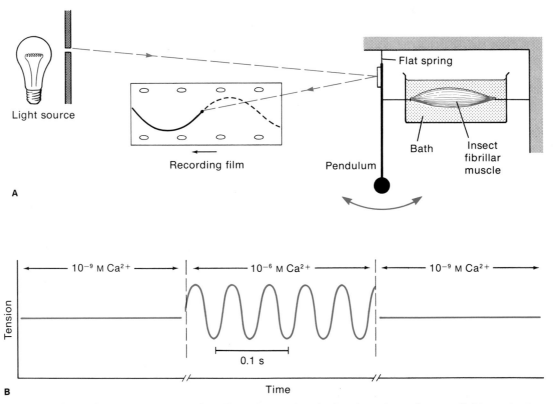

9-41    Calcium dependence of oscillatory contractions in glycerinated asynchronous flight muscle of a water beetle. (A) Experimental method. The pendulum provides the mechanical resonance in place of the thoracic exoskeleton and wings of beetle. Contraction of muscle causes movement of the pendulum to the right. As the pendulum continues its movement owing to momentum, tension in the muscle drops, inactivating it. Muscle is reactivated when the pendulum swings back to left and stretches the muscle. Movements are recorded on film as shown. (B) Ability of extracted flight muscle to produce oscillations is dependent on calcium concentration of reactivation solution. [Jewell and Rüegg, 1966.]

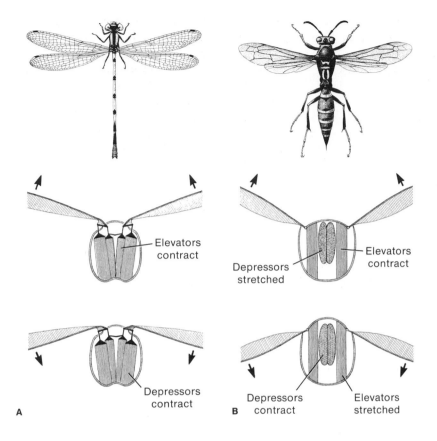

9-42 Two kinds of insect flight muscles. (A) A damselfly with synchronous flight muscles arranged so both elevators and depressors work vertically to raise or lower the wings. (B) A wasp with asynchronous flight muscles arranged so that contraction of the elevators pulls the roof of the thorax down, raising the wings, whereas contraction of the longitudinal depressors causes the roof of the thorax to arch upward, lowering the wings. The thorax tends toward either position, passing through an intermediate configuration of low stability. [From "The Flight Muscles of Insects" by D. S. Smith. Copyright © 1965 by Scientific American, Inc. All rights reserved.]

roof of the thorax to snap down past the "click" point (much as in a clicker toy). This does three things: (1) elevates the wings; (2) stretches, and thereby activates, the *depressors* (muscles that lower the wings by shortening the thoracic exoskeleton front to back so as to expand the thorax dorsoventrally); and (3) slackens the elevator muscles suddenly, thereby inactivating them. The cycle is completed in reverse as the stretch-activated depressors produce an upward deformation of the roof of the thorax until it "clicks" back to the raised (wing-depressing) position (Figure 9-42B, bottom).

When nervous input to these muscles ceases, the muscle membrane repolarizes and the sarcoplasmic calcium levels drop, with the result that the cross bridges are unable to attach to the actin filaments. Externally applied stretch then no longer produces an active state, and the flight movements stop. Thus,

motor input to fibrillar muscle acts largely as an ON–OFF switch. The frequency of contraction depends on the mechanical properties of the muscle and the mechanical resonance of the flight apparatus (thorax, muscles, wings). Thus, if the wings are clipped short, the wing-beat frequency will increase, even though the nerve impulses occur with unchanged frequency.

## Cardiac Muscle

The contractile mechanism of vertebrate ventricular muscle (Figure 9-43B) fundamentally resembles that of skeletal twitch muscle. The major specializations are in the action potential, which differs from that of nerve and skeletal muscle in the following ways:

*1.* It has a plateau hundreds of milliseconds long following the upstroke (Figure 9-44B). The long duration of the action potential relative to the velocity of

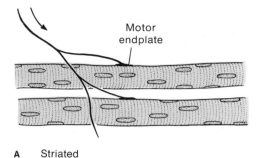

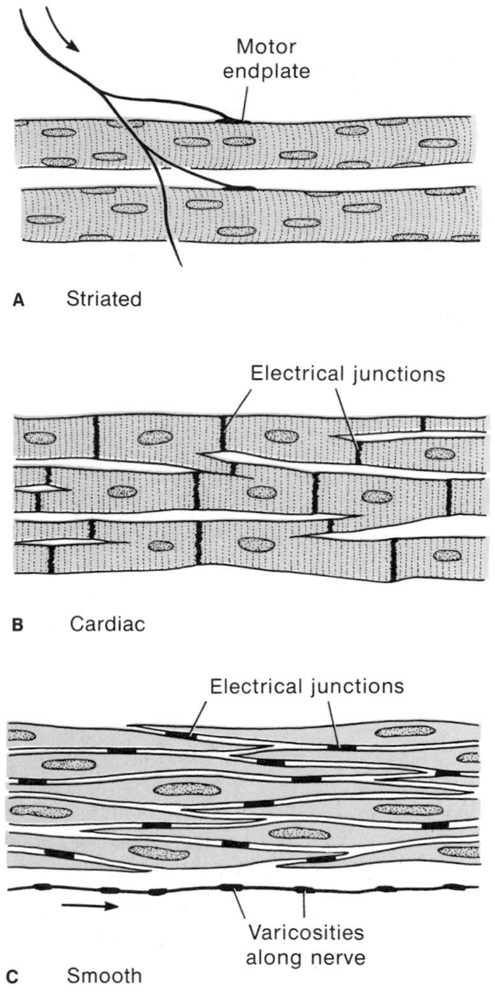

**A    Striated**

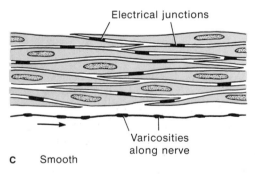

**B    Cardiac**

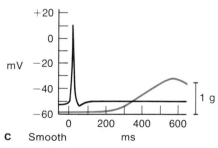

**C    Smooth**

**9-43**  Gross morphology of three major classes of vertebrate muscle. (A) Striated muscle fibers are cylindrical and multi-nucleate. (B) Cardiac muscle cells are joined end to end by electrical junctions and are therefore electrically continuous. (C) Smooth-muscle fibers are small and spindle-shaped, each with a single nucleus. They are joined laterally into electrically coupled groups. The innervation is diffuse, the transmitter arising from varicosities along the motor nerve.

*2.* Each cardiac action potential is followed by a refractory period of several hundred milliseconds. This long period of refractoriness prevents tetanic contraction, requiring the muscle to relax and permitting the ventricle to fill with blood between action potentials. As a result of regularly paced, prolonged action potentials, the heart contracts and relaxes at a rate suitable for its function as a pump.

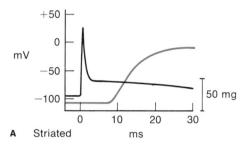

**A    Striated**

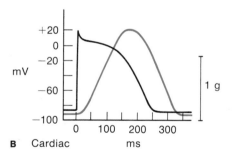

**B    Cardiac**

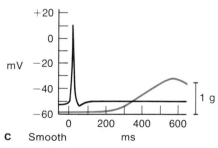

**C    Smooth**

**9-44**  Relationships between electrical (black traces) and mechanical (colored traces) activity in three major classes of vertebrate muscle. (A) Striated muscle. [Hodgkin and Horowicz, 1957.] (B) Cardiac muscle. [Brooks et al., 1955.] (C) Smooth muscle. [Marshall, 1962.]

conduction ensures that all the cells of the ventricle are excited approximately in unison, causing the ventricular muscle to contract as a unit. This concerted contraction is essential for the efficient pumping action of the heart (p. 552).

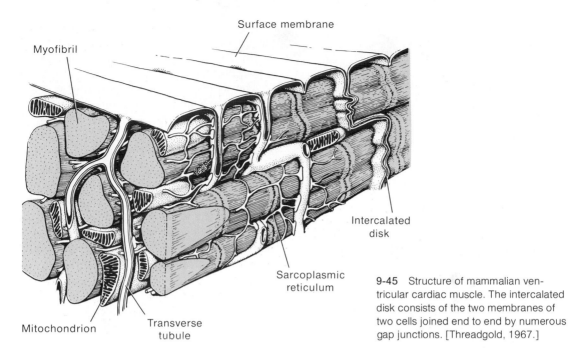

Surface membrane

Myofibril

Intercalated disk

Sarcoplasmic reticulum

Mitochondrion

Transverse tubule

**9-45** Structure of mammalian ventricular cardiac muscle. The intercalated disk consists of the two membranes of two cells joined end to end by numerous gap junctions. [Threadgold, 1967.]

The cells of mammalian cardiac muscle possess an elaborate SR and system of T tubules (Figure 9-45). Like skeletal twitch muscle, cardiac muscle is activated by the release of $Ca^{2+}$ from the SR. The cardiac muscle of amphibians is more simply organized than that of higher vertebrates and has therefore proved useful in elementary studies of how contraction is regulated by the electrical activity of the cell membrane. Cardiac muscle of the frog has only a rudimentary reticulum and tubular system. Like those of smooth muscle, the cells of the frog heart are much smaller than skeletal muscle fibers. The relatively large surface/volume ratios of these small cells reduces the need for an elaborate intracellular reticulum for the storage, release, and uptake of $Ca^{2+}$. Instead, much of the calcium supplied for contraction in smooth muscle and in the amphibian heart enters these cells through the surface membrane as a result of the membrane's increased calcium permeability during depolarization (see discussion of calcium action potentials, p. 167).

The relation between membrane potential and tension in small strips of frog ventricle is shown in Figure 9-46B. As the cell is depolarized (Figure 9-46A), $Ca^{2+}$ diffuses into the cell because of the increased calcium conductance of the surface membrane. Because the influx of $Ca^{2+}$ is voltage-dependent, tension develops as a function of depolarization, with greater depolarization producing greater tension. Reduction of the extracellular $Ca^{2+}$ concentration results in a weaker contraction for a given depolarization (Figure 9-46B), presumably because less $Ca^{2+}$ enters the cell when the concentration difference between cell interior and exterior is reduced.

## Smooth Muscle

Muscles are called "smooth" if they lack the characteristic striations produced by the organized groups of actin and myosin filaments that form sarcomeres. The filaments of smooth muscle are distributed somewhat randomly within the myoplasm. There are several types of smooth muscle, distinguished on the basis of cell morphology. For example, the byssus retractor muscle and the adductor muscles of bivalve mollusks have long cylindrical cells, and the smooth muscle that

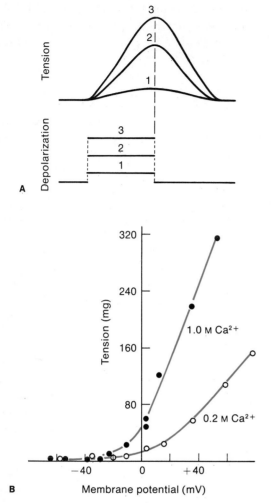

**9-46** Relations between depolarization and tension in frog ventricular muscle. **(A)** Tension (upper traces) at three voltage steps (lower traces). **(B)** The amount of tension depends on the amount of $Ca^{2+}$ in the extracellular medium as well as on the amount of depolarization. The ordinate is tension recorded at the end of the voltage step. The membrane potential during the voltage step is plotted along the abscissa. [Morad and Orkand, 1971.]

forms the walls of vertebrate visceral organs (e.g., alimentary canal, urinary bladder, ureters, arteries, and arterioles) consists of small, mononucleate, spindle-shaped cells 2–20 μm in diameter and 10 to 100 times as long as they are wide (Figure 9-43C). Groups of these cells are connected to each other by gap junctions that

permit the electrotonic spread of current from cell to cell. Such cells are thus electrically coupled in bundles about 100 μm across and several millimeters long. Such coupled groups of cells form functional units.

The innervation of smooth muscle differs significantly from that of skeletal muscle, which has discrete and intimate synaptic junctions between the motor terminal and the muscle fiber. In vertebrate smooth muscle, the transmitter is released from many swellings, or varicosities, along the length of autonomic axons that travel along the smooth muscle tissue. These do not form intimate junctions with differentiated postsynaptic specializations. Instead, the transmitter released at a given varicosity diffuses over some distance, encountering a number of the small, spindle-shaped smooth muscle cells along the way. The postsynaptic receptor molecules in smooth muscle appear to be distributed diffusely over the cell surface. Smooth muscle of vertebrates is entirely under autonomic and hormonal control and is not "voluntary" as is skeletal muscle. It contracts and relaxes far more slowly than striated muscle and is generally capable of more sustained contractions.

In the cells of smooth muscle, the SR characteristic of striated muscle is at most rudimentary in form, being present only as smooth flat vesicles close to the inner surface of the cell membrane. A highly developed SR like that of striated muscle fibers is unnecessary because the cells of smooth muscle are small and therefore have large surface/volume ratios. No point in the cytoplasm is more than a few micrometers distant from the surface membrane. Thus, the surface membrane of smooth-muscle cells can perform calcium-regulating functions similar to those ascribed to the membranes of the SR of striated muscle. Calcium is constantly pumped outward across the surface membrane so as to keep internal levels of that ion very low. Whenever the membrane becomes depolarized, it becomes more permeable to calcium ions. This permits an influx of $Ca^{2+}$, which activates contraction. Relaxation, according to this hypothesis, occurs when the calcium permeability returns to the low resting level while the membrane pumps out calcium. Large depolarizations lead to action potentials in which calcium carries the inward current (Figure 9-47A). Action potentials produce the greatest $Ca^{2+}$ influx, and thus

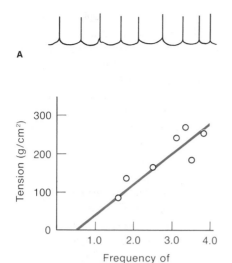

**9-47** Relations between electrical activity and tension in mammalian smooth muscle. (A) Slow depolarizations of smooth muscle produce calcium spikes. (B) Tension produced by the muscle is related to the frequency of action potentials. [Diamond and Marshall, 1969.]

evoke the largest contractions. Tension is graded with membrane depolarization (Figure 9-47B).†

Calcium ions in smooth muscle have a regulatory function that differs considerably from their role in regulating contraction in striated muscle. Instead of binding to troponin, which is absent in smooth muscle, $Ca^{2+}$ binds to *calmodulin,* an important regulatory molecule very similar to troponin C (p. 356). Calcium–calmodulin, in turn, complexes with and activates a protein kinase that phosphorylates a myosin light chain, which forms part of the myosin head. The phosphorylated form of the myosin is activated by actin, and the two molecules interact (in a poorly understood manner) to cause contraction.

An interesting feature of smooth muscle is the sensitivity of the membrane to mechanical stimuli. Stretching the muscle produces some depolarization, which in turn produces some contraction. As a result, muscle tension is maintained over a large range of muscle

———

†For an account of metabolic mechanisms augmenting the electrical regulation of smooth-muscle activity, see page 475.

length. Because there is no sarcomere structure with which to gauge the overlap of actin and myosin filaments, it is difficult to interpret the length–tension relations of smooth muscle.

## Musculoskeletal Mechanics

To do work, a muscle must be able to transfer the energy of its contraction to a load. This energy transfer can be achieved with a mechanical system such as the vertebrate skeleton, with its levers, joints, pulleys, and muscle attachments.

The arthropod exoskeleton also consists of rigid elements connected by joints, but it is arranged as a protective armor enclosing the muscles and other tissues. It functions on the same mechanical principles as an endoskeleton. Many invertebrates have no rigid skeletons, but even these animals are generally able to move, change shape, and propel themselves by muscle contractions. Some animals, such as coelenterate polyps and annelid worms, utilize a hydrostatic "skeleton" (Figure 9-48) in which opposing circularly and longitudinally arranged layers of muscle contract against an incompressible aqueous fluid contained within a lumen. Contraction of longitudinal muscles shortens the organism; contraction of circular muscles elongates it. Movements along the vertebrate digestive tract arise in a similar manner.‡

A number of factors determine the speed with which contracting muscles can move a distal part of the body (e.g., the foot) relative to a proximal part (e.g., the knee). These factors include the relative distances between joints and muscle attachments, limb proportions, and muscle load, length, and strength. The variations in basic skeletal architecture among various animals and the limbs of a given animal are the result of adaptations for different mechanical specializations. The simple principles of lever mechanics will be recalled from elementary physics. A comparison of lever

———

‡The mammalian penis and clitoris become erect as the result of inflation of spongy tissue by blood under hydrostatic pressure. The stiffness of the erect organ arises in the same way as that of any inflated object and does not involve muscle tissue other than those smooth muscles that control the flow of blood into and out of the organ.

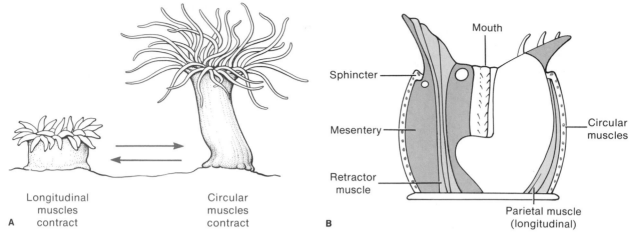

**9-48** A hydrostatic skeleton. (A) The sea anemone has no rigid skeleton, but antagonistic muscles can act against each other through the incompressibility of the fluids contained within the animal, causing shortening or elongation. (B) Schematic of the various major circular and longitudinal muscle systems of *Metridium*. [Batham and Pantin, 1950.]

mechanics in the leg of a burrowing animal like a badger with those of a high-speed runner like a deer (Figure 9-49) reveals that the ratio of distances—toe to pivot/heel to pivot—is higher for the deer. This ratio, of course, conveys greater speed (but less force) to the deer's toe for a given velocity and force of gastrocnemius contraction. Figure 9-50 illustrates three basic lever systems in vertebrate skeletons. In each case, the muscle works against a load by pivoting one skeletal member against another at a point of articulation that serves as a fulcrum, or pivot.

The orientation of the muscle fibers relative to the tendon is an important factor in the function of the musculoskeletal system. Thus, the velocity and force exerted on the tendons that connect a muscle to the bone depend on both the length and the cross-sectional area of the muscle. For a given sarcomere length and contractile characteristics, the longer the muscle, the faster the velocity of overall shortening. The greater the cross-sectional area (i.e., the more myofilaments in parallel), the greater the force it can develop. In the course of evolution of specific muscles, either speed or mechanical advantage (force capability) has been favored by the way in which the muscle and skeletal elements have become organized. An interesting example is the *pinnate* or semipinnate arrangement of

muscle fibers (Figure 9-51 B and C) characteristic of the gastrocnemius muscle as opposed to a *parallel* arrangement (Figure 9-51A) characteristic of the sartorius muscle. For a given size and proportion, the pinnate muscle is capable of producing greater tension on the tendon than is a parallel-fiber muscle. This capability derives from the greater number of fibers between the *origin* and *insertion* of fibers in a pinnate muscle. The trade-off, of course, is that in a pinnate muscle the muscle fibers will be shorter than in a parallel-fiber muscle of similar mass and proportions and therefore will not be capable of producing as rapid an overall contraction. Thus, parallel-fiber muscles tend to occur where speed of contraction is important, whereas pinnate arrangements are favored where force is at a premium.

In addition to the principles of skeletal mechanics and the architecture and properties of the muscles, the elastic properties of the tendons are important for musculoskeletal function. Until recently, the considerable elasticity and extensibility of tendons were largely overlooked. It has become evident that long tendons, in particular, undergo considerable stretching during muscle contraction and contribute significantly not only to the performance of the muscle but to the mechanics of locomotion. A tendon stores kinetic energy when it is stretched. This potential energy is imparted

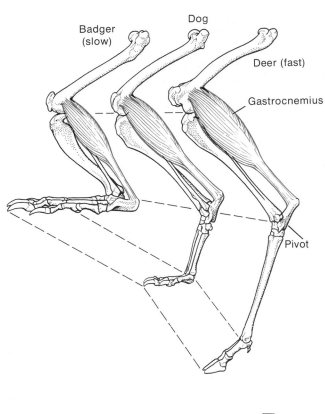

**9-49** Adaptations for strength and speed in skeleton and musculature of three mammals. Each is scaled so that the femurs are all of the same length. The lengthened metatarsals of the deer and dog result in a longer foot and altered ankle posture, which afford these two species a much higher running speed than the badger. The legs of the badger have evolved an efficient structure for the strength required to dig burrows. Note the differences in the ratio of the length of the foot to the length of the heel bone. The greater this ratio, the higher the speed of movement of the toes in response to a given contraction of the gastrocnemius. [Hildebrand, 1960.]

**9-50** Three basic arrangements by which muscles work against a load by moving one skeletal member against a pivot provided by its articulation with another member. [Goldspink, 1977.]

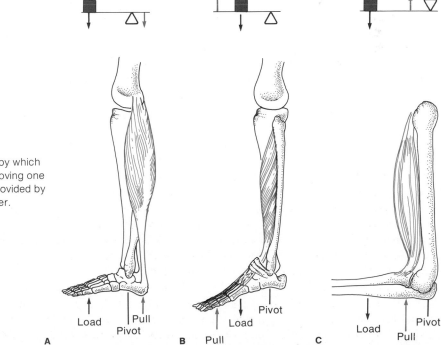

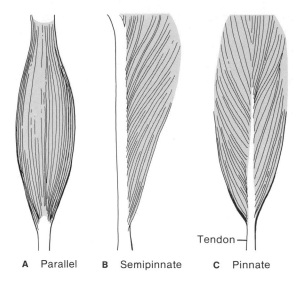

**A** Parallel   **B** Semipinnate   Tendon—   **C** Pinnate

to the tendon by the contraction of its muscle and by outside forces such as the momentum of the body acting to bend a joint. For example, when a biped runs, the energy stored in the stretched tendon is used to propel the heel away from the ground (Figure 9-52). An extreme form of this occurs in the bounding of a kangaroo.

9-51   Variations in muscle-fiber arrangements. Assuming similar overall length and mass, the pinnate configuration will produce greater force on the tendon, while the parallel arrangement will produce higher velocity of contraction. [Goldspink, 1977.]

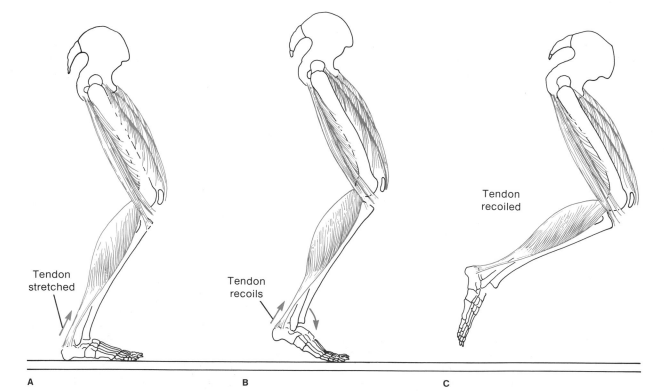

Tendon stretched

Tendon recoils

Tendon recoiled

**A**          **B**          **C**

9-52   Mechanical-energy storage and release in tendons during locomotion. The Achilles tendon is stretched (A) by combination of muscle contraction and planting of the foot during running. The tendon recoils (B) while the muscle remains contracted, propelling the foot downward. The recoil is completed (C) as the foot is lifted clear. [Courtesy of G. Goldspink.]

Another example of energy storage in elastic elements occurs in the ballistic jump of some insects such as fleas, grasshoppers, and crickets. For instance, before a grasshopper jumps, the leg is fully flexed when the body is close to the ground. The leg extensors and flexors are then *coactivated* (i.e., excited to contract against one another) while the leg remains in the same position. In this way a great deal of energy is stored in the tendons, muscles, and exoskeleton of the "cocked" leg. In response to sensory alarm input through visual, auditory, and tactile organs, a set of inhibitory interneurons to the flexor motoneurons is activated, causing a sudden relaxation of the flexor muscles of the leg. This relaxation allows the stored tension in the extensors and associated elastic structures to extend the leg with great speed and force, propelling the insect into a ballistic arc.

## Summary

Muscles are classified into two major types: striated and smooth. Because striated muscle has been so intensively studied, its structure is perhaps better understood than that of any other tissue. Its striated appearance is the result of the regular arrangement of parallel myofilaments, which produce the banded sarcomeres. The sarcomere consists of myosin and actin filaments. Myosin filaments, which make up the A band, interdigitate and slide between the thinner actin filaments, which make up the I bands. During muscle activity, active sliding is produced by interactions between the actin filaments and cross bridges that project from the myosin filaments. The head of the myosin cross bridge exhibits ATPase activity. Hydrolysis of ATP, which requires $Mg^{2+}$, produces a conformational change that permits the cross bridge to undergo a cycle of attachment, rotation, and detachment on the actin filament. Following each cycle, the myosin head reattaches farther along the actin filament. The force-generating rotation of the myosin head against the actin is believed to result from successive attachment of several sites on the myosin head to the actin filament. As successive sites attach and detach, the rotation of the head against the actin filament pulls the cross-bridge link. This link, in turn, pulls on the thick filament from which the cross bridge arises, causing the myosin filament to slide past the thin filament toward the end of the sarcomere. Because this happens symmetrically at both ends of the thick filaments, the sarcomere shortens. A number of classic muscle properties can now be understood in the light of this sliding-filament model of muscle contraction.

At rest, attachment of the myosin heads is prevented by steric hindrance of the actin sites by tropomyosin, a long protein molecule associated with the actin filament. When the muscle cell is depolarized, as by an action potential, calcium ions are released from within the sarcoplasmic reticulum. Calcium ions entering the sarcoplasm then bind to troponin C, a part of a group of globular protein molecules attached to both the actin and the tropomyosin. The $Ca^{2+}$ effects a conformational change in the troponin, and this change then moves the tropomyosin molecule so that it no longer hinders the binding of the myosin cross-bridge heads to the actin filament. In that way, calcium regulates contraction in vertebrate striated muscle. As the surface membrane repolarizes, the sarcoplasmic reticulum (SR) begins pumping up the $Ca^{2+}$ again, removing it from the troponin, terminating the active state of the muscle, and causing the muscle to relax.

The electrical activity of the surface membrane is carried into the cell interior by transverse (or T) tubules, which are thin invaginating projections of the surface membrane. These come into close apposition with the lateral cisternae of the collars of SR that wrap around each sarcomere of each myofibril. Although the nature of the signal transmitted from T tubules to lateral cisternae is not known, it is evident that depolarization of the T tubules leads to the release of $Ca^{2+}$ from the SR. The rise in free $Ca^{2+}$ in the sarcoplasm is the final signal that couples activation of the contractile mechanism to depolarization of the surface membrane.

The control of muscle tension by the nervous system has evolved in several ways in different animal groups. Vertebrate striated muscle responds to single motoneuron impulses with all-or-none twitches, because it contracts in response to all-or-none action potentials. Many arthropod striated muscle fibers give graded contractions in response to graded, nonconducted depolarizations at synapses distributed along each muscle fiber. Most arthropod muscle fibers have an inhibitory innervation, in addition to being innervated by several kinds of excitatory motor axons.

Smooth-muscle cells are usually spindle-shaped, and they are electrically coupled to one another in groups. They show both slow-wave electrical activity and action potentials in response to stretch and to diffuse transmitter release by innervating nerves. Actin and myosin fibers are present, but not in the organized manner characteristic of striated muscle. The calcium that activates contraction enters primarily from the cell exterior during depolarization; this entry is feasible in smooth muscle because the contractions are slow and because the small cells have a large surface/volume ratio.

Vertebrate cardiac muscle is organized at the ultrastructural level much like skeletal striated muscle. It differs in the following ways:

*1.* The fibers of cardiac muscle consist of many short individual cells that are electrically coupled to each other for through-conduction of the action potential. In skeletal mus-

cle, the embryonic cells fuse into long cylindrical fibers, losing their integrity as individual cells.

*2.* The ionic mechanisms of cardiac muscle are specialized for pacemaker activity in the atrial wall and for prolonged action potentials in the ventricular wall.

The mechanical properties of a musculoskeletal system depend on a number of factors. These include:

*1.* The innate contractile properties of the muscle, which depend on the ATPase activity of the myosin head, the length and overlap of thick and thin filaments, the rate of $Ca^{2+}$ release and uptake by the SR, and so on. These factors are under the developmental control of the motoneuron that innervates the muscle fiber.

*2.* The architecture of the muscle and the skeletal elements will determine the mechanics of movement. Thus, a long muscle will shorten at its end faster than will a short muscle. The positions of muscle attachment relative to the pivot point and load of the bone it moves will determine the speed and force of movement of the appendage.

## Exercises

*1.* Distinguish each of the following levels of muscle organization: myofilaments, myofibrils, muscle fibers, and muscle.

*2.* What kinds of evidence led to the sliding-filament hypothesis?

*3.* Draw and label from memory the various components of the sarcomere.

*4.* Discuss the functions of myosin, actin, troponin, and tropomyosin.

*5.* Why do muscles become stiff several hours after an animal dies?

*6.* According to prevailing theory, how is the force responsible for sliding produced by the cross bridges?

*7.* Explain how calcium regulates muscle contraction.

*8.* List the steps involved in muscle contraction and relaxation.

*9.* Outline the evidence for the coupling function of the sarcotubular system.

*10.* How is depolarization of the surface membrane of a muscle fiber communicated to the calcium-sequestering reticulum?

*11.* A muscle attached to a 50 g weight requires more time before it lifts the weight off a table than it requires to lift a 5 g weight. Explain why.

*12.* Discuss the factors that determine the production of heat during contraction.

*13.* What limits the tension attainable by a myofibril? by a muscle?

*14.* Why does velocity of shortening decrease as heavier loads are placed on a muscle?

*15.* Explain the effect of the series elastic components on tension developed in muscle.

*16.* Define the active state of muscle.

*17.* How is a vertebrate twitch muscle, the fibers of which are activated all-or-none by single motor impulses, controlled so as to give graded contractions?

**18.** How do arthropods achieve graded muscular contraction with only a very limited number of motoneurons?

**19.** What factors determine frequency of contraction in insect fibrillar muscle?

**20.** Discuss the major functional differences between skeletal, smooth, and cardiac muscle.

**21.** What factors contribute to the intrinsic speed of shortening and the endurance of a muscle fiber?

**22.** All things being equal, a pinnate muscle develops higher tension at the tendon than a parallel-fiber muscle. Explain why.

**23.** The grasshopper jump is initiated by the relaxation of a muscle. Explain how.

## Suggested Reading

Alexander, R. McN., and G. Goldspink, eds. 1977. *Mechanics and Energetics of Animal Locomotion.* London: Chapman and Hall.

Bendall, J. R. 1969. *Muscles, Molecules, and Movement.* New York: Elsevier.

Carlson, F. D., and D. R. Wilkie. 1974. *Muscle Physiology.* Englewood Cliffs, N.J.: Prentice-Hall.

Cohen, C. 1975. The protein switch of muscle contraction. *Scientific American* 233(5): 36–45.

Duncan, C. J., ed. 1976. *Calcium in Biological Systems.* New York: Cambridge University Press.

Ebashi, S. 1980. Contractile proteins of cardiac muscle. *Proc. Roy. Soc.* (London) Ser. B. 207:259–286.

Ebashi, S., K. Maruyama, and M. Endo, eds. 1980. *Muscle Contraction: Its Regulatory Mechanisms.* New York: Springer-Verlag.

Hanson, J., and H. E. Huxley. 1955. The structural basis of contraction in striated muscle. *Symp. Soc. Exp. Biol.* 9:228–264.

Huddart, H. 1975. *The Comparative Structure and Function of Muscle.* New York: Pergamon.

Huxley, A. F., and R. E. Taylor. 1958. Local activation of striated muscle fibres. *J. Physiol.* 144:426–441.

Huxley, H. E. 1969. The mechanism of muscular contraction. *Science* 164:1356–1365.

Margaria, R. 1976. *Biomechanics and Energetics of Muscular Exercise.* Oxford: Clarendon.

*The Mechanism of Muscle Contraction.* 1973. *Cold Spring Harbor Symp. Quant. Biol.* Vol. 37.

Tonomura, Y. 1973. *Muscle, Proteins, Muscle Contraction, and Cation Transport.* Baltimore: University Park Press.

Trueman, E. R. 1975. *The Locomotion of Soft-bodied Animals.* London: Arnold.

Weber, A., and J. M. Murray. 1973. Molecular control mechanisms in muscle contraction. *Physiol. Rev.* 53:612–673.

Wessells, N. K., ed. 1968. *Vertebrate Adaptations: Readings from Scientific American.* San Francisco: W. H. Freeman and Company.

## References Cited

Ashley, C. C. 1971. Calcium and the activation of skeletal muscle. *Endeavor* 30(109):18–25.

Ashley, C. C., and E. B. Ridgway. 1970. On the relationships between membrane potential, calcium transient, and tension in single barnacle muscle fibres. *J. Physiol.* 209:105–130.

Aubert, X. 1956. La relation entre la force et la vitesse d'allongement et de raccourcissement du muscle strié. *Arch. S. Int. Physiol. Biochem.* 64:121.

Batham, E. J., and C. F. A. Pantin. 1950. Muscular and hydrostatic action in the sea anemone *Metridium senile* (L.). *J. Exp. Biol.* 27:264–289.

Bloom, W., and D. W. Fawcett. 1968. *A Textbook of Histology.* 9th ed. Philadelphia: Saunders.

Brooks, C. M. C., B. F. Hoffman, E. E. Suckling, and O. Orias. 1955. *Excitability of the Heart.* New York: Grune and Stratton.

Close, R. 1971. Neural influences on physiological properties of fast and slow limb muscles. In R. J. Podolsky, ed., *Contractility of Muscle Cells and Related Processes.* Englewood Cliffs, N.J.: Prentice-Hall.

Davson, H. 1964. *A Textbook of General Physiology.* 3rd ed. Boston: Little, Brown.

Diamond, J., and J. M. Marshall. 1969. A comparison of the effects of various muscle relaxants on the electrical and mechanical activity of rat uterus. *J. Pharma. Exp. Ther.* 168:21–30.

Ebashi, S., M. Endo, and I. Ohtsuki. 1969. Control of muscle contraction. *Quart. Rev. Biophys.* 2:351–384.

Goldspink, G. 1977. Design of muscles in relation to locomotion. In R. McN. Alexander and G. Goldspink, eds., *Mechanics and Energetics of Animal Locomotion.* London: Chapman and Hall.

Gordon, A. M., A. F. Huxley, and F. J. Julian. 1966. The variation in isometric tension with sarcomere length in vertebrate muscle fibres. *J. Physiol.* 184:170–192.

Heilbrunn, L. V. 1940. The action of calcium on muscle protoplasm. *Physiol. Zool.* 13:88–94.

Heilbrunn, L. V., and F. J. Wierczinski. 1947. The action of various cations on muscle protoplasm. *J. Cell. Comp. Physiol.* 29:15–32.

Hellam, D. C., and R. J. Podolsky. 1967. Force measurements in skinned muscle fibres. *J. Physiol.* 200:807–819.

Hildebrand, M. 1960. How animals run. *Scientific American* 202(5):148–157.

Hill, A. V. 1938. The heat of shortening and the dynamic constants of muscle. *Proc. Roy. Soc.* (London) Ser. B. 126:136–195.

Hill, A. V. 1948. On the time required for diffusion and its relation to processes in muscle. *Proc. Roy. Soc.* (London) Ser. B. 135:446–453.

Hodgkin, A. L., and P. Horowicz. 1957. The differential action of hypertonic solutions on the twitch and action potential of a muscle fibre. *J. Physiol.* 136:17–18.

Hodgkin, A. L., and P. Horowicz. 1960. Potassium contractures in single muscle fibres. *J. Physiol.* 153:386–403.

Hoyle, G. 1967. Specificity of muscle. In C. A. G. Wiersma, ed., *Invertebrate Nervous Systems.* Chicago: University of Chicago Press.

Huxley, A. F., and R. Niedergerke. 1954. Structural changes in muscle during contraction: Interference microscopy of living muscle fibres. *Nature* 173:971–973.

Huxley, A. F., and R. M. Simmons. 1971. Proposed mechanism of force generation in striated muscle. *Nature* 233:533–538.

Huxley, H. E. 1963. Electron microscope studies on the structure of material and synthetic protein filaments from striated muscle. *J. Molec. Biol.* 7:281–308.

Huxley, H. E., and J. Hanson. 1954. Changes in the cross-striations of muscle during contraction and stretch and their structural interpretation. *Nature* 173:973–976.

Jewell, R. R., and J. C. Rüegg. 1966. Oscillatory contraction of insect fibrillar muscle after glycerol extraction. *Proc. Roy. Soc.* (London) Ser. B. 164:428–459.

Kamada, T., and H. Kinosita. 1943. Disturbances initiated from naked surface of muscle protoplasm. *Jap. J. Physiol.* 10:469–493.

Keynes, R. D., and D. J. Aidley. 1981. *Nerve and Muscle.* Cambridge: Cambridge University Press.

Lehninger, A. L. 1975. *Biochemistry.* 2nd ed. New York: Worth.

Marshall, J. M. 1962. Regulation of activity in uterine smooth muscle. *Physiol. Rev.* 42:213–227.

Miller, A., and R. T. Tregear. 1971. X-ray studies on the structure and function of vertebrate and invertebrate muscle. In R. J. Podolsky, ed., *Contractility of Muscle Cells and Related Processes.* Englewood Cliffs, N.J.: Prentice-Hall.

Morad, M., and R. Orkand. 1971. Excitation–contraction coupling in frog ventricle: Evidence from voltage clamp studies. *J. Physiol.* 219:167–189.

Peachey, L. D. 1965. Transverse tubules in excitation–contraction coupling. *Fed. Proc.* 24:1124–1134.

Ringer, S., and D. W. Buxton. 1887. Upon the similarity and dissimilarity of the behavior of cardiac and skeletal muscle when brought into relation with solutions containing sodium, calcium, and potassium salts. *J. Physiol.* 8:288–295.

Smith, D. S. 1965. The flight muscles of insects. *Scientific American* 212(6):76–88.

Szent-Györgyi, A. 1947. *Chemistry of Muscular Contraction.* New York: Academic.

Threadgold, L. J. 1967. *Ultra-structure of the Animal Cell.* New York: Pergamon.

Vander, A. J., J. H. Sherman, and D. S. Luciano. 1975. *Human Physiology.* 2nd ed. New York: McGraw-Hill.

Wilkie, D. R. 1968. *Muscle.* London: Arnold.

# CHAPTER *10*

# *Motility of Cells*

*L*iving matter is constantly in motion: cytoplasm streams; cells divide; chromosomes separate; vesicles undergo exocytosis; embryonic cells creep in choreographed migration, giving rise to the morphogenetic changes of embryogenesis; and unicellular organisms swim about through the beating of miniature hairlike appendages. Most or all of these movements, together with muscle contraction, are closely related to one another at the molecular level. One needs only consider the swimming of a sperm cell and the separation of chromosomes during metaphase to recognize the fundamental importance of such so-called primitive types of cell motility to the survival of all eukaryotic species.

Cell movement other than muscle contraction, in the form of swimming sperm cells, was first observed through a crude microscope by Anton van Leuwenhoek over 300 years ago, but only recently have methods been developed that help us understand the mechanisms of cell movements. We shall begin with the molecules involved in cell motility, and then go on to mechanisms of cytoplasmic motility and amoeboid movement and finally to ciliary and flagellar movement.

## *The Molecules of Motility*

To our knowledge, all active cell movement originates in the interactions of only a few different classes of protein molecules. The unit proteins are assembled into long filamentous polymers. Those that take an active and direct part in generating movement are collectively termed *contractile proteins,* even though the molecules themselves appear not to shorten during contraction. Contraction or contractionlike forces are produced when two or more species of contractile proteins interact and move relative to one another. It has been suggested that cell movement can also result when a single species of monomers undergoes rapid polymerization into filaments. How such polymerization might produce contractile force is not understood.

In addition to the contractile proteins, there are *regulatory proteins* and, frequently, smaller molecules such as $Ca^{2+}$ or cyclic nucleotides (p. 463) involved in the activation or control of most contractile processes. Cell motility also requires a source of chemical energy such as ATP or GTP. We will briefly consider each category of contractile protein and regulatory protein, keeping in mind that each category may be represented by molecules of slightly different structure in different species and in different kinds of cells in a given animal species.

## CONTRACTILE PROTEINS

### Actin

The organization of striated muscle actin was discussed in Chapter 9. In 1960, S. Hatano and F. Osawa first demonstrated the presence of actin in a nonmuscle tissue, in this case the slime mold *Physarum polycephalum*. Since then, several forms of actin, similar to but not precisely identical with the actin found in muscle, have been found in different cell types throughout the animal and plant kingdoms. Amoebas and motile amoeboid cells such as fibroblasts and blood platelets have especially large amounts of this protein—up to 15% of their total protein. Actin filaments, similar in most respects to the actin filaments of the muscle sarcomere, make up the ubiquitous *microfilaments,* a major class of subcellular building blocks, characterized by a diameter of about 6 nm.

The presence of actin in a cell can be determined in several ways. Antibodies to muscle actin react against microfilaments in other cells. The isolated actin monomers exhibit amino acid composition, electrophoretic migration, and sedimentation properties similar to those of muscle actin. In addition, monomers of heavy meromyosin (p. 346), when added to actin filaments, "decorate" the filaments by attaching in an orderly and characteristic angled and spiral manner to myosin-selective binding sites on the actin filaments, as seen in the electron microscope (Figure 10-1). Those sites presumably correspond to the actin sites in muscle to which the myosin heads attach when the cross bridges are active during muscle contraction. H. E. Huxley had found that the actin filaments of the sarcomere have a polarized organization that causes the myosin heads on the decorated actin filament to attach so as to produce arrowhead profiles all pointing away from the Z line. Subsequent workers found that nonmuscle actin, when decorated with myosin, shows a similar polarized organization of arrowhead profiles along the actin filament. This organization can be seen in the microvilli that project into the lumen of

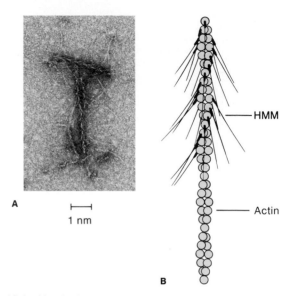

**10-1** Myosin-decorated actin filaments. (A) Actin filaments, isolated from the slime mold *Physarum polycephalum,* show arrowhead figures produced by addition of heavy meromyosin (HMM) isolated from muscle. [Courtesy of V. Nachmias and D. Kessler.] (B) Diagram of HMM attached to actin filaments so as to produce the arrowhead appearance. Note the polarized (i.e., unidirectional) nature of the HMM orientation on the actin. [Holberton, 1977.]

the vertebrate intestine from epithelial absorptive cells (Figures 10-2 and 15-18). The interior of each microvillus contains large numbers of parallel actin filaments that are anchored to the surface membrane of the apical end of the microvillus. When isolated myosin heads are allowed to react with the actin filaments, the myosin decorates the filaments so that the arrowheads point toward the cell interior (Figure 10-2A). Thus, all the actin filaments in the microvillus are structurally polarized in the same direction. As we will see later, this polarization enables actin filaments to produce contractions of the microvilli.

Actin filaments can run through the cytoplasm over long distances as bundles (Figure 10-3). Bundles of actin filaments originate at the inner surface of the cell membrane at specializations termed *adhesion plaques.* In addition, many individual actin filaments are attached to the inner surface of the membrane throughout many types of cells. In some cells, actin filaments

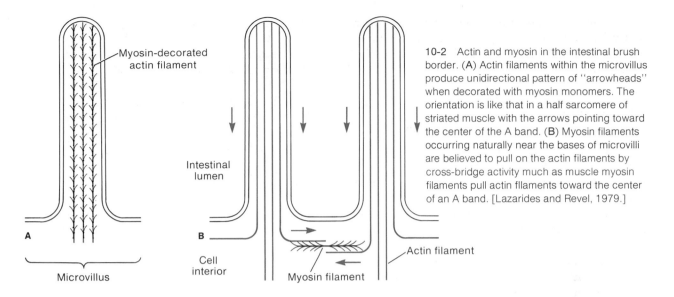

**10-2** Actin and myosin in the intestinal brush border. (A) Actin filaments within the microvillus produce unidirectional pattern of "arrowheads" when decorated with myosin monomers. The orientation is like that in a half sarcomere of striated muscle with the arrows pointing toward the center of the A band. (B) Myosin filaments occurring naturally near the bases of microvilli are believed to pull on the actin filaments by cross-bridge activity much as muscle myosin filaments pull actin filaments toward the center of an A band. [Lazarides and Revel, 1979.]

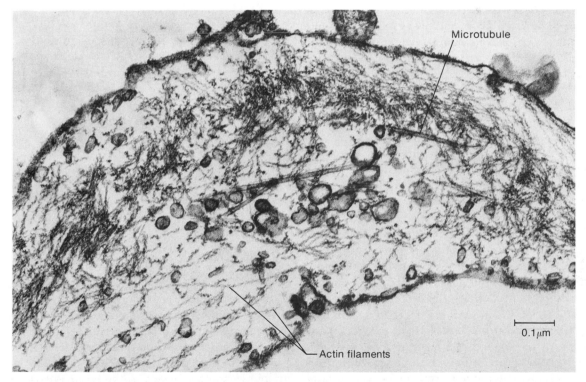

**10-3** Electron micrograph of mammalian blood platelet, showing cytoplasmic microfilaments and microtubules. The preparation was labeled with heavy meromyosin prepared from rabbit muscle (see Figure 10-1). Because these microtubules selectively bound the myosin, a behavior that is an important diagnostic characteristic of actin, the microfilaments were identified as actin filaments. The microtubules are fewer, thicker, and stiffer in appearance than the microfilaments. [Courtesy of J. L. Travis.]

form amorphous meshworks beneath the cell membrane. Microfilament bundles can, in fact, rapidly dissociate into their constituent actin monomers, which then reorganize into the fibrous meshwork form, and vice versa. Different species of actin have different ionic requirements for polymerization and depolymerization, some polymerizing and others depolymerizing with increases in the free $Ca^{2+}$ concentration.

*Myosin*

Long after its discovery in muscle, myosin was first identified in nonmuscle cells (slime mold, again) by S. Hatano and M. Tazawa at Osaka University in the late 1960s. Myosins capable of reacting with actin were subsequently described in a variety of cell types. The different myosin molecules exhibit differences in their ATPase activities that suggest parallels to their functional roles. Thus, the rate of ATP hydrolysis by amoeba myosin is slow compared to that of muscle myosin, reflecting that the amoeba "contracts" more slowly than does striated muscle. Several types of both myosin and actin have been found to coexist in the same cell type, and some of these have different regulatory ion requirements for polymerization and ATPase activity. This variety allows for a number of actin–myosin combinations, possibly for different contractile functions in the same cell.

The filaments formed when nonmuscle myosin polymerizes are thinner and shorter than the myosin filaments of muscle. This makes them hard to distinguish from actin filaments in the electron microscope. The technique of *immunofluorescence* has therefore been especially useful in studying the distribution of contractile proteins. In this immunological approach, molecules of a known protein are used to produce antibodies against that protein. The antibodies are tagged with a fluorescent "label," a molecule that fluoresces (i.e., emits light when excited by shorter wavelength illumination). With this technique, it was possible to identify myosin filaments in a number of nonmuscle cell types viewed under the light microscope. In a related approach, useful for the electron-microscopic identification of these molecules, the antibodies are tagged with an electron-opaque molecule such as *ferritin*, a large iron-containing protein. The association of the ferritin with the antigen-containing organelles

can be seen in electron micrographs. This method has recently been used to show that myosin molecules are present near the bases of intestinal microvilli, but not within the microvilli themselves.

As in muscle, the nonmuscle myosin filaments are assembled from myosin monomers, with the myosin heads pointing away from the filament midregion toward each end so as to produce the bipolar arrangement that allows each half of the myosin filament to cross-link with and slide against actin filaments running in opposite directions. This arrangement provides a basis in nonmuscle cells for a sliding-filament mechanism of contraction similar to that of striated muscle.

*Tubulin*

Tubulin *heterodimers* (units consisting of two dissimilar monomers joined together) assemble into *microtubules* in the cytoplasm, using guanosine triphosphate (GTP) as an energy source. Microtubules have a diameter of 25 nm and form the mitotic spindle "fibers," axonal microtubules, presynaptic microtubules, parts of the centrioles, parts of the ciliary and flagellar apparatus, and numerous other intracellular structures. The plant-derived drug *colchicine* interferes with the assembly of tubulin into microtubules and, by shifting the dimer–polymer equilibrium, can cause microtubules (other than those of cilia and flagella) to disassemble. Whereas *actin forms filaments that are able to resist tension* (which enables them to "pull"), *tubulin builds cylindrical structures that can resist compression* (and are thus able to "push") as well as resist tension.

## REGULATORY PROTEINS

*Tropomyosin*

This short protein filament is closely associated with actin in muscles ranging from those of mollusks to those of mammals. As explained in Chapter 9, tropomyosin, along with the troponins, plays a key role in regulating the interaction of myosin and actin filaments in striated muscle of vertebrates. Tropomyosin has not, however, been found in conjunction with actin in motile nonmuscle cells, although it does make its appearance when the same cells become immobile. It is therefore speculated that tropomyosin in nonmuscle cells is associated with actin only in its cross-linked "skeletal" (i.e., cell-shape maintaining) state.

*Troponin*

This group of proteins, which has not been found outside of muscle, is discussed on page 356.

*Calmodulin*

This troponin C–like calcium-binding protein is found in all animal cells. In its calcium-complexed form, it regulates the activity of many calcium-dependent enzymes and other proteins. (It is discussed at greater length in Chapter 11.)

*Alpha-actinin*

This protein was first discovered in the meshwork structure of the Z lines of the muscle sarcomere, where its function appears to be purely structural. Fluorescent antibodies made against muscle α-actinin have been used to localize actinin in close association with actin in nonmuscle cells such as fibroblasts and epithelial cells, especially at the adhesion plaques. It appears to play a role in the attachment of actin filaments to the cell membrane. There is no evidence that actinin is involved directly in the contractile process.

*Filamin*

This actin-binding protein occurs in smooth muscle and nonmuscle cells localized in a periodic manner along actin-filament bundles. It is involved in the aggregation of actin filaments into a gel that is important in giving rigidity to the cytoplasm.

*Profilin*

In spleen cells (and perhaps in other cells as well), monomeric actin exists free in the cytosol in association with a small protein molecule called *profilin*. Removal of the profilin unit from the monomeric actin allows the actin to polymerize into filaments.

## CHEMISTRY OF CYTOPLASMIC MOTILITY

The actins and myosins isolated from muscle and from various nonmuscle cells cross-react with one another. That is, muscle myosin will decorate actin filaments isolated from other cell types, and vice versa. This observation, together with biochemical characterization, indicates that the myosins and actins found in a variety of cells are basically similar. However, there are some differences in their behavior. The contraction

of actin and myosin in skeletal muscle is *actin-regulated.* That is, $Ca^{2+}$ regulates the actin–myosin interaction through regulatory proteins (i.e., troponin and tropomyosin) associated with the actin filament. In vertebrate smooth muscle and many invertebrate muscles, as well as in amoeboid cells, regulation of the actin–myosin interaction has been found to depend on the binding of $Ca^{2+}$ to the "light chain" of the myosin head, and so these systems are said to be *myosin-regulated.*

The interactions of contractile and regulatory proteins in producing cytoplasmic movement are poorly understood, and their elucidation is hampered by the existence of several different mechanisms. One of the most easily understood is the actin–myosin motility of intestinal microvilli, which appears to be similar to the sliding-filament mechanism of muscle contraction. The actin filaments of the microvillus (Figure 10-2B) are believed to slide along myosin filaments located near the bases of the microvilli, in a manner very much like the sliding of actin filaments along myosin filaments in a muscle sarcomere as a result of cross-bridge activity. Since the actin filaments are anchored at the apical end of the microvillus, actin–myosin countersliding causes the microvilli to shorten. Contractions of the microvilli occur rhythmically, helping mix the luminal contents of the small intestine at the absorptive surface of the epithelial lining.

There is evidence that amoeboid movements may arise by similar actin–myosin interactions. However, there is little structural stability in the rather amorphous organization of a locomoting amoeba. Rapid changes in the localization of contracting regions are possible owing to rapid polymerizations and depolymerizations of the contractile proteins in amoeboid cells. Thus, the actin filaments in a region of cytoplasm that at one moment is an actively contracting tail region may at a later time depolymerize into actin monomers as that portion of cytoplasm is displaced and becomes more fluid as it is squeezed forward into an advancing *pseudopodium* ("false foot"—an extension of the amoeboid cell).

The regulation of actin–myosin interactions in motile nonmuscle cells as presently understood is outlined in Figure 10-4. The actin-filament mesh or gel interacts with phosphorylated myosin in the presence

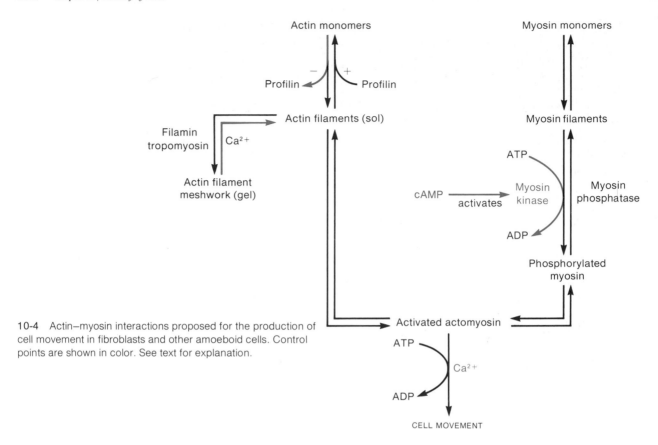

**10-4**   Actin–myosin interactions proposed for the production of cell movement in fibroblasts and other amoeboid cells. Control points are shown in color. See text for explanation.

of calcium ions to produce contraction. The mechanical details (i.e., filament sliding and cross-bridge activity) of contraction are poorly understood in non-muscle cells. It is noteworthy, however, that there are several possible points of control in the scheme shown in Figure 10-4. On the actin side, there is the polymerization of monomeric actin into actin filaments, with the filaments organizing into a meshwork or gel. On the myosin side, there is the phosphorylation of myosin by a cAMP-activated kinase (p. 467). Finally, contraction of the actomyosin complex depends on ATP as an energy source and on $Ca^{2+}$ as a regulatory agent.

### AMOEBOID LOCOMOTION

Simple microscopic observation of locomoting amoebas, such as the large free-living protozoan amoebas *Amoeba proteus* and *Chaos carolinensis,* reveals one kind of amoeboid movement. The vacuolated core of cyto-plasm, termed *endoplasm,* streams forward in the direction of overall movement and into the pseudopodia, which grow forward as the cell gradually flows into them (Figure 10-5). This flow takes place within a peripheral tube of relatively clear, stiffer gel-like cyto-plasm called *ectoplasm.* The latter forms at the tips of the pseudopodia and acts as a jacket within which the less viscous endoplasm streams forward. The endo-plasm becomes gradually more viscous as it advances. As the cell moves about, its direction of locomotion often changes: its flow into one of the pseudopodia stops and reverses direction when the tip of that pseudopodium retracts, and streaming then continues into another pseudopodium, or a new pseudopodium develops.

One long-standing question about amoeboid move-ment concerns the site at which the motile force is gen-erated within the advancing cell. Is the endoplasm pushed forward by a contraction of the ectoplasm at

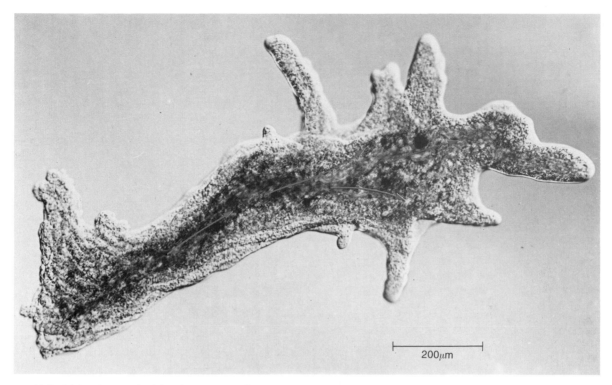

200μm

**10-5**  Light micrograph of the giant amoeba *Chaos carolinensis*. Endoplasm was flowing toward the pseudopodia, as shown by the arrows. Note the ruffled appearance of the retracting trailing parts at the left as compared to the smooth surface appearance of the advancing pseudopodia. [Courtesy of R. D. Allen.]

the rear, like toothpaste squeezed forward in its tube, or is the endoplasm *pulled* forward by forces that arise within the pseudopodium? The distribution of forces within a locomoting amoeboid cell are still not entirely clear and may, in fact, differ among amoeboid cell types. Recent optical experiments by D. L. Taylor and colleagues (1978) on *A. proteus* and *C. carolinensis* have led to a model for the molecular events that give rise to amoeboid movement in these cells. The essence of this model is that actin fibrils, cross-linked by gelation factors such as filamin, form a highly viscous *gel* (Figure 10-6A, left). To interact with myosin and undergo sliding-filament contraction, the gel-state actin must lose its cross-links so as to free the individual filaments. This low-viscosity, unlinked condition is the *sol* state. The individual actin filaments can then interact with myosin filaments to produce sliding-filament contrac-

tion (Figure 10-6B, right). The forces of the contraction are transferred to the surrounding areas through filaments connected to the surrounding gelled actin. Solation of the actin gel requires elevation of free cytoplasmic $Ca^{2+}$ or $H^+$. In the presence of myosin, the actin–myosin sliding occurs at elevated levels of cytoplasmic $Ca^{2+}$, with ATP as the energy source. According to this scheme, contraction cannot occur in the gel state, because the actin filaments are cross-linked and cannot slide against the myosin to produce force. Contraction requires that the actin cross-links first be broken in response to elevated $Ca^{2+}$.

This *solation–contraction hypothesis* needs further study, but is supported by the following observations:

*1.* Cytoplasm isolated from amoebas remains nonmotile, gelled, and nonbirefringent (does not polarize

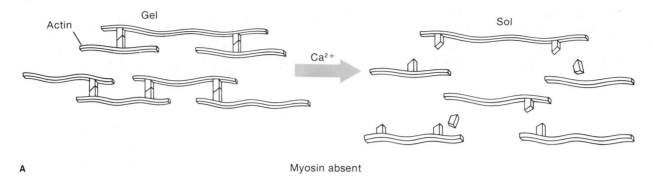

10-6    Roles of actin, cross-linking factors, $Ca^{2+}$, and myosin in sol–gel-contraction transitions. (A) At low cytoplasmic free $Ca^{2+}$, actin is cross-linked by gelation factors (perhaps filamin). Elevation of $Ca^{2+}$ causes gel to break down. (B) With myosin present, contraction due to actin–myosin sliding can occur. [Taylor et al., 1978.]

light, indicating random fibrillar orientation) if the free $Ca^{2+}$ concentration is below $10^{-6}$ M. Elevation of $Ca^{2+}$ to levels of about $10^{-6}$ M causes contraction of the cytoplasm.

*2.* In cytoplasm from which myosin has first been extracted, the organization of the actin is converted from the gel state to the sol state in response to $Ca^{2+}$ elevation, and no contraction takes place (Figure 10-6A). Contraction will, however, occur if myosin is added back to the system (Figure 10-6B).

*3.* When $Ca^{2+}$ is injected into living amoebas, the cytoplasm around the injection pipette contracts.

*4.* Observations on living amoebas injected with the calcium-sensitive photoprotein aequorin indicate that locomotion is associated with local elevations in cytoplasmic free $Ca^{2+}$, especially in the tail regions of the amoebas and in the tips of the pseudopodia.

*5.* Similar optical studies with fluorescent antibodies against actin indicate that elevated concentrations of actin occur preferentially in the tail region and tips of the pseudopodia.

At this time there is no satisfying hypothesis that explains just how the calcium- and myosin-dependent interactions between gel state, sol, state, and contracting state of the actin are coordinated within the cell so as to produce amoeboid movement. It does seem likely from evidence such as items 4 and 5, above, that the local transitions between these states depend, in part, on the local control of the level of cytoplasmic free $Ca^{2+}$.

## Cilia and Flagella

The cilia and flagella of eukaryotic cells arise from cellular organelles homologous with centrioles. They are developmentally and structurally unrelated to bacterial flagella and the nonmotile stereocilia found in eighth-nerve mechanoreceptors of vertebrates. The organelles we will consider typically have a characteristic organization of "9 + 2" tubular substructures. These organelles are generally termed *flagella* (singular, *flagellum*) if relatively long, or *cilia* (singular, *cilium*) if short. There appears to be no fundamental difference in structure or mechanism between a cilium and a flagellum. Generally, cilia occur in large numbers on a cell, whereas flagella occur singly or in pairs.

Flagella are familiar as the motile organelles that propel spermatozoa and flagellates by means of traveling waves (Figure 10-7). Several examples of flagella are also found in epithelia, such as the flame cells of flatworms and the collar cells of sponges. Flagella are about 0.2 $\mu$m in diameter and attain lengths up to 100–200 $\mu$m. Cilia have the same diameter but

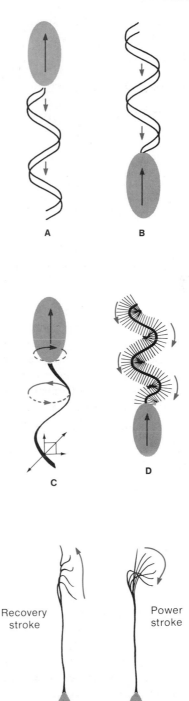

**10-7** Examples of flagellar motion in different protozoa. (A–B) Planar waves. Organism moves in direction opposite to direction of wave propagation. In some species, the flagellum leads (as in B). (C) Helical waves produce more complex force vectors, adding a counterturning spin to the body. (D) Mastigonemes, hairlike filaments projecting normal to the surface of the flagellum of *Mastigamoeba setosa,* reverse the direction of thrust, causing the cell to be propelled in the same direction as flagellar wave propagation. This reversal occurs because the mastigonemes projecting on the outside of the curvature of a traveling wave swing in the direction opposite to the movement of the wave (curved arrows). (E) *Peranema trichophorum* exhibits beating movements restricted to the terminal portion of its flagellum. The power stroke is shown at the right, propelling the cell forward. The recovery stroke, which exerts less force on the medium, is shown at left. [A after Jahn et al., 1964; B–C, Grell, 1973; D, Jahn and Votta, 1972; E, Jahn and Bovee, 1967.]

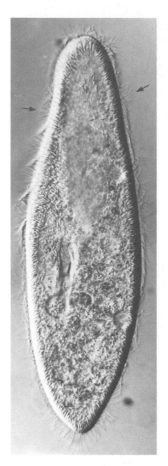

10-8   Photomicrograph of a swimming *Paramecium multimicronucleatum* in interference contrast, showing metachronal waves of ciliary activity. Arrows indicate cilia performing power stroke toward rear (bottom of figure), thereby propelling specimen forward. [Machemer, 1972.]

up and out of the respiratory system. Tobacco smoke paralyzes the cilia, preventing this function when most needed. The cilia of the mammalian fallopian tubes are important in transporting the ovum.

## TYPES OF MOVEMENTS

There are several types of beating patterns, all of which can generally be ascribed to bending or propagated bending. The most easily described are those in which all the bending takes place in two dimensions rather than in three. This kind of movement has been analyzed in cilia with the aid of high-speed photography (Figure 10-9). The beating cycle can be divided into two parts, a *power* (or *effective*) *stroke* and a *recovery stroke*. During the power stroke, the cilium remains relatively straight and develops a bend near its base as it sweeps through the water. During the recovery stroke, the bend formed near the base enlarges and propagates along the cilium, as the basal portion swings in the recovery direction. The result is that the cilium applies a minimum of force to the extracellular medium on its return to the starting position for the next power stroke, which is executed at a higher velocity than the recovery stroke. These movements result in a vector of net force applied to the extracellular medium.

Planar, or two-dimensional, movement is characteristic of epithelial cilia in metazoans, especially those in which many long, closely grouped cilia function together to form a *cirrus* (found in clam gills) or *comb plate* (in ctenophores). The cilia of protozoa and some epithelia beat in a more complex three-dimensional pattern. The power stroke is nearly planar, but during the recovery stroke the cilium departs from that plane of movement, leaning far over toward one side, close to the cell surface (Figure 10-10).

Flagella show several forms of movement (Figure 10-7), some with planar waves and some with helical movements. At first glance, one might suppose that the undulatory movements of flagella are due to passive traveling waves set up by active movements of the basal end of the flagellum—analogous to the motion of a whip. But this is apparently not so. Instead, bending forces are generated along the entire length of the organelle or along whatever part undergoes movement. Several lines of evidence indicate that this is the case:

are generally less than 15 $\mu$m in length. Cilia on the surface of certain groups of protozoa and invertebrate larvae provide the propulsive energy for locomotion (Figure 10-8), and those on certain epithelial cells of metazoans move mucus-trapped particles over the surface or serve to circulate water to facilitate either feeding or respiration. The cilia of the mammalian trachea and bronchioles serve to move mucus-trapped particles

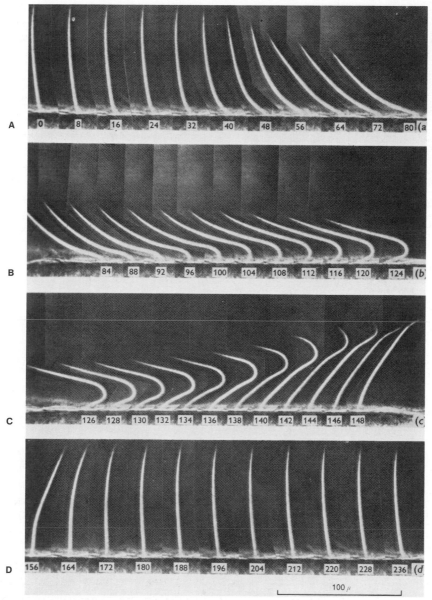

**10-9** High-speed (450 frame/s) cinematographic analysis of the movement of the giant compound cilium found on mussel gills. Frames are numbered below. Frames were cut out and arranged in this montage. Every eighth frame is shown in parts A and D, every fourth in B, and every second in C. The power stroke extends from frame 40 to about frame 84; the recovery stroke extends from about frame 84 to frame 148. Following the recovery stroke, the cilium stands upright until the next power stroke. The movement of this specialized compound giant cilium is largely two-dimensional and thus simpler than the three-dimensional movements characteristic of protozoan cilia. [Baba and Hiramoto, 1970.]

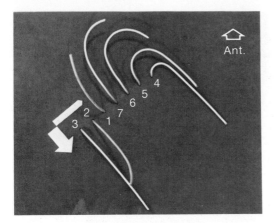

**10-10** Reconstructed stages in the ciliary cycle of *Paramecium*. Seven cilia along a line perpendicular to the metachronal wave are numbered in the sequence exhibited by a single cilium. The power stroke (steps 1–3) is planar in form; the recovery stroke (steps 4–7) takes place close to the cell surface, with the cilium leaning to one side. The small arrow shows the direction of the metachronal wave. The anterior of the ciliate is indicated by the arrow marked ''Ant.'' Based on photomicrographs of beating cilia. [Machemer, 1972.]

*1.* If passive waves were set up by active movement near the base, they would show progressively diminishing amplitude as they move toward the free end. The traveling waves seen in flagella of sea urchin sperm show no such decrease in their amplitude (Figure 10-11).

*2.* Isolated flagella can move independently of their normal attachments.

*3.* In some flagella, the waves begin at the free end and travel toward the base (Figure 10-7B).

*4.* The movement of some flagella is restricted to a small portion of the free end; the proximal region remains immobile and straight (Figure 10-7E).

### FINE STRUCTURE

The major structures in a cilium or a flagellum are microtubules, which extend continuously from one end of the organelle to the other. The hierarchy of microtubular organization is shown in Figure 10-12. In the shaft of the typical cilium or flagellum, nine microtubular doublets form a ring around a central pair of microtubules, which originate above the basal plate. The nine doublets and two central microtubules, together with their associated structures, form the *axoneme*, which is fully enclosed in an evaginated extension of the cell membrane.

The microtubules of the axoneme have a substructure of globular heterodimers, each consisting of one $\alpha$ and one $\beta$ monomer. In the central pair, the tubulin dimers are organized into 13 longitudinal protofilaments per tubule. The outer nine doublets have a similar substructure of tubulin. The dimers are assembled into one complete tubule, *tubule A,* and an attached incomplete *tubule B* (Figure 10-13).

Tubule A of each doublet bears two *arms* that point toward tubule B of the next doublet clockwise around the ring (as viewed from the base toward the tip of

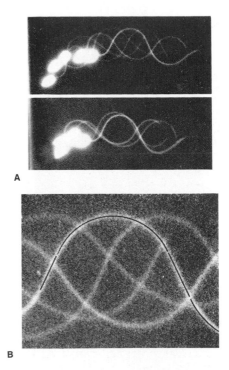

**A**

**B**

**10-11** The beating form of sea urchin sperm tails. (A) The traveling waves in the flagellum remain constant in amplitude throughout the length of the flagellum, as seen in multiple stroboscopic exposures. (B) Straight sections alternate with regions of circular arcs. [Brokaw, 1965.]

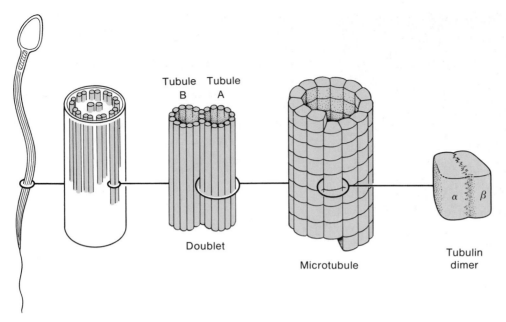

**10-12** Hierarchy of axonemal substructure. The building blocks of the microtubules are tubulin heterodimers arranged in protofilaments. The axoneme, surrounded by the surface membrane, makes up the major portion of the sperm tail. [From ''How Living Cells Change Shape'' by N. Wessells. Copyright © 1971 by Scientific American, Inc. All rights reserved.]

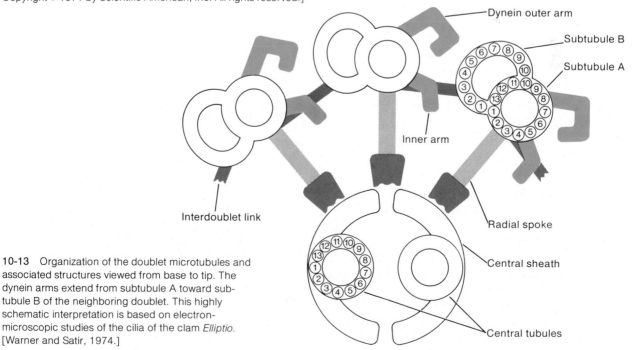

**10-13** Organization of the doublet microtubules and associated structures viewed from base to tip. The dynein arms extend from subtubule A toward subtubule B of the neighboring doublet. This highly schematic interpretation is based on electron-microscopic studies of the cilia of the clam *Elliptio*. [Warner and Satir, 1974.]

the shaft). The arms, which repeat at intervals along tubule A, consist of a protein, *dynein,* which exhibits magnesium-activated ATPase activity. Thus far, dynein is the only component of the axoneme that has been shown to have ATPase activity, but other ATPases in the axoneme have not been ruled out.

Each peripheral doublet also bears a series of *radial spokes.* These have been seen in some flagella to occur in pairs along the entire length of each doublet, as shown in Figure 10-13. Each radial spoke can be seen to make contact with a corresponding protuberance on the *central sheath.* The arrangement of radial spokes along the length of each microtubular doublet is regularly displaced from one doublet to the next. The result is a helical progression of spokes reminiscent of the arrangement of cross bridges emerging from the myosin filaments in striated muscle.

At the base of each cilium or flagellum is a *basal body* (also called *kinetosome*) homologous with the centrioles, organelles that give rise to the microfilaments and tubules (e.g., the spindle fibers of dividing cells). The basal body is a short squat cylinder with an outer wall of nine triplet tubules.

Although 9 + 2 is by far the most prevalent microtubular organization found in cilia and flagella, a number of variations on this fundamental scheme have been reported, including 3 + 0, 6 + 0, 9 + 0, 9 + 1, 9 + 3, and 9 + 7 in different species. There are also instances in which there are more than nine doublets in the peripheral array.

Electron-microscopic studies both on clam-gill cilia, which beat in a single plane, and on cilia of the parasitic protozoan *Opalina,* which beat with a complex three-dimensional movement, indicate that bending of the shaft consistently occurs in a plane passing between the two central tubules and perpendicular to a line passing through both the central tubules. This finding is especially interesting in the cilia of *Opalina,* because they bend in a three-dimensional pattern, which is shifted when the effective stroke changes its direction in response to membrane depolarization (described later). The perpendicular relation between the central pair and the plane of bending holds true near the base as well as near the tip of the cilium. It was recently discovered that in *Paramecium*

the central pair actually undergoes a full 360° rotation during each beat cycle. It remains uncertain whether the central pair determines the direction of bending of the shaft or simply complies passively so as to give the least resistance to bending.

### CHEMISTRY OF CILIA AND FLAGELLA

The localization of the different protein components and ATPase in the axoneme was made possible by the technique of chemical dissection, applied to the cilia of *Tetrahymena* by I. R. Gibbons in 1965. The first step after isolation of the cilia is extraction with a detergent that removes the cell membrane and the amorphous matrix of the ciliary shaft. The remaining axoneme is then dialyzed against a solution of low ionic strength containing EDTA, an agent that chelates both $Ca^{2+}$ and $Mg^{2+}$. This extraction procedure removes the two central tubules and the arms from the outer nine doublets. The ring of nine outer doublets otherwise remains intact. Ultracentrifugation of the dialysate (extract obtained by dialysis) produces two principal protein fractions. One, which sediments slowly, contains the proteins of the two central fibrils; the other fraction consists of the ATPase dynein, which occurs as a monomer and as more rapidly sedimenting polymers of various lengths. Gibbons showed that the enzymatic activity of dynein requires a divalent ion, such as $Mg^{2+}$, and liberates inorganic phosphate from ATP (Figure 13-9). The dynein monomer is a globular protein about 140 Å long and 80 Å in diameter. It polymerizes to form a unit about 400–5,000 Å long.

When the purified dynein fraction is mixed with the extracted axonemes, the dynein reattaches to the A tubules of the outer nine doublets to reconstitute the extracted arms. Because of their location and their ATPase activity, the dynein arms appear to serve a function analogous to that of the cross bridges of striated muscle, utilizing the energy released by hydrolysis of ATP to produce sliding forces between adjacent tubules, as described next.

## Mechanisms of Flagellar Bending

Whereas the contraction of a muscle fiber is simply a shortening in one dimension, the active movements of cilia and flagella are two- or three-dimensional, which makes the mechanism of ciliary movement much more difficult to study than that of muscle. In addition, these organelles are very small and cannot be attached to a mechanical transducer for simple tension measurements. As a result, we do not yet have a thoroughly tested theory for the mechanism of ciliary and flagellar movement. Nevertheless, recent findings suggest a sliding-tubule mechanism as the basis for ciliary and flagellar motility.

### SLIDING-TUBULE HYPOTHESIS

There are two ways in which the ring of nine doublets might behave during bending of the axoneme: (1) The doublets on the inside curvature of the axoneme might *contract*, as illustrated in Figure 10-14A (or the tubules on the outside curvature might expand); or (2) the doublets on the inside curvature of the bending axo-

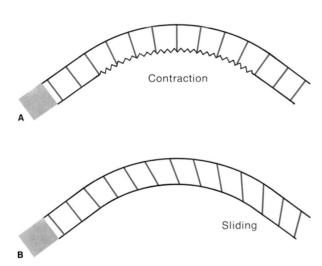

**10-14** Two hypothetical mechanisms to explain the active bending of cilia. **(A)** If the peripheral doublets on one side contract, the axoneme should bend toward the side undergoing contraction. This process would work on the same principle as the bending of a bimetal strip, which changes shape as temperature changes cause differential contraction of the metals. There is no evidence for such differential contraction. **(B)** If the doublets are fixed at the base and the doublets on one side of the axoneme slide actively in relation to the others, the axoneme should bend toward the doublets that tend to slide away from the base. This appears to be the way cilia bend. [Brokaw, 1968.]

neme might *slide* relative to the others while all the doublets maintain a constant length (Figure 10-14B). It now appears that the second of these two possibilities is the basis for ciliary and flagellar mobility. The evidence for this conclusion follows.

P. Satir (1973) has shown that the microtubules of the axoneme slide past one another during bending of the ciliary shaft. This behavior was demonstrated by careful electron-microscopic examination of the terminations of the outer nine doublets in cilia fixed instantaneously with osmium while in various phases of the beating cycle. The microtubules on the inside curvature of the cilium always project beyond those that happen to be located on the outer arc of the curved cilium, regardless of the part of the ciliary cycle in which the cilium is stopped by the fixative (Figure 10-15). Such a projection is precisely what would be expected if the microtubules maintain a constant length while sliding past one another. But this evidence does not in itself distinguish between (1) active sliding between microtubules, which would *cause* bending of the shaft, and (2) passive sliding, which might occur as the *result* of bending produced by another process.

Active sliding would require a physical means for the development of force between apposed or adjacent doublets or between doublets and the central sheath. In muscle, this force is developed with cross bridges from the myosin acting on the actin filaments. The axoneme contains at least two structures that might behave in a manner analogous to that of the cross bridges of muscle. These structures are the dynein arms that project from the A tubules of the outer nine doublets toward the B tubules and the radial spokes that project inward from the A tubules toward the central sheath (Figure 10-13). No ATPase activity has as yet been traced to the radial spokes; therefore, unless such an ATPase can be demonstated, it is unlikely that the radial spokes perform a function analogous to the role of myosin cross bridges in muscle contraction. On the other hand, the ATPase activity of the dynein arms suggests that they are somehow involved in the process of mechanochemical transduction.

K. Summers and Gibbons (1971) demonstrated active sliding of the outer doublets energized by ATP.

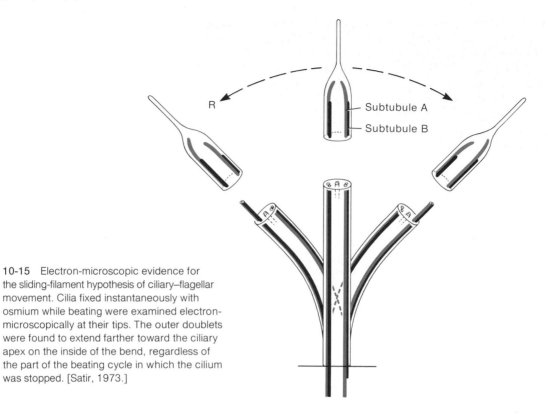

R

Subtubule A

Subtubule B

**10-15** Electron-microscopic evidence for the sliding-filament hypothesis of ciliary–flagellar movement. Cilia fixed instantaneously with osmium while beating were examined electron-microscopically at their tips. The outer doublets were found to extend farther toward the ciliary apex on the inside of the bend, regardless of the part of the beating cycle in which the cilium was stopped. [Satir, 1973.]

Flagella of sea urchin sperm were extracted with the detergent Triton X-100 to remove the surface membrane and the soluble components in the shaft (Box 9-1). The extracted axonemes were then briefly exposed to the enzyme trypsin, which attacks certain peptide bonds. The trypsin treatment modifies the axonemes so that subsequent exposure to ATP and $Mg^{2+}$ causes them initially to elongate and then to dissociate into individual tubules and groups of tubules. Examination with dark-field microscopy revealed that the elongation results from a longitudinal sliding of the nine outer doublets with respect to one another (Figure 10-16), so that the total length of overlapping tubules increases to as much as five times the length of individual tubules. These results may be interpreted as follows. In the living cilium, ATP energizes a sliding action between the dynein arms and the B tubule of the next doublet. This active movement is opposed by an elastic restraint, perhaps the attachment of the nine outer doublets to the central sheath

via the radial spokes. The limited sliding motion produced in this way is responsible for the bending of the flagellum (or cilium). When the trypsin has digested certain structures, unrestrained sliding takes place after addition of $Mg^{2+}$–ATP, producing a progressive elongation of the total structure as one doublet slides along upon another, which in turn slides along upon still another, and so forth (Figure 10-16B). This sliding has the same energetic and ionic requirements as reactivated beating in extracted models.

More recently, W. S. Sales and P. Satir (1977) examined the unrestrained sliding of digested cilia of *Tetrahymena* by electron-microscopic methods. They found that the relative movement generated between adjacent doublets is such that the A tubule, bearing the dynein arms, slides toward the base of the B tubule with which it interacts. Thus, the dynein arms must produce their force by an active swinging movement made in the direction of the ciliary tip, away from the ciliary base.

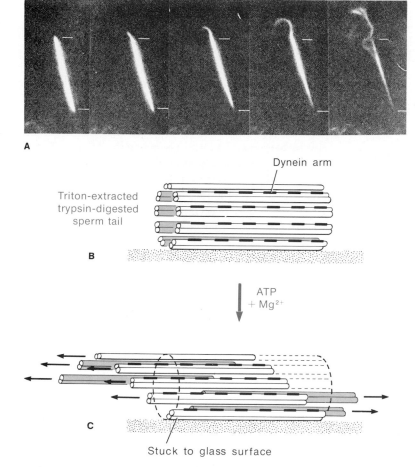

**A**

Dynein arm

Triton-extracted
trypsin-digested
sperm tail

**B**

ATP
+ Mg²⁺

**10-16** Direct evidence for ATP-energized sliding of doublets. The addition of $Mg^{2+}$–ATP to Triton-extracted, trypsin-treated sea urchin flagella causes tubules to slide past one another. (A) Dark-field light micrographs of such a preparation taken at 10–30 s intervals. The two white lines indicate the initial position of each end of the fragment. [Summers and Gibbons, 1971.] (B–C) Schematic interpretation of these findings.

**C**

Stuck to glass surface

It appears, then, that ciliary and flagellar movements are produced by a sliding-tubule mechanism somewhat similar in principle to the sliding-filament mechanism of muscle. A major question that remains at this time is how the radially arrayed set of nine doublets is regulated to undergo sliding movements that produce the various kinds of asymmetrical movements characteristic of beating cilia and flagella.

## PROPAGATION OF BENDING ALONG THE AXONEME

It has been suggested that the cell membrane, which is continuous over the surface of the ciliary shaft, conducts an electric impulse that travels along the shaft and triggers active bending along the way. However, bend propagation by impulses has been ruled out by experiments with glycerinated or detergent-extracted preparations in which the membrane is either physically or functionally eliminated. In such models, bend propagation and other movements of individual cilia are fully retained. A plausible hypothesis that accounts for the propagation of the bend along the shaft of a cilium or flagellum has emerged from the sliding-tubule hypothesis. C. J. Brokaw (1971) has suggested that the propagation of bending along the axoneme of a flagellum can be attributed to a sliding mechanism. If it is assumed that in the inactive segments of the axoneme there is some resistance to

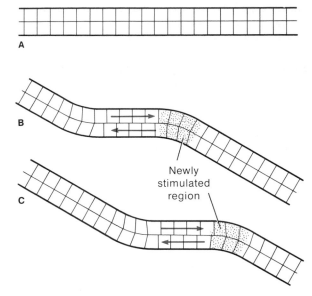

10-17  Diagram illustrating the idea that active sliding in a straight section of the axoneme of a flagellum produces passive bending in opposite directions in adjacent inactive regions. If passive bending activates the sliding mechanism (**B**), the sequence of alternating straight and curved regions will propagate along the axoneme (**C**). See text and Figure 10-18.

relative sliding of the tubules, it will be apparent that active sliding within one section of the axoneme will produce passive bending in opposite directions of the inactive sections on either side of the actively sliding section (Figure 10-17B).

Longitudinal propagation of activity along the axoneme can be explained if it is assumed that the bending of an inactive region stimulates it to become active (Figure 10-17). Sliding forces produced by chemomechanical transduction in the active region will be transferred longitudinally to adjacent inactive regions, causing these to bend passively. The passive bending initiated by nearby active sliding initiates active sliding that bends the next section, and so on, along the length of the axoneme. A mathematical model based on this hypothesis has been used for computer simulation of movement patterns that closely resemble real flagellar movements (Figure 10-18). Sensitivity to mechanical strain is seen in the "giant cilium" found on the gill filament of clams. This cilium (actually a closely packed bundle of very long individual cilia) remains poised indefinitely at the end of its effective stroke under certain ionic conditions (Figure 10-19A). If then given a small passive displacement with a microneedle, the cirrus initiates a complete cycle of active bending (Figure 10-19D–G). Intracellular electrical recordings show no potential changes across the membrane associated with the mechanical stimulus or the active mechanical response. Apparently the mechanical stimulus acts directly on the mechanism responsible for active sliding.

Evidence that ATPase activity is stimulated by mechanical forces associated with flagellar movements was obtained by slowing the frequency of movements of glycerinated sperm tails by increasing the viscosity of the medium in which they were suspended. This procedure produced a reduction in ATP hydrolysis. Thus, it appears that the rate of ATP hydrolysis depends on the frequency with which inactive regions are stimulated by regions undergoing active bending. A relation between passive tension on cross bridges and ATP hydrolysis is observed in insect fibrillar muscle, in which ATPase activity and tension production of the cross bridges are stimulated by externally applied tension.

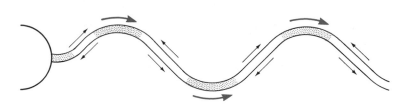

10-18.  Propagation of alternating linear and bent regions of axoneme along sperm-tail flagellum. Small arrows show proposed active countersliding of axoneural tubules in active straight segments. This countersliding is believed to produce passive bending of the inactive neighboring segment.

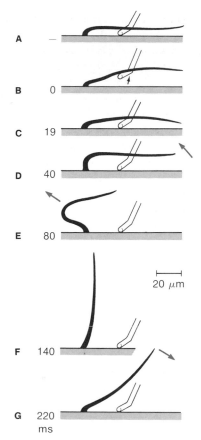

**10-19** Mechanical activation of the giant cilium of the clam. A crystal-driven microstylus is used to give the stationary cilium an abrupt displacement (**B**). This displacement initiates an entire beat cycle (**B–G**). [Thurm, 1968.]

## Coordination of Ciliary Movement

The questions of control and coordination of movement to be considered here exist at three principal levels. First, how is the beat initiated and the beating frequency regulated? Second, what causes the individual cilia of a population to beat with a timing that produces *metachronal waves* that sweep steadily and repeatedly over the ciliary field (Figure 10-8)? Third, how are the cilia of protozoa and of certain metazoan epithelia controlled so as to undergo changes in the

spatial organization of the beat, most evident as a change in the direction of the power stroke and in the direction of the metachronal waves?

### SPONTANEITY AND BEATING FREQUENCY

Cilia and flagella exhibit endogenous rhythmicity. The organelles can beat spontaneously after isolation from a cell, and they will beat after extraction of the cell membrane and cytoplasmic metabolites by glycerination or treatment with detergents (Box 9-1) if provided with an external supply of ATP and essential ions. In certain species, the flagellum continues to beat when it is severed near the base and separated from the basal body (Figure 10-20). It must be concluded that, in at least some species, the beat can originate in the shaft itself.

The cilia of certain vertebrate epithelia (e.g., frog

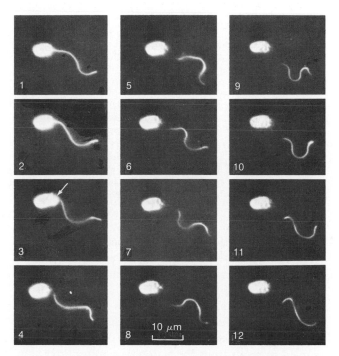

**10-20** Flagellar beat is independent of basal body. If the flagellum is severed from the cell body by a laser beam, the flagellum continues to beat for several cycles until its ATP is exhausted. [Goldstein et al., 1970.]

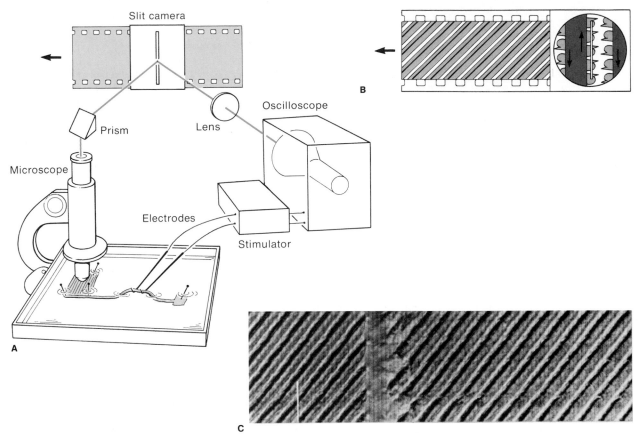

**10-21** Method of studying the neural control of ciliary beating. Electrical stimulation of the nerve that innervates the gills of the mussel *Mytilus* stops the ciliary beat. (**A**) The experimental apparatus. A steadily moving strip of film travels past a slit while the image of the metachronal waves of the beating cilia moves along the longitudinal axis of the slit. (**B**) This movement produces a diagonal pattern on the film. Alterations in the activity of the cilia interrupt this pattern. (**C**) Records showing the stoppage of the beat of cilia in response to nerve stimulation. [Takahashi and Murakami, 1968.]

palate) increase their frequency of beating in response to ACh and slow down in response to epinephrine. They also increase beating frequency when nerves that release ACh in the tissue are stimulated. Brief electrical stimulation of the nerve innervating the ciliated gills of the mussel *Mytilus* results in a transient cessation of ciliary activity (Figure 10-21). This is presumed to result from the action of a transmitter substance released by the nerve terminals. The same arrest of beating can be produced by experimental procedures that cause a rise in intracellular $Ca^{2+}$ in the ciliated cells of the gill. Thus, it has been postulated

that the arresting action of the neurotransmitter is mediated by a rise in intracellular $Ca^{2+}$. The neurotransmitter serotonin has the opposite effect, accelerating the cilia of clam and mussel gills and thus mimicking the excitatory effect of stimulating the branchial nerve.

### CILIARY REVERSAL

Most epithelial cilia beat in a fixed direction, but the cilia of protozoa and of the epithelium of tunicate larvae can alter the direction of the power stroke.

The most extensive investigations of ciliary reversal

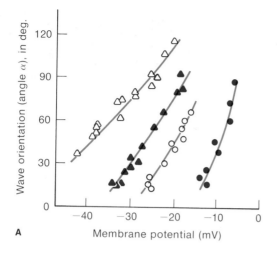

A

**10-22** Orientation of the metachronal waves as a function of membrane potential. The ciliate *Opalina* was immersed in various concentrations of KCl to set the membrane to different potentials. The potentials were recorded with a microelectrode, and the orientation of metachronal waves (angle α) was plotted against the membrane potential (**A**). The more depolarized the membrane, the more anteriorly directed are the waves. The different symbols in the graph refer to different specimens. (**B**) Direction of wave propagation on the surface of *Opalina,* with depolarization progressing from left to right. [Kinosita, 1954.]

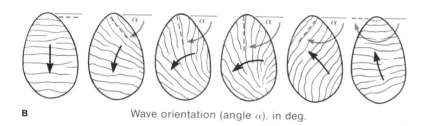

B
                          Wave orientation (angle α), in deg.

have been made on ciliated protozoa. There is an intimate relation between change in the direction of ciliary beat and changes in the direction of metachronal waves. A shift in the direction of the effective stroke is accompanied by a similar shift in the direction of metachronism. This relation was used to correlate direction of ciliary beating with the membrane potential in *Opalina.* The membrane potential was altered by changes in the KCl concentration of the external medium, and the orientation of the metachronal waves was plotted against membrane potential (Figure 10-22). As the membrane is depolarized, the direction of the power stroke shifts in a clockwise direction until the beat is directed toward the anterior. In the free-swimming cell, this redirection produces backward locomotion. This shift in the direction of the power stroke is termed *ciliary reversal.* The orientation of the beat is graded with membrane depolarization in somewhat the same way that muscle contraction is graded with membrane potential.

The mechanism that produces reversal of the direction of effective stroke, like the mechanism for muscle contraction, requires a certain concentration of $Ca^{2+}$. If $Ca^{2+}$ is absent from the extracellular fluid, the cilia do not reverse their direction of beating in response to depolarization. The high calcium sensitivity of the ciliary apparatus was demonstrated by Y. Naitoh and H. Kaneko (1972) in detergent-extracted specimens of *Paramecium.* After the membrane is destroyed as a diffusion barrier with Triton X-100 (a nonionic detergent), the cells can be reactivated to swim in a solution of ATP and $Mg^{2+}$. If the concentration of $Ca^{2+}$ in the reactivation solution is held at less than $10^{-6}$ mol/L, the ciliated cells swim forward; if it is higher, the cilia beat in reverse, causing the ciliated cells to swim backward (Figure 10-23). Because the ciliary

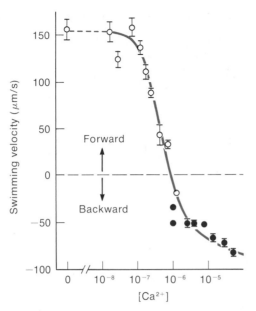

10-23   Relation between $Ca^{2+}$ concentration and direction of ciliary beat in Triton-extracted models of *Paramecium*. Extracted cells swim if reactivated with Mg–ATP. Swimming is forward if $Ca^{2+}$ is below $10^{-6}$ M and backward if above that concentration. [Naitoh and Kaneko, 1972.]

meability of the membrane to $Ca^{2+}$ (much as depolarization increases sodium conductance in squid axon, as described in Chapter 5). The influx of $Ca^{2+}$ that occurs in response to depolarization is seen in electrical recordings as an impulselike $Ca^{2+}$ response (Figure 10-24) that is graded with stimulus intensity rather than being all-or-none like a nerve impulse. The influx of $Ca^{2+}$ through channels in the membrane covering the cilia leads to a rise in the $Ca^{2+}$ concentration within the cilia. The intraciliary $Ca^{2+}$ alters the activity of the sliding tubules so as to produce a reversal in the orientation of the ciliary movements. Thus, it is the influx of $Ca^{2+}$ due to increased calcium conductance that couples a membrane depolarization produced by a sensory stimulus to the reversal of ciliary beating in ciliates such as *Paramecium* (Figure 10-25).

apparatus of the extracted *Paramecium* is exposed directly to the reactivation medium (ATP + $MgCl_2$ + $CaCl_2$), these results indicate that the direction of the ciliary power stroke is normally controlled by the internal concentration of $Ca^{2+}$. ATP is required for the calcium-activated reversal. If $Mg^{2+}$ is omitted from the reactivation medium, the cilia fail to beat, but they will shift toward the anterior-pointing orientation as the concentration of $Ca^{2+}$ is increased, provided that ATP is present.

In the living ciliate, the concentration of free $Ca^{2+}$ in the cytoplasm is determined by a balance between efflux of $Ca^{2+}$ produced by a metabolically energized calcium pump and influx of $Ca^{2+}$ through calcium channels or carriers in the surface membrane. If the permeability of the membrane to $Ca^{2+}$ increases in response to a stimulus, the balance shifts and the intracellular concentration of $Ca^{2+}$ rises. Electrophysiological experiments have shown that depolarization of the cell membrane in *Paramecium* increases the per-

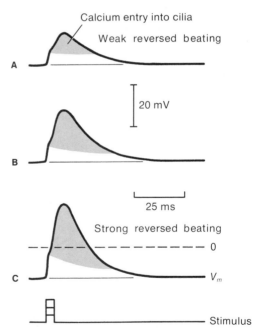

10-24   Calcium response of *Paramecium*. Injection of brief current pulses of three different intensities (bottom trace) produces responses of increasing intensity (A–C). The colored areas indicate approximately the depolarization contributed by the influx of $Ca^{2+}$ into the cell. The cilia respond to the graded influx of $Ca^{2+}$ with reversed beating of graded strength and duration. [Eckert, 1972.]

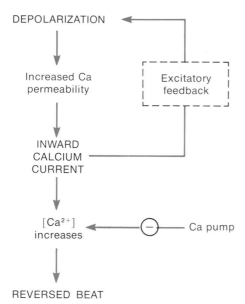

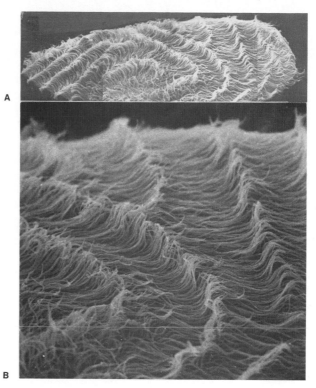

**10-25** Coupling of ciliary reversal to membrane depolarization. Depolarization produced by stimulation of the membrane causes calcium channels to open. The large electrochemical gradient of $Ca^{2+}$ causes it to flow through these channels into the cell, primarily into the cilia, as most channels are located in the portions of membrane covering the cilia. This inward $Ca^{2+}$ flux is seen as a calcium response, a weakly regenerative, spikelike depolarization. The increase in ciliary free $Ca^{2+}$ causes the beat to undergo a change in the direction of the power stroke. Removal of the $Ca^{2+}$ by membrane-transport mechanisms or by diffusion into the rest of the cell is followed by restoration of normal beating. [Eckert, 1972.]

**10-26** Metachronal waves in *Opalina*. The specimen was fixed instantaneously with osmium and then shadowed with gold for examination by scanning electron microscope. Part B is an enlargement from part A and measures 17 $\mu$m across the lower edge. [Tamm and Horridge, 1970.]

## METACHRONISM

The term *metachronism* refers to the waves of in-phase activity that spread over a population of beating cilia. In general, the direction of the waves is at right angles to the planes of the effective stroke. Figure 10-26 shows an instantaneously fixed specimen of the ciliate *Opalina*. As can be seen, all cilia are in phase along one axis—the axis of the wave front. At right angles to this, there is a phase shift from one cilium to the next, producing the sequence of waves. As the cilia produce their cyclic movements, the waves move smoothly at right angles to the axis of the wavefront.

Two hypotheses have been devised to explain the coordination of cilia in metachronal waves. The *neuroid hypothesis* proposes that repetitive signals carried by a cellular epithelium or cell membrane arise at a pacemaker region in one part of the epithelium or cell membrane and sweep over the surface, triggering or entraining each cilium as the signals pass. There are various versions of the neuroid hypothesis, some involving the surface membrane and others the intracellular fibrils connecting the basal bodies. The idea common to all versions is that the progression of the metachronal wave results directly from propagated coordinating signals of some kind. The *coupled-oscillator hypothesis* holds that metachronal waves arise as a consequence of viscous coupling between neighboring cilia through their fluid medium, and perhaps as a consequence of the sensitivity of the cilium to passive bending, already described.

There is no convincing evidence in support of the neuroid hypothesis, with the possible exception of the comb-plate movements of ctenophores, which may be coordinated by nerve-borne signals. No electrical signals recorded across the cell membrane can be correlated with ciliary beating, either in protozoa or in vertebrate epithelium. The evidence to the contrary is most convincing. In extracted models of *Paramecium* lacking a functional cell membrane, well-coordinated metachronism has been demonstrated following $Mg^{2+}$–ATP reactivation (Figure 10-23). Microsurgical experiments have demonstrated that intracellular fibrils are not involved in coordination between cilia; instead, the coordinated *reversal* of beating depends on a depolarization of the membrane, which occurs uniformly over the cell by electrotonic spread.

There is considerable evidence that the metachronal waves arise as a consequence of viscous coupling between the organelles, which is caused by the viscosity of the extracellular medium. The role of viscous coupling was first suggested in 1922 by J. Gray, who studied a population of independently undulating spirochetes (a type of bacteria) and found that, as they were crowded more closely together, they fell into a common rhythm, each beating in harmony with its neighbors, apparently as a result of mutual interference. Mechanical interaction does not require that cilia be in direct contact, because they are loosely coupled to their neighbors through the viscous drag of the extracellular fluid. Although each cilium has an inherent spontaneous beating frequency, it becomes entrained to beat nearly in phase with its closest neighbors along the axis of greatest physical interaction, because that requires less energy than does beating out of phase against its neighbors. This axis is determined by the form and orientation of the beat of individual cilia and by their spacing along the cell surface.

## Summary

The chapter considers two categories of cell motility: (1) cytoplasmic contraction and streaming (e.g., amoeboid movement), and (2) ciliary and flagellar motion. The former depends on interactions of cytoplasmic filaments for force generation, whereas the latter arises out of relative sliding between microtubules.

The contractile proteins responsible for amoeboid and other forms of cytoplasmic motility include actin and myosin molecules, which are similar to but not identical with the actin and myosin of striated muscle. Regulatory proteins include α-actinin, which is involved in the attachment of actin to the cell membrane; tropomyosin and filamin, which play a role in cross-linking actin filaments into a gel; and profilin, which associates with the actin monomers to prevent their polymerization into actin filaments. Actin and myosin filaments can interact to produce mutual sliding and thus force production, providing the myosin is first phosphorylated and that $Ca^{2+}$ levels are sufficiently high. There is evidence that such localized actomyosin contraction occurs to produce the forces behind amoeboid movement.

The cilia and flagella of eukaryotic cells arise from basal bodies homologous with centrioles and have a basic "9 + 2" organization of tubules. The two central tubules are surrounded by a ring of nine doublet tubules. Like other cellular microtubules, these tubules form by polymerization of tubulin dimers. The nine doublet tubules of cilia and flagella have arms positioned at intervals along the length of tubule A and project toward tubule B of the next doublet. These arms are composed of dynein, a protein that exhibits ATPase activity. It is believed that the dynein arms are involved in producing an ATP-energized sliding movement between adjacent doublets.

Spokes radiate from the sheath that surrounds the central pair of tubules to the nine doublet tubules. Although the function of these spokes is not clear, they appear to become displaced longitudinally along their respective doublets as the shaft of the cilium or flagellum is bent. When extracted and trypsinized flagella are supplied with ATP, the doublets slide past one another, suggesting that the movements of cilia and flagella depend on a sliding-tubule mechanism similar in principle to the sliding-filament mechanism of muscle. When cilia bend, the doublets on the inside of the bend are seen to project out beyond the other doublets near the tip of the cilium. This also suggests that bending is produced not by selective contraction of the doublets, but by relative sliding between doublets.

Analysis of undulating movements of flagella show that straight regions alternate with bending regions. The bends propagate along the shaft without diminishing in amplitude, indicating that—unlike the traveling wave of a whip, which is produced by active movement at one end only—force is generated along the entire length of the flagellum. It has been proposed that the bends propagate by a regenerative mechanical process, with the strain produced by an actively sliding region being transferred to the inactive portion ahead to activate the force-generating mechanism that causes the tubules in the previously inactive region to undergo active sliding.

In vertebrate epithelial cilia, frequency appears to be determined by cytoplasmic concentrations of ATP and $Ca^{2+}$, which are controlled by membrane permeability regulated by neurotransmitters. In protozoa, $Ca^{2+}$ influx due to cell-membrane depolarization induces a reversal of the direction of the power stroke.

Although individual cilia exhibit endogenous rhythmicity, populations of cilia exhibit a coordinated wave of motion called metachronism. No electrical signals corresponding to phases of the beating cycle occur over the cell membrane. Instead, the entrainment of neighboring cilia is a hydrodynamic phenomenon; the viscous drag of the medium between cilia brings about coherent bending of all the cilia along the paths of greatest force production.

## Exercises

*1.* Describe three contractile proteins and four regulatory proteins and discuss their function in cell motility.

*2.* What is meant by the gel and sol states of actin, and how are they believed to be regulated?

*3.* Compare the structures and properties of microfilaments and microtubules.

*4.* Discuss the regulatory steps present in the pathways to contracted actomyosin in amoeboid cells.

*5.* Discuss evidence that flagellar movements are not generated solely at the base and simply propagated in whiplike fashion toward the tip.

*6.* Describe the properties and functions of the two main protein fractions found in the ciliary axoneme.

*7.* What is the evidence that ATP supplies the energy for ciliary and flagellar bending?

*8.* Discuss the evidence in favor of the sliding-tubule hypothesis.

*9.* Describe an experiment that suggests that the cilia can respond to passive bending by generating active bending.

*10.* How do nerves regulate the activity of epithelial cilia?

*11.* How is reversal of ciliary beat regulated by membrane depolarization in the protozoa?

*12.* Discuss evidence for the viscous-coupling hypothesis of metachronism.

*13.* What evidence indicates that the cycle of ciliary beating is independent of cyclic regulation by the cell membrane?

## Suggested Reading

Allen, R. D., and N. S. Allen. 1978. Cytoplasmic streaming in amoeboid movement. *Ann. Rev. Biophys. Bioeng.* 7:469–495.

Blum, J. J., and M. Hines. 1979. Biophysics of flagellar motility. *Quart. Rev. Biophys.* 12:103–180.

Eckert, R. 1972. Bioelectric control of ciliary activity. *Science* 176:473–481.

Goldman, R., T. Pollard, and J. Rosenbaum, eds. 1976. *Cell Motility.* Cold Spring Harbor, N.Y.: Cold Spring Harbor Laboratory.

Holberton, D. V. 1977. Locomotion of protozoa and single cells. In R. McN. Alexander and G. Goldspink, eds., *Mechanics and Energetics of Animal Locomotion.* London: Chapman and Hall.

Lazarides, E., and J. P. Revel. 1979. The molecular basis of cell movement. *Scientific American* 240(5):100–113.

Pepe, F. A., J. W. Sanger, and V. T. Nachmias, eds. 1979. *Motility in Cell Function.* New York: Academic.

Satir, P. 1974. How cilia move. *Scientific American* 231(4):44–52. Also available as Offprint 1304.

Sleigh, M. A., ed. 1974. *Cilia and Flagella.* New York: Academic.

Sugino, K., and Y. Naitoh. 1982. Simulated cross-bridge patterns corresponding to ciliary beating in *Paramecium. Nature* 295:609–611.

Taylor, D. L., and J. S. Condeelis. 1979. Cytoplasmic structure and contractility in amoeboid cells. *Int. Rev. Cytol.* 56:57–144.

Warner, F. D., and P. Satir. The structure basis of ciliary bend formation. *J. Cell. Biol.* 63:35–63.

Wessells, N. 1971. How living cells change shape. *Scientific American* 225(4):76–82.

## References Cited

Baba, S. A., and Y. Hiramoto. 1970. A quantitative analysis of ciliary movement by means of high speed microcinematography. *J. Exp. Biol.* 52:645–690.

Brokaw, C. J. 1965. Non-sinusoidal bending waves of sperm flagella. *J. Exp. Biol.* 43:155–169.

Brokaw, C. J. 1968. Mechanisms of sperm movement. *Symp. Soc. Exp. Biol.* 22:101–117.

Brokaw, C. J. 1971. Bend propagation by sliding filament model for flagella. *J. Exp. Biol.* 55:289–304.

Gibbons, I. R. 1965. Chemical dissection of cilia. *Arch. Biol.* 76:317–352.

Gibbons, I. R. 1966. Studies on the adenosine triphosphate activity of 14S and 30S dynein from cilia of *Tetrahymena. J. Biol. Chem.* 241:5590–5596.

Goldstein, S. F., M. E. J. Holwill, and N. R. Silvester. 1970. High speed recording of laser damage. *Vision Res.* 8:245–248.

Gray, J. 1922. The mechanism of ciliary movement. *Proc. Roy. Soc.* (London) Ser. B. 93:104–121.

Grell, K. G. 1973. *Protozoology.* New York: Springer-Verlag.

Jahn, T. L., and E. C. Bovee, 1967. Motile behavior of protozoa. In T. T. Chen, ed., *Research in Protozoology.* Vol. 1. New York: Pergamon.

Jahn, T. L., M. D. Landman, and J. R. Fonseca. 1964. The mechanism of locomotion of flagellates, II: Function of mastigonemes of *Ochromonas. J. Protozool.* 11:291–296.

Jahn, T. L., and J. J. Votta. 1972. Locomotion of protozoa. *Ann. Rev. Fluid Mech.* 4:93–116.

Kinosita, H. 1954. Electric potentials and ciliary response in *Opalina. J. Fac. Sci.* (Tokyo University, Sec. 4) 7:1–14.

Machemer, H. 1972. Temperature influences on ciliary beat and metachronal coordination in *Paramecium. J. Mechanochem. Cell Motil.* 1:57–66.

Naitoh, Y., and H. Kaneko. 1972. Reactivated Triton-extracted models of *Paramecium:* Modification of ciliary movement by calcium ions. *Science* 176:523–524.

Sales, W. S., and P. Satir. 1977. Direction of active sliding of microtubules in *Tetrahymena* cilia. *Proc. Nat. Acad. Sci.* 74:2045–2049.

Satir, P. 1973. Structural basis of ciliary activity. In A. Perez-Miravette, ed., *Behavior of Microorganisms.* New York: Plenum.

Summers, K., and I. R. Gibbons. 1971. Adenosine triphosphate–induced sliding of tubules in trypsin-treated flagella of sea-urchin sperm. *Proc. Nat. Acad. Sci.* 68:3092–3096.

Takahashi, K., and A. Murakami. 1968. Nervous inhibition of ciliary motion in the gill of the mussel *Mytilus edulis. J. Fac. Sci.* (Tokyo University, Sec. 4) 2:359–372.

Tamm, S. L., and G. A. Horridge. 1970. The relation between the orientation of the central fibrils and the direction of beat in cilia of *Opalina. Proc. Roy. Soc.* (London) Ser. B. 175:219–233.

Taylor, D. L., S. B. Hellewell, H. W. Virgin, and J. Heiple. 1978. The solation–contraction coupling hypothesis of cell movements. In S. Hatano, H. Ishikawa, and H. Sato, eds., *Cell Motility: Molecules and Organization.* Baltimore: University Park Press.

Thurm, U. 1968. Steps in transducer process of mechanoreceptors. *Symp. Zool. Soc. London* 23:199–216.

# CHAPTER *11*

# *Chemical Messengers and Regulators*

*T*his chapter is about chemical agents employed to control the activities of the cells, tissues, and organs in the animal body. Without molecular regulatory agents, the functions of living tissue would be chaotic. Chemical regulation of cellular processes makes its appearance in some of the most primitive plant and animal species. For example, individual amoebas of the cellular slime molds exhibit aggregating behavior in response to a substance liberated by the individual amoebas. Before its identity was known, this substance was given the name "acrasin" but now is known to be the ubiquitous regulatory molecule cyclic adenosine 3′,5′-monophosphate (cAMP). An example of a still more primitive kind of chemical regulation is seen in the freshwater coelenterate *Hydra*. The water from a crowded culture of *Hydra* induces the differentiation of reproductive tissues in that animal. This effect is now ascribed to the high $CO_2$ concentration built up by a dense culture.

Chemical regulatory agents include relatively nonspecific molecules, such as $CO_2$, $H^+$, $O_2$, and $Ca^{2+}$, and those, like cAMP, that are more complex and are produced specifically as regulators, or *messengers.*

Chemical messengers operate at all levels of biological organization from subcellular to interorganismal. Some animal species, for example, utilize chemical messengers as means of communication between individuals of that species. A familiar example of such a *pheromone* is bombykol, the powerful female sex attractant of the commercial silkworm moth (p. 242). Pheromones occur in a wide variety of animal groups, but are most familiar in the insects, in which they function not only as sex attractants but also as a means of identifying the members of a colony or caste. In certain marine invertebrates, such as clams and starfish, spawning of eggs and sperm is triggered by pheromones liberated along with the gametes. Thus, the spawning of one individual triggers spawning in others of both sexes. The adaptive value of such *epidemic spawning* is that it enhances the probability that sperm and egg will meet and that fertilization will occur. *Crustecdysone,* a steroid that induces molting in crabs, also serves as the female sex attractant, producing behavioral responses in males at concentrations in the seawater as low as $10^{-13}$ M. Major categories of chemical messengers are listed in Table 11-1.

**TABLE 11-1**    Some classes of chemical messengers.

| Messenger type | Origin | Mode of action | Examples |
|---|---|---|---|
| Calcium | Extracellular or intracellular stores | Intracellular messenger action; coupling agent between membrane and internal effectors; "second messenger" for some hormones | |
| Cyclic nucleotides | Most cells | Intracellular messenger action; "second messenger" for many hormones | cAMP<br>cGMP |
| Neurotransmitters | Nerve cells | Synaptic transmission; distances transported are short, duration of activity brief | Acetylcholine<br>Serotonin<br>Norepinephrine |
| Neurohormones | Nerve cells | Endocrine function; transported by circulation | Vertebrate neurohypophysial hormones<br>Arthropod brain hormone |
| Glandular hormones | Nonneural endocrine tissues | Endocrine function; transported throughout body to distant target organs | Epinephrine<br>Ecdysone<br>Juvenile hormone<br>Insulin |
| Pheromones | Glands opening to the outside | Intraspecific communication between individuals | Bombykol |

## Hormones as Messengers

Both invertebrates and vertebrates have specialized tissues that secrete regulatory molecules into the blood and act on *target cells* within the same organism. These tissues constitute the various *endocrine glands.* The messenger molecules they secrete are called *hormones.* This term, introduced by E. H. Starling in 1905 in first describing secretin (page 678), comes from the Greek for "I arouse." We now consider many hormones to act as the first messenger in a series of two or more sequential messengers that lead to a specific response in the target tissue. In the course of their circuit in the blood and interstitial fluid, the hormone molecules encounter *receptor molecules,* which are specific for the hormone and which reside at the surface of the target cells or within them. By an interaction of hormone molecule with receptor molecule, there is initiated in the target cells a series of steps that influence one or more aspects of the physiology or metabolism of those cells. Although hormone molecules come into contact with all the tissues in the body, only cells that contain receptors specific for the hormone are affected by the hormone.

Because the amount of hormone produced by an endocrine gland is generally very small, its concentration in the blood remains very low. The target cells, however, are extraordinarily sensitive to the hormone. Thus, some target cells respond to plasma concentrations of hormone as low as $10^{-12}$ M. If the human taste buds were that sensitive to sugar, we would be able to detect a pinch of sugar dissolved in a large swimming pool full of coffee.

Because the secretion and circulation of hormones both take place relatively slowly, endocrine systems are best suited for regulatory functions that are sustained for minutes, hours, or days, such as maintenance of blood osmolarity, blood sugar and metabolic levels (antidiuretic hormone, insulin, growth hormone, thyroxine), for control of sexual activity and the reproductive cycles (sex hormones), and for the modification of behavior (various hormones). It seems apparent, then, that the slow, steady actions of the endocrines complement the more rapid performance of the nervous system in coordinating bodily functions.

Hormones are characterized by the following:

*1.* Hormones are produced and secreted into the circulation by endocrine cells in trace amounts.

*2.* They are carried in the blood to the target tissue.

*3.* They react with specific receptor molecules present in certain target cells.

*4.* They act in catalytic quantities, frequently by causing activation of specific enzymes.

*5.* A single hormone may have multiple effects on a single target tissue or on several different target tissues.

### STRUCTURAL CLASSIFICATION OF HORMONES

It must be assumed that some hormones remain undiscovered. Thus, no classification is certain to be complete. Nevertheless, we can place most of the known hormones into a few molecular groups: (1) *amines,* (2) *prostaglandins,* (3) *steroids,* and (4) *polypeptides* and proteins (Figure 11-1). The simplest are the amines. They are among the first to have been discovered, epinephrine having been described, and its function recognized, at the beginning of this century. Next in complexity are the prostaglandins, which are cyclic unsaturated fatty acids. The steroid hormones are cyclic hydrocarbon derivatives synthesized in all instances from the precursor steroid cholesterol. The largest and most complex are the polypeptides and proteins. The peptide hormones are particularly interesting as models of molecular evolution. The sequence of amino acid residues in a given polypeptide hormone (e.g., vasopressin, Table 11-2) is, of course, genetically determined; within different animal groups, substitutions of individual amino acid residues produce different analogs of the polypeptide. From inspection of Table 11-2, it is evident that some residues are highly conserved (never undergo substitution) and are presumably necessary for function; those that are not conserved (positions 3, 4, and 8) seem to be functionally neutral and probably serve only to place the essential residues in the positions appropriate for the activity of the polypeptide.

11-1 Representative structures of the four major classes of hormones: amines, prostaglandins, steroids, and polypeptides.

A Amine    Epinephrine

B Prostaglandin    Prostaglandin PGE$_2$

C Steroid    Testosterone

D Polypeptide    Insulin (bovine)

Chain A  Gly Ile Val Glu Glu Cys Cys Ala Ser Val Cys Ser Leu Tyr Glu Leu Glu Asp Tyr Cys Asp
         1   2   3   4   5   6   7   8   9  10  11  12  13  14  15  16  17  18  19  20  21

Chain B  Phe Val Asp Glu His Leu Cys Gly Ser His Leu Val Glu Ala Leu Tyr Leu Val Cys Gly Glu Arg Gly Phe Phe Tyr Thr Pro Lys Ala
         1   2   3   4   5   6   7   8   9  10  11  12  13  14  15  16  17  18  19  20  21  22  23  24  25  26  27  28  29  30

**TABLE 11-2**   Some forms of the neurohypophysial hormones in different animal groups.

| Name | Positions of amino acid residues | | | | | | | | | Animal group |
|------|----|----|----|----|----|----|----|----|----|--------------|
| | 1 | 2 | 3 | 4 | 5 | 6 | 7 | 8 | 9 | |
| Parent molecule | Cys—Tyr— · · · | | | · · ·—Asn—Cys—Pro— · · ·—Gly—($NH_2$) | | | | | | |
| Oxytocin | Cys—Tyr—Ile | —Gln—Asn—Cys—Pro—Leu—Gly—($NH_2$) | | | | | | | | Mammals |
| Arginine vasopressin | Cys—Tyr—Phe | —Gln—Asn—Cys—Pro—Arg—Gly—($NH_2$) | | | | | | | | Mammals |
| Lysine vasopressin | Cys—Tyr—Phe | —Gln—Asn—Cys—Pro—Lys—Gly—($NH_2$) | | | | | | | | Pigs and relatives |
| Arginine vasotocin | Cys—Tyr—Ile | —Gln—Asn—Cys—Pro—Arg—Gly—($NH_2$) | | | | | | | | All vertebrate classes |
| Isotocin | Cys—Tyr—Ile | —Ser—Asn—Cys—Pro—Ile —Gly—($NH_2$) | | | | | | | | Some teleost fishes |
| Mesotocin | Cys—Tyr—Ile | —Gln—Asn—Cys—Pro—Ile —Gly—($NH_2$) | | | | | | | | Reptiles, amphibians, and lungfish |
| Glumitocin | Cys—Tyr—Ile | —Ser—Asn—Cys—Pro—Gln—Gly—($NH_2$) | | | | | | | | Some cartilaginous fishes |

*Source:* Frieden and Lipner, 1971.

## FUNCTIONAL CLASSIFICATION

The actions of hormones on their target tissues are diverse and not easily subject to generalization. It is possible, however, to recognize four major classes of endocrine-mediated effects (Table 11-3). *Kinetic effects* include pigment migration, muscle contraction, and glandular secretion. *Metabolic effects* consist mainly of changes in the rate and balance of reactions and concentrations of tissue constituents. *Morphogenetic effects* have to do with growth and differentiation. *Behavioral effects* result from hormonal influences on the functioning of the nervous system. Hormones often have multiple effects, and some in fact cut across at least two of these classes of effects. For example, the thyroid hormones have metabolic effects on cells as well as morphogenetic effects on certain tissues. This suggests that the hormones do not produce their end effects directly, but instead activate intermediate processes, which may differ in different tissues or cell types. In fact, this indirect effect is borne out by recent studies of how hormones act at the cellular level.

The mechanisms of hormone action are considered in the last section of this chapter. In the meantime, we will note that there are two major classes of hormones, categorized according to the nature of their primary actions (Figure 11-2). Hormones of the first class act on the genetic machinery of the cell. They readily penetrate the surface membranes of their target cells and react or combine with internal receptors to form active complexes that in turn influence the genetic machinery to alter the production of proteins,

including some enzymes. These effects are relatively slow and long lasting. Examples of this class are the steroids and thyroid hormones, whose lipid solubility allows them to pass readily across membranes. Hormones of the second class cannot readily enter the cell because of their low lipid solubility; instead, they interact with receptor proteins located in the surface membrane. These receptor molecules, when activated by binding the hormone, become instrumental in eliciting the cellular response to the hormone. Lipid-insoluble hormones such as the catecholamines and the peptides exemplify this class.

## IDENTIFICATION OF ENDOCRINE GLANDS AND HORMONES

The various endocrine tissues are structurally and chemically diverse. Some contain more than one kind of secretory cell, each elaborating a different hormone. Unlike *exocrine glands* (e.g., salivary glands, enzyme-secreting cells of the pancreas, prostate, mammary glands), which empty their secretions into ducts, endocrine tissues liberate their secretions (i.e., hormones) directly into the extracellular space, from which the hormone diffuses into the blood. Since there is no common morphological plan for endocrine glands, nor any distinctive gross morphological feature (other than a rich vascularization), such as a secretory duct, it has proved difficult in some cases to establish unequivocally whether tissues suspected of having an endocrine function actually do. No single characteristic can be used for this purpose, so the following set of criteria has been used to establish whether a tissue has an endocrine function.

*1.* Ablation of the suspected tissue should produce deficiency symptoms in the subject.

TABLE 11-3  The main types of effects of vertebrate hormones.

| Class of effect | Effect on target tissues | Hormones |
|---|---|---|
| Kinetic | Contraction of muscle | Epinephrine, oxytocin |
| | Concentration and dispersion of pigment | Melatonin |
| | Secretion of exocrine glands | Secretin, gastrin |
| | Secretion of endocrine glands | ACTH, TSH, LH, releasing hormones |
| Metabolic | Control of respiration rate | Thyroxins |
| | Carbohydrate and protein balance | Insulin, growth hormone, glucocorticoids, glucagon |
| | Electrolyte and water balance | ADH, aldosterone |
| | Calcium and phosphorus balance | Parathormone, calcitonin, vitamin $D_3$ |
| Morphogenetic | General growth | Growth hormone |
| | Molting | Thyroxins, corticosteroids |
| | Metamorphosis | Thyroxins |
| | Regeneration | GH |
| | Gonad maturation | FSH |
| | Gamete release | LH |
| | Differentiation of genital ducts | Androgens |
| | Development of secondary sexual characteristics | Estrogens and androgens |
| Behavioral | Trophic effects on developing nervous system | Estrogens Progesterone Androgens |
| | Sensitization to specific stimuli | Prolactin |
| | Release of behavior patterns | |

*2.* Deficiency symptoms should be relieved by replacing the suspected hormone by injection. In practice, this is first done with a crude tissue extract of the suspected endocrine gland; if this works, the extract is subsequently fractionated and eventually purified to a single molecular species. Successful replacement is the most important criterion for identification of a suspected endocrine tissue and its hormone. It is also the basis of *replacement therapy,* which is so important for patients who for medical reasons have lost the function of an endocrine gland. For example, the effects of surgical removal of the thyroid gland because of a malignancy can be compensated by daily administration of thyroxine. Likewise, replacement therapy successfully maintains normal function when it is necessary to remove the gonads or when the insulin-secreting cells of the pancreas fail.

*3.* Replacement (i.e., reimplantation) of the ablated tissue elsewhere in the body should prevent or reverse these deficiency symptoms. If the effects produced by removal of the tissue are due to the absence of a blood-borne substance produced by that tissue, replacement of the ablated tissue should restore the missing hormone. Misleading results may be expected, however, when ablation-and-reimplantation experiments are done with tissues closely associated with the nervous system because of the interruption of neural connections.

*4.* Following purification, the molecular structure of the active substance is determined. When technology permits, the molecule is synthesized and tested for biological potency.

These four classic criteria for the identification of endocrine glands and hormones are now supplemented by a variety of modern approaches. For example, in peptide- or protein-secreting glands, electron-microscopic investigation of the tissue may be employed to locate the membrane-limited *secretory granules* (secretory vesicles),† 1000–4000 Å in diameter, that contain the hormone. Once a hormone has been isolated

†These are similar in many respects to synaptic vesicles, which are about 400 Å in diameter. We will use the terms *secretory granule* and *secretory vesicle* interchangeably, depending on whether the emphasis is on its contents (granule) or its limiting membrane (vesicle).

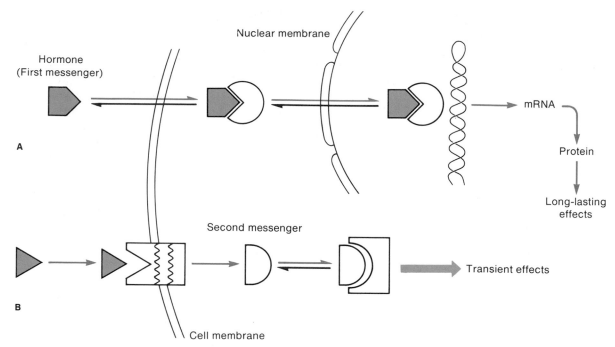

**11-2** Two classes of hormones based on primary mode of action. (**A**) Lipid-soluble hormone penetrates the membrane and combines with intracellular receptor molecule to form an active complex that acts on genetic machinery. (**B**) Lipid-insoluble hormone binds to the surface receptor, activating enzymatic production of the second messenger, which in turn combines with another molecule to produce a metabolically active complex. [Concept suggested by M. J. Berridge.]

in purified form, its molecular composition can be determined. Such efforts have opened the way for artificial synthesis of hormones important for replacement therapy.

## Regulation of Hormone Secretion

### FEEDBACK CONTROL

The secretory activities of endocrine tissues are modulated by negative feedback (Box 1-1). That is, the concentration of the hormone itself, or a response to the hormone by a target tissue, will have an inhibitory effect on the synthetic or secretory processes responsible for the elaboration of the hormone in question. Feedback can be either *short-loop* or *long-loop* (Figure 11-3).

In *short-loop* feedback, either a product of the target tissue or an effect produced by it acts directly back on the endocrine tissue that elaborates the hormone.

This direct action keeps hormone secretion in check. Negative feedback, as the name implies, requires a sign inversion somewhere in the loop, usually some inhibitory effect that the product of the target tissue has on the endocrine tissue. As the concentration of the hormone rises, the hormone indirectly inhibits any further increase in its own plasma concentration.

A long loop operates on the same principle, but it includes more elements (Figure 11-3). Regardless of the number of elements in the loop, there must be an odd number of sign inversions to produce the regulatory effect. For example, two sign inversions (i.e., inhibitions) in the same loop will not do, for two negatives would cancel each other, producing an overall positive feedback.

Negative-feedback control is illustrated by several of the pituitary hormones that stimulate secretion by glandular tissues found in other parts of the body.

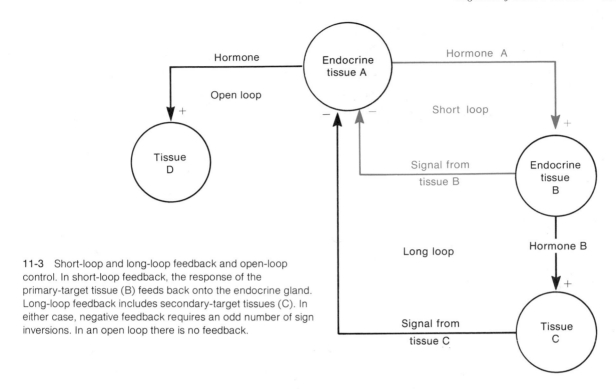

**11-3** Short-loop and long-loop feedback and open-loop control. In short-loop feedback, the response of the primary-target tissue (B) feeds back onto the endocrine gland. Long-loop feedback includes secondary-target tissues (C). In either case, negative feedback requires an odd number of sign inversions. In an open loop there is no feedback.

For most of these, discussed later, their release from the pituitary gland is inhibited by those hormones whose release they stimulate. Such feedback control is important in preventing runaway secretion of these very potent hormones.

The negative-feedback signal acting to slow the output of an endocrine tissue may be a physiological response to the hormone secreted by that tissue (e.g., reduced blood glucose levels in the insulin loop), or it may be the rise in concentration of a second hormone, the secretion of which is stimulated by the first hormone (e.g., high testosterone levels inhibit production of the gonadotropic hormone that stimulates testosterone production). Thus, the negative-feedback signal may be simply the reduction of a stimulus to the endocrine cells, or it may be an active inhibition of secretion of the hormone.

### INTRACELLULAR PACKAGING AND STORAGE OF HORMONES

Endocrine cells, like other secretory cells, generally show a morphological polarity, with the synthesis and packaging of the hormone taking place at one part of

the cell and its secretion taking place at another part. This polarity is illustrated in Figure 11-4. The nature of synthesis and storage varies from one class of hormones to another. For example, the steroid hormones, all derived from cholesterol, appear to be secreted in molecular (i.e., unpackaged) form. Most other hormones are packaged in membrane-bound vesicles within the secretory cell, later to be liberated into the extracellular space.

The protein and peptide hormones are synthesized on messenger RNA templates of the polyribosomes (polysomes) of the *rough endoplasmic reticulum (RER)* and accumulate within the reticulum. Pulse-chase radiography—a tracer technique in which radioactive labeled amino acid is incorporated for a brief period into newly synthesized proteins—permits observation of the intracellular travels of secretory proteins. From the RER, the polypeptides pass into the smooth, polysome-free portions of the endoplasmic reticulum, termed *transitional elements* (Figure 11-4), where the membrane of the reticulum buds off, encapsulating the secretory products. These vesicles then migrate to the Golgi complex, which consists of a stacked set of slightly

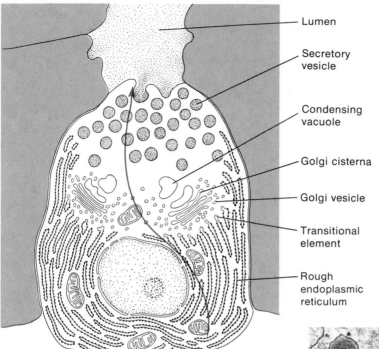

Lumen

Secretory
vesicle

Condensing
vacuole

Golgi cisterna

Golgi vesicle

Transitional
element

Rough
endoplasmic
reticulum

**11-4**  Representative epithelial endocrine cell. The colored arrow represents the general pathway of hormone production and secretion, beginning with the rough endoplasmic reticulum and ending with the exocytosis of secretory granules. [Jamieson, 1975.]

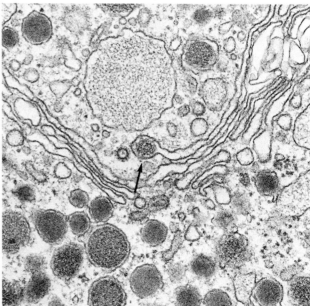

**11-5**  Electron micrograph showing a Golgi apparatus in an anterior pituitary gonadotroph cell. Notice the relatively dense free secretory granules and the one budding off from the Golgi cisterna (arrow). Magnification 20,000×. [Farquhar, 1971.]

concave, nearly flat membrane saccules with closely associated free vesicles and vacuoles (Figure 11-5). In an as yet unknown way, the protein is delivered to the Golgi saccules. Microscopic studies suggest that the membranes of the vesicle and saccule become fused. Within the Golgi complex, which contains enzymes bound to the luminal membrane surfaces, some proteins are altered by the addition of sugar residues, others (e.g., insulin) by the excision of fragments joining two polypeptide chains. It is believed that in a process that begins in the *Golgi cisternae,* but which takes place mainly in the *condensing vacuoles* (Figure 11-4), water is drawn out of the future secretory vesicle osmotically, with the result that the protein becomes concentrated 20–25 times. The mature secretory vesicle eventually reaches the plasma membrane to await the appropriate signal to release its inclusions to the cell exterior.

Retention of the hormone within the secretory vesicle is accomplished by a variety of means. Hormones of large molecular weight, such as proteins, are retained simply because of their size, which renders them incapable of crossing the membrane of the vesicle. Some small hormone molecules are bound to larger accessory molecules, usually proteins. There is evidence that the catecholamines norepinephrine and epinephrine are kept in their secretory vesicles at least in part by continual active uptake into the vesicles from the cytosol. The tranquilizing drug *reserpine* interferes with this uptake, thereby causing the catecholamines to leak out of their secretory vesicles and out of the secretory cells.

The duration of storage of a hormone within a secretory tissue varies widely. The steroid hormones appear to diffuse out of the cell across the surface membrane in a matter of minutes after synthesis, as they are not packaged in vesicles and, being lipid soluble, readily cross membranes. Secretory residues of most endocrine cells are held until their release is stimulated by mechanisms discussed in the next subsection. The thyroid hormones are secreted into the extracellular storage spaces of spherical clusters of cells termed *follicles* and are stored there for up to several months. Once secreted into the circulation, a hormone can still be considered stored in the bloodstream, the length of this phase of storage depending on the rate of degradation or uptake by cells. The hydrophobic hormones (steroids and thyroid) are, in fact, carried in the circulation by *binding proteins,* and remain inactive until they dissociate from the protein.

## SECRETORY MECHANISMS

There are several conceivable mechanisms by which hormones stored within the cell in secretory vesicles might find their way to the cell exterior. For most hormones (except the steroids, which appear not to be stored in vesicles), the most widely accepted theory is that the entire contents of a vesicle are delivered to the cell exterior by the process of exocytosis. A current view of the formation, transport, release, and reconstitution of secretory vesicles is summarized in Figure 11-6.

Secretion occurs in response to appropriate stimulation of the endocrine cell. The stimulus may be the arrival of another hormone or neurotransmitter at the membrane of the secreting cell (e.g., the release of acetylcholine in the chromaffin tissue of the adrenal medulla), or secretion may result from a nonhumoral stimulus (e.g., stimulation of the ADH-secreting neurons by an increase in plasma osmolarity). In neurosecretory nerve cells (next section), the stimulus sets up action potentials that travel to the axon terminals and there elicit the release of hormone from the endings. It is clear that depolarization in the form of the impulse leads to secretion in these cells. This has been demonstrated by stimulating such cells electrically at a distance from their endings, so as to set up impulses, while monitoring the release of hormone from the endings. The rate of hormone secretion increases with increased frequency of impulses (Figure 11-7). Depolarization *without* the production of action potentials—for example, by experimental increase of the extracellular $K^+$ concentration—is also accompanied by a rise in the rate of hormone secretion. Secretion rises to a maximum with increasing $[K^+]_o$ and hence with increasing depolarization (Figure 11-8). At still higher concentrations, secretion decreases. The stimulation of secretion by depolarization suggests that the action potential also evokes secretion by virtue of its depolarization. In either case, there must be a means by which the depolarization of the surface membrane stimulates the process of secretion. Because of the well-known role of $Ca^{2+}$ in regulating neurotransmitter release (described in Chapter 6), it comes as no surprise that calcium has also been implicated in the coupling of hormone secretion to membrane stimulation.

The evidence that $Ca^{2+}$ is the *secretogogue* that couples secretion to membrane stimulation comes from experiments on several kinds of endocrine tissues. Any stimulus that leads to an influx of $Ca^{2+}$ in the output portion of the cell (colored in Figure 11-9) is followed by an increase in secretory activity. The stimulus is

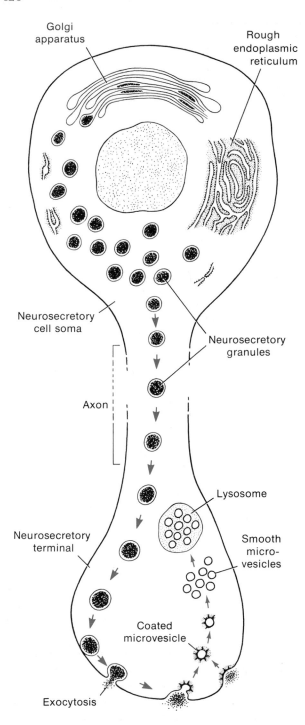

Golgi apparatus

Rough endoplasmic reticulum

Neurosecretory cell soma

Neurosecretory granules

Axon

Neurosecretory terminal

Lysosome

Smooth micro-vesicles

Coated microvesicle

Exocytosis

**11-6** Formation and fate of secretory vesicles. After their formation by the Golgi apparatus (top), secretory vesicles are transported to the site of release. In a secretory neuron, as shown here, this entails transport of the vesicles down the axon to the terminals (bottom), where release occurs by exocytosis. Vesicles may be reformed by a pinocytotic process, possibly using the same membrane that made up the original vesicle. The microvesicles thus formed then aggregate and eventually become incorporated into new secretory vesicles. [Douglas, 1974.]

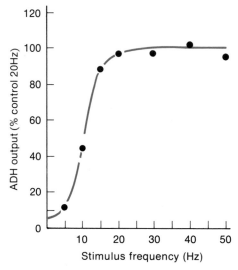

**11-7** Relation between electrical stimulation and the release of ADH from the rat neurohypophysis. The stimulus pulses at each frequency were continued for 5 min. [Mikiten, 1967.]

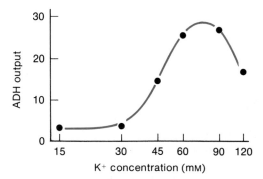

**11-8** ADH release (arbitrary units) as a function of potassium depolarization. Freshly dissected neurohypophyses were placed in incubation media of different $K^+$ content for 10 min, after which the released ADH was assayed. [Douglas, 1974.]

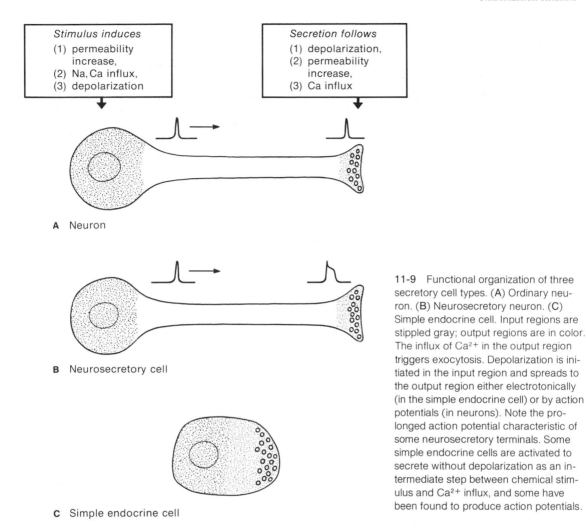

*Stimulus induces*
(1) permeability increase,
(2) Na,Ca influx,
(3) depolarization

*Secretion follows*
(1) depolarization,
(2) permeability increase,
(3) Ca influx

**A** Neuron

**B** Neurosecretory cell

**C** Simple endocrine cell

**11-9** Functional organization of three secretory cell types. (A) Ordinary neuron. (B) Neurosecretory neuron. (C) Simple endocrine cell. Input regions are stippled gray; output regions are in color. The influx of $Ca^{2+}$ in the output region triggers exocytosis. Depolarization is initiated in the input region and spreads to the output region either electrotonically (in the simple endocrine cell) or by action potentials (in neurons). Note the prolonged action potential characteristic of some neurosecretory terminals. Some simple endocrine cells are activated to secrete without depolarization as an intermediate step between chemical stimulus and $Ca^{2+}$ influx, and some have been found to produce action potentials.

sensed by the receptors in the input region of the cell (shaded gray). In neurosecretory cells (discussed in the next section) and in ordinary nerve cells, the input and output regions are separated by a conducting region. Incoming stimuli (synaptic input, physical or chemical changes in the plasma) produce a depolarizing permeability change that, in turn, leads to an increased frequency of impulse firing in the axon. By invading and depolarizing the terminal membranes, impulses (action potentials) cause calcium-permeable channels to open. The influx of $Ca^{2+}$ then triggers exocytosis by mechanisms presently not known.

## Neuroendocrine Relations

The endocrine system is closely associated with the nervous system. This is most obvious in those endocrine tissues that are made up of nerve cells or have a neural embryological origin. For example, the *chromaffin cells* of the *adrenal medulla,* which secrete the catecholamines epinephrine and norepinephrine, are embryologically as well as functionally distinct from the steroid-secreting tissue of the *adrenal cortex.* The chromaffin cells are derived from the embryonic tissue that also gives rise to postganglionic sympathetic neurons of the autonomic ganglia (Figure 8-13). They are, in fact, the postganglionic cells of the preganglionic sympathetic neurons that arise in the thoracolumbar segments of the autonomic ganglia. Their secretory activity is stimulated by ACh released from the preganglionic sympathetic fibers that terminate in the chromaffin tissue.

Another important example of neuroendocrine affinity is seen in the secretory neurons of the vertebrate hypothalamus, which will be discussed shortly.

## NEUROSECRETION

Nerve cells that produce hormones and secrete them into the bloodstream belong to a special class of neurons termed *neurosecretory cells.* These are at the same time nerve cells and endocrine cells. *Neurosecretion* occurs in nearly all metazoans but has been most extensively studied in the insects, crustaceans, annelids, mollusks, and vertebrates.

In principle, there is little difference between an ordinary nerve cell and most neurosecretory cells. The release of a hormone from the terminals of a neurosecretory cell is similar to the release of a transmitter substance from a simple nerve cell. The major distinctions are morphological. Ordinary nerve cells form synapses with other cells at their terminations, whereas neurosecretory axons terminate close to capillaries and release their secretions into the interstitial space (Figures 11-10 and 11-11), whence they diffuse into the bloodstream and are carried away to target tissues elsewhere in the organism. The neurosecretory hormone is packaged within the neuron soma into vesicles that are 1000–4000 Å in diameter (Figure 11-11), in contrast to the 300–600 Å vesicles that contain the neurotransmitter substances of ordinary nerve cells.

The *axonal transport* of secretory products can be demonstrated in neurosecretory cells by ligating the bundle of axons with a fine silk strand and then examining the proximal (toward the cell body) and distal (away from the cell body) sides of the ligature at different times. When the neurosecretory material is stained with a specific dye, it can be seen to accumulate on the proximal side of the ligature. The distal side of the ligature shows a depletion of the material. The neurosecretory material is produced in the cell body and is then transported distally within the axon toward the terminal, where it is eventually released. Since the

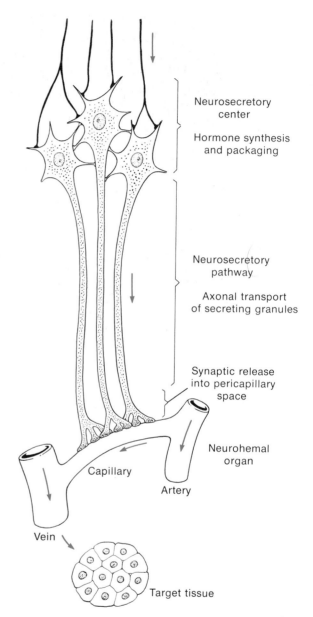

**Neurosecretory center**

**Hormone synthesis and packaging**

**Neurosecretory pathway**

**Axonal transport of secreting granules**

**Synaptic release into pericapillary space**

**Neurohemal organ**

**Capillary**

**Artery**

**Vein**

**Target tissue**

**11-10**    Organization of neurosecretory system.

discovery of axonal transport in the 1930s, it has been found that there are both slow and fast transport systems in a given axon. Neurosecretory granules have been reported to be transported by the fast system at rates up to 2800 mm a day.

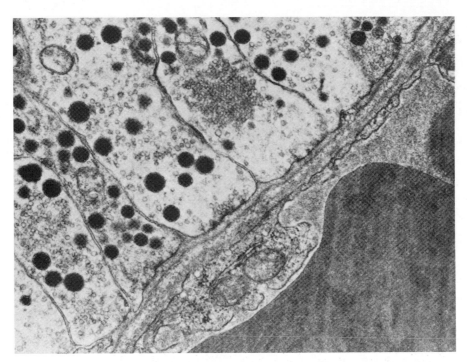

**11-11** Terminals of neurosecretory axons in the neurohypophysis of the hamster, as seen with the electron microscope. The large dark bodies are the secretory vesicles (or granules). The terminals end on an endothelial basement membrane that separates the terminals from a fenestrated capillary wall. The large dark object at the bottom is a red blood cell in the capillary. Magnification 27,000×. [Douglas et al., 1971.]

## The Mammalian Endocrine System

To discuss in any detail all the vertebrate endocrine systems, their glands, their hormones, and their interrelations would require many chapters or even volumes. We will limit this section to a brief overview of some of the more prominent components of the mammalian endocrine system. Descriptions of those endocrines involved in the regulation of renal function and digestion are deferred to Chapters 12 and 15, respectively. The major vertebrate hormones—their origin, structure, target tissues, actions, and regulation—are summarized in Tables 11-4 to 11-10.

Much of this section will be devoted to the hormones of the *hypophysis,* or *pituitary gland,* a small but complex appendage at the base of the hypothalamus in the brain stem. This organ secretes at least nine hormones, many of which regulate the function of other endocrine tissues. For this reason, the hypophysis has been called the "master gland." Recently, however, it has been discovered that the pituitary tissue is, in fact, itself under the control of neurosecretory cells located in the hypothalamus.

## Hypothalamus and Pituitary

### NEUROHYPOPHYSIAL HORMONES

The *posterior lobe* of the pituitary gland (Figure 11-12) is also called the *pars nervosa,* or the *neurohypophysis.* The neurohypophysis consists of neurosecretory axons and their terminals (Figure 11-11). The cell bodies of these

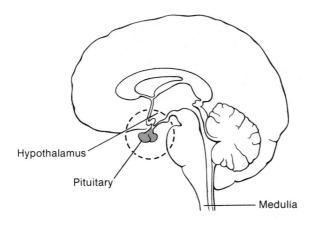

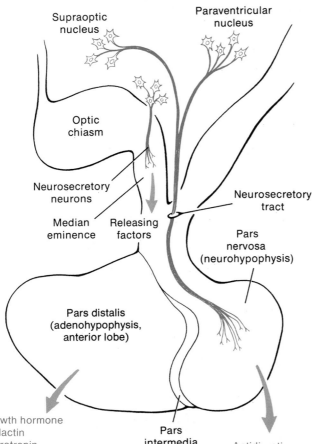

**11-12** The primate hypophysis, or pituitary gland, situated at the base of the hypothalamus. The major components are the anterior lobe (adenohypophysis) and the posterior lobe (neurohypophysis). The neurohypophysis is an extension of the brain. The adenohypophysis consists of nonneural glandular tissue.

axons reside in the anterior hypothalamus in two structures called the *supraoptic* and *paraventricular* nuclei. The secretory products, synthesized and packaged in the cell bodies, are transported within the axons to the nerve terminals in the posterior lobe, within which they are released into a capillary bed. This system was the first example of neurosecretion discovered in the vertebrates.

The neurohypophysial secretory endings release two octapeptides: *vasopressin,* also termed antidiuretic hormone (ADH), and *oxytocin* (Table 11-4). Both are mildly effective in fostering contractions of the smooth-muscle tissue in arterioles and the uterus. In mammals, oxytocin is best known for stimulating the uterine contractions during parturition and for stimulating the release of milk from the mammary gland; in birds, it stimulates motility of the oviduct. The foremost function of ADH is the stimulation of water retention in the kidney (p. 523). These functions are summarized in in Figure 11-13.

These neurosecretory hormones occur in several molecular forms, differing from one vertebrate group to another in the identity of the amino acid residues at three loci in the peptide chain (Table 11-2). There is no clear structural distinction between the oxytocins and the vasopressins.

TABLE 11-4  Octapeptide hormones of the mammalian neurohypophysis.

| Hormone | Structure | Target tissue | Primary action | Regulation |
|---|---|---|---|---|
| Antidiuretic hormone (ADH) | Peptide | Kidney | Increased water resorption | Increases in plasma osmotic pressure or decreased blood volume cause increased release |
| Oxytocin | Peptide | Uterus Mammary glands | Smooth muscle contraction Milk ejection | In mammals, suckling stimulus and cervical distention cause release of hormone; increased progesterone inhibits release |

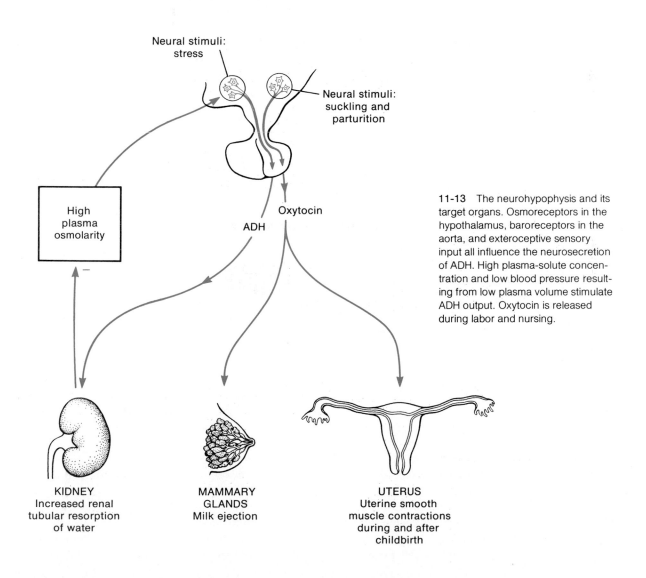

11-13  The neurohypophysis and its target organs. Osmoreceptors in the hypothalamus, baroreceptors in the aorta, and exteroceptive sensory input all influence the neurosecretion of ADH. High plasma-solute concentration and low blood pressure resulting from low plasma volume stimulate ADH output. Oxytocin is released during labor and nursing.

**TABLE 11-5**  Hormones of the adenohypophysis.

| Hormone | Structure | Target tissue | Primary action | Regulation |
|---------|-----------|---------------|----------------|------------|
| Adrenocorticotropin (ACTH) | Peptide | Adrenal cortex | Increases steroidogenesis and secretion in adrenal cortex | CRH secretion stimulates release; ACTH inhibits release |
| Thyroid-stimulating hormone (TSH) | Glycoprotein | Thyroid gland | Increases thyroid hormone synthesis and secretion | TRH induces secretion; thyroid hormones block release |
| Growth hormone (GH) | Peptide | (see Table 11-7) | | |
| Follicle-stimulating hormone (FSH) | Glycoprotein | Seminiferous tubules (male); ovarian follicles (female) | In male, increases sperm production; in female, stimulates follicle maturation | FSH/LH-RH stimulates release; inhibin interferes with release |
| Luteinizing hormone (LH) | Glycoprotein | Ovarian interstitial cells (female); testicular interstitial cells (male) | Induces final maturation of follicle, estrogen secretion, ovulation, corpus luteum formation and progesterone secretion; in males, increases synthesis and secretion of androgens | FSH/LH-RH stimulates release |
| Prolactin | Peptide | Mammary glands (alveolar cells) | Increases synthesis of milk proteins and growth of mammary glands | Secretion is continuously blocked by PIH secretion; increased estrogen and a decreased PIH secretion permit release |
| Melanocyte-stimulating hormone (MSH) | Peptide | Melanophores (ectotherm pigment cells); melanocytes (endotherm pigment cells) | Increases melanin synthesis in melanophores and melanocytes; increases dispersal in melanophores (skin darkening) | MIH secretion inhibits release |

Within their respective neurosecretory cells, these two neurohypophysial peptides are associated with cysteine-rich protein molecules termed *neurophysins*. Within the secretory granules, the hormone molecules appear to be noncovalently complexed in the ratio of 1:1 with the neurophysin molecules, of which there are two major fractions, neurophysin I and neurophysin II. Oxytocin is associated with I and vasopressin with II. The role of these proteins is not clear, since they have no hormone activity, even though they are secreted along with the two hormones. It is conjectured that a parent protein molecule is enzymatically cleaved, producing the eight-residue peptide and the neurophysin moiety, both of which are secreted during exocytosis. One possible role of the neurophysin is that of a *storage protein,* which serves to retain the hormone in the secretory granules until release. A storage protein called *chromagranin* is also found associated with epinephrine and norepinephrine in the chromaffin granules of the adrenal medulla.

### ADENOHYPOPHYSIAL HORMONES

The *anterior lobe* of the pituitary, also termed the *adenohypophysis* (Figure 11-12), secretes at least seven peptide hormones in mammals (Table 11-5). In contrast to the hormones of the neurohypophysis, the hormones of the adenohypophysis do not originate in neurosecretory cells, but instead arise from simple secretory cells that reside entirely within the anterior lobe. These cells occur in three histochemically distinct types but are otherwise difficult to distinguish. It appears that the acidophils (cells that take up acidic stains) secrete *growth hormone (GH;* also termed *somatotropin)* and *pro-lactin,* and that the basophils secrete *thyroid-stimulating hormone (TSH)* and two *gonadotropins—luteinizing hormone (LH)* and *follicle-stimulating hormone (FSH).*

### HYPOTHALAMIC CONTROL OF THE ADENOHYPOPHYSIS

In recent years, evidence has been accumulating that the secretory activity of the adenohypophysial endocrine cells is regulated by at least seven *hypothalamic hormones* of neurosecretory origin. Four of these are *releasing hormones* (or *releasing factors),* and three are *release-inhibiting hormones.* They are produced by neurosecretory cells that are located in the hypothalamus and have endings in the *median eminence* at the floor of the hypothalamus. These hormones, which are nearly all polypeptides, are named according to their actions on adenohypophysial secretion (Table 11-6). Only in recent years has the elaborate neuroendocrine organization of the hypothalamus and pituitary become known, and much more undoubtedly will be learned.

As early as the 1930s, studies revealed that capillaries in the median eminence converge to form a series of *portal vessels* that carry blood from the neurosecretory

**TABLE 11-6** Hypothalamic releasing hormones and releasing factors.

| Hormone | Structure | Target tissue | Primary action | Regulation |
|---|---|---|---|---|
| Corticotropin-releasing hormone (CRH) | Polypeptide | Adenohypophysis | Stimulates ACTH release | Stressful neural input increases secretion; ACTH inhibits secretion |
| TSH-releasing hormone (TRH) | Polypeptide | Adenohypophysis | Stimulates TSH release and prolactin secretion | Low body temperatures induce secretion; thyroid hormone inhibits secretion |
| GH-releasing hormone (GRH) | Polypeptide | Adenohypophysis | Stimulates GH release | Hypoglycemia stimulates secretion |
| FSH- and LH-releasing hormone (FSH/LH-RH) | Polypeptide | Adenohypophysis | Stimulates release of FSH and LH | In the male, low blood testosterone levels stimulate secretion; in the female, neural input and decreased estrogen levels stimulate secretion; high blood FSH or LH inhibits secretion |
| GH-inhibiting hormone (GIH) (or somatostatin) | Polypeptide | Adenohypophysis | Inhibits GH release; interferes with TSH release | Exercise induces secretion; hormone is rapidly inactivated in body tissues |
| Prolactin-release-inhibiting hormone (PIH)† | Dopamine | Adenohypophysis | Inhibits prolactin release | High levels of prolactin increase secretion; estrogen, testosterone, and neural stimuli (suckling) inhibit secretion |
| MSH-release-inhibiting hormone (MIH) | Polypeptide | Adenohypophysis | Inhibits MSH release | Melatonin stimulates secretion |

†This releasing hormone and its actions are still tentative.

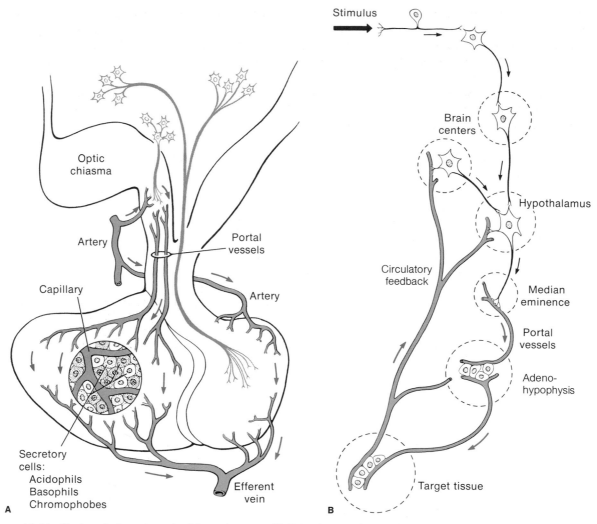

**11-14** The hypothalamo-hypophysial portal system. (A) Releasing or release-inhibiting hormones secreted by neurosecretory endings in the median eminence are carried via the portal vessels to the adenohypophysis, where they stimulate (or inhibit) secretory activity. (B) Input from various neural sources to the hypothalamus elicits the secretion of releasing and release-inhibiting hormones in the median eminence. Compare with Figure 11-15.

tissue to the secretory tissue of the anterior pituitary. There they break up again into a capillary bed before finally converging into the venous system (Figure 11-14). The portal system enhances chemical communication from the hypothalamus to the adenohypophysis by carrying the hypothalamic hormones directly to the interstitium of the adenohypophysis. In the adenohypophysis, the hypothalamic hormones come into contact with the endocrine cells that secrete the seven anterior pituitary hormones, either stimulating or inhibiting their secretory activity. The discovery of the hypothalamic releasing hormones has proved to be one of the more important recent developments in vertebrate endocrinology, opening an entire new area of investigation into the orchestration of vertebrate endocrine function.

The first physiological evidence for the neurohumoral control of the anterior pituitary gland came in the late 1950s with the discovery of a substance that stimulates the release of *adrenocorticotropic hormone* (*ACTH*) from the anterior pituitary. This substance, obtained by the extraction of the hypothalamus from thousands of pigs, was given the name *corticotropin-releasing hormone* (*CRH*). Minute amounts of CRH are liberated from neurosecretory cells in the hypothalamus when they are activated by neural input in response to a variety of stressful stimuli to the organism (e.g., cold, fright, sustained pain). The axons of these cells terminate close to capillaries in the median eminence, from which their secretions are carried directly

via the portal vessels to the capillary bed of the anterior pituitary (Figure 11-14). Because there is a direct portal connection from the hypothalamus to the anterior pituitary, the amount of releasing hormone (in this case CRH) required to produce effective concentrations in the anterior pituitary tissue need not be large. Once the releasing hormones enter the total circulation, they are diluted to ineffective concentrations and are enzymatically degraded within several minutes.

The three release-inhibiting hormones of the hypothalamo-hypophysial relay system suppress the release from the adenohypophysis of *melanocyte-stimulating hormone* (*MSH*), prolactin, and GH. Of these, GH is also under the control of a releasing hormone (Table 11-6). These relations are summarized in Figure 11-15, which shows (1) the closed, short and long feedback loops of the hypothalamo-hypophysial relay system, controlled by ACTH, TSH, FSH, and LH, and (2) the open-loop control of nonendocrine tissues by GH, prolactin and MSH.

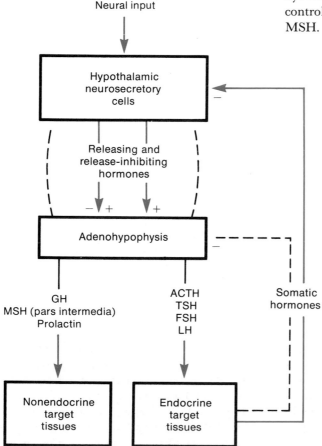

11-15  Regulatory pathways for the adenohypophysial hormones. Production of GH, MSH, and prolactin is regulated by releasing hormones and release-inhibiting hormones carried by the portal vessels from the median eminence. In contrast ACTH, TSH, FSH, and LH all stimulate the release of hormones from somatic (i.e., nonpituitary) target tissues, and those hormones exert negative feedback on the hypothalamic neurosecretory cells and perhaps also on the corresponding adenohypophysial cells themselves.

**TABLE 11-7** Hormones that regulate energy metabolism.

| Hormone | Tissue of origin | Structure | Target tissue | Primary action | Regulation |
|---|---|---|---|---|---|
| Insulin | Pancreas $\beta$-cells | Peptide | All tissues (except most neural tissue) | Increases glucose and amino acid uptake by cells | High plasma glucose and amino acids, presence of glucagon increase secretion; somatostatin inhibits secretion |
| Glucagon | Pancreas $\alpha$-cells | Peptide | Liver; fat | Stimulates glycogenolysis and releases glucose from liver; lipolysis | Low serum glucose increases secretion |
| Thyroxin | Thyroid | Amino acid derivatives | Most cells, but especially those of muscle, heart, liver, and kidney | Increases metabolic rate, thermogenesis, growth, and development; promotes amphibian metamorphosis | TSH secretion induces release |
| Norepinephrine and epinephrine | Adrenal medulla, chromaffin cells | Amino acid derivative (catecholamine) | Most cells | Increases cardiac activity; induces vasoconstriction; increases glycolysis, hyperglycemia, and lipolysis | Sympathetic stimulation via splanchnic nerves increases secretion |
| Growth hormone (GH) | Adenohypophysis | Peptide | All tissues | Stimulates RNA synthesis, protein synthesis, and tissue growth; increases transport of glucose and amino acids into cells; increases lipolysis and antibody formation | Insulin and elevated amino acid levels stimulate release via GRH and GIH; somatostatin inhibits release |
| Glucocorticoids | Adrenal cortex | Steroid | Most cells | Stimulates mobilization of amino acids from muscle and gluconeogenesis in liver to produce elevated blood glucose; exhibits anti-inflammatory action | Physiological stress; biological clock via CRH and ACTH |

# *Metabolic and Developmental Hormones*

### *GLUCOCORTICOIDS*

The adrenal gland, which is situated close to the kidney, is actually composed of two functionally unrelated glandular tissues—an outer cortex and an inner medulla. Cells of the adrenal cortex are stimulated by ACTH to synthesize and secrete a family of steroids termed *glucocorticoids* (Table 11-7) derived from cholesterol. These include *cortisone, cortisol, corticosterone,* and *11-deoxycorticosterone* (Figure 11-16). The basal level of secretion is maintained by a feedback action of glucocorticoids on the CRH-secreting neurons of the hypothalamus and on the adenohypophysis (Figure 11-17). This basal level of glucocorticoid secretion undergoes a diurnal rhythm resulting from cyclically varying CRH-secretion, which appears to be influenced by an endogenous biological clock. Basal glucocorticoid levels in humans are maximal during the early hours of the morning prior to waking. This is adaptively useful because of the energy-mobilizing consequences of these hormones. In addition to the endogenous cycle of secretion, the adrenal cortex is stimulated to secrete glucocorticoids in response to stress of various types (including starvation). Stress, acting through the nervous system, causes an elevation in ACTH and hence stimulation of the adrenal cortex.

Il-Deoxycorticosterone

Corticosterone

Cortisol (Hydrocortisone)

Cortisone

Aldosterone

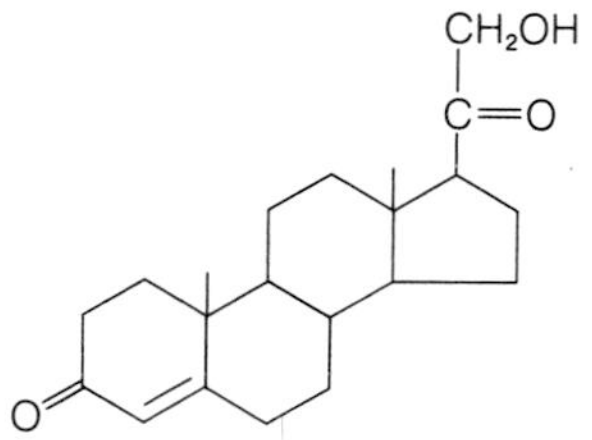

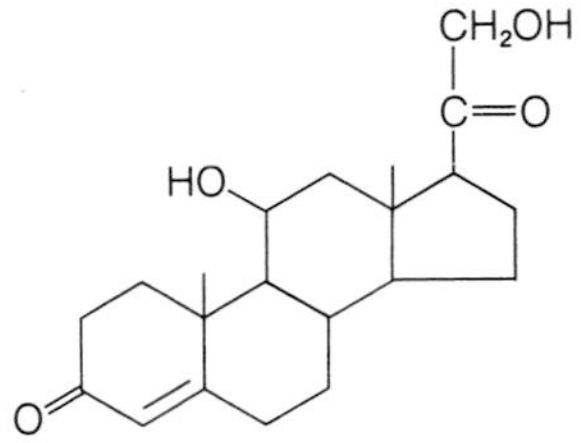

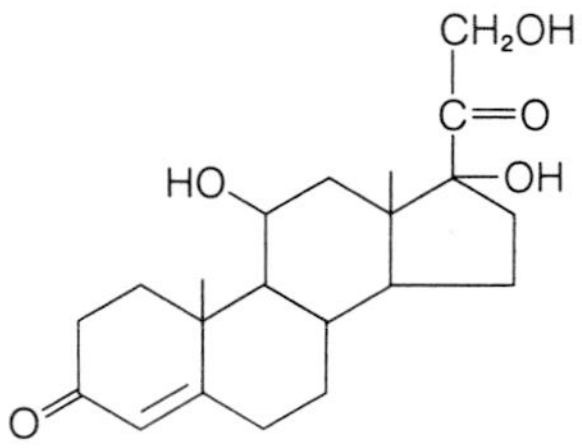

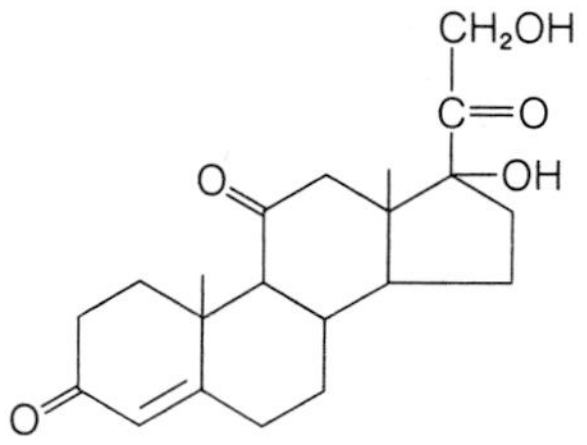

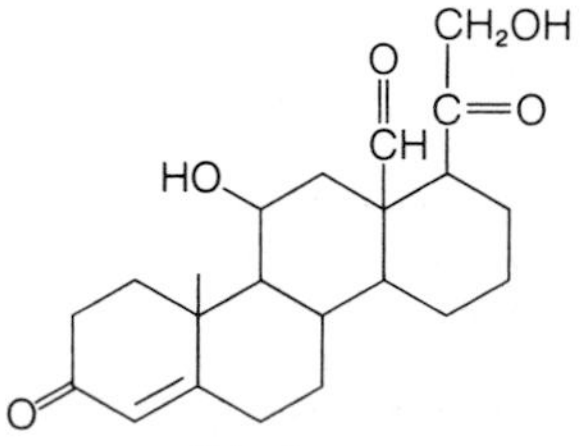

**11-16** Structures of some adrenal steroids.

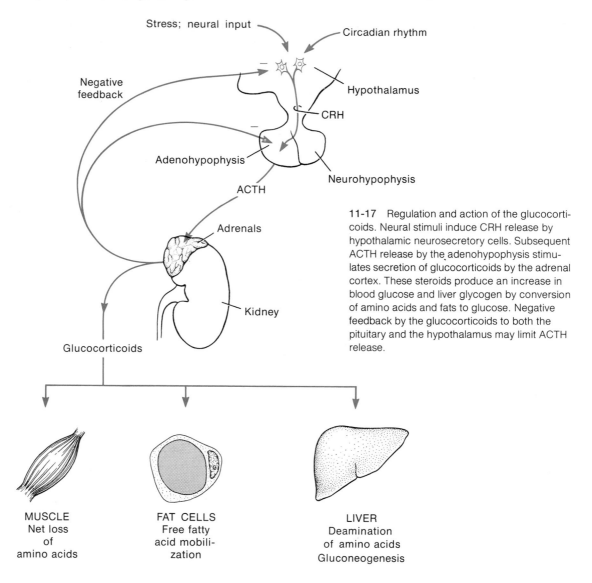

Stress; neural input — Circadian rhythm

Negative feedback

Hypothalamus

CRH

Adenohypophysis

Neurohypophysis

ACTH

Adrenals

Kidney

Glucocorticoids

11-17 Regulation and action of the glucocorticoids. Neural stimuli induce CRH release by hypothalamic neurosecretory cells. Subsequent ACTH release by the adenohypophysis stimulates secretion of glucocorticoids by the adrenal cortex. These steroids produce an increase in blood glucose and liver glycogen by conversion of amino acids and fats to glucose. Negative feedback by the glucocorticoids to both the pituitary and the hypothalamus may limit ACTH release.

MUSCLE
Net loss
of
amino acids

FAT CELLS
Free fatty
acid mobili-
zation

LIVER
Deamination
of amino acids
Gluconeogenesis

The glucocorticoids act on the liver, increasing the synthesis of enzymes that promote *gluconeogenesis* (synthesis of glucose from substances other than carbohydrates). The glucose is released into the circulation, causing a rise in blood glucose levels. Some of the glucose is stored as glycogen in the liver. While glucose is mobilized by the liver, the uptake of amino acids by muscle tissues is decreased, and amino acids are released from muscle cells into the circulation. This release increases the quantity of amino acids available for deamination and conversion into glucose in the liver under glucocorticoid stimulation. Another action of the glucocorticoids is the mobilization of fatty acids from stores of fat in adipose tissue. Again, this has the effect of increasing the available substrates for gluconeogenesis in the liver. All these actions tend to produce *hyperglycemia* (i.e., increased blood glucose levels). The mobilization of glucose produced from noncarbohydrate sources increases the availability of quick energy to muscle and nervous tissue.

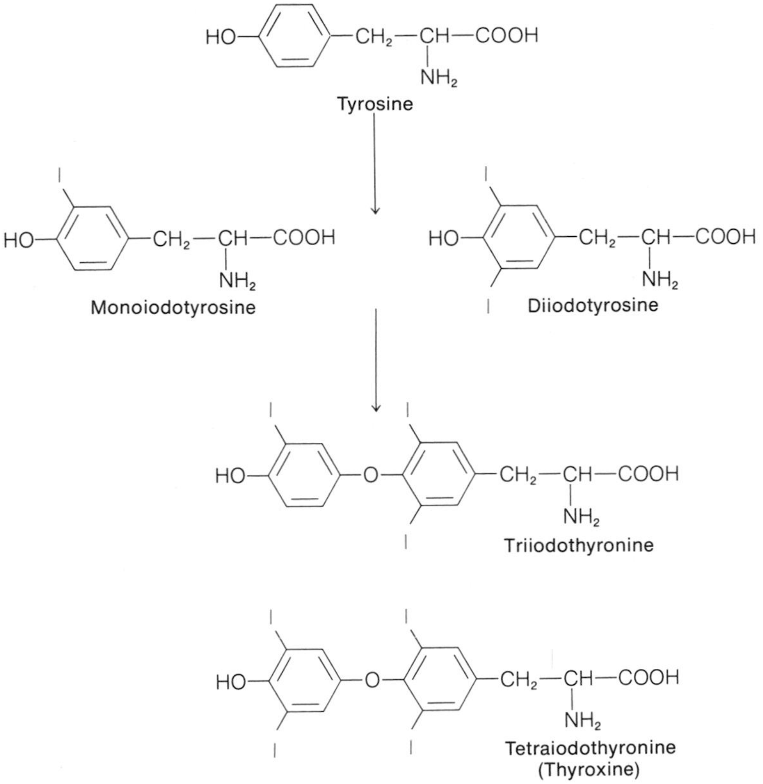

**11-18** Formation of the thyroid hormones DIT, $T_3$, and thyroxine by the condensation of iodinated molecules of the amino acid tyrosine in a peptide linkage. $T_3$ is also produced by removal of one iodide from $T_4$.

## THYROID HORMONES

Thyroid-stimulating hormone, released from the adenohypophysis, maintains the volume, weight, and secretory activity of the thyroid gland. As its name implies, it stimulates the release of thyroid hormones. The two major thyroid hormones, *thyroxine* and *3,5,3-triiodothyronine,* are synthesized in the follicles of the thyroid tissue from two iodinated tyrosine molecules (Figure 11-18). Iodine is actively accumulated by the thyroid tissue from the blood.

The release of TSH is regulated by the secretion of *TSH-releasing hormone (TRH)* from the median eminence. Both the hypothalamic neurosecretory cells, which produce TRH, and the TSH-secreting cells of the adenohypophysis are inhibited by a rise in the circulating levels of thyroid hormones (Figure 11-19). Superimposed on this chemical regulation, however, is neural stimulation of the hypothalamus; a low skin temperature, for example, will reflexly stimulate the release of hypothalamic TRH.

The thyroid hormones act on the liver, kidney, heart, nervous system, and skeletal muscle, sensitizing these tissues to epinephrine and stimulating cellular respiration, oxygen consumption, and metabolic rate. The effects of thyroid develop slowly; after a rise in thyroid hormone concentration, up to 48 hours may pass before the effects are seen.

The acceleration of metabolism stimulated by thyroid hormones leads to a rise in heat production. This is of major importance in the *thermoregulation* of homeotherms (birds and mammals). The catecholamines (discussed in the next subsection) also participate in thermoregulatory responses.

The thyroid hormones have only a limited thermogenic effect in poikilotherms ("cold-blooded" lower vertebrates), although these hormones do have important metabolic and developmental effects. Thyroid hormones have been reported to play a role in physiological adaptation to changes in environmental salinity associated with the migrations of euryhaline teleosts

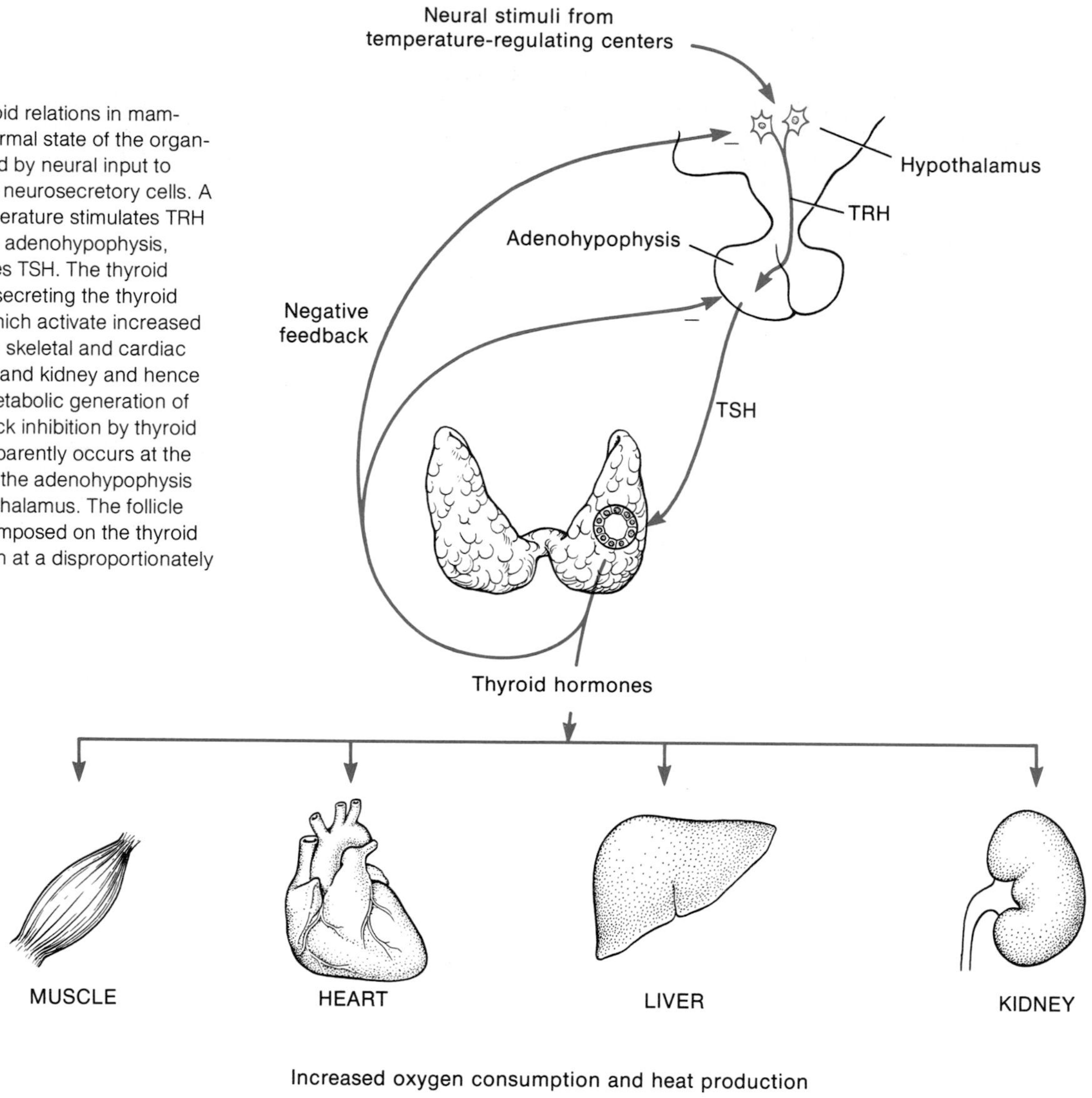

**11-19**  Thyroid relations in mammals. The thermal state of the organism is signaled by neural input to hypothalamic neurosecretory cells. A low skin temperature stimulates TRH release to the adenohypophysis, which secretes TSH. The thyroid responds by secreting the thyroid hormones, which activate increased metabolism in skeletal and cardiac muscle, liver, and kidney and hence lead to the metabolic generation of heat. Feedback inhibition by thyroid hormones apparently occurs at the levels of both the adenohypophysis and the hypothalamus. The follicle shown superimposed on the thyroid gland is drawn at a disproportionately large scale.

Neural stimuli from temperature-regulating centers

Hypothalamus

TRH

Adenohypophysis

Negative feedback

TSH

Thyroid hormones

MUSCLE          HEART          LIVER          KIDNEY

Increased oxygen consumption and heat production

(such as salmon) from fresh water into seawater. In some teleosts, increased thyroid secretion induces a behavioral preference for salt water; in others, a preference for fresh water.

The thyroid hormones play important and varied roles in the development and maturation of various vertebrate groups. The developmental effects of thyroid hormones occur only in the presence of GH, and vice versa. Both GH and the thyroid hormones acting together promote protein synthesis during development. *Hypothyroidism* during early stages of development in fish, birds, and mammals results in a deficiency disease (*cretinism* in humans) in which somatic, neural, and sexual development are severely retarded, the metabolic rate is reduced to as little as about half the normal rate, and resistance to infection is reduced. In young humans, the lack of dietary iodine causes cretinism. It also causes goiter, a malady character-

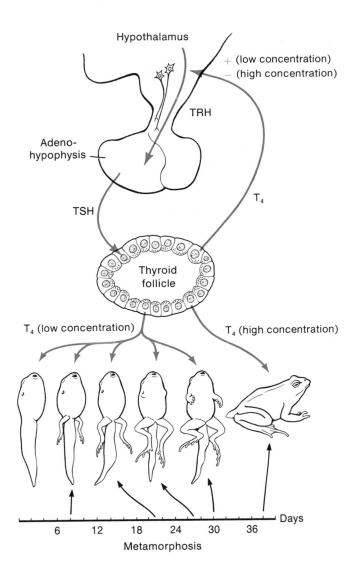

Hypothalamus

+ (low concentration)
− (high concentration)

TRH

Adeno-
hypophysis

$T_4$

TSH

Thyroid
follicle

$T_4$ (low concentration)

$T_4$ (high concentration)

Days

6    12    18    24    30    36

Metamorphosis

**11-20** Role of thyroid hormones in the control of metamorphosis in the grass frog. During the first 20 days (premetamorphosis), the median eminence is undifferentiated, and TRH and TSH secretion is low. The thyroid gland is also poorly developed and inactive except for iodine binding and hormone synthesis. The next 20 days (prometamorphosis) are characterized by maturation of the thyroid gland, with increasing iodine uptake and secretory activity. Tissue sensitivity to the thyroid hormone ($T_4$) is age-dependent and thus determines in part the timetable of morphological transformations during amphibian metamorphosis. [Spratt, 1971.]

ized by an enlargement of the throat due to hypertrophy of the thyroid tissue. Hypertrophy is caused by excessive TSH production that stems from the absence of negative feedback to the hypothalamus in the form of thyroid hormones. Both cretinism and goiter have been reduced in areas where table salt is routinely "iodized" and the populace is no longer dependent on natural trace amounts of iodine in food.

The role of thyroid hormones in development is especially dramatic in the metamorphosis of amphibians. In the absence of thyroxine or triiodothyronine, tadpoles fail to metamorphose into frogs. The role of thyroid hormones in the metamorphosis of the common grass frog is outlined in Figure 11-20. The development of tadpole into frog occurs in three stages:

*1.* During *premetamorphosis* (about 20 days), the immature thyroid binds iodine and synthesizes thyroxine.

*2.* The first part of metamorphosis, *prometamorphosis,* is characterized by slow morphological changes, growth of the thyroid gland, iodine concentration and binding, increased secretory activity of the thyroid tissue, and differentiation of the median eminence and hypothalamus.

*3.* In the final stage, the *metamorphic climax,* in which the adult form emerges, the median eminence undergoes its final differentiation and becomes highly vascularized.

Sensitivity of metamorphosing tissues to thyroid hormones increases with development, but postmetamorphic growth in the frog occurs under the direction of growth hormone.

### THE CATECHOLAMINES

The catecholamines are secreted by specialized cells derived from the neural crest and located in the central region (medulla) of the adrenal gland. These cells are called chromaffin cells because they stain readily with chromium-containing dyes. The chromaffin tissue of the adrenal medulla secretes catecholamines in response to activation of the preganglionic neurons of the sympathetic nervous system (Figure 8-13). Whereas postganglionic sympathetic nerve cells secrete norepinephrine almost exclusively, the chromaffin cells secrete mostly epinephrine (Figure 6-34), norepinephrine amounting to about one-fourth the total secretion. Cells that contain catecholamines can be conveniently identified in appropriately fixed preparations by a characteristic yellow fluorescence when viewed in ultraviolet light, the fluorescence arising from the catecholamine molecules stored in the secretory granules of these cells. Fluorescence and electron microscopy both indicate that epinephrine and norepinephrine are produced in two different sets of cells intermingled in the chromaffin tissue of the adrenal medulla.

The two adrenal catecholamines have a large number of actions, most of which contribute to the sympathetic fight-or-flight response to emergencies. They have a strong glycogenolytic effect in skeletal and cardiac muscle, as will be discussed later. The effect is to mobilize glucose in those tissues (e.g., muscle and cardiovascular system tissues) that are stimulated during the response to emergencies. In addition, these hormones stimulate the strength and rate of the heartbeat and the contraction of vascular smooth muscle, thereby raising the blood pressure.

Epinephrine and norepinephrine are not equivalent in their actions (Table 11-8), although in some tissues they exert similar effects. Differences in their actions are due to differences in the membrane receptor molecules with which they interact. Epinephrine binds most effectively to $\beta$-adrenergic receptors, and norepinephrine binds to $\alpha$-adrenergic receptors (p. 296).

### INSULIN AND GLUCAGON

The level of glucose in the blood is enhanced by uptake from the gut and release into the circulation from storage within cells. The level drops as a result of uptake into various tissues for metabolism or storage. The uptake of glucose into cells is stimulated by *insulin,* which is secreted by the $\beta$-cells of the *pancreatic islets,* small patches of endocrine tissue scattered throughout the exocrine tissue of the *pancreas.* Insulin facilitates the movement of glucose from its higher concentration in the plasma to its lower concentration in the cells of tissues and organs. It does this by increasing the facilitated transport of glucose across cell membranes.

**TABLE 11-8** Physiological responses to epinephrine and norepinephrine.

|  | Epinephrine | Norepinephrine |
| --- | --- | --- |
| Heart rate | Increase | Decrease† |
| Cardiac output | Increase | Variable |
| Total peripheral resistance | Decrease | Increase |
| Blood pressure | Rise | Greater rise |
| Respiration | Stimulation | Stimulation |
| Skin vessels | Constriction | Constriction |
| Muscle vessels | Dilatation | Constriction |
| Bronchus | Dilatation | Less dilatation |
| Metabolism | Increase | Slight increase |
| Oxygen consumption | Increase | No effect |
| Blood sugar | Increase | Slight increase |
| Kidney | Vasoconstriction | Vasoconstriction |

†This effect is secondary to peripheral vasoconstriction that raises blood pressure. In the isolated heart, norepinephrine increases the rate.
*Source:* Bell et al., 1972.

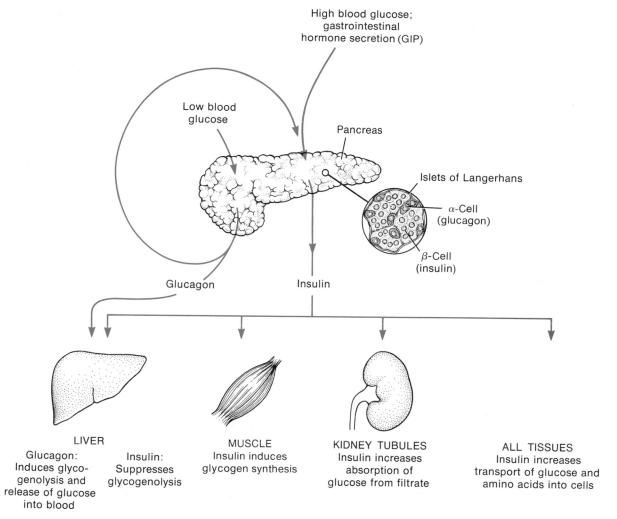

High blood glucose;
gastrointestinal
hormone secretion (GIP)

Low blood
glucose

Pancreas

Islets of Langerhans

α-Cell
(glucagon)

β-Cell
(insulin)

Glucagon

Insulin

**LIVER**
Glucagon:
Induces glyco-
genolysis and
release of glucose
into blood

Insulin:
Suppresses
glycogenolysis

**MUSCLE**
Insulin induces
glycogen synthesis

**KIDNEY TUBULES**
Insulin increases
absorption of
glucose from filtrate

**ALL TISSUES**
Insulin increases
transport of glucose and
amino acids into cells

**11-21** Relationships of insulin and glucagon in the regulation of glucose metabolism. High levels of blood glucose and glucagon and/or gastrointestinal hormones signaling food ingestion, such as gastrointestinal inhibitory peptide (GIP), stimulate the pancreatic β-cells to secrete insulin, which stimulates glucose uptake in all tissues. Glucagon, secreted by pancreatic α-cells, exerts an action that is antagonistic to that of insulin in the liver, where it stimulates glycogenolysis and glucose release.

Once glucose has entered a cell, it is immediately phosphorylated to prevent its uncontrolled escape. In muscle, glucose is either metabolized or converted into glycogen for storage. In the cells of the liver, insulin increases energy storage by stimulating glycogenesis (polymerization of glucose into glycogen) and lipogenesis. Adipose cells also respond to insulin by increasing glucose uptake and lipogenesis. Insulin and GH share the property of promoting the uptake and incorporation of amino acids into protein; but insulin also inhibits the conversion of amino acids into glucose by gluconeogenesis, an action counter to that of the glucocorticoids.

High blood glucose acts as the major secretory stimulus to the β-cells of the pancreas, which respond by discharging insulin (Figure 11-21). The release of insulin is also stimulated by glucagon, by *gastric inhibitory peptide* (*GIP,* also known as *glucose-dependent insulin-releasing peptide*), by epinephrine, and by elevated levels of amino acids.

*Diabetes mellitus* in humans is usually due to a disorder of the pancreatic $\beta$-cells, in which insulin production is either lost or impaired, or from which a defective form of insulin is secreted. The symptoms can be reversed by administration of insulin or an agent with similar pharmacological properties and by diet, or even, in certain cases, by diet alone. Insulin deficiency causes severe hyperglycemia, *glycosuria* (spillover of excess glucose into the urine), reduced ability to metabolize carbohydrates or convert them into fat, and loss of protein, which is broken down for energy in place of glucose. In addition, mobilized fat particles that cannot be rapidly metabolized accumulate in the blood as ketone bodies. These are excreted in the urine but can also interfere with liver function.

*Glucagon* is secreted by the $\alpha$-cells of the pancreatic islets in response to *hypoglycemia* (low levels of blood glucose). This hormone stimulates glycogenolysis, gluconeogenesis, and glucose release by the liver as other defenses against glucose starvation of tissues. The molecular basis of the action of glucagon on cell metabolism is understood perhaps better than that of any other hormone, and we will return to that subject later.

### GROWTH HORMONE

Growth hormone is dependent on the thyroid hormones for its synthesis in the adenohypophysis. The growth-enhancing effects of GH and the thyroid hormones are difficult to separate because the action of each relies on the presence of the other hormone. An important action ascribed to GH is tissue growth—in particular, the proliferation of cartilage and the consequent growth of bone. GH-stimulated tissue growth occurs by an increase in cell number. However, GH apparently does not stimulate growth of cells directly. Instead, it acts on the liver to cause the production of growth-promoting factors called *somatomedins,* which act directly on cells to promote growth. The importance of GH in the regulation of growth is seen in humans in the extreme cases of *gigantism,* caused by infantile GH hypersecretion; *acromegaly* (enlargement of bones of the head and of the extremities), caused

by late onset of GH hypersecretion; and *dwarfism,* caused by hyposecretion of GH.

The metabolic effects of GH are diverse (Figure 11-22). It strongly promotes protein synthesis, thereby stimulating growth. It depresses glycogenolysis but initiates the mobilization of stored fat for energy metabolism. The fatty acids released from adipose tissue into the bloodstream in response to GH are converted in the liver to glucose and released into the circulation. GH-stimulated fatty acid uptake in muscle further promotes the utilization of fatty acids as an energy source and thereby helps conserve glycogen stores.

GH has an "anti-insulin" action. Whereas insulin is released in response to elevated blood glucose levels, the release of GH is stimulated by a drop in plasma glucose levels. Moreover, GH has an action on glucose metabolism opposite that of insulin—namely, it blocks glucose uptake by tissues, in addition to raising blood glucose by promoting the conversion of fatty acids to glucose. Thus, the release of GH in response to insulin-induced hypoglycemia serves to counteract the low glucose state and maintain a balanced glucose level in the blood. GH reaches its peak plasma level several hours after a meal, when immediate energy supplies (e.g., blood glucose, amino acids, and fatty acids) are less abundant. In most tissues other than those of the central nervous system, GH suppresses glycolysis, thereby conserving glucose for use by the CNS.

GH stimulates insulin secretion both directly, through its action on the pancreatic $\beta$-cells, and indirectly, through its effect in elevating plasma glucose levels by stimulating gluconeogenesis from fats. The insulin released "corrects" the GH-induced hyperglycemia by promoting the cellular uptake of glucose. Thus, GH and insulin have antagonistic roles, each responding to opposite extremes of blood glucose concentration, with GH counteracting hypoglycemia and insulin counteracting hyperglycemia.

## Hormonal Regulation of Electrolyte Balance

The hormones that regulate electrolyte and water balance are:

*1.* Antidiuretic hormone, which increases the water permeability of the collecting ducts of the kidney, the

**11-22** Actions of GH and insulin. Output of insulin from the pancreatic β-cells occurs in response to high blood glucose, as after a meal. GH is released, usually several hours after a meal or after prolonged exercise, in response to insulin-induced hyperglycemia. GH causes lipolysis and fatty acid uptake by muscle tissue for energy and by the liver for gluconeogenesis. There is a general depression of glucose uptake, except in the central nervous system. The rise in plasma glucose stimulates insulin secretion, which then stimulates glucose uptake into cells and thus counteracts GH-induced hyperglycemia.

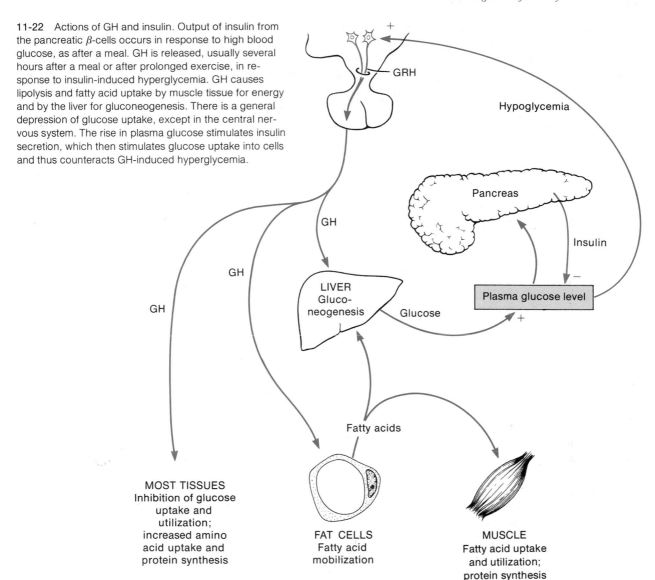

GRH

Hypoglycemia

Pancreas

GH

Insulin

GH

Plasma glucose level

GH

LIVER
Gluco-
neogenesis

Glucose

−

+

Fatty acids

MOST TISSUES
Inhibition of glucose
uptake and
utilization;
increased amino
acid uptake and
protein synthesis

FAT CELLS
Fatty acid
mobilization

MUSCLE
Fatty acid uptake
and utilization;
protein synthesis

end effect of which is a reabsorption and conservation of urinary water.

*2.* The *mineralocorticoids,* in particular *aldosterone,* which enhance retention of sodium (and, indirectly, chloride) by the kidney.

*3. Calcitonin, parathyroid hormone (PTH, or parathormone), vitamin* $D_3$, *and prolactin, which are important in the metabolism of calcium in bone, plasma, kidney, and other tissues (Table 11-9).*

Antidiuretic hormone and the mineralocorticoids

are discussed in the section on renal control in Chapter 12, so we will limit ourselves here to the third category.

Millimolar changes in the concentration of calcium ion in the plasma and other extracellular tissue fluids are more critical than changes in the concentration of most other common metal ions because of calcium's role as a regulatory agent in many cellular processes. This ion is actively absorbed through the intestinal wall into the plasma and is deposited in bone, the major depot for storage of calcium. Elimination of

Ca$^{2+}$ from the body occurs through the kidney. PTH, calcitonin, and vitamin D all participate in the regulation of Ca$^{2+}$ and PO$_4^{3-}$ exchange between the skeleton and the blood.

PTH is secreted from the paired parathyroid glands (Figure 11-23). Release occurs in response to a drop in plasma Ca$^{2+}$ levels. In its short span of activity (a half-life of about 20 min), it promotes calcium mobilization from bone, increases renal calcium uptake from kidney tubules, increases renal phosphate excretion,

**TABLE 11-9**   Other vertebrate hormones.

| Hormone | Tissue of origin | Structure | Target tissue | Primary action | Regulation |
|---|---|---|---|---|---|
| Melatonin | Pineal gland | Amino acid derivative | Melanophores (ectotherms) | Induces aggregation of melanin (skin blanching) | Darkness (or blinding) increases synthesis and secretion; synthesis and secretion under photoperiodic control |
| | | | Melanocytes (endotherms); nervous system | May regulate secretion of the gonadotropin-releasing hormones and thus seasonal reproductive cycles | |
| Calcitonin (CT) | Thyroid | Peptide | Bones, kidney | Decreases bone calcium release; decreases blood calcium levels; increases calcium and phosphorus excretion | Elevation of plasma Ca$^{2+}$ increases secretion |
| Parathyroid hormone (PTH) | Parathyroid | Peptide | Bones, kidney, intestine | Increases blood calcium and phosphorus levels; increases mobilization of bone calcium; decreases calcium excretion from kidney; increases intestinal calcium absorption | Decline in plasma Ca$^{2+}$ increases secretion |
| Erythropoietin | Kidney | Glycoprotein | Bone marrow | Causes hyperplasia of the bone marrow; increases production and release of erythrocytes | Low atmospheric O$_2$ and anemia increase secretion |
| Relaxin | Ovary | Peptide | Pelvic ligaments | Induces relaxation of pelvic ligaments and cervix | Increased progesterone and estrogen levels late in pregnancy increase secretion |
| Renin | Kidney | Protein | Blood | Activates angiotensin | Increased renal sympathetic activity, decreased plasma Na$^+$, decreased renal arteriole distention and decreased blood volume or pressure increase secretion |

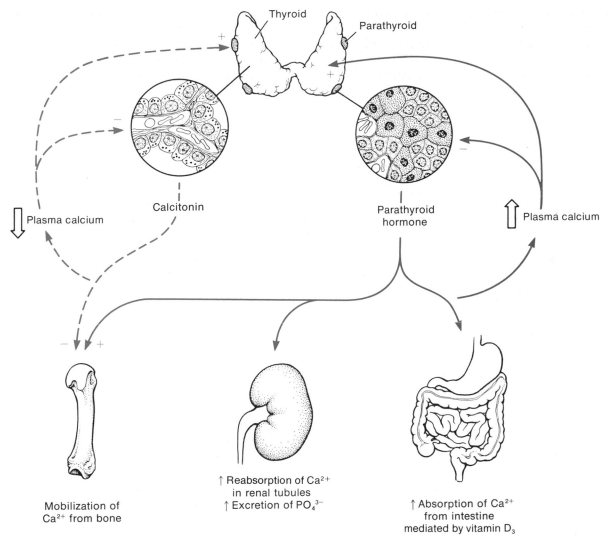

**11-23** Plasma calcium regulation by calcitonin and parathyroid hormone in mammals. Low levels of plasma calcium stimulate the cells of the parathyroid glands to release PTH, which initiates liberation of $Ca^{2+}$ from bone, enhances $Ca^{2+}$ absorption from food, and increases $Ca^{2+}$ reabsorption from the urine by the renal tubules. High concentrations of $Ca^{2+}$ in the blood induce release of calcitonin from thyroid parafollicular cells, thus promoting deposition of $Ca^{2+}$ in the bone.

and, with the help of vitamin D, enhances intestinal calcium absorption (Figure 11-23).

Calcitonin is secreted from the *parafollicular* (or *C*) *cells* in the thyroid gland in response to high calcium levels in the plasma. It rapidly suppresses calcium loss from bone. Calcitonin acts more rapidly on bone than

PTH and produces the dominating effect. Although the two hormones act opposingly, there is no feedback interaction between them. The dominance of calcitonin prevents hypercalcemia and extensive dissolution of the skeleton. Essentially, then, bone acts as a large reservoir and buffer for $Ca^{2+}$ and $PO_4^{3-}$. Plasma

TABLE 11-10  Prostaglandins.

| Tissue of origin | Structure | Target tissues | Primary action | Regulation |
|---|---|---|---|---|
| Seminal vesicles, uterus, ovaries | Cyclical unsaturated fatty acids | Uterus, ovaries, fallopian tubes | Potentiates smooth muscle contraction and possibly luteolysis; may mediate LH stimulation of estrogen; progesterone synthesis | Introduced during coitus with semen |
| Kidney | | Blood vessels (especially renal) | Regulates vasodilation or constriction | Increased angiotensin II and increased epinephrine stimulate secretion; inactivated in lungs and liver |
| Neural tissue | | Adrenergic terminals | Blocks norepinephrine-sensitive adenylate cyclase | Neural activity increases level |
| Most cells | | Most cells | Alters cAMP levels | |

$Ca^{2+}$ and $PO_4^{3-}$ are held within narrow limits by PTH and calcitonin, which have opposite actions, regulating the flux of these minerals between plasma and bone.

Vitamin D is a steroidlike compound that is synthesized in the skin in response to sunlight and is ingested along with some dietary oils. It is converted in the liver and kidneys into *1,25-dihydroxycholecalciferol,* which stimulates intestinal calcium absorption.

## Prostaglandins

The cyclic unsaturated fatty acids called prostaglandins (Figure 11-1B) were first discovered in the 1930s in seminal fluid. They were thought to be produced by the prostate gland—hence the name. Seminal prostaglandins have been shown, instead, to be produced by the seminal vesicles. They have now been found in all mammalian tissues that have been studied. There are several prostaglandins: PGA, $PGE_1$, $PGE_1\alpha$, and $PGE_2\alpha$. These are converted to other prostaglandins that are also biologically active. The prostaglandins undergo rapid oxidative degradation to inactive products in the liver and lungs.

The prostaglandins are numerous and have diverse actions on a variety of tissues (Table 11-10), making generalization difficult, although many of the effects involve smooth muscle. They are produced by many tissues, in some cases acting locally and in others acting on distant target tissues.

## Sex Hormones

The sex hormones of vertebrates—steroids derived from cholesterol (Figure 11-24) in the gonads and adrenal cortex of both sexes—include the *estrogens,* the *androgens,* and *progesterone.* Apparently the only function of progesterone in males is as a biosynthetic intermediate. The estrogens and androgens are important in various aspects of growth, development, and morphological differentiation, as well as in the development and regulation of sexual and reproductive behavior and cycles. Estrogens predominate in the female, and androgens predominate in the male. The production and secretion of these steroids is under the control of follicle-stimulating hormone (FSH) and luteinizing hormone (LH) (Table 11-5). Both of these gonadotropins are present in both sexes.

11-24 Synthesis of sex hormones from cholesterol.

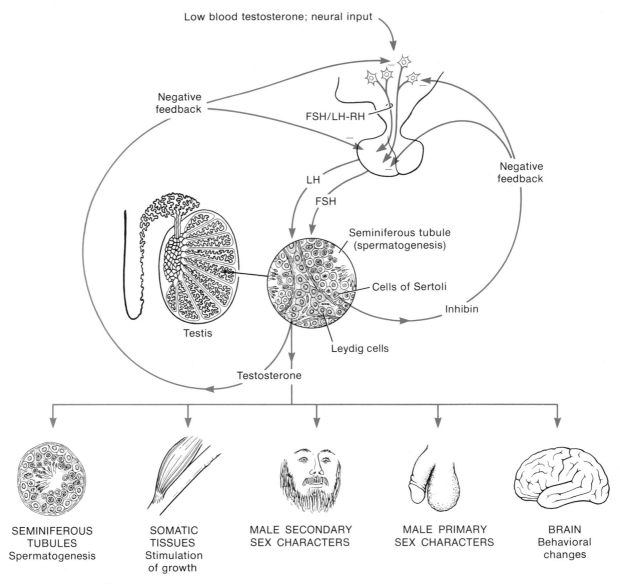

**11-25**   Regulation and actions of testosterone. A decrease in blood testosterone stimulates the secretion of the releasing hormone FSH/LS-RH, which promotes the release of FSH and LH. Some of the actions of testosterone are indicated at the bottom of the figure. Inhibin, which is secreted by the testes along with testosterone, is inhibitory to the FSH-secreting cells of the adenohypophysis.

The mammalian testes consist primarily of tubules lined with germ cells, the *seminiferous tubules* (Figure 11-25). Under stimulation by FSH, spermatogenesis occurs in these tubules after sexual maturity, either con-tinuously or seasonally, depending on the species. The *cells of Sertoli,* which also line the tubules, support development of the sperm and are responsible for the synthesis of an androgen-binding protein. The *interstitial*

**TABLE 11-11** Steroid hormones.

| Hormone | Tissue of origin | Target tissue | Primary action | Regulation |
|---|---|---|---|---|
| Testosterone (androgen) | Cells of Leydig | Most tissues | Promotes development and maintenance of masculine characteristics and behavior; spermatogenesis | Increased LH level stimulates secretion |
| Estradiol-17$\beta$ (estrogen) | Ovarian follicle, corpus luteum | Most tissues | Promotes development and maintenance of female characteristics and behavior; oocyte maturation; uterine proliferation | Increased FSH and LH levels stimulate secretion |
| Progesterone (progestin) | Corpus luteum | Uterus, mammary glands | Maintains uterine secretion and stimulates mammary duct formation | Increased LH and prolactin levels stimulate secretion |
| Cortisol (glucocorticoid) | Adrenal cortex | Liver, adipose tissue, muscle | Stimulates the transfer of amino acids from muscle and fatty acids from adipose cells to liver; gluconeogenesis; antiinflammatory action | Increased ACTH level stimulates secretion |
| Aldosterone (mineralocorticoid) | Adrenal cortex | Distal kidney tubules | Promotes resorption of $Na^+$ from urinary filtrate | Angiotensin II stimulates secretion |

*cells,* or *Leydig cells,* which lie between the seminiferous tubules, are the main site of steroid synthesis in the testes, also promoted by LH. The testicular androgens, of which *testosterone* (Table 11-11) and *dihydrotestosterone* are the most important, trigger development of the penis, vas deferens, seminal vesicles, prostate gland, epididymis, and such secondary sex characteristics as the lion's mane, the rooster's comb and plumage, and facial hair in male humans. The androgens also contribute to general growth and protein synthesis—in particular, the synthesis of myofibrillar proteins, as evidenced by the enhanced muscularity of the males relative to the females of many vertebrate species.

The estrogens are a family of steroids produced in the vertebrate ovary, testis, and adrenal cortex. Cholesterol is converted to progesterone, which is then transformed to the androgens androstenedione and testosterone. The estrogens, of which *estradiol-17$\beta$* is the most potent, are made from these androgens. The female genitalia develop because of an absence of androgens in the female embryo. However, estrogens stimulate further growth and maturation of the female sex organs. The estrogens are also responsible for the development of the secondary sex characteristics and for the regulation of reproductive cycles (Figure 11-26), but their function in the male is not yet understood.

Prenatal sexual differentiation of the genital tract is dependent on the secretion or absence of androgens during a critical period in development. Without them, the female configuration develops with the Müllerian ducts retained and the Wolffian ducts eliminated. In the presence of androgens, the male tract develops (Wolffian ducts retained; Müllerian ducts eliminated). In normal male development, maternal estrogen is bound to a protein that serves to keep the circulating levels of estrogens within the male fetus low.

## REPRODUCTIVE CYCLES

Simultaneous reproduction within an entire population can be of obvious survival value to a species. The gathering of large numbers of individuals of both sexes for mating, bearing young, and tending to the needs of the young during their period of high vulnerability can be timed to coincide with favorable

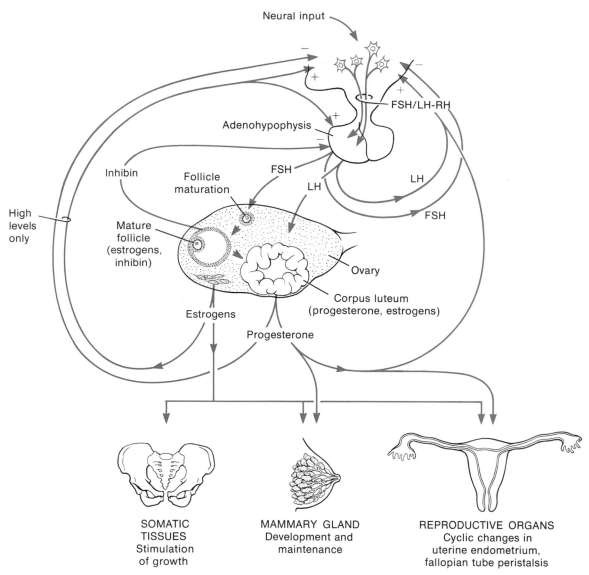

**11-26** Regulation and actions of the female sex hormones in mammals. Drops in progesterone and estrogen levels permit the release of FSH/LH-RH and hence FSH. The FSH stimulates development of the primordial follicles. Estrogens secreted by the follicles and by the interstitial cells eventually reach levels that inhibit FSH/LH-RH. The LH surge associated with ovulation and subsequent development of the corpus luteum is due to the stimulating effect of high estrogen levels on FSH/LH-RH at that time. The corpus luteum secretes primarily progesterone and estrogens, which function to maintain pregnancy.

weather and an adequate food supply. Moreover, the sudden appearance of large numbers of defenseless individuals of a species can have an overwhelming effect on even the most voracious of predators, per-

mitting the survival of enough individuals of the new generation to assure survival of the species. In general, reproductive cycles arise from within the animal, but in many species the inner cycles are entrained by environmental signals, such as the changes in day length that accompany the change of seasons.

In vertebrates, the neuroendocrine mechanisms of the hypothalamus and adenohypophysis play important roles in the sexual and reproductive cycles. The pituitary gonadotropins (LH and FSH) maintain the activity of the testes and ovaries. Steroid-secreting cells in the gonads and the adrenal cortex are responsible for secretion of sex hormones. Secretion of pituitary gonadotropin is modulated by the feedback of sex steroids to the hypothalamic neurons, which secrete the corresponding releasing hormones (Table 11-6), and possibly by direct action of these steroids on the corresponding endocrine cells of the adenohypophysis. The interactions between the gonadotropins and steroids provide the endocrine basis for the cyclicity most apparent in the mammalian female.

The female bird or mammal is born with her full complement of *oocytes* (each of which gives rise to one *ovum*), and each oocyte becomes embedded in a *follicle* within the ovary. Most of the follicles and their oocytes degenerate early, but even before puberty some develop just short of yolk formation or maturation. In humans, about 400 ova are available for release between *menarche* (onset of *menstruation*) and *menopause*. Oogenesis in lower vertebrates occurs throughout life.

In mammalian females, a chemical fugue begins with the first trickle of FSH (Table 11-5) during puberty. FSH stimulates development of the follicles to the ripe form, each a fluid-filled cavity enclosed by a membranous sac of several cell layers in which the ovum is embedded. One of these layers, the *theca interna*, is the ovarian site of androgen biosynthesis and secretion under the influence of LH. FSH stimulates the production of an enzyme in the cells of the *ovarian granulosa* that converts the androgens to estrogens. Thus, the granulosa cells are the actual site of estrogen synthesis and secretion. The adult ovarian cortex contains follicles in all stages of growth from *primary* to *Graafian follicles*. During the follicular phase of the ovarian cycle, the FSH- and LH-induced ripening of the follicle leads to increasing production and release of estrogen into the blood (Figure 11-27). At high concentrations of estrogen, characteristic of the time of ovulation, estrogen feeds positively back onto hypothalamic FSH/LH-RH secreting cells and onto the FSH- and

LH-secreting cells of the adenohypophysis, *stimulating* increased release of FSH and thereby accelerating follicle maturation. At other times, however, the estrogens *inhibit* the release of FSH, possibly by lowering the sensitivities of the adenohypophysial secretory cells. The follicular phase is also the time of proliferation of the *endometrium,* the tissue that lines the uterus.

A surge of LH (Table 11-5), along with the increased release of FSH, precipitates ovulation. The ovum bursts forth from the follicle and begins its journey down the ciliated fallopian tube. At that time, estrogen secretion declines, and under the influence of LH the follicle becomes transformed into a temporary endocrine tissue, the *corpus luteum.* During the *luteal phase,* the corpus luteum secretes estrogen and progesterone, the latter being responsible for the stimulation of the secretion of endometrial fluid by the endometrial tissue. In the absence of fertilization and implantation of an ovum, the corpus luteum degenerates after a $14 \pm 1$–day period (in humans), and secretions of estrogen and progesterone subside. FSH and LH secretions by the pituitary, which remain low during the luteal phase, are then able to increase again. In humans and some other primates, this precipitates the *menses,* or shedding of the uterine lining. With the reduction in estrogen concentration, FSH/LH-RH secretion resumes, and a new cycle is thereby initiated.

In the event that an ovum is fertilized and becomes implanted in the endometrium of a placental mammal, an endocrine signal—in the form of *chorionic gonadotropin (CG)* originating in the placenta—induces further growth of the active corpus luteum, so that progesterone and estrogen secretion continues. The CG is similar to the LH produced by the pituitary, but not identical. The placenta begins secreting CG within about a day of implantation of the ovum and effectively takes over the gonadotropic functions of the pituitary during pregnancy, in essence "rescuing" the corpus luteum. Pituitary FSH and LH are not secreted again until after parturition (birth of the fetus). In many mammals, including humans, the corpus luteum continues to grow and to secrete progesterone and some estrogen until the placenta fully takes over, at which time the corpus luteum degenerates. In other mammals, such as the rat, the secretions of the corpus luteum, which are stimulated by prolactin (Table

454

11-27 The primate menstrual cycle. Before ovulation, FSH promotes maturation of the follicle, which secretes estrogen. At *high* estrogen levels, a surge of LH is initiated. The LH triggers ovulation, promotes development of the corpus luteum, and induces the corpus luteum to secrete progesterone and some estrogen. If implantation and pregnancy occur, secretion of chorionic gonadotropin (CG) by the placenta "rescues" the corpus luteum, which maintains secretion of estrogen and progesterone for the first two to three months of pregnancy in the human. Thereafter, the placenta itself secretes estrogens and progesterone. In the absence of implantation, the progesterone and estrogen levels peak and then fall, initiating menstruation. The high progesterone level that prevails just before menstruation initiates the secretion of FSH, renewing the cycle. This rhythmicity is the result of the alternating dominances of the two steroids and the two gonadotropic hormones. [McNaught and Callander, 1975.]

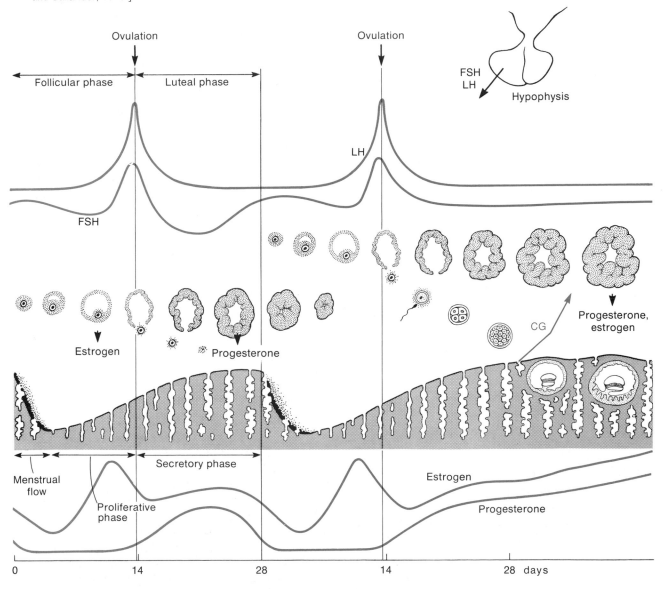

11-5), are essential to the maintenance of pregnancy throughout its term. Progesterone and estrogen prevent ovulation during gestation and initiate growth of the mammary gland ducts in preparation for lactation.

Birth control pills contain small amounts of progesterone and estradiol or of their synthetic analogs. Taken daily, these steroids mimic the earliest stages of pregnancy, preventing ovulation and also acting on the endometrium, thereby providing a highly effective (ca. 99%) means of avoiding conception.

The durations of the follicular and luteal phases of the reproductive cycle vary among different mammalian groups. They are about equal in the primate *menstrual cycle,* but in the *estrous cycle* characteristic of nonprimate mammals, the luteal phase is much shorter. The number of cycles per year also varies among species. The human menstrual cycle of approximately 28 days normally occurs 13 times a year. Among nonprimate mammals, some have only one estrous cycle per year (usually in spring) and others, such as the laboratory rat, are polyestrous throughout the year.

The estrous cycle is based on endocrine relations similar but not identical to those controlling the primate menstrual cycle. We can list four major points of difference between the menstrual and the estrous cycles:

*1.* In estrous mammals, the luteal phase is greatly reduced, so there is little or no proliferation and hence no precipitous shedding of the endometrium. As a result, estrous animals exhibit no menses.

*2.* Occurrence of the estrous cycle is often sensitive to environmental (especially seasonal) factors.

*3.* During that part of the estrous cycle in which the follicle is ready for ovulation, the female becomes receptive to the male, exhibiting pronounced behavioral changes that increase the probability of mating. This period is termed *estrus;* in the vernacular, the female is said to be "in heat."

*4.* In some estrous species (e.g., cat, rabbit), ovulation is normally triggered just before or during estrus by coitus. In the majority of estrous mammals, ovulation occurs independent of stimulation, as it does in menstrual mammals.

In the laboratory rat, in which the estrous cycle repeats itself every four days (these days are called *diestrus-1, diestrus-2, proestrus,* and *estrus*), ovulation is spontaneous, but, interestingly enough, activation of the corpora lutea is not. Instead, their activation depends on neural input from mechanical stimulation of the uterine cervix at about the time of ovulation. Such stimulation is normally associated with coitus, but it can be mimicked in the laboratory by stimulation of the cervix with a glass rod, which results in a hormonal and physiological condition known as *pseudopregnancy.* This condition lasts about 12 days, delaying the next estrous cycle by the same amount. Pseudopregnancy is interesting in the present context primarily because it demonstrates the neural influence that activates the hormonal mechanism in this species for rescuing the corpora lutea, and because pseudopregnancy resembles the early phase of pregnancy in endocrine behavior.

Hormonal levels determined in the laboratory rat during the four-day estrous cycle are shown in black in Figure 11-28. Except for a gradual buildup of estrogens released from the growing ovarian follicle, the sex hormone levels remain low during most of the cycle, exhibiting a large surge during proestrus. It is noteworthy that the estrous cycle is characterized by a greatly attenuated luteal phase, in contrast to the prominent luteal phase of the menstrual cycle.

The colored plots in Figure 11-28 show the hormonal changes induced by mechanical stimulation of the uterine cervix. These changes are characteristic of the state of pseudopregnancy as well as the initial phase of true pregnancy in the rat. Both states are characterized by repeating large surges of prolactin initiated by neural input to the hypothalamus. Interestingly, it is prolactin rather than LH that mediates the rescue of the corpora lutea in the rat, causing them to sustain the secretion of progesterone responsible for the 12-day pseudopregnancy. In the event of true pregnancy (i.e., fertilization and implantation of the ova), the corpora lutea are eventually sustained by secretion of gonadotropins by the placenta.

In mammals, prolactin and a hormone from the placenta, *placental lactogen,* aid in preparing the mammary glands for lactation. After parturition, a decrease in progesterone levels relieves inhibition of the milk-synthesizing machinery, permitting lactation to begin.

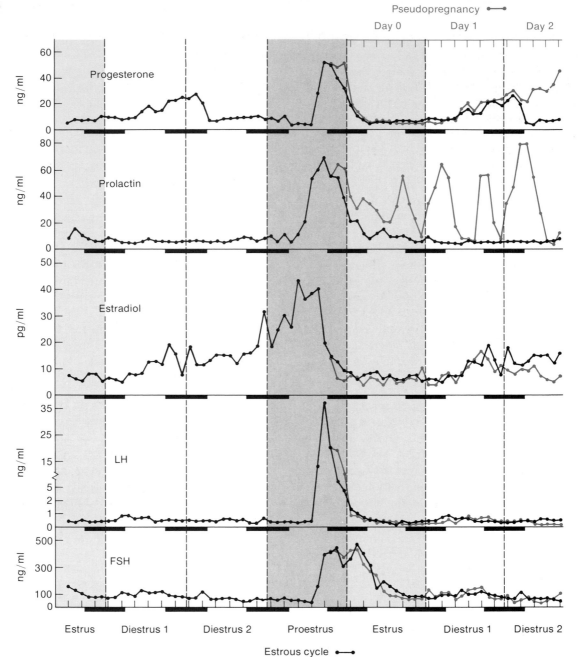

**11-28** Sex hormone levels determined in the blood of the common laboratory rat during its four-day estrous cycle and during three days of pseudopregnancy. The estrous cycle levels are plotted in black; those during pseudopregnancy in color. Pseudopregnancy was evoked by mechanical stimulation of the cervix during proestrus. By neural reflex, this stimulation leads to surges in prolactin levels, the immediate effect of which is to sustain the corpora lutea for several days, delaying the next estrous cycle by the same amount. In the event of coitus and impregnation, these events would generally be accompanied by fertilization, implantation, and pregnancy, as a consequence of which the corpora lutea would be sustained by secretion of placental gonadotropins. Note the steady climb in estrogen levels during diestrus and proestrus, correlated with the ripening of the ovarian follicles. [Smith et al., 1975.]

Milk production is mediated by prolactin along with the glucocorticoids, and the release of the milk is induced by oxytocin (Table 11-4). Both prolactin and oxytocin are released during suckling as a result of neural input to the hypothalamus from the nipple.

In both males and females of some species of birds (e.g., pigeons), prolactin stimulates secretion of "crop milk," a nutritive substance fed to chicks by regurgitation. In many birds, prolactin stimulates the development of brood patches—highly vascularized bald regions on the undersides for incubating eggs.

### SEX HORMONES AND BEHAVIOR

There are numerous examples of behavior whose development or expression is under hormonal control. The most familiar examples are seen in the sexual behavior of animals. In temperate-zone birds, the progressive increase in day length with approaching springtime induces growth of the gonads, testicular weight in some species increasing up to 500 times; this is accompanied by a large rise in the levels of sex hormones, which in turn promotes the development of various aspects of sexual behavior, such as selection of territory and singing in the male and nest-building in the female. Although hormones play a direct role in such behavior, they also play indirect roles by promoting the differentiation of sex-related morphological features, such as plumage, color, and voice. The behavior of one sex also releases behavioral responses (p. 336) in the other sex. An interesting example is the stimulation of nest-building behavior in female canaries by the singing of courting males. The poorer the repertoire of the male, the less intense the nest-building. Conversely, the more elaborate the serenade, the more intense the efforts of the female. Although the nest-building behavior itself is promoted by increased progesterone levels in the female, it is nevertheless dependent on the appropriate sensory input. Similarly, the singing of the male, though promoted by testosterone, depends for development of quantity and quality on exposure to the songs of the older, more experienced males.

Some behavioral differences between male and female mammals are due to differences in levels of sex hormones. Increased levels of androgens circulating in the blood lead to greater sexual receptivity in female primates and to greater sexual aggressiveness in all mammalian males—in short, increased sexual arousability in both sexes. The more general behavioral consequences of castration in male domestic animals and humans have been known for centuries. Thus, an aggressive bull is readily converted into a placid ox. More recently, it has been shown that aggressive behavior can be restored in castrated mice by administering either estradiol or testosterone. In contrast, simultaneous progesterone administration interferes with the induction of aggressiveness.

The modification of behavior by hormones implies that certain hormones can modify the properties of certain nerve cells. Direct evidence for endocrine influences on single nerve cells has been obtained in several studies. Recordings from single units in the lateral areas of the hypothalamus of rats at different times during the estrous cycle show that the percentage of neurons synaptically inhibited by such stimuli as pain, cold, or mechanical stimulation of the uterine cervix is higher during estrus than at other times. They found a similar augmentation of synaptic inhibition in that part of the hypothalamus when the estrogen was administered to the rat. Neurons in the septal region of the rat brain were found to respond differently to estrogen. These cells exhibit a decrease in reflex inhibition in response to cervical stimulation, pain, and cold during estrus or when estrogen is administered. Thus, an increase in the plasma level of estrogen, whether due to ovarian secretion during estrus or to experimental administration of the hormone, increases the inhibitory effects of sensory input in some neurons while decreasing it in others. Specific changes in neural excitability in various parts of the hypothalamus are also seen in response to progesterone and to prolactin.

Hormones not only influence behavior temporarily by influencing neural events, but also have, as recently proven, effects on the development and anatomy of neural tissue. Enlarged clusters of neurons associated with one sex or the other in various vertebrates have been shown to develop in the young animal in response to the presence of sex hormones of one kind or another. The implication is that certain sexually differentiated behaviors are structurally programmed into the brain under the influence of sex hormones during development.

# The Endocrine System of Insects

Endocrine cells—in particular, neurosecretory cells—have been identified in all invertebrate groups, including the primitive hydroid coelenterates. In *Hydra*, for example, neurons secrete what is believed to be a growth-promoting hormone during budding, regeneration, and growth. In most of the invertebrate groups investigated so far, the presence of hormones has been mainly inferred from morphological and histochemical observations. The endocrinology of development and metamorphosis in insects has been widely studied, however, and is appropriate for a brief overview here.

### REGULATION OF INSECT DEVELOPMENT

The life cycle of *hemimetabolous* insects—that is, those that exhibit incomplete metamorphosis, such as the Hemiptera (bugs), Orthoptera (locusts, crickets), and Dictyoptera (roaches, mantids)—begins with the development of the egg into an immature *nymphal* stage. The nymph eats and grows and undergoes several *molts*, replacing its old exoskeleton with a soft new one that expands to a larger size before hardening. The stages between molts are termed *instars*. The final nymphal instar gives rise to the *adult* stage. The development of the *holometabolous* orders—that is, those that exhibit complete metamorphosis, such as the Diptera (flies), Lepidoptera (butterflies, moths), and Coleoptera (beetles)—is more complex. The egg develops into a *larva* (e.g., maggot, "worm," caterpillar), which grows through several instars. The last larval instar undergoes an especially pronounced tanning and stiffening of its cuticle when it molts into a *pupa*. This stage is an outwardly dormant one in which extensive internal reorganization takes place. The pupa gives rise to the adult form, which shows little if any morphological resemblance to the previous stages. The larval stage of a holometabolous insect is specialized for eating and is therefore the one that causes the major damage to many agricultural crops. The adult is the reproductive stage, and in some species is not even equipped to feed.

The first experiments to demonstrate a probable endocrine control of insect development were done between 1917 and 1922 by S. Kopeč, who ligated the last larval instars of a moth at various positions along their length. He found that when the ligature was tied

before a certain critical period, the larva would pupate anteriorly to the ligature but remain larval posteriorly. Cutting the nerve cord had no effect, so he concluded that a circulating, pupa-inducing substance had its origin in a tissue located in the anterior portion of the larva. By testing various tissues, Kopeč found that removal of the brain prevents pupation and that reimplantation of the brain allows it to proceed again. It was subsequently found that a hormone secreted by cells in the brain stimulates the activity of the *prothoracic (thoracic) glands,* the tissue that elaborates the hormone that induces molting. Thus, ligating posteriorly to the prothoracic glands after their activation by the *prothoracicotropic hormone (PTTH)*, also called *brain hormone,* prevents pupation of the abdomen, whereas ligating anteriorly allows pupation to proceed. In the former case, pupation can be initiated by implanting activated prothoracic glands into the isolated abdomen.

The hardiness of insects makes them ideal subjects for the kind of experiments that were to demonstrate the humoral control of molting and metamorphosis. It is possible to carry out extended parabiosis experiments (Figure 11-29) in which two insects or two parts

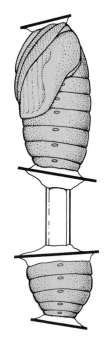

**11-29** Parabiosis as a method in insect endocrinology. Insect tissues readily survive in the face of such radical surgery as transection and decapitation. In this experiment by Williams (1974), the abdomen of one pupa is joined to another pupa through a glass tube. Glass windows at either end permit visual inspection of the developing tissues.

TABLE 11-12   Insect developmental hormones.

| Hormone | Tissue of origin | Structure | Target tissue | Primary action | Regulation |
|---|---|---|---|---|---|
| Prothoracicotropin, PTTH (brain hormone) | Neurosecretory cells in brain | Polypeptide | Prothoracic gland | Stimulates ecdysone release | Various environmental and internal cues including photoperiod, temperature, crowding, abdominal stretch; inhibited by juvenile hormone in some species |
| Ecdysone (molting hormone) | Prothoracic glands, ovarian follicle | Steroid | Epidermis, fat body, imaginal disks | Increases synthesis of RNA, protein, mitochondria, endoplasmic reticulum; stimulates secretion of new cuticle | Secretion stimulated by brain hormone |
| Juvenile hormone (JH) | Corpora allata | Terpene derivative | Epidermis, ovarian follicles, sex accessory glands, fat body | In larva: promotes synthesis of larval structures; inhibits metamorphosis; in adult: stimulates yolk protein synthesis and uptake, activates ovarian follicles and sex accessory glands | Inhibitory and stimulatory factors from the brain |
| Bursicon | Neurosecretory cells of CNS | Polypeptide | Epidermis | Promotes cuticle development; induces tanning of cuticle of newly molted adults | Stimuli associated with molting stimulate secretion |
| Diapause hormone (silk moth, *Bombyx mori*) | Neurosecretory cells in sub-esophageal ganglion | Polypeptide | Ovaries, eggs | Induces diapause of egg | Release is controlled by CNS |
| Eclosion hormone | Neurosecretory cells of brain | Polypeptide | Nervous system | Induces emergence of adult from puparium | Endogenous "clock" |

of one insect are joined so that they share a common circulation, exchanging body fluid. Windows made of thin glass make it possible to observe developmental changes.

The endocrine control of development and molting in insects depends on several major hormones (Table 11-12). The insect endocrine system responsible for development and metabolic regulation is characterized by the following cells and functions:

*1.* Neurosecretory cells, which have their cell bodies in the *pars intercerebralis* of the brain (Figure 11-30), manufacture PTTH. The chemical structure of this hormone is not known, but it appears to be a small protein of about 5000 daltons.

*2.* PTTH is shipped via axoplasmic transport to storage depots, or neurohemal organs. The *corpora cardiaca* (Figure 11-30) have been considered the neurohemal organs that store and release PTTH, but more

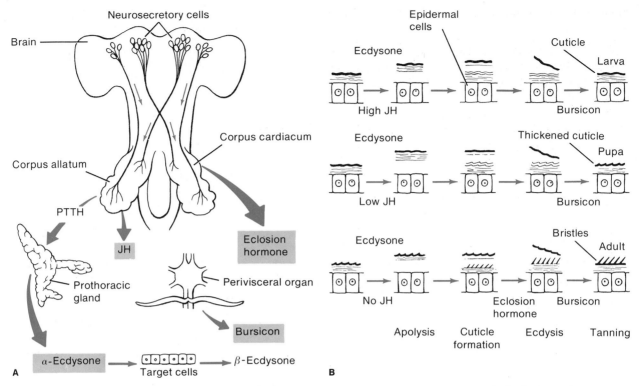

**11-30** (A) Insect endocrine system and its role in the molting process. Neurosecretory cells in the brain convey hormones via their axonal transport to the corpora cardiaca and corpora allata, which are neurohemal organs. The neurosecretory products PTTH and eclosion hormone are stored in the nerve endings until their release into blood sinus spaces surrounding the endings. The corpora allata also contain nonneural cells that elaborate juvenile hormone. (B) The cuticular changes involved in molts leading (top to bottom) to larval, pupal, and adult stages. See text for explanation. [Riddiford and Truman, 1978.]

recent evidence in the tobacco hornworm moth, *Manduca sexta,* indicates that the axons of the PTTH-producing neurosecretory brain cells actually pass through the corpora cardiaca and end within the *corpora allata,* which are located at the posterior ends of the corpora cardiaca. Thus, the corpora allata appear to be the site at which the neurosecretory endings release PTTH into the blood.† It remains to be determined if this is true of all insects.

---

†The transport via neurosecretory axons of hormone from cells in the brain to an appendage of neural tissue where it is stored and released into the circulation is reminiscent of the vertebrate hypothalamo-hypophysial system. Since most vertebrate and invertebrate neuroendocrine systems are not phylogenetically homologous, they offer remarkable examples of convergent evolution in widely divergent animal groups.

*3.* PTTH, released into the blood, activates the prothoracic gland to secrete the molt-inducing factor, *α-ecdysone* (Figure 11-31), a steroid that closely resembles cholesterol and the vertebrate steroid hormones, differing from these primarily in the presence of a number of hydroxyl groups, which greatly increase the water solubility of the molecule. Insects require cholesterol in their diets to synthesize α-ecdysone. The latter is now thought to be a prohormone that is converted to the physiologically active form, 20-hydroxyecdysone (*β-ecdysone*), in several peripheral target tissues.

*4.* Other neurosecretory cells in the brain and nerve cord produce *bursicon,* a hormone that influences certain aspects of cuticle development, including the process of *tanning* (i.e., hardening and darkening), which is completed several hours after each molt. This hormone is a small protein with a molecular weight of about 40,000.

**5.** The corpora allata (Figure 11-30) contain nonneural endocrine tissue that produces and secretes *juvenile hormone (JH)* (Figure 11-32), a modified hydrocarbon chain. This substance, acting in association with β-ecdysone, promotes the retention of the immature ("juvenile") characteristics of the larva, thereby postponing metamorphosis until the adult stage.

The presence of JH in the early nymphal instar was demonstrated in the mid-1930s in experiments by V. B. Wigglesworth in which parabiotic coupling (Figure 11-29) of the early instar to a final instar prevented the latter from becoming an adult. The circulating concentration of JH is at its highest early in larval life, dropping to a minimum at the end of the pupal period (Figure 11-33). Metamorphosis to the

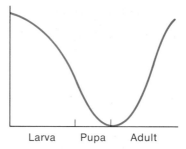

**11-33** Changes in the level of juvenile hormone during the insect life cycle. Metamorphosis to the adult form can occur only if its concentration is very low. After the adult insect has fed, secretion of JH recurs, regulating ovarian activity and stimulating development of male accessory organs. [Spratt, 1971.]

### Ecdysone

**11-31** α-Ecdysone. This steroid prohormone is synthesized in and liberated by the prothoracic glands of insects. Subsequent to release, α-ecdysone is converted in certain target tissues into the physiologically active hormone β-ecdysone by hydroxylation at C-20 (*).

### Juvenile hormone I

**11-32** A juvenile hormone extracted from the cecropia moth *Hyalophora cecropia*. This is one of several homologs that occur naturally in insects, promoting the retention of juvenile characteristics in larvae and inducing reproductive maturation in adults.

adult stage occurs when JH disappears from the circulation. The concentration then rises again in the reproductively active adult. In the males of some insect species, JH promotes development of the accessory sexual organs; in many female insects, it induces yolk synthesis and promotes maturation of the eggs.

The surface cuticle of the insect is an important structure that undergoes conspicuous changes during insect growth and development. Therefore, considerable attention has been given to the production of new cuticle and shedding of the old cuticle during molting. The interaction of hormones regulating the molting cycle is illustrated in Figure 11-30. PTTH, JH, and β-ecdysone are all involved in the initiation of molting. Ecdysone, secreted by the prothoracic glands in response to stimulation by PTTH, acts on the cuticle to initiate the steps for production of the new cuticle, which begins with *apolysis,* the detachment of the old cuticle from the underlying epidermal cells. This is followed by synthetic activity that lays down the new cuticle, while the old cuticle is partially digested from within by enzymes in the *molting fluid* secreted by the epidermis. If JH is present in high concentrations, a larval-type new cuticle is formed. At low levels of JH, an adult-type cuticle is produced, along with the other events of metamorphosis.

Two additional hormones, *eclosion hormone* and *bursicon,* regulate the terminal phase of the molting process. The actual *ecdysis* (i.e., shedding) of the cuticle of

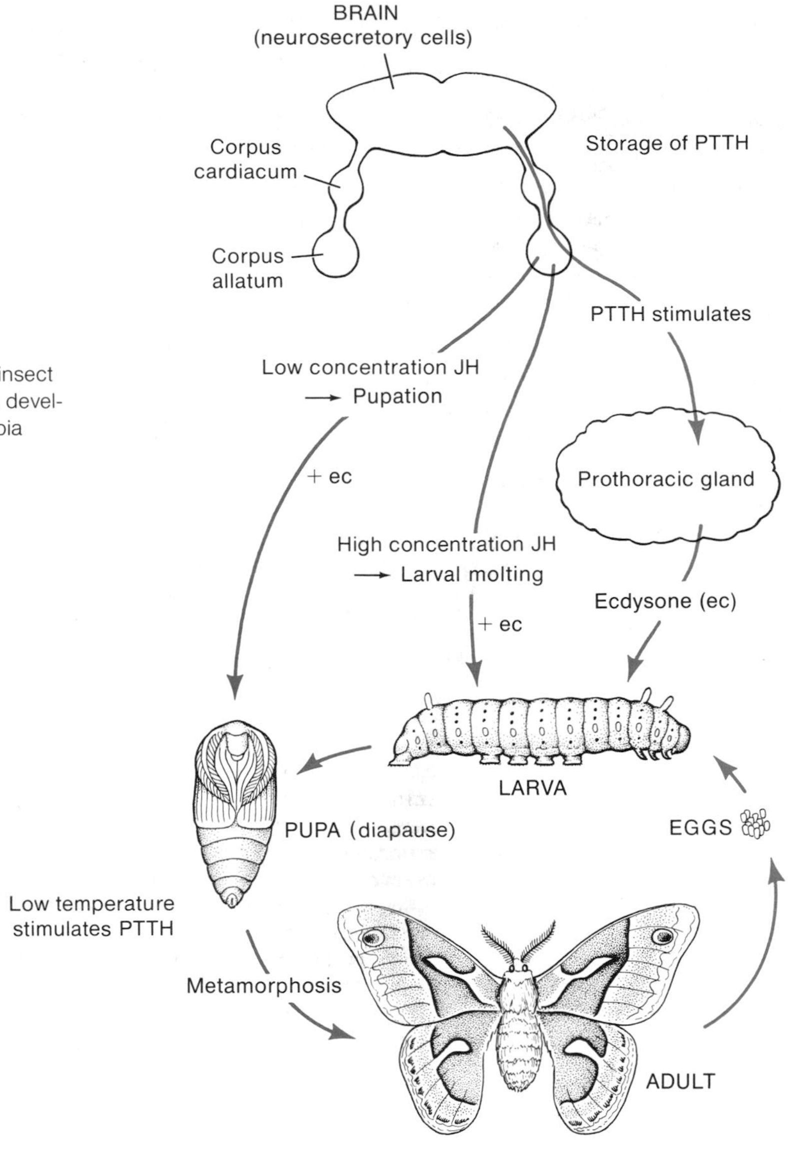

BRAIN
(neurosecretory cells)

Corpus
cardiacum

Corpus
allatum

Storage of PTTH

PTTH stimulates

Low concentration JH
⟶ Pupation

+ ec

High concentration JH
⟶ Larval molting

+ ec

Prothoracic gland

Ecdysone (ec)

PUPA (diapause)

LARVA

EGGS

Low temperature
stimulates PTTH

Metamorphosis

ADULT

**11-34**  Hormone interactions in insect metamorphosis, illustrated by the developmental sequence of the cecropia moth. [Spratt, 1971.]

the pupae of at least some holometabolous species is triggered by the eclosion hormone. The newly molted insect has a pale, soft cuticle that is expanded to the next size before it hardens, or tans, under the regulation of bursicon (Figure 11-30).

The interaction of the insect developmental hormones in the life cycle of the cecropia moth is illustrated in Figure 11-34. Larval molts are initiated by release of PTTH, which in turn stimulates the prothoracic gland to secrete ecdysone. Growth continues through a series of instars, which remain larval as long as the concentration of JH remains above a minimum. This process of growth and molting is usually completed in four or five instars, during which there is a progressive decline in JH. Low concentration of JH permits the molt to the pupal stage, which is the overwintering stage of the cecropia moth, causing it to undergo obligatory diapause. After prolonged exposure to cold, release of PTTH is stimulated, inducing the release of ecdysone, which, in the absence of JH, gives rise to the adult moth.

Thus, the normal development of an insect depends on precisely adjusted concentrations of JH at each stage. The role of JH is somewhat analogous to that of thyroid hormones in the regulation of amphibian development. In both cases, disturbance of the relations between hormone concentration and developmental stage leads to abnormal development. Because of its potency in preventing maturation in insects, JH or synthetic analogs have been proposed as potential nontoxic, ecologically sound means of combating insect pests.

## The Second-Messenger Concept

Now that we have surveyed some of the endocrines and their hormones, it is appropriate to ask how the hormones bring about such diverse effects in their target tissues. Recall that we can divide the hormones into two groups: (1) those like the steroids and other lipophilic hormones that readily penetrate the surface membranes of cells, and (2) all the rest that penetrate poorly or not at all (Figure 11-2). The former appear to act directly on the genetic apparatus to influence the synthesis of enzymes; they thereby produce slower but longer-lasting effects, some of a permanent, morphogenetic nature. The majority of hormones belong to the nonpenetrating group, which produce quicker, shorter-lived responses, generally of a more strictly metabolic nature (Figure 11-2B). The direct action of a nonpenetrating hormone ceases after binding to and activating receptor protein molecules situated in the surface membranes of its target cells. The consequence of the hormone–receptor protein interaction is the appearance within the cytosol of the target cell of an intracellular regulatory molecule. In this scheme, the hormone is termed the *first messenger*. The intracellular regulatory molecule or complex that appears as the direct result of the hormone–receptor interaction is termed the *second messenger*. The primary action of many hormones, then, is to stimulate the elevation of second-messenger concentration within the target cell.

It is the second messenger that acts as the immediate regulatory agent for the appropriate response in the target cells.

The number of hormones and of the responses they elicit is considerable, the number rising with increased complexity of the organism. In contrast to the large number of hormones and hormone effects, only a very few second messengers serve to translate the hormone's signal into cellular action. In fact, only two distinctly different kinds of second-messenger molecule have been identified in a large variety of tissues throughout the animal kingdom.

One class of these second messengers is represented by the calcium ion, which we have seen in earlier chapters to act in a regulatory capacity in a number of physiological responses. The other class of second messengers consists of the *cyclic nucleotide monophosphates* (cNMPs), of which there are at least two kinds—*adenosine 3',5'-monophosphate* (cyclic AMP, or cAMP) and the closely related *guanosine 3',5'-monophosphate* (cGMP). The overall relations between these first and second messengers and between both types of messenger and physiological responses are schematized in Figure 11-35.

## cAMP as Second Messenger

The advance of science generally depends on two forms of progress. There is the everyday growth of scientific knowledge by the slow but steady accumulation of data in thousands of laboratories. These small-scale but essential increments of progress represent by far the major effort expended by the entire community of scientists. They generally build upon another kind of advance—the major leap forward, generally unanticipated and providing the revolutionary new insight or point of departure. Such breakthroughs open new paths of inquiry, which are then explored by the step-by-step everyday mode of progress until the next major breakthrough provides new insight and again alters the course of daily investigation.

The discovery of cAMP—and the investigations, started in the mid-1950s by the late E. W. Sutherland and associates, of its role as a cellular regulatory agent—is an example of a fundamental advance in biological science. The discovery of the regulatory role of cAMP ranks at least equal in importance to the discovery,

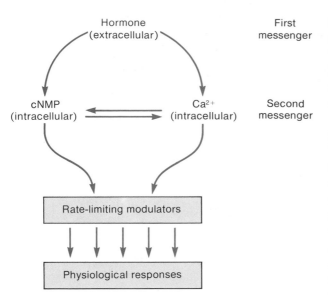

**11-35**   Concepts of first and second messenger in the regulation of cell function by hormones. The levels of cyclic nucleotides (cNMPs) and intracellular $Ca^{2+}$ levels influence one another through their complex interactions. By combining with modulators, the second messengers regulate physiological responses.

several decades before, of the role of its parent molecule, ATP, as the major energy currency of the cell. Sutherland's discovery has led to a vast accumulation of data from many types of cells, confirming the importance and ubiquitousness of cAMP and cGMP as cellular regulatory agents.

In his initial studies on cAMP, Sutherland noted that the activity of *adenylate cyclase* (the enzyme that catalyzes conversion of ATP to cAMP) in cell-free homogenates of liver is activated by hormones that stimulate it when applied to the exterior of intact cells. He then examined the various fractions of the cell-free homogenate of liver tissue and found that adenylate cyclase activity disappears if the cell membrane fragments of the homogenate are removed. It was subsequently found that the enzyme is intimately associated with a *hormone receptor* in the membrane. In this context, it is interesting to note that neither ATP nor cAMP readily penetrates the surface membrane if placed in the extracellular fluid (analogs, such as dibutyryl cAMP, enter as a result of their enhanced lipid solu-

bility). Furthermore, hormones that stimulate adenylate cyclase activity do so without entering the cell. The hormone acts on the outer surface of the cell membrane, whereas cAMP is derived from ATP at the inner surface of the membrane. The hormone must therefore convey its signal through the surface membrane.

The discovery of the membrane-bound enzyme adenylate cyclase provided the first evidence for a link between extracellular hormones and intracellular messenger molecules and led to the *second-messenger hypothesis*. In Figure 11-36 we see that a hormone-receptor molecule projecting from the outer surface of the target-cell membrane interacts with a hormone molecule for which the receptor is specific. This interaction causes a conformational change in the receptor, leading to an allosteric activation of adenylate cyclase, which faces the cytoplasm. The receptor molecule interacts through a *transducer subunit* with the adenylate cyclase, activating the enzyme so that it can catalyze the hydrolysis of adenosine triphosphate to cAMP. The cAMP then stimulates or inhibits enzymes or processes that are specific for the type of target cell. The conversion of ATP to cAMP (Figure 11-37) requires $Mg^{2+}$ and a trace of $Ca^{2+}$.

The intracellular level of cAMP depends not only on the rate at which it is synthesized, but also on the rate at which it is inactivated by conversion to ordinary adenosine 5'-monophosphate (AMP), a reaction catalyzed by a *phosphodiesterase* (Figure 11-37). Thus, there exists a cycle of

$$ATP \xrightarrow{\ 1\ } cAMP \xrightarrow{\ 2\ } AMP$$
$$\underset{3}{\underleftarrow{\hspace{4cm}}}$$

The last of these steps, regeneration of ATP from AMP, is energized by intermediary metabolism as described in Chapter 3. The level of cAMP is determined by the balance between steps 1 and 2, with step 1 in many endocrine target tissues under the control of hormones that modulate the activity of adenylate cyclase. Step 2, dependent on phosphodiesterase activity, can be slowed by the addition of the methyl xanthines caffeine or theophylline, which inhibit phosphodiesterase activity and can thereby increase the intracellular concentration of cAMP. The basal concentration of cAMP within the cells ranges from as low as $10^{-12}$ M to above $10^{-7}$ M.

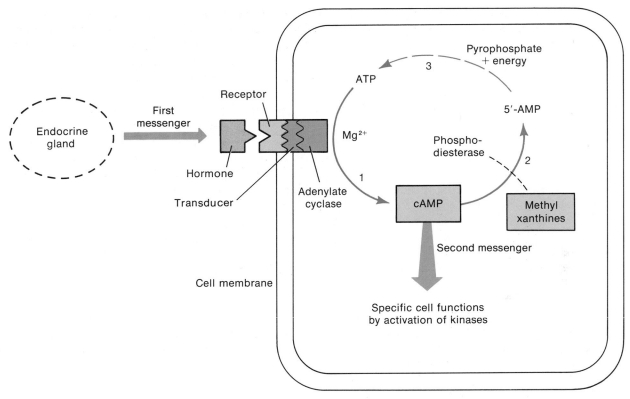

**11-36** The first and second messengers. The first messenger is the circulating hormone. The hormone–receptor interaction activates adenylate cyclase, which catalyzes the synthesis of cAMP from ATP. The cAMP acts as an intracellular second messenger, leading to the stimulation of certain cellular reactions. It does this by removing inhibition from protein kinases. Degradation of cAMP to AMP can be inhibited by addition of methyl xanthines, such as caffeine and theophylline, and in many cells by an elevation in intracellular $Ca^{2+}$ concentration.

## MULTIPLE ACTIONS OF cAMP

Since the discovery by Sutherland of cAMP as the messenger linking glucose mobilization to hormone action in liver and muscle, cAMP production has been found to be stimulated in a large number of target tissues by the appropriate hormones. Furthermore, when a permeant and less readily destroyed analog such as dibutyryl cAMP is applied to those tissues, the effects normally induced by the hormones are elicited in the absence of the hormone. These effects can often also be produced by blocking the phosphodiesterase with methyl xanthines, thereby elevating cAMP levels (Figure 11-36). These lines of evidence are all taken to indicate that cAMP acts as the intracellular messenger in a large variety of target tissues. Just a few examples of cAMP-mediated hormone actions on target cells are:

*1.* Enhanced synthesis and release of hormone in various endocrine tissues, including the secretory response of anterior pituitary to hypothalamic releasing hormone and responses of endocrine tissues to various tropic hormones, such as ACTH, FSH, and TSH.

*2.* ADH-stimulated increases in water permeability of the renal collecting duct.

*3.* Stimulation of lipid mobilization and inhibition of fat storage in adipose (fat) tissue by epinephrine, glucagon, and TSH.

*4.* Epinephrine-evoked increased rate and force of beating in cardiac muscle.

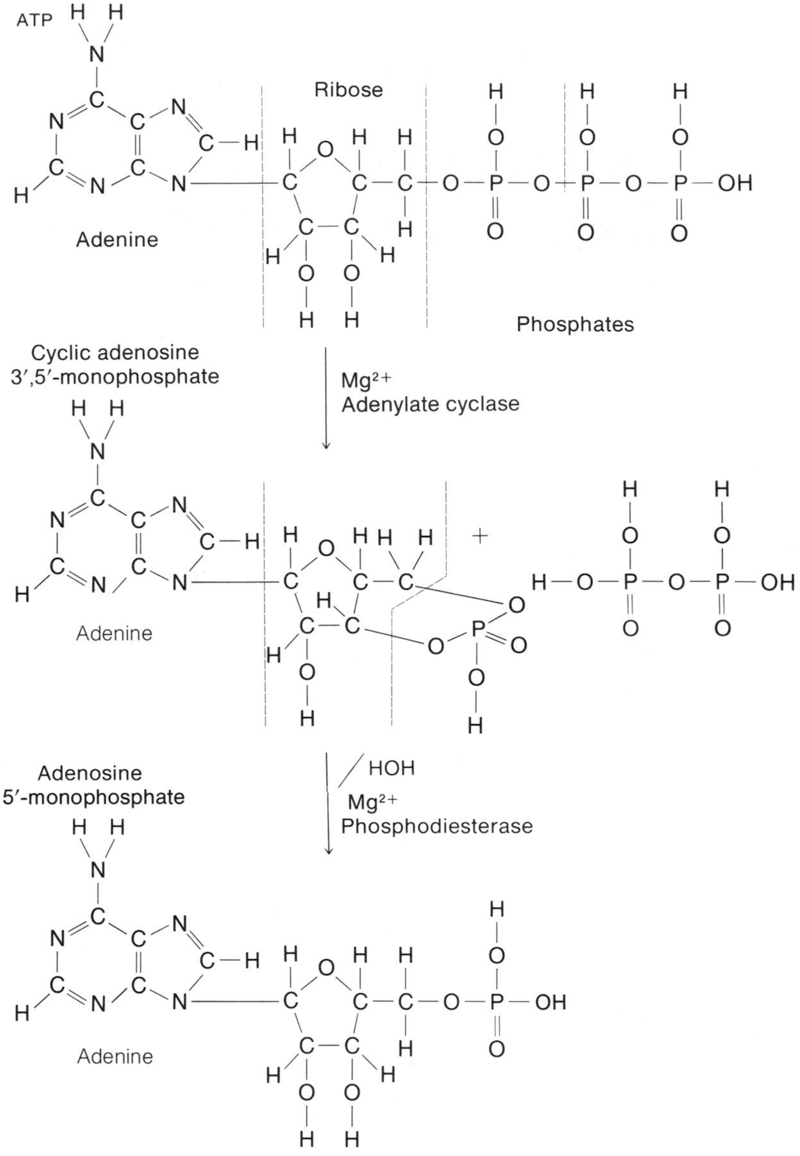

**11-37** Synthesis and degradation of cAMP. Membrane-bound adenylate cyclase catalyzes the conversion of intracellular ATP to cAMP in response to hormone–receptor interaction at the outer cell surface. Removal of cAMP from the system occurs via the phosphodiesterase-catalyzed hydrolysis of the ribose–phosphate bond, converting the cAMP to adenosine 5′-monophosphate (AMP), which is recycled by metabolic phosphorylation to ATP.

**5.** Postsynaptic changes in membrane permeability in response to the neural transmitter dopamine.

How, one might ask, can the same messenger molecule be responsible for mediating such a diversity of biochemical and physiological responses? In all systems that have been studied so far, the step that follows hormone-stimulated cAMP production is the activation of a *protein kinase* by removal of an *inhibitory subunit* by the cAMP (Figure 11-38). The key to the diversity of effects that cAMP exerts lies in the different forms of protein kinase, each one effective in phosphorylating a different substrate protein. There are various protein kinases in a given type of cell. Thus,

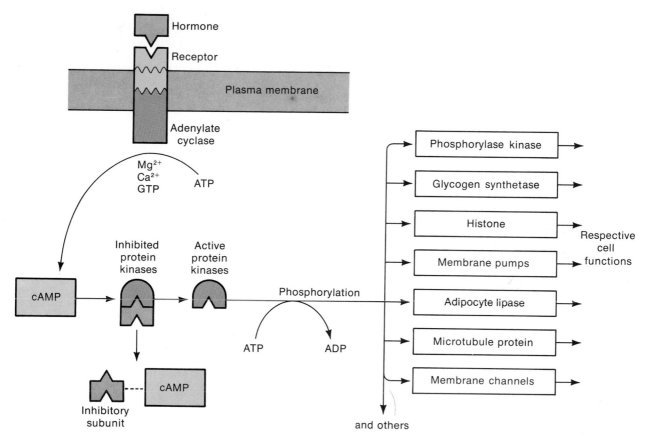

**11-38** Mediation of various hormone actions by cAMP. Hormone association with the membrane receptor leads to the synthesis of cAMP from ATP. The cAMP then removes the inhibition of a protein kinase by removing an inhibitory subunit from the kinase. The active protein kinase can then phosphorylate another intracellular enzyme, thereby either activating or repressing it, depending on the enzyme. Since there are various forms of kinase with different substrates that they phosphorylate, the hormone-induced increase in cAMP concentration can lead to any of a variety of effects, depending on the macromolecular specializations of the target cell. [Goldberg, 1975.]

in Figure 11-38, for each protein or enzyme listed at the right (and for many others) there is a different cAMP-activated protein kinase. This diversity, of course, is the reason why a hormone can exert multiple effects in one or more cell types, using cAMP as second messenger in evoking each response. The action of cAMP in any given cell is determined by the particular protein kinases that it encounters and activates as it is produced in response to the hormone.

It is apparent, then, that the relations between hormones and the responses of their target tissues are determined by genetic and developmental mechanisms. First, there are hormone-receptor specificities that determine which cells will recognize a given hormone. Second, there is molecular specificity between cAMP-activated protein kinase and the protein (often an enzyme) that is phosphorylated and thereby activated by the kinase.

## Box 11-1   *Amplification of Hormone Action*

The sequence of steps between the binding of glucagon or epinephrine and the breakdown of glycogen is rather complex (part A of figure). It would be simpler, of course, for cAMP to activate the final enzyme in the sequence directly and thereby save several steps. On the other hand, the large number of steps makes sense if one considers the need for

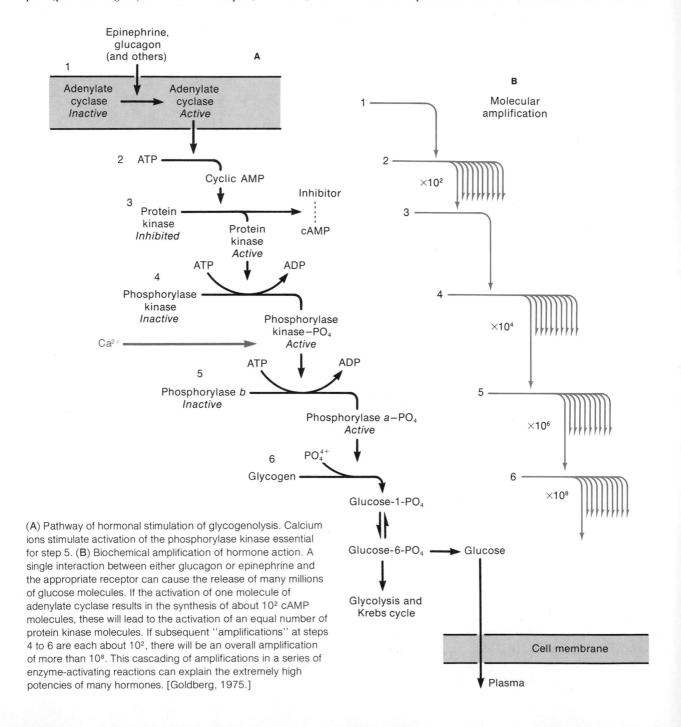

(A) Pathway of hormonal stimulation of glycogenolysis. Calcium ions stimulate activation of the phosphorylase kinase essential for step 5. (B) Biochemical amplification of hormone action. A single interaction between either glucagon or epinephrine and the appropriate receptor can cause the release of many millions of glucose molecules. If the activation of one molecule of adenylate cyclase results in the synthesis of about $10^2$ cAMP molecules, these will lead to the activation of an equal number of protein kinase molecules. If subsequent "amplifications" at steps 4 to 6 are each about $10^2$, there will be an overall amplification of more than $10^8$. This cascading of amplifications in a series of enzyme-activating reactions can explain the extremely high potencies of many hormones. [Goldberg, 1975.]

amplification—that is, for producing a large effect from a few hormone molecules. The concept of *biochemical amplification* is illustrated in part B of the figure. Each enzymatic step, beginning with number 1 and ending with number 6 (with the exception of step 2), is an *activating reaction* that converts a catalytically inactive (or weakly active) molecule into an active enzyme. The result is a cascading of amplification through four steps. If it is assumed, conservatively, that each activated enzyme molecule catalyzes the activation of 100 molecules in the next step, then four steps would produce an overall amplification of $10^8$. Thus, the interaction of one molecule of glucagon or epinephrine with the membrane receptor of liver or muscle can result in the mobilization of about 100,000,000 or more molecules of glucose. Since the basal intracellular concentration of cAMP is very low ($10^{-12}$ to $10^{-8}$ M), a small increase in the number of cAMP molecules can be equivalent to a large percentage change in cAMP concentration, and a few hormone molecules can produce significant changes in cAMP levels.

## HORMONE-INDUCED MOBILIZATION OF GLUCOSE: A BIOCHEMICAL MODEL FOR cAMP AS SECOND MESSENGER

It will be useful to review the sequence in which the roles of cAMP and $Ca^{2+}$ as second messengers have been most completely worked out—namely, the endocrine-stimulated mobilization of glucose from glycogen (Box 11-1).

It had been known for some time that the hormone glucagon causes glycogenolysis in the liver and that epinephrine does the same in skeletal and cardiac muscle. Sutherland and his associates, working with liver tissue, found that an increase in cAMP levels due to activation of adenylate cyclase occurs in the presence of the hormone glucagon. The cAMP was found to promote breakdown of glycogen into glucose, inhibit the synthesis of glycogen from glucose, and stimulate the formation of glucose from lactate and amino acids. The end effect is a rise in blood glucose.

The interaction of the hormone (glucagon in liver and epinephrine in skeletal and cardiac muscle) with the membrane-bound receptor activates the adenylate cyclase. The activation takes place through an intermediate molecule within the membrane, the transducer (not shown in the figures), that is now known to be a guanosine-triphosphate (GTP)–binding protein. In the absence of GTP, transduction is blocked. In any event, the result of the hormone–receptor interaction is an increased rate of cAMP synthesis from ATP (Box 11-1, steps 1 and 2). The immediate action of cAMP—one that appears to be the common step in most if not all cAMP-regulated systems—is the activation of a protein kinase (step 3). The cAMP molecule does this by complexing with the inhibitory subunit associated with the protein kinase (Figure 11-38). Once activated, the protein kinase is free to catalyze the phosphorylation (by ATP) of another enzyme, *phosphorylase kinase* to *phosphorylase kinase-$PO_4$*, which in turn catalyzes the phosphorylation of *phosphorylase b* to *phosphorylase a-$PO_4$* (steps 4 and 5). It is the latter enzyme that then phosphorylates glycogen residues to glucose-1-$PO_4$ (step 6), making the glucose available for use within the cell in the glycolytic pathway or for release from the cell as unphosphorylated glucose (Figure 3-41).

It is interesting that the pathway leading to the *activation* of phosphorylase *a*, the enzyme that breaks down glycogen into glucose, concomitantly produces an *inhibition* of *glycogen synthetase*, the enzyme that catalyzes the polymerization of glucose into glycogen. The inhibition results from a phosphorylation of the glycogen synthetase by a protein kinase. Thus, the hormone-stimulated rise in cAMP stimulates (by phosphorylation) the enzyme that breaks down glycogen and inhibits (also by phosphorylation) the enzyme that resynthesizes it from glucose. This synergistic effect is important, for it keeps the rise in glucose from driving by mass action a resynthesis of glycogen from glucose. It is also an example of the multiplicity of cAMP-mediated effects that take place simultaneously within a single cell.

## CYCLIC GMP AS SECOND MESSENGER

The study of cAMP and calcium ions as second messengers is complicated by the existence of the other regulatory nucleotide monophosphate, cGMP (Figure 11-39). This nucleotide, present in cells at concentrations even lower than those of cAMP, is produced from GTP (guanosine triphosphate) by the enzyme

Guanine          Ribose          Phosphate

**Cyclic GMP**

**11-39**  Cyclic guanosine 3′, 5′-monophosphate (cGMP). This cyclic nucleotide differs from cAMP only in one portion (shown shaded) of its purine group. Compare with Figure 11-37. The cGMP is synthesized from GTP by guanylate cyclase, is active in extremely low concentrations, and appears to act in an opposite way to cAMP in many tissues.

*guanylate cyclase,* which occurs free in the cytoplasm. In some processes at least, cGMP promotes responses opposite to those promoted by cAMP. For example, in heart muscle, as we noted earlier, epinephrine stimulates the production of cAMP. Acetylcholine, on the other hand, stimulates the production of cGMP. ACh and the catecholamines have opposing effects on the rate and strength of the heartbeat, the former decreasing them and the latter increasing them. This difference suggests that cAMP and cGMP have opposing intracellular actions. Activation of the β-adrenergic receptors of cardiac muscle, brain, smooth muscle, and lymphocytes simultaneously produces a rise in the level of cAMP and a drop in the level of cGMP. Conversely, ACh causes a drop in the level of cAMP while producing a rise in the level of cGMP.

The production of cGMP is particularly sensitive to $Ca^{2+}$. In every system in which GMP is the intermediary for a hormone-elicited effect, the hormone is ineffective if $Ca^{2+}$ is absent from the extracellular fluid. Studies on isolated guanylate cyclase indicate that the enzyme is inactive in the absence of $Ca^{2+}$ and becomes progressively more active with increases in free $Ca^{2+}$. In contrast, isolated preparations of adenylate cyclase are stimulated by $Ca^{2+}$ at low concentrations, but are inhibited at high concentrations as well as at zero $Ca^{2+}$. Thus, the optimum $Ca^{2+}$ concentration for

this enzyme is lower than for guanylate cyclase. In view of the differences in the calcium sensitivities of these two enzymes, the relative concentrations of cAMP and cGMP can in principle be influenced by intracellular free $Ca^{2+}$. In addition, the greater calcium responsiveness of cGMP synthesis suggests that in some systems $Ca^{2+}$ may act as a second messenger to stimulate the production of cGMP, which would then act as a third messenger.

## Calcium as Intracellular Messenger

In recent years, $Ca^{2+}$ has come into prominence as a very important and ubiquitous intracellular regulatory agent, as well as one of the second messengers linking intracellular responses to the presence of an extracellular first messenger. We have already noted its role in regulating muscle contraction and ciliary activity and in triggering synaptic release and exocytosis in general. A few of the many other functions regulated or triggered by elevations in intracellular levels of $Ca^{2+}$ include cell cleavage, various reactions in intermediary metabolism, oxidative phosphorylation, microtubule polymerization, amoeboid movement, and DNA replication.

There are two sources for elevations of $Ca^{2+}$ in the cytosol. In certain cells, such as muscle fibers, $Ca^{2+}$ is released from intracellular stores in response to stimulation. More commonly, the stimulus causes the opening of channels specific for $Ca^{2+}$, allowing these ions to flow into the cell. This influx occurs, for example, in the mammalian liver and salivary gland when epinephrine activates the α-adrenergic receptor, as will be discussed further on.

How is $Ca^{2+}$ adapted to its ubiquitous role of intracellular regulatory agent? First, we note that although the *total* $Ca^{2+}$ content of most cells is not particularly low, the activity, or free ionized concentration of $Ca^{2+}$, $[Ca^{2+}]_i$, in the cytosol is maintained at extraordinarily low levels, usually well below $10^{-6}$ M. Extracellular $Ca^{2+}$ is typically $10^{-2}$ M. The discrepancy between total and free intracellular concentrations of $Ca^{2+}$ arises in part from the sequestering of $Ca^{2+}$ within such membrane-bound organelles as mitochondria and the reticulum. The low intracellular concentration of free $Ca^{2+}$ is possible, of course, only because of uphill transport of $Ca^{2+}$ against a very large concentration gradient.

The advantage of the very low $[Ca^{2+}]_i$ is quite simple. Numerous intracellular functions are highly sensitive to changes in the concentration of $Ca^{2+}$. With $[Ca^{2+}]_i$ very low, small increments in the *amount* of free $Ca^{2+}$ result in large *percentage* increments in $Ca^{2+}$. This percentage increase is illustrated by comparing the relative changes in $[Ca^{2+}]_i$ and $[Na^+]_i$ that result from the entrance of equal quantities of these two ion species into the cell in response to a transient increase in the permeability of the membrane to both ions (Figure 11-40).

Agents or stimuli (i.e., hormones, potential changes) that act on the membrane to increase its permeability to $Ca^{2+}$ by this means produce an increase in $[Ca^{2+}]_i$ and stimulate the calcium-sensitive reactions in the cell. Calcium therefore acts as a *coupling agent,* or messenger, between stimuli acting on the cell membrane and processes that take place inside the cell.

Calcium that enters the cell is eventually pumped out. In the short course of time, free $Ca^{2+}$ entering the cell is kept from reaching high levels by two additional mechanisms of control. First, the calcium is rapidly bound by molecules that are therefore said to "buffer"

the $Ca^{2+}$ concentration, limiting perturbations in free $Ca^{2+}$ levels. The buffering makes it hard for calcium to diffuse from one part of the cell interior to another, permitting local transients in $Ca^{2+}$ concentration—for instance, near the membrane where the transient elevation in $Ca^{2+}$ can have a local regulatory function. Second, free calcium levels are also controlled by the calcium-uptake mechanism of mitochondria, which liberate $H^+$ in exchange for $Ca^{2+}$.

Certain technical advances have been essential in studies of the physiological functions of $Ca^{2+}$. One advance makes it possible to adjust $Ca^{2+}$ at concentrations below the levels of normal calcium contamination of distilled water (i.e., less than $10^{-5}$ M $Ca^{2+}$). This adjustment is made by adding known amounts of both $CaCl_2$ and a strong calcium chelating agent, such as EGTA or EDTA, to a solution. This produces a *pCa buffer system* analogous to a pH buffer system. The $CaCl_2$ behaves like the strong acid, and the calcium–EGTA or calcium–EDTA (both of which dissociate extremely weakly) acts like the weak acid. By this means, free $Ca^{2+}$ can be buffered at, for example, $10^{-8}$ M in the presence of much larger quantities of total $Ca^{2+}$.

Another important methodological advance was the discovery in 1963 of the jellyfish protein aequorin,

$$\text{A} \qquad \underbrace{10^{-8}\,\text{M}}_{[Ca^{2+}]_{init}} + \underbrace{10^{-6}\,\text{M}}_{\Delta[Ca^{2+}]} = \underbrace{1.01 \times 10^{-6}\,\text{M}}_{[Ca^{2+}]_{final}} \qquad \frac{[Ca^{2+}]_{final}}{[Ca^{2+}]_{init}} = \frac{1.01 \times 10^{-6}}{10^{-8}} \approx 100 \times \text{initial } [Ca^{2+}]$$

(with $Ca^{2+}$ influx arrow above)

$$\text{B} \qquad \underbrace{10^{-2}\,\text{M}}_{[Na^+]_{init}} + \underbrace{10^{-6}\,\text{M}}_{\Delta[Na^+]} = \underbrace{1.0001 \times 10^{-2}\,\text{M}}_{[Na^+]_{final}} \qquad \frac{[Na^+]_{final}}{[Na^+]_{init}} = \frac{1.0001 \times 10^{-2}}{10^{-2}} \approx 1 \times \text{initial } [Na^+]$$

(with $Na^+$ influx arrow above)

**11-40**  Elevation of intracellular free $Ca^{2+}$ by transient influx. (**A**) In this example, the low initial concentration, $[Ca^{2+}]_{init}$, is raised 100 times by an increment of $10^{-6}$ M produced by a transient influx, $\Delta[Ca^{2+}]$. (**B**) Since $[Na^+]_{init}$ is already $10^{-2}$ M, a similar increment, $\Delta[Na^+]$, produces virtually no change in the intracellular sodium concentration.

11-41 Calcium–calmodulin-regulated processes or enzymes, established or presumed. [Cheung, 1979.]

Adenylate cyclase

Myosin light chain kinase

Phosphodiesterase

Phosphorylase kinase

Phospholipase

Guanylate cyclase ← Calcium–Calmodulin → $Ca^{2+}$–ATPase

$Ca^{2+}$-dependent protein kinase?

Microtubule disassembly?

Neurotransmitter release?

Others?

Membrane phosphorylation

which emits light when it complexes with $Ca^{2+}$. Since light can be measured with very sensitive instruments, injection of aequorin into cells has provided a means of detecting minute changes in the intracellular free $Ca^{2+}$ levels. More recently, calcium-sensitive dyes such as arsenazo III have been successfully employed to make spectrophotometric measurements of calcium levels within single cells.

### CALCIUM-BINDING PROTEINS: CALMODULIN

A number of protein molecules selectively bind calcium ions, and some of these serve as intracellular receptors for messenger $Ca^{2+}$. In Chapter 9 we encountered the first $Ca^{2+}$ receptor protein to be discovered—the striated-muscle protein troponin C, which together with three other subunits, troponin $T_1$, $T_2$, and troponin I, makes up the troponin complex that regulates the contraction of vertebrate skeletal muscle. When the troponin-C molecule binds $Ca^{2+}$, the helical conformation of its peptide chain increases. This conformational change, brought about by the forces associated with the binding of $Ca^{2+}$, is transmitted to the other troponin subunits and thus to the regulatory muscle protein, tropomyosin. The result is that the latter is displaced from the position in which it blocks the contractile interaction of actin and myosin. In this example, the first messenger is the depolarizing signal sent to the sarcoplasmic reticulum from the surface membrane along the transverse tubules of the muscle fiber. $Ca^{2+}$, released from the SR, acts as the second messenger, and troponin acts as the calcium receptor.

Troponin C is found only in striated muscle. However, every eukaryotic tissue examined thus far contains relatively large amounts of a closely related $Ca^{2+}$-binding protein named *calmodulin*. This protein is inactive in its native state but binds $Ca^{2+}$ to form an active complex. The $Ca^{2+}$–calmodulin complex, rather than $Ca^{2+}$ alone, appears to be the actual regulatory agent responsible for influencing the activity of numerous enzymes and effector proteins (Figure 11-41).

The $Ca^{2+}$-binding properties of both calmodulin and troponin C (as well as *parvalbumen*, a poorly understood $Ca^{2+}$-binding protein of the striated-muscle cytosol), reside in their amino acid sequences. These $Ca^{2+}$-binding proteins have very similar peptide structure—indeed, calmodulin isolated from cow brains has been found to substitute effectively for troponin C in activating the actomyosin ATPase of rabbit muscle. About 70% of the amino acid sequences of these different calcium-binding regulatory proteins are homologous. The amino acid sequence of calmodulin appears to be especially similar throughout a wide range of species, and for that reason this regulatory protein is believed to have a very ancient origin. Thus, calmodulin from plants and from invertebrates can cross-react immunologically with calmodulin extracted from mammalian tissue.

Calmodulin and troponin C each have four $Ca^{2+}$-binding sites. These consist of acidic amino acid residues that are carboxylate-bearing and rich in oxygen atoms. The oxygen atoms, carrying full or partial negative charges, occur in a loop of the peptide chain so that 6–8 oxygen atoms form a cavity of just the right size to harbor the positively charged calcium ion (crystal radius 0.099 nm, in contrast to 0.065 nm for $Mg^{2+}$). When the oxygen-rich receptor regions of troponin C or calmodulin bind $Ca^{2+}$, the molecules undergo changes in tertiary structure that give rise to their active conformations.

When a hormone or other stimulus leads to an increase in cytoplasmic free $Ca^{2+}$, the binding of the calcium ion to unoccupied molecules of calmodulin produces the active $Ca^{2+}$–calmodulin complex:

$$Ca^{2+} + (\text{calmodulin})_{\text{inactive}}$$
$$\Longrightarrow (Ca^{2+}\text{–calmodulin})_{\text{active}} \qquad (11\text{-}1)$$

The active complex can then bind to the regulatory unit of an enzyme $(E)$, increasing the activity of the enzyme:

$$(E)_{\text{less active}} + (Ca^{2+}\text{–calmodulin})_{\text{active}}$$
$$\Longrightarrow (Ca^{2+}\text{–calmodulin–}E)_{\text{active}} \qquad (11\text{-}2)$$

The role of calmodulin in cellular regulation is just beginning to be understood, but this ubiquitous regulatory protein has already been implicated in the activation of a number of enzymes. Prominent among these are the enzymes responsible for synthesis of cAMP and for the hydrolysis of both cAMP and cGMP. It is also interesting that calmodulin is the calcium receptor regulating contraction in vertebrate smooth muscle, a function somewhat analogous to that of troponin C in vertebrate striated muscle.

## MULTIPLE RECEPTORS: CONVERGENT AND DIVERGENT PATHWAYS

A number of hormones elicit different sets of responses in cells of different tissues, or even in the same cells. A familiar example of this is seen in cells that have two kinds of adrenergic receptors (i.e., receptors binding epinephrine or norepinephrine) in their membranes. They are termed the $\alpha$-adrenergic receptor and $\beta$-adrenergic receptor. Figure 11-42 presents examples of two kinds of cells, from the mammalian liver and salivary gland, that both have $\alpha$ and $\beta$ pathways. In liver (Figure 11-42A), we see an example of *convergent pathways* leading to the same end effect. The immediate effect of the binding of epinephrine by the $\alpha$-receptor is an elevation of intracellular $Ca^{2+}$, which leads to activation of the phosphorylase kinase. The immediate

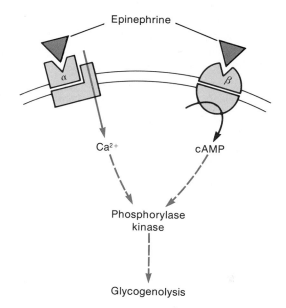

**A** Mammalian liver

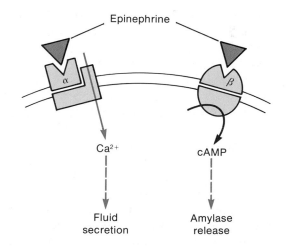

**B** Mammalian salivary gland

**11-42** The hormone epinephrine acts on both $\alpha$- and $\beta$-receptors in various vertebrate tissues. (**A**) In some tissues, activation of the two receptors by the hormone leads to the same end result via "parallel pathways." (**B**) In other tissues, activation of $\alpha$- and $\beta$-receptors leads to different end effects. [Suggested by M. J. Berridge.]

effect of the binding of epinephrine by the $\beta$-receptor is an elevation of cAMP levels. This increase in cAMP also leads to activation of the phosphorylase kinase, which in turn leads to glycogenolysis and the release of glucose into the bloodstream. Thus, the one hormone in this case leads to elevated concentrations of two different second messengers that both evoke the same end response.

In the salivary gland (Figure 11-42B), we see an example of *divergent pathways*. Here $Ca^{2+}$ elevation and cAMP synthesis, again stimulated respectively through the $\alpha$- and $\beta$-adrenergic receptors, lead to different, independent, end effects—fluid secretion and amylase secretion by the mammalian salivary gland cells. In this case, the pathways initiated by the binding of epinephrine to $\alpha$- and $\beta$-receptors diverge rather than converge.

Of interest in these examples is the occurrence of both $Ca^{2+}$ and cAMP acting as second messengers on behalf of the same first messenger. There is growing evidence that epinephrine is not unique in having multiple receptors in the same animal species, or even in the same cell.

Divergent pathways have also been identified for intracellular second messengers. The multiple actions of cAMP already noted (Figure 11-38) can, of course, occur in the same cell in response to a first-messenger signal. An interesting example of divergence in the action of messenger calcium occurs in muscle (Figure 11-43). The elevation of free cytoplasmic $Ca^{2+}$ by electrical stimulation (Figure 9-26) activates two separate pathways, one leading to ATP hydrolysis and energizing the production of actin–myosin force (as described in Chapter 9), the other leading to enhanced phosphorylation of ADP to ATP. The effect of the latter, of course, is to replenish the ATP converted to ADP during muscle contraction.

### INTERACTIONS BETWEEN $Ca^{2+}$ AND CYCLIC NUCLEOTIDES

Not long after the discovery of the second-messenger role of cAMP, it became apparent that most cAMP-mediated cell responses also depend in some way on the presence of $Ca^{2+}$. In some instances, a cAMP-mediated response can be evoked by elevation of $Ca^{2+}$ in the cytoplasm or can be diminished by lowering the $Ca^{2+}$

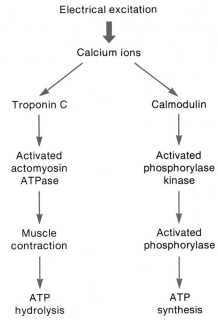

11-43 Divergence of second-messenger calcium actions in mammalian skeletal muscle, mediated by two intracellular calcium receptors—troponin C and calmodulin. The end effect of the troponin-C pathway is muscle contraction and hydrolysis of ATP to ADP. The end effect of the calmodulin pathway is the stimulation of regeneration of ATP by the transphosphorylation of ADP from creatine phosphate. [Cohen, 1982.]

level. As it turns out, $Ca^{2+}$ and cyclic nucleotide metabolism are intimately intertwined. A general appreciation of the interdependence between $Ca^{2+}$ and cAMP can be gained from the following relations:

*1.* Calcium ion (actually $Ca^{2+}$–calmodulin) in low concentrations is essential for activation of adenylate cyclase. However, at high concentrations, $Ca^{2+}$ inhibits the activity of the cyclase and thus suppresses cAMP production.

*2.* $Ca^{2+}$ also stimulates the enzyme guanylate cyclase and thus the production of cGMP, whose actions in many cases oppose those of cAMP. However, cGMP is also required for the production of cAMP, as it is essential in the activation of adenylate cyclase.

*3.* $Ca^{2+}$–calmodulin activates the cyclic nucleotide phosphodiesterase, increasing the rate of cAMP and cGMP breakdown. The overall effect of $[Ca^{2+}]_i$ on the levels of cAMP and cGMP in a given cell type apparently depends in part on the relative calcium dependence of the destruction versus the synthesis of the cyclic nucleotides.

**4.** cAMP stimulates uptake of $Ca^{2+}$ by intracellular (i.e., endoplasmic, sarcoplasmic) reticula on the one hand, but facilitates release of $Ca^{2+}$ from mitochondria on the other hand. The net effect of a rise in cAMP on $[Ca^{2+}]_i$ will depend, then, on which effect is greater—the release of $Ca^{2+}$ from mitochondria or the enhanced uptake of $Ca^{2+}$ into the reticulum.

Because of these multiple effects of $Ca^{2+}$ on cyclic nucleotide metabolism on the one hand, and of cyclic nucleotides on intracellular calcium metabolism on the other hand, it is not yet possible to predict for any given tissue what effect a change in the concentration of one of these agents will have on the concentration of the other. Following are some examples of how $Ca^{2+}$ and cAMP can participate in regulating cell processes.

Two regulatory processes control the rate of glycogenolysis in skeletal muscle. Skeletal muscle responds to epinephrine with the sequence of enzyme activations—already familiar from Box 11-1—that leads to the breakdown of glycogen to glucose-6-phosphate. In that sequence, phosphorylase kinase is sensitized to *resting* calcium levels (i.e., lower than $10^{-7}$ M) by cAMP-induced phosphorylation. In addition to this epinephrine-activated, cAMP-regulated pathway, there is a calcium-dependent pathway, which is activated during muscle contraction by the increase in the concentration of intracellular free $Ca^{2+}$ that results from release of $Ca^{2+}$ by the sarcoplasmic reticulum. The increased $[Ca^{2+}]_i$ enhances phosphorylase kinase activity independent of a hormone-stimulated rise in cAMP (see colored arrow, part A of figure, Box 11-1).

These two controls owe their characteristics to the differences in the relative activities of phosphorylase kinase and phosphorylase kinase–$PO_4$ at different $Ca^{2+}$ concentrations. The phosphorylase kinase requires an intracellular $Ca^{2+}$ concentration of $10^{-6}$–$10^{-5}$ M, characteristic of the contracting muscle, before it will catalyze the conversion of phosphorylase $b$ to phosphorylase $a$ (Figure 11-44). In contrast, phosphorylase kinase–$PO_4$ (produced in response to epinephrine) catalyzes

this reaction at intracellular calcium concentrations below $10^{-6}$ M, which is the calcium concentration characteristic of the relaxed muscle. Thus, glucose is mobilized via phosphorylase kinase as needed in response to increases in $[Ca^{2+}]_i$ during contraction. The epinephrine-stimulated mobilization of glucose via phosphorylase kinase–$PO_4$, however, anticipates energy needs and augments the calcium-activated mobilization in the fight-or-flight syndrome elicited by the sympathetic division of the autonomic nervous system (Figure 8-13A), through secretion of the catecholamines epinephrine and norepinephrine.

Another example of $Ca^{2+}$–cAMP interaction is seen in the smooth muscle of the mammalian intestine. Acetylcholine released from parasympathetic postganglionic neurons (Figure 8-13B) causes a depolarization and concomitant influx of $Ca^{2+}$ in the spindle-shaped cells of smooth muscle (Figure 11-45A). As the $[Ca^{2+}]_i$ rises, the cells contract. (Recall that in smooth muscle, the calcium receptor for regulation of contraction is calmodulin.) At the same time, the elevated $[Ca^{2+}]_i$ (also in the form of $Ca^{2+}$–calmodulin) stimulates the activity of both the phosphodiesterase and the adenylate cyclase, the effect on the latter

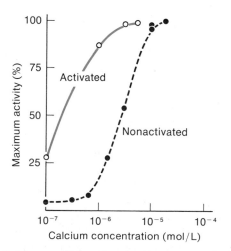

**11-44** Calcium sensitivity of the phosphorylated ("activated") and unphosphorylated form of phosphorylase kinase. The phosphorylated form shows greater activity at lower calcium concentrations. [Rasmussen, 1975.]

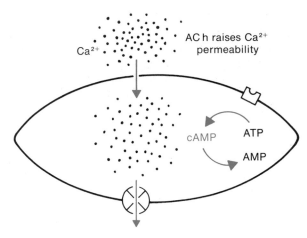

**A** ACh induces contraction

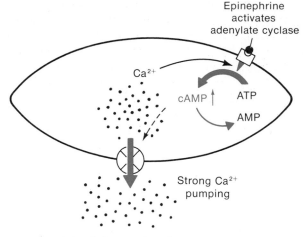

**A** Epinephrine induces relaxation

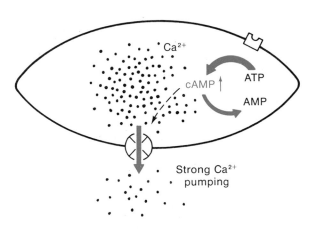

**B** Feedback limits duration of contraction

**11-45** Calcium-mediated and cAMP-mediated smooth muscle responses to acetylcholine. (A) Contraction is produced by an influx of $Ca^{2+}$ in response to the depolarization caused by ACh. (B) The increased $[Ca^{2+}]_i$ interferes with phosphodiesterase activity, leading to an increase in the cAMP level. The cAMP stimulates the calcium pump, lowering $[Ca^{2+}]_i$ and thus causing smooth muscle to relax. [Rasmussen, 1975.]

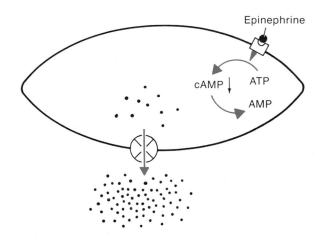

**B** Feedback stabilizes relaxation

**11-46** cAMP-mediated relaxation of smooth muscle in response to epinephrine. (A) Epinephrine activates adenylate cyclase. The cAMP then rises and stimulates the $Ca^{2+}$ pump, causing $[Ca^{2+}]_i$ to fall. (B) The drop in $[Ca^{2+}]_i$ removes $Ca^{2+}$ inhibition of phosphodiesterase activity, and the cAMP levels tend toward their steady-state resting levels. [Rasmussen, 1975.]

outweighing that of the former, thereby producing an elevation in cAMP concentration (Figure 11-45B). The rise in cAMP stimulates the calcium pump, which transports $Ca^{2+}$ out of the cell and into the $Ca^{2+}$-sequestering reticulum (not shown). The result of this negative-feedback regulation of the intracellular $Ca^{2+}$

concentrations is to reestablish the resting $[Ca^{2+}]_i$ and the resting muscle tension when stimulation by ACh ceases.

The cAMP also mediates the relaxation of intestinal smooth muscle in response to the secretion of epinephrine by the adrenal medulla and the liberation of norepinephrine from sympathetic postganglionic neurons. The catecholamines stimulate cAMP production in smooth muscle by activation of adenylate cyclase (Figure 11-46A). The rise in cAMP concentration stimulates the pumping of $Ca^{2+}$ from the cytosol, as we have already noted, fostering relaxation of the contractile mechanism. Concomitantly, the drop in $[Ca^{2+}]_i$ leads to a reduction in adenylate cyclase activity and hence a reduction in cAMP synthesis (Figure 11-46B). A steady state of $Ca^{2+}$ and cAMP levels is thereby restored, ensuring relaxation of the contractile mechanism.

The dynamic nature of cAMP metabolism is illustrated by the discovery that in cardiac muscle the concentration of cAMP undergoes fluctuations correlated with the cycle of beating. Thus, the concentration of cAMP is highest in samples of heart tissue rapidly frozen just before or just at the beginning of contraction. The cAMP concentrations at that time are about 50% higher than during the relaxed phase. It may be that the increase in $[Ca^{2+}]_i$ during contraction raises cAMP levels by stimulating adenylate cyclase activity. By analogy with smooth muscle, a rise in cAMP may stimulate the calcium pump of the sarcoplasmic reticulum, helping restore the low $[Ca^{2+}]_i$ required for relaxation of the contractile apparatus.

## Hormone Action on Genetic Mechanisms

Steroid hormones appear to penetrate the surface membrane and act on internal receptor sites. Some, it appears, exert their cellular actions by modulating the expression of the genetic program. Presumptive evidence for this comes from several sources. Injection of the insect growth and molting hormone ecdysone into larvae of the fly *Chironomus* is followed after a short delay (3–5 min) by the appearance of *puffs* at specific loci on the chromosomes of the salivary gland (Figure 11-47). The larval salivary gland is used because the chromosomes are polytene (composed of many chromatin strands) and are therefore "giant" sized, with easily observed bands. The puffing consists of a swelling of one or more of the bands and indicates the elaboration of a specific messenger RNA. Other tissues in the larval fly respond to ecdysone by puffing of other bands, suggesting that the programmed re-

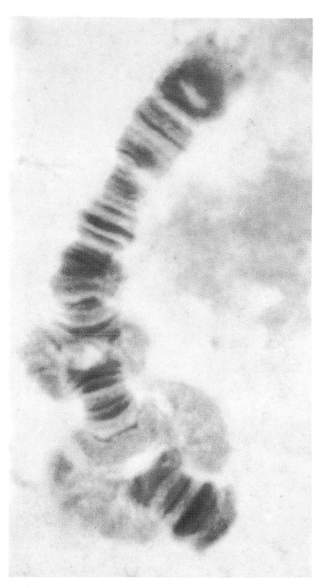

**11-47** Puffing induced by ecdysone in a giant chromosome of the fly *Chironomus*. Two puffs are seen in the bottom portion of the chromosome. [From "Chromosome Puffs" by W. Beerman and U. Clever. Copyright © 1964 by Scientific American, Inc. All rights reserved.]

sponse to ecdysone is tissue specific and results in different synthetic activity in different tissues. It is still not clear how ecdysone influences the transcription of DNA to RNA.

*478*

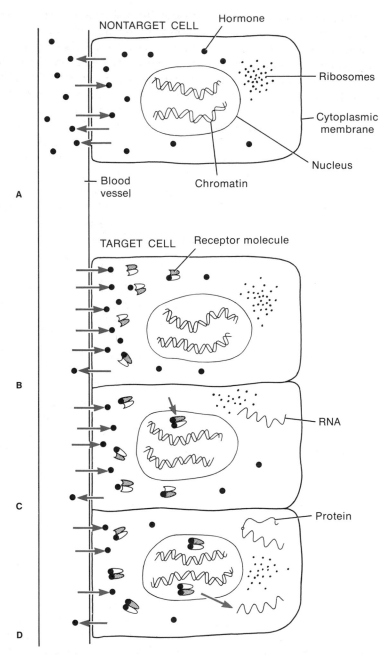

**11-48** Postulated mechanism of steroid hormone action. (**A**) Steroids diffuse randomly into and out of non-target cells without interaction or binding. (**B**) Target cells, however, preferentially retain specific steroid hormone molecules, which complex in the cytoplasm with receptor protein molecules consisting of two subunits. (**C**) These complexes then accumulate in the nucleus, where they bind to a specific fraction of the group of heterogeneous histone proteins associated with the DNA in the chromatin. It has been proposed that this association makes a segment of the DNA available for gene transcription. (**D**) The resulting messenger RNA is then translated into protein by the ribosomes. [O'Mally and Schrader, 1976.]

The steroid hormones of vertebrates also exert powerful effects on the genetic machinery. This is evidenced by the stimulation of synthesis of certain proteins in restricted tissues in response to steroid hormones in the general circulation. For example, the androgens stimulate the synthesis of myofibrillar proteins, but only in striated-muscle cells, not in other tissues.

Because of their low molecular weight (ca. 300; similar to sucrose) and their lipid solubility, steroid hormones can readily enter and leave cells by diffusing across membranes. It was therefore interesting to find by autoradiography that steroid hormones accumulate in the nuclei of their target cells, but not in the nuclei of other cells. This specific accumulation occurs very rapidly and persists for some time after the labeled steroid is removed from the circulation. These findings, made in the 1960s, suggested that these must be steroid-hormone-specific binding molecules within target cells. Such receptor molecules were found by fractionating target tissue incubated with labeled hormone and separating components of different molecular weight by sucrose-density-gradient centrifugation. R. Gorski (1979) and associates did this using estradiol as the hormone and rat uterus as the target tissue. The receptor–hormone complex was identified by the radioactivity of the tritiated estradiol; the receptor turned out to be a protein molecule with a molecular weight of about 200,000. This protein, which binds estradiol very strongly, was present in uterine tissues but not in other tissues. Most significant was the observation that substances that mimic the hormonal action of estradiol in the uterus were all bound by this protein. It therefore appears that the formation of the hormone–protein complex is an intermediate in the chain of events that leads to the end effect of estradiol action in the uterus. Similar receptor proteins have since been identified in the target tissues of the other steroid hormones. Conversely, the receptor molecules were found only in the target tissues of the respective hormones.

The steroid hormone is initially bound by the receptor protein in the cytoplasm before the complex enters the nucleus (Figure 11-48), as seen in labeling and fractionation experiments. Initially, nearly all the label appears in the cytoplasm, and, with time, increasing amounts appear in the nucleus. In the absence of the hormone, the receptor protein is distributed more or less uniformly between the nucleus and the cytoplasm. With the addition of hormone, the hormone–receptor complex accumulates in the nucleus, suggesting that the complex binds selectively to some component of the nucleus. This component is the *chromatin,* the substance of chromosomes, which consists of DNA, the fairly uniform basic protein *histone,* and a group of heterogeneous acid proteins collectively termed the *nonhistone chromosomal proteins* (*NCPs*). As it turns out, the hormone–receptor complex binds to a fraction of the NCPs, which, unlike the histones, are specific for each tissue.

## Summary

Physiological and biochemical processes are controlled and coordinated in the animal body in large part by special blood-borne messenger molecules termed hormones, which are liberated by the cells of endocrine secretory tissues. In vertebrates, these messenger molecules belong to four categories: (1) amines, (2) prostaglandins, (3) steroids, and (4) polypeptides and proteins. After liberation from the endocrine tissues in which they originate, the molecules of a hormone are distributed by the circulation in low concentrations throughout the body, where they come in contact with target cells that exhibit a specific and high sensitivity to the hormone molecules.

The peptide and amine hormones bind to receptors on the surface membranes of the target cells. This interaction produces a signal (perhaps a conformational change in the receptor molecule) that is transmitted to the active site of an associated membrane-bound enzyme facing the cytoplasm. The membrane-bound enzyme adenylate cyclase participates in the initial response to many of these hormones, mediating many and diverse cell responses. This enzyme catalyzes the conversion of ATP to cyclic adenosine 3',5'-monophosphate (cAMP), which acts as a second messenger, conveying the influence of the extracellular first messenger, the hormone, to the cell interior. The cAMP does this by activating a protein kinase by removal of an inhibitory subunit from the enzyme. In the cAMP-mediated action, the target cell's ultimate chemical or metabolic responses to the hormone depend on the nature of the protein kinases present in the cell and are therefore subject to activation by the cAMP. A given kinase will phosphorylate a certain protein, generally an enzyme, activating it and thereby determining which cell reactions or functions are stimulated. Cyclic GMP, derived from guanosine triphosphate by the action of a cytoplasmic enzyme, also acts as a second messenger, as does the calcium ion. The metabolism of cyclic nucleotides and $Ca^{2+}$ are intimately interrelated in the control of various intracellular enzymes and processes.

The steroid hormones, being lipid-soluble, enter cells freely and appear at least in some cases to influence the transcription of certain genes. The steroid hormone molecules bind to free receptor-protein molecules present in the cytoplasm and nucleus. The hormone—receptor complex binds to chromatin in the nucleus in such a way that transcription of one or more codons into RNA is stimulated.

Secretion of most hormones occurs by exocytotic expulsion of hormone molecules contained within minute secretory vesicles, a process requiring the intervention of $Ca^{2+}$, which enters the secretory cell through the surface membrane. The secretory activities of many endocrine tissues are modulated by the action of hormones that arise in other tissues. The hypothalamus, for example, is the site of neurosecretory cells that secrete polypeptide hormones that are carried into the anterior pituitary gland by a system of portal vessels. There they stimulate (or inhibit) the secretion of other polypeptide hormones. The hormones of the anterior pituitary in turn stimulate secretion of hormones in various other endocrine tissues throughout the body. Feedback from the responses of target glands and other target tissues plays a major role in preventing runaway activity by endocrine tissues. Thus, the secretion of a hormone becomes reduced when the secreting cells sense a signal associated with the increased activity of the target tissue.

The nervous system is intimately involved in the endocrine physiology of both vertebrates and invertebrates. This is especially evident in neurosecretory systems, made up of neurons that secrete hormones into the circulation. Both neural and nonneural endocrine secretions are important in the regulation of diverse kinetic, metabolic, and morphogenetic events.

## Exercises

*1.* Give three examples of chemical regulation that do not involve hormones.

*2.* What criteria must be met before a tissue can be unequivocally identified as having an endocrine function?

*3.* Give an example of negative feedback in the control of hormone secretion.

*4.* Discuss the evidence that calcium plays an important role in hormone secretion.

*5.* Give two examples that illustrate the intimate functional association of the nervous and endocrine systems.

*6.* How do the glucocorticoids (also growth hormone and glucagon) combat hypoglycemia?

*7.* What role do the thyroid hormones play in mammalian thermoregulation?

*8.* How does insulin produce its hypoglycemic effects?

*9.* What factors influence secretion of growth hormone?

*10.* Compare the mechanisms and sites of action of aldosterone and ADH in the mammalian kidney.

*11.* Discuss the endocrine regulation of plasma calcium levels.

*12.* What biochemical reason would explain why the average man has larger muscles than the average woman?

*13.* Discuss the endocrine control of the menstrual cycle.

*14.* How is conception prevented by the use of birth control pills?

*15.* Give some examples of endocrine influence on animal behavior.

*16.* Discuss the role of juvenile hormone in the development and metamorphosis of an insect.

*17.* Explain how it can be that the actions of epinephrine and glucagon are similar but are specific to different tissues.

*18.* How does cAMP activate a protein kinase?

*19.* If cAMP is the second messenger in many different target tissues, how do you explain the multitude of specific and distinct responses?

**20.** Explain how a small number of hormone molecules can elicit responses involving millions of times as many molecules.

**21.** How can a working muscle mobilize glycogen stores without stimulation of glycogenolysis by epinephrine?

**22.** Describe the interrelations between $Ca^{2+}$ and cAMP in the mammalian salivary gland and liver to illustrate convergent and divergent second-messenger pathways for a single first messenger such as epinephrine.

## Suggested Reading

Barrington, E. J. W. 1975. *An Introduction to General and Comparative Endocrinology.* 2nd ed. Oxford: Clarendon.

Catt, K. J. 1971. *An ABC of Endocrinology.* Boston: Little, Brown.

Cheung, W. Y. 1979. Calmodulin plays a pivotal role in cellular regulation. *Science* 207:17–27.

Cheung, W. Y. 1982. Calmodulin. *Scientific American* 246(6):60–70.

Cohen, P. 1982. The role of protein phosphorylation in neural and hormonal control of cellular activity. *Nature* 296:613–620.

Frieden, E. H. 1976. *Chemical Endocrinology.* New York: Academic.

Frieden, E. H., and H. Lipner. 1971. *Biochemical Endocrinology of the Vertebrates.* Englewood Cliffs, N.J.: Prentice-Hall.

Krieger, D. T., and J. C. Hughes, eds. 1980. *Neuroendocrinology.* Sunderland, Mass.: Sinauer.

Novak, V. J. A. 1975. *Insect Hormones.* New York: Wiley.

O'Mally, B. W., and W. T. Schrader. 1976. The receptors of steroid hormones. *Scientific American* 234(2):32–43.

Rasmussen, H., and D. B. P. Goodman. 1977. Relationships between calcium and cyclic nucleotides in cell activation. *Physiol. Rev.* 57:421–509.

Riddiford, L. M., and J. W. Truman. 1978. Biochemistry of insect hormones and insect growth regulators. In Morris Rockstein, ed., *Biochemistry of Insects.* New York: Academic.

Schulster, D., and A. Levitzki, eds. 1980. *Cellular Receptors for Hormones and Neurotransmitters.* New York: Wiley.

Tepperman, J. 1973. *Metabolic and Endocrine Physiology.* 3rd ed. Chicago: Year Book Medical Publishers.

Turner, C. D., and J. T. Bagnara. 1976. *General Endocrinology.* Philadelphia: Saunders.

Weissman, G., and R. Claiborne, eds. 1975. *Cell Membranes: Biochemistry, Cell Biology and Pathology.* Chaps. 18, 19, 20. New York: Hospital Practice Publishing Co.

Wigglesworth, V. B. 1970. *Insect Hormones.* San Francisco: W. H. Freeman and Company.

Wigglesworth, V. B. 1972. *The Principles of Insect Physiology.* London: Chapman and Hall.

## References Cited

Beerman, W., and U. Clever. 1964. Chromosome puffs. *Scientific American* 210(4):50–58.

Bell, G. H., J. N. Davidson, and H. Scarborough. 1972. *Textbook of Physiology and Biochemistry.* 8th ed. Edinburgh: Churchill Livingstone.

Douglas, W. W. 1974. Mechanism of release of neurohypophysial hormones: Stimulus–secretion coupling. In R. O. Greep, ed., *Handbook of Physiology.* Section 7. Endocrinology (Vol. 4, Part 1, Pituitary Gland). Washington, D.C.: American Physiological Society.

Douglas, W. W., J. Nagasawa, and R. Schulz. 1971. Electron microscopic studies on the mechanism of secretion of posterior pituitary hormones and significance of microvesicles ("synaptic vesicles"): Evidence of secretion by exocytosis and formation of microvesicles as a by-product of this process. In H. Heller and K. Lederis, eds., *Subcellular Organization and Function in Endocrine Tissues. Mem. Soc. Endocrin.* No. 19. New York: Cambridge University Press.

Farquhar, M. G. 1971. Processing of secretory products by cells of the anterior pituitary gland. In H. Heller and K. Lederis, eds., *Subcellular Organization and Function in Endocrine Tissues. Mem. Soc. Endocrin.* No. 19. New York: Cambridge University Press.

Goldberg, N. D. 1975. Cyclic nucleotides and cell function. In G. Weissman and R. Claiborne, eds., *Cell Membranes: Biochemistry, Cell Biology and Pathology.* New York: Hospital Practice Publishing Co.

Gorski, R. A. 1979. Long-term hormonal modulation of neuronal structure and function. In F. O. Schmitt and F. G. Worden, eds. *The Neurosciences: Fourth Study Program.* Cambridge, Mass.: MIT Press.

Jamieson, J. D. 1975. Membranes and secretion. In G. Weissman and R. Claiborne, eds., *Cell Membranes: Biochemistry, Cell Biology and Pathology.* New York: Hospital Practice Publishing Co.

Kopeč, S. 1917. Experiments on metamorphosis in insects. *Bull. Acad. Sci. Cracovie.* 57–60.

Kopeč, S. 1922. Studies on the necessity of the brain for the inception of insect metamorphosis. *Biol. Bull.* 42:323–342.

McNaught, A. B., and R. Callander. 1975. *Illustrated Physiology.* New York: Churchill Livingstone.

Mikiten, T. M. 1967. *Electrically Stimulated Release of Vasopressin from Rat Neurohypophyses* in Vitro. Ph.D. dissertation, Yeshiva University, New York.

Rasmussen, H. 1975. Ions as "second messengers." In G. Weissman and R. Claiborne, eds. *Cell Membranes: Biochemistry, Cell Biology, and Pathology.* New York: Hospital Practice Publishing Co.

Smith, M. C., M. E. Freeman, and J. D. Neill. 1975. The control of progesterone secretion during the estrous cycle and early pseudopregnancy in the rat: Prolactin, gonadotropin and steroid levels associated with rescue of the corpus luteum of pseudopregnancy. *Endocrinology* 96:219–226.

Spratt, N. T., Jr. 1971. *Developmental Biology.* Belmont, Calif.: Wadsworth.

Williams, C. M. 1947. Physiology of insect diapause: II, Interaction between the pupal brain and prothoracic glands in the metamorphosis of the giant silkworm *Platysamia cecropia. Biol. Bull.* 93:89–98.

# CHAPTER *12*

# *Osmoregulation and Excretion*

The unique physical and chemical properties of water have played a major role in the origin of living organisms, and water thus is indispensable for all biochemical and physiological processes. These properties of water, described in Chapter 2, made it possible for life as we know it to arise several billion years ago in a shallow, salty sea. Living cells to this day carry this aqueous heritage in their intracellular milieu and, moreover, are generally dependent on the immediate presence of extracellular water, even if merely a very thin layer. The macromolecular machinery of living cells also requires certain inorganic molecules and ions, which play a variety of important roles in both the intra- and extracellular fluids (Table 12-1).

The ability to survive in an osmotically unfavorable environment was achieved in the more advanced animal groups by the evolution of a *stable internal environment,* which acts to buffer the internal tissues against the vagaries and extremes of the external environment (Figure 12-1).

## *General Considerations*

In this chapter, we will consider the osmotic environment and the mechanisms used by various animals to cope with environmental osmotic extremes. The movement of water and solutes across cell membranes and multicellular epithelial layers has been covered in Chapter 4 and forms an essential background for an understanding of osmoregulatory organs such as the kidney, gill, and salt gland covered in this chapter. Toward the chapter's end we will consider the problem of eliminating toxic nitrogenous wastes produced during the metabolism of proteins.

One of the requirements in the regulation of the internal environment is that appropriate quantities of water be retained. Another major requirement for cell survival is the presence, in appropriate concentrations, of various solutes, such as salts and nutrient molecules. Some tissues require an extracellular ionic environment that is more or less an approximation of seawater—namely, fluid high in sodium and chloride and relatively low in the other major ions, such as potassium and the divalent cations (Tables 12-2 and 12-3). For the simpler forms of marine invertebrates, the seawater itself acts as the extracellular medium; for most of the more complex forms, the internal fluids are in near ionic equilibrium with the seawater. In most multicellular animals, there are mechanisms for some degree of regulation (between broad or narrow limits) of extracellular solutes (Figure 12-1). Finally, the cellular environment must be freed of toxic wastes that accumulate as by-products of metabolism. In the simplest and smallest marine and aquatic organisms, this purification happens simply by diffusion of the wastes into the surrounding water. In animals that

**TABLE 12-1**  Major inorganic ions of tissues.

| Ion | Distribution | Some functions |
|---|---|---|
| $Na^+$ | Main extracellular cation | Contributes major osmotic pressure of extracellular fluid<br><br>Concentration gradient set up by sodium pump provides potential energy for transport of substances across cell membrane<br><br>Carries inward current for membrane excitation |
| $K^+$ | Main cytoplasmic cation | Contributes to osmotic pressure of cytoplasm<br><br>Establishes the resting potential<br><br>Activates some enzymes<br><br>Carries outward current for membrane repolarization |
| $Ca^{2+}$ | Low concentration in cells | Stabilizes membranes<br><br>Regulates exocytosis and muscle contraction<br><br>Involved in "cementing" cells together<br><br>Carries inward current in some excitable membranes<br><br>Regulates many enzymes and processes: acts as second messenger or coupling agent |
| $Mg^{2+}$ | Intra- and extracellular | Stabilizes membranes<br><br>Antagonist of calcium action in many functions<br><br>Acts as cofactor for many enzymes (e.g., myosin ATPase) |
| $HPO_4^{2-}$<br>$HCO_3^-$ | Intra- and extracellular | Buffers $H^+$ concentration |
| $Cl^-$ | Main extracellular anion of tissues | Counterion for inorganic cations |

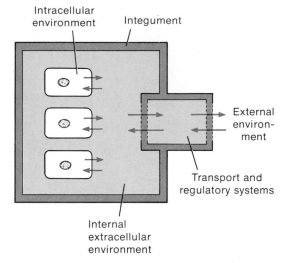

**12-1**  Regulatory systems act as buffers between the external and the internal environments of the more advanced groups of animals. The cells and tissues of these animals are protected from large fluctuations in the external environment, such as osmotic extremes, since the composition of the internal extracellular fluids is maintained within narrow limits.

**TABLE 12-2**  Electrolyte composition of the human body fluids.

| Electrolytes | Serum (meq/kg $H_2O$) | Interstitial fluid (meq/kg $H_2O$) | Intracellular fluid (muscle) (meq/kg $H_2O$) |
|---|---|---|---|
| **Cations** | | | |
| $Na^+$ | 142 | 145 | 10 |
| $K^+$ | 4 | 4 | 156 |
| $Ca^{2+}$ | 5 | | 3 |
| $Mg^{2+}$ | 2 | | 26 |
| *Totals* | 153 | 149 | 195 |
| **Anions** | | | |
| $Cl^-$ | 104 | 114 | 2 |
| $HCO_3^-$ | 27 | 31 | 8 |
| $HPO_4^{2-}$ | 2 | | 95 |
| $SO_4^{2-}$ | 1 | | 20 |
| Organic acids | 6 | | |
| Protein | 13 | | 55 |
| *Totals* | 153 | 145 | 180 |

*Note:* Some of the ions contained within cells are not all freely dissolved in the cytoplasm, but may also be sequestered within cytoplasmic organelles. Thus, the true free $Ca^{2+}$ concentration in the cytoplasm is typically below $10^{-6}$ meq/kg $H_2O$ rather than the overall value given in the table. Failure of anion and cation totals to agree reflects incomplete tabulation.

TABLE 12-3  Composition of extracellular fluids of representative animals (concentrations in millimoles per liter of $H_2O$).

| | Habitat† | Milliosmoles | [Na$^+$] | [K$^+$] | [Ca$^{2+}$] | [Mg$^{2+}$] | [Cl$^-$] | [SO$_4{}^{2-}$] | [HPO$_4{}^{2-}$] | Urea |
|---|---|---|---|---|---|---|---|---|---|---|
| Seawater‡ | | 1000 | 460 | 10 | 10 | 53 | 540 | 27 | | |
| Coelenterata | | | | | | | | | | |
| *Aurelia* (jellyfish) | SW | | 454 | 10.2 | 9.7 | 51.0 | 554 | 14.6 | | |
| Echinodermata | | | | | | | | | | |
| *Asterias* (starfish) | SW | | 428 | 9.5 | 11.7 | 49.2 | 487 | 26.7 | | |
| Annelida | | | | | | | | | | |
| *Arenicola* (lugworm) | SW | | 459 | 10.1 | 10.0 | 52.4 | 537 | 24.4 | | |
| *Lumbricus* | | | | | | | | | | |
| (earthworm) | Ter. | | 76 | 4.0 | 2.9 | — | 43 | — | | |
| Mollusca | | | | | | | | | | |
| *Aplysia* (sea slug) | SW | | 492 | 9.7 | 13.3 | 49 | 543 | 28.2 | | |
| *Loligo* (squid) | SW | | 419 | 20.6 | 11.3 | 51.6 | 522 | 6.9 | | |
| *Anodonta* (clam) | FW | | 15.6 | 0.49 | 8.4 | 0.19 | 11.7 | 0.73 | | |
| Crustacea | | | | | | | | | | |
| *Cambarus* (crayfish) | FW | | 146 | 3.9 | 8.1 | 4.3 | 139 | — | | |
| *Homarus* (lobster) | SW | | 472 | 10.0 | 15.6 | 6.7 | 470 | | | |
| Insecta | | | | | | | | | | |
| *Locusta* | Ter. | | 60 | 12 | 17 | 25 | | | | |
| *Periplaneta* | | | | | | | | | | |
| (cockroach) | Ter. | | 161 | 7.9 | 4.0 | 5.6 | 144 | | | |
| Cyclostomata | | | | | | | | | | |
| *Eptatretus* (hagfish) | SW | 1002 | 554 | 6.8 | 8.8 | 23.4 | 532 | 1.7 | 2.1 | 3 |
| *Lampetra* (lamprey) | FW | 248 | 120 | 3.2 | 1.9 | 2.1 | 96 | 2.7 | — | 0.4 |
| Chondrichthyes | | | | | | | | | | |
| Dogfish shark | SW | 1075 | 269 | 4.3 | 3.2 | 1.1 | 258 | 1 | 1.1 | 376 |
| *Carcharhinus* | | | | | | | | | | |
| (freshwater shark) | FW | — | 200 | 8 | 3 | 2 | 180 | 0.5 | 4.0 | 132 |
| Coelacantha | | | | | | | | | | |
| *Latimeria* | SW | — | 181 | 51.3 | 6.9 | 28.7 | 199 | — | — | 355 |
| Teleostei | | | | | | | | | | |
| *Paralichthys* | | | | | | | | | | |
| (flounder) | SW | 337 | 180 | 4 | 3 | 1 | 160 | 0.2 | — | — |
| *Carassius* (goldfish) | FW | 293 | 142 | 2 | 6 | 3 | 107 | — | — | |
| Amphibia | | | | | | | | | | |
| *Rana esculenta* (frog) | FW | 210 | 92 | 3 | 2.3 | 1.6 | 70 | — | — | 2 |
| *Rana cancrivora* | FW | 290 | 125 | 9 | — | — | 98 | — | — | 40 |
| | 80% SW | 830 | 252 | 14 | — | — | 227 | — | — | 350 |
| Reptilia | | | | | | | | | | |
| *Alligator* | FW | 278 | 140 | 3.6 | 5.1 | 3.0 | 111 | — | — | — |
| Aves | | | | | | | | | | |
| *Anas* (duck) | FW | 294 | 138 | 3.1 | 2.4 | | 103 | — | 1.6 | — |
| Mammalia | | | | | | | | | | |
| *Homo sapiens* | Ter. | | 142 | 4.0 | 5.0 | 2.0 | 104 | 1 | | |
| Lab rat | Ter. | | 145 | 6.2 | 3.1 | 1.6 | 116 | — | | |

†SW = seawater; FW = fresh water; Ter. = terrestrial.

‡The osmolarity and composition of seawater vary, and the values given here are not intended to be absolute. The composition of body fluids of osmoconformers will also vary, depending on the composition of the seawater they are tested in.

*Sources:* Schmidt-Nielsen and Mackay, 1972; Prosser, 1973.

have evolved circulatory systems, the blood typically passes through excretory organs, generally termed *kidneys*. In terrestrial animals, the kidneys not only play an important role in the removal of organic wastes but are the primary organs of osmoregulation. In fishes and aquatic invertebrates, a major portion of the osmoregulatory functions are carried out by other organs, such as the gills, the integument, and even the intestines.

The ability to maintain a suitable internal environment in the face of osmotic stress has played a most important role in animal evolution. There are two main reasons. First, animals are restricted in their geographic distribution by environmental factors, one of the most important being the osmotic nature of the environment. Second, geographic dispersal followed by genetic isolation is an important mechanism for the divergence of species in the process of evolution. If, for example, the arthropods and the vertebrates had not evolved means of regulating their extracellular compartments, they would have been far less successful in their invasion of the osmotically hostile freshwater and terrestrial environments. As a consequence, the number and diversity of arthropod and vertebrate species would now be limited largely to those found in the oceans and fresh waters of the world, which support far less than half the living species of animals. In the absence of terrestrial arthropods and vertebrates, other groups would have evolved with greater diversity to fill the vacant terrestrial niches.

Thus, the animal groups that have been most successful in colonizing extreme osmotic environments have evolved means of regulating their water content and the concentration of solutes dissolved in their body water. There is always a difference between the optimal intracellular and extracellular concentrations of ions. A number of mechanisms are employed to handle osmotic problems and regulate the differences (1) between intracellular and extracellular compartments and (2) between the extracellular compartment and the external environment. These are collectively termed *osmoregulatory mechanisms,* a term coined in 1902 by Rudolf Höber to refer to the regulation of osmotic pressure and ionic concentrations in the extracellular compartment of the animal body.

The evolution of efficient osmoregulatory mechanisms had extraordinarily far-reaching effects on other aspects of animal speciation and diversification. The various adaptations and physiological mechanisms evolved by animals to cope with the rigors of the osmotic environment form especially fascinating examples of the resourcefulness of evolutionary adaptation. This is the theme of an excellent book by the late Homer Smith (1953) entitled *From Fish to Philosopher.*

## Problems of Osmoregulation

Although there may be hourly and daily variations in osmotic balance, an animal is generally in an osmotic steady state over the long term. That is, on the average, the input–output balance sheet over an extended period sums up to zero (Figure 12-2). Water enters with food and drink, and in a freshwater environment it enters primarily through the respiratory epithelia—the gill surfaces of fish and invertebrates,

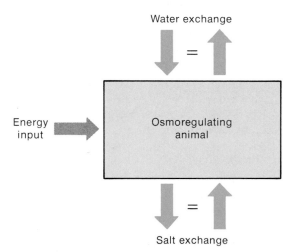

**12-2** In a strictly osmoregulating animal, the amounts of internal salt and water are held relatively constant in the face of environmental changes. This constancy requires that intake and outflow of water and salts be equal over an extended period of time. Such osmotic homeostasis is maintained at the cost of metabolic energy.

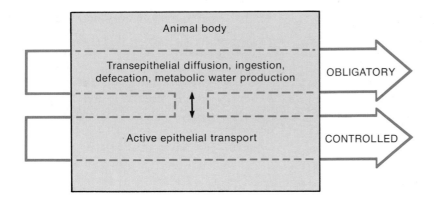

12-3  Two major classes of osmotic exchange between an animal and its environment. Obligatory exchanges are those that occur in response to physical factors over which the animal has little short-term physiological control. Controlled exchanges are those that the animal can vary physiologically to maintain internal homeostasis. Substances entering the animal by either path can leave by the other path.

and the integument of amphibians and many invertebrates. Water leaves the body in the urine, in the feces, and by evaporation through the integument and lungs.

The problem of osmotic regulation does not end with the intake and output of water. If that were so, osmoregulation would be a relatively simple matter: A frog sitting in fresh water far more dilute than its body fluids would merely have to eliminate the same amount of water as leaked in through its skin, and a camel would just stop urine production between oases. Osmoregulation also includes the requirement of maintaining favorable solute concentrations in the extracellular compartment. Thus, the frog immersed in hypotonic pond water is faced not only with the need to eliminate excess water, but also with the problem of retaining salts, which tend to leak out through the skin, because the skin in amphibians is much more permeable than that in the other vertebrate classes. The camel is faced with a rather different set of problems. In the face of a limited water supply, it must make a compromise: On the one hand, it must conserve water and eliminate toxic end products of metabolism, such as urea; on the other hand, it must regulate the salt concentrations of its extracellular fluids as it loses water through evaporation and through the production of urine.

The osmotic exchanges that take place between an animal and its environment can be divided into two classes (Figure 12-3): (1) *obligatory exchanges*—namely, those that occur mainly in response to physical factors over which the animal has little or no physiological control; and (2) *regulated exchanges*, which, as the name indicates, are physiologically controlled and serve to aid in maintaining internal homeostasis. Regulation of osmotic exchange forms the major theme of this chapter. The various factors that contribute to the obligatory exchange have been reviewed by Bentley (1971); these are outlined next.

### FACTORS INFLUENCING OBLIGATORY EXCHANGES

*1. Gradients between the extracellular compartment and the environment.* The greater the difference between the concentration of a substance in the external medium and that in the body fluids, the greater the tendency for net diffusion in the direction of low concentration. Thus, although a frog immersed in a pond tends to take up water from its hypotonic environment, a bony fish in seawater is faced with the problem of losing water into the hypertonic seawater.

*2. Surface/volume ratio.* The volume of an animal varies with the cube of its linear dimensions, whereas its surface area varies with the square of its linear dimensions. That is, the surface/volume ratio is greater for small animals than for large animals. It follows that the surface area of the integument, through which water or a solute can exchange with the environment, is greater relative to the water content of a small animal than for a large animal. This means that for

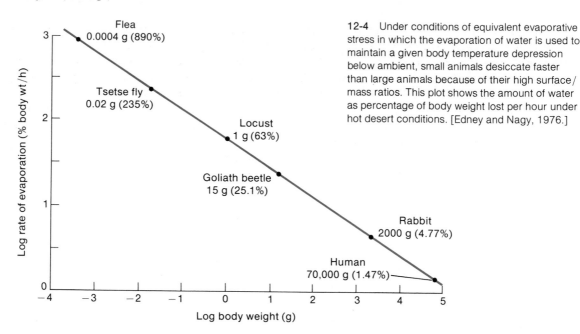

12-4 Under conditions of equivalent evaporative stress in which the evaporation of water is used to maintain a given body temperature depression below ambient, small animals desiccate faster than large animals because of their high surface/mass ratios. This plot shows the amount of water as percentage of body weight lost per hour under hot desert conditions. [Edney and Nagy, 1976.]

a given net rate of exchange across the integument (in moles per second per square centimeter), a small animal will desiccate (Figure 12-4) (or hydrate) more rapidly than a larger animal of the same shape.

*3. Permeability of the integument.* The integument acts as a barrier between the extracellular compartment and the environment. The permeability of the integument to water and solutes varies with animal groups. Amphibians have moist, highly permeable skins, through which they exchange $O_2$ and $CO_2$ and through which water and ions move by passive diffusion. Amphibian skin compensates for loss of electrolytes by active transport of salts from the aquatic environment into the animal. Fish gills are necessarily permeable, since they engage in the exchange of $O_2$ and $CO_2$ between the blood and the aqueous environment. The gills also engage in active transport of salts. Terrestrial animals (i.e., reptiles, birds, mammals, insects) have integuments with much lower water permeability and thus lose relatively little water through this route. The low integumental permeability of these groups is maintained in those species that have secondarily become marine or aquatic, such as pond insects and marine mammals.

In general, the integumental permeability to water is lower in osmoregulators than in osmoconformers. This reduced permeability is often correlated with osmotic demand. In euryhaline fishes (those that can tolerate a wide range of osmolarity), adaptation to salt water is accompanied by a drop in permeability to water. The blood circulation to the gill epithelium in some fishes has been shown to decrease as respiratory demand drops and to rise in response to increased $O_2$ need. This reduces unnecessary perfusion of the epithelium by blood and thus effectively limits unnecessary osmotic transfer through the gill epithelium.

*4. Feeding.* Water and solutes are taken in during feeding. The diet may include excess water or excess salts. A gull feeding on seashore invertebrates ingests a relatively high quantity of salt relative to water and must therefore have special means of excreting the excess salt. A freshwater fish or amphibian, on the other hand, ingests large quantities of water relative to salts and requires special means of conserving salts.

*5. Temperature, exercise, and respiration.* Because of its high heat of vaporization, water is ideally suited for the elimination of body heat by evaporation from epithelial surfaces. During evaporation, those water molecules with the highest energy content enter the gaseous phase and thus take with them their thermal energy. As a result, the water left behind becomes cooler. The importance of water in temperature regulation leads to conflicts and compromises between physiological adaptation to environmental temperatures and osmotic stresses in terrestrial animals. Desert animals, faced with both high temperatures and a meager water supply, are especially hard-pressed, since they must avoid becoming overheated and yet avoid losing large quantities of body water. Conversely, strenuous exercise generates high body temperatures owing to muscle metabolism and must be compensated by a high rate of heat dissipation. This compensation can be accomplished best by evaporative cooling over respiratory surfaces, such as the lungs, air passages, and tongue, or by evaporative water loss through the skin. Even during basal conditions (no exercise beyond breathing), the nature of the respiratory mechanism of many terrestrial animals leads to the loss of water through the respiratory surfaces. We will return to problems of temperature regulation in Chapter 16.

*6. Metabolic factors.* Those end products of metabolism that cannot be used by the organism must be eliminated. Carbon dioxide diffuses into the environment from the respiratory surfaces. Water, the other major end product of cellular respiration, is produced in small enough quantities so that its elimination is no problem (Table 12-4). In fact, this so-called *metabolic water* must be conserved in some animals; it is the major source of water for many desert dwellers. The kangaroo rat, for example, minimizes water loss by being nocturnal; it can therefore exist exclusively on

dry seeds. Osmotic problems are posed by the production of nitrogenous end products of metabolism, such as ammonia, urea, and uric acid. Ammonia, which is rather toxic, is highly soluble, so it poses no problem for aquatic or marine invertebrates, from which it merely diffuses into the aqueous environment. Uric acid requires little water for its elimination, since it is relatively insoluble in water and is therefore the preferred form of nitrogenous waste in certain terrestrial groups, such as lizards and birds.

## OSMOREGULATORS AND OSMOCONFORMERS

Animals that maintain an internal osmolarity different from the medium in which they are immersed have been termed *osmoregulators*. An animal that does not actively control the osmotic condition of its body fluids and instead conforms to the osmolarity of the ambient medium is termed an *osmoconformer*. Table 12-3 reveals these two extremes of adaptation. Most vertebrates, with the notable exception of the hagfish, are strict osmoregulators, maintaining the composition of the body fluids within a small osmotic range. Although some osmotic differences do exist between vertebrate species, the blood in all vertebrates is hypoosmotic to seawater and hyperosmotic to fresh water. This is true, as well, of fishes that migrate between fresh and saltwater environments, employing endocrine mechanisms to meet the changing osmotic stress accompanying environmental change.

Many terrestrial invertebrates also osmoregulate to a large degree. Aquatic, brackish-water, and marine invertebrates are, of course, exposed to various environmental osmolarities. Marine invertebrates, as a

**TABLE 12-4** Typical relations between metabolic water production, energy conversion, and $O_2$ consumption when various foodstuffs are oxidized to $CO_2$.

| Foodstuff | Grams of water per gram of food | Liters of $O_2$ per gram of food | Liters of $O_2$ per gram of water | Kilojoules per gram of food | Grams of water per kilojoule |
|---|---|---|---|---|---|
| Carbohydrates | 0.56 | 0.83 | 1.49 | 17.58 | 0.032 |
| Fats | 1.07 | 2.02 | 1.89 | 39.94 | 0.027 |
| Proteins | 0.40 | 0.97 | 2.44 | 17.54 | 0.023 |

*Note:* In the case of proteins, catabolism goes as far as urea. Note that oxidation of fats provides the most water per gram of foodstuff, whereas carbohydrates provide the most metabolic water for the energy expended.
*Source:* Edney and Nagy, 1976.

rule, are in osmotic balance with seawater. Table 12-3 shows that the concentrations of $Na^+$, $K^+$, $Ca^{2+}$, $Mg^{2+}$, and $Cl^-$ of the body fluids of marine invertebrates are close to the concentrations of the seawater in which the species live. This similarity has allowed the use of seawater as a physiological saline by physiologists studying the tissues of marine species. Ionic concentrations vary far more in the body fluids of different freshwater and terrestrial invertebrates, in which the body fluids are invariably more dilute than seawater but considerably more concentrated than fresh water.

Some aquatic invertebrates are *strict osmoregulators* like the vertebrates, some are *limited osmoregulators,* and some are *strict osmoconformers*. These classes are illustrated in Figure 12-5, in which the osmolarity of the extracellular compartment is plotted against the osmolarity of the aqueous environment. As the osmolarity of the environment changes, the osmolarity of a strict osmoconformer changes by an equal amount, paralleling the line that describes internal–external osmolar equality. In contrast, a strict osmoregulator maintains a constant internal osmolarity over a large range of external osmolarities, so as to produce a horizontal plot parallel with the abscissa. Limited osmoregulators regulate over a limited range of osmolarities and conform at other environmental osmolarities.

It is evident, then, that there are two ways in which animals cope with large differences in environmental osmolarity. First, they may osmoconform and display a high degree of *cellular osmotic tolerance*. Second, they may osmoregulate and maintain strict *extracellular osmotic homeostasis* in the face of the large environmental differences in electrolyte concentration. In osmoregulating animals, the internal tissues are generally not able to cope with more than minor changes in extracellular osmolarity and must depend entirely on the osmotic regulation of the extracellular fluid. The cells of osmoconformers, on the other hand, can cope with high plasma osmolarities by increasing their intracellular osmolarities. They do this by increasing the concentration of intracellular *organic osmolytes,* molecules that by their presence in high concentrations act to increase intracellular osmolarity.

In certain marine vertebrates and invertebrates, organic osmolytes are also found in the blood and interstitial fluids, so that extracellular osmolarity is brought close to that of seawater. The best-known examples of such osmolytes are urea and trimethylamine oxide, both utilized by various marine elasmobranchs, the primitive coelacanth fish *Latimeria,* and the crab-eating, brackish-water frog *Rana cancrivora* of Southeast Asia. Urea, a simple by-product (or waste product) of nitrogen metabolism (p. 535), is toxic in such high concentrations to most other animals.

### EPITHELIUM AS THE OSMOREGULATORY TISSUE
The osmoregulatory capabilities of metazoans depend to a great extent on one class of tissue made up of highly adapted cells. The tissues involved here are *transport epithelia* of gills, skin, kidneys, and gut, and the cells are epithelial cells (Figure 4-43). These cells differ from all other kinds of cells in being anatomically and functionally polarized so that one side, the *apical end,* faces a space that is continuous with the external world (the sea, the pond, the lumen of the gut, the lumen of a kidney tubule, etc.). The other side of the epithelial cell, the *basal end,* faces the internal compartment filled with extracellular fluid. This internal compartment is the one that contains all the other cells of the remaining body tissues. These have, so to speak, their own private "pond," comprised of the extracellular fluid in which they are bathed, thanks to the osmoregulatory work and barrier functions carried

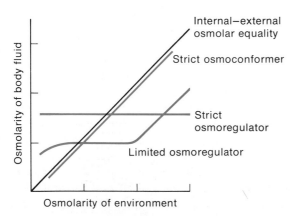

**12-5** Relation between the osmolarity of body fluids and that of the environment for three classes of *aquatic* animals.

out by epithelial cells. The mechanisms of solute and water movements across epithelial tissues have been considered in Chapter 4.

## *Osmoregulation in Aqueous Environments*

### *FRESHWATER ANIMALS*

Bodies of water range in osmolarity from several milliosmoles (mosm) per liter to about 1000 mosm/L in ordinary seawater, or more in landlocked salt seas. In between, there is a complete range of intermediate salinities, as in brackish bogs, marshes, and estuaries. As a rule, the body fluids (i.e., interstitial fluids and blood) tend away from the environmental osmotic extremes. The fluids of freshwater animals, including invertebrates, fishes, amphibians, reptiles, and mammals, are generally hyperosmotic to their aqueous surroundings (Table 12-3). Thus, freshwater animals are faced with two kinds of osmoregulatory problems: (1) Because their body fluids are hyperosmotic to the surrounding water, they are subject to swelling by movement of water into their bodies owing to the osmotic gradient; and (2) because the surrounding medium is low in salt content, they are faced with the disappearance of their body salts by continual loss to the environment. Thus, freshwater animals must prevent net gain of water and net loss of salts. They do so by several means.

Net gain of water is prevented in part by removal of excess water, as when the kidney produces a dilute urine. Thus, in closely related fishes, those that live in fresh water produce a more copious (i.e., plentiful and hence dilute) urine than their saltwater relatives (Figures 12-6 and 12-7). The useful salts are largely retained by reabsorption into the blood from the pre-urine in the tubules of the kidney (p. 509), and thus a dilute urine is excreted. Nonetheless, some salts pass out in the urine, so there is a potential problem of gradually washing out biologically important salts such as KCl, NaCl, $CaCl_2$, and $MgCl_2$. Lost salts are replaced, in part, from ingested food. An important specialization for salt replacement in freshwater animals is active transport of salt from the external dilute medium across the epithelium into the interstitial fluid and blood. This activity is accomplished across trans-porting epithelia such as those in the skin of amphibians and in the gills of fishes. In fishes and many aquatic invertebrates, the gills act as the major osmoregulatory organs, exceeding the kidneys in this function.

Freshwater animals have remarkable abilities to take up salts from their dilute environment. Freshwater fishes are able, for example, to extract $Na^+$ and $Cl^-$ with their gills (Figure 12-7) from water containing less than 1 mM/L NaCl, even though the plasma concentration of the salt exceeds 100 mM/L NaCl. Thus, the active transport of NaCl in the gills takes place against a concentration gradient in excess of 100-fold. In some freshwater animals, including fishes, reptiles, birds, and mammals, water uptake and salt loss are minimized by an integument having low permeabilities to both salts and water. As a general rule, freshwater animals other than reptiles, birds, and mammals, all of which have relatively impermeable integuments, refrain from drinking fresh water, reducing the need to expel excess water.

### *MARINE ANIMALS*

In general, the body fluids of marine invertebrates, including the ascidians (primitive chordates), are close to seawater both in osmolarity and in the individual plasma concentrations of the major inorganic salts (Table 12-3). Such animals therefore need not expend much energy in regulating the osmolarity of their body fluids. A rare example of a vertebrate whose plasma is isosmotic to the environment is the hagfish *Myxine*. It differs from most marine invertebrates, however, in that it does regulate the concentrations of individual ions. In particular, $Ca^{2+}$, $Mg^{2+}$, and $SO_4^{2-}$ are maintained significantly lower than they are in seawater, whereas $Na^+$ and $Cl^-$ are higher. Since various functions of excitable tissues such as nerve and muscle of vertebrates are especially sensitive to the concentrations of $Ca^{2+}$ and $Mg^{2+}$, the regulation of these divalent cations may have evolved to accommodate the idiosyncrasies of neuromuscular function.

Like the hagfish, the cartilaginous fishes such as sharks, rays, and skates, as well as the primitive coelacanth *Latimeria*, have plasma that is isosmotic to seawater. They differ, however, in that they maintain far lower electrolyte (e.g., inorganic ion) concentrations, making up the difference with organic osmolytes such

| | Blood concentration relative to environment | Urine concentration relative to blood | |
|---|---|---|---|
| Marine elasmobranch | Isotonic | Isotonic | Does not drink seawater / Hypertonic NaCl from rectal gland |
| Marine teleost | Hypotonic | Isotonic | Drinks seawater / Secretes salt from gills |
| Freshwater teleost | Hypertonic | Strongly hypotonic | Drinks no water / Absorbs salt with gills |
| Amphibian | Hypertonic | Strongly hypotonic | Absorbs salts through skin |
| Marine reptile | Hypotonic | Isotonic | Drinks seawater / Hypertonic salt-gland secretion |
| Desert mammal | — | Strongly hypertonic | Drinks no water / Depends on metabolic water |
| Marine mammal | Hypotonic | Strongly hypertonic | Does not drink seawater |
| Marine bird | — | Weakly hypertonic | Drinks seawater / Hypertonic salt-gland secretion / Weakly hypertonic urine |
| Terrestrial bird | — | Weakly hypertonic | Drinks fresh water |

**12-6** (facing page) Exchange of water and salt in some vertebrates. Only active exchange is indicated; passive loss of water through the skin, lungs, and alimentary tract is not indicated.

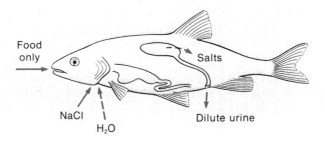

**A**  Freshwater teleost

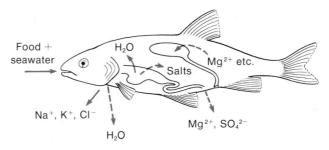

**B**  Marine teleost

**12-7**  Salt and water exchange in (**A**) freshwater and (**B**) marine teleosts. Solid arrows indicate active transport; broken arrows, passive transport. Note the active role of the gills in salt transport in both groups. [Prosser, 1973.]

as urea and trimethylamine oxide. In the elasmobranchs and coelacanths, excess inorganic electrolytes such as NaCl are excreted via the kidneys and also by means of a special excretory organ, the *rectal gland,* located at the end of the alimentary canal.

The body fluids of marine teleosts (modern bony fishes), like those of most higher vertebrates, are hypotonic to seawater, so there is a tendency for these fishes to lose water to the environment, especially across the gill epithelium. To replace the lost volume of water, they drink salt water (Figure 12-7). By absorption across the intestinal epithelium, 70–80% of the ingested seawater enters the bloodstream, along with most of the NaCl and KCl in the seawater. Left behind in the gut and expelled through the anus are most of the divalent cations such as $Ca^{2+}$, $Mg^{2+}$, and $SO_4^{2-}$. The excess salt absorbed along with the water is subsequently eliminated from the blood by active transport of $Na^+$, $Cl^-$, and perhaps $K^+$ across the gill epithelium into the seawater, and by secretion of divalent salts by the kidney (Figure 12-7). The urine is isotonic to the blood, but rich in those salts, especially $Mg^{2+}$, $Ca^{2+}$, and $SO_4^{2-}$, that are not secreted by the gills. The net result of the combined osmotic work of gills and kidneys in the marine teleost is a net retention of water that is hypotonic both to the ingested seawater and to the urine. Using those means, some teleost species—for example, the salmon of the Pacific Northwest—are able to maintain a more or less constant plasma osmolarity even though they migrate between marine and freshwater environments.

Like marine teleosts, marine reptiles (e.g., iguanas, estuarine sea turtles, crocodiles, sea snakes) and also marine birds are unable to produce a urine that is significantly hyperosmotic to their body fluids (Figure 12-6). Instead, they are endowed with cranial organs specialized for the secretion of salts in a strong hyper-osmotic fluid. These are the so-called *salt glands* (discussed in more detail later), which are generally located in birds on the bill below the eyes and in lizards near the nose or eyes. Recently, salt glands were discovered in the tongue of brackish-water crocodiles, which had been suspected of using extrarenal means of excreting salts. Although neither reptilian nor avian kidneys are capable of producing a very hypertonic urine, the salt glands of marine reptiles and birds secrete enough salt to enable them to drink salt water (Figure 12-8). Salt glands, along with the gills of marine teleosts, compensate in these groups for the inability of the submammalian kidney to produce a urine that is strongly hypertonic relative to body fluids.

Marine mammals such as pinnipeds (e.g., sea lions, seals) and cetaceans (e.g., porpoises, whales) have no extrarenal salt-secreting organs like the salt glands of birds and reptiles or the gills of fishes. But they are endowed, as are other mammals, with highly efficient kidneys capable of producing a very hypertonic urine. Nonetheless, marine mammals do not imbibe seawater, but only ingest the water that is present in their food. In this way, one might say, they make use of the past osmoregulatory efforts of their prey, avoiding much of the osmotic work that befalls teleosts, reptiles, and birds that drink seawater (Figure 12-6). As mentioned earlier, another source of water for some desert mammals as well as for some marine mammals is the metabolic water obtained from the metabolism of food molecules, during which hydrogen atoms combine with oxygen atoms to produce $H_2O$ (Figure 3-40; Table 12-4).

Human beings, like other mammals, are not equipped to drink seawater. The human kidney can remove up to about 6 g of $Na^+$ from the bloodstream per liter of

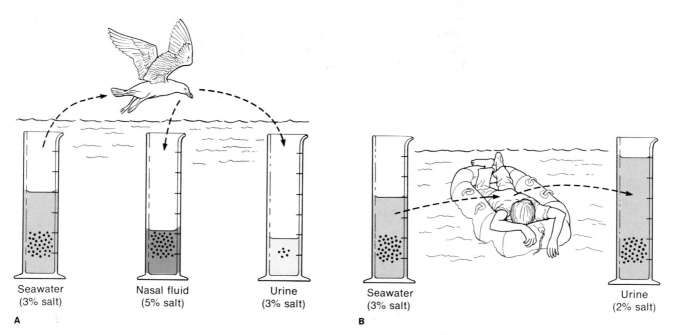

Seawater
(3% salt)

Nasal fluid
(5% salt)

Urine
(3% salt)

A

Seawater
(3% salt)

Urine
(2% salt)

B

12-8    (A) Specialization of the gull for salt and water balance at sea. The bird can drink seawater because it is able to secrete, via the salt glands, 80% of the ingested salt along with only 50% of the ingested water. As a result, it can produce a hypotonic urine without desiccating. (B) Humans and most other terrestrial mammals cannot survive on seawater because the urine cannot be concentrated sufficiently to conserve water while eliminating the ingested salt. [From ''Salt Glands'' by K. Schmidt-Nielsen. Copyright © 1959 by Scientific American, Inc. All rights reserved.]

urine produced. Seawater contains about 12 g of Na$^+$. Thus, imbibing seawater will cause the subject to accumulate salt without adding a physiologically equivalent amount of water (Figure 12-8). Stated differently, to excrete the salt ingested with a given volume of seawater, the human kidney must pass more water than is contained in that volume, which, of course, will rapidly lead to dehydration.

To generalize, then, freshwater animals tend to take on water passively and to remove it actively through the osmotic work of kidneys (vertebrates) or kidneylike nephridial organs (invertebrates). They lose salts to the dilute environment and replace them by actively absorbing ions from the surrounding fluids into their bodies through skin, gills, or other actively transporting epithelia. Marine vertebrates, on the other hand, lose water osmotically through the gills or through the integument, if it is permeable. To replace lost water, they drink seawater and actively secrete back into the environment the excess salt ingested with the seawater. This process takes place through active transport in extrarenal osmoregulatory organs such as gills and salt glands. Marine mammals, which lack salt glands or similar specializations, avoid drinking seawater, get their water entirely from their food, and depend primarily on their kidneys for maintaining osmotic balance. Marine invertebrates are generally osmoconformers.

## Osmoregulation in Terrestrial Environments

In a terrestrial environment, animals can be thought of as submerged in an ocean of air rather than water. Unless the humidity of the air is close to saturation, animals having a water-permeable epithelium will be subject to dehydration very much as if they were submerged in a hypertonic medium such as seawater. Dehydration would be avoided if all epithelial surfaces exposed to air were totally impermeable to water. The evolutionary process has not found this to be a feasible solution to the problem of desiccation, since an epithelium that is impermeable to water (and thus dry) will not have adequate permeability to respiratory gases ($O_2$ and $CO_2$) to meet the respiratory needs of an animal. As a consequence, air-breathing animals

are subject to dehydration through their respiratory epithelia. Various means have evolved to minimize water loss into the air by this route and others.

### WATER LOSS AND GAIN THROUGH THE INTEGUMENT

The integument of some terrestrial animals is relatively impermeable to water. In such animals, there is little water lost through the skin. Insects, for example, have a waxy cuticle that is highly impermeable to water, and so they lose very little moisture through that pathway. The wax is deposited on the surface of the exoskeleton through fine canals that penetrate the cuticle (Figure 12-9). The importance of the waxy layer for water retention by insects has been demonstrated by measuring the rate of water loss at different temperatures. In Figure 12-10, we see that there is a sudden jump in rate of water loss coincident with the melting point of the wax coating.

The water permeability of vertebrate integuments varies widely (Table 12-5). Reptiles, birds, and many mammals have relatively impermeable skins. However, amphibians and many invertebrates, as well as mammals that perspire, can become dehydrated at low humidity because of water loss through the integument. Animals with highly permeable skin are simply not able to tolerate very hot, dry environments. Most frogs stay close to water. Toads and salamanders can venture a bit farther, but they also are limited to moist woods or meadows not far from puddles, streams, or bodies of water in which they can replenish their supply of body water. These animals also minimize loss by behavioral strategies, avoiding desiccation by staying in cool, damp microenvironments during hot, dry times of day. Frogs, and in particular toads, which may temporarily wander from a body of water or may have to wait for periods between rains, are endowed with oversized urinary bladders in which they store water until needed. At such times, water will move osmotically from the lumen of the bladder into the partially dehydrated interstitial fluid and blood. The epithelium of the bladder, like the amphibian skin, is capable of actively transporting Na$^+$ and Cl$^-$ from the bladder lumen into the body to compensate for loss of salts that accompanies excessive hydration during times of plentiful water. Thus, the anuran bladder

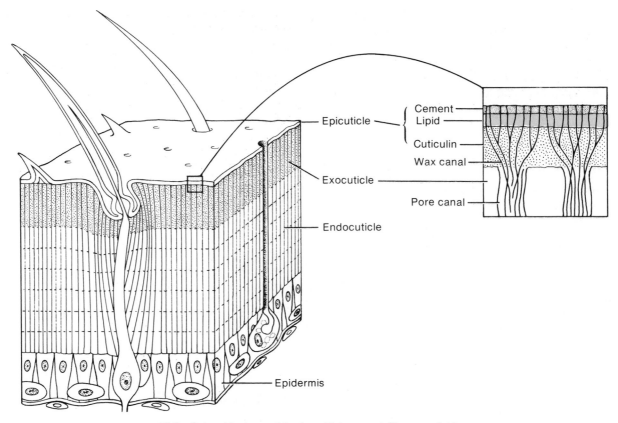

Cement
Lipid
Cuticulin
Wax canal
Pore canal

Epicuticle
Exocuticle
Endocuticle
Epidermis

**12-9** General features of the insect integument. The waxy lipid layer deposited through minute canals serves as the major water barrier of the integument. [Edney, 1974.]

serves a dual function as a water reservoir in times of desiccation and as a source of salts during times of excessive hydration.

The high water permeability of amphibian skin is used to advantage to take up water from hypoosmotic sources such as puddles. Many anurans have *seat patches,* specialized regions of skin on the abdomen and thighs, that when immersed can take up water at a rate of $3\frac{1}{2}$ times the body weight per day. The permeability of amphibian skin is controlled by arginine vasotocin, which enhances water permeability. The outer layers of toad skin contain minute channels that draw water by capillary action to moisten the skin, conserving internal water during cutaneous evaporation.

## WATER LOSS DURING AIR BREATHING

As we have noted, respiratory surfaces are, by their very nature, a major avenue for water loss in air-breathing animals. By having the respiratory surfaces internalized in a body cavity (i.e., lung), terrestrial vertebrates reduce evaporative loss. Even within the lung, however, the ventilation of the respiratory epithelium by unsaturated air will cause evaporation of the moisture wetting the epithelial surface. The moisture lost in this way from the surface of the respiratory epithelium is replaced by osmotic movement of water from inside the respiratory tissue. Such evaporative loss of water is enhanced in birds and mammals by virtue of the difference between body temperature and ambient

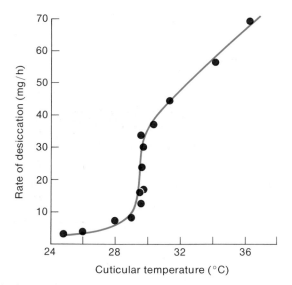

**12-10** Effect of cuticle temperature on the loss of water from a cockroach. As the temperature reaches 30°C, there is a dramatic jump in rate of desiccation. This transition temperature corresponds to the melting point of the waxy layer covering the surface of the cuticle. [Beament, 1958.]

**TABLE 12-5** Evaporative water loss of representative animals under desert conditions.

| Species | Water loss (mg/cm²/h) | Remarks |
|---|---|---|
| Arthropods | | |
| *Eleodes armata* (beetle) | 0.20 | 30°C; 0% r.h. |
| *Hadrurus arizonensis* (scorpion) | 0.02 | 30°C; 0% r.h. |
| *Locusta migratoria* (locust) | 0.70 | 30°C; 0% r.h. |
| Amphibian | | |
| *Cyclorana alboguttatus* (frog) | 4.90 | 25°C; 100% r.h. |
| Reptiles | | |
| *Gehydra variegata* (gecko) | 0.22 | 30°C; dry air |
| *Uta stansburiana* (lizard) | 0.10 | 30°C |
| Birds | | |
| *Amphispiza belli* (sparrow) | 1.48 | 30°C |
| *Phalaenptilus nuttallii* (poorwill) | 0.86 | 35°C |
| Mammals | | |
| *Peromyscus eremicus* (cactus mouse) | 0.66 | 30°C |
| *Oryx beisa* (African oryx) | 3.24 | 22°C |
| *Homo sapiens* | 22.32 | 70 kg; nude, sitting in sun; 35°C |

*Note:* Compared with the others, all of which are desert animals, the human exhibited the greatest evaporative loss.

*Source:* Hadley, 1972.

temperature. The same holds for those reptiles and amphibians that raise their body temperature by behavioral mechanisms. Warm air can hold more moisture when saturated than cool air can (Figure 12-11), and so the warmer expired air tends to contain more water than the inspired air.

In a number of vertebrates, this respiratory loss of water is minimized through a mechanism first discovered in the nose of the desert-dwelling kangaroo rat, *Dipodomys merriami,* by Knut Schmidt-Nielsen (1964) and co-workers (Schmidt-Nielsen et al., 1970). This mechanism, which they called a *temporal countercurrent system* (Figure 12-12B), retains most of the respiratory vapor by condensing it on cooled nasal passages during exhalation. Cool external air entering the lungs via the nasal passages is warmed and humidified by the loss of heat and moisture from the tissues of the nasal passages. The inspired air then unavoidably takes on more moisture from the respiratory epithelium of the lungs, as it is warmed further to about the core temperature (37°–38°C). During exhalation, the process of heat exchange between air and nasal tissues is reversed, and the warm air is cooled to somewhat above ambient as

it passes back out through the nasal passages that had been cooled by the same air during inhalation. As the expired air gives up some of its heat to the tissues of the nasal pasages, its relative humidity rises (Figure 12-11), and most of the acquired moisture condenses on the nasal epithelium (Figure 12-13). With the next inhalation, this condensed moisture again contributes to the humidification of the inspired air, and the cycle is repeated, most of the vapor being recycled within the respiratory tract.

Thus, mammals employing this mechanism have cold noses. Such temporal countercurrent mechanisms have been identified in the nasal passages of a number of mammals, including the camel and the elephant seal. These animals all have a characteristic labyrinth-like proliferation of epithelial surfaces in the nasal passages. These surfaces appear to serve the purpose of increasing the rate of heat exchange between air and tissues in the nose. A similar mechanism for trapping exhaled moisture occurs in numerous birds and lizards.

Where salt glands drain into the nasal passages, as they do in lizards, the water in the excreted salt solution enters this cycle during inspiration, and is largely conserved by recycling as it is recondensed on exhalation.

A remarkable example of water retention in a marine mammal faced with desertlike problems of water conservation occurs in the recently weaned baby elephant seal. The baby seal has to go for 8–10 weeks without food or water, during which time its only source of water is the water it derives from the oxidation of its body fat. It weighs about 140 kg at the time of weaning and loses only about 800 g of water per day, of which less than 500 g is lost through respiration. This economy is ascribed both to its nasal countercurrent heat exchanger and to a periodic slowing of metabolic rate, which allows it to stop breathing for 40 min and then alternate with 5 min of deep breath-

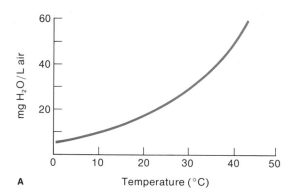

**A**

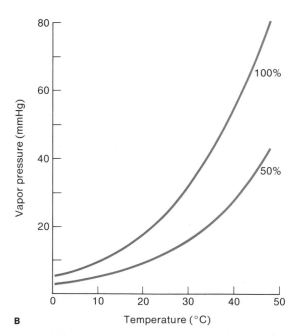

**B**

12-11   Relations between air temperature and water content of the air. (A) Quantity of water at saturation, that is, at 100% relative humidity (*r.h.*). [Schmidt-Nielsen, 1972.] (B) The partial pressure of water vapor at different temperatures for 100% and 50% *r.h.* [Buxton, 1931.]

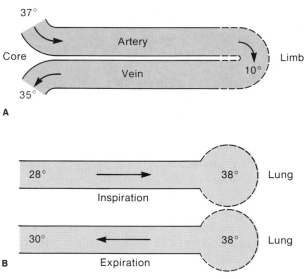

12-12   Spatial and temporal countercurrent exchange. (A) Spatial countercurrent system that acts to conserve the heat content of blood perfusing the extremities of cold-climate birds and mammals. This system is described more fully in Box 12-2 and in Chapter 16. (B) Temporal countercurrent exchange in the respiratory systems of many vertebrates acts to conserve body heat and body water. During inspiration (top), cool air (e.g., 28°C) is warmed as it flows toward lungs, removing heat from nasal passages. During expiration (bottom), the same air loses most of the heat it gained earlier as it warms the cool nasal passages on its way out. [Schmidt-Nielsen et al., 1970.]

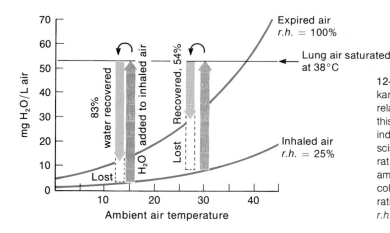

12-13  Recovery of water from exhaled air in the kangaroo rat. When air is inhaled at 15°C and 25% relative humidity, the amount of water needed to bring this air to saturation at body temperature (38°C) is indicated by the gray bar placed at 15°C on the abscissa. Under these climatic conditions, the kangaroo rat exhales air at 13°C (lower than ambient!). The amount of recondensed water is represented by the colored column. The columns to the right give evaporation and recovery at ambient air of 30°C and 25% *r.h.* [Schmidt-Nielsen et al., 1970.]

ing. The ability to suspend breathing is, of course, not uncommon for marine mammals such as the elephant seal, which can dive for prolonged periods.

The major route of water loss in terrestrial insects is via the *tracheal system,* which consists of air-filled *tracheoles* that penetrate the tissues. As long as the tracheoles are open to the air, water vapor can diffuse out while $O_2$ and $CO_2$ diffuse down their respective gradients. The entrances to the tracheoles are guarded by valvelike *spiracles* that are closed periodically by the spiracular muscles, conserving water. As the $CO_2$ concentration in the hemolymph rises, the muscle relaxes to allow respiratory exchange. The level of $CO_2$ required to cause spiracular opening increases, however, if the insect is undergoing osmotic stress due to desiccation, thereby conserving respiratory water vapor when appropriate.

### WATER ABSORPTION FROM THE AIR

Certain terrestrial arthropods have the ability to extract water vapor directly from the air, even, in some species, when the relative humidity is as low as 50% (Table 12-6). To date, this poorly understood ability has been demonstrated only in certain arachnids (ticks, mites) and in a number of wingless forms of insects, primarily larvae. Those species that exhibit this ability live in habitats devoid or nearly devoid of free water. The ability to remove water from the air is all the more remarkable in these arthropods because it normally occurs against an osmotic gradient, that is, when the

vapor pressure of the hemolymph exceeds that of the air, which it does at all values of relative humidity below 99%. In insects that extract water from air, the site of entry appears to be the rectum, which reduces the water content of fecal matter to remarkably low levels. As water is removed from the feces, the latter can take on new water from the air, if the water vapor pressure is high enough and if the air has access to the rectal lumen. In ticks, tissues in the mouth have been implicated in the uptake of water vapor.

TABLE 12-6   The critical equilibrium humidities for some arthropods that are able to extract water from the vapor phase.

|  | Relative humidity |
|---|---|
| Arachnida | |
| *Ixodes ricinus* | 92·0 |
| *Rhipicephalus sanguineus* | 84·0–90·0 |
| *Acarus siro* | 75·0 |
| *Laelaps echidnina* | 90·0 |
| Insecta | |
| *Thermobia domestica* | 45·0 |
| *Liposcellis bostrychophilus* | 60·0 |
| *Xenopsylla brasilinensis* | 50·0 |
| *Tenebrio molitor* larvae | 88·0 |
| *Arenivaga investigata* | 82·5 |

*Note:* At relative humidities below critical, the animal is unable to extract moisture from the air.

*Source:* Edney and Nagy, 1976.

## WATER LOSS DURING EXCRETION AND ION REGULATION

Unavoidable loss of body water in terrestrial animals accompanies processes for the regulation of ion concentrations and for the excretion of nitrogenous wastes (discussed later). A number of physiological adaptations address themselves to the task of minimizing the loss of water associated with these important functions of excretory systems.

Among terrestrial invertebrates, insects are highly effective in conserving water in the course of eliminating nitrogenous and inorganic waste. Before leaving the rectum, most of the water is removed from the pellet that consists of feces and uric acid. The pre-urine, formed in the Malpighian tubules (described later) and containing nitrogen in the form of potassium urate, passes into the rectum, where the relatively soluble urate is converted to uric acid as $K^+$ is actively transported out of the rectum into the blood. As noted earlier, uric acid has a very low water solubility and thus readily precipitates as crystals, no longer contributing significantly to the osmotic pressure of the rectal contents. By means described further on, much of the fecal water is transported out of the rectum back into the hemolymph, so that relatively little water is lost during elimination of the fecal pellet.

The extent to which ions are reabsorbed in the insect rectum or eliminated with the feces is regulated according to the osmotic condition of the insect. This is seen in Table 12-7. Locusts were allowed to drink either pure water or a concentrated (450 mosm) saline containing NaCl and KCl. Whereas the salt concentration of the hemolymph differed by a factor of about 0.5, the salt concentration of the feces was several hundred times higher after the insect drank saline than when it drank pure water.

The kidney is the chief organ of osmoregulation and nitrogen excretion in most terrestrial vertebrates, and most especially in mammals, which have no other provision for the excretion of salts or nitrogen. Because of this, the vertebrate kidney is considered in a special section later. It will suffice here to mention that the kidneys of birds and mammals have specializations that utilize countercurrent multiplication (described further on) as a means to produce a urine more concentrated than the plasma. These specializations, centered

**TABLE 12-7**   Ionic regulation in locusts.

| Fluid | Concentration (mean values in meq/L) | | |
|---|---|---|---|
| | Na | K | Cl |
| Saline for drinking | 300 | 150 | 450 |
| Hemolymph | | | |
|   With water to drink | 108 | 11 | 115 |
|   With saline to drink | 158 | 19 | 163 |
| Rectal fluid | | | |
|   With water to drink | 1 | 22 | 5 |
|   With saline to drink | 405 | 241 | 569 |

*Note:* Desert locusts were given strong saline or pure water to drink. When they drank saline, the ionic concentrations in the hemolymph rose, but not to the level of the saline. Ionic concentrations in their rectal fluid became higher than those in the saline.
*Source:* Edney and Nagy, 1976.

on a hairpinlike loop made by the kidney tubule, have undoubtedly been of major importance in allowing birds and mammals to exploit dry environments. The so-called *loop of Henle* reaches its highest degree of specialization in desert animals such as the kangaroo rat and Australian hopping mice, which can produce a urine of up to 9000 mosm/L. In birds, the countercurrent organization of the loop of Henle is less efficient, and concentrations of only up to 2000 mosm/L have been recorded (in the salt-marsh savannah sparrow). Reptiles and amphibians, whose kidneys are not organized for countercurrent multiplication, are unable to produce a hyperosmotic urine. As an adaptive consequence, some amphibians, when faced with desiccation, are able to cease urine production entirely during the period of osmotic stress.

### THE KANGAROO RAT: INTEGRATED SPECIALIZATION FOR DESERT LIFE

*Dipodomys merriami,* a native of the American Southwest, has become a classic example of how small mammals survive in the desert. A brief review of survival strategies practiced by the kangaroo rat will provide an overview of a variety of osmoregulatory adaptations to desert life exemplified in this appealing little rodent (Figure 12-14), which survives arid conditions without ingesting any free water.

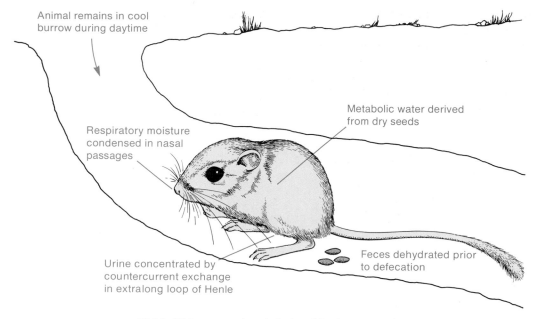

Animal remains in cool burrow during daytime

Respiratory moisture condensed in nasal passages

Metabolic water derived from dry seeds

Urine concentrated by countercurrent exchange in extralong loop of Henle

Feces dehydrated prior to defecation

**12-14**   Water-conserving strategies of the kangaroo rat.

Desert animals such as the kangaroo rat are faced with physiological double jeopardy—excessive heat and near-absence of free water. Water regulation and temperature regulation are, of course, closely related, since one important means of channeling excess heat out of the body into the environment is by evaporative cooling. Since evaporative cooling is at odds with water conservation, most desert animals cannot afford this method and have devised means of circumventing it. The kangaroo rat, like many desert mammals, avoids much of the daytime heat through a nocturnal life-style, keeping cool during daylight hours by remaining in a burrow. The nocturnal habit is an important and widespread *behavioral adaptation* to desert life.

Not only does the cool burrow reduce the animal's temperature load, but it reduces respiratory water loss. The nasal countercurrent mechanism for conserving respiratory moisture in the kangaroo rat and many other arid land mammals and birds depends, of course, on the ambient temperature being significantly lower than the 37°–40°C characteristic of the core temperatures of birds and mammals. If the rodent ventures out of its cool burrow into air close to its own temperature,

its respiratory water loss will rise abruptly, since the cooling properties of the nasal epithelium will be absent. Also important for the desert mammal is the avoidance of heat-generating exercise during the day, when removal of excess heat from the body is slowed by the higher ambient temperature.

Because of its efficient kidneys, the kangaroo rat excretes a highly concentrated urine, and rectal absorption of water from the feces results in essentially dry fecal pellets.

Using all these adaptations for desert survival, the kangaroo rat achieves a balance of water gain equaling water loss. In spite of extreme osmotic economy, the small amount of lost water must, of course, be replaced, or the animal will eventually dry up. Since the kangaroo rat eats dry seeds, is not known to drink, and in fact survives quite well in the near absence of free water, it must have a cryptic source of water. This, it turns out, is the metabolic water noted earlier. The exquisite conservation of water by the kangaroo rat (Table 12-8) allows it to survive primarily on the water produced during the oxidation of foodstuff hydrogen. The dry seeds it eats contain only a trace of free water.

**TABLE 12-8**   Water balance in a desert mammal, the kangaroo rat.

| Gains | | Losses | |
|---|---|---|---|
| Metabolic water | 90% | Evaporation + respiration | 70% |
| Free water in "dry" food | 10% | Urine | 25% |
| Drinking | 0% | Feces | 5% |
| | 100% | | 100% |

*Source:* Schmidt-Nielsen, 1972.

## Osmoregulatory Organs

Transport epithelia, such as frog skin and the mammalian gallbladder (discussed in Chapter 4), incorporate the basic cellular mechanisms of all excretory or osmoregulatory organs. The capabilities of transport epithelia are greatly enhanced in osmoregulatory organs by their anatomical organization, as is exquisitely evident in the kidneys of higher vertebrates. Here, in addition to a high degree of cellular differentiation for transepithelial transport, there is the added dimension of "plumbing"; that is, the epithelium is organized into tubules that are arranged so as to enhance the transport capabilities of the tubular epithelium. This combination of cell function and tissue organization has produced a marvelously efficient osmoregulatory and excretory organ. We will now examine the mammalian kidney in some detail before going on to consider briefly the osmoregulatory organs of some other animals.

## The Vertebrate Kidney

To speak of the vertebrate kidney would be misleading unless we note that it is organized somewhat differently in different groups of vertebrates. Comparisons can best be made after we first consider the mammalian kidney. The mammalian kidney performs certain functions that in lower vertebrates are shared by such organs as the skin and bladder of amphibians, the gills of fishes, and the salt glands of reptiles and

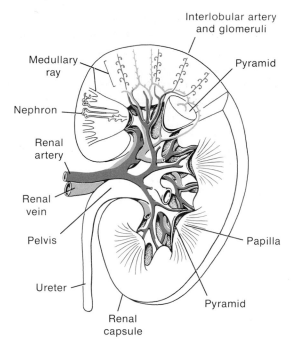

12-15   Anatomy of the mammalian kidney. The nephrons (see Figure 12-16) are arranged in parallel, with their collecting ducts opening through the papillae into a central cavity termed the renal pelvis. The urine passes from the pelvis into the ureter, which takes it to the urinary bladder. [Smith, 1956.]

birds. It is also the osmoregulatory organ of which we have the most complete understanding, thanks to intensive research over the past four to five decades.

### ANATOMY OF THE MAMMALIAN KIDNEY

The gross anatomy of the mammalian kidney is shown in Figure 12-15. The organs occur in pairs, one located on each side against the dorsal inner surface of the lower back, outside the peritoneum. In view of their small size (about 1% of total body weight in humans), they receive a remarkably large amount of blood, carrying about 20–25% of the total cardiac output. Surrounding the hollow *pelvis* is the *medulla*, which sends *papillae* projecting into the pelvis. The outer functional layer, the *cortex*, is covered by a tough *capsule* of connective tissue. The pelvis gives rise to the *ureters*, which empty into the *urinary bladder*. The urine leaves the bladder during *micturition* (urination) via the *urethra,* which leads to the end of the penis in males and into the vulva in females. The formation of urine is completed as it reaches the pelvic cavity of the kidney. The urine is then carried to the bladder and expelled without further modification. For this reason, the portions of the urinary tract below the kidney might be considered mere plumbing. Fortunately, this plumbing is equipped to allow occasional release of stored urine rather than a continual dribble. This release is accomplished by the neural control of a skeletal muscle sphincter around the opening of the bladder,

which leads into the urethra. As the bladder wall is stretched by gradual filling of the bladder, stretch receptors in the wall of the bladder generate nerve impulses that are carried by sensing neurons to the spinal cord and brain, producing the associated sensation. The sphincter can then be relaxed by inhibition of motor impulses, allowing the smooth muscle of the bladder wall to contract under autonomic control and empty the contents.

The functional unit of the mammalian kidney is the *nephron* (Figure 12-16), an intricate epithelial tube that is closed at one end and that opens into the renal

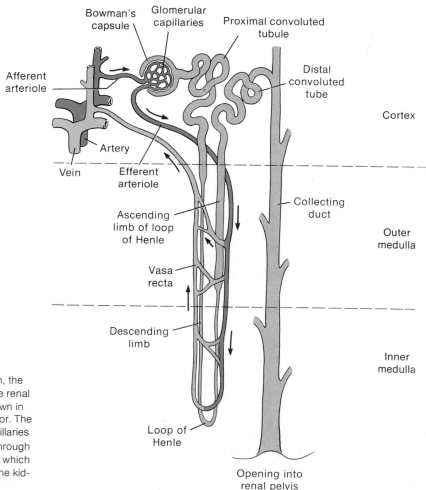

**12-16** The unit of kidney function, the nephron, with its blood supply. The renal tubule and collecting duct are shown in gray; the vascular elements, in color. The blood first passes through the capillaries of the glomerulus and then flows through the hairpin loops of the vasa recta, which plunges deep into the medulla of the kidney along with the loop of Henle.

pelvis through collecting ducts at the other end. At the closed end, the nephron is expanded—somewhat like a balloon that has been pushed in from one end toward its neck—to form the cup-shaped *Bowman's capsule.* The lumen of the capsule is continuous with the narrow lumen that extends through the *renal tubule.* Associated with the capsule is the tuft of capillaries that forms the *glomerulus* inside Bowman's capsule. This remarkable structure is responsible for the first step in urine formation. A *filtrate* of the blood passes through the single-cell layer of the capillary walls, through a basement membrane, and finally through another single-cell layer of epithelium that forms the wall of Bowman's capsule. The filtrate accumulates in the lumen of the capsule to begin its trip through the various segments of the renal tubule, finally descending the *collecting duct* into the renal pelvis. The wall of the renal tubule is one cell layer thick; this epithelium separates the lumen, which contains the urinary filtrate, from the interstitial fluid. In some portions of the nephron, these epithelial cells are morphologically specialized for transport, bearing a dense pile of *microvilli* on their luminal surfaces (Figure 12-17). The epithelial cells are tied together by leaky tight junctions, which permit paracellular diffusion between the lumen and interstitial space surrounding the renal tubule.

The nephron can be divided into three main regions: the *proximal nephron,* the *loop of Henle,* and the *distal nephron.* The proximal nephron consists of Bowman's capsule, the *proximal convoluted tubule,* and the *proximal straight tubule.* The hairpin loop of Henle comprises a *descending limb* and an *ascending limb.* The latter merges into the distal nephron, which consists of *distal convoluted* and *straight tubules,* and these join a collecting duct serving several nephrons. The number of nephrons per kidney varies from several hundred in lower vertebrates to many thousands in small mammals, and a million or more in humans and other large species.

Before continuing, we might make several observations about the loop of Henle. This feature is known only in birds and mammals and is believed to be of central importance in concentrating the urine. Vertebrates that lack the loop of Henle are incapable of producing a urine that is hyperosmotic relative to

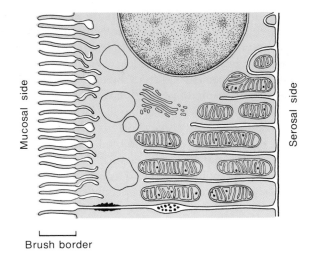

Mucosal side · Serosal side · Brush border

**12-17** Diagram of a proximal tubular cell. The surface membrane facing the lumen of the proximal tubule of the nephron is thrown into fingerlike projections (microvilli), greatly increasing its surface area. Mitochondria are concentrated near the serosal surface. [Rhodin, 1954.]

the blood. In mammals, the nephron is so oriented that the loop of Henle and the collecting duct lie parallel to each other and in a radial orientation in the kidney. The glomeruli are found in the renal cortex, and the loops of Henle reach down into the papillae of the medulla.

The anatomy of the renal circulatory system also appears to be important in the function of the nephron. The vascular plumbing leading to the capillaries of the glomerulus is designed for maximum pressure difference between inflowing and outflowing blood. As we will see in the next subsection, this aids in the filtration of plasma into the lumen of Bowman's capsule. The pressure in an artery drops linearly with distance, assuming constant diameter. The glomerulus and the artery that supplies it are connected by a very short length of arteriole, minimizing the pressure drop (Figure 12-16). The capillary tuft then recombines to form a second segment of arteriole. This formation is unusual, for in most other tissues the capillaries empty into venules, which in turn merge into veins. The blood, on leaving the glomerulus, located in the renal cortex, enters the second arteriole and is carried into the medulla in a descending and subsequently ascending loop of anastomosing (interconnecting) capillaries before leaving the kidney via a vein. We will return later to the functional significance of this circulatory loop, which lies in parallel with the loop of Henle.

## AN OVERVIEW OF URINE FORMATION

The initial stage of urine formation is the filtration of plasma† and the accumulation of the ultrafiltrate in the lumen of Bowman's capsule. The glomerular filtrate contains essentially all the constituents of the blood except for the blood cells and proteins. Filtration in the glomerulus is so extensive that 15–25% of the water and solutes are removed from the plasma that flows through it. The glomerular filtrate is produced at the rate of 125 ml/min, or about 200 L/day, in human kidneys. When this number is compared to the normal intake of water, it is evident that unless most of the glomerular filtrate is subsequently reabsorbed into the bloodstream, the body would be quickly dehydrated. *Reabsorption* is one of the important functions of the nephron. As we will see later, 99% of the water, and nearly as large a proportion of the salts, is reabsorbed before formation of the urine is completed. Some substances, however, appear in the urine in concentrations higher than can be accounted for by the rate of plasma filtration, which means that those substances are *secreted* into the lumen of the nephron during formation of the urine.

It is apparent, then, that there are three processes that contribute to the ultimate composition of the urine (Figure 12-18). These are listed here and then discussed in more detail.

*1. Glomerular filtration* of water and nonprotein solutes (crystalloids, such as $Na^+$, $K^+$, $Cl^-$, glucose, and urea) in approximately the proportions in which they occur in the plasma. These proportions have been determined by direct sampling of the glomerular filtrate by micropuncture (Figure 12-19). With the exception that blood cells and large proteins are ex-

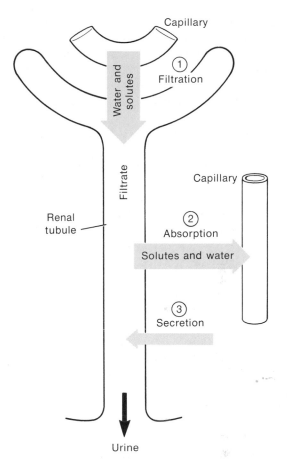

**12-18** The three processes involved in urine formation in the glomerular nephron. Filtration, the initial step, takes place in Bowman's capsule. Absorption and secretion take place along the renal tubule. The final product of these three processes is the urine.

cluded, the process of glomerular filtration is completely nonselective and based entirely on molecular size (Table 12-9).

*2. Tubular reabsorption* of approximately 99% of the water and most of the salts. The reabsorption is somewhat selective, resulting in the concentration (relative to water and salt) of waste products such as urea. Some substances, such as NaCl, are reabsorbed from the glomerular filtrate by active transport, while other

---

†Plasma is the fluid portion of the blood that remains when the various blood cells are removed.

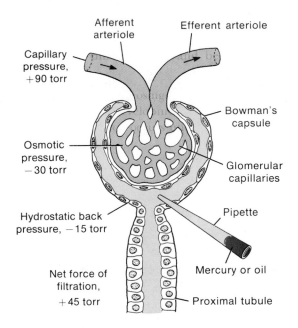

**12-19**   Filtration mechanism. The various forces affecting filtration rate are indicated at the left. At the right, a micropipette is inserted to sample the glomerular filtrate. The mercury in the pipette is pushed to the tip by pressure before penetration of the capsule. A sample is then sucked into the calibrated tip for subsequent microanalysis. [Merck, Sharp, and Dohme, in Hoar, 1975.]

substances, like water, are reabsorbed through passive diffusion down osmotic gradients.

*3. Tubular secretion* of a number of substances. This is relatively selective and is responsible for the regulation of blood concentrations of $K^+$, $H^+$, and bicarbonate as well as for the elimination of foreign substances, such as drugs. Tubular secretion is accomplished through active transport in nearly all instances.

The kidney may be considered a "black box" in which the final concentration of any substance in the urine is a function of one, two, or all three of the processes listed above. This is best illustrated by some simple quantitative comparisons of the amounts of various substances appearing in the urine and the amounts of those substances appearing in the glomerular filtrate in Bowman's capsule. Such comparisons require an understanding of the concept of *renal clearance* (Box 12-1).

The *clearance* of a plasma-borne substance is the volume of blood plasma from which that substance is "cleared" (i.e., completely removed) per unit time by the kidneys. The total renal clearance is the quantity of ultrafiltrate in milliliters produced by the kidney in 1 min. A substance that is freely filtrated along with

**TABLE 12-9**   Relation between molecular size and the ratio of the concentration of the substance in the filtrate appearing in Bowman's capsule to its concentration in the plasma, [filtrate]/[filtrand].

| Substance | Mol. wt. (g) | Radius from diffusion coefficient | Dimensions from X-ray diffraction | [filtrate] / [filtrand] |
|---|---|---|---|---|
| Water | 18 | 1.10 | | 1.0 |
| Urea | 60 | 1.6 | | 1.0 |
| Glucose | 180 | 3.6 | | 1.0 |
| Sucrose | 342 | 4.4 | | 1.0 |
| Inulin | 5,500 | 14.8 | | 0.98 |
| Myoglobin | 17,000 | 19.5 | 54 / 8 | 0.75 |
| Egg albumin | 43,500 | 28.5 | 88 / 22 | 0.22 |
| Hemoglobin | 68,000 | 32.5 | 54 / 32 | 0.03 |
| Serum albumin | 69,000 | 35.5 | 150 / 36 | <0.01 |

*Source:* Pitts, 1968.

## Box 12-1  *Renal Clearance*

A substance that passes freely from the blood into Bowman's capsule during formation of the glomerular filtrate, and subsequently is neither secreted nor reabsorbed by the renal tubule, can be very useful in renal physiology because it exhibits a renal clearance equal to the glomerular filtration rate. Thus, inulin, a substance that has been found to meet these criteria, has been used routinely to determine the glomerular filtration rate of intact kidneys. In such studies, the substance is first injected into the subject's circulation and allowed to mix to uniform concentration in the bloodstream. A sample of blood is removed, and the plasma concentration of inulin, $P$, is determined from the sample. The rate of appearance of inulin in the urine is determined by multiplying the concentration, $U$, of inulin in the urine by the volume of urine, $V$, produced per minute. The amount of inulin appearing in the urine per minute ($VU$) must equal the rate of plasma filtration (GFR) multiplied by the plasma concentration of inulin:

$$\frac{VU}{(GFR)P} = \frac{\text{amt. inulin appearing in urine/min}}{\text{amt. inulin removed from blood/min}} = 1$$

In this special case, the substance in question, inulin, is freely filtered and unchanged by tubular absorption or secretion. Therefore, the GFR and the clearance, $C$, of the substance are equal. Substituting $C$ for GFR gives, for inulin,

$$\frac{VU}{CP} = 1$$

Therefore,

$$\frac{VU}{P} = C = \text{plasma clearance (ml/min)}$$

If the amount of substance $x$ appearing in the urine per minute deviates from the amount of $x$ present in the volume of plasma that is filtered per minute, this will be reflected in a value of $C_x$ that differs from that of the inulin plasma clearance $C$. For example, if the inulin clearance of a subject, and hence the GFR, is 125 ml/min and substance $x$ exhibits a clearance of 62.5 ml/min,

$$\frac{VU_x}{P_x} = C_x = 62.5 \text{ ml/min}$$
$$= 0.5(\text{GFR})$$

In this case, a volume of plasma *equivalent* to half that filtered each minute is cleared of substance $x$. Stated differently, only half the amount of substance $x$ present in a volume of blood plasma equal to the volume filtered each minute actually appears in the urine per minute. There are two possible reasons why the plasma clearance for a substance would be less than the GFR. First, it may not be freely filterable: it may be hindered in the filtration process by binding to serum proteins, by its large molecular size, which would interfere with filtration, or by some other factor. Second, it may be freely filtered, but the amount of $x$ that appears in the urine might be reduced by tubular reabsorption. As a matter of fact, most molecules below a molecular weight of about 5000 are freely filtered (Table 12-9). Many of these are either partially reabsorbed or partially secreted, and the extent of reabsorption or secretion can be gauged by the plasma clearance of a substance. Reabsorption reduces the plasma clearance to below the GFR. Tubular secretion, however, will cause more of a substance to appear in the urine than is carried into the tubule by glomerular filtration.

water into the nephron, but which is neither reabsorbed nor secreted, permits the calculation of total plasma clearance merely by dividing the amount of the nontransported substance appearing in the urine by the concentration of that substance in the plasma. One such molecule is inulin (*not* insulin), a small starch-like carbohydrate (molecular weight 5000). Since the inulin molecule is neither reabsorbed nor secreted by the renal tubule, the *inulin clearance* is identical with the rate at which the glomerular filtrate is produced—

the *glomerular filtration rate* (GFR), generally given in milliliters per minute. Having determined the GFR, and knowing the concentration of a substance in the plasma (thus also its concentration in the ultrafiltrate), it is easy to calculate whether it undergoes a net reabsorption or net secretion during the passage of the ultrafiltrate along the renal tubule (Box 12-1). Thus, if less of a substance appears in the urine than was filtered in the glomerulus, it must have undergone some reabsorption in the tubule. This is true for water,

NaCl, glucose, and many other essential constituents of the blood. If, however, the quantity of a substance appearing in the urine over a period of time is greater than the amount that passed into the nephron because of glomerular filtration, it can be concluded that this substance is actively secreted into the lumen of the tubule. The *clearance technique* is of limited usefulness in studies of renal function, since it indicates only the *net* output of the kidney relative to input and fails to provide insight into the physiological details. Nevertheless, it is useful clinically, and it indicates that some substances are reabsorbed while others are secreted. It does not indicate if a substance is both reabsorbed and secreted by different portions of the nephron, as is true for potassium ions.

Glucose clearance offers a simple example of a clinical application. A healthy mammal exhibits a plasma glucose clearance of 0 ml/min. That is, even though the glucose molecule is small and is freely filtered by the glomerulus, it is completely reabsorbed (up to certain concentrations) by the epithelium of the renal tubule. The appearance of glucose in the urine is usually due to its presence in excessive concentration in the blood plasma. In diabetes mellitus, appearance of glucose in the urine is due to saturation of carriers responsible for glucose transport in the epithelial cells of the tubular epithelium. Figure 12-20 reveals that there is a maximum rate (milligrams per minute) at which glucose can be removed from the tubular urine by reabsorption. This *transfer maximum,* or Tm, is about 375 mg/min in humans. Below plasma glucose levels of 200 mg %, all the glucose appearing in the glomerular filtrate is reabsorbed. At about 400 mg %, the carrier mechanism is fully saturated, so that any additional amount of glucose appearing in the filtrate will be passed out in the urine. The arterial plasma glucose concentration in humans is normally held at about 100 mg % by an endocrine feedback loop involving insulin. Since this is well below the Tm for glucose, normal urine contains essentially no glucose.

### *GLOMERULAR FILTRATION*

The process of ultrafiltration in the glomerulus (Figure 12-19) depends on three factors: (1) the net hydrostatic pressure difference between the lumen of the

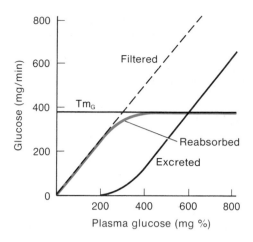

**12-20**   Relation of urine glucose to blood glucose levels. The concentration of glucose in the glomerular filtrate (broken line) is proportional to the plasma glucose concentration. The renal tubules are capable of reabsorbing the glucose by active transport at rates up to 375 mg/min ($Tm_G$). Glucose entering the filtrate in excess of this rate is necessarily excreted in the urine. [Pitts, 1974.]

capillary and the lumen of Bowman's capsule, which favors filtration; (2) the colloid osmotic pressure, which opposes filtration; and (3) the hydraulic permeability (sievelike properties) of the three-layered tissue separating these two compartments. The net pressure gradient results from the sum of the hydrostatic pressure difference between the two compartments and the colloid osmotic-pressure difference. The latter arises because of the separation of proteins during the filtration process. In humans, the proteins remaining in the capillary plasma produce an osmotic pressure difference of about $-30$ torr (mmHg), and the hydrostatic pressure difference (capillary blood pressure minus intracapsular back pressure in the renal tubule) is about $+75$ torr (Table 12-10). The result is a net filtration pressure of about $+45$ torr. It is this pressure acting on the high water and crystalloid permeability of the glomerular ultrasieve (about 100 times higher than the permeability of systemic capillaries) that produces the phenomenal rate of filtrate formation by the millions of glomeruli in each human kidney—125 ml/min, or 200 L/day. It is important to note that the filtration process in the kidney is entirely passive, deriving its energy from the hydrostatic pressure of the

**TABLE 12-10** Balance sheet of pressures (in torr) involved in glomerular ultrafiltration as illustrated in Figure 12-19.

|  | Salamander | Man |
|---|---|---|
| Glomerular capillary pressure | 17.7 | 90 |
| Intracapsular pressure | − 1.5 | −15 |
| Net hydrostatic pressure | 16.2 | 75 |
| Osmotic pressure | − 10.4 | −30 |
| Net filtration pressure | 5.8 | 45 |

*Source:* Pitts, 1968.

blood, which is maintained by the contractions of the heart.

The kidneys are perfused by 500–600 ml of plasma per minute, or 20–25% of the cardiac output, yet constitute less than 1% of the body weight. This preferential perfusion takes place in a relatively low-resistance vascular bed within the kidney. A high renal blood pressure is the result of the relatively direct arterial supply; since arteries and arterioles are large in diameter and short in length, the loss of pressure due to friction is minimized. The efferent arterioles (those taking the blood away from the glomeruli) are of smaller diameter, providing greater resistance and hence maintaining a higher pressure within the glomeruli.

*TUBULAR REABSORPTION*

Glomerular filtration is only the first step in urine formation; as the filtrate makes its way through the nephron, its original composition is quickly modified by reabsorption of some substances from the glomerular filtrate and by the secretion of some substances into the pre-urine formed from the filtrate.

The human kidneys produce about 200 L of filtrate per day, but the final volume of urine is only about 1½ L. Of the 1800 g of NaCl in the original filtrate, only 10 g appears in the urine. Essentially all the glucose filtered is reabsorbed (Figure 12-20), but varying amounts of other filtered solutes are reabsorbed from the tubular lumen or secreted into the lumen.

The details of tubular function vary from species to species. Our knowledge of the changes in urinary composition along different portions of the nephron

is based to a large extent on the technique of micropuncture, first developed by Alfred Richards (1935) and his co-workers in the 1920s. A glass capillary micropipette is used to remove a minute sample of the tubular fluid from the lumen of the nephron. The osmolarity of the sample is then determined by measuring its freezing point. The lower the freezing point, the higher its osmolarity. Subsequent modifications of this technique include the injection of oil into the tubule to isolate a portion of the lumen and analyze the action of that section of the tubule on injected samples of defined solution (Figure 12-21). Microchemical methods are now used to determine the concentrations of individual ion species in the sample. Techniques have recently been developed by which a given segment of renal tubule can be steadily perfused

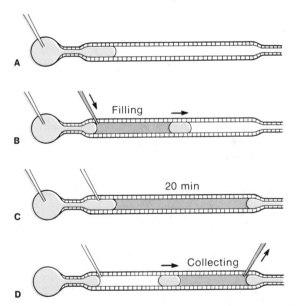

**12-21** The stopped-flow microperfusion technique. (**A**) A micropipette is inserted into Bowman's capsule (left), and oil is injected until it enters the proximal tubule. (**B**) Perfusion fluid is injected through a second pipette into the middle of the column of oil, forcing a droplet ahead of it. (**C**) The tubule is full when the droplet reaches the far end of the tubule. (**D**) After about 20 min, the fluid is collected for microanalysis by the injection of a second liquid behind the oil remaining near the glomerulus. [From ''Pumps in the Living Cell'' by Arthur K. Solomon. Copyright © 1962 by Scientific American, Inc. All rights reserved.]

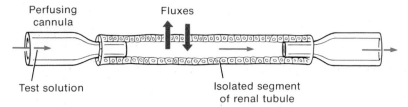

**12-22**    Perfusion of a cannulated segment of renal tubule. The perfusate is subjected to chemical and radiotracer analysis to determine the fluxes of ions across the tubular wall.

and the perfusate analyzed (Figure 12-22). The major findings on tubular transport, summarized in Figure 12-23, are listed below:

*1.* The *proximal convoluted tubule* initiates the process of concentrating the glomerular filtrate. In this segment, about 75% of the $Na^+$ is removed from the lumen by active transport, and a nearly proportional amount of water and certain other solutes, such as $Cl^-$, follow passively. So about 70% of the filtrate is reabsorbed before the filtrate reaches the loop of Henle. The result is a tubular fluid that is isosmotic with respect to the plasma and interstitial fluids.

Using the stopped-flow perfusion technique (Figure 12-21), A. K. Solomon and co-workers found that when the NaCl concentration inside the tubule is decreased, the movement of water also decreases. This result is just the opposite of the one that should be expected if the water moves by simple osmotic diffusion, and it indicates that water transport is coupled to active sodium transport. The coupling is presumably due to a standing-gradient mechanism (described in Chapter 4). The actual pumping of sodium ions takes place at the serosal surface of the epithelial cells of the proximal tubule, just as it does in frog skin and gallbladder epithelia. In amphibians, this active transport leaves the tubular lumen about 20 mV negative relative to the fluid surrounding the nephron. This potential difference accounts for the passive net diffusion of $Cl^-$ out of the proximal tubule as the counter-ion for sodium. At the most distal portion of the proximal tubule (where it joins the thin descending limb of the loop of Henle), the glomerular filtrate is already reduced to one-fourth its original volume. As a result

of the reduction in the volume of tubular fluid, substances that are not transported across the tubule or that do not passively diffuse across it become four times as concentrated toward the end of the proximal tubule as in the original filtrate. In spite of this great reduction in the volume of tubular fluid, the fluid at this point is isosmotic relative to the fluid outside the nephron, having an osmolarity of about 300 mosm (Figure 12-23). It is interesting to note that the active transport of NaCl alone can account for the major changes in volume of the fluid along the proximal tubule and for the increased concentration of urea and many other filtered substances.

The proximal tubule is ideally structured for the massive reabsorption of salt and water. Numerous microvilli at the luminal border of the tubular epithelial cells form the so-called *brush border* (Figure 12-17). This absorptive surface greatly increases the area of membrane and thereby reduces the barrier to diffusion of salt and water from the tubular lumen into the epithelial cell. The serosal border has a smaller surface area and hence offers a greater resistance to the back leakage of $Na^+$ pumped out of the cell across that membrane.

*2.* The *descending limb* of the loop of Henle and the inner medullary ascending portion are made up of very thin cells containing few mitochondria and no brush border. The morphology of the thin segments suggests that they do not actively transport solutes across the tubular wall. They have proven too small for micropuncture studies, but success has recently been obtained in perfusing segments of the renal tubule in situ by using tracers and determining the alterations produced in the perfusate during its passage through the

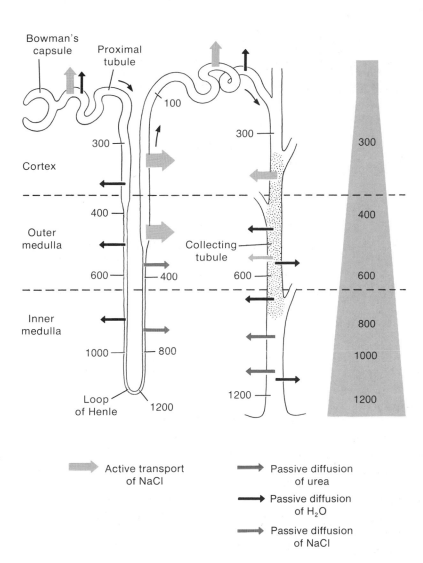

**12-23** Overview of ion and water fluxes along the renal tubule of a mammal. Numbers indicate milliosmoles per liter. The stippled portion of the collecting duct is ADH-sensitive. The osmotic gradient in the extracellular fluid is indicated by the gray wedge. [Pitts, 1959.]

Bowman's capsule

Proximal tubule

Cortex

Outer medulla

Inner medulla

Loop of Henle

Collecting tubule

Active transport of NaCl

Passive diffusion of urea

Passive diffusion of $H_2O$

Passive diffusion of NaCl

tubule (Figure 12-22). Using this approach, Maurice Burg and his co-workers at the National Institutes of Health have demonstrated that there is no active salt transport in the descending loop. Moreover, this segment exhibits very low permeability to NaCl and low permeability to urea but is permeable to water. As will be described later, the descending loop plays an entirely passive, but important, role in the urine-concentrating system of the nephron.

*3.* The *thin segment* of the ascending limb has also been shown by perfusion experiments to be inactive in salt transport. Nevertheless, it is highly permeable to $Na^+$ and $Cl^-$. Its permeability to urea is low, and to water, very low.

*4.* The *medullary thick ascending limb* differs from the rest of the loop of Henle in that it exhibits active transport of $Na^+$ and $Cl^-$ outward from the lumen to the interstitial space. This portion, along with the rest of the ascending limb, has a very low permeability to water. As a result of NaCl reabsorption, the fluid reaching the distal convoluted tubule is somewhat hypoosmotic relative to the interstitial fluid. The importance of salt reabsorption by the ascending tubule will be discussed in the section on the countercurrent mechanism.

**5.** The movement of salt and water across the *distal convoluted tubule* is complex; the distal tubule is made up of several morphologically and functionally different segments in series. The distal tubule is important in the transport of $K^+$, $H^+$, and $NH_3$ into the lumen and of $Na^+$, $Cl^-$, and $HCO_3^-$ out of the lumen and back into the interstitial fluid. As salts are pumped out of the tubule, water follows passively.

**6.** The *collecting duct* is permeable to water, and thus it allows water to flow from the dilute urine into the more concentrated interstitial fluid of the renal medulla. This is the final step in the production of a hyperosmotic urine. The duct also reabsorbs NaCl by active transport of $Na^+$; it is impermeable to salts and exhibits a variable permeability to water. An important feature of the inner medullary segment of the collecting duct, toward its distal end, is its high permeability to urea. The significance of this will be evident in the discussion, later, on the countercurrent mechanism that concentrates the urine in the collecting duct. The rate at which water is reabsorbed in the collecting duct is under delicate feedback control via antidiuretic hormone.

### TUBULAR SECRETION

The nephron has several distinct systems that secrete substances by transporting them from the plasma into the tubular lumen. Most thoroughly investigated are the systems for secretion of $K^+$, $H^+$, $NH_3$, organic acids, and organic bases. The number of secretory pathways is small and limited; nevertheless, the nephron secretes innumerable "new" substances such as drugs and toxins, as well as endogenous and natural molecules. How is the nephron able to recognize and to transport all these substances? The answer seems to reside in the role of the vertebrate liver in modifying many such molecules so that they can react with the transport systems located in the wall of the nephron. These secretory mechanisms are highly important because they remove potentially dangerous substances from the blood. In the liver, many of these substances, along with endogenous, normal metabolites, are con-

jugated with glucuronic acid or sulfate. Both these classes of conjugated molecules are actively transported by the system that secretes organic acids. Since they are highly polar, these conjugated molecules, once deposited in the lumen of the nephron, cannot readily diffuse back into the peritubular space or into the blood and are therefore excreted in the urine.

### Secretion of $K^+$

Normally, most of the potassium ions that are freely filtered at the glomerulus and appear in the pre-urine are reabsorbed, so that little $K^+$ is passed out in the urine. The rate of active reabsorption in the proximal tubule and loop of Henle continues unabated even when the level of $K^+$ in the blood and filtrate rises to excessive levels in response to excessive intake of this ion. However, the distal tubules and collecting ducts are able to secrete as well as reabsorb $K^+$, so they are able to produce a net secretion of $K^+$ to achieve homeostasis in the face of a $K^+$ load due to abnormally high levels of ingested $K^+$. The transport of $K^+$ appears to depend on $K^+$ uptake into the tubular cell from the interstitial fluid by the usual $Na^+–K^+$ pump, with leakage of cytoplasmic $K^+$ into the luminal fluid. The latter is electronegative with respect to the cytoplasm, so $K^+$ can simply flow from inside the renal tubular cell into the lumen. The rate of $K^+$ secretion by these mechanisms is stimulated by the adrenocortical hormone aldosterone, which is released in response to elevated plasma $K^+$.

### Secretion of $H^+$

The membrane facing the lumen of the tubule contains an active transport system that pumps $H^+$ from the cytoplasm of the epithelial cell into the lumen. It removes $H^+$ from HOH, leaving $OH^-$ in the cytoplasm to react with $H^+$ to form HOH again. Although $H^+$ carries a positive charge, the lumen does not develop excessive electropositivity, for $Na^+$ moves in the opposite direction into the renal tubular cell owing to active transport by the $Na^+–K^+$ pump located on membrane facing the peritubular space. The secreted potassium ions, as already noted, are also reabsorbed. As we will see next, the secretion of $H^+$ is important in the regulation of the extracellular pH of the body.

## pH REGULATION BY THE KIDNEY

Several buffering systems in the body fluids regulate the extracellular pH. All are in equilibrium with one another insofar as compartmentalization of the tissues permits. Thus, a change in the buffering activity of one of these systems will affect the $H^+$ concentration of the body fluids to the extent that all the other buffering systems will allow. However, the most important pH buffering system by far, and the one primarily responsible for controlling the pH of the vertebrate extracellular space, is the *carbon dioxide–bicarbonate buffering system* in which

$$H_2O + CO_2 \rightleftharpoons H_2CO_3 \rightleftharpoons H^+ + HCO_3^-$$

This system is regulated primarily by (1) respiratory exchange, which largely determines the $CO_2$ concentration of the body (if, for example, respiratory exchange is reduced and $CO_2$ levels increase, there will be a drop in blood pH as bicarbonate ions accumulate; p. 631), and by (2) the kidney, which is ultimately responsible for maintaining the plasma $H_2CO_3$ concentration.

Plasma bicarbonate is regulated at the level of the individual nephron in two ways: first, by the variable rate of reabsorption of $H_2CO_3$ from the glomerular filtrate; and second, by the conversion of $CO_2$ and $H_2O$ into bicarbonate, which then passes into the plasma.

Both these mechanisms require the active transport of $H^+$ from the renal tubular cells into the lumen of the tubule, as we shall see next.

### Bicarbonate Reabsorption

Since the bicarbonate ion, $HCO_3^-$, like all small molecules, freely passes through the glomerulus into the tubular fluid, it must be reabsorbed to prevent a severe case of *acidosis* resulting from excess acid left behind in the body fluids. In fact, nearly all the bicarbonate, which has alkaline properties, is normally reabsorbed, and almost none appears in the urine. The reabsorption, however, is not accomplished by an active pumping of $HCO_3^-$ out of the lumen. Instead, the $H^+$ that, as we have noted, is secreted into the lumen combines with the bicarbonate ion to form $H_2CO_3$ (Figure 12-24); and $CO_2$ in the renal tubular cell combines with $H_2O$ to form $H_2CO_3$ also. The latter dissociates immediately, with catalytic assistance from the enzyme *carbonic anhydrase,* into $H^+$ and $HCO_3^-$, and the anion diffuses into the peritubular space. In the meantime, the $H^+$ combines with the $^-OH$ liberated at the membrane by the hydrolysis of HOH that occurs when $H^+$ is actively transported out across the membrane into the lumen. The $H^+$ that is pumped into the lumen replaces $Na^+$ diffusing passively from the lumen into the cell and

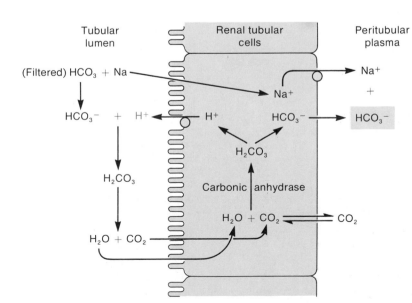

12-24  Reabsorption of filtered bicarbonate and sodium. The intracellular enzyme carbonic anhydrase catalyzes the production of $H_2CO_3$ from $CO_2$ and water. The bicarbonate reacts with $OH^-$ (not shown) to form HOH and $HCO_3^-$, which diffuses into the plasma. The free $H^+$ is actively transported into the tubular lumen, where it combines with $HCO_3^-$, filtered from plasma at the glomerulus, to form $H_2O$ and $CO_2$, both of which diffuse back into the tubular cells to repeat the cycle. Sodium leaks from the lumen into the cell down its concentration gradient and is then pumped out of the cell on the serosal side. The net effect is to recover bicarbonate and sodium that otherwise would be excreted in the urine. [Pitts, 1974; Vander, 1980.]

combines with $HCO_3^-$ to form $CO_2$ and $H_2O$. The luminal $CO_2$, which is freely permeant across the cell membrane, diffuses into the cell and then passes into the plasma as $CO_2$ or as $HCO_3^-$. The overall effect, then, is removal of $HCO_3^-$ and $Na^+$ from the urine and its addition to the plasma, where it replaces those bicarbonate ions lost from the plasma during glomerular filtration. Note that the $H^+$ that was actively secreted into the lumen does not remain in the urine, but is incorporated into water, most of which is reabsorbed in the nephron before formation of the final urine.

### Excretion of Acid

The removal of excess $H^+$ from the plasma is an important function of the kidney. The way the kidney does this is by secretion of $H^+$, which is then "trapped" in the urine in an impermeant form. At the same time, new $HCO_3^-$ is added to the plasma. Such a system is seen in the excretion of $H^+$ tied up in phosphate (Figure 12-25). Here, also, $H_2CO_3$ derived from $CO_2$ and $H_2O$ dissociates, with the $HCO_3^-$ passing into the plasma while $H^+$ is actively transported into the lumen. Some of the $H^+$ that is secreted combines with $HPO_4^{2-}$, which was freely filtered from the plasma at the glomerulus, forming $H_2PO_4^-$. The latter is highly impermeant across the tubular membranes and passes along the tubule without reabsorption, to be excreted finally in the urine. It is interesting that the same cell mechanisms come into play here as in the bicarbonate reabsorption seen in Figure 12-24. The difference lies in which anion the $H^+$ reacts with after being secreted into the lumen.

Secreted $H^+$ also combines in the tubular lumen with $NH_3$ to form $NH_4^+$ (Figure 12-26). Like $H_2PO_4^-$, $NH_4^+$ is highly polar and impermeant, so it passes out of the nephron in the urine without reabsorption, carrying with it the secreted proton. Unlike phosphate, which appears in the tubular fluid because of glomerular phosphate filtration, ammonia is produced within the renal tubular cells by enzymatic deamination of amino acids. In its nonpolar, un-ionized form, ammonia freely diffuses across the cell membrane into the lumen. It is trapped there by being converted to the impermeant ammonium ion in a reaction with the $H^+$ that was actively secreted into the lumen. In this way, $NH_4^+$ appears in the urine as a vehicle for excretion of both nitrogen atoms and hydrogen ions.

## THE CONCENTRATING MECHANISM OF THE NEPHRON

The urine of birds and mammals becomes concentrated by osmotic removal of water from the collecting ducts as they plunge through the renal medulla. In the medulla, the osmotic gradient that is responsible for the

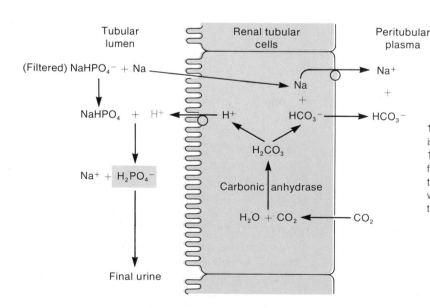

12-25  Excretion of hydrogen ions. $H^+$ that is produced and secreted, as in Figure 12-24, replaces $Na^+$ that leaks into the cell from the filtrate. The $H^+$ combines with filtered phosphate ions to produce $H_2PO_4^-$, which is impermeant and thus remains in the urine. [Pitts, 1974; Vander, 1980.]

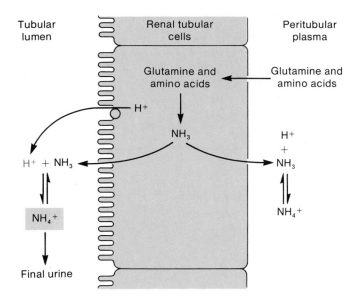

**Tubular lumen**

**Renal tubular cells**

**Peritubular plasma**

Glutamine and amino acids ← Glutamine and amino acids

H⁺

NH₃

H⁺ + NH₃ ←

H⁺ + NH₃

NH₄⁺

Final urine

**12-26** Secretion of ammonia. NH₃ produced in renal cells by deamination of glutamine and amino acids diffuses freely out of the cells into the plasma and the tubular lumen. In the latter, it reacts with H⁺ secreted by the renal cells, producing NH₄⁺, which is impermeant and thus is trapped in the urine, unable to diffuse back into the cell. The trapping of NH₃ in this way sets up a steep gradient for ammonium, facilitating its diffusion into the lumen. [Vander, 1980.]

osmotic removal of water from the collecting duct is established by a remarkable combination of cellular specialization and anatomical organization. There is a clear-cut correlation between the architecture of the vertebrate kidney and its ability to manufacture a urine that is hypertonic relative to the body fluids. Kidneys capable of producing a hypertonic urine (i.e., those of mammals and birds) all have nephrons featuring the loop of Henle. In general, the longer the loop and the deeper it extends into the renal medulla, the greater the concentrating power of the nephron. Thus, desert mammals have both the longest loops of Henle and the most hypertonic urine. This fact, together with the progressive increase in tonicity of intra- and extratubular fluids toward the deeper regions of the renal medulla, led B. Hargitay and W. Kuhn to propose in 1951 that the loop of Henle acts as a *countercurrent multiplier* (Box 12-2). Though a very attractive and plausible hypothesis, it was initially hard to test because of the difficulty of sampling the intratubular fluid in the thin loop of Henle. Determinations of the melting point of the fluid in slices of frozen kidney revealed that the osmolarity of both the intratubular and the extratubular fluid increases in gradient from the upper to the lower ends of the loop of Henle. This *corticomedullary concentration gradient* is represented by the gray wedge in Figures 12-23 and 12-27. More recently, the counter-

current hypothesis of urine concentration has been extended by the perfusion, in situ, of segments of the loop noted earlier (Table 12-11). The fluid that enters the descending limb from the proximal tubule is isosmotic with respect to the extracellular fluid at that point (in the outer portion of the renal medulla), having a concentration of about 300 mosm. The concentration of the fluid gradually increases as it makes its way down the descending limb toward the hairpin turn in the loop, where its concentration reaches 1000–3000 mosm in most mammals. At this point, too, it is nearly isosmotic relative to the surrounding extracellular fluid in the deep portion of the renal medulla. We will see shortly why the interstitial fluid in the deep parts of the medulla has a high osmolarity.

The tubular fluid flowing down the descending limb comes into osmotic equilibrium with the interstitial fluid because the walls of the descending limb are relatively permeable to water, but far less permeable to NaCl or urea. Thus, the concentration of the tubular fluid, initially about 300 mosm, approaches the 1000–3000 mosm (or greater in some desert mammals) concentration of the interstitial fluid around the hairpin turn of the loop (Figure 12-23). As the fluid flows up in the thick segment of the ascending limb, it undergoes a progressive loss of NaCl (but not water) as the result of an active transport of salt out of the tubule. Both the

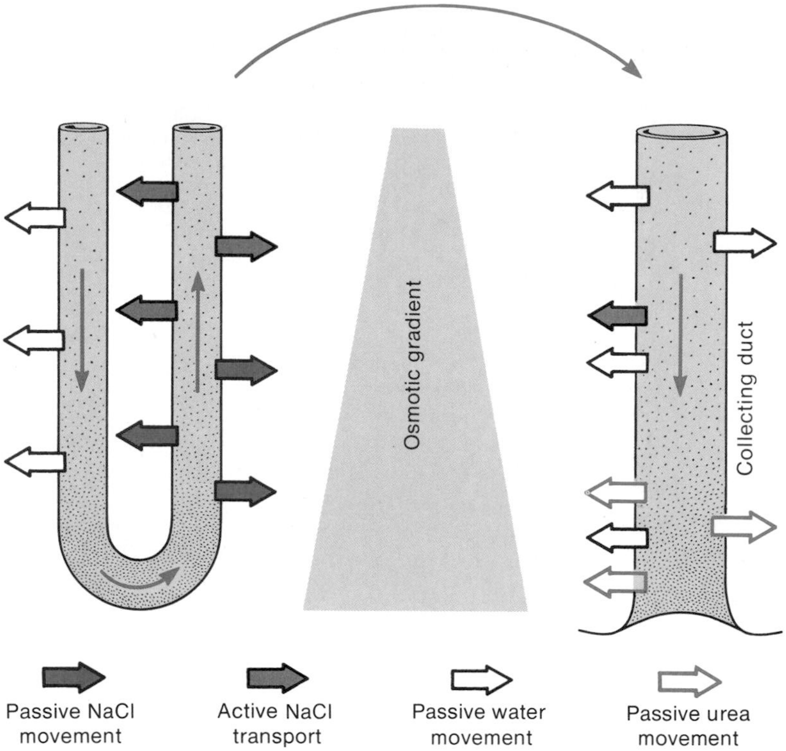

12-27 Countercurrent model of the loop of Henle. The solid colored arrows indicate the active transport of NaCl. The gradient of extracellular salt concentration is indicated by the colored wedge. The osmotic movement of water into the interstitial fluid is shown by open arrows; passive diffusion of urea by open colored arrows. See text for explanation.

Passive NaCl movement · Active NaCl transport · Passive water movement · Passive urea movement

TABLE 12-11  Active transport and passive permeability of different segments of rabbit nephron as determined in perfused tubules.

| | Active salt transport | Relative permeability | | |
| --- | --- | --- | --- | --- |
| | | $H_2O$ | NaCl | Urea |
| Thin descending limb | 0 | + + + + | 0 | + |
| Thin ascending limb | 0 | 0 | + + + | + |
| Thick ascending limb | + + + + | 0 | 0 | 0 |
| Cortical and outer medullary collecting duct | + | + + +† | 0 | 0 |
| Inner medullary collecting duct | + | + + +† | 0 | + + + |

*Note:* The zeros indicate negligible or unmeasurable values.
†Collecting duct water permeability assumes ADH present.
*Source;* Modified from Kokko and Rector, 1972.

thin and the thick segments of the ascending limb are relatively impermeable to water.

The functional asymmetry between the descending and the ascending limbs of the loop of Henle, together with the countercurrent principle, accounts for the corticomedullary concentration gradient of NaCl and urea shown in Figure 12-28. The osmotic loss of water from the descending limb allows the lumen of that limb to approach osmotic equilibrium with the interstitial fluid. Since the fluid is moving along within the lumen, the loss of water from the lumen of the tubule to the progressively higher concentrations present in the interstitium leads to a progressively higher concentration of solute as the fluid approaches the bend in the loop of Henle (Figure 12-27).

The interstitial osmotic gradient (Figure 12-28) is believed to be established by a combination of features that include the active transport of NaCl from the ascending thick segment and selective passive permeabilities to water, salt, and urea along specific segments of the nephron (Figure 12-27). Recall that the descend-

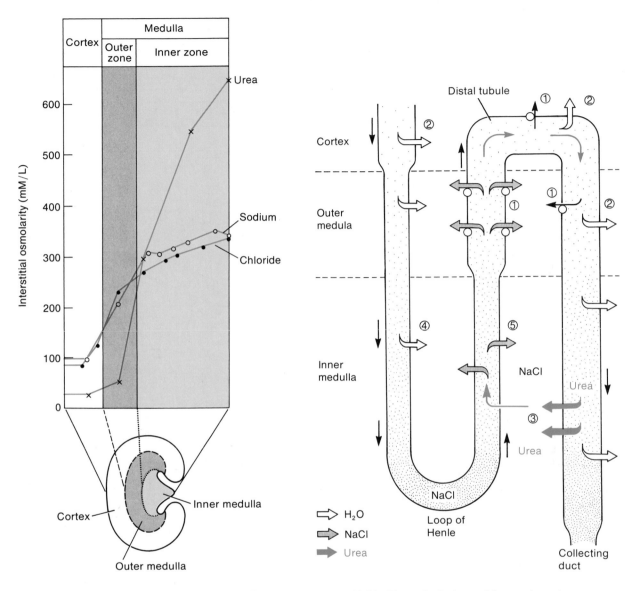

**12-28** Profile of solute concentrations along the corticomedullary axis in a mammalian kidney. In progressing from the cortex into the depths of the medulla, the concentrations of urea, $Na^+$, and $Cl^-$ increase. Note that most of the increase in urea concentration occurs across the inner medulla, whereas most of the increase in NaCl concentration occurs across the outer medulla. Since the osmotic contributions of $Na^+$ and $Cl^-$ sum, the total osmotic contributions of NaCl and urea are about equal deep within the medulla. [Ullrich et al., 1961.]

**12-29** The major features of the renal counter-current multiplier model. The corticomedullary osmotic gradient is set up as the result of active NaCl accumulation and passive urea accumulation in the medulla. The steady-state gradient (Figure 12-28) depends on differing permeabilities and active salt transport in different segments of the nephron, as well as on the anatomical layout of the nephron and circulatory supply (the vasa recta, not shown). The size of the arrows indicates relative flux rates. See text for explanation. [Based on Jamison and Maffly, 1976.]

### Box 12-2 Countercurrent Systems

In 1944, L. C. Craig published a method of concentrating chemical compounds based on the countercurrent principle. This method has proved useful in many industrial and laboratory applications. As in many other instances, man's ingenuity turns out to be a latter-day reflection of Nature's inventiveness; countercurrent mechanisms have since been found to operate in a variety of biological systems, including the vertebrate kidney, the gas-secreting organ of swim bladders and gills of fishes, and in the limbs of various birds and mammals that live in cold climates.

The principle can be illustrated with a hypothetical countercurrent multiplier that employs an active-transport mechanism much like the one that operates in the mammalian kidney. The model system shown in part A of the figure consists of a tube bent into a loop with a common dividing wall between the two limbs. A sodium chloride solution flows in one limb of the tube and then out the other. Let us assume that, within the common wall separating the two limbs of the tube, there is a mechanism that actively transports NaCl from the outflow limb to the inflow limb of the tube, without any accompanying movement of water. As bulk flow carries the fluid along the inflow limb, the effect of NaCl transport is cumulative, and the salt concentration becomes progressively higher. As the fluid rounds the bend and begins flowing through the other limb, its salt concentration progressively falls as a result of the cumulative effect of outward NaCl transport along the length of the outflow limb. By the time it reaches the end of that limb, its osmolarity is slightly lower than that of the fresh fluid beginning its inward flow in the other limb.

This example resembles the loop of Henle in principle but not in detail. The loop of Henle has no common wall dividing the two limbs; nevertheless, the limbs are coupled functionally through the interstitial fluid, so that the NaCl pumped out of the ascending limb can diffuse the short distance toward the descending limb and cause osmotic reabsorption of water from that limb.

Of special interest are the following general points about the countercurrent principle:

*1.* The standing concentration gradient set up in both limbs is due to both the continual movement of fluid through the system and the cumulative effect of transfer from the outflow limb to the inflow limb. The gradient would disappear if either fluid movement or transport across the partition were to cease.

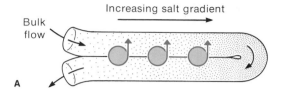

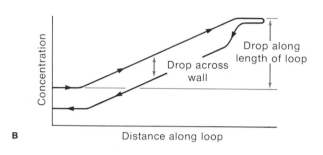

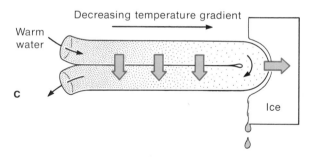

Active and passive models illustrating the countercurrent principle. (A) Active system. A salt solution flows through a U-shaped tube with a common dividing wall that pumps salt from the outflow to the inflow limb. (B) A plot of salt concentration along the two limbs, illustrating the principle that the concentration difference across the wall at any point is small relative to the total concentration difference along the length of the loop. The length of the loop as well as the efficiency of transport across the wall will determine the overall difference along the entire length of the loop. (C) Passive system. Warm water flows down the input limb and gives up part of its heat to cooler water flowing in the opposite direction in the outflow limb. Some heat is lost to the heat sink represented by the ice, but much of the heat is conserved by passive transfer from the inflow to the outflow limbs.

**2.** The difference in concentration from left to right and from right to left of the two limbs of the countercurrent multiplier is far greater than the difference across the partition at any one point (part B of the figure). As a consequence, the countercurrent multiplier can produce greater concentration changes than would be attained by a simple transport epithelium without the configuration of a countercurrent system. The longer the multiplier, the greater the concentration differences that can be attained.

**3.** The multiplier system can work only if it contains an asymmetry. In this case, there is an active net transport of salt in one direction across the partition. Countercurrent systems are also used to conserve heat (part C of the figure). In the extremities of birds and mammals that inhabit cold climates, there is a temperature differential between the arterial and venous flow of blood, because the blood is cooled as it descends into the leg. The arterial blood gives up some of its heat to the venous blood leaving the leg, thereby reducing the amount lost to the environment.

---

ing limb of the loop of Henle has high water, low salt, and low urea permeability, whereas the ascending limb has low water, low urea, and high salt permeability. Beginning with step 1 in Figure 12-29, we see that sodium and chloride are transported actively out of the thick segment of the ascending limb of the loop of Henle and out of the distal tubule. The resulting addition of NaCl to the interstitium of the cortex and outer medulla leads to osmotic loss of water (step 2) from those same segments and from the salt-impermeant descending limb in the cortex and outer medulla. Because of the net loss of water and salt from the loop of Henle and the further loss of both in the distal tubule, a high concentration of urea is left behind in the collecting duct. An important feature of this process is that the collecting duct becomes highly permeable to urea only as it passes into the depths of the medulla, where the urea leaks out down its concentration gradient (step 3), raising the interstitial osmolarity of the inner medulla. The resulting high interstitial osmolarity draws water from the descending limb (step 4), producing a very high intratubular solute concentration at the bottom of the loop. As the highly concentrated tubular fluid then flows up the highly salt-permeable thin segment of the ascending limb, NaCl leaks out (step 5) down its concentration gradient. Further loss of NaCl occurs in the thick segment by active transport (step 1). The lower collecting duct is the only section of the nephron with a high urea permeability; the small amount of urea that leaks back into more proximal parts of the nephron is simply recycled. The high osmolarity of the inner medullary interstitium (Figure 12-28) is based largely on the passive accumulation there of urea by the countercurrent mechanism of the nephron. If the ascending limb were as permeable to urea as is the collecting duct, this accumulation would not occur. If NaCl were not actively removed (with water following passively), urea would not become concentrated in the collecting duct, and the high medullary accumulation of urea would not take place either. It is interesting that the medullary urea gradient is established largely by passive means, although the active transport of NaCl is an essential component of the system and accounts for most of the metabolic energy expenditure necessary to set up the NaCl and urea gradient. The result of this combination of cellular specialization and tissue organization is a standing corticomedullary gradient of urea and salt in which the osmolarity becomes progressively higher with distance into the depths of the renal medulla, both inside the tubule and in the peritubular interstitium. This gradient is responsible for the final osmotic loss of water from the collecting ducts into the interstitium and the consequent production of a hyperosmotic urine.

A countercurrent feature in the organization of the blood vessels around the nephron, the *vasa recta*, is essential in maintaining the standing concentration gradient in the interstitium. Blood descends from the cortex into the deeper portions of the medulla in capillaries that form looplike networks around each nephron (Figure 12-16). It then ascends toward the cortex. In this circuit, the blood takes up salt and gives up water osmotically, and the surrounding interstitial fluid becomes increasingly hyperosmotic during the descent of the blood into the medullary depths. The reverse occurs as the blood returns back up toward the cortex

12-30 Renin—angiotensin feedback loops. The juxtaglomerular apparatus consists of renin-secreting cells located primarily in the wall of the afferent arteriole and of osmotically sensitive cells in the closely apposed wall of the distal segment of the renal tubule. Renin, liberated by such stimuli as afferent arteriolar constriction due to decreased pressure in the vessel and to low salt concentration in the distal tubule, leads to an increase in the titre of angiotensin II and aldosterone. Aldosterone stimulates salt transport out of the renal tubule.

and encounters an interstitium of progressively lower osmolarity. Thus, there is not much change in blood osmolarity during the circuit through the vasa recta, although the water and solutes removed from the glomerular filtrate in its passage through the nephron is carried away by the blood. However, this represents only a small percentage of the large volume of blood that perfuses the kidney.

An important consequence of the countercurrent organization of the vasa recta is that it allows this high rate of perfusion without disrupting the corticomedullary standing gradient of salt and urea concentration. The rate of blood circulation through the kidney is, in fact, among the highest through any vertebrate tissue. As the blood leaves the glomerulus and moves down

the vasa recta into the medulla, it passively accepts interstitial NaCl and urea as it encounters ever-increasing interstitial osmolarities. NaCl and urea in the blood reach their peak concentrations as the blood traverses the loop of the vasa recta in the depths of the medulla. On ascending back toward the cortex, the excess NaCl and urea diffuse back into the interstitium, staying behind as the blood leaves the kidney. But before leaving the kidney, the blood, in fact, regains some of the water it lost during glomerular filtration. This happens because of colloid osmotic pressure having been elevated during ultrafiltration.

To summarize, the formation of urine begins with the concentration of the glomerular filtrate into a hyperosmotic fluid in the proximal tubule, where 75%

of the salt and water are removed from the pre-urine in osmotically equivalent amounts, leaving urea and certain other substances behind. In this fluid's course through the loop of Henle and the distal tubule, there is little overall change in osmolarity of the pre-urine, but the countercurrent mechanism sets up a standing concentration gradient in the medullary interstitium along the length of the loop. This gradient provides the basis for the osmotic removal of water from the urine as it makes its way down the collecting duct within the medulla. Interestingly, this process takes place without active transport of water at any place along the nephron.

## RENAL REGULATORY MECHANISMS

An animal can experience osmotic stress owing to changes in temperature or salinity and to the ingestion of food and drink. Perturbations in the osmotic state of the body fluids are minimized through feedback mechanisms by which the osmoregulatory organs adjust their activity so as to maintain the internal status quo. These feedback control mechanisms may be neural, endocrine, or a combination of the two. There are several means by which the formation of urine is regulated in response to osmotic stress and other signals. These include the regulation of (1) the glomerular filtration rate, (2) the rate at which salts are absorbed from the lumen of the renal tubule, and (3) the rate at which water is osmotically drawn out of the pre-urine.

### Control of Glomerular Filtration Rate

The control of the glomerular filtration rate (GFR) is not entirely clear. Since the amount of salt and water passing into Bowman's capsule is more than 100 times greater than the amount appearing in the urine, large changes in the GFR could result in the disruption of salt and water balance, unless there were heroic corrective action by mechanisms of reabsorption and secretion. In the human kidney, the renal blood flow remains quite constant in spite of large changes in systemic blood pressure. This continuity, of course, aids in maintaining a constant GFR. In lower vertebrates, altered GFR is one of the variables that the renal system uses to cope with changes in osmotic

stress. Thus, in trout the GFR is high in fresh water, but it drops by about 90% when the trout is transferred to seawater.

### Control of Tubular Reabsorption of Na+

The kidneys receive a nerve supply from the sympathetic system, which innervates the *juxtaglomerular apparatus (JGA)*, shown in Figure 12-30 and 12-31. The JGA consists of the *macula densa*, an extratubular tissue close to the glomerulus, and of specialized receptor and secretory cells surrounding a segment of the *afferent arteriole* closely apposed to the distal segment of the renal tubule. Activation of the sympathetic innervation induces the release of a proteolytic enzyme, *renin*, from the secretory cells located in the wall of the afferent arteriole that carries blood into the glomerular capillaries in Bowman's capsule. Renin is also liberated from those cells in response to a drop in the plasma Na+ concentration and to local stimuli such as loss of distention of the arteriolar walls due to decreased blood pressure and circulating catecholamines. Salt concentrations are believed to be sensed by cells of the macula densa, which is in close communication with the renin-secreting cells of the arteriole.

As indicated in Figure 12-31, renin is released by the cells of the JGA into the afferent arteriole. There it acts on $\alpha_2$-globulin, a glycoprotein molecule that is manufactured in the liver and is present in the plasma. Renin cleaves a leucine–leucine bond in the globulin molecule, releasing a 10-residue peptide, *angiotensin I*. A *converting enzyme* then removes two additional amino acids to form the 8-residue peptide *angiotensin II*. This hormone has several actions, one of which is to cause general vasoconstriction (constriction of arterioles), which raises the blood pressure, thereby increasing both renal blood flow and the rate of glomerular filtration. Another important action of angiotensin II is to stimulate the release, from the adrenal cortex, of the steroid hormone *aldosterone* (Table 11-11), one of the so-called *mineralocorticoids*, which are important in promoting the retention of salts. Aldosterone acts on the distal tubule of the nephron, increasing the rate of sodium reabsorption (and, secondarily, osmotic reabsorption of water). It increases the uptake of Na+ from the renal tubule by stimulating the active exchange of plasma K+ and H+ in the glomerular filtrate. Rather than

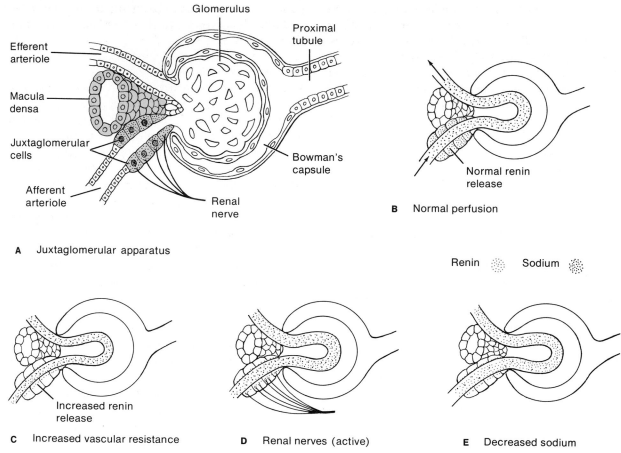

**A** Juxtaglomerular apparatus

**B** Normal perfusion

**C** Increased vascular resistance

**D** Renal nerves (active)

**E** Decreased sodium

**12-31** Juxtaglomerular apparatus. The JGA, which is the source of renin released into the blood, has two receptor tissues: a vascular receptor in the afferent arteriole, which monitors vascular distension; and the macula densa, which monitors the sodium concentration. The renal nerve also regulates the rate of renin secretion. The overall relations of renin secretion are shown in Figure 12-30. [J. O. Davis, "Control of Renin Release," *Hospital Practice,* 9, (4), April 1974. Illustration (p. 57) by Alan Iselin.]

being directly stimulated by a sodium pump, it appears that increased sodium transport arises from an aldosterone-induced increase in *sodium permeability* of the renal cell in the membrane on the *mucosal side* (i.e., the side facing the lumen) of the tubular epithelium. As a consequence of this increase in sodium permeability of the luminal membrane, there is an increased influx of $Na^+$ from the tubular lumen into the epithelial cells (followed by $Cl^-$). The rise in intracellular $Na^+$ concentration provides the pump on the *serosal side* (i.e., the side facing the blood) of the cells with a higher $Na^+$ concentration and thus increases the rate of sodium transport from the kidney tubule across the epithelial cell layer into the interstitial fluid and blood.†

---

†Aldosterone works the same way in certain other transport epithelia, such as the bladder wall and integument of amphibians (discussed in Chapter 4). In mammals, this hormone also promotes $Na^+$ reabsorption in the salivary glands, sweat glands, and colon.

Thus, one ultimate effect of renin secretion into the afferent arteriole is an increased recovery of NaCl from the glomerular filtrate. This recovery has the obvious regulatory effect of compensating for any reduction in plasma volume or Na$^+$ levels (recall that renin is released in response to a drop in plasma Na$^+$ and to loss of distention of the arteriolar wall).

### Control of Osmotic Water Retention

The tubular fluid is concentrated by the osmotic removal of water as it passes down the collecting duct into the hyperosmotic depths of the renal medulla (Figure 12-27). This concentration provides a means of regulating the amount of water passed in the urine. The rate at which water is osmotically drawn out across the wall of the collecting duct from the urine into the interstitial fluid depends on the water permeability of the epithelium that forms the wall of the collecting duct. The antidiuretic hormone (ADH), released from the neurohypophysis (the posterior lobe of the pituitary gland), regulates the water permeability of the collecting duct and thereby controls the amount of water leaving the animal via the urine. The higher the level of ADH in the blood, the more permeable the epithelial wall of the collecting duct, and hence the more water is drawn out of the urine as

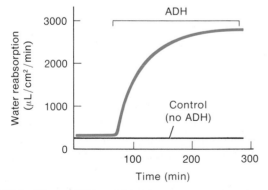

**12-32** Effect of ADH on osmotic reabsorption of water from the collecting duct into the interstitial fluid. The osmolarity of the fluid perfused through the duct was 125 mosm/L; that of the external bath, 290 mosm/L. The hormone was applied during the period indicated. [Grantham, 1971.]

it passes down the duct toward the renal pelvis (Figure 12-32). The level of ADH in the blood is a function of the osmotic pressure of the plasma. Osmotically sensitive neurons with cell bodies located in the hypothalamus respond to increased plasma osmolarity by an increased rate of firing. These are neurosecretory cells that send their axons to the neurohypophysis. The increased neural activity of these cells increases the rate at which ADH is released into the bloodstream from their axon terminals and thus increases the blood level of ADH. The regulatory effect of this neuroendocrine mechanism is illustrated in Figure 12-33. If, for example, the osmolarity of the blood is increased as a result of desiccation, the activity of the neurosecretory neurons is increased, more ADH is released, the collecting ducts become more permeable, and water is osmotically drawn from the urine at a higher rate. This process results in the excretion of a more concentrated urine and the conservation of body water. The hypothalamic cells that produce and release ADH receive inhibitory input from receptors that respond to increases in blood pressure. These receptors are located in various parts of the circulatory system, but the major ones are located in the left atrium of the heart. Any factor that raises the arterial blood pressure (e.g., an increase in blood volume due to dilution by ingested water) will inhibit the ADH-releasing hypothalamic cells and thereby result in an increased loss of body water through the urine.

In humans, the ingestion of ethyl alcohol inhibits the release of ADH and therefore leads to copious urination and an increase of plasma osmolarity beyond the normal set-point level. Some degree of dehydration results, and this contributes to the uncomfortable feeling of a hangover.

The action of mammalian ADH and related peptides of nonmammalian species is not limited to the kidney. These antidiuretic hormones, applied to frog skin and toad bladder, also increase the water permeability of those epithelia.

## EVOLUTION OF THE VERTEBRATE KIDNEY

The evolution of the vertebrate nephron is summarized in Figure 12-34. A primitive forerunner is found in some prochordates, in which the proximal end of the

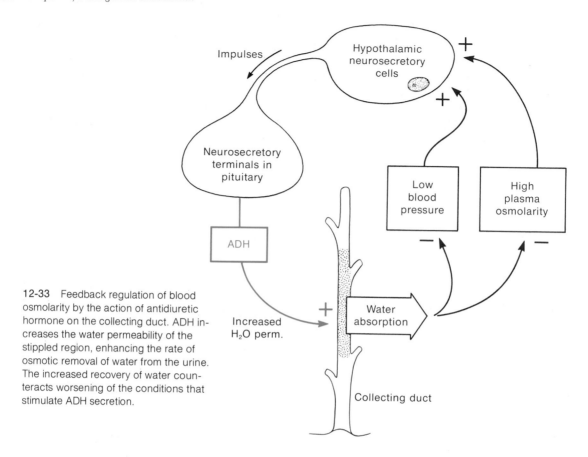

**12-33** Feedback regulation of blood osmolarity by the action of antidiuretic hormone on the collecting duct. ADH increases the water permeability of the stippled region, enhancing the rate of osmotic removal of water from the urine. The increased recovery of water counteracts worsening of the conditions that stimulate ADH secretion.

nephron opens into the coelom. The simplest vertebrate nephron occurs in certain marine teleosts that have neither glomeruli nor Bowman's capsules. In such *aglomerular* kidneys, the urine is formed entirely by secretion and perhaps reabsorption, since there is no specialized mechanism for the production of a filtrate. The opposite extreme, again, is seen in the marine hagfishes (class Cyclostomata); the glomeruli are present in these animals, but the tubules are absent and the Bowman's capsules empty directly into the collecting duct. We have little information on the physiology of these primitive nephrons, except that they actively secrete divalent ions such as $Ca^{2+}$, $Mg^{2+}$, and $SO_4^{2-}$. These kidneys carry out little or no osmoregulation, serving primarily to turn over a certain fraction of blood, water, and crystalloids per unit time as a means of preventing the toxic buildup of nitrogenous wastes.

Thus, the extracellular fluids (Table 12-3) of the most primitive living vertebrate, the hagfish, are relatively similar to seawater in concentration of major salts, and their plasma is essentially isotonic relative to seawater. The plasma of sharks and rays (class Chondrichthyes) is lower in the concentrations of the major ions but nonetheless essentially isosmotic relative to seawater. The balance of the osmolarity depends on the retention of large concentrations of urea in the plasma of these cartilaginous fishes. The teleosts represent the next level of osmoregulation, for they maintain a plasma that is hypoosmotic relative to seawater and hyperosmotic relative to fresh water. However, this depends not on their kidneys, which are incapable of producing a concentrated urine, but instead on salt secretion by their gills for maintenance of their body fluids hypoosmotic to seawater.

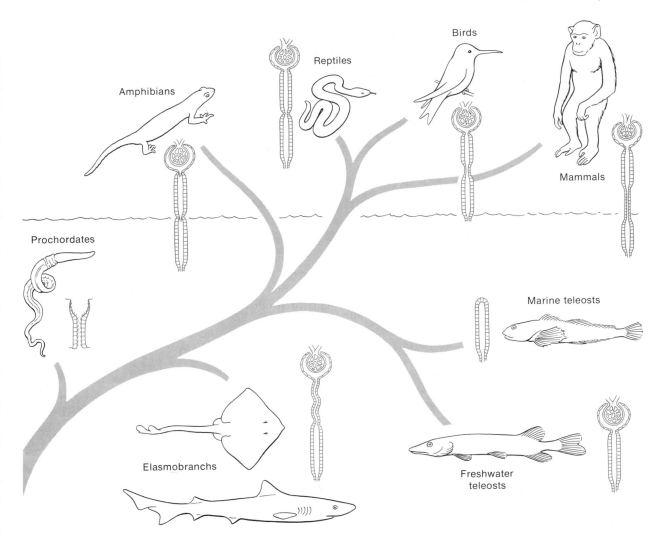

**12-34** Simplified outline of vertebrate renal phylogeny. (See text.) [Smith, 1939.]

As a general rule, freshwater teleosts have larger glomeruli and more of them than do their marine relatives. This is also true for all the higher vertebrates. In lower vertebrates up to the reptiles, the kidney is incapable of producing a hypertonic urine (i.e., of higher osmolarity than the plasma). The ability to produce a hypertonic urine is closely correlated with the organization of the nephron, for without a countercurrent system the kidney cannot produce a urine of significantly greater osmolarity than the plasma. Only birds and mammals have a loop of Henle, and thus only these animals have renal plumbing that is so organized as to allow osmotic countercurrent multiplication. The avian kidney resembles that of mammals, except that some of the nephrons are without loops of Henle, and in some birds the loop is oriented perpendicular to the collecting duct, producing a less efficient concentrating mechanism. In mammals, there is a nice correlation between the length of the loop and the ability to concentrate the urine. The loop of Henle is longest in desert dwellers, such as the kangaroo rat; these longer loops produce larger overall gradients in osmolarity from renal cortex to medulla, thus permitting more efficient osmotic extraction of water from the collecting duct.

## Extrarenal Osmoregulatory Organs of Vertebrates

Although the kidney is the most highly developed organ of osmoregulation and excretion among the vertebrates, several other vertebrate organs are equally important in the maintenance of osmotic homeostasis.

### OSMOREGULATORY FUNCTIONS OF FISH GILLS

The epithelial surface area of a gill must be large if it is to function efficiently as an organ for respiratory gas exchange (described in Chapter 14). Although this feature makes gills an osmotic liability for animals such as fishes, which are out of osmotic equilibrium with their aqueous environment, it does make gills suitable as organs for osmoregulation. Thus, the gills of numerous aquatic species, vertebrate and invertebrate, are active not only in gas exchange but also in such diverse functions as ion transport, excretion of nitrogenous wastes, and maintenance of acid–base balance. In teleost fishes, for example, it is the gills that play the central role in coping with osmotic stress.

The structure of a teleost gill is seen in Figure 12-35A. The blood transfusing the gill flows along two major pathways, the afferent and efferent vessels inside the filament and the lumen inside the lamellae. The epithelium separating the blood from the external water consists of two distinct cell types. The epithelium of the lamellae consists entirely of flat cells no more than 3–5 $\mu$m thick, containing few mitochondria. These are clearly best suited for respiratory exchange, acting as minimal barriers for diffusion of gases. The epithelium covering the gill filaments consists of these cells plus a second type that is more columnar in shape and several times thicker from base to apex than the flat cells. These so-called *chloride cells* are deeply invagi-

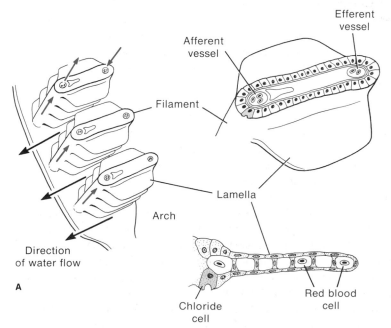

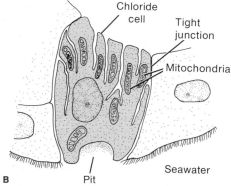

12-35   Teleost gill as an osmoregulatory organ. (A) Gill filaments extending up from the gill arch. Blood flows in capillaries, as shown by colored arrows. Flattened lamellae (right) extend out from the filaments. The lamellae contain extensive spaces through which blood circulates. (B) Chloride cell bordered by flat respiratory epithelial cells. Note its rich supply of mitochondria and its extensive invaginations and clefts. [Maetz, 1971.]

nated by infoldings of the surface membrane from the basal and lateral surfaces. In addition, they are heavily laden with mitochondria and with enzymes related to active salt transport. Glutamic dehydrogenase and other enzymes associated with nitrogen metabolism and excretion are also present in high concentrations, as is carbonic anhydrase, which is important for acid–base regulation.

The chloride cells were first described in 1932 by Ancel Keys and Edward Willmer, who ascribed to them the transport of chloride because they exhibit histochemical similarities to cells that secrete hydrochloric acid in the amphibian stomach, and because it had already been shown that the teleost gill is the site of extrarenal excretion of chloride (and sodium). More recent histochemical studies have confirmed the presence of high levels of chloride in these cells, especially near the pit that develops on the mucosal (external) border of these cells in fishes that have become adapted to high salinities. Relatively little is known about the mechanism of salt transport by the chloride cells, but the evidence suggests that chloride is pumped actively out of the cell and that $Na^+$ follows passively through a leakage pathway because of the potential difference set up across the cell membrane by the active transport of $Cl^-$.

It is interesting that the direction of salt transport across the gill epithelium changes so as to adapt to changes in environmental salinity in species that migrate between seawater and fresh water. Salts are actively taken up in fresh water and actively excreted in salt water. In experiments done to study the effects of transferring such fishes from low to high salinities, it was found that the physiological adaptation of the gills (i.e., change from inward pumping to outward pumping of NaCl) is a gradual process, involving the synthesis and/or destruction of molecular components of the epithelial transport systems and changes in the morphology and number of the chloride cells. When the external salinity is experimentally reduced, characteristic changes occur in the chloride cells, including the disappearance of the pits on the chloride cells. Conversely, exposure to high salinities results in (1) the appearance of the pits; (2) an increase in the number of chloride cells; (3) a corresponding increase in the activity of sodium- and potassium-activated ATPases and of carbonic anhydrase; and (4) the elaboration of

extracellular space by development of increased infoldings by the chloride cells. When *euryhaline* fishes (i.e., those able to tolerate a wide range of salinities) are transferred from fresh water to salt water, new ion-pump activity ensues after a delay of about a day. It is now known that the external salinity induces those osmoregulatory changes through endocrine mechanisms that influence epithelial differentiation and metabolism. The steroid hormone cortisol mediates the changes induced by transition to seawater, whereas prolactin is involved in the changes that accompany the transition to fresh water.

Since the chloride cells were first characterized and associated with the transport of chloride ions, they have been found to be responsible for the exchange of other ions as well, including $H^+$, $Na^+$, $K^+$, $NH_4^+$, and $HCO_3^-$. Although it is difficult to demonstrate rigorously that the chloride cells alone take part in the transport and exchange of those ions, a variety of evidence, including the changes that the chloride cells undergo during adaptation to high salinities, indicates that they are the major site of active exchange in the gills of teleost fishes.

## EXTRARENAL SECRETION BY AVIAN AND REPTILIAN SALT GLANDS

In 1957, Knut Schmidt-Nielsen and his co-workers, investigating the means by which marine birds maintain their osmotic balance without access to fresh water, discovered that the *nasal salt glands* secrete a hypertonic solution of NaCl (Figure 12-8). It was found in those early studies that if cormorants or gulls are administered seawater by intravenous injection or by stomach tube, the increase in the plasma salt concentration leads to a prolonged nasal secretion of fluid with an osmolarity two to three times that of the plasma. Salt glands have subsequently been described in many species of birds and reptiles, especially those subjected to the osmotic stress of a marine or desert environment. These species include nearly all marine birds, ostriches, the marine iguana, sea snakes, and marine turtles, as well as many terrestrial reptiles. Crocodilians have a similar salt-secreting gland in the tongue.

The salt glands of birds and some reptiles occupy shallow depressions in the skull above the eyes (Figure 12-36). In birds, the salt gland consists of many lobes about 1 mm in diameter, each of which drains via

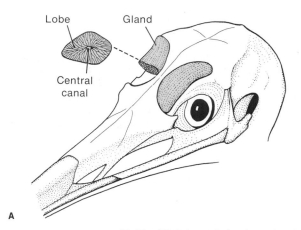

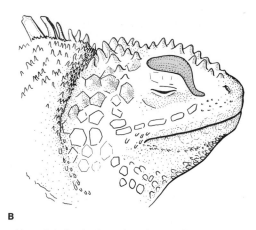

12-36 (A) Avian salt glands are located above the orbit and drain via ducts into the nasal region. The gland consists of a longitudinal arrangement of many lobes. Each lobe consists of tubules and capillaries (see Figure 12-37) arranged radially around a central canal. The secretion collects in a duct that empties into the nasal region. [Schmidt-Nielsen, 1960.] (B) Reptilian salt glands are found associated with eyes, nostrils, cheeks, and tongues in various groups. A marine iguana is shown here with the subdermal salt gland in color. [Dunson, 1969.]

branching secretory tubules and a central canal (Figure 12-37) into a duct that, in turn, runs through the beak and empties into the nostrils. Active secretion takes place across the epithelium of the secretory tubules and consists of characteristic secretory cells that have a profusion of deep infoldings at their basal ends and are heavily laden with mitochondria. As in other transport epithelia, adjacent cells are tied together by tight junctions, which preclude the massive leakage of water or solutes past the cells, from one side of the epithelium to the other.

The formation of nasal gland fluid does not include filtration, as does the formation of urine by the glomerular kidney. The absence of filtration can be deduced from the failure of small filterable molecules, such as insulin or sucrose, to appear in the nasal gland fluid after they have been injected into the bloodstream. Although our knowledge of cellular mechanisms in the salt gland is limited, there is good evidence that active transport takes place. A sodium- and potassium-activated ATPase has been localized at the basal membrane, and the application of ouabain (p. 113) to the basal surface of the epithelium blocks salt transport. Since this inhibitor does not pass across epithelia and can block

the pump only by direct contact with the transport mechanism, the sodium transport mechanism appears to operate in the basal membrane of the epithelial cell.

The avian salt gland appears to be organized as a countercurrent system that might aid in concentrating salts in the secreted fluid. The capillaries are so arranged that the flow of blood is parallel to the secretory tubules and occurs in the direction opposite to the flow of secretory fluid (Figure 12-37). This flow maintains a minimal concentration gradient from blood to tubular lumen along the entire length of a tubule; it thereby minimizes the concentration gradient for uphill transport from the plasma to the secretory fluid.

The secretory activity of the salt gland is controlled largely by direct neural control through the parasympathetic nerves, but it is also under neuroendocrine control (Figure 12-38). Osmoreceptors in the hypothalamus respond to an increase in plasma tonicity by a sensory discharge. This response, together with input from extracranial osmoreceptors, activates parasympathetic cholinergic neurons that innervate the salt gland. Acetylcholine liberated from the terminals of those neurons not only directly stimulates the secretion

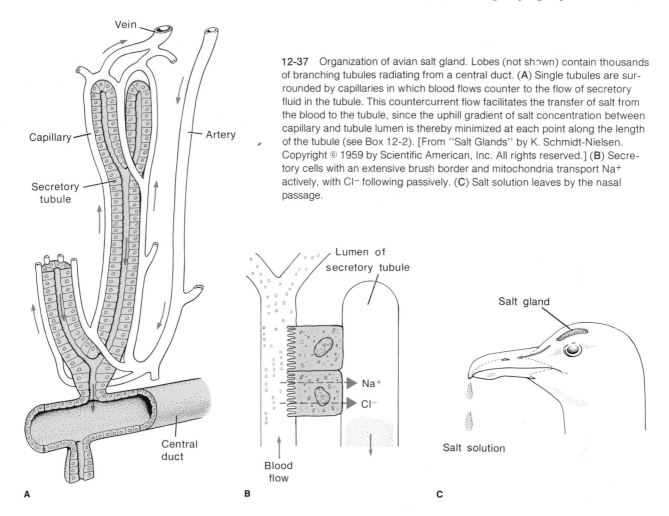

**12-37** Organization of avian salt gland. Lobes (not shown) contain thousands of branching tubules radiating from a central duct. (**A**) Single tubules are surrounded by capillaries in which blood flows counter to the flow of secretory fluid in the tubule. This countercurrent flow facilitates the transfer of salt from the blood to the tubule, since the uphill gradient of salt concentration between capillary and tubule lumen is thereby minimized at each point along the length of the tubule (see Box 12-2). [From "Salt Glands" by K. Schmidt-Nielsen. Copyright © 1959 by Scientific American, Inc. All rights reserved.] (**B**) Secretory cells with an extensive brush border and mitochondria transport $Na^+$ actively, with $Cl^-$ following passively. (**C**) Salt solution leaves by the nasal passage.

of salt, but further enhances secretion by causing vasodilation and thus increased perfusion by blood of the secretory tissue. Secretion is also stimulated by adrenocorticosteroids and by prolactin. Preliminary studies on the endocrine control of the avian salt gland suggest that several hormones, including corticosterone and arginine vasotocin (AVT), play a role in regulating secretion. Although direct neural control is most important in making short-term adjustments to osmotic stress, corticosterone is required for maintaining the responsiveness of the salt gland. For instance, when an animal's adrenal cortex, the source of corticosteroids, is removed, the infusion of a high-tonicity salt solution

is no longer effective in stimulating salt-gland secretion (Figure 12-39). But if corticosterone is then replaced by injection, salt-gland function is retained. AVT, an ADH-like neurohypophysial peptide that is secreted in response to elevated blood osmolarity, has been shown to exert a direct stimulatory action on the avian salt gland. The gland responds by producing a hypertonic secretion, lowering the tonicity of the blood toward the set point of this osmoregulatory feedback system. Simultaneously, the AVT acts to slow glomerular filtration in the bird's kidneys, which, as we noted earlier, are capable of producing only a slightly hypertonic urine.

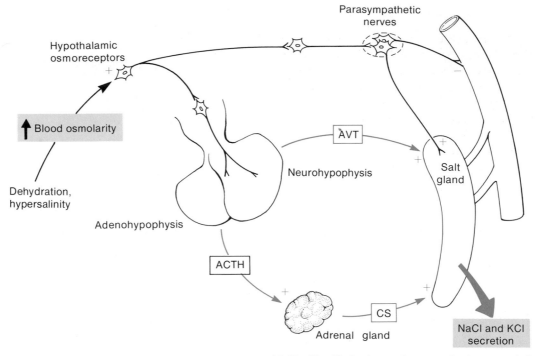

**12-38** Simplified scheme of neuroendocrine control of avian salt-gland secretion. Osmotically sensitive neurons in the hypothalamus, plus input from peripheral osmoreceptors, activate parasympathetic pathways to the salt gland directly and to the vessels supplying blood to the salt gland. Also activated is pituitary secretion of ACTH, arginine vasotocin, and prolactin (not shown). The ACTH stimulates release of corticosterone from the adrenal cortex. [Thomas and Phillips, 1978.]

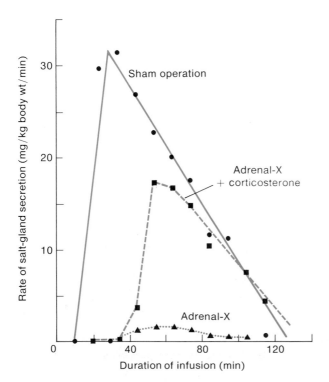

**12-39** Effect of adrenalectomy and corticosterone replacement therapy on salt-gland secretion in the duck. Two days after the operation the salt gland was challenged by an infusion of 10% NaCl into the blood. The plot of squares (■) was obtained when the NaCl was accompanied by corticosterone, a steroid hormone from the adrenal cortex. [Thomas and Phillips, 1975.]

## *Invertebrate Osmoregulatory Organs*

The invertebrates exhibit far greater evolutionary diversity than the vertebrates and have evolved a variety of osmoregulatory organs unrelated to the vertebrate kidney (Figure 12-40). In general, invertebrate osmoregulatory organs employ mechanisms of filtration, reabsorption, and secretion similar in principle to those of the vertebrate kidney to produce a urine that is significantly different in osmolarity and composition from the body fluids. That there has been convergent evolution of physiological mechanisms in nonhomologous organs underscores the utility of these mechanisms. In addition, there is some evidence, still controversial, that urine and feces may be dehydrated by active transport of water across the epithelium in the hindgut of some insects. This evidence may be verified in future studies, but there is as yet no evidence for such a mechanism of osmoregulation among the vertebrates. In them, water transport has been shown to occur only by passive osmosis from a compartment of lesser solute concentration to a compartment of greater solute concentration.

### *FILTRATION–REABSORPTION ORGANS*

Filtration of plasma, similar in principle to that which occurs in the Bowman's capsule of vertebrates, appears to underlie the formation of the primary urine in both mollusks and crustaceans. This conclusion is derived from several observations:

*1.* When polymers such as inulin are injected into the bloodstream or coelomic fluid, they appear in high concentrations in the urine. Since it is unlikely that they are actively secreted, they must enter the urine during a filtration process in which all those molecules below a certain size pass through a sievelike membrane of tissue. During the reabsorption of water and essential solutes, these polymers remain behind in the urine.

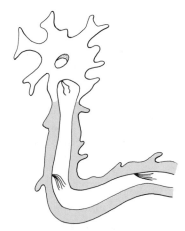

**A**  Protonephridium  (turbellarian)

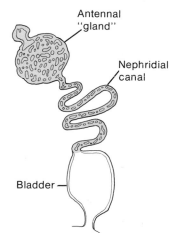

**B**  Crayfish antennal "gland"

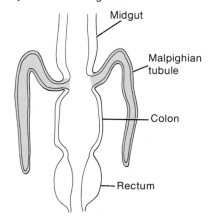

**C**  Generalized excretory apparatus of an insect

**12-40**  General organization of three representative invertebrate excretory and osmoregulatory systems. The tissues active in excretion are shown in color. [Phillips, 1975.]

**2.** The normal urine of some animals is found to contain little or no glucose, even though substantial levels occur in the blood. In several mollusks, for example, when the blood glucose is elevated by artificial means, such as by injection, glucose appears in the urine. In each species, glucose appears in the urine at a characteristic threshold concentration of blood glucose. As in the vertebrate kidney, the urine glucose concentration rises linearly with blood glucose concentration beyond that point (Figure 12-20). By analogy with the vertebrate kidney, this behavior is believed to result from a saturation of the transport system by which the glucose that entered the tubular fluid via filtration is reabsorbed from the filtrate into the blood. Once the transport system is saturated, the "spillover" of glucose in the urine is proportional to its concentration in the blood. More conclusive evidence is obtained with the drug *phlorizin,* which is known to block active glucose transport. When phlorizin is administered to mollusks and crustaceans, glucose appears in the urine even at normal blood glucose levels. The most reasonable explanation for this effect is that glucose enters the urine as part of a filtrate and remains in the urine when the reabsorption mechanism is blocked by phlorizin.

**3.** Analysis of tubular fluids near suspected sites of filtration indicate a crystalloid composition similar to that of the plasma.

**4.** The rate of urine formation has been found in some animals to depend on the blood pressure. This relationship is consistent with a filtration mechanism, but the change in blood pressure may also produce a change in the circulation to the osmoregulatory organ.

The site of primary urine formation by filtration is known in only a few invertebrates. In a number of marine and freshwater mollusks, filtration takes place across the wall of the heart into the pericardial cavity, and the filtrate is conducted to the "kidney" through a special canal. Glucose, amino acids, and essential electrolytes are reabsorbed in the kidney. In the crayfish, the major organ of osmoregulation is the so-called *antennal gland.* Part of that organ, the *coelomosac,* resembles the vertebrate glomerulus in ultrastructure. Micropuncture measurements have shown that the excretory fluid that collects in the coelomosac is produced by ultrafiltration of the blood.

Since the final urine in mollusks and crustaceans differs in composition from the initial filtrate, there must be either secretion of substances into the filtrate or reabsorption of substances from the filtrate. The reabsorption of electrolytes is well established in freshwater species, for the final urine has a lower salt concentration than either the plasma or the filtrate. Glucose must also be reabsorbed, since it is present in the plasma and in the filtrate but is either absent or very low in concentration in the final urine.

It is interesting that the filtration–reabsorption type of osmoregulatory system has appeared in at least three phyla (Mollusca, Arthropoda, Chordata) and perhaps more. This kind of system has the important advantage that all the low-molecular-weight constituents of the plasma are filtered into the ultrafiltrate in proportion to their concentration in the plasma. Such physiologically important molecules as glucose, and, in freshwater animals, such ions as $Na^+$, $K^+$, $Cl^-$, and $Ca^{2+}$, are subsequently removed from the filtrate by tubular reabsorption, leaving toxic substances or unimportant molecules behind to be excreted in the urine. This process avoids the need for active transport into the urine of toxic metabolites, or for that matter unnatural, man-made substances of a neutral or toxic nature encountered in the environment. A disadvantage of the filtration–reabsorption osmoregulatory system is its high energetic cost for the organism: The filtering of large quantities of plasma requires the active uptake of large quantities of salts, either in the excretory organ itself or in other organs, such as gills or skin. In frog skin, for example, it has been shown that 1 mol of oxygen must be reduced in the synthesis of ATP for every 16–18 mol of sodium ions transported. In freshwater clams, the cost of maintaining sodium balance amounts to about 20% of the total energy metabolism. In marine invertebrates, the filtration–reabsorption system proves metabolically less expensive, since salt conservation is much less of a problem.

### SECRETORY-TYPE OSMOREGULATORY ORGANS

In insects, the *Malpighian tubules* together with the *hindgut* form the major excretory–osmoregulatory system. In broad outline, this system (Figure 12-41) consists of the long, thin Malpighian tubules, which empty into the alimentary canal at the junction of the midgut and

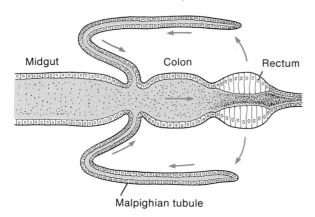

Midgut    Colon    Rectum

Malpighian tubule

**12-41** The insect excretory system in simplified outline. The pre-urine is produced by secretion into the lumen of the Malpighian tubules and flows into the rectum, where it is concentrated by the extraction of water. The decrease in volume of the urine in the rectum is due to reabsorption of water rather than to secretion of solutes. The arrows indicate the circular pathway of water and ions described in the text. There are numerous Malpighian tubules, even though only two are shown. [Wigglesworth, 1932.]

hindgut and whose closed ends lie in the hemocoel (body cavity containing blood). The secretion formed in the tubules passes into the hindgut, where it is dehydrated and passed into the rectum and voided as a concentrated urine through the anus. With the evolution of tracheolar respiration in insects (described in Chapter 14), the importance of an efficient circulatory system diminished. As a consequence, the Malpighian tubules do not receive a direct, pressurized arterial blood supply, as the mammalian nephron does. Instead, they are surrounded with blood, which is at a pressure essentially no greater than the pressure within the tubules. Since there is no significant pressure differential across the walls of the Malpighian tubules, filtration cannot play a role in urine formation in the insect. Instead, the urine must be formed entirely by secretion, with perhaps the subsequent reabsorption of some constituents of the secreted fluid. This process is analogous to the formation of urine by secretion in the aglomerular kidneys of marine teleosts. The serosal surface of the Malpighian tubule exhibits a profusion of microvilli and mitochondria (Figure 12-42), a specialization often associated with a highly active secretory epithelium.

The details of urine formation by tubular secretion differ among different insects, but some major features seem to be common throughout. Potassium and, to a lesser extent, sodium are secreted actively into the tubular lumen, along with such waste products of nitrogen metabolism as uric acid and allantoin. It appears that the transport of $K^+$ is the major driving force for the formation of the tubular urine, with most of the other substances following passively. This has been concluded from the following observations:

*1.* The tubular urine is more or less isotonic relative to the hemolymph.

*2.* The tubular urine has a high $K^+$ concentration in all insects.

*3.* The rate of tubular urine formation is a function of $K^+$ concentration in the fluid surrounding the tubule, higher potassium concentrations producing more rapid tubular urine accumulation.

*4.* The formation of tubular urine is largely independent of the sodium concentration of the surrounding fluid.

Although potassium is osmotically the most important substance actively transported, there is evidence that active transport plays an important role in the secretion of uric acid and other nitrogenous wastes.

The pre-urine formed in the Malpighian tubules is relatively uniform in composition from one species to

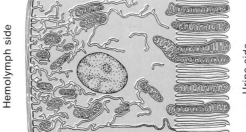

Hemolymph side

Urine side

**12-42** Secretory cell in the wall of a Malpighian tubule of a cockroach. The surface facing the urine is expanded into a dense pile of microvilli, some of which are distended by mitochondria. The side facing the hemolymph contains a profusion of branched invaginations. [Oschman and Berridge, 1971.]

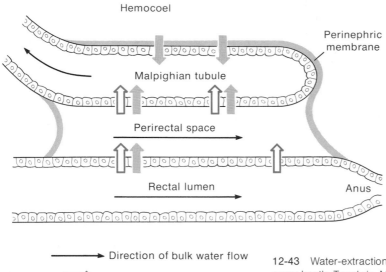

**12-43**   Water-extraction apparatus of the rectum of the mealworm beetle *Tenebrio*. Most of the water and KCl entering the rectal lumen is recycled into the Malpighian tubules. (See text.) [Phillips, 1970.]

another, and in each species it remains isotonic relative to the hemolymph under different osmoregulatory demands. The fluid formed in the Malpighian tubules passes into the hindgut, where several important changes in its composition occur. In the hindgut, water and ions are removed in amounts that maintain the proper composition of the hemolymph. Thus, it is in the hindgut that the composition of the final urine is determined.

The water and ions removed from the urine by the hindgut are transferred through intimate connections to the lumen of the Malpighian tubules. These substances are thereby retained and recycled in the Malpighian tubule–hindgut circuit.

The most complete study of the osmoregulatory behavior of the hindgut has been done with the desert locust *Schistocerca*. When a solution similar to hemolymph is injected into the hindgut of this insect, water, potassium, sodium, and chloride are absorbed into the surrounding hemolymph. Evidence from electrical measurements suggests that the ions are transported actively. The locust hindgut is capable of removing a large amount of water, producing a hypertonic urine, with urine/blood osmolarity ratios as high as 4. It has been argued that the uptake of water is active (i.e.,

directly coupled to the hydrolysis of an energy-donor molecule, such as ATP), since it takes place even against an osmotic gradient in which the osmolarity of the rectal contents is five times that of the hemolymph. In the mealworm *Tenebrio,* the urine/blood osmolarity ratio can be as high as 10, which is comparable to the concentrating ability of the most efficient mammalian kidneys. Alternatively, a purely physical mechanism may produce the net uphill movement of water without direct coupling of water transport to the release of chemical energy. It has been suggested, for example, that uphill transport of water in some species, such as *Tenebrio,* results from an arrangement of the Malpighian tubules, the perinephric space, and the rectum that acts as a countercurrent multiplier mechanism (Figure 12-43). According to Phillips (1970), water is drawn osmotically from the rectum into the Malpighian tubules because of the KCl gradient produced by active transport. The direction of bulk flow in these compartments is such that the osmotic gradient along the length of the complex is maximized, with the absolute osmolarities highest toward the anal end of the rectum. This gradient may allow the concentrations near the anal end to exceed those of the hemolymph by several times. Moreover, the recycling of KCl (Figure 12-41) in the Malpighian tubules and hindgut provides the osmotic pressure needed in the perinephric space and Malpighian tubular lumen to draw water from the rectum. Thus, a standing-gradient system appears to be responsible for solute-coupled uphill water transport in the hindgut of insects.

Very little is known about the feedback regulation of osmolarity among the invertebrates, but there is evidence for such regulation in insects. The bug *Rhodnius* becomes bloated after sucking blood from a mammalian host. Within 2–3 min, the Malpighian tubules increase their secretion of fluid by more than a thousand times, producing a copious urine. Artificially bloating the insect with a saline solution does not produce such diuresis in an unfed *Rhodnius*. It has also been found that isolated Malpighian tubules immersed in the hemolymph of unfed individuals remain quiescent, but if immersed in the hemolymph of a recently fed *Rhodnius*, they produce a copious secretion. A factor that stimulates the secretion of these tubules can be extracted from the neural tissue containing the cell bodies or axons of neurosecretory cells, primarily those of the metathoracic ganglion. Thus, it appears that these cells release a diuretic hormone in response to a factor present in the ingested blood. The only identified neurohumor that stimulates the diuretic action of the neurosecretory cells is serotonin. Similar findings in other insect species suggest that diuretic and antidiuretic hormones produced in the nervous system regulate the secretory activity of the Malpighian tubules or the reabsorptive activity of the rectum. In earthworms, the removal of the anterior ganglion results in the retention of water and a concomitant decrease in plasma osmolarity. Injection of homogenized brain tissue reverses these effects, suggesting a humoral mechanism.

## Excretion of Nitrogenous Wastes

When amino acids are catabolized, the amino group is released or transferred to another molecule for removal or for reuse. Unlike the atoms of the carbon skeleton of an amino acid, which can be oxidized to $CO_2$ and water, the amino group must either be salvaged for the resynthesis of amino acids or be excreted to avoid a toxic rise in the concentration of nitrogenous wastes. Thus, in most animals, there is a close link between the osmoregulatory functions and the processes for the

**12-44** Structures of the three common nitrogenous excretory products. Ammonia is both highly soluble and highly toxic. Uric acid is poorly soluble and nontoxic.

elimination of excess nitrogen. In those animals faced with a limited water supply, this relation gives rise to a serious conflict—namely, the conflict between conserving water, on the one hand, and preventing toxic accumulation of nitrogenous wastes, on the other. As we will see, animals have adapted excretory strategies appropriate to their water economies.

The waste amino groups are *excreted* in one of three forms: *ammonia, urea,* or *uric acid* (Figure 12-44). These three nitrogenous compounds differ in their properties, so that in the course of evolution some animal groups have found it more opportune to produce one or the other of these forms for excretory purposes during all or part of their life cycles.

Most teleosts and aquatic invertebrates produce little or no urea, but instead excrete their nitrogenous wastes primarily as ammonia. This approach is not feasible for most land animals, however, since ammonia is both highly toxic and highly soluble. For example, a blood concentration of $3 \times 10^{-5}$ mol/L of ammonia is fatal in rabbits. Large quantities of water would be required, therefore, to dissolve and carry off the ammonia, 300–500 ml for 1 g of nitrogen in the form of ammonia. Most land animals avoid this problem by converting nitrogenous wastes either to uric acid or to urea instead.†

Although urea is quite soluble, it is far less toxic than ammonia, and only about 50 ml of water is required to excrete 1 g of nitrogen. Moreover, urea contains only two hydrogen atoms for each nitrogen atom. Uric acid or guanine is the choice of birds, reptiles, and

---

†An interesting exception is the terrestrial isopod (sowbug), an arthropod that eliminates ammonia as a gas.

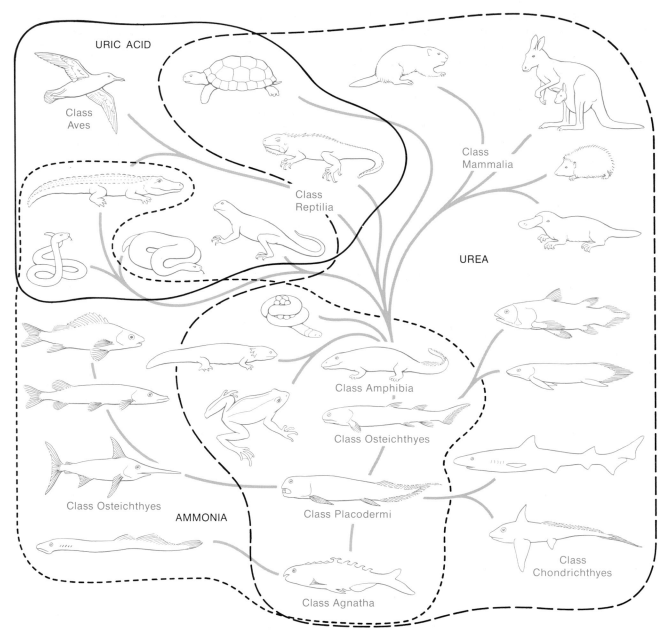

**12-45** Phylogenetic relations and nitrogen excretion among the vertebrates. The trunk of the tree includes extinct ancestral forms. Note the overlaps between ammonotelic, ureotelic, and uricotelic excretion in some of the groups. [Schmidt-Nielsen and Mackay, 1972.]

most terrestrial arthropods. Birds distinguish themselves with a *guano* containing white crystals of uric acid. Although highly toxic, uric acid is only slightly soluble in water, so it can be excreted as a pasty precipitate, with only 10 ml of water required per gram of the nitrogen.

Unusual among amphibians are two arid-land toads, *Chiromantis xerampelina* and *Phyllomedusa sauwagii,* which not only have an extremely low evaporative loss of water from their skin but, like reptiles, excrete nitrogen as uric acid rather than ammonia or urea as most other amphibians do. The low solubility of uric acid causes it to precipitate readily in the cloaca, and allows these toads, like reptiles and birds, to minimize the volume of urine necessary to eliminate their excess nitrogen.

Avian embryos produce ammonia for the first day or so, and then switch to uric acid, which is deposited within the egg as an insoluble solid and thus has no effect on the osmolarity of the precious little fluid contained in the egg. Lizards and snakes have various developmental schedules for switching from the production of ammonia and urea to the production of mainly uric acid. In species that lay their eggs in moist sand, the switch to uric acid production occurs late in development, but before hatching. The switch to uric acid production is a kind of biochemical metamorphosis that prepares the organism for a dry, terrestrial habitat.

The phylogenetic relations of nitrogen excretion among the vertebrates is illustrated in Figure 12-45. It is evident that there are overlaps of different excretory products in several groups. Thus, the adults of a given species may excrete part of their nitrogen as ammonia, urea, or uric acid, and lesser quantities in the form of such compounds as creatinine, creatine, or trimethylamine oxide. The type of excretion is generally related to habitat: Terrestrial birds excrete about 90% of their nitrogenous wastes as uric acid and 3–4% as ammonia, but semiaquatic birds, such as ducks, excrete only 50% of their nitrogenous wastes as uric acid and 30% as ammonia. Mammals excrete most of their waste as urea.

*Ureotelic* (urea-excreting) animals utilize one of two pathways for urea formation. In all vertebrates other than teleost fishes the synthesis of urea takes place primarily in the liver through the *ornithine–urea cycle*

(Figure 12-46). Two amino groups and a molecule of $CO_2$ are added to ornithine to form arginine. The urea molecule is removed with the aid of the enzyme arginase, which is present in relatively large quantities in these animals. The teleosts and many invertebrates utilize the so-called *uricolytic pathway* (Figure 12-47), in which urea is produced from uric acid that either arises from a transamination via aspartate or is produced during nucleic acid metabolism. Uric acid is converted first to *allantoin* and *allantoic acid* with the help of the enzymes *uricase* and *allantoinase,* respectively.

*Uricotelic* animals excrete nitrogen chiefly in the form of uric acid. The nitrogen atoms incorporated into uric acid ultimately arise from the breakdown of the amino acids glycine, aspartate, and glutamine (Figure 12-47). Since these animals do not have the enzyme uricase to break down the uric acid, the catalysis of nitrogenous molecules is terminated at this point and uric acid, which largely precipitates because of its low solubility, is excreted as the end product. It is adaptively significant that, in its precipitated form, uric acid contributes nothing to the tonicity of the "urine" or feces, and thus the excretion requires very little urinary water. In general, uricotelic animals are adapted to conditions of limited availability of water.

In *ammonotelic* (ammonia-excreting) animals, the amino groups of various amino acids are transferred with the aid of a transaminase enzyme, to glutamate, which is converted to *glutamine* (Figure 12-48). Finally, the glutamine is deaminated in the kidney tubules, and ammonia is liberated into the tubular fluid. The ammonia can take up a proton to form the ammonium ion, $NH_4^+$, and because the ion is very poorly diffusible across cell membranes, the ammonium cannot diffuse back into the tissues, so it leaves the body via the urine. Since ammonia in both its free and ionized forms is highly toxic, it makes good sense that glutamine, which is nontoxic, should act as the amino-group carrier through blood and tissues until its deamination in the ammonotelic kidney.

In freshwater teleosts, ammonium and $H^+$ are excreted by the epithelium of the gills in exchange for $Na^+$. Thus, in a single process of counterdiffusion, the gill epithelium of freshwater fishes takes up much needed $Na^+$ while ridding itself of toxic $NH_4$ and excess $H^+$. In marine teleosts, the same mechanism operates, causing a small concomitant increase in $Na^+$,

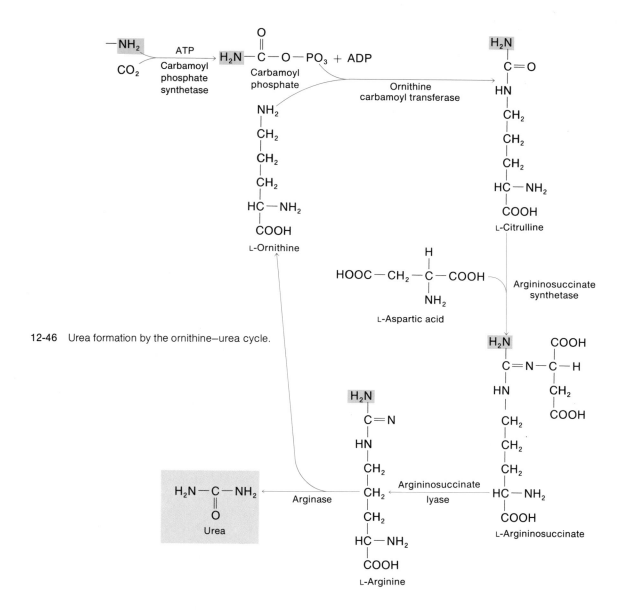

**12-46** Urea formation by the ornithine–urea cycle.

**12-47** Uric acid and urea production via the uricolytic pathway. Uric acid arises from a purine ring that is synthesized by a complex union of aspartic acid, formic acid, glycine, glutamate, and $CO_2$.

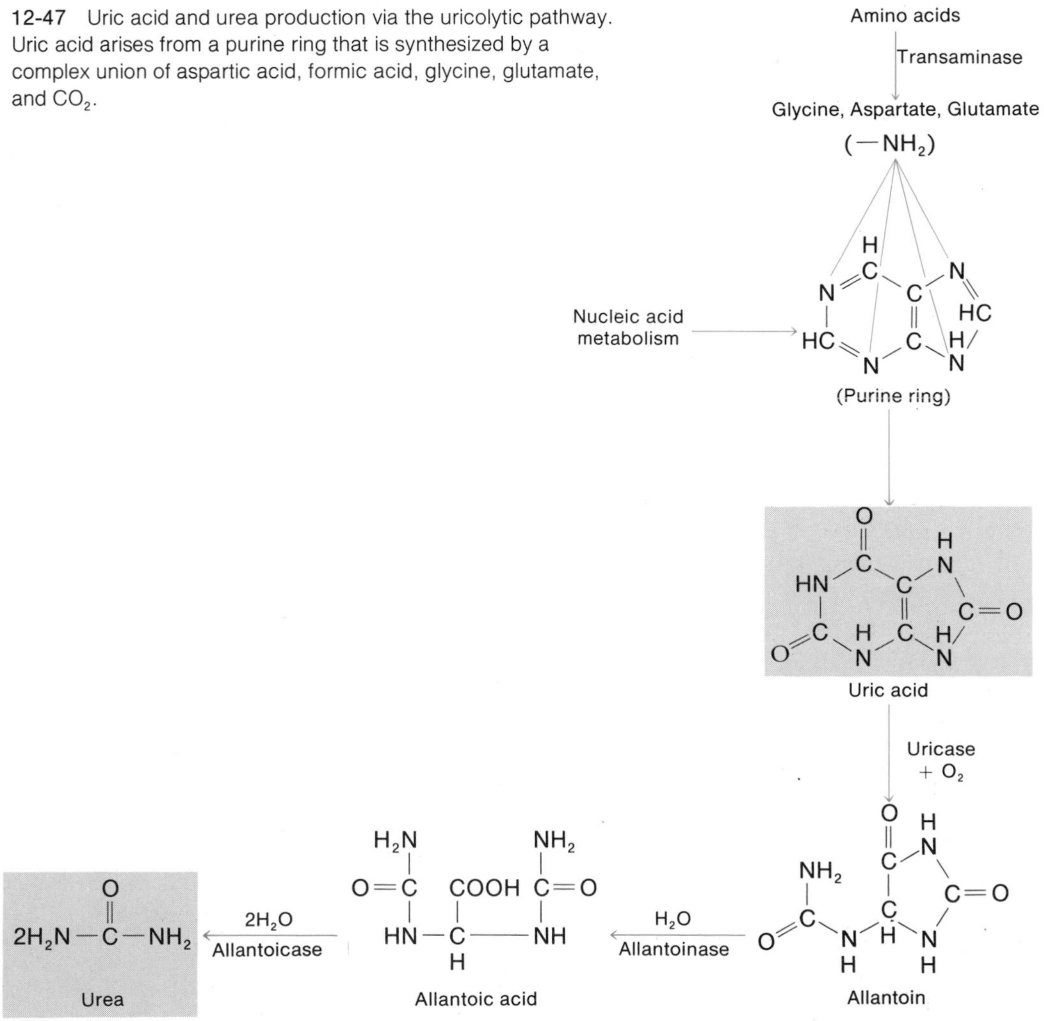

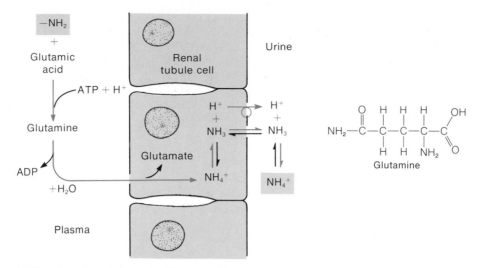

**12-48**  Formation of ammonia. Glutamic acid is transaminated to form glutamine, which serves as the amino-group carrier in the blood. Within the cells of the kidney tubule, the amino group is removed as $NH_4^+$. The uncharged lipid-soluble form, $NH_3$, crosses the cell membrane into the urine, by diffusion, while the $H^+$ is actively transported into the lumen, where it reacts to form $NH_4^+$.

which is actively transported out of the gill along with excess $Na^+$ from other sources.

Concentration of nitrogenous wastes in the urine of vertebrates takes place in two ways: (1) by glomerular filtration with subsequent removal of water and salts, leaving the waste molecules behind to be eliminated; and (2) by active tubular secretion of those molecules across the tubular epithelium into the urine. In the aglomerular kidneys of marine teleosts, secretion alone accounts for the passage of nitrogenous wastes into the urine. In the mammalian kidney, urea is passed into the urine by both filtration and secretion.

## Summary

The extracellular environment in many marine and non-marine animals broadly resembles dilute seawater. This similarity may have had its origin in the shallow, dilute primeval seas that are believed to have been the setting for the early evolution of animal life. The ability of many animals to regulate the composition of their internal environment is closely correlated with their ability to occupy ecological environments that are osmotically at odds with the osmotic requirements of their tissues. Osmoregulation requires the exchange of salts and water between the extracellular environment and the external environment to compensate for obligatory, or uncontrolled, losses and gains.

The transport of solutes and water across epithelial layers is fundamental to all osmoregulatory activity. The obligatory exchange of water depends on (1) the osmotic gradient that exists between the internal and external environments, (2) the surface/volume ratio of the animal, (3) the permeability of the integument, (4) the intake of food and water, (5) the evaporative losses required for thermoregulation, and (6) the disposal of digestive and metabolic waste through urine and feces.

Marine and terrestrial animals are faced with dehydration, whereas freshwater animals must prevent hydration by uncontrolled osmotic uptake of water. Marine birds, reptiles, and teleosts replace lost water by drinking seawater and actively secreting salt through secretory epithelia. Freshwater fishes do not drink water; they replace lost salts by active uptake. Birds and mammals are the only vertebrates to secrete a hypertonic urine. Many desert species in addition utilize mechanisms for minimizing respiratory water loss.

Mammalian and avian kidneys utilize filtration, reabsorption, secretion, and a countercurrent mechanism to produce a hypertonic urine. The filtration of the plasma in the glomerulus is dependent on arterial pressure. Crystalloids and small organic molecules are filtered, leaving blood cells and large molecules behind. Salts and organic molecules such as sugars are partially reabsorbed in the renal tubules from the glomerular filtrate, and certain substances are secreted into the tubules. A countercurrent multiplier system that includes the collecting duct and the loop of Henle sets up a steep extracellular concentration gradient of salt and urea that extends deep into the medulla of the kidney. Water is drawn osmotically out of the collecting duct as it passes through high medullary concentrations of salt and urea toward the renal pelvis. Endocrine control of the water permeability of the collecting duct determines the volume of water reabsorbed and retained in the circulation. The final urine, then, is the product of filtration, reabsorption, and secretion. These processes allow the urinary composition to depart strongly from the proportions of substances occurring in the blood.

The formation of urine follows the same major outline in all or most vertebrates and invertebrates. A pre-urine is formed that contains essentially all the small molecules and ions found in the blood. In most vertebrates and in the crustaceans and mollusks, this formation is accomplished by ultrafiltration; in insects, by the secretion through the epithelium of the Malpighian tubules of $K^+$, $Na^+$, and phosphate, with water and other small molecules, such as amino acids and sugars, following passively by osmosis and diffusion down their concentration gradients. The pre-urine is subsequently modified by the selective reabsorption of ions and water and, in some animals, by secretion of waste substances into the lumen of the nephron by the tubular epithelium.

The gills of teleost fishes and many invertebrates perform osmoregulation by active transport of salts, the direction of transport being inward in freshwater fish and outward in marine fish. Marine birds and reptiles both utilize the nasal salt gland for the active secretion and expulsion of concentrated salts via epithelial tubules. Secretory activity by the salt gland is stimulated by neuroendocrine circuits with osmosensory endings in the heart and brain.

The nitrogen produced in the catabolism of amino acids and proteins is concentrated into one of three forms of nitrogenous waste, depending on the osmotic environment of different animal groups. Ammonia, highly toxic and soluble, requires large volumes of water to dilute; it is produced by teleosts. Uric acid is less toxic and poorly soluble; it is excreted as a semisolid suspension by birds and reptiles. Urea is the least toxic and requires a moderately small amount of water. Mammals convert their nitrogen into urea, and excrete it via the urine; elasmobranchs use urea as an osmotic agent in their blood.

## Exercises

*1.* How has the development of osmoregulatory mechanisms affected animal evolution?

*2.* What factors influence obligatory osmotic exchange with the environment?

*3.* Explain why respiration, temperature regulation, and water balance in terrestrial animals are closely interrelated. Give examples.

*4.* Describe three anatomical or physiological mechanisms used by insects to minimize dehydration in dry environments.

*5.* How do marine and freshwater fishes maintain osmotic homeostasis?

*6.* Name and describe the three major processes used by the vertebrate kidney to achieve the final composition of the urine.

*7.* What factors determine the rate of ultrafiltration in the glomerulus?

*8.* What is meant by the renal clearance of a substance?

*9.* If the intratubular fluid in the loop of Henle remains nearly isosmotic relative to the extracellular fluid along its path, and is even slightly hypotonic on leaving the loop, in what way is the final urine made hypertonic?

*10.* Explain why the consumption of 1L of beer will lead to a greater urine production than an equal volume of water.

*11.* What role does the kidney play in the regulation of blood pressure?

*12.* Discuss the role of the kidney in the control of plasma pH.

*13.* Compare the function of aglomerular and glomerular vertebrate kidneys.

*14.* What evidence supports the existence of filtration–absorption organs of osmoregulation in mollusks and crustaceans?

*15.* Why is it more adaptive for animals to reabsorb essential physiological molecules or ions from urine than to secrete wastes actively into the urine?

*16.* How do insects produce concentrated, hypertonic urine and excrement?

*17.* In the course of evolution, terrestrial organisms have come to excrete mainly uric acid and urea rather than ammonia. What are the adaptive reasons for such a change?

*18.* Explain why gulls can drink seawater and survive but humans cannot.

*19.* After the injection of inulin into a small mammal, the plasma inulin concentration was found to be 1 mg/ml, the concentration in the urine 10 mg/ml, and the urine flow rate through the ureter 10 ml/h. What was the rate of plasma filtration and the clearance in milliliters per minute? How much water was reabsorbed in the tubules per hour?

*20.* What evidence is there that the mammalian nephron employs tubular secretion as one means of eliminating substances into the urine?

*21.* Countercurrent systems have evolved for gas exchange in the gills and swim bladder of fishes, for heat conservation in the vascular systems of the extremities of some homeotherms, and for establishing osmotic gradients in the kidneys of birds and mammals. Explain why the countercurrent system is more efficient in physical transport and transfer than a system in which fluids in opposed vessels flow in the same direction.

## Suggested Reading

Bentley, P. J. 1971. *Endocrines and Osmoregulation.* New York: Springer-Verlag.

Bentley, P. J. 1972. Comparative endocrinology and osmoregulation. *Fed. Proc.* 31:1583–1624.

Brenner, B. M. 1974. Renal handling of sodium. *Fed. Proc.* 33:13–36.

Conte, F. P., ed. 1980. Biology of the chloride cell. *Amer. J. Physiol.* 7:R139–R269.

Gupta, B. L., R. B. Moreton, J. L. Oschman, and B. J. Wall. 1977. *Transport of Ions and Water in Animals.* London: Academic.

Hadley, N. 1972. Desert species and adaptation. *American Scientist* 60:338–347.

Harvey, R. J. 1974. *The Kidneys and the Internal Environment.* New York: Wiley.

Koushanpour, E. 1976. *Renal Physiology: Principles and Functions.* Philadelphia: Saunders.

Krogh, A. 1939. *Osmotic Regulation in Aquatic Animals.* Cambridge: Cambridge University Press.

Maloiy, C. M. O., ed. 1979. *Comparative Physiology of Osmoregulation in Animals.* Vols. 1 and 2. New York: Academic.

Moffat, D. B. 1971. *The Control of Water Balance by the Kidney.* New York: Oxford University Press.

Pitts, R. F. 1974. *Physiology of the Kidney and Body Fluids.* 3rd ed. Chicago: Year Book Medical Publishers.

Potts, W. F. W., and G. Parry. 1964. *Osmotic and Ionic Regulation in Animals.* New York: Macmillan.

Schmidt-Nielsen, K. 1964. *Desert Animals: Physiological Problems of Heat and Water.* London: Oxford University Press.

Schmidt-Nielsen, K. 1972. *How Animals Work.* Cambridge: Cambridge University Press.

Smith, H. W. 1953. *From Fish to Philosopher.* Boston: Little, Brown.

Vander, A. J. 1980. *Renal Physiology.* 2nd ed. New York: McGraw-Hill.

Wessels, N. K., ed. 1968. *Vertebrate Adaptations: Readings from Scientific American.* Pt. 4. San Francisco: W. H. Freeman and Company.

# References Cited

Beament, J. W. L. 1958. The effect of temperature on the waterproofing mechanism of an insect. *J. Exp. Biol.* 35:494–519.

Burg, M., J. Grantham, M. Abramow, and J. Orloff. 1966. Preparation and study of fragments of single rabbit nephrons. *Amer. J. Physiol.* 210:1293–1298.

Buxton, P. A. 1931. The law governing the loss of water from an insect. *Proc. Entomol. Soc.* (London) 6:27–31.

Davis, J. O. 1974. Control of renin release. *Hospital Practice* 9:55–65.

Dunson, W. A. 1969. Electrolyte secretion by the salt gland of the Galápagos marine iguana. *Amer. J. Physiol.* 216:995–1002.

Edney, E. B. 1974. Desert arthropods. In G. W. Brown, ed., *Desert Biology.* Vol. 2. New York: Academic.

Edney, E. B., and K. A. Nagy. 1976. Water balance and excretion. In J. Bligh, J. L. Cloudsley-Thompson, and A. G. MacDonald, eds., *Environmental Physiology of Animals.* Oxford: Blackwell Scientific Publications.

Grantham, J. J. 1971. Mode of water transport in mammalian renal collecting tubules. *Fed. Proc.* 30:14–21.

Hargitay, B., and W. Kuhn. 1951. Das Multiplikationsprinzip als Grundlage der Harnkonzentrierung in der Niere. *Z. Electrochem.* 55:539–558.

Hoar, W. S. 1975. *General and Comparative Physiology,* 2nd ed. Englewood Cliffs, N.J.: Prentice-Hall.

Jamison, R. L., and R. H. Maffly. 1976. The urinary concentrating mechanism. *New England J. Med.* 295:1059–1067.

Kokko, J. P., and F. C. Rector, Jr. 1972. Countercurrent multiplication system without active transport in the inner medulla. *Kidney Int.* 2:214–223.

Maetz, J. 1971. Fish gills: Mechanisms of salt transfer in fresh water and sea water. *Phil. Trans. Roy. Soc.* (London) Ser. B. 262:209–249.

Oschman, J. L., and M. J. Berridge. 1971. The structural basis of fluid secretion. *Fed. Proc.* 30:49–56.

Phillips, J. E. 1970. Apparent transport of water in insect excretory system. *Amer. Zool.* 10:413–436.

Phillips, J. G. 1975. *Environmental Physiology.* New York: Wiley.

Pitts, R. F. 1959. *The Physiological Basis of Diuretic Therapy.* Springfield Ill.: Thomas.

Pitts, R. F. 1968. *Physiology of the Kidney and Body Fluids.* 2nd ed. Chicago: Year Book Medical Publishers.

Prosser, C. L. 1973. *Comparative Animal Physiology.* Vol. 1. Philadelphia: Saunders.

Rhodin, J. 1954. Correlation of ultrastructural organization and function in normal and experimentally changed proximal convoluted tubule cells of the mouse kidney. Thesis, Karolinska Institute, Stockholm. In R. F. Pitts, *Physiology of the Kidney and Body Fluids.* 2nd ed. Chicago: Year Book Medical Publishers.

Richards, A. N. 1935. Urine formation in the amphibian kidney. *Harvey Lect.* 30:93–118.

Schmidt-Nielsen, B. M., and W. C. Mackay. 1972. Comparative physiology of electrolyte and water regulation, with emphasis on sodium, potassium, chloride, urea, and osmotic pressure. In M. H. Maxwell and C. R. Kleeman, eds., *Clinical Disorders of Fluid and Electrolyte Metabolism.* New York: McGraw-Hill.

Schmidt-Nielsen, K. 1959. Salt glands. *Scientific American* 200(1):109–116.

Schmidt-Nielsen, K. 1960. The salt-secreting gland of marine birds. *Circulation* 21:955–967.

Schmidt-Nielsen, K., F. R. Hainsworth, and D. E. Murrish. 1970. Countercurrent heat exchange in the respiratory passages: Effect on water and heat balance. *Resp. Physiol.* 9:263–276.

Smith, H. W. 1939. *Physiology of the Kidney.* Lawrence: University of Kansas Press.

Smith, H. W. 1956. *Principles of Renal Physiology.* New York: Oxford University Press.

Solomon, A. K. 1962. Pumps in the living cell. *Scientific American* 207(2):100–108.

Thomas, D. H., and J. G. Phillips. 1975. Studies in avian adrenal steroid function, Pts. 4–5. *Gen. Comp. Endocr.* 26:427–450.

Thomas, D. H., and J. G. Phillips. 1978. The anatomy and physiology of the avian nasal salt glands. *Pavo* 16:89–104.

Ullrich, K. J., K. Kramer, and J. W. Boylan. 1961. Present knowledge of the countercurrent system in the mammalian kidney. *Prog. Cardiovasc. Dis.* 3:395–431.

Wigglesworth, V. B. 1932. On the function of the so-called rectal-glands of insects. *Quart. J. Microscop. Sci.* 75:131–150.

# CHAPTER 13

# Circulation of Blood

*M*etabolic processes utilize substrates and yield end products that, in animals less than about 1 mm thick, are delivered and removed by diffusion. This is a relatively slow process, and substrate levels will limit metabolism if the end products are delivered by diffusion alone over large distances. Circulatory systems—that is, blood and the vessels in which it moves—have evolved to overcome the limitations of diffusion by transporting materials among various regions of the body. These materials include respiratory gases, nutrients, waste products, hormones, antibodies, and salts. Blood is a complex tissue and contains many special cell types. It acts as a vehicle for most homeostatic processes and plays some role in nearly all physiological functions. This chapter is confined to the properties of blood associated with its movement through blood vessels. Other properties of blood, such as the clotting process, the production of erythrocytes, and the functions of leucocytes, are not discussed.

This chapter reviews the circulation of blood and how it is controlled to meet the requirements of the tissues. Most attention is given to the mammalian circulatory system, because it is the best known. Mammals are very active, predominantly aerobic, terrestrial animals, and their circulatory system evolved to meet their particular requirements. The mammalian system is only one of several types of circulation. All circulatory systems, however, can be divided into a number of parts with similar functions. These components consist of:

*1.* A main propulsive organ, usually a *heart,* which forces blood around in the body.

*2.* An *arterial system,* which can act both to distribute blood and as a pressure reservoir.

*3.* *Capillaries,* in which transfer of materials occurs between blood and tissues.

*4.* The *venous system,* which acts as a blood reservoir and as a system for returning blood to the heart.

## General Plan of the Circulatory System

The movement of blood through the body results from the forces imparted by rhythmic contractions of the heart, by the squeezing of blood vessels during body movements, and/or by the peristaltic contractions of smooth muscle surrounding blood vessels (Figure 13-1). The relative importance of each of these mechanisms in generating flow varies. In vertebrates, the heart plays the major role in blood circulation; in arthropods, movements of the limbs and contractions of the dorsal heart are equally important in generating blood flow; in the giant earthworm *Glossoscolex giganteus,* peristaltic contractions of the dorsal vessel are responsible for moving blood in an anterior direction and filling the five pairs of lateral hearts (Figure 13-2). In all animals, valves and/or septa determine the direction of flow; and sphincters surrounding blood vessels regulate the amount of blood that flows through a particular pathway and thus control the distribution of blood within the body.

Many invertebrates have an *open circulation*—that is, a system in which blood pumped by the heart empties via an artery into an open fluid space, the *hemocoel* (or *blastocoel*), which lies between the ectoderm and endoderm. The fluid contained within the hemocoel, referred to as *hemolymph* or blood, is not circulated through capillaries but bathes the tissues directly. Figure 13-3A and B illustrates the organization of the main vessels of open circulation from two groups of invertebrates. The hemocoel is not emphasized in this

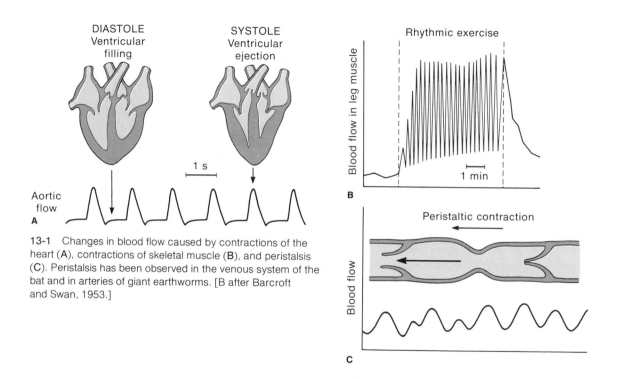

13-1  Changes in blood flow caused by contractions of the heart (**A**), contractions of skeletal muscle (**B**), and peristalsis (**C**). Peristalsis has been observed in the venous system of the bat and in arteries of giant earthworms. [B after Barcroft and Swan, 1953.]

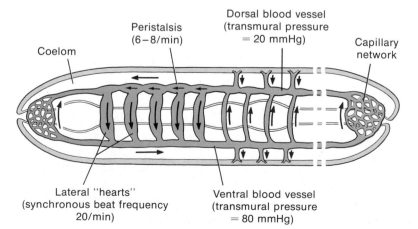

13-2  Circulation in the giant earthworm *Glossoscolex giganteus*. [Data from Johansen and Martin, 1965.]

diagram, but in many animals it is large and may constitute 20–40% of body volume. In some crabs, for instance, blood volume is about 30% of body volume. By contrast, vertebrates with a closed circulation have a blood volume that is typically about 5–10% of body volume. Open circulatory systems have low pressures, seldom exceeding 5–10 mmHg. Higher pressures have been recorded in portions of the open circulation of the terrestrial snail *Helix,* but these are exceptional. In snails, these high pressures are generated by contractions of the heart, whereas in some bivalve mollusks high pressures in the foot are generated by contractions of surrounding muscles rather than of the heart. The functional significance of these high pressures in an open circulation is that they appear to help the animal maintain posture.

Animals with an open circulation generally have a limited ability to alter the velocity and distribution of blood flow. As a result, in bivalve mollusks and other animals that have an open circulation and use blood for gas transport, changes in oxygen uptake are usually slow and maximum rates of oxygen transfer low per unit weight. Insects have avoided this problem by evolving a tracheal system in which direct gas transport to tissues occurs through air-filled tubes that bypass the blood. Consequently, although insects have an open circulation, they have a large capacity for aerobic metabolism. Animals with a low-pressure open circulation generally cannot produce an excretory fluid by ultrafiltration. In insects, for example, the primary excretory fluid is formed in the Malpighian tubules by secretion.

Some invertebrates, such as cephalopods (octopuses, squids) (Figure 13-3C), and all vertebrates have a closed circulation, with blood flowing in a continuous circuit of tubes from *arteries* to *veins* through *capillaries* (Figure 13-4). In general, there is a more complete separation of functions in closed circulatory systems than in open ones. In a closed circulation, the heart is the main propulsive organ, pumping blood into the *arterial system* and maintaining a high *blood pressure* in the arteries. The arterial system, in turn, acts as a pressure reservoir forcing blood through the capillaries. The capillary walls are thin, thus allowing high rates of transfer of material between blood and tissues. Each tissue has many capillaries, so that each cell is no more

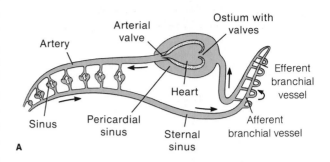

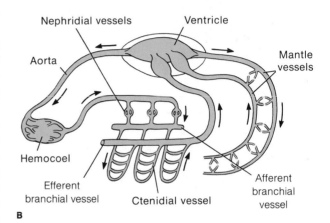

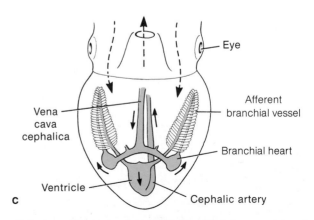

**13-3** Invertebrate circulations. (**A**) Simplified diagram of the circulation in the crayfish. (**B**) Simplified diagram of the circulation in a bivalve mollusk. (**C**) Heart and main blood vessels of a cephalopod. The broken arrows indicate the direction of water flow; other arrows, direction of blood flow.

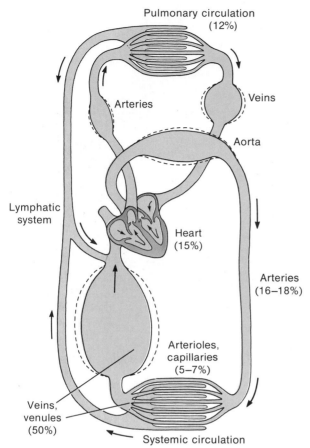

Pulmonary circulation
(12%)

Arteries

Veins

Aorta

Lymphatic
system

Heart
(15%)

Arteries
(16–18%)

Veins,
venules
(50%)

Arterioles,
capillaries
(5–7%)

Systemic circulation

**13-4**   The mammalian circulation. The percentages indicate the relative proportion of blood in different parts of the circulation.

than two or three cells away from a capillary. Capillary networks are in parallel, allowing fine control of blood distribution between tissues. Because the capillary walls are permeable and pressures high, fluid can slowly filter across the walls and into the space between cells. In addition, the blood is under sufficiently high pressure to permit the ultrafiltration of blood in the kidneys.

A *lymphatic system* has evolved in conjunction with the high-pressure, closed circulatory system of vertebrates to recover fluid lost to tissues from the blood. The problems of filtration across capillary walls are

ameliorated in some tissues by having less permeable walls. In addition, filtration in the mammalian lung is reduced because arterial pressures, and therefore capillary pressures, are lower than those of the arterial system serving the rest of the body. Different pressures can be maintained in the *systemic* (body) and *pulmonary* (lung) circulations of mammals because the mammalian circulatory system is equipped with a completely divided heart. The right side of the heart pumps blood in the pulmonary circulation, and the left side pumps blood in the systemic circulation.

The *venous system* collects blood from the capillaries and delivers it to the heart via the veins, which are typically low-pressure, compliant structures in which large changes in volume have little effect on venous pressure. Thus, the venous system contains most of the blood and acts as a large-volume reservoir. Blood donors give blood from this reservoir, and as there is little change in pressure as the venous volume decreases, the volumes and flows in other regions of the circulation are not markedly altered.

## *The Mammalian Heart*

The mammalian heart consists of four chambers (Figure 13-5). Blood returning from the lungs enters the *left atrium*, passes into the *left ventricle*, and is then ejected into the body circulation. Blood from the body collects in the right atrium, passes into the right ventricle, and is pumped to the lungs. Valves prevent backflow of blood from the aorta to the ventricle, the atrium, and the veins. These valves are passive and are opened and closed by pressure differences between heart chambers. The atrioventricular valves (bicuspid and tricuspid valves; Figure 13-5) are connected to the ventricular wall by fibrous strands. These strands prevent the valves from being everted into the atria when the ventricles contract and intraventricular pressures are much higher than those in the atria.

The *myocardium* (i.e., heart muscle) consists of three types of muscle fiber. Muscle cells found in the sinus node and the atrioventricular node are often smaller than other myocardial cells, are only weakly contractile, are autorhythmic, and exhibit very slow conduction between cells. The largest myocardial cells, found

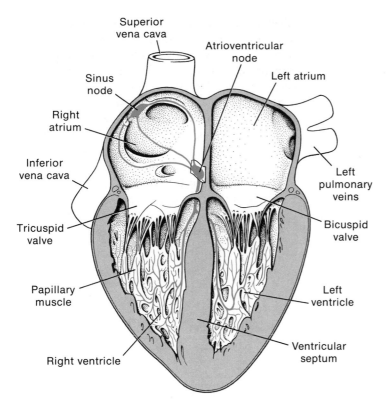

Superior
vena cava

Atrioventricular
node

Left atrium

Sinus
node

Right
atrium

Inferior
vena cava

Left
pulmonary
veins

Tricuspid
valve

Bicuspid
valve

Papillary
muscle

Left
ventricle

Right ventricle

Ventricular
septum

**13-5**    Cutaway view of the human heart, showing the rear portion, depicts the pacemaker and the path of its impulses. The pacemaker is the structure near the top of the heart; it is variously called the sinus node and the sinoatrial node. Impulses from the pacemaker spread along the paths indicated by the colored lines to the atrioventricular node, from which they are transmitted to the ventricles. Pacemaker cells differ from both nerve cells and muscle cells; they are usually described as modified muscle cells. [From "The Heart's Pacemaker" by E. F. Adolph. Copyright © 1967 by Scientific American, Inc. All rights reserved.]

in the ventricular endocardium, are also weakly contractile, but are specialized for fast conduction and constitute the system for spreading the excitation over the heart. The intermediate-sized myocardial cells are strongly contractile and constitute the bulk of the heart.

The walls of the ventricle, especially the left chamber, are thick and muscular; the inner portion of the wall, or *endocardium,* is generally more spongy than the outer region, or *epicardium.* The mammalian heart is contained within a fibrous but flexible bag, the *pericardium* (Figure 13-6).

## Electrical Activity of the Heart

Hearts are muscular pumps. Vertebrate cardiac and skeletal muscle fibers are similar in many respects, except that the T-tubule system (p. 360) is less extensive in myocardial (cardiac) cells of lower vertebrates and

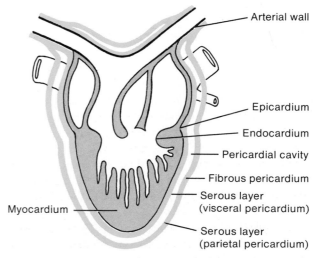

Arterial wall

Epicardium

Endocardium

Pericardial cavity

Fibrous pericardium

Serous layer
(visceral pericardium)

Myocardium

Serous layer
(parietal pericardium)

**13-6**    Diagram of the pericardium and its relation to the heart. [Skandalakis et al., 1980.]

cardiac muscle cells are electrically coupled. Except for differences in the uptake and release of $Ca^{2+}$, the mechanisms of contraction of vertebrate skeletal and cardiac muscle are generally considered to be alike.

A heartbeat consists of a rhythmic contraction and relaxation of the whole muscle mass. Contraction of each cell is associated with an action potential in that cell. The electrical activity is initiated in a pacemaker region of the heart, and electrical activity spreads over the heart from one cell to another because the cells are electrically coupled via membrane junctions. The nature and extent of coupling determines the pattern by which the electrical wave of excitation spreads over the heart and also influences the rate of conduction.

## PACEMAKER REGION

Pacemaker cells are capable of spontaneous activity and may be either neurons (as in many invertebrate hearts) or muscle cells (as in vertebrate and some invertebrate hearts). If the heartbeat is initiated in a neuron, the pacemaker is called a *neurogenic pacemaker;* if the beat is initiated in a modified muscle cell, the term *myogenic pacemaker* is used. Hearts are often categorized by the type of pacemaker and hence are called either neurogenic or myogenic hearts.

### Neurogenic Pacemakers

In many invertebrates, it is not clear whether the pacemaker is neurogenic or myogenic. Decapod crustaceans, however, do have neurogenic hearts. The cardiac ganglion, situated on the heart and consisting of nine or more neurons (depending on the species) acts as a pacemaker. If the cardiac ganglion is removed, the heart stops beating but the ganglion continues to be active and shows intrinsic rhythmicity. The cardiac ganglion consists of small and large cells. The small cells act as pacemakers and are connected to the large follower cells, which are all electrically coupled. Activity from the small pacemaker cells is fed into and integrated by the large follower cells and then distributed to the heart muscle. The crustacean cardiac ganglion is innervated by excitatory and inhibitory nerves originating in the central nervous system, and these nerves can alter the rate of firing of the ganglion and, therefore, the heart rate.

### Myogenic Pacemakers

Vertebrate, molluscan, and many other invertebrate hearts are driven by myogenic pacemakers. These tissues have been studied extensively in a variety of species. In the vertebrate heart, the pacemaker is situated in the sinus venosus (Figure 13-19) or in a remnant of it called the sinoatrial node and consists of small, weakly contractile, specialized muscle cells (Figure 13-5).

A heart may contain many cells capable of pacemaker activity, but because all cardiac cells are electrically coupled, the cell (or group of cells) with the fastest intrinsic rate is the one that stimulates the whole heart to contract and that determines the heart rate. These pacemaker cells will normally overshadow those with slower pacemaker activity, but if the normal pacemaker were to stop, the other pacemaker cells would determine a new, lower heart rate. Thus, cells with the capacity for spontaneous activity may be categorized as pacemakers and latent pacemakers. In the event that a latent pacemaker becomes uncoupled electrically from the pacemaker, it may beat and control a portion of cardiac muscle, generally an entire chamber, at a rate different from that of the normal pacemaker. Such an *ectopic pacemaker* is dangerous, as it will desynchronize the pumping action of the heart chambers.

### Cardiac Pacemaker Potentials

An important characteristic of pacemaker cells is the absence of a stable resting potential between action potentials. Consequently, the membranes of cells in pacemaker tissue undergo a steady depolarization, termed a pacemaker potential (p. 171), preceding each action potential (Figures 13-7, 13-8, and 13-9). As the pacemaker potential brings the membrane to the threshold potential, it gives rise to an all-or-none *cardiac action potential.* The interval between action potentials, which of course determines the heart rate, depends on the rate of depolarization of the pacemaker potential. A slower depolarization brings the membrane to a firing level later and thus decreases the frequency of firing.

Pacemaker activity has its origin in time-dependent changes in membrane conductance. In the frog sinus venosus (Figure 13-7), the pacemaker depolarization begins immediately after the previous action potential, when the potassium conductance of the membrane is very high. The potassium conductance then gradually drops, and the membrane shows a corresponding depolarization owing to a moderately high, steady conductance for sodium. The pacemaker depolarization

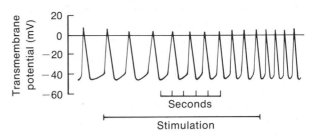

**13-9** The effect of sympathetic nerve stimulation on the action potential of the pacemaker cell. Several action potentials are shown. When the sympathetic nerves are stimulated, there is an increase in the rate of depolarization, which results in an increase in the frequency of firing of the pacemaker cell. [Hutter and Trautwein, 1956.]

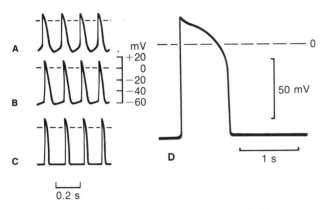

**13-7** Cardiac action potentials. (**A** and **B**) Recorded from different cells in the sinus venosus of the frog. Note the slowly depolarizing pacemaker potentials between the action potentials. (**C**) Recorded from the atrial muscle. [Hutter and Trautwein, 1956.] (**D**) Recorded from a cell in the ventricle of the frog. [Orkand, 1968.] Note the abrupt origin of the action potential in nonpacemaker tissue.

continues until it activates the sodium conductance. The Hodgkin cycle (p. 162) then predominates to produce the rapid regenerative upstroke of the cardiac action potential.

Acetylcholine, which slows the heart when released by activity of the vagus nerve (tenth cranial nerve), does so by increasing potassium conductance of the pacemaker cells. This greater conductance keeps the membrane potential near $E_K$ for a longer time, thereby slowing the pacemaker depolarization and delaying the onset of the next upstroke (Figure 13-8). Epinephrine, on the other hand, increases the slope of the pacemaker potential, thus increasing the heart rate (Figure 13-9). Epinephrine increases sodium and calcium conductance, but this is probably not the mechanism involved in speeding up the pacemaker rhythm. It is possible that epinephrine decreases the time-dependent potassium efflux during diastole and thereby increases the rate of pacemaker depolarization.

*Cardiac Action Potentials*

Action potentials precede contraction in all vertebrate cardiac muscle cells and are of long duration compared with those in skeletal muscle. The action potential in skeletal muscle is complete, and the membrane is in a nonrefractory stage before the onset of contraction; hence, repetitive stimulation and tetanic contraction are possible. In cardiac muscle, by contrast, the membrane remains in a refractory state until the heart

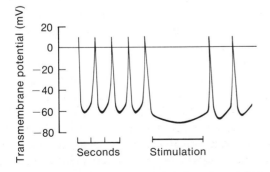

**13-8** The effect of vagus nerve stimulation on the pacemaker action potential. Several action potentials are shown. Stimulation of the vagus nerve produces a rise in diastolic transmembrane potential, a decrease in the rate of depolarization of the pacemaker potential, and a decrease in the duration of the action potential. [Hutter and Trautwein, 1956.]

has returned to a relaxed state. Thus, summation of contractions cannot occur in cardiac muscle.

Cardiac action potentials begin with a rapid depolarization that results from an increase in sodium conductance. Repolarization of the cell membrane is delayed while the membrane remains depolarized for hundreds of milliseconds (Figure 13-7). The long duration of the cardiac action potential produces a prolonged contraction, so that an entire chamber can fully contract before any portion begins to relax—a process that is essential for efficient pumping of blood.

The prolonged plateau results from a maintained high calcium conductance and a delay in the subsequent increase in potassium conductance (unlike the situation in skeletal muscle). A rapid repolarization terminates the plateau, apparently owing to a fall in calcium conductance and an increase in potassium conductance. The high $Ca^{2+}$ conductance during the plateau phase allows $Ca^{2+}$ to flow into the cell as the muscle contracts. This influx is especially important in lower vertebrates, in which much of the calcium essential for activation of contraction enters through the surface membrane. In birds and mammals, the surface/volume ratio of the larger cardiac cells is too small to allow sufficient entry of calcium to fully activate contraction. Therefore, most of the calcium is released during the plateau phase from the extensive sarcoplasmic reticulum characteristic of the hearts of higher vertebrates.

The duration of the plateau and the rates of depolarization and repolarization vary in different cells of the same heart (Figure 13-12). Atrial cells generally have an action potential of shorter duration than ventricular cells. The duration of the action potential in atrial or ventricular fibers from hearts of different species also varies. The duration of the action potential is one factor correlated with the maximum frequency of the heartbeat; in smaller mammals, heart rates are normally higher, and the duration of the ventricular action potential is shorter.

There is great diversity among the hearts of different invertebrate phyla; hence, little generalization can be made about the ionic mechanisms of invertebrate hearts other than that participation of $Ca^{2+}$ is widespread. For instance, bivalve mollusk hearts have a calcium action potential.

## TRANSMISSION OF EXCITATION OVER THE HEART

Activity initiated in the pacemaker region is conducted over the entire heart, depolarization in one cell resulting in the depolarization of neighboring cells by virtue of current flow through gap junctions. These junctions between cells are formed by the close apposition of large areas of neighboring cells. Area of contact is increased by folding and interdigitation of membranes. Gap junctions represent regions of low resistance between cells and allow current flow from one cell to the next across the *intercalated disks* (Figure 13-10). The extent of infolding and interdigitation increases during development of the heart and also varies among species (Figure 13-11).

Although the junctions between cells can conduct in both directions, transmission is usually unidirectional because it is initiated in and spreads only from the pacemaker region. There are usually several pathways for excitation of any single cardiac muscle fiber, as intercellular connections are numerous. If a portion of the heart becomes nonfunctional, the wave of excitation can easily flow around that portion, so that the remainder of the heart can still be excited. The prolonged nature of cardiac action potentials ensures that multiple connections do not result in multiple stimulation and a reverberation of activity in cardiac muscle. An action potential initiated in the pacemaker region results in a single action potential being con-

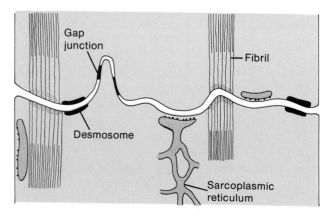

**13-10**  Diagram illustrating the features of a typical intercalated disk in the myocardium. [Navaratnam, 1980.]

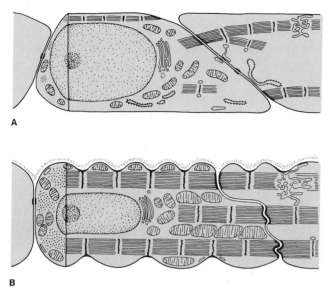

**13-11** Myocardial cells from (**A**) a mammalian embryonic heart and (**B**) a reptilian heart. [Hirakow, 1970.]

ducted through all the other cardiac cells, and another action potential from the pacemaker region is required for the next wave of excitation.

In the mammalian heart, the wave of excitation spreads from the sinus node over both atria in a concentric fashion at a velocity of about 0.8 m/s. The atria are connected electrically to the ventricles only through the atrioventricular node; other regions are joined by connective tissue that does not conduct the wave of excitation from the atria to the ventricles. Excitation spreads to the ventricle through small *junctional fibers,* in which the velocity of the wave of excitation is slowed to about 0.05 m/s. The junctional fibers are connected to nodal fibers, which in turn are connected via transitional fibers to the *bundle of His,* which consists of right and left bundles and covers the endocardium of the two ventricles. Conduction is slow through the nodal fibers (about 0.1 m/s) but rapid through the bundle of His (4–5 m/s). The bundle of His delivers the wave of excitation to all regions of the endocardium very rapidly, causing all the muscle fibers of the endocardium to contract synchronously. The wave of excitation is followed almost immediately by contractions of ventricular epicardial cells as the wave of excitation passes at a velocity of 0.5 m/s from

the endocardium to the epicardium. The functional significance of the electrical organization of the cardiac muscle cells is its ability to generate separate, synchronous contractions of the atria and the ventricles. Thus, slow conduction through the atrioventricular node allows atrial contractions to precede ventricular contractions and also allows time for blood to move from the atria into the ventricles.

Because of the large number of cells involved, the currents that flow during the synchronous activity of cardiac cells can be detected as small changes in potential from points all over the body. These potential changes—recorded as the *electrocardiogram*—are a reflection of electrical activity in the heart and can be easily monitored and then analyzed. A P-wave is associated with depolarization of the atrium, a QRS complex with depolarization of the ventricle, and a T-wave with repolarization of the ventricle (Figure 13-12). The exact form of the electrocardiogram is affected by the nature and position of recording electrodes as well as by the physiological state of the heart.

Various compounds alter the properties of cardiac muscle cells. As noted earlier, acetylcholine, released from cholinergic nerve fibers, increases the interval between action potentials in pacemaker cells and thus slows the heart rate (Figure 13-8). This process is sometimes referred to as a *negative chronotropic effect.* Parasympathetic cholinergic fibers in the vagus nerve innervate the sinus node and atrioventricular node of the verte-

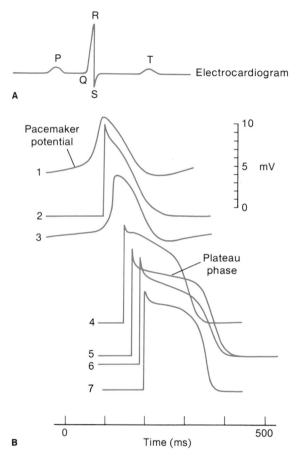

**13-12**    The electrocardiogram (**A**) and the underlying changes (**B**) recorded from individual cardiac muscle cells. P, atrial depolarization; QRS, ventricular depolarization; T, ventricular repolarization. The electrocardiogram arises from the electrical activity of the various parts of the heart represented by the intracellular signals shown in B. The potential changes were recorded from the following sites: (1) sinoatrial node, (2) atrium, (3) atrioventricular node, (4) bundle of His, (5) Purkinje fiber in a false tendon, (6) terminal Purkinje fiber, and (7) ventricular muscle fiber. Note the sequence of activation at the various sites, as well as the differences in the amplitude, configuration, and duration of the action potentials. [B from Hoffman and Cranefield, 1960.]

brate heart. As the heart rate slows, ACh also reduces the velocity of conduction from the atria to the ventricles through the atrioventricular node. High levels of ACh block transmission through the atrioventricular node, so that only every second or third wave of excitation is transmitted to the ventricle. Under these unusual conditions, the atrial rate will be two or three times that in the ventricle. Alternatively, high levels of ACh may block *conduction* through the atrioventricular node (*atrioventricular block*), giving rise to an ectopic pacemaker in the ventricle. The result is that the atria and ventricles are controlled by different pacemakers and contract at quite different rates, the two beats being uncoordinated.

Epinephrine and norepinephrine increase both the rate (termed a *positive chronotropic effect*) and force (termed a *positive inotropic effect*) of contraction of the heart. The effect of these catecholamines on rate of contraction is mediated via the pacemaker, whereas the increased strength of contraction is a general effect on all myocardial cells. Norepinephrine also increases conduction velocity through the atrioventricular node. It is released from adrenergic nerve fibers that innervate the sinus node, atria, atrioventricular node, and ventricle, so that sympathetic adrenergic stimulation has a direct effect on all portions of the heart. Metabolic actions of the catecholamines on cardiac cells are discussed in Chapter 11.

## Mechanical Properties of the Heart

The heart consists of a number of muscular chambers connected in series and guarded by valves or, in a few cases, sphincters (e.g., in some molluscan hearts), which allow blood to flow in only one direction. Contractions of the heart result in the ejection of blood into the circulatory system. A series of chambers permits step increases in pressure as blood passes from the venous to the arterial side of the circulation.

### CHANGES IN PRESSURE AND FLOW DURING A SINGLE HEARTBEAT

Contractions of the heart cause fluctuations in pressure and flow. Figure 13-13 shows the general plan of a mammalian heart. During diastole, closed aortic

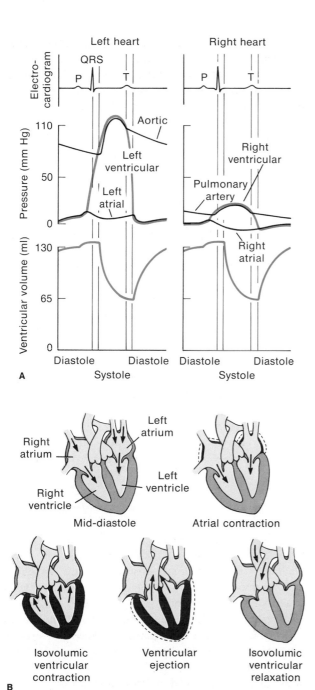

**13-13** (A) Changes in pressure and volume in the ventricles and aorta (left) and pulmonary artery (right) during a single cardiac cycle. [Vander et al., 1975.] (B) Black indicates the contracting ventricle; gray, the relaxed ventricle.

valves maintain large pressure differences between the relaxed ventricles and the systemic and pulmonary aortae. The atrioventricular valves are open, and blood flows directly from the venous system into the ventricles. When the atria contract, pressures rise in them, and blood is ejected from them into the ventricles. Then the ventricles begin to contract. Pressures rise in the ventricles and exceed those in the atria. The atrioventricular valves close and prevent the backflow of blood into the atria. Ventricular contraction proceeds; during this phase, both the atrioventricular and the aortic valves are closed, so that the ventricles form a sealed chamber and there is no volume change. That is, the ventricular contraction is "isometric." Pressure within the ventricle increases rapidly and eventually exceeds that in the aorta. The aortic valves open, and the volume of the ventricle decreases as blood is ejected into the aorta. The ventricle then begins to relax, intraventricular pressures fall below the pressure in the aorta, the aortic valves close, and there is an isometric relaxation of the ventricle. The atrioventricular valves open when ventricular pressure falls below that in the atria. Ventricular filling starts once more, and the cycle is repeated. In the mammalian heart, the volume of blood forced into the ventricle by atrial contraction is much less than the volume of blood ejected into the aorta by ventricular contraction. The ventricular filling is determined to a large extent by venous filling pressure; blood flows from the veins directly through the atria into the ventricles. Atrial contraction simply tops off the nearly full ventricles with blood.

Contraction of cardiac muscle is complex, and can be divided into two phases. The first is an isometric contraction during which tension in the muscle and pressure in the ventricle increase rapidly. The second phase is essentially isotonic; tension does not change very much, for as soon as the aortic valves open, blood is ejected rapidly from the ventricles into the arterial system with little increase in ventricular pressure. Thus, tension is generated first with almost no change in length, and then the muscle shortens with little change in tension; that is, during each contraction, cardiac muscle switches from an isometric to an isotonic contraction.

## WORK DONE BY THE HEART

It is a simple principle of physics that external work done is the product of force times distance moved. In the present context, work can be calculated as *pressure times flow*. Flow is directly related to the change in volume with each beat of the ventricle. Thus, a plot of pressure times volume for a single contraction of a ventricle yields a *pressure–volume loop* whose area (i.e., pressure × volume) is proportional to the external work done by that ventricle. In Figure 13-14, pressure–volume loops have been plotted for the right and left ventricles of a mammalian heart. The two ventricles eject equal volumes of blood, but the pressures generated in the pulmonary circuit are much lower; consequently, the external work done by the right ventricle is much less than that done by the left ventricle. Blood is ejected from the ventricle only when intraventricular pressures exceed the arterial pressure. If the arterial pressure is elevated, more external work must be done by the heart to raise the intraventricular pressure enough to maintain stroke volume at the original level. This, of course, means that there is an extra strain on the heart if blood pressure is high.

Not all energy expended by the heart will appear as changes in pressure and flow; some energy is expended to overcome frictional forces within the myocardium, and more will be dissipated as heat. The external work done by the heart, expressed as a fraction of the total energy expended, is termed the *efficiency of contraction*. The external work done can be determined from measurements of pressure and flow and converted into milliliters of $O_2$ consumed. This, in turn, can be expressed as a fraction of the total $O_2$ uptake by the heart in order to measure the efficiency of contraction. In fact, not more than 10–15% of the total energy expended by the heart appears as mechanical work.

Energy is expended to increase wall tension and raise blood pressure within the heart. According to *Laplace's Law,* the relationship between wall tension and pressure in a hollow structure is related to the radius of curvature of the wall. If the structure is a sphere, then

$$P = 2y/R$$

where $P$ is the transmural pressure (the pressure difference across the wall of the sphere), $y$ is the wall tension, and $R$ is the radius of the sphere. According to this relation, a large heart must generate twice the wall tension of a heart half its size to develop a similar pressure. Thus, more energy must be expended by larger hearts in developing pressure, and one might expect a larger ratio of muscle mass to total heart volume in these hearts. Hearts are not, of course, perfect spheres, but have a complex gross and microscopical morphology; nevertheless, Laplace's Law applies in general. The energy expended in ejecting a given quantity of blood from the heart will depend on

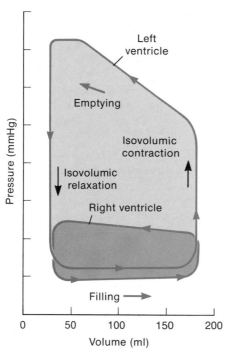

**13-14** Pressure–volume loops for the right and left ventricles of the mammalian heart. Once around a loop in a counterclockwise direction is one heartbeat. Ventricular filling occurs at low pressure; this pressure increases sharply only when the ventricles contract (the sharp upswing on the right-hand side of each loop). Ventricular volume decreases as blood flows into the arterial system, and ventricular pressure falls as the ventricle relaxes. Filling then begins again. Note that although the volume changes are similar, the pressure changes are much larger in the left ventricle than in the right one. Therefore, the left ventricle has a larger loop and hence does more external work than the right ventricle.

## Box 13-1   *Potential Energy, Pressure, and Kinetic Energy*

Contractions of the heart raise the potential energy of the blood in the ventricle. The rise in potential energy can be measured as an increase in blood pressure. Differences in pressure between two points in a flow path establish a pressure gradient and therefore the direction of flow—from high to low pressure. (An exception is a fluid at rest under gravity, where pressure increases uniformly with depth but flow does not occur.) Flowing blood has inertia, and energy is expended in setting the blood into motion; hence, fluids in motion possess kinetic energy. In static fluids, potential energy is measured in terms of pressure; in fluids in motion, however, potential energy is measured in terms of both pressure and kinetic energy. How does kinetic energy compare with pressure? The answer is that kinetic energy is usually negligible. The kinetic energy per milliliter of fluid is given by $\frac{1}{2}(QV^2)$, where $Q$ is the density of the fluid and $V$ the velocity of flow. If the velocity is measured in centimeters per second and density in grams per milliliter, then kinetic energy has the units of dynes per square centimeter, the same as pressure. The maximum velocity of blood flow occurs at the base of the aorta in mammals and is about 50 cm/s at the peak of ventricular ejection. The density of blood is about 1.055 g/ml. The kinetic energy of the blood in the aorta during peak ejection is $\frac{1}{2} \times 1.055 \times 50^2$, or 1320 dynes/cm$^2$. Since a pressure of 1 mmHg is equal to 1330 dynes/cm$^2$, the kinetic energy of the blood is close to 1 mmHg. This is small compared with peak systolic transmural pressures of around 120 mmHg. Blood velocity is low in the ventricle but accelerates as blood is ejected into the aorta; that is, blood gains kinetic energy as it leaves the ventricle. Pressure is converted into kinetic energy as blood is ejected from the heart, and this conversion accounts for most of the small drop in pressure that occurs between the ventricle and the aorta. Velocity (and therefore kinetic energy) is highest in the aorta. In the capillaries, the velocity is about 1 mm/s, and kinetic energy is virtually zero.

---

the efficiency of contraction, the pressures developed, and the size and shape of the heart. The cost of pumping blood in mammals is about 5–10% of the total energy they expend. Box 13-1 gives further consideration to the energetics of cardiovascular activity.

### STROKE VOLUME, HEART RATE, AND CARDIAC OUTPUT

*Cardiac output* is the volume of blood pumped per unit time from a ventricle. In mammals, it is defined as the volume ejected from the right or left ventricle, not the combined volume from both ventricles. The volume of blood ejected by each beat of the heart is termed the *stroke volume*. The mean stroke volume can be determined by dividing cardiac output by heart rate.

Stroke volume is the difference between the volume of the ventricle just before contraction (*end-diastolic volume*) and the volume of the ventricle at the end of a contraction (*end-systolic volume*). Changes in stroke volume may result from changes in either end-diastolic or end-systolic volume. The end-diastolic volume is determined by:

*1.* Venous filling pressure.

*2.* Pressures generated during atrial contraction.

*3.* Distensibility of the ventricular wall.

*4.* The time available for filling the ventricle.

The end-systolic volume is determined by:

*1.* The pressures generated during ventricular systole.

*2.* The pressure in the outflow channel from the heart (aortic or pulmonary artery pressure).

In 1914, E. H. Starling observed that increasing the venous filling pressure causes an increase in end-diastolic volume and results in an increased stroke volume from an isolated mammalian heart. End-systolic volume also increases, but not as much as the end-diastolic volume. Thus, cardiac muscle behaves in a way similar to that of skeletal muscle in that stretch of the relaxed muscle within a certain range of lengths results in the development of increased tension during a contraction. Starling also observed that increases in arterial pressure cause a rise in both end-diastolic and end-systolic volume with little change in stroke volume. In this instance, the increased mechanical work required to maintain stroke volume in the face of an elevated arterial pressure results from the increased stretch of cardiac muscle during diastole.

Otto Frank had previously derived a length-tension relationship for frog myocardium and had demonstrated that contractile tension increases with stretch up to a maximum and then decreases with further stretch. Although neither Starling nor Frank considered mechanical work, the increase in mechanical work from the ventricle caused by an increase in end-diastolic volume (or venous filling pressure) is termed the *Frank–Starling mechanism.* The curves derived from measuring work output from the ventricle at different venous filling pressures are described as *Starling curves* (Figure 13-15).

No single Starling curve, however, describes the relationship between venous filling pressure and work output from the ventricle. The mechanical, as well as electrical, properties of the heart are affected by a number of factors, including the level of activity in nerves innervating the heart and the composition of blood perfusing the myocardium. For instance, the relationship between ventricular work output and venous filling pressure is markedly affected by stimulation of sympathetic nerves innervating the heart (Figure 13-15).

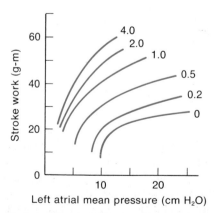

**13-15** Starling curves indicating the relationship between stroke work and venous filling pressure (left atrial mean pressure) at different levels of sympathetic nerve stimulation. The numbers indicate the stimulating frequency in hertz. [Sarnoff and Mitchell, 1962.]

Epinephrine and norepinephrine released from sympathetic nerves increase the force of contraction of the ventricle; hence both the rate and the extent of ventricular emptying are increased by these catecholamines. The effects of cholinergic (i.e., vagus) nerve activity on the rate and force of ventricular power output during each beat are much less marked than the effects of adrenergic sympathetic nerves. This difference in effects stems from a cholinergic innervation that is much smaller than the extensive adrenergic innervation of the ventricles.

The effects of sympathetic stimulation represent a series of integrated actions. Heart rate is increased by the action of sympathetic nerves on the membranes of pacemaker cells. Conduction velocity over the heart is increased to produce a more nearly synchronous beat of the ventricle. There are increases both in the rate of production of ATP and in the rate of conversion of chemical energy to mechanical energy, leading to an increase in ventricular work. As a result, the rate of ventricular emptying increases during systole, so that the same or a larger stroke volume is ejected in a much shorter time. Thus, when adrenergic nerve stimulation increases the heart rate, the same stroke volume is ejected from the heart in a shorter time. So, although the time available for emptying and filling the heart is reduced as heart rate increases, stroke volume remains relatively unchanged over a wide range of heart rates. For example, exercise in mammals is associated with a large increase in heart rate with little change in stroke volume; only at the highest heart rates does the stroke volume fall (Figure 13-16). This situation occurs because, over most of the range of heart rates, increased sympathetic activity ensures more rapid ventricular emptying, and elevated venous pressures result in more rapid filling as heart rate increases. There are limits, however, to which diastole can be shortened. These limits involve not only the maximum possible rate at which the ventricles can be filled and emptied but also the nature of the coronary circulation. Contractions of the heart occlude coronary capillaries, so the flow is very reduced during systole; but flow rises dramatically during diastole. A decrease in the diastolic period tends to reduce the period of coronary blood flow and therefore the nutrition of the heart.

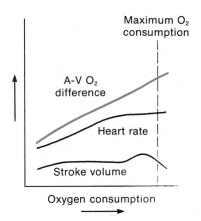

**13-16** Steady-state heart rate, stroke volume, and arterial—venous oxygen difference plotted against oxygen consumption associated with different levels of exertion in normal human subjects. Increases in cardiac output are associated with increases in heart rate rather than stroke volume except at very high levels of oxygen consumption, when heart rate levels off and stroke volume increases. [Rushmer, 1965b.]

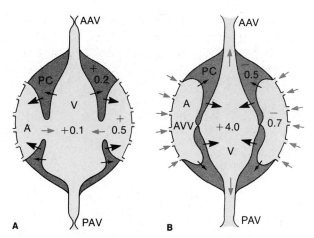

**13-17** The heart of the bivalve mollusk *Anodonta*, showing blood flow and pressure changes in the heart and pericardial cavity during (**A**) ventricular diastole and (**B**) ventricular systole. Pressures are in centimeters of seawater and are expressed relative to ambient pressures. Large black arrows indicate the movements of the walls of contracting chambers; small arrows indicate movements of walls of relaxing chambers. The colored arrows indicate the direction of blood flow. A, atrium; AAV, anterior aortic valve; AVV, atrioventricular valve; PC, pericardial cavity; PAV, posterior aortic valve; V, ventricle. [Brand, 1972.]

As we have noted, increases in cardiac output with exercise are associated with changes in heart rate rather than stroke volume in mammals (Figure 13-16). Following sympathetic denervation of the heart, exercise results in similar increases in cardiac output, but in this instance there are large changes in stroke volume rather than in heart rate. The increases in cardiac output are probably caused by an increase in venous return. The sympathetic nerves are not involved in increasing cardiac output per se but rather in raising heart rate and maintaining stroke volume, avoiding the large pressure oscillations associated with large stroke volumes and keeping the heart operating at or near its optimal stroke volume for efficiency of contraction. The sympathetic nerves therefore play an important role in determining the relationship between heart rate and stroke volume, but additional factors are involved in mediating the increase in cardiac output with exercise.

## The Pericardium

The heart is contained in a *pericardial cavity* and is surrounded by a *pericardial membrane* (or *pericardium*). The magnitude of the pressure changes within the *pericardial sac* depend on the rigidity of the pericardium and on the magnitude and rate of change of the heart volume. The membrane may be thin and flexible, as in mammals, in which case pressure changes that occur in the pericardial cavity during each heartbeat are negligible. The pericardium enveloping the mammalian heart is formed of two layers, an outer fibrous layer and an inner serous layer, the latter lining the pericardial cavity (Figure 13-6). In mammals, the serous layer secretes a pericardial fluid that acts as a lubricant, facilitating movement of the heart.

In elasmobranchs, lungfishes, and bivalve mollusks, contractions of the heart within a noncompliant (i.e., rigid) pericardium cause oscillations of pressure within the pericardial cavity. In the heart of bivalve mollusks (Figure 13-17) and crustaceans, contractions of the ventricle reduce pressure in the pericardial cavity and enhance flow into the atria from the venous system. Thus, tension generated in the ventricular wall

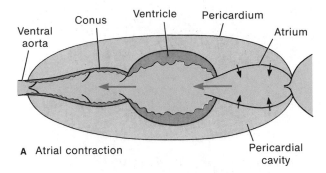

**A** Atrial contraction

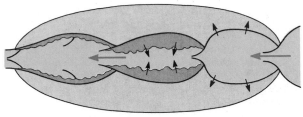

**B** Ventricular contraction

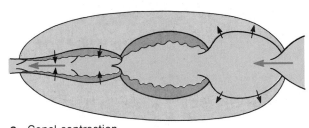

**C** Conal contraction

**13-18** The elasmobranch heart is contained in a noncompliant pericardium. Contractions of the ventricle create subatmospheric pressures in the pericardial cavity and assist atrial filling.

is utilized both to eject blood into the arterial system and to draw blood into the atria from the venous system.

The elasmobranch heart consists of three chambers—atrium, ventricle, and conus—all contained within a pericardium (Figure 13-18). The reduction in intrapericardial pressure during ventricular systole in elasmobranchs produces a suction that helps expand the atrium and thereby increases venous return to the heart. If the pericardial cavity is opened, cardiac output is reduced; hence, the increased venous

return to the atrium caused by reduced pericardial pressure is important in augmenting cardiac output. Contractions of the ventricle affect the conus as well as the atrium. The exit from the ventricle to the conus is guarded by a pair of flap valves, and there are from two to seven pairs of valves along the length of the conus. Just before ventricular systole, all valves except the set most distal to the ventricle are open. The conus and the ventricle are interconnected, but a closed valve at the exit of the conus maintains a pressure difference between the conus and the ventral aorta. During atrial systole, both the ventricle and the conus are filled with blood. Ventricular systole in elasmobranchs does not have an isovolumic phase, as in mammals, because at the onset of contraction blood is moved from the ventricle into the conus. Pressure rises in the ventricle and conus and eventually exceeds that in the ventral aorta. The distal valves open, and blood is ejected into the aorta. Conal systole begins after the onset of ventricular systole. During conal systole, the proximal valves close, preventing reflux of blood into the ventricle as it relaxes. Conal systole proceeds relatively slowly away from the heart toward the aorta; each set of valves closes, in turn, to prevent backflow of blood. Conus length is variable among species; in general, more valves are found in those species with a longer conus.

## Functional Morphology of the Vertebrate Heart

The structure of the heart varies in different vertebrates, and a comparative analysis of vertebrate circulatory systems produces insights into the relationships between heart structure and function.

The pulmonary (respiratory) circulation of birds and mammals is maintained by much lower pressures than is the systemic circulation because they have two series of heart chambers in parallel. The left side of the heart ejects blood into the systemic circuit, and the right side ejects blood into the pulmonary circulation (Figure 13-4). The advantage of a high blood pressure is that rapid transit times and sudden changes in flow can be readily achieved for blood passing through small-diameter capillaries. High transmural pressures, however, result in filtration of fluid across the capillary wall and therefore require an extensive lymphatic drainage of the tissues. In the mammalian lung, capillary flow can be maintained by relatively low input pressures, reducing the requirements for lymphatic drainage and avoiding the formation of large extracellular spaces that could increase diffusion distances between blood and air and impair the gas-transfer capacity of the lung. The advantage of a divided heart, like that of mammals, is that blood flow to the body and the lungs can be maintained by different input pressures. The disadvantage of a completely divided heart is that in order to avoid shifts in blood volume from the systemic to the pulmonary circuit, or vice versa, cardiac output must be the same in both sides of the heart, independent of the requirements in the two circuits. Lungfishes, amphibians, reptiles, bird embryos, and fetal mammals have either an undivided ventricle or some other mechanism that allows the shunting of blood from one circulation to the other. These shunts usually result in the movement of blood from the right (respiratory, pulmonary) to the left (systemic) side of the heart during periods of reduced gas transfer in the lung. Blood returning from the body, instead of being pumped to the lung, is shunted from the right to the left side of the heart and once again ejected into the systemic circuit, bypassing the lungs. A single undivided ventricle permits variations in the ratio of flows to the pulmonary and systemic circuits, but the same pressures must be developed on both sides of the heart.

### BONY FISHES

In fishes, the heart consists of four chambers in series (Figure 13-19). All chambers except the elastic bulbus are contractile. A unidirectional flow of blood through the heart is maintained by valves at the sinoatrial and atrioventricular junctions and at the exit of the ventricle. Blood pumped by the heart passes first through the gill (respiratory) circulation and then into a dorsal aorta that supplies the rest of the body (systemic circulation).

### AMPHIBIANS

The atria are completely subdivided in amphibians, but there is a single ventricle (Figure 13-20), whereas in reptiles there is either partial or complete separation

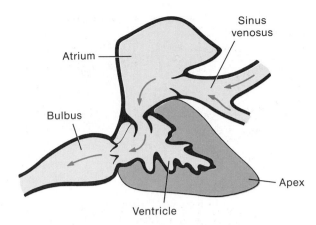

**13-19** The trout heart. [Randall, 1968.]

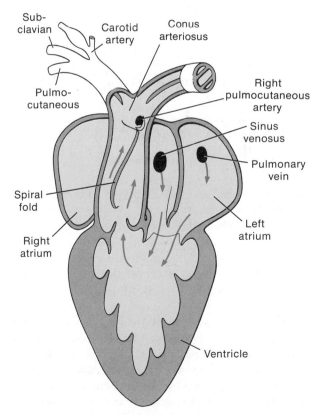

**13-20** Ventral view of the internal structure of the frog heart. [Goodrich, 1958.]

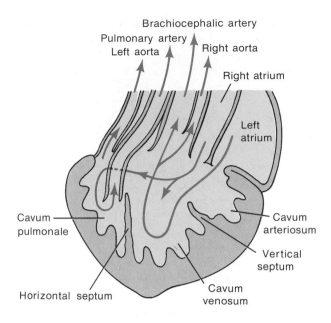

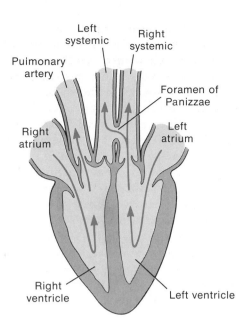

**13-21**  Diagrammatic illustration of a turtle heart seen from the ventral side. The cavum pulmonale lies ventral to the cavum venosum. The arrows indicate the pattern of blood flow through the various chambers in this incompletely divided ventricle. Blood entering the cavum pulmonale from the right atrium flows behind the outflow tract of the pulmonary artery. [Shelton and Burggren, 1976.]

of the ventricle into two chambers (Figures 13-21 and 13-22).

In the frog, there is a separation of blood within the heart even though the ventricle is undivided. Blood from the lungs and skin is preferentially directed toward the body, whereas deoxygenated blood from the body is directed toward the pulmocutaneous arch. This separation of oxygenated and deoxygenated blood is aided by a spiral fold within the conus arteriosus of the heart (Figure 13-20). Deoxygenated blood leaves the ventricle first during systole and enters the lung circulation. Pressures rise in the pulmocutaneous arch and become similar to those in the systemic arch. Blood flows into both arches, the spiral fold partially dividing the systemic and pulmocutaneous flows within the conus arteriosus.

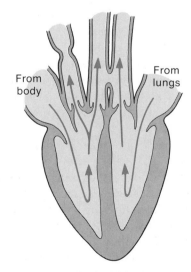

**13-22**  Diagrammatic view of the crocodilian heart. (**A**) Crocodile breathing air. The left systemic arch is closed off from the right ventricle, which ejects blood into the pulmonary artery. (**B**) Crocodile submerged. The right ventricle ejects blood into both the pulmonary artery and the left systemic arch. There is an increase in resistance to flow in the pulmonary circuit and a rise in pulmonary and right ventricular pressure.

The volume flow to the lungs or body is inversely related to the resistance of the two circuits to blood flow. Immediately following a breath, the resistance to blood flow through the lung is low and blood flow is high; between breaths, resistance gradually increases and is associated with a fall in blood flow. These oscillations in lung blood flow are possible because of the partial division of the heart, which, while directing deoxygenated blood toward the pulmocutaneous arch, allows adjustment of the ratio of lung to systemic blood flow. That is, when the animal is not breathing, blood flow to the lungs can be reduced and most of the blood pumped by the ventricle can be directed toward the body. When the animal is breathing, a more even distribution of flow to the lungs and body can be maintained. This distribution is possible only if the ventricle is not completely divided into right and left chambers (as in mammals).

## REPTILES

Noncrocodilian reptiles have a partially divided ventricle, whereas crocodilian reptiles have separate right and left ventricles. All reptiles have both right and left systemic arches.

In turtles, lizards, and snakes, the ventricle is partially subdivided by an incomplete muscular septum referred to as the *horizontal septum, Muskelleiste,* or *muscular ridge.* This horizontal septum separates the *cavum pulmonare* from the *cavum venosum* and *cavum arteriosum,* which in turn are partially separated by the *vertical septum* (Figure 13-21). The right atrium contracts slightly before the left atrium does and ejects deoxygenated blood into the cavum pulmonare across the free edge of the horizontal septum. Blood from the left atrium fills the cavum venosum and cavum arteriosum (Figure 13-21). Blood in the cavum pulmonare is ejected into the pulmonary artery, but the cavum venosum and cavum arteriosum empty into the systemic arches.

Measurements support the view that oxygenated blood from the left atrium passes into the systemic circuit, whereas deoxygenated blood from the right atrium passes into the pulmonary artery. Pulmonary artery diastolic pressure is often lower than systemic diastolic pressure. As a result, the pulmonary valves

open first when the ventricle contracts. Thus, flow occurs earlier in the pulmonary artery than in the systemic arches during each cardiac cycle. In turtles, there may be some recirculation of arterial blood in the lung circuit; that is, there is a left-to-right shunt within the heart. The ventricle remains functionally undivided throughout the cardiac cycle, and the relative flow to the lungs and systemic circuits is determined by the resistance to flow in each part of the circulatory system. When the turtle breathes, resistance to flow through the pulmonary circulation is low, and blood flow is high. When it does not breathe, as during a dive, pulmonary vascular resistance increases, but systemic vascular resistance decreases, resulting in a right-to-left shunt and a decrease in pulmonary blood flow. As in many other animals, there is a reduction in cardiac output associated with a marked slowing of the heart (bradycardia) during a dive.

Crocodilian reptiles have a completely divided ventricle. There are also two systemic arches, the left arising from the right ventricle, the right systemic from the left ventricle. The systemic arches are connected via the *foramen of Panizzae.* When the reptile is breathing normally, the impedance to blood flow through the lungs is low, and pressures generated by the right ventricle are lower than those generated by the left ventricle. Blood is pumped by the left ventricle into the right systemic arch and into the left systemic arch via the foramen of Panizzae (Figure 13-22). Pressures in the left systemic arch remain higher than the pressure in the right ventricle; consequently, the valves at the base of the left systemic arch remain closed throughout the cardiac cycle. All blood ejected from the right ventricle passes into the pulmonary artery and flows to the lungs. Thus, the crocodilian reptile is functionally the same as the mammal in that there is complete separation of systemic and pulmonary blood flow. The reptile, however, has the added capacity to shunt blood between the two circuits when it stops breathing and dives under water (Figure 13-22). During this period, there is an increase in the resistance to flow in the pulmonary circuit, which results in a rise in right ventricular pressure. The increase in pulmonary resistance is probably due to a constriction of the pulmonary artery and the pulmonary outflow tract and, in some species, to vasoconstriction of vessels within the lungs. Peak right ventricular pressure becomes equal to left ventricular pressure and exceeds left systemic pressure. As a result, the valves at the base of the left systemic arch open, and blood from the right ventricle is ejected into both the pulmonary and the systemic circulation (Figure 13-22). Thus, a portion of the deoxygenated blood returning to the heart from the body is recirculated in the systemic circuit during apnea (period of no breathing).

### MAMMALIAN FETUS

At birth, mammals shift from a placental to a pulmonary circulation, a process that involves several central cardiovascular readjustments. The lungs of the mammalian fetus are collapsed, presenting a high resistance to blood flow. In the fetus, the pulmonary artery is joined to the systemic arch via a short, but large-diameter blood vessel, the *ductus arteriosus*. This vessel closes at birth, separating the pulmonary and systemic circulations. In the fetus most of the blood ejected by the right ventricle is returned to the systemic circuit via the ductus arteriosus (Figure 13-23), blood flow through the pulmonary circulation is reduced, and there is a marked right-to-left shunt; that is, blood flows from the pulmonary to the systemic circuit. At birth, the lungs are inflated, reducing the resistance to flow in the pulmonary circuit. Blood ejected from the right ventricle passes into the pulmonary vessels, resulting in an increased venous return to the left side of the heart. At the same time, the placental circulation disappears, and the resistance to flow increases markedly in the systemic circuit. Pressures in the systemic circuit rise above those in the pulmonary circulation and, if the ductus arteriosus fails to close after birth, result in a left-to-right shunt; that is, blood flows from the systemic to the pulmonary circuit. Generally, the ductus arteriosus becomes occluded, and blood flow through the ductus does not persist. If the ductus remains open after birth, however, blood flow to the lungs exceeds systemic flow, because a portion of the left ventricular output passes via the ductus arteriosus into the pulmonary artery and to the lung. Under these circumstances, systemic flow is often normal, but blood flow to the lungs may be twice the systemic flow, and cardiac output from the left ventricle may exceed that from the right ventricle by a factor of two. The result is a

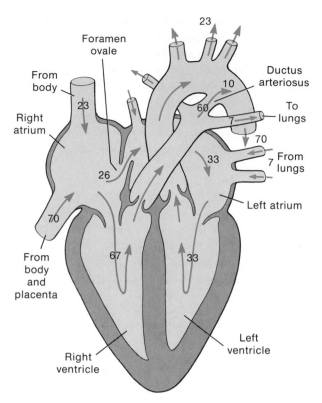

**13-23**  Blood flow in the mammalian fetal heart. Blood returning from the placenta is shunted from the right to the left atrium through the foramen ovale. The numbers refer to the relative flows of blood to and from different regions of the body.

marked hypertrophy of the left ventricle. The work done by the left ventricle during exercise is also much greater than normal, and the capacity to increase output is limited. As a result, the maximum level of exercise is much reduced if the ductus arteriosus remains open after birth. Furthermore, blood pressure in the lungs is increased, leading to a greater fluid loss across lung capillary walls and to possible pulmonary congestion.

Fetal blood is oxygenated in the placenta and mixed with the blood returning from the lower body via the inferior vena cava, a vein that in turn empties into the right atrium. In the interatrial septum is a hole, the foramen ovale, that is covered by a flap valve and permits flow from the right to the left atrium. The enlarged valve of the inferior vena cava directs blood into the left atrium through the foramen ovale. Oxygenated blood is then pumped from the left atrium into the left ventricle and ejected into the aorta, whence it flows to the head and upper limbs. Deoxygenated blood in the right atrium is preferentially directed toward the right ventricle, whence it flows into the systemic circuit via the ductus arteriosus. At birth, pressures in the left atrium exceed the pressure in the right atrium; as a result, the foramen ovale closes, and its position is later indicated by a permanent depression.

### BIRD EMBRYO

Oxygenated blood from the *chorioallantois* (a series of vessels containing blood that takes up oxygen diffusing through the eggshell) and deoxygenated blood from the head and body enter the right atrium of the bird embryo heart. Oxygenated blood from the chorioallantoic circulation passes from the right into the left atrium through several large and numerous small holes in the interatrial septum. The oxygenated blood is then pumped into the left ventricle and ejected into the aorta, whence it flows to the head and body. In the adult bird, the pulmonary and systemic circulations are completely separated, as in mammals.

## Hemodynamics

As we have noted, contractions of the heart generate blood flow through vessels—arteries, capillaries, and veins—that form the circulatory system. Before examining the properties of these vessels in detail, it is necessary to discuss the general patterns of blood flow in these vessels and the relationship between pressure and flow in the circulatory system. In vertebrates and other animals with a closed circulation, the blood flows in a continuous circuit. Since fluids are incompressible, blood pumped by the heart must cause flow of an equivalent volume in every other part of the circulation. That is, at any one time the same number of liters per minute flows through the arteries, the capillaries, and the veins. Furthermore, unless there is a change in total blood volume, a reduction in volume in one part of the circulation must lead to an increase in volume in another part. The velocity of flow at any point is related not to the proximity of the heart but

to the total cross-sectional area of that part of the circulation. The cross-sectional area refers not to that of a single artery or capillary but to the sum of the cross sections of all capillaries or arteries at that point in the circulation. Just as the velocity of water flow increases where a river narrows, so in the circulation the highest velocities of blood flow occur where the total cross-sectional area is smallest (and the lowest velocities occur where the cross-sectional area is largest). The arteries have the smallest total cross-sectional area, whereas the capillaries have by far the largest. Thus, the highest velocities occur in the aorta and pulmonary artery in mammals; then velocity falls markedly as blood flows through the capillaries, but it rises again as blood flows through the veins (Figure 13-24). The

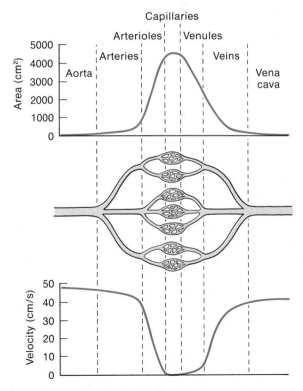

**13-24**  Blood velocity is inversely proportional to the cross-sectional area of the circulation at any given point. Blood velocity is highest in the arteries and veins and lowest in the capillaries; the converse is true for the cross-sectional area. [Feigl, 1974.]

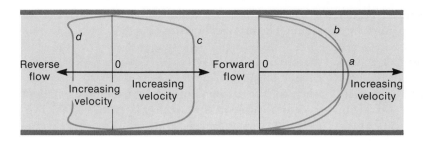

**13-25**  Velocity profiles for fluids flowing in tubes. Curve *a:* continuous laminar flow, plasma. Curve *b:* continuous laminar flow, blood. Curve *c:* pulsatile laminar flow, blood at peak of forward motion. Curve *d:* pulsatile laminar flow, blood at peak of flow reversal.

slow flow of blood in capillaries is of functional significance, because it is in capillaries that the time-consuming exchange of substances between blood and tissues takes place.

### LAMINAR AND TURBULENT FLOW

In many portions of the circulation, blood flow is streamlined, or laminar. *Laminar flow* is characterized by a parabolic velocity profile across the vessel (Figure 13-25). Flow is zero at the wall and maximum along the axis of the vessel. The blood adjacent to the vessel wall does not move, but the next layer of fluid slides over this layer, and so on, each successive layer moving at an increasingly higher velocity, with the maximum at the center of the vessel. A pressure difference supplies the force required to slide adjacent layers past each other, and viscosity is a measure of the resistance to sliding between adjacent layers of fluid. An increase in viscosity will require a larger pressure difference to maintain the same rate of flow.

Large arteries have a much more complex flow profile than do smaller vessels, in which continuous laminar flow dominates. In large arteries, blood is first accelerated and then slowed with each heartbeat; in addition, since the vessel walls are elastic, they expand and then relax as pressure oscillates with each heartbeat. The end result is that the velocity across large arteries has a much flatter profile (Figure 13-25) than the velocity across blood vessels with a less oscillating but laminar flow of blood.

*Turbulent flow* can occur at certain points in the circulation but is not common in the peripheral circulation. In turbulent flow, fluid moves in all directions at once, and much more energy is expended in moving fluid through a vessel under these conditions. Laminar flow is silent; turbulent flow noisy. In the bloodstream,

turbulence causes vibrations that produce the sounds of the circulation, and these sounds can be localized to indicate points of turbulence. Sounds can be heard in vessels when blood velocity exceeds a certain critical value and in heart valves when they open and close. The *Reynolds number (Re)*† is an empirically derived value that indicates whether flow will be laminar or turbulent under a particular set of conditions. In smooth vessels of the circulation, flow will be turbulent if the *Re* is greater than about 1000. Small back eddies may form at arterial branches and, like the back eddies in rivers, can become detached from the main flow regime and be carried downstream as small, discrete regions of turbulence. These eddies can form in the circulation at *Re*'s as low as 200. The *Re* is directly proportional to the flow rate, $\dot{Q}$ (in milliliters per second), and inversely proportional to the inside radius of the vessel, *r* (in centimeters), and the *kinematic viscosity, K,* of the blood:

$$Re = \frac{2\dot{Q}}{\pi r K}$$

The kinematic viscosity is the ratio of viscosity to density. The larger the kinematic viscosity, the less the likelihood that turbulence will occur. The relative viscosity, and therefore the kinematic viscosity, increases with *hematocrit* (volume of red blood cells per unit volume of blood), so that the presence of red blood cells decreases the occurrence of turbulence in the bloodstream. In general, blood velocity is seldom high

---

†In Chapter 16, we see that the Reynolds number is important in the energetics of locomotion through fluids. For that purpose, a somewhat different formulation of the *Re* will be necessary to deal with the movement of an object through an infinite volume of fluid.

enough to create turbulence, except during the high blood flows associated with maximum exercise. The highest flow rates in the mammalian circulation are in the proximal portions of the aorta and pulmonary artery, and turbulence may occur distal to the aortic and pulmonary valves at the peak of ventricular ejection or during backflow of blood as these valves close.

### PRESSURE AND FLOW: POISEUILLE'S EQUATION

Pressures generated by the heart are dissipated by the flow of blood, and they fall as the blood passes from the arterial to the venous side of the circulation (Figure 13-26).

The relationship between pressure and steady laminar flow of fluid in a rigid tube is described by *Poiseuille's Law,* which states that the flow rate of a fluid, $\dot{Q}$, is directly proportional to the pressure difference, $P_1 - P_2$, and the fourth power of the radius of the tube, $r$, and inversely proportional to the tube length, $L$, and fluid viscosity, $\eta$:

$$\dot{Q} = \frac{(P_1 - P_2)\pi r^4}{8L\eta}$$

As $\dot{Q}$ is proportional to $r^4$, very small changes in $r$ will have a profound effect on $\dot{Q}$. A doubling of vessel diameter will lead to a 16-fold increase in flow if the pressure difference ($P_1 - P_2$) along the vessel remains unchanged. Similarly, halving the vessel diameter will cause a 16-fold increase in resistance and a marked reduction in flow.

Poiseuille's equation applies to steady flows in straight rigid tubes. Blood pressure and flow are pulsatile, and blood is a complex fluid consisting of plasma and cells. Since the blood vessel walls are not rigid, the oscillations in pressure and flow are not in phase; consequently, the relationship between the two is no longer accurately described by Poiseuille's equation. The extent of the deviation of the relationship between pressure and flow from that predicted by Poiseuille's equation is indicated by the value of a non-dimensional constant $\alpha$:

$$\alpha = r\frac{\sqrt{2\pi n f \rho}}{\eta}$$

where $\rho$ and $\eta$ are the density and viscosity of the fluid, respectively, $f$ is the frequency of oscillation, $n$ is

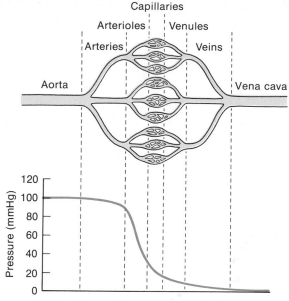

**13-26** Changes in mean pressure in different parts of the circulation. The major pressure decrease occurs in the arterioles. [Feigl, 1974.]

the order of the harmonic component, and $r$ is the radius of the vessel. If $\alpha$ is 0.5 or less, the relationship between pressure and flow is described by Poiseuille's equation. Values of $\alpha$ for the arterial systems of mammals and birds range from 1.3 to 16.7, depending on the species and the physiological state of the animal. Since most values of $\alpha$ for the arterial system are around 6, Poiseuille's equation is not applicable. The value for $\alpha$ in the small terminal arteries and veins is about 0.5, so this equation can be used to analyze the relationship between pressure and flow in this portion of the circulation. Flow in the terminal arteries therefore will be proportional to the pressure difference along the vessel and the fourth power of the radius and inversely proportional to the length of the vessel and the viscosity of blood.

### RESISTANCE TO FLOW

Since it is often difficult or impossible to measure the radii of all vessels in a vascular bed, we designate $8L\eta/\pi r^4$, the inverse of the term in Poiseuille's equation, as the resistance to flow, $R$, which is equal to the pressure difference $(P_1 - P_2)$ across a vascular bed divided by the flow rate, $\dot{Q}$:

$$R = \frac{P_1 - P_2}{\dot{Q}} = \frac{8L\eta}{\pi r^4}$$

The resistance to flow in the peripheral circulation is sometimes expressed in resistance units or PRUs (*peripheral resistance units*), 1 PRU being equal to the resistance in a vascular bed when a pressure difference of 1 mmHg results in a flow of 1 ml/s; that is, 1 PRU = 1 mmHg/cm³/s, and since 1 mmHg equals 1330 dynes/cm², then 1 PRU = 1330 dyne/s/cm⁵.

Blood flow is related not only to the pressure difference along a vessel but also to the resistance to flow, which is inversely proportional to the fourth power of the radius of the vessel. As pressure increases in an elastic vessel, so does the radius; as a result, flow increases as well. Let us consider a blood vessel with a constant pressure differential along its length but operating at different pressure levels—for example, input pressure 100 mmHg, outflow pressure 90 mmHg, and input pressure 20 mmHg, outflow pressure 10 mmHg. The flow rate in this vessel will be much greater at the higher pressure, if the vessel is distensible, simply because the radius will be increased and the resistance to flow reduced.

## VISCOSITY OF BLOOD

Changes in viscosity have a marked effect on the flow of blood. The presence of red blood cells alters the viscous properties of blood. Plasma has a viscosity relative to water of about 1.8; the addition of red blood cells increases the relative viscosity, so that mammalian and bird blood at 37°C has a relative viscosity between 3 and 4. Thus, owing largely to the presence of red blood cells, blood behaves as though it were three or four times more viscous than water. This characteristic means that larger pressure gradients are required to maintain the flow of blood through a vascular bed than would be needed if the vascular bed were perfused by plasma alone. Blood flowing through small tubes behaves as if the relative viscosity were much reduced. In fact, in tubes less than 1 mm in diameter

(i.e., capillaries), the relative viscosity of blood decreases with diameter and approaches the viscosity of plasma.

The velocity profile across a vessel with steady laminar fluid flow is parabolic (Figure 13-25). Maximum velocity is twice the mean velocity, which can be determined by dividing the flow rate by the cross-sectional area of the tube. The rate of change in velocity is maximal near the walls and decreases toward the center of the vessel. In flowing blood, red cells tend to accumulate in the center of the vessel, where velocity is highest but the rate of change in velocity between adjacent layers smallest. This accumulation leaves the walls relatively free of cells, so that fluid flowing from this area into small side vessels will have a low level of red blood cells and consist almost entirely of plasma. Such a process is referred to as *plasma skimming*.

The accumulation of red blood cells in the center of a bloodstream means that blood viscosity is highest in the center and decreases toward the walls, because viscosity of blood increases with hematocrit (volume percentage of red blood cells per unit volume of blood). Flow is inversely related to viscosity, and a change in viscosity between the center and the walls of the bloodstream will alter the velocity profile. In blood whose viscosity is lowest at the walls, the effect will be to increase flow slightly at the walls and reduce flow slightly in the center—that is, slightly flatten the parabolic shape of the velocity profile (Figure 13-25).

As noted earlier, the velocity profile is a parabola if flow is laminar and steady. If flow is turbulent, there is little change in velocity across the vessel. If flow is laminar but pulsatile, as in arteries, the velocity profile is also flattened, so that blood velocity is similar across the core of the bloodstream (Figure 13-25).

## COMPLIANCE IN THE CIRCULATORY SYSTEM

A further complication in analyzing the relationship between pressure and flow in the circulation is that blood vessels are not straight, rigid tubes, but contain elastic fibers that enable them to distend. As pressure increases, the walls are stretched and the volume of the vessel is enlarged. The ratio of change in volume to change in pressure is termed the *compliance*, or *capacitance*, of the system. The compliance of a system is related to its size and the elasticity of its walls. The larger the initial volume and the elasticity of the walls, the greater will be the compliance of the system.

The venous system is very compliant; small changes in pressure can produce large changes in volume. The venous system can therefore act as a *volume reservoir,* because large changes in volume have little effect on venous pressure (and therefore the filling of the heart during diastole or capillary blood flow). The arterial system is less compliant and therefore acts as a *pressure reservoir* in order to maintain capillary blood flow. Nevertheless, the portions of the arterial system near the heart *are* elastic in order both to dampen the pressure pulse generated by contractions of the heart and to maintain flow in distal arteries during diastole.

## The Arterial System

The arterial system consists of a series of branching vessels well suited to deliver blood from the heart to the fine capillaries that carry blood through the tissues. Arterial walls are thick, elastic, and muscular. All blood vessels are lined by *endothelial cells.* Capillaries consist of only an endothelial layer; the larger vessels have a surrounding layer of elastic and collagenous fibers. Circular and longitudinal smooth muscle fibers may intermingle with or surround the elastic and collagenous fibers. Blood vessels are covered by a limiting fibrous outer coat, the *tunica adventitia* (Figure 13-27). The endothelial layer and the elastic fibers form the *tunica intima,* and the circular and longitudinal smooth muscle form the *tunica media.* The boundary between the tunica intima and the tunica media is not well defined; the tissues blend into one another. Owing to increased muscularization, arteries have a thickened tunica media, and the larger arteries close to the heart are more elastic, with a wide tunica intima. The thick walls of blood vessels require their own capillary circulation, termed the *vasa vasorum.*

In general, arteries have thicker walls and much more smooth muscle than veins of similar outside diameter. In some veins, muscular tissue is absent.

Elastic tubes are unstable and tend to balloon. In blood vessels, this instability is prevented by a collagen sheath that limits the expansion of the vessel. Balloon-

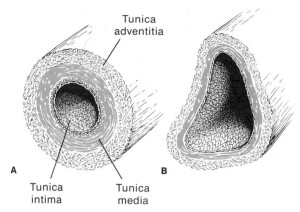

13-27 A diagram constructed to emphasize schematically the contrast between the cross-sectional appearance of an artery (**A**) and its corresponding vein (**B**). [Le Gros Clark, 1952.]

ing of blood vessels (aneurisms) can occur if the collagen sheath breaks down. This situation is possible because, according to Laplace's Law, the wall tension required to maintain a given pressure within a tube increases with increasing radius.

The arteries serve four main functions (Figure 13-28):

*1.* To act as a conduit for blood between the heart and capillaries.

*2.* To act as a pressure reservoir for forcing blood into the small-diameter capillaries.

*3.* To dampen oscillations in pressure and flow generated by the heart and produce a more even flow of blood into the capillaries.

*4.* To control distribution of blood to different capillary networks via selective constriction of the terminal branches of the arterial tree, the arterioles (to be described later).

Arterial blood pressure is finely controlled. The pressure is determined by the volume of blood the arterial system contains and the properties of its walls. If either is altered, the pressure will change. The volume of blood in the arteries is the resultant of filling by contractions of the heart and emptying via arterioles into capillaries. If cardiac output increases, arterial blood pressure will rise, but if capillary flow is increased,

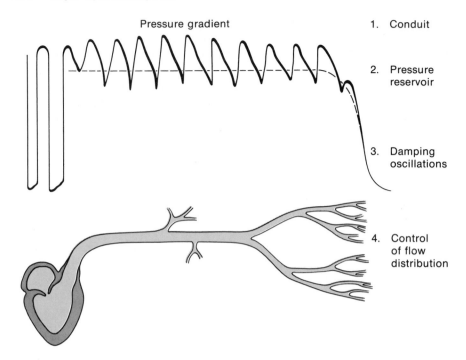

Pressure gradient

1. Conduit

2. Pressure reservoir

3. Damping oscillations

4. Control of flow distribution

**13-28**  Four functions of the systemic arterial system. (1) The conduit function is served by the vascular channels along which blood flows toward the periphery with minimal frictional loss of pressure. (2) The combination of distensible walls and high outflow resistance accounts for the pressure–reservoir function, which also allows (3) damping of oscillations in pressure and flow. (4) Controlled hydraulic resistance in the peripheral vascular beds controls the distribution of blood to the various tissues. [Rushmer, 1965a.]

blood pressure will fall. Normally, however, arterial blood pressure varies little, because the rates of filling and emptying are evenly matched by virtue of the fact that cardiac output and capillary flow are evenly matched. Capillary blood flow is proportional to the pressure difference between the arterial and venous systems. As venous pressure is low and changes little, arterial pressure exerts primary control over the rate of capillary blood flow and is responsible for maintaining adequate perfusion of the tissues. Arterial pressure varies among species, generally ranging from 50 to 150 mmHg. Pressure differences are small along large arteries (less than 1 mmHg), but there may be considerable drops in pressure along the small arteries and arterioles because of increasing resistance to flow with decreasing diameter of the vessel.

The elastic properties of arterial walls vary, and in general the elasticity as well as the magnitude of the muscular layer decreases with increasing distance from the heart. Close to the heart, the arteries are elastic and dampen the oscillations in pressure and flow generated by contractions of the heart. As blood is ejected into the arterial system, pressure rises and the vessels

expand. As the heart relaxes, blood flow to the periphery is maintained by the elastic recoil of the vessel walls, which causes a reduction in arterial volume. If the arteries were rigid tubes, pressures and flow in the periphery would exhibit the same large oscillations that are observed as blood leaves the ventricle. Further from the heart, the arteries do become more rigid and serve as blood conduits.

Thus, the extent of elastic tissue in arteries varies depending on the particular function of each vessel. In fishes, for example, blood pumped by the heart is forced into an elastic bulbus and a ventral aorta (Figure 13-29). The blood then flows through the gills and passes into a dorsal aorta, the main conduit for the distribution of blood to the rest of the body. A smooth, continuous flow of blood is required in the gill capillaries for efficient gas transfer. The bulbus, the ventral aorta, and the arteries leading to the gills are very compliant and act to smooth and maintain flow in the gills in the face of large oscillations produced by contractions of the heart (Figure 13-30). The dorsal aorta, which receives blood from the gills, is much less elastic than the ventral aorta. If the dorsal aorta were elastic,

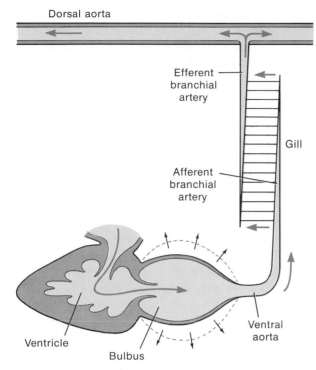

Dorsal aorta

Efferent
branchial
artery

Gill

Afferent
branchial
artery

Ventricle

Bulbus

Ventral
aorta

**13-29**  The teleost heart ejects blood into a short ventral aorta. Blood flows through the gills into a stiff, long dorsal aorta.

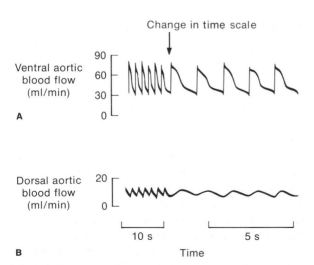

Change in time scale

Ventral aortic
blood flow
(ml/min)

**A**

90
60
30
0

Dorsal aortic
blood flow
(ml/min)

**B**

20

0

10 s        5 s

Time

**13-30**  Blood flow in the ventral (**A**) and dorsal (**B**) aortae of cod. Flow is more pulsatile in the ventral aorta than in the dorsal aorta. [Jones et al., 1974.]

there would be a rapid rush of blood through the gills during each heartbeat, when the dorsal aorta is filled with blood. This rush would increase rather than decrease the oscillations in flow through the gills. In this example, it can be seen that to ensure a steady flow through the gill capillaries, the major compliance must be placed *before, not after,* the gills to dampen the oscillations in flow through the gills. The ventral aorta must be elastic and the large-volume dorsal aorta stiff to achieve a smoothing of flow through the gills.

### BLOOD PRESSURE

Blood pressures reported for the arterial system are usually *transmural pressures* (i.e., the difference in pressure between the inside and outside, across the wall of the blood vessel). The maximum pressure is referred to as *systolic pressure,* the minimum as *diastolic pressure,* and the difference is the *pulse pressure.* Transmural pressures are typically given in millimeters of mercury and as systolic/diastolic, that is, 120/80 mmHg. Blood is 12.9 times less dense than mercury, so a blood pressure of 120 mmHg is equal to $120 \times 12.9 = 1550$ mm, or 155 cm of blood. In other words, if the blood vessel were suddenly opened, the blood would squirt out to a maximum height of 155 cm above the cut. (To convert to kilopascals, multiply the blood pressure in mmHg by 0.1333 kPa.)

The oscillations in pressure produced by the contractions and relaxations of the ventricle are reduced at the entrance to capillary beds and nonexistent in the venous system. Heart contractions cause small oscillations in pressure within capillaries. The pressure pulse travels at a velocity of 3–5 m/s. The velocity of the pressure pulse increases with decrease in artery diameter and increasing stiffness of the arterial wall. In the mammalian aorta, the pressure pulse travels at 3–5 m/s and reaches 15–35 m/s in small arteries.

Peak blood pressure and the size of the pressure pulse within the mammalian and avian aorta both *increase with distance* from the heart (Figure 13-31). This pulse amplification can be large during exercise. There are three possible explanations for this rather odd phenomenon. First, pressure waves are reflected from peripheral branches of the arterial tree; the initial and reflected waves summate; and, where peaks coincide, the pressure pulse and peak pressure are greater than

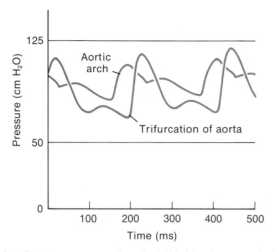

**13-31**   Simultaneous recording of rabbit's blood pressure in the aortic arch 2 cm from the heart, and at trifurcation of the aorta, 24 cm from the heart. Notice that peak and pulse pressures are much larger farther from the heart at the trifurcation of the aorta. Mean pressure, however, is slightly less than that from the aortic arch near the heart. [Langille, 1975.]

where they are out of phase. If the initial and reflected waves are 180° out of phase, the oscillations in pressure will be reduced. It has been suggested that the heart is situated at a point where initial and reflected waves are out of phase, thus reducing peak arterial pressure in the aorta close to the ventricle. As distance from the heart increases, the initial and reflected pressure waves move into phase, and a peaking of pressures is observed in vessels of the periphery. Second, the properties of the arterial walls change with distance from the heart; they become less elastic and the vessels taper, being of smaller diameter in the periphery. It is possible that these properties cause an increase in the magnitude of the pressure pulse with distance from the heart. Third, the pressure pulse is a complex wave form, consisting of several harmonics. Higher frequencies travel at higher velocities, and it has been suggested that the change in wave form of the pressure pulse with distance is due to summation of different harmonics. This third explanation is open to question, as the distances are too small to allow summation of harmonics.

## GRAVITY AND BODY POSITION

When a person is lying down, the heart is at the same level as the feet and head, and pressures will be similar in arteries in the head, chest, and limbs. Once a person moves to a sitting or standing position, the relationship between the head, heart, and limbs changes with respect to gravity and the heart is now a meter above the lower limbs. The result is an increase in arterial pressure in the lower limbs and a decrease in arterial pressure in the head.

Gravity has little effect on capillary flow, which is determined by the arterial–venous pressure difference rather than the absolute pressure. That is, gravity raises arterial and venous pressure by the same amount and therefore does not greatly affect the pressure gradient across a capillary bed. As the vascular system is elastic, however, an increase in absolute pressure expands blood vessels, particularly the compliant veins. Thus, pooling tends to occur, particularly in veins, in different regions of the body as the animal changes position with respect to gravity. This problem is related solely to the elasticity of blood vessels and would not occur if the blood flowed in rigid tubes.

*Vasoconstriction* in the limbs is required to prevent the pooling of blood as position changes with respect to gravity. This problem is more acute in some species than in others. When the giraffe is standing with its head raised, its brain is placed about 6 m above the ground and 1.6 m above the heart. If the arterial pressure of blood perfusing the brain is to be maintained at around 100 mmHg, aortic blood pressure must be between 200 and 300 mmHg near the heart. Aortic blood pressures greater than 200 mmHg have been recorded from an anesthetized giraffe with its head raised. As the giraffe lowers its head to the ground, arterial blood pressure at the level of the heart will be reduced considerably if brain blood flow is to remain constant. The probable means of control is extensive dilation and constriction in arterioles leading to capillary beds other than those in the head. As the giraffe raises its head, there is probably an extensive vasoconstriction of peripheral vessels other than those of the head. Flow to these capillary beds, however, cannot be varied without consequences. For example, there must be a fairly large, variable resistance to flow between the heart and the kidney in order to regulate the appropriate blood pressure at the level of the nephron. If the kidney tubule were subjected to the enormous changes in blood pressure associated with the raising and lowering of the giraffe's head, the rate of glomerular filtration would be chaotic. The rate of formation of an ultrafiltrate would be very high and would require that fluid be reabsorbed at an equally high rate each time arterial blood pressure were raised as the giraffe lifted its head. Thus, the giraffe must have mechanisms for adjusting peripheral resistance to flow in various capillary beds as it swings its head from ground level to a height of 6 m to eat. Similar problems must have been or are faced by a number of other animals with long necks, like dinosaurs and camels. Pooling of blood, with changes of position with respect to gravity, is not a problem in water, because the density of the environment is only slightly less than that of blood, whereas there are large differences in density between blood and air. In water, the hydrostatic pressure increases with depth, and this prevents blood pooling.

### BLOOD FLOW

Oscillations in flow vary with distance from the heart (Figure 13-32). They are largest at the exit to the ventricle and decrease with distance from the heart. At the base of the aorta, flow is turbulent and reverses during diastole as the aortic valves are closed by vortices created as blood is ejected into the aorta during systole. In most other parts of the circulation, flow is laminar, and oscillations are damped by the compliance of the aorta and proximal arteries.

The velocity of flow in a closed circulation is inversely proportional to the cross-sectional area of the circulation at that point. The total cross-sectional area of capillaries is much larger than that of the aorta, and therefore flow velocity is much lower in the capillaries. Blood velocity is maximal in the arteries, falls rapidly in the capillaries, but increases once more in the veins (Figure 13-24). Mean velocity in the aorta—the point of maximum blood velocity—is about 33 cm/s in humans. The cross-sectional area of the aorta (about 2.5 cm²) and cardiac output (about 5 L/min) were used to estimate this mean velocity. If we assume that

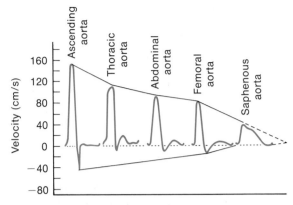

**13-32** Changes in form in the pulsatile flow recorded in dog's arteries ranging from the heart toward the periphery. There is a progressive decrease in amplitude until oscillatory flow is damped out in the capillaries (see Figure 13-35). A backflow phase is observed in the large arteries; in the ascending aorta, it is probably related to a brief reflux of blood through the aortic valves. [McDonald, 1960.]

maximum velocity in the vessel is twice the mean velocity (valid only if the velocity profile is a parabola), maximum velocity of blood flow will be 66 cm/s. If cardiac output is increased by a factor of 6 during heavy exercise, maximum velocity is raised to 3.96 m/s. These maximum values for blood flow velocity can be compared with pressure-pulse velocities of 3–35 m/s; that is, the pressure pulse travels faster than the flow pulse.

## The Venous System

The venous system acts as a conduit for the return of blood from the capillaries to the heart. It is a large-volume, low-pressure system with vessels of large inside diameter. In mammals, 50% of the total blood volume is contained in veins, and pressures seldom exceed 10 mmHg. The walls of veins are much thinner and less elastic than those of arteries. The venous compliance, unlike the arterial compliance, is related to its large volume rather than to the elasticity of its walls. The venous system acts as a storage reservoir for blood.

In the event of blood loss, venous and not arterial volume decreases in order to maintain arterial pressure and capillary blood flow.

Blood flow in veins is affected by a number of factors other than contractions of the heart. Activity of limb muscles and pressure exerted by the diaphragm on the gut both result in the squeezing of veins in those parts of the body. This squeezing, in the presence of venous *pocket valves* that allow flow only toward the heart, augments the return of blood to the heart. In the absence of skeletal muscle contraction, there may be considerable pooling of blood in the venous system of the limbs.

Breathing in mammals also contributes to the return of venous blood to the heart. The heart and large veins are contained within the thoracic cage. Expansion of the thoracic cage reduces pressure within the chest, with the result that blood is sucked from the veins of the head and abdominal cavity.

Peristaltic activity has been observed in the *venules* (small vessels joining capillaries to veins) of the bat wing, and peristaltic contractions of smooth muscle may be important in some instances in causing venous blood flow.

When bleeding depletes the venous blood reservoir, the loss is compensated for by a reduction in venous volume. The walls of many veins are covered by smooth muscle innervated by sympathetic adrenergic fibers. Stimulation of these nerves causes vasoconstriction and a reduction in the size of the venous reservoir. This reflex allows some bleeding to take place without a drop in venous pressure. Blood donors actually lose part of their venous reservoir; the loss is temporary, however, and the venous system gradually expands as blood is replaced.

Venous smooth muscle also aids in regulating the distribution of blood in the venous system. When a person shifts from a sitting position to a standing position, the change in the relative positions of heart and brain with respect to gravity activates sympathetic adrenergic fibers that innervate limb veins, causing contraction of venous smooth muscle and thereby promoting the redistribution of pooled blood. Such venoconstriction is inadequate, however, to maintain good circulation if the standing position is held for long periods in the absence of limb movements, as when soldiers stand immobile during a review. Under such circumstances, venous return to the heart, cardiac output, arterial pressure, and flow of blood to the brain are all reduced, which can result in fainting. Similar problems affect bedridden patients who attempt to stand after several days of inactivity and space pilots returning to Earth after a long period of weightlessness. It is possible that in these instances other control systems involving baroreceptors (pressure receptors) and arterioles may be disrupted as well. In the absence of body changes that shift the relative positions of the heart and brain with respect to gravity, the system of corrections falls into disrepair, and the result is the pooling of blood. The reflex control of venous volume is normally reestablished with use.

### RETE MIRABILE

Some vertebrates have a countercurrent arrangement (discussed in Chapter 12) of small arterioles and venules that form a *rete mirabile*. Before entering a tissue, the artery divides into a large number of small capillaries that parallel a series of venous capillaries leaving the tissue. The "arterial" capillaries are surrounded by "venous" capillaries, and vice versa, forming an extensive exchange surface between inflowing and outflowing blood. Retial capillaries serve to transfer heat or gases between arterial blood entering a tissue and venous blood leaving it (as elaborated in Chapter 16).

### CAPILLARIES

Most tissues have such an extensive network of capillaries that any single cell is not more than three or four cells away from a capillary. This is important for the transfer of gases, nutrients, and waste products, because diffusion is an exceedingly slow process. The capillary walls consist of a single layer of endothelial cells, which, although thin-walled, do not stretch or break easily when blood pressure is increased, because they also have a small diameter. The capillaries are usually about 1 mm long and 3–10 $\mu$m in diameter, just large enough for red blood cells to squeeze through. Large leucocytes, however, may become lodged in capillaries, stopping blood flow. The leucocytes are either dislodged by a rise in blood pressure or migrate slowly along the vessel wall until they reach a larger vessel and are swept into the bloodstream.

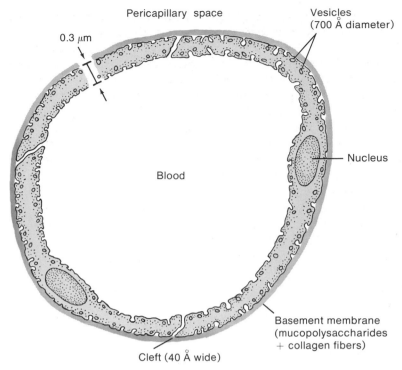

Pericapillary space

Vesicles
(700 Å diameter)

0.3 μm

Nucleus

Blood

Basement membrane
(mucopolysaccharides
+ collagen fibers)

Cleft (40 Å wide)

**13-33**   Cross section through a portion of the endothelial wall of a muscle capillary.

The small terminal arteries subdivide to form *arterioles,* which in turn subdivide to form *metarterioles* and subsequently capillaries, which then rejoin to form venules and veins. The arterioles are invested with smooth muscle that becomes discontinuous in the metarterioles and ends in a smooth muscle ring, the *precapillary sphincter.* The innervated smooth muscle of the arterioles and, in particular, the smooth muscle sphincter at the junction of arteries and arterioles control blood distribution to each capillary bed. Different tissues have varying numbers of capillaries and show some variation in the nature of the control of blood flow through the capillary bed. In some tissues, the precapillary sphincters, which appear to be under local control, twinkle open and closed and alter blood distribution within the capillary bed. In other tissues—for example, the brain—most, if not all, of the capillaries will be open, whereas in tissues such as the skin, capillaries may be closed for considerable periods. All capillaries combined have a potential volume of about 14% of an animal's total blood volume. At any one moment, however, only 30–50% of all capillaries are open, and thus only 5–7% of the total blood volume is contained in the capillaries.

Capillaries are completely devoid of connective tissue and smooth muscle, and the wall consists of only a single layer of endothelial cells surrounded by a *basement membrane* of collagen and mucopolysaccharides (Figure 13-33). The capillaries are often categorized as *arterial, middle,* or *venous capillaries,* the last being a little wider than the other two types. The venous capillaries empty into *pericytic venules* (small vessels outside the basement membrane), which in turn join the muscular venules and veins. The venules and veins are valved, and the muscle sheath appears after the first postcapillary valve.

Transfer of substances between blood and tissues occurs across the walls of capillaries, pericytic venules, and, to a lesser extent, metarterioles. The wall is specialized in that it is several orders of magnitude more permeable than epithelial cell layers, allowing substances to move with relative ease across the endothelium. In some tissues (e.g., the brain), capillary walls are much less permeable than those in bone and in the liver. The differences in permeability are associated with marked changes in structure. The capillaries of skeletal muscle have been studied extensively, and the endothelial walls have been found to be about 0.2–0.4 μm thick (Figure 13-34), the cells being separated by clefts that are about 40 Å wide at the narrowest point. The endothelial cells contain large numbers of pinocytotic vesicles of about 700 Å diameter.

Substances can move across the wall either through or between the endothelial cells. In moving across the endothelial cells, lipid-soluble substances diffuse through the membrane, whereas water and ions diffuse through water-filled pores. In addition, at least in brain capillaries, there are transport mechanisms for glucose and some amino acids. Large macromolecules can move across many capillary walls, but exactly how they are transferred is not always clear. Most endothelial cells contain large numbers of vesicles, most of them in association with the inner and outer surfaces of the endothelial cell, the remainder in the cell matrix. Under the electron microscope, the opaque substance horseradish peroxidase, when placed in the lumen of a muscle capillary, first appears in vesicles near the lumen, and then in vesicles close to the outer membrane but never in the surrounding cytoplasm. It has been con-

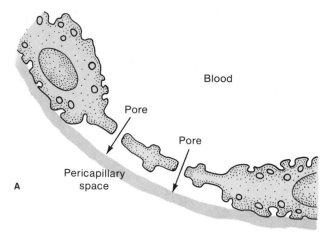

A

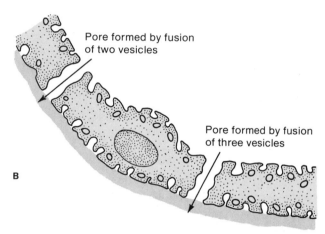

B

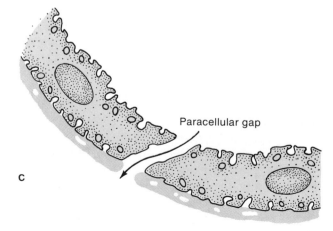

C

13-34   Portions of the endothelial wall of various types of capillaries. (**A**) A fenestrated endothelium. Note the thinning of the wall with large pores and a complete basement membrane. These capillaries are found in the renal glomerulus and the gut. (**B**) Pores formed in an endothelium by fusion of a row of vesicles. These pores have been observed in a number of capillary walls, including those of the rat diaphragm. (**C**) A capillary wall with large paracellular gaps and a broken basement membrane. These capillaries are found in bone and liver and are very permeable.

cluded that material is packaged in vesicles and shuttled through the cell. In support of this conclusion, it has been observed that endothelial cells of brain capillaries, which are less permeable, contain fewer vesicles than endothelial cells from other capillary beds. The reduced permeability of brain capillaries, however, is also considered to result from the tight junctions between endothelial cells. The movement of vesicles across endothelial cells is thought to be nondirectional, the vesicles forming and fusing at either surface at random. If this is true, vesicular transport should be nonselective over a certain range of molecular sizes, should transfer molecules of different sizes at the same rate, and, in addition, should require energy for fusion and fission of the vesicles. It has been observed, however, that different substances packaged in vesicles are transferred at widely differing rates and that anoxia and metabolic poisons do not affect the labeling of vesicles. So the notion of vesicular transport has been questioned. There are, however, large numbers of vesicles in endothelial cells, and they presumably serve some function. Another theory is that vesicles are not mobile within the cells but can coalesce to form pores through which substances can diffuse across the endothelium. In the capillaries of the rat diaphragm, vesicles have been observed to coalesce to form pores through the cell (Figure 13-34).

In the capillaries of the renal glomerulus and gut, the inner and outer plasma membranes of endothelial cells are opposed in some regions, and holes are formed through the cell. These capillary walls are referred to as *fenestrated endothelia.* Not surprisingly, these capillaries are permeable to nearly everything except large proteins and red blood cells. The kidney ultrafiltrate is formed across such an endothelial barrier. The basement layer of fenestrated endothelia is, however, normally complete, and it may constitute an important barrier to the movement of substances across fenestrated capillaries.

The paracellular gaps between endothelial cells are about 40 Å wide, but only molecules much smaller than 40 Å can get through, indicating the presence of some further sieving material in the cleft. Paracellular gaps vary within a single capillary network and are usually larger in the pericytic venules than in the

arterial capillaries. This is of functional significance because blood pressure, which is the filtration force for moving fluid across the wall, will decrease from the arterial to the venous end of the capillary network. Inflammation or treatment with a variety of substances, including histamine, bradykinin, and prostaglandins, increases the size of the paracellular gap at the venous end of the capillary network, making it very permeable. In liver and bone capillaries, there are always large paracellular gaps. Moreover, unlike fenestrated endothelia, the basement membrane is broken, with holes appearing in the region of the cell gaps (Figure 13-34C). Clearly, in these cells with large paracellular gaps, most of the exchange will occur between cells. As a result, the fluid surrounding the capillaries in liver has much the same composition as plasma.

In some less permeable endothelia, like those of lung capillaries, pulse pressure may play a role in augmenting movement of substances (e.g., oxygen) through the endothelium. As pressure rises, fluid is forced into the capillary wall; but as pressure drops, fluid will return to the blood. This tidal flushing of the capillary wall should enhance mixing in the endothelial barrier and effectively augment transfer.

The arrangement of arterioles and venules is such that all capillaries are only a short distance from an arteriole, and, as a result, a fairly uniform distribution of pressure and flow is possible within the capillary bed. Transmural pressures of about 10 mmHg have been recorded in capillaries by C. A. Wiederhielm and others (Figures 13-35 and 13-36). High pressures inside the capillary result in the filtration of fluid from the plasma into the interstitial space. This filtration pressure is opposed by the plasma colloid osmotic pressure. In the kidney, capillary pressure is high and filtration pressures exceed colloid osmotic pressures; hence, an ultrafiltrate is formed in the kidney tubule. In most other tissues, filtration pressure is balanced by osmotic pressure, so that there is little *net* movement of fluid across capillary walls. Pressures are higher, however, at the arterial end than at the venous end of the capillary. At the arterial end of the capillary, where filtration pressure exceeds colloid osmotic pressure, there is a net loss of fluid; at the venous end of the capillary,

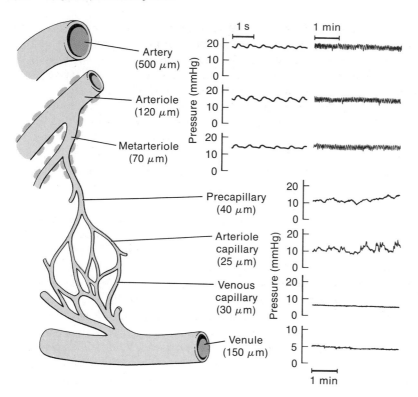

**13-35**   Capillary bed of the frog mesentery. Blood pressure is smoothed and falls from about 20 mmHg to 5 mmHg as it flows from the artery to the venule through the capillaries of the frog mesentery. [Wiederhielm et al., 1964.]

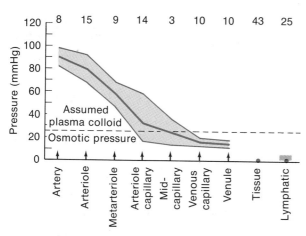

**13-36**   Pressure in the blood circulation, lymphatics, and tissues in the subcutaneous layers of the bat wing. The shaded area represents $\bar{x} \pm 1$ SE (standard error). [Wiederhielm and Weston, 1973.]

where osmotic pressure exceeds filtration pressure (Figure 13-36), there is a net uptake of fluid. Thus, there is a circulation of fluid from the arterial end of the capillary into the interstitial space and back to the venous end of the capillary. If the loss and uptake processes are balanced, there will be no net loss of fluid. If, however, there is a rise in capillary pressure, owing to a rise in either arterial or venous pressure, it will result in increased filtration into the interstitial space and a net loss of fluid from the blood. In general, though, arterial pressure remains fairly constant to prevent large oscillations in tissue volume. A drop in colloid osmotic pressure, resulting from a loss of protein from the plasma either by excretion or by increased capillary wall permeability and the movement of proteins into the interstitial space, will reduce the osmotic pressure difference between plasma and interstitial fluid. If filtration pressure remains constant, this will also result in an increase in net fluid loss to the tissue spaces. Thus, the volume and composition of interstitial fluid are closely correlated with the conditions for capillary blood flow. This correlation varies in different tissues; in general, liver capillaries are fairly permeable, brain capillaries much less permeable. Brain blood flow changes little, but blood flow in skeletal muscle changes enormously between rest and exercise. Thus, one might expect the interstitial fluid of the brain to vary less than that of the liver and skeletal muscle. In general, there are small net movements of fluid from the circulation into the interstitial spaces. This fluid does not collect in tissues, but is drained away via a low-pressure lymphatic system.

## The Lymphatic System

The lymphatic system begins with blind-ending capillaries that drain the interstitial spaces. These *lymphatic capillaries* join to form a treelike structure with branches reaching to all tissues. The larger lymphatic vessels resemble veins and empty via a duct into the blood circulation at a point of low pressure. In mammals and many other vertebrates, the lymph vessels drain via a *thoracic duct* into the anterior cardinal vein.

The lymphatic system serves to return to the blood the excess fluid and proteins that have been filtered across capillary walls. Large molecules, particularly fat absorbed from the gut and probably high-molecular-weight hormones, reach the blood via the lymphatic system. In addition, the lymphatic system is involved in the body's defense against infection by producing *lymphocytes* and *monocytes* in the *lymph nodes*. The *lymph* is a transparent, usually slightly yellow fluid within the lymphatic vessels, containing large numbers of lymphocytes. It is collected from all parts of the body and is returned to the blood via the lymphatic system by osmosis and the pumping action of muscular contraction.

The walls of lymphatic capillaries consist of a single layer of endothelial cells. The basement membrane is absent or discontinuous, and there are large pores between adjoining cells. This feature has been demonstrated by observing the passage of such substances as horseradish peroxidase or china ink particles through lymphatic capillary walls.

Lymphatic pressures are often as much as 1 mmHg less than the surrounding tissue pressures. Interstitial

fluid passes easily into the lymphatic channels. The vessels are valved and permit flow only away from the lymphatic capillaries. The larger lymphatic vessels are surrounded by smooth muscle and, in some instances, contract rhythmically, creating pressures of up to 10 mmHg and driving fluid away from the tissues (Figure 13-37). The vessels are also squeezed by contractions of the gut and skeletal muscle and by movements of the body generally, causing lymph flow. Lymph vessels are innervated by nerves, but it is not clear what type of innervation exists nor what function these nerves have.

Many cyclostomes, fishes, and amphibians have *lymph hearts,* which aid in the movement of fluid. A lymph heart located in the tail of the hagfish and many teleosts (Figure 13-38) aids in propelling fluid toward the heart. Lymph hearts are present in reptiles but absent in mammals. Bird embryos have a pair of lymph hearts located in the region of the pelvis; these hearts persist in the adult bird of a few species. Frogs have very large-volume lymph spaces, the function of which is not clear. They could act as a reservoir for water or ions and/or as a fluid buffer between the skin and underlying tissues.

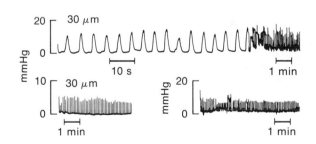

13-37   Pressures in lymphatic trunks (**A**) and capillaries (**B**). [Wiederhielm and Weston, 1973.]

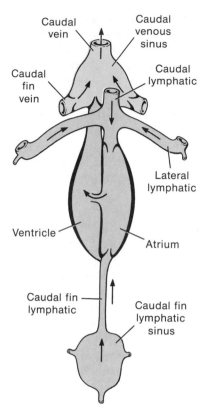

Caudal vein

Caudal venous sinus

Caudal fin vein

Caudal lymphatic

Lateral lymphatic

Ventricle

Atrium

Caudal fin lymphatic

Caudal fin lymphatic sinus

**13-38**  Schematic of the lymph heart located in the eel tail. The walls of the lymph heart contain skeletal muscle and beat rhythmically. Lymph is pumped into the caudal vein. [Kampmeier, 1969.]

Lymph flow is variable, ranging from 4 to 900 ml/kg/h in the human thoracic duct, 11 ml/h being an average value for resting individuals. This is 1/3000 of the cardiac output during the same time period. Nevertheless, although it is small, lymphatic flow is important in draining tissues of excess interstitial fluid. Lymph production in excess of lymph flow will cause retention of fluid in tissue spaces, a condition termed *edema,* which can be severe. In the tropical disease filariasis, larval nematodes, transmitted by mosquitoes to humans, invade the lymphatic system and cause the blockage of lymph channels and, in some cases, totally block lymphatic drainage from certain parts of the body. The consequent edema can cause parts of the body to become so severely swollen that the condition has come to be called *elephantiasis* because of the resemblance of the swollen, hardened tissues to the hide of an elephant.

The lymphatic vessels are not easily seen, as lymph is colorless, or almost so. As a result, even though the lymphatic system was first described about 400 years ago, it has not been nearly as extensively studied as the cardiovascular system.

## Regulation of Capillary Blood Flow

Capillary blood flow adjusts to meet the demands of the tissues. If the requirements change suddenly, as in skeletal muscle during exercise, then capillary flow also changes. If requirements for nutrients vary little with time, as in the brain, then capillary flow also varies little.

The regulation of capillary flow can be divided into two main types, local control and nervous control.

### NERVOUS CONTROL OF CAPILLARY BLOOD FLOW

*Nervous control* serves to maintain arterial pressure and thus to ensure that total flow in all capillaries is the same as cardiac output. The vertebrate brain and heart must be perfused with blood at all times. A failure in the perfusion of the human brain will rapidly result in damage. Nervous control of arterioles ensures that only a limited number of capillaries will be open at any moment, for if all capillaries were open, there would be a rapid drop in arterial pressure and blood flow to the brain would be reduced. The nervous control of capillary flow operates under a priority system. If arterial pressure falls, blood flow to the gut, liver, and muscles is reduced to maintain flow to the brain and heart.

Arterioles are normally innervated by sympathetic nerves, but some are innervated by parasympathetic nerves. Sympathetic nerves usually release norepinephrine, which binds to adrenergic receptors in vascular smooth muscle and usually causes a vasoconstriction of the arterioles.

A generalized effect of sympathetic stimulation is

to cause peripheral vasoconstriction and a rise in arterial blood pressure. This overall response is related to the release of norepinephrine from nerve endings; norepinephrine preferentially reacts with α-receptors in smooth muscle and results in an increase in smooth-muscle tension. Stimulation of a β-receptor, however, often results in vasodilation. The β-receptors are rarely innervated and usually respond to changing levels of circulating catecholamines. Catecholamines are released into the bloodstream from adrenergic neurons of the autonomic nervous system and from chromaffin cells in the adrenal medulla. Circulating catecholamines are dominated by epinephrine released from the adrenal medulla. Epinephrine preferentially reacts with β-receptors and causes vasodilation and, therefore, a decrease in peripheral resistance. Epinephrine, like norepinephrine, still causes a rise in arterial blood pressure because it also stimulates β-receptors in the heart and causes a marked increase in cardiac output. Thus, in general, stimulation of sympathetic nerves causes a peripheral vasoconstriction and a rise in arterial blood pressure, whereas an increase in circulating catecholamines causes a decrease in peripheral resistance, with a rise in arterial pressure because of concomitant stimulation of the heart and a rise in cardiac output. The response in any vascular bed depends on the type of catecholamine and the nature of the receptors involved and on the relationship between the catecholamine–receptor interaction and the change in muscle tone. Although stimulation of α-receptors is usually associated with a vasoconstriction and that of β-receptors with vasodilation, this is not invariably the case. An additional complicating factor is that *not all* sympathetic fibers are adrenergic. In some instances, they may be cholinergic, releasing acetylcholine from their nerve endings. The effect of sympathetic cholinergic nerve activity is to cause vasodilation in the vasculature of skeletal muscle.

Some arterioles, as noted earlier, are innervated by parasympathetic nerves—for example, in the circulation to the brain and the lungs. Parasympathetic nerves contain cholinergic fibers that release ACh from their nerve endings when stimulated (Figure 6-30). In mammals, parasympathetic nerve stimulation causes vasodilation in arterioles.

Some parasympathetic neurons may release ATP and other purines from nerve endings. ATP causes vasodilation, and it is possible (though not demonstrated) that some purinergic neurons participate in the control of capillary blood flow.

## LOCAL CONTROL OF CAPILLARY BLOOD FLOW

*Local control* of capillary flow ensures that the most active tissue has the most dilated vessels and, therefore, the most blood flow. The degree of dilation is dependent on local conditions in the tissue, and, in general, it is those conditions associated with high levels of activity that cause vasodilation.

The term *hyperemia* means increased blood flow to a tissue; *ischemia* means the cessation of flow. *Active hyperemia* refers to the increase in blood flow that follows increased activity in an organ, particularly skeletal muscle. Active tissues, metabolizing aerobically, utilize $O_2$ and produce $CO_2$, $H^+$, and various other metabolites. There are also other ionic changes—for example, a rise in extracellular $K^+$ in skeletal muscle following exercise. Increases in $CO_2$, $H^+$, $K^+$ (in muscle), and various other metabolites, or a decrease in $O_2$ in the tissues, have all been shown to cause vasodilation and a local increase in capillary blood flow. That is, the most active tissue has the most dilated vessels and therefore the highest blood flow. The lung capillary bed is an exception: low $O_2$ causes local vasoconstriction rather than vasodilation. The reason for the difference is that in the lung, $O_2$ is being taken up by the blood, and the blood must therefore flow to the regions of high $O_2$; in tissues, however, $O_2$ is being delivered in the blood flowing to them, and the highest blood flow should be to the area of greatest need, which is indicated by regions of low $O_2$.

If blood flow to an organ is stopped by clamping the artery or by a powerful vasoconstriction, there will be a much higher blood flow to that organ when the occlusion is removed than there was before the occlusion. This phenomenon is termed *reactive hyperemia*. Presumably during the ischemic period (a period of no blood flow), $O_2$ levels are reduced, and $CO_2$, $H^+$, and other metabolites build up and cause a local vasodilation. The result is that when the occlusion is removed, blood flow is much higher than normal.

Local injury in mammals and perhaps other vertebrates is accompanied by a marked vasodilation of vessels in the region of the injury. Histamine is released when cells are damaged and may mediate the vasodilation associated with local inflammation. Antihistamines ameliorate, but do not completely remove, this inflammation response. It is possible that another group of potent vasodilators, *plasma kinins,* may also be activated in damaged tissues. Kinins are formed by the action of many factors, including hypoxia, and by the action of proteolytic enzymes on *kininogen,* an $\alpha_2$-globulin. Tissue damage results in the release of proteolytic enzymes that split kinins from kininogen, and the kinins then cause local vasodilation. The prostaglandins found in most tissues may also be involved in the inflammation response. These compounds cause vasoconstriction of some vessels and dilation of others and a marked increase in capillary permeability.

Serotonin, found in high concentration in the gut and blood platelets, acts as a vasoconstrictor or vasodilator, depending on the vascular bed and on the dose level.

Angiotensin II, like norepinephrine, has a vasoconstrictor effect on arterioles. As noted in Chapter 12, it is formed by the action of renin on angiotensinogen, which circulates in the blood.

Histamine, bradykinin, and serotonin cause an increase in capillary permeability, and large proteins and other macromolecules tend to distribute themselves more evenly between plasma and interstitial spaces. As a result, the colloid osmotic pressure difference across the capillary wall is reduced, filtration is increased, and tissue edema occurs. On the other hand, norepinephrine, angiotensin II, and vasopressin tend to promote absorption of fluid from the interstitial fluid into the blood. This absorption could be achieved by reducing filtration pressure and/or the permeability of the capillaries.

## Cardiovascular Control by the Central Nervous System

Central control of the cardiovascular system hinges on the regulation of arterial blood pressure, which is adjusted to maintain an adequate level of capillary blood flow. The central priorities appear to be (1) to maintain arterial pressure in order to deliver an adequate supply of blood to the brain and heart and, (2) once these requirements are satisfied, to supply blood to other organs of the body. Arterial pressure is also maintained within narrow limits to control capillary pressure and, therefore, tissue volume and the composition of the interstitial fluid.

Receptors monitor blood pressure at various sites in the cardiovascular system. Information from baroreceptors, along with that from chemoreceptors monitoring $O_2$, $CO_2$, and pH in the blood, is transmitted to the brain. The integration of these inputs occurs in the mammalian brain at the level of the medulla oblongata and pons. This center of integration consists of a collection of neurons and is referred to as the medullary cardiovascular center. It is influenced by inputs from other regions of the brain, including the hypothalamus, the amygdala nucleus, and the cortex. The output from the medullary cardiovascular center is fed into autonomic motor neurons that innervate the heart and smooth muscle of arterioles and veins.

### BARORECEPTORS

There is a wide distribution of baroreceptors in the arterial system. The mammalian *carotid sinus baroreceptors* have been much more extensively studied than those of the aortic arch, subclavian, common carotid, or pulmonary arteries. In mammals, there appear to be only minor quantitative differences between the baroreceptors of the carotid sinus and the aortic arch. Birds have aortic arch baroreceptors that are sensitive to intravascular pressure changes.

The *carotid sinus* is a dilation of the internal carotid at its origin, where the walls are somewhat thinner than in other portions of the artery. The baroreceptors are finely branched nerve endings buried in the walls of the carotid sinus. Under normal physiological conditions, there is a resting discharge from the baroreceptors, and they increase their rate when the vessel wall is stretched (Figure 13-39). An increase in blood pressure stretches the wall of the carotid sinus and causes an increase in discharge frequency from the baroreceptors. The relationship between blood pressure and baroreceptor impulse frequency is sigmoidal, the system being most sensitive over the physiological range of blood pressures. The baroreceptor impulse frequency is higher when pressure is pulsatile than when it is constant (Figure 13-40). The carotid sinus baroreceptors are most sensitive to frequencies of pressure oscillation between 1 and 10 cycles/s—that is, within the normal physiological range. Sympathetic efferent fibers ter-

minate in the arterial wall near the baroreceptors; stimulation of these sympathetic fibers increases discharge of the carotid sinus baroreceptor. Under normal physiological conditions, these efferent neurons may be utilized by the central nervous system to control sensitivity of the receptors.

Information from baroreceptors is relayed through the *medullary cardiovascular center* in the medulla oblongata. An increase in blood pressure results in an increase in discharge frequency from the carotid sinus baroreceptors, which in turn causes a reflex reduction in both cardiac output and peripheral vascular resistance, thereby tending to reduce arterial blood pressure (Table 13-1). The reduction in cardiac output results from both a drop in the heart rate and the force of cardiac contraction. A reduction in arterial pressure reduces baroreceptor impulse frequency and causes a reflex increase in both cardiac output and peripheral resistance, which tends to increase arterial pressure. The baroreceptor reflex of the carotid sinus is therefore a negative feedback loop that tends to stabilize arterial

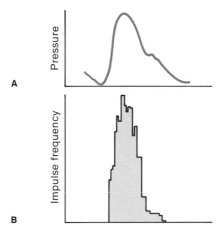

**13-39**   The relationship between the pressure pulse (**A**) and the impulse firing frequency (**B**) in a single carotid baroreceptor unit. [Korner, 1971.]

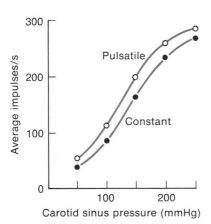

**13-40**   Average impulse frequency, recorded from a multifiber preparation of the carotid sinus nerve in relation to mean carotid sinus pressure while the pressure in the sinus is either constant or pulsatile. [Korner, 1971.]

**TABLE 13-1**   Open-loop reflex effects observed during changes in carotid sinus pressure.

| | Carotid sinus pressure | |
|---|---|---|
| Autonomic effector | Increased | Decreased |
| Cardiac vagus | + + + + | − |
| Cardiac sympathetic | − | + + + |
| Splanchnic bed | | |
|   Resistance vessels | − − | + + |
|   Capacitance vessels | − − | + + |
| Renal bed | ~0 | + |
| Muscle bed | | |
|   Resistance vessels | − − − | + + + + |
|   Capacitance vessels | − | + |
| Skin | | |
|   Resistance vessels | − | + + |
|   Capacitance vessels | ?0 | 0 |
| Adrenal catecholamines | ~0 | + + |
| Antidiuretic hormone | ? | + + |

*Note:* + = increased autonomic effect; − = decreased autonomic effect; 0 = no autonomic effect.
*Source:* Korner, 1971.

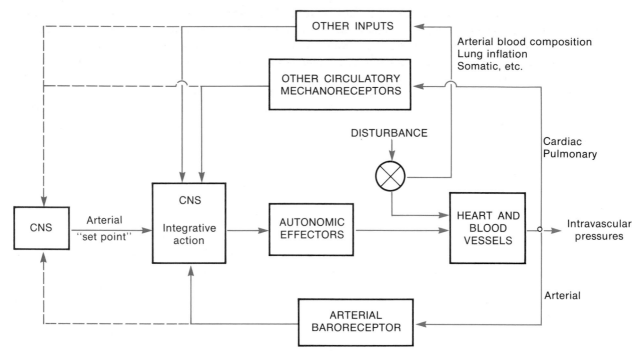

**13-41**   Circulatory control system in mammals. Changes in circulatory state are monitored by baroreceptors through changes in arterial pressure. Other mechanoreceptors and chemoreceptors provide collateral information about the physiological state of the body. The arterial set point is altered by inputs from other areas of the brain, which are in turn influenced by a variety of peripheral inputs. [Korner, 1971.]

blood pressure at a particular set point. The set point may be altered by interaction with other receptor inputs or be reset centrally within the medullary cardiovascular center by inputs from other regions of the brain (Figure 13-41).

Mechanoreceptors in the wall of the atrium monitor both the rate of atrial filling and the force of atrial contraction. Mechanoreceptors embedded in the ventricular wall are stimulated by stretch of the ventricle when end-diastolic volume is large. They may also be activated by ventricular contractions, which produce very small end-systolic volumes—that is, a high heart rate and a very low end-diastolic volume, a situation that may follow extensive blood loss. Stimulation of these receptors tends to inhibit the sympathetic system and activate the parasympathetic system, resulting in an increase in renal blood flow (which will tend to reduce blood volume) and a decrease in heart rate.

### ARTERIAL CHEMORECEPTORS

Arterial chemoreceptors, located in the *carotid* and *aortic bodies* (described in Chapter 14) not only have important reflex effects on ventilation, but also have some effect on the cardiovascular system. They respond with an increase in discharge frequency to an increase in $CO_2$ or to decreases in oxygen and pH of blood perfusing the carotid and aortic bodies. An increase in discharge frequency results in peripheral vasoconstriction and a slowing of the heart rate. But these reflex effects are seen only if the animal is not breathing. Peripheral vasoconstriction can cause a rise in arterial pressure, which then evokes reflex slowing of the heart by stimulation of the systemic baroreceptors. Nevertheless, chemoreceptor stimulation results in a slowing of the heartbeat even when arterial pressure is regulated at a constant level. Chemoreceptor stimulation therefore has a direct effect on heart rate, as well as an indirect action via changes in arterial pressure, resulting from peripheral vasoconstriction. Not surprisingly, there are many interactions between the control systems associated with both the respiratory and the cardiovascular system. The discharge pattern from stretch receptors in the lungs has a marked effect on the nature of the cardiovascular changes caused by chemoreceptor stimulation. That is, if the animal is breathing normally, changes in gas levels of the blood will cause one set of reflex changes; if, however, the animal is not breathing, chemoreceptor stimulation results in quite a different series of cardiovascular changes, as we will see.

A variety of inputs converge on the medullary cardiovascular center and cause reflex changes in circulation (Figure 13-41). The output from the medullary cardiovascular center is fed into sympathetic and parasympathetic autonomic motoneurons that innervate the heart and blood vessels. Stimulation of sympathetic nerves increases the rate and force of contraction of the heart and causes vasoconstriction; the result is a marked increase in arterial blood pressure and cardiac output. In general, the reverse effects occur upon parasympathetic nerve stimulation, the end result being a drop in arterial blood pressure and cardiac output. Regions have been located within the medulla that, when stimulated, result in activation of either the sympathetic or the parasympathetic system. Stimulation of one region, the *pressor center,* results in sympathetic activation and a rise in blood pressure, whereas stimulation of another region, the *depressor center,* results in activation of the parasympathetic system and a drop in blood pressure. In general terms, various inputs affect the balance between pressor and depressor activity; some activate the pressor center and inhibit the depressor center, and others have the reverse effect. Thus, the various inputs that converge on the medullary cardiovascular center are modified and integrated. The result is in an output that activates the pressor or the depressor center and produces cardiovascular changes in response to changing requirements of the body or disturbances to the cardiovascular system.

## Cardiovascular Responses to Exercise and Diving

During exercise, blood flow to skeletal muscle is increased in proportion to the level of activity of the muscle. This is achieved by an increase in cardiac output and a reduction in flow to the gut and kidney (Figure 13-42). Cardiac output can increase by up to an order of magnitude above the resting level owing to large increases in heart rate and small changes in stroke volume. Much of the increase in cardiac output can be accounted for by a decrease in peripheral resistance of about 50% of the resting value and by an increase in venous return due to pumping action of skeletal muscles on veins. There is increased sympathetic, but decreased parasympathetic, activity in nerves innervating the heart. The role of the cardiac nerves is to adjust heart rate to maintain stroke volume at a constant level rather than to cause the increase in cardiac output per se. There are only small changes in arterial blood pressure, pH, and gas tensions. The oscillations in $P_{CO_2}$ and $P_{O_2}$ with breathing are somewhat larger, as is the arterial pulse pressure. The increased oscillations in pulse pressure are dampened to some extent because of

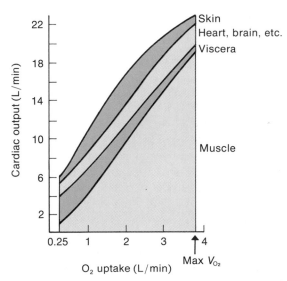

**13-42** Approximate distribution of cardiac output at rest and at different levels of exercise up to the maximal oxygen consumption (Max $V_{O_2}$) in a normal young man. The section labeled "viscera" reveals a progressive reduction in the absolute blood flow and percentage of cardiac output distributed to the splanchnic region and kidneys to augment muscle blood flow. Even skin is constricted during brief periods of exercise at high oxygen consumption. [Rowell, 1974.]

increased elasticity of the arterial walls due to a rise in circulating catecholamines. It is probable that arterial chemoreceptors and baroreceptors play only a minor role in the cardiovascular changes associated with exercise.

Motoneurons that innervate skeletal muscle are activated by higher brain centers at the onset of exercise; it is possible that this activating system also initiates changes in lung ventilation and blood flow. Proprioreceptive feedback from muscles may also play a role in increasing lung ventilation and cardiac output. Thus, exercise is responsible for a complex series of integrated changes that maintain blood flow to the exercising muscle. Active hyperemia is primarily responsible for increasing blood flow to skeletal muscle, and increases in cardiac output are a direct result of the decrease in peripheral resistance. The exact role of a variety of other control systems is not clear.

Many air-breathing vertebrates can remain submerged for prolonged periods. During submersion, there are marked respiratory and cardiovascular changes. Breathing stops, and the cardiovascular system is adjusted to meter out the limited oxygen store to those

organs that can least withstand anoxia (brain, heart, and some endocrine structures).

Breathing stops in all diving vertebrates because of stimulation of sensory receptors located in the nasal passages and pharynx (Figure 13-43). These sensory receptors respond to cold and submersion and reflexly cause apnea. In mammals, but not in other vertebrates, stimulation of these facial receptors also causes a marked bradycardia. In all vertebrates, the continued utilization of oxygen by the brain and heart in the absence of breathing results in the gradual fall in blood oxygen and rise in blood carbon dioxide levels, stimulating the arterial chemoreceptors and causing peripheral vasoconstrictions and a reduction both in cardiac output associated with a profound bradycardia and in blood supply to all tissues except the brain, heart, and some endocrine organs. The increase in peripheral resistance results from a marked rise in sympathetic output and involves constriction of fairly large arteries. Blood flow to skeletal muscle is reduced almost to zero in some diving mammals (e.g., seals), and there is enlargement of the venous reservoir. Most of the blood pumped by the heart into the systemic

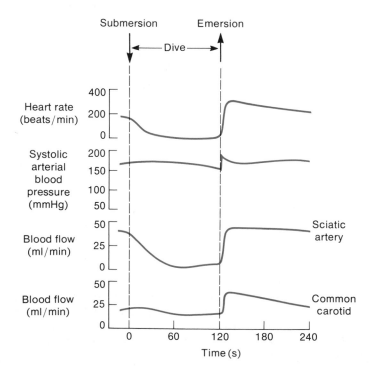

13-43    The effects of diving on the blood circulation in ducks (*Anas platyrhynchus*). Heart rate decreases following submersion, causing a fall in cardiac output. Systolic blood pressure does not fall because of peripheral vasoconstriction, which leads to a fall in blood flow in the sciatic artery and, therefore, to the legs. Blood flow in the common carotid does not decrease markedly, thus ensuring adequate flow to the brain. There is a large increase in cardiac output and sciatic artery and common carotid blood flow with emersion of the duck and the restoration of breathing. [Butler and Jones, 1971.]

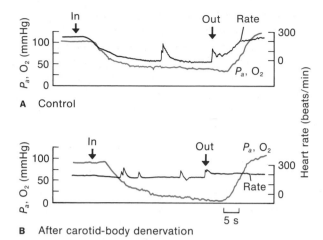

**13-44**  Changes in heart rate and in brachiocephalic artery's oxygen tension (0–100 mmHg) during a period of submergences of the head in water indicated by the in–out arrows. Six-week-old duck. **(A)** Control with all nerves intact. **(B)** The same duck three weeks after denervation of the carotid bodies. [Jones and Purves, 1970.]

circuit flows through the capillaries of the brain and heart. In some instances, there is a rise in arterial pressure, which results in arterial baroreceptor stimulation, and the bradycardia is maintained by a rise in both chemoreceptor and baroreceptor discharge frequency. The bradycardia is caused by an increase in parasympathetic and possibly a decrease in sympathetic activity in fibers innervating the heart.

Since the cardiovascular responses to diving can be elicited in decerebrate ducks, the medulla is presumed to be the site of integration of peripheral inputs. It has been shown in the seal, however, that the generation of the diving bradycardia can also involve some form of associative learning. In some trained seals, brady-cardia occurs before the onset of the dive and therefore before the stimulation of any peripheral receptors.

In birds, the "water" receptors are not directly involved in the cardiovascular changes associated with submersion. A decrease in heart rate is not observed either in submerged ducks breathing air through a tracheal cannula or in submerged ducks following carotid body denervation (Figure 13-44). Thus, the "water" receptors cause apnea, and the subsequent drop in blood $P_{O_2}$ and pH and the rise in $P_{CO_2}$ result in stimulation of the chemoreceptors, which then reflexly cause the cardiovascular changes.

Stimulation of lung stretch receptors modifies the reflex response initiated by chemoreceptor stimulation. In the absence of breathing, and hence stimulation of lung stretch receptors, different reflex responses are elicited by chemoreceptor stimulation than when the animal is breathing. Lung inflation tends to suppress reflex cardio-inhibition and peripheral vasoconstriction caused by stimulation of arterial chemoreceptors. The cardiovascular responses to hypoxia and hypercapnia (increased $CO_2$ levels) are therefore different when the animal is breathing than during a dive.

When the animal is breathing, stimulation of arterial chemoreceptors results in a marked increase in lung ventilation. Low blood oxygen and/or high blood carbon dioxide levels cause peripheral vasodilation. This vasodilation leads to an increase in cardiac output to maintain arterial pressure in the face of increased peripheral blood flow. Thus, hypoxia (low oxygen) during a dive (because the animal is not breathing) is associated with a bradycardia and a reduction in cardiac output. If hypoxia occurs when the animal is breathing (as, for instance after ascent to a high altitude), hypoxia is associated with an increase in heart rate and cardiac output.

## Summary

Circulatory systems can be divided into two broad categories—those with open and those with closed circulations. In open circulatory systems, transmural pressures are low, and blood pumped by the heart empties into a space in which blood bathes the cells directly. In closed circulatory systems, blood passes via capillaries from the arterial to the venous circulation. Transmural pressures are high, and fluid that has slowly leaked across capillary walls into the extracellular spaces is subsequently returned to the circulation via a lymphatic system.

The heart is a muscular pump that ejects blood into the arterial system. Excitation of the heart is initiated in a pacemaker, and the pattern of excitation of the rest of the muscle mass is determined by the nature of the contact between cells. The junctions between muscle fibers in the heart are of low resistance and allow the transfer of electrical activity from one cell to the next.

The initial phase of each heart contraction is isometric, and is followed by an isotonic phase in which blood is ejected into the arterial system. Cardiac output is dependent on venous inflow and, in mammals, changes in cardiac output are associated with changes in heart rate rather than in stroke volume.

The arterial system acts as a pressure reservoir and a conduit for blood between heart and capillaries. The elastic arteries dampen oscillations in pressure and flow caused by contractions of the heart, and the muscular arterioles control the distribution of blood to the capillaries.

Blood flow is generally streamlined, but because the relationship between pressure and flow is complex, Poiseuille's equation applies only to flow in smaller arteries and capillaries.

Capillaries are the site of transfer of material between the blood and tissues. Only 30–50% of all capillaries are open to blood flow at any particular time, but no capillary remains closed for long, because they all open and close continuously. Capillary blood flow is controlled by nerves that innervate smooth muscle around arterioles. Changes in the composition of blood and extracellular fluid in the region of the capillary causes the vessels either to constrict or to dilate, thereby altering blood flow.

The walls of capillaries are generally an order of magnitude more permeable than other cell layers. Material is transferred between blood and tissues by passing either through or between the endothelial cells that form the capillary wall. Endothelial cells contain large numbers of vesicles that may coalesce to form channels for the movement of material through the cell. In addition, some endothelial cells have specific carrier mechanisms for transferring glucose and amino acids. The size of the gaps between cells varies between capillary beds; brain capillaries have tight junctions, whereas liver capillaries have large gaps between cells.

The venous system acts both as a conduit for blood between capillaries and the heart and as a blood reservoir. In mammals, 50% of the total blood volume is contained in veins.

Capillary blood flow is adjusted to meet the requirements of the tissues, and arterial pressure is adjusted to maintain capillary blood flow. Arterial baroreceptors monitor blood pressure and reflexly alter cardiac output and peripheral resistance to maintain arterial pressure. Atrial and ventricular mechanoreceptors monitor venous pressure and derivatives of the heart contraction to ensure that activity of the heart is correlated to blood inflow from the venous system and blood outflow into the arterial system. Arterial chemoreceptors and other sensory receptors feed information into the medullary cardiovascular center, where the inputs are integrated to ensure an appropriate response of the circulatory system to changing requirements of the animal, as during exercise.

## Exercises

*1.* Describe the properties of myogenic pacemakers.

*2.* Describe the transmission of excitation over the mammalian heart.

*3.* Describe the changes in pressure and flow during a single beat of the mammalian heart.

*4.* Discuss the factors that influence stroke volume of the heart.

*5.* What are the nature and function of the nervous innervation of the mammalian heart?

*6.* What is the effect on cardiac function of a rigid versus a compliant pericardium?

*7.* What is the functional significance of a partially divided ventricle in some reptiles?

*8.* Discuss the changes in circulation that occur at birth in the mammalian fetus.

*9.* Discuss the applicability of Poiseuille's equation to the relationship between pressure and flow in the circulation.

*10.* What are the functions served by the arterial system?

*11.* Describe the factors that determine capillary blood flow.

*12.* Describe the location and role in cardiovascular control of various baroreceptors and/or mechanoreceptors in the mammalian circulatory system.

*13.* Compare and contrast the cardiovascular responses to breathing air low in oxygen with those associated with diving in mammals.

*14.* Describe the cardiovascular changes associated with exercise in mammals.

*15.* What are the consequences of raising or lowering arterial blood pressure for cardiac function and for exchange across capillary walls?

*16.* Discuss the relationship between capillary structure and organ function, comparing that found in different organs of the body.

*17.* Describe the ways in which substances are transferred between blood and tissues across capillary walls.

*18.* What are the functions served by the venous system?

*19.* Describe the effects of gravity on blood circulation in a terrestrial mammal. How are these effects altered if the animal is in water?

*20.* Define Laplace's Law. Discuss the law in the context of the structure of the cardiovascular system.

*21.* Discuss the role of the lymphatic system in fluid circulation. Discuss how and why this role may vary in different parts of the body.

## Suggested Reading

Bourne, G. H., ed. 1980. *Hearts and Heart-like Organs.* Vol. 1. New York: Academic.

Bundgaard, M. 1980. Transport pathways in capillaries: In search of pores. *Ann. Rev. Physiol.* 42:325–326.

Crone, C. 1980. Ariadne's thread: An autobiographical essay on capillary permeability. *Microvasc. Res.* 20:133–149.

Farner, D. S., and J. R. King., eds. 1972. *Avian Biology.* Vol. 2. New York: Academic.

Hecht, H. H. 1965. Comparative physiological and morphological aspects of pacemaker tissues. *Ann. N.Y. Acad. Sci.* 127:49–83.

Hoar, W. S., and D. J. Randall, eds. 1978. *Fish Physiology.* Vol. 7. New York: Academic.

Hoffman, B. F., and P. F. Cranefield. 1960. *Electrophysiology of the Heart.* New York: McGraw-Hill.

Korner, P. I. 1971. Integrative neural cardiovascular control. *Physiol. Rev.* 51:312–367.

Lewis, D. H., ed. 1979. Lymph circulation. *Acta Physiol. Scand.* Suppl. 463.

McDonald, D. A. 1960. *Blood Flow in Arteries.* Baltimore: Williams and Wilkins.

Paintal, A. S. 1973. Vagal sensory receptors and their reflex effects. *Physiol. Rev.* 53:159–227.

Robb, J. S. 1965. *Comparative Basic Cardiology.* New York: Grune and Stratton.

Schmidt-Nielsen, K. 1972. *How Animals Work.* New York: Cambridge University Press.

Stevens, E. D., H. M. Lam, and J. Kendall. 1974. Vascular anatomy of the countercurrent heat exchanger of skipjack tuna. *J. Exp. Biol.* 61:145–153.

## References Cited

Adolph, E. F. 1967. The heart's pacemaker. *Scientific American* 216(3):32–37.

Barcroft, H., and H. J. C. Swan. 1953. *Sympathetic Control of Human Blood Vessels.* London: Arnold.

Brand, A. R. 1972. The mechanisms of blood circulation in *Anodonta anatina* L. (Bivalvia Unionidae). *J. Exp. Biol.* 56:362–379.

Butler, P. J., and D. R. Bones. 1971. The effect of variations in heart rate and regional distribution of blood flow on the normal pressor response to diving in ducks. *J. Physiol.* 214:457–479.

Feigl, E. O. 1974. Physics of the cardiovascular system. In T. C. Ruch and H. D. Patton, eds., *Physiology and Biophysics.* 20th ed. Vol. 2. Philadelphia: Saunders.

Goodrich, E. S. 1958. *Studies on the Structure and Development of Vertebrates.* Vol. 2. New York: Dover.

Hirakow, R. 1970. Ultrastructural characteristics of the mammalian and sauropsidan heart. *Amer. J. Cardiol.* 25: 195–203.

Hutter, O. F., and W. Trautwein. 1956. Vagal and sympathetic effects on the pacemaker fibers in the sinus venosus of the heart. *J. Gen. Physiol.* 39:715–733.

Johansen, K., and A. W. Martin. 1965. Circulation in a giant earthworm, *Glossoscolex giganteus,* I: Contractile processes and pressure gradients in the large blood vessels. *J. Exp. Biol.* 43:333–347.

Jones, D. R., B. L. Langille, D. J. Randall, and G. Shelton. 1974. Blood flow in dorsal and ventral aortas of the cod, *Gadus morhua. Amer. J. Physiol.* 226:90–95.

Jones, D. R., and M. J. Purves. 1970. The effect of carotid body denervation upon the respiratory response to hypoxia and hypercapnia in the duck. *J. Physiol.* 211:295–309.

Kampmeier, O. F. 1969. *Evolution and Comparative Morphology of the Lymphatic System.* Springfield, Ill.: Thomas.

Langille, B. J. 1975. A comparative study of central cardiovascular dynamics in vertebrates. Ph.D. dissertation. University of British Columbia, Vancouver, Canada.

Le Gros Clark, W. E. 1952. *The Tissues of the Body.* 3rd ed. Oxford: Clarendon.

Navaratnam, V. 1980. Anatomy of the mammalian heart. In G. H. Bourne, ed., *Hearts and Heart-like Organs.* Vol. 1. New York: Academic.

Orkand, R. K. 1968. Facilitation of heart muscle contraction and its dependence on external $Ca^{2+}$ and $Na^{2+}$. *J. Physiol.* 196:311–325.

Randall, D. J. 1968. Functional morphology of the heart in fishes. *Amer. Zool.* 8:179–189.

Rowell, L. B. 1974. Circulation to skeletal muscle. In T. C. Ruch and H. D. Patton, eds., *Physiology and Biophysics.* 20th ed. Vol. 2. Philadelphia: Saunders.

Rushmer, R. F. 1965a. The arterial system: Arteries and arterioles. In T. C. Ruch and H. D. Patton, eds., *Physiology and Biophysics.* 19th ed. Philadelphia: Saunders.

Rushmer, R. F. 1965b. Control of cardiac output. In T. C. Ruch and H. D. Patton, eds., *Physiology and Biophysics.* 19th ed. Philadelphia: Saunders.

Sarnoff, S. J., and J. H. Mitchell. 1962. The control of the function of the heart. In H. W. Magoun, ed., *Handbook of Physiology.* Sec. 1, Vol. 1. Baltimore: Williams and Wilkins.

Shelton, G., and W. Burggren. 1976. Cardiovascular dynamics of the Chelonia during apnoea and lung ventilation. *J. Exp. Biol.* 64:323–343.

Skandalakis, J. E., S. W. Gray, and J. S. Rowe. 1980. The anatomy of the human pericardium and heart. In G. H. Bourne, ed., *Hearts and Heart-like Organs.* Vol. 1. New York: Academic.

Vander, A. J., J. H. Sherman, and D. S. Luciano. 1975. *Human Physiology: The Mechanisms of Body Function.* 2nd ed. New York: McGraw-Hill.

Wiederhielm, C. A., and B. U. Weston. 1973. Microvascular lymphatic and tissue pressures in the unanesthetized mammal. *Amer. J. Physiol.* 225:992–996.

Wiederhielm, C. A., J. W. Woodbury, S. Kirk, and R. F. Rushmer. 1964. Pulsatile pressures in the microcirculation of frog's mesentery. *Amer. J. Physiol.* 207:173–176.

# CHAPTER *14*

# *Exchange of Gases*

*A*nimals utilize oxygen and produce carbon dioxide during the course of cellular respiration, a process whose reactions take place at the level of the mitochondria, as described in Chapter 3. Oxygen is obtained from the environment, and carbon dioxide liberated into the environment. For cellular respiration to proceed, there must be a steady supply of $O_2$, and the waste product $CO_2$ must be continually removed.

This chapter reviews the transport of $O_2$ and $CO_2$ in the blood and the systems that have evolved in animals primarily to facilitate the movement of these two gases both between the environment and the blood and between the blood and the tissues. The main focus is on systems found in vertebrates, particularly mammals, because these have been investigated most thoroughly. Other interesting systems occur in animals for transferring gases besides the system that transports $O_2$ and $CO_2$ between the lungs and tissues. Three of these systems are discussed briefly toward the end of this chapter to demonstrate a few of the many intriguing problems of gas transfer in animals.

## *General Considerations*

Oxygen and carbon dioxide are transferred passively across the body surface (i.e., skin or special respiratory epithelium) by diffusion. If we consider a mass of gas, $M$, transferred across an epithelium, the rate of transfer, $M/dt$, is determined by the area available for diffusion, $A$, the diffusion distance, $x$, the diffusion coefficient, $D$, and the concentration difference across the respiratory surface, $(a_1 - a_2)$:

$$M = \frac{DA(a_1 - a_2)}{x}$$

Therefore, to facilitate the rate of gas transfer for a given concentration difference, the surface area of the respiratory epithelium should be as large as possible and diffusion distances as small as possible.

The $O_2$ requirements and $CO_2$ production of an animal increase in proportion to its mass, whereas the rate of gas transfer across the body surface is related to a large extent to surface area. It will be recalled that the area of a sphere increases as the square of its diameter, whereas the volume increases as the cube of its diameter. In very small animals, the distances for diffusion are small, and the ratio of surface area to volume is large. For this reason, diffusion alone is sufficient for the transfer of gases in small animals, such as rotifers and protozoa, which are less than 0.5 mm in diameter. Increases in size result in increases in diffusion distances and reductions in the ratio of surface area to volume. Large surface-area/volume ratios are maintained in larger animals by the elaboration of special areas for the exchange of gases. In some animals, the whole body surface participates in gas transfer, but in large, active

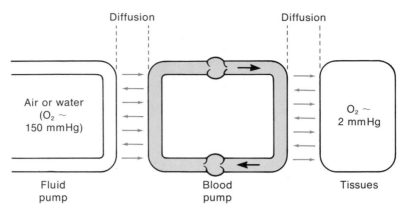

**14-1**  Overall gas-transport system of a vertebrate, consisting of two pumps and two diffusion barriers alternating in series between the external environment and the tissues. [Rahn, 1967.]

animals there is a specialized respiratory surface. This surface is made up of a thin monolayer of cells, the *respiratory epithelium,* which is 0.5–5 $\mu$m thick. This surface comprises the major portion of the total body surface. In humans, for instance, the respiratory surface area of the lung is between 50 and 100 m², varying with age and lung inflation, whereas the area of the rest of the body surface is less than 2 m².

Stagnation of the medium close to the surface of the respiratory epithelium is avoided in most animals by the exchange of air or water through breathing movements. A circulatory system has evolved in larger animals to transfer oxygen and carbon dioxide by the flow of blood between the tissues and the respiratory epithelium. Blood flows through an extensive capillary network and is spread in a thin film just beneath the respiratory surface, thereby reducing diffusion distances required to distribute the contained gases. The gases are transported between the respiratory surface and the tissues by bulk flow of blood. Gases diffuse between blood and tissues across the capillary wall. Once again, to facilitate gas transfer, the area for diffusion is large, and the diffusion distance between any cell and the nearest capillary is small.

*Graham's Law* states that the rate of diffusion of a substance along a given gradient is inversely proportional to the square root of its molecular weight (or density). Oxygen and carbon dioxide molecules are of similar size and therefore will diffuse at similar rates, and they are also utilized ($O_2$) and produced ($CO_2$) at approximately the same rates. It can therefore be expected that a transfer system that meets the oxygen requirements of an animal will also ensure adequate rates of carbon dioxide removal.

Thus, the transfer of gases in many animals takes place in several stages (Figure 14-1):

*1.* Breathing movements, which assure a continual supply of air or water to the respiratory surface (e.g., lungs or gills).

*2.* Diffusion of $O_2$ and $CO_2$ across the respiratory epithelium.

*3.* Bulk transport of gases by the blood.

*4.* Diffusion of $O_2$ and $CO_2$ across capillary walls between blood and mitochondria in tissue cells.

## Oxygen and Carbon Dioxide in Blood

In considering the movement of oxygen and carbon dioxide between the environment and the cells, we will first discuss transport in the blood, because the way in which these gases are carried in the blood affects their transfer between the environment and the tissues.

Relevant physical laws regarding the behavior of gases, along with some of the terminology used in respiratory physiology, are reviewed in Box 14-1.

## Box 14-1 The Gas Laws

Over 300 years ago, Robert Boyle determined that at a given temperature the product of pressure and volume is constant for a given number of molecules of gas. *Gay-Lussac's Law* states that either the pressure or the volume of a gas is directly proportional to absolute temperature if the other is held constant. Combined, these laws are expressed in the equation of state for a gas:

$$PV = nR\,\text{K}$$

where $P$ is pressure, $V$ is volume, $n$ is number of molecules of a gas, $R$ is universal gas constant (0.08205 L-atm/K-mol, or $8.314 \times 10^7$ ergs/K-mol, or 1.987 cal/K-mol), and K is absolute temperature. For accurate use, the equation should be modified using Van der Waals' constants.

The equation of state for a gas indicates that equal volumes of different gases at the same temperature and pressure contain equal numbers of molecules (Avogadro's Law). One mole of gas occupies approximately 22.414 L at 0°C and 760 mmHg. Because the number of molecules per unit volume is dependent on pressure and temperature, the conditions should always be stated along with the volume of gas. Gas volumes in physiology are usually reported as being at body temperature, atmospheric pressure, and saturated with water vapor (BTPS); or at ambient temperature and pressure, saturated with water vapor (ATPS); or at standard temperature and pressure (0°C, 760 mmHg) and dry (zero water-vapor pressure) (STPD).

Gas volumes measured at BTPS can be converted to ATPS or STPD volumes by using the equation of state for a gas. For example, the volume of air expired from a mammalian lung at body temperature (37°C, or 273 + 37 = 310 K) is often measured at room temperature (e.g., 20°C, or 273 + 20 = 293 K). The drop in temperature will reduce the expired gas volume. A gas in contact with water will be saturated with water vapor. The water vapor pressure at 100% saturation varies with temperature. Expired air is saturated with water, but as temperature decreases, water will condense, and this condensation will also reduce the expired gas

volume. If the measured expired volume at 20°C is 500 ml, then, if the barometric pressure is 760 mmHg and the water vapor pressure at 37°C and 20°C is 47.1 mmHg and 17.5 mmHg, respectively, the BTPS expired volume can be calculated as follows:

$$500 \times \frac{(760 - 17.5)}{(760 - 47.1)} \times \frac{(273 + 37)}{(273 + 20)} = 551 \text{ ml}$$

Thus, under the conditions stated above, a gas volume of 551 ml within the lung is reduced to 500 ml following exhalation because of the drop in gas temperature and the condensation of water.

*Dalton's Law* of partial pressure states that the partial pressure of each gas in a mixture is independent of other gases present, so that the total pressure equals the sum of the partial pressures of all gases present. The partial pressure of a gas in a mixture will depend on the number of molecules present in a given volume at a given temperature. Usually, oxygen accounts for 20.94% of all gas molecules present in dry air; thus, if the total pressure is 760 mmHg, the partial pressure of oxygen, $P_{O_2}$, will be $760 \times 0.2094 = 159$ mmHg. But air usually contains water vapor, which contributes to the total pressure. If the air is 50% saturated with water vapor at 22°C, the water vapor pressure is 10 mmHg. If the total pressure is 760 mmHg, the partial pressure of oxygen will be $(760 - 10) \times 0.2094 = 157$ mmHg. If the partial pressure of $CO_2$ in a gas mixture is 7.6 mmHg and the total pressure is 760 mmHg, then 1% of the molecules in air are $CO_2$.

Gases are soluble in liquids. The quantity of gas that dissolves at a given temperature is proportional to the partial pressure of that gas in the gas phase (*Henry's Law*). The quantity of gas in solution equals $\alpha P$, where $P$ is the partial pressure of the gas and $\alpha$ is the *Bunsen solubility coefficient*, which is independent of $P$ but varies with temperature. Thus, the amount of any gas in solution may be expressed either as a partial pressure or as a gas content per unit volume, the gas content being expressed in grams, moles, or milliliters per unit volume.

## RESPIRATORY PIGMENTS

Oxygen diffuses across the respiratory epithelium into the blood and combines with a *respiratory pigment* that gives the characteristic color to the blood. The best-known respiratory pigment, *hemoglobin,* is red. By binding oxygen, the respiratory pigment increases the $O_2$ content of blood.

In the absence of a respiratory pigment, the $O_2$ content of blood would be low. The *Bunsen solubility coefficient* of oxygen in blood at 37°C is 2.4 ml $O_2$/100 ml blood/atm of oxygen pressure. Therefore, the concentration of $O_2$ in physical solution (i.e., not bound to a respiratory pigment) in human blood at a $P_{O_2}$ of 95 mmHg will be $2.4 \times 95/760 = 0.3$ ml $O_2$/100 ml

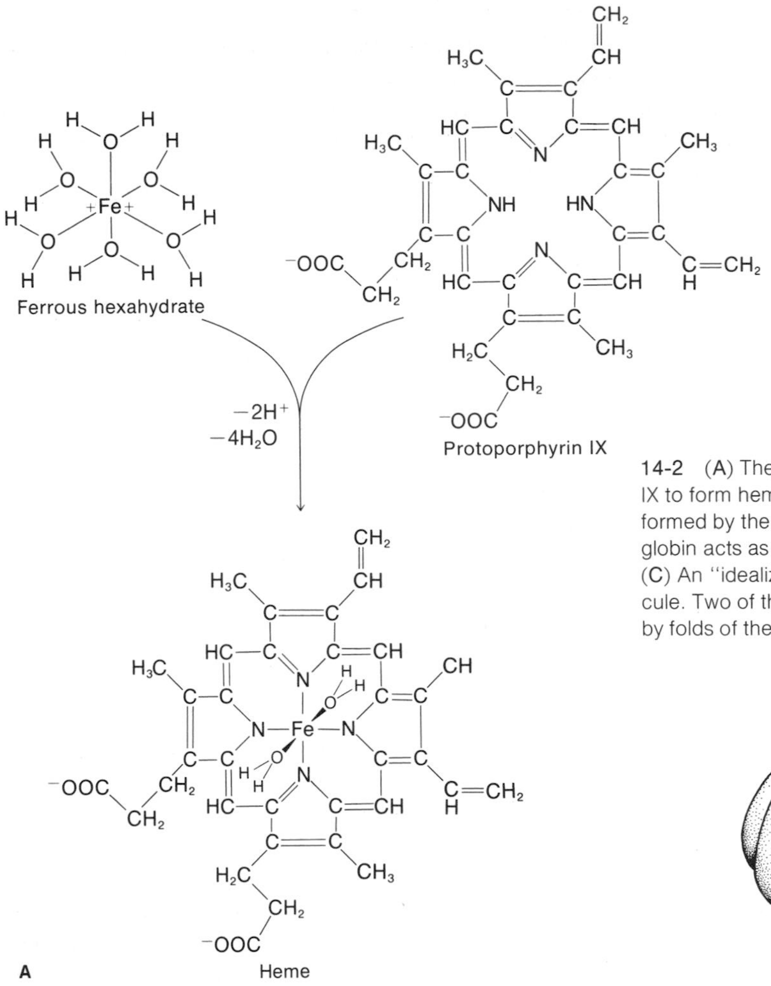

CH₂
‖
CH
|
H₃C—C ═ C—CH
   |       ‖
HC—C     C—CH
   ‖   N   ‖
H₃C—C       C—C—CH₃
   |   NH HN   |
   C—C       C—C
   |   ‖   N   ‖   |
⁻OOC  CH₂  HC—C  C═CH  C═CH₂
       |       ‖       |
       CH₂    C—C      H
              |  ‖
           H₂C  CH₃
              |
              CH₂
              |
             ⁻OOC

Ferrous hexahydrate

Protoporphyrin IX

−2H⁺
−4H₂O

CH₂
‖
CH
|
H₃C—C ═ C—CH
   |       ‖
HC—C     C—CH
   ‖   N   O H
H₃C—C     H
   |   N—Fe—N
   C—C   H O  C—C
⁻OOC CH₂  H H   C═CH₂
       |  HC—C  H
       CH₂  ‖ CH
           C—C
           |  ‖
        H₂C  CH₃
           |
           CH₂
           |
          ⁻OOC

**A**  Heme

**14-2** (A) The combination of ferrous ion and protoporphyrin IX to form heme. (**B**) Schematic diagram of heme in a pocket formed by the globin molecule. A histidyl side chain from the globin acts as an additional ligand for the iron atom in heme. (**C**) An "idealized," exploded diagram of the hemoglobin molecule. Two of the four heme units are visible in pockets formed by folds of the peptide chains. [McGilvery, 1970.]

L-Histidine
(His)

COO⁻
|
⁺H₃N—C—H
|
CH₂
|

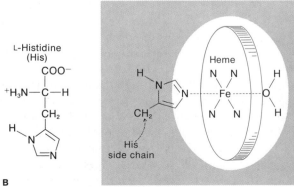

**B**

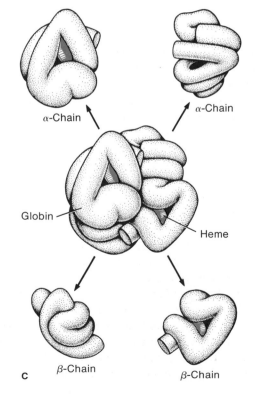

α-Chain

α-Chain

Globin

Heme

β-Chain

β-Chain

**C**

blood, or 0.3 vol % $O_2$. The total $O_2$ content of human arterial blood at a similar $P_{O_2}$ is, in fact, 20 vol %. The 70-fold increase in content is due to combination of oxygen with hemoglobin. The antarctic icefish is an exception among the vertebrates; the blood of this fish lacks a respiratory pigment and therefore has a low $O_2$ content. It compensates for the absence of hemoglobin with an increased blood volume and cardiac output, but its rate of oxygen uptake is reduced compared with that of species from the same habitat that have hemoglobin. Low temperatures are probably a factor in the evolution of fishes lacking hemoglobin. Low temperatures are associated with low metabolic rates in poikilotherms, and oxygen, like all gases, has a higher solubility at low temperatures.

Respiratory pigments are complexes of proteins and metallic ions, and each one has a characteristic color. The color of the respiratory pigment changes with its $O_2$ content. Thus, hemoglobin, which is bright red when it is loaded with $O_2$, becomes a dark, maroon-red when deoxygenated.

Vertebrate hemoglobin, except that of cyclostomes, has a molecular weight of 68,000 and contains four iron porphyrin prosthetic groups—*heme*—associated with the protein *globin* (Figure 14-2), which in turn is made up of two equal parts, each consisting of two polypeptide chains (an α-chain and a β-, γ-, or ε-chain). Hemoglobin will dissociate into four subunits of approximately equal weight, each containing one polypeptide chain and one heme group. Iron in the ferrous state is bound into the porphyrin ring of the heme, forming coordinate links with the four pyrrole nitrogens. The two remaining coordinate linkages are used to bind the heme group to an imidazole ring of the globin and to bind oxygen. If $O_2$ is bound, the molecule is referred to as *oxyhemoglobin;* if $O_2$ is absent, the term *deoxyhemoglobin* is used. The oxygen-binding characteristics of hemoglobins vary. These differences are related to variations in the structure of the globin molecule.

*Methemoglobin,* which does not bind oxygen and is therefore nonfunctional, is formed if ferrous iron of heme is oxidized to the ferric state. This formation occurs normally, but red blood cells contain the enzyme methemoglobin reductase, which reduces methemoglobin to the functional ferrous form. Certain compounds (nitrates or chlorates) act either to oxidize hemoglobin or to inactivate methemoglobin reductase, thereby increasing the level of methemoglobin and impairing oxygen transport.

The affinity of hemoglobin is about 200 times as great for carbon monoxide as for oxygen. As a result, carbon monoxide will displace oxygen and saturate hemoglobin, even at very low partial pressures of carbon monoxide, causing a marked reduction in oxygen transport to the tissues. Hemoglobin saturated with carbon monoxide is called *carboxyhemoglobin.* The effect of such saturation on oxidative metabolism is similar to that of oxygen deprivation, which is why the carbon monoxide produced by cars or improperly stoked coal or wood stoves is so extremely toxic. Even the levels found in city traffic can impair brain function owing to partial anoxia.

Different groups of invertebrates have different respiratory pigments, including *hemerythrin* (Priapulida, Brachiopoda, Annelida), *chlorocruorin* (Annelida), *hemocyanin* (Mollusca, Arthropoda), and hemoglobin (many invertebrate groups). Many invertebrates do not have a respiratory pigment.

Hemocyanin, a large, copper-containing respiratory pigment, has many properties similar to those of hemoglobin. It binds oxygen when the partial pressure is high and releases it when the partial pressure is low (Figure 14-3). It binds oxygen in the ratio of 1 mol of $O_2$ to approximately 75,000 g of the respiratory pigment. Unlike hemoglobin, it is not packaged in cells and is not associated with high levels of carbonic anhydrase in the blood. In its oxygenated form, it is light blue; in its unoxygenated form, it is colorless.

### OXYGEN TRANSPORT IN BLOOD

Each hemoglobin molecule can combine with four oxygen molecules, each heme combining with one molecule of oxygen. The extent to which $O_2$ is bound to hemoglobin varies with the partial pressure of the gas. If all sites on the hemoglobin molecule are occupied by $O_2$, the blood is 100% saturated and the oxygen *content* of the blood is equal to its oxygen *capacity.* A millimole of heme can bind a millimole of $O_2$, which represents a volume of 22.4 ml of $O_2$. Human blood contains about 0.9 mmol of heme per 100 ml of blood. The oxygen capacity is therefore $0.9 \times 22.4 = 20.2$ vol %. The oxygen content of a unit volume of blood includes the $O_2$ in physical solution as well as that combined with hemoglobin, but in most cases the $O_2$ in physical solution is only a small fraction of the total $O_2$ content.

The oxygen capacity increases in proportion to the hemoglobin concentration; consequently, in order to compare blood of different hemoglobin content, we use the term *percent saturation,* expressing $O_2$ content as a percentage of $O_2$ capacity. *Oxygen dissociation curves* (Figures 14-3 to 14-5) describe the relationship between percent saturation and the partial pressure of oxygen.

The oxygen dissociation curves of *myoglobin* (a respiratory pigment that stores $O_2$ in vertebrate muscles) and lamprey hemoglobin are hyperbolic, whereas other vertebrate hemoglobin oxygen dissociation curves are sigmoid (Figure 14-4). This difference in their dissociation curves is correlated with the fact that myoglobin and lamprey hemoglobin have a single heme group, whereas other hemoglobins have several (usually four) heme groups. The sigmoid shape of the curves exhibited by hemoglobins having several heme groups is explained in terms of *subunit cooperativity.* Oxygena-

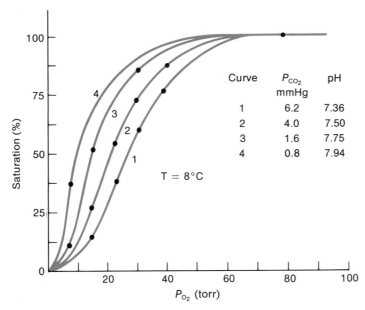

| Curve | $P_{CO_2}$ mmHg | pH |
|-------|------|------|
| 1 | 6.2 | 7.36 |
| 2 | 4.0 | 7.50 |
| 3 | 1.6 | 7.75 |
| 4 | 0.8 | 7.94 |

T = 8°C

**14-3**   Examples of hemocyanin–oxygen dissociation curves for the crab *Cancer magister.* The $P_{CO_2}$ and pH conditions for the different curves are given on the right. Note that hemocyanin, like hemoglobin, shows a Bohr shift. [Unpublished data supplied by D. G. McDonald.]

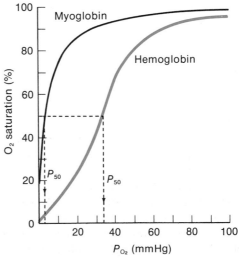

**14-4**   Characteristic oxygen dissociation curves for hemoglobin and myoglobin. Note the hyperbolic shape of the myoglobin curve and the sigmoid shape of the hemoglobin curve. Lamprey hemoglobin, with a single heme group, has a dissociation curve similar to myoglobin.

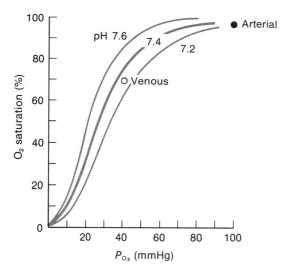

**14-5**  Bohr effect. The effect of changes in pH on examples of blood–oxygen dissociation curves in humans. [Bartels, 1971.] The $P_{O_2}$ of mixed venous and arterial blood is indicated.

tion of the first heme groups facilitates oxygenation of subsequent heme groups, owing to conformational changes in the protein globin. The steep portion of the curve corresponds to oxygen levels at which at least one heme group is already occupied by an oxygen molecule, increasing the affinity of the remaining heme groups for oxygen.

An important property of respiratory pigments is that they combine reversibly with $O_2$ over the range of partial pressures normally encountered in the animal. At low $P_{O_2}$, only a small amount of $O_2$ binds to the respiratory pigment; at high $P_{O_2}$, however, a large amount of $O_2$ is bound. Because of this property, the respiratory pigment can act as an oxygen carrier, loading at the respiratory surface (a region of high $P_{O_2}$) and unloading at tissues (a region of low $P_{O_2}$). In some animals, the predominant role of a respiratory pigment may be to serve as an oxygen reservoir, releasing $O_2$ to the tissues only when $O_2$ is relatively unavailable. For instance, seals and other diving mammals have high levels of myoglobin in their muscles. Myoglobin, which is related in structure to hemoglobin, acts as an oxygen store, releasing $O_2$ only during periods when oxygen levels in the muscle decrease, as they do during a dive.

The presence of a respiratory pigment increases the transfer of oxygen through a solution, because the

oxygenated pigment co-diffuses with oxygen down the concentration gradient. That is, a gradient exists for both oxygen and the oxygenated pigment in the same direction through the solution; whereas the gradient for the deoxygenated pigment is in the reverse direction from that for oxygen and the oxygenated pigment. Hence, the oxygenated pigment diffuses in the same direction as oxygen; whereas the deoxygenated pigment diffuses in the reverse direction. Thus, a pigment such as hemoglobin may facilitate the mixing of gases in the blood, and myoglobin may play a similar role within tissues.

Hemoglobins that have high oxygen affinities are saturated at low partial pressures of oxygen, whereas hemoglobins with low oxygen affinities are completely saturated only at relatively high partial pressures of oxygen. This difference in oxygen affinity is not due to differences in the heme group; rather, it is related to differences in the properties of the protein globin. The affinity is expressed in terms of the $P_{50}$, the partial pressure of oxygen at which the hemoglobin is 50% saturated with oxygen.

The rate of oxygen uptake increases in proportion to the difference in partial pressure across an epithelium. A hemoglobin with a high oxygen affinity facilitates the movement of $O_2$ into the blood from the environment because $O_2$ is bound to hemoglobin at low $P_{O_2}$. Thus, a large difference in $P_{O_2}$ across the respiratory epithelium—and therefore a high rate of oxygen transfer into the blood—is maintained until hemoglobin is fully saturated. Only then does blood $P_{O_2}$ rise. Hemoglobin with a high oxygen affinity, however, will not release $O_2$ to the tissues until the $P_{O_2}$ of the tissues is very low. In contrast, a hemoglobin with a low oxygen affinity will facilitate the release of $O_2$ to the tissues, maintaining large differences in $P_{O_2}$ between blood and tissues and a high rate of oxygen transfer to the tissues. Thus, a hemoglobin of high oxygen affinity favors the uptake of $O_2$ by the blood, whereas a hemoglobin of low oxygen affinity facilitates the release of $O_2$ to the tissues. From a functional viewpoint, therefore, hemoglobin should have a low $O_2$ affinity in the tissues and a high $O_2$ affinity at the respiratory surface. In light of this, it is highly significant that the hemoglobin–$O_2$ affinity is affected by changes in chemical and physical factors in the blood that favor oxygen binding at the respiratory epithelium and oxygen release in the tissues.

The hemoglobin–oxygen affinity is labile and is reduced either by increases in $P_{CO_2}$, temperature, and organic phosphate ligands—for example, 2,3-diphosphoglycerate (DPG) or ATP—or by a decrease in pH. The term *Bohr effect,* or *Bohr shift,* is used to describe the effect of pH on the hemoglobin–oxygen affinity. Increases in H⁺ (decrease in pH) cause a reduction in hemoglobin–oxygen affinity (Figure 14-5). Conversely, oxygenation of hemoglobin results in the release of protons from hemoglobin and therefore a decrease in pH. Carbon dioxide reacts with water to form carbonic acid and with $-NH_2$ groups on plasma proteins and hemoglobin to form carbamino compounds (Figure 14-9). An increase in $P_{CO_2}$ results in a decrease in blood pH and an increase in carbamino compounds in the blood. Increases in $CO_2$ cause a reduction in the oxygen affinity of hemoglobin, both by reducing pH (Bohr effect) and by the direct combination of $CO_2$ with hemoglobin to form carbamino compounds. Carbon dioxide enters the blood at the tissues and facilitates the unloading of $O_2$ from hemoglobin, whereas $CO_2$ leaves the blood at the lung or gill and facilitates the uptake of $O_2$ by the blood.

The oxygen dissociation curve for myoglobin, unlike that for hemoglobin, is relatively insensitive to changes in pH.

In some fishes, cephalopods, and crustaceans, an increase in $CO_2$ or a decrease in pH causes not only a reduction in the oxygen affinity of hemoglobin but also a reduction in oxygen capacity, which is termed the *Root effect,* or *Root shift* (Figure 14-6).

A decrease in pH results in an increase in oxygen affinity in hemocyanins from several gastropods and from the horseshoe crab *Limulus.* This greater oxygen affinity is referred to as a *reverse Bohr effect* and may serve to facilitate oxygen uptake during periods of low oxygen availability when reductions in blood pH are maintained in these animals.

An increase in temperature exacerbates problems of oxygen delivery in poikilothermic aquatic animals such as fishes. A rise in temperature not only reduces oxygen solubility in water but also decreases hemoglobin–oxygen affinity, making oxygen transfer between water and blood more difficult. Unfortunately, this occurs

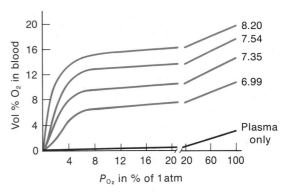

**14-6**  Examples of oxygen equilibrium curves of eel blood at 14°C and pH from 6.99 to 8.20. Notice the effect of pH on the $O_2$ capacity (Root effect). Bottom line describes the $O_2$ content of plasma. Hematocrit of whole blood, 40 vol %. [Steen, 1963.]

at a time when tissue oxygen requirements are increasing, also as a result of the rise in temperature.

Mammalian red blood cells contain high levels of 2,3-diphosphoglycerate. Hemoglobin and DPG are nearly equimolar in human red blood cells. DPG binds to the β-chains of deoxyhemoglobin and reduces oxygen affinity. Increases in DPG levels accompany reductions in blood $O_2$ levels, increases in pH, and reductions in hemoglobin concentrations in blood. Low blood $O_2$ levels may result from a climb to a higher altitude. Barometric pressure and the partial pressure of oxygen in air both decrease with altitude. The resultant DPG rise in response to high altitude is completed in 24 h with a half-time of about 6 h. At elevations of 3000 m, there is a 10% increase in DPG over the concentration at sea level. The resultant reduced $O_2$ affinity enhances oxygen transfer to the tissues but adversely affects oxygen uptake at the lungs. In some vertebrates, other phosphorylated compounds, such as adenosine triphosphate in fishes or inositol phosphate in birds, are in higher concentration in the red cell and have a greater effect on the oxygen affinity of hemoglobin than does DPG. Phosphorylated compounds in the red cell not only affect the oxygen affinity of hemoglobin but also increase the magnitude of the Bohr effect and may affect subunit interaction. It appears that the functional significance of increased

DPG levels is that they favor the release of $O_2$ to the tissues under conditions of low $O_2$, as at high altitude.

It is generally assumed that a particular hemoglobin has evolved to meet the special gas-transfer and H+-buffering requirements of the animal. The properties of hemoglobin (and red cells) vary among species and may change during development (Figure 14-7). For instance, fetal hemoglobins often have a higher $O_2$ affinity than the postnatal and adult hemoglobin (Figure 14-8). The higher $O_2$ affinity of fetal hemoglobin enhances oxygen transfer from mother to fetus.

It is important to remember that in most animals hemoglobin is contained within red blood cells. Values of blood parameters usually refer to conditions in the plasma, not in the red blood cell. There are differences between the inside and outside of cells, and the red blood cell is no exception. For example, mammalian arterial blood pH at 37°C is usually 7.4. This is the pH of arterial blood *plasma;* the pH inside the red blood cell is less, about 7.2 at 37°C.

### CARBON DIOXIDE TRANSPORT IN BLOOD

Carbon dioxide diffuses into the blood from the tissues, is transported in the blood, and diffuses across the respiratory surface into the environment. Figure 14-9 outlines the sequence of reactions involved in the exchange of $CO_2$ and $O_2$ and the relative distribution

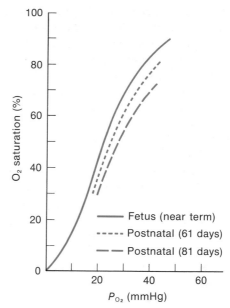

14-8 Changes in blood–oxygen dissociation curves in humans. Blood pH = 7.40. [Bartels, 1971.]

of $CO_2$ entering the blood from the tissues of a typical mammal.

Carbon dioxide reacts with water to form carbonic acid, a weak acid, which dissociates into bicarbonate and carbonate ions:

$$CO_2 + H_2O \rightleftharpoons H_2CO_3 \rightleftharpoons$$
$$H^+ + HCO_3^- \rightleftharpoons H^+ + CO_3^{2-}$$

The proportion of $CO_2$, $HCO_3^-$, and $CO_3^{2-}$ in solution is dependent on pH, temperature, and the ionic strength of the solution. In blood, the ratio of $CO_2$ to $H_2CO_3$ is approximately 1000:1, and the ratio of $CO_2$ to bicarbonate ions is about 1:20. Bicarbonate is therefore the predominant form of $CO_2$ in the blood at normal blood pH. The carbonate content is usually negligible in birds and mammals; in poikilotherms, however, with their low temperature and high blood pH, the carbonate content may approach 5% of the total $CO_2$ content of the blood, but bicarbonate is still the predominant form of $CO_2$.

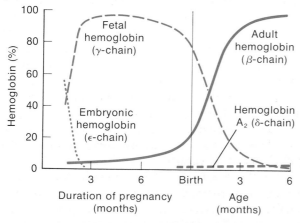

14-7 Changes in hemoglobins during development in humans. [Young, 1971.]

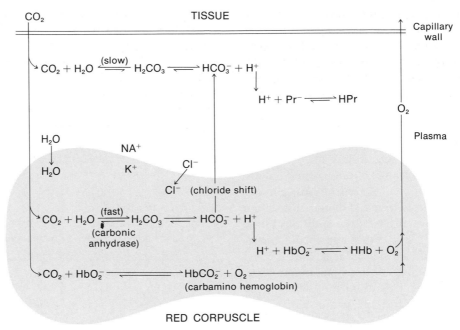

**14-9** Transfer of carbon dioxide between blood and tissues. The reverse process occurs in the lungs. Pr = protein; Hb = hemoglobin.

Carbon dioxide also reacts with $-NH_2$ groups on proteins and, in particular, hemoglobin to form carbamino compounds.

$$Hb-N\begin{array}{c}H\\ \\H\end{array} + CO_2 \rightleftharpoons H^+ + Hb-N\begin{array}{c}H\\ \\COO^-\end{array}$$

$$protein-N\begin{array}{c}H\\ \\H\end{array} + CO_2 \rightleftharpoons H^+ + protein-N\begin{array}{c}H\\ \\COO^-\end{array}$$

The extent of carbamino formation will depend on the number of available $-NH_2$ groups, as well as on blood pH and $P_{CO_2}$.

The sum of all forms of $CO_2$ in the blood—that is, molecular $CO_2$, $H_2CO_3$, $HCO_3^-$, $CO_3^{2-}$, and carbamino compounds—is referred to as the total $CO_2$ content of the blood. The $CO_2$ content varies with $P_{CO_2}$,

and the relationship can be described graphically in the form of a $CO_2$ dissociation curve (Figure 14-10). As $P_{CO_2}$ increases, the major change is in the bicarbonate content of the blood. The formation of bicarbonate is, of course, pH-dependent. The relationship between these parameters can be described in graphical form by a "Davenport" diagram (Figure 14-11). A decrease in pH at constant $P_{CO_2}$ is associated with a fall in bicarbonate. The pH of red blood cells is less than that of plasma, but $P_{CO_2}$ is in equilibrium across the cell membrane. Therefore, bicarbonate levels are lower in erythrocytes than in plasma.

Carbon dioxide is added to the blood in the tissues and removed at the respiratory surface, and levels of $CO_2$, $HCO_3^-$, and carbamino compounds all change during this transfer. Carbon dioxide both enters and leaves the blood as molecular $CO_2$ rather than as bicarbonate ion because $CO_2$ molecules are nonpolar and diffuse through membranes much more rapidly than $HCO_3^-$ ions. In the tissues, $CO_2$ enters the blood and either is hydrated to form bicarbonate ions or reacts with $-NH_2$ groups of hemoglobin and other proteins to form carbamino compounds. The reverse process occurs when $CO_2$ is unloaded from the blood.

The $CO_2$ hydration–dehydration reaction is slow and has an uncatalyzed time course of several seconds.

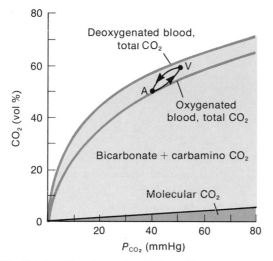

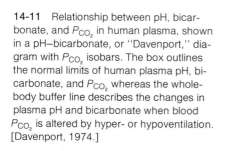

**14-10** Relationship between total $CO_2$ and $P_{CO_2}$ in oxygenated and deoxygenated blood. The volume of molecular $CO_2$ increases linearly with $P_{CO_2}$. A and V refer to arterial and venous blood levels, respectively.

In the presence of the enzyme *carbonic anhydrase,* found in red blood cells, the reaction approaches equilibrium in much less than a second. Although plasma has a higher total $CO_2$ content than red blood cells, most of the $CO_2$ entering and leaving the plasma does so via the red blood cell. The reason for this is that carbonic anhydrase is present in the red blood cell and not in the plasma. Therefore, bicarbonate or $CO_2$ formation occurs much more rapidly in the red blood cell, and bicarbonate ions subsequently diffuse either from or into the plasma (Figure 14-9). A second reason why most of the $CO_2$ entering or leaving the blood passes through the red blood cell is that deoxygenated hemoglobin (Hb) acts as a proton acceptor, facilitating the formation of bicarbonate:

$$CO_2 + H_2O \rightleftharpoons H_2CO_3 \rightleftharpoons$$
$$HCO_3^- + H^+ + HbO_2 \rightleftharpoons HHb + O_2$$

Thus, changes in pH caused by the hydration of $CO_2$ are minimized because of the binding of protons with the formation of deoxyhemoglobin as oxygen is released to the tissues. For every $HCO_3^-$ or carbamino compound formed, a proton is liberated; as deoxygenation proceeds, however, more proton acceptors become available on the hemoglobin molecule. In fact, complete deoxygenation of saturated hemoglobin, releasing 1 mol $O_2$, results in the binding of 0.7 mol of

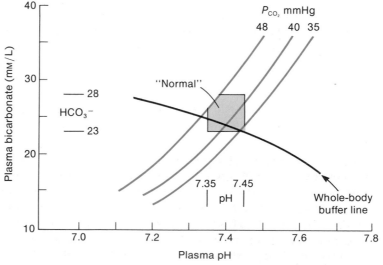

**14-11** Relationship between pH, bicarbonate, and $P_{CO_2}$ in human plasma, shown in a pH–bicarbonate, or "Davenport," diagram with $P_{CO_2}$ isobars. The box outlines the normal limits of human plasma pH, bicarbonate, and $P_{CO_2}$ whereas the whole-body buffer line describes the changes in plasma pH and bicarbonate when blood $P_{CO_2}$ is altered by hyper- or hypoventilation. [Davenport, 1974.]

hydrogen ions. At a respiratory quotient of 0.7 (pure fat diet), the transport of $CO_2$ can proceed without any change in blood pH. If the respiratory quotient is 1, the additional 0.3 mol $H^+$ is buffered by blood proteins, including hemoglobin, and blood undergoes only a small change in pH.

At a given $P_{CO_2}$, deoxyhemoglobin binds more protons and carbamino $CO_2$ than does oxyhemoglobin; the total $CO_2$ content of deoxygenated blood at a given $P_{CO_2}$ is therefore higher than that of oxygenated blood (Figure 14-10). Thus, deoxygenation of hemoglobin in the tissues reduces the change in $P_{CO_2}$ and pH as $CO_2$ enters the blood; this is termed the *Haldane effect.*

On entering the blood, $CO_2$ diffuses into red blood cells, and bicarbonate is formed rapidly in the presence of carbonic anhydrase. This results in a rise in bicarbonate levels, and as a result, $HCO_3^-$ leaves the red blood cells. Electrical balance within the cells is maintained by anion exchange; as $HCO_3^-$ moves from the red blood cells into the plasma, there is a net influx of $Cl^-$ ions into the erythrocyte. This process is called the *chloride shift.*

The rate of movement of $CO_2$, $HCO_3^-$, $Cl^-$, and $O_2$ into or out of the red blood cell will be determined by the surface/volume ratio of the cell and the diffusion coefficient of these substances through the red cell, as well as by the transport capacity of the $HCO_3^-/Cl^-$ exchange mechanism. The size of erythrocytes varies considerably among species. Hence it is not surprising that small erythrocytes, with their larger surface/volume ratio, are oxygenated faster than larger cells (Figure 14-12). The size of erythrocytes may also be important in determining the rates of $HCO_3^-$ and $Cl^-$ movement between the red blood cell and the plasma. Nevertheless, it is not clear why different species have evolved red blood cells of different sizes.

Since gas transfer is a dynamic process that takes place as blood moves rapidly through capillaries, rates of diffusion, reaction velocities, and steady-state conditions for gases in blood must all be taken into account. For instance, a Bohr shift would have little importance if it occurred after the blood had left the capillaries that supply an active tissue. The Bohr shift, in fact, occurs very rapidly, having a half-time at 37°C of 0.12 s in human red blood cells. A reduction in temperature always decreases reaction velocities

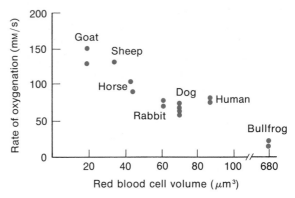

**14-12**   Relationship between the rate of oxygenation and volume of red blood cells. Small cells are oxygenated faster than large cells. [Holland and Forster, 1966.]

in any one species. Nevertheless, eel red blood cells at 15°C have the same half-time for the Bohr shift as human red blood cells at 37°C. These two species have evolved similar reaction velocities for the Bohr shift at very different temperatures. In most instances, reaction velocities of $O_2$ and $CO_2$ in blood are too fast to be important rate-limiting steps in gas transfer, and animals do not adjust reaction velocities in order to regulate rates of gas transfer. Variations in gas-transfer rates in mammals are achieved either by adjusting breathing rate and volume or by adjusting the flow rate and distribution of blood in both tissues and the respiratory surface.

## The Vertebrate Lung: Air Breathing

The previous section considered the properties of oxygen and carbon dioxide and described how these gases are carried in the blood. This section examines the ways in which $O_2$ and $CO_2$ move from the environment into the blood—for example, between air and blood across the *respiratory epithelium* of the *lung.*

The transfer of $O_2$ and $CO_2$ between air and blood will be considered first, followed in a later section by a discussion of gas transfer between water and blood across gills. This order of presentation was chosen because the structure of a system for gas exchange is influenced by the properties of the media, as well as the requirements of the animal. For example, the lungs of mammals have a very different structure from the gills of fish and are ventilated in a different manner. This dissimilarity exists because the density and viscosity of water are both approximately 1000 times greater than that of air, but water contains only one-thirtieth as much molecular oxygen. Moreover, gas molecules diffuse 10,000 times more rapidly in air than in water. Thus, in general, air breathing consists of the reciprocal movement of air into and out of the lungs (Figure 14-13A), whereas water breathing consists of a unidirectional flow of water over the gills (Figure 14-13B). These variations in the environment, in the structure of the respiratory apparatus, and in the nature of ventilation result in differences in the partial pressures of gases in the blood and tissues of air-breathing and water-breathing animals, particularly in $P_{CO_2}$.

## FUNCTIONAL ANATOMY OF THE LUNG

The vertebrate lung develops as a diverticulum of the gut and consists of a complex network of tubes and sacs, the actual structure varying considerably among species. If one compares the lungs of amphibians, reptiles, and mammals in that order, the sizes of terminal air spaces are seen to become smaller, but the total number per unit volume of lung is seen to become greater. The structure of the amphibian lung is variable, ranging from a smooth-walled pouch in some urodeles to a lung subdivided by septa and folds into numerous interconnected air sacs in frogs and toads. The degree of subdivision is increased in reptiles, and increases even more in mammals, the total effect being an increase in respiratory surface area per unit volume of lung. There is considerable variation in area of the respiratory surface in mammals; in general, the area

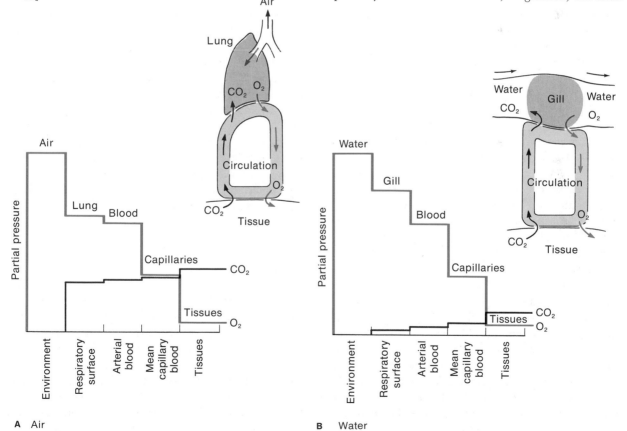

**14-13** Changes in $P_{O_2}$ and $P_{CO_2}$ between (**A**) air and (**B**) water and tissues in air-breathing and water-breathing animals.

increases with body weight (Figure 14-14A) and oxygen uptake (Figure 14-14B).

The mammalian lung consists of millions of blind-ending, interconnected sacs, termed *alveoli.* The *trachea* subdivides to form *bronchi,* which branch repeatedly, leading eventually to *terminal* and then *respiratory bron-* chioles, each of which is connected to a terminal spray of alveolar ducts and sacs (Figure 14-15). The total cross-sectional area of the airways increases rapidly as a result of extensive branching, although the diameter of individual air ducts decreases from the trachea to the terminal bronchioles. Gases are transferred across the thin-walled alveoli (Figure 14-16A, B) found throughout the portion of the lung distal to the terminal bronchiole, termed the *acinus.* The acinus, made up of the

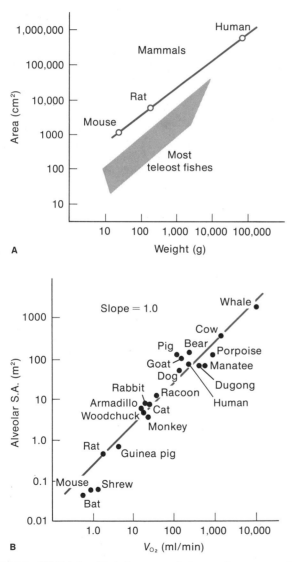

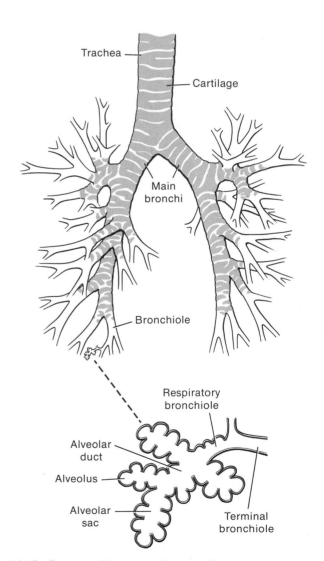

**14-14** (A) Relationship between respiratory surface area and body weight for a number of vertebrates. [Randall, 1970.] (B) Relationship between alveolar surface area (S.A.) and oxygen uptake in mammals. [Tenney and Temmers, 1963.]

**14-15** Structure of the mammalian lung. The respiratory epithelium is shown in color.

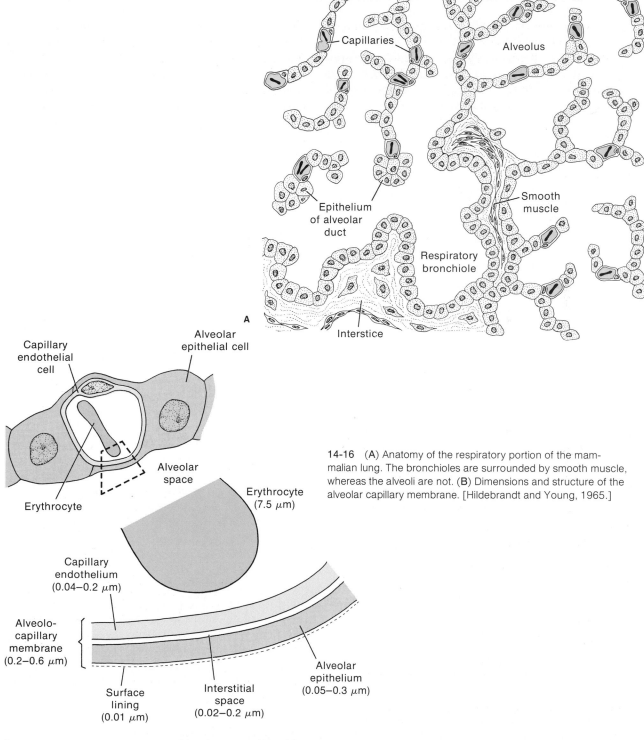

Capillaries

Alveolus

Epithelium
of alveolar
duct

Smooth
muscle

Respiratory
bronchiole

Interstice

**A**

Capillary
endothelial
cell

Alveolar
epithelial cell

Erythrocyte

Alveolar
space

Erythrocyte
(7.5 $\mu$m)

Capillary
endothelium
(0.04–0.2 $\mu$m)

Alveolo-
capillary
membrane
(0.2–0.6 $\mu$m)

Surface
lining
(0.01 $\mu$m)

Interstitial
space
(0.02–0.2 $\mu$m)

Alveolar
epithelium
(0.05–0.3 $\mu$m)

**B**

Alveolar space (50–300 $\mu$m)

14-16    (A) Anatomy of the respiratory portion of the mammalian lung. The bronchioles are surrounded by smooth muscle,
whereas the alveoli are not. (B) Dimensions and structure of the
alveolar capillary membrane. [Hildebrandt and Young, 1965.]

respiratory bronchioles, the alveolar ducts, and the alveolar sacs, constitutes the respiratory portion of the lung. The airways leading to the acinus constitute the nonrespiratory portion of the lung. Alveoli in adjoining acini are interconnected by a series of holes, the *pores of Kohn,* allowing the collateral movement of air, which may be a significant factor in gas distribution during lung ventilation.

Air ducts leading to the respiratory portion of the lung contain cartilage and a little smooth muscle and are lined with cilia, which move particles toward the mouth. Cartilage is absent, being replaced by smooth muscle in the respiratory portions of the lung. Activity in this smooth muscle can have a marked effect on the dimensions of the airways in the lungs.

Small mammals have a higher resting $O_2$ uptake per unit body weight than large mammals; this uptake is associated with their greater alveolar surface area per unit body weight. The increase in area is achieved by a reduction in the size but an increase in the number of alveoli per unit volume of lung. In the human lung, most alveoli develop after birth. There is a rapid increase in number of alveoli, the adult complement of about 300 million being attained by the age of eight years; subsequent increases in respiratory area are achieved by increases in the volume of each alveolus. The resting $O_2$ uptake per unit weight is higher in children than in adults, and once again there is a correlation between uptake per unit weight and alveolar surface area per unit body weight.

The diffusion barrier in mammals is made up of an aqueous surface film, the epithelial cells of the alveolus, the interstitial layer, endothelial cells of the blood capillaries, plasma, and the wall of the red blood cell (Figure 14-16). It is generally assumed, but not demonstrated, that the coefficient of diffusion for gases does not vary in tissues obtained from the lungs of different animals, the only structural variables being lung area and diffusion distance between air and blood.

Various terms are used to describe different types of breathing. *Eupnea* refers to the normal, quiet breathing typical of an animal at rest. *Hyperventilation* and *hypoventilation* refer, respectively, to an increase and a decrease in the amount of air moved in or out of the lung by changes in the rate and/or depth of breathing, such that ventilation no longer matches $CO_2$ production and blood $CO_2$ levels change. When breathing increases with $CO_2$ production, as during exercise, the increased lung ventilation is called *hyperpnea. Apnea* refers to the absence of breathing. *Dyspnea* denotes labored breathing associated with the unpleasant sensation of breathlessness, whereas *polypnea* indicates an increase in breathing rate without an increase in the depth of breathing.

Air exchanged between the alveoli and the environment must pass through a series of tubes (trachea, bronchi, nonrespiratory bronchioles) not directly involved in gas transfer. The volume of air contained in these nonrespiratory tubes is referred to as the *anatomical dead space.* Some air may be supplied to nonfunctional alveoli, or certain alveoli may be ventilated at too high a rate, increasing the volume of air not directly involved in gas exchange. This volume of air, termed the *physiological dead space,* is equal to or greater than, but includes, the anatomical dead space. The amount of air moved in or out of the lungs with each breath is referred to as the *tidal volume.* The amount of air moving in and out of the alveolar air sacs is the tidal volume minus the anatomical dead-space volume, and is referred to as the *alveolar ventilation volume.* Only this gas volume is directly involved in gas transfer. The lungs are not completely emptied even at maximal expiration, leaving a *residual volume* of air in the lungs. The maximum volume of air that can be moved in or out of the lungs is referred to as the *vital capacity* of the lungs (Figure 14-17).

Oxygen content is lower and $CO_2$ content higher in alveolar gas than in ambient air because only a portion of the lungs' gas volume is changed with each breath. There are differences in the movement of $O_2$ and $CO_2$ along the air ducts, $O_2$ diffusing toward the alveoli and $CO_2$ away from them. Alveolar ventilation in humans is about 350 ml per breath, whereas functional residual volume of the lungs exceeds 2000 ml. The volume of alveoli increases during inspiration by elongation and widening of the ducts leading to the alveoli. During breathing, air moves in and out of the acinus and may also move between adjacent alveoli through the pores of Kohn. Mixing of gases in the ducts and alveoli occurs by diffusion and by convection currents caused by breathing (Figure 14-18). Partial pressures of $O_2$ and $CO_2$ are probably fairly uniform across the

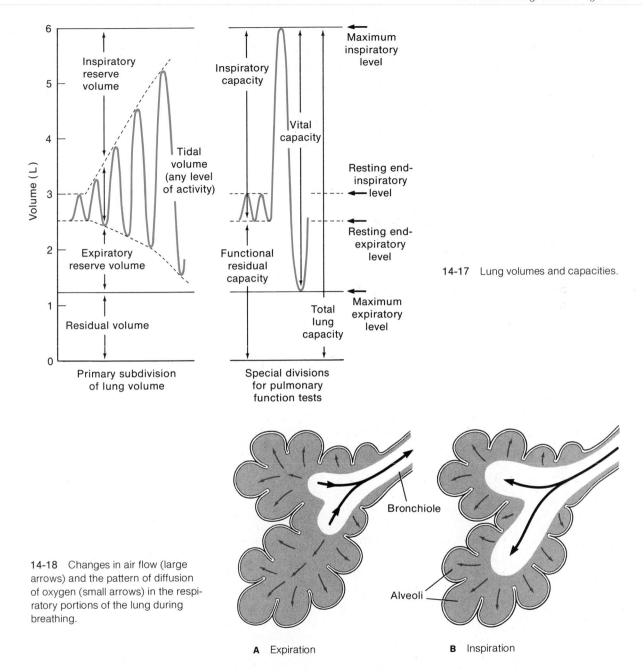

14-17   Lung volumes and capacities.

14-18   Changes in air flow (large arrows) and the pattern of diffusion of oxygen (small arrows) in the respiratory portions of the lung during breathing.

**A** Expiration   **B** Inspiration

alveoli, because diffusion is rapid in air and the distances involved are small. The partial pressures of gases within the alveoli oscillate in phase with the breathing movements, the magnitude depending on the extent of tidal ventilation.

The $O_2$ and $CO_2$ levels in alveolar gas will be determined by both the rate of gas transfer across the respiratory epithelium and the rate of alveolar ventilation. Alveolar ventilation will depend on breathing rate, tidal volume, and anatomical dead-space volume.

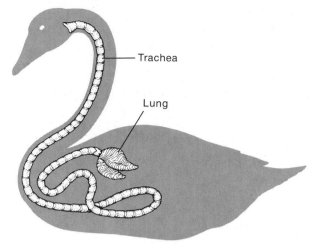

**14-19**  The extremely long trachea of the trumpeter swan results in a large anatomical dead-space volume. [Banko, 1960.]

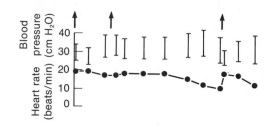

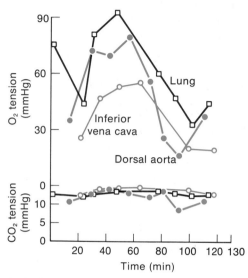

**14-20**  Changes in $O_2$ and $CO_2$ tensions in the lung, dorsal aorta, and inferior vena cava of a 515 g *Amphiuma* during two breathing–diving cycles. Heart rate and systolic and diastolic blood pressures in the dorsal aorta are also plotted. Vertical arrows indicate times when the animal surfaced and ventilated its lungs. [Toews et al., 1971.]

Variations in the magnitude of the anatomical dead space will alter gas tensions in the alveolus in the absence of changes in tidal volume. Thus, artificial increases in anatomical dead space, produced in human subjects breathing through a length of hose, result in a rise in $CO_2$ and a fall in $O_2$ in the lungs and a subsequent increase in tidal volume. Many animals, such as the giraffe and the swan, have long necks, which increase the tracheal length and therefore anatomical dead space. The trumpeter swan is an extreme example of tracheal elongation (Figure 14-19). Without concomitant increases in tidal volume, gas tensions in the lungs and blood would be adversely affected.

Breathing rate and tidal volume vary considerably in animals. Humans breathe about 12 times per minute and have a tidal volume at rest of about one-tenth of total lung volume. The exclusively aquatic but air-breathing amphibian *Amphiuma*, which lives in swamp water, rises to the surface of the water about once each hour to breathe; its tidal volume, however, is more than 50% of its lung volume. This large tidal volume produces large, slow oscillations in $P_{O_2}$ in the lung and blood; compared with the rather small oscillations in the more rapidly breathing human, these oscillations are more or less in phase with the breathing movements (Figure 14-20). *Amphiuma* is preyed on by snakes and is most vulnerable when it rises to

breathe. As it lives in water of low oxygen content, aquatic respiration is not a suitable alternative. The hazard of being eaten while surfacing to breathe may have influenced the evolution of its very low breathing rate, its large tidal and lung volume, and the cardiovascular adjustments it is capable of for maintaining $O_2$ delivery to the tissues in the face of widely oscillating blood gas levels.

In summary, $O_2$ and $CO_2$ levels in alveolar gas are determined by ventilation and the rate of gas transfer. Ventilation of the respiratory epithelium is determined by breathing rate, tidal volume, and anatomical dead-space volume. The nature and extent of ventilation also influences the magnitude of oscillations in $O_2$ and $CO_2$ in blood during a breathing cycle.

## BLOOD CIRCULATION TO THE LUNG

The lung, like the heart, receives blood from two sources. The major flow is of deoxygenated blood from the pulmonary artery that perfuses the lung, taking up $O_2$ and giving up $CO_2$. A second, smaller supply, the *bronchial circulation,* comes from the systemic (body) circulation, and supplies the lung tissues with $O_2$ and other substrates for growth and maintenance. Our discussion here is confined to the *pulmonary circulation.*

In mammals, the lung is perfused at lower pressures than is the systemic circulation. Pulmonary blood pressure in birds and mammals is low to reduce filtration of fluid into the lung, because any fluid that collects at the lung surface increases the diffusion distance between blood and air and reduces gas transfer. To further ensure that no fluid collects in the lung, there is an extensive lymphatic drainage of lung tissues.

The mean arterial pressure in the human lung is about 17 cm $H_2O$, oscillating between 10 and 30 cm $H_2O$ with each contraction of the heart. In the vertical (upright) human lung, arterial pressure is just sufficient to raise blood to the apex of the lung; hence flow is minimal at the top and increases toward the base of the lung (Figure 14-21). Blood is distributed more evenly to different parts of the horizontal lung.

The pulmonary vessels are very distensible and are subject to distortion by breathing movements. Small vessels within the interalveolar septa are particularly sensitive to changes in alveolar pressure. The diameter of these thin-walled collapsible capillaries is determined by the transmural pressure (blood pressure within capillaries minus alveolar pressure), and if alveolar pressure exceeds blood pressure within the capillaries, they collapse and blood flow ceases. This collapse may occur at the apex of the vertical human lung, where blood pressures are low (Figure 14-21). If pulmonary arterial pressure is greater than alveolar pressure, which in turn is greater than pulmonary venous pressure, then the difference between arterial and alveolar pressure will determine the diameter of capillaries in the interalveolar septa and, in the manner of a sluice gate, control blood flow through the capillaries. Venous pressure will not affect flow into the venous reservoir as long as alveolar pressure exceeds venous pressure. Flow in the upper portion of the vertical lung is probably determined in this way by the difference between arterial blood pressure and alveolar pressure. Arterial blood pressure (and therefore blood flow) increases with distance from the apex of the lung. In the bottom half of the vertical lung,

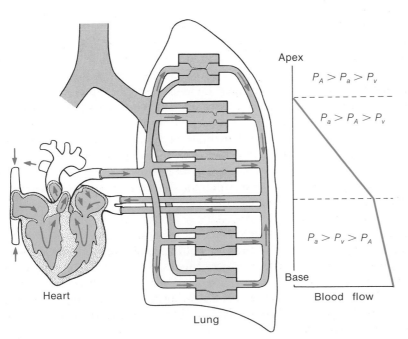

Apex

$P_A > P_a > P_v$

$P_a > P_A > P_v$

$P_a > P_v > P_A$

Base

Blood flow

Heart

Lung

**14-21** Pattern of blood flow in the vertical human lung. The boxes within the lung represent the condition of vessels in the interalveolar septum in different portions of the lung. $P_A$ = alveolar pressure; $P_a$ = arterial pressure; $P_v$ = venous pressure. [West, 1970.]

where venous pressure exceeds alveolar pressure, blood flow is determined by the difference between arterial and venous blood pressures. This pressure difference does not vary, but both the arterial and venous pressures increase toward the base of the lung. This increase in absolute pressure results in an expansion of vessels and, therefore, a decrease in resistance to flow. Thus, flow increases toward the base of the lung, even though the arterial–venous pressure difference does not change (Figure 14-21). The position of the lungs with respect to the heart is therefore an important determinant of pulmonary blood flow. The lungs surround the heart, thus minimizing the effect of gravity on pulmonary blood flow as an animal changes from a horizontal to vertical position. This close proximity of lungs and heart within the thorax also has significance for cardiac function: the reduced pressures within the thorax during inhalation aid venous return to the heart.

There is an absence of well-defined arterioles in the mammalian pulmonary circulation, but there are sympathetic adrenergic and parasympathetic cholinergic innervations of the smooth muscle around blood vessels and bronchioles. The pulmonary circulation has much less innervation than does the systemic circulation, and is relatively unresponsive to nerve stimulation or injected drugs. Sympathetic nerve stimulation or the injection of norepinephrine causes a slight increase in resistance to blood flow, whereas parasympathetic nerve stimulation or acetylcholine has the opposite effect.

Reductions in either oxygen levels or pH have a marked direct effect on pulmonary blood vessels: both cause a local vasoconstriction. The response to low oxygen is considered to be important in distributing blood to well-ventilated portions of the lung. Those regions of the lung that are poorly ventilated will have low alveolar $O_2$ levels, and this will cause a local vasoconstriction and therefore a reduction in blood flow to that area of the lung. Alternatively, a well-ventilated area of the lung will have high alveolar oxygen levels, local blood vessels will be dilated, and blood flow will be high. This vasoconstrictor response to low oxygen, which is the opposite to that observed in systemic capillary networks, ensures that blood flows to where the air is distributed within the lung.

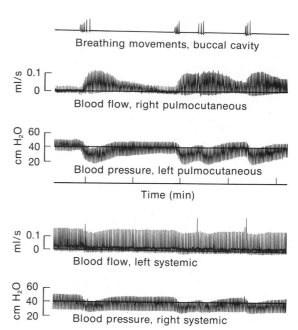

**14-22** Pressures and flows in the arterial arches of *Xenopus*, 85 g. Pressure changes in the buccal cavity produced by movements of the buccal floor are recorded on the upper trace. Each of the bursts of movement recorded was of the lung-ventilating type. The effect of breathing movements on individual flow and pressure pulses can be seen. [Shelton, 1970.]

Cardiac output to the pulmonary circuit is identical to cardiac output in the systemic circuit in mammals and birds. In amphibians and reptiles, with a single or partially divided ventricle that ejects blood into both the pulmonary and the systemic circulation, the ratio of pulmonary to systemic blood flow can be altered. In turtles and frogs, there is a marked increase in blood flow to the lung following a breath. During periods between breaths in the frog *Xenopus*, pulmonary blood flow decreases, but systemic blood flow is hardly changed (Figure 14-22).

## MECHANISMS OF VENTILATION

The lungs of mammals are elastic, multichambered bags suspended within the *pleural cavity* and opening to the exterior via a single tube, the trachea (Figure 14-23A). The walls of the pleural cavity, often referred to as the *thoracic cage*, are formed by the ribs and the

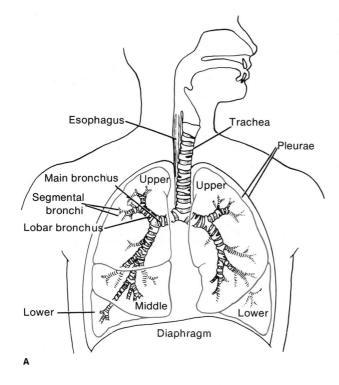

**A**

*diaphragm.* When isolated, lungs are somewhat smaller than they are in the thoracic cage because they are elastic. In situ, this elasticity creates a pressure several mmHg below atmospheric in the fluid-filled pleural space between the lungs and the wall of the thoracic cage. This pleural space is of low volume and is sealed; the lungs fill most of the thoracic cage. If the thoracic cage is punctured, air is drawn into the pleural cavity and the lungs collapse—a state known as *pneumothorax.* Normally, the space between the lungs and thorax is small and fluid-filled. Fluids are essentially incompressible, so when the thoracic cage changes volume, the gas-filled lungs do too.

If intact lungs are filled to various volumes and the entrance is closed with the muscles relaxed, then, as expected, alveolar pressure is found to increase with lung volume. At low pulmonary volumes, alveolar pressure is less than ambient owing to the resistance of the thorax to collapse. At large pulmonary volumes, alveolar pressure exceeds ambient because of the forces required to expand the thoracic cage. If lung volume is large, then once the mouth and glottis are opened, air will flow out of the lungs because the weight of the ribs will reduce pulmonary volume. At some intermediate volume, $V_r$, alveolar pressure in the relaxed thorax is equal to ambient pressure (Figure 14-23B).

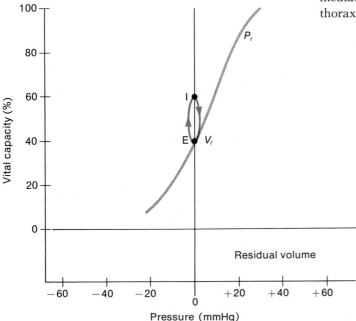

**B**

**14-23** Breathing in humans. (**A**) The thoracic cavity and lungs. The right lung has three lobes; the left lung only two. (**B**) The effect of lung volume on pressure ($P_r$) within the thorax when muscles are relaxed, but the glottis is closed. $V_r$ is the lung volume when alveolar pressure is the same as ambient pressure and the lung–chest system is relaxed. The points I and E represent the pressure and volume of the system following inspiration (I) and expiration (E) during quiet breathing.

During normal breathing, the thoracic cage is expanded and contracted by a series of skeletal muscles, the diaphragm, and the external and internal intercostal muscles (Figure 14-24). Contractions of these muscles are determined by activity of motoneurons controlled by the respiratory center within the medulla oblongata (p. 292). The volume of the thorax is increased as the ribs are raised and moved outward by the contraction of the external intercostals and by the contraction (and therefore the lowering) of the diaphragm. Contractions of the diaphragm account for up to two-thirds of the increase in pulmonary volume. The increase in volume reduces alveolar pressure, and air is drawn into the lungs. Relaxation of the diaphragm and external intercostal muscles reduces thoracic volume, thereby raising alveolar pressure and forcing air out of the lungs. During quiet breathing, pulmonary volume is at $V_r$ between breaths (Figure 14-23), and exhalation is often passive and is simply due to relaxation of the diaphragm and external intercostals. With increased tidal volume, expiration becomes active, owing to contractions of the internal

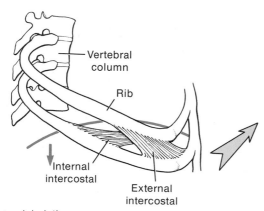

**A**    Inhalation

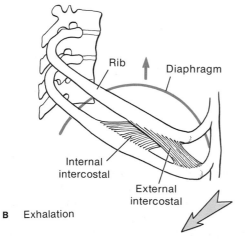

**B**    Exhalation

**14-24**  Changes in the position of the ribs and diaphragm during (**A**) inspiration and (**B**) expiration in mammals.

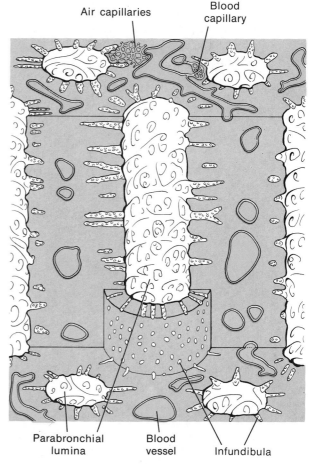

**14-25**  Semischematic drawing of the parabronchi characteristic of songbirds. Oxygen and carbon dioxide are exchanged between air and blood in the air capillaries branching from the parabronchi. [Duncker, 1972.]

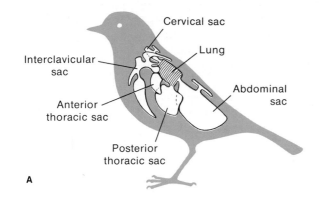

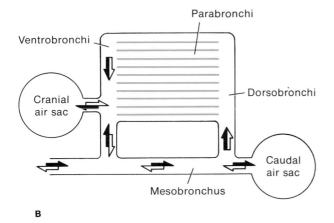

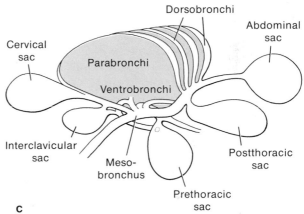

**14-26**   **(A)** Position of air sacs and lungs in a bird. [Salt, 1964.] **(B)** Diagram of a bird lung. The arrows indicate the qualitative flow direction, but not the flow rates. **(C)** Schematic diagram of the bronchial tree, with the connections to the air sacs. The air sacs of the cranial group (cervical, interclavicular, and prethoracic sacs) depart from the three cranial ventrobronchi, whereas the air sacs of the caudal group (postthoracic and abdominal sacs) are connected directly to the mesobronchus. [B and C after Scheid et al., 1972.]

intercostal muscles, and further reduces thoracic volume until it drops below $V_r$ at the end of expiration.

The mechanism of lung ventilation varies considerably in vertebrates. In birds, gas transfer takes place in small air capillaries (10 μm in diameter) that branch from tubes—the *parabronchi* (Figure 14-25), the functional equivalent of the mammalian alveolar sac. The parabronchi are a series of small tubes extending between large *dorsobronchi* and *ventrobronchi,* both of which are connected to an even larger tube, the *mesobronchus,* which joins the trachea anteriorly (Figure 14-26B). The parabronchi and connecting tubes form the lung, which is contained within a thoracic cavity. A tight horizontal septum closes the caudal end of the thoracic cage. The ribs are curved to prevent lateral compression and move forward only slightly during respiration; the volume of the thoracic cage and lung

changes little during breathing. The large flight muscles of birds are attached to the sternum and have little influence on breathing. Although there is no mechanical relation between flight and respiratory movements in birds, "in phase" flight and breathing movements may result from synchronous neural activation of the two groups of muscles involved.

How, then, is the avian lung ventilated? The answer lies in the associated air-sac system connected to the lungs (Figure 14-26A and B). As these air sacs are squeezed, air is forced through the parabronchi. The system of air sacs extends as diverticula and penetrates into adjacent bones and between organs, reducing the density of the bird. Of the many air sacs, only the thoracic (cranial) and abdominal (caudal) sacs show marked changes in volume during breathing. Volume changes in the air sacs are achieved by a rocking motion of the sternum against the vertebral column and by lateral movements of the posterior ribs. Air flow is two-directional in the mesobronchus, but unidirectional through the parabronchi (Figure 14-26B). During inspiration, air flows into the caudal air sacs through the mesobronchus; air also moves into the cranial air sacs via the ventrobronchus, the dorsobronchus, and the parabronchi. During expiration, air leaving the caudal air sacs passes through the parabronchi and, to a lesser extent, through the mesobronchus to the trachea. The cranial air sacs, whose volume changes less than that of the caudal air sacs, are reduced somewhat in volume by air moving from the cranial sac via the ventrobronchi and the mesobronchus to the trachea. Thus, there is a unidirectional flow of air through the parabronchi during both phases of breathing. This flow is achieved not by mechanical valves but by *aerodynamical valving*. The openings of the ventrobronchi and dorsobronchi into the mesobronchus show a variable, direction-dependent resistance to air flow. The structure of the openings is such that eddy formation, and therefore resistance to flow, varies with the direction of air flow.

The ribs of reptiles, like those of mammals, form a thoracic cage around the lungs. During inhalation, the ribs are moved cranially and ventrally, enlarging the thoracic cage. As this expansion reduces the pressure within the cage below atmospheric pressure, the nares and glottis open, and air flows into the lungs. Relaxation of muscles that enlarge the thoracic cage releases energy stored in stretching the elastic component of the lung and body wall, allowing passive exhalation. Reptiles do not possess a diaphragm, but pressure differences between the thoracic and abdominal cavities have been recorded, indicating at least a functional separation of these cavities. In tortoises and turtles, the ribs are fused to a rigid shell. The lungs are filled by outward movements of the limb flanks and/or the plastron (ventral part of the shell) and by forward movement of the shoulders. The reverse process results in lung deflation. As a result, the retraction of limbs and head into the shell occurs at the expense of pulmonary volume.

In frogs, the nose opens into a buccal cavity, which is connected via the glottis to paired lungs. The frog can open and close its nares and glottis. Air is drawn into the buccal cavity and then forced into the lungs by raising the buccal floor with the nares closed and the glottis open. This lung-filling process may be repeated several times in sequence. Reducing pulmonary volume may also be a step process, the lungs releasing air in portions to the buccal cavity (Figure 14-27). There may be alternate filling and emptying move-

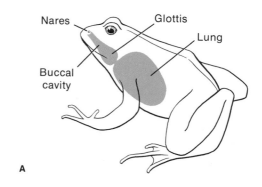

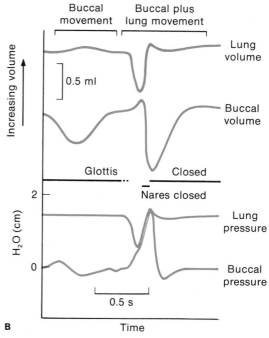

**14-27** Changes in pressure and volume in the buccal cavity and lung of a frog during buccal movements alone and during buccal and lung movements. [West and Jones, 1975.]

ments of the lungs during which some air is exhaled but the remainder, mixed with air in the buccal cavity, is pumped back into the lungs. That is, a mixture of pulmonary air, presumably low in $O_2$ and high in $CO_2$, is mixed with fresh air in the buccal cavity and returned to the lungs. The reason for this complex method of lung ventilation is not clear. Ventilation may be directed toward reducing oscillations in $CO_2$ levels in the lungs in order to stabilize and regulate blood $P_{CO_2}$ and control blood pH.

There is also variety among the lungs of invertebrates; some are ventilated, and others are not, relying only on diffusion of gases between the lung and the environment. Spiders have paired, nonventilated lungs on the abdomen. The respiratory surface consists of a series of thin, blood-filled plates that extend like the leaves of a book into a cavity guarded by an opening, or *spiracle*. The spiracle can be opened or closed to regulate the rate of water loss from these *book lungs*. Snails and slugs have ventilated lungs that are well-vascularized invaginations of the body surface, the mantle cavity. The volume change that the snail lung is capable of undergoing enables the animal to emerge from and withdraw into its rigid shell. When the snail retracts into its shell, the lungs empty, a situation similar to that seen in tortoises. In aquatic snails, the lungs serve to reduce the animal's density.

### THE PROBLEM OF ALVEOLAR COLLAPSE

The small dimensions of the fragile alveolar sacs create mechanical problems that might cause them to collapse. This problem can be illustrated by a tiny bubble that is alternately inflated and then deflated. According to *Laplace's Law*, the pressure differential between the inside and outside of a bubble is proportional to $2y/r$, where $y$ is the wall tension per unit length and $r$ the radius of the bubble. If two bubbles have a similar wall tension but a different radius, the pressure in the small bubble will be higher than that in the large bubble. As a result, if the bubbles are joined, the small bubble will empty into the large bubble (Figure 14-28).

A somewhat similar situation exists in the lung. We can consider the alveoli as a number of interconnected bubbles. If the wall tension is similar in alveoli of different size, the small alveoli will tend to collapse,

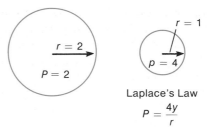

**A**   Two bubbles in air

Laplace's Law

$$P = \frac{4y}{r}$$

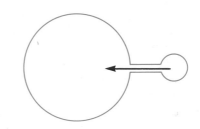

**B**   Bubbles joined

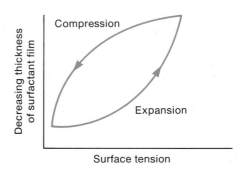

**C**   Properties of surfactant

**14-28** (A) Laplace's Law states that the pressure ($P$) in a bubble decreases with increased radius ($r$) if the wall tension ($y$) remains constant. The pressure in the small bubble is two times that of the large bubble, which is double the radius but has the same wall tension as the small bubble. The equation is written $4y/r$ rather than $2y/r$ because the bubble in air has an inner and an outer surface. (B) If the bubbles are joined, the small bubble (higher pressure) collapses into the large bubble (lower pressure). (C) The problem of alveolar collapse is ameliorated by a surfactant lining. As the surfactant film expands with the alveolus, the thickness of the film decreases and the surface tension increases, tending to stabilize the alveoli.

emptying into the larger alveoli. There is evidence, however, that wall tension increases with expansion, thereby stabilizing the alveoli and reducing the probability of alveolar collapse. In addition, surrounding tissue helps prevent overexpansion of alveoli. The wall tension depends on the properties of the alveolar wall and the surface tension at the liquid–air interface. The lungs are lined with a *surfactant,* a lipoprotein complex with a preponderance of dipalmitoyl lecithin. This lipoprotein complex bestows a very low surface tension on the liquid–air interface, which increases if the surface film is expanded and decreases if it is compressed. Because the film is expanded as the alveolar volume increases, the surfactant is spread out, causing it to have a smaller effect in lowering surface tension. As a result, the surface tension of the alveolus increases as it expands. The effect will be to minimize pressure differences between large and small alveoli and thus reduce the chance of collapse and permit easy inflation of any collapsed alveoli.

Surfactants are found in the lungs of amphibians, reptiles, birds, and mammals, and they may exist in some fishes that build bubble nests. Surfactants appear in the lung prior to birth in mammals. Their presence reduces the forces required to inflate the lung at birth. The lipoprotein film is stable, the lipid probably forming an outer monolayer firmly associated with a protein layer underneath. Surfactants are produced rapidly by specialized cells within the alveolar lining and have a half-life of about half a day.

### HEAT AND WATER LOSS WITH RESPIRATION

Increases in lung ventilation not only increase gas transfer but also result in more loss of heat and water. Thus, the evolution of lungs has involved some compromises. Air in contact with the respiratory surface becomes saturated with water vapor and comes into thermal equilibrium with the blood. Cool, dry air entering the lung of mammals is humidified and heated. Exhalation of this hot, humid air results in considerable loss of heat and water, which will be proportional to the rate of ventilation of the lung surface. Many air-breathing animals live in very dry environments, where water conservation is of paramount importance. It is therefore not surprising that these animals in particular have evolved means of minimizing the loss of water (discussed in Chapter 12).

The rates of heat and water loss from the lung are intimately related. As air is inhaled, it is warmed and humidified by evaporation of water from the nasal mucosa. Because the evaporation of water cools the nasal mucosa, a temperature gradient exists along the nasal passages. As the moist air leaves the lung, it is cooled, so that water condenses on the nasal mucosa, since the water-vapor pressure for 100% saturation decreases with temperature. Thus the cooling of exhalant air in the nasal passages results in the conservation of both heat and water. The blood circulation to the nasal mucosa is capable of supplying water to saturate the inhalant air, but the temperature gradients established by water evaporation and air movement are not destroyed by the circulation.

The structure of the nasal passages in vertebrates is variable, and to some extent it can be correlated with the ability of animals to regulate heat and water loss. Humans have only a limited ability to cool exhaled air, which is saturated with water vapor and is at a temperature only a few degrees below core body temperature.

Reptiles and amphibians, whose body temperatures adjust to the ambient temperature, exhale air saturated with water at temperatures about 0.5–1.0°C below body temperature. Pulmonary air temperatures and body surface temperatures are often slightly below ambient because of the continual evaporation of water. In some reptiles, however, body temperature is maintained above ambient. In the iguana, heat and water loss is controlled in a manner similar to that observed in mammals. In addition, this lizard conserves water by humidifying air with water evaporated from the excretory fluid of the nasal salt glands.

## The Vertebrate Gill: Water Breathing

Gills are ventilated with a unidirectional flow of water in most species (Figure 14-13B). Tidal flow of water, similar to that of air in the lung, would be costly because of the high density and viscosity of water. Blood flows through the gills, and the possible arrangements of the flow of blood relative to the flow of water are either concurrent or countercurrent, or some combination of these two arrangements (Figure 14-29). The advantage of a countercurrent over a concurrent flow

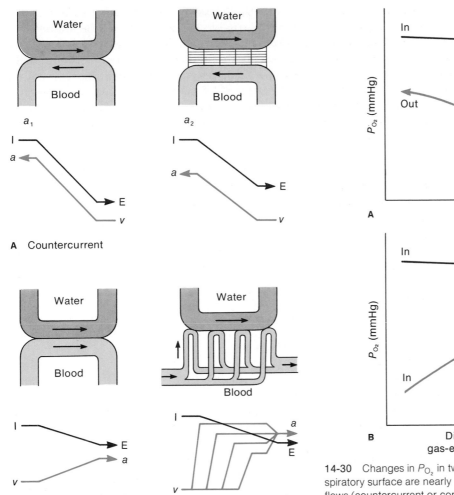

**A** Countercurrent

**B** Concurrent   **C** Multicapillary

**14-29** Various arrangements for the flows of water and blood at the respiratory surface in aquatic animals. Relative changes in $P_{O_2}$ in water and blood are indicated below each diagram. I = inhalant; E = exhalant; a = arterial blood; v = venous blood.

**14-30** Changes in $P_{O_2}$ in two fluids on either side of a respiratory surface are nearly independent of the arrangement of flows (countercurrent or concurrent), because fluid *a* has a much higher flow rate and/or oxygen solubility coefficient than does fluid *b*. A countercurrent arrangement of flows is most advantageous if the flows and oxygen solubility coefficients of the two fluids are equal.

of blood and water is that a larger difference in partial pressure of oxygen can be maintained across the exchange surface to allow more transfer of gas. A countercurrent flow is most advantageous if the capacity rate for $O_2$ (flow times $O_2$ content) is similar in blood perfusing and water flowing over the gills. If the capacity/rate ratio of blood and water is far from unity, there is little advantage of countercurrent over concurrent flow. For example, if water flow were very high

in relation to blood flow, there would be little change in $P_{O_2}$ in the water as it flowed over the gills, and the mean $P_{O_2}$ difference across the gills would be similar in concurrent and countercurrent arrangements of flow (Figure 14-30). But the capacity/rate ratio of oxygen of blood perfusing the gills and of oxygen in water flowing over the gills is close to unity, and the arrangement of water to blood flow is countercurrent in most fishes. The $O_2$ content of fish blood is generally higher than that of water, but the capacity/ rate ratio is maintained close to 1 by a flow rate of water that is much higher than the flow rate of blood.

Flow of water over the gills of teleost fishes is maintained by the action of skeletal muscle pumps in the buccal and opercular cavities. Water is drawn into the mouth, passes over the gills, and exits through the opercular (gill-covering) clefts (Figure 14-31). Valves guard the entrance to the buccal cavity and opercular clefts, maintaining a unidirectional flow of water over the gills. The buccal cavity changes volume by raising and lowering the floor of the mouth. The operculum swings in and out, enlarging and reducing the size of the opercular cavities. Changes in volume in the two cavities are nearly in phase, but a pressure differential is maintained across the gills throughout most of each breathing cycle. The pressure in the opercular cavity is a few mmHg below that in the buccal cavity, result-

ing in a unidirectional flow of water across the gills throughout most, if not all, of the breathing cycle (Figure 14-31).

Many active fish "ram-ventilate" their gills, opening their mouths so as to ventilate the gills by the forward motion of the body. The remora, a fish that attaches itself to the body of a shark, ventilates its own gills only when the shark stops swimming; normally, it relies on the motion of its host to ventilate its gills.

The lamprey is an exception to the rule that gill water flow is unidirectional. The mouth of this para-

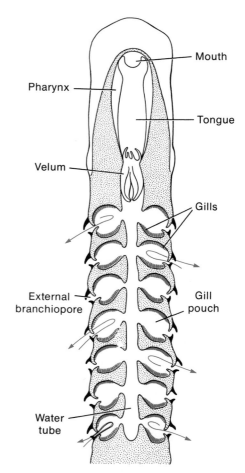

**14-32** Longitudinal transverse section through the head of an adult lamprey. Water is moved in and out of each gill pouch via the external branchiopore. Arrows mark the direction of water flow. The valves of the external branchiopore move in and out with the oscillating water flow.

**14-31** Schematic diagram of the process of gill ventilation in teleost fish.

sitic animal is often blocked by attachment to a host. The gill pouches, although connected internally to the pharyngeal and mouth cavities, are ventilated by tidal movements of water through a single external opening to each pouch (Figure 14-32). This unusual method of gill ventilation is clearly associated with a parasitic mode of life. The ammocoete larvae of lampreys are not parasitic and maintain a unidirectional flow of water over their gills, typical of aquatic animals in general.

The details of gill structure vary among species, but the general plan is similar. The gills of teleost fishes are taken to be representative of an aquatic respiratory surface. The four gill arches on either side of the head separate the opercular and buccal cavities (Figure 14-33A). Each arch has two rows of *filaments,* and each

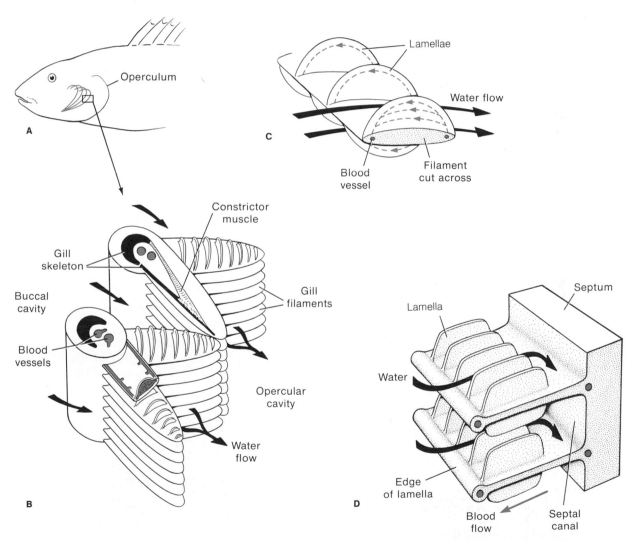

**14-33** (**A**) Position of the four gill arches beneath the operculum on the left side of a fish. (**B**) Part of two of these gill arches shown with the filaments of adjacent rows touching at their tips. Also shown are the blood vessels that carry the blood before and after its passage over the gills. (**C**) Part of a single filament with three secondary folds on each side. The flow of blood is in the opposite direction to that of the water. (**D**) Part of the dogfish gill. As in teleost fish, the flow of blood is in the opposite direction to that of the water. [A–C after Hughes, 1964; D after Grigg, 1970.]

filament, flattened dorsoventrally, has an upper and a lower row of *lamellae* (Figure 14-33B and C). The lamellae of successive filaments in a row are in close contact. The tips of filaments of adjacent arches are juxtaposed so that the whole gill forms a sieve-like structure in the path of water flow. Water flows in slitlike channels between neighboring lamellae. These channels are about 0.02–0.05 mm wide and about 0.2–1.6 mm long; the lamellae are about 0.1–0.5 mm high (Figure 14-34). As a result, the water flows in thin sheets between the lamellae, which represent the respiratory portion of the gill, and diffusion distances in water are reduced to a maximum of 0.01–0.025 mm (half the distance between adjacent lamellae on the same filament).

The lamellae are covered by thin sheets of epithelial cells (Figure 14-34), the internal walls of which are held together by *pillar cells,* which occupy about 20% of the internal volume of the lamella. Blood flows in the spaces between the pillar cells. The diffusion distance between the center of the red blood cell and the water is between 3 and 8 $\mu$m. The total area of the lamellae is large, varying 1.5–15 cm$^2$/g of body weight, depending on the size of the fish and on whether it is generally active or sluggish. Blood and water flow in a countercurrent arrangement (Figure 14-35).

## Regulation of Gas Transfer and Respiration

The regulation of the rate of $O_2$ and $CO_2$ transfer has been studied extensively only in mammals. This section is therefore confined mainly to a discussion of the regulation of gas transfer in mammals. The movement of $O_2$ and $CO_2$ between the environment and mitochondria in mammals is regulated by altering lung ventilation and the flow and distribution of blood within the body. Here we place emphasis on the control of breathing; Chapter 13 presents details of the control of the cardiovascular system.

### VENTILATION/PERFUSION RATIOS
Energy is expended in ventilating the respiratory surface with air or water and in perfusing the respiratory epithelium with blood. The total cost of these two processes is difficult to assess, but probably amounts to 4–20% of the total aerobic energy output of the

animal, depending on the species in question and the physiological state of the animal. Thus, gas transfer between environment and cell accounts for a considerable proportion of the total energy output of the animal and represents a significant selective pressure

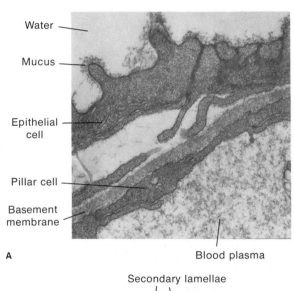

Water —
Mucus —
Epithelial cell —
Pillar cell —
Basement membrane —
Blood plasma

**A**

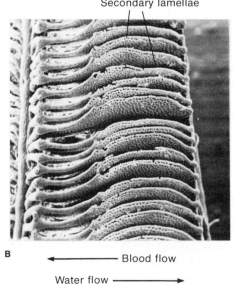

Secondary lamellae

**B**

← Blood flow
Water flow →

**14-34**   (A) Transverse section through gill lamella of trout, showing the water–blood barrier. Magnification 40,000×. (B) Scanning electron micrograph of a plastic cast of the vasculature of a trout gill filament, showing several lamellae. Magnification 160×. [Courtesy of B. J. Gannon.]

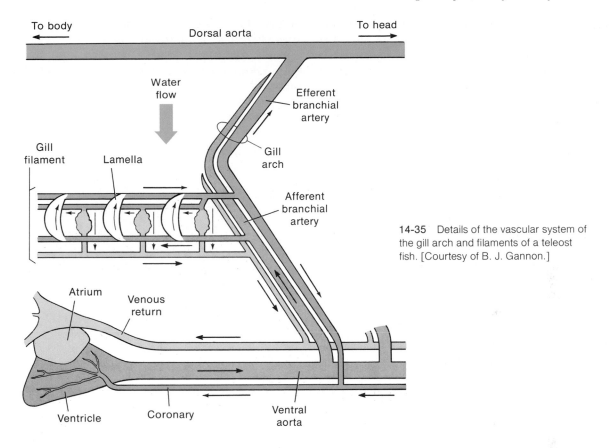

To body

Dorsal aorta

To head

Water flow

Efferent branchial artery

Gill arch

Gill filament

Lamella

Afferent branchial artery

**14-35** Details of the vascular system of the gill arch and filaments of a teleost fish. [Courtesy of B. J. Gannon.]

Atrium

Venous return

Ventricle

Coronary

Ventral aorta

in favor of the evolution of mechanisms for the close regulation of ventilation and perfusion in order to conserve energy.

The rate of blood perfusion of the respiratory surface is related to the requirements of the tissues for gas transfer and to the gas-transport capacities of the blood. To ensure that sufficient oxygen is delivered to the respiratory surface to saturate the blood with oxygen, the rate of ventilation must be adjusted in accord with the perfusion rate.

The ventilation/perfusion ratio at the respiratory surface depends on the difference in gas content between arterial and venous blood and between inhalant and exhalant air or water (Box 14-2). This difference, in turn, will depend on the partial pressure of oxygen and the gas content per unit volume per mmHg of $P_{O_2}$ in the blood and in the medium. The oxygen content of arterial blood in humans is similar to that of air. The $\dot{V}_A/\dot{Q}$ ratio, therefore, is about 1 in

humans (Figure 14-36). Water, however, contains only about one-thirtieth as much dissolved oxygen as an equivalent volume of air at the same $P_{O_2}$ and temperature. Thus, in fishes, the ratio between water flow ($\dot{V}_G$) over and blood flow through the gills is between 10:1 and 20:1, much higher than the $\dot{V}_A/\dot{Q}$ ratio in air-breathing mammals. The ratio of ventilation to perfusion in fishes is not 30:1 (as might be expected in comparing the dissolved oxygen content in water with that in air) because the oxygen capacity of the blood of many lower vertebrates is only half that of mammalian blood.

Changes in $P_{O_2}$ also affect the $\dot{V}_A/\dot{Q}$ ratio. A reduction in $P_{O_2}$ in inhalant air or water must be compensated by an increase in ventilation, and hence an increase in the ventilation/perfusion ratio, if the rate of oxygen uptake is to be maintained.

The ventilation/perfusion ratio must be maintained over each portion of the respiratory surface as

## Box 14-2  Ventilation/Perfusion Ratios

The actual ratio of ventilation to perfusion is affected by a number of factors. The oxygen uptake by the blood as it passes through the respiratory surface is given by

$$\dot{V}_{O_2} = \dot{Q}(C_{aO_2} - C_{vO_2}) \qquad (1)$$

where $\dot{V}_{O_2}$ is oxygen uptake per unit time, $\dot{Q}$ is blood flow per unit time, and $C_{aO_2}$ and $C_{vO_2}$ are, respectively, the oxygen content of blood leaving (oxygenated) and entering (deoxygenated) the respiratory epithelium.

Similarly, the amount of oxygen leaving the medium is given by

$$\dot{V}_{O_2} = \dot{V}_A(C_{IO_2} - C_{EO_2}) \qquad (2)$$

where $\dot{V}_A$ is alveolar ventilar volume and $C_{IO_2}$ and $C_{EO_2}$ are, respectively, the $O_2$ content in inhalant and exhalant medium. Equation 2 is true only if inspired and expired volumes are equal, which they generally are not in air-breathing animals. Therefore, mean values from replicates of samples taken during several breathing movements should be used. From Equations 1 and 2, we obtain

$$\dot{Q}(C_{aO_2} - C_{vO_2}) = \dot{V}_A(C_{IO_2} - C_{EO_2})$$

and

$$\dot{V}_A/\dot{Q} = (C_{aO_2} - C_{vO_2})/(C_{IO_2} - C_{EO_2})$$

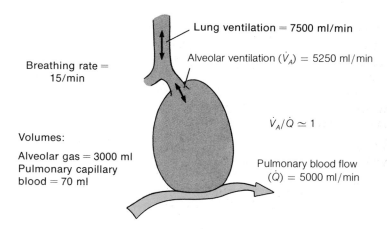

Lung ventilation = 7500 ml/min

Alveolar ventilation ($\dot{V}_A$) = 5250 ml/min

Breathing rate = 15/min

Volumes:

Alveolar gas = 3000 ml
Pulmonary capillary blood = 70 ml

$\dot{V}_A/\dot{Q} \simeq 1$

Pulmonary blood flow ($\dot{Q}$) = 5000 ml/min

**A**  Lung (human)

14-36  An approximation of volumes and flows in (A) lung (human) and (B) gill (trout). Actual values may vary considerably. Note that the ventilation/perfusion ratio is about 1 in the lung and about 10 in the gill.

$P_{O_2}$ = 150 mmHg

Water

Flow ($\dot{V}_G$) = 40 ml/min

Breathing rate = 75/min

$P_{O_2}$ = 80 mmHg

$P_{aO_2}$ = 100 mmHg

Blood

Flow ($\dot{Q}$) = 4 ml/min

Oxygen uptake = 0.13 ml/min

Heart rate = 50/min

$P_{vO_2}$ = 30 mmHg

$\dot{V}_G/\dot{Q}$ = 10

**B**  Gill (trout, body weight 200 g, 8°C)

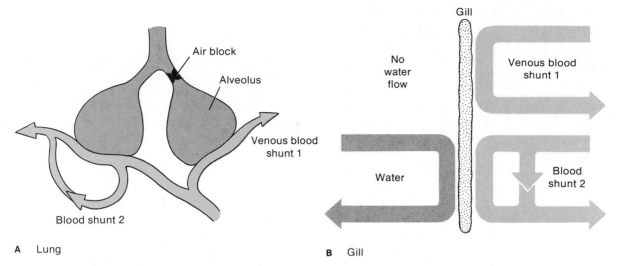

**A** Lung        **B** Gill

**14-37** Problems of inefficient gas transfer can arise because blood flows to a portion of the respiratory surface without adequate ventilation (shunt 1) or because blood does not flow in proximity to the respiratory epithelium (shunt 2). Blood flow is regulated to avoid the development of blood shunts in the lung and gills.

well as over the whole surface. The pattern of capillary blood flow changes in both gills and lungs, changing the distribution of blood over the respiratory surface. The distribution of air or water must reflect the blood distribution. Perfusion of an alveolus without ventilation is as pointless as ventilating an alveolus without blood perfusion. This is an extreme example; the maintenance of too high or too low a blood flow or ventilation rate will also result in a reduction in gas transfer per unit of energy expended. For efficient gas transfer, the optimal ventilation/perfusion ratio should be maintained over the whole respiratory surface. This optimal maintenance does not preclude differential rates of blood perfusion over the respiratory surface, but requires only that the flows of blood and medium be matched. In fact, as discussed in the section on lung blood flow, hypoxic vasoconstriction in the pulmonary circulation plays an important role in ensuring that more blood flows to the well-ventilated portions of the lung.

The efficiency of gas exchange is diminished if some of the blood entering the lungs either bypasses the respiratory surface or perfuses a portion of the respiratory surface that is inadequately ventilated (Figure 14-37). The magnitude of this *venous shunt* can be expressed as a percentage of total flow to the respiratory epithelium, and it can be calculated from arterial and venous $O_2$ content, assuming an ideal arterial $O_2$ content. In the lung, for instance, blood is almost in equilibrium with alveolar gas tensions. If these tensions and the blood–oxygen dissociation curves are known, the expected ideal $O_2$ content of arterial blood can be determined. Let us assume that this ideal content is 20 ml of $O_2$ per 100 ml of blood (20 vol %) and the measured values for arterial and venous blood are 17 and 5 vol %, respectively. This reduction in measured arterial $O_2$ content from the ideal situation can be explained in terms of a venous shunt, oxygenated arterial blood (20 vol %) being mixed with venous blood (5 vol %) in the ratio of 4:1 to give a final $O_2$ content of 17 vol %; that is, 20% of the blood perfusing the lung is passing through a venous shunt. This is an extreme example to illustrate a point; in most cases, venous shunts are very small.

Flows of blood and air or water are regulated to maintain the optimal ventilation/perfusion ratio over the surface of the respiratory epithelium under a variety of conditions. In general terms, $\dot{Q}$ is regulated to meet the requirements of the tissues, $\dot{V}_A$ to maintain adequate rates of $O_2$ and $CO_2$ transfer. Such mechanisms as hypoxic vasoconstriction of blood vessels help to maintain optimal $\dot{V}_A/\dot{Q}$ ratios in different parts of the respiratory surface. Low oxygen levels cause a vasoconstriction in lung vessels, tending to reduce blood flow to poorly ventilated, and therefore hypoxic, regions and to increase blood flow to well-ventilated regions of the respiratory surface.

### NEURAL REGULATION OF BREATHING

The lung is ventilated by the action of the diaphragm and muscles between the ribs. These muscles are activated by spinal motoneurons and the phrenic nerve, which receive inputs from groups of neurons that constitute the *medullary respiratory centers.* The control of respiratory muscles can be very precise, allowing extremely fine control of air flow, as is required for such complex actions in humans as singing, whistling, and talking, as well as simply breathing.

Brain sections indicate that rhythmic activity is maintained by neurons in the pons (*pneumotaxic center*) and in the medulla, and that some neurons just anterior to the medulla cause prolonged inspiration in the absence of rhythmic drive from the pneumotaxic center and reflex vagal inhibition from *pulmonary stretch receptors.* Thus, the respiratory centers contain a *central rhythm generator* that maintains breathing. This generator is dependent on neuronal inputs from *peripheral* and *central chemoreceptors* and from pulmonary stretch receptors.

The medullary respiratory center contains *inspiratory neurons,* whose activity coincides with inspiration, and *expiratory neurons,* whose activity coincides with expiration. The respiratory rhythm was once considered to arise from reciprocal inhibition between inspiratory and expiratory neurons, with reexcitation and accommodation within each set of neurons. But this model of the central rhythm generator is not tenable, because normal breathing can be maintained in the absence of any expiratory activity and reciprocal inhibition requires expiratory neuronal activity. In addition, respiratory neurons show little evidence of accommodation, and although expiratory neurons are inhibited by activity in inspiratory neurons, expiratory neuronal activity does not inhibit inspiratory neurons within the medulla. What, then, is the nature of the central control mechanism for breathing in mammals? Inspiratory neuronal activity, recorded from either the phrenic nerve or some individual neurons in the medulla, shows a rapid onset, a gradual increase, and then a sharp cutoff with each burst of activity asso-

ciated with inhalation. This neuronal activity results in a contraction of the inspiratory muscles and a decrease in intrapleural pressure (Figure 14-38A). Increased blood $CO_2$ levels cause the progressive growth of inspiratory activity to increase more rapidly (Figure 14-38B). Thus, the rate of rise of inspiratory activity is increased by inputs from chemoreceptors, and this results in a more powerful inspiratory phase. The "OFF-switch" of inspiratory neurons occurs once activity in the neuron has reached a threshold level. Expansion of the lung stimulates pulmonary stretch receptors, and activity from these receptors reduces the threshold for the inspiratory "OFF-switch" (Figure 14-38C). The pulmonary stretch receptors, through their action on inspiratory neurons, prevent overexpansion of the lung.

The interval between breaths is determined by the interval between bursts of inspiratory neuronal activity. The interval between bursts of activity is related to the level of activity in the previous burst and in afferent nerves from pulmonary stretch receptors. In general, the greater the level of inspiratory activity, the longer the pause between inspirations. The result is that the ratio of inspiratory to expiratory duration remains constant in spite of changes in the length of each breathing cycle. This ratio is affected by the level of activity of the pulmonary stretch receptors. If, for example, the lung empties only slowly during expiration, the pulmonary stretch receptors will remain active while the lung remains inflated. This continued pulmonary receptor activity prolongs the duration of expiration and increases the time available for exhalation.

The neuronal mechanisms causing phasic activation of inspiratory neurons are poorly understood, as is the nature of the central rhythm generator located in the pontine–medullary region of the brain. These neurons may be either pacemaker cells capable of intrinsic rhythmicity or a neuronal network whose integrity is required to maintain rhythmicity. Expiratory neurons are not always active, and exhalation may be a largely passive process. This is especially true during quiet, normal breathing. Expiratory neurons are active only when inspiratory neurons are quiescent, and then they show a burst pattern somewhat similar to, but out of phase with, that of inspira-

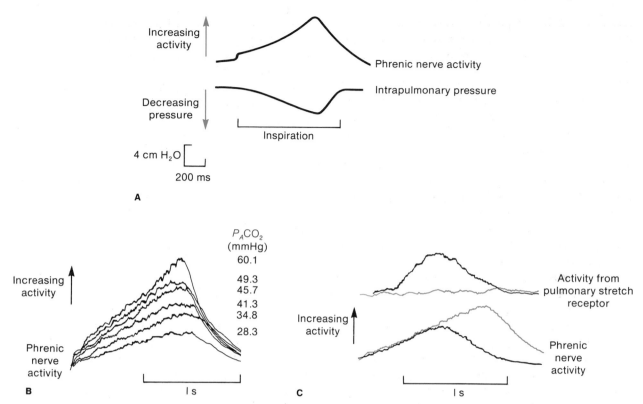

**14-38**  Pattern of phrenic nerve activity during inspiration. (**A**) Relationship between phrenic nerve activity and intrapulmonary pressure. Note the sudden onset, gradual rise, and then "OFF-switch," or termination, of inspiratory activity. (**B**) Effect of increasing alveolar $P_{CO_2}$ levels ($P_ACO_2$) on discharge in the phrenic nerve. Increasing $P_{CO_2}$ results in a more rapid rise of activity in the phrenic nerve during inspiration. (**C**) Effect of increasing activity from pulmonary stretch receptors on activity in the phrenic nerve. Increased activity in the receptors results in an earlier OFF-switch of activity in the phrenic nerve. The rate of increase in activity before the OFF-switch is unaffected.

tory neurons. Inspiratory neuronal activity inhibits expiratory activity, showing the dominance of inspiratory neurons in the generation of rhythmic breathing. In the absence of inspiratory activity, expiratory neurons are continually active. Inspiratory neuronal activity, however, imposes a rhythm, via inhibition, on expiratory neurons. The rate and depth of breathing are influenced by changes in $O_2$, $CO_2$, and pH; by the emotions; by sleep; by lung inflation and deflation and lung irritation; by variations in light and temperature; and by the requirements for speech. These influences are integrated by the medullary respiratory centers. Breathing can also, of course, be controlled by conscious volition.

Most, if not all, animals respond to changes in $O_2$ and $CO_2$ with changes in ventilation. The receptors involved have been localized in only a few groups of animals. Chemoreceptors monitor changes in $O_2$ and $CO_2$ in arterial blood in the *carotid and aortic bodies* of mammals, in the carotid body of birds, and in the carotid labyrinth of amphibians. In all cases, the chemoreceptors are innervated by branches of the ninth (glossopharyngeal) or tenth (vagus) cranial nerve.

The carotid and aortic bodies of mammals (Figure 14-39) receive a generous blood supply and have a high oxygen uptake per unit weight. Sensory fibers innervating the carotid body show an increase in activity when $P_{O_2}$ is decreased or $P_{CO_2}$ increased in the carotid

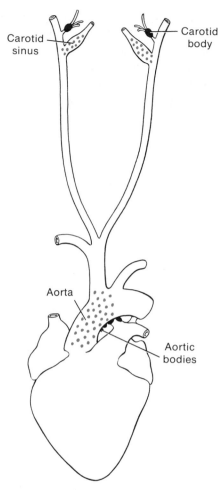

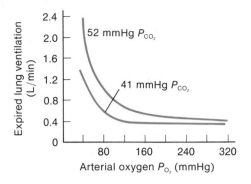

**14-40**   Effect of alterations in arterial $P_{O_2}$ on lung ventilation in the duck. This relationship is affected by changes in arterial $P_{CO_2}$. [Jones and Purves, 1970.]

**14-39**   Carotid and aortic bodies (dog). The approximate location of the carotid and aortic chemoreceptors and stretch receptors is shown. [Comroe, 1962.]

body perfusate. It has been reported that changes in the levels of arterial $P_{O_2}$ and, especially, arterial $P_{CO_2}$ in the intact animal cause reflex changes in ventilation via the carotid body. The response to a change in $P_{O_2}$ depends on the $CO_2$ level, and vice versa (Figure 14-40). Electrical stimulation of efferent sympathetic nerve fibers innervating the carotid body reduces blood flow, oxygen uptake of the carotid body, and the activity in its afferent sensory fibers, and therefore the ventilatory response to carotid-body stimulation.

Increases in blood $CO_2$ affect $CO_2$ levels and pH in the cerebral spinal fluid (CSF) of mammals and, via $H^+$ receptors in the brain, cause an increase in breathing. The cerebral spinal fluid of mammals, and possibly of other vertebrates, is produced by the choroid plexus of the brain, modified by exchange with the brain and glial cells, and finally absorbed by the arachnoid plexus of the brain. The production rate is extremely variable, ranging from 2 to 164 $\mu l/min$ in a variety of mammalian species. The CSF is very low in protein and is essentially a solution of NaCl and $NaHCO_3$, with low but closely regulated levels of $K^+$, $Mg^{2+}$, and $Ca^{2+}$. Changes in blood $P_{CO_2}$ cause corresponding changes in the $P_{CO_2}$ of the CSF, and these in turn result in changes in the pH of the CSF. A decrease in pH stimulates receptors that are possibly located in the region of the medullary respiratory center, causing the reflex increases in breathing (Figure 14-41). Prolonged changes in $P_{CO_2}$ result in the adjustment of the pH of the CSF by changes in $HCO_3^-$ levels.

Many reflexes participate in regulating inflation of the lung and preventing mechanical or chemical irritation of the respiratory surface. In 1868, Ewald Hering and Josef Breuer observed that inflation of the lungs decreases the frequency of breathing.† The *Hering–*

---

†Later, Breuer became one of the first proponents of psychoanalysis and collaborated with Sigmund Freud in producing the book *Studien über Hysterie.*

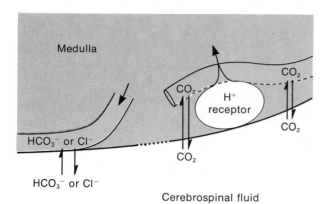

**14-41** Central $H^+$ receptors are influenced by the pH of cerebrospinal fluid (CSF) and by arterial $P_{CO_2}$. Carbon dioxide molecules diffuse readily across the walls of the brain capillaries and alter CSF pH, but there is a barrier to other molecules. Across some capillary walls, exchange of $HCO_3^-$ and $Cl^-$ helps to maintain a constant pH of the CSF in the face of a maintained change in $P_{CO_2}$.

*Breuer reflex* is abolished by cutting the vagus nerve. Inflation stimulates pulmonary stretch receptors in the bronchi and/or bronchioles, which have a reflex inhibitory effect, via the vagus, on the medullary inspiratory center (nucleus tractus solitarius) and therefore on inspiration. $CO_2$-sensitive mechanoreceptors have been localized in the lungs of rabbits. Increased $CO_2$ levels reduce the inhibitory effects of these pulmonary stretch receptors on the respiratory center, thereby increasing depth of breathing and lung ventilation. It is not clear if the $CO_2$-sensitive receptors investigated in the lungs of birds are pure $CO_2$ receptors or $CO_2$-sensitive mechanoreceptors, as observed in mammals. Increased $CO_2$ in the lungs of birds, however, has a greater effect on sensory discharge from the lung than that observed in mammals.

In addition to pulmonary stretch receptors, there are a variety of irritant receptors in the lung that are stimulated by mucus and dust or other irritant particles and cause reflex *bronchioconstriction* and *coughing*. A third group of receptors is positioned close to the pulmonary capillaries in interstitial spaces; these are called *juxtapulmonary capillary receptors* or, more simply, *type J receptors*. These receptors were previously termed "deflation receptors," but their natural stimulus appears not to be lung deflation but an increase in interstitial volume. Stimulation of the type J receptors elicits a sensation of breathlessness. Violent exercise probably results in a rise in pulmonary capillary pressure and an increase in interstitial volume, which could cause stimulation of type J receptors and therefore breathlessness.

## RESPONSES TO CHANGES IN RESPIRATORY GAS LEVELS

### Reduced Availability of Oxygen

Aquatic animals are subjected to more frequent and rapid changes in oxygen levels than are air-breathing animals. The changes in oxygen levels in water may or may not be accompanied by changes in carbon dioxide. In air, the $O_2$ and $CO_2$ levels are relatively stable, and local environments of low $O_2$ or high $CO_2$ are rare and more easily avoided. There is, of course, a gradual reduction in $P_{O_2}$ with altitude, and animals vary in their capacity to climb to high altitudes and withstand the reduction in oxygen levels in the air. The highest permanent human habitation is at about 5800 m, where the $P_{O_2}$ is 79 mmHg. Many birds migrate over long distances at altitudes above 6000 m, where atmospheric pressures would cause severe respiratory distress in many mammals. High altitudes are associated with low temperatures as well as low pressures, and this also has a marked effect on animal distribution. Many aquatic animals can withstand very long periods of hypoxia. Some fishes—for example, carp—overwinter in the bottom mud of lakes, where the $P_{O_2}$ is very low. Many invertebrates also bury themselves in mud of low $P_{O_2}$ but high nutritive content. Some parasites live in hypoxic regions during one or more phases of their life cycle. Limpets and bivalve mollusks close their shells during exposure at low tide to avoid desiccation, but as a consequence are subject to a period of hypoxia. Many of these animals utilize a variety of anaerobic metabolic pathways to survive the period of reduced $O_2$ availability. Others also adjust the respiratory and cardiovascular systems to maintain $O_2$ delivery in the face of reduced $O_2$ availability.

A reduction in oxygen in air results in a decrease in blood $P_{O_2}$ and causes an increase in lung ventilation in mammals as a result of hypoxic stimulation of the carotid and aortic body chemoreceptors. The rise in lung ventilation results in an increase in $CO_2$ elimination and a decrease in blood $P_{CO_2}$. The decrease in blood $P_{CO_2}$ causes a reduction in $P_{CO_2}$ and therefore an increase in pH of the CSF. Decreases in blood $P_{CO_2}$ and increases in cerebral spinal fluid pH tend to reduce ventilation and can result in attenuation of the increase in lung ventilation because of hypoxia. If, however, the hypoxic conditions are maintained, as occurs when animals move to high altitude, both blood and cerebral spinal fluid pH are returned to normal levels by the elimination of bicarbonate. This process takes about one week in humans. Thus, as cerebral spinal fluid pH returns to normal, the reflex effects of hypoxia on ventilation predominate, and the result is a gradual increase in ventilation as the animal acclimatizes to altitude. This response to prolonged hypoxia may also involve modulation of the effects of $CO_2$ on the carotid and aortic bodies to reset these chemoreceptors to the new lower $CO_2$ level at high altitude.

Low oxygen levels cause a local vasoconstriction in the pulmonary capillaries in mammals, producing a rise in pulmonary arterial blood pressure. This response normally has some importance in redistributing blood away from portions of the lung that are not adequately ventilated and are therefore hypoxic. When hypoxia is general, however, the increase in the resistance to flow through the whole lung can have detrimental effects. In some mammals that live at high altitudes, there is a reduced local pulmonary vasoconstrictor response to hypoxia that is probably genetically determined. Humans residing at high altitudes are usually small, are barrel-chested, and have large lung volumes. Pulmonary blood pressures in these people are high, and there is often hypertrophy of the right ventricle. High pulmonary pressures produce more even distribution of blood in the lung, and so augment the diffusing capacity for oxygen.

Long-term adaptations also occur during prolonged exposure to hypoxia. Most vertebrates respond by increasing the number of red blood cells and the blood hemoglobin content—and therefore the oxygen capacity of the blood. The level of phosphorylated compounds, such as 2,3-diphosphoglycerate, adjusts to alter the oxygen affinity of hemoglobin. In humans, a climb to high altitude is accompanied by an increase in 2,3-diphosphoglycerate levels and by an adaptive reduction in the hemoglobin–oxygen affinity. The increasing 2,3-diphosphoglycerate levels offset the effects of high blood pH on hemoglobin–oxygen affinity. The high blood pH is the result of hyperventilation in response to low $O_2$ supplies. Hypoxia due to travel to high altitude also results in systemic vasodilation and an increase in cardiac output. The higher cardiac output lasts only a few days and returns to normal or drops below normal as $O_2$ supplies to tissues are restored by the compensatory increases in ventilation and blood hemoglobin levels. Exposure to hypoxia stimulates a proliferation of capillaries in tissues, ensuring a more adequate oxygen delivery to the tissues.

The gills of fishes and amphibians are larger in species exposed to prolonged periods of hypoxia. Similar enlargement of the respiratory surface apparently does not occur in mammals. These processes augment the transfer of oxygen, its transport in the blood, and its delivery to the tissues, but they take from several hours to days or weeks to reach completion.

*Increased $P_{CO_2}$ Levels*

In many animals, an increase in $P_{CO_2}$ results in an increase in ventilation. In mammals, the increase is proportional to the rise in the $CO_2$ level in the blood. The effect is mediated by modulation of the activity of several receptors that send messages to the medullary respiratory center. These receptors include the chemoreceptors of the aortic and carotid bodies, the $H^+$ receptors within the medulla, and the mechanoreceptors in the lungs. A marked increase in ventilation occurs almost immediately in response to a rise in $CO_2$. The increase is maintained for long periods in the presence of increased $CO_2$, but ventilation eventually returns to a level slightly above the volume that prevailed before hypercapnia. This return to a value only slightly greater than the initial ventilation level is related to increases in levels of plasma bicarbonate and CSF bicarbonate, with the result that pH returns to normal even though the raised $CO_2$ levels are maintained.

## RESPIRATORY ADAPTATIONS FOR DIVING

Diving mammals and birds are, of course, subjected to periods of hypoxia during submergence. The vertebrate central nervous system cannot withstand anoxia and must be supplied with oxygen throughout the dive.

Diving animals solve the problem by utilizing oxygen stores in the body. To minimize depletion of available stores, oxygen is utilized only by the brain and the heart during a dive, blood flow to other organs is reduced, and these tissues adopt anaerobic metabolic pathways. There is a marked slowing of heart rate and a reduction of cardiac output. The blood is recirculated through the brain and the heart (see also p. 586).

Many diving animals exhale before diving. During a deep dive, the increase in hydrostatic pressure results in lung compression. As the animal reduces lung volume before a dive, air is forced out of the alveoli as the lungs collapse and is contained within the trachea and bronchi, which are more rigid but less permeable to gases. If gases remained in the alveoli, they would diffuse into the blood as pressure increased. At the end of the dive, the partial pressure of nitrogen in the blood would be high, and a rapid ascent would result in the formation of bubbles in the blood—the equivalent of decompression sickness, or the "bends," in humans.

Situated near the glottis and near the mouth and nose (depending on the species) are receptors that detect the presence of water and that inhibit inspiration during a dive; blood $O_2$ levels fall and $CO_2$ levels rise, but these changes do not stimulate ventilation via the chemoreceptors of the carotid and aortic bodies. The information from these receptors is ignored by the respiratory neurons during the dive.

During birth, a mammal emerges from an aqueous environment into air and survives a short period of anoxia between the time the placental circulation stops and the time air is first inhaled. The respiratory and circulatory responses of the fetus during this period are similar in several respects to those of a diving mammal.

### RESPIRATORY RESPONSES DURING EXERCISE

Exercise increases $O_2$ utilization, $CO_2$ production, and metabolic acid production. Cardiac output increases to meet the higher demands of the tissues. Transit time for blood through the lung capillaries is reduced, but there is still adequate time for gas transfer (Figure 14-42). Ventilation volume increases in order to maintain gas tensions in arterial blood in the face of increased blood flow. The increase in ventilation in mammals is rapid, coinciding with the onset of exercise. This sudden increase in ventilation volume is followed by a more gradual rise until a steady state is obtained both

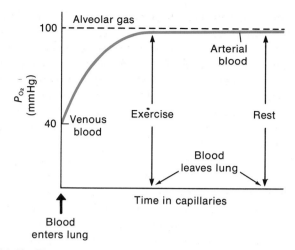

**14-42** Diagram of the way in which $P_{O_2}$ rises as blood flows along the pulmonary capillaries. Note that blood $P_{O_2}$ rapidly reaches near-equilibration with alveolar $P_{O_2}$. During exercise, blood flow increases; blood spends less time in the lung capillaries, but equilibration is still almost complete. [West, 1970.]

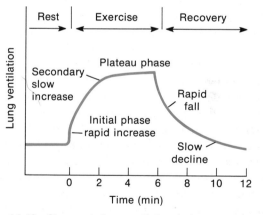

**14-43** Changes in lung ventilation during exercise and recovery in the human.

for ventilation volume and oxygen uptake (Figure 14-43). When exercise is terminated, there is a sudden decrease in breathing, followed by a gradual decline in ventilation volume. During exercise, $O_2$ levels are reduced and $CO_2$ and $H^+$ levels raised in venous blood, but the mean partial pressures of $O_2$ and $CO_2$ in arterial blood do not vary markedly, except during severe exercise. The oscillations in arterial blood $P_{O_2}$ and $P_{CO_2}$

associated with each breath increase in magnitude, although the mean level is unaltered.

There appear to be a number of receptor systems involved in the responses to exercise, not all of which may have been identified. In the absence of exercise, large changes in $CO_2$ and $O_2$ are required to produce equivalent changes in ventilation. It would seem that the chemoreceptors in both the aortic and the carotid body and in the medulla are probably not directly involved in the ventilatory responses to exercise, because mean $P_{O_2}$ and $P_{CO_2}$ levels in arterial blood do not change very much during exercise. It is possible, however, that the sensitivity of the receptors is increased during exercise, so that only small changes cause the increase in ventilation. In this regard, it is significant that catecholamines, which are released in increased quantities during exercise, increase the sensitivity of medullary receptors to changes in $CO_2$.

Contractions of muscles stimulate the stretch, acceleration, and position mechanoreceptors in muscles, joints, and tendons. Activity in these receptors reflexly stimulates ventilation, and this system probably causes the sudden changes in ventilation that occur at the beginning and end of a period of exercise. It has also been suggested that changes in neural activity in the brain and spinal cord leading to muscle contraction may also affect the medullary respiratory center, causing an increase in ventilation.

Muscle contraction generates heat and raises body temperature, thereby increasing ventilation via action on temperature receptors in the hypothalamus. The exact response elicited by stimulation of the hypothalamus depends on the ambient temperature. The increase in ventilation is more pronounced in a hot environment. Since the rise and fall of temperature that follow exercise and subsequent rest are gradual, they would appear to account for only slow changes in ventilation during exercise.

## Regulation of Body pH

Gas transfer, in particular the rate of carbon dioxide excretion, can have a marked effect on body pH. Changes in body pH alter the ionization of proteins and other weak acids. The net charge on proteins determines enzyme activity and subunit aggregation, influences membrane characteristics, and contributes

to the osmotic pressure of body compartments. Animals therefore regulate their internal pH in the face of a continual metabolic release of hydrogen ions.

Hydrogen and hydroxyl ions are in very low concentration because water is only weakly dissociated (as discussed in Chapter 2). Human blood plasma at $37°C$ has a pH of 7.4 or a hydrogen ion concentration of 40 nanomoles per liter (1 nM $= 10^{-9}$M). Normal function can be maintained in mammals at $37°C$ over a blood plasma pH range of 7.0–7.8, that is, between 100 and 16 nM $H^+$/L. This is, in fact, a rather large deviation in terms of percent change from the normal value of 40 nM/L, when compared with an animal's tolerance of variations in sodium or potassium levels in the body. It is important, however, to bear in mind that the changes in concentration are small, as are the actual concentrations of hydrogen ions in the body.

### HYDROGEN ION PRODUCTION AND EXCRETION
Hydrogen ions are produced and excreted on a continuing basis. The largest pool and flux is due, usually, to the metabolic production of $CO_2$, which reacts with water to form protons and bicarbonate ions. Hydrogen ions are also produced as a result of the formation of organic acids—for example, fatty acids and, in particular, lactic acid (as an end product of anaerobic metabolism)—which are almost completely dissociated at body pH. Some inorganic acids, such as phosphoric and sulfuric acids, are produced through the catabolism of proteins.

The relationship between pH and the extent of dissociation of a weak acid, such as carbonic acid, is described by the Henderson–Hasselbalch equation (given in Chapter 2). The equation can be rewritten for the $CO_2$–bicarbonate system in the form

$$pH = pK' + \log \frac{[HCO_3^-]}{\alpha P_{CO_2}}$$

where $P_{CO_2}$ is the partial pressure of $CO_2$ in blood, $\alpha$ is the Bunsen solubility coefficient for $CO_2$, and $[HCO_3^-]$ is the concentration of bicarbonate. Changes in pH will affect the $HCO_3^-/P_{CO_2}$ ratio, and vice versa.

At the pH of the body, nearly all dissolved $CO_2$ is in the form of $HCO_3^-$. Weak acids have their greatest buffering action when pH $= pK'$. The $pK'$ of plasma proteins and hemoglobin is close to the pH of blood. These compounds are therefore important physical buffers in the blood. The p$K'$ of the reaction $H_2CO_3^-$ $\rightleftharpoons$ $HCO_3^-$ + $H^+$ is about 6.2, and the p$K'$ of the reaction $HCO_3^-$ $\rightleftharpoons$ $CO_3^{2-}$ + $H^+$ is around 9.4. Thus, the $CO_2$ hydration–dehydration reaction is of less importance than either hemoglobins or proteins in providing a physical buffer system. The importance of the $CO_2$–bicarbonate system is that an increase in breathing can rapidly increase pH by lowering $CO_2$ levels in the blood, and that $HCO_3^-$ can be excreted via the kidney to decrease blood pH. That is, the $CO_2$/bicarbonate ratio can be adjusted by excretion in order to regulate pH.

Clearly, if lung ventilation is reduced so that $CO_2$ excretion drops below $CO_2$ production, body $CO_2$ levels will rise and pH will fall. This decrease in body pH is referred to as *respiratory acidosis*. The reverse effect— that is, a rise in pH due to hyperventilation—is termed a *respiratory alkalosis*. The word "respiratory" is used to differentiate these pH changes from those caused by changes in metabolism or kidney function. For example, anaerobic metabolism results in lactic acid production, which reduces body pH; such a change is referred to as a *metabolic acidosis*.

## HYDROGEN ION DISTRIBUTION
## BETWEEN COMPARTMENTS

Cell membranes separating the intracellular and extracellular compartments and layers of cells between two body compartments are much more permeable to carbon dioxide than to either hydrogen or bicarbonate ions. Membrane permeability to hydrogen ions is often greater than that for bicarbonate ions, a notable exception being the red blood cell membrane, which is very permeable to bicarbonate ions. Red blood cell membranes are able to exchange bicarbonate for chloride ions at high rates, as occurs when bicarbonate is excreted as $CO_2$, via erythrocytes, as blood perfuses the lungs. In general, the membrane permeability for hydrogen ions is greater than that for $K^+$, $Cl^-$, or $HCO_3^-$.

An increase in $P_{CO_2}$ in extracellular fluid will cause an increase in both bicarbonate and hydrogen ion concentration as well. The increase in extracellular $P_{CO_2}$ will create gradients for $CO_2$, bicarbonate, and hydrogen ions between the cell and extracellular fluid. $CO_2$ will move rapidly into the cell and cause a sharp fall in intracellular pH. This results in the slow extrusion of hydrogen ions from the cell (Figure 14-44A).

The loss of hydrogen ions from the cell is such that if $P_{CO_2}$ levels are returned to the original value, cell pH is higher than the initial value; that is, there is a small overshoot in pH.

A similar situation exists for weak bases like ammonia. Cell membranes are much more permeable to molecular ammonia than to ammonium ions. If $NH_4Cl$ levels in the extracellular fluid are raised, ammonia penetrates the cell much more rapidly than ammonium ions. The result is, of course, that ammonia levels in the cell are increased much more rapidly too. The ammonia combines with hydrogen ions to form ammonium ions within the cell, thus raising cell pH (Figure 14-44B). After reaching a maximum, pH starts to fall during $NH_4Cl$ exposure. $NH_3$ levels equilibrate rapidly, but the slow passive influx of $NH_4^+$ continues causing this fall in pH. The return of external $NH_4Cl$ levels to the original value results in a sharp fall in intracellular pH as $NH_3$ diffuses out of the cell; but because of the accumulation of $NH_4^+$, pH is reduced below the initial level (Figure 14-44B). It slowly returns to the initial level as $NH_4^+$ diffuses out of the cell.

If an acid is injected into a cell, it is extruded from the cell at rates that increase in proportion to the decrease in cell pH. Although a portion of the hydrogen ion efflux may be related to $H^+$ diffusion out of the cell, some of the efflux is coupled to sodium influx. Acid extrusion is also accompanied by chloride efflux, presumably in exchange for $HCO_3^-$, because external $HCO_3^-$ has been shown to be a requirement for pH regulation by cells. In addition, the drug SITS (4-acetamido-4'-isothiocyanostilbene-2,2'-disulfonic acid), which has been shown to block anion transport in erythrocytes, also inhibits pH regulation in other cells. Thus, intracellular pH is adjusted via $Na^+$–$H^+$ and bicarbonate–$Cl^-$ coupled exchanges in the cell membrane. An acid load in the cell is accompanied by $H^+$ efflux and $Na^+$ influx and by bicarbonate influx coupled to chloride efflux. The movement of $HCO_3^-$ into the cell is equivalent to $H^+$ efflux. This is because bicarbonate ions that enter the cell are then dehydrated, consuming protons, and leave the cell as $CO_2$ (Figure 14-45).

These mechanisms of pH adjustment are activated by either a reduction in intracellular pH or an increase in extracellular pH. Acid extrusion is reduced to low levels if extracellular pH falls below 7.0 or intracellular pH rises above 7.4. There is no evidence that cells

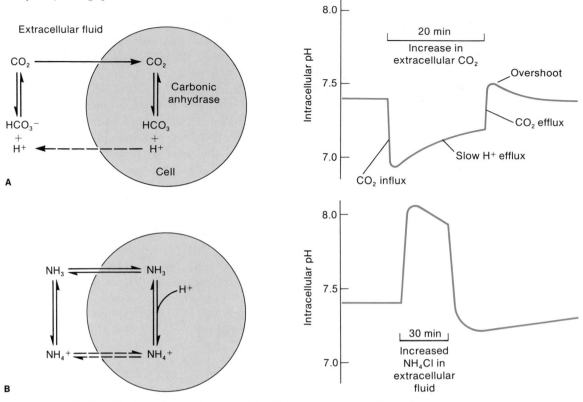

**14-44** Effects of changes in extracellular $CO_2$ and ammonium chloride levels on intracellular pH. (**A**) If $CO_2$ levels in the extracellular fluid are suddenly increased, a sharp fall in intracellular pH occurs as $CO_2$ rapidly enters the cell. (**B**) A sudden increase in extracellular $NH_4Cl$ levels results in a sharp increase in intracellular pH because $NH_3$ diffuses rapidly into the cell and combines with hydrogen ions to form ammonium ions, which do not diffuse across the cell membrane readily.

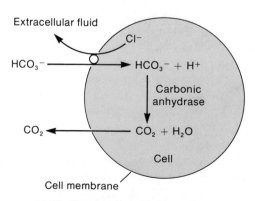

**14-45** Bicarbonate–chloride exchange across cell membranes is an important mechanism for adjusting intracellular pH, because bicarbonate influx results in the formation of $CO_2$, which diffuses out of the cell, in effect removing acid.

possess any specialized transport mechanism to deal with alkaline loads. Elevations in cell pH are probably rare, as there is usually a net release of hydrogen ions by metabolic processes.

*FACTORS INFLUENCING INTRACELLULAR pH*

Intracellular pH will be stable if the rate of acid loading, from metabolism or from influx into the cell, is equal to the rate of acid removal. Any sudden increase in cell acidity will be buffered by physical buffers (e.g., proteins and phosphates) that are located within the cell and bind protons as pH falls. Bicarbonate can also act as a buffer by combining with protons, so that the $CO_2$ thus formed diffuses out of the cell. An acid load is removed from the cell by diffusion protons from the cell and by activation of $Na^+$–$H^+$ and/or bicarbonate–$Cl^-$ exchange processes. Finally, the production of protons through metabolism is pH-sensitive, so the source of hydrogen ions may be regulated. Many enzymes are inhibited by low pH, and it is possible that the inhibition of glycolysis (and possibly some other metabolic pathways) at low pH may serve to regulate intracellular pH by reducing the production of protons during periods of increased acidity in cells.

### FACTORS INFLUENCING BODY pH

A stable body pH requires that acid production be matched to acid excretion. In mammals, this symmetry is achieved by adjusting the excretion of $CO_2$ via the lungs and acid or bicarbonate via the kidneys, so that excretion balances production, which is largely determined by the metabolic requirements of the animal. In aquatic animals, the external surfaces have the capacity to extrude acid in ways similar to those seen in single cells. Fish gills, for example, have transport mechanisms for $Na^+$–$H^+$ and bicarbonate–$Cl^-$ exchange. If these mechanisms are inhibited by appropriate drugs, body pH is affected.

Temperature can have a marked effect on body pH. The dissociation of water varies with temperature, and the pH of neutrality is 7.00 only at 25°C. The dissociation of water decreases, and the pH of neutrality therefore increases, with a decrease in temperature (Figure 14-46). At 37°C, neutral pH is 6.8, whereas at 0°C it

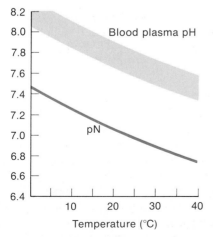

**14-46** Blood pH in turtles, frogs, and fishes at various temperatures, together with the change in neutrality of water, pN. [Rahn, 1967.]

is 7.46. Human plasma at 37°C has a pH of 7.4, so it is slightly alkaline. Most animals maintain the same alkalinity of many of their tissues relative to neutral pH independent of the temperature of their bodies (Figure 14-46). Fishes at 5°C have a plasma pH of 7.9–8.0, turtles at 20°C a plasma pH of about 7.6, and mammals at 37°C a plasma pH of 7.4. Thus, all have the same relative alkalinity and therefore the same $OH^-$/$H^+$ ratio of around 20 in plasma. Tissues are generally less alkaline than plasma, erythrocytes have an intracellular pH about 0.2 pH units less than plasma, and muscle cells have an $OH^-$/$H^+$ ratio of approximately 1.

It can be seen from the Henderson–Hasselbalch equation that changes in p$K'$ will cause changes in pH or in the dissociation of weak acids. Temperature has a marked effect on p$K'$ of plasma proteins, the p$K'$ increasing as temperature decreases. As temperature drops, the pH of arterial blood plasma increases (Figure 14-46) and offsets changes in the p$K'$ of plasma proteins (as a result of the temperature change), so that the extent to which the plasma proteins dissociate remains constant. Because the p$K'$ of the $CO_2$ hydration–dehydration reaction changes less with temperature than does blood pH, animals must adjust the $CO_2$/bicarbonate ratio in the blood. In general, it appears that as temperature falls, air-breathing, poikilothermic vertebrates keep bicarbonate levels constant but decrease molecular $CO_2$ levels. In aquatic animals, on the other hand, $CO_2$ levels remain the same and bicarbonate levels increase as temperature drops. This process results in the same adjustment of the $CO_2$/bicarbonate ratio and hence pH in both aquatic and air-breathing vertebrates. The important point is that if body pH changes with temperature in the same way as the p$K'$ of proteins, then, considering the Henderson–Hasselbalch equation, the charge on proteins should remain unchanged. If there is little or no change in the net charge on proteins, function will be retained over a wide range of temperatures.

Changes in blood pH in mammals can occur in response to acid movement between compartments. For example, following a heavy meal, the production of large volumes of acid in the stomach can produce an *alkaline tide* in the blood owing to transfer of acid from the blood into the stomach. In a similar manner, the production of large volumes of alkaline pancreatic juices can result in an *acid tide* in the blood. In general, acid can be redistributed between body compartments, and this is of functional significance because some tissues are more adversely affected by changes in pH than others. The brain is particularly sensitive, whereas muscle can and does tolerate much larger oscillations in pH. As a result, the brain has extensive, if poorly understood, mechanisms for regulating the pH of the cerebrospinal fluid. In the face of a sudden acid load in the blood, hydrogen ions are taken up by the muscles, reducing oscillations in the blood and protecting the brain and other more sensitive tissues. Hydrogen ions are then slowly released into the blood from muscle and excreted either via the lungs as $CO_2$ or via the kidney in acid urine. Thus, when there is a sudden acid load in the body, the muscle can act as a temporary $H^+$ reservoir by reducing the magnitude of the oscillations in pH in other regions of the body.

## Other Gas-Transfer Systems

### THE INSECT TRACHEAL SYSTEM

The system that insects have evolved for transferring gases between the tissues and the environment differs fundamentally from that found in air-breathing vertebrates. The insect tracheal system takes advantage of the fact that oxygen and carbon dioxide diffuse 10,000 times more rapidly in air than in water, blood, or tis-

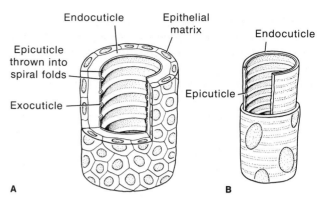

**14-48**   Structure of tracheae. (**A**) Tracheal branch close to spiracle. (**B**) Small tracheal branch, more highly magnified. [Wigglesworth, 1965.]

sues. Consisting of a series of air-filled tubes that penetrate from the body surface to the cells (Figure 14-47), this system acts as a pathway for the rapid movement of $O_2$ and $CO_2$, thereby avoiding the need for a circulatory system to transport gases between the respiratory surface and the tissues. The tracheae (Figure 14-48), having a wall structure similar to that of the cuticle, are invaginations of the body surface that branch everywhere in the tissues. The smallest branches, or tracheoles, are blind-ending and poke between and into individual cells (without disrupting the cell membrane). At various intervals, the system of branching tubes may have a number of air sacs (Figure 14-47B), which enlarge tracheal volume and therefore oxygen stores. In some cases, these air sacs reduce the specific gravity of organs, either for buoyancy or for balance.

Except in a few primitive forms, the tracheal entrances are guarded by spiracles, which control air flow into the tracheae, regulate water loss, and keep out dust. The bug *Rhodnius,* for example, dies in three days if its spiracles are kept open in a dry environment.

The air sacs and tracheae are often compressible, allowing changes in tracheal volume. Some insects ventilate the larger tubes and air sacs of the tracheal system by alternate compression and expansion of the body wall, particularly the abdomen. Different spiracles may open and close during different phases of the breathing cycle so as to control the direction of air flow. In the locust, air enters through the thoracic spiracles but leaves through more posterior openings.

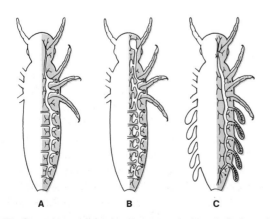

**14-47**   Some types of respiratory system in insects (schematic). (**A**) Simple anastomosing tracheae, with sphincters in the spiracles. (**B**) Mechanically ventilated air sacs developed. (**C**) Tracheal system entirely closed with abdominal gills, formed as evaginations of the tracheoles. [Wigglesworth, 1965.]

Tracheal volume in insects is highly variable; it is 40% of body volume in the beetle *Melolontha* but only 6–10% of body volume in the larva of the diving beetle *Dytiscus*. Each ventilation results in a maximum of 30% of tracheal volume being exchanged in *Melolontha* and 60% in *Dytiscus*. Not all insects ventilate their tracheal system; in fact, many calculations have shown that diffusion of gases in air is rapid enough to supply tissue demands in many species. To augment gas transfer, ventilation occurs in larger insects and, during high levels of activity, in some smaller insects.

Gases are transferred between air and tissues across the walls of the tracheoles. The walls are very thin, with an approximate thickness of only 40–70 nm. The tracheolar area is very large, and only rarely are insect cells more than three cells away from a tracheole. The tips of the tracheoles, except in a few species, are filled with fluid, so that gases diffusing from the tracheoles to the tissues move through the fluid in the tracheoles, the tracheolar wall, the extracellular space (often negligible), and the cell membrane to the mitochondria. This diffusion distance can be altered in active tissues either by an increase in tissue osmolarity, which causes water to move out of the tracheoles and into tissues (Figure 14-49) or by changes in the activity of an ion pump, which results in the net flow of ions and water out of the tracheoles. The fluid is replaced by air, so that oxygen can more rapidly diffuse into the tissues. Insect flight muscle has the highest recorded $O_2$ uptake rate of any tissue, with $O_2$ uptake increasing 10- to 100-fold above the resting value during flight. In general, more active tissues have more tracheoles, and in larger insects the tracheal system is more adequately ventilated.

There are many modifications of the generalized tracheal system just described. Some larval insects rely on cutaneous respiration, the tracheal system being closed off and filled with fluid. Other aquatic insects have a closed, air-filled tracheal system in which gases are transferred between water and air across tracheal gills. The gills are evaginations of the body that are filled with tracheae, the air of which is separated from the water by a membrane 1 $\mu$m thick. As this tracheal system is not readily compressible, it allows the insect to change depth under water without impairment of gas transfer.

Many aquatic insects, such as mosquito larvae,

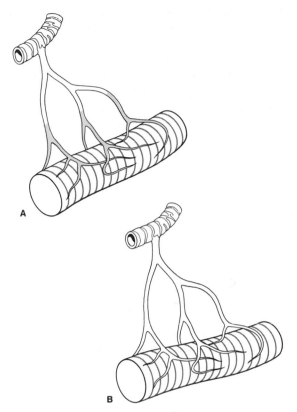

**14-49** Tracheoles running to a muscle fiber (semischematic). (**A**) Muscle at rest; terminal parts of tracheoles (shown in color) contain fluid. (**B**) Muscle fatigued; air extends far into tracheoles. [Wigglesworth, 1965.]

breathe through a hydrofuge (water-repellent) siphon that protrudes above the surface of the water; others take bubbles of air beneath the surface with them. The water bug *Notonecta* carries air bubbles that cling to hydrofuge velvetlike hairs on its ventral surface when the water beetle *Dytiscus* dives with air bubbles beneath its wings or attached to its rear end. When such insects dive, gases are transferred between the bubble and the tissues via the tracheal system; gases can also diffuse however, between the bubble and the water. The rates of $O_2$ transfer from water to bubble will depend on the gradients established and the area of the air–water interface.

The oxygen content of the bubble will decrease owing to uptake by the animal; this will establish an

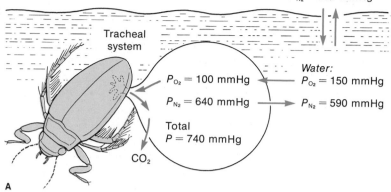

*Air:*    $P_{O_2} = 150$ mmHg
          $P_{N_2} = 590$ mmHg

Tracheal
system

$P_{O_2} = 100$ mmHg  ←  *Water:*  $P_{O_2} = 150$ mmHg

$P_{N_2} = 640$ mmHg  →  $P_{N_2} = 590$ mmHg

Total
$P = 740$ mmHg

$CO_2$

**A**

**14-50**  Partial pressures of $O_2$, $CO_2$, and $N_2$ and total pressure ($P$) in bubbles under water. Arrows indicate diffusion of gas molecules. Note that the sum of the gas partial pressures for the water phase (and the atmosphere) always equals 740 mmHg.

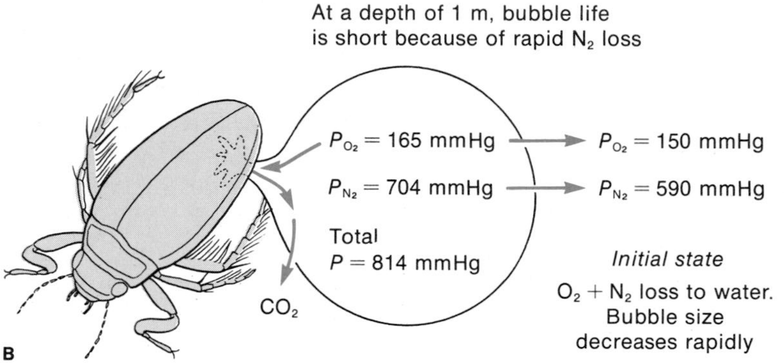

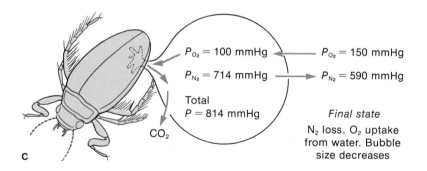

At a depth of 1 m, bubble life is short because of rapid $N_2$ loss

$P_{O_2} = 165$ mmHg  →  $P_{O_2} = 150$ mmHg

$P_{N_2} = 704$ mmHg  →  $P_{N_2} = 590$ mmHg

Total
$P = 814$ mmHg

$CO_2$

*Initial state*
$O_2 + N_2$ loss to water.
Bubble size
decreases rapidly

**B**

$P_{O_2} = 100$ mmHg  ←  $P_{O_2} = 150$ mmHg

$P_{N_2} = 714$ mmHg  →  $P_{N_2} = 590$ mmHg

Total
$P = 814$ mmHg

$CO_2$

*Final state*
$N_2$ loss, $O_2$ uptake
from water. Bubble
size decreases

**C**

$O_2$ gradient between the bubble and the water (assuming the water is in gaseous equilibrium with air), and oxygen will diffuse into the bubble from the water. As $P_{O_2}$ in the bubble is reduced, $P_{N_2}$ will increase; if the bubble is just below the surface, the pressure will be maintained at approximately atmospheric pressure. Nitrogen will therefore diffuse slowly from the bubble into the water (Figure 14-50). (Because of the high solubility of $CO_2$ in water, $CO_2$ levels in the bubble are always negligible.) If the bubble is taken to depth, however, the pressure will increase by 0.1 atm for every meter of depth, increasing both $P_{O_2}$ and $P_{N_2}$ and speeding the diffusion of both $N_2$ and $O_2$ from the bubble into the water. The bubble will gradually get smaller and eventually disappear as nitrogen leaves the bubble. Thus, the life of the bubble depends on the insect's

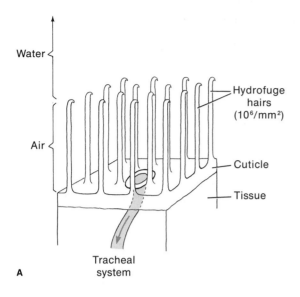

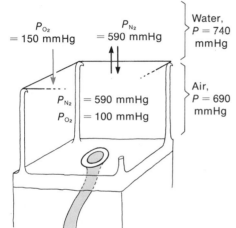

**14-51** (A) Hydrofuge hairs on the surface of some insects and insect eggs have an incompressible air space that acts as a gill under water. Oxygen diffuses from water into the air space contained within the plastron and then into the animal via the tracheal system. (B) Partial pressures of oxygen and nitrogen in the air and water phases. Most of the hairs have been removed for simplicity.

metabolic rate, the initial size of the bubble, and the depth to which it is taken. It collapses because nitrogen is lost from the bubble as the insect uses the oxygen. It has been calculated that up to seven times the initial $O_2$ content diffuses into the bubble from the water and

is therefore available to the insect before the bubble disappears.

If the bubble were noncollapsible, the animal would not need to surface; oxygen would diffuse from the water via the bubble into the tracheal system and thence to the tissues. The insect *Aphelocheirus* has such a structure (Figure 14-51); it consists of a plastron of hydrofuge hairs ($10^6/mm^2$) that contain a small volume of air. The plastron can withstand pressures of several atmospheres before collapsing. In the small air space, $N_2$ is presumably in equilibrium with the water, $P_{O_2}$ is low, and oxygen therefore diffuses from water into the plastron, which is continuous with the tracheal system.

### GAS TRANSFER IN BIRD EGGS

The shells of bird eggs have fixed dimensions but contain an embryo whose gas-transfer requirements increase by a factor of $10^3$ between laying and hatching. Thus, the transfer of $O_2$ and $CO_2$ must take place across the shell at ever-increasing rates during development while the dimensions of the transfer surface (eggshell) do not change. Gases diffuse through small air-filled pores in the eggshell and then through underlying membranes, including the allantoic membrane (Figure 14-52A). The allantoic circulation is in close apposition to the eggshell and increases with the development of the embryo. If the egg is kept in a hypoxic environment, more capillaries develop in the allantoic membrane. The key to increasing rates of gas transfer during development in the bird's egg is the development of an underlying circulation in the allantoic membrane and an increase in the $P_{CO_2}$ and $P_{O_2}$ difference across the eggshell, both of which increase transfer of $O_2$ and $CO_2$ between the environment and the embryo (Figure 14-52B and C).

Water is lost from the egg during development, and this loss results in the gradual enlargement of an air space within the egg, which is as much as 12 ml at hatching in the chicken egg. Just before they hatch, birds ventilate their lungs by poking their beaks into the air space. Blood $P_{CO_2}$ is initially low, but it rises to 40 mmHg just before hatching (Figure 14-52C). This pressure is maintained after hatching, thus avoiding any marked acid–base changes when the bird switches from the shell to its lungs for gas exchange.

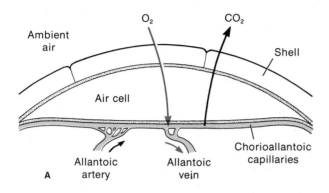

**A**

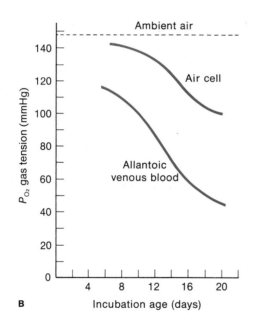

**B**

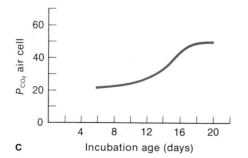

**C**

### ACCUMULATION OF OXYGEN IN SWIM BLADDERS

Many aquatic animals maintain a neutral buoyancy, compensating for a dense skeletal structure by the incorporation of lighter materials in specialized organs. These "buoyancy tanks" may be $NH_4Cl$ solutions (squids), lipid layers (many animals, including sharks), or air-filled swim bladders (many fishes). $NH_4Cl$ and lipid floats have the advantage of being incompressible, not changing volume with changes in hydrostatic pressure that accompany vertical movement in water. Swim bladders are less dense and can be much smaller than $NH_4Cl$ and lipid floats, but they are compressible and change in volume, thus changing the buoyancy of the animal with changes in depth. Hydrostatic pressure increases by approximately 1 atm for every 10 m of depth. If a fish is swimming just below the surface and suddenly dives to a depth of 10 m, the total pressure in its swim bladder doubles from 1 to 2 atm and the bladder volume is reduced by one half, thus increasing the density of the fish. The fish will now continue to sink because it is more dense than water. Similarly, if the fish rises to a shallower depth, its swim bladder will expand, decreasing the fish's density, so that it continues to rise. One advantage of swim bladders is that they are of low density. But they are essentially unstable because of the volume changes they undergo with changes in depth. One means of preventing volume changes is for gas to be added or removed as the fish descends or ascends. Many fishes do have mechanisms for increasing or decreasing the amount of gas in the swim bladder in order to maintain a constant volume over a wide range of pressures.

**14-52**   (A) Diagram of the diffusion pathway between air and chick embryo blood across the eggshell in the region of the air cell. (B) Plot of oxygen tension versus incubation age, comparing measurements of air cell and allantoic venous blood $P_{O_2}$ (C) Plot of carbon dioxide tension of air cell and allantoic venous blood versus incubation age. In contrast to data for oxygen, there is no $P_{CO_2}$ difference between air-cell gas and allantoic venous blood during a chick embryo's development. [Wangensteen, 1972.]

Fishes with swim bladders spend most of their time in the upper 200 m of lakes, seas, and oceans. The pressure in the bladder will range from 1 atm at the surface to about 20 atm at 200 m. Water is in equilibrium with air, and the gas content in water will not vary much with depth (Figure 14-53). The swim-bladder gas in most fishes consists of $O_2$, but those of some species are filled with $CO_2$ or $N_2$. If the fish dives from a depth of 100 m, $O_2$ is added to the swim bladder. The environment is the source, and $O_2$ is moved from the environment to the swim bladder against a pressure difference—in this example, a difference of nearly 10 atm (water $P_{O_2} = 0.228$ atm, bladder $P_{O_2} = 10$ atm). To understand how this occurs, let us review the structure of the swim bladder.

The teleost swim bladder is a pouch of the foregut (Figure 14-54). In some fishes, there is a duct between gut and bladder; in others, the duct is absent in the adult. The bladder wall is tough and impermeable to gases, with very little leakage even at very high pressures, but the wall expands easily if pressures inside the bladder exceed those surrounding the fish. Those animals capable of moving oxygen into the bladder at high pressure have a rete mirabile. The rete consists of several bundles of capillaries (both arterial and venous) in close apposition, the blood flow being countercurrent between arterial and venous blood. It has been calculated that eel retia have 88,000 venous capillaries and 116,000 arterial capillaries containing about 0.4 ml of blood. The surface area of contact between the venous and arterial capillaries is about 100 cm². Blood passes first through the arterial capillaries of the rete, then through a secretory epithelium in the bladder wall, and finally back through the venous capillaries in the rete. The arterial blood and the venous blood in the rete are separated by a distance of about 1.5 $\mu$m.

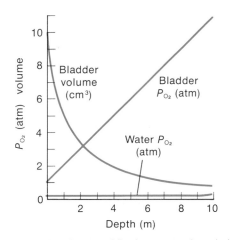

**14-53** Changes in volume and $P_{O_2}$ in water and gas bubble with depth in water. Hydrostatic pressure increases by approximately 1 atm every 10 m. In this example, oxygen is assumed to be the only gas present and is neither added nor removed from the bubble. Fishes can maintain constant density only by maintaining constant bladder volume. This volume can be achieved by adding $O_2$ to the bladder with increasing depth. Note the increasing $P_{O_2}$ difference between water and bladder with depth. Oxygen must be moved from water into the swim bladder even though this $P_{O_2}$ difference occurs.

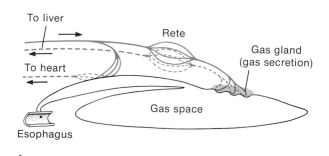

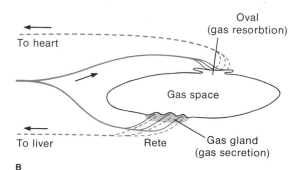

**14-54** The two main types of swim bladders. **(A)** A physostome bladder (from the eel *Anguilla vulgaris*). **(B)** A physoclist bladder (from the perch *Perca fluviatilis*). [Denton, 1961.]

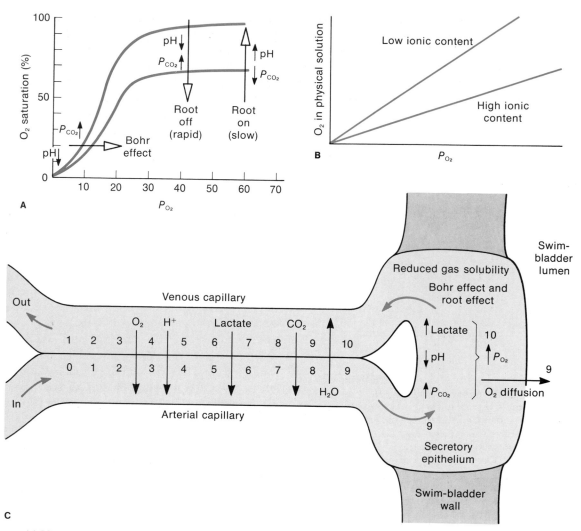

**14-55**   (A) Effect of changes in $P_{CO_2}$ and pH on the blood oxygen dissociation curve in fish. (B) Relationship between $P_{O_2}$ and oxygen content in solutions of different ionic content. Increasing ionic content reduces oxygen solubility, often referred to as a "salting-out effect." (C) Schematic diagram summarizing events in the rete mirabile and the gas gland during the secretion of oxygen into the swim bladder. Small black arrows refer to increasing or decreasing levels; large black arrows, diffusion gradients; colored arrows, direction of blood flow.

The rete structure allows blood to flow into the bladder wall without a concomitant large loss of gas from the swim bladder. Blood leaving the secretory epithelium at high $P_{O_2}$ passes into the venous capillaries, and oxygen diffuses into the arterial capillaries within the rete. $P_{O_2}$ decreases in both arterial and venous capillaries with distance from the secretory epithelium. The $P_{O_2}$ difference between arterial and venous blood at the end of the rete distal to the swim bladder is small compared with the $P_{O_2}$ difference between the environment and the swim bladder. The rete therefore acts as a countercurrent exchanger (Box 12-2), reducing the loss of oxygen from the swim bladder.

The rete structure reduces gas loss from the swim bladder, but *how is oxygen secreted into the swim bladder?* First, consider the relationship between $P_{O_2}$, oxygen solubility, and oxygen content. Oxygen is carried in blood bound to hemoglobin and in physical solution. If oxygen is released from hemoglobin into physical solution, $P_{O_2}$ will increase. The release of oxygen from hemoglobin can be caused by a reduction in pH via the Bohr, or Root, effect (Figure 14-55). An increase in ionic concentration reduces oxygen solubility and also results in an increase in $P_{O_2}$, as long as oxygen content in physical solution remains unchanged. Thus, an increase in blood $P_{O_2}$ can be achieved by releasing oxygen from hemoglobin or increasing the ionic concentration of the blood.

Glycolysis occurs in the secretory epithelium of the swim bladder, even in an oxygen atmosphere, resulting in the production of two lactate molecules for each glucose molecule. This process results in (1) a decrease in pH, which causes the release of oxygen from hemoglobin (Root OFF-shift) and (2) an increase in ionic concentration and therefore a reduction in oxygen solubility (sometimes termed the "salting-out effect"). Both changes cause the $P_{O_2}$ in the secretory epithelium to increase more than that in the swim bladder, so that oxygen diffuses from blood into the gas space of the swim bladder. The salting-out effect will also reduce the solubility of other gases, such as nitrogen and carbon dioxide, and may explain the high levels of these gases sometimes observed in swim bladders.

## Summary

At the level of the mitochondria, the number of oxygen molecules that an animal extracts from the environment and utilizes is approximately the same as the number of carbon dioxide molecules it produces and releases into the environment. In very small animals, gases are transferred between the surface and the mitochondria by diffusion alone, but in larger animals a circulatory system has evolved for the bulk transfer of gases between the respiratory surface and the tissues.

Large surface/volume ratios are maintained in animals by the elaboration of a respiratory surface characterized by large surface areas and small distances for diffusion between the medium and the blood to facilitate gas transfer. Breathing movements assure a continual supply of oxygen and prevent stagnation of the medium close to the respiratory epithelium. The design of the respiratory surface and the mechanism of breathing are related to the nature of the medium (i.e., gills in water, lungs in air).

Bulk transport of $O_2$ and $CO_2$ in the blood is augmented by the presence of a respiratory pigment (e.g., hemoglobin); the pigment not only increases the oxygen-carrying capacity of the blood, but also aids the uptake and release of $O_2$ and $CO_2$ at the lungs and tissues.

The rate of gas transfer across a respiratory surface depends on the ratio of ventilation of the respiratory surface to blood flow, as well as on absolute ventilation volume and cardiac output. These factors are closely regulated to maintain adequate rates of gas transfer to meet the requirements of the tissues. The control system has been studied extensively only in mammals. It consists of a number of mechano- and chemoreceptors that feed information into a central integrating region, the medullary respiratory center. This center, through a variety of effectors, causes appropriate changes in breathing and blood flow to maintain rates of $O_2$ and $CO_2$ transfer at a level sufficient to meet the requirements of metabolism.

Animals regulate $H^+$ concentration in the body in the face of a continual rate of production and excretion. Production varies with the metabolic requirements of the animal, and excretion rates via the lungs and kidney are adjusted to match production. Buffers, particularly proteins and phosphates, ameliorate any oscillations in body $H^+$ due to imbalance between production and excretion. The muscle tissues are utilized as a temporary hydrogen ion reservoir to further protect more sensitive tissues such as the brain from wide swings in pH until the excess $H^+$ can be excreted from the body. Intracellular pH is adjusted by modulation of $Na^+ - H^+$ and bicarbonate$-Cl^-$ exchange mechanisms located in the cell membrane.

Insects have evolved a tracheal system that takes advantage of the rapid diffusion of gases in air and avoids the necessity of transporting gases in the blood. The tracheal system consists of a series of air-filled, thin-walled tubes that extend throughout the body and serve as diffusion pathways for $O_2$ and $CO_2$ between the environment and the cells. In some large active insects, the tracheal system is ventilated.

Bird eggs and the swim bladder of the fish present interesting problems in gas transfer. A bird's egg contains an embryo whose oxygen must be transferred across a shell of fixed dimensions, the transfer requirements increasing by $10^3$ between laying and hatching. Gas tensions in the fish swim bladder often exceed that in the blood by several orders of magnitude, but the design of the blood supply and gas gland is such that gases move from the blood into the swim bladder.

## Exercises

*1.* Calculate the percent change in volume when dry air at $20°C$ is inhaled into the human lung ($T = 37°C$).

*2.* Define the following terms: (a) oxygen capacity, (b) oxygen content, (c) percent saturation, (d) methemoglobin, (e) Bohr effect, (f) Haldane effect.

*3.* Describe the role of hemoglobin in the transfer of oxygen *and* carbon dioxide.

*4.* Describe the effects of gravity on the distribution of blood in the human lung. What effect does alveolar pressure have on lung blood flow?

*5.* Compare and contrast ventilation of the mammalian lung and the bird lung.

*6.* What is the functional significance of the presence of surfactants in the lung?

*7.* How have insects avoided the necessity of transporting gases in the blood?

*8.* The number and dimensions of air pores in eggshells are constant for a given species. What effect would doubling the number of pores have on the transfer of oxygen, carbon dioxide, and water across the eggshell?

*9.* Discuss the role of the rete mirabile in the maintenance of high gas pressures in the fish swim bladder.

*10.* How is oxygen moved into the swim bladder of teleost fishes?

*11.* Describe the structural and functional differences between gills and lungs.

*12.* Why is the ventilation/perfusion ratio much higher in water-breathing than in air-breathing animals?

*13.* Describe the role of central chemoreceptors in the control of carbon dioxide excretion.

*14.* What is the importance of the Hering–Breuer reflex in the control of breathing?

*15.* Describe the processes involved in the acclimation of mammals to high altitude.

*16.* What is the effect on intracellular pH of elevating extracellular $NH_4Cl$ levels at either low or high extracellular pH?

*17.* Describe the role of the $CO_2$/bicarbonate systems in pH regulation in mammals.

*18.* Explain the significance of the localization of the enzyme carbonic anhydrase within the red blood cell and not in the plasma.

*19.* Describe the possible mode of operation of the medullary respiratory center.

*20.* Discuss the interaction between gas transfer and heat and water loss in air-breathing vertebrates.

## Suggested Reading

Bolis, L., K. Schmidt-Nielsen, and S. H. P. Maddrell. 1973. *Comparative Physiology.* New York: American Elsevier.

Cohen, M. I. 1979. Neurogenesis of respiratory rhythm in the mammal. *Physiol. Rev.* 59:1105–1173.

Davenport, H. W. 1974. *The A.B.C. of Acid–Base Chemistry.* 6th rev. ed. Chicago: University of Chicago Press.

Dejours, P. 1975. *Principles of Comparative Respiratory Physiology.* New York: American Elsevier.

Dejours, P., ed. 1972. Comparative physiology of respiration in vertebrates. *Resp. Physiol.* 14:1–236.

Diamond, J. 1982. How eggs breathe while avoiding desiccation and drowning. *Nature* (London) 295:10–11.

Euler, C. von. 1980. Central pattern generation during breathing. *Trends Neurosci.* 3:275–277.

Farner, D. S., and J. R. King., eds. 1972. *Avian Biology.* Vol. 2. New York: Academic.

Hoar, W. S., and D. J. Randall, eds. *Fish Physiology.* Vol. 4. New York: Academic.

Hochachka, P. W., and G. N. Somero. 1973. *Strategies of Biochemical Adaptation.* Philadelphia: Saunders.

Johansen, K. 1971. Comparative physiology: Gas exchange and circulation in fishes. *Ann. Rev. Physiol.* 33:569–612.

Jones, J. D. 1972. *Comparative Physiology of Respiration.* London: Arnold.

Krogh, A. 1968. *The Comparative Physiology of Respiratory Mechanisms.* New York: Dover.

Leusen, I. 1972. Regulation of cerebral spinal fluid composition with reference to breathing. *Physiol. Rev.* 52:1–56.

Mitchell, R. A. 1977. Location and function of respiratory neurons. *Amer. Rev. Resp. Dis.* 115:209–216.

Rahn, H. 1966. Aquatic gas exchange theory. *Resp. Physiol.* 1:1–12.

Reuck, A. V. S. de, and R. Porter, eds. 1967. *Development of the Lung.* London: Churchill.

Roos, A., and W. F. Boron. 1981. Intracellular pH. *Physiol. Rev.* 61:296–434.

Ruch, T. C., and H. D. Patton, eds. 1974. *Physiology and Biophysics.* 20th ed. Vol. 2. Philadelphia: Saunders.

Schmidt-Nielsen, K. 1972. *How Animals Work.* New York: Cambridge University Press.

Steen, J. B. 1971. *Comparative Physiology of Respiratory Mechanisms.* New York: Academic.

West, J. B. 1974. *Respiratory Physiology: The Essentials.* Baltimore: Williams and Wilkins.

## References Cited

Banko, W. E. 1960. *The Trumpeter Swan.* North American Fauna, No. 63. Washington, D.C.: U.S. Dept. of the Interior, Fish and Wildlife Service.

Bartels, H. 1971. Blood oxygen dissociation curves: Mammals. In P. L. Altman and S. W. Dittmer, eds., *Respiration and Circulation.* Bethesda, Md.: Federation of American Societies for Experimental Biology.

Comroe, J. H. 1962. *Physiology of Respiration.* Chicago: Year Book Medical Publishers.

Dejours, P. 1966. *Respiration.* New York: Oxford University Press.

Denton, E. J. 1961. The buoyancy of fish and cephalopods. *Prog. Biophys.* 11:178–234.

Duncker, H. R. 1972. The lung air sac system of birds: A contribution to the functional anatomy of the respiratory apparatus. *Ergebn. Anat. Entwickl. Ges.* 45(6):1–171.

Grigg, G. C. 1970. Water flow through the gills of Port Jackson sharks. *J. Exp. Biol.* 52:565–568.

Hildebrandt, J., and A. C. Young. 1965. Anatomy and physics of respiration. In T. C. Ruch and H. D. Patton, eds., *Physiology and Biophysics.* 19th ed. Philadelphia: Saunders.

Holland, R. A. B., and R. E. Forster. 1966. The effect of size of red cells on the kinetics of their oxygen uptake. *J. Gen. Physiol.* 49:727–742.

Hughes, C. M. 1964. How a fish extracts oxygen from water. *New Scientist* 11:346–348.

Jones, D. R., and M. J. Purves. 1970. The carotid body in the duck and the consequences of its denervation upon the cardiac response to immersion. *J. Physiol.* 211:279–294.

McGilvery, R. W. 1970. *Biochemistry: A Functional Approach.* Philadelphia: Saunders.

Rahn, H. 1967. Gas transport from the external environment to the cell. In A. V. S. de Reuck and R. Porter, eds., *Development of the Lung.* London: Churchill.

Randall, D. J. 1970. Gas exchange in fish. In W. S. Hoar and D. J. Randall, eds., *Fish Physiology.* Vol. 4. New York: Academic.

Salt, G. W. 1964. Respiratory evaporation in birds. *Biol. Rev.* 39:113–136.

Scheid, P., H. Slama, and J. Piiper. 1972. Mechanisms of unidirectional flow in parabronchi of avian lungs: Measurements in duck lung preparations. *Resp. Physiol.* 14:83–95.

Shelton, G. 1970. The effect of lung ventilation on blood flow to the lungs and body of the amphibian *Xenopus laevis. Resp. Physiol.* 9:183–196.

Steen, J. B. 1963. The physiology of the swimbladder of the eel *Anguilla vulgaris,* I: The solubility of gases and the buffer capacity of the blood. *Acta Physiol. Scand.* 58:124–137.

Tenney, S. M., and J. E. Temmers. 1963. Comparative quantitative morphology of the mammalian lung: Diffusing area. *Nature* (London): 197:54–57.

Toews, D. P., G. Shelton, and D. J. Randall. 1971. Gas tensions in the lungs and major blood vessels of the urodele amphibian, *Amphiuma tridactylum. J. Exp. Biol.* 55:47–61.

Wangensteen, O. D. 1972. Gas exchange by a bird's embryo. *Resp. Physiol.* 14:64–74.

West, J. B. 1970. *Ventilation/Blood Flow and Gas Exchange.* 2nd ed. Oxford: Blackwell Scientific Publications.

West, N. H., and D. R. Jones. 1975. Breathing movements in the frog *Rana pipiens,* I: The mechanical events associated with lung and buccal ventilation. *Canad. J. Zool.* 52:332–334.

Wigglesworth, V. B. 1965. *The Principles of Insect Physiology.* 6th ed. London: Methuen.

Young, M. 1971. Changes in human hemoglobins with development. In P. L. Altman and D. W. Dittmer, eds., *Respiration and Circulation.* Bethesda, Md.: Federation of American Societies for Experimental Biology.

# CHAPTER 15

# Feeding, Digestion, and Absorption

Living organisms depend on external sources for raw materials and energy needed for growth, maintenance, and functioning. The sum total of material and energetic intake must, of course, equal output plus the material incorporated into new tissue plus all energy expended. Food, used to provide material for production of new tissue and the repair of old tissue and for use as an energy source, is obtained from a variety of plant, animal, and inorganic sources. The chemical energy for fueling all processes in the animal body comes ultimately, however, from a single source, the sun (Figure 3-3). Solar energy is unavailable (except in the form of heat) to organisms other than the photosynthetic *autotrophic,* or self-nourishing, organisms (i.e., chlorophyll-containing plants), which harness radiant energy to synthesize low-entropy carbon compounds from high-entropy precursors—$CO_2$ and $H_2O$. These compounds are repositories of chemical energy that can be released and utilized through coupled reactions (p. 51) to drive energy-consuming reactions in living tissue. Thus, all animals are *heterotrophic:* they derive all their energy-yielding carbon compounds from ingested foodstuffs that are ultimately derived from autotrophic organisms.

A simplified overview of the flow of energy from the sun to a molecule of ATP is given in Figure 15-1. Monosaccharides such as glucose are synthesized by green plants from $CO_2$ and $H_2O$. These elementary carbon compounds occur at the beginning of the *food chain* represented by the broken arrows in Figure 15-1. In a short food chain, green plants are eaten by a large heterotroph such as an elephant. This heterotroph, having no natural enemies except humans, is at the end of that chain until it dies and is consumed by bacteria and carrion-eating scavengers. In a longer chain, the succession is: phytoplankton $\longrightarrow$ zooplankton $\longrightarrow$ small fish $\longrightarrow$ medium fish $\longrightarrow$ large fish. But nutrient flow is generally more complex (Figure 3-3).

At each stage of conversion, passing from one trophic level to another, there is a loss of usable material and of free energy. The grain produced by an acre of wheat contains more material and energy directly available for human consumption than it does if used as cattle feed and thus converted to beef. For example, on the average, an acre of cereals produces 5 times more (and an acre of legumes produces 10 times more) protein than does an acre devoted to meat production; and to produce 1 lb of protein for human consumption, a cow

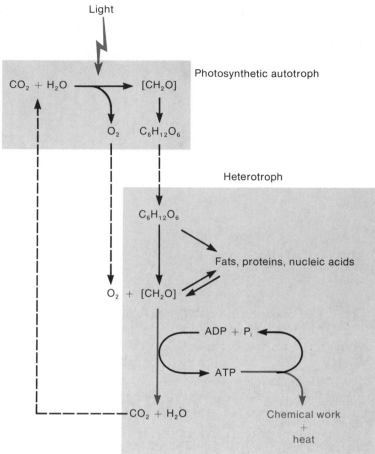

Light

CO$_2$ + H$_2$O ⟶ [CH$_2$O]

Photosynthetic autotroph

O$_2$  C$_6$H$_{12}$O$_6$

Heterotroph

C$_6$H$_{12}$O$_6$

Fats, proteins, nucleic acids

O$_2$ + [CH$_2$O]

ADP + P$_i$

ATP

CO$_2$ + H$_2$O

Chemical work
+
heat

**15-1** Generalized flow diagram of chemical energy in the biotic world. The flow chart begins with the photosynthetic formation of high-energy-content molecules (sugars) from low-energy-content raw materials (CO$_2$ and H$_2$O). Oxidation of carbon compounds yields free energy coupled to the synthesis of high-energy compounds, such as ATP, used as common energy currency in metabolism. Chemical energy content is at its peak following the photosynthetic production of sugars. With each subsequent transaction, some chemical energy is converted to heat and thus is lost as a source of energy for driving biological processes.

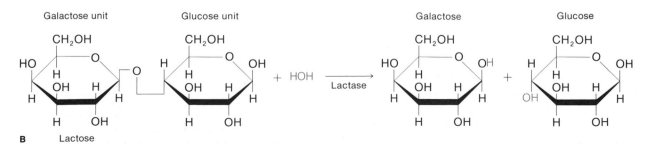

**A** Peptide      Acid    Amine

Galactose unit    Glucose unit      Galactose      Glucose

**B**   Lactose

**15-2** Hydrolysis of (A) peptides and (B) disaccharides. Under enzyme catalysis, a molecule of water is added to the two residues as shown, breaking the covalent bond holding the residues together.

must be fed more than 20 lb of protein. At each level of feeding, digestion, and incorporation along the food chain, there is loss due to the energetic cost of tissue maintenance and of breaking down food molecules into small ones and then rebuilding these small molecules into complex tissue molecules. It follows that, all else being equal, a shorter food chain conserves greater amounts of photosynthetically captured energy.

## Digestive Hydrolysis

To utilize food for tissue maintenance and growth and for the release of chemical energy, the animal must first digest it. *Digestion* is primarily a complex chemical process in which special enzymes catalyze the breakdown of large foodstuff molecules into simpler compounds that are small enough to cross cell membranes readily for incorporation into tissue. For example, starch, a long-chain polysaccharide, is degraded to much smaller disaccharides and monosaccharides; proteins are broken down into polypeptides and then into tripeptides, dipeptides, and amino acids.

The chemical processes we call digestion all involve *hydrolysis*—the utilization of water during bond cleavage, so that $H^+$ is added to one residue and $OH^-$ to the other (Figure 15-2). Hydrolysis of the anhydrous bonds frees the constituent residues (e.g., monosaccharides, amino acids, monoglycerides) from which the polymer is formed, making them small enough for absorption from the alimentary canal into the circulating body fluids and for subsequent entry into cells to be metabolized.

Chemical energy liberated during hydrolysis in the digestive tract cannot be utilized by the body except as heat. For this reason, it is of great adaptive significance that the enzymes catalyzing digestive processes do not hydrolyze those bonds in food molecules containing large amounts of energy. Thus, digestive hydrolysis releases only a small fraction of the total chemical energy stored in a polymeric molecule. Most of the chemical energy of the foodstuff molecules resides in bonds located within the individual residues, from which it can be subsequently liberated in discrete increments during intracellular metabolic steps, as described in Chapter 3.

In metazoans, digestion takes place primarily in the lumen of the *alimentary canal* (Figure 15-3), which is

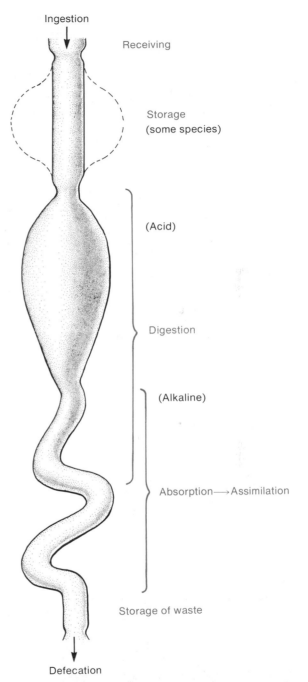

**15-3** Generalized digestive tract. One-way passage of food allows simultaneous operation of sequential stages in the processing of food and reduces mixing of digested and undigested matter. Dashed outline represents crop for storage found in some animals.

supplied with the necessary enzymes and other chemicals. The digested material enters the bloodstream from the alimentary canal by the process of *absorption*. *Assimilation* then takes place as digested molecules from the blood enter the tissues. *Nutrition* is a broad term for the sum of processes that the animal uses to obtain, ingest, digest, absorb, and assimilate foodstuffs.

## Feeding Strategies

Obtaining nutritional essentials is clearly a key to the success of any species. Much of the routine functioning of an animal is directed toward this end. The complexity and sophistication of the nervous system, for example, are due largely to the selective pressure brought to bear on the obtaining of sufficient food and on the avoidance of becoming someone else's meal. Animals use various strategies to feed. Some search, stalk, pounce, capture, and kill. Sessile species, unable to move about, resort to more subtle means, such as surface absorption, filter feeding, or trapping.

### ABSORPTION OF FOOD THROUGH BODY SURFACE

Certain protozoa, endoparasites, and aquatic invertebrates are able to take up nutrient molecules through their integument from the medium in which they live. The endoparasites (animals that live within other animals) include parasitic protozoa, tapeworms, flukes, and certain molluscans and crustaceans. All are surrounded by host tissues or by alimentary canal fluids, both of which are high in nutrients. Tapeworms make no pretense of having even a rudimentary digestive system. Apparently they did not lose it, but rather evolved from a primitive flatworm that was acoelomic (i.e., without a body cavity). Parasitic crustaceans, which belong to the cirripeds (barnacle group), also lack an alimentary canal, but they appear to have evolved from nonparasitic ancestors possessing a gut. Some free-living protozoa and invertebrates have also been found to utilize surface uptake of nutrients from the surrounding medium. Small molecules such as amino acids are taken up from dilute solution, against a concentration gradient, by transport mechanisms (described in Chapter 4), and larger molecules or particles are taken up by a bulk process described next.

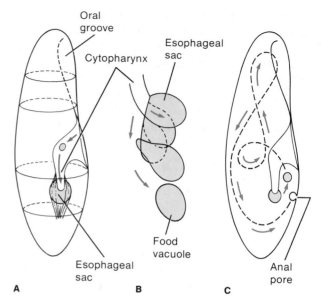

**15-4** Food vacuole in *Paramecium*. (**A**) Food particles are swept into cytopharynx by action of cilia in oral groove. (**B**) Particles enter membrane-limited esophageal sac, which pinches off to form the food vacuole. (**C**) Food vacuole circulates with cyclosis of cytoplasm. Wastes are ejected by exocytosis at the anal pore after digestion of food particles. [Mast, 1947; Jennings, 1972.]

### ENDOCYTOSIS

*Endocytosis* comprises both pinocytosis (cell drinking) and phagocytosis (cell eating). It occurs when a molecule or small particle binds to the cell surface and the membrane invaginates under it, forming an *endocytotic tube*. The morsel is engulfed in a vesicle that pinches off the bottom of the tube. The vesicle (or *food vacuole* in protozoa) floats free in the cytoplasm, where it fuses with *lysosomes*, organelles containing intracellular digestive enzymes. After digestion of the contents, the vacuolar membrane dissolves, and the digestion products enter the cytoplasm. This method of feeding is familiar in protozoa such as *Paramecium* (Figure 15-4) but also occurs in the lining of the alimentary canals and other digestive tissues of many metazoans.

### FILTER FEEDING

Animals that are sessile must wait for food to come to them. Marine and freshwater environments contain diverse groups of *filter feeders*, or *suspension feeders*, most of which depend on the water currents to carry along small organisms, which they capture by a variety of clever means. Most filter feeders are small sessile animals such as sponges, brachiopods, lamellibranchs, and tunicates.

The brachiopods actually rotate on their pedal stalks so as to present the most efficient hydrodynamic orientation for capturing the water current. A number of other sessile animals in moving water make use of *Bernoulli's effect* (i.e., fluid pressure drops as fluid velocity increases) to increase the rate of feeding flow at no energy cost to themselves. An example of such

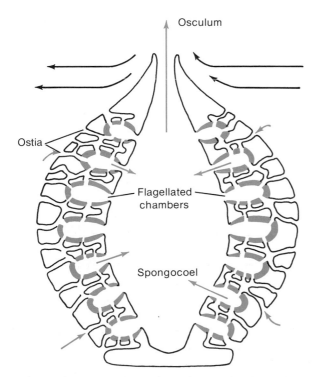

**15-5** Diagrammatic section through a syconoid sponge, with colored arrows indicating water flow. This flow results in part from the activity of flagellated choanocytes in the chambers, and in part from the reduction in hydrostatic pressure at the osculum due to the Bernoulli effect produced by the transverse water currents (black arrows) flowing over the osculum at elevated velocity. Nutrients are taken up by endocytosis. [Hyman, 1940; and from "Organisms That Capture Currents" by S. Vogel. Copyright © 1978 by Scientific American, Inc. All rights reserved.]

passively assisted filter feeding is seen in sponges (Figure 15-5), which also utilize the flagella of the *choanocytes,* cells that line the body cavity, to create internal water currents. Some sponges living in moving water "pump" water up to 20,000 times their body volume per day. The flow of water across the conical *osculum,* the large terminal opening, causes a drop in pressure (Bernoulli's effect) outside the osculum. As a result, water is drawn out of the sponge through the osculum, replenished through the numerous *ostia* (mouthlike openings) in the body wall. The drop in pressure is facilitated by the shape of the sponge's exterior, which causes the water over the osculum to flow with greater velocity than the water flowing past the ostia. A plastic model of a sponge has been shown by Steven Vogel (1978) to also exhibit such a passive flow of water when placed in a water current. Food particles, swept into the ostia of the sponge along with the water, are engulfed by the flagellated choanocytes.

*Mucus,* a sticky mixture of mucopolysaccharides, plays an important role in many filter feeders. Water-borne microorganisms and food particles are trapped in a layer of mucus that covers a ciliated epithelium. The mucus is then transported to the oral parts by beating cilia. The cilia carry water through sessile animals not only to capture suspended food but also to aid in respiration. This is of greatest importance in still water. In the mussel *Mytilus* (Figure 15-6), the lateral cilia draw a stream of water through the inhalant siphon, passing the water between the gill filaments. The frontal cilia are responsible for keeping the mucus traveling down along the filaments (i.e., 90° to water flow) to the tip of the gill, where it travels under ciliary power toward the mouth in a ropelike string of mucus in a special groove. Sand particles and other inedibles are rejected and passed out with the water leaving the exhalant siphon.

The largest filter feeders are the baleen whales, such as the right whale. In their upper jaws are horny *baleen plates* bearing a fringe of parallel filaments of hairlike keratin that hang down between upper and lower jaws and act as strainers (Figure 15-7A). These whales swim into schools of pelagic crustaceans such as krill, engulfing vast numbers suspended in tons of water. As the jaws close, the water is squeezed out through the baleen strainers with the help of the large tongue, and the

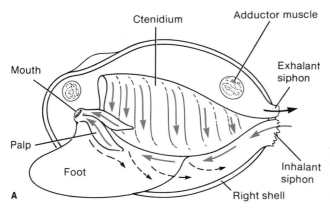

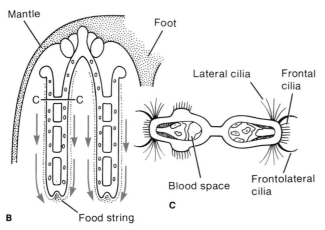

**15-6**  Ciliary feeding in lamellibranch mollusks. (**A**) Side view of generalized lamellibranch with left valve and ctenidium (gill) removed. Colored arrows show paths of mucus and food particles from gill surface to mouth. Broken arrows show paths of rejected particles. (**B**) Vertical section through ctenidium of the mussel *Mytilus.* Mucus and food are shown in color. (**C**) Section taken at level C–C in part B, showing arrangement of the cilia. [Jennings, 1972.]

crustaceans, left behind inside the mouth, are swallowed. It is noteworthy that feeding on this scale supports some of the largest animals that have ever existed.

On a somewhat smaller scale, the flamingo uses a similar apparatus (Figure 15-7B) to filter small animals and other morsels it finds in the muddy bottoms of its freshwater habitat. The flamingo and the right whale exhibit remarkable convergent evolution: they both have a deep-sided lower jaw, a recurved rostrum,

fibrous-fringe filters suspended from the upper jaw, and a large fleshy tongue. The major difference is that the flamingo feeds with its head upside down. Filter feeders also include some teleosts and sharks that strain food organisms from water passing through their gills. Long gill rakers of teleosts serve the same function as the baleen plate in the right whale.

### PIERCING AND SUCKING

Feeding by piercing and sucking occurs among the platyhelminthes, nematodes, annelids, and arthropods. Among the chordates, blood-sucking has been ascribed to vampire bats, but their feeding is more akin to lapping blood that oozes from an incision made with their sharp teeth. Leeches, among the annelids, are true bloodsuckers, using an anticoagulant in their saliva

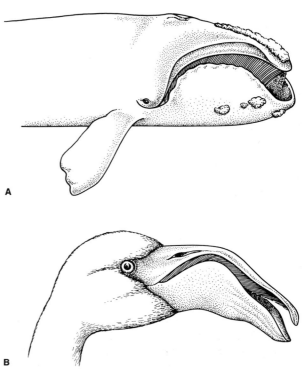

**15-7**  Convergence in filter-feeding mechanisms in (**A**) the black right whale, *Eubalaena glacialis,* and (**B**) the lesser flamingo, *Phoeniconaias minor.* Both the baleen in the whale's mouth and the fringe along the edge of the flamingo's bill act as strainers. [Milner, 1981.]

to prevent blood from clotting as it leaves the blood vessels ruptured by their rasping jaws. Some free-living flatworms seize their invertebrate prey by wrapping themselves around it. They then penetrate the body wall with a protrusible pharynx that sucks out the victim's body fluids and viscera. Penetration by the pharynx and liquefaction of the victim's insides are facilitated by proteolytic enzymes secreted by the muscular pharynx.

A variety of arthropods feed by piercing and sucking. Most familiar and irksome of these to humans are mosquitoes, bedbugs, and lice. Although the majority of sucking arthropods victimize animal hosts, some, especially among the hemiptera (true bugs), pierce and suck plants, from which they draw sap. Sucking insects like the hemipterans possess exquisitely fine piercing mouthparts in the form of a *proboscis.* In plant-sucking bugs, for example, a jointed needlelike tube is formed from the *labium,* which supports and guides the other mouth parts (Figure 15-8). The two *maxillae* are shaped so that they make up two canals that run to the tip of the proboscis. One of these, the *dorsal canal,* is the passage for juices sucked from the host; the other, the *ventral canal,* carries saliva, containing anticoagulants and enzymes, from the salivary glands into the host. Sucking occurs by the action of a muscular pharynx. After feeding, the proboscis in some hemipterans is folded back so as to be out of the way.

## JAWS, BEAKS, AND TEETH

A rasplike structure termed a *radula* is used by gastropods to scrape algae from rock surfaces or to rasp through vegetation (Figure 15-9). Although no true teeth occur among the invertebrates, various invertebrates have beaklike or toothlike chitinous structures for biting or feeding. The octopus, for example, has a sharp, tearing beak. The lower vertebrates—cyclostomes, elasmobranchs, teleosts, amphibians, and reptiles—have pointed teeth, mounted on the jaws or palate, that aid in holding, tearing, and/or swallowing prey. Among the lower vertebrates it is common to swallow prey whole.

Birds have no teeth, but instead have horny beaks that exemplify adaptive radiation suited to a gastronomic life-style. For instance, beaks may have finely serrated edges, or sharp, hooklike upper bills, or sharp,

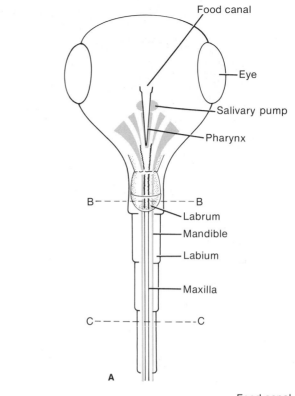

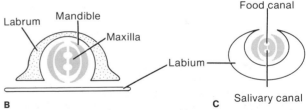

**15-8** Sucking apparatus of plant-feeding bug. (**A**) Anterior view of head and proboscis. (**B**) Transverse section through B–B. (**C**) Transverse section through C–C. The separate parts fit together to form channels for delivery of saliva and for sucking. Since they are free to slide past each other, the parts are readily folded between feedings. [From J. B. Jennings, *Feeding, Digestion and Assimilation in Animals.* By permission of Macmillan, London and Basingstoke.]

wood-pecking points (Figure 15-10). Raptorial birds capture prey with their talons or beak. Seed-eating birds eat their food whole, but may subject it to grinding in a muscular *gizzard* containing pebbles that aid the grinding process.

Submammalian teeth are generally poorly differentiated from one another. One exception is found

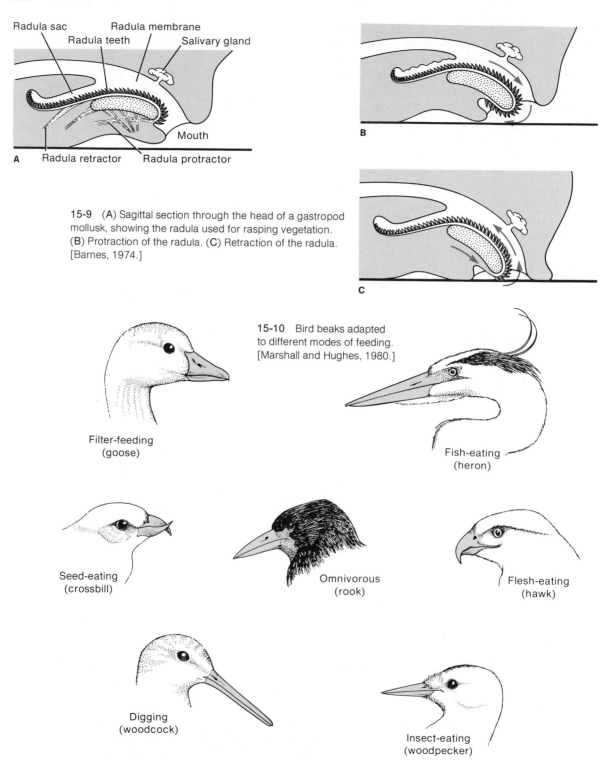

Radula sac   Radula membrane
Radula teeth   Salivary gland

Mouth

Radula retractor   Radula protractor

**A**

**B**

**C**

**15-9** (A) Sagittal section through the head of a gastropod mollusk, showing the radula used for rasping vegetation. (B) Protraction of the radula. (C) Retraction of the radula. [Barnes, 1974.]

**15-10** Bird beaks adapted to different modes of feeding. [Marshall and Hughes, 1980.]

Filter-feeding
(goose)

Fish-eating
(heron)

Seed-eating
(crossbill)

Omnivorous
(rook)

Flesh-eating
(hawk)

Digging
(woodcock)

Insect-eating
(woodpecker)

among the poisonous snakes, such as vipers, cobras, and rattlesnakes, which have modified teeth, called *fangs,* that they use to inject venom (Figure 15-11). These fangs either are equipped with a groove that guides the venom or are hollow, very much like a syringe needle. In rattlesnakes, the fangs fold back against the roof of the mouth, but extend perpendicularly when the mouth is opened to strike. A snake's lower jaws are held together with an elastic ligament that allows them to spread apart during swallowing. This enables the snake to swallow animals larger than the diameter of its head.

Mammals make extensive use of their teeth for killing prey and then cutting and grinding it up, and the teeth have evolved very different shapes for these purposes (Figure 15-12). Chisellike *incisors* are used especially by rodents and lagomorphs (hares,

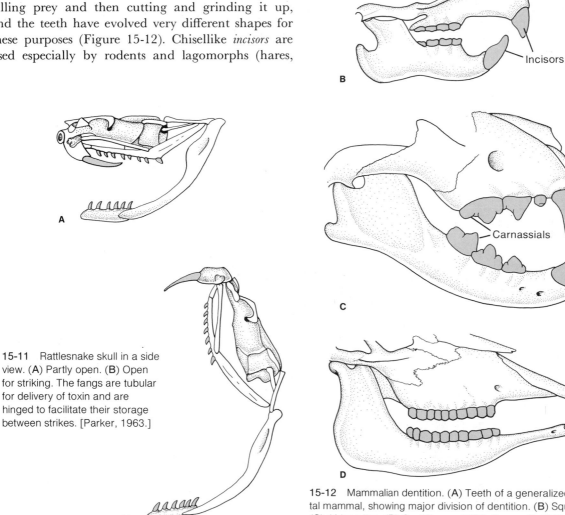

**15-11**  Rattlesnake skull in a side view. (**A**) Partly open. (**B**) Open for striking. The fangs are tubular for delivery of toxin and are hinged to facilitate their storage between strikes. [Parker, 1963.]

**15-12**  Mammalian dentition. (**A**) Teeth of a generalized placental mammal, showing major division of dentition. (**B**) Squirrel. (**C**) African lion. (**D**) Ox. [Romer, 1962; Cornwall, 1956.]

rabbits) for gnawing. In the Proboscidea (elephants, mammoths), the incisors are modified as a pair of tusks. Pointy, daggerlike *canines* are used by the carnivores, insectivores, and primates for piercing and tearing. In some groups like the wild pigs and pinnipeds, the canines are elongated as tusks, which are used for prying and as weapons. *Carnassials* are knifelike molars that, along with sharp incisors, are used by the carnivores to cut flesh into pieces small enough to swallow. Most complex and interesting in their form are the molars of some herbivorous groups such as the Artiodactyla (cattle, pigs, hippopotamuses), the Perissodactyla (horses, zebras), and the Proboscidea. These teeth, which are used in a side-to-side grinding motion, are comprised of folded layers of enamel, cement, and dentine which all differ in hardness and in rate of wearing. Because the softer dentine wears rather quickly, the harder enamel and cement layers form ridges that enhance the effectiveness of the molars in the mastication of grass and other tough vegetation.

### PREY CAPTURE TOXINS

Toxins, most of them acting on the nervous system, are utilized in subduing prey and in defense by members of various phyla. Among the coelenterates, there is extensive use of *nematocysts* (stinging cells). Concentrated in large numbers on the tentacles, the nematocysts inject paralytic toxins into prey and immobilize it while the tentacles carry it to the mouth. Many nemertine worms paralyze their prey by injecting venom through a stilettolike proboscis. Poisons are also used by annelids, gastropod mollusks, and arthropods. Among the last, scorpions and spiders are most notorious for their toxins. After grabbing its prey with its large chelae (pincerlike organs), a scorpion will arch its tail to bring its sting to bear (Figure 15-13), injecting the victim with a poison containing a *neurotoxin* that interferes with the inactivation of the sodium current and thus with proper firing of nerve impulses. Spider poisons, too, contain neurotoxins. The venom from the black widow spider contains a substance that induces massive release of neurotransmitter at the motor endplate in muscle. A neurotoxin, α-bungarotoxin (Box 6-3), found in the venom of the cobralike krait, binds

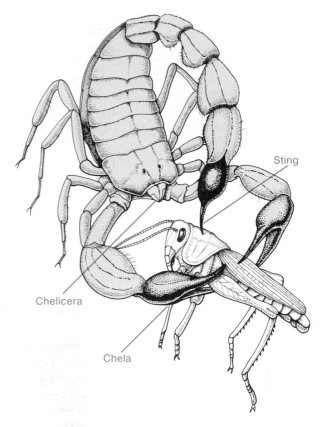

**15-13** Prey capture and poisoning by the scorpion *Androctonus*. Its tail is arched over the head at the moment the prey is captured, bringing the sting into position. [From J. B. Jennings, *Feeding, Digestion and Assimilation in Animals*. By permission of Macmillan, London and Basingstoke.]

to the acetylcholine receptor, thereby blocking neuromuscular transmission in vertebrates. The venoms of various species of rattlesnake contain hemolytic (blood-cell-destroying) substances. These toxins are generally proteins and thus are rendered harmless by proteolytic enzymes when the predator ingests its poisoned prey.

## Nutritional Requirements

*Nutrients* are substances that serve as sources of metabolic energy, raw material for growth or repair of tissues, and general maintenance of body functions. Included among nutrients are essential trace elements such as iodine, zinc, and other metals. Animals vary widely in their specific nutritional needs depending on species. Within a species, those needs vary according to differences in genetic makeup, body size and composition, activity, age, sex, and reproductive functions. Thus, an egg-producing or pregnant female will require more nutrients than a male. As discussed in Chapter 16, a small animal requires more food for energy per gram of body weight than does a larger animal, because its metabolic rate per unit body weight is higher. Similarly, an animal with a high body temperature requires more food for greater energy needs than does a cooler animal.

For an animal to be in a *balanced nutritional state,* it must be supplied with:

*1.* Sufficient energy to power all body processes.

*2.* Enough protein and amino acids to maintain a positive nitrogen balance (i.e., to avoid net loss of body proteins).

*3.* Enough water and minerals to compensate for losses or incorporation.

*4.* Those essential vitamins not synthesized within the body.

*Energy balance* requires that caloric intake over a period of time equal the number of calories consumed for tissue maintenance and repair and for work (metabolic and otherwise) plus the production of body heat in birds and mammals. Insufficient intake of calories can be temporarily offset by calling on tissue fat, carbohydrates, or proteins, with a resultant loss of body weight. Conversely, excessive caloric intake will result in increased storage of body fat (often seen in *Homo sapiens*).

Differences in specific (nonenergetic) nutritional requirements are known to be genetically determined in some instances. Thus, for a given animal species, certain cofactors ($Zn^{2+}$, $I^-$, etc.) or building blocks (amino acids, etc.) essential for important biochemical reactions or for the production of tissue molecules may be required from food sources simply because those cofactors cannot be produced by the animal itself. Such an item is known as an *essential nutrient* in the diet of that species.

## NUTRIENT MOLECULES

### Proteins and Amino Acids

Proteins are used as structural components of soft tissues and as enzymes. They can also be utilized as energy sources if first broken down to amino acids. The proteins of animal tissues are composed of about 20 or so different amino acids. The ability to synthesize amino acids differs among species. Those amino acids that cannot be synthesized by an animal, but are required for synthesis of essential proteins, are the so-called *essential amino acids* for that animal. Recognition of this principle has been of enormous economic significance in the poultry industry. The rate of growth of chickens at one time was limited by too small a proportion of one essential amino acid in the grain diet they were provided. Supplementing the diet with this amino acid allowed full utilization of the other amino acids present in the feed, greatly increasing the rate of protein synthesis and hence the rate of poultry growth and egg laying.

### Carbohydrates

These are used primarily as immediate (glucose-6-phosphate) or stored (glycogen) sources of chemical energy; but they may also be converted to metabolic intermediates or to fats. Conversely, proteins and fats can be converted by most animals into carbohydrates. The major sources of carbohydrate are the sugars, starches, and cellulose found in plants and the glycogen stored in animal tissues.

### Fats and Lipids

These are especially suitable as concentrated energy reserves, for each gram of fat provides over two times as much caloric energy as a gram of protein or carbohydrate and need not be dissolved in water. Thus, fat is commonly stored by animals for times of *caloric deficit,* when energy expenditure exceeds energy intake. Fats and lipids are important in such tissue components as plasma membranes and other membrane-based organelles of the cell and the myelin sheath of axons. The fatty molecules or lipids include fatty acids, monoglycerides, triglycerides, sterols and phospholipids.

TABLE 15-1 Some mammalian vitamins.

| Designation letter and name | Major sources and solubility | Metabolism | Function | Deficiency symptoms |
|---|---|---|---|---|
| A—Carotene | Egg yolk, green or yellow vegetables and fruits; FS† | Absorbed from gut; bile aids, in liver | Formation of visual pigments; maintenance of epithelial structure | Night blindness, skin lesions |
| D$_3$—Calciferol | Fish oils, liver; FS | Absorbed from gut; little storage | Increases calcium absorption from gut; bone and tooth formation | Rickets (defective bone formation) |
| E—Tocopherol | Green leafy vegetables; FS | Absorbed from gut; stored in adipose and muscle tissue | Humans—maintains red cells | Increased fragility of red blood cells |
| | | | Other mammals—maintains pregnancy | Abortion, muscular wastage |
| K—Naphthoquinone | Synthesis by intestinal flora; liver; FS | Absorbed from gut; little storage; excreted in feces | Enables prothrombin synthesis by liver | Failure of coagulation |
| B$_1$—Thiamine | Brain, liver, kidney, heart; whole grains; WS† | Absorbed from gut; stored in liver, brain, kidney, heart | Formation of cocarboxylase enzyme involved in decarboxylation (Krebs cycle) | Stoppage of CH$_2$O metabolism at pyruvate, beriberi, neuritis, heart failure |
| B$_2$—Riboflavin | Milk, eggs, liver, whole cereals; WS | Absorbed from gut; stored in kidney, liver, heart | Flavoproteins in oxidative phosphorylation | Photophobia; fissuring of skin |
| Niacin | Whole grains; WS | Absorbed from gut; distributed to all tissues | Coenzyme in hydrogen transport, (NAD, NADP) | Pellagra; skin lesions; digestive disturbances, dementia |
| B$_{12}$—Cyanocobalamin | Liver, kidney, brain. Bacterial synthesis in gut; WS | Absorbed from gut; stored in liver, kidney, brain | Nucleoprotein synthesis, prevents pernicious anemia | Pernicious anemia; malformed erythrocytes |
| Folic acid (folacin, pteroylglutamate acid) | Meats; WS | Absorbed from gut; utilized as taken in | Nucleoprotein synthesis; formation of erythrocytes | Failure of erythrocytes to mature; anemia |
| Pyridoxine (B$_6$) | Whole grains; WS | Absorbed from gut; one-half appears in urine | Coenzyme for amino acid and fatty acid metabolism | Dermatitis; nervous disorders |
| Pantothenic acid | WS | Absorbed from gut; stored in all tissues | Forms part of coenzyme A (CoA) | Neuromotor, cardiovascular disorders |

TABLE 15-1 (continued) *657*

| Designation letter and name | Major sources and solubility | Metabolism | Function | Deficiency symptoms |
|---|---|---|---|---|
| Biotin | Egg white; synthesis by flora of GI tract; WS | Absorbed from gut | Protein synthesis, $CO_2$ fixation and transamination | Scaly dermatitis; muscle pains, weakness |
| Para-amino benzoic acid (PABA) | WS | Absorbed from gut; little storage | Essential nutrient for bacteria; aids in folic acid synthesis | No symptoms established for humans |
| Ascorbic acid (Vitamin C) | Citrus; WS | Absorbed from gut; stored | Vital to collagen and ground substance | Scurvy—failure to form connective tissue |

†FS = fat-soluble; WS = water-soluble.
*Source:* McClintic, 1978.

### Nucleic Acids

Although nucleic acids are essential for the genetic machinery of the cell, all animal cells appear to be capable of synthesizing them from simple precursors.

### Inorganic Salts

Some chloride, sulfate, phosphate, and carbonate salts of the metals calcium, potassium, sodium, and magnesium are important constituents of intra- and extracellular fluids. Calcium phosphate occurs as *hydroxyapatite* $(Ca_{10}(PO_4)_6(OH)_2)$, a crystalline material that lends hardness and rigidity to the bones of vertebrates and the shells of mollusks. Iron, copper, and other metals are required for redox reactions and for oxygen transport and binding. Many enzymes require certain metal atoms for their catalytic functions. Animal tissues need moderate quantities of some metal ions (Ca, P, K, Na, Mg, and Cl) and trace amounts of others (Mn, Fe, S, I, Co, Cu, Zn, and Se).

### Vitamins

These diverse and chemically unrelated organic substances generally function in small quantities. Some vitamins important in human nutrition, along with their diverse functions, are listed in Table 15-1. The ability to synthesize different vitamins differs among species, and those essential vitamins that an animal cannot produce itself must be obtained from other sources, primarily from plants but also from dietary animal flesh or from intestinal microbes. Thus, ascorbic acid, also known as vitamin C, is synthesized by many animals, but not by humans. Humans also depend on intestinal bacteria for production of vitamins K and $B_{12}$. Fat-soluble vitamins such as A, D, E, and K can be stored in body fat deposits, but water-soluble vitamins such as ascorbic acid are not stored and are carried out in the urine. These must be ingested or produced continually if adequate levels are to be maintained.

Certain vitamin deficiences are manifested by characteristic symptoms. For example, scurvy, a condition of ascorbic acid deficiency, was common on board ships before the British admiralty instituted the use of citrus fruit to supplement the diet of the crews. Albert Szent-Györgyi was awarded the Nobel Prize in 1937 largely for his work in characterizing vitamin C, which he isolated from Hungarian peppers and potatoes. This vitamin has been found to be important as, among other things, an oxidizing agent, and in this capacity it is believed to have the additional helpful property of eliminating harmful free radicals in tissue.

### Water

Of all the constituents of animal tissue, none is more pervasively important to living tissue than water. This marvelous substance comprises up to 95% or more of the weight of some animal tissues. It is replenished in most animals by drinking, by ingestion with food, and, to a modest extent, by its metabolic production during the oxidation of fats and carbohydrates. Some marine and desert animals depend almost entirely on "metabolic water" to replace water lost by evaporation, defecation and urination, as discussed in Chapter 12.

## An Overview of Alimentary Systems

In some of the simpler forms of animal life, food particles are engulfed undigested by endocytosis directly into cells themselves, where they undergo *intracellular digestion* by acids and enzymes. More complex animals rely primarily on *extracellular digestion* which takes place in a tubular cavity that extends through the organism (Figure 15-3). The lumen of this *alimentary canal* is topologically external to the body, although sphincters and other devices prevent uncontrolled exchange between the lumen and the external world. Food is exposed to various mechanical, chemical, and bacterial treatment as it passes through this canal, and *digestive* juices are mixed with the food at appropriate stages. As the food is digested, absorbable nutrients are transported into the circulatory system, while undigested material is stored briefly until it, along with bacterial remains, is expelled as feces by *defecation.*

The development of extracellular digestion in an alimentary canal was an important evolutionary innovation. It freed many animals from feeding continuously, for they could now quickly ingest a few large chunks of food rather than slowly obtaining many particles small enough to enter cells and undergo intracellular digestion. The overall tubular organization of alimentary canals (Figures 15-14 and 15-15) is efficient because it allows food to travel in one direction, passing through different regions of digestive specialization. Thus, both acidic and alkaline phases occur in the tract of vertebrates, and both are active at the same time and provide different types of digestive action. All phyla above the flatworms have this tubular alimentary organization. Below these phyla, the coelenterates, for example, have a blind cavity, the *coelenteron,* which opens only at a "mouth" that serves also for the expulsion of undigested remains.

In general, alimentary canals have four major divisions, the functions of which are (1) receiving, (2) conducting and storing, (3) digesting and absorbing nutrients, and (4) absorbing water and defecating. Representative alimentary canals from the different vertebrate classes are illustrated in Figure 15-15.

### RECEIVING REGION

The beginning of the alimentary canal consists of organs and devices for feeding and swallowing, including the mouthparts, *buccal cavity, pharynx,* and associated structure such as bills, teeth, tongue, and salivary glands.

Salivary glands occur in most metazoans other than small-particle feeders such as coelenterates, flatworms, and sponges. The primary function of the salivary secretion, *saliva,* is lubrication to assist swallowing. The lubrication is provided in many cases by a slippery mucus in which the chief constituent is a type of mucopolysaccharide named *mucin.* The saliva often contains additional agents such as digestive enzymes, toxins, and anticoagulants (in blood-lapping or blood-sucking animals such as vampire bats and leeches).

Tongues, an innovation of the chordates, assist in the swallowing process. In some animals they are used to grasp food. They are also used in chemoreception, bearing gustatory receptors called taste buds (Figure 7-16A). Snakes use their forked tongues to take olfactory samples from the air and the substratum, retracting the tongue to wipe the samples in Jacobson's organ, which consists of a pair of richly innervated chemosensory pits located in the roof of the buccal cavity.

15-14 (facing page)   Some invertebrate digestive systems. (A) Section through body wall of *Hydra,* a coelenterate. The epithelial lining of coelenteron includes phagocytosing cells (called nutritive muscle cells) and gland cells that secrete digestive enzymes. (B) Digestive system of a polyclad flatworm. (C) Digestive system of a prosobranch gastropod mollusk. Arrows show ciliary currents and the rotation of the mucous mass within the style sac. [Barnes, 1974.] (D) Digestive system of the cockroach *Periplaneta.* The proventriculus (or gizzard) contains chitinous teeth for grinding food. [Imms, 1949.]

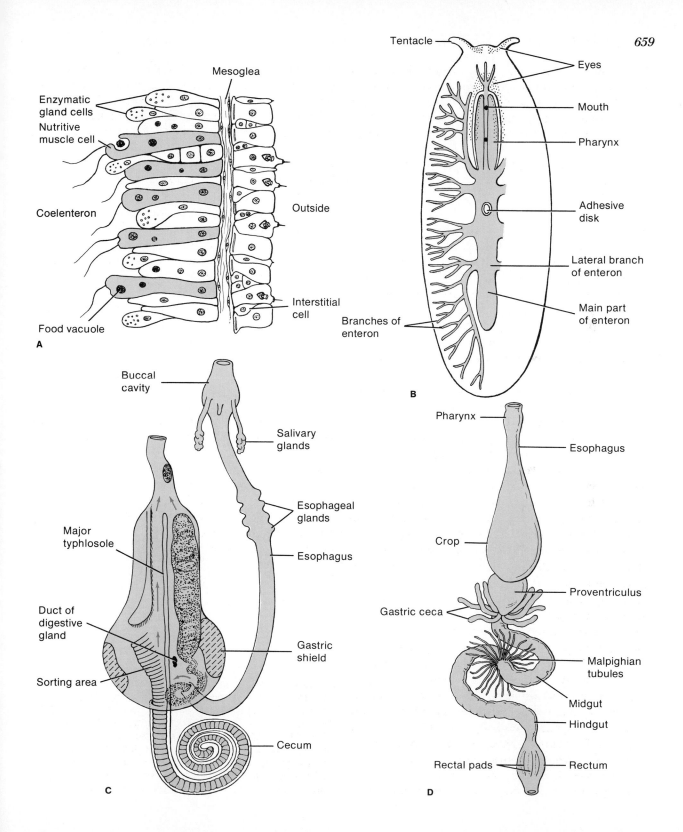

660

Cyclostome
(hagfish)

Elasmobranchs
(shark)

Teleost fishes
(bass)

Amphibians
(frog)

Aves
(pigeon)

Mammals
(rabbit)

Mammals
(human)

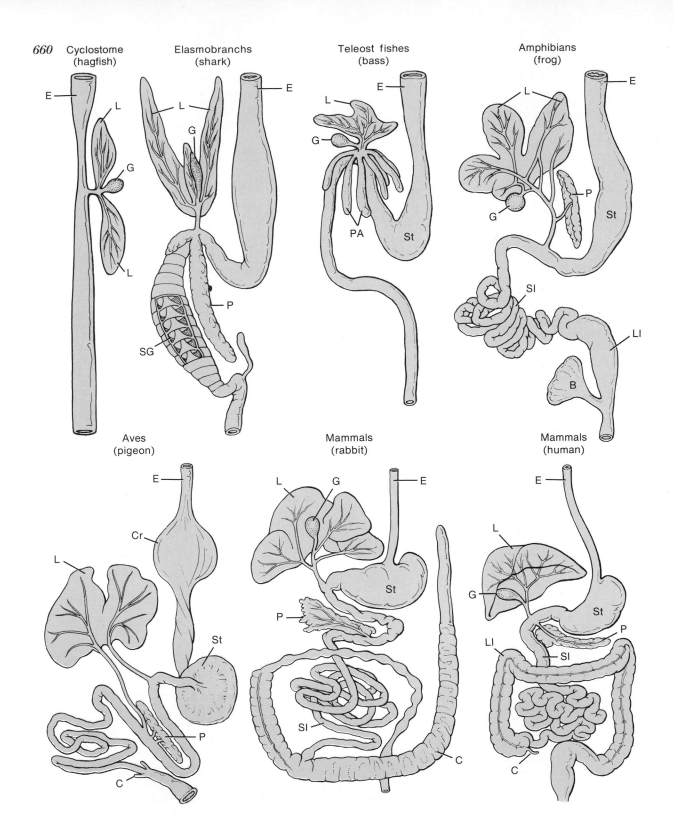

**15-15** (facing page)   Digestive tracts of vertebrates. B, bladder; C, cecum; Cr, crop; E, esophagus; G, gallbladder; L, liver; Ll, large intestine; SG, spiral gut; P, pancreas; PA, pyloric appendices; SI, small intestine; St, stomach. [Florey, 1966, after Stempell, 1926.]

## CONDUCTING AND STORAGE REGION

The *esophagus* of chordates and some invertebrates serves to conduct the *bolus* (mass of chewed food) by *peristaltic movement* from the buccal cavity or pharynx. In some animals, this region contains a saclike expanded section, the *crop,* used for food storage prior to digestion. This feature, generally related to a lifestyle that includes infrequent feeding, allows quantities of food to be stored for delayed digestion. One example of this is seen in leeches, which take very infrequent large blood meals, storing the blood for many weeks and digesting it in small amounts between their rare feedings. Crops are also used to ferment mildly or digest foods for purposes other than their immediate utilization. Parent birds prepare food in this way to be regurgitated for their nestlings.

## DIGESTIVE REGION

Most digestive processes take place in a region of the alimentary canal that is divided into two large divisions, *stomach* and *intestine,* in vertebrates and some invertebrates. The stomach provides the initial stages of digestion, which often require an acidic environment. In vertebrates and some invertebrates, it also continues the mechanical mixing begun during ingestion.

### Monogastric Stomach

Mechanical *trituration* is achieved by a variety of means. A strong muscular tube or sac is characteristic of vertebrates that are carnivorous or omnivorous (Figure 15-16). This single chamber contracts so as to mix the contents with digestive juices. Instead of a stomach, some invertebrates, including insects (Figure 15-14), have outpouchings termed *gastric ceca* (singular: *cecum*),

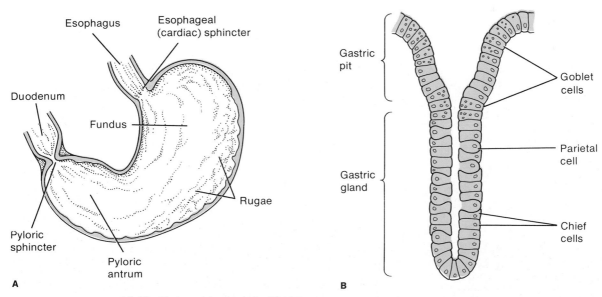

**15-16**   Monogastric stomach. (A) Major parts of the mammalian stomach. (B) Fundic, or gastric, gland. The epithelium contains chief (pepsinogen-secreting) and parietal (HCl-secreting) cells as well as goblet cells. [Madge, 1975, from Passmore and Robson, 1968.]

which are lined with enzyme-secreting cells and phago-cytic cells that continue the processes of digestion after engulfing partially digested food. In these alimentary systems, the processes of digestion and absorption are completed in the ceca, and the remainder of the alimentary canal is concerned primarily with water and electrolyte balance and nitrogen excretion.

Some birds have a tough, muscular gizzard (Figure 15-15). Sand, pebbles, or stones are swallowed and then lodge in the gizzard, where they aid the grinding of seeds and grains. The *proventriculus* of insects and the stomach of decapod crustaceans contain grinding apparatuses for chewing swallowed food.

*Digastric Stomach*

Multichambered *digastric stomachs* (Figure 15-17) are found in the suborder Ruminantia (deer, elk, giraffe, bison, sheep, cattle, etc.). Somewhat similar digastric stomachs occur outside this suborder, in particular in the suborder Tylopoda (camel, llama, alpaca, vicuña). All these groups engage in *rumination,* a process in which partially digested food, initially swallowed without chewing, is *regurgitated* for *remastication* after

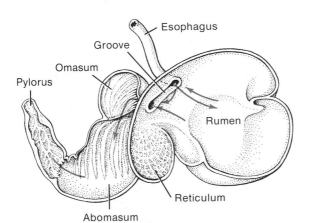

**15-17** Digastric stomach. This sheep stomach, characteristic of the Ruminidae, has two divisions made up of four chambers. The rumen and reticulum make up the fermentative division. The omasum and abomasum (true stomach) make up the digestive division. [Romer, 1962.]

being fermented by microorganisms in the first division of the stomach. This procedure allows the ruminant to swallow food hastily while grazing and then to chew it more thoroughly later when at leisure in a place of relative safety. After the regurgitated food is chewed, it is swallowed again. This time it passes into the second division of the digastric stomach and begins the second stage of digestion, in which hydrolysis takes place, assisted by digestive enzymes secreted by the stomach lining.

The digastric stomach of the Ruminantia (Figure 15-17) has four chambers, separated into two divisions. The first division consists of the *rumen* and *reticulum* chambers; the second division comprises the *omasum* and the *abomasum* (true stomach). The rumen and reticulum act as a fermentation vat that receives unchewed vegetation as the animal grazes. Bacteria and protozoa in these chambers thrive on the vegetation, causing extensive digestive breakdown by fermentation of carbohydrates to butyrate, lactate, acetate, and propionate. These products of fermentation, along with some peptides, amino acids, and short-chain fatty acids, are absorbed into the bloodstream from the rumen fluid. Symbiotic microorganisms grown in the rumen, along with undigested particles, are passed into the omasum (missing in the Tylopoda) and then into the abomasum. Only the latter secretes digestive enzymes and is homologous to the monogastric stomach of nonruminants.

Fermentation in the stomach is not limited to ruminating animals. It is found in other animals in which the passage of food through the stomach is delayed, allowing the growth of symbiotic microorganisms. Examples are the stomach of the kangaroo and the crops of galliform (chickenlike) birds.

*Small Intestine*

As food is ready to pass on from the vertebrate stomach, it is released into the intestine through the *pyloric sphincter,* which relaxes as the peristaltic movements of the stomach squeeze the acidic contents into the initial segment of the *small intestine.* In the small intestine, digestion continues, generally in an alkaline environment.

The intestinal region varies widely among animal groups. In animals that have extensive ceca and diverticula (blind pouches or sacs arising from a main passage), as do many invertebrates, the intestine serves no digestive function. Among the vertebrates, carnivores have shorter and simpler intestines than do herbivores. For example, a tadpole, which is herbivorous, is reputed to have a longer intestine than the much larger adult frog, which is carnivorous.

The vertebrate small intestine is typically divided into three distinct portions. The first, rather short, section is the *duodenum,* the lining of which secretes digestive enzymes and receives secretions carried by ducts from the *liver* and *pancreas.* Next is the *jejunum,* which also secretes digestive juices. The more posterior section, the *ileum,* acts primarily to absorb digested nutrients. Some digestion, begun in the duodenum and jejunum, continues in the ileum.

The secretory functions of the duodenal epithelium are supplemented by secretions from the liver and pancreas. As will be discussed later, the cells of the liver produce *bile salts,* which are carried in the *bile fluid* to the duodenum through the *bile duct.* Bile fluid is important, both in emulsifying fats and in neutralizing acidity introduced into the duodenum from the stomach. The pancreas, an important exocrine organ, produces and releases into the *pancreatic duct* the *pancreatic juice* that contains many of the proteases, lipases, and carbohydrases essential for intestinal digestion in vertebrates. The pancreatic juice is also important in neutralizing gastric acid in the intestine.

The intestine of most animals contains large numbers of bacteria, protozoa, and fungi. These multiply, contributing enzymatically to digestion, and usually are in turn digested themselves. An important function of some intestinal symbionts is the synthesis of essential vitamins.

### INTESTINAL EPITHELIUM

Lining the small intestine are columnar *absorptive cells* (Figure 15-18). Forming a monocellular epithelium, these cells cover the *villi,* which stand about 1 mm tall and are each surrounded by a circular depression known as the *crypt of Lieberkühn* (Figure 15-19C). Within each villus is a network of blood vessels—capillaries and venules—and a network of *lymph vessels,* including the *central lacteal.* It is into these blood and lymph vessels that the nutrients find their way. The

absorptive cells proliferate at the base of the villus and steadily migrate toward its tip, whence they are sloughed off at the rate of about $2 \times 10^{10}$ cells per day within the human intestine. The villi themselves are protrusions from extensive *folds* of the *mucosa* (Figures 15-19 and 15-20).

Each absorptive cell bears a striated structure at its apical surface. This is the *brush border* that is made up of densely arrayed *microvilli* (Figure 15-18A and D), which number several thousand per cell (about $2 \times 10^5$ per square millimeter), and each stands 0.5–1.5 $\mu$m tall and about 0.1 $\mu$m wide. The microvilli are enclosed in the plasma membrane and contain actin filaments that interact with myosin filaments present at the base of each microvillus. This interaction is responsible for the rhythmic motions of the microvilli that may help mix and exchange the *intestinal chyme* (semifluid mass of partially digested food) near the absorptive surface.

The hierarchy of folds, villi, and microvilli results in an enormous elaboration of absorptive surface area. In humans, the lumen of the small intestine has a gross cylindrical surface area of about 0.4 m². The folds, villi, and microvilli increase the area at least 500 times, to a total of 200–300 m². This increase, of course, is important for the process of absorption, for the rate of absorption should be proportional to the area of the major diffusional barrier, which is the apical surface membrane of the absorptive cells.

Covering the surface of the microvilli is the *glycocalyx,* a meshwork up to 0.3 $\mu$m thick, made up of acid mucopolysaccharide and glycoprotein. Within the interstices of the glycocalyx, water and mucus are trapped in an "unstirred layer." The mucus is secreted by *mucous cells,* or *goblet cells* (so called because of their shape) that occur among the absorptive cells.

The absorptive cells are held to one another by desmosomes. Near the apex, the zonula occludens encircles each cell, making a tight junction with its neighbors. The tight junctions are especially tight in this epithelium, and so the apical membranes of the absorptive cells are joined into a continuous sheet of apical membrane. All nutrients must pass through this sheet and through the absorptive cell cytoplasm to get from the lumen to the blood and lymph vessels within the villi.

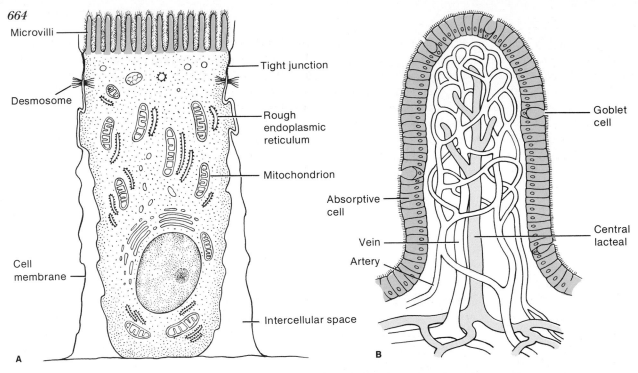

Microvilli

Tight junction

Desmosome

Rough endoplasmic reticulum

Mitochondrion

Cell membrane

Intercellular space

**A**

Goblet cell

Absorptive cell

Vein

Artery

Central lacteal

**B**

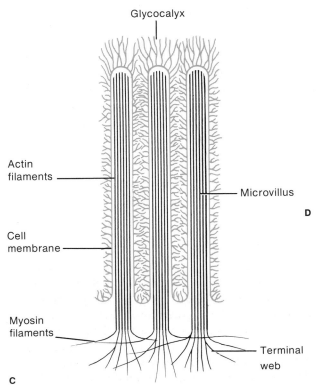

Glycocalyx

Actin filaments

Cell membrane

Myosin filaments

Microvillus

Terminal web

**C**

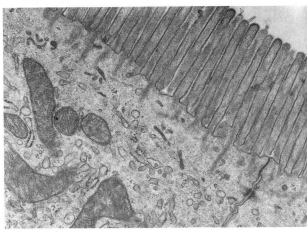

**D**

**15-18** Lining of the mammalian small intestine. The luminal surface is shown in color. (**A**) Villus covered with the digestive epithelium consisting primarily of absorptive cells (**B**) and occasional goblet cells. The luminal or apical surface of the absorptive cell bears a brush border of microvilli (**C**), which consist of evaginations of the surface membrane (color) enclosing bundles of actin filaments. [From "The Lining of the Small Intestine" by F. Moog. Copyright © 1981 by Scientific American, Inc. All rights reserved.] (**D**) Electron micrograph of portion of absorptive cells from the small intestine of the rat, showing brush border. Magnification 32,000×. [Courtesy of K. R. Porter.]

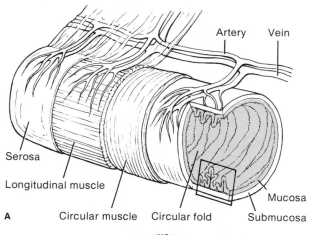

Serosa

Longitudinal muscle

**A**  Circular muscle   Circular fold

Artery  Vein

Mucosa

Submucosa

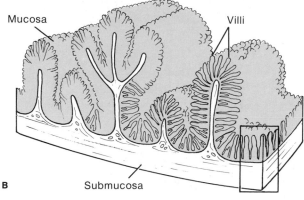

Mucosa

Villi

**B**  Submucosa

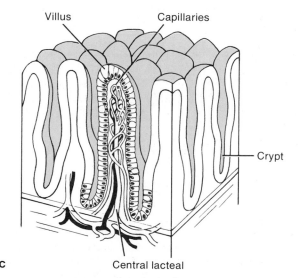

Villus  Capillaries

Crypt

**C**  Central lacteal

15-19  Anatomy of the small intestine. (**A**) Overall plan.
(**B**) Fingerlike villi of the mucosa (**C**) covering intestinal folds.
[From "The Lining of the Small Intestine" by F. Moog. Copy-
right © 1981 by Scientific American, Inc. All rights reserved.]

## WATER-ABSORBING AND ELIMINATION REGION

The final section of an alimentary canal is generally concerned with the further removal of excess water from the intestinal contents and the consolidation of the undigested material into feces, before the latter are expelled through the anus. In vertebrates, this function is carried out primarily in the latter portion of the small intestine and in the large intestine. In some insects, the feces within the rectum are rendered almost dry by a specialized mechanism for removing water from the rectal contents, as was described on page 534.

## Motility of the Alimentary Canal

Motility is important to alimentary function for (1) translocation of food along the canal and the final expulsion of fecal material, (2) mechanical treatment by grinding and kneading to help mix in digestive juices and convert food to a soluble form, and (3) mixing of the contents so that there is continual renewal of material in contact with the absorbing surfaces of the epithelial lining. Muscular mechanisms of motility alone are utilized by arthropods and chordates. Ciliary mechanisms alone are used to translocate food along the alimentary canals of annelids, lamellibranch mollusks, tunicates, and cephalochordates, but they are used in conjunction with muscular mechanisms in the echinoderms and most mollusks. Muscular mechanisms permit handling of harder and larger pieces of food.

### PERISTALSIS

In groups other than arthropods, the alimentary musculature is made up of smooth muscle tissue. In arthropods and vertebrates, the arrangement of the musculature consists of an *inner circular layer* and an *outer longitudinal layer* (Figure 15-20). The coordinated contraction of the circular layer with relaxation of the longitudinal layer produces an active constriction with passive elongation; or active shortening with relaxation of the circular layer produces passive distention. *Peristalsis* occurs as a traveling wave of constriction produced by contraction of circular muscle and is preceded by a simultaneous contraction of the longitudinal muscle and relaxation of the circular

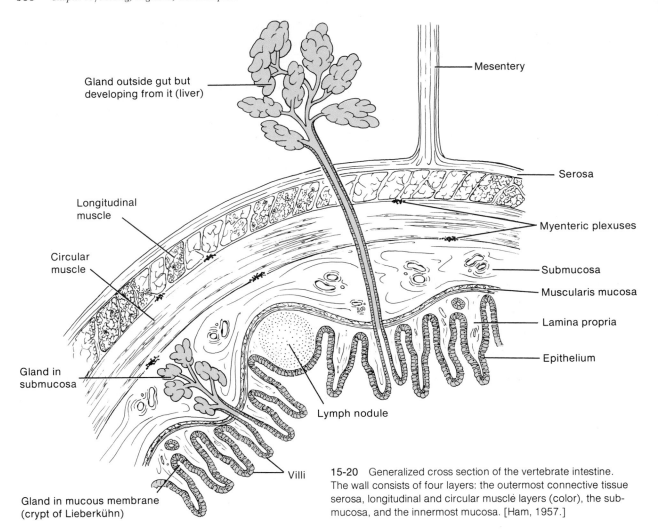

Gland outside gut but
developing from it (liver)

Mesentery

Serosa

Longitudinal
muscle

Myenteric plexuses

Circular
muscle

Submucosa

Muscularis mucosa

Lamina propria

Epithelium

Gland in
submucosa

Lymph nodule

Gland in mucous membrane
(crypt of Lieberkühn)

Villi

**15-20**   Generalized cross section of the vertebrate intestine.
The wall consists of four layers: the outermost connective tissue
serosa, longitudinal and circular muscle layers (color), the submucosa, and the innermost mucosa. [Ham, 1957.]

muscle (Figure 15-21). This pattern of contraction
produces a longitudinal displacement of the luminal
contents in the direction of the peristaltic wave. Mixing
of the luminal contents is achieved primarily by a
process called *segmentation,* which consists of rhythmic
contractions of the circular muscle layer that occur
asynchronously along the intestine at various points
without participation of the longitudinal muscle. Regurgitation occurs when peristalsis takes place in the
reverse direction, moving the luminal contents into the
buccal cavity. Ruminants regularly use regurgitation

to bring up the unchewed food for further chewing,
and other vertebrates use it during vomiting.

*Swallowing* in vertebrates involves the integrated
movements of muscles in the tongue and pharynx,
as well as peristalsis of the esophagus. These actions
propel the bolus to the stomach. Normal peristalsis
in the vertebrate stomach occurs with the ring of contraction only partially closed. There is thus a mixing
action in which the contents are squeezed backward
(opposite the direction of the wave) centrally through
the partially open ring and forward peripherally in the

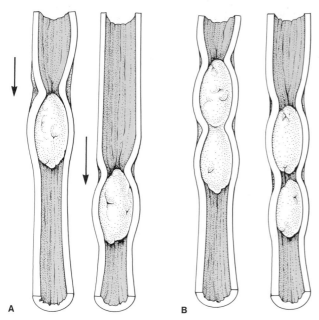

15-21 Motility of the gastrointestinal tract. (A) Peristalsis occurs as a traveling wave of contraction of circular muscle is preceded by its relaxation. This produces longitudinal movement of the bolus. (B) Segmentation occurs as alternating relaxations and contractions, primarily of circular muscle. The result is a kneading and mixing of the intestinal contents.

direction of peristalsis as the partially closed ring of contraction moves from the cardiac end to the pyloric end of the stomach.

### CONTROL OF MOTILITY

The motility of the alimentary canal of vertebrates depends on the coordinated contractions of circular and longitudinal smooth muscle layers and is regulated by a combination of three separate mechanisms.

First, the smooth muscle tissue is myogenic—namely, capable of producing an intrinsic cycle of electrical activity that leads to contraction. This cycle occurs as rhythmic depolarizations and repolarizations called the *basic electric rhythm* (*BER*) and consists of spontaneous slow waves of depolarization (Figure 15-22). Some of these slow waves give rise to action potentials produced by an inward current carried by calcium ions. It is these calcium "spikes" that give rise to contractions of the smooth muscle cells in which they occur.

Second, the smooth muscle is diffusely innervated by an intrinsic network of cholinergic neurons, the *myenteric plexus,* located entirely within the muscular layers of the gut (Figure 15-23). This network, in turn,

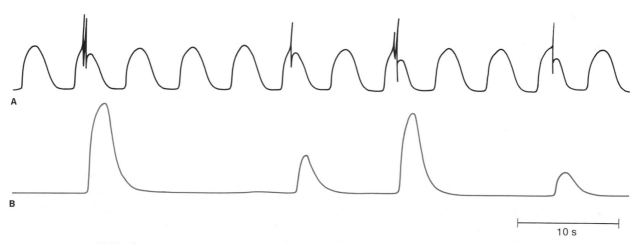

15-22 Correlation between electrical activity and contractions in the cat jejunum. (A) The slow basic electric rhythm occasionally gives rise to calcium action potentials at their peaks. (B) These, in turn, elicit contractions of the smooth muscle in which they occur. [Bortoff and Davis, 1976.]

is innervated by efferent axons from the parasympathetic (primarily excitatory) division carried by the vagus, pelvic, and splanchnic nerves. Postganglionic axons from the sympathetic (primarily inhibitory) division of the autonomic nervous system directly innervate all the tissues of the gut wall as well as the neurons of the myenteric and submucous plexus. Activity in the parasympathetic system excites motility via the intrinsic cholinergic network, whereas activity of the sympathetic efferents inhibits the motility of the stomach and intestine.

The smooth muscle cells are inhibited (i.e., prevented from spiking) by norepinephrine, liberated by the sympathetic innervation, and are excited by ace-

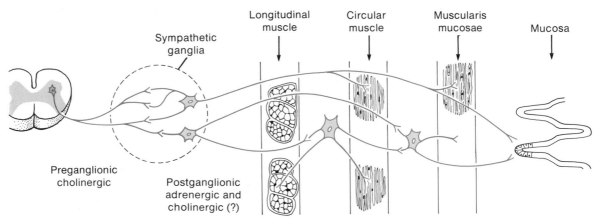

**A**  Efferent sympathetic innervation

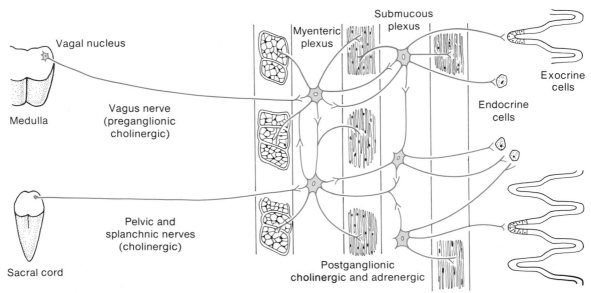

**B**  Parasympathetic innervation

15-23   Innervation of the gastrointestinal tract. (**A**) Efferent sympathetic innervation. (**B**) Parasympathetic innervation. All nerve endings on the gastrointestinal target tissues (muscle, glands) are postganglionic. [Davenport, 1977.]

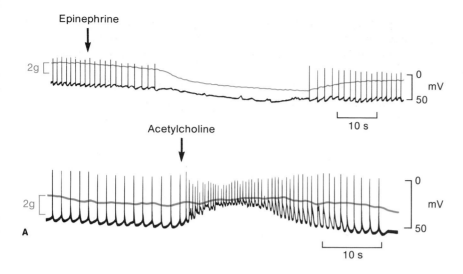

Epinephrine

Acetylcholine

15-24　Membrane potential and tension in the tenia coli, a longitudinal muscle along the colon. (A) Effects of topically applied epinephrine and acetylcholine. (B) Time correlation between action potentials (black) and tension (color). [Bülbring and Kuriyama, 1963; Bülbring, 1959.]

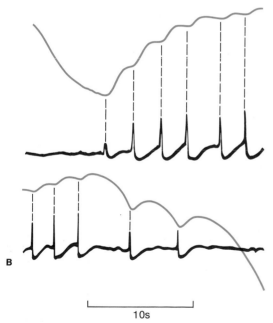

The third means of controlling motility is by a group of *peptide hormones* secreted by cells in the stomach and intestine. These hormones influence the motility of the gallbladder, stomach, and intestine (Table 15-2) and will be discussed shortly.

## Gastrointestinal Secretions

The alimentary canal produces both endocrine and exocrine secretions. As explained in Chapter 11, hormones are produced by cells of ductless endocrine glands and are liberated into the bloodstream, acting as messengers to receptor molecules in target tissues. Exocrine secretions are quite different.

### EXOCRINE GLANDS

Unlike endocrine secretions, the output of an *exocrine gland* does not ooze into the circulation but generally flows through a duct into a body cavity such as the mouth, gut, nasal passage, or urinary tract that is in continuity with the exterior. Exocrine secretions consist of aqueous mixtures rather than single species of molecule (as in the case of endocrine secretions). In the alimentary canal, these mixtures typically consist of water, ions, enzymes, and mucus. Exocrine tissues of the alimentary canal include the salivary glands, secretory cells in the stomach and intestinal epithelium, and secretory cells of the liver and pancreas.

tylcholine, liberated in response to activity of the parasympathetic innervation (Figure 15-24A). Each impulse associated with excitation produces an increment of tension, which subsides with cessation of impulses (Figure 15-24B). It now appears that there are, in addition to ACh and norepinephrine, other intestinal neurotransmitters, among them some neuropeptides and perhaps the purine nucleoside adenosine.

TABLE 15-2 Some gastrointestinal hormones (all peptides).

| Hormone | Tissue of origin | Target tissue | Primary action | Stimulus to secretion |
|---|---|---|---|---|
| Gastrin | Stomach and duodenum | Secretory cells and muscles of stomach | HCl production and secretion; stimulates gastric motility | Vagus nerve activity; peptides and proteins in stomach |
| Cholecystokinin-pancreozymin (CCK-PZ)† | Upper small intestine | Gallbladder | Contraction of gallbladder | Fatty acids and amino acids in duodenum |
| | | Pancreas | Pancreatic juice secretion | |
| Secretin† | Duodenum | Pancreas; secretory cells and muscles of stomach | Water and $NaHCO_3$ secretion; inhibits gastric motility | Food and strong acid in stomach and small intestine |
| Gastric inhibitory peptide (GIP) | Upper small intestine | Gastric mucosa and musculature | Inhibits gastric secretion and motility; stimulates Brunner's glands | Monosaccharides and fats in duodenum |
| Bulbogastrone | Upper small intestine | Stomach | Inhibits gastric secretion and motility | Acid in duodenum |
| Vasoactive intestinal peptide (VIP)† | Duodenum | | Increases blood flow; secretion of thin pancreatic fluid; inhibits gastric secretion | Fats in duodenum |
| Enteroglucagon | Duodenum | Jejunum, pancreas | Inhibits motility and secretion | Carbohydrates in duodenum |
| Enkephalin† | Small intestine | Stomach, pancreas, intestine | Stimulates HCl secretion; inhibits pancreatic enzyme secretion and intestinal motility | |
| Somatostatin† | Small intestine | Stomach, pancreas, intestines, splanchnic arterioles | Inhibits HCl secretion, pancreatic secretion, intestinal motility, and splanchnic blood flow | |

†Peptides marked with a dagger are also found in central nervous tissue as neuropeptides. Neuropeptides not listed here, but identified in both brain and gut tissue, include substance P, neurotensin, bombesin, insulin, pancreatic polypeptide, and ACTH.

*Sources:* Madge, 1975; Dockray and Gregory, 1980.

An exocrine gland typically consists of an invaginated epithelium of closely packed secretory cells lining a blind cavity called the *acinus* (Figure 15-25A). The acinus connects to a small duct that, in turn, along with other acini, connects to a larger duct leading to the lumen of the digestive canal. The basal surfaces of the epithelial cells are usually in close contact with

the circulation. Secretory granules (Figure 15-26) are formed within the acinar cells by the synthetic activity of ribosomes attached to the rough endoplasmic reticulum (Figure 15-25B). The newly synthesized products are then packaged into granules in the Golgi apparatus. The granules are membrane-bound vesicles distinguished from smaller synaptic vesicles of neural

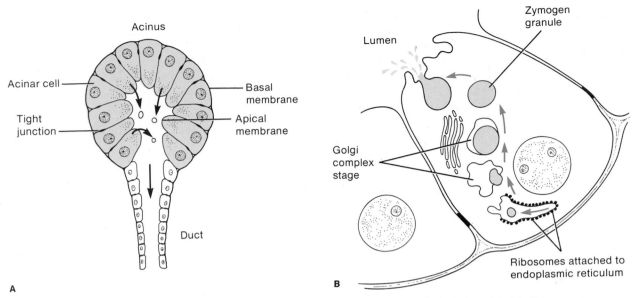

**15-25** Exocrine secretion. (A) Generalized excretory gland. Acinar cells (color) secrete into the acinus, which leads into the exocrine duct, whose epithelium further modifies the secretory fluid. (B) Formation and secretion of zymogen granules from an acinar cell. [B from Stryer, 1981, after G. Palade.]

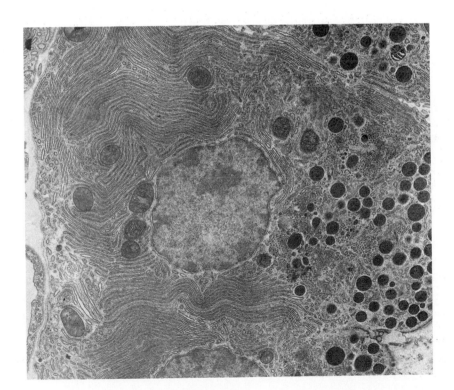

**15-26** Zymogen granules in an acinar cell of the pancreas of a bat. Magnification 10,600×. [Courtesy of K. R. Porter.]

tissue by their larger and less uniform size. Enzyme-containing secretory granules are known as *zymogen granules.* The way in which the granules are released into the acinus depends on the animal group and the tissue. There are four known means of secretion:

*1. Exocytosis* entails the fusion of the membranes surrounding secretory vesicles with the cell membrane so that the contents are expelled from the cell. In every case investigated, this process is regulated by the level of intracellular free $Ca^{2+}$. Exocytosis appears to be the mechanism for secretion in all exocrine and endocrine cells in which the secretory product is stored in vesicles.

*2.* In *apocrine secretion,* the apical portion of the acinar cell, which contains the secretory material, is sloughed off, and the cell then reseals at its apex. This occurs in mollusks.

*3.* In *merocrine secretion,* the apical portion of the acinar cell pinches off, and this portion, containing the secretory products, breaks open in the acinus. Arthropod and annelid exocrine glands utilize this method.

*4.* In *holocrine secretion,* the entire acinar cell is cast off and lyses to release its contents. This occurs in some insect and molluscan exocrine tissues.

Once the primary secretory products are free in the acinar lumen, they generally become secondarily modified in the secretory duct. This modification can involve further transport of water and electrolytes into or out of the duct to produce the final secretory juice. *Primary secretion* and *secondary modification* are illustrated in the salivary gland (Figure 15-27).

### WATER AND ELECTROLYTES

The exocrine glands of the alimentary canal secrete large quantities of fluid, most of which is reabsorbed in the distal portions of the gut. The secretions contain various quantities of enzymes, electrolytes, *mucin,* water, and special products, such as bile constituents in the case of the vertebrate liver (Figure 15-28).

In aqueous solution, the mucus produced in the

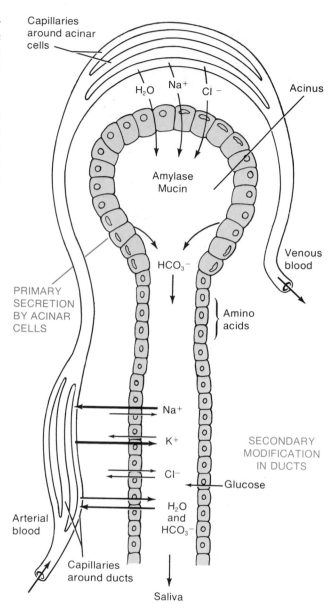

**15-27** Secretory activity in the mammalian salivary gland. The acinar cells transport electrolytes from their basal sides into the acinus and secrete mucin and amylase by exocytosis, with water flowing osmotically. As the salivary fluid moves down the duct, it undergoes modification by active transport across the epithelium of the duct. [Davenport, 1977.]

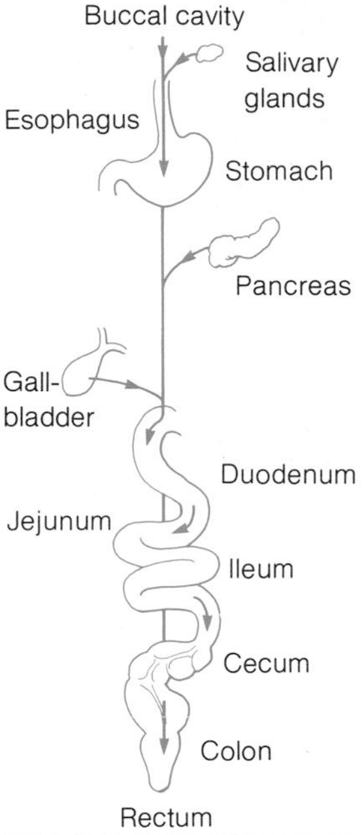

| Region | Secretion | Composition† | Daily amount (L) | pH |
|---|---|---|---|---|
| Buccal cavity / Salivary glands | Saliva | Amylase, bicarbonate | 1+ | c. 6.5 |
| Stomach | Gastric juice | Pepsinogen, HCl, rennin in infants, "intrinsic factor" | 1–3 | c. 1.5 |
| Pancreas | Pancreatic juice | Trypsinogen, chymotrypsinogen, carboxy- and aminopeptidase, lipase, amylase, maltase, nucleases, bicarbonate | 1 | 7–8 |
| | Bile | Fats and fatty acids, bile salts and pigments, cholesterol | c. 1 | 7–8 |
| Duodenum | 'Succus entericus' | Enterokinase, carboxy- and aminopeptidases, maltase, lactase, sucrase, lipase, nucleases | c. 1 | 7–8 |

(Diagram at left labeled: Buccal cavity, Esophagus, Stomach, Pancreas, Gall-bladder, Duodenum, Jejunum, Ileum, Cecum, Colon, Rectum)

†Excluding mucus and water, which together make up some 95% of the actual secretion.

**15-28** Secretions of the mammalian alimentary canal, shown in generalized outline at the left. The figures to the right are representative of the volume and pH of the secretions in the human.

goblet cells of the stomach and intestine (Figures 15-16 and 15-18) provides a slippery, thick lubricant that helps prevent mechanical and enzymatic injury to the lining of the gut. The salivary glands and pancreas secrete a thinner mucoid solution.

Secretion of inorganic constituents of digestive fluids generally occurs in two steps. First, water and ions are secreted either by passive ultrafiltration owing to hydrostatic pressure differences or by active processes from the interstitial fluid bathing the basal portions of the acinar cells. The latter is believed to usually entail active transport of ions by these cells, followed by osmotic flow of water into the acinus. There is subsequent modification of this fluid by active or passive transport across the epithelium lining the ducts, as the fluid passes along the exocrine ducts toward the alimentary canal.

### BILE AND BILE SALTS

The vertebrate liver does not produce digestive enzymes, but provides *bile,* a fluid essential for digestion of fats. *Bile* consists of water and a weakly basic mixture of cholesterol, lecithin, inorganic salts (in concentrations similar to that of plasma), *bile salts,* and *bile pigments.* The bile salts are organic salts composed of *bile acids* that are manufactured by the liver from cholesterol and are conjugated with amino acids complexed with sodium (Figure 15-29). The bile pigments derive from the hemoglobin metabolites *biliverdin* and *bilirubin.*

Bile is produced in the liver and transported via the hepatic duct to the *gallbladder* for storage. In the gallbladder, bile is effectively concentrated by the active removal of $Na^+$ and $Cl^-$ through the gallbladder epithelium, with water following osmotically.

**15-29** Sodium glycholate, a mammalian bile salt. Cholic acid (colored area) is conjugated with the amino acid glycine and sodium.

Bile serves three functions. First, its high alkalinity is important for the terminal stages of digestion after the high acidity provided by the gastric juice. Second, the bile salts help disperse fat for digestion. Third, the bile carries with it water-insoluble waste substances removed from the bloodstream by the liver, such as hemoglobin pigments, steroids, and drugs.

Because the bile-salt molecule contains the lipid-soluble bile acid together with a water-soluble amino acid, it acts as a detergent to emulsify fat droplets, dispersing them in aqueous solution for more effective attack by digestive enzymes. After entering the upper intestine from the bile duct and traveling down to the lower intestine, the bile salts are removed from the intestine by a highly efficient active transport system. Back in the bloodstream, the bile salts become bound to a plasma carrier protein and are returned to the liver to be recycled.

### DIGESTIVE ENZYMES

Digestive enzymes, like all enzymes, exhibit substrate specificity and are sensitive to temperature, pH, and certain ions. Corresponding to the three major types of foodstuffs are three major groups of digestive enzymes: *proteases, carbohydrases,* and *lipases* (Table 15-3).

**TABLE 15-3** Action of the enzymes of the mammalian alimentary tract.

| Enzyme | Site of secretion | Site of action | Substrate acted upon | Products of action |
|---|---|---|---|---|
| Salivary α-amylase | Mouth | Mouth | Starch | Disaccharides (few) |
| Pepsinogen→pepsin | Stomach | Stomach | Proteins | Large peptides |
| Pancreatic α-amylase | Pancreas | Small intestine | Starch | Disaccharides |
| Trypsinogen→trypsin | Pancreas | Small intestine | Proteins | Large peptides |
| Chymotrypsin | Pancreas | Small intestine | Proteins | Large peptides |
| Elastase | Pancreas | Small intestine | Elastin | Large peptides |
| Carboxypeptidases | Pancreas | Small intestine | Large peptide | Small peptides (oligopeptides) |
| Aminopeptidases | Pancreas | Small intestine | Large peptides | Oligopeptides |
| Lipase | Pancreas | Small intestine | Triglycerides | Monoglycerides, fatty acids, glycerol |
| Nucleases | Pancreas | Small intestine | Nucleic acids | Nucleotides |
| (Enterokinase | Small intestine | Small intestine | Trypsinogen | Trypsin) |
| Disaccharidases | Small intestine | Small intestine† | Disaccharides | Monosaccharides |
| Peptidases | Small intestine | Small intestine† | Oligopeptides | Amino acids |
| Nucleotidases | Small intestine | Small intestine† | Nucleotides | Nucleosidases, phosphoric acid |
| Nucleosidases | Small intestine | Small intestine† | Nucleosides | Sugars, purines, pyrimidines |

†Intracellular.

*Source:* Madge, 1975.

## Proteases

These proteolytic enzymes consist of two major kinds, the *endopeptidases* and *exopeptidases,* both of which attack peptide bonds of proteins and polypeptides (Figure 15-2A). They differ in that the endopeptidases confine their attacks to bonds well within the protein molecule, breaking large peptide chains into shorter polypeptide segments and thus providing many sites of action for the exopeptidases; the exopeptidases attack only peptide bonds near the end of a peptide chain, providing free amino acids, plus dipeptides and tripeptides. Some proteases exhibit marked specificity for the amino acid residues located on either side of the bonds they attack. Thus, the endopeptidase *trypsin* attacks only those peptide bonds in which the carboxyl group is provided by arginine or lysine, regardless of where they occur within the peptide chain. The endopeptidase *chymotrypsin* attacks peptide bonds containing the carboxyl groups of tyrosine, phenylalanine, tryptophan, leucine, and methionine.

In mammals, protein digestion usually begins in the stomach by the action of the gastric enzyme *pepsin,* which has a low optimum pH value. Its action is aided by secretion of gastric HCl and results in the hydrolysis of proteins into polypeptides and some free amino acids. In the mammalian intestine, several proteases produced by the pancreas continue the proteolytic process, yielding a mixture of free amino acids and small peptide chains. Finally, proteolytic enzymes intimately associated with the epithelium of the intestinal wall hydrolyze the polypeptides into *oligopeptides,* which consist of residues of two or three amino acids, and into individual amino acids.

## Carbohydrases

These enzymes also can be divided into two categories, the *polysaccharidases* and *glycosidases.* The former hydrolize the glycosidic bonds of long-chain carbohydrates such as cellulose, glycogen, and starch, whereas the latter act on disaccharides such as sucrose, fructose, maltose, and lactose, breaking them down into their constituent monosaccharides (Figure 15-2B). The most common polysaccharidases are the *amylases,* which hydrolize all but the terminal glycosidic bonds within starch and glycogen, producing disaccharides and oligosaccharides. Amylases are secreted in vertebrates by the salivary glands and pancreas, and in most invertebrates by salivary glands and intestinal epithelium. *Cellulase* is produced by symbiotic microorganisms in the gut of host animals as diverse as cattle and termites, which themselves are incapable of producing the enzyme required for the digestion of cellulose. In termites, the cellulose is liberated into the intestinal lumen by the symbiont and functions extracellularly to digest the ingested wood. In cattle, the symbiotic microbes take up cellulose molecules (from ingested grass, etc.), digesting them intracellularly and passing some digested fragments into the surrounding fluid. The bacteria, in turn, multiply and are themselves subsequently digested. Were it not for these symbiotic microorganisms, cellulose, the major nutritional constituent of grass, hay, and leaves, would be unavailable as food for grazing and browsing animals. Only a few animals, such as the shipworm (a wood-boring clam), *Limnoria* (an isopod), and the silverfish (a primitive insect), can produce cellulase on their own without the help of symbionts.

## Lipases

Fats present the special problem of being water-insoluble. They must therefore undergo special treatment before they can be processed in the aqueous contents of the digestive tract. This treatment is accomplished in two stages. First, they are *emulsified* (dispersed into small droplets) by churning of the intestinal contents, aided by detergents such as bile salts and *lecithin* (a polysaccharide) under conditions of neutral or alkaline pH. The effect is somewhat like the dispersal of salad oil in vinegar and egg yolk when mayonnaise is produced. Second, there is an enzymatic attack by intestinal lipases (in invertebrates) or pancreatic lipases (in vertebrates), producing fatty acids plus monoglycerides and diglycerides.

The next step, in vertebrates, is the formation of *micelles* (Figure 2-16), aided by bile salts. Micelles have polar hydrophilic groups at one end and nonpolar hydrophobic groups at the other end, so that their polar ends face outward into the aqueous solution. The lipid core of each micelle is about $10^6$ times smaller than the original emulsified fat droplets.

### Proenzymes

Certain digestive enzymes, in particular proteolytic enzymes, are synthesized, stored, and released in an inert molecular form, so they require activation before they are functional. This inertia prevents self-digestion of the enzyme and its container while it is stored in zymogen granules. In its inactive form, an enzyme is known as a *proenzyme* or *zymogen.* The proenzyme is activated by the removal of a portion of the molecule, either by the action of another enzyme specific for this purpose and/or through a rise in acidity. Trypsin and chymotrypsin are good examples. The proenzyme *trypsinogen,* a 249-residue polypeptide, is inert until a six-residue segment is cleaved from the $NH_2$-terminal end, either by the action of another trypsin molecule or by *enterokinase,* an intestinal proteolytic enzyme. Trypsin is also instrumental in converting inactive *chymotrypsinogen* into active *chymotrypsin* by way of three proteolytic steps.

### Other Digestive Enzymes

In addition to the major classes just described, there are a number of enzymes that play a less important role in digestion. *Nucleases, nucleotidases,* and *nucleosidases,* as their names imply, hydrolyze nucleic acids and their residues. *Esterases* hydrolyze *esters,* which include those fruity-smelling compounds that are so important in rendering ripe fruit nearly irresistible to birds, apes, and humans. These and other minor digestive enzymes are not essential for nutrition, but they make the utilization of ingested food more efficient.

## Control of Digestive Secretions

Very little is known about the control of digestive secretions in the invertebrates. Filter feeders evidently maintain a steady secretion while they continuously feed. Other invertebrates secrete enzymes in response to the presence of food in the alimentary canal, but precise control mechanisms remain obscure owing to the lack of intensive investigation and to the formidable variety of invertebrate types.

Among vertebrates, the primary stimulus for secretion of digestive juices in a given part of the digestive tract is the presence of food there or, in some instances, elsewhere in the tract. The presence of food molecules

stimulates sensory endings, which leads to the reflex activation of autonomic efferents that activate or inhibit motility and exocrine secretion. Appropriate food molecules also directly stimulate epithelial endocrine cells by contact with receptors of endocrine cells, causing reflex secretion of gastrointestinal hormones into the local circulation (Table 15-2). These reflexes permit secretory organs separate from the alimentary tract proper, such as the liver and the pancreas, to be properly coordinated with the digestive needs of food passing along the digestive tract. None of these mechanisms is under simple voluntary control.

The properties of the two types of control, hormonal and neural, employed in each part of the alimentary canal are related to the length of time food is normally present there. For example, salivary secretion is very rapid and entirely under involuntary neural control; gastric secretions are under hormonal as well as neural control; and intestinal secretions are slower and are primarily under hormonal control. As in other systems, neural control predominates in rapid reflexes, whereas endocrine mechanisms are involved in reflexes that transpire over minutes or hours.

Gastrointestinal secretion is largely under the control of *gastrointestinal peptide hormones* secreted by endocrine cells of the gastric and intestinal mucosa. Several of these hormones turn out to be identical to neuropeptides that act as transmitters in the central nervous system. This suggests that the genetic machinery for producing these biologically active peptides has been put to use by cells of both the CNS and the gastrointestinal tract. Some gastrointestinal hormones are listed in Table 15-2.

### SALIVARY AND GASTRIC SECRETIONS

The saliva of mammals contains water, electrolytes, mucin, and amylase (Figure 15-27). In the absence of food, the salivary glands produce a slow flow of watery saliva. Secretion is stimulated by psychic factors (Box 15-1) via cholinergic parasympathetic nerves to the salivary glands and by the presence of food in the mouth. The amylase mixed in with the food during chewing digests starches. The mucin and watery fluid conditions the bolus to help it slide smoothly toward the stomach by the peristaltic movements of the esophagus.

## Box 15-1    *Behavioral Conditioning in Feeding and Digestion*

Illustrious in the histories of both psychology and physiology are the experiments of I. Pavlov nearly a century ago, demonstrating reflexive secretion of saliva in dogs. A dog was given food together with the sounding of a bell. A dog normally salivates in response to the sight or taste of food, but not in response to a bell. However, after several presentations of the bell (conditioned stimulus) together with food (unconditioned stimulus), the bell alone elicited salivation. This was the first recognition of a *conditioned reflex*. These experiments became important for the development of theories of animal behavior and psychology. In the context of this chapter, Pavlov's experiments demonstrated that some secretions of the digestive tract are under *cephalic* (i.e., brain) control. Thus, in vertebrates, neural control of digestive secretions consists of two categories: in the first, *secretomotor output* to gland tissue occurs by an unconditioned reflex elicited directly by food in contact with chemoreceptors; in

the second, secretomotor output is evoked indirectly by *association* of a conditioned stimulus with an unconditioned stimulus.

Another example of the cephalic control of secretions is the reflexive activation of secretion of salivary and gastric (i.e., stomach) fluids evoked by the sight, smell, or anticipation of food. This reaction is based on past experience (i.e., associative learning). Closely related to this is the discovery that some animals exhibit one-trial avoidance learning of noxious foods. Thus, a meal will be rejected even before it is tasted if it looks or smells like something previously sampled that proved to be noxious. Insect-eating birds have been found to avoid a particular species of bad-tasting insect prey on the basis of a one-trial experience with that prey. Examples of avoidance of noxious foods by one-trial learning have also been described in several mammalian species.

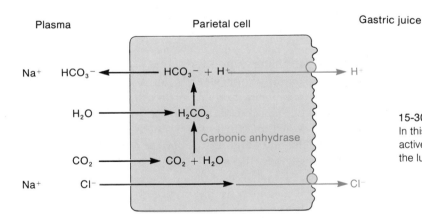

15-30    Secretion of HCl by parietal cells. In this scheme, H+ and Cl− are transported actively across the apical membrane into the lumen of the stomach.

A major secretion of the stomach lining is hydrochloric acid, which is produced by the *parietal* or *oxyntic cells* located in the gastric mucosa. The secretion of HCl is stimulated by motor activity in the vagus nerve and by the action of gastrin. The HCl helps break the peptide bonds of proteins, activates some gastric enzymes, and kills bacteria that enter with the food.

The parietal cells produce a concentration of hydrogen ions in the gastric juice $10^6$ times that of the plasma. They do this with the aid of the enzyme *carbonic anhydrase,* which catalyzes the reaction

$$CO_2 + H_2O \longrightarrow H_2CO_3 \rightleftharpoons HCO_3^- + H^+$$

The $HCO_3^-$ diffuses into the plasma while $H^+$ is actively secreted by the apical cell membrane along with $Cl^-$ into the lumen of the stomach, with water following passively (Figure 15-30).

The major gastric enzyme is *pepsin.* This proteolytic

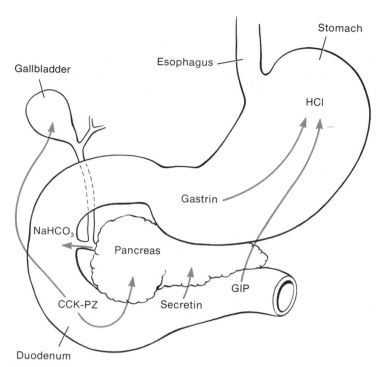

**15-31** Actions of several gastrointestinal hormones. Gastrin from the lower stomach stimulates the flow of HCl and pepsin from stomach secretory cells and the churning action of the muscular walls. Gastrin is secreted in response to intragastric protein, stomach distention, and input from the vagus nerve. GIP, liberated from the small intestine in response to high levels of fatty acids, inhibits these activities. Neutralization and digestion of the chyme is accomplished by pancreatic secretions stimulated by CCK-PZ, which also induces contraction of the gallbladder, liberating the fat-emulsifying bile into the small intestine. CCK-PZ is secreted in response to the presence of amino and fatty acids in the duodenum.

enzyme is secreted in the form of *pepsinogen* by the *chief cells* (Figure 15-16B). These exocrine cells are under vagal control and are also stimulated by the hormone *gastrin,* which arises from the gastric antrum (Figure 15-31). The inactive proenzyme pepsinogen, of which there are several variants, is converted to the active pepsin by a low-pH-dependent cleavage of a part of the peptide chain. An endopeptidase, pepsin selectively cleaves inner peptide bonds that occur adjacent to carboxylic side groups of large protein molecules.

In some young mammals, including bovine calves and human infants, the stomach secretes *rennin,* an endopeptidase that clots milk by promoting the formation of calcium caseinate from the milk protein casein. The curdled milk is then digested by proteolytic enzymes, including rennin.

Goblet cells in the lining of the stomach secrete a gastric mucus containing various mucopolysaccharides. The mucus coats the gastric epithelium, pro-

tecting it from digestion by pepsin and HCl. HCl can penetrate the layer of mucus but is neutralized by alkaline electrolytes trapped within the mucus.

The control of gastric secretion in mammals is divided into three phases: the cephalic, the gastric, and the intestinal. In the *cephalic phase,* secretion occurs in response to the sight, smell, and/or taste of food or in response to conditioned reflexes (Box 15-1). This phase is mediated by the brain and is abolished by section of the vagus nerves. In the *gastric phase,* mediated by the hormone gastrin and the compound *histamine,* the secretion of HCl and pepsin is stimulated directly by the presence of food in the stomach. The *intestinal phase* is controlled by hormones such as gastrin, *secretin, vasoactive intestinal peptide (VIP),* and the *gastric inhibitory peptide (GIP)* (Table 15-2). GIP, for instance, is liberated by endocrine cells in the mucosa of the upper small intestine in response to the entry of fats and sugars into the duodenum (Table 15-2; Figure 15-31).

The control of gastric-phase secretion was determined by use of the "Heidenhain pouch," which is a denervated pouch surgically constructed within the animal from a part of the stomach so as to open to the exterior. Its only contact with the rest of the stomach is indirect, through the circulation. Since it is not innervated, the pouch can exhibit no cephalic-phase secretion. However, it does secrete gastric juice in response to food placed into the stomach proper. This secretion suggested to researchers that a messenger is released into the bloodstream when food is in the stomach. Such a hormone was indeed discovered. It was named gastrin (Table 15-2) and was later found to be a polypeptide. Gastrin is secreted from endocrine cells of the pyloric mucosa of the stomach in response to gastric chyme containing protein and distending the stomach. It stimulates stomach motility and induces a strong secretion of HCl and a moderate secretion of pepsin. When the pH of the gastric chyme drops to 3.5 or below, gastrin secretion slows, and at pH 1.5 it stops. As already noted, secretion of histamine by the gastric mucosa also stimulates secretion of HCl, as does mechanical distention of the stomach.

The intestinal phase of gastric secretion is more complex and is poorly understood. Depending on its composition, stomach chyme entering the duodenum causes either stimulation or inhibition of gastric secretion. Protein digestion products entering the duodenum lead to a secretion of gastric acid, but the mechanism of this control is not clear. Gastrin is released from the duodenal mucosa in response to the presence of polypeptides and proteins in the duodenal chyme and acts back on the stomach. As we have mentioned, several hormones, including GIP, inhibit gastric secretions.

### INTESTINAL AND PANCREATIC SECRETIONS
The epithelium of the mammalian small intestine secretes the *intestinal juice,* also termed the *succus entericus.* This consists of a viscous, largely enzyme-free, alkaline mucoid fluid secreted by *Brunner's glands,* and a thinner, enzyme-containing alkaline fluid arising in the crypts of Lieberkühn. The secretion of these juices is under the control of several hormones, including secretin, GIP, and gastrin, and under neural control as well. Distention of the wall of the small intestine elicits a local secretory reflex. The vagal innervation also stimulates secretion. The large intestine secretes no enzymes, but merely a thin alkaline fluid containing bicarbonate and potassium ions plus some mucus that binds the fecal matter together.

In addition to its endocrine secretion of insulin from the islets of Langerhans, the pancreas contains exocrine tissue that produces several digestive secretions that enter the small intestine through the pancreatic duct. The pancreatic enzymes, including *alpha-amylase, trypsin, chymotrypsin, elastase, carboxypeptidases, aminopeptidases, lipases,* and *nucleases,* are delivered in an alkaline, bicarbonate-rich fluid that helps neutralize the acid chyme formed in the stomach. This buffering is essential since the pancreatic enzymes require a neutral or slightly alkaline pH.

Exocrine secretion by the pancreas is controlled by the peptide hormones produced in the upper small intestine. Acid chyme reaching the small intestine from the stomach stimulates the release of secretin and VIP, both produced by endocrine cells in the upper small intestine (Table 15-2). On entering the bloodstream, these peptides stimulate the pancreas to produce its thin bicarbonate fluid, but stimulate little secretion of pancreatic enzymes. Gastrin secreted from the stomach lining also elicits a small flow of pancreatic juice in anticipation of the food that will enter the duodenum.

Secretion of pancreatic enzymes is elicited by another upper intestinal hormone secreted from epithelial endocrine cells in response to fatty acids and amino acids in the intestinal chyme. This is the peptide *cholecystokinin* (Table 15-2), now known to be identical with pancreozymin and thus also called cholecystokinin-pancreozymin (CCK-PZ). It stimulates pancreatic secretion of enzymes as well as contraction of the smooth muscular wall of the gallbladder, forcing bile into the duodenum (Figure 15-31).

The neuropeptides somatostatin and enkephalin have also been identified in endocrine cells of the upper intestinal mucosa. Both have a variety of actions on gastrointestinal function. Somatostatin inhibits gastric acid secretion, pancreatic secretion, and intestinal motility as well as blood flow. The enkephalins stimulate gastric acid secretion and inhibit pancreatic enzyme secretion and intestinal motility.

## Absorption

To be useful to the organism, the products of digestion must find their way into all its tissues and cells. In a unicellular organism, this simply means that the products of digestion leave the food vacuole to enter the surrounding cytoplasm. In a metazoan alimentary canal, these products must first be transported across the *absorptive epithelium* into the circulation, and then from the body fluids across cell membranes of body tissues.

As noted earlier, the translocation of digested foodstuff molecules from the lumen of the alimentary canal into the absorptive cells takes place across the apical membrane of the absorptive cell—primarily through those portions of the membrane that cover the microvilli. This membrane carries specializations related to this function. These specializations include the glycocalyx, digestive enzymes intimately associated with the membrane, membrane systems for the transport of substances out of the lumen and into the absorptive cell, and systems to transport these substances out of the absorptive cell into the circulation through its basolateral membranes.

### INTESTINAL TRANSPORT

The carbohydrate-rich filaments composing the glycocalyx arise from, and are continuous with, the surface membrane itself. The filaments of the glycocalyx appear to be the carbohydrate side chains of glycoproteins imbedded in the membrane. Further, the brush border (microvilli plus glycocalyx) has been found to contain digestive enzymes for the *terminal digestion* of various small foodstuff molecules. These enzymes are membrane-associated glycoproteins having carbohydrate side chains protruding into the lumen. The enzymes found associated with the brush border include disaccharidases, aminopeptidases, and phosphatases. Thus, some of the terminal stages of digestion are carried out at the absorptive cell membrane, close to the sites of uptake from the lumen into the absorptive cells.

Several categories of transport (described in Chapter 4) are involved in the absorptive process, including cotransport, countertransport, active transport, passive diffusion, and endocytosis.

The absorption of monosaccharides and amino acids presents two problems. First, they are hydrophilic owing to $-OH$ groups and/or ionization; and second, they are too large to be carried through water-filled pores by solvent drag or simple diffusion. These problems are overcome by carrier-mediated transport across the absorptive cell membrane. Thus, some sugars (e.g., fructose) are carried down their concentration gradient by facilitated transport, a process in which a hydrophobic, lipid-insoluble substance diffuses down the gradient with the help of lipid-soluble carrier molecules located in the membrane. This process requires no energy other than that provided by the concentration gradient of the diffusing substance. Other sugars, such as glucose and galactose, as well as amino acids, are carried across the membrane by cotransport mechanisms in which the motive force is derived from the electrochemical gradient of the sodium ion. $Na^+$, flowing down its gradient from the lumen into the absorptive cell, provides the energy for moving the organic molecules against their concentration gradients. The $Na^+$ and sugar or amino acid molecule share a carrier molecule in passing across the membrane. The $Na^+$ leaking into the cell through this mechanism is continually removed by ATP-energized active transport by the $Na^+$–$K^+$ pump located in the basolateral membranes of the absorptive cell. Thus, sodium ions that leak into the cytoplasm of the absorptive cell from the intestinal lumen are pumped out by active transport across the basal membrane into the interstitial fluid of the villus (Figure 15-32). Upon the distribution of sugars and amino acids by the circulatory system, similar transport mechanisms utilize the sodium gradient once more to transport the sugars and amino acids from the interstitial fluid into the cells of various body tissues.

In the mammalian intestine, the sodium-driven transport of amino acids into the absorptive cells takes place via four separate and noncompeting cotransport systems, one for each of four categories of amino acids: the 15 neutral amino acids; the 3 dibasic amino acids having two amino groups each; the 2 diacidic amino acids glutamate and aspartate, having two carboxylic

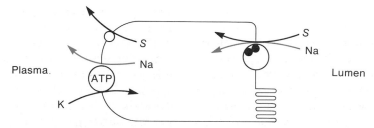

**15-32** Transport of amino acids and sugars from lumen to plasma in the small intestine. Coupled sodium–substrate transport driven by the transmembrane Na+ gradient carries the substrate, *S,* into the absorptive cell. The substrate leaves the cell by facilitated diffusion (upper left), entering the plasma. The common active Na+–K+ pump (lower left) maintains the sodium concentration gradients in the face of passive sodium entry into the cell.

groups each; and a fourth class consisting of glycine, proline, and hydroxyproline. Another separate transport system exists for dipeptides and tripeptides. Once inside the cell, these oligopeptides are cleaved into their constituent amino acids by intracellular peptidases. This has the advantage that there is no concentration buildup within the cell of the oligopeptides, and so there is always a large inwardly directed gradient promoting their inward transport.

Some monosaccharides are taken up by a third mechanism, *hydrolase transport.* An enzyme in the membrane hydrolyzing the parent disaccharide (e.g., maltose) also acts as, or is coupled to, the carrier transporting the monosaccharide into the absorptive cell, and it does so without using the Na+ gradient.

Simple diffusion can also take place across the lipid bilayer (providing the substance has a high lipid solubility) or through water-filled pores. Substances that diffuse across the bilayer include fatty acids, monoglycerides, cholesterol, and other fat-soluble substances. Substances that pass through water-filled pores include water, certain sugars, alcohols, and other small, water-soluble molecules. For nonelectrolytes, the rate of passive non-carrier-mediated net diffusion is proportional to the concentration gradient. For electrolytes, it is proportional to the electrochemical gra-

dient. For passive diffusion, net transport is always "downhill" in the direction of the gradient.

Some oligopeptides may be taken up by absorptive cells through endocytosis. This process is responsible for the uptake, by the suckling's intestine, of undigested immunoglobulin molecules derived from the mother's milk. Once inside the absorptive cell, nutrients pass through the laterobasal membranes of the absorptive cell (Figure 15-32) into the interior of the villus and then move from the interstitial fluid into the circulatory system. Transport of sugars and amino acids across the laterobasal membranes occurs by facilitated transport, as noted earlier.

The digestion products of fats, monoglycerides, fatty acids, and glycerol are reconstructed within the absorptive cell into triglycerides and are collected together with phospholipids and cholesterol into tiny droplets termed *chylomicrons,* about 1500 µm in diameter (Figure 15-33). They are coated with a layer of protein, are loosely contained in vesicles formed by the Golgi apparatus and are expelled by exocytosis through fusion of these vesicles with the laterobasal membrane of the absorptive cell.

From the interstitial fluid of the villus, digestion products enter the blood or the *lymphatic circulation.* About 80% of the chylomicrons, for example, enter the bloodstream via the lymphatic system (Figure 15-33), while the rest enter the bloodstream directly. The pathway into the lymphatic system begins with the blind *central lacteal* of the villus (Figure 15-18B), from which the *lymph* is carried, in higher mammals, to the *left thoracic lymph duct*. Sugars and amino acids enter primarily the capillaries of the villus, which are drained by venules that lead into the *hepatic portal vein*. This vein takes the blood from the intestine directly to the liver. There, much of the glucose is taken up into *hepatocytes,* and in these cells it is converted to glycogen granules for storage and subsequent release into the circulation after being converted back to glucose. The hormonal regulation of glycogen breakdown, sugar metabolism, fat metabolism, and amino acid metabolism is discussed in Chapter 11.

### WATER AND ELECTROLYTE BALANCE IN THE GUT

In the process of producing and secreting their various digestive juices, the exocrine tissues of the alimentary canal and its accessory organs pass a great deal of water and electrolytes into the lumen of the alimentary canal. In humans, this can normally amount to over 8 L per day (Figure 15-34). Clearly, it would be osmotically disastrous if this quantity of water and electrolytes were passed out of the body with the feces. In fact, nearly all of it, along with ingested water, is recovered by uptake of electrolytes and water by the intestine. Although water is reabsorbed throughout the intestine, most of the reabsorption takes place in the lower part of the small intestine.

Since the absorptive cells are bound together by tight junctions near their apical borders (Figure 15-18),

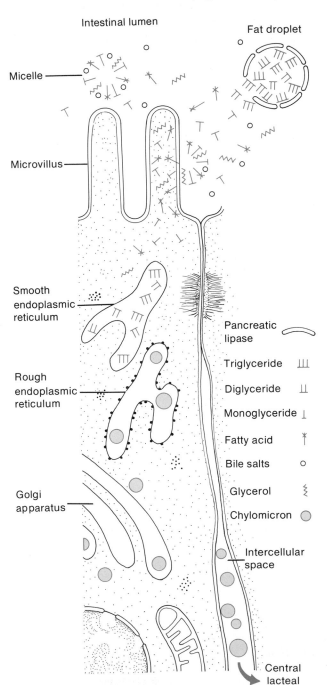

15-33 Transport of lipids from intestinal lumen through absorptive cells and into the interstitial space. Hydrolytic products of triglyceride digestion—monoglycerides, fatty acids, and glycerol—form micelles with the bile salts in solution. They enter the absorptive cell by passive, lipid-soluble diffusion across the microvillous membrane. Within the cell, they accumulate in the smooth endoplasmic reticulum, from which they are passed to the rough endoplasmic reticulum. There, they are resynthesized into triglycerides and, together with a smaller amount of phospholipids and cholesterol, are stored in the Golgi apparatus as chylomicrons—droplets about 1500 Å diameter. These then leave the laterobasal portions of the cell by exocytosis. [Madge, 1975, from Porter, 1969.]

nearly obliterating free paracellular pathways, absorbed water must pass through the apical and basal membranes or between cells through the minute intercellular spaces of the tight junctions. Tracer studies using deuterium oxide, $D_2O$, indicate that water leaves the intestinal lumen through channels that occupy only 0.1% of the epithelial surface. Flux studies employing isotopically labeled solutes indicate that these unseen channels will not pass water-soluble molecules of molecular weights exceeding 200. Solutes smaller than that are carried passively along by *solvent drag* with water as it flows down its osmotic gradient through hydrated channels.

The motive force that leads to net movement of water from the intestinal lumen to the interior of the villus is osmotic pressure. Net water movement, therefore, is entirely passive. No active transport of water has been found in any vertebrate alimentary canal. The osmotic gradient driving water from the lumen into the villus is set up primarily by the metabolically energized transport of substances from the lumen into the villus, in particular the transport of salt, sugars, and amino acids. The elevated osmotic pressure, especially in the lateral clefts of the epithelium (Figures 4-48 and 4-49) draws water osmotically from the absorptive cell, which is replenished by water entering osmotically across the apical membrane from the lumen.

Most of the absorption of water and electrolytes across the absorptive cell epithelium occurs at or near the tips of the villi. The greater absorption of water at the villous tip results from an elevated concentration of $Na^+$ near the upper end of the villous lumen, decreasing with distance from the villous tip (Figure 15-35). The reasons for this concentration gradient are twofold. First, most of the active absorption of $Na^+$ takes place across absorptive cells located at the

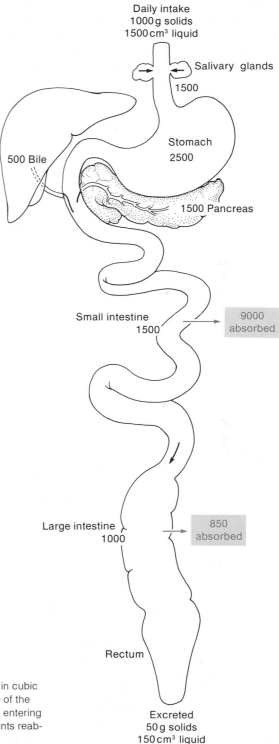

Daily intake
1000 g solids
1500 cm³ liquid

Salivary glands

1500

Stomach
2500

500 Bile

1500 Pancreas

Small intestine
1500

9000 absorbed

Large intestine
1000

850 absorbed

Rectum

Excreted
50 g solids
150 cm³ liquid

**15-34** Water fluxes in the human alimentary canal, in cubic centimeters. Figures vary with the condition and size of the subject. The figures in black are the amounts of fluid entering the alimentary tract, and those in color are the amounts reabsorbed from the lumen. [Madge, 1975.]

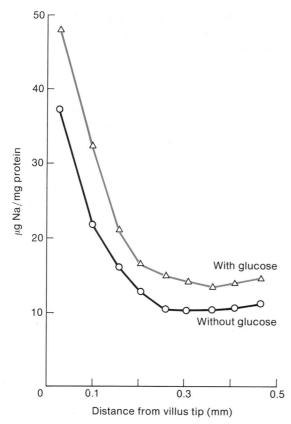

**15-35** Concentration of sodium within the villus, measured from the villous tip during perfusion of a cat's intestinal lumen with isotonic NaCl solution (black plot). Addition of glucose to the salt solution in the lumen caused the sodium concentration within the villus to go up (colored plot). [Haljamae et al., 1973.]

The absorption of $Na^+$ into the villus is enhanced by high concentrations of glucose and certain other hexose sugars in the intestinal lumen. Enhanced absorption of $Na^+$ results from stimulation of sodium–sugar cotransport. The effect of elevated glucose on the profile of intravillous $Na^+$ concentration is shown in Figure 15-35.

Excessive uptake of water, leading to a semidry lumen (and hence constipation), is prevented by an inhibitory action on electrolyte and water uptake by

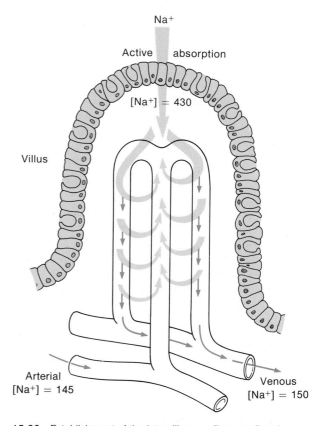

**15-36** Establishment of the intravillous sodium gradient by a countercurrent mechanism. Active sodium transport from the lumen into the villus is greatest at the villous tip. Some of the $Na^+$ carried away by venous flow "short circuits" by diffusion to the lower sodium concentration of the arterial blood entering the villus. This causes a buildup of $Na^+$ concentration in the villus far above the plasma levels, a factor that is important in osmotically drawing water from the lumen into the villus. [Davenport, 1977.]

tip of each villus. $Cl^-$ follows, and NaCl accumulation is therefore greatest at the blind upper end of the villous lumen. Second, the organization of the circulation within the villus leads to a further concentration of NaCl at the upper end of the villous lumen because of a countercurrent mechanism (Box 12-2), as illustrated in Figure 15-36. Arterial blood flowing toward the tip of the villus picks up $Na^+$ from sodium-enriched blood leaving the villus in a descending venule. The "short-circuiting" of sodium in this manner recirculates and concentrates it in the villous tip, promoting osmotic flow of water from the intestinal lumen into the villus.

some of the gastrointestinal hormones. Gastrin inhibits water absorption from the small intestine, while secretin and CCK-PZ reduce the uptake of $Na^+$, $K^+$, and $Cl^-$ in the upper jejunum. Bile acids and fatty acids also inhibit the absorption of water and electrolytes.

A special transport mechanism exists for the active transport of $Ca^{2+}$. Calcium is first bound to a *calcium-binding protein* found in the microvillous membrane and is transported into the absorptive cell by an energy-consuming process; it then passes into the blood. The presence of calcium-binding protein is regulated by 1,25-dihydroxy-vitamin $D_3$. The release of $Ca^{2+}$ from the absorptive cell into the villous lumen is accelerated by parathyroid hormone.

The largest water-soluble essential nutrient taken up intact is vitamin $B_{12}$, which has a molecular weight of 1357. This highly charged cobalt-containing compound is associated in the ingested food with protein to which it is bound as a coenzyme. In the process of absorption, $B_{12}$ transfers from the dietary protein to a mucoprotein known as *intrinsic factor* (or *hemopoietic factor*) which is produced by the $H^+$-secreting parietal cells of the stomach. Since $B_{12}$ is essential for the synthesis and maturation of red blood corpuscles, *pernicious anemia* occurs in persons in whom $B_{12}$ absorption is prevented by interference with its binding to intrinsic factor. Some tapeworms "steal" $B_{12}$ in the intestine of the host by producing a compound that removes $B_{12}$ from intrinsic factor, making it unavailable to the host but available to the worm.

## Summary

All heterotrophic organisms acquire carbon compounds of moderate to high energy content from the tissues of other organisms. The chemical energy contained in these compounds originally was converted from radiant energy into chemical energy trapped in sugar molecules by photosynthesizing autotrophs. Subsequent synthetic activity by autotrophs and heterotrophs convert these simple carbon compounds into more complex carbohydrates, fats, and proteins. After such foodstuff molecules are ingested by an animal, they must first be digested before the food matter can be assimilated, that is, incorporated into the animal. Digestion consists of the enzymatic hydrolysis of large molecules into their monomeric building blocks. In metazoans, this takes place extracellularly in an alimentary canal. Digestive hydrolysis occurs only at low-energy bonds, most of the chemical energy of foodstuffs being conserved for intracellular energy metabolism once the products of digestion have been assimilated into the animal's tissues. Stepwise intracellular oxidations by coupled reactions then lead to a controlled release of chemical-bond energy and material for cell growth and functioning.

Food is obtained in different ways, including absorption through the body surface in some aquatic or marine species, endocytosis in microorganisms, filter feeding, mucus trapping, sucking, biting, and chewing. Once ingested, the food may be temporarily stored, as in a crop or rumen, or immediately subjected to digestion. Digestion in vertebrates begins in a region of low pH, the stomach, and proceeds to a region of higher pH, the small intestine. Proteolytic enzymes are released as proenzymes, or zymogens, which are inactive until a portion of the peptide chain is removed by digestion. This procedure avoids the problem of proteolytic destruction of the enzyme-producing cells that store and secrete the zymogen granules containing the proenzyme. Other exocrine cells secrete digestive enzymes (e.g., carbohydrases and lipases), mucin, or electrolytes such as HCl or $NaHCO_3$.

The motility of the vertebrate digestive tract depends on the coordinated activity of longitudinal and circular layers of smooth muscle. Peristalsis occurs when a ring of circular contraction proceeds along the gut preceded by a region in which the circular muscles are relaxed. The parasympathetic innervation stimulates motility, whereas the sympathetic innervation inhibits motility.

Secretion of digestive juices as well as the motility of smooth muscle are under neural and endocrine control. All gastrointestinal hormones are peptides, and many of them are also known as neuropeptides in the central nervous system, where they act as transmitters or short-range neurohormones. Both direct activation by food in the gut and neural activation stimulate the endocrine cells of the gastrointestinal mucosa that secrete peptide hormones. These hormones are active in either stimulating or inhibiting the activity of the various kinds of exocrine cells in the gut that produce digestive enzymes and juices.

The products of digestion are taken up by the absorptive cells of the intestinal mucosa and transferred to the lymphatic and circulatory systems. The absorptive surface, in effect consisting of a continuous membrane sheet formed by the apical membranes of the myriad absorptive cells joined by tight junctions, is greatly increased in area by virtue of microvilli, the microscopic evaginations of the apical membrane. In addition, the absorptive cells cover larger fingerlike villi that reside on convoluted folds and ridges in the wall of the intestine.

The process of terminal digestion takes place in the brush border formed by the microvilli and the glycocalyx that covers the apical membrane. Here short-chain sugars and peptides are hydrolyzed into monomeric residues before membrane transport takes place. Transport of some sugars can occur by facilitated diffusion, which requires a membrane carrier but no metabolic energy. Most sugars and amino acids require energy expenditure for adequate rates of absorption. An important transport mechanism for these substances is cotransport with $Na^+$, utilizing a common carrier and the potential energy of the electrochemical gradient driving $Na^+$ from the lumen into the cytoplasm of the absorptive cell. Endocytosis plays a role in the uptake of small polypeptides and, rarely, of larger proteins, such as immunoglobulin in infants. Fatty substances enter the absorptive cell by simple diffusion across the cell membrane.

Large quantities of water and electrolytes enter the alimentary canal as constituents of digestive juices, but these quantities are nearly all recovered by active uptake of salts by the intestinal mucosa. Active transport of electrolytes out of the intestinal lumen results in the passive osmotic movement of water out of the lumen back into the bloodstream. Without such recycling of electrolytes and water, the digestive system would impose an awesome osmotic load on the animal.

## Exercises

*1.* Define the terms *digestion, absorption, assimilation,* and *nutrition.*

*2.* In what way is Bernoulli's effect significant to the feeding of a sponge?

*3.* Cite two unrelated examples of proteins produced specifically for the purpose of obtaining and utilizing food.

*4.* What is an essential amino acid?

*5.* Explain why it would be inadvisable for the digestive system to fragment amino acids, hexose sugars, and fatty acids into still smaller molecular fragments, even though this might facilitate absorption.

*6.* Explain why proteolytic enzymes fail to digest the exocrine cells in which they are produced and stored before release.

*7.* Give several examples of symbiotic microorganisms in alimentary canals, and explain how they benefit the host.

*8.* State two adaptive advantages of the digastric stomach. Since it has three or four chambers, why is it referred to as "digastric"?

*9.* Explain how bile aids the digestive process even though it contains little or no enzyme.

*10.* Outline the autonomic innervation of the intestinal wall, explaining the organization and functions of sympathetic and parasympathetic innervation.

*11.* How is HCl produced and secreted into the stomach by parietal cells?

*12.* Compare and contrast endocrine and exocrine systems. What do they have in common?

*13.* Describe four mechanisms, found in exocrine cells of different animal groups, by which the secretory products are liberated into the exocrine duct. Which of these mechanisms also occurs in all endocrine secretion?

*14.* What is meant by secondary modification of an exocrine secretion?

*15.* Describe the roles of gastrin, secretin, and cholecystokinin in mammalian digestion.

*16.* Why are some gastrointestinal hormones also classified as neuropeptides? Give examples.

**17.** Explain what is meant by the cephalic, the gastric, and the intestinal phases of gastric secretion. How are they regulated?

**18.** How are amino acids and some sugars transported against a concentration gradient from intestinal lumen into absorptive cells?

**19.** Why is the countercurrent principle important in the removal of water from the intestinal lumen?

**20.** How is pernicious anemia related to intestinal function?

## Suggested Reading

Becker, H. D., D. D. Reeder, and J. C. Thompson. 1975. Vagal control of gastrin release. In J. C. Thompson, ed., *Gastrointestinal Hormones: A Symposium.* Austin: University of Texas Press.

Davenport, H. W. 1978. *Digest of Digestion.* 2nd ed. Chicago: Year Book Medical Publishers.

Davenport, H. W. 1977. *Physiology of the Digestive Tract.* 4th ed. Chicago: Year Book Medical Publishers.

Jennings, J. B. 1972. *Feeding, Digestion and Assimilation in Animals.* New York: St. Martin's Press.

Madge, D. S. 1975. *The Mammalian Alimentary System.* London: Arnold.

Moog, F. 1981. The lining of the small intestine. *Scientific American* 245(5):154–176.

Smyth, D. H., ed. 1974. *Intestinal Absorption.* 2 vols. London: Plenum.

Swenson, M. J. 1977. *Duke's Physiology of Domestic Animals.* Ithaca, N.Y.: Cornell University Press.

Vogel, S. 1978. Organisms that capture currents. *Scientific American* 239(2):128–139.

## References Cited

Barnes, R. D. 1974. *Invertebrate Zoology.* Philadelphia: Saunders.

Bortoff, A., and R. S. Davis. 1968. Myogenic transmission of antral slow waves across the gastroduodenal junction *in situ. Amer. J. Physiol.* 215:889–897.

Bülbring, E. 1959. *Lectures on the Scientific Basis of Medicine.* Vol. 7. London: Athlone.

Bülbring, E., and H. Kuriyama. 1963. Effects of changes in ionic environment on the action of acetylcholine and adrenaline on smooth muscle cells of guinea pig taenia coli. *J. Physiol.* 166:59–74.

Cornwall, I. W. 1956. *Bones for the Archeologist.* London: Phoenix House.

Dockray, G. J., and R. A. Gregory. 1980. Relations between neuropeptides and gut hormones. *Proc. Roy. Soc.* (London) Ser. B. 210:151–164.

Florey, E. 1966. *General and Comparative Animal Physiology.* Philadelphia: Saunders.

Haljamae, H., M. Jodal, and O. Lundgren. 1973. Countercurrent multiplication of sodium in intestinal villi during absorption of sodium chloride. *Acta Physiol. Scand.* 89:580–593.

Ham, A. W. 1957. *Histology.* Philadelphia: Lippincott.

Hyman, L. H. 1940. *The Invertebrates: Protozoa Through Ctenophora.* New York: McGraw-Hill.

Imms, A. D. 1949. *Outlines of Entomology.* London: Methuen.

McClintic, J. R. 1978. *Physiology of the Human Body.* New York: Wiley.

Marshall, P. T., and G. M. Hughes. 1980. *Physiology of Mammals and Other Vertebrates.* Cambridge: Cambridge University Press.

Mast, S. O. 1947. The food vacuole in *Paramecium. Biol. Bull.* 92:31–72.

Milner, A. 1981. Flamingos, stilts and whales. *Nature* (London) 289:347.

Parker, H. W. 1963. *Snakes.* London: Hale.

Passmore, R., and J. S. Robson. 1968. *A Companion to Medical Studies.* Oxford: Blackwell Scientific Publications.

Porter, K. R. 1969. Independence of fat absorption and pinocytosis. *Fed. Proc.* 28:35–40.

Romer, A. S. 1962. *The Vertebrate Body.* 3rd ed. Philadelphia: Saunders.

Stempell, W. 1926. *Zoologie im Grundriss.* Berlin: G. Borntraeger.

Stryer, L. 1981. *Biochemistry.* 2nd ed. San Francisco: W. H. Freeman and Company.

# CHAPTER 16

# Animal Energetics and Temperature

Animals can be thought of as living machines. Like some industrial machines, they require chemical energy to do work. In addition, animals require energy to maintain their molecular and structural integrity. The more effectively they utilize the energy resources of their environment, the better they are able to compete with other species and the more "fit" they are for survival. In Chapters 3 and 15, we noted that animals degrade large, organic compounds in a way that permits the transfer of some of their inherent chemical energy to special "energy-rich" molecules (e.g., ATP), which are used to drive endergonic reactions. Thus, living cells eventually convert chemical energy of foodstuff molecules into other forms, such as electrical energy, osmotic energy, and contractile energy.

Like all other energy-converting machines, animals are far less than 100% efficient in their energy conversions. They therefore lose a large fraction of metabolic energy as heat, a low-grade energy by-product of the conversion of chemical energy to other biologically useful forms of energy. This metabolic heat is akin to the unused, waste heat produced by a gasoline engine in converting chemical energy to mechanical work. In some animals, metabolic heat production is sufficient to raise the temperature of the tissues to levels that significantly enhance the rate of chemical reactions. This heat thus serves the same useful purpose as does the waste heat that warms a gasoline engine to a more efficient temperature. In some instances, however, and especially in hot climates, heat production can create problems of excessive body temperature. In cold climates immoderate heat loss can lower body temperature excessively.

Clearly, the metabolic activity of an animal is closely linked to body temperature. Low body temperatures preclude high metabolic rates because of the temperature dependence of enzymatic reactions, while high metabolic rates can lead to overheating and attending deleterious effects on tissue function. Moreover, in animals that elevate their body temperatures above that of the environment, the temperature is regulated in part by their metabolic rates. Therefore, in this chapter we will consider those factors that influence and are influenced by both metabolic rate and body temperature.

## The Concept of Energy Metabolism

The term *metabolism,* in its broadest sense, embodies the sum total of all the chemical reactions occurring in an organism. Metabolic pathways fall into two major categories: *anabolism,* in which simple substances are built up into more complex molecules necessary for the organism; and *catabolism,* in which complex molecules are broken down into simpler ones. Anabolism requires energy and is associated with repair, regeneration, and growth. Although it is difficult to measure anabolic metabolism quantitatively, a positive nitrogen balance (i.e., net incorporation of nitrogen) in an organism is indicative of anabolism. That is, anabolic activity leads to net incorporation through protein synthesis, rather than net loss of nitrogen-containing molecules through excretion. In catabolism, the degradation of complex molecules into simpler ones is accompanied by the release of chemical energy. Some of this energy is stored in the form of high-energy phosphate compounds, which can subsequently be used·to drive cellular activities.

In the absence of external work or storage of chemical energy, all the energy released during metabolic processes appears inexorably as heat. This simple fact makes it possible to equate heat production with *energy metabolism*. The conversion of chemical energy to heat is measured as the *metabolic rate*—heat energy released per unit time. Measurements of metabolic rates are useful for several reasons. For example, they can be used to calculate the energy requirements of an animal. To survive, an animal must take in as much energy in the form of energy-yielding foodstuff molecules as the total of all the energy that it releases and stores. Measurements of metabolic rate at different ambient (environmental) temperatures will also provide information about the heat-conserving or heat-dissipating mechanisms of an animal. Measurements of metabolic rate during different types of exercise help us understand the energy cost of such activities. How energy-expensive, for example, is it simply to stay alive, or to fly, swim, run, or walk a given distance?

The metabolic rate of an animal will, of course, vary with the number and intensity of the activities it performs. These include tissue growth and repair; chemical osmotic, electrical, and mechanical internal work; external work for locomotion, communication; and loss through secretions, urine, and feces (Figure 16-1). Among the other factors that influence the metabolic rate are environment temperature, time of day,

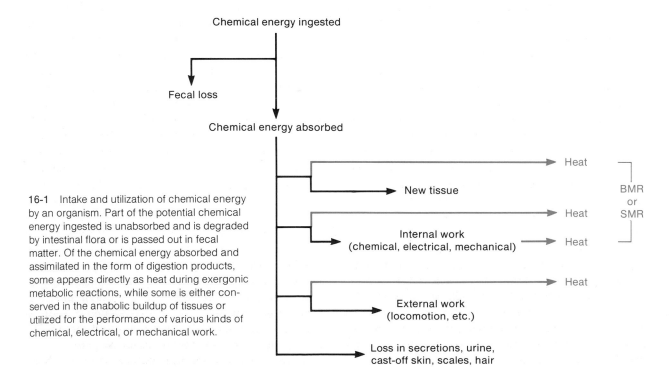

**16-1** Intake and utilization of chemical energy by an organism. Part of the potential chemical energy ingested is unabsorbed and is degraded by intestinal flora or is passed out in fecal matter. Of the chemical energy absorbed and assimilated in the form of digestion products, some appears directly as heat during exergonic metabolic reactions, while some is either conserved in the anabolic buildup of tissues or utilized for the performance of various kinds of chemical, electrical, or mechanical work.

season of year, age, sex, weight, size, stress, type of food being metabolized, and pregnancy. Consequently, metabolic rates of different animals can be meaningfully compared only under carefully chosen and closely controlled conditions, which we will consider later. First, some essential definitions.

*Endotherms* are animals that generate their own body heat, typically elevating their body temperatures above ambient temperatures. They produce heat metabolically at high rates, and by virtue of their low heat conductivity (i.e., good insulation) they lose heat at the same high rate at elevated body temperatures. Mammals and birds regulate their temperatures within rather narrow limits and are therefore also said to be *homeothermic* endotherms.

*Ectotherms* gain most of their body heat from the environment. In most of these animals, the body temperature fluctuates passively with the ambient temperature. The invertebrates and lower vertebrates are ectotherms, although at certain times some of them do raise their temperatures above ambient to varying degrees by metabolic means and may even effectively regulate their temperatures to some degree at those times.

The *basal metabolic rate* (*BMR*) is the stable rate of energy metabolism measured in mammals and birds under conditions of minimum environmental and physiological stress, after fasting has temporarily halted digestive and absorptive processes.

Since the body temperature of ectotherms varies depending on the ambient temperature, and since the minimum metabolic rate varies with the body temperature, it is necessary to measure the equivalent of the BMR at a controlled temperature. For that reason, the *standard metabolic rate* (*SMR*) is defined as an animal's resting and fasting metabolism *at a given body temperature*. Interestingly, the SMR of some ectotherms depends on their previous temperature history, owing to metabolic compensation, or thermal acclimation, which is described further on. Both the BMR and the SMR are typically determined with the animal in an unnaturally controlled and quiet state. These measurements are simply baselines against which comparisons can be made both between species and for differing physiological conditions, within a given species.

The *active metabolic rate* measures the rate of energy metabolism when the organism is performing one or more of its normal activities, such as running, flying, or swimming. Under these circumstances, the energy requirements of the activity are added to the basal or the standard metabolism. These topics are discussed in greater detail later. The *net metabolic rate* is the sum of energy conversions attributable to all factors for the time and conditions under consideration.

Although the BMR and SMR are useful measurements for baseline comparisons of metabolic rates in different animals, they give little information about the metabolic costs of normal activities carried out by the animals, because the conditions under which the BMR and the SMR are measured differ greatly from natural conditions. The quantity that best describes the metabolic behavior of an animal in its natural state is called its *field metabolic rate*, and represents the average rate of energy utilization as the animal goes about its daily activities. The range of metabolic rates of which an animal is capable is called its *metabolic scope*. This is defined as the ratio of the maximum sustained metabolic rate to the BMR or the SMR determined under controlled resting conditions. Metabolic scope thus indicates the increase in the amount of energy an animal can expend over and above the amount it consumes under resting conditions. Among different animals, it is common for the metabolic rate to increase 10 to 15 times with activity. But studies on metabolic scope have to contend with certain difficulties. Thus, there may be a significant contribution from anaerobic metabolism (buildup of "oxygen debt," p. 83), especially during short periods of exertion (Figure 16-2). This component may not be detected during short-term measurements, since the aerobic breakdown of anaerobic products can be delayed. For this reason, it is best to make measurements of metabolic scope for sustained activity only. White muscles in some vertebrates are specially adapted to develop an oxygen debt through anaerobic metabolism and are therefore particularly suitable for short-term spurts of intense activity. Another practical problem is that any animal's cooperation and motivation are essential for obtaining maximum exertion.

A great deal of work on metabolic scope has been

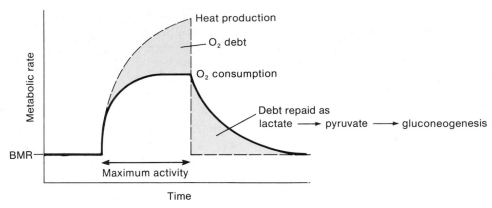

**16-2**   Muscle tissue with anaerobic capabilities can build up an "oxygen debt" that is paid off in the form of delayed oxidation of the anaerobic product, lactic acid, following a period of sustained activity. As a result, an elevated metabolic rate continues after cessation of activity.

done on fishes in flow tanks, in which faster swimming can be stimulated by faster rates of water flow, and the results indicate that the metabolic scope varies with specimen size. The ratio of active to standard metabolism in salmon, for example, increases from less than 5 in specimens weighing 5 g to more than 16 in those weighing 2.5 kg. Flying insects, especially those that sustain high body temperatures during flight, can exhibit ratios of up to 100, probably the highest in the animal kingdom.

## Measurement of Metabolic Rates

### MEASUREMENT FROM FOOD INTAKE AND WASTE REMOVAL

In principle, metabolic rate may be determined from a balance sheet of energy gain and loss. Living organisms obey the laws of energy conservation and transformations that were initially derived for chemical and physical nonliving systems. We could, therefore, in principle determine the metabolic rate of an animal maintaining constant mass using the formulation

energy intake − energy loss = metabolic energy

Total energy intake approximates the chemical energy content of ingested food over a given period. Energy loss is the chemical energy that remains in the feces and urine produced by the animal over the same period. The energy content of food and wastes can be obtained from the heat of combustion of these materials in a *bomb calorimeter*. In this method, the material to be tested is dried and placed inside an ignition chamber that is enveloped in a jacket of water. The material is burned to an ash under its own heat of burning with the aid of oxygen. The heat produced, which is captured in the surrounding water jacket, is determined from the increase in temperature of the known quantity of water.

Not all the energy extracted from food is, however, available for the metabolic needs of the animal. A variable fraction, depending on the type of food, is used up in the process of assimilation, and a correction for this fraction must be made in measurements of energy intake. Measurement errors also occur because (1) energy can be obtained during the period of measurement from an increased utilization of an animal's tissue reserves, and (2) some fraction of the organism's food is generally degraded by microorganisms in the gut. Thus, the balance-sheet approach to energy metabolism not only is cumbersome but must contend with variables that are difficult to control. Modern studies on energy metabolism generally use the more direct approaches discussed next.

## DIRECT CALORIMETRY

When no physical work is being performed and no new molecules are being synthesized, all the chemical energy released by an animal in carrying out its metabolic functions appears finally as heat. According to *Hess' Law* (1840), the total energy released in the breakdown of a fuel to a given set of end products is always the same irrespective of the intermediate chemical steps or pathways utilized. The metabolic rate of an organism is therefore effectively determined by measuring the amount of energy released as heat over a given period. Such measurements are made in a calorimeter, and the method is called *direct calorimetry*. The animal is placed in a well-insulated chamber, and the heat lost by the animal is determined from the rise in temperature of a known mass of water used to trap that heat. The earliest and simplest calorimeter was that devised in the 1780s by Antoine Lavoisier and Pierre de Laplace, in which the heat given off by the animal melted ice packed around the chamber (Figure 16-3). This heat loss was calculated from the mass of collected water and the latent heat of melting ice. Figure 16-4 shows the essential elements of a modern calorimeter. Water flows through coiled copper pipes in the measuring chamber. The total heat lost by the subject is the sum of the heat gained by the water plus the latent heat present in the water vapor of the expired air and of evaporated skin moisture. To measure this latent heat, the mass of the water vapor is determined by passing the air through sulfuric acid, which absorbs the water. The energy content of each gram of water absorbed is 0.585 kcal, the latent heat of vaporization of water at 20°C. The results are generally reported in calories or kilocalories per hour (Box 16-1).

Direct calorimetry has been used most often with birds and small mammals that have high metabolic rates. With large animals and small animals having low metabolic rates, the precision of the technique is inadequate. Another disadvantage of direct calorimetry is that the behavior (and therefore metabolism) of the animal is unavoidably altered because of the restrictions imposed by the conditions of the measurement. Though simple in principle, direct calorimetry is rather cumbersome in practice.

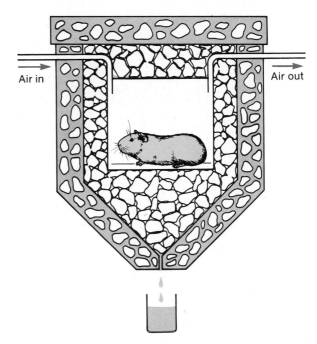

**16-3** Lavoisier's ice-jacket calorimeter. The animal's heat production was determined from the quantity of water melted from ice, since 80 kcal of heat melts 1 kg of ice. [Kleiber, 1961.]

## INDIRECT CALORIMETRY

*Indirect calorimetry* depends on the measurement of some factor(s) other than heat production related to energy utilization. Oxygen uptake and carbon dioxide production (milliliters or liters per hour) are the factors commonly used. The energy contained in food molecules becomes available for use by an animal when those molecules or their products are subjected to oxidation, as described in Chapter 3. In aerobic oxidation, the amount of heat produced is related to the quantity of oxygen consumed. *Spirometry* is the measurement of an animal's respiratory exchange. In the study of aquatic animals, a closed system is commonly used. The animal is confined to a closed chamber in which the amount of oxygen removed and carbon dioxide introduced is monitored (Figure 16-4). Oxygen consumption is revealed by serial determinations of the decreasing concentration of oxygen dissolved in the water contained in the chamber. These

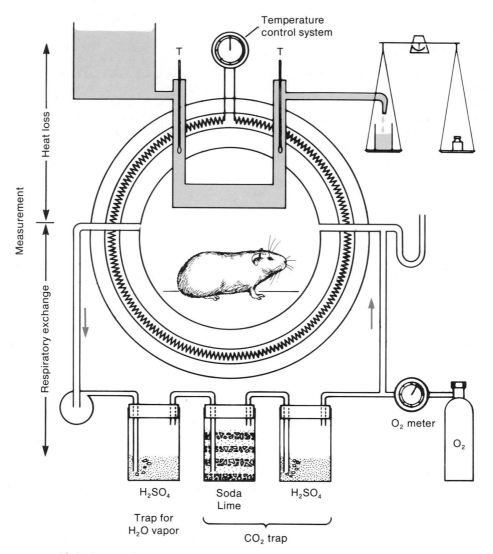

16-4   Atwater–Rosa respiration calorimeter. The animal is confined to an insulated chamber that is kept at a constant temperature. The heat produced by the animal is determined from the heat absorbed by the coolant water (top). T = thermometers. Respiratory exchange is determined by measuring $CO_2$ and $O_2$. [Kleiber, 1961.]

measurements can be obtained very conveniently with the aid of an oxygen electrode and the appropriate electronic circuitry. The oxygen dissolved in the water determines the electrical signal produced in the electrode. In air-breathing animals, $O_2$ and $CO_2$ can be measured in the gaseous phase by paramagnetic and infrared devices, respectively. These modern methods of gas analysis allow mass-flow analytical techniques in which the flow of gas into and out of a chamber is monitored, and the difference in concentrations is used to calculate respiratory exchange. Spirometry is also carried out on animals fitted with breathing masks, a

## Box 16-1   *Energy Units*

The most commonly used unit of measurement of heat is the *calorie,* defined as the quantity of heat required to raise the temperature of 1 g of water 1°C. This quantity varies slightly with temperature, and so the calorie is more precisely the amount of heat required to raise the temperature of 1 g of water from 14.5 to 15.5°C. A more practical unit is the *kilocalorie* (1 kcal = 1000 cal). The use of calorie and kilocalorie has persisted largely because of their familiarity. However, according to the International System of Units (SI), heat is defined in terms of work, and the unit of measurement is the *joule,* the more practical unit being the *kilojoule* (1 kJ = 1000 J). Thus, 1 cal = 4.184 J; 1 kcal = 4.184 kJ. If we assume a respiratory quotient (RQ) of 0.79, which is a typical figure, 1 L of oxygen used in the oxidation of substrate will release 4.8 kcal, or 20.1 kJ, of energy.

Power is energy expended per unit time and is given in *watts* (*W*), with 1 W = 1 J/s. Conversion tables are given in Appendix 3.

method primarily useful in wind-tunnel tests of flying animals, because mass air flow would make inflow–outflow comparisons of $O_2$ and $CO_2$ inaccurate.

The determination of metabolic rate from $O_2$ consumption rests on two important assumptions. First, the relevant chemical reactions are assumed to be aerobic. This assumption usually holds for most animals at rest, since energy available from anaerobic reactions is relatively minor except during vigorous activity. However, anaerobiosis is important in animals that live in oxygen-poor environments, as do gut parasites and invertebrates that dwell in deep lake-bottom muds. Oxygen consumption would, of course, be an unreliable index of metabolic rate in such animals. Second, the amount of heat produced (i.e., energy released) when a given volume of oxygen is consumed is assumed to be constant irrespective of the metabolic substrate. This assumption is not precisely true—more heat is produced when 1 L of $O_2$ is used in the breakdown of carbohydrate than when fats or proteins are the substrate. However, the error involved in this assumption is no greater than about 10%, as can be seen in Table 16-1. Unfortunately, it is generally difficult to identify the substrate(s) being oxidized with any precision in correcting for differences in caloric yield.

### RESPIRATORY QUOTIENT

To translate the amount of oxygen consumed in the oxidation of food into equivalent heat production, one must know the relative amounts of carbon and hydrogen oxidized. The oxidation of hydrogen atoms is hard to determine, however, since metabolic water (i.e., that produced by oxidation of hydrogen atoms available in foodstuffs), together with other water, is lost in the urine and from a variety of body surfaces at a rate that is irregular and determined by unrelated factors such as osmotic stress. So it is more practical

TABLE 16-1   Heat production and respiratory quotient for the three major foodstuff types.

| | Heat production (kcal) | | | RQ |
| --- | --- | --- | --- | --- |
| | Per gram of foodstuff | Per liter of $O_2$ consumed | Per liter of $CO_2$ produced | $\dfrac{\text{Mol } CO_2}{\text{Mol } O_2}$ |
| Carbohydrates | 4.1 | 5.05 | 5.05 | 1.00 |
| Fats | 9.3 | 4.74 | 6.67 | 0.71 |
| Proteins (to urea) | 4.2 | 4.46 | 5.57 | 0.80 |

to measure, along with the $O_2$ consumed, the amount of carbon converted to $CO_2$, as explained earlier. From such measurements, the amount of water produced by oxidative metabolism can in fact be readily calculated from the difference between total oxygen consumed and that used to oxidize carbon to $CO_2$. The ratio of the volume of $CO_2$ expired to the volume of $O_2$ consumed within a given time is called the *respiratory quotient (RQ)*:

$$RQ = \frac{\text{rate of } CO_2 \text{ production}}{\text{rate of } O_2 \text{ consumption}} \quad (16\text{-}1)$$

The RQ is characteristic of the type of foodstuff catabolized—carbohydrate, fat, or protein—reflecting the proportions of carbon and hydrogen in the food molecules. The following examples illustrate how the RQ of the major food types may be calculated from a formulation of their oxidation reactions.

*Carbohydrates*

The general formula of carbohydrates is $(CH_2O)_n$. During complete oxidation of a carbohydrate, respiratory oxygen is utilized *in effect* only to oxidize the carbon, forming $CO_2$. Each mole of a carbohydrate upon complete oxidation produces $n$ mol of both $H_2O$ and $CO_2$ and consumes $n$ mol of $O_2$. The RQ for carbohydrate oxidation is thus 1. The overall catabolism of glucose, for example, may be formulated as

$$C_6H_{12}O_6 + 6O_2 \longrightarrow 6CO_2 + 6H_2O$$

$$RQ = \frac{6 \text{ volumes of } CO_2}{6 \text{ volumes of } O_2}$$

$$= 1.0$$

*Fats*

The RQ characteristic of the oxidation of a fat such as tripalmitin may be calculated as follows:

$$2C_{51}H_{98}O_6 + 145O_2 \longrightarrow 102CO_2 + 98H_2O$$

$$RQ = \frac{102 \text{ volumes of } CO_2}{145 \text{ volumes of } O_2}$$

$$= 0.703$$

Since different fats contain different ratios of carbon, hydrogen, and oxygen, they differ slightly in their RQs.

The relations between the caloric value of 1 L of

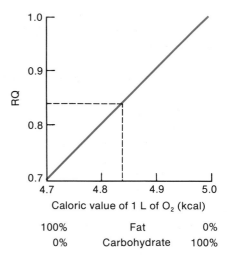

16-5    Relation between respiratory quotient and caloric value of oxygen. See text for details. [Hardy, 1979.]

$O_2$ consumed and the proportions of fat and carbohydrate oxidized are illustrated in Figure 16-5. As the proportion of fat and carbohydrate increases from 0 to 100%, the RQ rises from 0.7 to 1.0, and the caloric value of 1 L of $O_2$ increases from 4.7 to 5.0 kcal.

*Proteins*

The RQ characteristic of protein catabolism presents a special problem because proteins are not completely broken down during oxidative metabolism. Some of the oxygen and carbon of the constituent amino acid residues remain combined with nitrogen and are excreted as nitrogenous wastes in urine and feces. In mammals, the excreted end product is urea, $(NH_2)_2CO$, and in birds, primarily uric acid, $C_5H_4N_4O_2$. To obtain the RQ, it is therefore necessary to know the amount of ingested protein as well as the amount and kind of nitrogenous wastes excreted. The carbon and hydrogen oxidized during the catabolism of protein typically require 96.7 volumes of $O_2$ for the production of 77.5 volumes of $CO_2$, giving an RQ of about 0.80.

It is routinely assumed in making deductions from RQ that (1) the only substances metabolized are carbohydrates, fats, and proteins, (2) no synthesis takes place alongside breakdown, and (3) the amount of $CO_2$ exhaled in a given time equals the $CO_2$ produced by the tissues in that interval. These assumptions are not strictly true, so caution must be exercised in using RQ values at rest and in postabsorptive (fasting) states. Under such conditions, protein utilization is negligible, and the animal is considered to be metabolizing only fat and carbohydrate. In Table 16-1, it is seen that the oxidation of 1 g of mixed carbohydrate releases about 4.1 kcal as heat. Further, when 1 L of $O_2$ is used to oxidize carbohydrate, 5.05 kcal is obtained; the value for fats is 4.75 kcal and for protein (to urea), 4.46 kcal. A fasting aerobic animal presumed to be metabolizing only carbohydrates and fats produces between 4.75 kcal and 5.05 kcal of heat for every liter of oxygen consumed, depending on the proportion of fat to carbohydrate metabolized.

### SPECIFIC DYNAMIC ACTION

A marked increase in metabolism accompanies the processes of digestion and assimilation of food. About 1 h after a meal is eaten, the metabolic rate increases, reaching a peak some 3 h later and remaining above the basal value for several hours. Thus, following food intake, an animal's oxygen consumption and heat production increase independently of other activities. Max Rubner reported in 1885 that this increase in heat production differed for each of the classes of foodstuffs, and he gave it the rather awkward name *specific dynamic action (SDA)*. Its mechanism is not clearly understood, but apparently the work of digestion (and the concomitant increase in metabolism of the tissues of the gastrointestinal tract) is responsible for only a small part of the elevated metabolism. A more likely explanation for this rise in metabolic rate may be that certain organs such as the liver expend extra energy to prepare the products of digestion for entry into metabolic pathways. The extra energy consumed by such processes is lost as heat. The magnitude of the increased metabolic rate ranges from 5% to 10% of total energy of ingested carbohydrates and fats and from 25% to 30% for proteins. It is quite important, therefore, that basal metabolism be measured during the postabsorptive state so as to minimize any contribution of the SDA.

### ENERGY STORAGE

Most animals can not strike a moment-to-moment balance between food intake and energy expenditure, since they cannot ingest food continually as they continually expend metabolic energy. As food is taken in, it exceeds immediate energy requirements, and the excess is stored for later use, primarily as fats and carbohydrates. Protein is not an ideal storage material for energy reserves because nitrogen is a relatively scarce commodity and is generally the limiting factor in growth and reproduction; so it would be wasteful to tie up valuable nitrogen in energy reserves. Fat is the most efficient form for energy storage, since oxidation of fat yields 9.56 kcals/g, nearly twice the yield per gram for carbohydrate or protein. This efficiency is of great importance in animals such as migrating birds, in which economy of weight and volume is of the essence. Not only is the energy yield per gram of carbohydrate lower than that of fats, but carbohydrates are stored in a bulky hydrated form, with as much as 4–5 g of water required per gram of carbohydrate. Nonetheless, some carbohydrates are important in energy storage. Glycogen, a branched, starchlike carbohydrate polymer, is stored as granules in skeletal muscle fibers and liver cells of vertebrates. Muscle glycogen is converted to glucose for oxidation within the muscle cells during intense activity, and liver glycogen is utilized to maintain blood glucose levels. Glycogen is broken down directly into glucose-6-phosphate, providing fuel for carbohydrate metabolism more directly than does fat.

## Body Size and Metabolic Rate

Although both whales and water shrews normally submerge, a whale can remain under water, holding its breath, far longer than a shrew. This observation illustrates the general principle that small animals respire at higher rates than large animals. An inverse relationship, in fact, exists between the rate of $O_2$ consumption per gram of body weight and the total mass of the animal. Thus, a 100 g mammal will consume far more than $\frac{1}{10}$ the energy per unit time of a 1000 g mammal. The nonproportionality that exists for BMRs of mammals ranging from very small to very large

TABLE 16-2  Oxygen consumption in mammals of various body sizes.

| Animal | Body mass (g) | Total O₂ consumption (ml/h) | O₂ consumption per gram (ml/g/h)† |
|---|---|---|---|
| Shrew | 4.8 | 35.5 | 7.40 |
| Harvest mouse | 9.0 | 22.5 | 1.50 |
| Kangaroo mouse | 15.2 | 27.3 | 1.80 |
| Mouse | 25 | 41.0 | 1.65 |
| Ground squirrel | 96 | 98.8 | 1.03 |
| Rat | 290 | 250 | 0.87 |
| Cat | 2,500 | 1,700 | 0.68 |
| Dog | 11,700 | 3,870 | 0.33 |
| Sheep | 42,700 | 9,590 | 0.22 |
| Human | 70,000 | 14,760 | 0.21 |
| Horse | 650,000 | 71,100 | 0.11 |
| Elephant | 3,833,000 | 268,000 | 0.07 |

†The figures in this column are proportional to metabolic intensity.
*Source:* Schmidt-Nielsen, 1975.

is illustrated by the "mouse-to-elephant" curve (Table 16-2; Figure 16-6A). A similar relation holds for other vertebrate groups. The log of whole-body metabolic rate has been found to be linearly related to log of body weight:

$$M = aW^b \qquad (16\text{-}2)$$

where $M$ is the BMR or SMR, $W$ is the body weight, $a$ is a proportionality coefficient that differs in different species, and $b$ is an empirically determined exponent equal to the slope of the log $M$ plot (Figure 16-6C).

*Weight-specific metabolic rate, $M/W$, also termed metabolic intensity,* is the metabolic rate of a unit mass of tissue. It is determined by dividing both sides of Equation 16-2 by $W$:

$$\frac{M}{W} = \frac{aW^b}{W} = aW^{(b-1)} \qquad (16\text{-}3)$$

This relation is shown in color in Figure 16-6B. To work with straight-line rather than curved plots, it is customary to make logarithmic transformations of

Equations 16-2 and 16-3. Thus, Equation 16-2 becomes

$$\log M = \log a + b\,(\log W) \qquad (16\text{-}4)$$

and Equation 16-3 becomes

$$\log \frac{M}{W} = \log a + (b - 1)\log W \qquad (16\text{-}5)$$

These are plotted in Figure 16-6C.

In a large sample of vertebrates, the value for $b$ lies between 0.65 and 0.8. The same holds true for other animals and even unicellular organisms (Figure 16-7). This essentially universal exponential relationship between body size and metabolic rate has attracted the attention of physiologists since it was first recognized, and a rational "explanation" for the logarithmic relation between body mass and metabolism has been earnestly, but unsuccessfully, sought. A plausible but short-lived theory was the *surface hypothesis,* or *surface "law."* It was reasoned that the metabolic rate of endotherms should be proportional to body surface area because the rate of heat transfer between two compartments (i.e., warm animal interior and cooler environment) is proportional, all else being equal, to their area of mutual contact. The surface area of an object of given density and shape is proportional to the $\frac{2}{3}$ power of its weight, since weight increases as the cube of linear size, whereas surface area increases only as the square. The surface hypothesis predicts that because heat loss to the environment occurs through the body surface, and lost heat is replenished by metabolic activity, the metabolic rates of endotherms of different sizes should vary with body surface and hence as the $\frac{2}{3}$ power of their body weight.

The surface hypothesis quickly gained support from findings that heat production in mammals is proportional to approximately the 0.70 power of body weight (i.e., $M = aW^{0.7}$), or even closer to the predicted value of 0.67 (Heusner, 1982). Additional studies revealed that metabolic rate per unit area of surface is similar for mammals and birds of varying sizes. The differences in metabolic intensity of animals of differing sizes are correlated with the number of mitochondria per unit volume of tissue. Thus, the cells of a small mammal contain more mitochondria in a given volume of

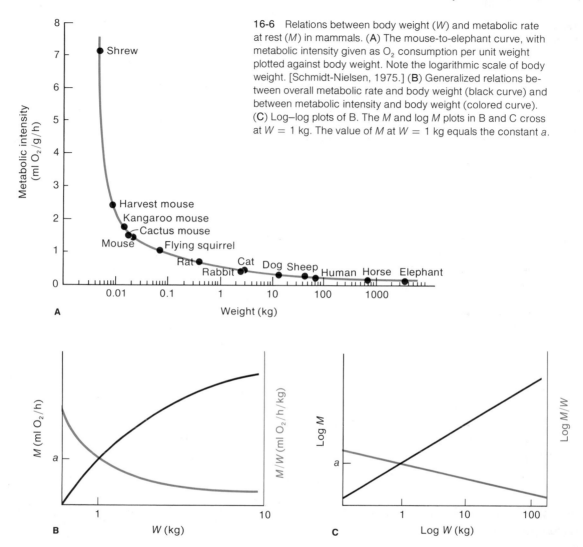

**16-6** Relations between body weight (*W*) and metabolic rate at rest (*M*) in mammals. (**A**) The mouse-to-elephant curve, with metabolic intensity given as $O_2$ consumption per unit weight plotted against body weight. Note the logarithmic scale of body weight. [Schmidt-Nielsen, 1975.] (**B**) Generalized relations between overall metabolic rate and body weight (black curve) and between metabolic intensity and body weight (colored curve). (**C**) Log–log plots of B. The *M* and log *M* plots in B and C cross at *W* = 1 kg. The value of *M* at *W* = 1 kg equals the constant *a*.

cytoplasm than do the cells of a large mammal. Since the mitochondria are the site of oxidative respiration, this correlation comes as no surprise.

In spite of the logical attractiveness of the surface hypothesis, it appears that the approximate proportionality of metabolic intensity to surface area, or $\frac{2}{3}$ power of body weight, may be nothing more than an incidental relationship, for the SMRs of ectotherms exhibit nearly the same relations to body weight as the BMRs of endotherms (Figure 16-7). There is no self-evident reason why the metabolic rate of an ectotherm should be causally related through heat loss to body surface area, since relatively little or no metabolic energy is expended to warm an ectotherm above the ambient temperature, and no net heat flux occurs across the surface of an ectotherm that is in temperature equilibrium with its environment. The significance of the proportionality of metabolic intensity to a 0.65–0.8 power of body weight has not been satisfactorily explained.

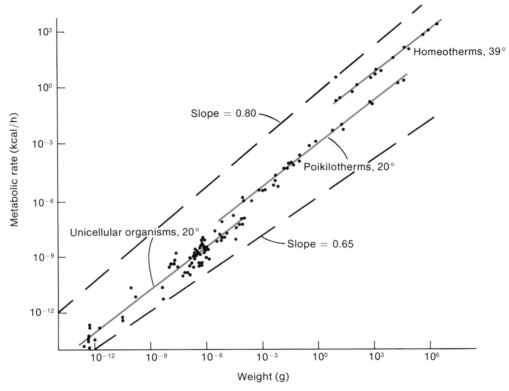

**16-7**  Minimal metabolic rates of unicellular organisms, poikilotherms, and homeotherms are related to body weight by similar exponents. The solid lines all represent exponents of 0.75. The vertical position of each group on the graph is related to the coefficient *a* in Equation 16-2. [Hemmingsen, 1969.]

## Energetic Cost of Locomotion

### ANIMAL SIZE, VELOCITY, AND COST OF LOCOMOTION

The metabolic cost of muscular activity is calculated as the additional cost above the BMR or SMR and is usually given as kilocalories per kilogram per kilometer. Measurements of $O_2$ consumption and $CO_2$ production associated with locomotion are generally made while the animal is running on a treadmill, swimming in a flow tank, or flying in a wind tunnel, and the rate of gas exchange is translated into rate of energy conversion.

Relations between the *net* work done in the locomotion of an animal and the *gross* energy conversion powering the underlying muscle activity are compli-

cated by several factors, not all of which are sufficiently well understood to be discussed here. We might note, nonetheless, that a significant percentage of muscular effort during locomotion does not contribute directly to the production of forward motion. Some muscle contraction is devoted to holding limb joints in their proper articulating positions. Another large percentage of muscle work (to be discussed at greater length later) is performed in the elongating muscle to counteract gravity, absorb shocks, and finely tune the movements of limbs during the contraction of antagonists. The comparative energetics of animal locomotion are further complicated by the inverse relation that has long been known to exist between the force produced by a contracting muscle and its rate of shortening (i.e., muscle lengths or sarcomere length

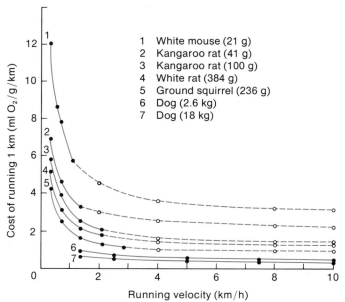

1  White mouse (21 g)
2  Kangaroo rat (41 g)
3  Kangaroo rat (100 g)
4  White rat (384 g)
5  Ground squirrel (236 g)
6  Dog (2.6 kg)
7  Dog (18 kg)

**16-8**  Energetic cost of locomotion at different velocities of running in mammals of differing sizes. The cost of running 1 km drops and levels off with increasing velocity. Dashed portions are extrapolated. [Taylor et al., 1970.]

per second; Figure 9-30). Owing to scaling principles, small animals employ higher rates of muscle shortening to achieve a given velocity of locomotion than do larger animals, and for that reason they must convert correspondingly larger amounts of metabolic energy to produce a given amount of force per unit cross section of contractile tissue in moving their limbs.

In relating overall energetic cost of locomotion to size and velocity of an animal, several generalizations become apparent. As velocity of running increases, for example, the metabolic cost of traveling a given distance decreases (Figure 16-8). On the other hand, the rate of oxygen consumption in excess of the BMR increases linearly with the velocity (Figure 16-9A). It is noteworthy, however, that the increase in energy utilization per unit weight for a given increment in speed is less for larger than for smaller animals. This can be seen in Figure 16-8 and in the different slopes of the plots in Figure 16-9A. When the cost of locomotion is plotted as energy utilization per gram of tissue per kilometer against body weight, it is again apparent that larger animals expend less energy to move a given mass a given distance (Figure 16-9B). The lower energetic efficiency of small animals during locomotion may, to a limited degree, be due to the

greater drag (to be discussed shortly) they experience, but this explanation certainly does not suffice for terrestrial animals moving at low and moderate speeds through air, where drag is negligible. More likely, the lower energetic efficiency is related to the lower efficiency of rapidly contracting muscle mentioned earlier.

### PHYSICAL FACTORS INFLUENCING LOCOMOTION
The metabolic cost of moving a given mass of animal tissue over a given distance depends also on certain purely physical factors, including the following two.

*1. Inertia.* Every object possesses inertia proportional to its mass. The larger the animal, the greater its inertia and the greater its momentum once it is in motion. The high *inertial forces* that must be overcome during the acceleration of a large animal account for a significant utilization of energy during the period of acceleration (Figure 16-10A). Small animals, like small cars, require less energy to accelerate to a given velocity; likewise, they need less energy to decelerate. Therefore, the small animal starts and stops abruptly at the beginning and end of a locomotory effort, whereas a large animal accelerates more slowly after locomotion begins and slows down more gradually as

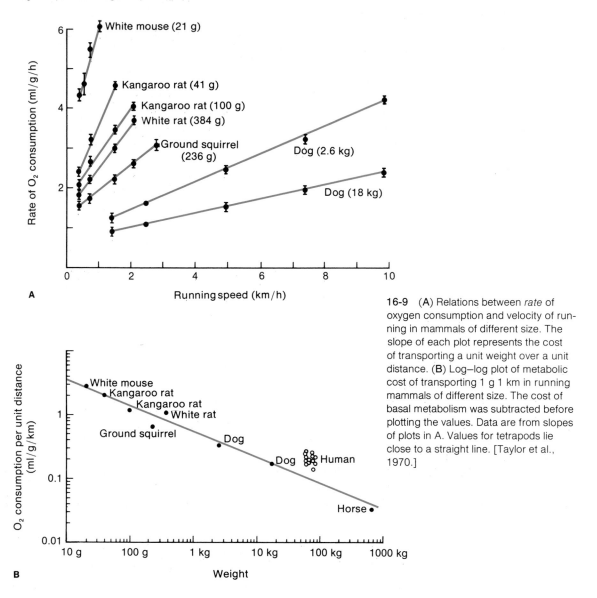

16-9    (A) Relations between *rate* of oxygen consumption and velocity of running in mammals of different size. The slope of each plot represents the cost of transporting a unit weight over a unit distance. (B) Log–log plot of metabolic cost of transporting 1 g 1 km in running mammals of different size. The cost of basal metabolism was subtracted before plotting the values. Data are from slopes of plots in A. Values for tetrapods lie close to a straight line. [Taylor et al., 1970.]

locomotion ends (Figure 16-10B). Similarly, in terrestrial animals, the limbs engaged in back-and-forth movements during running are subject to inertial forces related to their mass, the limbs of a large animal exhibiting greater inertia and momentum than those of a small animal.

*2. Drag.* Since animals do not exist within a vacuum, the energetics of sustained locomotion are influenced by the physical properties of the gas or liquid through which they move. The viscosity and density of the gas or liquid environment produces a *drag force* opposite in direction to the animal's movement. The drag produced in a given medium depends on the surface area and the shape of an object. For an object of a given shape, drag is proportional to the surface area. Since larger animals have lower surface/mass

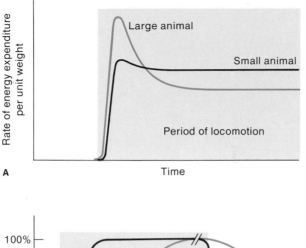

**16-10** Effect of body weight on energetics of locomotion. (A) Rate at which energy is utilized per unit weight during onset and maintenance of locomotion in a large and a small animal of similar type. (B) Velocity during acceleration and deceleration of a small and a large animal.

ratios, they experience less fluid drag per unit mass than do smaller animals, for whom overcoming drag is energetically more costly. Once it is under way, the larger animal expends less energy per unit mass to propel itself at a given velocity than does a smaller animal of similar type (Figure 16-10A). These distinctions are more pronounced, of course, in water than in air, since water, having the higher viscosity and density, produces far more drag on a moving object than air does.

The ease with which an object moves through water depends in part on the *flow pattern* set up in the medium through which it moves. The fluid at the immediate surface of the object moves at the same velocity as the object, whereas the fluid at a great distance is undisturbed. If the transition in fluid velocity is smoothly continuous as the fluid progresses away from the object's surface, the flow at the boundary layer is said to be *laminar* (Figure 16-11A). *Turbulent flow* occurs as a result of sharp gradients in the velocity of flow.

Because of conservation of energy, pressure and velocity are reciprocally related in a given fluid system, and the higher the velocity of fluid at a given locus, the lower the pressure. Thus, strongly differing flow rates around an object will cause *eddy currents* owing to secondary flow patterns set up between regions of high and regions of low pressure. Moreover, the higher the viscosity of the medium or the relative motion between the object and the surrounding fluid, the greater the shear forces produced and hence the greater the tendency toward turbulence. Because its production requires energy, turbulence retards the efficient conversion of metabolic energy into propulsive movement.

Long, streamlined shapes promote laminar flow, minimizing eddy current formation. Fishes and marine mammals such as seals, porpoises, and whales are admirably streamlined, exhibiting nearly turbulence-free passage through water even at high speeds. An additional factor reducing turbulence in these animals is the compliance (resiliency) of the body surface.

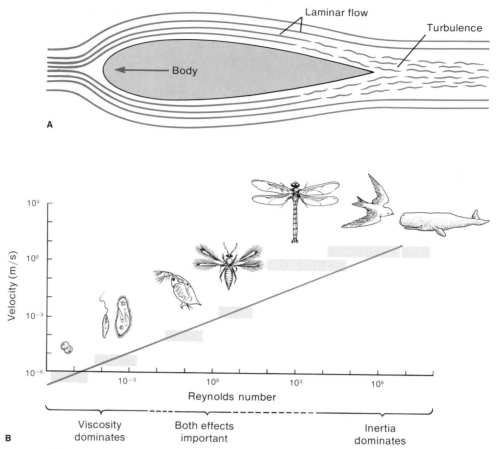

**16-11** (**A**) Flow around a moving body. Movement through a fluid can create turbulence owing to uneven fluid pressures. Laminar flow occurs where pressure gradients are minimal. The larger the body and the less viscous the fluid, the higher the velocity before turbulence occurs. (**B**) Log of animal size plotted against the log of the respective Reynolds numbers at cruising velocities. Small animals that move slowly have small *Re*'s because viscous forces predominate at small dimensions. Larger animals move rapidly with high *Re*'s because inertial forces predominate at large dimensions. [B from Nachtigall, 1977.]

This compliance damps small perturbations in the pressure of water flowing over the body surface and thereby lessens the local variations in water pressure that give rise to energy-consuming turbulence.

### EFFECT OF REYNOLDS NUMBER

The flow pattern set up around an object moving in a fluid is determined not only by the density and viscosity of the fluid but also by the dimensions and velocity of the object. Osborne Reynolds lumped these four factors together in a dimensionless ratio comparing inertial forces (proportional to density, size, and velocity) to viscous forces. This is the so-called *Reynolds number* (*Re*), calculated as

$$Re = \frac{\rho V L}{\mu} \qquad (16\text{-}6)$$

in which $\rho$ is the density of the medium, $V$ is the velocity of the body, $L$ is an appropriate linear dimension, and $\mu$ is the viscosity of the medium. Thus, when a body moves through a fluid such as water or air, the flow pattern depends on the $Re$. The larger the object or the higher its speed in water, the higher the $Re$. The same object moving at the same speed in air as in water would be characterized by a lower $Re$ in air (about 15 times lower) because of air's much lower density.

An $Re$ below about 1.0 characterizes movement in which the object produces a purely laminar pattern of flow by the fluid passing over its surface. Above an $Re$ of about 40 turbulence begins to appear in the wake of the object. As $Re$ exceeds about $10^6$, the fluid in contact with the object becomes turbulent. At this point, the energy needed to further increase the velocity of a streamlined body rises steeply. The upper limit is somewhat higher for streamlined objects like dolphins than for unstreamlined objects like human scuba divers.

For a small organism such as a bacterium, spermatozoan, or ciliate, the watery medium appears far more viscous than it does to a human. The viscosity encountered by a paramecium swimming through water has been compared to the viscosity that would be experienced by a human swimming through a pool of warm tar (which, indeed, is difficult to comprehend). Because of such *scaling factors,* the movements of small organisms are dominated by *viscous effects,* while those of large animals are dominated by *inertial effects* (Figure 16-11B).

The relative importance of *viscous drag* on the "coasting" of objects of different size can be demonstrated by pushing a matchstick (low $Re$) floating on water to a given velocity, say 0.1 m/s, and then doing the same to a 100 kg floating log (high $Re$) of similar physical proportions but of larger size. The little matchstick abruptly comes to rest owing to the drag exerted on it by the viscosity of the water, whereas the massive log coasts for many seconds because its greater inertia when in motion overcomes the drag. Similarly, a paramecium comes to an abrupt halt if it stops beating its cilia, whereas a whale coasts smoothly between thrusts of its flukes with little loss of velocity. Perhaps because of these differences between the locomotion of small and large aquatic animals, small teleosts tend to have predominantly oxidative-type locomotor muscles, whereas larger teleosts tend to have predominantly white, anaerobic-type locomotor muscles, capable of building up an oxygen debt. The muscles of larger marine animals, used in a more discontinuous fashion, can briefly build up an oxygen debt that is quickly repaid during periods of coasting. The muscles of small swimming animals, in contrast, are kept more nearly continuously active so as to overcome viscous drag, for they cannot rely on the anaerobic "borrow now, pay later" principle.

## AQUATIC, AERIAL, AND TERRESTRIAL LOCOMOTION

### Aquatic

Animals that swim in water, such as fishes, need not support their own weight. Most are equipped with flotation bladders that enable them to suspend themselves at a given depth with little expenditure of energy. However, though the high density of water allows them to float, it also produces high drag. This hindrance to objects moving through a fluid has led to a convergence of body forms among marine mammals and fishes. The streamlined, fusiform body shape is wonderfully developed in most sharks, teleost fishes, and dolphins. The reasons are evident enough on intuitive grounds, but they can be understood more precisely in terms of flow pattern.

The speed of a swimming or flying animal is proportional to the *power/drag ratio.* Power is directly proportional to muscle mass, and, assuming that muscle mass increases in proportion to total body mass, power rises in proportion to body mass. On the other hand, for a large swimming animal, drag per unit mass decreases with body mass because, if the shape remains constant, the frontal area (which determines the drag) increases as a function of some linear dimension squared, whereas body mass (which determines the power available) increases as a function of that linear dimension cubed. Thus, a large aquatic animal can develop power out of proportion to its drag forces, and it therefore will be able to attain higher swimming velocities than a smaller animal of the same shape. Large fishes and mammals swim faster than their smaller counterparts. Because of the high drag forces developed in water, aquatic animals can reach only the speeds of a bird in flight if they are much larger than the bird.

Interestingly, the ratio of power output to fluid drag allows all aquatic animals, whatever their size, to achieve velocities of about 10–15 body lengths per second. So a paramecium achieves about 0.01 km/h, a tuna about 75 km/h.

### Aerial

Unlike water, air offers little buoyant support, so all flyers must overcome gravity by utilizing principles of aerodynamic *lift*. Although the effects of drag increase with speed, there is still less need for streamlining among birds than among fishes because of the low density of air. Thus, thanks to the relatively low drag forces developed, birds can achieve much higher speeds than fishes. The production of propulsive force, to drive the bird forward, and of lift, to keep it aloft, is accomplished simultaneously during the downstroke of the bird wing, as shown in Figure 16-12. The wing is driven downward and forward with an angle of attack, $\alpha$, that pushes the air both downward and backward, creating an upward and forward *thrust*. This thrust resolves into both lift and forward propulsion, which are required to overcome the bird's weight and drag, respectively.

The body shapes of fishes and birds demonstrate the great differences that exist between the physical properties of water and those of air, as well as the biological divergence that results from adaptation to these two dissimilar media. When a bird is gliding, its elongated, extended wings produce excellent lift, but if placed in water they would obviously generate far too much drag. Thus, the wings of penguins are modified as short paddles and are folded against the body while coasting under water. Since drag forces are much higher in water than in air, only medium-sized to large animals can coast in water. In contrast, only very small flying animals, smaller than a dragonfly, are unable to coast (glide) in air. Small insects

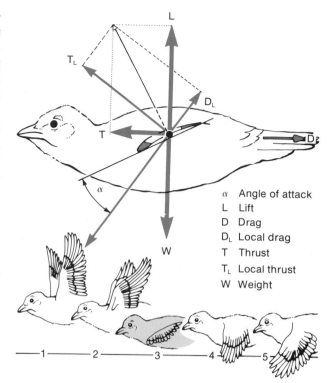

| | |
|---|---|
| $\alpha$ | Angle of attack |
| L | Lift |
| D | Drag |
| $D_L$ | Local drag |
| T | Thrust |
| $T_L$ | Local thrust |
| W | Weight |

**16-12** Forces developed by downstroke of bird wing. Shown is the wing in stage 3 (below) of the flight cycle. Forces relating to the wing beat are shown as colored arrows, and those relating to the body as black arrows. $D_L$ (also known as induced drag) equals the drag produced as a consequence of lift production. $T_L$ (also known as induced thrust) is complementary to $D_L$. [Nachtigall, 1977.]

such as flies and mosquitoes must continually beat their wings to maintain headway, since they have very little momentum.

### Terrestrial

When swimming, flying, and running are compared with respect to the energetic cost of moving a given body mass at a given velocity (Figure 16-13), it is apparent that terrestrial locomotion (i.e., running) is the most costly, while swimming is the least expensive. It is perhaps not surprising that a fish swimming in water is more efficient in its locomotion than a bird flying through the air, for, as we have noted, the fish

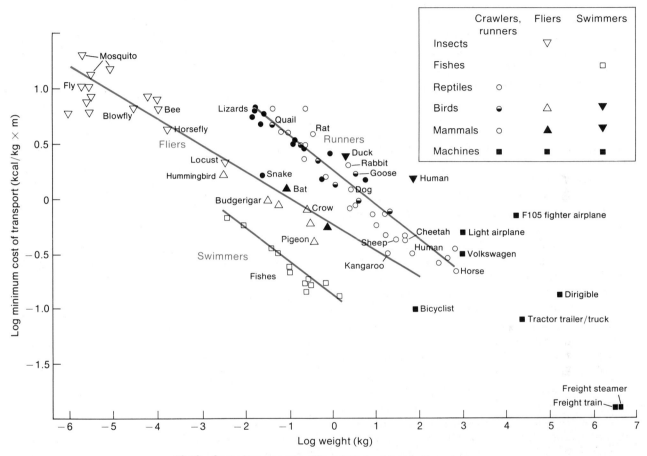

**16-13** Cost of transport for different kinds of locomotion, given in kilocalories times meters per second. The cost is more closely related to the kind of locomotion than it is to the kind of organism. [Tucker, 1975.]

is close to neutral buoyancy, whereas a bird must expend energy to stay aloft. But why is running less efficient than either flying or swimming? An excellent discourse on this subject is given by V. A. Tucker (1975), who also explains why riding a bicycle over a given distance is more energy-efficient than running.

Running differs from cycling, fish swimming, and bird flying in the way the muscles are used, and this difference accounts for the low work efficiency of running. When a biped or quadruped animal runs, its center of mass (CM) necessarily rises and falls cyclically with the gait. The rise in the CM occurs when the foot and leg extensors push the body up and forward, and the fall occurs as gravity inexorably tugs at the body, bringing it back to earth between locomotory extensions. Efficiency is lost because the antigravity extensor muscles that contract to propel the CM upward and forward must also break the fall of the CM that occurs before the next stride. To control the fall, the extensor muscles must expend energy to resist lengthening as they slow the rate of descent of the body in preparation for the next cycle. This technically unproductive use of muscle energy to counteract the pull of gravity is said to produce "negative work." Such negative work is also required of the extensor muscles of hikers' legs during descent of a steep trail.

In short, running is less efficient than flying, swimming, or bicycling because muscles are used for deceleration (negative work) as well as acceleration (positive work). During bike riding as well as wing or fin locomotion, muscle energy need not be expended to slow the momentum of the body mass as it is during running or walking.

Energy storage in elastic elements appears to be especially important in running and hopping animals. The concept of *elastic energy storage* can be illustrated by a child hopping with a pogo stick. The higher the hop, the greater the speed of descent, the more energy transferred to the spring, and the greater the force of elastic recoil of the spring.

By changing gait at appropriate speeds, land animals optimize their locomotory efficiency. For example, say a pony is made to trot on a treadmill at a speed at which it would normally gallop, or to gallop when it would normally trot, or to trot when it would normally walk. In all cases, it would expend more energy than if allowed to change its gait naturally. It is speculated that optimum gaits result from the relative amounts of energy stored in the elastic elements of the body, such as tendons, when performing the different gaits. For instance, little energy is stored when an animal walks, somewhat more when it trots. When the animal is galloping, its entire trunk is involved in elastic storage. At least half the negative work done in absorbing kinetic energy during the stretching of an active muscle appears as heat, while the remainder is stored in stretched elastic elements such as muscle cross bridges and tendons. Only the elastically stored energy is available for recovery on rebound, and only 60–80% of it is in fact recovered on its release. The energy converted to heat is, of course, not available for conversion to mechanical work in living tissue.

On simple energetic grounds, it might be conjectured that terrestrial endotherms and ectotherms of the same size will expend the same net metabolic energy to run at a given velocity. This reasonable assumption is almost, but not entirely, correct. When $O_2$ consumption is plotted against running velocity for a lizard and a mammal of similar size (Figure 16-14), the plots for both exhibit rather similar slopes. Thus, once under way, increasing velocity results in similar increments in metabolic energy expenditure for both the mammal and the lizard. The difference between the two lies in the lower $y$-intercept of the lizard's plot relative to its SMR while at rest. The reason for the differences between the resting metabolic rates and the $y$-intercepts of the "running" plots (arrows) is not absolutely certain, but it has been suggested that these differences represent the "postural cost" of locomotion—for example, the energy required to move the limbs up and down. It appears that this cost is higher for a mammal than for a lizard.

## Temperature Classifications of Animals

The most fundamental temperature classification for animals is based on the source of body heat, endotherms generating it on their own and ectotherms depending almost entirely on their surroundings to supply it. As endotherms (all birds and mammals plus some lower vertebrates and a few insects) maintain

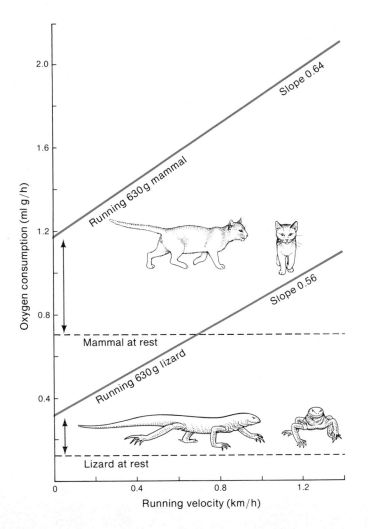

**16-14** Comparison of oxygen consumption in a mammal and a lizard of comparable size. Both show similar increments with increasing velocity (i.e., similar slopes). Extrapolated to zero velocity, the plot for the mammal exhibits a larger background (i.e., velocity-independent) increment (arrows) than the lizard above the resting metabolic rate. [Taylor, 1973.]

their body temperatures well above ambient in cold climates, they have access to habitats too cold for most ectotherms. Although most endotherms are well insulated with feathers or fur, they keep warm at considerable metabolic cost, the metabolic rate of an endotherm at rest being at least five times that of an ectotherm of equal size and body temperature.

Ectotherms have low rates of metabolic heat production and high *thermal conductances*—that is, they are poorly insulated. As a result, the heat derived from metabolic processes is quickly lost to the surroundings. On the other hand, the high thermal conductance allows ectotherms to absorb heat readily from their surroundings. Thus, body temperature in these animals adjusts largely passively to ambient temperatures. *Behavioral temperature regulation* is, for the most part, the only type of thermoregulation of which the true ectotherm is capable. Thus, a lizard or a snake may bask in the sun with its body oriented for maximal warming until it achieves a temperature suitable for efficient muscular function. In general, the most effective thermoregulatory actions taken by ectotherms is the selection of a suitable *microclimate* in the environment—one that is close to the *optimal body temperature*.

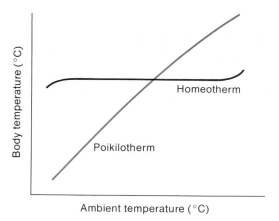

**16-15** Generalized relations between body temperature and ambient temperatures in homeotherms and poikilotherms.

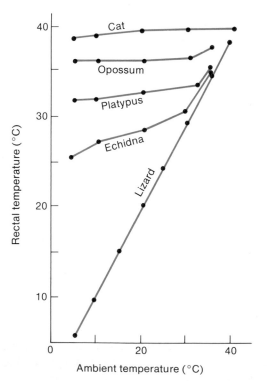

**16-16** Relations between body temperature and ambient temperatures in different types of mammals and a lizard. The marsupials and monotremes exhibit temporal heterothermy behavior. [Marshall and Hughes, 1980.]

Another way of classifying animals is based on whether they regulate their body temperatures within narrow limits independently of environmental temperatures or whether they allow it to fluctuate more or less passively with that of the environment. These two extremes are illustrated in Figure 16-15. Thus, *homeotherms* (or *homoiotherms*) are those animals that regulate their temperatures close to a set-point value by controlling heat production and heat loss. In mammals, the set point for core body temperature is typically 37°–38°C, while in birds it is closer to 40°C. *Poikilotherms* (from the Greek root for "variable") are those animals in which body temperature fluctuates more or less with the ambient temperature. This grouping includes essentially all lower vertebrates and all invertebrates. As we will see, intermediate conditions exist in some vertebrates between homeothermy and poikilothermy (Figure 16-16). Note that all homeotherms are endotherms as well.

The colloquial terms *warm-blooded* for homeotherms and *cold-blooded* for poikilotherms are unsatisfactory, since the blood of some poikilotherms can become quite warm under certain conditions. For example, a locust sustaining flight in the equatorial sun or a lizard exercising at midday in a hot desert may have blood temperatures exceeding those of "hot-blooded" mammals.

*Heterotherms* are those animals generally capable of varying degrees of endothermic heat production, but they generally do not regulate body temperature within a narrow range. They may be divided into two groups, regional and temporal heterotherms.

*Temporal heterotherms* constitute a broad category of animals whose temperatures vary widely over time. Examples are many flying insects, pythons and mackerels, all of which raise their body temperatures well above ambient temperature by virtue of heat generated as a by-product of intense muscular activity. Some insects prepare for flight by exercising their flight muscles for a time to raise their temperatures before takeoff. Monotremes such as *Echidna* are temporal heterotherms.

Certain small mammals and birds, while possessing accurate temperature control mechanisms and thus being basically homeothermic, behave as temporal heterotherms because they allow their body temperatures to undergo daily cyclical fluctuations, having endothermic temperatures during periods of activity and lower temperatures during periods of rest. In hot environments, this flexibility gives certain large animals such as camels the ability to absorb great quantities of heat during the day and to give it off again during the cooler night. Certain tiny endotherms such as hummingbirds must eat frequently to support their high daytime metabolic rate. To avoid running out of energy stores at night when they cannot feed, they enter into a sleeplike state of *torpor* in which they allow the body temperature to drop toward ambient.

*Regional heterotherms* are poikilotherms, such as certain large teleosts, that can achieve high *core* (i.e., deep-tissue) *temperatures* through muscular activity, while their peripheral tissues and extremities approach the ambient temperature. Examples include mako sharks, tuna, and many flying insects.

### GEOGRAPHIC FACTORS

Endothermy and ectothermy offer animals differing advantages in different climates. In the tropics, ectotherms such as reptiles compete successfully with, or even outcompete, mammals both in the abundance of species and in numbers of individuals. This success is thought to be due in part to the greater energy economy enjoyed by the ectotherm, since it need not expend energy to elevate its body temperature. The energy saved by tropical ectotherms in this way can be diverted to reproduction and to other uses that promote species survival. In moderate and cold climates, ectotherms are necessarily more sluggish, are thus less competent as predators, and are generally less successful than mammals are in those climates. Endotherms have a significant competitive edge over ectotherms in the cold because their tissues are kept warm. The farther terrestrial animals are from the equator, the higher the prevalence of endothermy. In polar regions, for example, there are no reptiles or insects, although a few genera of amphibians and insects may occupy subpolar arctic environments.

Endotherms in moderate and cold climates do, however, lose heat to the environment. As the ambient temperature drops, the rate of heat loss increases. Various measures have evolved to minimize this heat loss, as will be discussed later.

## Effects of Temperature on Animals

### TEMPERATURE-DEPENDENCE OF METABOLIC RATE

Enzymatic reaction rates are highly temperature-dependent. Therefore, tissue metabolism and ultimately the life of an organism depend on maintenance of the internal environment at suitable temperatures. The temperature-dependence of a reaction is described by the *Arrhenius equation*:

$$k = Ae^{-E_q/RT} \qquad (16\text{-}7)$$

in which $k$ is the velocity constant of the reaction, $A$ is a constant related to collision frequency of molecules, $E_q$ is the activation energy, and $e$ is the base of natural log, 2.72. When considering the influence of temperature on the rate of a reaction, it is useful to obtain a *temperature quotient* by comparing the rate of two different temperatures. A temperature difference of $10°C$ has become a standard, if arbitrary, span over which to determine the temperature sensitivity of a biological function. The so-called $Q_{10}$ is calculated by using the *Van't Hoff equation*:

$$Q_{10} = \frac{k_2^{10/(t_2-t_1)}}{k_1} \qquad (16\text{-}8)$$

where $k_1$ and $k_2$ are velocity constants at temperatures $t_1$ and $t_2$, respectively. For our present discussion, the following form of the Van't Hoff equation is more appropriate:

$$Q_{10} = \frac{M_2^{10/(t_2-t_1)}}{M_1} \qquad (16\text{-}9)$$

in which $M_1$ and $M_2$ are the metabolic rates at temperature $t_1$ and $t_2$, respectively. For temperature intervals of precisely $10°C$, the following form of the equation is simpler:

$$Q_{10} = \frac{M_{(t+10)}}{M_t} \qquad (16\text{-}10)$$

where $M_t$ is the metabolic rate at the lower temperature, and $M_{(t+10)}$ is the metabolic rate at the higher temperature.

It is important to note that $Q_{10}$ has no theoretical basis but instead is an entirely empirical value. Moreover, for a given reaction, the $Q_{10}$ differs over different

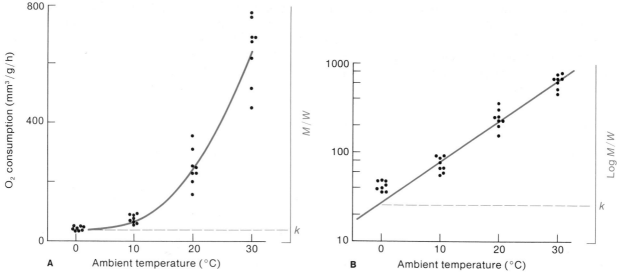

**16-17**  Oxygen consumption of an ectotherm, a tiger moth caterpillar, at different temperatures. **(A)** Rectangular coordinates. **(B)** Semilog coordinates. The generalized ordinates are drawn in color at the right in reference to Equations 16-11 and 16-12. The constant $k$ is obtained by extrapolating the metabolic rate to a body temperature of zero and is the proportionality factor in Equations 16-11 and 16-12. [Scholander et al., 1953.]

temperature ranges, so it is important when citing a $Q_{10}$ value that the range of temperatures (i.e., $t_1$ and $t_2$) for which it was determined be indicated. In general, chemical reactions have $Q_{10}$ values of about 2 to 3, while purely physical processes, such as diffusion, have lower temperature sensitivities (i.e., closer to 1.0).

Aquatic environments normally have a range of temperatures similar to that in which proteins remain stable and in which enzymatic reactions can proceed optimally ($-2°$ to $+40°C$). So ambient temperature is not generally a serious problem for aquatic ectotherms. Terrestrial ectotherms, on the other hand, must cope with temperatures that can be as extreme as $-65°C$ or $+70°C$.

The general relationship between metabolic intensity and body temperature in ectotherms, illustrated in the tiger moth caterpillar in Figure 16-17A, is described by the equation

$$\frac{M}{W} = k10^{b_1 t} \qquad (16\text{-}11)$$

where $M/W$ is the metabolic intensity (kilocalories

per kilogram per hour), $k$ and $b_1$ are constants, and $t$ is temperature.

Again, it is useful to transform the relation into a logarithmic one so as to produce a linear plot, from which it is easier to extrapolate. Thus, Equation 16-11 becomes

$$\log \frac{M}{W} = \log k + b_1 t \qquad (16\text{-}12)$$

In this case, coefficient $b_1$ gives the slope of the line, namely, the rate of increase in $\log M/W$ per degree. The $Q_{10}$ of $M/W$ can be determined by multiplying $b_1$ by 10 and taking the antilog. The data of Figure 16-17A are replotted on semilog coordinates in Figure 16-17B to conform with Equation 16-12.

Although metabolic rates in most ectotherms are at the mercy of variable body temperature, being two to three times higher for every $10°C$ increase in ambient temperature, the metabolic rates of some ectotherms exhibit a remarkable temperature independence. Thus, some intertidal invertebrates that experience large swings in ambient temperature with the ebb and flow of the tides have metabolic rates with a $Q_{10}$ of about 1.0, so that the rate shows no change over ranges of temperature as great as 20°C. These animals appear to possess enzyme systems with extremely broad temperature optima. Such systems may result from a spread in the temperature optima of sequential enzymes in a reaction, so that a drop in the rate of one step in the reaction chain "compensates" for an increased rate of other steps in the chain.

## THERMAL ACCLIMATION

In many species, naturally occurring stress from environmental heat or cold elicits compensatory changes in physiology or morphology that help the individual cope with the stress. For example, an ectotherm that cannot escape the winter cold, such as a pond-dwelling teleost fish, will gradually undergo, in the course of several weeks, a constellation of compensatory biochemical adaptations to low temperature. The overall change that the animal undergoes in the natural setting is termed *acclimatization.* We will confine ourselves here to a more restricted concept, *acclimation,* which refers to the specific physiological change(s) developed over time in the laboratory in response to a specific stress.

An example of *thermal acclimation* is illustrated in Figure 16-18. Specimens of the grass frog *Rana pipiens*

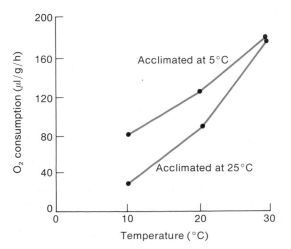

**16-18** Oxygen consumption at different temperatures of frogs acclimated at 5°C and at 25°C. [Rieck et al., 1960.]

were acclimated for several weeks to temperatures of 5°C (abnormally low) and 25°C (somewhat above their normal temperature). The rates of oxygen consumption of both groups were then measured at 10°, 20°, and 30°C. The frogs acclimated at low temperature exhibited a higher rate of $O_2$ consumption at temperatures intermediate between those to which the two groups were acclimated. Thus, a compensatory change must have taken place in one or both groups to cause them to metabolize at different rates at the same test temperatures. Such compensatory changes are adaptively useful. For example, the enhanced metabolic rate at the lower temperature will compensate, at least in part, for the normal reduction in reaction rates of enzymatic function that accompanies reduced temperature.

Acclimation has also been examined in individual tissues. For instance, muscles of winter frogs and summer frogs were found to differ at a given test temperature in their contractile properties. Similarly, nerve conduction persists at low temperatures in cold-acclimated fishes, but it is blocked at these same temperatures in warm-acclimated ones.

How are such changes brought about by exposure to different temperatures? It is reasonable to suppose that enzymatic reactions have been affected. In Figure 16-18, the plots of respiration against temperature for frogs acclimated to 5°C and 25°C show differing slopes. That is, the net respiratory processes in the two acclimation groups exhibit different $Q_{10}$'s. This suggests that there has been a modification in the temperature sensitivity of enzymatic activity. A change in enzymatic behavior can indicate a change in either the molecular structure of one or more enzymes, or in some other factor that influences the behavior.

More direct evidence for quantitative changes in enzymes during temperature acclimation has been obtained by isolating certain enzymes from animals acclimated at different temperatures and then comparing their kinetics. Thus, acetylcholinesterase extracted from trout brain exhibits a $K_M$ temperature curve (Figure 16-19) that depends on the temperature of acclimation. The enzyme isolated from the trout experiencing the lower acclimation temperature exhibits a curve displaced to lower temperatures than that of the high-temperature trout. Clearly, there must be a difference

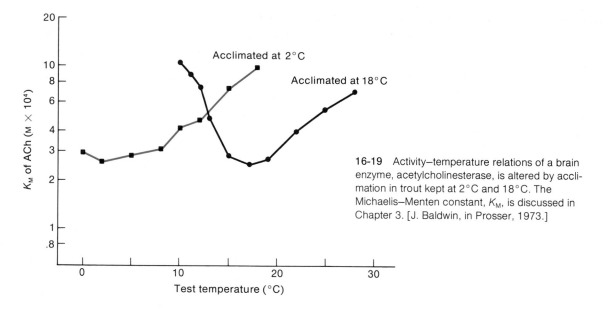

16-19   Activity–temperature relations of a brain enzyme, acetylcholinesterase, is altered by acclimation in trout kept at 2°C and 18°C. The Michaelis–Menten constant, $K_M$, is discussed in Chapter 3. [J. Baldwin, in Prosser, 1973.]

at the molecular level between the two enzyme samples. Such changes have been ascribed to changes in the proportions of two or more *isozymes,* slightly different forms of the same basic molecule, all encoded in the genome and all exhibiting more or less the same enzyme properties. At different body temperatures, one or another of the isozymes predominates. Since the different isozymes of an enzyme exhibit different temperature optima, the change in their proportions causes a shift in the overall temperature optimum of the reaction catalyzed by the enzyme.

In some instances of acclimation, however, thermal compensation appears to result simply from a change in the quantity of an enzyme. This is indicated in experiments in which the plot relating a metabolic function to the test temperature exhibits displacement without a change in slope (Figure 16-20). Since the $Q_{10}$ of the process remains unchanged, but the activity is higher at every temperature in the cold-acclimated group, the acclimation appears to have led to an increase in the number of enzyme molecules without any change in the kinetics of the enzymes involved.

## Determinants of Body Heat and Temperature

The temperature of an animal, endotherm or ectotherm, depends on the amount of heat (calories) per unit mass of tissue (Box 16-1). Since tissues consist primarily of water, the *heat capacity* of tissues between 0°C and +40°C approximates 1.0 cal/1°C/g. It follows that the larger the animal, the greater its body heat at a given temperature. The rate of change of body heat depends on: (1) the rate of *heat production*

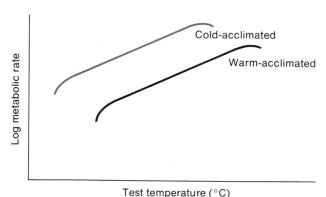

16-20   Generalized plot of log metabolic rate against test temperature in a cold-acclimated individual and in a warm-acclimated one. The similarity of slope in the two plots indicates identical $Q_{10}$'s.

through metabolic means, (2) the rate of *heat gain,* or (3) the rate of *heat loss* to the environment (Figure 16-21). We can state that

$$\text{body heat} = \text{heat produced} + \text{gained} - \text{heat lost}$$
$$= \text{heat produced} + \text{heat transfer}$$

Thus, body heat, and hence the body temperature of an animal, can be regulated by changes in the rate of heat production and heat exchange (i.e., heat gained minus heat lost), as outlined by Bligh (1973).

### Heat Production

The processes that influence the rate of body heat production can be classified as follows: (1) *behavioral mechanisms* such as simple exercise, (2) *autonomic mechanisms* such as accelerated metabolism of energy reserves, and (3) *adaptive mechanisms* or acclimatization, which are slower than the other two processes, producing an elevation in basal metabolism.

### Heat Transfer

The rate of heat transfer (kilocalories per hour) into or out of an animal depends on three factors:

*1. Surface area.* As we have noted, surface area per gram of tissue decreases with increases in body mass, providing small animals with a high heat flux per unit of body weight.

*2. Temperature difference,* $T_a - T_b$. The closer an animal maintains its temperature to the ambient temperature, the less heat will flow into or out of its body.

*3. Weight-specific heat conductance* of the animal's surface. Surface tissues of poikilotherms have high heat conductances, and therefore these animals are typically at the same temperature as their surroundings. Exceptions occur, such as the elevation of body temperature during basking in sunlight. Homeotherms have evolved feathers, fur, or blubber to decrease the

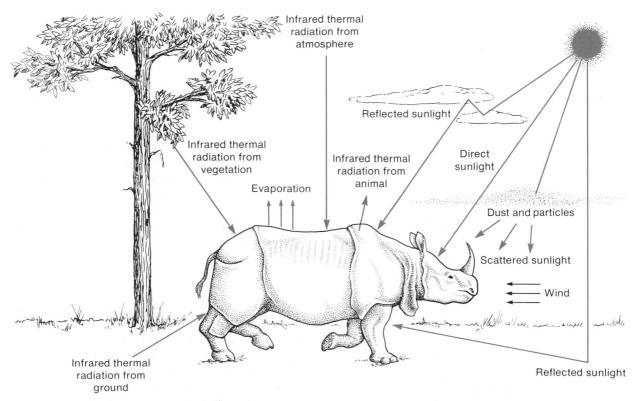

**16-21**   Channels of heat transfer between an animal and its environment. See Box 16-2 for discussion. [Porter and Gates, 1969.]

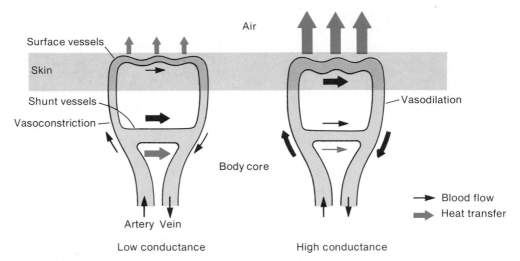

**16-22**    Role of blood flow to the skin in regulating the heat conductance of the body surface. Vasomotor control of peripheral arterioles shunts the arterial blood either to or away from the skin. In response to environmental cold, blood is shunted away from the surface of an endotherm; in response to high temperatures, the blood is diverted to the skin, where it approaches temperature equilibrium with the environment. In ectotherms, cutaneous blood flow is often increased in order to absorb heat from the environment.

heat conductance of their body surfaces. One effect of such insulation is to spread out the temperature difference over several millimeters or centimeters so that the *temperature gradient* (i.e., change in temperature per unit of distance) is less steep. Another feature of fur and feathers is that they trap and hold air, which has a very low thermal conductivity and therefore retards the transfer of heat. The rate of heat transfer is proportional to each of these factors. Increasing the value of any one of them enhances heat flow in the direction of the temperature gradient.

Among the mechanisms used by animals to regulate the exchange of heat between themselves and the environment are the following:

*1. Behavioral* control includes moving to an environment in which the temperature is closer to the optimum. For instance, a desert ground squirrel retires to its burrow during the midday heat; a lizard suns itself to warm up. Animals also control the amount of surface area available for heat exchange by adjusting their postures.

*2. Autonomic* control of blood flow to the vertebrate skin influences the temperature gradient, and hence the heat flux, at the body surface (Figure 16-22). Activation of piloerector muscles determines the extent of fluffing of pelage and plumage, which determines the effectiveness of insulation. Sweating and salivation during panting cause evaporative cooling.

*3. Adaptive* control includes long-term changes in pelage or subdermal fatty layer insulation, probably hormonally determined, and changes in the capacity for autonomic control of evaporative heat loss through sweating.

## Temperature Relations of Ectotherms

### ECTOTHERMS IN COLD ENVIRONMENTS

As long as they remain well submerged in the sea, most marine invertebrates will remain above the freezing point of their body fluids. On the other hand, many invertebrates that live in less moderate areas of subpolar regions survive subfreezing temperatures. The muscles and organs of these animals may become distorted on freezing as ice crystals form in their tissues. The formation of ice crystals within cells is usually lethal, for as the crystals grow in size, they rupture and destroy the cells. This problem is minimized in some animals (e.g., certain beetles) that can withstand freezing temperature because the extracellular fluid contains a substance that accelerates nucleation (the process of crystal formation). Consequently, the extracellular fluids freeze more readily than the intracellular fluids. As water leaves the extracellular fluid to form ice crystals, the fluid becomes more concentrated. This process draws water out of the cells, lowering the intracellular freezing point. As the temperature drops further, the process continues and produces further depression of the intracellular freezing point. The adaptiveness of this process lies in the failure of crystals in the extracellular fluid to do much tissue damage. The freshwater larvae of the midge *Chironomus,* which survive repeated freezing, will yield some unfrozen liquid at temperatures as low as $-32°C$. No animal has yet been proven to withstand the total freezing of all its cell water.

Some animals can undergo "supercooling," wherein the body fluids can be cooled to below their freezing temperature, yet remain unfrozen because of the absence of ice crystals. Crystals will fail to form as long as they have no nuclei (mechanical "seeds," so to speak) to initiate crystal formation. Thus, certain fishes dwelling at the bottom of arctic fjords live in a continually supercooled state and will not freeze unless they come into contact with ice nuclei, whereupon ice crystallizes rapidly throughout their bodies, so that they die almost immediately. To survive, they must remain well below the surface, where ice is absent.

The body fluids of some cold-climate ectotherms contain antifreeze substances. The parasitic wasp *Brachon cephi* contains glycerol, the concentration of which increases during the winter. The glycerol, acting as an antifreeze solute, lowers the freezing point to as low as $-17°C$. The tissues of *Brachon* larvae can withstand even lower temperatures; they have been supercooled to $-47°C$ without ice crystal formation. The blood of the antarctic ice fish *Trematomus* contains a glycoprotein antifreeze that is 200 to 500 times more effective than an equivalent concentration of sodium chloride in preventing formation of ice. The glycoprotein lowers the temperature at which the ice crystals enlarge, but it does not lower the temperature at which they melt. Although this glycoprotein has been isolated and its chemical structure determined, the mechanism by which it retards crystal formation is not fully understood.

### ENDOTHERMY IN HETEROTHERMS

The ectotherms include species that can also be classified as heterotherms. The best-studied examples of heterothermy are found among certain insects. These can be considered as both temporal and regional heterotherms because, at times when they prepare to fly, they will raise the core temperatures in their thoracic parts to more or less regulated levels, while at other times they behave as strict ectotherms. At moderate ambient temperatures, these insects are unable to take off and fly without prior warmup because their flight muscles contract too slowly to produce sufficient power for flight at temperatures much below $40°C$. Once the insect is aloft, its flight muscles produce enough heat to maintain adequately elevated muscle temperatures, and the insect even employs heat-dissipating mechanisms to prevent overheating. These insects, which include locusts, beetles, cicadas, and arctic flies, generally have relatively large mass; and some, like bumblebees, butterflies, and moths, are covered with heat-insulating "hairs" or scales. To warm up, these insects activate their large thoracic flight muscles, which are among the metabolically most active tissues known. The activation is such that antagonistic muscles work against one another, producing heat without much wing movement other than small, rapid vibrations akin to shivering. Flight is finally initiated when the thoracic temperature has reached the temperature that is maintained during flight, about $40°C$ (Figure 16-23). At ambient temperatures approaching $0°C$, convective heat loss is so rapid that flight temperatures cannot be maintained. High ambient temperatures, on the other hand, place the insect in danger of overheating. At ambient temperatures above $20°C$, the hovering sphinx moth *Manduca sexta* prevents thoracic overheating by regulating the flow of warm blood to the abdomen (Figure 16-24). The flow of heat from the active thorax to the relatively inactive and poorly insulated abdomen increases the loss of heat to the environment through the body surface and through the tracheal system.

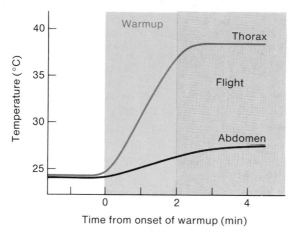

**16-23** Preflight thermogenesis in the sphinx moth *Manduca sexta*. Shivering of the thoracic flight muscles causes a steep increase in thoracic temperature prior to flight. [Heinrich, 1974.]

Small and medium-sized teleosts are typically cold-bodied (i.e., strictly ectothermic). Large fishes such as the tuna have enough mass to generate and retain sufficient heat through muscular activity to raise the core temperature 10°C or more above their surroundings; therefore they are regional heterotherms. The large mass (and hence small surface/volume ratio) of these fishes helps retain endothermic heat. In warm-bodied fishes, the retention of heat in the body core is further enhanced by the organization of the vascular system. Unlike cold-bodied fishes, which have a centrally located aorta and postcardinal vein, heterothermic fishes, such as the tuna and mako shark, have these major vessels located under the skin (Figure 16-25). Blood is delivered to the deep, red muscles via a rete that acts as a countercurrent heat-exchange system (Figure 16-26). Arterial blood, cooled during passage

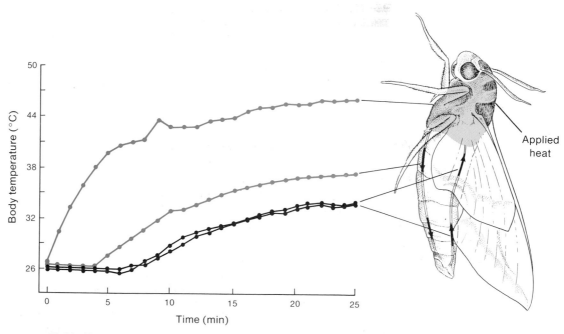

**16-24** Temperature regulation in the immobilized sphinx moth during 25 min of external heat application to the thorax (colored shading). The thoracic temperature rises steeply with onset of heating. When the temperature reaches about 39°C, blood flow from the thorax into the abdomen begins to heat the latter, while cool abdominal blood flows in the dorsal vessel from the abdomen into the thorax, as reflected in a lowered rate of heating in the thorax. [Heinrich, 1974.]

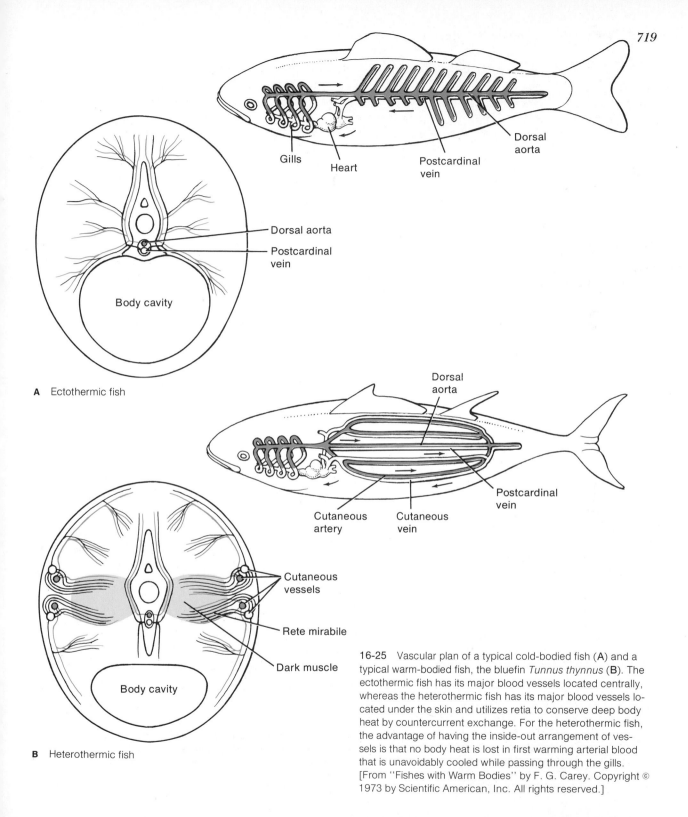

**A** Ectothermic fish

Gills

Heart

Postcardinal vein

Dorsal aorta

Dorsal aorta

Postcardinal vein

Body cavity

**B** Heterothermic fish

Dorsal aorta

Cutaneous artery

Cutaneous vein

Postcardinal vein

Cutaneous vessels

Rete mirabile

Dark muscle

Body cavity

**16-25** Vascular plan of a typical cold-bodied fish (A) and a typical warm-bodied fish, the bluefin *Tunnus thynnus* (B). The ectothermic fish has its major blood vessels located centrally, whereas the heterothermic fish has its major blood vessels located under the skin and utilizes retia to conserve deep body heat by countercurrent exchange. For the heterothermic fish, the advantage of having the inside-out arrangement of vessels is that no body heat is lost in first warming arterial blood that is unavoidably cooled while passing through the gills. [From ''Fishes with Warm Bodies'' by F. G. Carey. Copyright © 1973 by Scientific American, Inc. All rights reserved.]

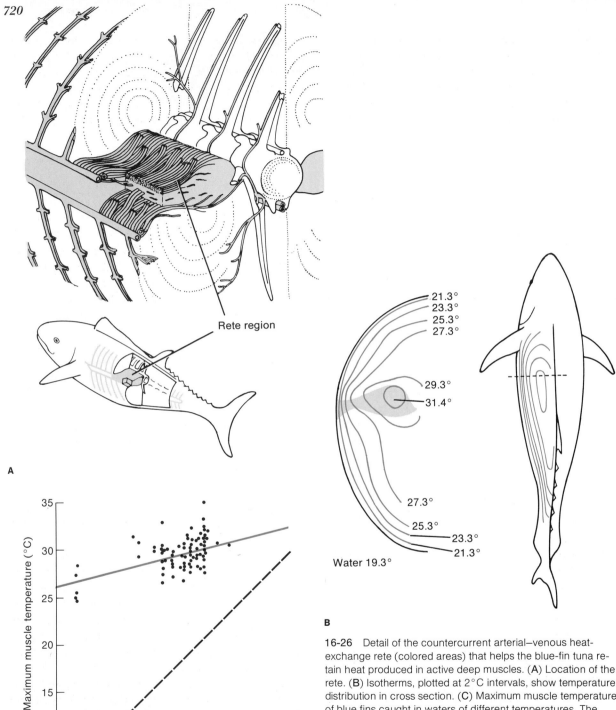

Rete region

Water 19.3°

**16-26** Detail of the countercurrent arterial–venous heat-exchange rete (colored areas) that helps the blue-fin tuna retain heat produced in active deep muscles. (**A**) Location of the rete. (**B**) Isotherms, plotted at 2°C intervals, show temperature distribution in cross section. (**C**) Maximum muscle temperatures of blue fins caught in waters of different temperatures. The broken line indicates equality between body temperatures and water temperatures. [Carey and Teal, 1966.]

through the respiratory tissues of the gills and through the surface vessels, passes from the cool periphery into the warm deeper muscle tissue through a network of fine arteries (rete) that intermingle with small veins carrying warm blood away from the muscles. This process causes the cool arterial blood to pick up heat from the warm venous blood leaving the muscle tissue, retaining the heat in the deep, red muscle tissue. In this way, heterothermic fishes maintain their swimming muscles at a temperature suitable for vigorous muscular activity, while the temperature of the surface tissues approaches that of the surrounding water.

### ECTOTHERMS IN HOT ENVIRONMENTS

Since heat exchange with the environment is closely related to body surface area, the temperature of a small ectotherm will rise and fall rapidly as environmental temperatures undergo diurnal fluctuations. Although some reptiles can tolerate a broad range of body temperatures, for most ectotherms tissue functions are handicapped by a decreased affinity of hemoglobin for oxygen at the upper limit of body temperature. At 50°C, the blood of a chuckwalla *(Sauromalus)* cannot bind more $O_2$ than 50% saturation at atmospheric pressure and thus prevents vigorous activity by the animal. At slightly lower temperatures (47°–48°C), the desert iguana *Dipsosaurus* continues to be active. Above 43°C, the iguana pants, much as a dog does, to minimize further heating.

A common reptilian tactic is to expose the body to sun or to shade in order to absorb more or less heat, respectively, from the environment. The effectiveness of such behavioral thermoregulation is enhanced by the high heat conductance of ectotherms. Certain reptiles, however, use nonbehavioral means to control the rates at which their bodies heat and cool. For example, the diving Galápagos marine iguana *Amblyrhynchus* (Figure 16-27) can permit its body temperature to rise at about twice the rate at which it drops. It does so by regulating both heart rate and flow of blood to its surface tissues. By diverting blood to the surface when basking in the sun to warm up, the iguana increases its heat conductance and speeds its absorption of heat. Increased pumping of blood accelerates the removal of heat from surface tissues to deeper tissues. During the iguana's prolonged feeding dives in the cool ocean, loss of body heat is slowed by a reduction in blood flow to the surface tissues and by a general slowing of the circulation. This is apparent in experiments that demonstrate a hysteresis (asymmetrical behavior) of the heart rate relative to body temperature during a rise and fall in body temperature (Figure 16-27C).

## Temperature Relations of Endotherms

In homeothermic endotherms (most mammals and birds), body temperature is closely regulated by exquisite homeostatic mechanisms that alter rates of heat production and heat loss so as to maintain a relatively constant body temperature. Core temperatures are maintained nearly constant at 37°–40°C in mammals and 37°–41°C in birds. The temperatures of peripheral tissues and extremities are held less constant and are sometimes allowed to approach environmental temperatures. Basal heat production for different homeotherms of a given size is about the same (Figures 16-6 and 16-7), and this BMR is 3 to 10 times as high as the SMR of ectotherms of comparable size measured at similar body temperatures. It is this elevated basal metabolism, in conjunction with heat-conserving and heat-dissipating mechanisms, that allows homeotherms to maintain a constant body temperature.

### THERMAL NEUTRAL ZONE

The degree of thermoregulatory activity that homeotherms require to maintain a constant core temperature increases with increasing extremes of environmental temperature. Minimum regulatory effort is needed when the environmental temperature is within what is termed the *thermal neutral zone* for an animal. Within this temperature zone, the animal compensates for differences in ambient temperature entirely through minor alterations of the thermal conductance of the body surfaces and thereby controls the rate at which heat is lost to the environment. These adjustments include such responses as vasomotor changes (Figure 16-22), postural changes to alter exposed areas of surface, and pilomotor regulation of the insulating effectiveness of the pelage. Thus, within this range, fur or feathers are fluffed by pilomotor muscles in the skin to provide a thicker layer of stagnant air, and at the upper end of this range, fur or feathers are applied

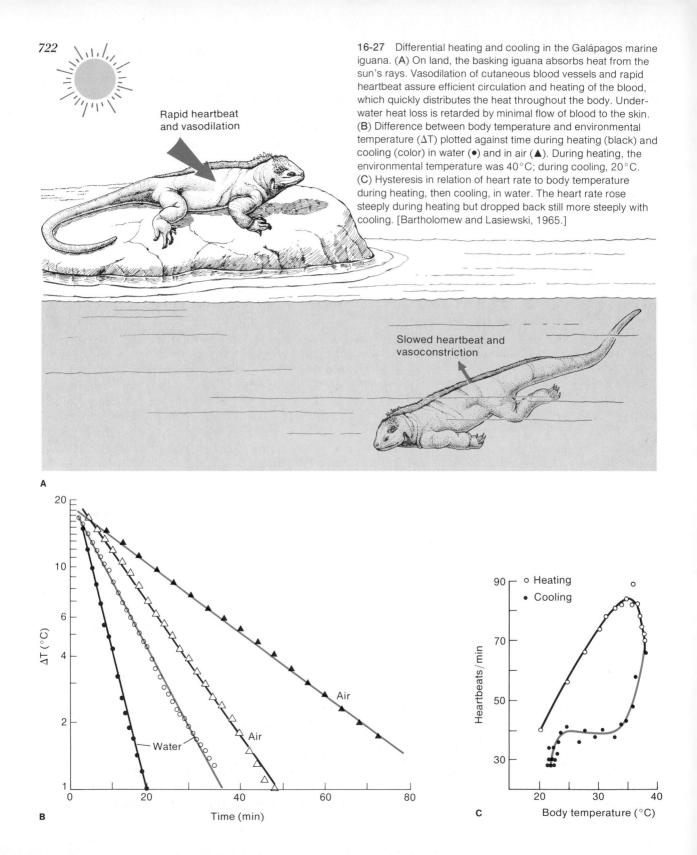

**16-27** Differential heating and cooling in the Galápagos marine iguana. (A) On land, the basking iguana absorbs heat from the sun's rays. Vasodilation of cutaneous blood vessels and rapid heartbeat assure efficient circulation and heating of the blood, which quickly distributes the heat throughout the body. Underwater heat loss is retarded by minimal flow of blood to the skin. (B) Difference between body temperature and environmental temperature ($\Delta T$) plotted against time during heating (black) and cooling (color) in water (•) and in air (▲). During heating, the environmental temperature was 40°C; during cooling, 20°C. (C) Hysteresis in relation of heart rate to body temperature during heating, then cooling, in water. The heart rate rose steeply during heating but dropped back still more steeply with cooling. [Bartholomew and Lasiewski, 1965.]

Rapid heartbeat and vasodilation

Slowed heartbeat and vasoconstriction

**A**

**B**

$\Delta T$ (°C)

Water

Air

Air

Time (min)

**C**

○ Heating
● Cooling

Heartbeats / min

Body temperature (°C)

closer to the skin. Humans exhibit "goose bumps" as a vestige of the pilomotor control of a long-lost pelage.

The lowest ambient temperature at which body temperature can be regulated without increased metabolic heat production is termed the *lower critical temperature (LCT)*. When environmental temperatures drop below the LCT, homeotherms thermoregulate by increased metabolic activity (i.e., thermogenesis, described in the next subsection). Heat production then rises in a linear manner with continuing drop in temperature (Figure 16-28). This is the *zone of metabolic regulation*. If the environmental temperature drops below the zone of metabolic regulation, compensating mechanisms fail, the body cools, and the metabolic rate drops. The animal now enters the state of *hypothermia*. If this condition persists, the animal grows progressively cooler; and since cooling only further lowers the metabolic rate, the animal soon dies.

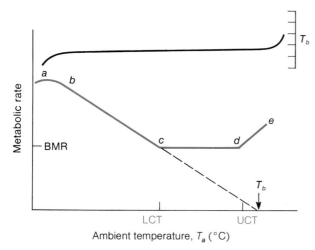

**16-28** Metabolic rate (colored plot) and body temperature (black plot) of a homeotherm at different ambient temperatures. The thermal neutral zone extends from the lower to the upper critical temperature, points *c* and *d*, respectively. Above and below this range, the metabolic rate must rise to either increase thermogenesis (segment *b–c*) or increase active dissipation of heat by evaporative cooling (segment *d–e*) if body temperature is to remain constant. Within the thermal neutral zone, body temperature is regulated entirely by changing the heat conductance of the body surface, which requires essentially no change in metabolic effort. At ambient temperatures below point *b*, thermogenesis is unable to replace body heat at the rate it is lost to the environment.

Note that in the plot in Figure 16-28, the thermal neutral zone lies entirely below the normal body temperature, $T_b$ (37°–40°C). Why is this? Consider that, at the *upper critical temperature (UCT)*, heat loss by passive mechanisms cannot be increased further because the surface insulation is minimal (i.e., it cannot be made lower) at that temperature. Any further increase in ambient temperature, $T_a$, above that temperature will therefore cause a rise in body temperature, unless active heat-dissipating mechanisms such as sweating or panting are brought into play. Without evaporative heat loss, temperatures above the thermal neutral zone lead to *hyperthermia*, because the heat produced by basal metabolism cannot escape passively from the body as fast as it is being produced. Hot-tub enthusiasts would be wise to bear this in mind.

Getting back to Figure 16-28, we might ask why the metabolic rate should rise linearly with temperature below the LCT, along a line that extrapolates to zero at an ambient temperature equal to body temperature. This is explained by considering *Fourier's law of heat flow*:

$$\dot{Q} = C(T_b - T_a) \qquad (16\text{-}13)$$

in which $\dot{Q}$ is the rate of heat loss from the body (in calories per minute), and $C$ is the thermal conductance (Box 16-2). As $T_b$ is constant, $\dot{Q}$ varies linearly with the ambient temperature. $C$ determines the slope of the plot below the neutral zone; the better the insulation (i.e., the lower $C$ is), the shallower the slope, and the less heat must be produced metabolically at low temperatures.

The extrapolated intercept with zero occurs at $T_b$, because if $T_a = T_b$, $C(T_b - T_a) = 0$. With $\dot{Q} = 0$, there will be no net heat loss. We know, of course, that the metabolic rate does not normally drop below the BMR. When $T_a = T_b$, the body temperature must be above the neutral zone, and thus the animal must cool by some means other than heat conduction. As we have noted, the only means of cooling when $T_a$ lies above the UCT is by evaporation.

### THERMOGENESIS

When the ambient temperature drops below the LCT, the endothermic animal responds by generating additional heat from energy stores, thereby preventing

## Box 16-2   *Physical Concepts of Heat and Temperature*

The total heat content of an animal is determined by the metabolic production of heat and the *thermal flux* between the animal and its terrestrial surroundings, as shown in Figure 16-21. This can be represented as follows (Schmidt-Nielsen, 1975):

$$H_{\text{tot}} = H_m + H_c + H_r + H_e + H_s$$

in which $H_{\text{tot}}$ is the total heat, $H_m$ the heat produced metabolically, $H_c$ heat lost or gained by conduction and convection, $H_r$ the net heat transfer by radiation, $H_e$ the heat lost by evaporation, and $H_s$ the heat stored in the body.

Heat leaving the animal has a negative ($-$) value, whereas heat entering the body from the environment has a positive ($+$) value. The key terms are defined as follows:

*Conduction* is the transfer of heat between objects and substances that are in contact with each other. It results from the direct transfer of kinetic energy of the motion from molecule to molecule, with the net flow of energy being from the warmer to the cooler region. The rate of heat transfer through a plate conductor of uniform properties can be expressed as

$$\dot{Q} = kA \frac{t_2 - t_1}{l}$$

in which $\dot{Q}$ is the rate of heat transfer by conduction; $k$ is the *thermal conductivity* of the conductor; $A$ is the surface area; and $l$ is the distance between points 1 and 2 normal to the surface that are at temperatures $t_1$ and $t_2$, respectively.

*Convection* is the mass transfer of heat due to mass movement of a gas or liquid. It accelerates heat transfer by conduction between a solid and a fluid, since continuous replacement of the fluid (e.g., air, water, blood) in contact with the solid of different temperature maximizes the temperature difference between the two phases (i.e., $t_2 - t_1$) and hence the conductive transfer of heat between the solid and the fluid.

*Radiation* is the transfer of heat by electromagnetic radiation, and it takes place without direct contact between objects. All physical bodies at a temperature above absolute zero emit electromagnetic radiation in proportion to the fourth power of the absolute temperature of the surface. As an example of how radiation works, the sun's rays may warm a black body to a temperature well above the temperature of the air surrounding the body. A dark body radiates and absorbs more strongly than a more reflective body having a lower *emissivity*. For temperature differences between the surfaces of two bodies of about 20°C or less, the net radiant heat exchange is approximately proportional to the temperature difference.

*Evaporation.* Every liquid has its own *latent heat of vaporization,* the amount of energy required to change it from a liquid to a gas of the same temperature. The energy required to convert 1 g of water to water vapor is relatively high, approximately 585 cal. Many animals dissipate heat by allowing water to be evaporated from body surfaces.

*Heat storage* leads to an increase in temperature of the heat-storing mass. The larger the mass, or the higher its specific heat, the smaller its rise in temperature (°C) for a given quantity of heat (calories).

---

lowering of the core temperature. There are two primary means of extra heat production other than exercise: *shivering* and *nonshivering thermogenesis.* In both of these, a normal energy-converting metabolic mechanism is rendered highly inefficient, so essentially all the chemical-bond energy released in the process is fully degraded to heat rather than to a chemically or mechanically useful form of energy. In other words, thermogenesis involves the burning of energy stores with a purposeful inefficiency, causing all or most of the chemical energy to be converted to heat.

Shivering is a means of using muscle contraction to liberate heat. The nervous system activates motor units of groups of antagonistic skeletal muscles so that little net movement results other than shivering. The activation of muscle causes ATP to be hydrolyzed to provide energy for contraction. Since the muscle contractions are inefficiently timed and mutually opposed, they produce no useful physical work, but the chemical energy released during contraction appears as heat. Shivering thermogenesis is practiced by both insects and vertebrates.

In nonshivering thermogenesis, enzyme systems for the metabolism of fats are activated throughout the body, so that fats are broken down and oxidized to produce heat. Very little of the energy released is conserved in the form of newly synthesized ATP. A specialization found in some mammals for fat-fueled thermogenesis is *brown fat*. Generally deposited in the neck and between the shoulders (Figure 16-29), it is an adaptation for rapid, massive heat production. This fat

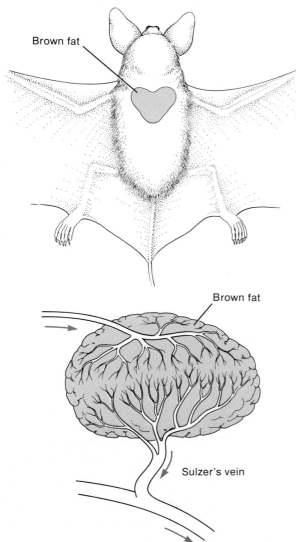

Brown fat

Brown fat

Sulzer's vein

16-29 Brown-fat deposits are found between the scapulae in bats and many other mammals. During brown-fat oxidation, this tissue is detectable as a warm region by its infrared emission.

contains such extensive vascularization and so many mitochondria that it is brown (owing largely to mitochondrial cytochrome oxidase) rather than white. Oxidation of brown fat takes place within the fat cells, which are richly endowed with fat-metabolizing enzyme systems, whereas ordinary fat must first be reduced to fatty acids, which enter the circulation and eventually are taken up by other tissues, where they are oxidized.

Thermogenesis in brown fat is activated by the sympathetic nervous system through the release of norepinephrine, which binds to receptors on the adipose cells of brown-fat tissue. Through a second-messenger mechanism, described in Chapter 11, this signal leads to thermogenesis by two mechanisms. In the first of these, normal ATP utilization for cellular processes rises in these fat cells in response to the sympathetic signal, accounting for part of increased heat production. Through processes such as ion pumping by the cell membrane, ATP is hydrolyzed to produce work and heat. In the second mechanism, ATP production is uncoupled during respiratory chain oxidation. ATP resynthesis from ADP and $P_i$ is normally coupled to movement of protons ($H^+$) down their electrochemical gradient from cytoplasm into mitochondria across the inner mitochondrial membrane. Thermogenesis in brown fat is characterized by the appearance of a separate pathway by which protons leak across this membrane without the energy of their downhill movement being harnessed for the phosphorylation of ADP to ATP. Once inside the mitochondrion, the protons oxidize substrate in the mitochondrion to produce water and heat or require further substrate oxidation and energy utilization to pump them out again (Figure 16-30).

During thermogenesis, brown fat heats up significantly, and this newly produced heat is rapidly dispersed to other parts of the body by blood flowing through the extensive vascularization of the brown-fat tissue. This intense form of thermogenesis is especially pronounced in hibernating or torpid mammals during arousal, when it supplements shivering to facilitate rapid warming. One consequence of acclimation to cold by mammals is an increase in brown-fat deposits, which makes for a gradual changeover from shivering to nonshivering thermogenesis at low ambient temperatures.

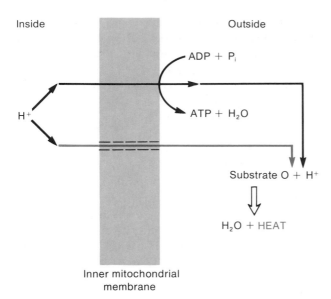

Inside

Outside

ADP + P$_i$

H$^+$

ATP + H$_2$O

Substrate O + H$^+$

H$_2$O + HEAT

Inner mitochondrial
membrane

**16-30**   Mechanism of thermogenesis during brown-fat metabolism. Protons, driven by their electrochemical gradient, enter the inner mitochondrial space across the inner membrane. Normally, the energy of this downhill transport drives the phosphorylation of ADP to form ATP. An alternate pathway (color) allows H$^+$ to enter without coupled ATP synthesis. Once inside the mitochondrion, the H$^+$ oxidizes substrate molecules to produce water. Heat is given off during the oxidation. [Horowitz, 1978.]

## ENDOTHERMY IN COLD ENVIRONMENTS

Endotherms adapted to cold environments have necessarily evolved a number of mechanisms, both temporary and permanent, that help them retain body heat. For example, an animal sensing heat loss in a windy place will fluff its pelage and move to a more sheltered area. This reduces convection, the dissipation of body heat to the wind.

More enduring responses to cold include the thick layers of insulation in many arctic animals, in the form of subcutaneous fat or a thicker pelage or plumage. The insulating effectiveness of pelages in arctic and subarctic animals changes adaptively with both season and latitude. In addition, animals living in the temperate zone exhibit seasonal variations by shedding old fur and growing new fur, thereby providing thick insulation during the winter, yet not producing undue heat stress during the summer.

The specific conductances of homeotherms vary over a large range and are related to body size (Figure 16-31). Smaller animals have higher specific conductances owing to their thinner coats of fur or feathers. In addition, they face greater heat loss in cold climates because of their relatively large surface areas. Thus, one adaptation of endotherms to cold latitudes is an increase in body size. According to *Bergmann's rule,* cold-climate forms of an endothermic animal type (e.g., foxes, coyotes, deer) tend to be larger than their warm-climate relatives. All else being equal, large endotherms also have thicker layers of pelage or plumage. As pelage becomes thicker, and conductance decreases, the LCT of a homeotherm decreases (Figure 16-32).

*Blubber,* a fatty tissue, is a good insulator because it, like air, has a lower thermal conductivity than water, which is the main constituent of nonfatty tissues. In addition, fatty tissues are metabolically very inactive and require little perfusion by blood, which would ordinarily carry heat to be lost at the body surface. Cetaceans have a thick layer of blubber, the outer side of which is nearly always at a temperature near that of the surrounding water.

An important means of controlling heat loss from the surface tissues of both vertebrate endotherms and ectotherms is the diversion of blood flow to or away from the

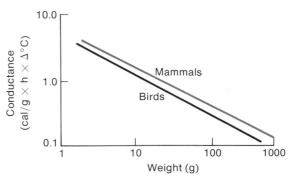

**16-31**   Thermal conductance as a function of body weight. [Herried and Kessel, 1967.]

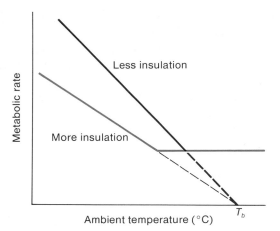

**16-32** Effect of thermal conductance on metabolic rate at different ambient temperatures. A decrease in insulation (i.e., an increase in conductance) raises the lower critical temperature and makes the slope of increasing metabolism steeper. The slope, however, still extrapolates to body temperature at zero metabolic rate.

skin (Figure 16-22). Vasoconstriction of arterioles leading to the skin keeps warm blood from perfusing cold skin and conserves the heat of the body core.

### COUNTERCURRENT HEAT EXCHANGE

To avoid mechanical hindrance during locomotion, the limbs of endotherms are better off without a massive layer of insulation. The flukes and flippers of cetaceans and seals and the legs of wading birds, arctic wolves, caribou, and other cold-weather homeotherms require blood to nourish cutaneous tissue and muscles used in locomotion. Such limbs are potential major avenues of body heat loss, for they are thin and have large surface areas.

Excessive heat loss from these appendages is drastically reduced by *countercurrent heat exchange*. Arterial blood, originating in the animal's core, is warm. Conversely, the venous blood returning from peripheral tissues may be very cold. As blood flows from the core, it enters arteries in the limb that lie next to veins that carry blood returning from the extremity. As the arteries and veins pass each other, the warm arterial blood

gives up heat to the returning venous blood and thus becomes successively cooler as it enters the extremity. By the time it reaches the periphery, the arterial blood is precooled to within a few degrees of the ambient temperature, and little heat is lost. Conversely, the venous blood returning is warmed by the arterial blood, so it is nearly up to core temperature as it flows into the core. A highly evolved example of this has been described by Schmidt-Nielsen (1973) in the flipper of the porpoise (Figure 16-33). Here, the artery carrying warm blood flowing toward the extremity is completely encased in a circlet of veins carrying cold blood back from the extremity. Birds and arctic land mammals also utilize countercurrent exchange to minimize heat loss from their extremities in cold climates. As a result, the extremities of cold-climate endotherms are maintained at temperatures far below the core temperature and often approaching the ambient temperature (Figure 16-34).

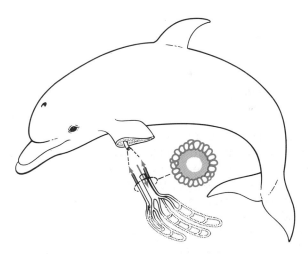

**16-33** Countercurrent system for conserving core heat in the porpoise flipper. The cross section shows the artery completely surrounded by small veins. Heat flows from the warm arterial blood to the cool venous blood. This process keeps the flipper cool and minimizes the difference between the temperature of the flipper and the temperature of the surrounding water. [Schmidt-Nielsen, 1975.]

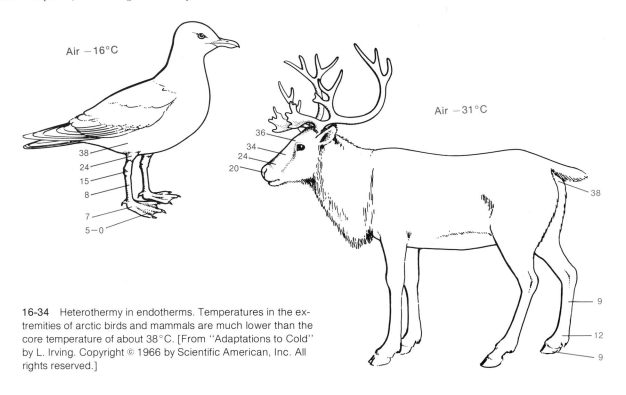

**16-34**   Heterothermy in endotherms. Temperatures in the extremities of arctic birds and mammals are much lower than the core temperature of about 38°C. [From "Adaptations to Cold" by L. Irving. Copyright © 1966 by Scientific American, Inc. All rights reserved.]

### LOW-TEMPERATURE LIPIDS

Because of the low ambient temperature and the presence of countercurrent heat exchange in the extremities of arctic and subarctic mammals, the tissues in their legs and feet must tolerate temperatures close to freezing. This situation could, in principle, create serious problems, because lipids become highly viscous as their temperatures fall below their melting points, and they therefore could, for example, alter the behavior of the lipid bilayer of the cell membrane. We can imagine the effect of temperature on lipid viscosity by recalling that room temperatures are below the melting point of a cooking grease like Crisco but are above the melting point of a cooking oil like Wesson oil. The difference between the oil and the grease lies in the degree of hydrogenation of the carbon backbone. The greater the proportion of unsaturated (i.e., double unhydrogenated) carbon–carbon bonds of a lipid's fatty acid molecules, the lower its melting point. At temperatures above the melting point, the lipid is an "oil," and below, it is a "fat."

A consequence of acclimatization of poikilotherms to cold or hot environments is that the membrane lipids exhibit an adaptive change, becoming more saturated during acclimatization to warmth and less saturated during acclimatization to cold.

There is an interesting correlation in mammals between the tissue temperature and the melting point of lipids. In the cool extremities of mammals, the tissue lipids are less saturated than the fats of the body core and thus exhibit lower melting points. At 37°C, they are much "oilier" than the waxier fats of the warmer body parts. Humans have taken advantage of these low-viscosity oils of mammalian limbs by extracting them from slaughtered cattle as neat's-foot oil, useful as a penetrating preservative and softener of leather.

### ENDOTHERMY IN HOT ENVIRONMENTS

In very hot and dry climates, large animals have the advantage of relatively low surface/mass ratios and large heat capacities. Camels, well known for their ability to tolerate heat, have, in addition to large mass, a thick pelage that helps insulate them from external heat. Low surface area and thick pelage retard the absorption of heat from the surroundings. Furthermore, because of its large mass and the high specific heat of tissue water, the camel, as well as other large mammals, can absorb relatively large quantities of heat for a given rise in body temperature. The dehydrated camel can also tolerate an elevation of its core temperature by several degrees, further increasing its heat-absorbing capacity. Large amounts of heat gradually accumulated during daytime hours are subsequently dissipated in the cool of the night. In preparation for the next onslaught of daytime heat, the dehydrated camel allows its core temperature to drop several degrees below normal during the night. Starting the day with a heat deficit allows the camel to absorb additional heat during the hot part of the day without reaching harmful temperatures. By practicing limited hetero-thermy in letting its temperature fluctuate in this manner, the camel manages to tolerate the extreme daytime desert heat without using much water for evaporative cooling.

The antelope ground squirrel (*Ammospermophilus leucurus*), a diurnal desert mammal, also survives in environments of extreme heat, although it is much smaller than the camel. It cannot continuously gain heat for several hours in the hot sun because its small surface/mass ratio leads to rapid heating. Instead, it exposes itself to high environmental temperatures for only about 8 min at a time. It then returns to its burrow, where its stored heat escapes into the cool underground air. By allowing its temperature to drop a bit below normal before returning to the hot desert floor, it is able to extend its stay without overheating.

An important factor influencing the loss of heat to the environment is the temperature of the body surface, since that determines the temperature difference $T_b - T_a$. Heat can be lost by conduction, convection, and/or radiation (Box 16-2) as long as the ambient temperature is below the body surface temperature. Thus, the closer the surface temperature is to the core temperature in an endotherm, the higher the rate of heat loss through the surface to cooler surroundings. Heat is transferred from the core to the surface primarily by the circulation; thus, the rate of heat loss to the environment is a function of the flow of blood to surface vessels (Figure 16-22). An example of this is seen in rabbits, whose thin, membranous, and lightly furred ears, with their extensively anastomosing arterioles and venules, are used to dissipate excess heat.

## EVAPORATIVE COOLING

Evaporation, which requires 585 cal/g of water, is the most effective means of removing excess heat, providing there is sufficient water available. Certain reptiles and birds and some mammals take available water (saliva, urine, or standing water) and spread it on various body surfaces, allowing it to evaporate at the expense of body heat. Other vertebrates, however, use sweating or panting for evaporative cooling.

In *sweating*, found in some mammals, sweat glands in the skin actively extrude water through pores onto the surface of the skin. Sweating is under autonomic control. Although it is a mechanism for evaporative cooling, it may persist in the absence of evaporation, and water will continue to be extruded even if the humidity is too high for evaporation to keep up with the rate of sweating.

Mammals and birds also utilize the respiratory system to lose heat by evaporative cooling. To increase heat loss, mammals breathe through the mouth instead of through the nose. Heat is carried away in exhalant air because the dimensions of the mouth are such that exhalant air retains the heat absorbed in the lungs. As noted earlier (p. 497), the nasal passages and their vascularization are effective in many mammalian species in retaining both water and body heat. Mammals also hyperventilate to increase heat loss. A change in alveolar ventilation, however, will result in a change in blood $P_{CO_2}$ and blood pH. This situation is avoided by increasing dead-space ventilation (i.e., flow through the mouth and trachea) *without* increasing ventilation of the alveolar respiratory surface (Figure 16-35). Breathing rate is increased, but tidal volume is reduced. This process is, of course, the familiar *panting*. Over-heated dogs pant, inhaling through the nose and exhaling through the mouth, exposing the tongue to encourage further water evaporation and therefore heat loss (Figure 16-36). This strategy produces an evaporatively effective one-way air flow over the non-respiratory surfaces of the nose, trachea, bronchi, and mouth, avoiding stagnation of saturated air in these

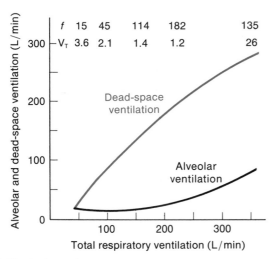

16-35   As the total respiratory ventilation (abscissa) increases in the panting ox, the dead-space ventilation increases steadily. The alveolar ventilation, however, does not increase until the total ventilation exceeds about 200 L/min. In extreme panting, the respiratory frequency ($f$) is decreased as tidal volume ($V_T$) is increased (figures at top of graph). [Hales, 1966.]

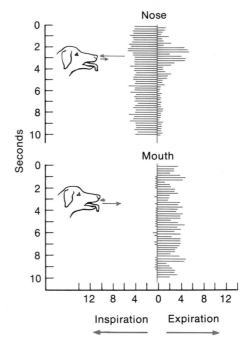

16-36   (**A**) Air flow through the nose of a panting dog. Horizontal lines extending to the left of the vertical midline indicate inspiration; to the right, expiration. Mean inhaled and exhaled volumes are indicated by vectors placed adjacent to the dog's nose. (**B**) Air flow through the mouth of a panting dog. Inspiration through the mouth is virtually zero; expiration through the mouth carries most of the air taken in through the nose. [Schmidt-Nielsen et al., 1970.]

passages. The respiratory work involved in panting is lower than it might seem, because a panting animal will cause its respiratory system to oscillate at its resonant frequency, thereby minimizing muscular effort. Panting is accompanied by increased secretions from the salivary glands and the glands of the nose, secretions that are under autonomic control. Most of the water that is not evaporated by panting is swallowed and conserved.

Because evaporation from skin or respiratory epithelium is the most effective means of ridding the body of excess heat, there is a close link between water balance and temperature control in hot environments. In hot, dry, desertlike environments, animals can be faced with the choice of either overheating or desiccation. Dehydrated mammals will conserve water by reducing evaporation caused by panting or sweating and will thus allow the body temperature to rise. Having a small heat capacity, a small mammal exposed to the desert heat will, in the absence of thermoregulatory water, undergo a rise in temperature that is far more rapid and more threatening than it is for a larger mammal. To survive, such smaller mammals must either ingest water or stay out of the heat.

The *reciprocity of water conservation and heat dissipation* in a small desert animal can be illustrated by contemplating water balance and temperature control in the kangaroo rat. To conserve water, this animal uses a temporal countercurrent heat-exchange system in which the nasal epithelium is cooled during inspiration by the inhalant air. During exhalation, most of the moisture picked up by the air in the warm, humid respiratory passages is conserved by its condensation on the cool nasal epithelium (Figures 12-12 and 12-13). However, this mechanism also recycles body heat and requires that the inhaled air be cooler than the body core. As a consequence, the rat is confined to its cool burrow during the hot times of day. If inhaled air were at or above body temperature, the rat's loss of respiratory moisture would increase. Although the evaporative loss of water would help cool the animal, it would also destroy its water balance.

The importance of water in the control of body temperature in a large desert mammal has been illustrated by simple observations carried out on camels (Figure 16-37). The camels were either allowed to drink ad libitum (freely) or subjected to periods of dehydration during which water was withheld for several days. Rectal temperatures were found to fluctuate between highs (daytime) and lows (nighttime). These fluctuations were minimal when the camels were allowed to drink, but still large in comparison with a water-drinking human. The temperature swings became even more exaggerated during periods of desiccation, when reserves of body water dwindled, leaving less for heat storage and for thermoregulation by sweating.

## Thermostatic Regulation of Body Temperature

Homeothermic endotherms utilize a system of body temperature control that has similarities to the mechanized thermostatic control found in a laboratory temperature bath (Figure 1-4) or a home heating system. In the water bath, a *temperature sensor* compares the water temperature, $T_w$, with a *set-point temperature*, $T_{sp}$. If $T_w$ is below $T_{sp}$, the *thermostat* closes the circuit that activates the production of additional heat until $T_w = T_{sp}$, after which the thermostat contacts open and heat production ceases. The cycle is repeated as $T_w$ drops again.

Homeostatic regulation of body temperature, $T_b$, in homeotherms is not fully understood, but it appears to work along similar principles of negative feedback (Box 1-1), although details differ from the physical

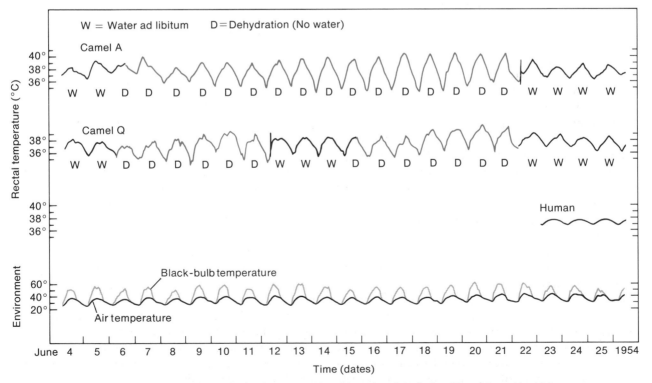

**16-37**  Core-temperature fluctuations in camels subjected to dehydration (D) or allowed to drink water ad libitum (W). The black line in the bottom trace gives air temperature; the gray line gives black-bulb temperature indicating radiant heat in the environment. [Schmidt-Nielsen et al., 1957.]

model. The homeotherm, for example, has not one but many temperature sensors in various parts of the body. Furthermore, to maintain $T_b$ at about $T_{sp}$, the animal can call on several heat-producing and heat-exchanging mechanisms, so that the thermostat controls heat-conserving and heat-loss mechanisms as well as heat production. That would be analogous to a hypothetical home heating and cooling system (perhaps yet to come) in which the thermostat, in addition to cycling the furnace and air conditioner, controls the position of window shades, window opening and closing, the conductance of the wall and roof insulation, and so on. Furthermore, control of thermogenesis in the homeotherm is not all-or-none as the turning ON and OFF of a furnace. Instead, the rate of heat production by metabolic means is graded according to need. The colder the temperature sensors become (within limits, of course), the higher the rate of thermogenesis. In engineering jargon, this is termed *proportional control,* because heat production and conservation are more or less proportional to the difference $T_b - T_{sp}$.

## THE MAMMALIAN THERMOSTAT

Mammalian body temperature can vary widely (up to 30°C) between the periphery and the body core, with the extremities undergoing far more variation than the core. Temperature-sensitive neurons or nerve endings occur in the brain, the spinal cord, the skin, and sites in the body core, providing input to thermostatic centers in the brain. Although there may be several thermoregulatory centers in mammals, the most important one, considered to be the body's "thermostat," is located in the hypothalamus (Figure 11-12). This was discovered by H. G. Barbour in 1912 during a series of experiments in which a small temperature-controlled probe was implanted into different parts of the rabbit brain. The probe evoked strong thermoresponses only in response to heating or cooling the hypothalamus. Cooling the hypothalamus produced an increase in metabolic rate and a rise in $T_b$, whereas heating it evoked panting and a drop in $T_b$. This experiment is analogous to changing the temperature of a home thermostat by holding a lighted match nearby. As the thermostat is warmed above its set-point temperature, it shuts down the furnace, allowing the room temperature to drop below the set point. A modern apparatus

for controlling hypothalamic temperature and measuring a homeotherm's response to changes in that temperature is shown in Figure 16-38. Such experimental procedures have shown that the *hypothalamic thermostat* is highly sensitive to temperature.

Variation of mammalian brain temperature of only a few degrees centigrade seriously affects the functioning of the brain, so it is not surprising to find the major thermoregulatory center of mammals located there. Neurons that are highly temperature-sensitive are located in the anterior part of the hypothalamic thermostat. Some of these neurons show a sharply defined increase in firing frequency with increased hypothalamic temperature (Figure 16-39A), whereas others show a decrease in firing frequency with increase in temperature above a certain value (Figure 16-39B).

Still other neurons increase in firing frequency when the brain temperature drops below the set-point temperature. The latter appears to control the activation of heat-producing responses (e.g., shivering, brown-fat metabolism) and heat conserving (e.g., pilomotor) responses. In addition, there are neurons that increase their firing rate in response to increased brain temperature. These are believed to activate heat-dissipating responses such as vasodilation and sweating.

In addition to the information about its own temperature generated by these thermosensitive neurons, the hypothalamus receives neural input from thermoreceptors in other parts of the body. All this thermal information is integrated and used to control the output of the thermostat. Neural pathways leaving the hypothalamus make connections with other parts of the nervous system that regulate heat production and heat loss.

A generalized schematic of the neural circuitry believed to underlie mammalian thermoregulation is given in Figure 16-40. The pathways shown in color are activated by high temperatures signaled from peripheral and spinal thermoreceptors and by the hypothalamic temperature-sensitive neurons. The efferent pathways activate increased sweating or panting, as well as a lowered peripheral vasomotor tone that produces increased blood flow to the skin. Conversely, body cooling leads to thermogenesis and increased peripheral vasomotor tone.

The experimental lowering of hypothalamic tem-

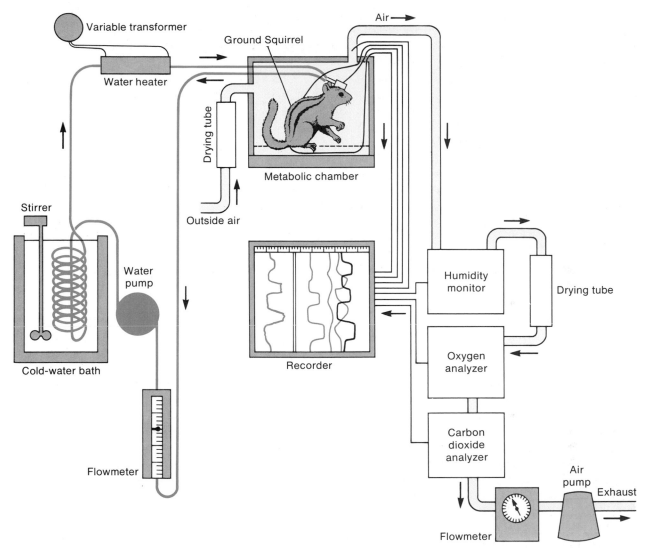

**16-38**  Apparatus for measuring temperature sensitivity of the hypothalamus and thermoregulatory responses to changes in hypothalamic temperature. Hypothalamic temperature is altered by means of a water-perfused thermode implanted in the hypothalamus. Metabolic rate and evaporative water loss are measured by analyzing the effluent air for water, $O_2$, and $CO_2$ content. The metabolic chamber is at constant temperature. [From "The Thermostat of Vertebrate Animals" by H. C. Heller, L. I. Crawshaw, and H. T. Hammel. Copyright © 1978 by Scientific American, Inc. All rights reserved.]

perature in a dog, produced as shown in Figure 16-38, elicited additional metabolic heat production by shivering, whereas warming of the dog's hypothalamus elicited the heat-dissipating response of panting (Figure 16-41).

The physical feedback circuit for mammalian thermoregulation is shown in Figure 16-42. The vascular system forms an important part of the circuit by carrying the heated or cooled blood from the thermoeffector tissues (brown fat, muscle, evaporative surfaces) back to the temperature sensors.

A rise in core temperature of only 0.5°C causes such profound peripheral vasodilation that the blood

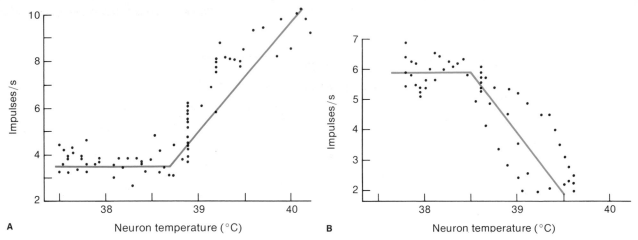

**16-39** Temperature–activity patterns of hypothalamic neurons in a rabbit. (**A**) Firing frequency of a neuron shows steep linear increase with temperature above 38.7°C. (**B**) Firing frequency of a different neuron in the same region exhibits linear decrease at temperatures above 38.4°C. [Hellon, 1967.]

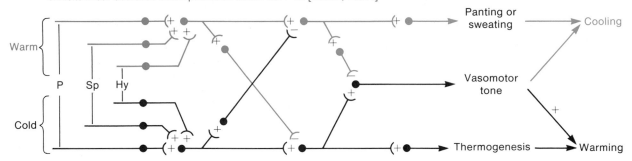

**16-40** Generalized circuitry proposed for neural regulation of body temperature. Peripheral (P), spinal (Sp), and hypothalamic (Hy) temperature sensors connect with neurons innervating networks that ultimately control heat-dissipating (colored pathways), heat-conserving, or heat-generating mechanisms. Pluses (+) and minuses (−) refer to excitatory and inhibitory inputs. [Bligh, 1973; and "The Thermostat of Vertebrate Animals" by H. C. Heller, L. I. Crawshaw, and H. T. Hammel. Copyright © 1978 by Scientific American, Inc. All rights reserved.]

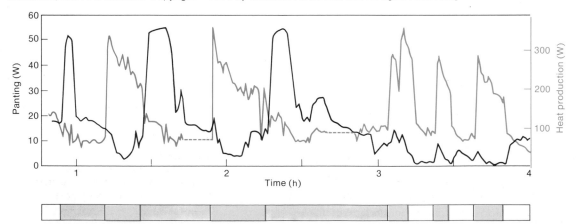

**16-41** Heat production (by shivering; in color) and heat dissipation (by panting; in black) in response to hypothalamic cooling and heating, respectively, in a dog. The apparatus used was similar to that in Figure 16-38. Cooling of the hypothalamus is shown on the time scale below in color, heating in gray. [From "The Thermostat of Vertebrate Animals" by H. C. Heller, L. I. Crawshaw, and H. T. Hammel. Copyright © 1978 by Scientific American, Inc. All rights reserved.]

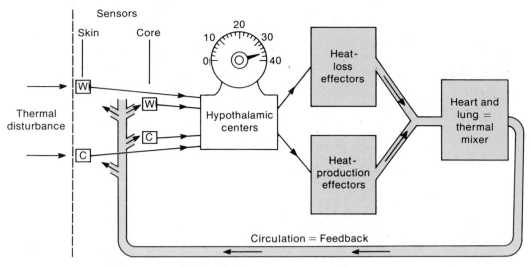

**16-42** Outline of components of mammalian thermoregulatory system. The hypothalamic centers integrate the information from distributed peripheral and core temperature receptors, compare it with the temperature set point, and accordingly regulate the heat-loss mechanism (to cool the blood) or heat-production mechanism (to heat the blood). The blood is mixed and distributed by the circulatory system, feeding back to the distributed sensors and to the hypothalamic centers. As the sensory inputs indicate a temperature approaching the set point, activation of thermoregulatory effectors is reduced. When the set-point temperature is achieved, activation of the effectors ceases. [Cremer and Bligh, 1969.]

flow to the skin can increase up to seven times normal, producing a flushed appearance to the skin. The effect of elevated core temperature on peripheral vasodilation and hence skin temperature is illustrated in Figure 16-43, in which it can be seen that the skin temperature of a rabbit ear rose very sharply to its maximum as the rabbit's core temperature exceeded 39.4°C. The influence of the hypothalamic thermostat on such peripheral heat-exchange mechanisms is about 20 times as great in some mammals as are reflexive adjustments initiated by peripheral temperature sensors. This hypothalamic "override" is significant in light of the importance of carefully regulated brain temperature. Without dominance of the hypothalamic thermostat, an internally overheated animal exercising in a cold environment would fail to activate heat-dissipating blood flow to the surface capillaries, and its core temperature would continue to rise dangerously.

In some homeotherms, especially small animals subject to rapid cooling at low ambient temperatures, the set-point temperature of the hypothalamic thermostat changes with ambient temperature, presumably

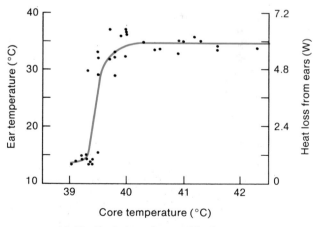

**16-43** Ear temperature and heat loss at different core temperatures in rabbits at an environmental temperature of 10°C. The core temperature was raised by forcing the rabbits to run on a treadmill. As temperature rose to above 39.5°C, blood flow to the ears increased, raising ear temperature and heat loss (given in watts). [Kluger, 1979.]

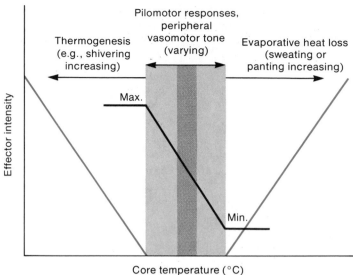

16-44 Generalized relation between degree of thermoregulatory response and core temperature. Within a small range (gray area) of the set-point temperature (dark gray area), regulation of body temperature entails only control of heat conductance to the environment by varying the peripheral blood flow or the insulating effectiveness of fur or feathers. Above and below this range, these passive measures are exhausted, and active thermogenesis (left) or evaporative heat loss (right) are brought into play. [Bligh et al., 1976.]

because ambient deviations are sensed by peripheral receptors. Thus, in the kangaroo rat, a sudden drop in ambient temperature is quickly followed by a rise in set-point temperature. This rise causes an increase in metabolic heat production in anticipation of increased heat loss to the environment.

The relations between thermoregulatory responses controlled by the hypothalamic centers and core temperature are illustrated in Figure 16-44. Small deviations in core temperature from the set point produce only peripheral vasomotor and pilomotor responses (black plot) that in effect alter the thermal conductivity of the body. These small deviations in core temperature usually result from moderate variations within a range of ambient core temperatures corresponding to the thermal neutral zone (Figure 16-28). When the core temperature is forced out of this range by more extreme ambient-temperature deviations or by exercise, passive thermoregulatory responses no longer suffice, and the hypothalamic centers institute active measures (colored plots in Figure 16-44), that is, thermogenesis or evaporative heat loss.

### NONMAMMALIAN THERMOREGULATORY CENTERS

Thermostatic control of body temperature has received less attention in birds than in mammals, perhaps because the manner of control seems to be more complex in birds. The hypothalamic thermoregulatory center in those birds tested (pigeons) is itself virtually insensitive to temperature changes. The spinal cord was found to be a site of central temperature sensing in pigeons, penguins, and ducks; but core receptors outside the CNS are the dominant central receptors in birds. The temperature sensors in the core presumably signal the avian hypothalamic thermostat, which in turn integrates the input and activates the thermoregulatory effectors.

Fishes and reptiles, though not able to regulate body temperature like homeotherms, do have a temperature-sensitive center in the hypothalamus. Heating the hypothalamus with an implanted thermode leads to hyperventilation in the scorpion fish; cooling leads to slower ventilatory movements. Peripheral cooling also produces similar ventilatory responses. Since the fish's rate of metabolism varies with body temperature, a rise in temperature leads to an increased need for oxygen. The temperature-determined adjustment in the rate of respiration is adaptive in that it anticipates changes in respiratory need and serves to minimize fluctuations in blood oxygen. The reptilian response to cooling of the hypothalamus is to engage in *thermophilic* (i.e., heat-seeking) *behavior,* whereas heating of the hypothalamus elicits *thermophobic* (i.e., heat-avoiding) *behavior.*

## FEVER

An interesting feature of the hypothalamic thermo-regulatory center is its sensitivity to certain chemicals collectively termed *pyrogens* (fever-producing substances). Two general categories of pyrogens have been recognized on the basis of their origins. The *exogenous pyrogens* are bacterial endotoxins produced by gram-negative bacteria. These heat-stable, high-molecular-weight polysaccharides are so potent that only $10^{-9}$ g of purified endotoxin injected into a large mammal will cause an elevation of body temperature. *Endogenous pyrogens,* on the other hand, arise from the animal's own tissues and, unlike those of bacterial origin, are heat-labile proteins. Leucocytes release endogenous pyrogens in response to circulating exogenous pyrogens produced by infectious bacteria. Thus, it appears that exogenous pyrogens cause a rise in body temperature indirectly by stimulating the release of endogenous pyrogens that act directly on the hypothalamic center. This idea is supported by evidence that the hypothalamus is more sensitive to direct application of endogenous pyrogens than to exogenous ones.

The sensitivity of the temperature-sensing neurons to these molecules leads to an elevation in the set point to a higher temperature than normal. The result is that the body temperature is allowed to rise several degrees, and the animal experiences a *fever*. Anesthetics and opiates such as morphine, in contrast to pyrogens, cause a lowering of the set-point temperature and hence a drop in body temperature. The adaptive significance of endogenous pyrogens and of their production of fever in homeotherms is not clear. Although elevated body temperature has bacteriostatic effects, and might therefore be considered useful in combating infection, the fever itself has deleterious effects on body tissues and is often fatal to the host.

## THERMOREGULATION DURING EXERCISE

Muscle contraction has an energy efficiency of about 25%. That is, for every calorie of chemical energy converted into mechanical work, 3 cal is degraded to heat. During exercise, this extra heat, added to that produced by basal metabolism, will cause a rise in body temperature unless it is dissipated to the environment at the same rate as it is produced. Most of the excess heat is indeed transferred to the environment, but a rise in core temperature of homeotherms occurs during exercise, indicating incomplete removal of the excess heat. The rise in temperature is moderately useful in two respects: (1) It increases the difference $T_b - T_a$ and thereby increases the effectiveness of the heat-loss processes, and (2) it leads to an increased rate of chemical reactions, including those that support physical activity. However, the core temperature can rise to dangerously high levels during heavy exercise in warm environments.

The level to which the core temperature rises in homeotherms is proportional to the rate of muscular work. During light or moderate exercise in cool environments, body temperature rises to a new level and is regulated at that level as long as the exercise continues. Thus, temperature appears to remain under the control of the body's thermostat. The rise in temperature proportional to the level of exercise appears to be a consequence primarily of an increase in the error signal, $T_b - T_{sp}$, of the thermostatic feedback control by the hypothalamus. The error signal is the difference between the thermostat's set point and the actual core temperature. The greater this difference (i.e., the greater the error signal), the greater the activation of heat-loss mechanisms. Thus, the rate of heat dissipation increases as the core temperature rises above the set point, and a new equilibrium becomes established between heat production and heat loss. During heavy exercise, especially in warm environments, the heat-dissipating mechanisms are not able to balance heat production until body temperature rises several degrees, increasing the $T_b - T_a$ difference. Thus, elevations of $4°-5°C$ in core temperature are commonly observed in humans after strenuous sustained running and in sled dogs after a race.

The error signal is kept small by the high sensitivity of the feedback control of heat-dissipating mechanisms. For example, a small increase in $T_b$ above the set-point temperature produces a strong and steep increase in the rate of sweating (Figure 16-45). The heat-loss mechanisms are initiated by vigorous exercise even before peripheral body temperature has undergone any significant increase. For example, in humans, an increased rate of sweating begins within 2 s after onset of heavy physical work, even though there is no detectable increase in skin temperature during that

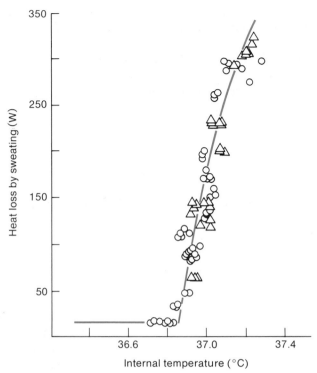

16-45   Rate of sweating at different body temperatures in a human. Core temperature was elevated by exercise (△) or by elevating the ambient temperature (○). [Benzinger, 1961.]

time. However, core blood temperature has in fact been found to exhibit a detectable rise in temperature within 1 s after exercise has begun. Apparently, the onset of sweating, nearly concurrent with the onset of neural activity underlying exercise, results from the reflex activation of sweating by central temperature receptors.

Certain groups of hoofed mammals (e.g., sheep, goats, and gazelles) and carnivores (e.g., cats and dogs) employ a special countercurrent heat exchanger to prevent overheating of the brain during strenuous exercise such as running. This system, the *carotid rete* (Figure 16-46A), uses cool venous blood returning from respiratory passages to remove heat from hot arterial blood supplying the brain. In these animals, most of the blood to the brain flows through the

external carotid artery. At the base of the skull the carotid anastomoses to form hundreds of small arteries into a vascular rete, the vessels of which rejoin just prior to passage into the brain (Figure 16-47). These arteries pass through a large sinus of venous blood, the *cavernous sinus*. This venous blood is significantly cooler than the arterial blood because it has come from the walls of the nasal passages, where it was cooled by respiratory air flow. Thus, the hot arterial blood flowing through the rete gives up some of its heat to the cooler venous blood before it enters the skull. As a result, brain temperature may be 2°–3° lower than the core temperature of the body. Although sustained running in hot surroundings inevitably places a heat load on these animals, the most serious and acute consequence of overheating—spastic brain function—is thereby avoided. This system of cooling is most effective when the animal is breathing hard during strenuous exercise. Some mammals, such as rodents and primates, have no countercurrent mechanism for cooling carotid arterial blood (Figure 16-46B).

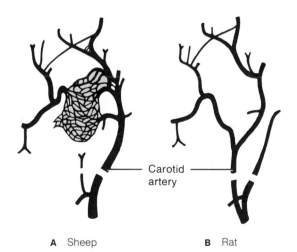

**A**  Sheep               **B**  Rat

16-46   The carotid rete found in some mammals is shown in color. A network of small arteries acts as a heat exchanger for blood supplying the brain. Loss of heat from the rete into the venous blood of the cavernous sinus that surrounds the rete helps prevent overheating of the brain. The rete is extensive in sheep (**A**) and absent in rats (**B**). [Daniel et al., 1953.]

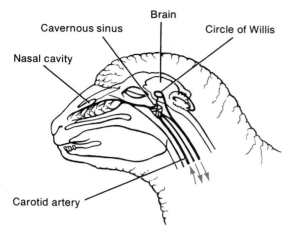

**16-47** Carotid rete and countercurrent cooling of carotid blood in the sheep. Cool venous blood returning from the nasal cavity bathes the carotid rete contained in the cavernous sinus, removing heat from arterial blood heading for the circle of Willis and on to the brain. [Hayward and Baker, 1969.]

## Dormancy

Five forms of animal dormancy are recognized: *sleep, torpor, hibernation, winter sleep,* and *estivation.*

### Sleep

Sleep is poorly understood, even though it has been studied intensively in humans and other mammals, and has been found to involve extensive adjustments in brain function. In mammals, sleep is associated with a drop in both body temperature and hypothalamic temperature sensitivity. There is evidence of sleep-inducing substances that build up during wakefulness, accumulating in extracellular fluids of the CNS. The identity and mode of action of these substances are not known.

The remaining four categories of dormancy are even less well understood than sleep, but in homeotherms all appear to be manifestations of physiologically related processes.

### Torpor

The closer the body temperature is to the air temperature, the less heat is lost to the environment (Equation 16-13). The lower the $T_b$, the lower the rate of conversion of energy stores, such as fatty tissues, into body heat. Thus, except when an animal is under heat stress, it is advantageous for it to allow its temperature to drop to a low level during periods of nonfeeding. Small endotherms, because of their high rates of metabolism, are subject to starvation during periods of inactivity when they are not feeding. Some of these enter a state of torpor during those periods, in which temperature and metabolic rate subside. Then, before the animal becomes active, its body temperature rises as a result of a burst of metabolic activity, especially through oxidation of brown-fat stores. Daily torpor is practiced by many terrestrial birds and small mammals. There is, of course, some degree of risk associated with torpor, since a torpid animal stands less chance of escaping if detected by a predator. So an animal generally resorts to torpor only when energy reserves are very low.

### Hibernation and "Winter Sleep"

Some mammals, primarily those in the orders Rodentia, Insectivora, and Chiroptera, store enough energy reserves to go into hibernation, a period of deep torpor or winter dormancy, which lasts up to weeks or even several months in cold climates. Hibernators arouse periodically to tend to their needs, such as emptying the bladder. During hibernation, the hypothalamic thermostat is reset to as low as 20°C or more below normal. At ambient temperatures between 5°C and 15°C, many hibernators keep their temperature as little as 1°C above ambient. If the air temperature falls to dangerously low levels, the animal increases its metabolic rate to maintain a constant low $T_b$ or becomes aroused. During torpor and hibernation, thermoregulatory control is evidently not suspended, but continues with a lowered set point. In the hibernating marmot, for example, experimental cooling of the anterior hypothalamus with an electronically controlled implanted probe resulted in an increased metabolic production of body heat (Figure 16-48). The increase in heat production was proportional to the difference between the set-point temperature and the actual hypothalamic temperature. The set-point temperature dropped about 2.5°C within one day, as the animal entered a deeper state of hibernation.

As one might expect, body functions are greatly

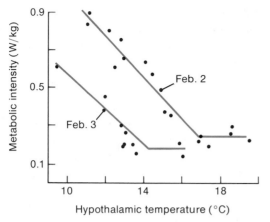

**16-48** Hypothalamic thermosensitivity during hibernation in *Marmota*. The hypothalamic temperature was slowly lowered on two consecutive days during the period in which the marmot entered a progressively deeper state of hibernation. Below the lower critical temperature, the metabolic rate increased in proportion to the drop in temperature. The critical temperature decreased from about 16.3°C on February 2 to about 14.0°C on February 3. [Florant et al., 1978.]

slowed with the lowered body temperatures characteristic of torpor and hibernation. Total blood flow is typically reduced to about 10% of normal, although the head and brown-fat tissue receive a much higher blood flow than other tissues. The cardiac output slows considerably, pumping blood at a rate that is only a small percentage of the normal rate. This retardation is accomplished by a drastic slowing of the rate of heartbeat, while the stroke volume remains essentially normal. The metabolic rate in the ground squirrel *Citellus lateralis,* for example, drops to about 7% of normal when the body temperature has subsided to 8°C. As the squirrel's hibernation begins, the transition to the torpid state is completed within 12–18 h (Figure 16-49), whereas arousal requires less than 3 h. The speed of arousal in this mid-sized mammal depends on the rapid heating that is initiated by intensive oxidation of brown fat, and then accompanied by shivering, which leads to the large overshoot in metabolic rate seen in the figure.

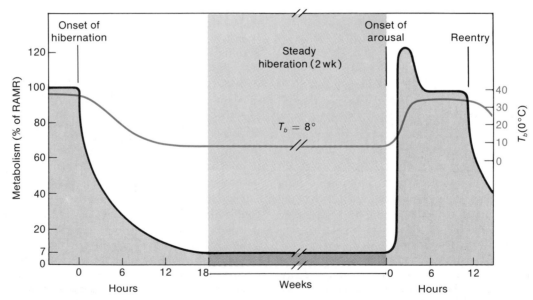

**16-49** Metabolic relations during entry into and arousal from hibernation in a ground squirrel kept in a chamber having a temperature of 4°C. The period of steady-state hibernation is shaded in color, and the body temperature, $T_b$, is graphed in color. Metabolism is graphed in black. At onset of hibernation, the set point for body temperature is depressed. Metabolism decreases, allowing $T_b$ to drop to 1°–3°C above ambient throughout hibernation. Arousal occurs when the set-point temperature climbs to 38°C, and a strong surge of metabolic heat production raises $T_b$ to the new set-point level. RAMR = resting average metabolic rate. [Swan, 1974.]

Although many small endotherms undergo a daily cycle of torpor, their high rate of metabolism precludes extended periods of torpor (i.e., hibernation), since even in the hibernating state they would quickly consume stored energy reserves. All true hibernators are mid-sized mammals, as they are large enough to store sufficient reserves for extended hibernation. But there are no true hibernators among large mammals. Bears, which were once thought to hibernate, in fact simply enter a "winter sleep" without a substantial drop in body temperature, remaining curled up in a protected microhabitat such as a cave or hollow log. They are able to wake and become active quickly now and then during the long winter.

Why are there no large hibernators? First, they have less need to save fuel, since their normal BMRs are low relative to their fuel stores. Second, because of their large mass and relatively low rate of metabolism, it would require a prolonged metabolic effort to raise body temperature from a low level near ambient to normal body temperature. It has been calculated, for example, that a large bear would require at least a day or two to warm up to 37°C from a hibernating temperature of 5°C.

*Estivation*

This poorly defined term, which means "summer sleep," applies to both vertebrates and invertebrates and refers to a dormancy that some species enter into in response to high ambient temperatures and/or danger of desiccation. Thus, land snails such as *Helix* and *Otala* will become dormant during long periods of low humidity after sealing the entrance to the shell by secreting a diaphragmlike operculum that retards loss of water by evaporation. Some small mammals, such as the Columbian ground squirrel, spend the hot late summer inactive in their burrows, with their core temperature approaching the ambient temperature. This state is probably similar physiologically to hibernation, but it differs in seasonal timing.

## Summary

The utilization of chemical energy in tissue metabolism is accompanied by the inexorable production of heat as a low-grade energy by-product. The total energy liberated in going from compound A to a lower-energy end product, compound X, is independent of the chemical route taken from A to X. In addition, a given class of food molecules will consistently liberate the same amount of heat and require the same amount of $O_2$ when oxidized to $H_2O$ and $CO_2$. These characteristics of energy metabolism make it possible to use either the rate of heat production or the rate of $O_2$ consumption (and $CO_2$ production) as a measure of metabolic rate. The respiratory quotient—the ratio of $CO_2$ production to $O_2$ consumption—is useful in determining the proportions of carbohydrates, proteins, and fats metabolized, each of which has a different characteristic energy yield per liter of $O_2$ consumed.

The basal metabolic rate (BMR) and the standard metabolic rate (SMR) are related to body size: the smaller the animal, the greater its metabolic intensity—namely, the higher the metabolic rate of a unit mass of tissue. Although there is a correlation of metabolic intensity with body surface/volume ratio, this correlation appears to be incidental, for similar correlations are seen in ectotherms, which are in temperature equilibrium with the environment when the SMR is determined.

The energetics of locomotion is also related to body size. The smaller the animal, the higher the metabolic cost of transporting a unit mass of body tissue over a given distance. The Reynolds number (*Re*) of a body moving through a liquid or gaseous medium is the ratio of the relative importance of inertial and viscous forces in the medium. Small animals swim with a low *Re* and large animals with a high *Re* because, with increasing size, viscosity plays a lesser role and inertia a greater one.

The rate of energy utilization during different kinds of locomotion typically increases with velocity. By changing gait from walking to running to hopping or trotting, and so on, terrestrial animals optimize efficiency. Increased efficiency occurs when the energy of falling at the end of the stride is elastically stored for release during the next stride, as in a hopping kangaroo.

Dependence of metabolic rate on tissue temperature is described by the $Q_{10}$, the ratio of metabolic rate at a given temperature to the metabolic rate at a temperature of 10°C less. This ratio typically lies between 2 and 3. Exposure of an ectotherm to high or low temperatures for extended periods leads to acclimation, in which the metabolic rate at a given test temperature is elevated in cold-acclimated animals and depressed in warm-acclimated ones.

Ectotherms have evolved a variety of strategies for survival in temperature extremes. Some species cope with subzero temperatures by using antifreeze substances or by supercooling without ice crystal formation, but no animals have been shown to survive freezing of water within the cells. Other ectotherms elevate body temperature by shivering or by ordinary muscle contraction at certain times or in certain parts of their bodies. Such heterothermy is utilized by certain insects and large fishes to warm locomotor muscles to optimal operating temperatures. Heat absorption or heat loss to the environment is regulated in some ectothermic species by changes in blood flow to the skin. In this way, heat absorbed from the sun's rays can be quickly transferred by the blood from the body surface to the body core during heating, or conversely, core heat can be conserved in a cold environment by restricted circulation to the skin.

Endotherms subjected to cold environments conserve body heat by increasing the effectiveness of their surface insulation. They do this by decreasing peripheral circulation, increasing fluffiness or thickness of pelage or plumage or by adding fatty insulating tissue. In cold-climate endotherms, heat is also conserved by countercurrent heat-exchange mechanisms in the circulation to the limbs. Within the thermal neutral zone of ambient temperatures, changes in surface conductance compensate for changes in ambient temperature. Below this temperature zone, thermogenesis compensates for heat loss to the environment. Thermogenesis occurs by shivering or by nonshivering oxidation of substrates.

At ambient temperatures above the thermal neutral zone, endotherms actively dissipate heat by means of evaporative cooling, either by sweating or by panting. The use of water places an osmotic burden on desert dwellers. Most small desert inhabitants, subject to rapid changes in body temperature, minimize such changes by usually remaining in cool microenvironments to avoid daytime heat. Large desert mammals, buffered against rapid temperature changes by more favorable surface/volume ratios and large thermal inertia, can conserve water that they would otherwise use for cooling by slowly absorbing heat during the day without reaching lethal body temperatures; they can then rid themselves of heat during the cool night. The brain is specially protected from overheating in some mammals by a highly developed carotid rete in which cool venous blood from the nasal epithelium removes heat from arterial blood heading toward the brain.

Body temperature in homeotherms is regulated by a neural thermostat sensitive to differences between the actual temperature of neural sensors and the thermostatic set-point temperature. Differences result in a neural output to thermoregulatory effectors for corrective heat loss or heat gain. Fever occurs when the set-point temperature is raised by the cellular action of endogenous pyrogens, which are protein molecules released by leucocytes in response to exogenous pyrogens produced by infectious bacteria.

Ordinary sleep, torpor, hibernation, "winter sleep," and estivation are all neurophysiologically and metabolically related forms of dormancy. During periods when food intake is necessarily absent or restricted, small- to medium-sized homeotherms allow their temperatures to drop in accord with a lowered set-point temperature of the body thermostat. By lowering body temperature to within a few degrees of ambient air, the homeotherm minimizes heat production and conserves energy stores. Oxidation of brown fat and shivering thermogenesis are utilized to produce rapid warming at the termination of torpor or hibernation.

## Exercises

*1.* Define ectotherm, endotherm, poikilotherm, homeotherm, basal metabolic rate, standard metabolic rate, and respiratory quotient.

*2.* Explain why the rate of heat production can be used to measure metabolic rate accurately.

*3.* Explain why respiratory gas exchange can also be used to measure metabolic rate.

*4.* Why is the surface hypothesis untenable as an explanation for the high metabolic intensity of small animals?

*5.* Why is the locomotion of a small animal influenced more strongly by the viscosity of the medium than that of a large animal?

*6.* What factors influence the flow pattern of fluid around a swimming animal? What factors minimize turbulence?

*7.* Why does riding a bicycle at 10 km/h for 1 h require less energy than running the same distance at the same speed?

*8.* Give examples of the influence of body size on the metabolism and locomotion of animals.

*9.* The potency of some medications depends on metabolic factors. Explain why it would be risky to give a 100 kg human 100 times the dose of a drug proven effective in a 1.0 kg guinea pig.

*10.* Give two examples, with explanations, of north–south geographic influence on the distribution of homeotherms and poikilotherms.

*11.* Give examples of low-temperature adaptations of some ectotherms.

*12.* Define thermal acclimation and give two molecular mechanisms that appear to come into play during acclimation.

*13.* What determines the limits of the thermal neutral zone of a homeotherm?

*14.* What thermoregulatory mechanisms are available to a homeotherm at temperatures below and above the thermoregulatory zone?

*15.* Explain and give examples of the relations that exist between water balance and temperature regulation in a desert animal.

*16.* Why are small animals in greater danger of temperature extremes than large animals?

*17.* Discuss the integration of peripheral and core temperatures in the thermostatic control of temperature in a mammal.

*18.* Discuss two naturally occurring situations in which the set-point temperature of the hypothalamic thermostat is changed and the body temperature correspondingly changes.

*19.* Explain the mechanism of heat production in two different kinds of thermogenesis.

*20.* Discuss the role of countercurrent heat exchange in porpoises, arctic mammals, tuna fishes, and sheep.

*21.* What are the sphinx moth's two major means of regulating thoracic temperature?

*22.* What means does the marine iguana use to speed the elevation of its body temperature and then retard cooling during diving?

## Suggested Reading

Alexander, R. McN. 1982. *Locomotion of Animals.* London: Chapman and Hall.

Alexander, R. McN., and G. Goldspink. 1977. *Mechanics and Energetics of Animal Locomotion.* London: Chapman and Hall.

Bligh, J. 1973. *Temperature Regulation in Mammals and Other Vertebrates.* Amsterdam: North-Holland.

Carey, F. G. 1973. Fishes with warm bodies. *Scientific American* 228(2):36–44.

Childress, S. 1981. *Mechanics of Swimming and Flying.* Cambridge: Cambridge University Press.

Crawshaw, L. I., B. P. Moffitt, D. E. Lemons, and J. A. Downey. 1981. The evolutionary development of vertebrate thermoregulation. *American Scientist* 69:543–550.

Elder, H. Y., and E. R. Trueman. 1980. *Aspects of Animal Movement,* Cambridge: Cambridge University Press.

Florant, G. L., B. M. Turner, and H. C. Heller. 1978. Temperature regulation during wakefulness, sleep, and hibernation in marmots. *Amer. J. Physiol.* 235:282–286.

Hainsworth, F. R. 1981. Energy regulation in hummingbirds. *American Scientist* 69:420–429.

Hardy, R. N. 1979. *Temperature and Animal Life.* Baltimore: University Park Press.

Heglund, N. C., M. A. Febak, C. R. Taylor, and G. A. Cavagna. 1982. Energetics and mechanics of terrestrial locomotion. *J. Exp. Biol.* 97:57–66.

Heinrich, B. 1974. Thermoregulation in endothermic insects. *Science* 185:747–756.

Heller, H. C., L. I. Crawshaw, and H. T. Hammel. 1978. The thermostat of vertebrate animals. *Scientific American* 239(2):102–113.

Kleiber, M. 1961. *The Fire of Life: An Introduction to Animal Energetics.* New York: Wiley.

Kluger, M. J. 1979. *Fever: Its Biology, Evolution, Function.* Princeton: Princeton University Press.

Porter, W. P., and D. M. Gates. 1969. Thermodynamic equilibria of animals with environment. *Ecol. Monographs* 39:227–244.

Satinoff, E., ed. 1980. *Thermoregulation.* Stroudsburg, Pa.: Dowden, Hutchinson and Ross.

Schmidt-Nielsen, K. 1973. *How Animals Work.* New York: Cambridge University Press.

Swan, H. 1974. *Thermoregulation and Bioenergetics: Patterns for Vertebrate Survival.* New York: American Elsevier.

Tucker, V. A. 1975. The energetic cost of moving about. *American Scientist* 63:413–419.

Vernberg, F. J., ed. 1975. *Physiological Adaptation to the Environment.* New York: Crowell.

Wang, L. C. H., and J. Hudson, eds. 1978. *Strategies in Cold: Natural Torpidity and Thermogenesis.* New York: Academic.

## References Cited

Bartholomew, G. A., and R. C. Lasiewski. 1965. Heating and cooling rates, heart rate and simulated diving in the Galapagos marine iguana. *Comp. Biochem. Physiol.* 16:573–582.

Benzinger, T. H. 1961. The diminution of thermoregulatory sweating during cold reception at the skin. *Proc. Nat. Acad. Sci.* 47:1683–1688.

Bligh, J., J. L. Cloudsey-Thompson, and A. G. MacDonald. 1976. *Environmental Physiology of Animals.* Oxford: Blackwell Scientific Publications.

Carey, F. G., and J. M. Teal. 1966. Heat conservation in tuna fish muscle. *Proc. Nat. Acad. Sci.* 56:1464–1469.

Cremer, J. E., and J. Bligh. 1969. Body temperature and responses to drugs. *Br. Med. Bull.* 25:299–306.

Daniel, P. M., J. D. K. Dawes, and M. M. L. Prichard. 1953. Studies of the carotid rete and its associated arteries. *Phil. Trans. Roy. Soc.* (London) 237:173–208.

Hales, J. R. S. 1966. The partition of respiratory ventilation of the panting ox. *J. Physiol.* 188:45–68.

Hayward, J. N., and M. A. Baker. 1969. A comparative study of the role of the cerebral arterial blood in the regulation of brain temperature in five mammals. *Brain Res.* 16:417–440.

Hellon, R. F. 1967. Thermal stimulation of hypothalamic neurones in unanaesthetized rabbits. *J. Physiol.* 193: 381–395.

Hemmingsen, A. M. 1969. Energy metabolism as related to body size and respiratory surfaces, and its evolution. *Rep. Steno. Mem. Hosp. Nordisk Insulinlaboratorium* 9:1–110.

Herried, C. F., and B. Kessel. 1967. Thermal conductance of birds and mammals. *Comp. Biochem. Physiol.* 21:405–414.

Heusner, A. A. 1982. Energy metabolism and body size. *Resp. Physiol.* 48:1–25.

Horowitz, B. A. 1978. Neurohumoral regulation of non-shivering thermogenesis in mammals. In L. C. H. Wang and J. Hudson, eds., *Strategies in Cold: Natural Torpidity and Thermogenesis.* New York: Academic.

Irving, L. 1966. Adaptations to cold. *Scientific American* 214(15):94–101.

Marshall, P. T., and G. M. Hughes. 1980. *Physiology of Mammals and Other Vertebrates.* 2nd ed. Cambridge: Cambridge University Press.

Nachtigall, W. 1977. On the significance of Reynolds number and the fluid mechanical phenomena connected to it in swimming physiology and flight biophysics. In W. Nachtigall, ed., *Physiology of Movement-Biomechanics.* Stuttgart: G. Fischer Verlag.

Prosser, C. L. 1973. *Comparative Animal Physiology.* Philadelphia: Saunders.

Rieck, A. F., J. A. Belli, and M. E. Blaskovics. 1960. Oxygen consumption of whole animals and tissues in temperature acclimated amphibians. *Proc. Soc. Exp. Biol. Med.* 103: 436–439.

Schmidt-Nielsen, K. 1975. *Animal Physiology, Adaptation and Environment.* New York: Cambridge University Press.

Schmidt-Nielsen, K., B. Schmidt-Nielsen, S. A. Jarnum, and T. R. Houpt. 1957. Body temperature of the camel and its relation to water economy. *Am. J. Physiol.* 188: 103–118.

Schmidt-Nielsen, K., W. L. Bretz, and C. R. Taylor. 1970. Panting in dogs: Unidirectional air flow over evaporative surfaces. *Science* 169:1102–1104.

Scholander, P. F., W. Flagg, V. Walters, and L. Irving. 1953. Climatic adaptation in arctic and tropical poikilotherms. *Physiol. Zool.* 26:67–92.

Taylor, C. R., K. Schmidt-Nielsen, and J. L. Raab. 1970. Scaling of energetic costs of running to body size in mammals. *Am. J. Physiol.* 219:1104–1107.

# *Appendixes*

## *Appendix 1: SI Units*

Basic SI units.

| Physical quantity | Name of unit | Symbol for unit |
|---|---|---|
| Length | meter | m |
| Mass | kilogram | kg |
| Time | second | s |
| Electric current | ampere | A |
| Temperature | kelvin | K |
| Luminous intensity | candela | cd |

SI multipliers.

| Multiplier | Prefix | Symbol |
|---|---|---|
| $10^9$ | giga | G |
| $10^6$ | mega | M |
| $10^3$ | kilo | k |
| $10^2$ | hecto | h |
| 10 | deka | da |
| $10^{-1}$ | deci | d |
| $10^{-2}$ | centi | c |
| $10^{-3}$ | milli | m |
| $10^{-6}$ | micro | $\mu$ |
| $10^{-9}$ | nano | n |
| $10^{-12}$ | pico | p |

Derived SI units.

| Physical quantity | Name of unit | Symbol for unit | Definition of unit |
|---|---|---|---|
| Acceleration | meter per second squared | $m/s^2$ | |
| Activity | 1 per second | $s^{-1}$ | |
| Electric capacitance | farad | fd | $A\text{-}s/V$ |
| Electric charge | coulomb | C | $A\text{-}s$ |
| Electric field strength | volt per meter | $V/m$ | |
| Electrical resistance | ohm | $\Omega$ | $V/A$ |
| Entropy | joule per kelvin | $J/K$ | |
| Force | newton | N | $kg\text{-}m/s^2$ |
| Frequency | hertz | Hz | $s^{-1}$ |
| Illumination | lux | lx | $lm/m^2$ |
| Luminance | candela per square meter | $cd/m^2$ | |
| Luminous flux | lumen | lm | $cd\text{-}sr$ |
| Power | watt | W | $J/s$ |
| Pressure | newton per square meter | $N/m^2$ | |
| Voltage, potential difference | volt | V | $W/A$ |
| Work, energy, heat | joule | J | $N\text{-}m$ |

# Appendix 2: Selected Atomic Weights

Atomic weights.

| Element | Symbol | Atomic number | Atomic weight | Valence |
|---|---|---|---|---|
| Aluminum | Al | 13 | 26.9815 | 3 |
| Barium | Ba | 56 | 137.34 | 2 |
| Boron | B | 5 | 10.811 | 3 |
| Cadmium | Cd | 48 | 112.40 | 2 |
| Calcium | Ca | 20 | 40.08 | 2 |
| Carbon | C | 6 | 12.01115 | 2, 4 |
| Cesium | Cs | 55 | 132.905 | 1 |
| Chlorine | Cl | 17 | 35.453 | 1, 3, 5, 7 |
| Chromium | Cr | 24 | 51.996 | 2, 3, 6 |
| Cobalt | Co | 27 | 58.9332 | 2, 3 |
| Copper | Cu | 29 | 63.54 | 1, 2 |
| Fluorine | F | 9 | 18.9984 | 1 |
| Gold | Au | 79 | 196.967 | 1, 3 |
| Helium | He | 2 | 4.0026 | 0 |
| Hydrogen | H | 1 | 1.00797 | 1 |
| Iodine | I | 53 | 126.9044 | 1, 3, 5, 7 |
| Iron | Fe | 26 | 55.847 | 2, 3 |
| Lanthanum | La | 57 | 138.91 | 3 |
| Lead | Pb | 82 | 207.19 | 2, 4 |
| Lithium | Li | 3 | 6.939 | 1 |
| Magnesium | Mg | 12 | 24.312 | 2 |
| Manganese | Mn | 25 | 54.9380 | 2, 3, 4, 6, 7 |
| Mercury | Hg | 80 | 200.59 | 1, 2 |
| Nickel | Ni | 28 | 58.71 | 2, 3 |
| Nitrogen | N | 7 | 14.0067 | 3, 5 |
| Osmium | Os | 76 | 190.2 | 2, 3, 4, 8 |
| Oxygen | O | 8 | 15.9994 | 2 |
| Palladium | Pd | 46 | 106.4 | 2, 4, 6 |
| Phosphorus | P | 15 | 30.9738 | 3, 5 |
| Platinum | Pt | 78 | 195.09 | 2, 4 |
| Plutonium | Pu | 94 | 244 | 3, 4, 5, 6 |
| Potassium | K | 19 | 39.102 | 1 |
| Radium | Ra | 88 | 226 | 2 |
| Rubidium | Rb | 37 | 85.47 | 1 |
| Selenium | Se | 34 | 78.96 | 2, 4, 6 |
| Silicon | Si | 14 | 28.086 | 4 |
| Silver | Ag | 47 | 107.870 | 1 |
| Sodium | Na | 11 | 22.9898 | 1 |
| Strontium | Sr | 38 | 87.62 | 2 |
| Sulfur | S | 16 | 32.064 | 2, 4, 6 |
| Thallium | Tl | 81 | 204.37 | 1, 3 |
| Zinc | Zn | 30 | 67.37 | 2 |

# Appendix 3:

## Conversions, Formulas, Physical and Chemical Constants, Definitions

Units and conversion factors.

| To convert from | to | Multiply by |
|---|---|---|
| angstroms | inches | $3.937 \times 10^{-9}$ |
| | meters | $1 \times 10^{-10}$ |
| | micrometers ($\mu$m) | $1 \times 10^{-4}$ |
| atmospheres | bars | 1.01325 |
| | dynes per square centimeter | $1.01325 \times 10^6$ |
| | grams per square centimeter | 1033.23 |
| | torr ($=$ mm Hg; 0°C) | 760 |
| | pounds per square inch | 14.696 |
| bars | atmospheres | 0.9869 |
| | dynes per square centimeter | $1 \times 10^6$ |
| | grams per square centimeter | 1019.716 |
| | pounds per square inch | 14.5038 |
| | millimeters of mercury (0°C) | 750.062 |
| calories | British thermal units | $3.968 \times 10^{-3}$ |
| | ergs | $4.184 \times 10^7$ |
| | foot-pounds | 3.08596 |
| | kilocalories | $10^{-3}$ |
| | horsepower-hours | $1.55857 \times 10^{-6}$ |
| | joules | 4.184 |
| | watt-hours | $1.1622 \times 10^{-3}$ |
| | watt-seconds | 4.184 |
| ergs | British thermal units | $9.48451 \times 10^{-11}$ |
| | calories | $2.39 \times 10^{-8}$ |
| | dynes per centimeter | 1 |
| | foot-pounds | $7.37562 \times 10^{-8}$ |
| | gram-centimeters | $1.0197 \times 10^{-3}$ |
| | joules | $1 \times 10^{-7}$ |
| | watt-seconds | $1 \times 10^{-7}$ |
| grams | daltons | $6.024 \times 10^{23}$ |
| | grains | 15.432358 |
| | ounces (avdp) | $3.52739 \times 10^{-2}$ |
| | pounds (avdp) | $2.2046 \times 10^{-3}$ |

Units and conversion factors (continued).

| To convert from | to | Multiply by |
|---|---|---|
| inches | angstroms | $2.54 \times 10^8$ |
| | centimeters | 2.54 |
| | feet | $8.333 \times 10^{-2}$ |
| | meters | $2.54 \times 10^{-2}$ |
| joules | calories | 0.239 |
| | ergs | $1 \times 10^7$ |
| | foot-pounds | 0.73756 |
| | watt-hours | $2.777 \times 10^{-4}$ |
| | watt-seconds | 1 |
| liters | cubic centimeters | $10^3$ |
| | gallons (US,liq) | 0.2641794 |
| | pints (US,liq) | 2.113436 |
| | quarts (US,liq) | 1.056718 |
| lumens | candle power | $7.9577 \times 10^{-2}$ |
| lux | lumens per square meter | 1 |
| meters | angstroms | $1 \times 10^{10}$ |
| | micrometers ($\mu$m) | $1 \times 10^6$ |
| | centimeters | 100 |
| | feet | 3.2808 |
| | inches | 39.37 |
| | kilometers | $1 \times 10^{-3}$ |
| | miles (statute) | $6.2137 \times 10^{-4}$ |
| | millimeters | 1000 |
| | yards | 1.0936 |
| newtons | dynes | $10^5$ |
| pints (US,liq) | cubic centimeters | 473.17647 |
| | gallons | 0.125 |
| | liters | 0.47316 |
| | ounces | 16 |
| | quarts | 0.5 |
| watts | British thermal units per second | $9.4845 \times 10^{-4}$ |
| | calories per minute | 14.3197 |
| | ergs per second | $1 \times 10^7$ |
| | foot-pounds per minute | 44.2537 |
| | horsepower | $1.341 \times 10^{-3}$ |
| | joules per second | 1 |

Temperature conversion.

$$°C = 5/9(°F - 32)$$
$$°F = 9/5(°C) + 32$$
$$0°K = -273.15°C = -459.67°F$$
$$0°C = 273.15°K = 32°F$$

Useful formulas.

| | | | |
|---|---|---|---|
| Electric potential | $E = IR = q/C$ | Electrostatic force of attraction | $F = \dfrac{q_1\, q_2}{\epsilon r^2}$ |
| | $E$ = electric potential (voltage) | | |
| | $I$ = current | | $r$ = distance separating $q_1$ and $q_2$ |
| | $R$ = resistance | | $\epsilon$ = dielectric constant |
| | $q$ = charge | | |
| | $C$ = capacitance | Potential energy | $E = mgh$ |
| | | | $h$ = height of mass above surface of Earth |
| Power | $p = w/t$ | | |
| | $w$ = work | | |
| | $t$ = time | Kinetic energy | $E = 1/2mv^2$ |
| | | | $v$ = velocity of mass |
| Electric power | $p = RI^2 = EI$ | | |
| | $E$ = electric potential | Energy of a charge | $E = 1/2qV$ |
| | | | $q$ = charge |
| Work | $W = RI^2t = EIt = Pt$ | | $V$ = electric potential |
| Pressure | $P$ = force (f)/unit area | Perfect Gas Law | $PV = nRT$ |
| | | | $P$ = pressure |
| Weight | $W = mg$ | | $V$ = volume |
| | $m$ = mass | | $n$ = number of moles |
| | $g$ = acceleration of gravity | | $R$ = gas constant |
| | | | $T$ = absolute temperature |
| Force | $f = ma$ | | |
| | $m$ = mass | Hooke's Law of Elasticity | $F = kT$ |
| | $a$ = acceleration | | $k$ = spring constant |
| | | | $F$ = force |
| Dalton's Law of Partial Pressures | $PV = V(p_1 + p_2 + p_3 + \ldots + p_n)$ | | $T$ = tension |
| | $P$ = pressure of gas mixture | | |
| | $V$ = volume | Energy of a photon | $E = h\nu$ |
| | $p$ = pressure of each gas alone | | $h$ = Planck's constant |
| | | | $\nu$ = frequency |

Physical and chemical constants.

| | |
|---|---|
| Avogadro's number | $N_A = 6.022 \times 10^{23}$ |
| Faraday constant | $F = 96,487$ C/mol |
| Gas constant | $R = 8.314$ J/K-mol |
| | $= 1.98$ cal/K-mol |
| | $= 0.082$ L-atm/K-mol |
| Planck's constant | $h = 6.62 \times 10^{-27}$ ergs/s |
| | $= 1.58 \times 10^{-34}$ cal/s |
| Speed of light in a vacuum | $c = 2.997 \times 10^{10}$ cm/s |
| | $= 186,000$ mi/s |

Dimensions of plane and solid figures.

Area of a square $= l^2$
Surface area of a cube $= 6l^2$
Volume of a cube $= l^3$
Circumference of a circle $= 2\pi r$
Area of a circle $= \pi r^2$
Surface area of a sphere $= 4\pi r^2$
Volume of a sphere $= 4/3\pi r^3$
Surface area of a cylinder $= 2\pi rh$
Volume of a cylinder $= \pi r^2 h$

Chemical definitions.

1 mol = the mass in grams of a substance equal to its molecular or atomic weight; this mass contains Avogadro's number ($N_A$) of molecules or atoms

Molar volume = the volume occupied by a mole of gas at standard temperature and pressure ($25°$C, 1 atm) $= 22.414$ L

1 molal solution = 1 mol per 1000 g of solvent

1 molar solution = 1 mol of solute in 1 L of solution

1 equivalent = 1 mol of 1 unit charge

1 einstein = 1 mol of photons

# *Glossary*

**A band**  A region of a muscle sarcomere that corresponds to the myosin thick filaments.

**Abomasum**  The true digestive stomach of the ruminant digastric stomach.

**Absolute temperature**  Temperature measured from absolute zero, the state of no atomic or molecular thermal agitation. The absolute scale is divided into kelvin units (K), with 1 K having the same size as 1°C. Thus, 0 K is equal to $-273.15°C$ or $-459.67°F$.

**Acclimation**  The persisting change in a specific function due to prolonged exposure to an environmental condition such as high or low temperature.

**Acclimatization**  The persisting spectrum of changes due to prolonged exposure to environmental conditions such as high or low temperature.

**Accommodation**  The temporary increase in threshold that develops during the course of a subthreshold stimulus.

**Acetylcholine (ACh)**  An acetic acid ester of choline $(CH_3—CO—O—CH_2—CH_2—N(CH_3)_3—OH)$, important as a synaptic transmitter.

**Acetylcholinesterase**  An enzyme that hydrolyzes ACh and resides on postsynaptic membrane surface.

**Acid**  Proton donor.

**Acidosis**  Excessive body acidity.

**Acinus**  A small sac or alveolus lined with exocrine-secreting cells.

**Acromegaly**  Hypersecretion of growth hormone in adulthood, causing enlargement of the skeletal extremities and facial structures.

**Actin**  A muscle protein. G-actin is the globular monomer that polymerizes to form F-actin, the backbone of the thin filaments of the sarcomere.

**Action potential (nerve impulse; spike)**  Transient all-or-none reversal of a membrane potential produced by regenerative inward current in excitable membranes.

**Action spectrum**  The degree of response to incident light of a given energy as a function of wavelength.

**Activating reaction**  A reaction that changes an inactive enzyme into an active catalyst.

**Activation energy**  The energy required to bring reactant molecules to velocities sufficiently high to break or make chemical bonds.

**Activation heat**  Heat produced during excitation and activation of muscle tissue, but independent of shortening.

**Active site**  Catalytic region of an enzyme molecule.

**Active state**  Condition of relative inextensibility of muscle, before and during contraction, due to attachment of myosin cross bridges to actin filaments.

**Active transport**  Energy-requiring translocation of a substance across a membrane, usually against its concentration or electrochemical gradient.

**Activity**  Capacity of a substance to react with another; the effective concentration of an ionic species in the free state.

**Activity coefficient**  A proportionality factor obtained by dividing the effective reactive concentration of an ion (as indicated by its properties in a solution) by its molar concentration.

**Actomyosin**   A complex of the muscle proteins actin and myosin.

**Acuity**   Resolving power.

**Adaptation**   Decrease in sensitivity during sustained presentation of a stimulus.

**Adductor muscle**   One that brings a limb toward the median plane of the body.

**Adenine**   A white, crystalline base 6-amino-purine, $C_5H_5N_5$; purine base constituent of DNA and RNA.

**Adenohypophysis (anterior pituitary gland; anterior lobe)**   The glandular anterior lobe of the hypophysis, consisting of the pars infundibularis and pars distalis.

**Adenosine diphosphate (ADP)**   A nucleotide formed by hydrolysis of ATP, with the release of one high-energy bond.

**Adenosine triphosphatase**   *See* ATPase.

**Adenosine triphosphate (ATP)**   A nucleotide used as a common energy currency by all cells.

**Adenylate cyclase (adenyl cyclase)**   A membrane-bound enzyme that catalyzes the conversion of ATP to cAMP.

**Adipose**   Fatty.

**ADP**   *See* Adenosine diphosphate.

**Adrenalin**   Trade name for epinephrine.

**Adrenergic**   Relating to neurons or synapses that release epinephrine, norepinephrine, and other catecholamines when stimulated.

**α-Adrenergic receptors**   Receptors on cell surfaces that bind norepinephrine and, less effectively, epinephrine; the binding leads to enzymatically mediated responses by the cells.

**β-Adrenergic receptors**   Epinephrine-binding sites.

**Adrenocorticotrophic hormone (ACTH; adrenocorticotropin; corticotropin)**   A hormone released by cells in the adenohypophysis that acts mainly on the adrenal cortex, stimulating growth and corticosteroid production and secretion in that organ.

**Aequorin**   A protein extracted from the jellyfish *Aequorea;* on combining with $Ca^{2+}$, it emits a photon of blue-green light.

**Aerobic**   Utilizing molecular oxygen.

**Afferent**   Transporting or conducting toward a central region; centripetal.

**Afferent fiber**   An axon that relays impulses from a sensory receptor to the central nervous system.

**1a Afferent fiber**   An axon with a peripheral sensory ending innervating a muscle spindle organ and responding to stretch of the organ; its central terminals synapse directly onto alpha motoneurons of the homonymous muscle.

**1b Afferent fiber**   An axon whose sensory terminals innervate the tendons of skeletal muscle and respond to tension.

**Affinity sequence (selectivity sequence)**   The order of preference with which an electrostatic site will bind different species of counterions.

**Aldehydes**   A large class of substances derived from the primary alcohols by oxidation and containing the —CHO group.

**Aldosterone**   A mineralocorticoid secreted by the adrenal cortex; the most important electrolyte-controlling steroid, acting on the renal tubules to increase the reabsorption of sodium.

**Alkali earth metals**   A group of grayish-white, malleable metals easily oxidized in air, comprising Be, Mg, Ca, Sr, Ba, and Ra.

**Alkaline tide**   A period of increased body and urinary alkalinity associated with excessive gastric HCl secretion during digestion.

**Alkaloids**   A large group of organic nitrogenous bases found in plant tissues, many of which are pharmaceutically active (e.g., codeine, morphine).

**Alkalosis**   Excessive body alkalinity.

**Allantoic membrane**   One of the membranes within a bird eggshell; important in the respiration of the unhatched chick.

**Allantoin**   Waste product of purine metabolism.

**All-or-none**   Pertaining to the independence of response magnitude from the strength of the stimulus; response is "all" if the stimulus achieves threshold and "none" if the stimulus fails to achieve threshold.

**Allosteric site**   Area of an enzyme that binds a substance other than the substrate, changing the conformation of the protein so as to alter the catalytic effectiveness of the active site.

**Alpha helix**   Helical secondary structure of many proteins in which each NH group is hydrogen-bonded to a CO group at a distance equivalent to three amino acid residues; the helix makes a complete turn for each 3.6 residues.

**Alpha motoneurons**   Large spinal neurons that innervate extrafusal skeletal muscle fibers of vertebrates.

**Alpha receptors**   The class of adrenergic membrane receptors that are blocked by the drug phenoxybenzamine; their activation is highly sensitive to norepinephrine.

**Alveolar ventilation volume**   The volume of fresh atmospheric air entering the alveoli during each inspiration.

**Alveoli**   Small cavities, especially those microscopic cavities that are the functional units of the lung.

**Amacrine cells**   Neurons without axons, found in the inner plexiform layer of the vertebrate retina.

**Ambient**   Surrounding, prevailing.

**Amide**   An organic derivative of ammonia in which a hydrogen atom is replaced by an acyl group.

**Amine** Derivative of ammonia in which at least one hydrogen atom is replaced by an organic group.

**Amino acids** Class of organic compounds containing at least one carboxyl group and one amino group; the alpha-amino acids, $RCH(NH_2)COOH$, make up proteins.

**Amino group** $-NH_2$.

**Ammonia** $NH_3$; toxic, water-soluble, alkaline waste product of deamination of amino acids and uric acid.

**Ampere (A)** MKS unit of electric current; equal to the current produced through a 1 ohm ($\Omega$) resistance by a potential difference of 1 volt (V); the movement of 1 coulomb (C) of charge per second.

**Amphipathic** Pertaining to molecules bearing groups with different properties, such as hydrophilic or hydrophobic groups.

**Amphoteric** Having opposite characteristics; behaving as either an acid or a base.

**Anabolism** Synthesis by living cells of complex substances from simple substances.

**Anastomose** To interconnect.

**Anatomical dead space** The nonrespiratory conducting pathways in the lung.

**Androgens** Hormones having masculinizing activity.

**Aneurism** Localized dilation of an artery wall.

**Angiotensin** A protein in the blood, converted from angiotensinogen by the action of renin; it first exists as a decapeptide (angiotensin I) that is acted upon by a peptidase, which cleaves it into an octapeptide (angiotensin II), a potent vasopressor and stimulator of aldosterone secretion.

**Anion** Negatively charged ion; attracted to the anode or positive pole.

**Anode** Positive electrode or pole to which negatively charged ions are attracted.

**Anoxemia** A lack of oxygen in the blood.

**Anoxia** A lack of oxygen.

**Antagonist muscle** A muscle acting in opposition to the movement of another muscle.

**Antennal gland** Crustacean osmoregulatory organ.

**Antibody** An immunoglobulin, a 4-chain protein molecule of a specific amino acid sequence; an antibody will interact only with the antigen that brought about its production or one very similar to it.

**Antidiuretic hormone (ADH)** A hormone made in the hypothalamus and liberated from storage in the neurohypophysis; acts on the epithelium of the renal collecting duct by stimulating osmotic reabsorption of water, thereby producing a more concentrated urine; also acts as a vasopressor.

**Antigen** A substance capable of bringing about the production of antibodies and then to react with them specifically.

**Antimycin** An antibiotic that is isolated from a *Streptomyces* strain; acts to block electron transport from cytochrome *b* to cytochrome *c* in the electron-transport chain.

**Aorta** The main artery leaving the heart.

**Aortic body** A nodule on the aortic arch containing chemoreceptors that sense the chemical composition of the blood.

**Apical** Pertaining to the apex; opposite the base.

**Apnea** The suspension or absence of breathing.

**Apoenzyme** The protein portion of an enzyme; the apoenzyme and coenzyme form the functioning holoenzyme.

**Apolysis** Release; loosening from.

**Aporepressor** A repressor gene product that, in combination with a corepressor, reduces the activity of particular structural genes.

**Arteriole** A tiny branch of an artery; in particular, one nearest a capillary.

**Arteriosclerosis** A class of diseases marked by an increase in thickness and a reduction in elasticity of the arterial walls.

**Asynchronous muscle** A type of flight muscle found in the thorax of some insects; contracts without any one-to-one relation to motor impulses. *See also* Fibrillar muscle.

**ATP** *See* Adenosine triphosphate.

**ATPase (adenosine triphosphatase)** A class of enzymes that catalyze the hydrolysis of ATP.

**ATPS** Ambient temperature and pressure, saturated with water vapor; referring to gas volume measurements.

**Atrioventricular node** Specialized conduction tissue in the heart, which, along with Purkinje tissue, forms a bridge for electrical conduction of the impulse from atria to ventricles.

**Autonomic nervous system** The efferent nerves that control involuntary visceral functions; classically subdivided into the sympathetic and parasympathetic sections.

**Autoradiography** The process of making a photographic record of the internal structures of a tissue by utilizing the radiation emitted from incorporated radioactive material.

**Autorhythmicity** The generation of rhythmic activity without extrinsic control.

**Autotrophic** Pertaining to the ability to synthesize food from inorganic substances by utilizing the energy of the sun or of inorganic compounds.

**Avogadro's Law** Equal volumes of different gases at the same temperature and pressure contain equal numbers of molecules. One mole (mol) of an ideal gas at $0°C$ and 1 standard atmosphere (atm) occupies 22.414 liters (L).

Avogadro's number equals $6.02252 \times 10^{23}$ molecules/mol.

**Axon**    The elongated cylindrical process of a nerve cell along which action potentials are conducted; a nerve fiber.

**Axoneme**    Complex of microtubules and associated structures within the flagellar or ciliary shaft.

**Axon hillock**    The transitional region between an axon and the nerve cell body.

**Axoplasm**    The cytoplasm within an axon.

**Azide**    Any compound bearing the $N_3^-$ group.

**Basal body (kinetosome)**    Microtubular structure from which a cilium or flagellum arises; homologous with centriole.

**Basal metabolic rate (BMR)**    The rate of energy conversion in a homeotherm while it is resting quietly within the thermal neutral zone without food in the intestine.

**Base**    Proton acceptor.

**Basilar membrane**    The delicate ribbon of tissue bearing the auditory hair cells in the cochlea of the vertebrate ear.

**Bell–Magendie rule**    The dorsal root of the spinal cord contains only sensory axons, whereas the ventral root contains only motor axons.

**Beta keratin**    Insoluble, sulfur-rich scleroprotein; constituent of epidermis, horns, hair, feathers, nails, and tooth enamel. *Beta* refers to the protein's secondary structure, which is in pleated sheets.

**Beta receptors**    The class of adrenergic membrane receptors that are blocked by the drug propranolol; their activation is less sensitive to norepinephrine than is that of the alpha receptors.

**Bile**    Viscous yellow or greenish alkaline fluid produced by the liver and stored in the gallbladder; containing bile salts, bile pigments, and certain lipids, it is essential for digestion of fats.

**Bile salt**    Bile acid such as cholic acid conjugated with glycine or taurine, promoting emulsification and solubilization of intestinal fats.

**Bipolar cell**    A neuron with two axons emerging from opposite sides of the soma; one class of such neurons is found in the vertebrate retina, where they transmit signals from the visual receptor cells to the ganglion cells of the optic nerve.

**Birefringence**    Double refraction; the ability to pass preferentially light that is polarized in one plane.

**Bleaching**    Fading of photopigment color upon absorption of light.

**Bohr effect (Bohr shift)**    A change in hemoglobin–oxygen affinity due to a change in pH.

**Bombykol**    Sex attractant pheromone of the female silkworm moth (*Bombyx mori*).

**Book lungs**    The respiratory surface in spiders.

**Bowman's capsule (glomerular capsule)**    A globular expansion at the beginning of a renal tubule and surrounding the glomerulus.

**Boyle's Law**    At a given temperature, the product of pressure and volume of a given mass of gas is constant.

**Bradycardia**    A reduction in heart rate from the normal level.

**Bradykinin**    A hormone formed from a precursor normally circulating in the blood; a very potent cutaneous vasodilator.

**Brain hormone (prothoracotropin; activating hormone)**    A hormone synthesized by the neurosecretory cells of the pars intercerebralis and released by the corpora cardiaca of insects; activates the prothoracic glands to secrete ecdysone.

**Bronchi**    Conducting airways in the lung; branches of the tracheae.

**Bronchioles**    Small conducting airways in the lung; branches of the bronchi.

**Brood spot**    A prolactin-induced bald area on the ventral surface of some brooding birds that receives a rich supply of blood for the incubation of eggs.

**Brunner's glands**    Exocrine glands that are located in the intestinal mucosa and secrete an alkaline mucoid fluid.

**Brush border**    A free epithelial cell surface bearing numerous microvilli.

**BTPS**    Body temperature, atmospheric pressure, saturated with water vapor.

**Buccal**    Pertaining to the mouth cavity.

**Buffer**    A chemical system that stabilizes the concentration of a substance; acid–base systems serve as pH buffers, preventing large changes in hydrogen ion concentration.

**Bundle of His**    The conducting tissue within the interventricular septum of the mammalian heart.

**Bungarotoxin (BuTX)**    A blocking agent comprised of a group of neurotoxins isolated from the venom of members of the snake genus *Bungarus* (the krait) of the cobra family; binds selectively and irreversibly to acetylcholine receptors.

**Bunsen solubility coefficient**    The quantity of gas at STPD that will dissolve in a given volume of liquid per unit partial pressure of the gas in the gas phase. This coefficient is used only for gases that do not react chemically with the solvent.

**Bursicon**    A hormone secreted by neurosecretory cells of the insect central nervous system; tans and hardens the cuticle of freshly molted insects.

**Cable properties** Passive resistive and capacitive electrical properties of a cell, akin to those exhibited by a submarine cable.

**Calcitonin (thyrocalcitonin)** A protein hormone secreted by the mammalian parafollicular cells of the thyroid in response to elevated plasma calcium levels.

**Calcium response** A graded depolarization due to a weakly regenerative inward calcium current.

**Calmodulin** A troponinlike calcium-binding regulatory protein found in essentially all tissues.

**Calorie (cal)** The quantity of heat required to raise the temperature of 1 g of water from 14.5° to 15.5°C; most commonly used as kilocalorie (kcal) = 1000 cal.

**Calorimetry** Measurement of heat production in an animal.

**Capacitance** The property of storing electric charge by electrostatic means.

**Capacitive current** Current entering and leaving a capacitor.

**Capacity** The ability of a capacitor or other body to store electric charge. The unit of measure is the farad (fd), which describes the proportionality between charge stored and potential for a given voltage, $C = q/V =$ coulombs/volts.

**Carbohydrate** Aldehyde or ketone derivative of alcohol; utilized by animal cells primarily for the storage and supply of chemical energy; most important are the sugars and starches.

**Carbonic anhydrase** An enzyme reversibly catalyzing the degradation of carbonic acid to carbon dioxide and water.

**Carbonyl** The organic radical $-C=O$, which occurs in such compounds as aldehydes, ketones, carboxylic acids, and esters.

**Carboxyhemoglobin** The compound formed when carbon monoxide combines with hemoglobin; carbon monoxide competes successfully with oxygen for combination with hemoglobin, producing tissue anoxia.

**Carboxylates** $R-COO-$, salts or esters of carboxylic acids.

**Carboxyl group** The radical $-COOH$, which occurs in the carboxylic acids.

**Cardiac output** The total volume of blood pumped by the heart per unit of time; cardiac output equals heart rate times stroke volume.

**Carotid body** A nodule on the occipital artery just above the carotid sinus, containing chemoreceptors that sense the chemical composition of arterial blood.

**Carotid sinus baroreceptors** Receptors that sense arterial blood pressure; located in the carotid sinus, a dilatation of the internal carotid artery at its origin.

**Carrier molecules** Lipid-soluble molecules that act within biological membranes as carriers for certain molecules that have lower mobility in the membrane.

**Catabolism** Disassembly of complex molecules into simpler ones.

**Catalysis** An increase in the rate of a chemical reaction promoted by a substance—the catalyst—not consumed by the reaction.

**Catalyst** A substance that increases the rate of a reaction without being used up in the reaction.

**Catecholamines** A group of related compounds exerting a sympathomimetic action on nervous tissue; examples are epinephrine, norepinephrine, and dopamine.

**Cathode** The negative electrode, so called because it is the electrode to which cations are attracted.

**Cation** A positively charged ion; attracted to a negatively charged electrode.

**Caudal** Pertaining to the tail end.

**Cecum** A blind pouch in the alimentary canal.

**Central sulcus** A deep, almost vertical furrow on the cerebrum, dividing the frontal and parietal lobes.

**Cephalic** Pertaining to the head.

**Cerebellum** The part of the hind brain that is involved in coordination of motor output.

**Cerebral hemispheres** The large paired structures of the cerebrum, connected by the corpus callosum.

**Cerebrospinal fluid** A clear fluid that fills the cavities (ventricles) within the brain; it is a complex filtrate of blood plasma and is modified by brain cells before returning to the venous system.

**Charge, electric (q)** Measured in units of coulombs (C). To convert 1 g-equiv weight of a monovalent ion to its elemental form (or vice versa) requires a charge of 96,500 C (1 faraday, *F*). Thus, in loose terms, a coulomb is equivalent to 1/96,500 g-equiv of electrons. The charge on one electron is $-1.6 \times 10^{-19}$ C. If this is multiplied by Avogadro's number, the total charge is 1 *F*, or $-96,487$ C/mol.

**Chemoreceptor** A sensory receptor specifically sensitive to certain molecules.

**Chief cells** Epithelial cells of the gastric epithelium that release pepsin.

**Chitin** A structural polymer of D-glucosamine that serves as the primary constituent of arthropod exoskeletons.

**Chloride cells** Epithelial cells of fish gills that engage in active transport of salts.

**Chloride shift** The movement of chloride ions across the red blood cell membrane to compensate for the movement of bicarbonate ions.

**Chlorocruorin** A green respiratory pigment found in some marine polychaetes; similar to hemoglobin.

**Cholecystokinin (CCK; pancreozymin; CCK-PZ)** A hormone liberated by the upper intestinal mucosa that induces gallbladder contraction and release of pancreatic enzymes.

**Cholesterol** A natural sterol; precursor to the steroid hormones.

**Cholinergic** Relating to acetylcholine or substances with actions similar to ACh.

**Choroid plexus** Highly vascularized, furrowed projections into the brain ventricles that secrete cerebrospinal fluid.

**Choroid rete** A countercurrent arrangement of arterioles and venules behind the retina in the eyes of teleost fish.

**Chromaffin cells** Epinephrine-secreting cells of the adrenal medulla; named for their high affinity for chromium salt stains.

**Chromophore** A chemical group that lends a distinct color to a compound containing it.

**Chronotropic** Pertaining to rate or frequency, especially in reference to the heartbeat.

**Chyme** The mixture of partially digested food and digestive juices found in the stomach and the intestine.

**Ciliary body** A thick region of the anterior vascular tunic of the eye; joins the choroid and the iris.

**Ciliary muscle** A muscle of the ciliary body of the vertebrate eye; influences the shape of the lens in visual accommodation.

**Ciliary reversal** A change in the direction of the power stroke of a cilium, causing it to beat in reverse.

*Cis* A configuration with similar atoms or groups on the same side of the molecular backbone.

*Cis–trans* **isomerization** Conversion of a *cis* isomer into a *trans* isomer.

**Cochlear microphonics** Electrical signals recorded from the cochlea, having a frequency identical to that of the sound stimulus.

**Coelom** The body cavity of higher metazoans, situated between the gut and the body wall and lined with mesodermal epithelium.

**Coenzyme** An organic molecule that combines with an apoenzyme to form the functioning holoenzyme.

**Coenzyme A (CoA)** A derivative of pantothenic acid to which acetate becomes attached to form acetyl CoA.

**Cofactor** An atom, ion, or molecule that combines with an enzyme to activate it.

**Coitus** Sexual intercourse.

**Colchicine** An agent that disrupts microtubules by interfering with the polymerization of tubulin monomers.

**Collar cells** Flagellated cells lining the internal chambers of sponges (Porifera).

**Collaterals** Minor side branches of a nerve or blood vessel.

**Collecting duct** The portion of the renal tubules in which the final concentration of urine occurs.

**Colligative properties** Characteristics of a solution that depend on the number of molecules in a given volume.

**Colloid** A system in which fine solid particles are suspended in a liquid.

**Competitive inhibition** Reversible inhibition of enzyme activity due to competition between a substrate and an inhibitor for the active site of the enzyme.

**Compliance** The change in dimension per unit change in the applied force.

**Compound eye** The multifaceted arthropod eye; the functional unit is the ommatidium.

**Condensation** A reaction between two or more organic molecules leading to the formation of a larger molecule and the elimination of a simple molecule, such as water or alcohol.

**Conductance, electrical (G)** A measure of the ease with which a conductor carries an electric current; the unit is the siemens (S), reciprocal of the ohm ($\Omega$).

**Conductance, thermal** A quantity describing the ease with which heat flows by conduction under a temperature gradient across a substance or an object.

**Conductivity** The intrinsic property of a substance to conduct electric current; reciprocal of resistivity.

**Conductor** A material that carries electrical current.

**Cone** A vertebrate visual receptor cell having a tapered outer segment in which the lamellar membranes remain continuous with the surface membrane.

**Conjugate acid–base pair** Two substances related by gain or loss of an $H^+$ ion (proton).

**Contracture** A more or less sustained contraction in response to an abnormal stimulus.

**Contralateral** Pertaining to the opposite side.

**Conus** A chamber invested with cardiac muscle and found in series with and downstream from the ventricle in elasmobranchs.

**Corepressor** A low-molecular-weight molecule that unites with an aporepressor to form a substance that inhibits the synthesis of an enzyme.

**Corpora allata** Nonneural insect glands existing as paired organs or groups of cells dorsal and posterior to the corpus cardiaca; the corpora allata secrete juvenile hormone (JH).

**Corpora cardiaca** Major insect neurohemal organs existing as paired structures immediately posterior to the brain; they liberate brain hormone.

**Corpus luteum** The yellow ovarian glandular body that arises from a mature follicle that has released its ovum; it secretes progesterone. If the ovum released has been fertilized, the corpus luteum grows and secretes during gestation; if not, it atrophies and disappears.

**Cortex**   External or surface layer of an organ.

**Cortisol**   A steroid hormone secreted by the adrenal cortex.

**Cotransport**   Carrier-mediated transport in which two dissimilar molecules bind to two specific sites on the carrier molecule for transport in the same direction.

**Coulomb (C)**   MKS unit of electric charge; equal to the amount of charge transferred in 1 second (s) by 1 ampere (A) of current. *See also* Charge, electric.

**Countercurrent multiplier**   A pair of opposed channels containing fluids flowing in opposed directions and having an energetic gradient directed transversely from one of the channels into the other. Since exchange due to the gradient is cumulative with distance, the exchange per unit distance will be multiplied, so to speak, as a function of the total distance over which exchange takes place.

**Counterion**   An ion associated with, and of opposite charge to, an ion or an ionized group of a molecule.

**Countertransport**   The uphill membrane transport of one substance driven by the downhill diffusion of another substance.

**Coupled transport**   Uptake of one substance into a cell that depends on the downhill diffusion exit of another substance into the cell.

**Covalent bond**   A bond formed by electron-pair sharing between two atoms.

**Creatine phosphate**   A phosphorylated nitrogenous compound found primarily in muscle; acts as a storage form of high-energy phosphate for the rapid phosphorylation of ADP to ATP.

**Creatinine**   Nitrogenous waste product of muscle creatine.

**Cretinism**   A chronic condition caused by hypothyroidism in childhood; characterized by arrested physical and mental development.

**Critical fusion frequency**   The number of light flashes per second at which the light is perceived to be continuous.

**Crop milk**   A nutrient-rich substance fed by regurgitation to pigeon chicks by both parents.

**Cross-bridge link**   The thin connection between the globular head of the myosin cross bridge and the myosin thick filament.

**Cross bridges**   Spirally arranged projections from the myosin thick filaments that interact with the actin thin filaments during muscle contraction.

**Crustecdysone**   A steroid hormone that promotes molting in crabs.

**Cupula**   A small upside-down cup or domelike cap housing another structure; in lateral-line and equilibrium organs, the cupula covers hair cells in a gelatinous matrix.

**Curare (*d*-tubocurarine)**   South American arrow poison; blocks synaptic transmission at the motor endplate by competitive inhibition of acetylcholine receptors.

**Current, electric (*I*)**   The flow of electric charge. A current of 1 coulomb (C) per second is called an ampere (A). By convention the direction of current flow is the direction that a positive charge moves (i.e., from the anode to the cathode).

**Cyanide**   A compound containing cyanogen and one other element; blocks transfer of electrons from the terminal cytochrome $a$ and $a_3$ to oxygen in the respiratory chain.

**Cybernetics**   The science of information communication and control in animals and machines.

**Cyclic AMP (cAMP)**   A ubiquitous cyclic nucleotide (3′,5′-cyclic adenosine monophosphate) produced from ATP by the enzymatic action of adenylate cyclase; important cellular regulatory agent that acts as the second messenger for many hormones and transmitters.

**Cyclic GMP (cGMP)**   A cyclic nucleotide (guanosine 3′,5′-monophosphate) analogous to cAMP but present in cells at a far lower concentration and producing target cell responses that are usually opposite to those of cAMP.

**Cyclostomes**   A group of jawless vertebrates, including lampreys and hagfishes.

**Cytochalasin**   A drug that disrupts cytoplasmic microfilaments.

**Cytochromes**   A group of iron-containing proteins that function in the electron-transport chain in aerobic cells; they accept and pass on electrons.

**Cytoplasm**   The semifluid substance within a cell, exclusive of the nucleus but including other organelles.

**Cytosine**   Oxyamino-pyrimidine, $C_4H_5N_3O$; base component of nucleic acid.

**Cytosol**   The unstructured aqueous phase of the cytoplasm between the structured organelles.

**Dalton's Law**   The partial pressure of a gas in a mixture is independent of other gases present. The total pressure is the sum of the partial pressures of all gases present.

**Dark current**   Steady sodium current leaking into a visual receptor cell at the outer segment. The sodium is actively pumped out at the inner segment, completing the circuit. The dark current is reduced by photoexcitation.

**Decerebration**   Experimental elimination of cerebral activity by section of the brain stem or by interruption of the blood supply to the brain.

**Decussation**   Crossing over from one side to the other.

**Dehydrogenase**   An enzyme that "loosens" the hydrogen of a substrate in preparation for passage to a hydrogen receptor.

**Dehydroretinal**　An aldehyde of dehydroretinol.

**Dehydroretinol (retinol 2; vitamin A₂)**　The form of vitamin A occurring in the liver and retina of freshwater fishes, some invertebrates, and amphibians.

**Delayed outward current (late outward current)**　Current carried by K⁺ through channels that open with a time lag after onset of a depolarization; responsible for repolarization of the action potential.

**Denaturation**　Alteration or destruction of the normal nature of a substance by chemical or physical means.

**Dendrites**　Fine processes of a neuron, often providing the main receptive area of the cell for synaptic contacts.

**Deoxyhemoglobin**　Hemoglobin in which oxygen is not combined to the $Fe^{3+}$ of the heme moiety.

**Deoxyribonucleic acid**　*See* DNA.

**Depolarization**　The reduction or reversal of the potential difference that exists across the cell membrane at rest.

**Diabetes mellitus**　A metabolic malady in which there is a partial or complete loss of activity in the pancreatic islets; the concomitant insulin insufficiency leads to inadequate uptake of glucose into cells and loss of blood glucose in the urine.

**Dialysis**　The process by which crystalloids and macromolecules are separated by utilizing the differences in their diffusion rates through a semipermeable membrane.

**Diastole**　The phase in the heartbeat during which the myocardium is relaxed and the chambers are filling with blood.

**Dielectric constant**　A measure of the degree to which a substance is able to store electric charge under an applied voltage; depends on charge distribution within molecules.

**Diffusion**　Dispersion of atoms, molecules, or ions as a result of random thermal motion.

**1,25-Dihydroxycholecalciferol**　A substance that is converted from vitamin D in the liver and increases $Ca^{2+}$ absorption by the kidney.

**Dimer**　A molecule made by the joining of two molecules (i.e., monomers) of the same kind.

**Dinitrophenol (DNP)**　Any of a group of six isomers, $C_6H_3(OH)(NO_2)_2$, that act as aerobic metabolic inhibitors by virtue of their ability to uncouple oxidation from phosphorylation in mitochondrial electron transport.

**Dipole**　A molecule with separate regions of net negative and net positive charge, so that one end acts as a positive pole and the other as a negative pole.

**Dipole moment**　The electrostatic force required to align a dipolar molecule parallel to the electrostatic field; the force required increases as the separation of the molecular charges decreases.

**Dissociation**　Separation; resolution by thermal agitation or solvation of a substance into simpler constituents.

**Dissociation constant**　$K' = [H^+][A^-]/[HA]$. The empirical measure of the degree of dissociation of a conjugate acid–base pair in solution.

**Distal**　More distant from a point of reference in the centrifugal direction; i.e., away from the center.

**Distal convoluted tubule**　The renal tubules located in the renal cortex leading from (and continuous with) the ascending limb of the loop of Henle to the collecting duct.

**Distance of closest approach**　Shortest possible span between the centers of two atoms.

**Diuretic**　An agent that increases urine secretion.

**Divalent**　Carrying an electric charge of two units; a valence of two.

**DNA (deoxyribonucleic acid)**　The class of nucleic acids responsible for hereditary transmission and for the coding of amino acid sequences of proteins.

**Donnan equilibrium**　Electrochemical equilibrium that develops when two solutions are separated by a membrane permeable to only some of the ions of the solutions.

**Dopamine (hydroxytyramine)**　A product of the decarboxylation of dopa, an intermediate in norepinephrine synthesis; possibly a central nervous system transmitter.

**Dorsal root**　A nerve trunk that enters the spinal cord near the dorsal surface; contains sensory axons only.

**D600**　α-Isopropyl-α-[(*N*-methyl-*N*-homoveratryl)-α-aminopropyl]-3,4,5-trimethoxyphenylacetonitrile; an organic drug that blocks calcium influx through cell membranes.

**Duodenum**　The initial section of the small intestine, situated between the pylorus of the stomach and the jejunum.

**Dwarfism**　An abnormally small size in humans; a result of insufficient growth hormone secretion during childhood and adolescence.

**Dynein**　A ciliary protein with magnesium-activated ATPase activity.

**Dynein arms**　Projections from tubule A of one microtubule doublet toward tubule B of the next, composed of a protein exhibiting ATPase activity.

**Dyspnea**　Labored, difficult breathing.

**Early inward current**　Depolarizing current of excitable tissues, carried by Na⁺ or $Ca^{2+}$; responsible for the upstroke of the action potential.

**Early receptor potential (ERP)**　An almost instantaneous potential change recorded from the retina in response to a short flash of light that probably corresponds to a movement of charge that occurs as the photopigment undergoes conformational change.

**Eccentric cell**   In the ommatidium of *Limulus,* the afferent neuron that is surrounded by retinular cells.

**Ecdysis**   Shedding of the outer shell; molting in an arthropod.

**Ecdysone**   A steroid hormone secreted by the thoracic gland of arthropods that induces molting.

**Ectopic pacemaker**   A pacemaker situated outside the area where it is normally found.

**Ectoplasm**   The gel-state cytoplasm surrounding the endoplasm.

**Edema**   Retention of interstitial fluid in organs or tissues.

**EDTA**   Ethylenediamine tetraacetic acid; a calcium- and magnesium-chelating agent.

**Effector**   A tissue or organ that produces a response (i.e., contraction, secretion, etc.) to nerve signals or to a hormone.

**Efferent**   Centrifugal; carrying away from a center.

**EGTA**   Ethyleneglycol-*bis*(β-aminoethylether)-*N*, *N*′-tetraacetic acid; a calcium-chelating agent.

**Elastic**   Capable of being distorted, stretched, or compressed with subsequent spontaneous return to original shape; resilient.

**Electric potential**   Electrostatic pressure, analogous to water pressure; a potential difference (i.e., voltage) across a resistance is required for the flow of current.

**Electrocardiogram (ECG)**   The record of electrical events associated with contractions of the heart; obtained with electrodes placed on other portions of the body.

**Electrochemical equilibrium**   The state at which the concentration gradient of an ion across a membrane is precisely balanced by the electric potential.

**Electrode**   An electrical circuit element used to make contact with a solution, a tissue, or a cell interior; used either to measure potential or to carry current.

**Electrogenic**   Giving rise to an electric current or voltage.

**Electrolyte**   A compound that dissociates into ions when dissolved in water.

**Electromotive force (emf)**   The potential difference across the terminals of a battery or other source of electric energy.

**Electronegativity**   Affinity for electrons.

**Electroneutrality rule**   For a net potential of zero, the positive and negative charges must add up to zero; a solution must contain essentially as many anionic as cationic charges.

**Electron shells**   Energy levels of electrons surrounding the nucleus.

**Electron-transport chain (respiratory chain)**   A series of enzymes that transfer electrons from substrate molecules to molecular oxygen.

**Electro-osmosis**   Movement of water through a membrane of fixed charge in response to a potential gradient.

**Electroretinogram (ERG)**   An electrical signal recorded from the cornea of the eye; it represents the total activity of the visual receptors and neurons.

**Electrotonic potential**   Potential generated locally by currents flowing across the membrane; not actively propagated and not all-or-none.

**Endergonic**   Characterized by a concomitant absorption of energy.

**Endocardium**   The internal lining of the heart chamber.

**Endocrine glands**   Ductless organs or tissues that secrete a hormone into the circulation.

**Endocytosis**   Bulk uptake of material into a cell by membrane inpocketing to form an internal vesicle.

**Endogenous**   Arising within the body.

**Endolymph**   Aqueous liquid with a high $K^+$ concentration found in the scala media of the cochlea.

**Endometrium**   An epithelium that lines the uterus.

**Endoplasm**   The sol-state cytoplasm that streams within the cell interior.

**Endorphins**   Neuropeptides that exhibit morphinelike actions, found in the central nervous system of vertebrates; different types consist of 16, 17, or 31 amino residues.

**Endplate**   The traditional name of the vertebrate neuromuscular synapse, where the motor axon ramifies into fine terminal branches over a specialized system of folds in the postsynaptic membrane of the muscle cell.

**End-product inhibition (feedback inhibition)**   Inhibition of a biosynthetic pathway by the end product of the pathway.

**Energy**   Capacity to perform work.

**Enkephalin**   Neuropeptide exhibiting morphinelike actions, found in the central nervous system of vertebrates; the peptide consists of five amino acid residues.

**Enterogastrone**   A hormone that is secreted from the duodenal mucosa in response to fat ingestion and that suppresses a gastric motility and secretion.

**Entropy**   Measure of that portion of energy not available for work in a closed system; measure of molecular randomness. Entropy increases with time in all irreversible processes.

**Enzyme**   A protein with catalytic properties.

**Enzyme activity**   A measure of the catalytic potency of an enzyme: the number of substrate molecules that react per minute per enzyme molecule.

**Enzyme induction**   Enzyme production stimulated by the specific substrate (inducer) of that enzyme or by a molecule structurally similar to the substrate.

**Epicardium**   The external covering of the heart wall.

**Epididymis**   A long, stringlike duct along the dorsal edge of the testis; its function is to store sperm.

**Epinephrine**   Generic name for the catecholamine released from the adrenal cortex; also known by the trade name Adrenalin.

**Equilibrium**   The state in which a system is in balance due to equal action by opposing forces arising from within the system.

**Equilibrium potential**   That voltage difference across a membrane at which the ionic species in question is in electrochemical equilibrium; it is dependent on the concentration gradient of the ions, as described by the Nernst equation.

**Equimolar**   Having the same molarity.

**Eserine (physostigmine)**   An alkaloid ($C_{15}H_{21}N_3O_2$) of plant origin that blocks the enzyme cholinesterase.

**Estradiol-17$\beta$**   The most active natural estrogen.

**Estrogens**   A family of female sex steroids responsible for producing estrus and the female secondary sex characteristics; also prepares the reproductive system for fertilization and implantation of the ovum; synthesized primarily in the ovary, although some is made in the adrenal cortex and male testis.

**Estrous cycle**   Periodic episodes of "heat," or estrus, marked by sexual receptivity in mature females of most mammalian species.

**Ethers**   A class of compounds in which two organic groups are joined by an oxygen atom, $R_1 - O - R_2$.

**Eupnea**   Normal breathing.

**Euryhaline**   Able to tolerate wide variations in salinity.

**Exchange diffusion**   A process by which the movement of one molecule across a membrane enhances the movement of another molecule in the opposite direction; most likely involves a common-carrier molecule.

**Excitability**   The property of altered membrane conductance (and often membrane potential) in response to stimulation.

**Excitation–contraction coupling**   The process by which electrical excitation of the surface membrane leads to activation of the contractile process in muscle.

**Excitation–secretion coupling**   The concept that depolarization brings about secretion of transmitter at nerve endings by one or more intermediate steps known to include the influx of extracellular $Ca^{2+}$.

**Excitatory**   In neurophysiology, pertaining to the enhanced probability of producing an action potential.

**Exergonic**   Characterized by a concomitant release of heat energy.

**Exocrine gland**   A gland that secretes a fluid via a duct.

**Exocytosis**   Fusion of the vesicle membrane to the surface membrane and subsequent expulsion of the vesicle contents to the cell exterior.

**Extensor**   A muscle that extends or straightens a limb or other extremity.

**Exteroceptors**   Sense organs that detect stimuli arriving at the surface of the body from a distance.

**Facilitated transport (carrier-mediated transport)**   Downhill transmembrane diffusion aided by a carrier molecule that enhances the mobility of the diffusing substance in the membrane.

**Facilitation**   An increase in the efficacy of a synapse as the result of a preceding activation of that synapse.

**Farad (fd)**   The unit of electrical capacitance.

**Faraday's constant ($F$)**   The equivalent charge of a mole of electrons, equal to $9.649 \times 10^4$ coulombs (C) per mole of electrons.

**Feedback**   The return of output to the input part of a system. In *negative feedback,* the sign of the output is inverted before being fed back to the input so as to stabilize the output. In *positive feedback,* the output is unstable because it is returned to the input without a sign inversion, and thus becomes self-reinforcing, or regenerative.

**Fermentation**   Enzymatic decomposition; anaerobic transformation of nutrients without net oxidation or electron transfer.

**Ferritin**   A large protein molecule opaque to electrons, used as a marker in electron microscopy; normally present in the spleen as a storage protein for iron.

**Fibrillar muscle**   Oscillatory insect flight muscle; also termed asynchronous muscle because contractions are not individually controlled by motor impulses.

**Fibroblast**   A connective-tissue cell that can differentiate into chondroblasts, collagenoblasts, and osteoblasts.

**Final common pathway**   The concept that the sum total of neural integrative activity expressed in motor output is channeled through the motoneurons to the muscles.

**Firing level**   Potential threshold for the generation of an action potential.

**First Law of Thermodynamics**   Net energy is conserved in any process.

**First-order enzyme kinetics**   Describes enzymatic reactions, the rates of which are directly proportional to one reactant's concentration (either substrate or product).

**Flagellum**   A motile, whiplike organelle similar in organization to a cilium, but longer and generally present on a cell only in small numbers.

**Flame cells**   Flagellated cells at the ending of the excretory collecting tubules of flatworms and nemerteans.

**Flavin adenine dinucleotide (FAD)**   A coenzyme formed by the condensation of riboflavin phosphate and adenylic acid; it performs an important function in electron transport and is a prosthetic group for some enzymes.

**Flavoproteins**   Proteins combined with flavin prosthetic groups that are important as intermediate carriers of electrons between the dehydrogenases and cytochromes in the respiratory chain.

**Flexor**   A muscle that flexes or bends an extremity.

**Fluorescence**   The property of emitting light upon molecular excitation by an incident light; the emitted light is always less energetic (has a longer wavelength) than the light producing the excitation.

**Flux**   The rate of flow of matter or energy across a unit area.

**Follicle-stimulating hormone (FSH)**   An anterior pituitary gonadotropin that stimulates the development of ovarian follicles in the female and testicular spermatogenesis in the male.

**Follicular phase**   That part of the estrous and menstrual cycles that is characterized by formation of and secretion by the Graafian follicles.

**Foramen**   An orifice or opening.

**Fourier's Law**   The rate of flow of heat in a conducting body is proportional to its thermal conductance and to the temperature gradient.

**Fovea**   A depression in the center of the retina containing closely packed cone cells.

**Free energy**   The energy available to do work at a given temperature and pressure.

**Fructose**   A ketohexose, $C_6H_{12}O_6$, found in honey and many fruits.

**Fusimotor system**   The gamma motoneurons and the intrafusal fibers that they innervate.

**Gain**   The increase in signal produced by amplification.

**Gamma aminobutyric acid (GABA)**   Inhibitory transmitter identified in crustacean motor synapses and in the vertebrate central nervous system.

**Gamma efferents**   The motor axons innervating the intrafusal muscle fibers of spindle organs.

**Gamma motoneurons**   Nerve cells of the ventral spinal cord that innervate the intrafusal muscle fibers.

**Gamma rays**   Electromagnetic radiation of very short wavelength ($10^{-12}$ cm) and very high energy.

**Ganglion**   An anatomically distinct concentration of neuron cell bodies.

**Ganglion cells**   A nonspecific term applied to some nerve cell bodies, especially those located in ganglia of invertebrates or outside the vertebrate central nervous system proper.

**Gap junctions**   Specializations for electrical coupling between cells, where intercellular spacing is only about 20 Å and tubular assemblies of particles connect the apposed membranes.

**Gastric**   Pertaining to the stomach.

**Gastric inhibitory peptide (GIP)**   A gastrointestinal hormone released into the bloodstream from the duodenal mucosa, inhibiting gastric secretion and motility.

**Gastric juice**   Fluid secreted by the cells of the gastric epithelium.

**Gastrin**   A protein hormone that is liberated by the gastrin cells of the pyloric gland and induces gastric secretion and motility.

**Gay-Lussac's Law**   Either the pressure or the volume of a gas is directly proportional to absolute temperature if the other is held constant.

**Geiger counter**   An instrument that detects the presence of ionizing radiation.

**Gel**   The stiff, high-viscosity state of cytoplasm.

**Generator potential**   A receptor potential that depolarizes the spike-initiating zone of an axon.

**Geniculate body**   A projection of the geniculate nucleus that relays incoming sensory information to the cortex; named for its kneelike shape in cross section.

**Gestation**   Pertaining to pregnancy.

**Gigantism**   Excessive growth due to hypersecretion of pituitary growth hormone from birth.

**Glial cells (neuroglia)**   Inexcitable supportive cells associated with neurons in nervous tissue.

**Glomerular filtration rate (GFR)**   The amount of total glomerular filtrate produced per minute by all nephrons of both kidneys; equal to the clearance of a freely filtered and nonreabsorbed substance such as inulin.

**Glomerulus**   A coiled mass of capillaries.

**Glucagon**   A protein hormone released by the alpha cells of the pancreatic islets; its secretion is induced by low blood sugar or by growth hormone; it stimulates glycogenolysis in the liver.

**Glucocorticoids**   Steroids synthesized in the adrenal cortex with wide-ranging metabolic activity; included are cortisone, cortisol, corticosterone, and 11-deoxycorticosterone.

**Gluconeogenesis**   Synthesis of carbohydrates from noncarbohydrate sources, such as fatty acids or amino acids.

**Glutamate**   A putative excitatory synaptic transmitter in the vertebrate central nervous system and in arthropod neuromuscular junctions.

**Glycocalyx**   A meshwork of acid mucopolysaccharide and glycoprotein filaments that arise from the membrane covering the microvilli of the intestinal brush border.

**Glycogenesis**   The synthesis of glycogen.

**Glycogenolysis**   The breakdown of glycogen to glucose-6-phosphate.

**Glycogen synthetase**   An enzyme that catalyzes the polymerization of glucose to glycogen.

**Glycolipid**   A lipid containing carbohydrate groups, in most cases galactose.

**Glycolysis (Embden–Meyerhof pathway)**   The metabolic pathway by which hexose and triose sugars are broken down to simpler substances, especially pyruvate or lactate.

**Glycosuria (glucosuria)**   The excretion of excessive amounts of glucose in the urine.

**Goiter**   An abnormal increase in size of the thyroid gland, usually due to a dietary lack of iodine.

**Golgi tendon organs**   Tension-sensing nerve endings of the 1b afferent fibers found in muscle tendons.

**Gonadotropic hormones (gonadotrophic hormones; gonadotropins)**   Hormones that influence the activity of the gonads; in particular, those secreted by the anterior pituitary.

**Graafian follicle**   A mature ovarian follicle in which fluid is accumulating.

**Graded response**   One that increases as a function of the energy applied; a membrane response that is not all-or-none.

**Graham's Law**   The rate of diffusion of a gas is proportional to the square root of the density of that gas.

**Gray matter**   Tissue of the central nervous system consisting of cell bodies, unmyelinated fibers, and glial cells.

**Growth hormone (GH; somatotropin)**   A protein hormone that is secreted by the anterior pituitary and stimulates growth; directly influences protein, fat, and carbohydrate metabolism and regulates growth rate.

**Guanine**   2-amino-6-oxypurine ($C_5H_5N_5O$); a white, crystalline base; a breakdown product of nucleic acids.

**Guano**   White, pasty waste product of birds and reptiles; high in uric acid content.

**Guanosine triphosphate (GTP)**   A high-energy molecule similar to ATP that participates in several energy-requiring processes, such as peptide bond formation.

**Guanylate cyclase**   An enzyme that converts GTP to cyclic GMP.

**Gustation**   The sense of taste; chemoreception of ions and molecules  in solution by specialized epithelial sensory receptors.

**Habituation**   The progressive loss of behavioral response probability with repetition of a stimulus.

**Hair cell**   A mechanosensory epithelial cell bearing stereocilia and in some cases a kinocilium.

**Halide**   A binary compound of a halogen and another element.

**Halogens**   A family of related elements that form similar saltlike compounds with most metals; they are fluorine, chlorine, bromine, and iodine.

**Heat**   Energy in the form of molecular or atomic vibration that is transferred by conduction, convection, and radiation down a thermal gradient.

**Heat of shortening**   Thermal energy associated with muscle contraction; it is proportional to the distance the muscle has shortened.

**Heat of vaporization**   Heat necessary per mass unit of a given liquid to convert all of the liquid to gas at its boiling point.

**Heavy meromyosin (HMM; H-meromyosin)**   The "head" and "neck" of the myosin molecule, the portion containing ATPase activity.

**Helicotrema**   The opening that connects the scala tympani and the scala vestibuli at the cochlear apex.

**Hematocrit**   The percentage of total blood volume occupied by red blood cells; in humans, the hematocrit is normally 40–50%.

**Heme**   $C_{34}H_{33}O_4N_4FeOH$; an iron protoporphyrin portion of many respiratory pigments.

**Hemerythrin**   An invertebrate respiratory pigment that is a protein but does not contain heme.

**Hemocyanin**   An invertebrate respiratory pigment that is a protein, contains copper, and is found in mollusks and crustaceans.

**Hemolymph**   The blood of invertebrates with open circulatory systems.

**Hemopoiesis (hematopoiesis)**   The processes leading to the production of blood cells.

**Henderson–Hasselbalch equation**   $pH = pK + \log([H^+ \text{ acceptor}]/[H^+ \text{ donor}])$. The formula for calculation of the pH of a buffer solution.

**Henry's Law**   The quantity of gas that dissolves in a liquid is nearly proportional to the partial pressure of that gas in the gas phase.

**Hepatocyte**   A liver cell.

**Hering–Breuer reflex**   A reflex in which lung inflation activates pulmonary stretch receptors that inhibit further inspiration during that cycle; activity from stretch receptors is carried in the vagus nerve.

**Hertz (Hz)**   Cycles per second.

**Heterotherm**   An animal that derives essentially all of its body heat from the environment.

**Heterotrophic**   Pertaining to nutritional dependency on other organisms.

**Hexose**   A six-carbon monosaccharide.

**Histamine**   The base formed from histidine by decarboxylation; responsible for dilation of blood vessels.

**Histaminergic**   Referring to nerves that release histamine.

**Histone**   A simple, repeating, basic protein that combines with DNA.

**Hodgkin cycle**   The regenerative or positive-feedback loop responsible for the upstroke of the action potential; depolarization causes an increase in the sodium permeability, permitting an increased influx of Na$^+$, which further depolarizes the membrane.

**Homeotherm**   An animal (mammal, bird) that regulates its own internal temperature within a narrow range, regardless of the ambient temperature.

**Homonymous**   Pertaining to the same origin.

**Horizontal cell**   A nerve cell whose fibers extend horizontally in the outer plexiform layer of the retina.

**Hormone**   A chemical compound synthesized and secreted by an endocrine tissue into the bloodstream; influences the activity of a target tissue.

**Horseradish peroxidase**   A large protein molecule that is opaque in the electron microscope.

**Hydration**   Combination with water.

**Hydride**   Any compound consisting of an element or a radical combined with hydrogen.

**Hydrofuge**   Pertaining to structures with nonwetting surfaces.

**Hydrogen bond**   A weak electrostatic attraction between a hydrogen atom bound to a highly electronegative element in a molecule and another highly electronegative atom in the same or a different molecule.

**Hydrolysis**   Fragmentation or splitting of a compound by the addition of water, whereupon the hydroxyl group joins one fragment and the hydrogen atom the other.

**Hydronium ion**   A hydrogen ion (H$^+$) combined with a water molecule; $H_3O^+$.

**Hydrophilic**   Having an affinity for water.

**Hydrophobic**   Lacking an affinity for water.

**Hydrostatic pressure**   Force exerted over an area due to pressure in a fluid.

**Hydroxyl group (radical)**   The $-OH^-$ group.

**Hypercalcemia**   Excessive plasma calcium levels.

**Hypercapnia**   Increased levels of carbon dioxide.

**Hyperemia**   Increased blood flow to a tissue or an organ.

**Hyperglycemia**   Excessive blood glucose levels.

**Hyperosmotic**   Containing a greater concentration of osmotically active constituents than the solution of reference.

**Hyperpnea**   Increased lung ventilation; hyperventilation.

**Hyperpolarization**   An increase in potential difference across a membrane, making the cell interior more negative than it is at rest.

**Hyperthermia**   A state of abnormally high body temperature.

**Hypertrophy**   Excessive growth or development of an organ or a tissue.

**Hyperventilation**   *See* Hyperpnea.

**Hypoglycemia**   Low blood glucose levels.

**Hypoosmotic**   Containing a lower concentration of osmotically active constituents than the solution of reference.

**Hypophysis**   The pituitary gland.

**Hypopnea**   Hypoventilation; decreased lung ventilation.

**Hypothalamico-hypophysial relay system**   Portal veins linking the capillaries of the hypothalamic median eminence with those of the adenohypophysis; these transport hypothalamic neurosecretions directly to the adenohypophysis.

**Hypothalamus**   The part of the diencephalon that forms the floor of the median ventricle of the brain; includes the optic chiasma, mammillary bodies, tuber cinereum, and infundibulum.

**Hypothermia**   A state of abnormally low body temperature.

**Hypothyroidism**   Reduced thyroid activity.

**Hypoxia**   Reduced oxygen levels.

**Hysteresis**   A nonlinear change in the physical state of a system, such that the state depends in part on the previous history of the system.

**H zone**   The light zone in the center of the muscle sarcomere, where the myosin filaments are not overlapped by actin filaments; the region between actin filaments.

**I band**   The region between the A band and Z disk of a muscle sarcomere that appears light when viewed microscopically; contains the actin thin filaments without overlap from the myosin filaments.

**Impedance**   The dynamic resistance to flow met by fluids moving in a pulsatile manner.

**Impulse-initiating region (spike-initiating zone)**   The proximal portion of the axon, which has a lower threshold for action potential generation than either the soma or the dendrites.

**Infrared**   Thermal radiation; electromagnetic radiation of wavelengths greater than $7.7 \times 10^{-5}$ cm and less than $12 \times 10^{-4}$ cm; the region between red light and radio waves.

**Inhibitory**   In neurophysiology, pertaining to a reduction in probability of generating an action potential.

**Initial segment**   The portion of axon and axon hillock proximal to the first myelinated segment; generally the site of impulse initiation.

**Inner segment**    The portion of vertebrate photoreceptor cell that contains the cell organelles and synaptic contacts.

**Inotropic**    Pertaining to the strength of contraction of the heart.

**Instinct**    A species-specific set of unlearned behaviors and responses.

**Insulin**    A protein hormone synthesized and secreted by the beta cells of the pancreatic islets; controls cellular uptake of carbohydrate and influences lipid and amino acid metabolism.

**Integration, neural**    Synthesis of an output based on the sum of inputs to a neuron or neural network.

**Interneuron**    A nerve cell connecting two or more other neurons.

**Interstitial**    Between cells or tissues.

**Interstitial cell-stimulating hormone (ICSH)**    Identical to luteinizing hormone but in the male.

**Interstitium**    The tissue space between cells.

**Intrafusal fibers**    The muscle fibers located within a muscle spindle organ.

**Inulin**    An indigestible vegetable starch; used in studies of kidney function because it is freely filtered and not actively transported.

**In vitro**    "In a glass"; in an artificial environment outside the body.

**In vivo**    Within the living organism or tissue.

**Iodoacetic acid**    An agent that poisons glycolysis by inhibiting glyceraldehyde phosphate dehydrogenase.

**Ion**    An atom bearing a net charge due to loss or gain of electrons.

**Ion battery**    The electromotive force capable of driving an ionic current across a membrane; results from unequal concentrations of an ion species in the two compartments separated by the membrane.

**Ion-exchanger site (ion-binding site)**    An electrostatically charged site that attracts ions of the opposite charge.

**Ionic bond**    Electrostatic bond.

**Ionization**    The dissociation into ions of a compound in solution.

**Ionophores**    Molecules or molecular aggregates that promote the permeation of ions across membranes; these may be carrier molecules or ion-permeable membrane channels.

**Ipsilateral**    Relating to the same side.

**Iris**    The pigmented circular diaphragm located behind the cornea of the vertebrate eye.

**Ischemia**    The absence of blood flow (to an organ or a tissue).

**Islets of Langerhans**    Microscopic endocrine structures dispersed throughout the pancreas. They consist of three cell types: the alpha cells, which secrete glucagon; the beta cells, which secrete insulin; and the delta cells, which secrete gastrin.

**Isoelectric point**    The pH of a solution at which an amphoteric molecule has a net charge of zero.

**Isomer**    A compound having the same chemical formula as another, but with a different arrangement of its atoms.

**Isometric contraction**    Contraction during which the muscle does not shorten significantly.

**Isosmotic**    Having the same osmotic pressure.

**Isoteric interaction**    Chemical interaction involving molecules with the same number of valence electrons in the same configuration, but made up of different types and numbers of atoms.

**Isotonic contraction**    Contraction in which force remains constant while the muscle shortens.

**Isotope**    Any of two or more forms of an element with the same number of protons (atomic number), but a different number of neutrons (atomic weight).

**Isovolumic**    Having the same volume.

**Isozymes**    Multiple forms of an enzyme found in the same animal species or even in the same cell.

**Jejunum**    The portion of small intestine between the duodenum and the ileum.

**Joule (J)**    SI unit of work equivalent to 0.239 calories (cal).

**Juvenile hormone (JH)**    A class of insect hormones that are secreted by the corpora allata and that promote retention of juvenile characteristics.

**Juxtaglomerular cells**    Specialized secretory cells situated in the afferent glomerular arterioles; act as receptors that respond to low blood pressure by secreting renin, which converts angiotensinogen to angiotensin, resulting in the stimulation of aldosterone secretion.

**Kelvin**    *See* Absolute temperature.

**Ketone**    Any compound having a carbonyl group (CO) attached (by the carbon) to hydrocarbon groups.

**Ketone bodies**    Acetone, acetoacetic acid, and $\beta$-hydroxybutyric acid; products of fat and pyruvate metabolism formed from acetyl CoA in the liver; oxidized in muscle and by the central nervous system during starvation.

**Kinematic viscosity**    Viscosity divided by density; gases of equal kinematic viscosity will become turbulent at equal flow rates in identical airways.

**Kinetic energy**    Energy inherent in the motion of a mass.

**Kininogen**    Precursor of bradykinin.

**Kinocilium**    A true "9 + 2" or "9 + 0" cilium present in sensory hair cells.

**Kirchhoff's Laws**    *First law:* The sum of the currents entering a junction in a circuit equals the sum of the currents leaving the junction. *Second law:* The sum of the potential changes encountered in any closed loop in a circuit is equal to zero.

**Labeled-lines concept**  Sensory modalities are determined by the stimulus sensitivity of peripheral sense organs and the anatomical specificities of their central connections.

**Lamella**  A thin sheet or leaf.

**Laminar flow**  Turbulence-free flow of fluid in a vessel or past a moving object; a gradient of relative velocity exists in which the fluid layers closest to the wall or body have the lowest relative velocity.

**Laplace's Law**  The transmural pressure in a thin-walled tube is proportional to the wall tension divided by the inner radius of the tube.

**Larva**  The immature, active feeding stage characteristic of many invertebrates.

**Lateral inhibition**  Reciprocal suppression of excitation by neighboring neurons in a sensory network; the effect is enhanced lateral contrast and an increase in dynamic range.

**Lateral-line system**  Series of hair cells (*see* Neuromast) on the medial wall of canals running the length of the head and body of fishes and many amphibians; these channels have openings to the outside; the system is sensitive to water movement.

**Lecithin**  Any of a group of phospholipids found in animal and plant tissues; composed of choline, phosphoric acid, fatty acids, and glycerol.

**Length constant ($\lambda$)**  The distance along a cell over which a potential change decays in amplitude by $1 - 1/e$ or 63%.

**Leucocytes**  White blood cells.

**Leydig cells (interstitial cells)**  Cells of the testes that are stimulated by luteinizing hormone to secrete testosterone.

**Light meromyosin (LMM)**  The rodlike fragment of the myosin molecule that constitutes most of the molecule's backbone.

**Lipid**  Any of the fatty acids, neutral fats, waxes, steroids, and phosphatides; lipids are hydrophobic and feel greasy.

**Lipogenesis**  The formation of fat from nonlipid sources.

**Lipophilic**  Having an affinity for lipids.

**Local circuit current**  The current that spreads electrotonically from the excited portion of an axon during conduction of the nerve impulse, flowing longitudinally along the axon, across the membrane, and back to the excited portion.

**Loop of Henle**  A U-shaped bend in the portion of a renal tubule that lies in the renal medulla.

**Lumen**  The interior of a cavity or duct.

**Luminosity**  Brightness; relative quantity of light reflected or emitted.

**Luteal phase**  The part of the estrous or menstrual cycle characterized by formation of and secretion by the corpus luteum.

**Luteinizing hormone (LH)**  A gonadotropin that is secreted by the adenohypophysis and that acts with follicle-stimulating hormone (FSH) to induce ovulation of the ripe ovum and liberation of estrogen from the ovary; also influences formation of the corpus luteum and stimulates growth in and secretion from the male testicular Leydig cells.

**Lymph**  Plasmalike fluid collected from interstitial fluid and returned to the bloodstream via the thoracic duct; contains white, but not red, blood cells.

**Lysolecithin**  A lecithin without the terminal acid group.

**Lysosomes**  Minute electron-opaque organelles that occur in many cell types, contain hydrolytic enzymes, and are normally involved in localized intracellular digestion.

**Magnetite**  A magnetic mineral composed of $Fe_3O_4$ and found in some animals; believed to play a role in geomagnetic orientation.

**Malpighian tubules**  Insect excretory-osmoregulatory organs responsible for the active secretion of waste products and the formation of urine.

**Mass Action Law**  The velocity of a chemical reaction is proportional to the active masses of the reactants.

**Mastication**  The chewing or grinding of food with the teeth.

**Mastoid bone**  The posterior process of the temporal bone, situated behind the ear and in front of the occipital bone.

**Mechanism**  A theory that proposes that life is based purely on the action of physical and chemical laws.

**Mechanoreceptor**  A sensory receptor sensitive to mechanical distortion or pressure.

**Median eminence**  A structure at the base of the hypothalamus that is continuous with the hypophysial stalk; contains the primary capillary plexus of the hypothalamico-hypophysial portal system.

**Medulla oblongata**  A cone-shaped neural mass connecting the pons and the spinal cord.

**Medullary cardiovascular center**  A group of neurons in the medulla involved in the integration of information used in the control and regulation of circulation.

**Melanocyte-stimulating hormone (melanophore-stimulating hormone)**  A peptide hormone released by the adenohypophysis that effects melanin distribution in mammals and creates skin-color changes in fishes, amphibians, and reptiles.

**Melting point**  The lowest temperature at which a solid will begin to liquefy.

**Membrane potential**  The electrical potential measured from within the cell relative to the potential of the extracellular fluid, which is by convention at zero potential; the potential difference between opposite sides of the membrane.

**Menarche**  The onset of menstruation during puberty.

**Menopause**  The cessation of the menstrual cycle in the mature female human.

**Menstrual cycle**  Recurring physiological changes that include menstruation.

**Menstruation**  The shedding of the endometrium, an event that usually occurs in the absence of conception throughout the fertile period of the female of certain primate species, including humans.

**Messenger RNA (mRNA)**  A fraction of RNA that is responsible for transmission of the informational base sequence of the DNA to the ribosomes.

**Metabolic intensity**  The metabolic rate of a unit mass of tissue.

**Metabolic pathway**  A sequence of enzymatic reactions involved in the alteration of one substance into another.

**Metabolic water**  Water evolved from cellular oxidation.

**Metabolism**  The totality of physical and chemical processes involved in anabolism, catabolism, and cell energetics.

**Metachronism**  The progression of in-phase activity in a wavelike manner over a population of organelles, such as cilia.

**Metamorphic climax**  The last stage of amphibian metamorphosis, in which the adult form is attained.

**Metamorphosis**  A change in morphology—in particular, from one stage of development to another, such as juvenile to adult.

**Metarhodopsin**  Product of the absorption of light by rhodopsin; decomposes to opsin and *trans*-retinal.

**Methemoglobin**  Hemoglobin in which the $Fe^{3+}$ of heme has been oxidized to $Fe^{2+}$.

**Micelle**  A microscopic particle made from an aggregation of amphipathic molecules in solution.

**Michaelis–Menten constant ($K_M$)**  The concentration (moles/liter) of the substrate at half the maximal velocity of an enzymatic reaction.

**Microfilaments**  Filaments within the cytoplasmic substance; diameter of less than 100 Å.

**Microtubules**  Cylindrical cytoplasmic structures made of polymerized tubulin and found in many cells, especially motile cells, as constituents of the mitotic spindle, cilia, and flagella.

**Microvilli**  Tiny cylindrical projections on a cell surface that function to increase surface area; frequently found on absorptive epithelia.

**Micturition**  Urination.

**Mineralocorticoids**  Steroid hormones that are synthesized and secreted by the adrenal cortex and that influence plasma electrolyte balance—in particular, by sodium and chloride reabsorption in the kidney tubules. *See also* Aldosterone.

**Miniature endplate potentials (m.e.p.p.'s)**  Tiny depolarizations (generally 1 mV or less) of the postsynaptic membrane of a motor endplate; produced by presynaptic release of single packets of transmitter.

**Miniature postsynaptic potentials (m.p.p.'s)**  Potentials produced in a postsynaptic neuron by presynaptic release of single vesicles of transmitter substance.

**Mobility, electrical**  A quantity proportional to the migration rate of an ion in an electric field.

**Mobility, mechanical**  A quantity proportional to the rate at which a molecule will diffuse in a liquid phase.

**Modulatory agent**  One that either increases or decreases the response of a tissue to a physical or chemical signal.

**Molality**  The number of moles of solute in a kilogram of a pure solvent.

**Molarity**  The number of moles of solute in a liter of solution.

**Mole**  Avogadro's number ($6.023 \times 10^{23}$) of molecules of an element or a compound; equal to the molecular weight in grams.

**Monomer**  A compound capable of combining in repeating units to form a dimer, trimer, or polymer.

**Monopole**  An object or a particle bearing a single unneutralized electric charge, as, for example, an ion.

**Monosaccharide sugar**  An unhydrolyzable carbohydrate; a simple sugar. Such sugars are sweet-tasting, colorless crystalline compounds with the formula $C_n(H_2O)_n$. *See also* Saccharide.

**Monosynaptic**  Relating to or transmitted through one synapse only.

**Monovalent**  Having a valence of one.

**Monozygotic**  Arising from one ovum or zygote.

**Motoneuron (motor neuron)**  A nerve cell that innervates muscle cells.

**Motor cortex**  The part of the cerebral cortex that controls motor function; situated anterior to the central sulcus, which separates the frontal and parietal lobes.

**Motor program**  An endogenous coordinated motor output of central neural origin and independent of sensory feedback.

**Motor unit**  The unit of motor activity consisting of a motoneuron and the muscle fibers it innervates.

**mRNA** *See* Messenger RNA.

**Mucosa** Mucous membrane facing a cavity or the exterior of the body.

**Müllerian ducts** Paired embryonic ducts originating from the peritoneum that connect with the urogenital sinus to develop into the uterus and fallopian tubes.

**Multineuronal innervation** Innervation of a muscle fiber by several motoneurons, as in many invertebrates, especially arthropods.

**Multiterminal innervation** Numerous synapses made by a single motoneuron along the length of a muscle fiber.

**Muscarinic** Pertaining to muscarine; refers to acetylcholine receptors that are sensitive to muscarine but not to nicotine.

**Muscle fiber** A skeletal muscle cell.

**Muscle spindle (stretch receptor)** A length-sensitive receptor organ located between and in parallel with extrafusal muscle fibers; gives rise to the myotatic or stretch reflex of vertebrates.

**Mutation** A transmissible alteration in genetic material.

**Myelin sheath** A sheath formed by many layers of Schwann cell membrane wrapped tightly around segments of axon in vertebrate nerve; serves as electrical insulation in saltatory conduction.

**Myoblast** Embryonic cell precursor for muscle.

**Myocardium** Heart muscle.

**Myofibril** A longitudinal unit of muscle fiber made up of sarcomeres and surrounded by sarcoplasmic reticulum.

**Myogenic pacemaker** A pacemaker that is a specialized muscle cell.

**Myoglobin** An iron-containing protoporphyrin–globin complex found in muscle; serves as a reservoir for oxygen and gives muscle its pink color.

**Myoplasm** The cytosol in a muscle cell.

**Myosin** The protein that makes up the thick filaments and cross bridges of the A band.

**Myotatic reflex (stretch reflex)** Reflex contraction of a muscle in response to stretch.

**Myotube** A developing muscle fiber.

**Naloxone** An analog of morphine that acts as an opioid antagonist.

**Nares** Nostrils.

**Nephron** The morphological and functional unit of the vertebrate kidney; composed of the glomerulus and Bowman's capsule, the proximal and distal tubules, the loop of Henle (birds, mammals), and the collecting duct.

**Nernst equation** Equation for calculating the potential difference across a membrane that will balance the osmotic gradient of an ion.

**Nerve** *As a noun:* A bundle of axons held together as a unit by connective tissue. *As an adjective:* Neural.

**Neural network** A system of interacting nerve cells.

**Neurilemma** Connective tissue sheath covering a bundle of axons.

**Neurin** A filamentous protein found attached to the inner surface of the plasmalemma of nerve endings in the brain; similar to muscle actin and may play a role in exocytosis. *See also* Neurostennin.

**Neurogenic pacemaker** A pacemaker that is a specialized nerve cell.

**Neuroglia (glia)** Inexcitable supporting tissue of the nervous system.

**Neurohemal organ** Organ for storage and discharge into the blood of the products of neurosecretion.

**Neurohumor** Synaptic transmitters and neurosecretory hormones.

**Neurohypophysis (pars nervosa)** A neurally derived reservoir for hormones with antidiuretic and oxytocic action; consists of the neural lobe, which makes up its bulk, and the neural stalk, which is connected to and passes neurosecretions from the hypothalamus.

**Neuromast** A collection of hair cells embedded in a cupula in lateral-line mechanoreceptors of the lower vertebrates.

**Neuron** Nerve cell.

**Neuropeptide** A peptide molecule identified as a neurotransmitter substance.

**Neurophysins** Proteins associated with neurohypophysial hormones stored in granules in the neurosecretory terminals; cleaved from the hormones before secretion.

**Neuropil** A dense mass of closely interwoven and synapsing nerve cell processes (axon collaterals and dendrites) and glial cells.

**Neurosecretory cells** Nerve cells that liberate neurohormones.

**Neurostennin** A complex of two proteins found in brain nerve cell endings, possibly associated with exocytosis. *See also* Neurin; Stennin.

**Neurotransmitter** A chemical mediator released by a presynaptic nerve ending that interacts with receptor molecules in the postsynaptic membrane. This process generally induces a permeability increase to an ion or ions and thereby influences the electrical activity of the postsynaptic cell.

**Nicotinamide adenine dinucleotide (NAD)** A coenzyme widely distributed in living organisms, participating in many enzymatic reactions; made up of adenine, nicotinamide, and two molecules each of *d*-ribose and phosphoric acid.

**Nicotinic** Nicotinelike or sensitive to nicotine; refers to a class of acetylcholine receptors that are sensitive to nicotine but not to muscarine.

**Node of Ranvier**   Regularly spaced interruption (about every millimeter) of the myelin sheath along an axon.

**Noncompetitive inhibition**   Enzyme inhibition due to alteration or destruction of the active site.

**Norepinephrine (noradrenalin)**   A neurohumor secreted by the peripheral sympathetic nerve terminals, some cells of the central nervous system, and the adrenal medulla.

**Nucleic acids**   Nucleotide polymers of high molecular weight. *See also* DNA *and* RNA.

**Nucleotide**   A product of enzymatic (nuclease) splitting of nucleic acids; made up of a purine or pyrimidine base, a ribose or deoxyribose sugar, and a phosphate group.

**Nucleus**   *Of an atom:* The central, positively charged mass surrounded by a cloud of electrons. *Of a cell:* The membrane-bound body within eukaryotic cells that houses the genetic material of the cell. *Of nerve cells:* A related group of neurons in the central nervous system.

**Nymph**   A juvenile developmental stage in some arthropods; morphology resembles the adult.

**Nystatin**   A rod-shaped, antibiotic molecule that creates channels through membranes that allow the passage of molecules of a diameter less than 4 Å.

**$O_{10}$**   The ratio of the rate of a reaction at a given temperature to its rate at a temperature $10°C$ lower.

**Obligatory osmotic exchange**   An exchange between an animal and its environment that is determined by physical factors beyond the animal's control.

**Occipital lobe**   The most posterior region of the cerebral hemisphere.

**Ohm ($\Omega$)**   MKS unit of electrical resistance, equivalent to the resistance of a column of mercury 1 mm square in cross section and 106 cm long.

**Ohm's Law**   $I = V/R$. The strength of an electric current, $I$, varies directly as the voltage, $V$, and inversely as the resistance, $R$.

**Olfaction**   The sense of smell; chemoreception of molecules suspended in air.

**Oligodendrocytes**   A class of glial cells with few processes.

**Oligosaccharides**   Carbohydrates made up of a small number of monosaccharide residues.

**Omasum**   The part of the ruminant stomach lying between the rumen and the abomasum.

**Ommatidium**   The functional unit of the invertebrate compound eye, consisting of an elongated structure with a lens, focusing cone, and photoreceptor cells.

**Oocyte**   A developing ovum.

**Operator gene**   A gene that regulates the synthetic activity of closely linked structural genes via its association with a regulator gene.

**Operon**   A segment of DNA consisting of an operator gene and its associated structural genes.

**Opiates**   Opium-derived narcotic substances.

**Opioids**   Substances that exert opiatelike effects.

**Opsin**   Protein moiety of visual pigments; it combines with 11-*cis*-retinal to become a visual pigment.

**Optic axis**   An imaginary straight line passing through the center of curvature of a simple lens.

**Optic chiasma**   A swelling under the hypothalamus, where the two optic nerves meet and the optic fibers from the nasal portions of the retinas cross to the contralateral sides of the brain.

**Organ of Corti**   The tissue in the cochlear channel of the inner ear that contains the hair cells.

**Ornithine cycle (urea cycle)**   A cyclic succession of reactions that eliminate ammonia and produce urea in the liver of ureotelic organisms.

**Osmoconformer**   An organism that exhibits little or no osmoregulation, so that the osmolarity of its body fluids follows changes in the osmolarity of the environment.

**Osmolarity**   The effective osmotic pressure.

**Osmole**   The standard unit of osmotic pressure.

**Osmolyte**   A substance that serves the special purpose of raising the osmotic pressure or lowering the freezing point of a body fluid.

**Osmometer**   An instrument for the measurement of the osmotic pressure of a solution.

**Osmoregulation**   Maintenance of internal osmolarity with respect to the environment.

**Osmoregulator**   An organism that controls its internal osmolarity in the face of changes in environmental osmolarity.

**Osmosis**   The movement of pure solvent from a solution of a low solute concentration to one of higher concentration through a semipermeable membrane separating the two solutions.

**Osmotic flow**   The solvent flux due to osmotic pressure.

**Osmotic pressure**   Pressure that can potentially be created by osmosis between two solutions separated by a semipermeable membrane; the amount of pressure necessary to prevent osmotic flow between the two solutions.

**Ossicles**   Little bones. Auditory ossicles are the tiny bones (malleus, incus, stapes) of the middle ear, which transmit sound vibrations from the tympanic membrane to the oval window.

**Ouabain**   Cardiac glycoside, a drug capable of blocking some sodium pumps.

**Outer segment**   That portion of a vertebrate visual receptor that contains the pigmented receptor membranes;

attached to the inner segment by a thin ciliumlike connection.

**Oval window** An opening in the inside wall of the middle ear, closed by the base of the stapes.

**Overshoot** The reversal of membrane potential during an action potential; the voltage above zero to the peak of the action potential.

**Ovulation** The release of an ovum from the ovarian follicle.

**Ovum** An egg cell; the reproductive cell (gamete) of the female.

**Oxidant** An electron acceptor in a reaction involving oxidation and reduction.

**Oxidation** Loss of electrons or increase in net positivity of an atom or a molecule. Biological oxidations are usually achieved by removal of a pair of hydrogen atoms from a molecule.

**Oxidative phosphorylation** Respiratory chain phosphorylation; the formation of high-energy phosphate bonds via phosphorylation of ADP to ATP, accompanied by the transport of electrons to oxygen from the substrate.

**Oxygen debt** The extra oxygen necessary to oxidize the products of anaerobic metabolism that accumulate in the muscle tissues during intense physical activity.

**Oxygen dissociation curves** Curves that describe the relationship between the extent of combination of oxygen with the respiratory pigment and the partial pressure of oxygen in the gas phase.

**Oxyhemoglobin** Hemoglobin with oxygen combined to the Fe atom of the heme group.

**Oxyntic cells (parietal cells)** HCl-secreting cells of the stomach lining.

**Oxytocin** An octapeptide hormone secreted by the neurohypophysis; stimulates contractions of the uterus in childbirth and the release of milk from mammary glands.

**Pacemaker** An excitable cell or tissue that fires spontaneously and rhythmically.

**Pacemaker potentials** Spontaneous and rhythmical depolarizations produced by pacemaker tissue.

**Pacinian corpuscles** Pressure receptors found in skin, muscle, joints, connective tissue of vertebrates; they consist of a nerve ending enclosed in a laminated capsule of connective tissue.

**Pancreozymin** *See* Cholecystokinin.

**Parabiosis** The experimental connection of two individuals to allow mixing of their body fluids.

**Parabronchi** Air-conduction pathways in the bird lung.

**Parafollicular cells (C cells)** Cells in the mammalian thyroid that secrete calcitonin.

**Parasympathetic nervous system** The craniosacral part of the autonomic nervous system.

**Parathyroid glands** Small tissue masses (usually two pairs) close to the thyroid gland that secrete parathormone (parathyroid hormone).

**Parathyroid hormone (PH; parathormone)** A polypeptide hormone of the parathyroid glands secreted in response to a low plasma calcium level; stimulates calcium release from bone and calcium absorption by the intestines while reducing calcium excretion by the kidneys.

**Paraventricular nucleus** A group of neurosecretory neurons in the supraoptic hypothalamus that send their axons into the neurohypophysis.

**Parietal cells** *See* Oxyntic cells.

**Pars intercerebralis** The dorsal part of the insect brain; contains the cell bodies of neurosecretory cells that secrete brain hormones from axon terminals in the corpora cardiaca.

**Partition coefficient** Ratio of the distribution of a substance between two different liquid phases (e.g., oil and water).

**Parturition** The process of giving birth.

**Parvalbumin** Calcium-binding protein found in vertebrate muscle.

**Pentose** A five-carbon monosaccharide sugar.

**Pepsin** A proteolytic enzyme secreted by the stomach lining.

**Peptide** A molecule consisting of a linear array of amino residues. Protein molecules are made of one or more peptides. Short chains are oligopeptides; long chains are polypeptides.

**Peptide bond** The center bond of the —CO—NH— group, created by the condensation of amino acids into peptides.

**Perfusion** The passage of fluid over or through an organ, a tissue, or a cell.

**Pericardium** The connective-tissue sac that encloses the heart.

**Perilymph** Aqueous liquid contained within the scala tympani and scala vestibuli of the cochlea.

**Peripheral resistance units (PRU)** The drop in pressure (in millimeters of mercury, mmHg) along a vascular bed divided by mean flow in milliliters per second.

**Peritoneum** The membrane that lines the abdominal and pelvic cavities.

**Permeability** The ease with which substances can pass through a membrane.

**Phagocyte** A cell that engulfs other cells, microorganisms, or foreign particulate matter.

**Phagocytosis** The ingestion of particles, cells, or microorganisms by a cell into its cytoplasmic vacuoles.

**Phasic** Transient.

**Pheromone** A species-specific substance released into the environment for the purpose of signaling between individuals of the same species.

**Phlorizin** A glycoside that inhibits active transport of glucose.

**Phosphagen** High-energy phosphate compounds (e.g., phosphoarginine and phosphocreatine) that serve as phosphate-group donors for rapid rephosphorylation of ADP to ATP.

**Phosphoarginine** A compound that has phosphagen properties similar to those of phosphocreatine and that occurs in the muscles of some invertebrates.

**Phosphocreatine** A compound that is broken down in muscle metabolism into inorganic phosphorus and creatine in the rephosphorylation of ADP to ATP.

**Phosphodiesterase** A hydrolytic cytoplasmic enzyme that degrades cAMP to AMP.

**Phosphodiester group** —O—P—O—.

**Phospholipid** A phosphorus-containing lipid that hydrolyzes to fatty acids, glycerin, and a nitrogenous compound.

**Phosphorylase *a.*** Activated (phosphorylated) form of phosphorylase that catalyzes the cleavage of glycogen to glucose-1-phosphate.

**Phosphorylase kinase** Enzyme that, when phosphorylated by a protein kinase, converts phosphorylase *b* to the more active phosphorylase *a.*

**Phosphorylation** The incorporation of a $PO^{3-}$ group into an organic molecule.

**Photopigments** Pigment molecules that can become excited by light.

**Photoreceptor** A sensory cell specifically sensitive to light energy.

**pH scale** Negative log scale (base 10) of hydrogen ion concentration of a solution. $pH = -\log[H^+]$.

**Physiological dead space** That portion of inhaled air not involved in gas transfer in the lung.

**Pilomotor** Pertaining to the autonomic control of smooth muscle for the erection of body hair.

**Pinnate** Resembling a feather, with similar parts arranged on opposite sides of the axis.

**Pinocytosis** Fluid intake by cells via surface invaginations that seal off to become vacuoles filled with liquid.

**Pituitary gland (hypophysis)** A complex endocrine organ situated at the base of the brain and connected to the hypothalamus by a stalk. It is of dual origin: the anterior lobe (adenohypophysis) is derived from embryonic buccal epithelium, whereas the posterior lobe is derived from the diencephalon.

**p*K′*** The negative log (base 10) of an ionization constant, $K'$. $pK' = -\log_{10} K'$.

**Placebo** A physiologically neutral substance that elicits curative or analgesic effects through psychological means.

**Plane-polarized light** Light vibrating in only one plane.

**Plasma kinins** Hormones carried in blood—for example, bradykinin.

**Plasmalemma** Cell membrane; surface membrane.

**Plasma skimming** The separation of plasma from blood within the circulation.

**Plasticity** Compliance to external influence.

**Plastron** The ventral shell of a tortoise or turtle; also a gas film held in place under water by hydrofuge hairs, creating a large air–water interface.

**Pleura** The membranes that line the pleural cavity.

**Pleural cavity** The cavity between the lungs and the wall of the thorax.

**Pneumotaxic center** A group of neurons in the pons, thought to be involved in the maintenance of rhythmic breathing in mammals.

**Pneumothorax** Collapse of the lung due to a puncture into the pleural cavity of the chest wall or the lung.

**Poikilotherm** An animal whose body temperature more or less follows the ambient temperature.

**Poiseuille's Law** In laminar flow, the flow is directly proportional to the driving pressure, and resistance is independent of flow.

**Polymer** A compound composed of a linear sequence of simple molecules or residues.

**Polypeptide chain** A linear arrangement of more than two amino acid residues.

**Polypnea** Increased breathing rate.

**Polysynaptic** Referring to transmission through a sequence of more than one synapse.

**Polytene** Having many duplicate chromatin strands.

**Pores of Kohn** Small holes between adjacent regions of the lung, permitting collateral air flow.

**Porphyrins** A group of cyclic tetrapyrrole derivatives.

**Porphyropsin** A purple photopigment present in the retinal rods of some freshwater fishes.

**Portal vessels** Blood vessels that carry blood directly from one capillary bed to another.

**Postsynaptic** Located distal to the synaptic cleft.

**Posttetanic depression** Reduced postsynaptic response following prolonged presynaptic stimulation at a high frequency; believed to be due to presynaptic depletion of transmitter.

**Posttetanic potentiation (PTP)** Increased efficacy of synaptic transmission following presynaptic stimulation at a high frequency; often occurs after posttetanic depression.

**Potassium activation** An increase in the conductance of a membrane to potassium in response to depolarization.

**Potential energy** Stored energy that can be released to do work.

**Premetamorphosis** The developmental stage just preceding amphibian metamorphosis, during which iodine binding and hormone synthesis occur in the thyroid gland.

**Presynaptic** Located proximal to the synaptic cleft.

**Primary follicle** An immature ovarian follicle.

**Primary structure** The sequence of amino acid residues of a polypeptide chain.

**Proboscis** An elongated, protruding mouth part.

**Procaine** 2-diethylaminoethyl-*p*-aminobenzoate; a local anesthetic that interferes with some of the ion conductances of excitable membranes.

**Proenzyme (zymogen)** The inactive form of an enzyme before it is activated by removal of a terminal segment of peptide.

**Progesterone** A hormone of the corpus luteum, adrenal cortex, and the placenta that promotes growth of a suitable uterine lining for implantation and development of the fertilized ovum.

**Prolactin** An adenohypophysial hormone that stimulates milk production and lactation after parturition in mammals.

**Prometamorphosis** The first stage of amphibian metamorphosis, during which there is increased development and activity in the thyroid gland and median eminence.

**Proprioceptors** Sensory receptors situated primarily in muscles, tendons, and the labyrinth that relay information about the position and motion of the body.

**Prostaglandins** A family of recently discovered natural fatty acids that arise in a variety of tissues and are able to induce contraction in uterine and other smooth muscle, lower blood pressure, and modify the actions of some hormones.

**Prostate gland** A gland located around the neck of the bladder and urethra in males that contributes to the seminal fluid.

**Prosthetic group** An organic compound essential to the function of an enzyme. Prosthetic groups differ from coenzymes in that they are more firmly attached to the enzyme protein.

**Protagonistic muscles** Muscles whose contractions cooperate to produce a movement.

**Protein kinase** Any enzyme that catalyzes the transfer of a phosphate group from ATP to a protein, creating a phosphoprotein.

**Proteins** Large molecules composed of one or more chains of alpha amino acid residues (i.e., polypeptide chains).

**Proteolysis** The splitting of proteins by hydrolysis of peptide bonds.

**Proteolytic** Protein-hydrolyzing.

**Prothoracic glands** Ecdysone-secreting tissues situated in the anterior thorax of insects.

**Proximal convoluted tubules** Coiled portions of the renal tubules located in the renal cortex, beginning at the glomerulus and leading to (and continuous with) the descending limb of the loop of Henle.

**Pseudopodium** Literally, false foot; a temporary projection of an amoeboid cell for engulfment of food or for locomotion.

**Psychophysics** The branch of psychology concerned with relationships between physical stimuli and perception.

**Pulmonary** Pertaining to or affecting the lungs.

**Pupa** A developmental stage of some insect groups; between the larva and the adult.

**Pupil** The opening at the center of the iris through which light passes.

**Purinergic** Referring to nerve endings that release purines or their derivatives as transmitter substances.

**Purines** A class of nitrogenous heterocyclic compounds, $C_5H_4N_4$, derivatives of which (purine bases) are found in nucleotides; they are colorless and crystalline.

**P-wave** That portion of the electrocardiogram associated with depolarization of the atria.

**Pyloric** Pertaining to the caudal portion of the vertebrate stomach where it joins the small intestine.

**Pylorus** The distal stomach opening, ringed by a sphincter, that releases the stomach contents into the duodenum.

**Pyramidal tract** A bundle of nerve fibers originating in the motor cortex and descending down the brain stem to the medulla oblongata and to the spinal cord; responsible for mediating control of voluntary muscle movements.

**Pyrimidine** A class of nitrogenous heterocyclic compounds, $C_4H_4N_2$, derivatives of which (pyrimidine bases) are found in nucleotides.

**Pyrogen** Substance that leads to a resetting of a homeotherm's body thermostat to a higher set point, thereby producing fever.

**QRS-wave** That portion of the electrocardiogram related to depolarization of the ventricle.

**Quantal transmission** The concept that neurotransmitter is released in multiples of discrete "packets." It is now apparent that the quantal packets represent individual presynaptic vesicles or groups of simultaneously released vesicles.

**Quaternary structure** The characteristic ways in which the subunits of a protein containing more than one polypeptide chain are combined.

**Radial links**   Extensions from peripheral doublets to the central sheath in cilia and flagella.

**Radioisotope**   A radioactive isotope.

**Rate constant (specific reaction rate)**   The proportionality factor by which the concentration of a reactant in an enzymatic reaction is related to the reaction rate.

**Receptive field**   That area of a sensory field (e.g., the retina) that when stimulated influences the activity of a given neuron is the receptive field of that neuron.

**Receptor current**   A stimulus-induced change in the movement of ions across a receptor cell membrane.

**Receptor molecules**   Molecules that are situated on the outer surface of the cell membrane and that interact specifically with messenger molecules, such as hormones or transmitters.

**Receptor potential**   A membrane potential change elicited in receptor cells by the flow of receptor current.

**Reciprocal inhibition**   Inhibition of the motoneurons innervating one set of muscles during the reflex excitation of their antagonists.

**Redox pair**   Two compounds, molecules, or atoms involved in mutual reduction and oxidation.

**Reductant**   Donor of electrons in a redox reaction.

**Reduction**   The addition of electrons to a substance.

**Reduction potential**   A measurement of the tendency of a reducer to yield electrons in a redox reaction, expressed in volts.

**Reflex**   An action that is reflected; an involuntary motor response mediated by a neural arc in response to sensory input.

**Reflex arc**   A neural pathway used in reflex action; consists of afferent nerve input to a nerve center that produces activity in efferent nerves to an effector organ.

**Refraction**   The bending of light rays as they pass from a medium of one density into a medium of another density.

**Refractive index**   The refractive power of a medium compared with that of air, designated 1.

**Refractory period**   The period of increased membrane threshold immediately following an action potential.

**Regenerative**   Self-reinforcing; utilizing positive feedback; autocatalytic.

**Release-inhibiting hormone (RIH; release-inhibiting factor, RIF)**   A hypothalamic neurosecretion carried by portal vessels to the adenohypophysis, where it restrains the release of a specific hormone.

**Releasing hormone (RH; releasing factor)**   A hypothalamic neurosecretion that stimulates the liberation of a specific hormone from the adenohypophysis.

**Renal clearance**   That volume of plasma containing the quantity of a freely filtered substance that appears in the glomerular filtrate per unit time. Total renal clearance is the amount of ultrafiltrate produced by the kidney per unit time.

**Renin**   A proteolytic enzyme produced by specialized cells in renal arterioles; converts angiotensinogen to angiotensin.

**Rennin**   An enzyme that coagulates milk, found especially in the gastric juice of young mammals.

**Renshaw cells**   Short inhibitory interneurons in the ventral horn excited by branches of the motoneuron axons that feed back on the motoneuron pool.

**Repolarization**   The return to resting polarity of a cell membrane that has been depolarized.

**Repressor gene (regulator gene)**   A gene that produces a substance (repressor) that shuts off the structural-gene activity of an operon by an interaction with its operator gene.

**Reserpine**   A botanically derived tranquilizing agent that interferes with the uptake of catecholamine from the cytosol by secretory vesicles; its effect is to deplete the catecholamine content of adrenergic cells.

**Residual volume**   The volume of air left in the lungs after maximal expiratory effort.

**Resistance *(R)***   The property that hinders the flow of current. The unit is the ohm ($\Omega$), defined as the resistance that allows exactly 1 ampere (A) of current to flow when a potential drop of 1 volt (V) exists across the resistance. It is equivalent to the resistance of a column of mercury 1 mm$^2$ in cross-sectional area and 106.3 cm long. $R = \rho \times$ length/cross-sectional area.

**Resistivity ($\rho$)**   The resistance of a conductor 1 cm in length and 1 cm$^2$ in cross-sectional area.

**Respiratory quotient (RQ)**   The ratio of $CO_2$ production to $O_2$ consumption; depends on type of food oxidized by the animal.

**Resting potential**   The normal, unstimulated cell membrane potential of a cell at rest.

**Rete mirabile**   An extensive countercurrent arrangement of arterial and venous capillaries.

**Reticulum**   A small network.

**Retinal**   The aldehyde of retinol obtained from the enzymatic oxidative cleavage of carotene; unites with opsin in the retina to form the visual pigment.

**Retinol**   Vitamin A ($C_{20}H_{30}O$), an alcohol of 20 carbons; converted reversibly to retinal by enzymatic dehydrogenation.

**Retinular cell**   A photoreceptor cell of the arthropod compound eye.

**Reversal potential**   The membrane potential at which current through an activated synapse or receptor membrane is null.

**Reynolds number** *(Re)*   A unitless number; the tendency of a flowing gas or liquid to become turbulent is proportional to its velocity and density and inversely proportional to its viscosity. Calculated from these parameters, the Reynolds number indicates whether flow will be turbulent or laminar under a particular set of conditions.

**Rhabdome**   The aggregate structure consisting of a longitudinal rosette of rhabdomeres located axially in the ommatidium.

**Rhabdomere**   That portion of a retinular cell in which the photopigment-bearing surface membrane is thrown into closely packed microvilli; this region faces the central axis of the ommatidium.

**Rheogenic**   Producing electric current.

**Rhodopsin (visual purple)**   A purplish-red, light-sensitive chromoprotein with 11-*cis*-retinal as its prosthetic group; found in the rods of the retina; bleached to visual yellow (all-*trans*-retinal) by incident light.

**Ribonucleic acid**   *See* RNA.

**Ribose**   A pentose monosaccharide with the chemical formula $HOCH_2(CHOH)_3CHO$; a constituent of RNA.

**Ribosome**   Ribonucleoprotein particles found within the cytoplasm; the sites of intersection of mRNA, tRNA, and the amino acids during the synthesis of polypeptide chains.

**Rigor mortis**   The rigidity that develops in dying muscle as ATP becomes depleted and cross bridges remain attached.

**Ringer solution**   Physiological saline solution.

**RNA (ribonucleic acid)**   A nucleic acid made up of adenine, guanine, cytosine, uracil, ribose, and phosphoric acid; responsible for the transcription of DNA and the translation into protein.

**Rods**   One class of vertebrate visual receptor cells, the cones being the other; highly sensitive, but color-blind in most species.

**Root effect (Root shift)**   A change in blood oxygen capacity as a result of a pH change.

**Round window**   A membrane-covered opening that is located in the inside wall of the middle ear and through which pressure waves leave the cochlea after entering via the oval window.

**Rumen**   The storage and fermentation chamber in the digastric stomach of ruminants.

**Rumination**   The chewing of partially digested food brought up by reverse peristalsis from the rumen in ungulate animals and in other ruminants.

**Saccharide**   A family of carbohydrates that includes the sugars; they are grouped as to the number of saccharide ($C_nH_{2n}O_{n-1}$) groups comprising them: the mono-, di-, tri-, and polysaccharides.

**Saltatory**   Jumping; discontinuous.

**Salt glands**   Osmoregulatory organs of many birds and reptiles that live in desert or marine environments. A hypertonic aqueous exudate is formed by active salt secretion into the small tubules situated above the eyes and is excreted via the nostrils.

**"Salting out"**   A decrease in Bunsen solubility coefficient as a result of increased ionic strength of the solvent.

**Sarcolemma**   The surface membrane of muscle fibers.

**Sarcomere**   The contractile unit of a myofibril, being the span between two Z lines.

**Sarcoplasm**   Cytosol of a muscle cell.

**Sarcoplasmic reticulum (SR)**   A smooth, membrane-limited network surrounding each myofibril. Calcium is stored in the SR and released in ionic form during muscle excitation.

**Sarcotubular system**   The sarcoplasmic reticulum plus the transverse tubules.

**Saturated**   In reference to fatty acid molecules, indicates that the carbon–carbon bonds are single, with each carbon atom bearing two hydrogens. Without free valence electrons.

**Scala media**   The cochlear duct, a membranous labyrinth containing the organ of Corti and the tectorial membrane; it is filled with endolymph.

**Scala tympani**   A cochlear chamber connected with the scala vestibuli through the helicotrema; it is filled with perilymph.

**Scala vestibuli**   A cochlear chamber beginning in the vestibule, connecting with the scala tympani through the helicotrema; it is filled with perilymph.

**Schwann cell**   A neuroglial cell that wraps its membrane around axons during development to produce a myelinated insulating sheath between nodes of Ranvier.

**SDA (specific dynamic action)**   The increment in metabolic energy cost that can be ascribed to the digestion and assimilation of food; it is highest for proteins.

**Secondary structure**   Refers to the straight or helical configuration of polypeptide chains.

**Second Law of Thermodynamics**   All natural or spontaneous processes are accompanied by an increase in entropy.

**Second messenger**   A term applied to cAMP, cGMP, $Ca^{2+}$, or any other intracellular regulatory agent that is itself under the control of an extracellular first messenger, such as a hormone.

**Second-order enzyme kinetics**   Describes enzymatic reactions whose rates are determined by the concentrations of two reactants multiplied together or of one reactant squared.

**Secretagogue**   A substance that stimulates or promotes secretion.

**Secretin**    A polypeptide hormone secreted by the duodenal and jejunal mucosa in response to the presence of acid chyme in the intestine; induces pancreatic secretion into the intestine and is chemically identical to enterogastrone.

**Secretory granules (secretory vesicles)**    Membrane-bound cytoplasmic granules containing secretory products of a cell.

**Seminal vesicles**    Paired sacs attached to the posterior urinary bladder that have tubes joining the vas deferens.

**Semipermeable membrane**    A membrane that allows certain molecules but not others to pass through it.

**Sensillum**    A chitinous, hollow, hairlike projection of the arthropod exoskeleton that serves as an auxiliary structure for sensory neurons.

**Sensory filter network**    Neural circuits that selectively transmit some features of a sensory input and ignore other features.

**Series elastic components (SEC)**    Elasticity in series with contractile elements in muscle.

**Serosal**    Pertaining to the side of an epithelial tissue facing the blood, as opposed to the mucosal side, which faces the exterior or luminal space.

**Serotonin**    5-hydroxytryptamine, 5-HT; a neurotransmitter, $C_{10}H_{12}N_2O$.

**Servomechanism**    A control system that utilizes negative feedback to correct deviations from a selected level, the set point.

**Set point**    In a negative feedback system, the state to which feedback tends to bring the system.

**Siemens (S)**    The unit of electrical conductance; reciprocal of the ohm.

**Sign stimulus**    The most basic essential pattern of sensory input required to release an instinctual pattern of behavior.

**Sinus**    A cavity or sac; a dilated part of a blood vessel.

**Sinus node (sinoatrial node, SA node)**    The junction between the right atrium and the vena cava, the location of the pacemaker.

**Sliding-filament theory**    Shortening of muscle sarcomeres occurs by active sliding of the actin thin filaments toward the midregion of the myosin thick filaments.

**Sliding-tubule hypothesis**    Bending movements of cilia and flagella are produced by active longitudinal sliding of the axonemal microtubules past one another.

**Smooth muscle**    Muscle without sarcomeres and hence without striations. Myofilaments are nonuniformly distributed within small, uninucleated, spindle-shaped cells.

**Sodium activation**    An increased conductance of excitable membranes to sodium ions in response to membrane depolarization; believed to result from an opening of sodium gates associated with membrane channels.

**Sodium hypothesis**    The upstroke of an action potential is due to an inward movement of $Na^+$ down its electrochemical gradient as a result of a transient increase in sodium permeability.

**Sodium inactivation**    Loss of responsiveness of sodium gates to depolarization; develops with time during a depolarization and persists for a short period after repolarization of the membrane.

**Sodium pump**    Membrane mechanism responsible for active extrusion of sodium from the cell at the expense of metabolic energy. In some sodium pumps, there is a 3:2 exchange of intracellular $Na^+$ for extracellular $K^+$.

**Sol**    The low-viscosity state of cytoplasm.

**Solvation**    The process of dissolving a solute in a solvent; hydration, or clustering, of water molecules around individual ions and polar molecules.

**Soma**    The nerve cell body, or perikaryon; in general, the body.

**Somatic**    Referring to the body tissues as distinct from the germ cells.

**Somatosensory cortex**    That region of the cerebral cortex devoted to sensory input from the body surface.

**Spatial summation**    Integration of simultaneous synaptic current by a neuron.

**Specific resistance ($R_m$)**    Resistance per unit area of a membrane in ohms per square centimeter.

**Spectrum**    Specific, charted bands of electromagnetic radiation wavelengths produced by refraction or diffraction.

**Sphincter**    A ring-shaped band of muscle fibers capable of constricting an opening or a passageway.

**Sphingolipid**    A lipid formed by a fatty acid attached to the nitrogen atom of sphingosine, a long-chain, oily amino alcohol ($C_{18}H_{37}O_2N$). The sphingolipids occur primarily in the membranes of brain and nerve cells.

**Spinal cord**    The portion of the vertebrate central nervous system that is encased in the vertebral column, extending from the medulla oblongata to the upper lumbar region; constructed of a core of gray matter and an outer layer of white matter.

**Spindle organ**    A stretch receptor of vertebrate skeletal muscle.

**Spiracle**    Surface opening of the tracheal system in insects.

**Spirometry**    Measurement of an animal's $O_2$ and $CO_2$ turnover.

**Standard metabolic rate (SMR)**    Similar to basal metabolic rate, but utilized for metabolic rate of a heterotherm maintained at a selected body temperature.

**Standard temperature and pressure (STP; dry, STPD)**    $25°C$, 1 atmosphere (atm).

**Standing wave**   A resonating wave with fixed nodes.

**Stapes (stirrup)**   The innermost auditory ossicle, which articulates at its apex with the incus and whose base is connected to the oval window.

**Starch**   A polysaccharide of plant origin, formula $(C_6H_{10}O_5)_n$.

**Starling curves**   Curves that describe the relationship between heart work and filling pressure.

**Statocysts**   Gravity-sensing sensory organs made up of mechanoreceptor cells and associated statoliths.

**Statolith**   A small, dense solid granule found in statocysts.

**Steady state**   Dynamic equilibrium.

**Stennin**   A protein found in brain nerve endings; exhibits ATPase activity and some similarities with muscle myosin. *See also* Neurin.

**Stenohaline**   Able to tolerate only a narrow range of salinities.

**Stereocilia**   Nonmotile filament-filled projections of the surface of eighth-nerve hair cells; not to be confused with true "9 + 2" cilia.

**Steric**   Pertaining to the spatial arrangement of atoms.

**Sterols**   A group of solid, primarily unsaturated polycyclic alcohols.

**Stimulus**   A substance, action, or other influence that when applied with sufficient intensity to a tissue causes a response.

**STP**   *See* Standard temperature and pressure.

**Striated muscle**   Characterized by sarcomeres aligned in register. Skeletal and cardiac muscle are striated.

**Stria vascularis**   Vascular tissue layer over the external wall of the scala media; secretes the endolymph.

**Stroke volume**   The volume of blood pumped by one ventricle during a single heartbeat.

**Structural gene**   A gene coding for the sequence of amino acids that make up a polypeptide chain.

**Strychnine**   A poisonous alkaloid $(C_{21}H_{22}N_2O_2)$ that blocks inhibitory synaptic transmission in the vertebrate central nervous system.

**Substrate**   A substance that is acted on by an enzyme.

**Succus entericus**   Digestive juice secreted by the glands of Lieberkühn in the small intestine.

**Sulfhydryl group**   The radical —SH.

**Supraoptic nucleus**   A distinct group of neurons in the hypothalamus, just above the optic chiasma; their neurosecretory endings terminate in the neurohypophysis.

**Surface charge**   Electric charge at the membrane surface, arising from fixed charged groups associated with the membrane surface.

**Surface tension**   The elasticity of the surface of a substance (particularly a fluid), which tends to reduce the surface area at each interface.

**Surfactant**   A surface-active substance that tends to reduce surface tension.

**Swim bladder**   A gas-filled bladder used for flotation; found in many teleost fishes.

**Sympathetic nervous system**   Thoracolumbar part of the autonomic nervous system.

**Synapse**   A conjunction between two directly interacting nerve cells, where impulses in the presynaptic cell influence the activity of the postsynaptic cell.

**Synaptic cleft**   The space separating the nerve cells at a synapse.

**Synaptic delay**   The characteristic time lag encountered from the time an impulse reaches a presynaptic nerve terminal to the time a postsynaptic potential change occurs.

**Synaptic efficacy**   Effectiveness of a presynaptic impulse in producing a postsynaptic potential change.

**Synaptic inhibition**   A change in a postsynaptic cell that reduces the probability of its generating an action potential; produced by a transmitter substance that elicits a postsynaptic current having a reversal potential more negative than the threshold for the action potential.

**Synaptic noise**   Irregular changes in postsynaptic membrane potential produced by irregular subthreshold synaptic input.

**Syncitium**   A network of cells that are connected by low-resistance intercellular pathways.

**Syneresis**   Contraction of a gelled mixture so that a liquid is squeezed out from molecular interstices.

**Systemic**   Pertaining to or affecting the body; for example, systemic circulation.

**Tachycardia**   An increase in heart rate above the normal level.

**Target cells**   Cells that preferentially bind and respond to specific hormones.

**Taxis**   Locomotion oriented with respect to a stimulus direction or gradient.

**Tectorial membrane**   A fine gelatinous sheet lying on the organ of Corti of the ear in contact with the cilia of the hair cells.

**Teleost**   Bony fish, of the infraclass Teleostei.

**Temporal**   Referring to the lateral areas of the head above the zygomatic arch. Also, relating to time; time-limited.

**Temporal lobe**   A lobe of the cerebral hemisphere, situated in the lower lateral area, at the temples.

**Temporal summation**   Summation of membrane potentials over time.

**Terminal cisternae**   The closed spaces that make up part of the sarcoplasmic reticulum on both sides of the Z line, making close contact with the T tubules.

**Tertiary structure**   Refers to the way a polypeptide chain is folded or bent to produce the overall conformation of the molecule.

**Testosterone**   A steroid androgen synthesized by the testicular interstitial cells of the male; responsible for the production and maintenance of male secondary sex characteristics.

**Tetanus**   An uninterrupted muscular contraction due to a high frequency of motor impulses. Also the name of a neurotoxin that exhibits retrograde (toward the cell body) transport in axons.

**Tetraethylammonium (TEA)**   A quaternary ammonium agent, $(C_2H_5)_4N$, that can be used to block some potassium channels in membrane.

**Tetrodotoxin (TTX)**   The pufferfish poison, which selectively blocks sodium ion channels in the membranes of excitable cells.

**Theca interna**   The internal vascular layer encasing an ovarian follicle; responsible for the biosynthesis and secretion of estrogen.

**Theophylline**   A crystalline alkaloid $(C_7H_8N_4H_2O)$ found in tea; inhibits the enzyme phosphodiesterase, thereby increasing the level of cAMP; also releases $Ca^{2+}$ from calcium-sequestering organelles.

**Thermal neutral zone**   That range of ambient temperatures within which a homeotherm can control its temperature by passive measures and without elevating its metabolic rate to maintain thermal homeostasis.

**Thermogenesis**   The production of body heat by metabolic means such as brown-fat metabolism or muscle contraction during shivering.

**Thermoreceptor**   Sensory nerve ending responsive to temperature changes.

**Thick filament**   A myofilament made of myosin.

**Thin filament**   A myofilament made of actin.

**Threshold potential**   The potential just large enough to produce the response (e.g., action potential, muscle twitch).

**Threshold stimulus**   The minimum strength of stimulation necessary to produce a detectable response or an all-or-none response.

**Thymine**   A pyrimidine base, 5-methyluracil $(C_5H_6N_2O)$, a constituent of DNA.

**Thyroid-stimulating hormone (TSH)**   An adenohypophysial hormone that stimulates the secretory activity of the thyroid gland.

**Thyroxine**   An iodine-bearing, tyrosine-derived hormone that is synthesized and secreted by the thyroid gland; raises cellular metabolic rate.

**Tidal volume**   The volume of air moved in or out of the lungs with each breath.

**Tight junction**   An area of membrane fusion between adjoining cells; prevents passage of extracellular material between the cells.

**Time constant (T)**   A measure of the rate of accumulation or decay of an exponential process; the time required for an exponential process to reach 63% completion. In electricity, it is proportional to the product of resistance and capacitance.

**Tonic**   Steady; slowly adapting.

**Tonicity (hyper-, hypo-, iso-)**   The relative osmotic pressure of a solution under given conditions (e.g., its osmotic effect on a cell relative to the osmotic effect of plasma on the cell).

**Tonus**   Sustained resting contraction of muscle, produced by basal neuromotor activity.

**Torpor**   A state of inactivity, often with lowered body temperature and reduced metabolism, that some homeotherms enter into so as to conserve energy stores.

**Trachea**   The large respiratory passageway that connects the pharynx and bronchioles in the vertebrates.

**Tracheal system**   Consists of air-filled tubules that carry respiratory gases between the tissues and the exterior in insects.

**Tracheoles**   Minute subdivisions of the tracheal system of insects.

**Train of impulses**   A rapid succession of action potentials propagated down a nerve fiber.

*Trans*   A configuration with particular atoms or groups on opposite sides.

**Transcription**   The formation of an RNA chain of a complementary base sequence from the informational base sequence of DNA.

**Transducer**   A mechanism that translates energy or signals of one form into a different kind of energy or signals.

**Transducer molecule**   Hypothesized intermediate molecule within the cell membrane that transmits a hormone-initiated signal from the externally facing hormone receptor to the internally facing adenylate cyclase.

**Transduction**   General term for the modulation of one kind of energy by another kind of energy. Thus, sense organs transduce sensory stimuli into nerve impulses.

**Transfer RNA (tRNA)**   A small RNA molecule that is responsible for the transfer of animo acids from their activating enzymes to the ribosomes; there are 20 tRNAs, one for each amino acid.

**Translation**   Utilization of the DNA base sequence for linear organization of amino acid residues on a polypeptide; carried out by mRNA.

**Transmitter substance**   A chemical mediator liberated from a presynaptic ending, producing a conductance change or other response in the membrane of the postsynaptic cell.

**Transmural blood pressure** The difference in pressure across the walls of a blood vessel.

**Transphosphorylation** The transfer of phosphate groups between organic molecules, bypassing the inorganic phosphate stage.

**Transverse tubules (T tubules)** Branching, membrane-limited, intercommunicating tubules that come into close apposition with the lateral cisternae of the sarcoplasmic reticulum and are continuous with the surface membrane.

**Traveling wave** A wave that traverses the propagating medium, as opposed to a standing wave.

**Tricarboxylic acid cycle (TCA cycle; Krebs cycle; citric acid cycle)** The metabolic cycle responsible for the complete oxidation of the acetyl portion of the acetyl coenzyme A molecule.

**Trichromacy theory** The postulate that three kinds of photoreceptor cells exist in the retina, each exhibiting maximal sensitivity to a different part of the color spectrum.

**Triglyceride** A neutral molecule composed of three fatty acid residues esterified to glycerol; formed in animals from carbohydrates.

**Triiodothyronine** An iodine-bearing tyrosine derivative synthesized in and secreted by the thyroid gland; raises cellular metabolic rate, as does thyroxine.

**Trimer** A compound made up of three simpler identical molecules.

**Trimethylamine oxide** A nitrogenous waste product, probably from choline decomposition.

**Tritium** A radioactive isotope of hydrogen with an atomic mass of three ($H^3$).

**Triton X-100** A nonionic detergent used in cell biology to solubilize lipids and certain cell proteins.

**Trituration** Grinding; mastication.

**tRNA** *See* Transfer RNA.

**Trophic substances** Hypothetical chemical substances believed to be released from neuron terminals and to influence the chemical and functional properties of the postsynaptic cell.

**Tropomyosin** A long protein molecule located in the grooves of the actin filament of muscle; inhibits contraction by blocking the interaction of myosin cross bridges with actin filaments.

**Troponin** A complex of globular calcium-binding proteins associated with the actin and tropomyosin of the thin filaments of muscle. When troponin is combined with $Ca^{2+}$, a conformational change in the troponin complex allows the tropomyosin to move out of its myosin-blocking position on the actin filament.

**Tubulin** An actinlike, 40 Å globular protein molecule that is the building block of microtubules.

**Turbulent flow** Flow in which the fluid moves in all directions.

**Turgor** Distention; swollenness.

**T-wave** That portion of the electrocardiogram associated with repolarization (and usually relaxation) of the ventricle.

**Twitch muscle (fast muscle)** The most common, striated vertebrate skeletal muscle type, usually pale in color because of its low myoglobin content. It has few mitochondria, and its fibers are constructed of many clearly defined fibrils. These contract rapidly and derive most of their energy from anaerobic metabolism.

**Tympanic membrane** The eardrum.

**Tympanum** The middle ear cavity just inside the tympanic membrane; houses the auditory ossicles.

**Ultraviolet light** Light of wavelengths between $1.8 \times 10^{-5}$ and $3.9 \times 10^{-5}$ cm.

**Unit membrane** The sandwichlike profile of biological membrane seen in electron micrographs and believed to represent the bimolecular leaflet with a hydrophobic center region between hydrophilic surfaces.

**Unsaturated** In reference to fatty acid molecules, indicates that some of the carbon–carbon bonds are double. Having free valence electrons.

**Uracil** A pyrimidine ($C_4H_4O_2N_2$) constituent of RNA.

**Urea** The primary nitrogenous waste product in the urine of mammals.

**Ureotelic** Pertaining to the excretion of nitrogen in the form of urea.

**Ureter** A muscular tube passing urine to the bladder from the kidney.

**Urethra** The channel passing urine from the bladder out of the body.

**Uric acid** A crystalline waste product of nitrogen metabolism found in the feces and urine of birds and reptiles; poorly soluble in water.

**Uricolytic** Pertaining to the excretion of nitrogen in the form of uric acid.

**Vacuole** A membrane-limited cavity in the cytoplasm of a cell.

**Vagus nerve (tenth cranial nerve)** A major cranial nerve that sends sensory fibers to the tongue, pharynx, larynx, and ear; motor fibers to the esophagus, larynx, and pharynx; and parasympathetic and afferent fibers to the viscera of the thoracic and abdominal regions.

**Valence** The number of missing or extra electrons of an atom or a molecule.

**Van der Waals forces** The close-ranging, relatively weak attraction exhibited between atoms and molecules with hydrophobic properties.

**Varicosities**   Swellings along the length of a vessel or fiber.

**Vas deferens**   A testicular duct that joins the excretory duct of the seminal vesicle to form the ejaculatory duct.

**Vasoconstriction**   Contraction of circular muscle of arterioles, decreasing their volume and increasing the vascular resistance.

**Vasomotor**   Pertaining to the autonomic control of arteriolar constriction or dilation by contraction or relaxation of circular muscle.

**Vasopressin**   *See* Antidiuretic hormone.

**Vasopressor**   A substance that induces arterial and capillary smooth-muscle contraction.

**Vector**   A carrier; an animal transferring an infection from host to host. Also, a mathematical term for a quantity with direction, magnitude, and sign.

**Venous shunt**   A direct connection between arterioles and venules, bypassing the capillary network.

**Ventral**   Toward the belly surface.

**Ventral horn**   The ventral portion of gray matter in the vertebrate spinal cord in which motor nerve cell bodies are situated.

**Ventral root**   A nerve trunk leaving the spinal cord near its ventral surface; contains only motor axons.

**Ventricle**   A small cavity. Also, a chamber of the vertebrate heart.

**Venule**   A small vessel that connects a capillary bed with a vein.

**Villi**   Small, fingerlike projections of the intestinal epithelium.

**Viscosity**   A physical property of fluids that determines the ease with which layers of a fluid move past each other.

**Visible light**   Light of wavelengths between $3.9 \times 10^{-5}$ and $7.4 \times 10^{-5}$ cm.

**Visual cortex**   The outermost thin layer of gray matter in the occipital region of the cerebrum; devoted to the processing of visual information.

**Vital capacity**   The maximum volume of air that can be inhaled into or exhaled from the lungs.

**Vitalism**   The theory that postulates that biological processes cannot adequately be explained by physical and chemical processes and laws.

**Volt (V)**   MKS unit of electromotive force; the force required to induce a 1 ampere (A) current to flow through a 1 ohm ($\Omega$) resistance.

**Voltage (*E* or *V*)**   The electromotive force, or electric potential, expressed in volts. When the work required to move 1 coulomb (C) of charge from one point to a point of higher potential is 1 joule (J), or 1/4.184 calories (cal), the potential difference between these points is said to be 1 volt (V).

**Voltage clamping**   An electronic method of imposing a selected membrane potential across a membrane by means of feedback control.

**Watt (W)**   A unit of electrical power; the work performed at 1 joule (J) per second.

**Weber–Fechner Law**   Sensation increases arithmetically as stimulus increases geometrically; the least perceptible change in stimulus intensity above any background is proportional to the intensity of the background.

**White matter**   Tissue of the central nervous system, consisting mainly of myelinated nerve fibers.

**Wolffian ducts**   The embryonic ducts that are associated with the primordial kidney and that become the excretory and reproductive ducts in the male.

**Work**   Force exerted upon an object over a distance; force times distance.

**X-ray diffraction**   The method of examining crystalline structure using the pattern of scattered X rays.

**Xylocaine**   The trade name for lidocaine, a local anesthetic related to procaine.

**Z disk (Z line, Z band)**   A narrow zone at either end of a muscle sarcomere, consisting of a latticework from which the actin thin filaments originate.

**Zero-order kinetics**   Kinetics in which the rate of the reaction is independent of the concentration of any of its reactants. This would occur if the enzyme concentration were the limiting factor.

**Zonula occludens**   Tight junctions between epithelial cells, usually having a ring-shaped configuration and serving to occlude transepithelial extracellular passages.

**Zwitterion**   A molecule carrying both negatively and positively ionized or ionizable sites.

**Zygote**   A fertilized ovum before first cleavage.

**Zymogen**   *See* Proenzyme.

# Index of Names

Page references to figures and tables are in italic type. The letter "b" after a page reference indicates a box.

# Index of Topics

Page references to figures and tables are in italic type. The letter "b" after a page reference indicates a box.

during dormancy, 739, 740, 741
effect of exercise on, 489, 737–738, *738*
and fever, 737
of homeotherms, 710
mammalian, 7, 260
and metabolic rate, 689, 691, 711–713, *712*
and obligatory exchanges, 488–489
optimal, 709
of poikilotherms, 710
sensors of, 732
thermostatic regulation of, 731–736, *733–736*
and ventilation, 630
*See also* Evaporative cooling; Thermoregulation
Body weight:
  and calories, 655
  drug to reduce, 77
  and metabolic cost of locomotion, 701–704, *701, 703*
  and metabolic rate, 655, 677–700, *698–700,* 715
  and thermal conductance, 726, *726*
  water as main component of, 657–658
Bohr effect (Bohr shift), *597,* 598, 602, 641
  reverse, 598
Bolus, 661, 666, 676
Bomb calorimeter, 692
Bombycol, 243, 417
*Bombyxmori:*
  diapause hormone of, *459*
  olfactory receptors of, 242–243
Bonds:
  anhydrous, 647
  carbon, 15, 16, *16,* 31, 32, 271, 728
  covalent, 15
  disulfide, 40, *40, 41*
  ester, 32
  glycosidic, 675
  high-energy, 53–54, *54*
  hydrogen: *See* Hydrogen bonds
  ionic, 15
  leucine–leucine, 521
  oxygen, 15, 16, 17, *17,* 30, *30*
  peptide, 37, *38,* 39–40, *40,* 58, *58,* 406, 675, 677, 678
  phosphate, 53
  phosphodiester, 41–42
  polar covalent, 17–18, *17*
  silicon, 16
  sulfur, 17
Bone:
  calcium storage in, 445, 446, 447, *447,* 448, 657
  capillaries in, 577
  and growth hormone, 555
Boron, 15
Botulinum toxin, 196b
Bowman's capsule, 504, 505, 508, 521, 524, 531
  concentration of substances in, 506, *506*

Brachiopods, as filter feeders, 649, *649*
*Brachon cephi,* antifreeze substance in, 717
Bradycardia, during diving, 586–587
Bradykinin, 582
Brain, 1, 586, 738, 739
  anoxia and, 595
  and baroreceptor information, 582
  bat, 333
  blood flow in, 1, 572, 574, 579, 580, 581, 582
  capillaries in, 576, 577, 579, 580
  and control of secretions, 677b, 678
  cow, calmodulin from, 473
  evolution of, 289, 290
  $H^+$ receptors in, 626
  human, *290, 293,* 580
  insect, hormone secretion in, 458, 460
  invertebrate, 289
  mammalian, 70–71, 121, 582
  and neurosecretion, 360, 360n
  pH sensitivity of, 634
  rabbit, 732
  rat, 457
  and sensation, 234, 297
  and regulation of respiration, 624, 626, 630
  thermostatic centers in, 732, 735
  vertebrate, 290, *290,* 292–293, *293*
  visual centers of: *See* Visual cortex
Brain hormone (prothoracicotropin), 458–463 passim, *459*
Brain stem, 318
Breathlessness, 627
Bronchi, 604, 627, 629
Bronchial circulation, 609
Bronchioles, 604, 610, 627
  cilia of, 400
Bronchoconstriction, 627
Brood patches, 457
Brown fat, *725*
  thermogenesis in, 725, 739, 740
Brunner's glands, 679
Brush border:
  intestinal, 392, *393,* 663, 680
  renal tubular, 510
Bubble nests, 616
Bubbles:
  alveoli as, 615–616, *616*
  and respiration of aquatic insects, 635–637, *636*
Buccal cavity, 658, 666
  and gill ventilation, 618, 619
  and lung ventilation in frog, 614–615, *614*
Buffer systems, 25–26, *26*
  in blood, 598–602, *601,* 630–631
  calcium, 471
*Bufo marinus,* receptor potentials in, *268*
Bugs:
  feeding apparatus and strategies of, 651, *651*
  flight muscles of, 377

Building blocks, essential, 655
Bulbogastrone, actions of, *670*
Bulbus, 561, 570
Bullfrog, sacculus hair-cell response in, *249*
Bundle of His, 553
α-Bungarotoxin (BuTX), 196b, 204
Bunsen solubility coefficient, 593b
  of oxygen in blood, 593
Bursicon, *459,* 460–461, *460*
Butterfly, as perceived by compound eye, *277*
BuTX (α-bungarotoxin), 196b, 204
Butyrate, 662

$Ca^{2+}$: *See* Calcium ion
Cadmium ion ($Cd^{2+}$), 167
Caffeine:
  and cAMP levels, 464
  release of intracellular calcium by, 270, 364
Calcitonin (CT), *446*
  actions of, 445, 446, *446,* 447–448, *447*
Calcium caseinate, 678
Calcium channel, 160, 470
  of mechanoreceptors, 247
  and membrane excitability, 167, *168,* 169, 171, 326, 427
  in *Paramecium,* 167, *168,* 326, 412, *412, 413*
Calcium chloride, *126,* 471
  and tonicity, 102
Calcium conductance, 167, *168,* 169–170, 233
  in ciliary reversal, 412
  and heart contraction, 381, 552
  and pacemaker potentials, 171–172, *171,* 551, 572
Calcium current:
  and contraction of smooth muscle, 667, *667*
  and membrane excitability, 167, *168*
  and potassium current, 169–170, *169*
Calcium hypothesis of visual excitation, 269–270, *270*
Calcium ion ($Ca^{2+}$), 626, 657
  active transport of, 105, 119, 685
  binding of calmodulin, 383, 395, 472–473, *472,* 474–475, *474*
  binding to intestinal protein, 685
  and cAMP, 464, *468b,* 469, 473–477, *473, 475*
  and cGMP, 470, 474, 475
  chelation of, 354, 404
  and ciliary-beat arrest, 410
  and ciliary reversal, 411–412, *412*
  as cofactor, 61, 68
  conductance of, 167, *168,* 169–172, *171,* 233, 381, 412
  and contraction in cardiac muscle, 381
  and contraction in nonmuscle tissue, 391, 394–399 passim, *396, 398*